Petroleum Systems in the Southern Gulf of Mexico

Edited by
C. Bartolini and J. R. Román Ramos

AAPG Memoir 90

Published by

The American Association of Petroleum Geologists
Tulsa, Oklahoma, U.S.A.

ISBN13: 978-0-89181-371-2

AAPG Editor: Gretchen M. Gillis
AAPG Geoscience Director: James B. Blankenship

ON THE COVER: Church of the Virgin of Guadalupe in Real de Catorce, San Luis Potosi State in Central Mexico. The church was built in 1775 and is part of the old cemetery. The lower part of the hill exposes continental redbeds (red colors) of Early and Middle Jurassic age, whereas the white, upper part, of the hill consists of Upper Jurassic marine carbonates. Photograph by Claudio Bartolini.

This publication is available from:

The AAPG Bookstore
P.O. Box 979
Tulsa, OK U.S.A. 74101-0979
Phone: 1-918-584-2555 or 1-800-364-AAPG (U.S.A. only)
E-mail: bookstore@aapg.org
www.aapg.org

Canadian Society of Petroleum Geologists
600, 640 – 8th Avenue S.W.
Calgary, Alberta T2P 1G7
Canada
Phone: 1-403-264-5610
Fax: 1-403-264-5898
E-mail: reception@cspg.org
www.cspg.org

Geological Society Publishing House
Unit 7, Brassmill Enterprise Centre
Brassmill Lane, Bath BA13JN
United Kingdom
Phone: +44-1225-445046
Fax: +44-1225-442836
E-mail: sales@geoloc.org.uk
www.geolsoc.org.uk

Affiliated East-West Press Private Ltd.
G-1/16 Ansari Road, Darya Gaaj
New Delhi 110-002
India
Phone: +91-11-23279113
Fax: +91-11-23260538
E-mail: affiliate@vsnl.com

Acknowledgments

AAPG wishes to thank the following
for their generous contribution to

Petroleum Systems in the Southern Gulf of Mexico

Contributions are applied toward the production cost of publication,
thus directly reducing the book's purchase price
and making the volume available to a larger readership.

About the Sponsors

Repsol is an international integrated oil and gas company, operating in more than 30 countries and the leader in Spain and Argentina. It is one of the ten largest private oil companies worldwide and the leading private energy company in Latin America in terms of assets. Repsol is the leading refining and marketing company in Spain and Argentina, and the third leading private company in the world engaging in liquefied petroleum gas (LPG) distribution activities through its alliance with Gas Natural.

From exploration and production, up to marketing, Repsol is present in all the business phases. Its oil and gas production totals more than one million barrels per day and its reserves are situated mainly in North Africa and Latin America. Repsol has a refining capacity of more than 1.2 million barrels per day and sells its products though a large network of more than 6,800 service stations. In the Chemical business, the company is the leading producer in Spain and Portugal.

Repsol is a company oriented towards the customer and the creation of value, with a firm belief in technological innovation and committed to the environment and the community.

Petróleos Mexicanos is the largest company in Mexico and Latin America, and the largest taxpayer in Mexico. It is one of the few petroleum companies in the world involved in the entire production chain, from exploration, through distribution, to commercialization of the final products. During 2007, production was 3.082 million barrels per day (Mbd). Exported liquids (crude and condensates) averaged 1.686 Mbd, with a value of $US 44,388,000,000. Total sales, including service income, increased 2.9% over 2006, reaching a historic maximum of 11,350,000,000 Mexican pesos, due mainly to the higher price of exported liquids. The company's business plan recognizes the necessity to grow through fortification of the production and operational infrastructure, integrally improving operational execution, and harmonizing the efforts of the different lines of business in order to maximize economic value.

PEMEX, the pride of México and its oil workers, has been converted into one of the world's most competitive state oil companies, since it now operates in an opportunistic, modern, transparent, and efficient manner by applying high standards of excellence and honesty. The nation's economy has turned around, thanks to PEMEX. Through its industrial alliances, it is positioned as a lever of national development by generating high employment levels. Its leading edge technology has led to increased reserves and exports of higher quality crude oil at better prices. It has also achieving self-sufficiency in natural gas. Pemex supplies basic materials, products and services of high quality at competitive prices, and is a modern and growing petrochemical industry.

About the Editors

Claudio Bartolini is a Senior Exploration Geologist with the Repsol Regional Studies Group in Madrid, Spain. He first was involved in the exploration of deep and ultra-deep waters Gulf of Mexico Basin, and subsequently transferred to the Alaska Exploration Team. Claudio worked three years as an exploration geologist after he received his B.S. degree in geology from the College of Engineering, University of Sonora, Mexico. He earned a M.S. degree in geology from the Geosciences Department at the University of Arizona, and after graduation he worked for four years as an exploration geologist for Gold Fields Mining Corporation in Arizona, California and Mexico. In 1992, he enrolled at the University of Texas at El Paso, and earned a Ph.D. degree in geology in 1997, with a major in tectonics and sedimentary basins. Claudio was the recipient of a NASA Scholarship during the five years at the doctoral program. While pursuing his doctoral studies, Claudio worked as a consultant geologist in Mexico and Central America for Canadian exploration companies. After finishing his doctoral program, he joined ARCO International Oil and Gas Company, Latin America exploration group. Later, he was a basin researcher for IHS Energy in Houston, Texas. Claudio is the senior editor of the 1999 Geological Society of America Special Paper 340, *Mesozoic Sedimentary and Tectonic History of North-Central Mexico*. He also edited (2001) AAPG Memoir 74, *The Western Gulf of Mexico Basin: Tectonics, Sedimentary Basins and Petroleum Systems*, and more recently, AAPG Memoir 79, *The Circum-Gulf of Mexico and the Caribbean: Hydrocarbon Habitats, Basin Formation, and Plate Tectonics*. Claudio Bartolini maintains a strong interest on the Paleozoic and Mesozoic geology and tectonics of Mexico, with emphasis on sedimentary basins and petroleum systems.

Juan Rogelio Román Ramos is a leader of the geological petroleum system modeling at the Mexican Petroleum Company (PEMEX). He earned his B.S., M.S., and Ph.D. degrees as a Petroleum Geologist in Moscow, Russia between 1976 and 1987. He began his professional career in 1988 with the Mexican Petroleum Institute as a research scientist of the Enhanced Oil Recovery from tight and naturally fractured carbonated reservoirs. Simultaneously, he was teaching the Sedimentary Petrology class at the National University (UNAM), and was a member of the National Research System. In 1992 he joined PEMEX Exploracion & Produccion where he developed and supervised geochemical studies and geological modeling for the northern Mexican hydrocarbon basins, and participated in technologic transfer projects dealing with basin modeling and exploratory risk evaluation. From 1999 to 2004, Juan Rogelio was a Project Leader for the western Deepwater Gulf of Mexico where he supervised sampling programs leading to the geochemical characterization of hydrocarbons seeping from Gulf floor, data acquisition programs and geological interpretation. This latter project led to the delimitation of plays and the definition of the first drill locations in the Mexican Ridges and Perdido Fold Belt Deepwater Provinces. Since 2005, he has coordinated basin and petroleum systems studies for the Northern Regional Exploration Asset, creating the foundation for a better understanding of inter-regional 3D geological relationships, by modeling the generation and migration of hydrocarbons in the Tampico-Misantla, Burgos and Sabinas Basins, and relating them to the deepwater Gulf of Mexico. These studies have defined new areas with economically interesting petroleum potential. In 2006, he was elected as the Leader of the Integrated Network for Geological Modeling for Petroleum Exploration. Juan R. Román Ramos is an active member of the Mexican Petroleum Geological, Geophysical and Engineers Associations, where he has served as a Delegate President; Director of the Membership, and Technical Studies Commissions. He was recognized in 2006 as Distinguished National Expert. He has promoted affiliation with the American Association of Petroleum Geologists, has coordinated the visits of AAPG Distinguished Lecturers to Mexico, and has also collaborated in the organization of Joint AMGP-AAPG Conferences.

Dedication

Teodoro Díaz González (deceased) was born in Cerralvo, Nuevo Leon, Mexico on November 9, 1917. He earned his bachelor's degree from the University of Texas at Austin in 1940. Teodoro was among the pioneering geologists at Petroleos Mexicanos (Pemex) where he began as a geologist in 1940. He was geological party chief from 1942 to 1950, and was promoted to Chief Geologist for Pemex's northeast region from 1950 to 1966. He was promoted again to Manager of Subsurface Geology for the central exploration division from 1966 to 1967, and Senior Exploration Advisor from 1967 to 1973, the year he retired from Pemex. Teodoro implemented the systematic application of petrophysics to evaluate the petroleum potential of Jurassic and Cretaceous sequences in all the Mexican basins. Upon retirement, he became a valued consultant for Pemex and the Mexican Commission of Electricity. Teodoro was a member of the Mexican Association of Petroleum Geologists (AMGP), the American Association of Petroleum Geologists (AAPG), the Geological Society of America (GSA), the Mexican Geological Society (SGM), and the College of Geological Engineers of Mexico (CIGM). During his career his technical capabilities were recognized by participation in prestigious industry efforts, including committee member of the Cretaceous Symposium in the Western Hemisphere in 1956; leader for field trips during the XX International Geological Congress (1956); committee member for the Basement Map of North America produced by the AAPG and the U.S. Geological Survey in 1967; honorary committee member of the American Commission on Stratigraphic Nomenclature from 1966 to 1969; and the Martillo de Plata (the Golden Hammer) Medal awarded by the Mexican College of Geological Engineers and the Mexican Geological Society in 1997. Teodoro Díaz González published important national and international papers related to the petroleum geology of Mexico, including "The Geology of Northeastern Mexico," the University of Texas at Austin, in 1959; "Exploration and Production Results from the Offshore Extension of the Mexican Cretaceous Golden Lane," AAPG Annual Meeting, 1969; "Litología y correlación estratigráfica del Paleozoico Superior en la región de Palomas, Chihuahua, México," National Convention of Petroleum Exploration and Exploitation Engineers, AAPG, El Paso, Texas, November, 1961; and "Posibilidades petrolíferas de la región de San José de las Rusias a Sabino Gordo, Tamaulipas," First Mexican Petroleum Technology Convention, Mexico City, February, 1950.

Rafael Sánchez Montes de Oca (deceased) was born in Tuxpan, Veracruz, Mexico on May 10, 1938 and passed away in the city of Veracruz, Mexico on February 22, 2007. Rafael was a mentor and friend to many Mexican and visiting international geologists. He was a person with great human qualities including a fine sense of humor, a passion for geology, a love of fieldwork, and one who took joy in sharing his knowledge. He is survived by his wife, Lucina del Pilar Ferreiro Pérez, and children Leonardo Rafael, Lucina del Pilar, and Daniel Arturo Sánchez Ferreiro. Rafael completed his geological studies at the National Polytechnic Institute in 1960. In 1961 he was an intern in the Council of Non-Renewable Natural Resources where he completed his thesis entitled "The Surface Geology of the Mining District of Huautla, Morelos" which he defended that year, earning the title of Geological Engineer with Honors. Rafael joined the Southern District of Pemex in 1962 as an assistant field geologist, subsurface geologist and geological interpreter. From 1967 to 1974 he was the chief of a surface geology survey team and pioneered the use of satellite imagery for regional interpretation as part of a petroleum evaluation of the geologically complex and logistically challenging Sierra de Chiapas. From 1975 to 1981 Rafael was superintendent of geological operations for the Southern District of Pemex where he actively participated in the discovery of the Reforma and Campeche Sound oil fields. Between 1982 and 1989 he served as the general superintendent of exploration in Pemex's Northern and Southeastern Districts. In 1989 Rafael moved to Pemex headquarters in Mexico City where he was a leader in geological evaluation and planning until retiring in

1996. After retiring he consulted for private companies, including holding the position of exploration manager for ICA in Colombia during 1998 and 1999. Between 2002 and 2004, in an assignment from which Rafael derived tremendous satisfaction, he was responsible for the Field Geology Training Program at the National Polytechnic Institute together with the Technical Exploration Group of Pemex. During this work, he authored five geological synopses of the Sierra de Chiapas, which encapsulated his unique understanding of this area. He continued as an advisor and mentor in the Technical Exploration Group of Pemex until late 2006. Rafael Sánchez Montes de Oca was a member of the Association of Petroleum Engineers of Mexico (AIPM), the Mexican Association of Petroleum Geologists (AMGP) where he served as National President from 1992 to 1994, and the American Association of Petroleum Geologists (AAPG). AAPG presented Rafael with its Special Commendation Award in recognition of his work benefiting the international community of petroleum geologists. In 2006, the Mexican Association of Petroleum Geologists recognized Rafael for his invaluable contribution and leadership in the formation of new geoscientists for the national petroleum industry, and who presently are the technical foundation of Pemex.

José Carrillo Bravo was born in Zapotitlan Salinas, Puebla, Mexico. After finishing high school he moved to Mexico City and enrolled in the National Polytechnic Institute where he earned a bachelor's degree in geology with honors in 1951, becoming one of the first geologists to graduate from that institution. José Carrillo worked for Petroleos Mexicanos (Pemex) from 1952 until his retirement in 1989 as exploration manager for integration and interpretation in Pemex's Subdivision for Primary Production. As a Pemex geologist, José first described the Paleozoic section of northeastern Mexico, particularly the Silurian, which was unknown until that time. He was the first to document and date Triassic and Jurassic continental redbeds in the Gulf of Mexico coastal region. He also proposed, and later proved, the existence of the Mesozoic Valles-San Luis Potosi platform, thereby confirming the existence of central Mexico's Mesozoic Basin. He was also the first to identify several Tertiary paleocanyons in eastern Mexico, and proposed and documented the existence of the regional Tertiary unconformity responsible for the migration and accumulation of oil along the western margin of the Gulf of Mexico. For many years, José Carrillo Bravo was responsible for reviewing, analyzing and documenting Pemex's exploratory wells. Additionally, for 26 years he was a part-time faculty member at the National Autonomous University of Mexico (UNAM) and the National Polytechnic Institute (POLI) where he is revered for his outsanding academic standards and distinguished lectures José Carrillo Bravo has received a number of honors and awards, including Distinguished Member and former president of the Mexican Academy of Engineering (1987), the Geological and Petroleum Engineering National Award (1982), Honorary Member of the National Commission of Science and Technology (Conacyt) (1976), and the American Association of Petroleum Geologists (AAPG) Distinguished Service Award. He has been a member of AAPG since 1960. Selected publications of José Carrillo Bravo include "Exploración Petrolera de la Cuenca Mesozoica del Centro de México," 1982, *Bol. Asoc. Mex. de Geól. Petroleros,* v. XXXIV, n. 1, p. 21-46; "Paleocañones Terciarios de la Planicie Costera del Golfo de México," 1980, *Bol. Asoc. Mex. de Geól. Petroleros,* v. XXXII, n. 1, p. 17-55; "La Plataforma de Valles – San Luis Potosí," 1971, *Bol. Asoc. Mex. de Geól. Petroleros,* Vol. XXIII, n. 1-6, p. 1-102; "Estudio Geológico de una parte del Anticlinorio de Huayacocotla," 1965, *Bol. Asoc. Mex. de Geól. Petroleros,* v. XVII, n. 5-6, p. 73-96; "Geología del Anticlinorio de Huizachal Peregrina, al NW de Cd. Victoria, Tamaulipas," 1961, *Bol. Asoc. Mex. de Geól. Petroleros,* v. XIII, n. 1 y 2, p. 1-98; "Notas sobre el Paleozoico de la Región de Cd. Victoria, Tamps.," 1959, *Bol. Asoc. Mex. de Geól. Petroleros,* v. XI. n. 11 y 12, p. 671-680.

Abelardo Cantú-Chapa has had a distinguished career as a paleontologist with experience in the Mesozoic paleontology and biostratigraphy of Mexico. He has made important contributions to an understanding of the origin and evolution of the Gulf of Mexico, and to offshore Mesozoic stratigraphy. Abelardo earned his bachelor's degree in Mexico in 1956 at the University of Nuevo León and the National Polytechnic Institute. After two years as a micropaleontologist for Petróleos Mexicanos (Pémex), he studied sedimentology and sedimentary petrography at the University of Marseille, the French Petroleum Institute in Paris, and the University of Paris. In 1962, he earned his doctorate Docteur ès-Sciences Naturelles from the Université de France at the Sorbonne with a dissertation on the biostratigraphy of Upper Jurassic and Cretaceous ammonites of Mexico. His dissertion was published in the Société Geólogique de France, Memoire No. 99 (1963). He has published more than 100 papers and abstracts, with numerous presentations in national and international congresses. From 1962 to 1966, Abelardo was in charge of Pemex's macropaleontology laboratory, holding the same position at the Mexican Petroleum Institute from 1966 to 1989. He was chairman of the Geology Graduate Program at the National Polytechnic Institute from 1992 to 1995, and from 1995 to 2001 he served as its chairman of Graduate Studies and Research. Abelardo is the coeditor of the American Association of Petroleum Geologists Memoir 75, which was published in 2001. During his tenure at the National Polytechnic Institute, Abelardo directed 23 master's theses and 10 bachelor's theses on petroleum geology, mining geology, and geological risk, and he continues to serve on thesis committees. Over the years, Abelardo Cantú-Chapa received major research grants from the National Commission of Science and Technology (Conacyt) and the National Polytechnic Institute Division of Graduate Studies and Research to support petroleum and mining related research programs under his leadership. Abelardo has been recognized for his active participation in the Mexican System of National Research. His research has also been recognized by Mexican and international geological institutions. His specific awards include an honors diploma and the medal from the National Polytechnic Institute in recognition of 30 years of academic contributions; the Juan de Dios Batiz Award for 30 years of service to the National Polytechnic Institute; a diploma and prize from the College of Mexican Geological Engineers for his contributions to geology during the Twentieth Century; and an honorary membership in the College of Mining, Metallurgical and Materials Engineers. Recently, Abelardo's initiative to create the National Center for Geosciences and Hydrocarbons Administration was approved by both chambers of the Mexican Legislature. Cantu's most recent contributions to the petroleum geology of Mexico are included in this AAPG memoir.

Jaime Rueda Gaxiola graduated in 1961 as a geological engineer from the National Polytechnic Institute and then worked for the Natural Resources Commission until 1963. He earned his Ph.D. in geology at Lille University, France in 1967, graduating with honors for his dissertation on the palynology and petrography of coal in the Sabinas Basin. Jaime has pioneered the application of palynology and organic geochemistry to hydrocarbons exploration in the Mexican sedimentary basins and has received many honors for his life-long dedication to the geosciences, including member of the Scientific Committee of CONACYT in 1997; president of the Geosciences, Sea, and Atmospheric Sciences Committee of CONACYT in 1993; member of the Earth Sciences Committee of CONACYT in 1991; Lazaro Cardenas Medal and Award in 1989; National Research Scientist of SNI-SEP in 1988; and selection as one of 20 final candidates to be the first Mexican astronaut. Jaime is presently a professor at the College of Superior Education at the National Polytechnic Institute in Mexico City, where he has been teaching since 1972. Jaime was director of the College of Engineering and Architecture from 1986 to 1987, subdirector of the Geosciences Department from 1980 to 1983, academic coordinator for the Geological Engineering Division from 1975 to 1976, and chairman of laboratories in the Geosciences Department from 1973 to 1974. He has been the academic advisor for 32 bachelor's theses and has also taught at the graduate level. He has trained petroleum professionals for Pemex, the Mexican Petroleum Institute, and the Venezuelan and Chinese petroleum institutes. Jaime has written three books on applied palynology, another on natural hazards, and 38 professional papers published in Mexican and international forums. He conducted 19 research projects for the Mexican Petroleum Institute between 1968 and 1999, and in 1978, he founded the Institute's Department of Palynostratigraphy and Organic Geochemistry. In 1979, he was a member of the scientific team aboard the Glomar Challenger studying the Gulf of California. Jaime Rueda Gaxiola has been a member of 13 professional societies, and, over the years, has served as vice president of the Association of Geological Engineers of the National Polytechnic Institute, vice president of the Mexican Association of Petroleum Geologists, vice president of the Mexican Geological Society, and vice president of the College of Geologists of the Mexican Republic. He continues to be a highly regarded geologist with expertise in Mesozoic palynology, palynostratigraphy, and organic geochemistry.

Table of Contents

Introduction

Bartolini, C., 2009, Introduction, *in* C. Bartolini and J. R. Román Ramos, eds., Petroleum systems in the southern Gulf of Mexico: AAPG Memoir 90, p. xiii–xiv.

Claudio Bartolini

The petroleum geology of Mexico remains relatively unknown to the international geoscience community because of the limited number of articles published in international forums by geoscientists from Petroleos Mexicanos (PEMEX) and the Mexican Petroleum Institute. Sensitive Pemex data have traditionally been treated with strict confidentiality. Thus, in the realm of Mexican petroleum exploration over the last 70 years, the publication of Memoir 90 is a watershed event as the first international publication that is devoted exclusively to Mexican authors from the petroleum sector. In consideration of the confidentiality issues, cultural differences, and the language barrier, this pioneering volume is a tremendous achievement. We strongly believe that it sets an important and encouraging precedent for future international publications by Mexican geoscientists.

The 20 articles that comprise this book cover the onshore and offshore Mexican basins of the Circum-Gulf of Mexico Basin. Most of them have a multidisciplinary approach, with special emphasis on hydrocarbon exploration and petroleum geology. General subjects include petroleum systems, salt tectonics, structural geology, petroleum geochemistry, sequence stratigraphy, sedimentology, sediment provenance, play fairways, regional geology, biostratigraphy, and more.

In order to expand the geographic coverage of the original manuscripts submitted, we included four extended abstracts and one paper on the petroleum geology of Mexico, which were recently published by AAPG. These publications are a great complement to the diverse petroleum exploration topics covered in this volume. The extended abstracts and paper are placed at the end of the book/CD as complimentary chapters. We decided to provide the readers with both a book and a CD. The book will be entirely in color, as well as the CD, which also contains complete color versions of the articles for the convenience of our readership.

A large number of individuals contributed to the great quality of AAPG Memoir 90 in the form of their very constructive and accurate manuscript reviews. Our special thanks go to Joshua Rosenfeld, Angel F. Callejon, Gary Prost, Horacio Parent, Carlos Rivero, Francois Roure, Jon Blickwede, Mark Rowan, Abhijit Basu, Louis Liro, Scott Krueger, Thomas Buerkert, Carl Steffensen, Joe Cuariale, Mark Gordon, Emilio Garciacaro, Marc Edwards, William Ambrose, Jaap Klein, Carl Fiduk, Fred Diegel, Thomas Ewing, Arthur Berman, Terry Poulton, Mark Jiang, Christopher Denison, Norman Rosen, Pari Rosen, Ricardo Veiga, John Bacheller III, Peter Cobbold, Charles Winker, and Andrew Horbury.

Memoir 90 reflects a high degree of cooperation and a strong relationship between PEMEX and Repsol. Both companies have made a great effort to produce a book that will not only yield new insights into the petroleum geology of Mexico, but will also mark the beginning of a new era for PEMEX geoscientists.

We thank PEMEX for allowing the publication of articles related to the petroleum geology of Mexico, which will greatly benefit our geoscience community. Our appreciation goes to Antonio Escalera Alcocer, PEMEX subdirector of exploration in the southern region, for his encouragement and the support provided to Mexican geoscientists during the publication stages of the memoir.

We, the editors and authors, are grateful to Repsol E&P U.S.A. for being the sole financial sponsor of our volume and making possible the production of an affordable book for the Panamerican geoscientific community. We are particularly indebted to Jose C. Vicente Bravo, Repsol's North America exploration manager, who greatly encouraged the senior editor and supported this project from its inception.

We thank AAPG for their continued support of the geosciences. This book is another example of AAPG's relevant contributions to the geoscientists of the world.

DOI:10.1306/13191141M903340

The volume's editors are grateful to the authors for their dedication, hard work, and the extra effort required during the seemingly endless time-consuming tasks associated with manuscript preparation.

This book is dedicated with great respect to Abelardo Cantú Chapa, Jaime Rueda Gaxiola, and José Carrillo Bravo for their lifelong commitment to the geosciences and for their important contributions to the petroleum geology, geological research, and academic advancement of Mexico. Our dedication also goes to Teodoro Díaz Gonzáles (deceased) and Rafael Sanchez Montes de Oca (deceased) for their contributions to the petroleum geology of Mexico and their pioneering exploration work at PEMEX.

Gómez-Cabrera, P. T., and M. P. A. Jackson, 2009, Regional Neogene salt tectonics in the offshore Salina del Istmo Basin, southeastern Mexico, *in* C. Bartolini and J. R. Román Ramos, eds., Petroleum systems in the southern Gulf of Mexico: AAPG Memoir 90, p. 1–28.

Regional Neogene Salt Tectonics in the Offshore Salina del Istmo Basin, Southeastern Mexico

Pedro T. Gómez-Cabrera

Petróleos Mexicanos (PEMEX), Activo Integral Bellota-Jujo, Comalcalco, Tabasco, Mexico

Martin P. A. Jackson

Bureau of Economic Geology, Jackson School of Geosciences, the University of Texas at Austin, Austin, Texas, U.S.A.

ABSTRACT

Compared with the northern Gulf of Mexico, little has been published on the tectonics of its southern equivalent in the Bay of Campeche, especially in English. A middle Miocene change in the Pacific plate kinematics created the southern Mexico Neogene belt. This is locally expressed as four main fold belts underlying most of the study area: the Agua Dulce, Marbella, Marbella Norte, and Catemaco fold belts. Each fold belt contains productive Upper Jurassic and Cretaceous sections and may be part of a much larger fold belt that merges with adjoining fold belts in the Chiapas and Campeche areas. During the final stages of folding, salt was squeezed from diapirs and extruded over the eroded and uplifted fold belts. The extruded salt coalesced to form the discontinuous Sal Somera canopy. The salt canopy or equivalent weld is roughly conformable with the underlying fold belts, indicating that the canopy extruded during the final stage of folding or soon after. Evacuation of the thickest parts of the salt canopy allowed the greatest Pliocene–Pleistocene subsidence and minibasin growth. Three counterregional minibasins formed by seaward evacuation of the underlying salt canopy in the Pliocene–Pleistocene. Counterregional systems dominate the study area, the largest of which forms the seaward boundary of the 100-km-long (62-mi-long) Pescadores minibasin. Some minibasins are floored by the allochthonous salt canopy or equivalent salt weld, others rest directly on the fold belts, and others are more deeply rooted.

DOI:10.1306/13191074M903329

INTRODUCTION

The Salina del Istmo Basin, southwestern Gulf of Mexico, is the southern equivalent of the northern Gulf of Mexico. Both conjugate margins of the Gulf are underlain by Middle Jurassic evaporites and are dominated by salt tectonics. However, much less has been published on salt tectonics in the southern Gulf, especially in English. Most publications on the Salina del Istmo Basin focused onshore (e.g., Contreras and Castillon, 1960, 1968; Viniegra, 1971; Wilson, 1993). Even fewer publications describe the offshore basin. Ricoy (1989) was the first to apply sequence-stratigraphic ideas to southeastern Mexico, recognizing four genetic sequences in the Tertiary. Sequence I includes Paleogene sediments derived from erosion of the Chiapas massif and the Sierra de Chiapas. The sequence consists mostly of shale onshore and dolomitized limestone breccia in the Bay of Campeche. Sequence II, which accompanied an early Miocene tectonic event, is mostly alternating sandstone and blue-gray fossiliferous shale, probably deposited as bathyal submarine fans. Pliocene sequence III comprises gray fossiliferous shale and abundant intercalations of micaceous quartz sandstones deposited on a platform. The Pleistocene fluvial-deltaic sequence IV comprises gradational sequences of shale and sandstone and minor conglomerate.

García-Molina (1994) used onshore and offshore seismic data to study the structural evolution in southeastern Mexico. He inferred that the main detachment level is the Middle Jurassic evaporite sequence and that the Villahermosa-Sonda de Campeche fold belt formed during the late Miocene. He inferred that salt flowed initially into the cores of compressional folds, as in the Zagros Mountains (Iran). Sediment loading was thought to be responsible for salt tectonics in the offshore part of the Salina del Istmo Basin (Marbella area).

A Petróleos Mexicanos (PEMEX)-British Petroleum (1994) team conducted the first comprehensive study of the Salina del Istmo Basin using modern concepts of salt tectonics. From this work, Oviedo-Pérez (1996) inferred a geologic evolution for the onshore and offshore Salina del Istmo Basin, using concepts of tectonically controlled depositional megasequences (Hubbard et al., 1985a, b). Oviedo-Pérez (1996) reported three main tectonosequences relevant here. First, a Paleogene foredeep tectonosequence accumulated until the early Miocene. Then, regional shortening reached the offshore area in the middle to late Miocene, creating a fold-belt tectonosequence. Combined with the sediment load on the autochthonous Jurassic salt, this compression drove the emplacement of the salt canopies. The last tectonosequence is the passive-margin fill from the Pliocene to Holocene. This featured onshore uplift, rapid progradation, and evacuation and redistribution of the allochthonous salt canopy.

These studies on the Salina del Istmo Basin point to a complex tectonic history, which included orogenically driven shortening, gravity-driven extension, diapirism, and extrusion and evacuation of allochthonous salt. Some aspects of this history are similar to those in the northern Gulf of Mexico. However, other aspects are different in the southern Gulf: for example, Cenozoic Los Tuxtlas volcanism, major uplift of nearby provenance areas, orogenic shortening, collision of the Chortis block, and rise of the Sierra de Chiapas and Sierra de Zongolica, all representing far-field effects of the Cocos plate convergence in a northeast direction.

Our chapter addresses the Neogene salt-related structures in a study area on the continental shelf of the Bay of Campeche, southeastern Mexico (Figure 1). The study area is in the western part of the much larger study area of García-Molina (1994), whose data density was necessarily low. Only four, low-quality, two-dimensional (2-D) seismic profiles from García-Molina (1994) overlapped with our study area. Our study is based on 26 seismic lines of 2-D data and some profiles extracted from a small three-dimensional (3-D) seismic survey (Figure 2). This 2-D seismic grid allows the structures to be mapped in greater detail than previously possible. A companion chapter in this volume explores a subarea in more detail, covered by a 3-D seismic survey (Gómez-Cabrera and Jackson, 2009).

None of the nine available wells in the study area reached salt, so we identified salt bodies only by their seismic character. The top of salt was identified by (1) a high-amplitude reflection or doublet at the top-salt contact, (2) low amplitude below this high-amplitude reflection, (3) multiple chaotic seismic reflections below the high-amplitude reflection, (4) correlation between some bathymetric highs with the crests of high-amplitude reflections, (5) extensional faults and grabens above high-amplitude reflections, (6) structural tilting and deformation of strata on the flanks of diapirs, and (7) irregular geometry of the high-amplitude reflection (Fiduk, 1994).

The next three sections describe the main tectonic features in order of age: the Miocene fold belts, the partly evacuated late Miocene salt canopy, and the Pliocene–Pleistocene minibasins and diapirs.

MIOCENE FOLD BELTS

The study area includes fold belts, which we assign to the southern Mexico Neogene belt (Prost and Aranda, 2001). Between 15 and 12 Ma, a change in Pacific plate kinematics began to compress the southwestern Gulf of Mexico region. This change may be related to the capture of the subducted Farallon slab by the Pacific plate, which caused the oceanic spreading ridge to jump inland in the Gulf of California (Bohannon and Parsons,

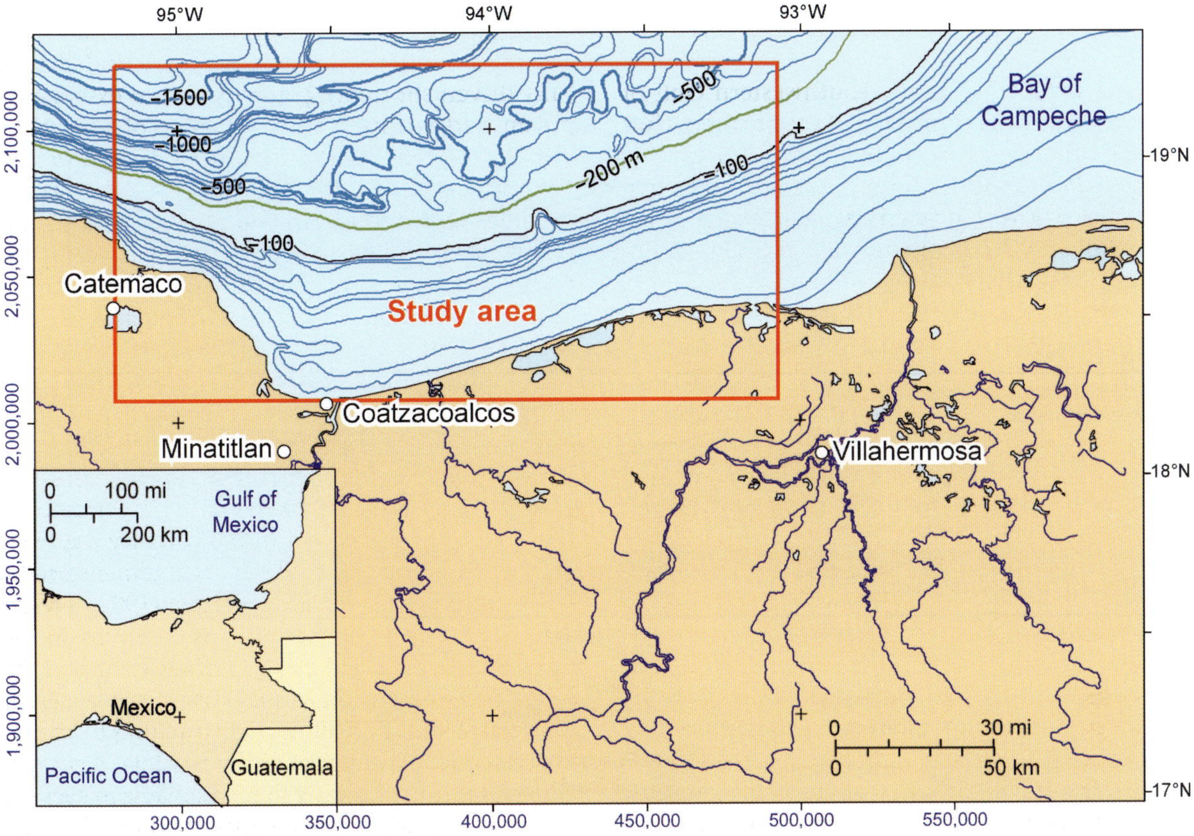

FIGURE 1. Location map of the offshore study area in the southern Gulf of Mexico. The contour interval of isobaths is 10 m (30 ft) from 0 to −100 m (−330 ft) and 100-m (330-ft) interval deeper. Universal Transverse Mercator (UTM) projection.

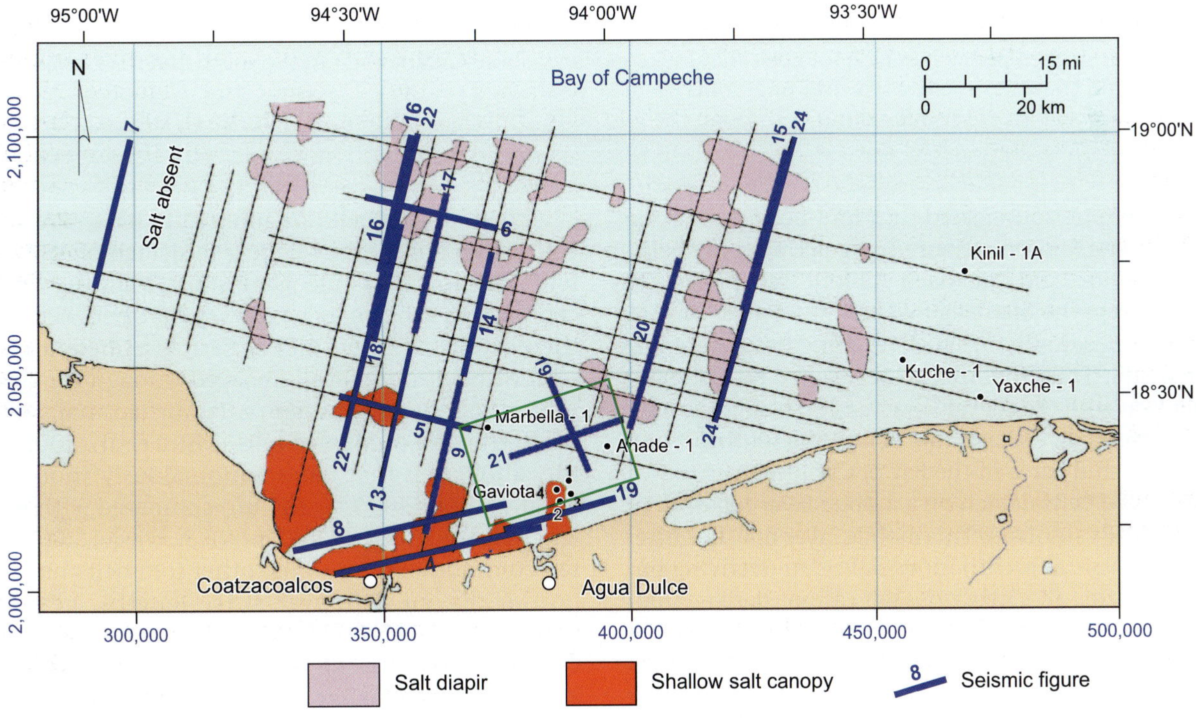

FIGURE 2. Map showing the subsurface database, location of seismic lines in other figures, and the main salt diapirs. Red areas depict shallow salt diapirs; pale-purple areas depict deeper salt diapirs. The green rectangle shows the border of the 3-D seismic survey. Diapiric planforms vary with depth and are only loosely constrained by the 2-D seismic grid and well data. Salt structures are not observed in the west. Universal Transverse Mercator (UTM) projection.

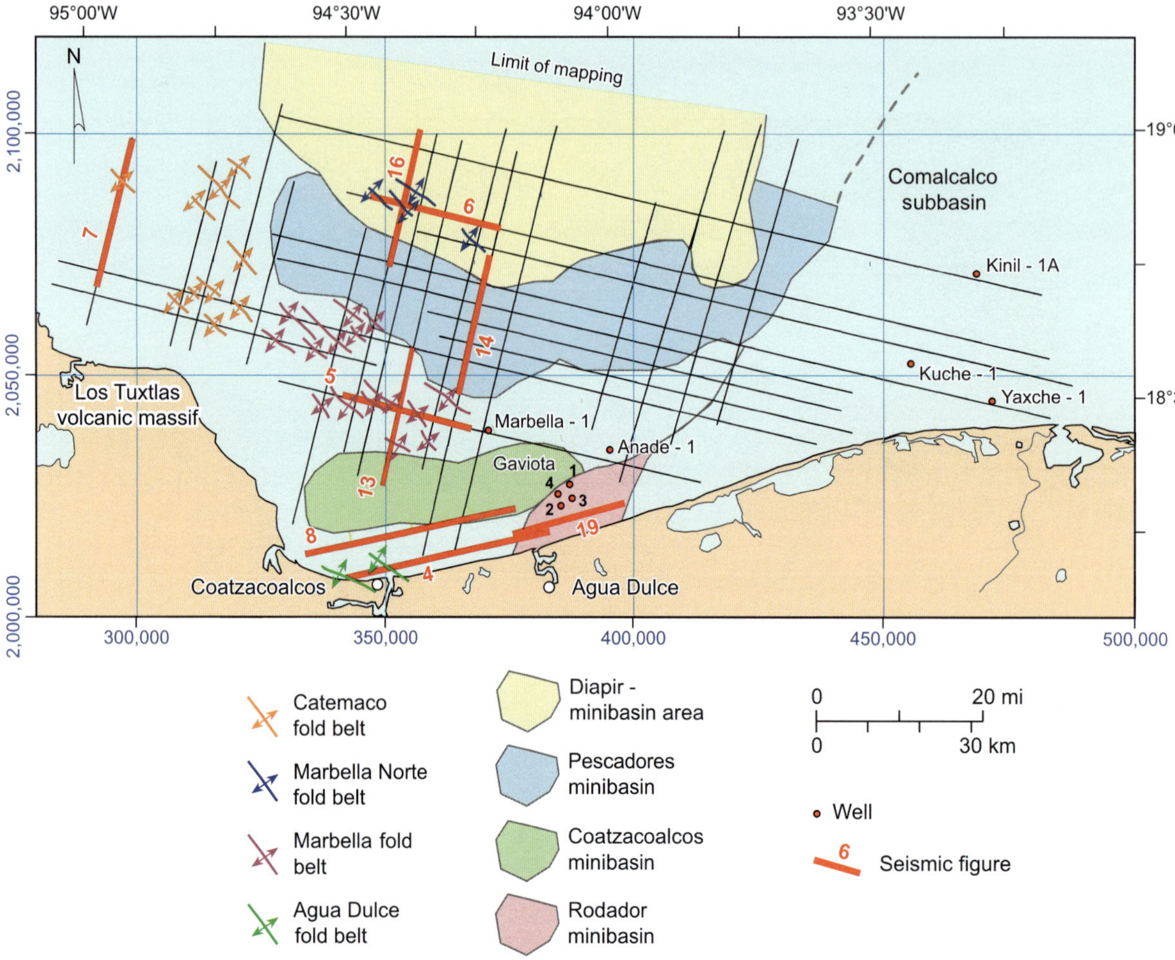

FIGURE 3. Map showing the named fold belts, minibasin provinces, and wells in the study area. These fold belts may all be merely better imaged parts of a single fold belt that includes the onshore Chiapas belt to the south and the Campeche Bay area. Fold trends are schematic and uncertain because of aliasing on the 2-D seismic grid. Numbered red lines represent the location of seismic examples through the fold belts. Universal Transverse Mercator (UTM) projection.

1995). Compression created four fold belts during the middle to late Miocene (Figure 3): Agua Dulce, Marbella, Marbella Norte, and Catemaco. All four were previously named except the Marbella fold belt. Inferences on fold trends are necessarily speculative, being based on a 2-D seismic grid. Thus, fold trends are inferred by assuming parallelism with the coeval Veracruz fold belt to the west and the Sierra de Chiapas fold belt to the southeast and east of the study area in the Bay of Campeche.

The fold belts detach on a mobile layer too deep to be seismically resolved. This mobile layer may be either Middle Jurassic autochthonous salt or an overlying but deep, Mesozoic allochthonous salt canopy derived from the Jurassic source layer. The offshore fold belts are close to coeval, orogenically driven fold belts of onshore southern Mexico and to the convergent Pacific margin. For these reasons, the regional shortening in the offshore Salina del Istmo Basin is interpreted to be orogenically driven from the Pacific plate margin and to involve the presalt basement.

The four fold belts are inferred to have several features in common: they appear to detach on salt, they fold Mesozoic and Paleogene strata, they are of Miocene age, and they may have a similar trend. Because of these similarities, they may be parts of a single large fold belt, which is only locally imaged through the overlying salt canopy and diapirs. Published maps of the southern Gulf of Mexico (e.g., Mitra et al., 2005) portray a fold belt that has an anomalously short strike length (about 380 km, 240 mi) compared with its dip length (about 300 km, 190 mi) (F. A. Diegel, 2007, written communication). Most other fold belts have proportionally much longer strike lengths, hence the term "belts." This anomaly suggests that the Chiapas-Campeche fold belt extends offshore into our study area and has a strike length of at least 530 km (330 mi). However, because of the fragmentary knowledge of these fold belts, we continue to refer to them by separate names, with the caveat that they may all be parts of the same large fold belt.

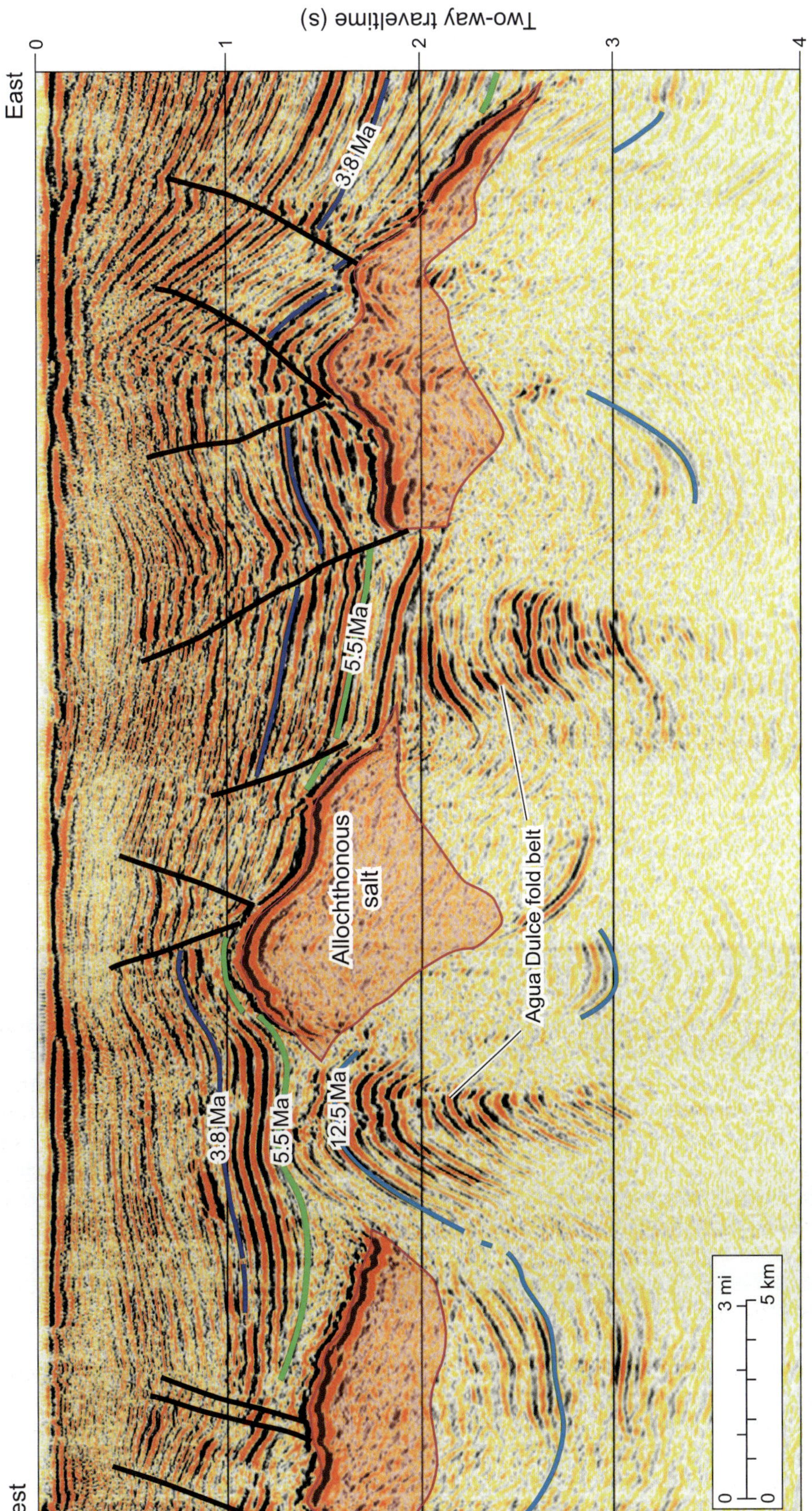

FIGURE 4. Interpreted 3-D inline seismic profile through the Agua Dulce fold belt, below allochthonous salt sheets (transparent pink fill) of the discontinuous Sal Somera canopy. Areas between salt sheets may be salt free or welded parts of the canopy. Figure 3 shows the location of the seismic section.

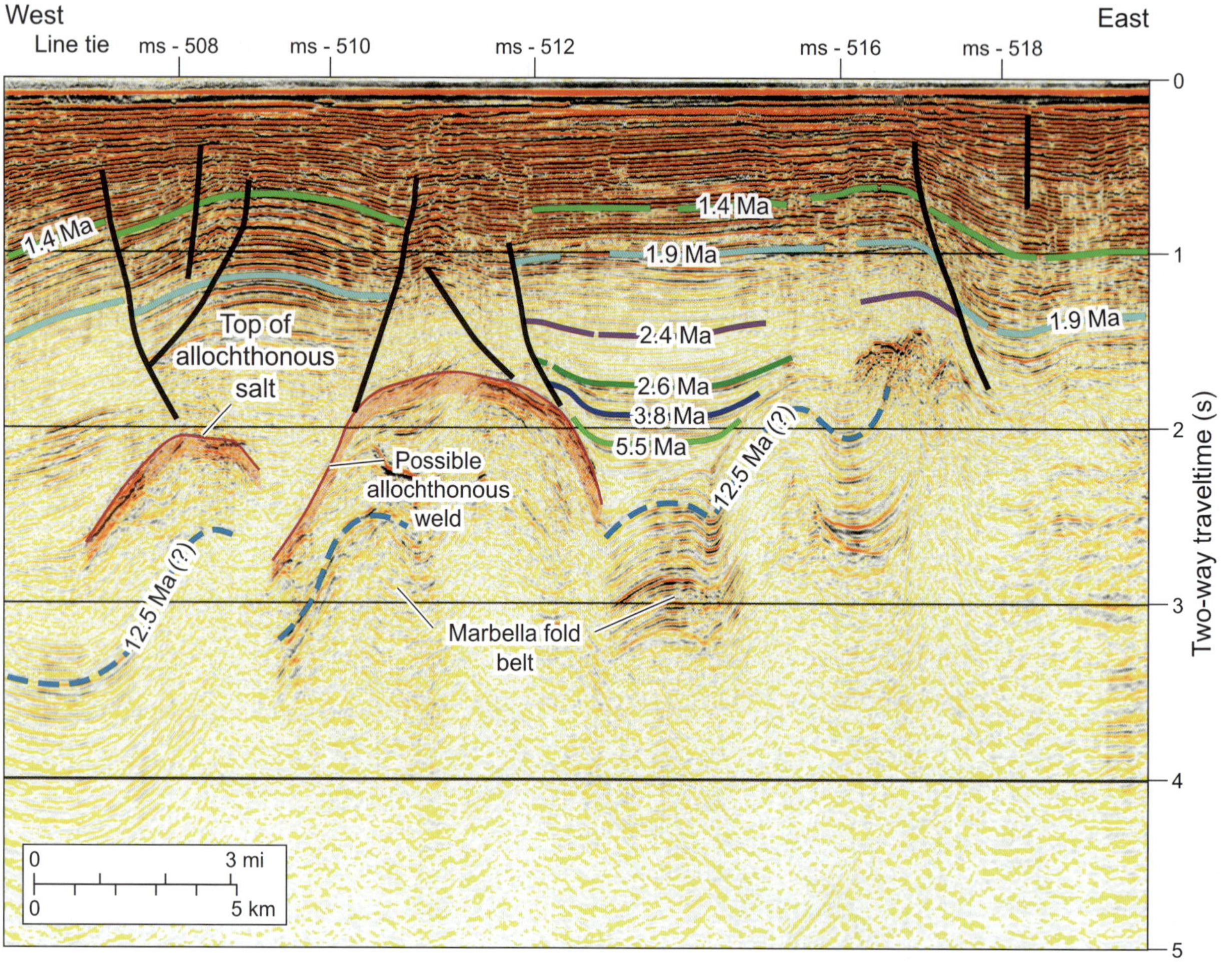

FIGURE 5. Interpreted 2-D seismic profile showing the Marbella fold belt. Figure 3 shows the location of the seismic section.

The Agua Dulce fold belt comprises thrusts in the coastal zone near the cities of Coatzacoalcos and Agua Dulce (Figure 3). Figure 4 is an inline section of a 3-D seismic survey over the fold belt. The high structural relief is chiefly the result of buckle folding. Allochthonous salt sourced from the basin farther east ponded in the synclines and thinly draped over parts of the crest of some anticlines.

The Marbella fold belt underlying the continental shelf between the Pescadores and the Coatzacoalcos minibasins (Figure 3) is named here after the nearby Marbella 1 well. The folds appear tighter than most in the study area (Figure 5).

The Marbella Norte fold belt underlies the upper continental slope (Figures 3, 6). Anticlinal crests in the Marbella Norte fold belt rose above sea level in the late Miocene, perhaps in reaction to shortening driven by the rise of the Chortis continental block farther south. These uplifted crests, which were variably eroded, created islands fringed by reefs (Oviedo-Pérez, 1996).

The Catemaco fold belt is in the west, offshore of the Los Tuxtlas volcanic massif (Figure 3). This fold belt comprises gentle folds (Figure 7). The strongly discordant onlap surface (best imaged on the limb of the most seaward anticline, just above 3 s) against parallel-bedded folded strata is compatible with a sudden onset of buckling after a long period of tectonic stability. The thrusts verge mostly seaward. The main shortening occurred during the late Miocene (Robles et al., 1998), coeval with the other three fold belts. Because Jurassic salt is thought to be absent here, the fold detachment level may be Paleogene shale.

LATE MIOCENE SALT CANOPY

During or just after the final stages of folding during the late Miocene, salt extruded over the eroded and uplifted fold belts. Estimates for when the Sal Somera canopy was emplaced vary: Paleogene (Oviedo et al., 1996; Oviedo-Pérez, 1996), Paleogene and Miocene (Oviedo et al., 1995), and base Miocene (García-Molina, 1994) to Miocene (Mora et al., 2001). Oviedo-Pérez's (1996) restoration of a mostly onshore cross section,

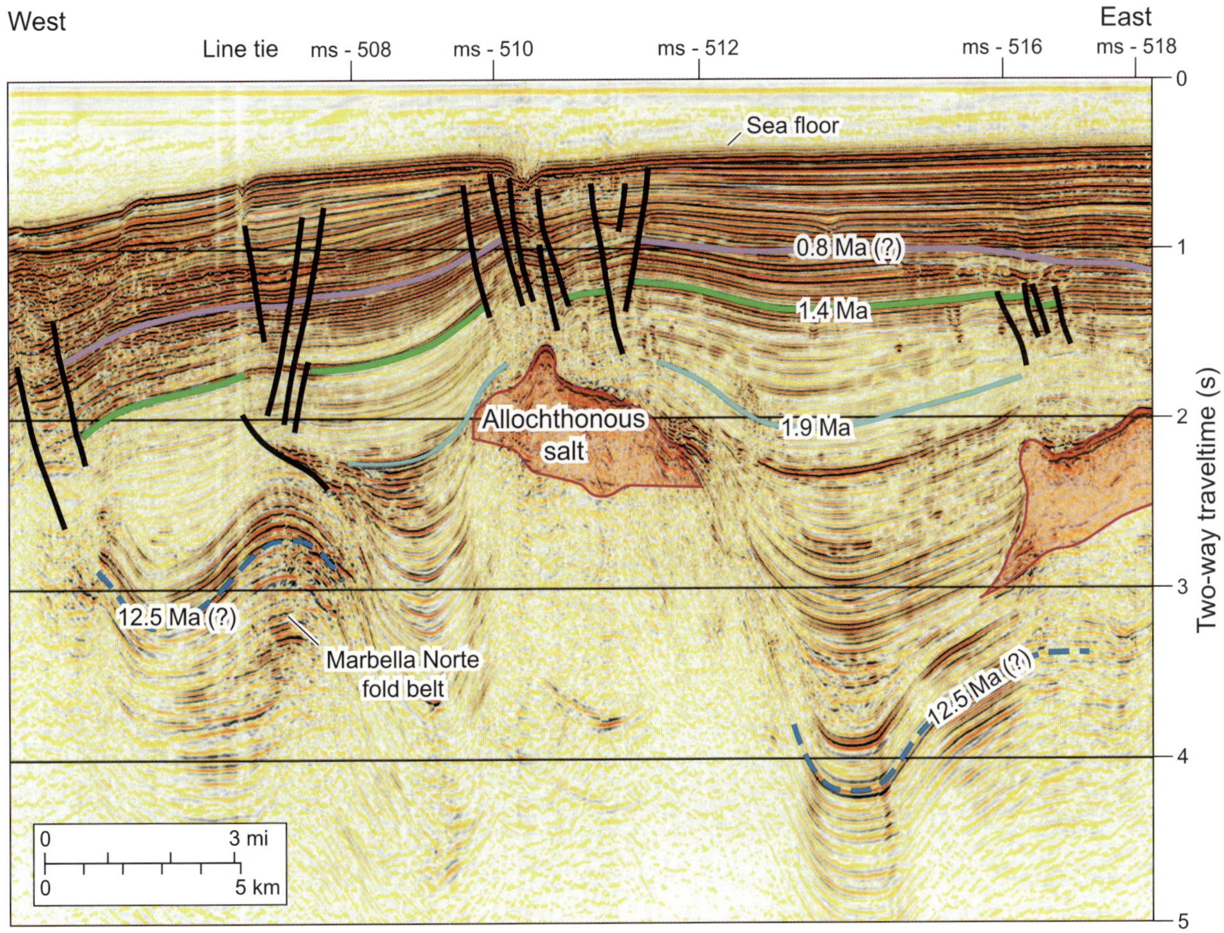

FIGURE 6. Interpreted 2-D seismic profile showing the Marbella Norte fold belt below allochthonous salt sheets (transparent pink fill). Figure 3 shows the location of the seismic section.

which extends into the center of our study area, suggests that much of the canopy was emplaced by 12 Ma and that it continued to extrude to reach maximum extent as recently as 3 Ma. For our study area, the wide age range from Oligocene to Pliocene can be narrowed because the canopy overlies an eroded fold belt, yet the top of the salt is mostly unfolded. Moreover, the salt canopy or equivalent weld is roughly conformable with the underlying fold belts. Thus, the salt canopy must have extruded during the late stages of folding or just after folding. The age of this fold belt is either middle Miocene (Oviedo et al., 1995; Soto-Cuervo et al., 2004) or late Miocene (García-Molina, 1994). Available well data are compatible with a middle to late Miocene age for canopy emplacement. The start of extrusion is poorly constrained because of uncertain stratigraphic ages just below the canopy. In most profiles, we have identified a possible 12.5-Ma horizon close to the early stage of canopy spreading. The canopy was mostly in place by 5.5. Ma (Soto-Cuervo et al., 2004). The canopy probably extruded diachronously, so different authors may have different but correct interpretations for their particular areas. For example, canopy emplacement may have become younger seaward for two reasons. First, progradation of the margin would have shifted the sedimentary load seaward. Second, the folding may have propagated seaward away from the causative orogenic shortening in the hinterland.

Similar ages between the late stages of folding and the canopy emplacement evoke a dynamic tectonic landscape. If salt extrusion overran and sealed the top of the fold belt, lateral salt flow into the synclines during the final stages of folding would reduce the surface relief, although the structural relief of the fold belt continued to grow. However, if salt sheets were emplaced soon after folding, the full structural relief of the fold belt would be expressed geomorphically in mountainous sea-floor anticlines, similar to those in the inner Perdido fold belt in the northern Gulf of Mexico. Such anticlines would be prone to erosion and slumping, and some evidence for this in the study area is seen (Figure 4, center).

Seismic data have not resolved the form or location of diapirs feeding the extrusions. However, by analogy to other areas (Jackson et al., 1990; Heaton et al., 1995; Nilsen et al., 1995; Vendeville and Nilsen, 1995), salt could have been squeezed upward out of the diapirs by

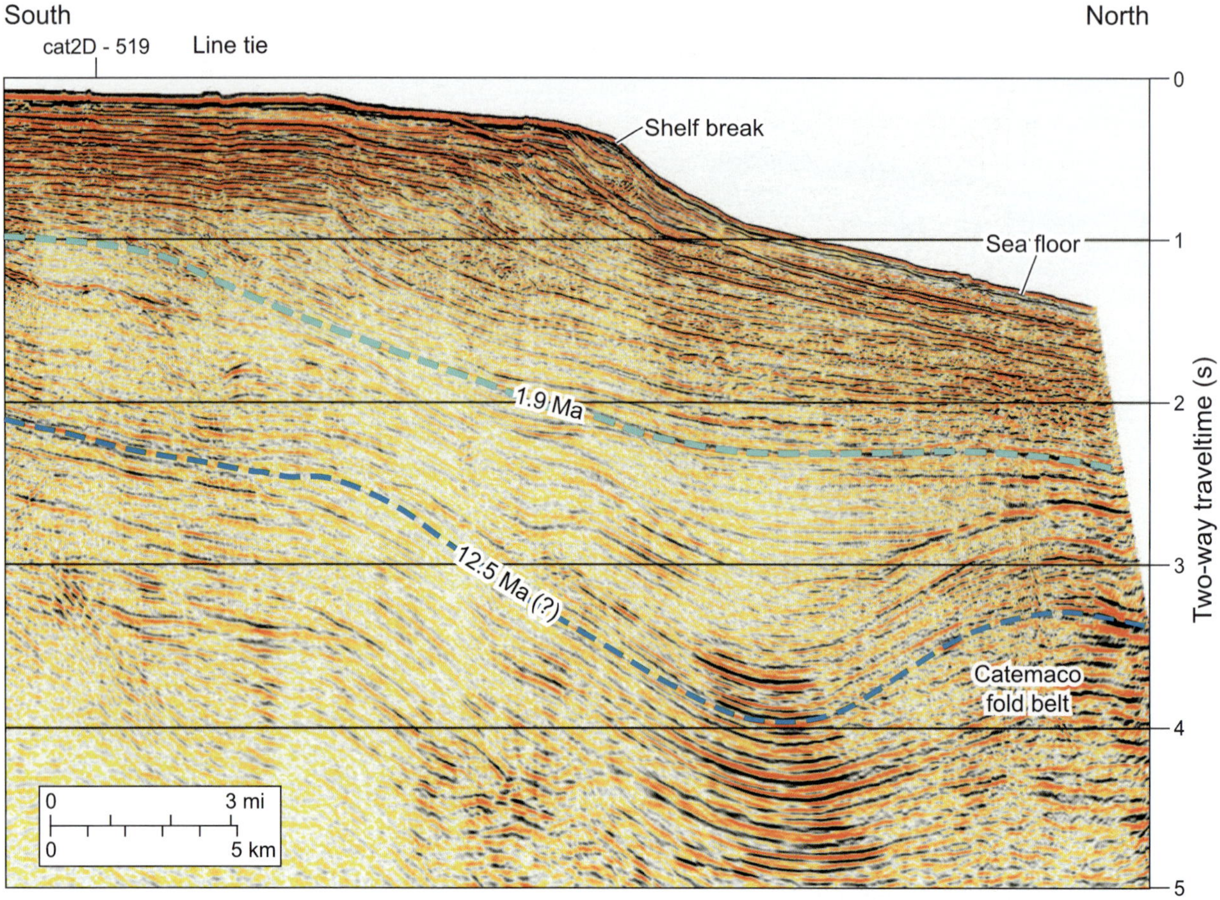

FIGURE 7. Interpreted 2-D seismic profile showing the Catemaco fold belt. Figure 3 shows the location of the seismic section.

the same regional shortening that formed the fold belts. In addition to the effects of lateral compression in expelling salt, a shortening-related uplift would have decreased the accommodation for sediments that would otherwise have buried the extruding salt glaciers. Both compression and uplift would therefore have encouraged salt extrusion. As the Mesozoic and Paleogene sections buckled, any preexisting salt stocks and walls would have been laterally compressed and would have shortened much more readily than the stronger encasing sediments. Diapir squeezing and the resulting salt extrusion enabled the further shortening of the fold belt by transferring diapiric salt to the surface.

Copious extrusions of salt coalesced at or near the sea floor to form one or more salt canopies. The allochthonous salt may have been partly protected from dissolution by a veneer of pelagic or hemipelagic sediments. Some salt sheets overlie anticlines (Figure 6) as might be expected if the diapirs feeding them predated the folding and acted as weak, elevated zones prone to rising during shortening. Elsewhere, salt sheets overlie synclines (Figure 4). These salt extrusions may have ponded in the low synclines although their feeders were actually in anticlines.

The Sal Somera salt canopy is historically significant as an early depiction of coalesced diapirs. Salt canopies were first named, modeled, and described in the Great Kavir of central Iran (Jackson and Cornelius, 1985; Jackson and Talbot, 1989; Jackson et al., 1990). However, even earlier, Correa-Perez and Gutierrez-y-Acosta (1983) inferred from gravity data that two salt sheets could have coalesced to form a laterally extensive salt sheet in the Salina del Istmo Basin. They speculated that this salt body with two stems could be more widely continuous in the offshore part of the Salina del Istmo Basin (the present study area). This speculation was confirmed when the Sal Somera canopy was recognized by seismic interpretation (Oviedo et al., 1996; Oviedo-Pérez, 1996).

The Sal Somera canopy, which extends throughout most of the study area, may comprise several smaller canopies. The top of the Sal Somera canopy is seismically identified by a high-amplitude reflection. The base of salt is unclear, and coherent subsalt reflections are sporadic. The salt may have extruded glacially over the sea floor (Fletcher et al., 1995) or it may have extruded by thrusting, along with its protective sedimentary roof (Jackson and Hudec, 2004). How the salt sheet was

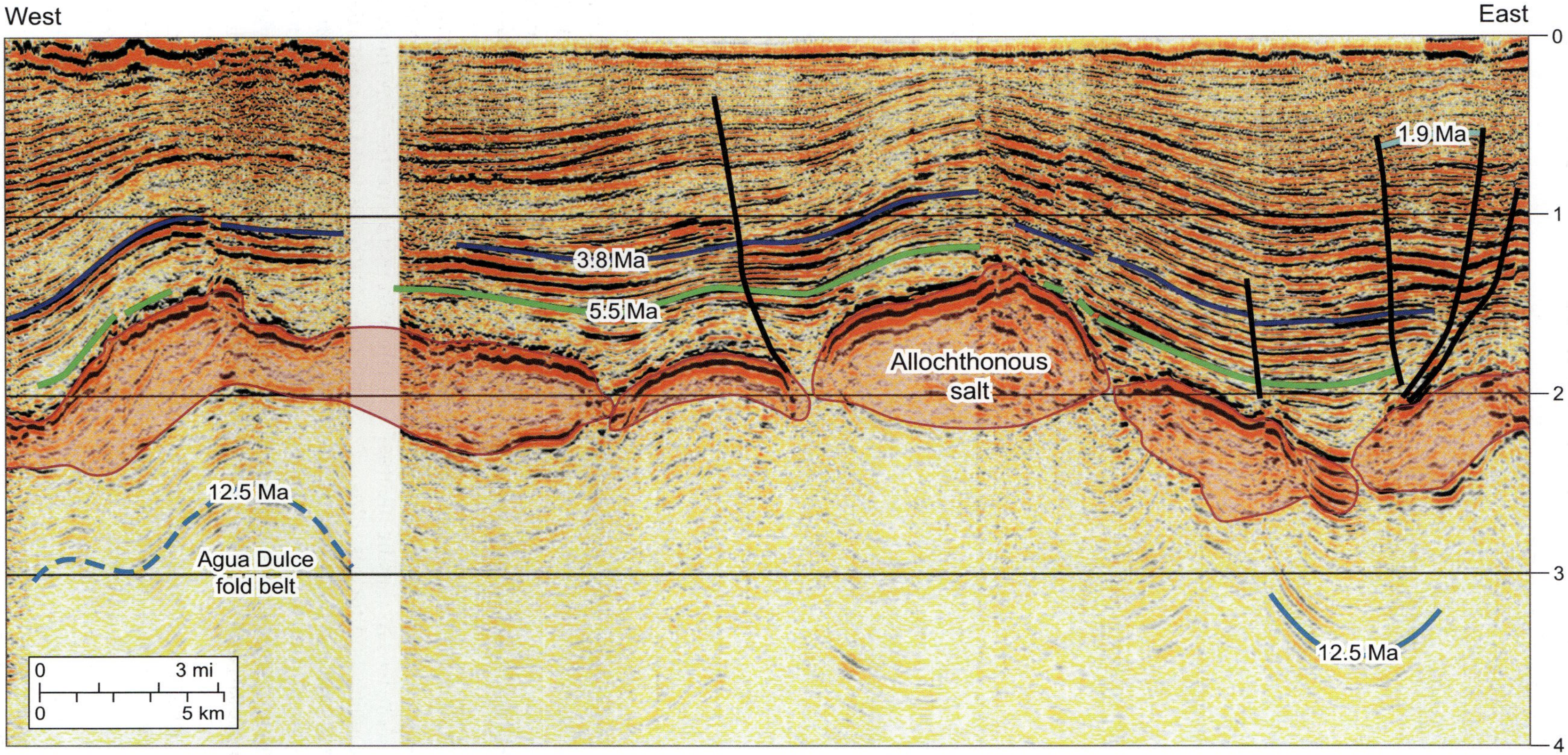

FIGURE 8. Interpreted 3-D inline seismic profile through the Sal Somera canopy (transparent pink fill) and Agua Dulce fold belt. Figure 3 shows the location of the seismic section.

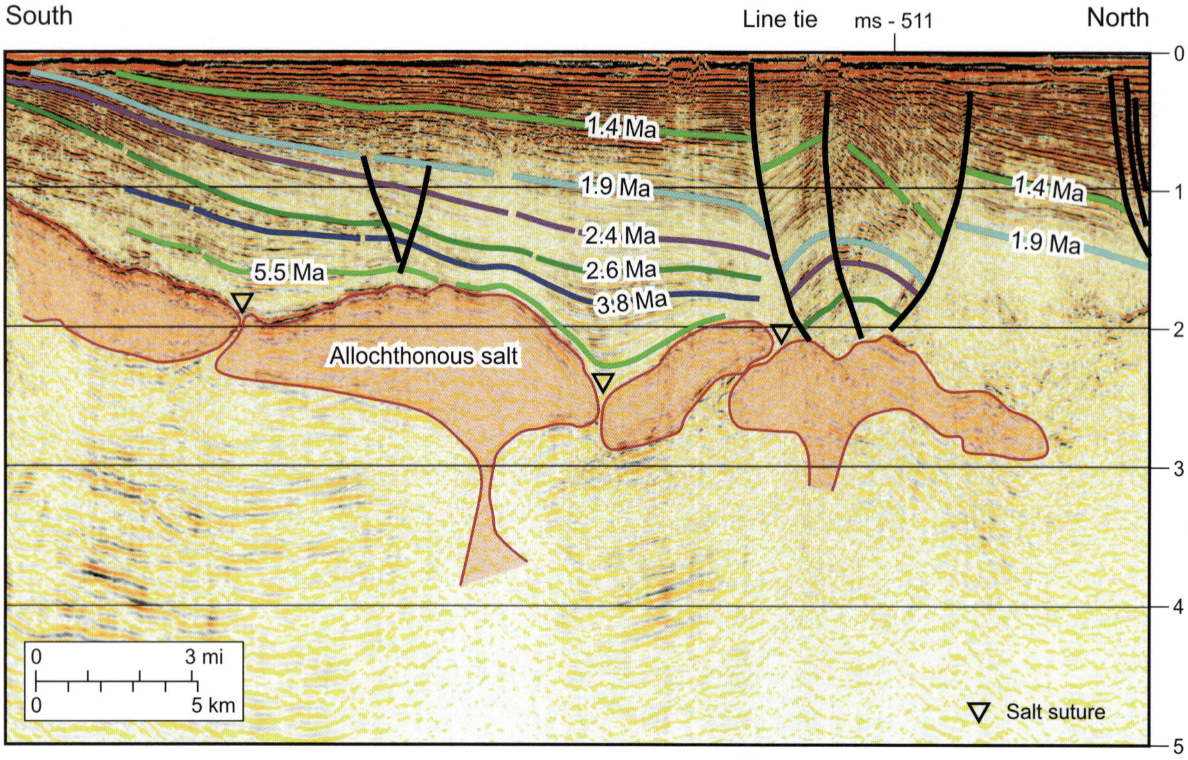

Figure 9. Interpreted 2-D seismic profile showing the Sal Somera canopy (transparent pink fill). Figure 2 shows the location of the seismic section.

emplaced is uncertain because critical evidence in the base-salt geometry is not seismically imaged.

Salt thickness was strongly controlled by the fold-belt paleotopography built by compressional uplift and degraded by erosion. Allochthonous salt is up to 700 ms thick in synclines and much thinner over anticlines. In rare places where the base of the canopy is imaged well (Figure 8), the base in places conforms to the underlying folds. Thus, either the salt extruded postorogenically over folded strata before they had eroded or the salt was emplaced and folded during the final stages of regional shortening.

Figure 9 shows part of the Sal Somera canopy. Some of the salt sheets may have their feeder stems out of the plane of the section, and some feeders may be pinched off. Salt sutures, where the allochthonous salt coalesced, form sharp-keeled synclines between the stocks. In the crestal graben indenting the northernmost salt stock, the tilted, parallel reflectors dip uniformly except near the surface. The persistence of parallel reflectors suggests that this local extension of the shelf was rapid and late. The possibility of transtension, forming a negative flower structure, cannot be eliminated given the spacing of the seismic lines. Seaward dip increased here in the Pleistocene, but farther seaward, the canopy dips landward. The salt canopy also dips gently eastward on a regional scale. In many places, the canopy has been thinned or welded (Figure 10), especially below counter-regional structures.

The tectonic setting of allochthonous salt emplacement during the late Miocene in the offshore part of the Salina del Istmo Basin has some similarities to the foreland fold and thrust belt of the Zagros Mountains in southern Iran (García-Molina, 1994). Collision of the Arabian and Iranian microplates began forming the Zagros fold belt in the Oligocene, culminated in the Miocene and Pliocene, and continues today. The main detachment level is the infra-Cambrian Hormuz salt, but shallower decollements contribute (Letouzey and Sherkati, 2004; Sepehr and Cosgrove, 2004). Of the more than 160 salt diapirs exposed in the Zagros fold belt, most have extruded evaporites, either as fringes or salt glaciers (Figure 11). All four visible salt extrusions originate from diapirs in the Zagros anticlines, although the two main glaciers have flowed into intervening lows across synorogenic Pliocene–Pleistocene deposits in the synclines. If much more allochthonous salt extruded, it could coalesce into canopies overlying the fold belt, similar to the situation in the Salina del Istmo Basin during the late Miocene.

PLIOCENE–PLEISTOCENE SUPRASALT MINIBASINS

The surface of the Sal Somera salt canopy (Figure 12) mostly reflects an interplay of two processes. The first

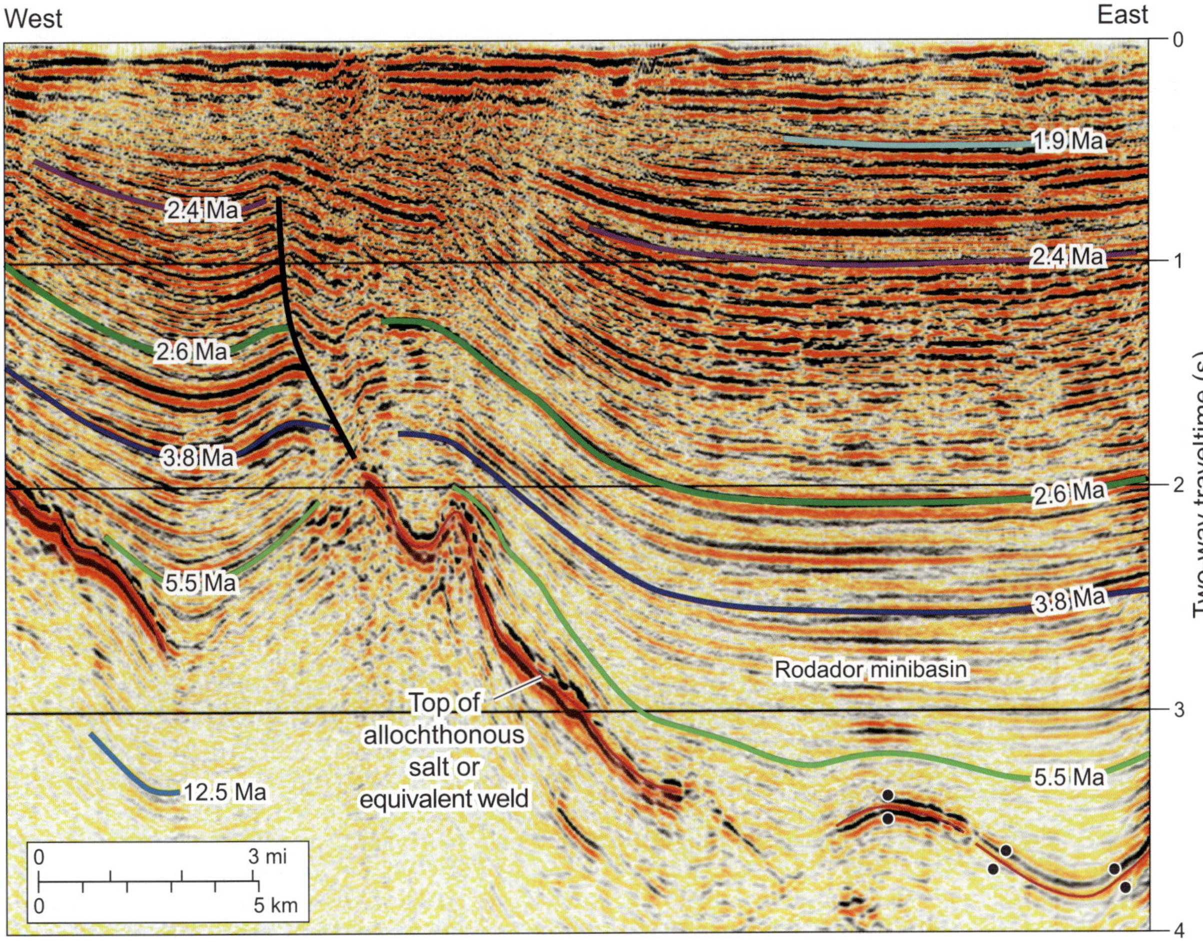

FIGURE 10. Strike-oriented, interpreted seismic profile showing the west part of the Rodador minibasin, underlain by the welded Sal Somera salt canopy (paired dots). Figure 3 shows the location of the seismic section.

was folding: areas having the shallowest salt or salt weld tend to overlie Miocene anticlines because the salt canopy or equivalent weld is roughly conformable with the underlying fold belts. Either the base of the salt was folded along with the subsalt strata or the salt infilled a previously folded sea floor. The second process was minibasin formation: areas originally having the thickest salt canopy allowed the greatest salt expulsion by overlying Pliocene–Pleistocene minibasins. This expulsion created the deepest depressions in the top of salt. The salt ends abruptly in the west (Figure 12). From south to north, three main provinces of minibasins formed by progressive seaward evacuation of the underlying salt canopy: (1) Coatzacoalcos-Rodador, (2) Pescadores, and (3) the diapir-minibasin area (Figure 3).

Coatzacoalcos-Rodador Minibasins

Below this region, the canopy is shallow (Figure 12). The Coatzacoalcos minibasin lies between the Agua Dulce fold belt to the south and the Marbella fold belt to the north from which it is separated by a counterregional structure (Figure 13). The Rodador minibasin lies on the fringe of the study area, but a small part of it appears on Figure 10.

Pescadores Minibasin

The Pescadores minibasin is a large minibasin at least 100 km (62 mi) long and up to 20 km (12 mi) wide. The Pescadores minibasin forms a prominent east-southeast–trending depression roughly parallel to the present coastline (Figures 3, 12). Major counterregional structures (diapirs, welds, and faults) bound this minibasin on its seaward flanks, marked by tight contours (Figure 12), forming the northern margin of this minibasin. The Pescadores minibasin is the westward extension of the Comalcalco minibasin (Figure 3), which is mostly outside the study area.

The Pescadores minibasin began to form during the early Pliocene when a clastic wedge prograded northward across the thick salt canopy, loading the salt with a thick turbidite section. The thickest parts of the salt canopy, which had ponded in synclines of the fold belts,

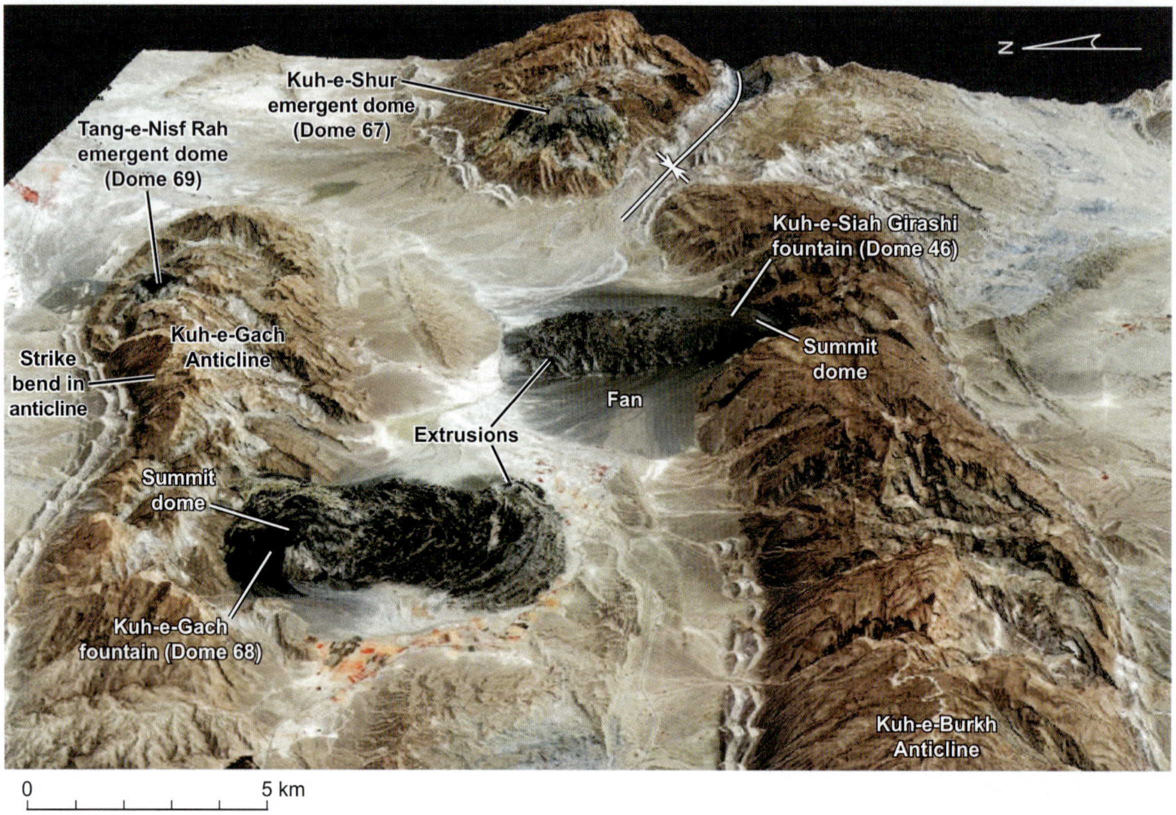

FIGURE 11. Possible analog in the central Zagros Mountains (Iran) for an early emplacement of extrusive salt in the Salina del Istmo Basin. Three salt diapirs comprising impure Hormuz salt emerge from the crests of three anticlines. Two diapirs (Kuh-e-Gach and Kuh-e-Siah Girashi) have extruded salt glaciers, choked with nonevaporitic residual breccia. This oblique view with a 2× vertical exaggeration was created by draping a band 3-2-1 (RGB) image over an ASTER-derived digital elevation model acquired on August 10, 2001. Image from National Aeronautics and Space Administration/Goddard Space Flight Center/Japan's Ministry of Economy, Trade and Industry/Earth Remote Sensing Data Analysis Center/JAROS and U.S./Japan Advanced Spaceborne Thermal Emission and Reflection Radiometer Science Team.

provided the greatest space for the Neogene overburden. Thus, these lows form the thickest parts of both the Coatzacoalcos and Pescadores minibasins (Figures 13, 14). The floor of the minibasin is mostly a salt weld created by the expulsion of the Sal Somera canopy under sediment loading. Figure 14 shows the deepest part of the Pescadores minibasin, from which most of the underlying salt was expelled.

Diapir-Minibasin Area

The diapir-minibasin area below the upper continental slope is just north of the large counterregional structure bounding the Pescadores minibasin (Figures 1, 3). The diapir-minibasin area has massive salt diapirs (mostly passive stocks, shown by rounded shallow contours), which also dominate other parts of the Salina del Istmo Basin in the Neogene (Figure 2). The bloblike shapes of salt diapirs in Figure 2 are interpolated between the 2-D seismic lines so are only approximate. Diapir crests average between 1 and 5 km (0.6 and 3 mi) in width. In general, these diapiric walls and stocks have steep sides surrounded by minibasins filled by deep-water sediments. Figure 15 shows three salt diapirs below the upper slope. Diapir A is triangular with well-defined flanks and crest. Only the crests of diapirs B and C are clear, but they are interpreted to have a narrow, roughly cylindrical stem. Converging, upswept reflections below the crest of diapir B suggest the possibility of a narrow feeder stem, probably detached from its source or a feeder off section. This geometry is compatible with diapiric pinch-off during regional shortening after a long history of passive rise. However, the proximity of these diapirs to extensional faults north of diapir A and their position on the upper slope suggest that any shortening must have occurred earlier when diapirs B and C were in deeper water. The depth of the base of these diapirs is unknown because reflections below the salt crest are diffuse, but the base could be at an allochthonous upper Miocene level between 4.0 and 4.5 s in the seismic section. A counterregional fault in the Pescadores minibasin dominates the south end of the seismic section.

Minibasins thicker than 2500 m (8200 ft) have compacted enough that they sink into the buoyant salt. This

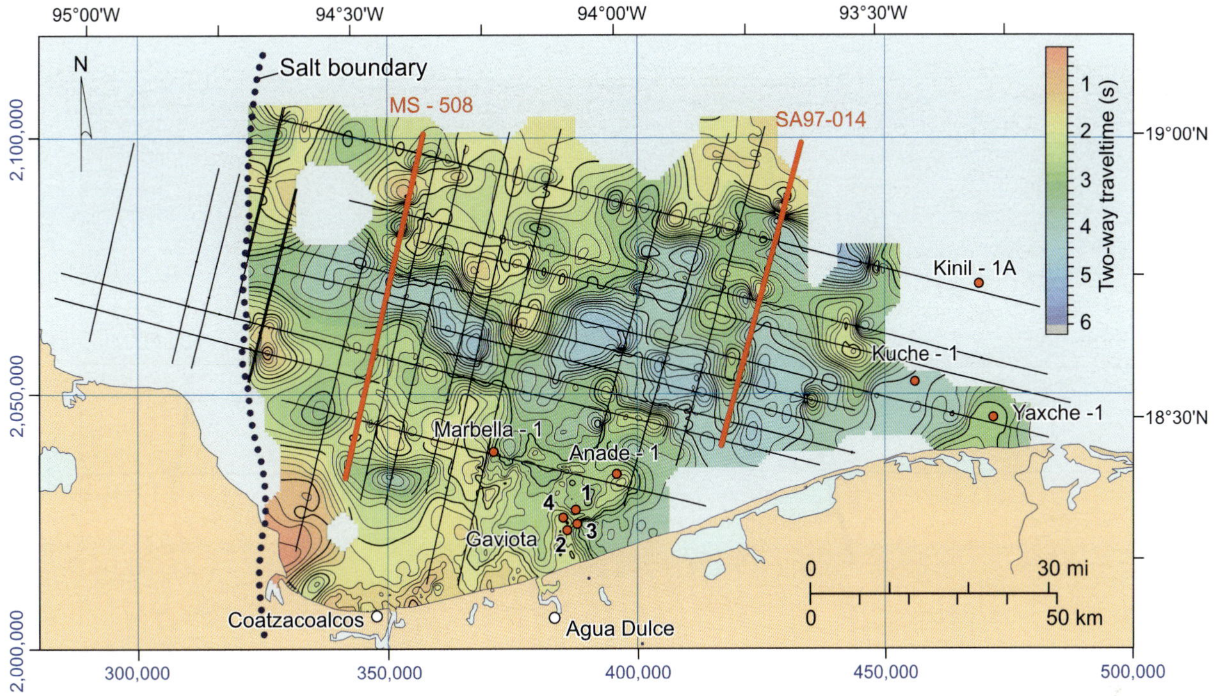

FIGURE 12. Structural contour map in two-way traveltime of the top of the salt or equivalent salt weld in the offshore Salina del Istmo Basin. The contour interval is 0.2 s. The map also shows the location of wells and the two restored cross sections (red lines).

density inversion can drive minibasin subsidence, which displaces underlying salt into the adjoining salt diapirs and accommodates additional sediments, further compacting the minibasin sediments, a process of positive feedback. In contrast, because minibasins containing less than 2500 m (<8200 ft) of siliciclastic sediments are overall less dense than salt, they float instead of sink to form a minibasin, a process of negative feedback that retards minibasin filling. Thus, how minibasins first form under thin sediment cover is problematic. Seismic evidence (Hudec et al., 2005) indicates a compressional origin for some minibasins in the northern Gulf of Mexico (as suggested by Rowan, 2002, and numerically modeled by Ings et al., 2004; Hudec et al., 2005). Shortening creates synclinal sags in which the basal units of the minibasins were thrusted. This process may be appropriate for initiating minibasins whose bulk density is less than that of salt. In contrast to shortening, regional extension drives the rise of reactive diapirs but is unfavorable for minibasins because the diapirs occupy structural lows, not highs.

Some minibasins are floored by the allochthonous salt canopy or equivalent salt weld. Other minibasins rest directly on synclines of the Miocene fold belts, with no intervening allochthonous weld visible (Figure 6). Still other minibasins may be more deeply rooted: for example, the northern minibasin in Figure 16 has the classic U-shaped profile of a deeply rooted primary minibasin, although poor seismic quality hides its base.

FAULT SYSTEMS

Salt-related fault families in the study area are keystone fault families, roller fault families, counterregional fault families, and rollover fault families, all of which are present in the northern Gulf of Mexico (Diegel et al., 1995; Schuster, 1995; Fox, 1998; Rowan et al., 1999).

The most spectacular fault systems in the study area are counterregional. Figure 17 shows a large counterregional structure along the seaward boundary of the Pescadores minibasin. The deepest part of the Pescadores minibasin near 4 s appears to be a salt weld. A modern submarine canyon is flanked by the slumped scarp of the counterregional fault. Counterregional systems form above and ahead of where salt is expelled upward and seaward from an allochthonous salt sheet. This salt expulsion accommodates seaward-expanding sediment wedges. The lower part of the counterregional structure is a salt weld, and its upper part is a true normal fault. The origin of landward-dipping faults in such settings is still not fully understood, although most workers interpret a dominance of salt expulsion in their formation after Schuster (1995) presented type examples in the northern Gulf of Mexico.

The keystone fault family comprises symmetric faults resulting from outer-arc bending above arched salt bodies. These faults form in the hinge zone of anticlines or monoclines where strata are locally stretched by high curvature (Rowan et al., 1999). The faults die out downward

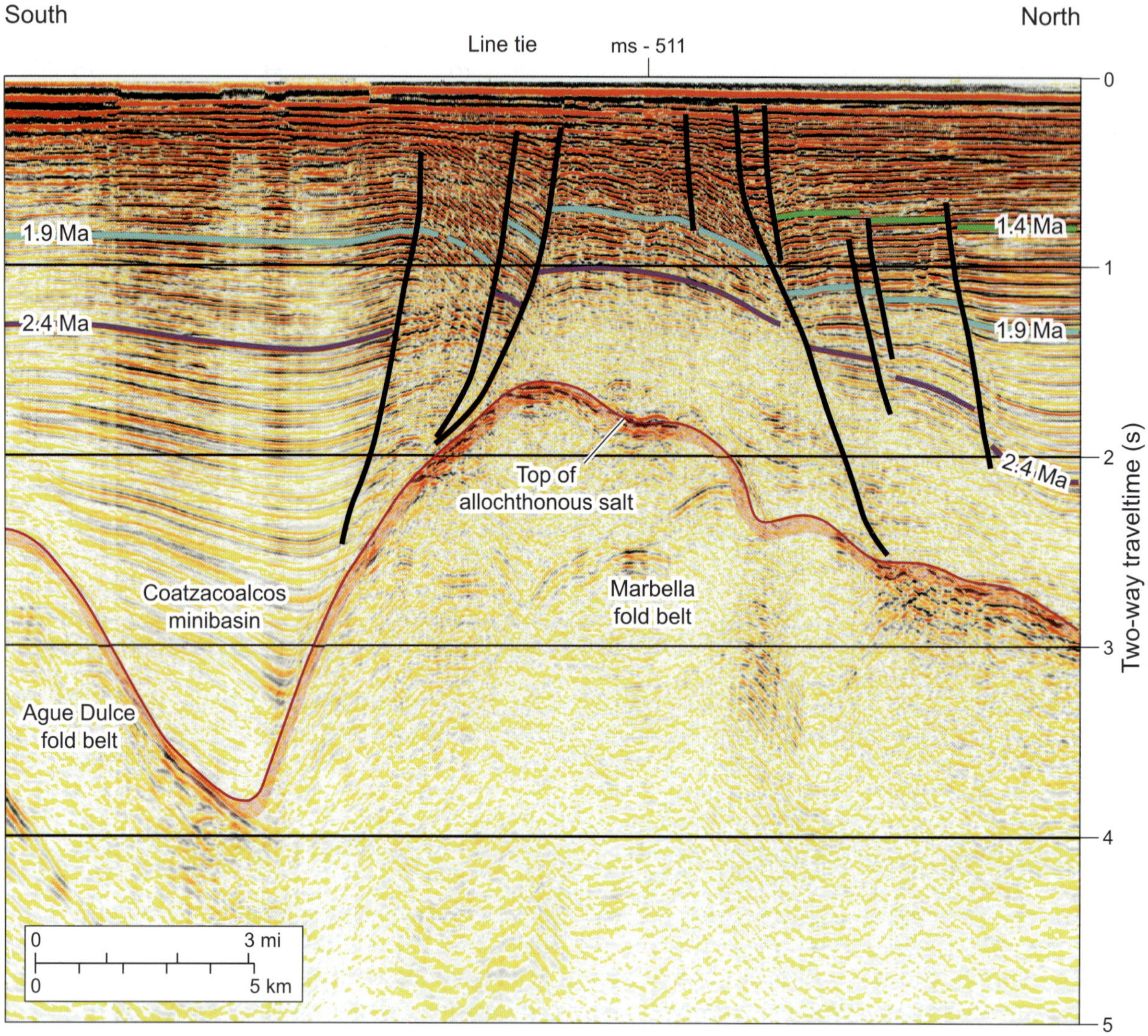

FIGURE 13. Interpreted seismic profile showing the Coatzacoalcos minibasin between the Agua Dulce and the Marbella fold belts. Figure 3 shows the location of the seismic section.

as extensional strain diminishes toward the core of the anticline. Major salt expulsion by sediment loading creates the stratal bending required to form the keystone fault family (Figures 18, 19). Fault strike cannot be confirmed by the 2-D seismic grid.

The roller fault family was defined by Rowan et al. (1999, p. 1461) as "dominantly basinward-dipping, listric growth faults that sole into a salt layer or its equivalent weld." In the study area, such faults are present only locally (e.g., Figure 13) and are much smaller than the landward-dipping faults.

Rollover faults form by bending or stretching when salt is evacuated (Figure 20). These planar faults dip basinward. Asymmetric rollover and symmetric keystone faults form in the hinge of monoclines where strata are most sharply curved. Keystone faults form preferentially over anticlines where flexure is bidirectional and rollover faults form over monoclines with a single direction of flexure (Rowan et al., 1999) as shown in Figure 20.

Ramp faults typically arc along the margin of the salt diapir. A ramp fault (sensu Rowan et al., 1999) is a slightly listric growth fault that is continuous with the basinward-dipping salt ramp along the updip base of the salt stock or the equivalent salt weld. Figure 21 shows a bird-shaped allochthonous salt body supporting a roller fault (soling into the top of salt) and a longer lived ramp fault (on trend with the base of salt). Both faults facilitated salt evacuation, mostly in the late Pliocene.

STRUCTURAL RESTORATION OF CROSS SECTIONS

Because many salt basins have fairly complete records of synkinematic deposition, sequential restoration is a powerful tool to show their structural evolution. Two regional cross sections illustrate the structural evolution. We depth converted them by vertical raypaths

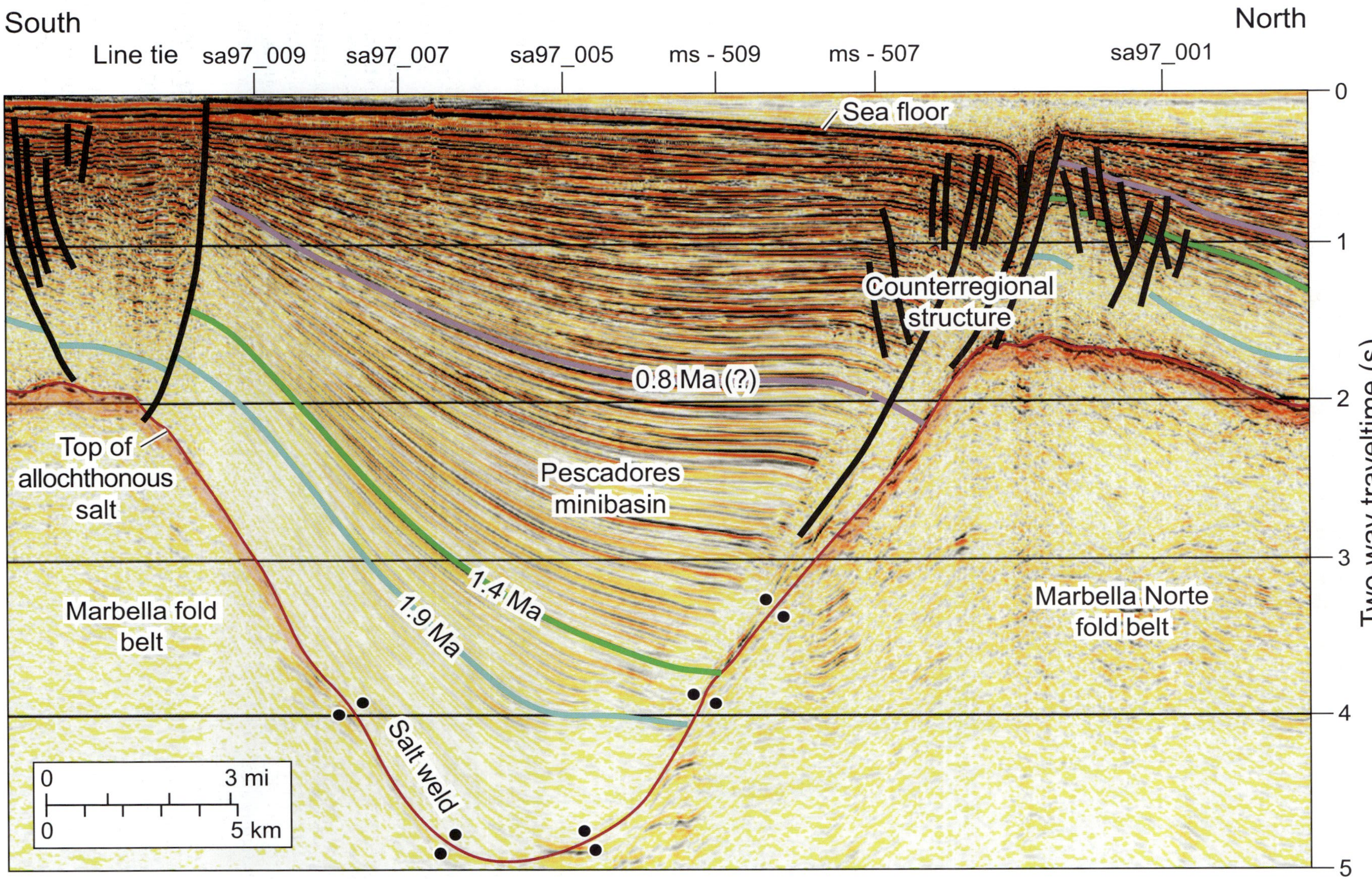

FIGURE 14. Interpreted 2-D seismic section showing the deepest part of the Pescadores minibasin between the Marbella and Marbella Norte fold belts. Figure 3 shows the location of the seismic section.

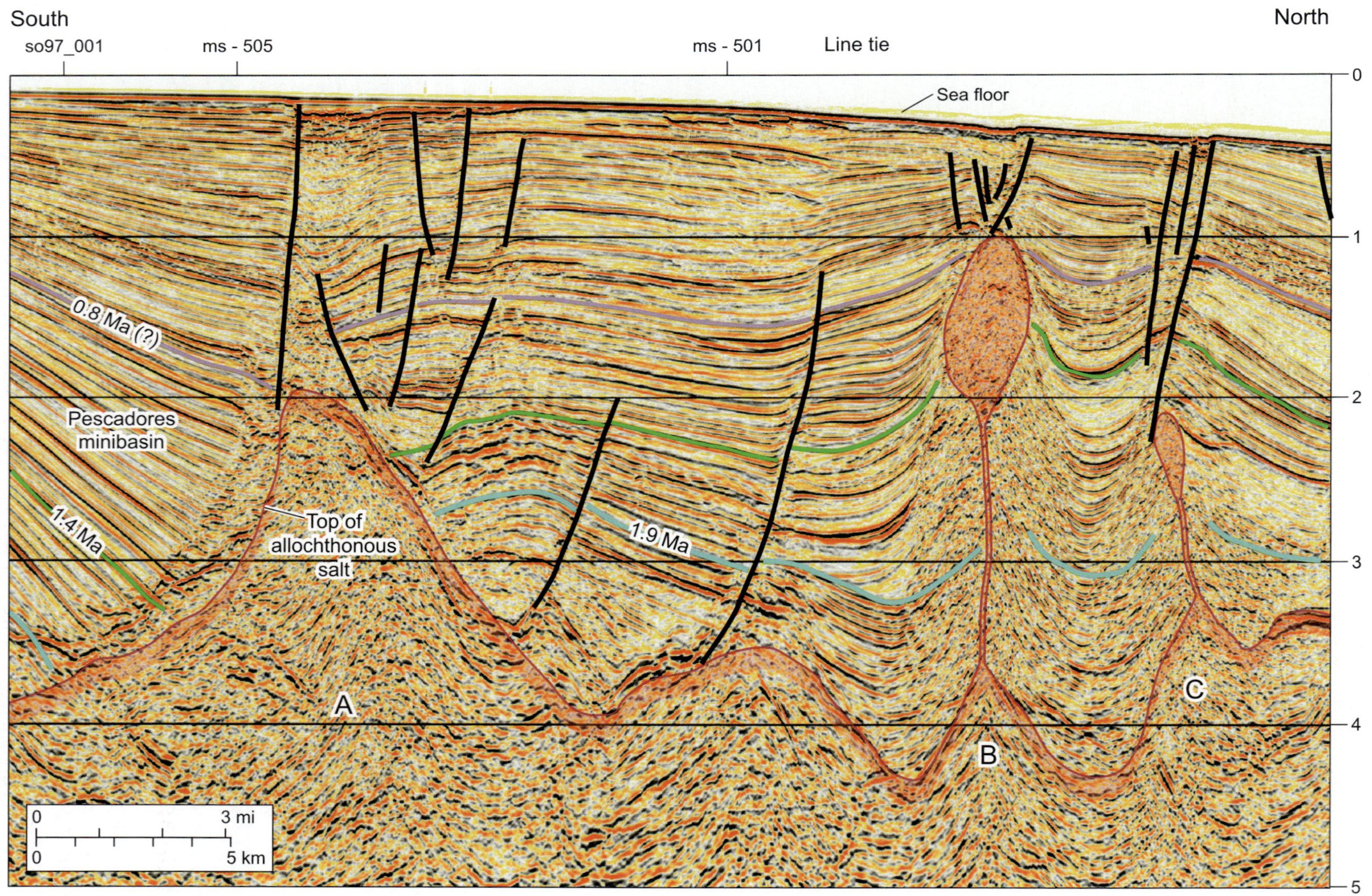

Figure 15. Interpreted seismic profile showing three salt diapirs in the Pescadores minibasin. Figure 2 shows the location of the seismic section.

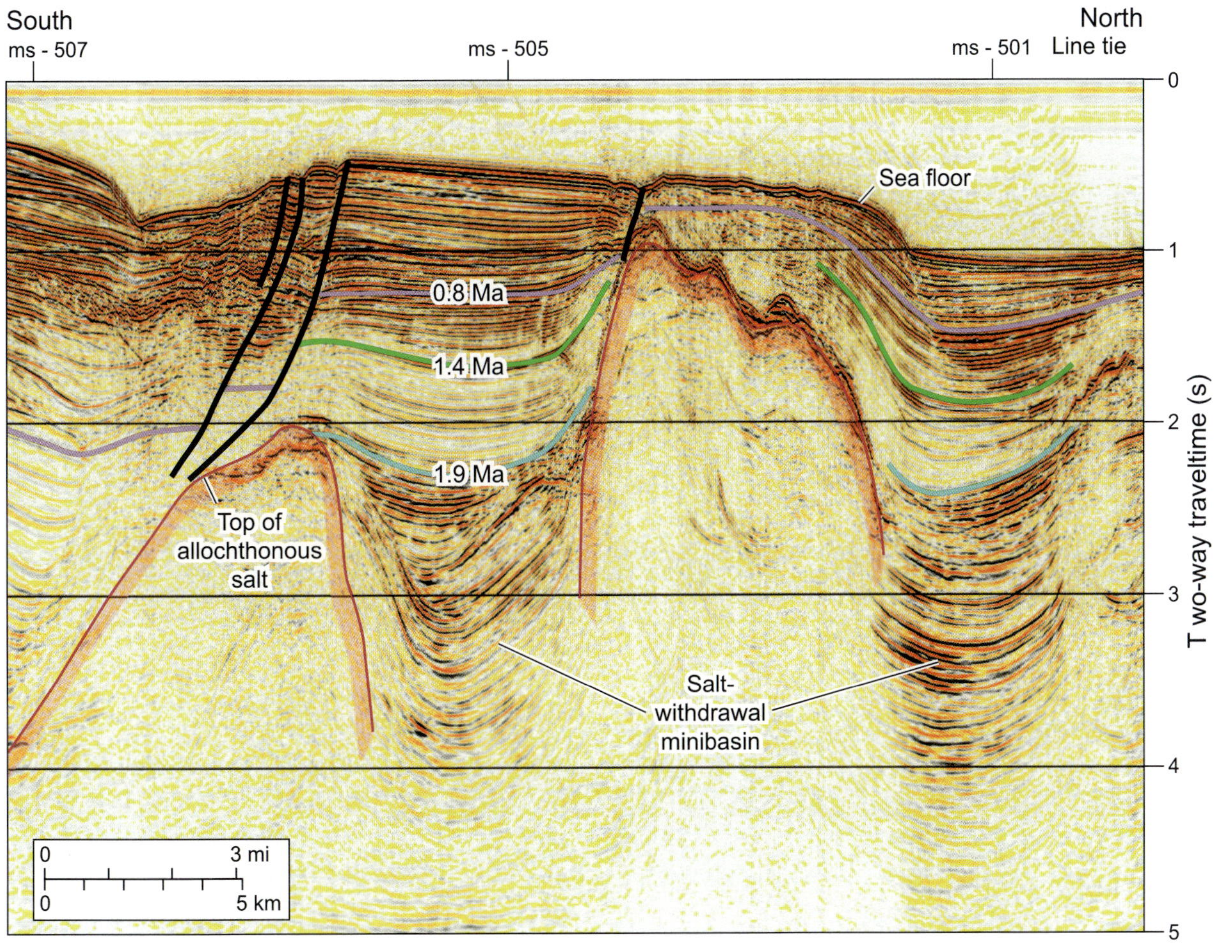

FIGURE 16. Interpreted 2-D seismic profile showing deeply rooted minibasins and a counterregional structure in the upper slope of the Salina del Istmo Basin. The magenta line depicts the interpreted top of allochthonous salt or equivalent salt weld. Figure 3 shows the location of the seismic section.

and restored them using Midland Valley's 2DMove software, following the methodology of Schultz-Ela (1992) and Hossack (1995). For simplicity, both restorations assumed rigid subsalt strata and a horizontal sea floor, although the paleobathymetry probably dipped seaward at about 0.5° by analogy to the present-day southern Gulf of Mexico margin (Smith and Sandwell, 1997; see also Pratson and Haxby, 1996, for the similar northern Gulf margin). We describe the structural evolution forward in time.

Cross Section MS-508

A seismic profile (Figure 22, located in Figure 12) in the west shows a large counterregional structure separating the landward Pescadores minibasin from the minibasin area in the upper slope. These minibasins are underlain by a variably evacuated, partly welded salt canopy, which overlies the Miocene fold belts. The deepest horizon interpreted is the lower Pleistocene sequence boundary (SB) 1.9 Ma (Figure 22). Deeper reflections are difficult to correlate because of structural complexity and poor imaging. Other interpreted horizons are at 1.4 Ma, an interpolated horizon around 0.85 Ma, and the sea floor, allowing four restoration stages (Figure 23).

By 1.9 Ma (Figure 23a), the salt canopy was being buried. The marked differential loading by the pre-1.9 Ma Pliocene overburden would have created the low, broad passive diapirs depicted as emergent salt bodies. During the Pliocene, the Pescadores minibasin began to form, whereas the prograding sedimentary wedge sank into the allochthonous salt and drove salt seaward. Restoration assumed only a few kilometers of regional extension. The lenses of the pre-1.9 Ma overburden could have been restored with smaller gaps between them, which would entail extension forward in time. However, we considered it more realistic if each lens started accumulating in the same base-of-salt low that they are in today, which is required to form the minibasins in the north.

At 1.4 Ma (Figure 23b), salt evacuation continued, and the Pescadores minibasin depocenter shifted slightly

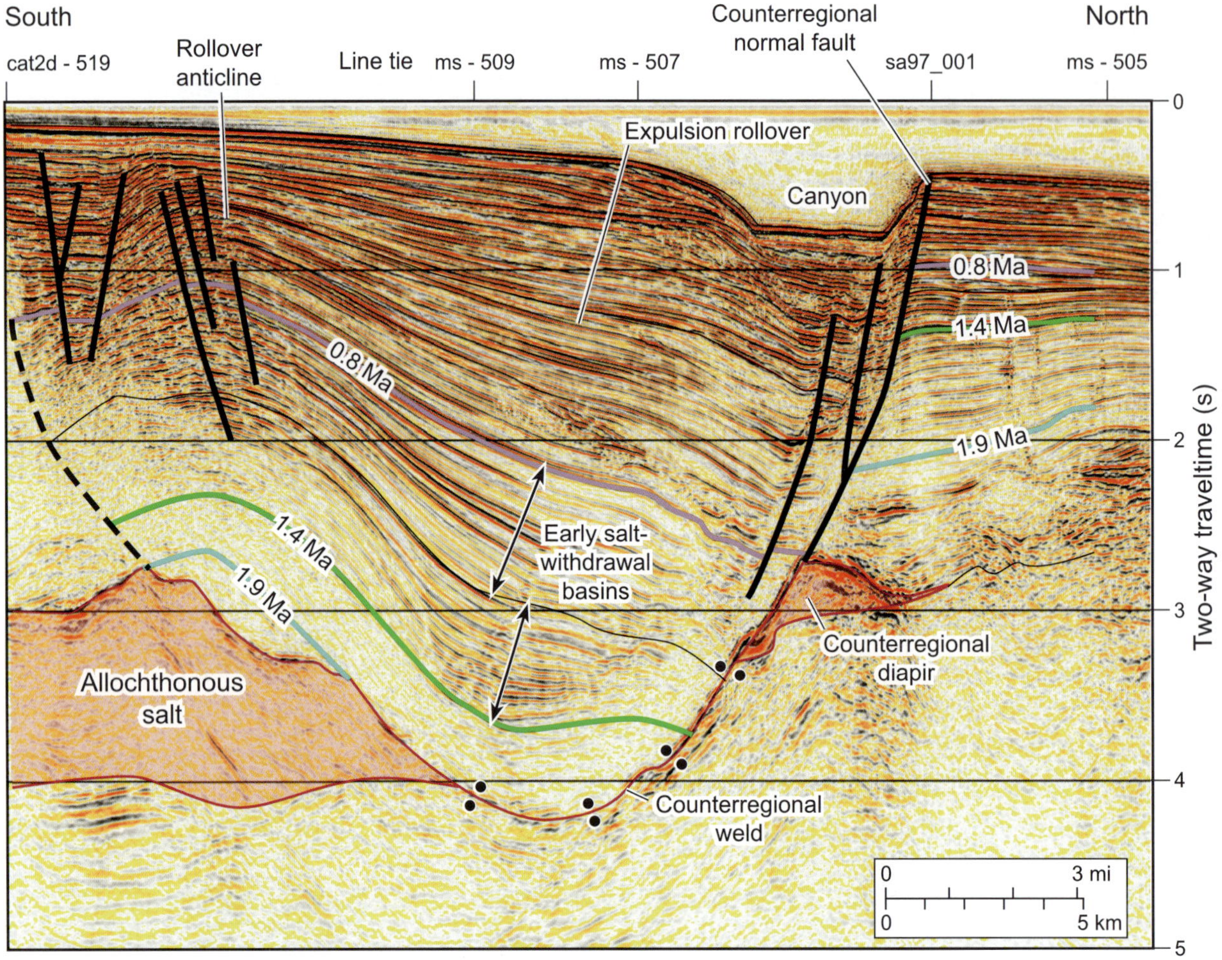

FIGURE 17. Interpreted seismic profile showing a counterregional structure (weld, diapir, and faults) overlain by a canyon in the sea floor. The magenta line depicts the interpreted top of allochthonous salt or equivalent salt weld. Figure 2 shows the location of the seismic section.

seaward because of progradation. Meanwhile, the two more-distal minibasins merely thickened.

At about 0.85 Ma (Figure 23c), the first counterregional growth faults (left side of section) formed as the adjoining Coatzacoalcos minibasin deepened. Pleistocene sedimentation was rapid enough to bury the largest diapir in the center as sediments ponded on its landward flank, although the restoration suggests that thick salt was still present to supply diapiric growth. Shallower passive diapirs remained sporadically emergent distally in the diapir-minibasin area, where sediments accumulated more slowly.

Continuous salt expulsion allowed minibasins to sink until eventually a discontinuous salt weld formed (Figure 23d). The restoration suggests that much of the salt cross-sectional area was lost over time. Some salt loss could be real because salt flowed out of the plane of section. Alternatively, some salt loss could be an artifact of restoration. Subsalt sediments were assumed to be rigid, so that they remained undeformed since 1.9 Ma, an assumption common to many restorations in the northern Gulf of Mexico. However, if restoration had allowed subsalt sediments to deform, then the initial state (1.9 Ma) could be restored with a smaller salt area, thus simulating less salt loss forward in time. In light of this ambiguity, the proportion of salt lost by dissolving directly into seawater or pore fluids is unknown.

Cross Section SA97-014

A seismic profile in Figure 24 (located in Figure 12) in the east shows the Pescadores minibasin, a triangular bounding passive diapir, other passive diapirs and minibasins basinward, and deep allochthonous salt pillows. These are all underlain by a sporadically welded salt canopy, which climbed upsection by advancing seaward in the upper slope.

At 1.9 Ma (Figure 25a), the wedge of prograding sediments had expelled the salt canopy seaward, forming a basal salt weld in the south. By this stage, only part

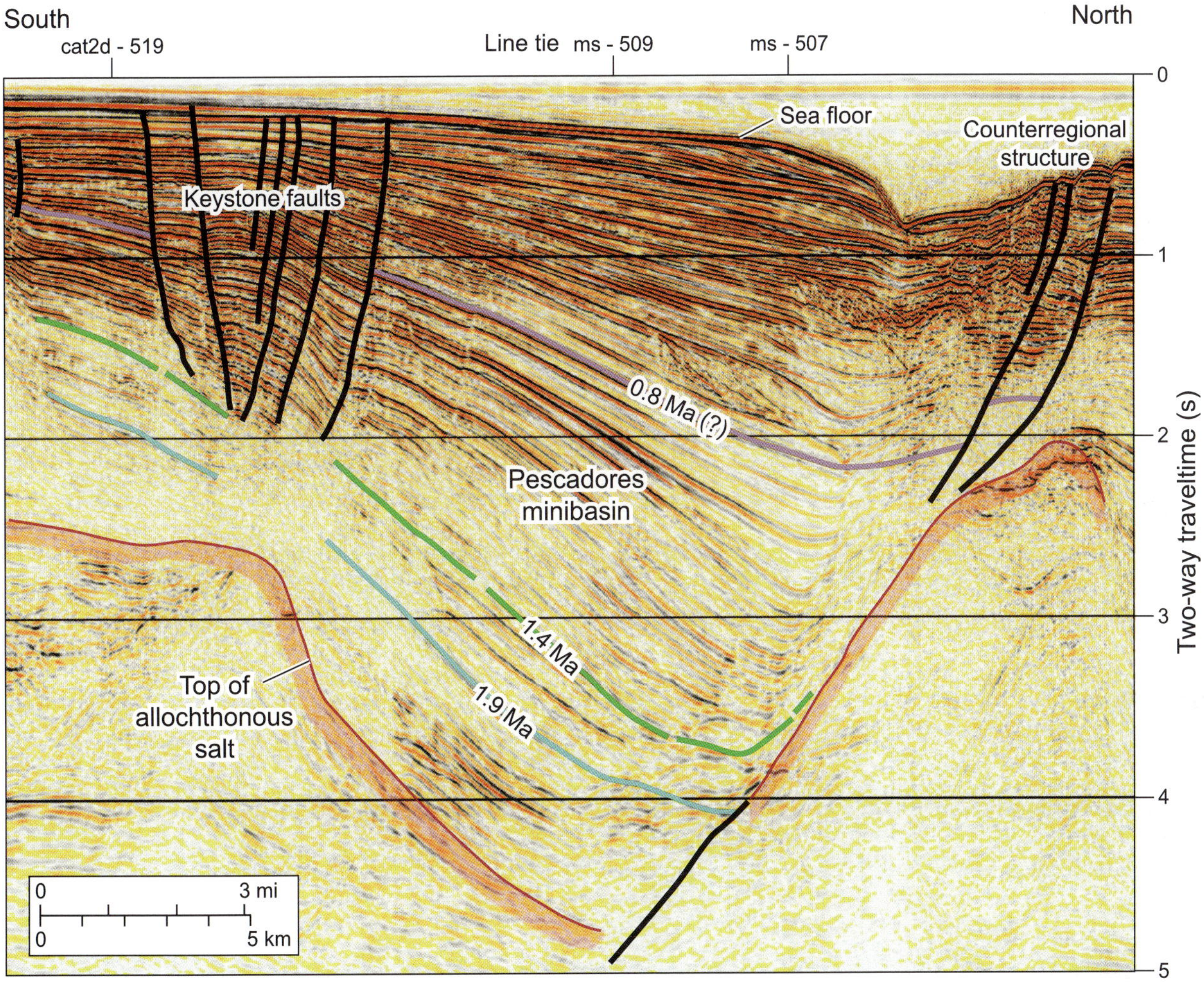

FIGURE 18. Interpreted seismic profile showing a keystone fault family in a counterregional system. The magenta line depicts the interpreted top of allochthonous salt or equivalent salt weld. Figure 2 shows the location of the seismic section.

of the extensive salt canopy remained emergent, probably covered by a condensed sedimentary veneer. Local extension affected the flank of this large, central diapir. Distally, more symmetric deep minibasins separated two passive diapirs.

At 1.4 Ma (Figure 25b), salt evacuation continued to accommodate the asymmetrically deepening Pescadores minibasin. All passive diapirs remained emergent, except for a northern one, which was buried and its crest began to be faulted.

By about 0.8 Ma (Figure 25c), seaward-thickening sediments had mostly filled the Pescadores minibasin. These prograding sediments trapped relict salt below them, forming a low diapir of residual salt. The largest passive diapir bounding the Pescadores minibasin continued to shrink laterally because it was onlapped by encroaching sediments and its flanks subsided.

By the present day (Figure 25d), all the allochthonous salt diapirs have been totally buried. Crestal extension of the newly buried salt diapirs was caused either by continued extensional sagging of the central diapir or by arching of the roof strata farther basinward. Meanwhile, the oldest faults became inactive.

As in the first restoration (Figure 23), little regional extension is shown in Figure 25. However, the second restoration could have been validly restored with more than 10 km (6 mi) of extension in the north by depicting the central emergent diapir in Figure 25a to be initially only a few kilometers wide at the surface instead of about 10 km (~6 mi) wide. We chose not to restore with major extension, however, to match the methodology for the first restoration.

HYDROCARBON PLAYS AND COMPARISON WITH THE NORTHERN GULF OF MEXICO

Hydrocarbon exploration has great potential in offshore Salina del Istmo Basin. The subsalt play, first

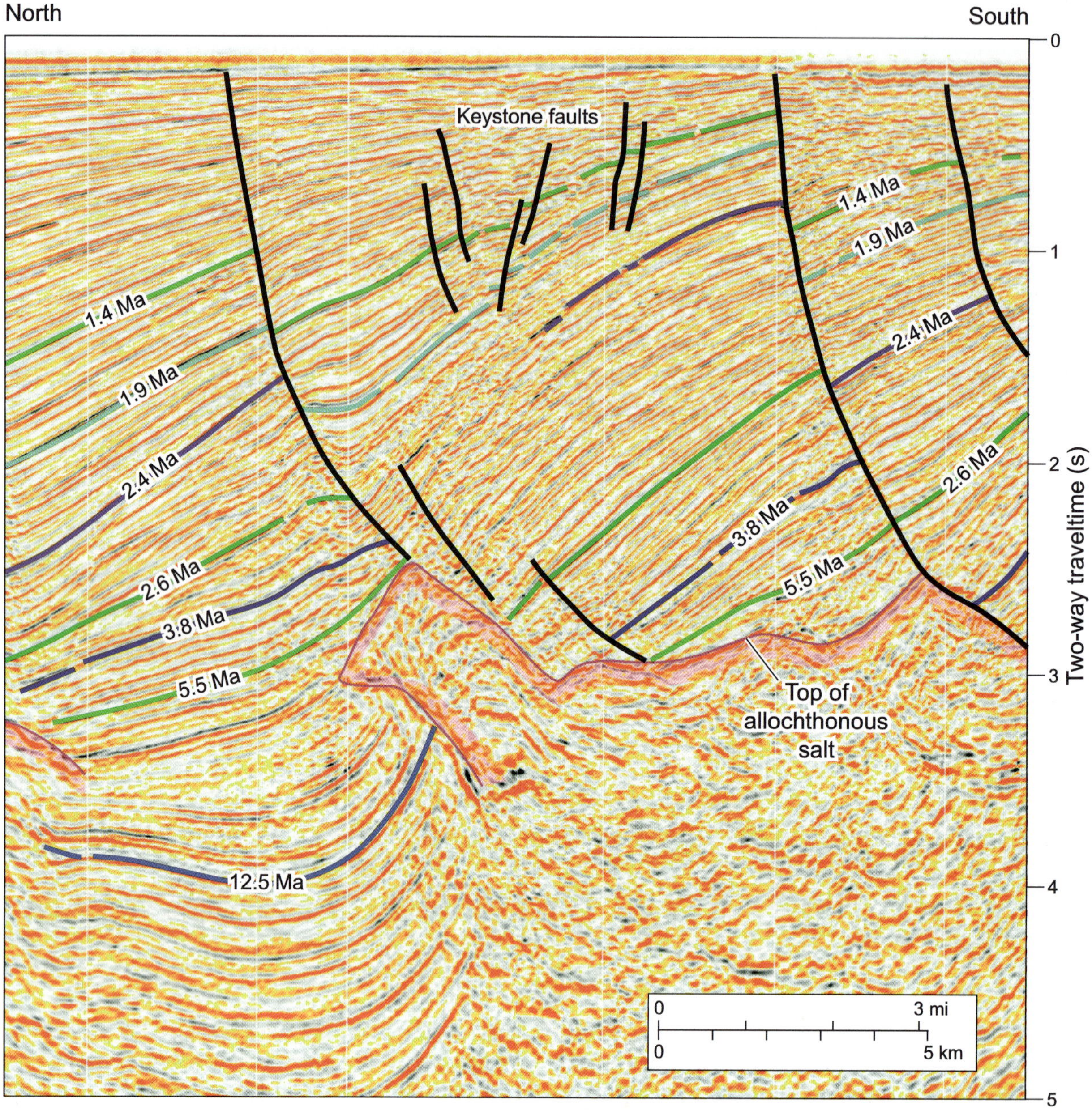

FIGURE 19. Interpreted seismic profile showing landward-dipping listric growth faults. The seaward divergence of reflectors indicates that the main phase of salt expulsion and extension began only in the Pleistocene. Figure 2 shows the location of the seismic section.

discovered in Miocene deep-water conglomerates in 1976 (Mora et al., 2001), can be extended much farther north to at least the upper continental slope. Three play types have been reported (Soto-Cuervo et al., 2004): (1) Neogene sandstones in structural and combined traps over salt bodies, (2) Paleogene sandstones in structural and combined traps under the salt canopy, and (3) Mesozoic carbonates in subsalt compressional anticlines. We add to this list the minibasin turbidite sands structurally enhanced by rollover into listric and counterregional faults.

The offshore part of the Salina del Istmo Basin in the southern Gulf of Mexico is both similar and dissimilar to its northern counterpart in United States waters. Although the study area is small compared with the northern Gulf of Mexico, a few similarities emerge.

1. Both areas have petroleum systems in which salt tectonics is an important factor in creating, changing, and destroying hydrocarbon traps and deepening the base of the oil window below thermally

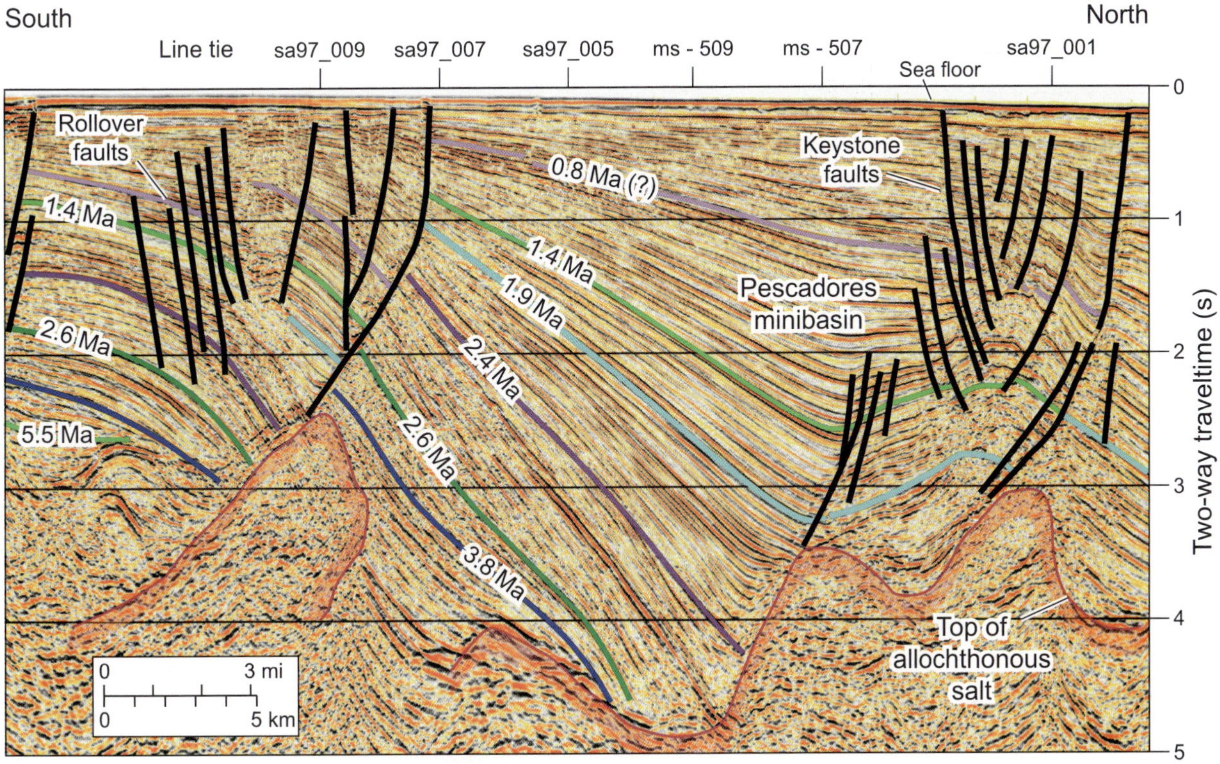

FIGURE 20. Interpreted seismic profile showing two major counterregional systems in the Pescadores minibasin. Seaward divergence of reflectors indicates that the main phase of salt expulsion extension began in the Pleistocene. Figure 2 shows the location of the seismic section.

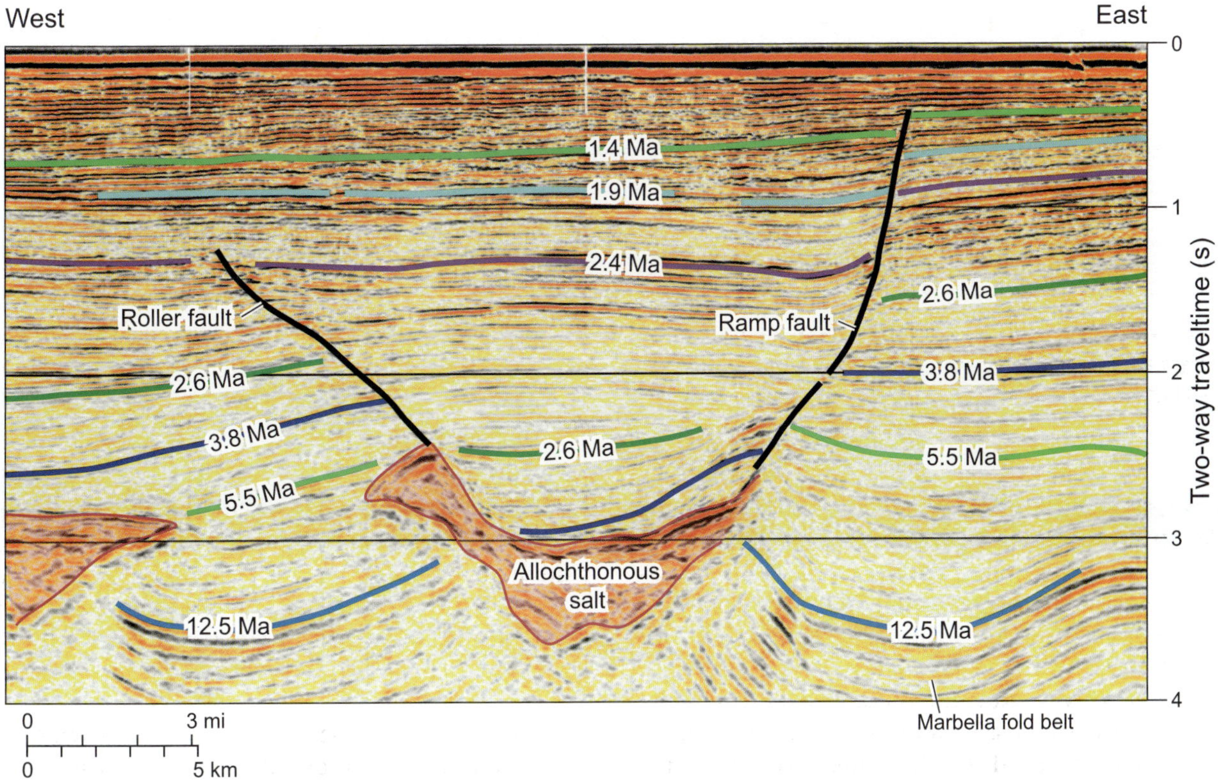

FIGURE 21. Interpreted 3-D seismic profile showing a roller fault and a ramp fault over an allochthonous bird-shaped salt sheet (transparent pink fill). Figure 2 shows the location of the seismic section.

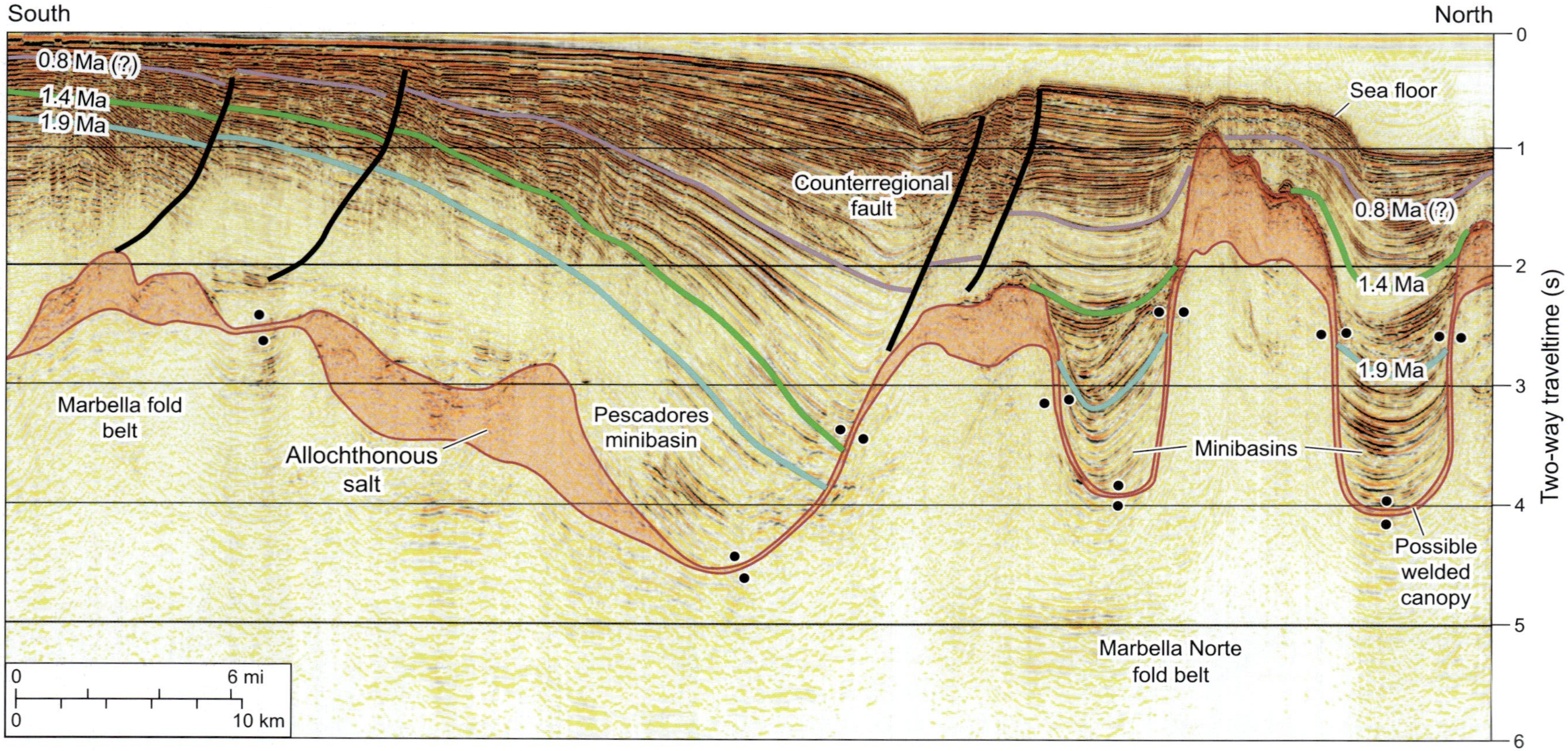

FIGURE 22. Interpreted 2-D regional seismic profile MS-508 in the west, restored in Figure 23. Allochthonous salt (transparent pink fill) or welded equivalent appears to underlie most of the profile. Crests of diapirs become shallower seaward on the upper slope. Figure 12 shows the location of the seismic section.

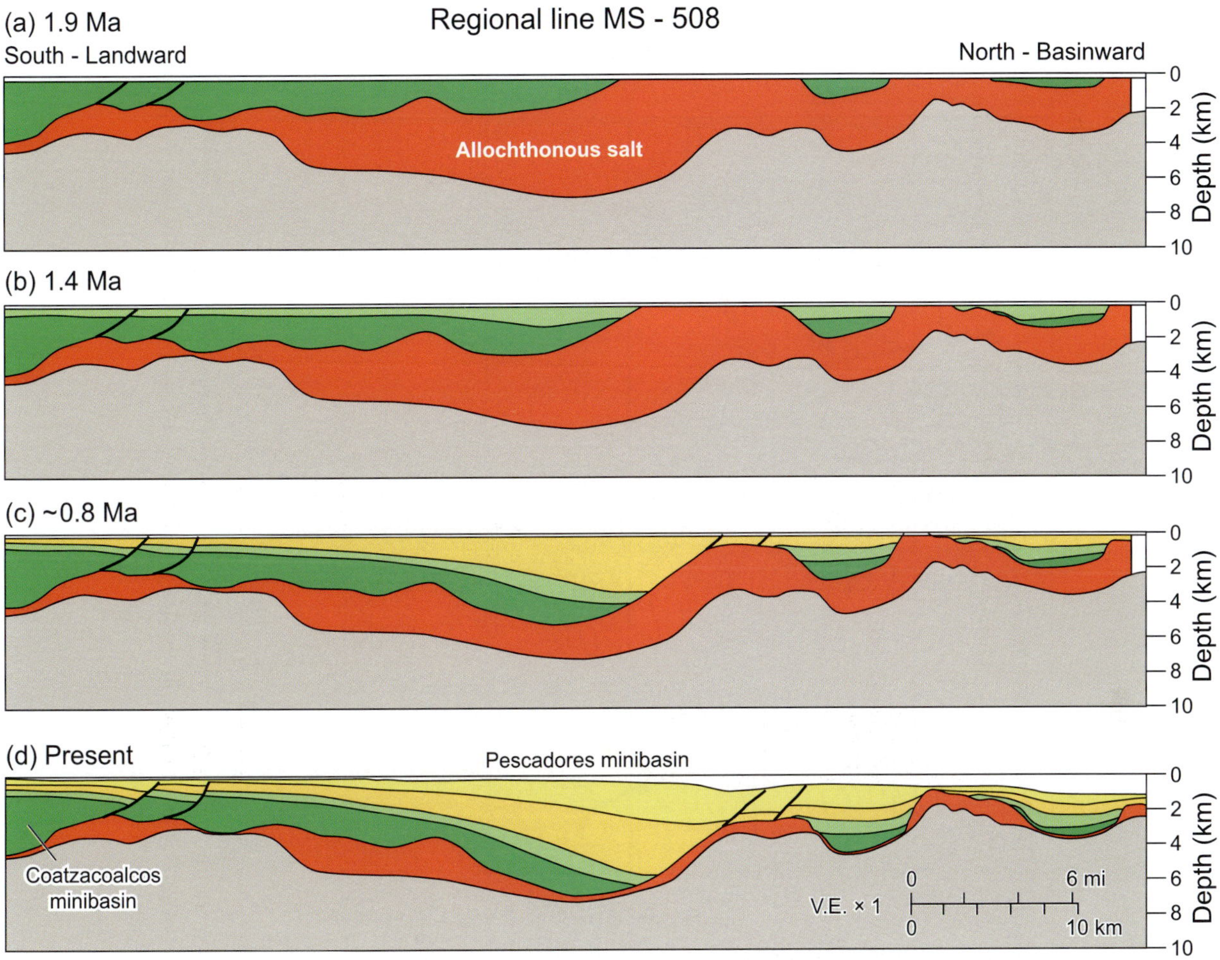

Figure 23. Pleistocene structural restoration of a depth section constructed from the seismic profile in Figure 22. V.E. = vertical exaggeration.

conductive allochthonous salt sheets and raising the top of the oil window above the sheets.

2. Both northern and southern margins of the Gulf contain large areas of mostly passive salt stocks and minibasins.
3. Most types of salt-related fault families and welds found in the northern Gulf of Mexico are also present in the Salina del Istmo Basin.

Differences are also apparent.

1. In the study area, Miocene fold belts are widespread instead of confined to specific areas. Although these folds are not well imaged, folding in the Salina del Istmo Basin appears to be more intense than the 6-km (4-mi) shortening estimated for the Mississippi Fan fold belt (Rowan et al., 2004).
2. Fold belts in the northern Gulf of Mexico originated purely or mostly by gravitational body forces (e.g., Mississippi Fan-Atwater and Perdido-Port Isabel fold belts, respectively). In contrast, the setting of Miocene fold belts in the Salina del Istmo Basin close to the convergent boundaries of the Pacific margin and the orogenic Sierra de Chiapas fold belt suggests that they originated mostly by plate convergence.
3. Among its diverse structural provinces, the northern Gulf of Mexico has highly extensional roho systems, consisting of seaward-dipping listric normal growth faults (Diegel et al., 1995). In contrast, counterregional systems dominate the Salina del Istmo. Landward-dipping faults in the study area have at least some component of extension; they are not simply welds formed by salt expulsion.

CONCLUSIONS

1. The study area in the offshore Salina del Istmo Basin in the southern Gulf of Mexico comprises Pliocene–Pleistocene diapir-minibasins overlying partly evacuated late Miocene allochthonous salt sheets, which in turn overlie middle to late Miocene fold belts.

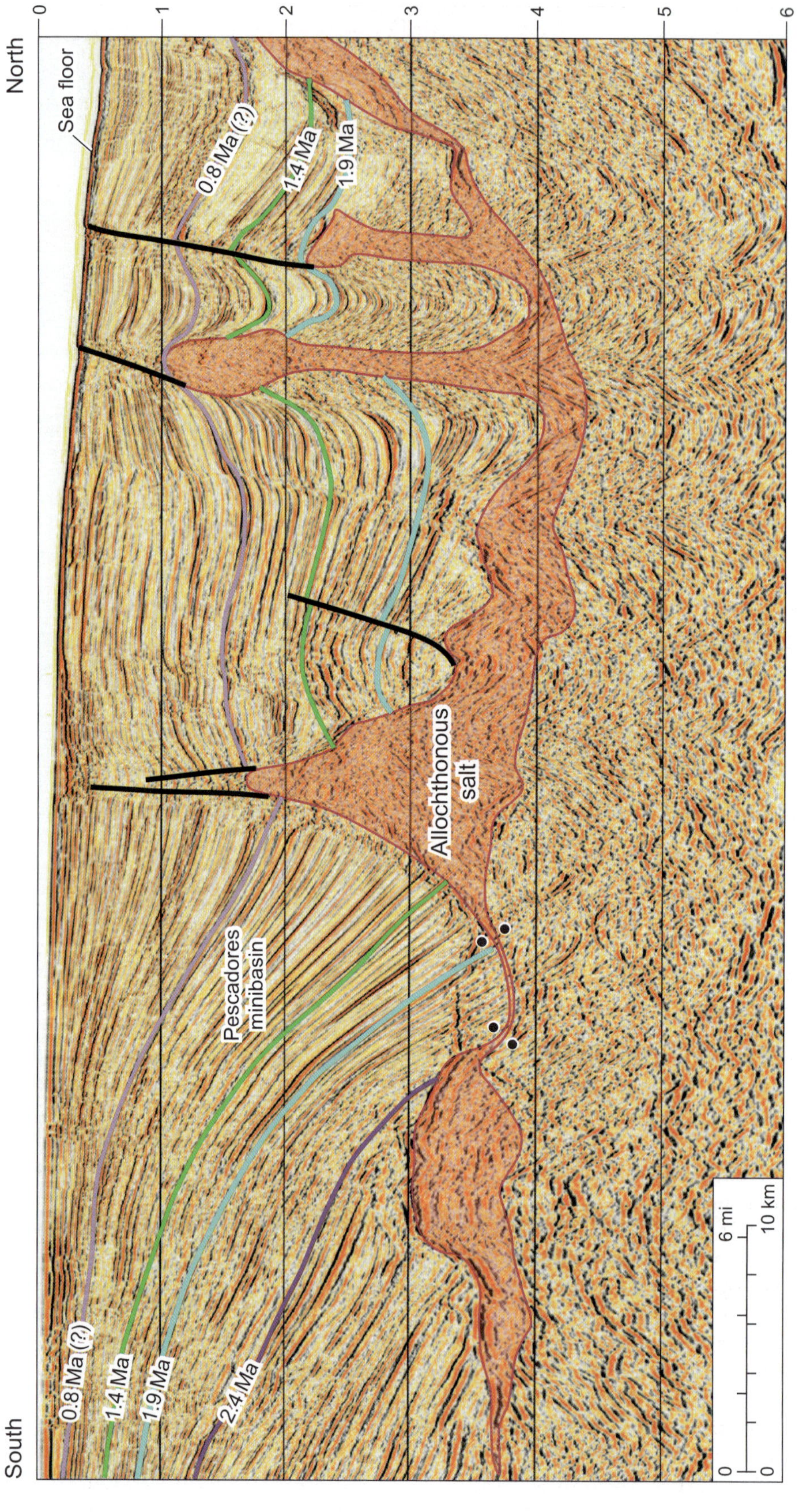

Figure 24. Interpreted 2-D regional seismic profile SA97-014 in the east, restored in Figure 25. Allochthonous salt (transparent pink fill) or welded equivalent appears to underlie most of the profile. Figure 12 shows the location of the seismic section.

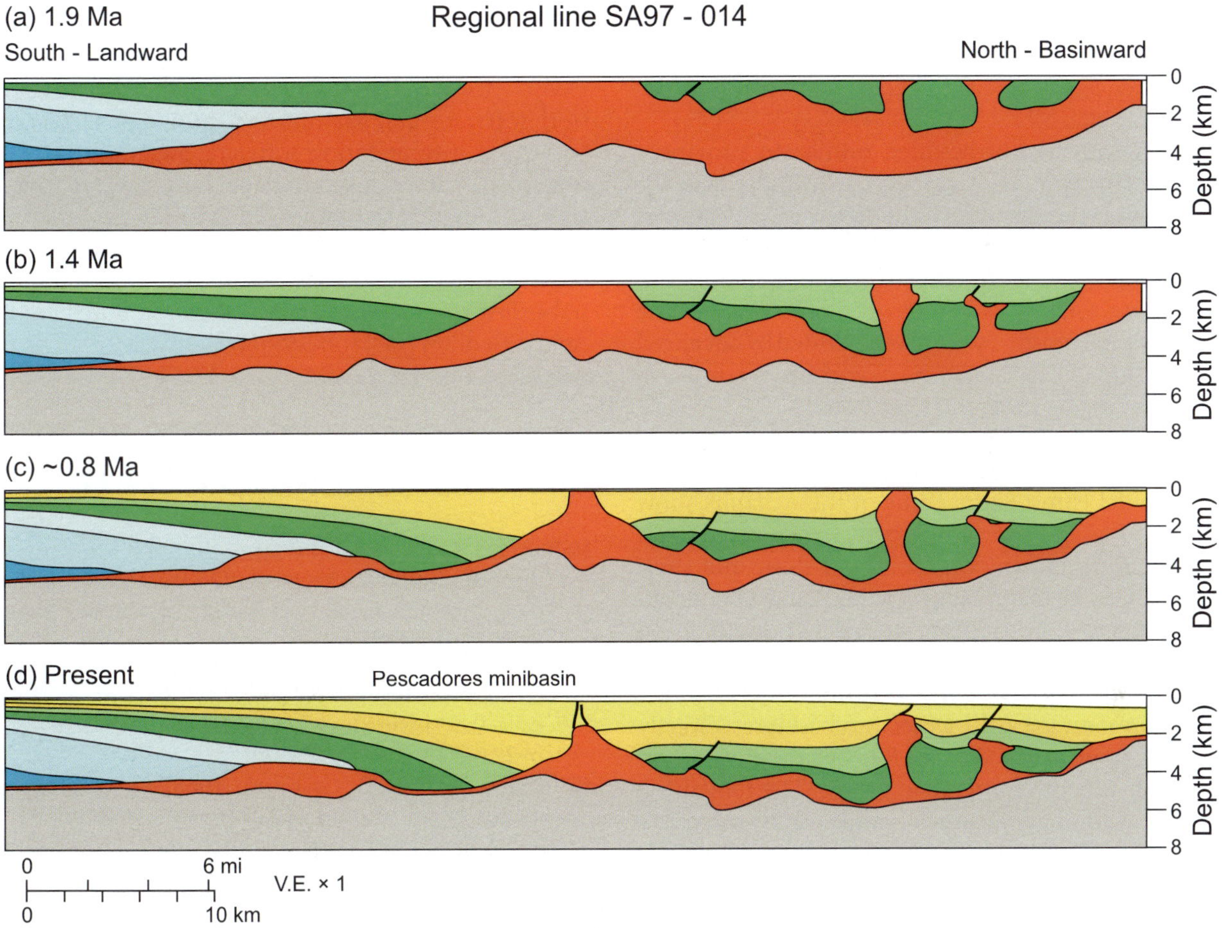

FIGURE 25. Pleistocene structural restoration of a depth section built from the seismic profile in Figure 24. The restoration shows the growth and burial of passive salt diapirs sourced in allochthonous salt (red). V.E. = vertical exaggeration.

2. Between 12 and 15 Ma, a change in Pacific plate kinematics created the southern Mexico Neogene belt. In the study area, this formed four main fold belts: Agua Dulce, Marbella, Marbella Norte, and Catemaco. However, these belts could be merely better-imaged patches of one giant Chiapas-Campeche fold belt, having a strike length of at least 530 km (330 mi) instead of the present estimate of about 380 km (240 mi). The first three mentioned fold belts detach on either Middle Jurassic autochthonous salt or a Mesozoic allochthonous salt canopy derived from this source layer. In contrast, the Catemaco fold belt, which lies west of the Jurassic salt basin, probably detaches on Paleogene shale. Given their proximity to the orogenic fold belts of southern Mexico and the convergent Pacific margin, shortening in all four Salina fold belts is likely to be orogenically driven and to involve the presalt basement. Folding affects the upper Mesozoic section (Upper Jurassic and Cretaceous), which can be reached by drilling.
3. During or just after the final stages of folding in the late Miocene, salt extrusion culminated as salt sheets flowed over the eroded and uplifted fold belts. When extrusion began is uncertain, but the process was mostly complete by the end of the Miocene. The salt was probably squeezed upward out of preexisting diapirs by the same regional shortening that formed the fold belts, as in the foreland fold and thrust belt of the Zagros Mountains of Iran. Both lateral compression and uplift caused by shortening would have enhanced salt extrusion. Copious salt sheets partly coalesced at or near the surface to form the discontinuous Sal Somera canopy. The number and style of diapirs supplying salt to the canopy are unknown because of poor seismic quality below the canopy, as was true for much of the northern Gulf of Mexico until recently. The present-day structure of the canopy mostly reflects the interplay of two factors. (1) The shallowest salt or salt welds tend to overlie Miocene anticlines. The salt canopy or equivalent weld is roughly conformable with the underlying fold belts, indicating that the canopy extruded either during the final stage of folding or soon after. (2) The originally thickest parts

of the salt canopy allowed the greatest salt evacuation by overlying Pliocene–Pleistocene minibasins and, hence, became the deepest depressions in the top of salt.

4. From south to north, three minibasin provinces formed by seaward evacuation of the underlying salt canopy: (1) Coatzacoalcos-Rodador, (2) Pescadores, and (3) the diapir-minibasin area. The Pescadores minibasin is a 100-km-long (62-mi-long) minibasin, trending roughly parallel to the present coastline and floored mostly by a salt weld created by advanced expulsion of the Sal Somera canopy under differential loading by prograding sediments.
5. The diapir-minibasin area has large salt diapirs (mostly passive), which also dominate other parts of the Salina del Istmo Basin. Some of these diapirs show signs of lateral shortening before they reached an extensional setting on the upper slope and shelf margin. Some diapirs appear to be sourced by allochthonous salt, but seismic imaging at depth is too poor to be certain. Some of the surrounding minibasins are floored by the allochthonous salt canopy or equivalent salt weld. Other minibasins rest directly on synclines of the Miocene fold belts, with no intervening allochthonous weld visible. The remaining minibasins may be much more deeply rooted in autochthonous salt or deep allochthonous salt.
6. The study area contains several salt-related fault families, such as keystone, roller, counterregional, and rollover, all of which have equivalents in the northern Gulf of Mexico. The most spectacular fault systems are counterregional. The largest of these forms the seaward boundary of the Pescadores minibasin.
7. Restorations suggest that allochthonous salt remained partly emergent until 1.4 or 0.8 Ma, after which the salt was completely buried by growth of large counterregional systems and salt-expulsion minibasins.
8. Although the study area in offshore Salina del Istmo Basin is broadly similar to the conjugate margin in the northern Gulf of Mexico, it differs chiefly in having widespread, instead of local, fold belts, which probably formed orogenically instead of by gravity spreading. The basin is dominated by counterregional fault systems, like in parts of offshore Louisiana (Diegel et al., 1995).

ACKNOWLEDGMENTS

We thank the following for making this study possible. Petróleos Mexicanos provided the data and gave the permission to publish. William L. Fisher cosupervised the Ph.D. dissertation (Gómez-Cabrera, 2003) on which this chapter was based and provided valuable stratigraphic insights. The second author (MPAJ) was funded by the Applied Geodynamics Laboratory consortium of oil companies and by the Jackson School of Geosciences. Midland Valley supplied 2DMove restoration software and Halliburton provided Landmark's SeisWorks interpretation software. Suggestions by peer reviewers Carl Fiduk and especially the creative insights of Fred Diegel improved the manuscript considerably. Nancy Cottington drew the illustrations, and Lana Dieterich edited the manuscript. This chapter is published by permission of the director, Bureau of Economic Geology, Jackson School of Geosciences, the University of Texas at Austin.

REFERENCES CITED

Bohannon, R. G., and T. Parsons, 1995, Tectonic implications of post-30 Ma Pacific and North American relative plate motions: Geological Society of America Bulletin, v. 107, p. 937–959.

Contreras, V., and B. Castillon, 1960, Morfología y origin de los domos salinos del Istmo de Tehuantepec: Boletín de la Asociación Mexicana de Geólogos Petroleros, v. 12, p. 221–242.

Contreras, V., and B. Castillon, 1968, Morphology and origin of salt domes of Isthmus of Tehuantepec, *in* J. Braunstein and G. D. O'Brien, eds., Diapirism and diapirs: AAPG Memoir 8, p. 244–260.

Correa-Perez, I., and J. Gutierrez-y-Acosta, 1983, Interpretacion gravimétrica y magnetométrica del occidente de la Cuenca Salina del Istmo: Revista del Instituto Mexicano del Petróleo, v. 15, no. 4, p. 5–25.

Diegel, F. A., J. F. Karlo, D. C. Schuster, R. C. Shoup, and P. R. Tauvers, 1995, Cenozoic structural evolution and tectono-stratigraphic framework of the northern Gulf Coast continental margin, *in* M. P. A. Jackson, D. G. Roberts, and S. Snelson, eds., Salt tectonics: A global perspective: AAPG Memoir 65, p. 109–151.

Fiduk, J. C., 1994, Plio-Pleistocene evolution of the upper continental slope, Garden Banks and East Breaks areas, northwestern Gulf of Mexico: Ph.D. dissertation, the University of Texas at Austin, 311 p.

Fletcher, R. C., M. R. Hudec, and I. A. Watson, 1995, Salt glacier and composite sediment-salt glacier models for the emplacement and early burial of allochthonous salt sheets, *in* M. P. A. Jackson, D. G. Roberts, and S. Snelson, eds., Salt tectonics: A global perspective: AAPG Memoir 65, p. 77–108.

Fox, J. F., 1998, Salt/sediment interaction: The Leading Edge, v. 17, no. 8, p. 1033–1041.

García-Molina, G., 1994, Structural evolution of SE Mexico (Chiapas-Tabasco-Campeche) offshore and onshore: Ph.D. dissertation, Rice University, 161 p.

Gómez-Cabrera, P. T., 2003, Stratigraphic and structural analysis of the Neogene sediments of the offshore part of the Salina del Istmo Basin, southeastern Mexico: Ph.D. dissertation, the University of Texas at Austin, 274 p.

Gómez-Cabrera, P. T., and M. P. A. Jackson, 2009, Neogene stratigraphy and salt tectonics of the Santa Ana area, offshore Salina del Istmo Basin, southeastern Mexico, *in* C. Bartolini and J. R. Román Ramos, eds., Petroleum systems in the southern Gulf of Mexico: AAPG Memoir 90, p. 237–255.

Heaton, R. C., M. P. A. Jackson, M. Bamahoud, and A. S. O. Nani, 1995, Superposed Neogene extension, contraction, and salt canopy emplacement in the Yemeni Red Sea, *in* M. P. A. Jackson, D. G. Roberts, and S. Snelson, eds., Salt tectonics: A global perspective: AAPG Memoir 65, p. 333–351.

Hossack, J. R., 1995, Geometric rules of section balancing for salt structures, *in* M. P. A. Jackson, D. G. Roberts, and S. Snelson, eds., Salt tectonics: A global perspective: AAPG Memoir 65, p. 29–40.

Hubbard, R. J., J. Pape, and D. G. Roberts, 1985a, Depositional sequence mapping as a technique to establish tectonic and stratigraphic framework and evaluate hydrocarbon potential on a passive continental margin, *in* O. R. Berg and D. Woolverton, eds., Seismic stratigraphy II: An integrated approach to hydrocarbon exploration: AAPG Memoir 39, p. 79–91.

Hubbard, R. J., J. Pape, and D. G. Roberts, 1985b, Depositional sequence mapping to illustrate the evolution of a passive continental margin, *in* O. R. Berg and D. Woolverton, eds., Seismic stratigraphy II: An integrated approach to hydrocarbon exploration: AAPG Memoir 39, p. 93–115.

Hudec, M. R., M. P. A. Jackson, and D. D. Schultz-Ela, 2005, A compressional origin for minibasins near the Sigsbee escarpment, Gulf of Mexico (abs.): AAPG Search and Discovery: http://www.searchanddiscovery.net /documents/abstracts/2005annual_calgary/abstracts /hudec.htm (accessed August 18, 2009).

Ings, S., C. Beaumont, and L. Gemmer, 2004, Numerical modeling of salt tectonics on passive continental margins: Preliminary assessment of the effects of sediment loading, buoyancy, margin tilt, and isostasy, *in* P. J. Post, D. L. Olson, K. T. Lyons, S. L. Palmes, P. F. Harrison, and N. C. Rosen, eds., Salt-sediment interactions and hydrocarbon prospectivity: Concepts, applications, and case studies for the 21st century: 24th Annual Research Conference, Gulf Coast Section of the Society of Economic Paleontologists and Mineralogists Foundation, Houston, Texas, p. 4.

Jackson, M. P. A., and R. R. Cornelius, 1985, Tertiary salt diapirs exposed at different structural levels in the Great Kavir (Dasht-e-Kavir) south of Semnan, north-central Iran: A remote-sensing study of their internal structure and shape: The University of Texas at Austin, Bureau of Economic Geology Open File Report OF-STSI-1985-22, 108 p.

Jackson, M. P. A., and M. R. Hudec, 2004, A new mechanism for advance of allochthonous salt sheets, *in* P. J. Post, D. L. Olson, K. T. Lyons, S. L. Palmes, P. F. Harrison, and N. C. Rosen, eds., Salt-sediment interactions and hydrocarbon prospectivity: Concepts, applications, and case studies for the 21st century: 24th Annual Research Conference, Gulf Coast Section of the Society of Economic Paleontologists and Mineralogists Foundation, Houston, Texas, p. 220–242.

Jackson, M. P. A., and C. J. Talbot, 1989, Salt canopies, *in* Gulf of Mexico salt tectonics, associated processes and exploration potential: Gulf Coast Section SEPM Foundation, 10th Annual Research Conference, Program and Abstracts, p. 72–78.

Jackson, M. P. A., R. R. Cornelius, C. H. Craig, A. Gansser, J. Stöcklin, and C. J. Talbot, 1990, Salt diapirs of the Great Kavir, central Iran: Geological Society of America Memoir 177, 139 p.

Letouzey, J., and S. Sherkati, 2004, Salt movement, tectonic events, and structural style in the central Zagros fold and thrust belt (Iran), *in* P. J. Post, D. L. Olson, K. T. Lyons, S. L. Palmes, P. F. Harrison, and N. C. Rosen, eds., Salt-sediment interactions and hydrocarbon prospectivity: concepts, applications, and case studies for the 21st century: 24th Annual Research Conference, Gulf Coast Section of the SEPM Foundation, Houston, Texas, p. 444–462.

Mitra, S., G. C. Figueroa, J. H. Garcia, and A. M. Alvarado, 2005, Three-dimensional structural model of the Cantarell and Sihil structures, Campeche Bay, Mexico: AAPG Bulletin, v. 89, p. 1–26.

Mora, G., A. Iparrea, and A. Oviedo, 2001, Petroleum geology of the Salina del Istmo Basin (abs.): AAPG Search and Discovery: http://www.searchanddiscovery.com /abstracts/html/2001/annual/abstracts/0543.htm (accessed August 18, 2009).

Nilsen, K. T., B. C. Vendeville, and J.-T. Johansen, 1995, Influence of regional tectonics on halokinesis in the Nordkapp Basin, Barents Sea, *in* M. P. A. Jackson, D. G. Roberts, and S. Snelson, eds., Salt tectonics: A global perspective: AAPG Memoir 65, p. 413–436.

Oviedo, A., G. Mora, and R. Herbert, 1995, Tertiary evolution and petroleum habitat of the Salina Basin, southeast Mexico: AAPG Bulletin, v. 79, no. 13, p. 73.

Oviedo, A., G. Mora, and R. Herbert, 1996, A new model for the role of salt in the Salina Basin of southeast Mexico: AAPG Bulletin, v. 80, p. 1320.

Oviedo-Pérez, A. E., 1996, Evolución neógena de las Cuencas Terciarias del Sureste: Tesis de Maestría, División de Estudios de Posgrado, Facultad de Ingeniería, Universidad Nacional Autónoma de México, 91 p.

Petróleos Mexicanos (PEMEX)-British Petroleum, 1994, Proyecto cuencas terciarias del sureste y area Marbella: Unpublished PEMEX Report, 297 p.

Pratson, L. F., and W. F. Haxby, 1996, What is the slope of the U.S. continental slope?: Geology, v. 24, p. 3–6.

Prost, G., and M. Aranda, 2001, Tectonics and hydrocarbon systems of the Veracruz Basin, Mexico, *in* C. Bartolini, R. T. Buffler, and A. Cantu-Chapa, eds., The western Gulf of Mexico Basin: Tectonics, sedimentary basins, and petroleum systems: AAPG Memoir 75, p. 271–291.

Ricoy, U. J., 1989, Tertiary terrigenous depositional systems of the Mexican Isthmus basins: Ph.D. dissertation, the University of Texas at Austin, 145 p.

Robles, J., C. Toledo, E. Pliego, and P. Boll, 1998, Interpretación Sismico-Estructural y de Secuencias en el área de Catemaco, Proyecto Coatzacoalcos Marino (SW del Golfo de México): Asociación Mexicana de Geofísicos de Exploración, Memorias Trabajos Técnicos, VIII Simposium de Geofísica y Exhibición de Equipo, Veracruz, Ver., México, October 7–9, 1998.

Rowan, M. G., 2002, Salt-related accommodation in the Gulf of Mexico deep water: Withdrawal or inflation, autochthonous or allochthonous?: Gulf Coast Association of Geological Societies Transactions, v. 52, p. 861–869.

Rowan, M. G., M. P. A. Jackson, and B. D. Trudgill, 1999, Salt-related fault families and fault welds in the northern Gulf of Mexico: AAPG Bulletin, v. 83, no. 9, p. 1454–1484.

Rowan, M. G., F. J. Peel, and B. C. Vendeville, 2004, Gravity-driven fold belts on passive margins, *in* K. R. McClay, ed., Thrust tectonics and hydrocarbon systems: AAPG Memoir 82, p. 157–182.

Schultz-Ela, D. D., 1992, Restoration of cross sections to constrain deformation processes of extensional terranes: Marine and Petroleum Geology, v. 9, no. 4, p. 372–388.

Schuster, D. C., 1995, Deformation of allochthonous salt and evolution of related salt-structural systems, eastern Louisiana Gulf Coast, *in* M. P. A. Jackson, D. G. Roberts, and S. Snelson, eds., Salt tectonics: A global perspective: AAPG Memoir 65, p. 177–198.

Sepehr, M., and J. W. Cosgrove, 2004, Structural framework of the Zagros fold-thrust belt, Iran: Marine and Petroleum Geology, v. 21, p. 829–843.

Smith, W. H. F., and D. T. Sandwell, 1997, Global sea floor topography from satellite altimetry and ship depth soundings: Science, v. 277, p. 1957–1962.

Soto-Cuervo, A.,V. Ortega-González, and G. Mora-Oropeza, 2004, Present and future of the Salina del Istmo Basin and its offshore extension into the Gulf of Mexico (abs.): AAPG Search and Discovery: http://www.searchanddiscovery.net/documents/abstracts/2004intl_cancun/index.htm?q=%2Btext%3A%22soto+cuervo%22#s (accessed August 18, 2009).

Vendeville, B. C., and K. T. Nilsen, 1995, Episodic growth of salt diapirs driven by horizontal shortening, *in* C. J. Travis, H. Harrison, M. R. Hudec, B. C. Vendeville, F. J. Peel, and B. F. Perkins, eds., Salt, sediment, and hydrocarbons: SEPM Gulf Coast Section 16th Annual Research Foundation Conference, p. 285–295.

Viniegra, O. F., 1971, Age and evolution of salt basins of southeastern Mexico: AAPG Bulletin, v. 55, p. 478–494.

Wilson, H. H., 1993, The age of salt in the Gulf of Mexico Basin: Journal of Petroleum Geology, v. 16, p. 125–152.

2

Sosa Patrón, A. A., J. G. Cárdenas López, C. Cárdenas Lara, O. Pinto Gómez, M. E. Guzzy Arredondo, M. E. Monroy Audelo, and R. Castellanos Calvo, 2009, Integrated geological interpretation and impact on the definition of Neogene plays in the Isthmus Saline Basin, Mexico, *in* C. Bartolini and J. R. Román Ramos, eds., Petroleum systems in the southern Gulf of Mexico: AAPG Memoir 90, p. 29–47.

Integrated Geological Interpretation and Impact on the Definition of Neogene Plays in the Isthmus Saline Basin, Mexico

Alejandro A. Sosa Patrón, Juan de G. Cárdenas López, Celia Cárdenas Lara, Oscar Pinto Gómez, Ma. Elena Guzzy Arredondo, Ma. Eugenia Monroy Audelo, and Reynaldo Castellanos Calvo

Petróleos Mexicanos (PEMEX) Exploracion and Produccion, Centro Técnico Administrativo, Activo Regional de Exploración Sur, Villahermosa, Tabasco C.P., Mexico

ABSTRACT

The Isthmus Saline Tertiary Basin covers 9800 km^2 (3784 mi^2)on the coastal plain of the Gulf of Mexico. This work concentrates mainly on updating the chronostratigraphic framework of the Neogene-producing plays in this economically important area by integrating data obtained from new three-dimensional (3-D) seismic surveys and recently drilled wells.

The interpretation of integrated data indicated that the basin began to evolve in the upper Miocene and also solved correlation problems in the upper Miocene and lower Pliocene sequences with a positive impact on updating the Neogene geological framework. The Neogene section is controlled by a system of extensional counterregional faults associated with minibasins formed by salt evacuation that governed the location of depocenters containing reservoir rocks in rollover and pinch-out traps. The Chiapaneca orogeny created additional pre-upper Miocene compressional traps. Paleoecological conditions when combined with seismostratigraphic analysis are relevant criteria for defining bioevents that equate to chronostratigraphic surfaces for plays. These surfaces are key for understanding the plays and for improving the economic evaluation of the basin.

Within this chronostratigraphic framework, the main reservoirs for the area's giant fields are turbiditic and shelfal deltaic sands in upper Miocene and lower to upper Pliocene sections. Additionally, as indicated by our analyses, lower and middle Miocene basinal sands have been tested and display exploration potential.

INTRODUCTION

Since the beginning of the last century, the geology of the southeastern Tertiary basins in the Gulf of Mexico has been studied using potential field methods, surface geology, wells, and seismic data (Sansores and Flores Covarrubias, 1972; Kohl, 1985; Sánchez, 1987; Petróleos Mexicanos [PEMEX]-British Petroleum, 1994; Oviedo,

DOI:10.1306/13191076M903330

1996; Gómez, 2003; Pereznegrón, 2003; PEMEX, 2004; Tarango and Sánchez, 2005; Martínez, 2006; Solís, 2006). Within this region, the Isthmus Saline Basin provides the setting for the current work. The Isthmus Saline Basin of the Gulf of Mexico coastal plain is bordered by the Comalcalco Basin, the Reforma-Akal fold belt, and the Macuspana Basin. The most relevant tectonic elements in this area are the Sierra de Chiapas to the south, the Yucatan Platform to the east, the Pescadores Basin in the Gulf of Mexico to the north, and the Veracruz Basin to the west (Figure 1).

To date, 40 fields have been discovered in the Isthmus Saline Basin with cumulative production of more than 1.5 billion bbl of oil and 2 tcf of gas. Four giant oil fields with reservoirs consisting mainly of turbidite slope and deltaic shelf sands of upper Miocene and lower Pliocene age are present. In addition, a middle Miocene and lower Miocene basin-floor fan turbidite play has been tested, which has exploration potential. The hydrocarbon source rocks are of Upper Jurassic age.

The goals of this chapter are to update the geological framework of the Isthmus Saline Basin through the integration of new three-dimensional (3-D) seismic data and recently drilled wells and to define the proven plays in the basin, understand their productivity, and evaluate the basin's remaining economic petroleum potential. Using 3-D seismic data, updating and detailing the geological framework of the Isthmus Saline Basin and determining the correct context for producing plays in the area were possible. The integrated interpretation of paleontologic, seismic, and lithostratigraphic data for the wells located in the southern part of the basin positively impacts the definition of the geological framework and, therefore, the economic evaluation of Neogene plays in the Isthmus Saline Basin.

Because of its particular economic importance, the main focus here is on the Neogene part of the basins, primarily the upper Miocene and lower Pliocene sections. To that end, we have applied a play fairway analysis methodology to characterize the lower Miocene, middle Miocene, upper Miocene, and lower Pliocene plays.

BIOSTRATIGRAPHY

Biostratigraphic analyses of 136 wells were applied mainly to second-order stratigraphic sequences and included both biochronostratigraphic and paleobathymetric determinations to establish the ranges and paleoecological behaviors of the different index fossil species.

In the years preceding this study, paleontological descriptions in the Isthmus Saline Basin were based on the identification of planktonic fossils; however, the geologic history of the basin has included major bathymetric variations that affected the presence and evolution of these fossils through time. Therefore, an integrated interpretation was undertaken here, including the analysis of the planktonic foraminiferal assemblages and their paleoecological relationships to benthonic foraminifera, thereby achieving a better understanding of the chronostratigraphic framework. Figure 2 is an example of the updated Neogene biostratigraphy.

The biochronostratigraphic determinations are supported by a seismic stratigraphic interpretation, which guided the correct interpretation of the paleontologic data and lead to significant advancements in the definition of the rock package. Through an updated correlation of high-resolution biostratigraphy for the wells, we have established a new chronostratigraphic framework that is tied to seismic data, as shown in Figure 3.

Biochronostratigraphic Analysis

This work required a very detailed analysis of species distribution patterns for both planktonic and benthonic foraminifera. The stratigraphic significance of a particular species was tested against a paleobathymetric framework to evaluate the overprint of paleoenvironmental factors on time as indicated. These analyses lead to the exclusion of data in some cases. The identification of benthonic foraminifera present in the samples allowed the determination of water-depth variation through geologic time. The resultant variation indicates that paleodepths transgressed time lines from south to north in the basin, a concept exemplified in Figure 4.

For planktonic species, it is also important to identify the geological time range over which the species existed to decide whether the occurrence of an index species in a well coincides with a bathymetric change. In such cases, the occurrence is considered to be an ecologically depressed top instead of a true extinction or datum (Armentrout and Clement, 1990) because it occurred deeper in the well than its true extinction, as determined on a worldwide basis (Stainforth et al., 1975; Bolli and Saunders, 1985; Hardenbol et al., 1998). True bioevents recognized by applying these concepts were correlated with seismic reflections to provide chronostratigraphic horizons, which were projected into the shallow-water environments (Figures 3, 5).

PALEOBATHYMETRIC MAPS

Paleobathymetric determinations made in this chapter employ the criteria of Tipsword et al. (1966) who defined seven benthonic foraminiferal zones with their corresponding bathymetries, as shown in Figure 6. The study of benthonic and planktonic foraminifera in 95 wells has allowed the determination of Neogene paleobathymetry.

The definition of lower bathyal (>500 m, >1640 ft) depths includes the presence of the following species

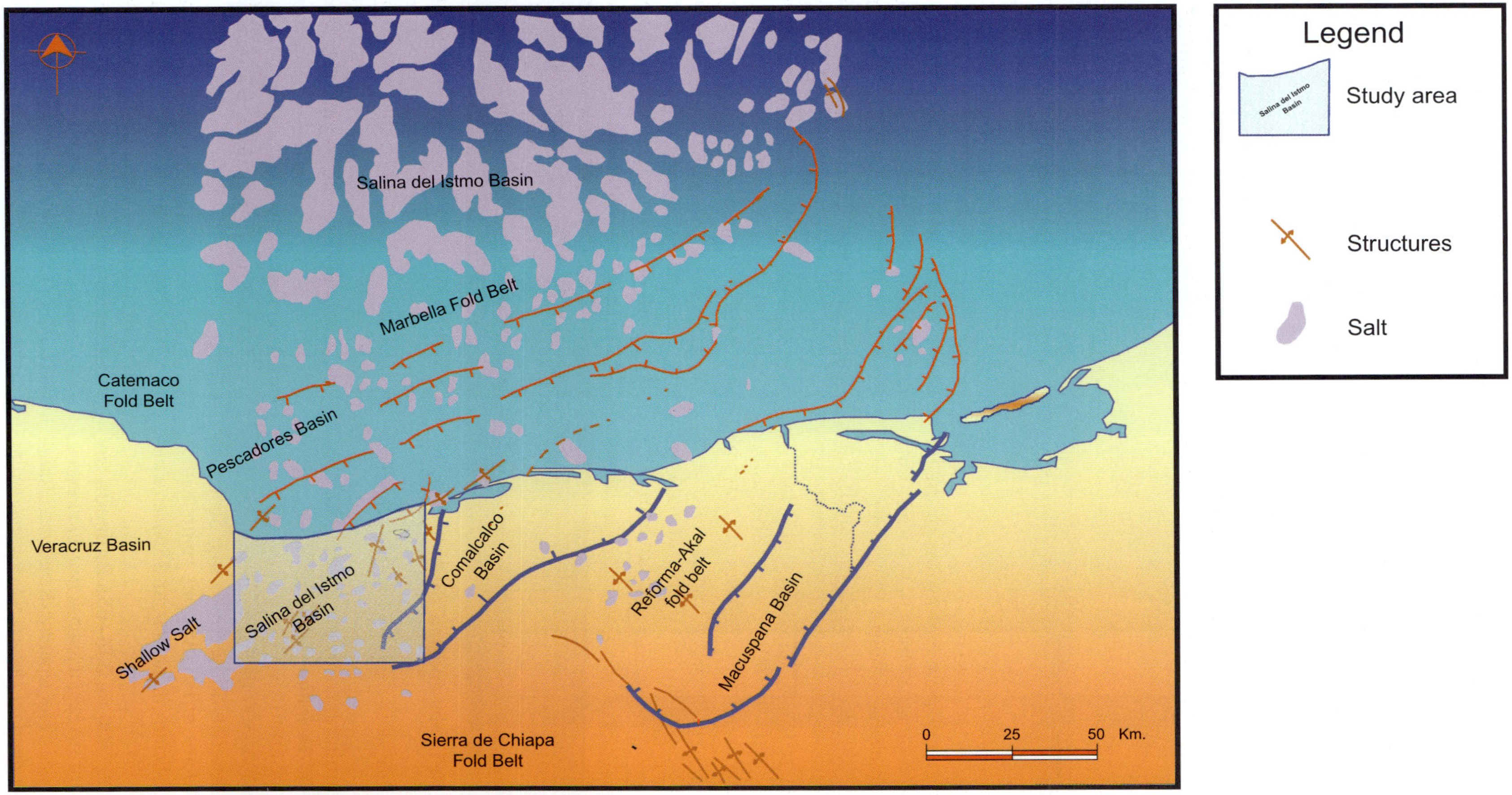

FIGURE 1. Location map of the southeast Tertiary basins showing the study area and principal tectonic elements.

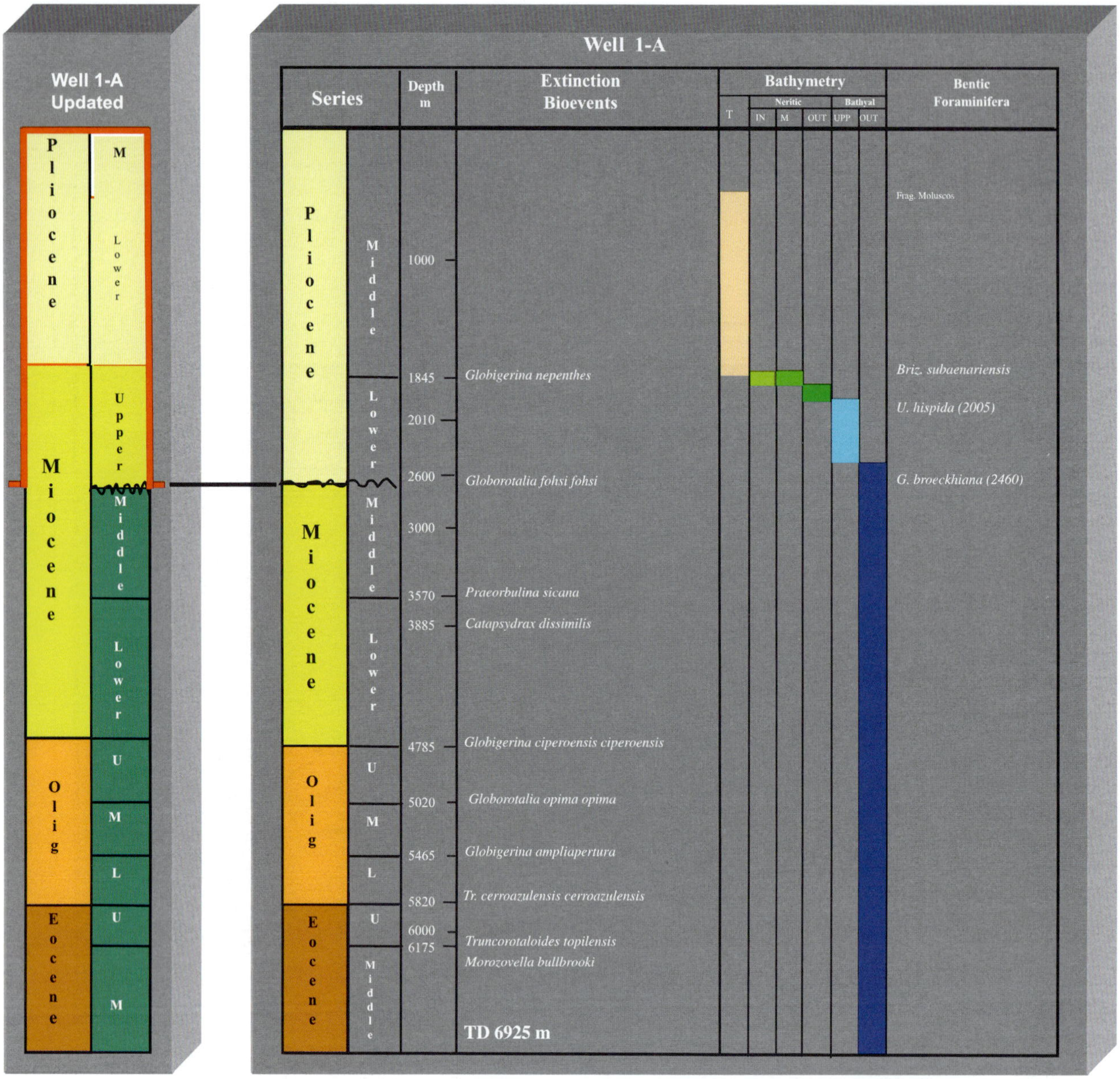

FIGURE 2. An example of the updated chronostratigraphic column for the wells in the study area. TD = total depth.

of benthonics foraminifera: *Gyroidinoides broeckhiana*, *Planulina wüellerstorfi*, *Anomalinoides cicatricosus*, *Cibicidoides robertsonianus*, and *Usbeskistania charoides*.

Upper bathyal (200–500 m, 656–1640 ft) depths are defined by the presence of the following: *Siphouvigerina auberiana*, *Melonis affinis*, *Melonis pompilioides*, *Karreriella caribaea*, *Pullenia qinqueloba*, and *Sphaerodina bulloides*.

Outer neritic (100–200 m, 328–656 ft) depths include the following forms: *Marginulinopsis marginulinoides*, *Marginulinopsis goajiraensis*, *Marginulinopsis messinae*, *Islandiella californica*, *Hoeglundina elegans*, *Plectofrondicularia californica*, *Rectuvigerina striata*, *Spiroplectammina mississipiensis*, and *Oridorsalis variapertura*.

Inner neritic (<30 m, <98 ft) depths include the following: *Amphistegina lessonii*, *Hanzawaia s.l.*, *Eponides antillarum*, and *Lenticulina americana*.

Transitional (intertidal) zones are defined by the presence of the following: *Ammonia beccarii*, *Elphidium s.l.*, and *Quinqueloculina lamarckiana*.

Paleobathymetric maps were then constructed and interpreted for the lower and middle Miocene based on the determination of benthonic foraminifera and generally indicate deep-water, lower bathyal environments (Figure 7a, b) (Tipsword et al., 1966; Sansores and Flores Covarrubias, 1972). During the upper Miocene accelerated shallowing of the depositional environment occurred, and at the end of the Miocene (5.73 Ma) we observe

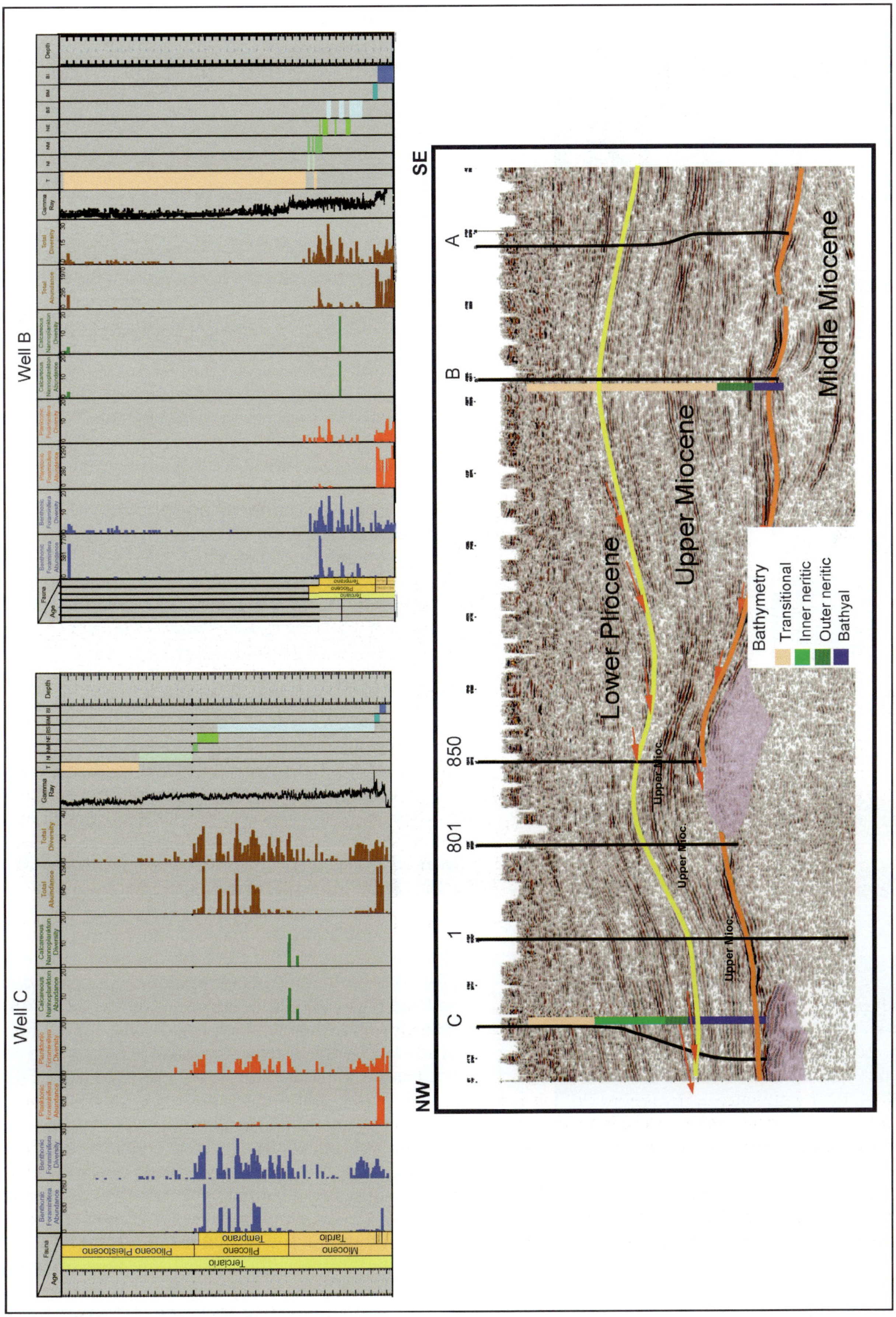

FIGURE 3. Correlation of chronostratigraphic horizons supported by high-resolution paleontologic studies tied to the seismic interpretation that allowed updating of the chronostratigraphic framework.

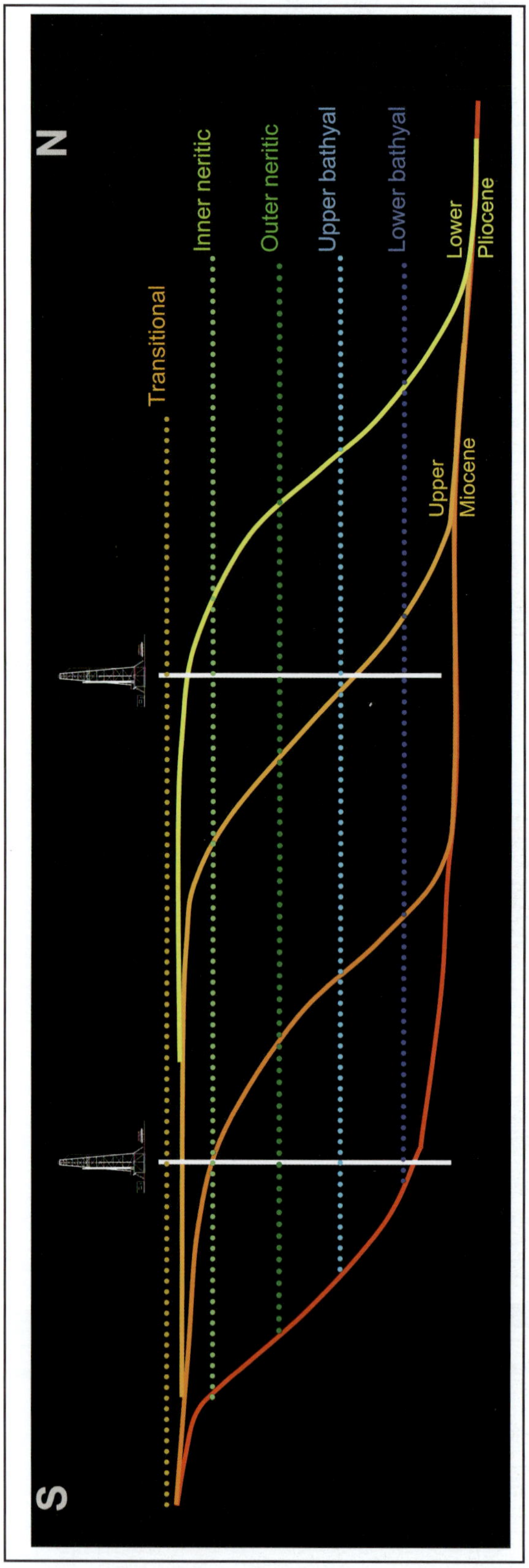

FIGURE 4. Chronostratigraphic horizon scheme (in yellow tones) reveals the progradation of the basin whereas paleobathymetric levels (dashed lines) are aligned in a parallel manner indicating that bathymetric variations transgressed the time lines.

a transition from shallow bathymetry (transitional-inner neritic) in the southern part of the basin to deep-water, upper and lower bathyal environments to the north. Finally, the same shallowing behavior continued during the lower Pliocene (Figures 7c, d; 8, 9).

STRATIGRAPHY

The stratigraphic sections with biostratigraphically correlated sequences have lower and middle Miocene sawtooth electrofacies from which high clay content is inferred. This section is interpreted to be composed of submarine fan deposits. A relevant feature is a regional unconformity that truncates the first two biozones (*Globorotalia mayeri* and *Globorotalia fohsi robusta*) and is commonly dated by *Globorotalia fohsi fohsi* (12.5 Ma). This middle Miocene unconformity is the evidence for a very important tectonic event locally known as the Chiapaneca orogeny. The Chiapaneca orogeny led to the deposition of a great quantity of sediment that caused the evacuation of allochthonous salt and the consequent generation of minibasins. These minibasins are the depocenters that contain the most important reservoir rocks in the Isthmus Saline Basin of upper Miocene and Pliocene age. These two significant regional unconformities are sometimes associated with salt welds (Figure 10).

Thirty-one wells that penetrated lower and middle Miocene strata have been drilled in the study area. The representative rocks in this sequence comprise alternating sandstones and shales deposited in slope environments and are characterized by sawtoothed electric log responses suggestive of high clay content.

Starting in the upper Miocene, the rate of sedimentation changed in the basin, leading to prograding sedimentary cycles that evolved from deep-water environments in the lower part of the upper Miocene to shallow environments in the top (Figures 8, 9).

The correlation of these distal environments along chronostratigraphic surfaces from the north is accomplished by integrating the paleontology with seismic data and logs (Figure 8). This is important for the upper Miocene sequence where well-log electrofacies show fining-upward sawtooth patterns suggestive of sand bodies deposited in submarine fans in the northern part of the section. From wells 5 to 10, the sawtooth electrofacies pattern continues with the addition of small cylindrical bodies on the upper Miocene base. These are indicative of slope channels eroded into preexisting sandy lobes. Furthermore, towards the upper Miocene top, a rapid shallowing of the depositional environment is noted in the central part of the section where the electrofacies display the transition from stacked, upward-coarsening, and cylindrical patterns typical of deltaic facies to swamp and fluvial transitional facies in the extreme southern

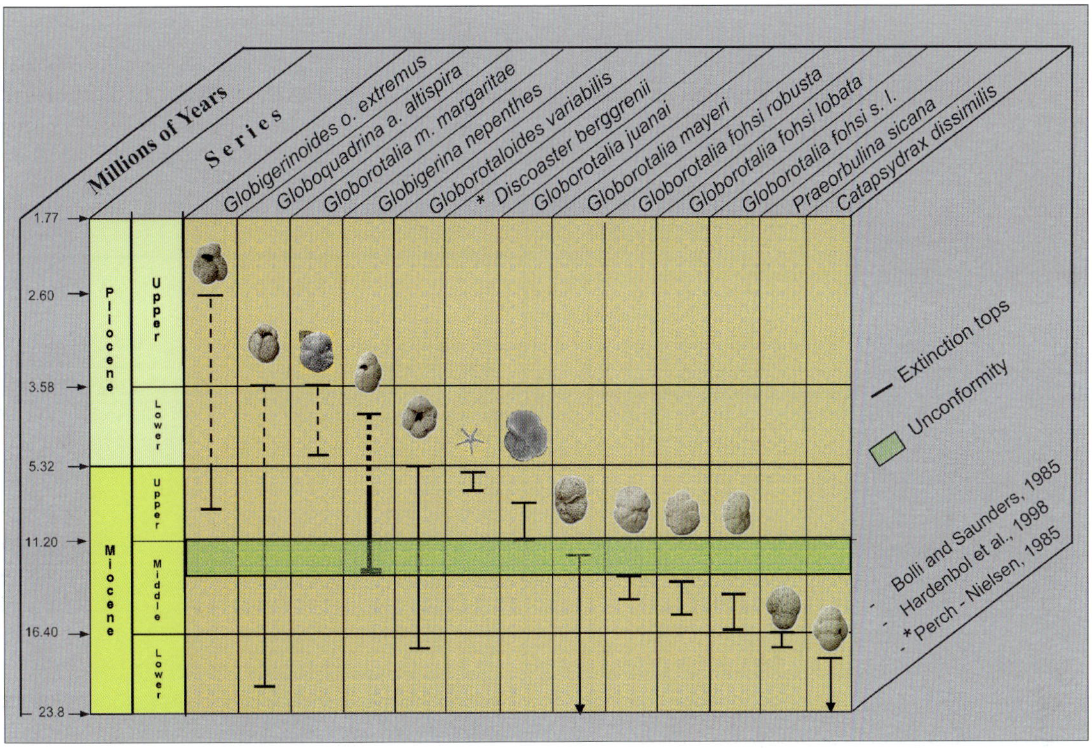

FIGURE 5. Stratigraphic ranges of bioevents that determine the age limits. Dashed lines are times that may represent the first appearance of the fossils, which do not always correspond to their extinction. In that case, they lose their character as a bioevent.

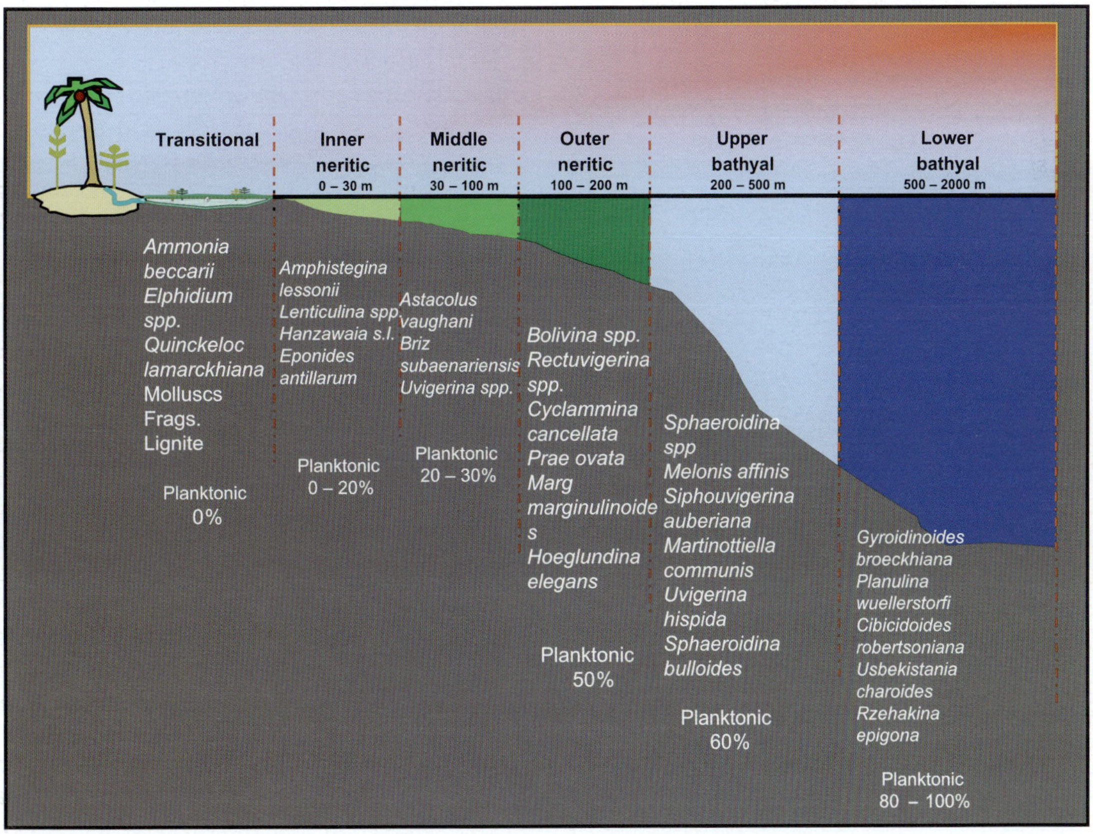

FIGURE 6. Scheme for the determination of the bathymetric levels associated with the benthonic faunal assemblage.

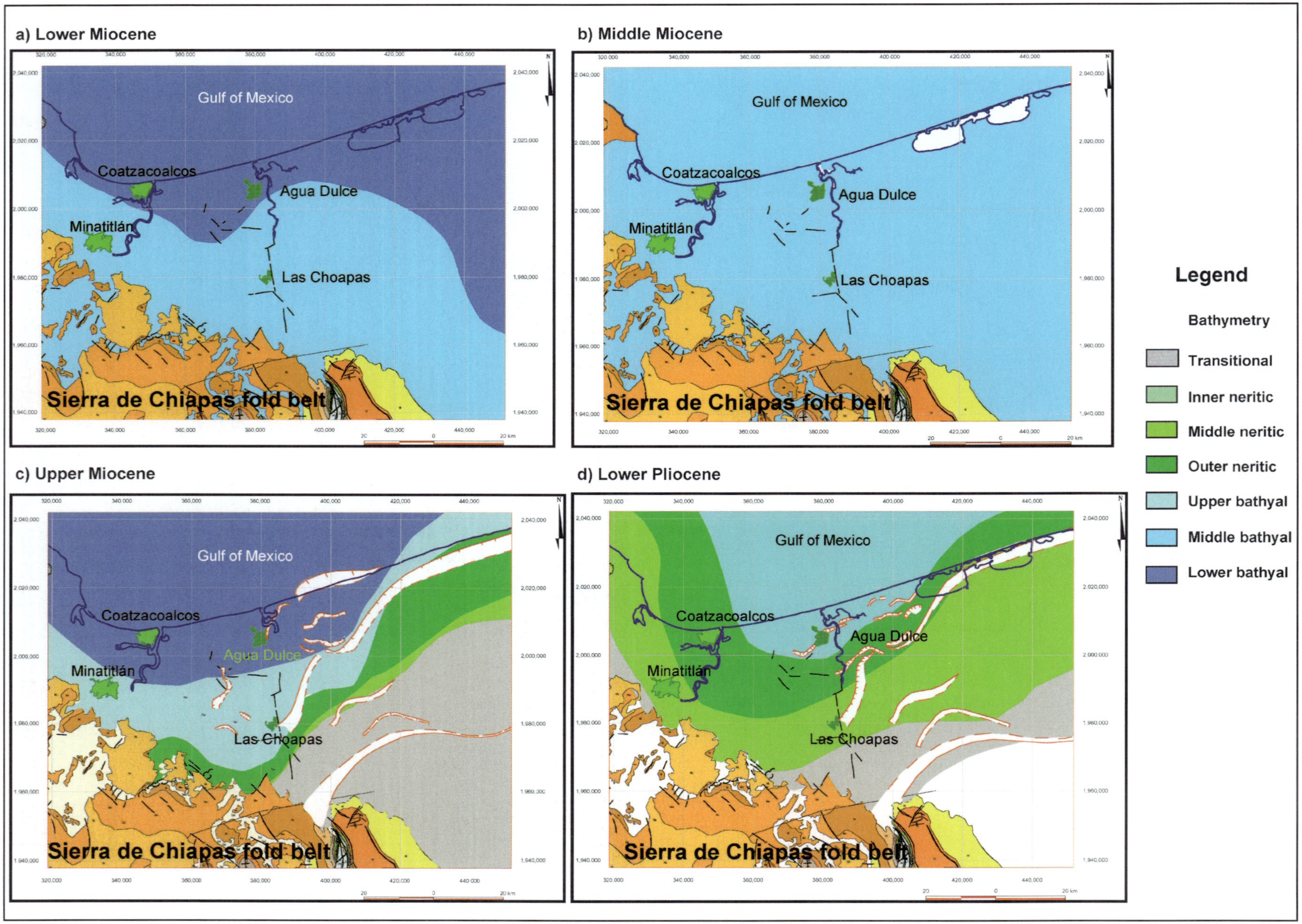

FIGURE 7. Maps showing the paleobathymetric evolution in (a) lower Miocene, (b) middle Miocene, (c) upper Miocene, and (d) lower Pliocene.

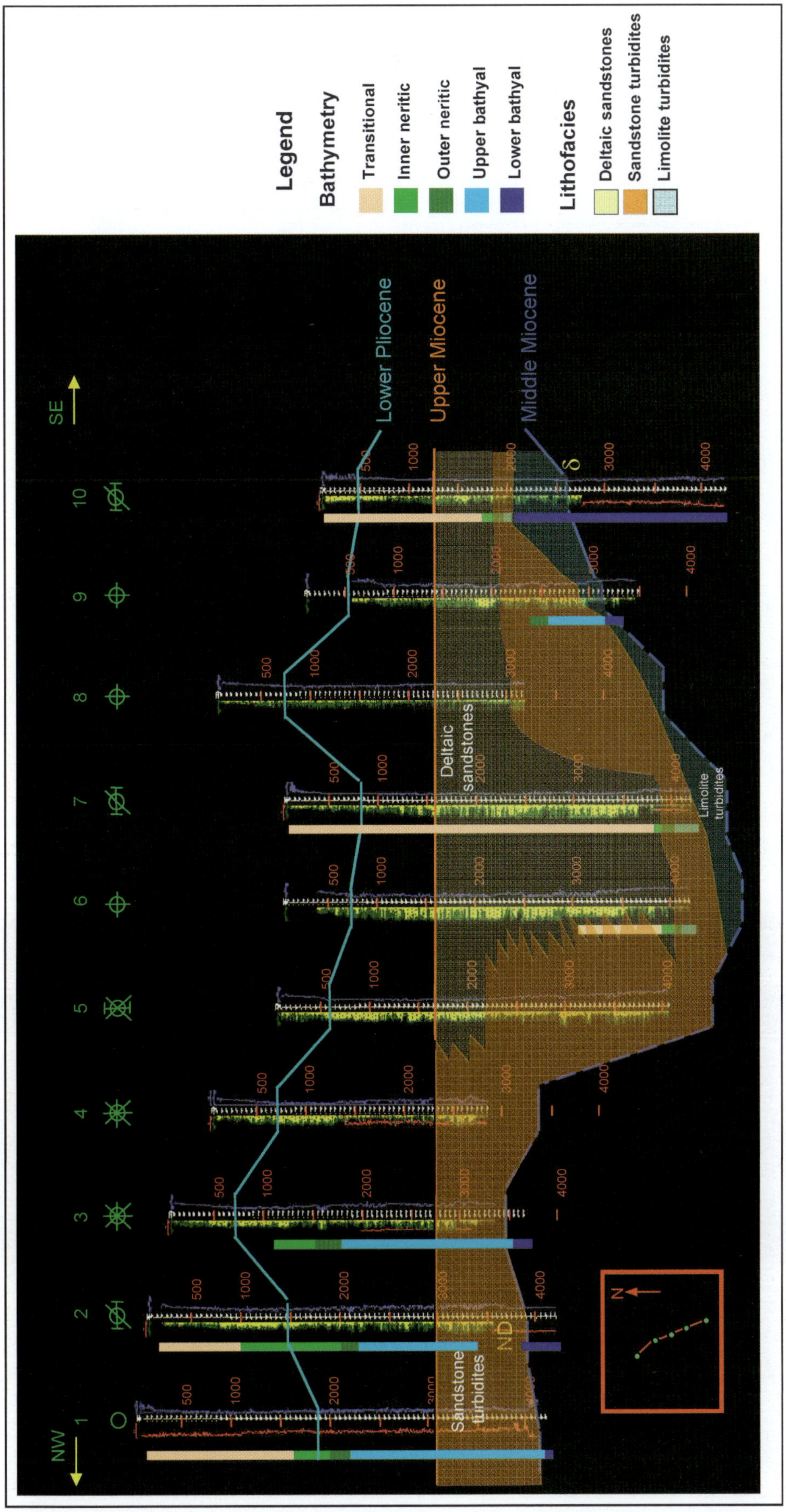

Figure 8. Neogene stratigraphic section for the Isthmus Saline Basin showing integrated data from geophysical logs, bathymetric analysis, relative ages, information interpreted from electrofacies, and distribution of depositional environments. ND = no data.

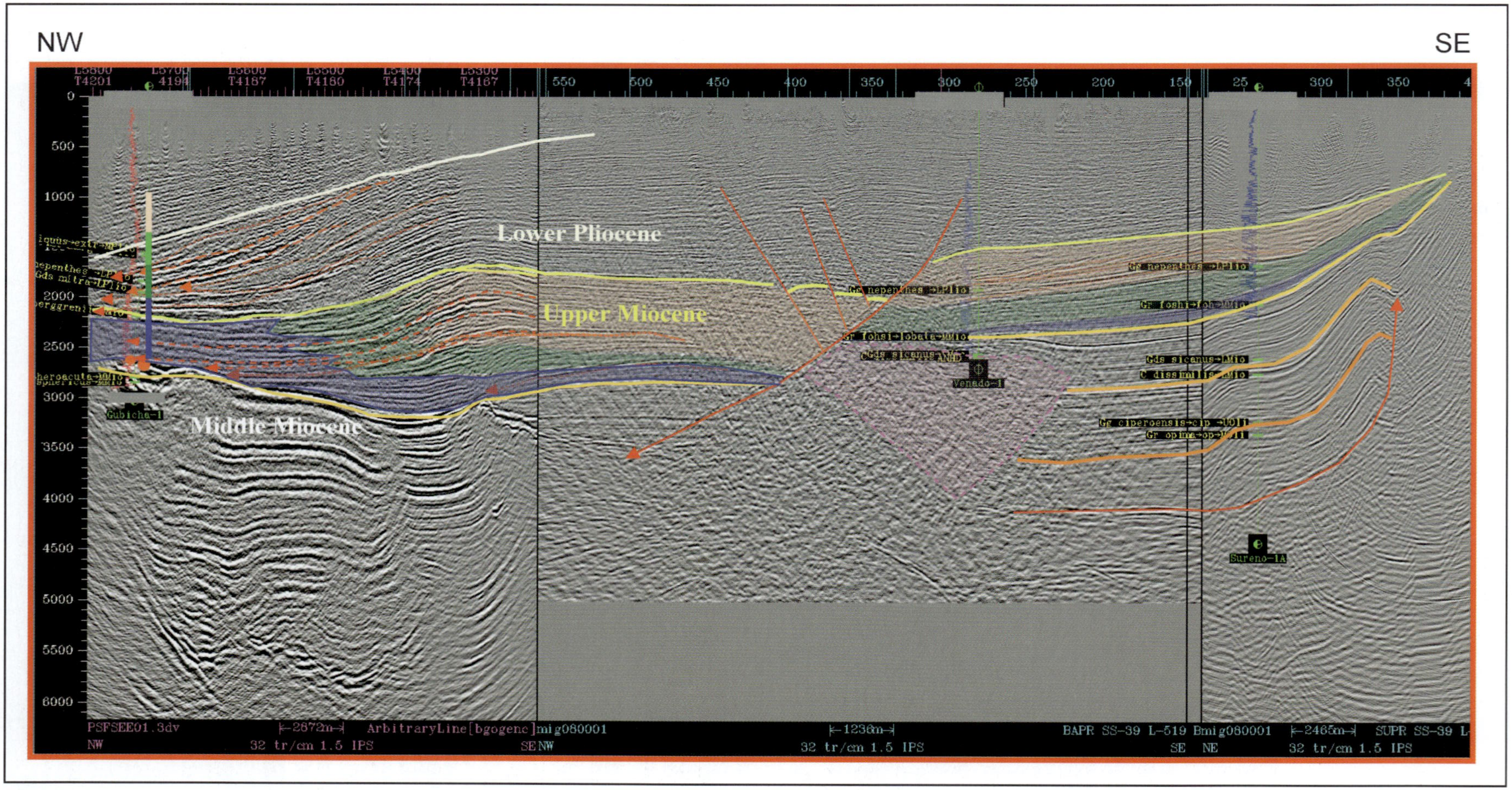

FIGURE 9. Composite seismic section showing the correlation of chronostratigraphic horizons where biostratigraphic resolution is nonexistent because of paleoecological conditions in the lower Pliocene and upper Miocene.

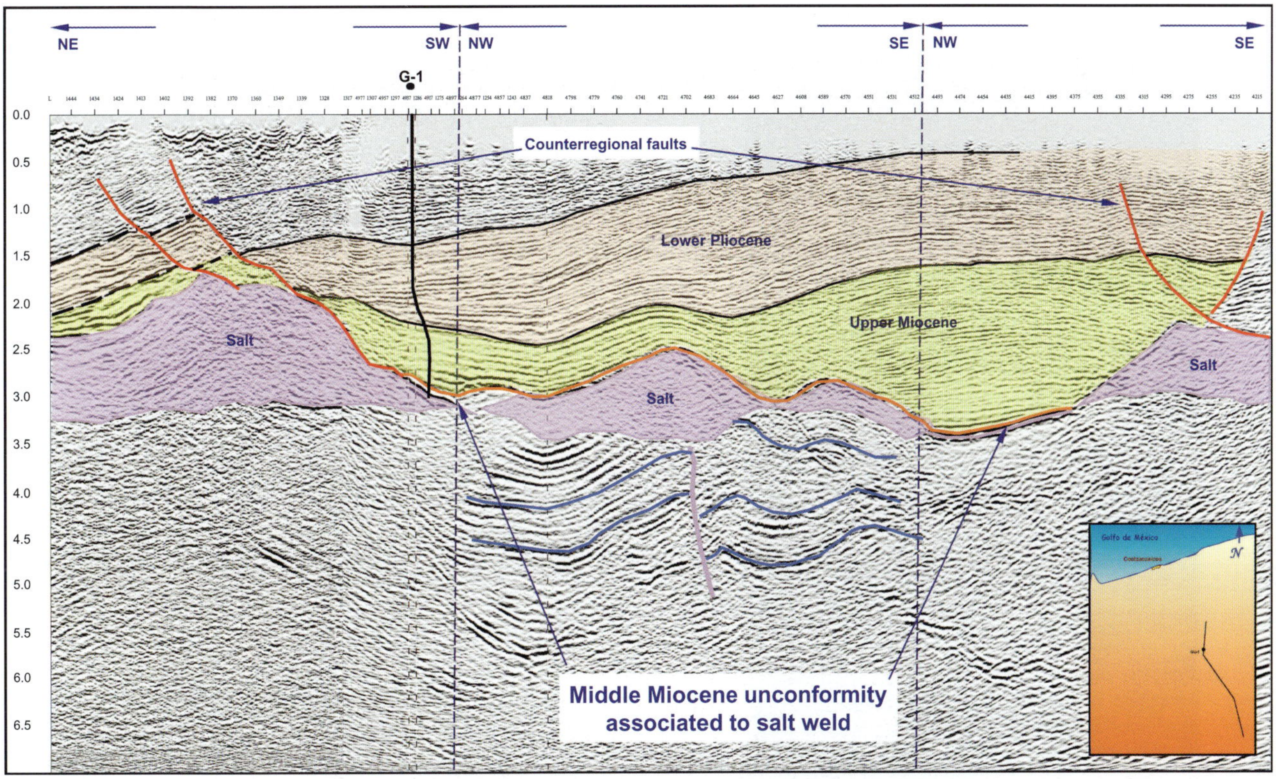

FIGURE 10. Seismic section showing thick sedimentation and changing structural style above the middle Miocene unconformity, which created the depocenters that formed the Isthmus Saline Basin.

part of the section. In this part of the section, the cylindrical electrofacies again become sawtoothed with both coarsening-upward and fining-upward patterns. The log facies characteristic of this environment of deposition occur in the area shown in Figure 11.

The biostratigraphy does not provide a sufficient resolution in the transitional facies tract where planktonic foraminifera are very scarce, or those that do exist may be reworked from older horizons or have other complications; therefore, the correlation of this surface would have been impossible without referring to the corresponding seismic chronostratigraphic surface.

The correlation of chronostratigraphic surfaces into areas where paleoecologic conditions do not allow dating by planktonic organisms is strongly supported by a seismic stratigraphic methodology that employs internal terminations, configurations, and patterns on seismic data to correctly determine biostratigraphic markers (Figure 9).

An analysis of the stratigraphic framework resulted in the determination that the Isthmus Saline Basin was created during the upper Miocene by the evacuation of a great volume of salt, whereas the prior interpretation was that the basin formed in the Pliocene. For example, Figure 10 shows where the middle Miocene unconformity marks the boundary between the compressional structural style (below) and an extensional structural style with counterregional growth faults (above). A thick upper Miocene sedimentation related to the mobilization and total evacuation of allochthonous salt is evident in the south central part of the section where it forms a turtle structure associated with the salt weld. This process created the minibasin depocenters of the Isthmus Saline Basin. As a result of this interpretation, the position of the most important Neogene productive plays has been updated.

DEPOSITIONAL ENVIRONMENTS

Mapping of the lithofacies distribution was accomplished by reviewing, integrating, and updating existing data and interpretations, and by incorporating the new seismic data and wells drilled in the last decade. The lower to middle Miocene throughout this area corresponds to deep-water environments in outcrops on the flanks of the Sierra de Chiapas and in the subsurface toward the north. These deposits consist of a variety of siliciclastic sediments. Examples of facies include alternating sandstones and shales in turbiditic deposits with classic features: deformation caused by gravity sliding, flame structures, loading structures, and channel fills with basal conglomerates and sandy clay matrix showing both normal and inverse graded bedding. Inner and middle fan slope facies can be found in outcrops of the Sierra de

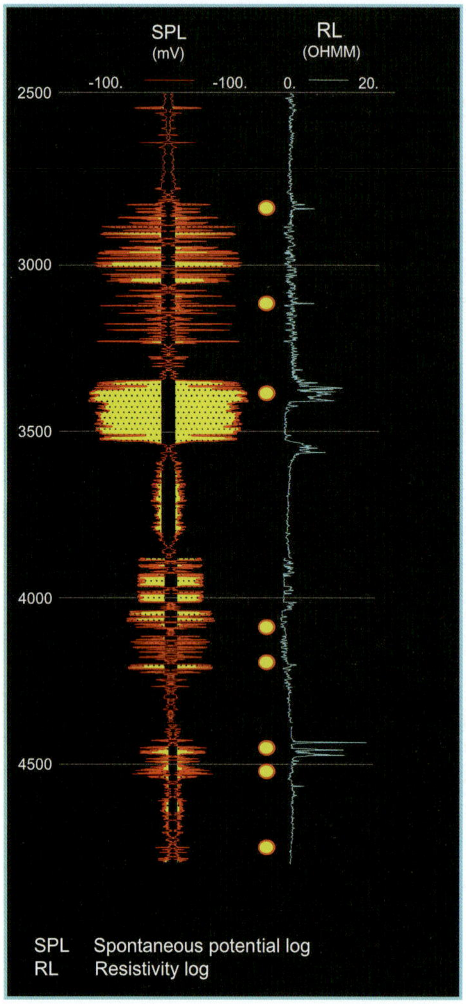

FIGURE 11. Electric log facies pattern characteristic of the Isthmus Saline Basin sedimentary environments. The fluvial environments show bell-shaped sawtoothed upward-fining patterns; the deltaic environments show upward-coarsening cylindrical- and funnel-shaped patterns. Finally, the turbiditic basinal facies show funnel-shaped sawtoothed patterns with some small box shapes.

Chiapas. These rocks occur in fans as channel, overbank, and levee deposits (Tarango and Sánchez, 2005) toward the north in our study area, where these sediments are found in the subsurface consisting mainly of sandy shale outer fan facies (Figure 12).

By the upper Miocene, based on sandy turbidite lithologies and electric well-log characteristics (Figure 11), the southern part of the area consisted of transitional deltaic and shelf sand facies, as shown on the map in Figure 13 and the cross section (shown in blue in the figure's seismic section). Toward the north, the environment is the slope base on which sandy turbidites were deposited in fan, amalgamated channel, overbank, and levee facies. The environment changes to the south with shallower conditions toward the top of the upper Miocene. The external neritic bathymetry coinciding with the shelf edge and the beginning of the slope is shown in green. Finally, shelf and subtidal sediments were deposited in the southernmost part of the study area into the upper Miocene (shown in orange in the figure). In general, the upper Miocene is characterized by facies variations, both laterally and vertically, that evolve from deeper to shallower water. The geometry of these deposits is controlled by minibasins caused by salt evacuation.

The robust and rapid facies progradation associated with counterregional growth faults and salt evacuation that began in the Miocene was accentuated in the lower Pliocene as the shelf margin advanced toward the northwest. Shallow-water facies continued to prograde in the same direction as the other facies, with deltaic facies to the south (Figure 14). The slope and shelf extended to the northwest across a wide area that was affected by salt diapirs, which in turn affected the trap formation in the sandy facies of anastomosing slope fans, amalgamated channels, overbank deposits, and levees.

PLAY ANALYSIS

A final map of petroleum system risk was generated (composite common risk segment [CCRS]) using the play fairway analysis methodology (Figure 15), which evaluates the presence and effectiveness of the petroleum system elements (source rock, reservoir, and seal) by mapping each risk element (common risk segment [CRS]) using green, yellow, and red for increasing levels of risk.

The presence and effectiveness of source rock, reservoir, and seal were described and defined to construct risk and composite risk maps for each of these play elements in the upper Miocene and lower Pliocene.

The range of geochemical analyses from wells and outcrops of the southeastern basins of Mexico, combined with basin modeling, provides information such as kerogen type, organic richness (total organic carbon [TOC]), maturity, oil-rock correlations, source rock age, and timing of hydrocarbon generation and expulsion. A compilation of this information is used to construct source-rock distribution and effectiveness maps to support the risk map (CRS) for this play element (Figure 16).

Net sand isopach maps were constructed to indicate the distribution of reservoir rock (Figure 17a). Petrophysical data, porosity, and permeability (Figure 18) were combined to predict minimum cutoff values for reservoir effectiveness in this area. This was followed by the compilation of petrographic data to determine the mineralogy of the sandstones and the diagenetic factors affecting porosity and permeability, yielding one more element for the evaluation of reservoir rock effectiveness. The combined risk map (CRS) for the presence and effectiveness of reservoir rock is shown in Figure 17c.

In an analysis similar to that of the reservoir rocks, seal rocks were evaluated through an isopach map combined

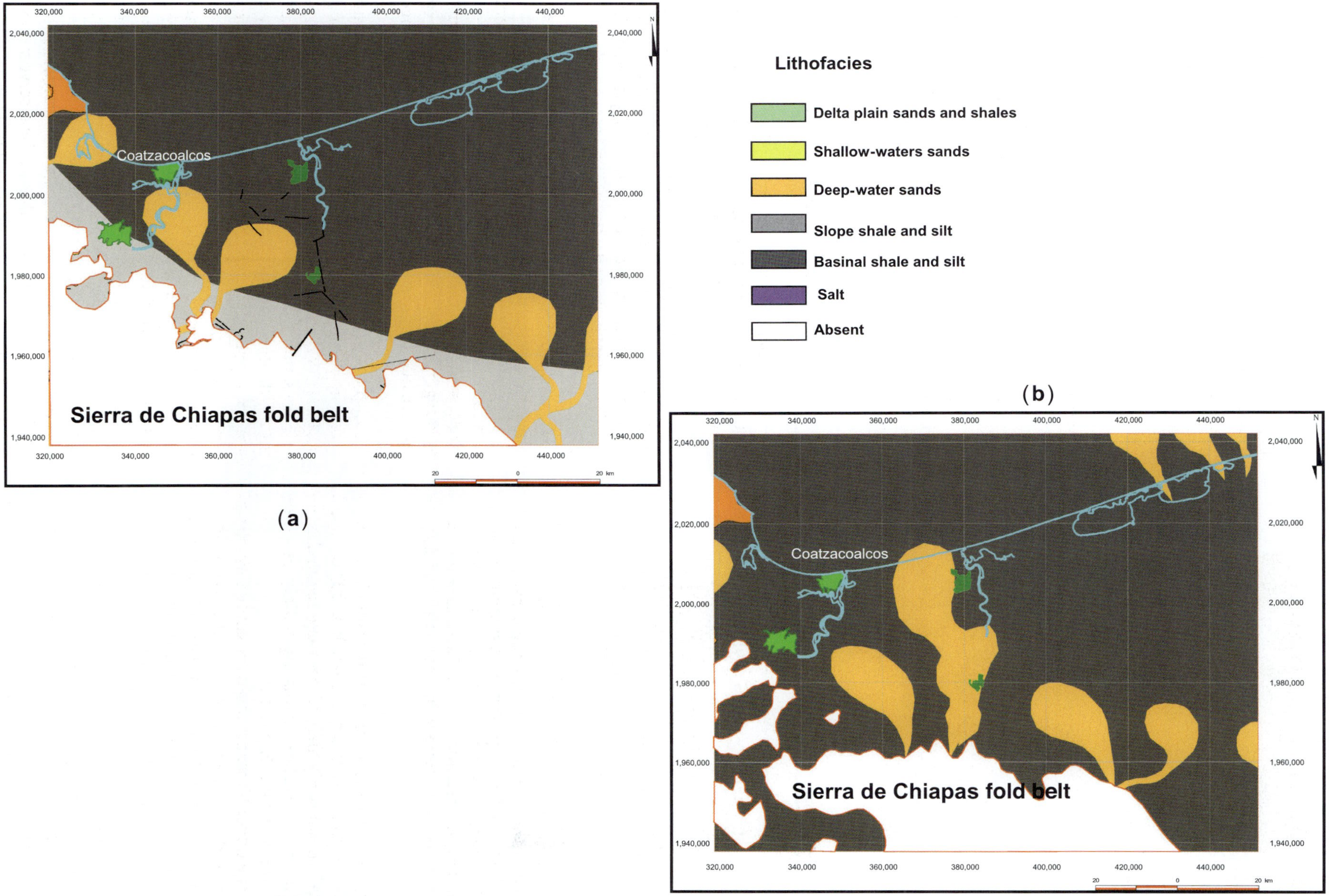

Figure 12. Maps of gross depositional environments (GDEs) in the lower Miocene and middle Miocene showing the presence of potential reservoir and seal rocks.

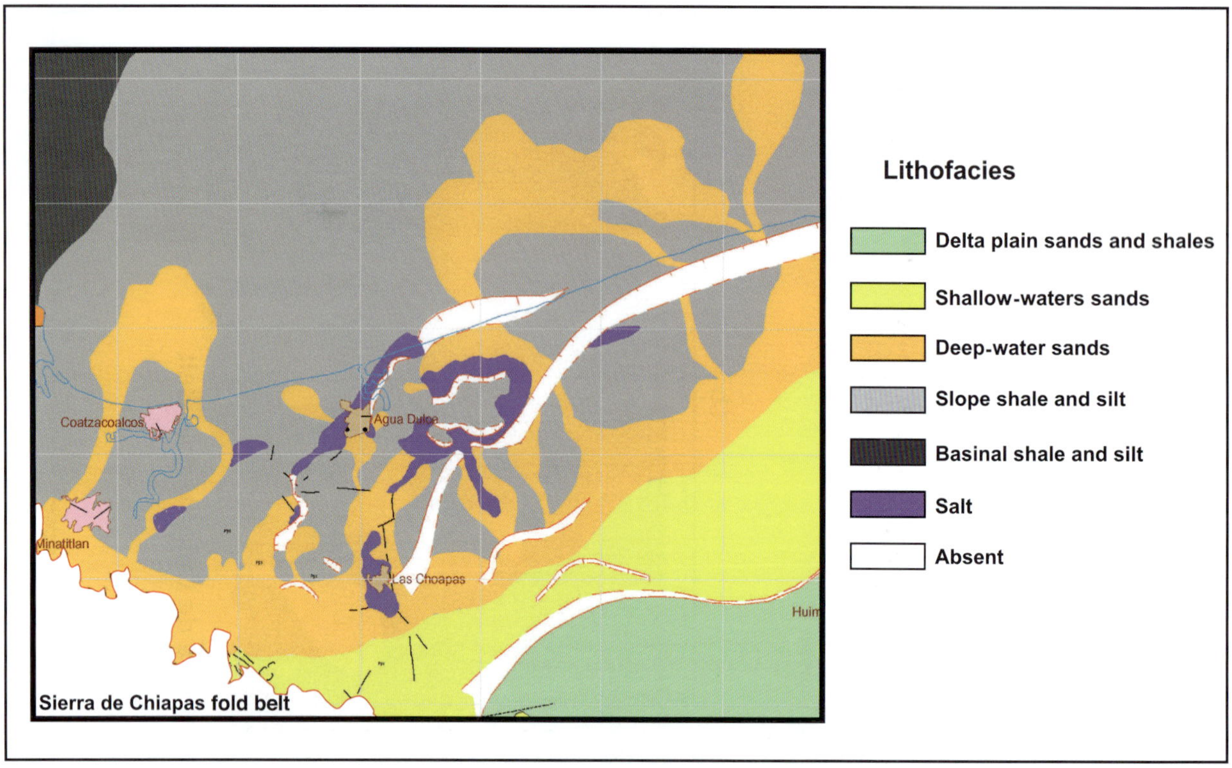

FIGURE 13. Map of depositional environments in the upper Miocene showing the presence of potential reservoir and seal rocks.

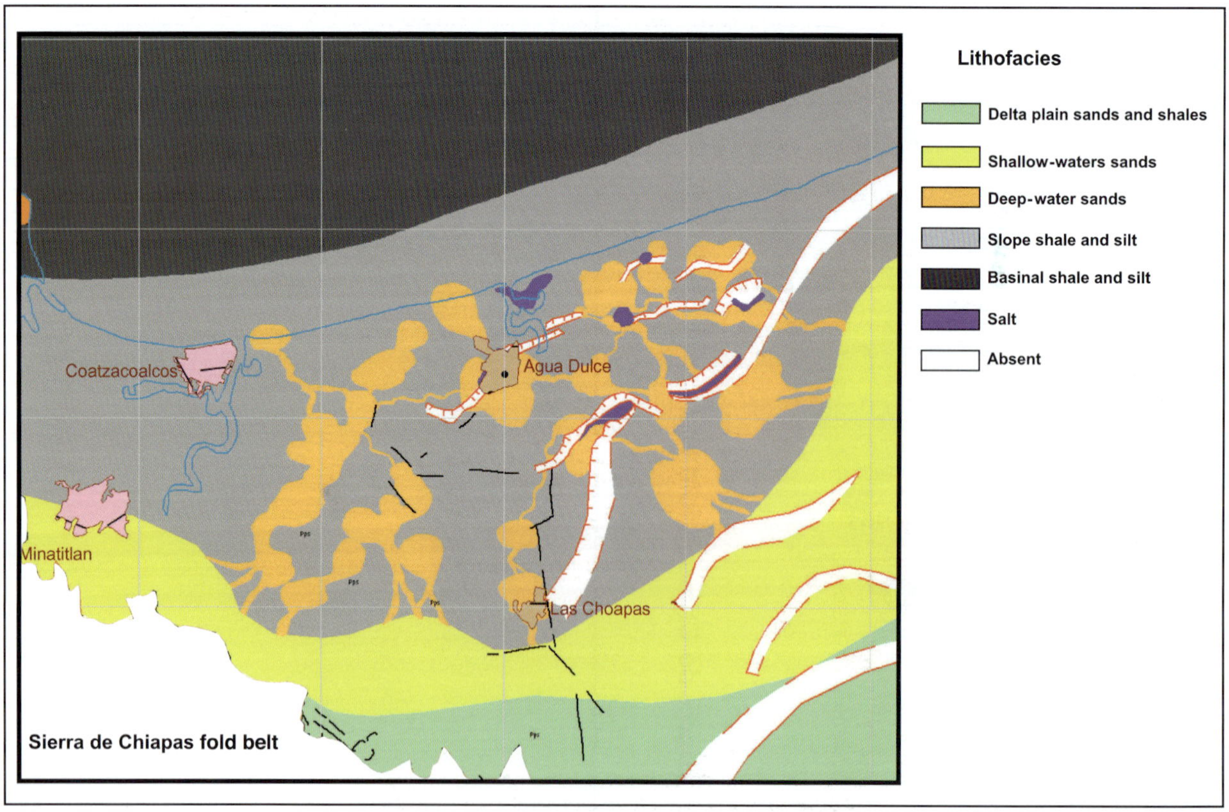

FIGURE 14. Map of gross depositional environments in the lower Pliocene showing the presence of potential reservoir and seal rocks.

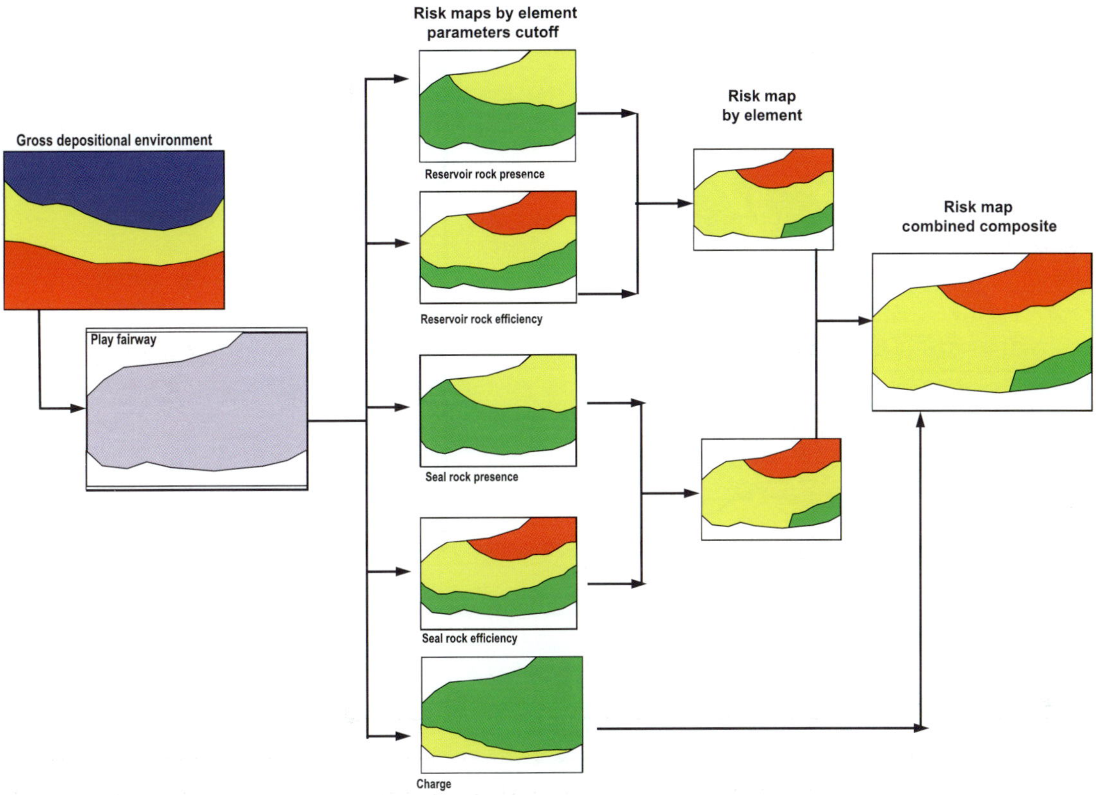

FIGURE 15. Work flow diagram showing the play analysis methodology employed in this study. Examples of mapped elements of the petroleum system and composite risk are shown for the upper Miocene.

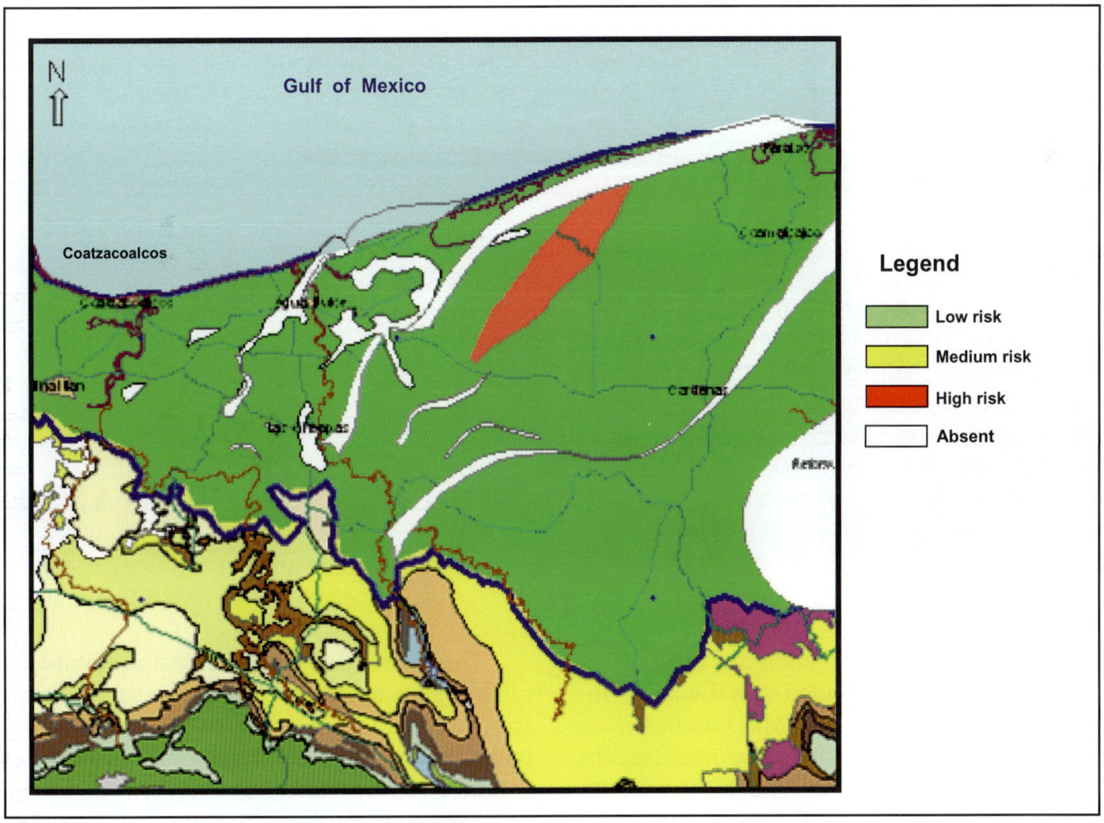

FIGURE 16. Source rock common risk segment (CRS) map.

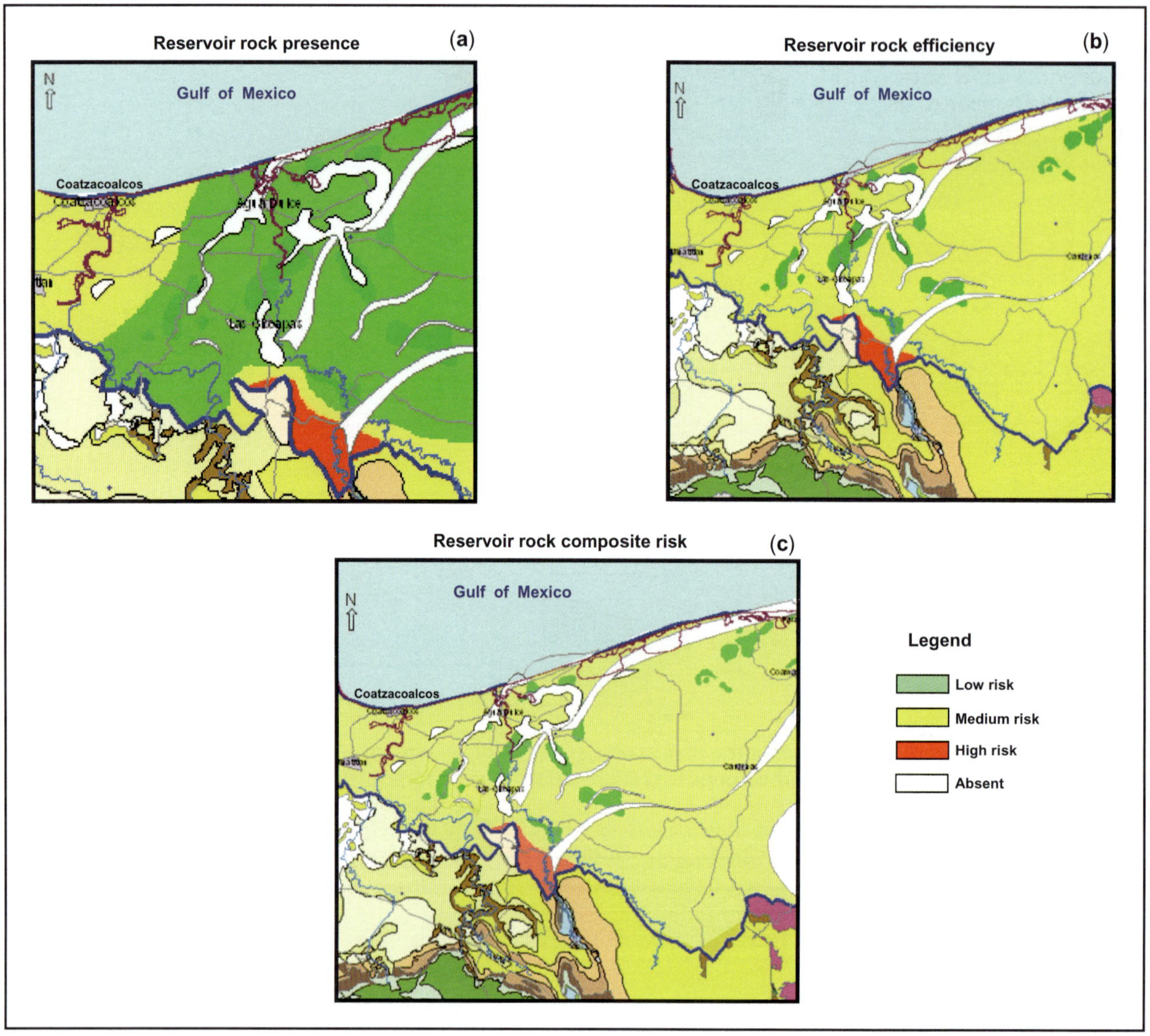

FIGURE 17. Maps of the presence, effectiveness, and common risk segment (CRS) for the reservoir rocks.

with an environmental deposition map (gross depositional environment [GDE]) to determine the presence of this element (Figure 19e). This was followed by mapping reservoir effectiveness (Figure 19b) and hydrocarbon retention capacity through the integration of reservoir pressure data, lithological descriptions to determine seal quality, the location of seeps, cumulative production, etc. The integrated seal risk map is shown in Figure 19c.

Neogene Plays

Sixteen principal fields were analyzed in the Isthmus Saline Basin, and we concluded that the most important established plays are in upper Miocene and lower Pliocene sandy turbidites. Plays were also identified in upper Miocene deltaic facies, albeit with lower cumulative production. An additional, mostly unexplored middle Miocene turbidite sand play was identified, which presents an additional exploration opportunity. As shown in Figure 20, some petroleum fields with production from more than one play are present.

VOLUMETRIC PLAY ANALYSIS

The volumetric analysis of Tertiary plays was performed using the following PEMEX Exploration and Production proprietary tools.

The BDOE (the Southern Region Database of Exploration Opportunities) is a proprietary database that contains the probabilistic analysis of each opportunity (prospect) based on objective horizons, areas, thicknesses, hydrocarbon type, expected hydrocarbon volumes, probability of commercial and geological success, and other economic considerations.

FIGURE 18. Graphs of the porosity and permeability behavior of the upper Miocene rocks that supported the determination of the reservoir rock efficiency.

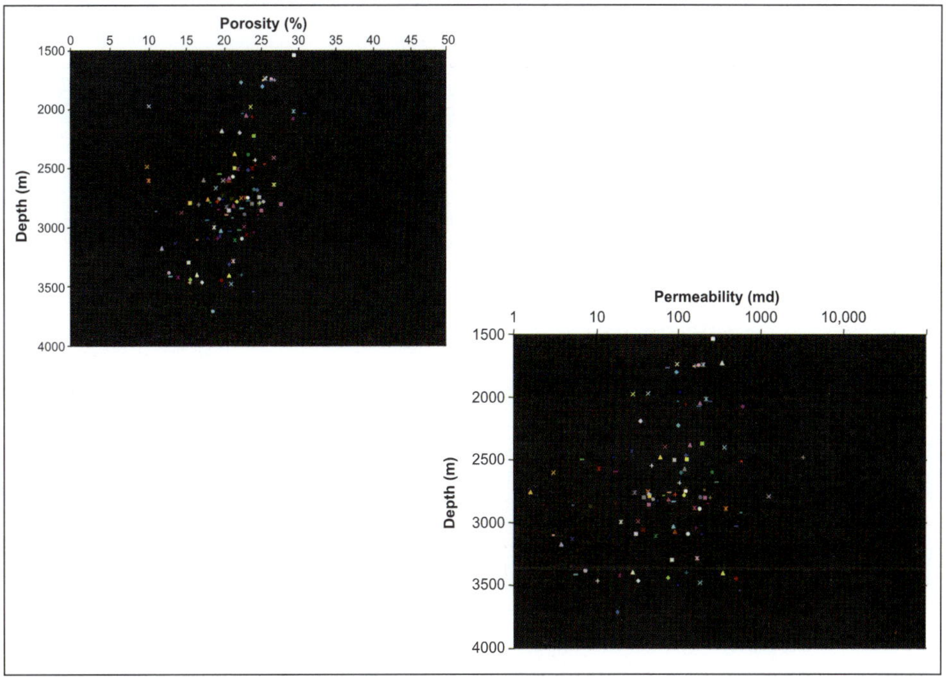

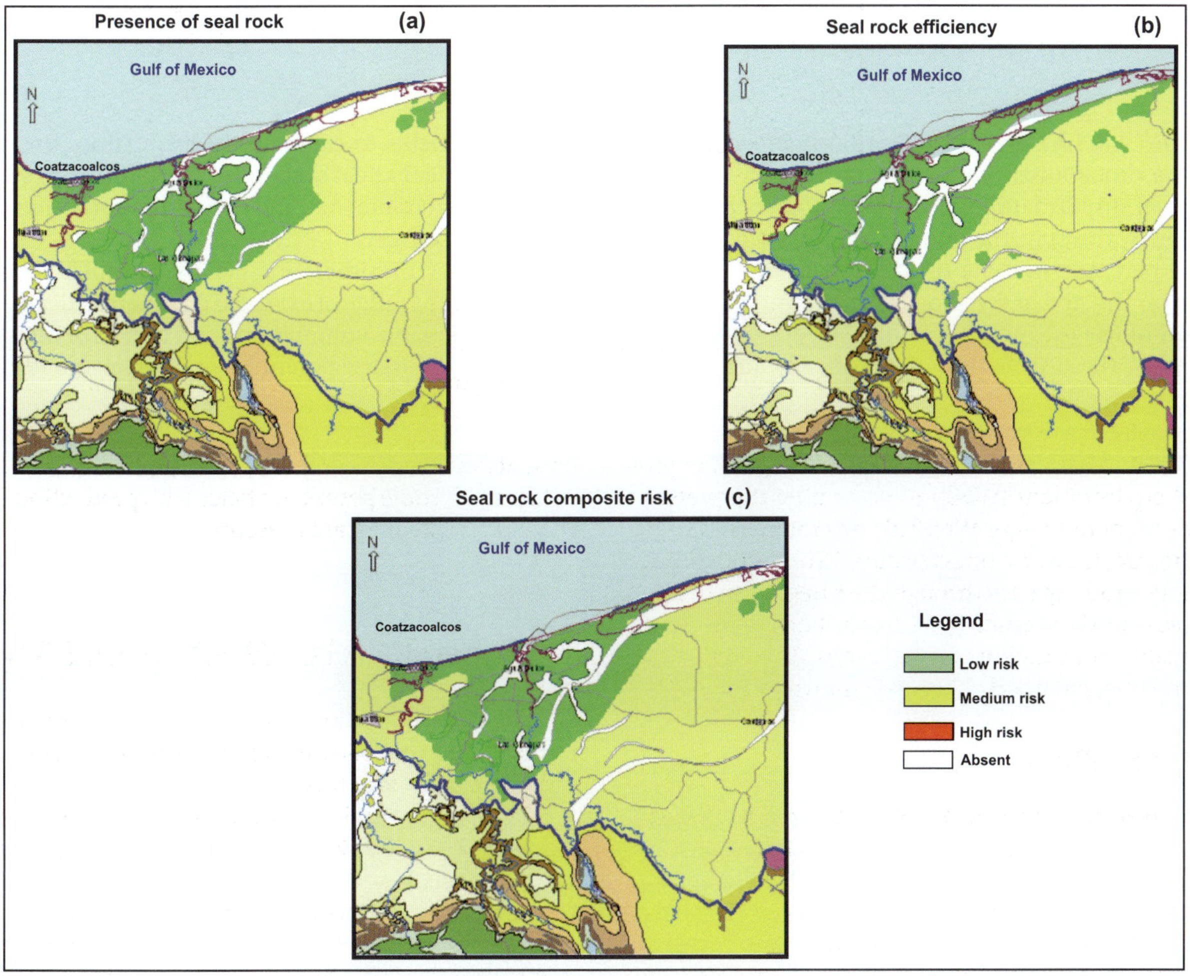

FIGURE 19. Maps of the presence, effectiveness, and common risk segment (CRS) for the seal rocks.

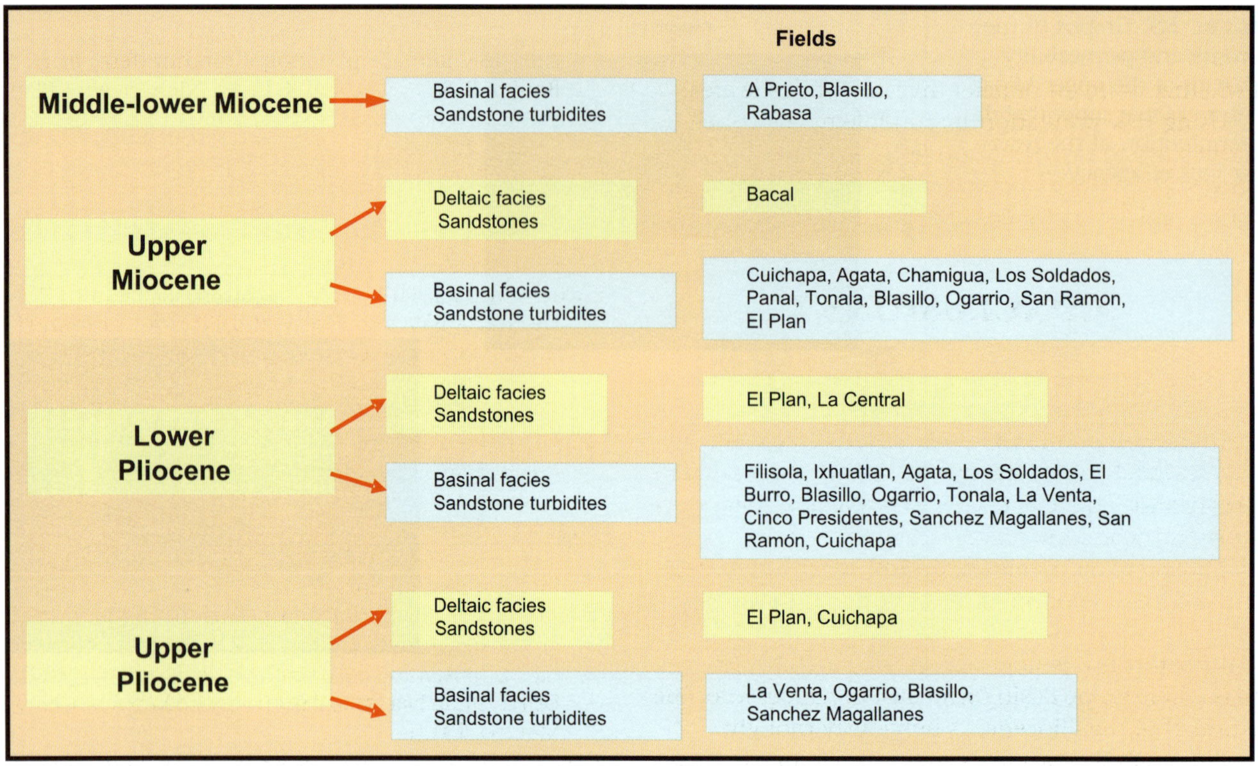

FIGURE 20. Stratigraphic location of the principal plays in the upper Miocene and lower Pliocene that produce hydrocarbons in different fields of the Isthmus Saline Basin.

The CERPLAY (Record of Play Evaluation) tool produces a probabilistic simulation of the play resource potential. Input data for the simulation include areas covered by the play, the number of visualized prospects in the play, the number of additional predicted leads based on the updated geological framework, the distribution of the estimated resource size for exploration prospects, the distribution of historically and recently discovered fields to calibrate the evaluation, the probabilities of geological success for the petroleum system and each of its elements, and the minimum commercial size for a discovery.

Through this simulation, we obtain probabilistic information for a range of expected commercial discoveries and their possible volumes (million barrels of oil equivalent [MMBOE]), the potential total petroleum resource

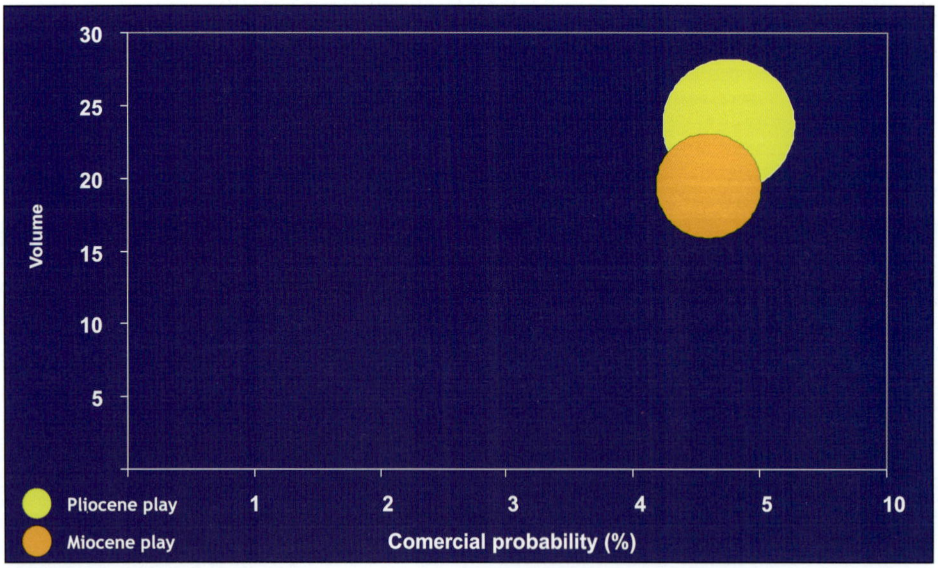

FIGURE 21. Volumetric evaluation of the remaining petroleum potential in Neogene plays of the Isthmus Saline Basin. We concluded that lower Pliocene plays have the highest potential and probability of commercial success.

of the play (MMBOE), and the probability that these will be discovered (percentage). This output gives a clear idea of whether to invest or defer investment in the area.

Using this simulation methodology, we conclude that, in comparison to the upper Miocene plays, the lower Pliocene plays have the most petroleum potential and highest probability for commercial success (Figure 21).

CONCLUSIONS

The integration of seismic stratigraphic interpretation with biostratigraphy was essential for the successful accomplishment of our congruent regional geologic correlation and analysis. This integrated geological-geophysical interpretation adds great value to stratigraphic play definition and economic evaluation in the basin.

This integrated framework highlights the upper Miocene and lower Pliocene turbidite sands as the most important Neogene reservoir rocks.

Updating the geological framework establishes that the Isthmus Saline Basin formed in the upper Miocene instead of in the Pliocene, as previously thought.

Play analysis is a successful methodology for compiling and assessing geologic risk in the different petroleum system elements, and it provides the ability to evaluate the economic probability of the remaining petroleum potential. Using this technique, we conclude that the Isthmus Saline Basin has greater petroleum potential and higher probability of success in the lower Pliocene plays compared to the upper Miocene plays.

ACKNOWLEDGMENTS

The authors thank the Subdirectorate of Technical Exploration Coordination (SCTER) of the Southern Region Exploration Asset Group of PEMEX for allowing the release of this technical document. We also give special recognition to Thomas Buerkert and Claudio Bartolini for their valuable commentary and corrections during the review of the original document.

REFERENCES CITED

Armentrout, J. M., and J. F. Clement, 1990, Biostratigraphic calibration of depositional cycles: A case study in High Island—Galveston East Breaks areas, offshore Texas: Gulf Coast Section SEPM Foundation, 11th Annual Research Conference, p. 21–51.

Bolli, H. M., and J. B. Saunders, 1985, Oligocene to Holocene low latitude planktonic foraminifera, *in* H. M. Bolli, J. B. Saunders, and K. Perch-Nielsen, eds., Plankton stratigraphy: United Kingdom, Cambridge University Press, v. 1, p. 155–262.

Gómez, C. P. T., 2003, Stratigraphic and structural analysis of the Neogene sediments of the offshore portion of the Salina Istmo Basin: Ph.D. dissertation, University of Texas, Austin, Texas, p. 22–35.

Hardenbol, J., J. Thierry, M. B., Farley, T. Jacquin, P. Ch. De Graciansky, and P. R. Vail, 1998, Mesozoic and Cenozoic sequence chronostratigraphic framework of European Basins: SEPM Special Publication 60, chart 2, Tulsa, Oklahoma, 764 p.

Kohl, H., 1985, Early Pliocene benthic foraminifers from the Salina Basin, southeastern Mexico: Bulletins of American Paleontology, v. 88, no. 322, p. 5–26.

Martínez, M. V., 2006, Reestudio Bioestratigráfico del Terciario del Pozo Tabaco-1: Reporte Inédito de Petróleos Mexicanos (PEMEX), 20 p.

Oviedo, P. A., 1996, Evolución Neógena de las Cuencas Terciarias del Sureste: Tesis de maestría, División de Estudios de postgrado de la Facultad de Ingeniería (DEPFI), Universidad Nacional Autonoma de México (UNAM), México, 231 p.

Pereznegrón, Z. I., 2003, Proyecto Encrucijada: Informe inédito Petróleos Mexicanos (PEMEX), p. 14–26.

Petróleos Mexicanos (PEMEX), 2004, Identificación, definición y delimitación de los plays Terciarios presentes en el activo de exploración reforma-Comalcalco: Informe inédito PEMEX, 431 p.

Petróleos Mexicanos (PEMEX)-British Petroleum, 1994, Cuencas Terciarias del Sureste y Área Marbella: Informe inédito PEMEX, p. 122–140, 176–190.

Sánchez, M. de O. R., 1987, Geología Petrolera del SE de México: Inédito de Petróleos Mexicanos (PEMEX), 230 p.

Sansores, J. C., and C. y Flores Covarrubias, 1972, Foraminíferos bentónicos del Terciario superior de la Cuenca Salina del Istmo de Tehuantepec: México, Instituto Mexicano del Petróleo, 528 p.

Stainforth, R. M., J. L. Lamb, H. Luterbacher, J. H. Beard, and R. M. Jeffords, 1975, Cenozoic. Planktonic foraminiferal zonation and characteristics of index forms: The University of Kansas Paleontological Contributions, Tables 1–8, p. 1–425.

Tarango, O. G., and M. de O. R. Sánchez, 2005, Entrenamiento de Campo 3: Libreto guía "Sierra de Chiapas", Brigada Escuela de Geociencias del Instituto Politécnico Nacional, 111 p.

Tipsword, H. L., F. M. Setzer, and F. L. Smith Jr., 1966, Interpretation of depositional environment in Gulf Coast exploration from paleoecology and related stratigraphy: Gulf Coast Association of Geological Societies Transaction, v. 16, p. 119–130.

3

Eguiluz-de Antuñano, S., 2009, The Yegua Formation: Gas play in the Burgos Basin, Mexico, *in* C. Bartolini and J. R. Román Ramos, eds., Petroleum systems in the southern Gulf of Mexico: AAPG Memoir 90, p. 49–77.

The Yegua Formation: Gas Play in the Burgos Basin, Mexico

Samuel Eguiluz-de Antuñano
Consultant, Manchester, United Kingdom

ABSTRACT

The Yegua Formation (Eocene) in the Burgos Basin is divided into three units. The lower Yegua represents a highstand systems tract (HST) whose top is truncated by a prominent regional sequence boundary at around 39.5 Ma. The middle Yegua is a facies complex that consists of basin-floor fan sandstones, channels, and unconformities corresponding to a lowstand systems tract and prograding wedge. The upper Yegua represents an HST underlying the Jackson Formation. The Yegua Formation probably reflects the final sediment influx toward the Gulf of Mexico Basin resulting from Laramide uplift and deformation.

The proposed sedimentary model and petroleum system allow for the increase of reserves and productivity in the Yegua gas play of the Burgos Basin.

INTRODUCTION

This is the first work published about the complex stratigraphy of the Yegua Formation in the subsurface of the Burgos Basin. Its purpose is to explain the regional stratigraphy and depositional environments of the Yegua Formation (Eocene Bartonian–early Priabonian). This study also contains data concerning the petroleum system and gas resources of the Yegua play. The results of this interpretation contribute to understanding the regional geology and permit an evaluation of the Yegua's hydrocarbon plays.

The study area is located in northeastern Mexico within the states of Tamaulipas and Nuevo León, between latitudes 25°20′ and 26°30′N, with the Rio Grande as its northern limit, and longitudes 98°23′ and 99°10′W. The area covers approximately 6000 km^2 (2317 mi^2) within the Burgos Basin, which is part of the Gulf of Mexico geological province (Figure 1).

The background for this chapter is the study of the Yegua and Jackson plays in the Burgos Basin conducted by the author for Petróleos Mexicanos (PEMEX) (unpublished data), in collaboration with Lawrence Meckel, Abel Octavio Pola, Gabriel Ramos, Javier Solano, and Angélica Tristán. The information included here only covers the Yegua interval.

The study involved the analysis of seismic three-dimensional (3-D) and two-dimensional (2-D) data, well logs and core, selected paleontologic data, and drill reports from more than 800 wells, including 120 exploratory wells and 680 development wells. Prior to the study, 61 wells in 12 gas fields had a cumulative production of 72 bcf (2.03 billion m^3) and a meager daily production of 3.8 million ft^3 (107,604 m^3) from the Yegua plays.

DOI:10.1306/13191077M902621

FIGURE 1. Location map of the study area (indicated by the red line) in the Burgos Basin, northeastern Mexico. The map shows subparallel sedimentary belts in a Tertiary regressive system that migrated through time from Paleocene in the west to Holocene in the east, and the facies changing from shallow to deep water in the same direction for each age.

EOCENE STRATIGRAPHY

The lithostratigraphic division of the Eocene in the Burgos Basin from base to top is represented by the Wilcox, Reklaw, Queen City, Weches, Cook Mountain, Yegua, and Jackson formations (Figure 2). The Claiborne Group and the Mount Selman subgroup nomenclatures have fallen into disuse, and the names Sparta, Pico, Laredo, and Crockett are not recognized. Prior to this study, the Yegua Formation was undivided, whereas the Jackson Group in the Burgos Basin is a formational unit informally divided into lower, middle, and upper members. Unfortunately, the imprecise definition of groups and formations in Texas has been carried into Mexico; therefore, difficulties exist in the correlations across both areas.

Lithostratigraphy

The lithostratigraphy of the Texas coastal plain and Burgos Basin was defined at the end of the 19th and beginning of the 20th centuries on the basis of outcrop studies. Eocene stratigraphic divisions in Texas (Figure 2) include the Wilcox Group at its base, the Claiborne Group in the middle part, and the Jackson Group at the top (Plummer, 1932). The Claiborne Group is subdivided into the Mount Selman and Cook Mountain subgroups, and the Yegua Formation. The Mount Selman includes, from base to top, the Reklaw, Queen City, and Weches formations; and the Cook Mountain subgroup includes the Sparta Formation at its base and the Crockett Formation at its top. The Carrizo Formation is the oldest stratigraphic unit in the Claiborne Group, whereas the Yegua Formation is the youngest. The Yegua Formation overlies the Crockett and underlies Jackson or Moody Branch strata. The Yegua Formation is equivalent to the Cockfield Ferry Formation in Louisiana.

The outcropping Claiborne Group in Texas was divided into two parts by Kennedy (1892); the lower part (Mount Selman) includes the base of the Carrizo through the top of the Weches formations and can readily be correlated from eastern to southwestern Texas. The upper part (Cook Mountain series) again included the Weches beds at the base and the Sparta and Crockett beds in the middle and at the top in central and eastern Texas, although it is difficult to follow toward the southwest on the surface because the basal sandy member (Sparta) is absent. The division of Kennedy (1892) lacks clarity, and Dumble (1894) supports renaming the Yegua beds contributing even more to the nomenclature problem.

Confusion among these definitions in eastern Texas and the difficulty of mapping them in southwestern Texas led to the imprecise subdivision of the Claiborne Group. Following the Claiborne subunits across southwestern Texas is difficult because the basal sandy member (Sparta) is absent; therefore, the Cook Mountain, as a group, is undifferentiated (Plummer, 1932). Eargle (1968) proposed to resolve the nomenclature problem of the Claiborne Group in southwestern Texas by naming the rocks under the Laredo Formation as the Pico Formation (Eargle, 1968), with both formations being equivalent to the Sparta Sand. The Pico and Laredo Formation names are not widely used.

In 1890, Penrose gave the name Fayette to the section above the Crockett Formation. In 1892, Kennedy gave the name Lufkin to the youngest beds of the Claiborne Group and in the same year, Dumble designated the same strata considered to be Fayette by Penrose as Yegua. Unfortunately, in 1894 Dumble included some of the marine beds of the Crockett Formation within the nonmarine Yegua, creating confusion that persists to this day. (Plummer, 1932) Adding to this confusion is the work of Kennedy (1896) who assigned the Crockett and Sparta beds to the Yegua. In that same year, Vaugham (1896) described the Cockfield Ferry beds in Louisiana as approximately equivalent to the original Yegua of Dumble (1892). Afterward, Deussen (1914) restricted the Yegua

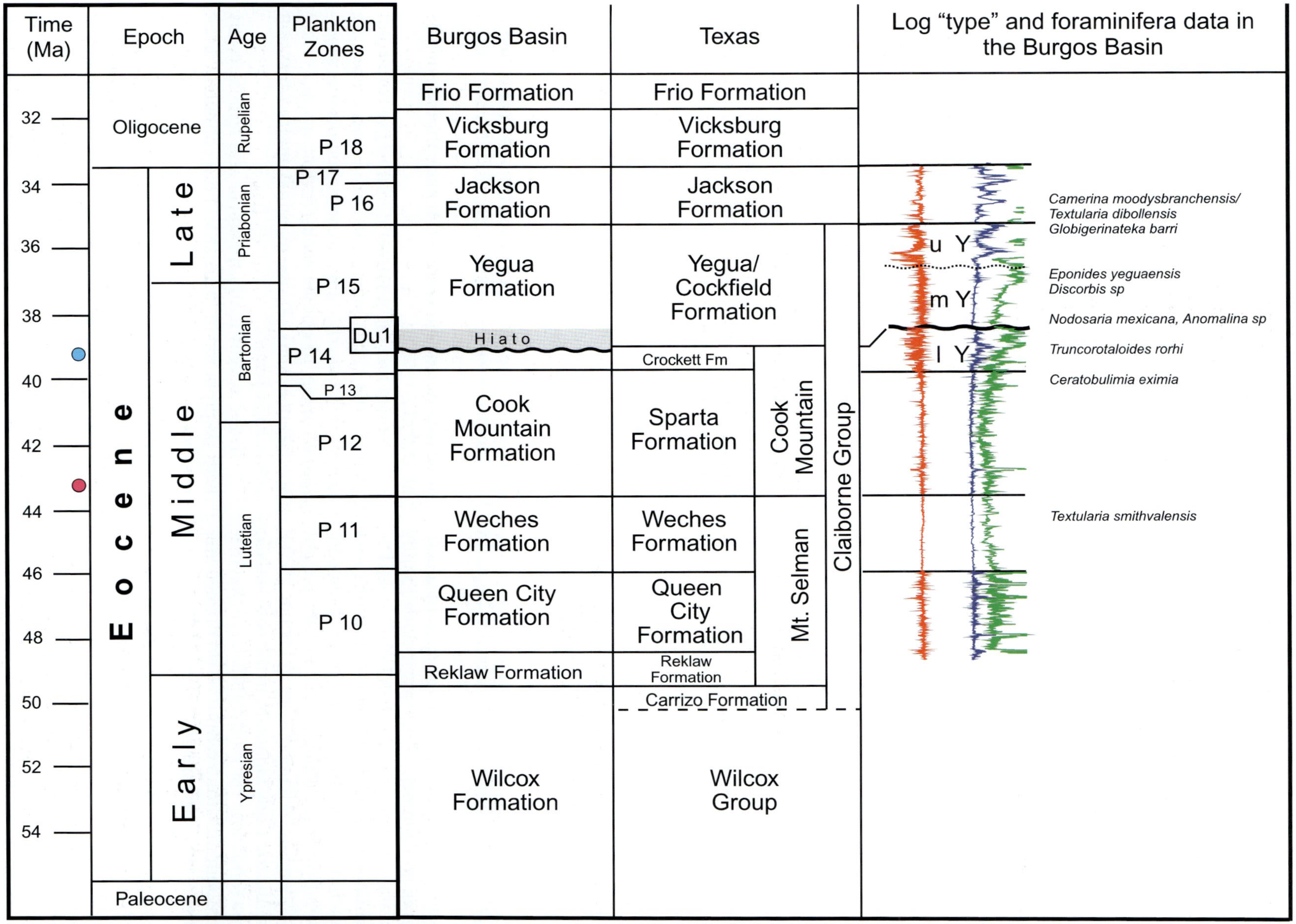

FIGURE 2. Stratigraphic column of Burgos Basin and Texas. Plankton zones, time, epoch, and age are from Berggren et al. (1995). Formation tops according to Rosen et al. (1994) based on plankton zones. Representative type log (gamma ray left and resistivity curves right) without scale showing the lower, middle, and upper updip Yegua. Foraminifera data from Cook Mountain to Jackson formations (PEMEX, 2000). The red dot is an isotopic age of rocks affected by Laramide deformation and the blue dot is an isotopic data of an undeformed intrusion, both data in the Sabinas Basin (Chávez-Cabello, 2005).

Formation, as initially proposed by Dumble (1892), to the nonmarine beds overlying the marine Claiborne and underlying the marine Jackson, and pointed out the equivalence of this unit to that defined by Vaugham (1896). Smaller divisions of the upper Claiborne section were not accepted by the U.S. Geological Survey (Plummer, 1932); the strata above Weches were named Cook Mountain, and the name Crockett, though fairly well established, is not used. The term Yegua Formation is currently used. This definition, accepted by the U.S. Geological Survey, became generally used for geological work in southwestern Texas, but in other parts of Texas, Louisiana, and Mississippi, the name Cockfield is synonymous with the Yegua and includes both the marine and nonmarine strata of the uppermost Claiborne Group.

The stratigraphy of southwestern Texas was extrapolated into the Burgos Basin, where it is applied with variations by subsurface geologists of the petroleum industry in northeastern Mexico. The stratigraphic division of the Eocene in the Burgos Basin (Figure 2) consists of the Reklaw, Queen City, Weches, Cook Mountain, Yegua, and Jackson formations. The Carrizo Formation is not used, and this unit is incorporated into the Wilcox Formation. The Claiborne Group and the Mount Selman subgroup have fallen into disuse. Cook Mountain is used as a formational name, whereas the names Pico, Sparta, and Laredo or Crockett are not recognized. The Yegua Formation, prior to this study, was undivided, whereas the Jackson Group in the Burgos Basin is a formational unit informally divided into lower, middle, and upper members. Unfortunately, the poor definitions of groups and formations in southwestern Texas have been applied to Mexican formations across the border, making the correlations across both areas very difficult.

Stratigraphic confusion concerning the Yegua in the Burgos Basin results from several factors. The first consideration was to follow Dumble's imprecise redefinition of Yegua (Plummer, 1932) that included sandy marine strata of Crockett Formation into Yegua. The second consideration was that Yegua Formation facies in both outcrop and subsurface of the Burgos Basin consist of littoral facies (continental facies is not present there) and therefore, the option was to include all the strata from the top Cook Mountain to the base Jackson in the Yegua Formation. The third consideration was that the Cook Mountain Group cannot be subdivided and is therefore relegated to formational status. The name Yegua was assigned by its stratigraphic position and not by lithological character.

Biostratigraphy

The Eocene biostratigraphy of the western Gulf of Mexico coastal plain is based on benthonic and planktonic foraminifera that were controlled by the paleoenvironments and facies distributions within the depositional systems. According to several authors (Edwards, 1990; Rosen et al., 1994; Ewing and Vincent, 1997a), the base of the Jackson Group, represented by the Moodys Branch Formation, contains *Globigerinatheka semiinvoluta* and *Globigerinatheka barri* that are associated, according to the environment, with *Textularia dibollensis*, *Camerina moodybranchensis*, *Glandulina laevigata*, *Anomalina hemisphaerica*, and *Anomalina perforata*. For Rosen et al. (1994), the biostratigraphic assemblage may be placed in the planktonic zone P 16 (Figure 2). In Berggren et al. (1995), the absolute age of the base zone P 16 corresponds to 35.2 Ma (late Eocene, Priabonian). In the Burgos Basin, the assemblage *G. semiinvoluta* (Kejizer), with benthic *C. moodybranchensis* and *T. dibollensis* Cushman, are placed in zone P 16 (Figure 2) in the lower Jackson member (PEMEX, 2000).

Rosen et al. (1994) divided the Yegua Formation into upper and lower units. The upper unit is represented by a sequence containing abundant *G. barri*, and by the presence of *Nonionella cockfieldensis*, and the scarce presence of *Discorbis yeguaensis* in the flooding zone, followed by *Eponides yeguaensis* and *Eponides mexicana* in the progradational wedge directly below this sequence. *Discorbis yeguaensis*, *Gyroidina octacamerata*, and *Bathysiphon eocenica* are found at the shelf margin. Rosen et al. (1994) indicated that *Nodosaria mexicana* normally appears below *E. yeguaensis* as the progradational wedge expands. In the upper Yegua unit, Rosen et al. (1994) considered that the abundance of sandy benthonic forms gives the impression of rapid deposition.

The faunal assemblage previously described by Rosen et al. (1994) assigns to this a late Eocene age in the planktonic P 15 zone. Berggren et al. (1995) considered the P 15 zone to range from 35.2 to 38.5 Ma, i.e., late to middle Eocene (Priabonian–Bartonian). New absolute time scale data (Gradstein et al., 2004) place the base of P 15 at 37.9 Ma.

Layman and Lagoe (1994) identified five major biofacies in the Yegua represented, from base to top, by outer shelf (*Uvigerina* spp.), middle shelf (*Eponides mexicanus*), inner to middle shelf (*Textularia* spp.), inner shelf (*Forilus bantkeni*), and marginal marine (*Ammobaculites hockleyensis*).

In the Burgos Basin, the upper and middle Yegua beds (as proposed in this article) both contain *G. semiinvoluta*, *E. yeguaensis* Cushman, *Globigerina yeguaensis* Applin, and *D. yeguaensis* and place the beds in P 15 zone. At the base of the middle Yegua strata, reworked foraminifera were frequently noted proceeding from the lower P 14 zone, and the mixed assemblage may be confused and provide unreliable information. The base of the middle Yegua strata is defined by a prominent regional erosion surface.

Swenson (1997) and Rosen et al. (1994) considered the lower part of the Yegua Formation to be represented by the presence of *Anomalina* spp. and *E. yeguaensis*; less commonly by *Eponides guayabalensis* and *N. mexicana*; and abundant *Anomalina umbonata* at the maximum flooding surface corresponding to this sequence, Rosen et al. (1994) mentioned that *Ceratobulimia eximia* may rarely occur in the lower part of this sequence and mention that all of these fauna correspond to the upper P 14 zone.

Rosen et al. (1994) calibrated the lower part of the Yegua Formation by the presence of *Truncorotaloides rohri*, *Globorotalia lehneri*, and *Globorotalia renzi* of the upper part of the P 14 zone. This work assumes that the *Anomalina*/*T. rohri* assemblage may be time equivalent to deposits of the Crockett Formation, the base of the marine Yegua Formation of Dumble (1894, in Plummer, 1932), and the base of the lower Yegua sequence of Rosen et al. (1994). In the Burgos Basin, the presence of *T. rohri* is at the base of the lower Yegua, and this stratum underlies the regional erosion surface dated in this article at about 39.5 Ma. Based on this assumption for the study area, the lower Yegua is assigned to the lower P 14 zone.

Rosen et al. (1994) calibrated the lower P 14 zone, containing *Clavulinoides guayabalensis* and *Truncorotaloides topilensis*, whose age is determined by the occurrence of *Globorotalia spinulosa* and *Globorotalia bullbrooki*. These species are included by Rosen et al. (1994) in the Cook Mountain sequence with *Orbitulinoides beckmanii* (P 13 zone). In the Burgos Basin, the top of the silty Cook Mountain Formation is determined by the presence of *C. eximia* (Figure 2), whereas *Orbitolinoides beckmanni* (P 13 zone) and *Asterigerina texana* (P 12 zone) are found below this unit. The Cook Mountain beds overly the Weches Formation assigned to the P 11 zone and characterized by the appearance of *Textularia smithvallensis* (Rodríguez Lozano, 1999).

According to the time scale of Berggren et al. (1995), the P 14 and P 13 zones range from 38.5 to 40.1 Ma, the base of the P 12 zone is at 43.5 Ma and the base of the P 11 zone is at 45.7 Ma. New data (Gradstein et al., 2004) give slightly younger ages placing the base of the P 13 zone at 39.8 Ma, the base P 12 at 42.6 Ma, and the base P 11 at 44.9 Ma.

In this study, the regional unconformity at the top of the lower Yegua is determined by its paleontologic assemblage and stratigraphic location. This unconformity is a prominent sequence boundary correlative with the sea level change proposed by Haq et al. (1988) at 39.5 Ma. In the Burgos Basin, the imprecise timing between the palentologic and lack of isotopic data impedes the precise dating of this unconformity placed around at 39.5 Ma.

In the Yegua Formation of Texas, Holroyd (2002) reported the presence of the vertebrate *Anthracotheriid artiodacty* (*Artiodactyla* Mammalia) genus *Heptacodon*. This new species appears to be the most primitive of the four species of *Heptacodon*. This terrestrial organism is placed in the late middle Eocene (Duchesnean, North American Land-Mammal Ages). Gradstein et al. (2004) correlated the Du 1 zone as being at the top of the P 14 and the base of the P 15 zones (Figure 2) from 38.1 to 39.2 Ma. The *Heptacodon*, in addition to the planktonic organisms, confirms the stratigraphic boundary between the marine and the regressive continental strata of the Yegua Formation.

Several authors have analyzed the biostratigraphy and sequences deposited in the Yegua interval of Texas (Edwards, 1990; Ewing and Fergeson, 1991; Hull, 1995; Luneau et al., 1995; Meckel and Galloway, 1996; Ewing and Vincent, 1997a, b). These authors recognize flooding surfaces within the Yegua as markers that can be followed for tens of kilometers across the Gulf coastal plain. Edwards (1990) described numerous high-frequency cycles within the Yegua and Cook Mountain that are separated by unconformities, and stated that these cycles may be similar to parasequences, some of which are interpreted to be localized effects of gravity slide deposition (C cycles) and other larger cycles are thought to be eustatically controlled (B cycles). This work proposes, however, that the major assemblage, which encompasses the previously described cycles, is tectonically controlled (A cycle). Edwards (1990) recognized that some of the unconformity surfaces within the Yegua were formed by canyon incision or gravity sliding on the continental slope (Jin-wook et al., 2003), and suggested that most of the unconformities are not continuous and, therefore, are of limited use as regional markers. Despite this, 8 flooding surfaces are identified in the Yegua-Cook Mountain formations of central Texas and 12 markers in southeastern Texas and Louisiana (Edwards, 1990; Meckel and Galloway, 1996), though Ewing and Vincent (1997a) recognized that the depositional systems change markedly around Live Oak County.

This sequence correlation aids in the recognition of stable shallow-water platform facies in the updip part of the basin that contain deltas separated by zones of bypass, with deltaic and delta-front facies displaced toward the deeper basin.

Meckel and Galloway (1996) agreed that several flooding surfaces within the Yegua are regional markers that separate fourth-order depositional cycles and, in Texas, propose that their intervals 1, 2 (basal Yegua), and 4 (middle part of Yegua) are sequences controlled mainly by sediment supply (sediment supply > accommodation space). Accommodation-dominated interval 6 (top Yegua) and pre-Yegua occurred under conditions dominated by low sediment supply and rising relative sea level (supply < accommodation space), and equilibrium conditions (intervals 3 and 5) underlie intervals 4 and 6.

Meckel and Galloway's (1996) correlations from Texas to the Burgos Basin are not clear. Preliminary sequence correlation in this article assumes that the lower Yegua may correspond to their intervals 1 and 2. The middle Yegua beds (a complex sequence) may correlate with their intervals 3 to 5, and therefore, the upper Yegua correlates to interval 6.

Ewing and Vincent (1997a) provided a logical regional correlation framework that indicates the relationship between multiple sequence boundaries and flooding surfaces in the Yegua. Within this framework, they show that unconformable sequence boundaries (SB) appear to be overlain by prograding wedges (PW), basin-floor fans (SF), and shelf margin systems tracts (SMST). Biofacies zonation and environmental variations shown by Ewing and Vincent (1997a) are observed to be similar in the Burgos Basin. The preliminary correlation of sequences from Texas and the Burgos Basin suggests that from the top *C. eximia* Y-50, adjacent to Y-40, the sequence possibly correlates with the lower Yegua beds of this article. Overlying the lower Yegua beds is a prominent sequence boundary that may correlate with the ~39.5-Ma unconformity of the Burgos Basin, and overlying that, the middle and the upper Yegua beds of this article could range from the Y-40 to near the Y-10 of Ewing and Vincent (1997a).

Hull (1995) indicated that the Yegua had an average sedimentation rate of 350 m (164 ft) per million years, which is very high when compared to the rates of deposition of the older Clairborne Group and the younger Jackson Formation, and are comparable to Wilcox deposition rates. In Burgos, the updip Yegua deposits are thinner whereas the downdip Yegua sequences are thick, and it is possible to derive an increasing average sedimentation rate of about 280 m (919 ft) per million years. Hull (1995) attributed high sedimentation rates in the Yegua to climatic factors. Galloway et al. (2000) believed that Yegua deposition was controlled by pulses of tectonic uplift in the Sierra Madre Oriental and based on the work of Yurewicz et al. (1997), cited the heating of the mantle and igneous activity as evidence.

Chávez-Cabello (2005) provided isotopic data demonstrating that Laramide deformation in the Sabinas Basin ended in the Eocene. Isotopic $^{40}Ar/^{39}Ar$ in hornblende and biotite shows that the emplacement and cooling of the magma occurred before 44 Ma. Structural data confirm that intrusion is contemporaneous with deformation and synorogenic sedimentation. Chávez-Cabello's (2005) data show that ~39.97 ± 0.04 Ma, stress in the Candela intrusive of the Sabinas Basin is no longer observed and this may signal the end of the Laramide deformation in this area (Figure 2). The proposed sequence boundary at ~39.5 Ma (this chapter) is an important regional unconformity and together with the complex sedimentation represented by the Yegua may be related to tectonic uplift during Laramide deformation (Eguiluz, 2004, 2007) with Yegua deposition being a possible consequence of tectonic activity and relative sea level change.

The greatest difficulties observed with the foregoing interpretations in the Burgos Basin consist of establishing the boundaries of the Yegua because of the confused definition of this stratigraphic unit, the lack of adequate paleontologic data, and the discrepancies in absolute ages that are dependent on which time scale is being used.

REGIONAL SECTIONS

Work for this study consisted of a workstation-based integration of digital 2-D and 3-D seismic data and well logs. Time and depth tables were prepared for 40 wells by integrating synthetic logs, sonic logs, and check shot surveys to obtain the velocity control within the area. Reports on paleontology, petrography, and well cores were analyzed and used to identify ages and depositional environments. A petrophysical evaluation of the stratigraphic intervals was conducted through the integration of drilling reports, well completions, and production histories.

Seismic lines selected across the sedimentary belts showed the principal growth faults that controlled sedimentation. The reference horizon used to identify the main growth faults was the top of the Queen City (using 3-D seismic data over most of the area). Only the most obvious faults were interpreted in the southern part of the area (from section V-V′ to section X-X′, Figure 3) because of the poor resolution of 2-D seismic data.

Well logs (gamma-ray, potential, and resistivity curves) were tied and correlated to more than 120 interpreted strike- and dip-oriented 3-D and 2-D seismic sections. A structural map was assembled on which structural blocks were defined (Figure 3). Furthermore, structural interpretation and stratigraphic correlation were derived from nine dip sections across the basin (sections I-I′ to IX-IX′) and seven strike sections lengthwise through the basin (sections 9-9′ to 14-14′). This chapter includes six of the dip sections and three of the strike sections (Figures 4a–f; 5a–c, respectively).

Representative type logs were identified for each structural block. These logs were then used to correlate the sandstone packages between the Yegua wells within each block which, in turn, served as the basis for mapping structural tops and sand thicknesses. Sand was determined based on a minimum spontaneous potential log (SP) cutoff of 1/3 of the total SP response, or in its absence, by a minimum resistivity of 5 ohms. A digital database was assembled from the tops, bases, sand thicknesses, and rotary table elevations to construct a sand thickness map (Figure 6) for the middle and upper Yegua interval.

The correlation of regional sections across the Burgos Basin (Figure 4a–f) demonstrates a very uniform regional

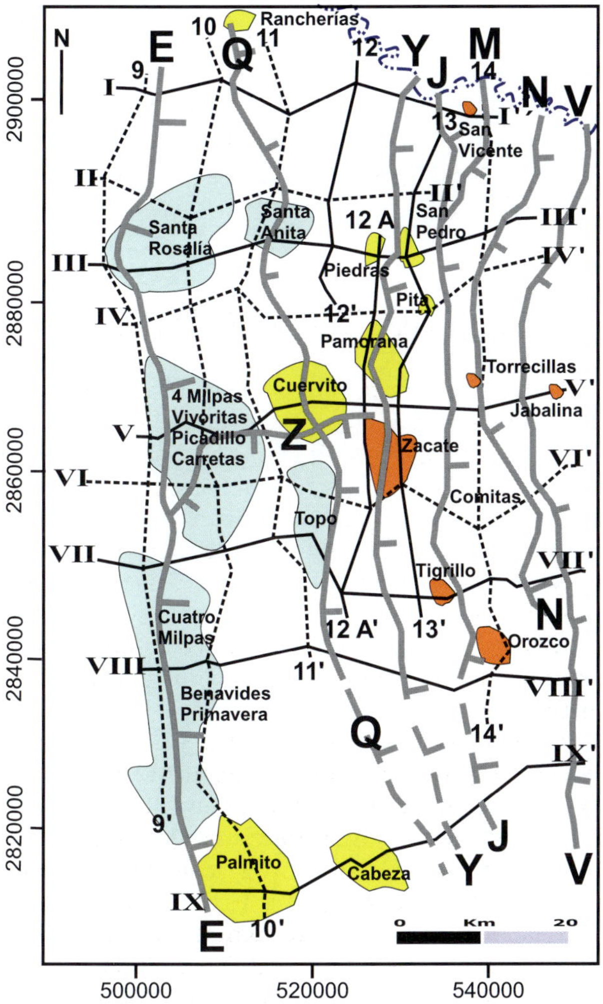

Figure 3. Structural map of the regional expansions faults that control deposits of the Eocene (E), Queen City (Q), Yegua (Y), Jackson (J), and Vicksburg (V); other important faults are the M and N with localities of the stratigraphic sections in the dip direction indicated by roman numerals (I-I′ to IX-IX′) and strike sections with arabic numerals (9-9′ to 13-13′). For both, the continuous lines are sections included in this text, and dashed lines are supporting sections not included in this text. Productive reservoirs in the Yegua are shown in yellow, Jackson reservoirs in orange, and Queen City fields with Yegua gas production in light blue. The coordinate system is Universal Transverse Mercator (UTM).

framework. To the west, the lower limit of the Yegua (in the new stratigraphic sense applied in the Burgos Basin) corresponds to a coarsening-upward sand lithology bounded at the top by a regional unconformity. The base of this sand body overlies Cook Mountain shales. This lithological assemblage corresponds to a prograding highstand systems tract (HST), whose upper sequence boundary corresponds to ~39.5 Ma based on the foraminiferal assemblage. This sandstone body is equivalent to the Crockett Formation and is named the lower Yegua in this chapter.

A depositional model was constructed based on paleontology, core and well logs, outcrops, and sand distribution within the structural blocks (Figure 7). The ~39.5-Ma unconformity toward the eastern end of the sections cuts older strata downdip into the basin and from west to east, gradually eliminates sediments of the lower Yegua (Crockett), Cook Mountain, Weches, and even the top of the Queen City, as seen in section III-III′ in the Santa Anita 101 well, section VII-VII′ in the Sansón 1 well, and in sections VIII-VIII′and IX-IX′, with uncertainty concerning the stratigraphic unit below the ~39.5-Ma unconformity, which may be pre-Queen City. The ~39.5-Ma unconformity is a regional marker and a very consistent sequence boundary throughout the area, although it can only be traced through the central and western parts of the area because the area beyond the faults to the east showing maximum growth (Y, J, and V) has no well penetrations deep enough to reach the unconformity.

The middle Yegua beds lie above the ~39.5-Ma unconformity. This depositional sequence corresponds to a thin wedge (250 m, 820 ft) west of the Marte R. Gómez 1, Santa Rosalía 1, Picadillo 14, Salitrillo 1, Quemado 1, and Esquina 2 wells, which thickens to the east of these wells near the Q expansion fault and the Y fault, attaining a thickness of more than 700 m (2296 ft) (Cuervito 1, Sierrita 1 A, Catarrín 1, Piedras 1, San Pedro 1, Clavel 1, Santa Gertrudis Sur 1 and 2 wells, etc.). This wedge consists mainly of shale with erratically distributed sand bodies along its base that are difficult to correlate. These are interpreted to be basin-floor fans and slides that onlap the ~39.5-Ma unconformity (Nauyaca 1, Santa Rosalía 14, Cuervito 1 and 21, Salitrillo 1, Topo 1, Sansón 1, Benavides 30, Palmito 19, Cabeza 1, Esquina 2 wells, etc.). The middle and upper parts of this middle Yegua interval contain numerous channels or incised valleys with multiple internal unconformities observed on seismic data (sections III-III′, V-V′, and VII-VII′), some of which show continuity more than tens of kilometers, although they are difficult to correlate with certainty between different seismic sections.

The upper Yegua overlies the middle Yegua, below the lower Jackson beds. The upper Yegua is predominantly sandy and coarsens upward (Tinta 1B, Nauyaca 1, Migueleño 1 wells in section I-I′; Santa Rosalía and Santa Anita fields, Santa Gertrudis Sur 1 and San Pedro 1 wells in section III-III′; Sierrita 1 A, Cascabel 1, Sansón 1, Comitas 1 and 5, and Chalupa 1 wells in sections V-V′ and VII-VII′; Orozco 2 in section VIII-VIII′, etc.). This sand body is about 250 m (820 ft) thick and is generally the gas-productive unit in downdip blocks controlled by regional growth faults (J or M).

Toward the north of the area (sections I-I′ and III-III′), the contact between Yegua and lower Jackson is conformable and is frequently marked by a lithologic

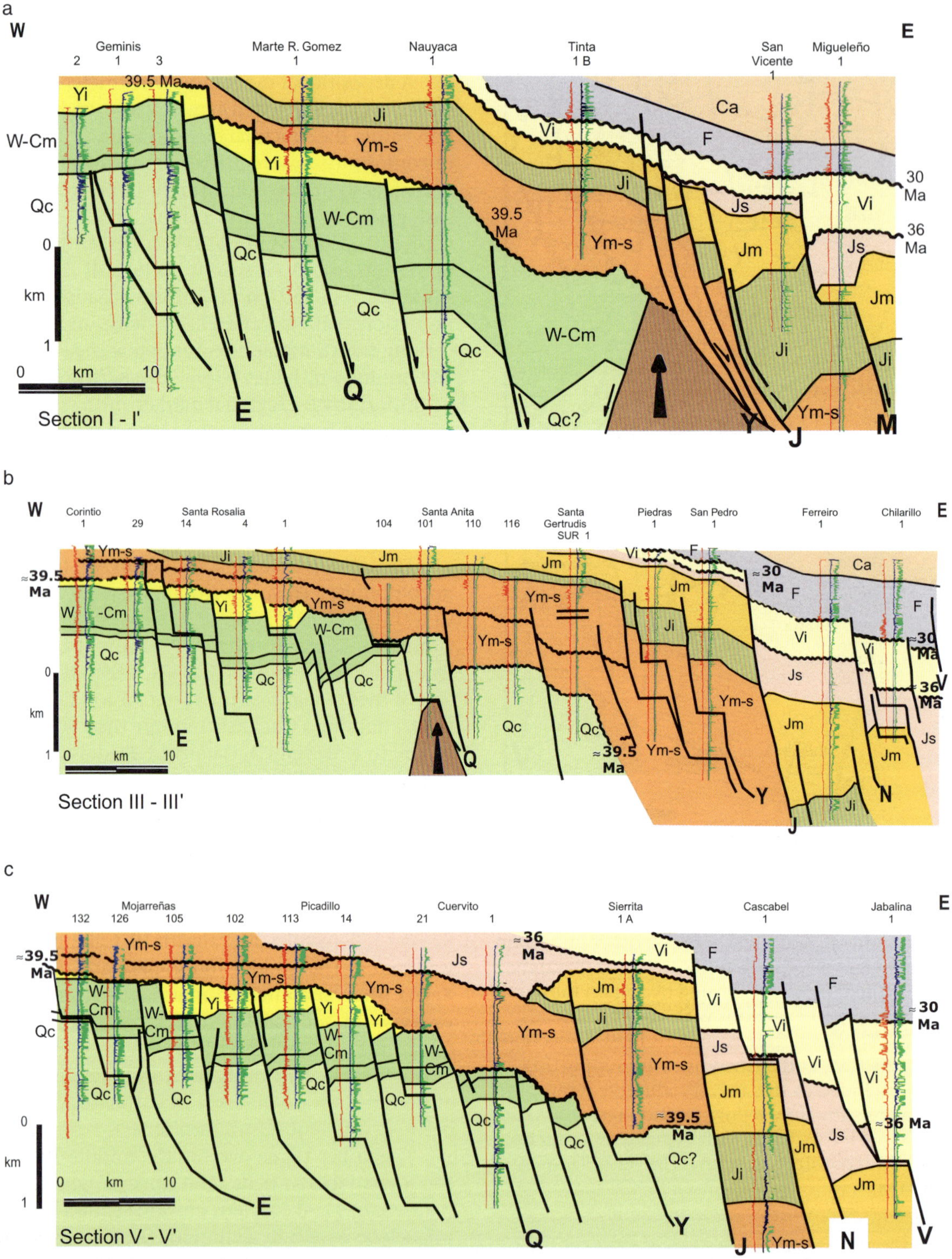

FIGURE 4. (a–f) Legend for the regional sections: east–west. Formations: Queen City (Qc); Weches and Cook Mountain (grouped as W-Cm); lower Yegua (Yi), middle and upper Yegua (Ym-s); lower Jackson (Ji), middle Jackson (Jm), and upper Jackson (Js); undifferentiated Vicksburg (Vi), undifferentiated Frio (F), and Catahoula (Ca). Expansion faults: Eocene (E), Yegua (Y), Jackson (J), Vicksburg (V), other important faults M and N. The thick arrows are shale diapirs. Log curves: spontaneous potential (SP) and/or gamma ray are to the left of the wells. Resistivity is to the right of the wells. Horizontal and vertical scales are indicated by bars. Major sequence boundaries: 39.5, 36, and 30 Ma. Several internal unconformities within the Yegua are observed. Between sections V-V′, VII-VII′, VIII-VIII′, and IX-IX′, the upper Jackson rests on middle to upper Yegua to the west; however, to the north and east within the structurally subsided blocks, the upper Jackson rests conformably on the middle Jackson. In the southern part of the area on sections VII-VII′, VIII-VIII′, and IX-IX′, the Vicksburg Formation rests unconformably (36 Ma) on the upper Jackson, and the unconformity even erodes into the middle Jackson. The 30 Ma unconformity invariably places the Frio on the Vicksburg formations.

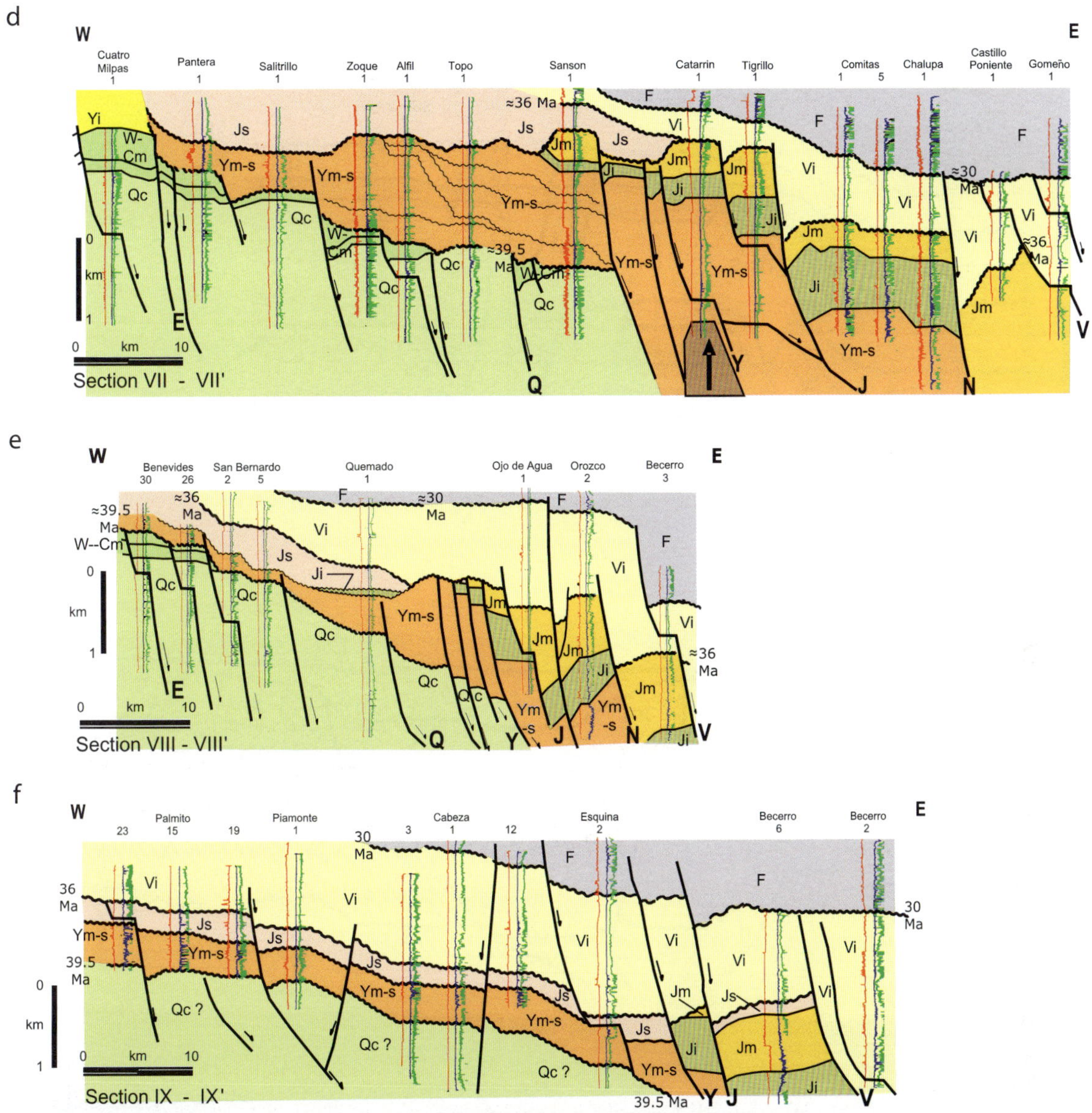

FIGURE 4. (cont.).

change from sandstone to shale and supported by foraminiferal data. When the sandy coarsening-upward section of the upper Yegua is absent, the middle and upper Yegua are hard to separate and are grouped together on the dip- and strike-oriented stratigraphic correlation sections. The middle and upper Yegua deepen eastward along growth faults.

Toward the southern part of the study area, both the base and top of the Yegua are deeply eroded. The top is covered unconformably by sediments of the upper Jackson Formation (sections V-V′ to IX-IX′), and the base of the Yegua rests on the ~39.5-Ma unconformity that erodes the Cook Mountain, Weches, Queen City, or even older units (sections VII-VII′ to IX-IX′). South of the area, the lower Yegua is eroded, and probably the upper Yegua as well, and is only represented by thin middle Yegua. The Yegua Formation in the southern part of the area is stratigraphically complex. Drastic sedimentary, stratigraphic, and structural changes, accompanied by deep erosion are recognized south of regional fault Z (Figure 3). Correlation is more difficult here because only poor quality 2-D seismic data are available.

Toward the west, both the well-log and the seismic data permit the good correlation of the Queen City, Weches, Cook Mountain, and lower Yegua between the Géminis 1, Corintio 1, Mojarreñas 132, Cuatro Milpas 1, and Benavides 26 wells. However, toward the east of fault E and south of fault Z, these stratigraphic units are deeply eroded in downstepping blocks with reduced thicknesses and difficult well-log and seismic correlations.

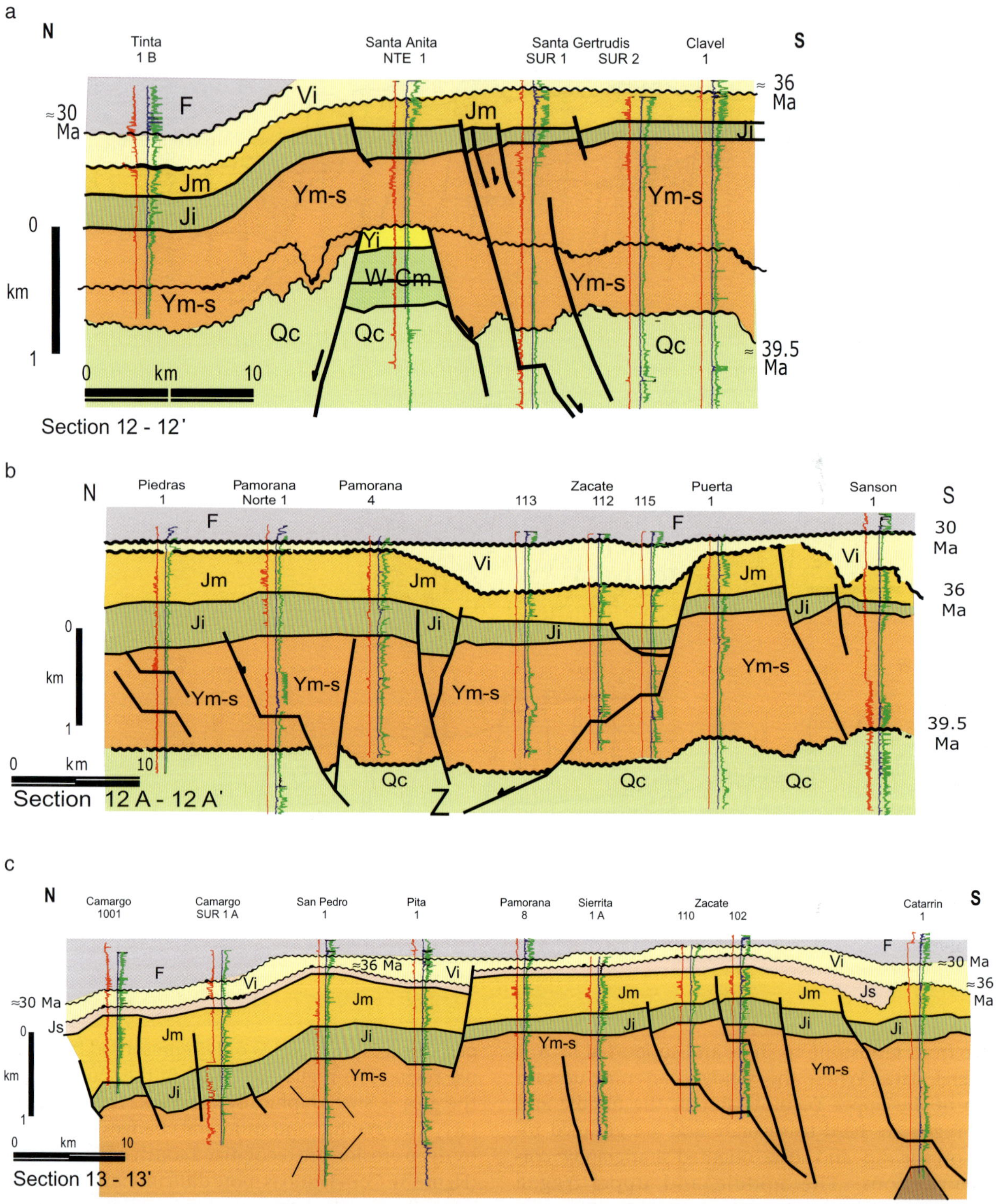

FIGURE 5. (a–c) Legend for the regional sections: north–south. Formations: Queen City (Qc); Weches and Cook Mountain (grouped as W-Cm); lower Yegua (Yi), middle and upper Yegua (Ym-s); lower Jackson (Ji), middle Jackson (Jm), and upper Jackson (Js); undifferentiated Vicksburg (Vi), undifferentiated Frio (F). Faults: Z and thick black lines. The thick arrows are shale diapirs. Log curves: spontaneous potential (SP) and/or gamma ray are to the left of the wells. Resistivity is to the right of the wells. Horizontal and vertical scales are indicated by bars. Major sequence boundaries: 39.5, 36, and 30 Ma. Several internal unconformities within the Yegua are observed. The lower Jackson overlies conformably on the upper Yegua.

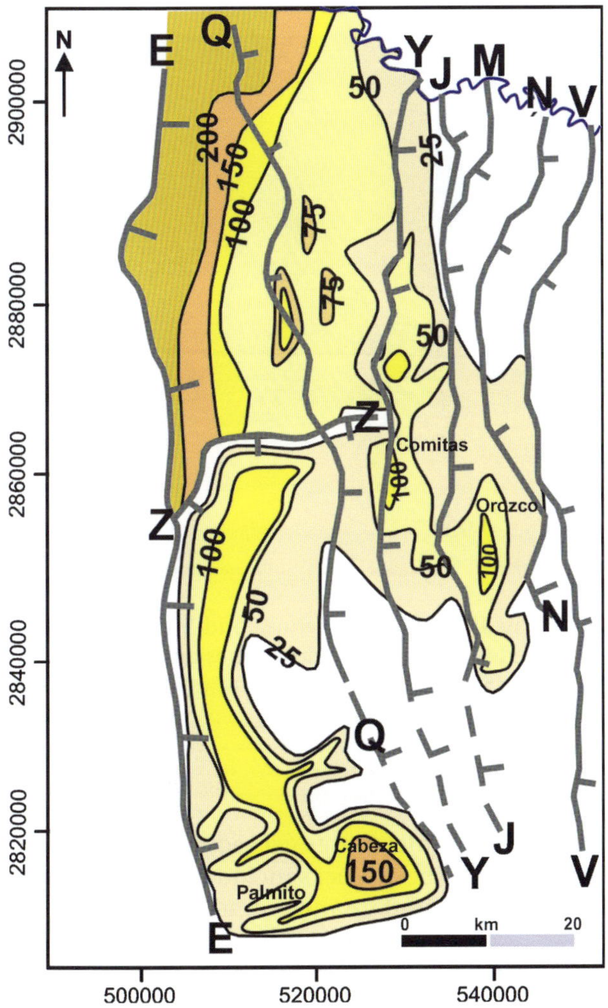

Figure 6. Net sand thicknesses in the Yegua play. The greatest thicknesses of sand in the Yegua Formation are found near expansion faults E, Y, and J and glide plane Z. Contours are in meters. Comitas and Orozco, Palmito, and Cabeza fields are shown in the map.

Toward the east and south of the sections VIII-VIII′ and IX-IX′ (Figure 4e–f), the lower Yegua is eroded. Yegua overlies undifferentiated strata of the Queen City-Reklaw in the Sanson 1, Quemado 1, San Bernardo 5, Piamonte 1, and Esquina 2 wells, and also in the Palmito and the Cabeza fields (sections VII-VII′, VIII-VIII′, and IX-IX′).

In the sections V-V′, VII-VII′, VIII-VIII′, and IX-IX′, the Yegua underlies the upper Jackson beneath an erosional unconformity. Based on paleontologic data from the Quemado 1 well on section VIII-VIII′, an erosional remnant of the lower Jackson may exist here, or lower Jackson fossils might have been recycled into the upper Jackson.

In the north–south direction, the Yegua interval in section 14-14′ (not included in this chapter) is very deep between the Migueleño 1 and Chalupa 1 wells (sections I-I′ and VII-VII′), whereas toward the Cascabel 1 well, it rises structurally and remains at depths of 2000 m (6562 ft). In Orozco and Comitas, great thicknesses of net sand are observed and that could be economic objectives in the Yegua play downdip toward the basin.

In sections 13-13′, 12-12′, and 12A-12A′, the Yegua becomes shallower as it climbs in a stepwise manner on each block, with depths ranging between 1000 and 1500 m (3280 and 4921 ft). Many gas shows in these blocks exist and a few wells produce gas.

A map was constructed to represent the regional sand accumulation in the Yegua play (Figure 6). The Yegua is thin in the northern part of the area, forming a wedge that thins toward the regional growth fault

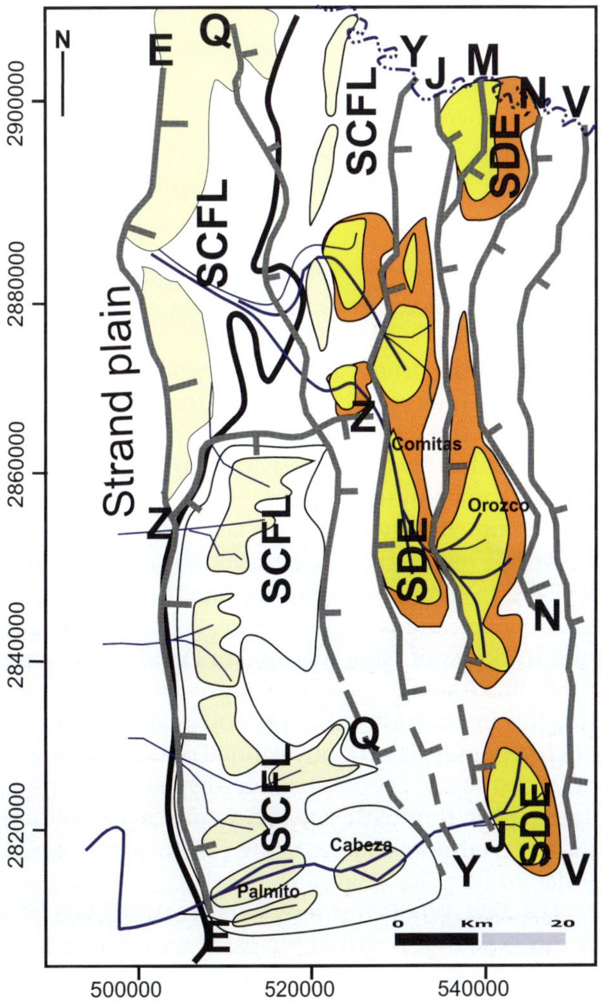

Figure 7. Distribution of Yegua sedimentary environments; yellow represents proximal deltaic sands, and orange represents distal sands. Faults Y and J control expanded deltaic environments (SDE). In the absence of expansion, the strata represent a complex of fluvial and shoreline environments (SCFL) in light yellow. Glide fault Z and fault E control estuarine and footwall deposition.

FIGURE 8. Arroyo Lobo locality, 143 km (88.86 mi), Federal Highway 40, between Reynosa and Monterrey. Lower Yegua has medium beds of fine sandstone and bioturbated, cross-bedded fine-grain sandstones and siltstones deposited in a shoreface environment. The notebook measures 15 cm (5.9 in.). The column shows lithologic details and the scale in meters.

(J). Thickness increases to the west and exceeds 200 m (656 ft).

Out of the study area, the Yegua Formation is eroded. The interval represented to the northwest corresponds to the lower Yegua sands.

Fault Z has a well-defined crescent-shaped boundary with a maximum stratigraphic thickness along the margin immediately east of the fault, forming a prism that wedges toward the Y fault suggesting a facies change. The Palmito and Cabeza fields contain the greatest gross thicknesses of sand averaging 150 m (492 ft), with a maximum thickness of more than 500 m (1640 ft). Net sand distribution (middle Yegua?) is elongated in the east–west direction in contrast to the previously described area, which generally has a poorly defined north–south sand body orientation. The area bounded by fault Z could be a stratigraphic trap with enhanced sand accumulation and reservoir potential, as shown in the depositional model for this play (Figure 7).

The best producing reservoirs in the Yegua play are in the southern part of the study area. The average gross thickness of the middle Yegua ranges from 400 to 500 m (1312 to 1640 ft) and reaches more than 800 m (2625 ft) in section VII-VII′ (Zoque 1, Alfil 1, Sansón 1, and Topo 1 wells). The reservoir sands are of lower quality in the northern part of the study area.

Regional faults Y, J, and M localized thicker accumulations of middle Yegua sand in semicircular fanlike patterns. Pamorana and Zacate fields appear to contain these sand concentrations, as may Comitas field, with net sand thicknesses exceeding 100 m (333 ft) (Figures 6, 7). These blocks include unexplored areas that may hold similar sand accumulations.

SEQUENCE ANALYSIS

Little biostratigraphic data are available for sequence control in the Burgos Basin (Rodríguez Lozano, 1999), and sequence identification is based primarily on 3-D seismic and vertical patterns expressed in well-log curves. The lower Yegua is interpreted as a prograding, coarsening-upward, highstand (HST) deposited in a nearshore environment (Figure 8) transitional with the Cook Mountain shales. The lower Yegua is a genetic unit whose top is truncated by a regional unconformity. Its stratigraphic position is based on paleontologic data and correlative age (Berggren et al., 1995). This unconformity (Figure 9) lies between the lower and middle Yegua and correlates with a second-order regional sequence boundary (39.5 Ma) and lowstand sea level change proposed by Haq et al. (1988). This unconformity in the Burgos Basin is assumed to have a submarine origin. Units as old as Reklaw and Queen City were eroded to the east, whereas in the west, the unconformity is less pronounced, eroding only the tops of the lower Yegua and Cook Mountain.

The Burgos Basin middle Yegua overlies the 39.5-Ma erosion surface proposed in this chapter. The complex

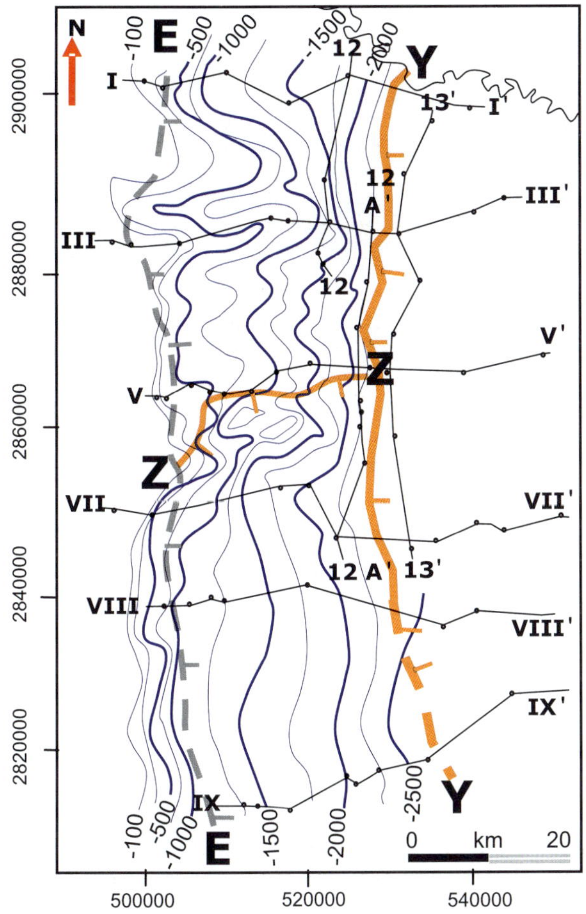

FIGURE 9. Contour map of unconformity around 39.5 Ma, referenced in meters below sea level.

middle Yegua sequence consists of down-lapping basin-floor sandstones and an upper section that is predominantly shaly with scattered sand bodies, including channels and deformation (slumps). Outcrops and well cores in the Burgos Basin show small-scale contorted strata, and some metric folds can also be observed in seismic sections. This complex sequence is interpreted as a lowstand systems tract (LST) at its base beneath a prograding wedge systems tract (PWST), both in the middle Yegua Formation. Middle Yegua may contain fourth-order sequences, as identified in Texas by Edwards (1990) and Meckel and Galloway (1996), although a detailed study has not yet been undertaken in the Burgos Basin.

The upper Yegua underlies the lower or upper Jackson. The shaly lower Jackson may represent a flooding surface. The upper Yegua generally coarsens upward with sandy facies possibly deposited in a nearshore environment. The systems tract is unclear and this work assumes that it may represent an HST updip. The downdip facies, when the sandy unit is not present, could be interpreted as the extension of the middle Yegua PWST, and a by-pass zone or erosion surface may have removed the sandy upper Yegua unit.

YEGUA PLAYS

Several plays were identified for the Yegua interval. The north-northwestern and western parts of the area are characterized by fluviolacustrine and coastal plain deltaic facies, with offshore bars of the inner and middle shelf. The southern part of the study area, at the western end of fault Z, has similar depositional environments. East of fault Z and south E fault, however, complex sandy fan deposits are found at the edge of the fault block. In the area of the Palmito and Cabeza fields, a series of dip-oriented channels and associated sandstones have been identified crossing the coastline trend. They possibly originated as fluvial channels to the west and estuarine bars to the east. More distal sand facies are interpreted when approaching regional fault Y, along with erratically distributed coast-parallel bars and lagoonal or low energy zones. The scarcity of seismic data and wells in this area prevents a conclusive interpretation conforming to the model (Figure 7) and, alternatively, these deposits may correspond to slump blocks downslope of fault Z.

Sand deposits are interpreted to be a series of deltaic lobes fed by channels coming from the west in the downthrown blocks of expansion faults Y and J. At the southeastern edge of the block bounded by fault Y and at the extreme north of the block limited by fault J, similar sand bodies may be present but are undrilled because of their depth or the early stage of exploration in this part of the basin. To the east of fault V, the Yegua sand facies may be more distal and were possibly deposited in a normal offshore environment.

Based on the foregoing, the author proposes that the Yegua contains at least three play types: a fluviolacustrine and nonexpanded coastal plain play in the lower Yegua, a downdip deltaic and deltaic-estuarine middle and upper Yegua play controlled by expansion faults, and a middle Yegua basin-floor fan or prograding wedge-complex play.

STRATIGRAPHIC TRAPS

The stratigraphic-structural sections (Figures 4, 5) and the sedimentary model (Figure 7) show facies changes that formed stratigraphic traps, mostly in delta plain and fluviolittoral environments of highstand sequences. Growth faults also produced thick sandy prisms that when combined with the structure of the downthrown block, form combination traps.

The Yegua and Jackson formations have thick shale packages that work as local and regional seals.

Where sand bodies wedge out against the ~39.5-Ma unconformity, stratigraphic traps may develop (Figure 10) from above (onlap) and below (truncation). The most

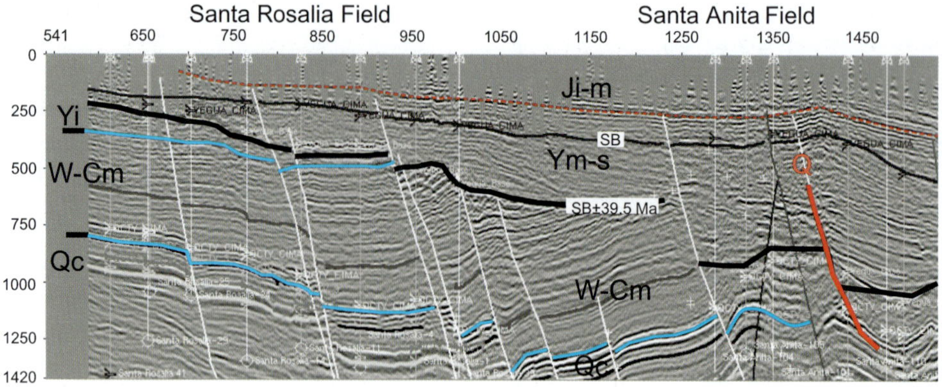

FIGURE 10. Seismic section between Santa Rosalía and Santa Anita fields. The lower Yegua (Yi) and the top the Cook Mountain-Weches (W-Cm) formations are truncated from west to east by the ~39.5-Ma unconformity in blocks that were broken and rotated by listric faults. The middle Yegua (Ym-s) contains basin-floor fans and a complex prograding sequence, with internal unconformities. The upper Yegua is combined with the middle Yegua, which continues toward the lower and middle Jackson (Ji-m) along a conformable contact. The location of this section is shown next to section III-III′ in Figure 3. SB = sequence boundaries.

economically important play occurs in sand bodies deposited above the ~39.5-Ma unconformity within basin-floor fans and slump deposits, as well as in mass flows that altered this erosion surface along a basal PWST. A transgressive systems tract (TST) developed later above these sands. Sand bodies in the middle Yegua wedge may be stratigraphically trapped within submarine erosional channels.

STRUCTURAL DEFORMATION

The structural style of the Gulf Coast Tertiary was described by Jackson and Galloway (1984). The deformation of the Tertiary sedimentary accumulation in the Burgos Basin was predominantly by extensional faults with rollover anticlines with a relatively small percentage of structures created by vertically rising shale diapirs. In a plate tectonic context, the Burgos Basin is a passive continental margin.

The structural framework of the area is characterized by north-south–trending listric faults that step down to the east and are detached at depth on the top of the Cretaceous (Méndez Formation). The eastern (basinward) parts of rotated fault blocks include deformed masses of ductile shale that form diapiric domes or walls. Distal parts of rotated blocks were uplifted and subjected to submarine erosion. At the same time, the western ends of fault blocks subsided to create accommodation space that filled with sediments.

The Z fault and south E fault (Figures 3, 11) are interpreted to be a subregional glide plane bounding a major sedimentary block. Yegua deposition may have been contemporaneous with the early movement of fault Z, which created instabilities in the downthrown fault block that, in turn, led to the deposition of slumped and contorted bed complexes related to a weak decollement. These folds may be confused on seismic sections with compressional folds. Testing this hypothesis will require new data and further study.

Numerous growth faults and associated folds (rollover anticlines) are observed in the area. Some growth faults became inactive over time, and others were reactivated leading to a complicated pattern of structural traps containing both expanded and nonexpanded section. Narrow fault-bounded structures that are elongated approximately north–south are observed within the regional blocks. The largest structures (defined by middle Jackson seismic isochrons) correspond to the Pamorana and Comitas fields (Figure 3) and the Becerro 6 well (Figure 4f). Seismic mapping of the Yegua interval shows that many structures are present, most of which are small. Some of these are near the Tinta 1B well and Camargo, Rancherías, San Pedro, and Piedras fields (Figures 3; 4a; 5a, b). Larger compartmentalized structures are found at Pamorana, Santa Gertrudis-Clavel, and Chalupa-Orozco fields.

In northern Mexico, the Upper Cretaceous–lower Tertiary is characterized by the continuous migration of siliciclastic facies in space and time (Tardy et al., 1974) from the western Central Mexican Basin (CMB) to the eastern Burgos Basin. The first appearance of detritus marks the pretectonic onset of the Laramide-age deformation moving toward the foreland basin.

In the CMB, the first siliciclastic appearance is caused by the pre-Laramide deformation of Turonian age (*Helvetoglobotruncana* zone and *Globotruncana fornicata* Plummer), whereas in the La Popa area (Giles et al., 1999), the siliciclastic deposits appear during the Maastrichtian (*Exogyra costata*, *Exogyra ponderosa*, and *Coahuilites sheltoni* zones). In the western Burgos Basin,

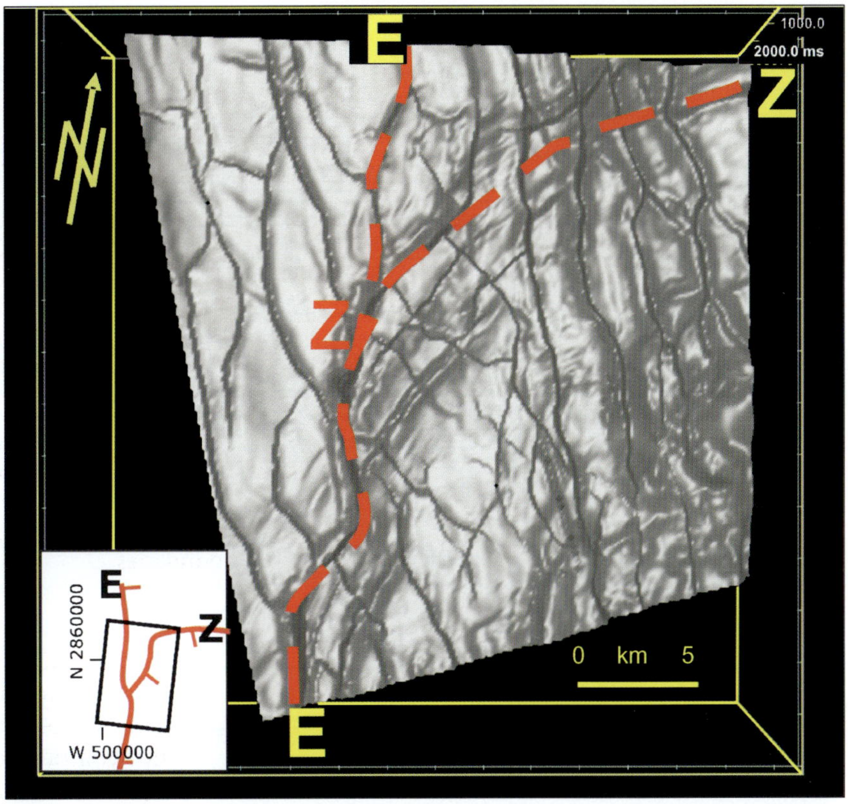

Figure 11. Time slice at the top of the Cook Mountain (Vivoritas area). Fault E is joined by fault Z and both are downthrown to the southeast.

sandstone facies do not appear until the Paleocene (*Globorotalia daubjergensis* and *Globorotalia pseudobulloides* zones), represented by the first Wilcox deposits and shift to the eastern Burgos Basin in the early Eocene (*Globorotalia velascoensis* and *Globorotalia soldadoensis* zones) within the Wilcox progradation (Eguiluz, 2004, 2007).

In the La Popa area, the Ypresian (Vega-Vera and Perrilliat, 1989) Adjuntas Formation and the post-Ypresian, Viento, and Carroza formations consist of fluviolacustrine facies more than 500 m (1640 ft) thick. Correlative formations in the Burgos Basin are the upper Wilcox, Reklaw, and Queen City, which represent nearshore marine facies, and it is conceivable that the La Popa and Burgos areas formed an extensive coastal plain with synorogenic deposition, whereas the Laramide-age deformation front was still far to the west of the La Popa area, and that Laramide-age deformation occurred here at a later time.

The Yegua sequences described in this chapter are similar to the Reklaw sequences (Figure 12), both having deep erosion surfaces, an LST, a prograding wedge, and complex facies architecture (Eguiluz, 2007). Both the Yegua and Reklaw reflect sea level change and voluminous mass-transport deposits into the deep Gulf of Mexico, perhaps derived from continental denudation (Galloway et al., 2000). Carrillo (1980) mentioned an erosion surface in the Gulf of Mexico related to the Chicontepec and La Laja paleocanyons, and both paleocanyons may be correlatives to the prominent erosion surface at the base of the middle Yegua in the Burgos Basin.

Additional satellite images and surface mapping, from Laredo to the General Bravo areas (28°00′N; 100°18′W to 25°31′N; 99°00′W), show outcrops of Paleocene and Eocene sedimentary beds in broad synclines oriented north-northwest to south-southeast. These structures seem to have a late tectonic deformation style (tarditectonique of Tardy, 1980). In seismic sections, the Burgos Basin basement steps down to the east, with Mesozoic rocks in the deep subsurface folded into broad synclines and narrow anticlines detached on Jurassic salt, whereas the lower Tertiary rocks are affected by listric faults detached on the Méndez Formation, thereby modifying the broad Laramide folds in the lower Tertiary rocks (Figure 13 and Geminis wells in Figure 4a).

Several authors (i.e., Yurewicz et al., 1997; Eguiluz et al., 2000; Gray et al., 2001; Chávez-Cabello, 2005) have proposed that the Laramide deformation and exhumation of the Mexican fold belt may have occurred in several stages, but that the deformation ended in about the middle Eocene, with Chávez-Cabello (2005) suggesting that the end occurred between 44 and 40 Ma.

The foregoing data support the idea that the ~39.5-Ma unconformity and sequences in the middle Yegua (lowstand and prograding wedge) were both caused mainly by the progressive tectonic uplift of the eastern Sierra

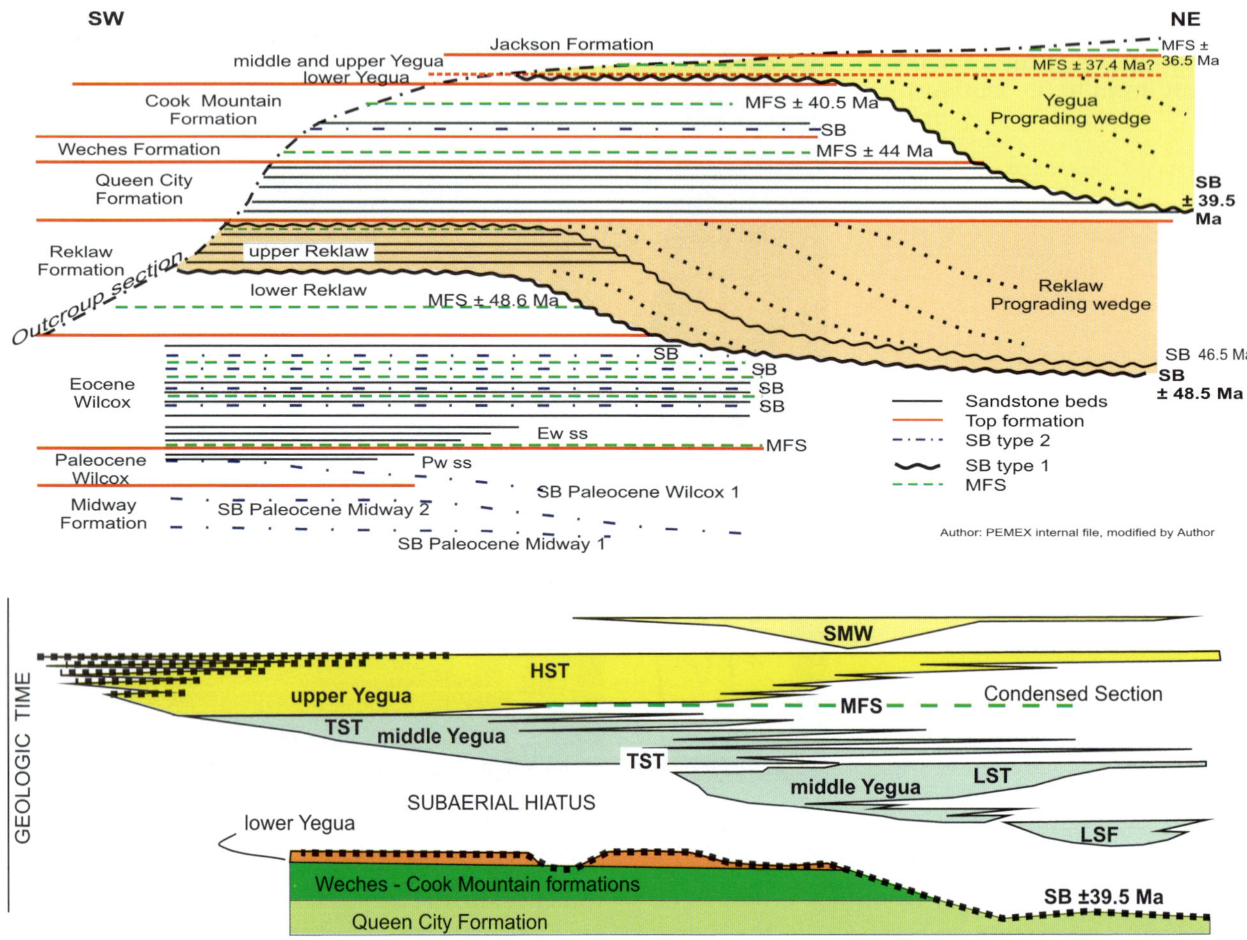

FIGURE 12. (a) The Paleogene in the Burgos Basin is integrated by numerous genetic sequences. The bases of the Reklaw and Yegua units both have a prominent sequence boundary at around 48.5 and 39.5 Ma, associated to a lowstand systems tract and following a transgressive and highstand systems and prograding wedge. MFS = maximum flood surface, SB 1 = unconformity sequence boundary, SB 2 = conformity sequence boundary. Age placed by paleontologic ensemble (PEMEX, unedited file). (b) Systems tract of the Yegua deposits in relation to geologic time from Cook Mountain to Jackson formations. Adapted by the author from Haq et al. (1988). LST = lowstand systems tract; TST = transgressive systems tract; HST = highstand systems tract; LSF = lowstand sea floor; MFS = maximum flooding surface; SMW = shell margin wedge; Ew = Eocene Wilcox sandstone; Pw = Paleocene Wilcox sandstone.

Madre Oriental and Sabinas fold belts and the denudation of the western continental areas. In a preliminary fashion, this chapter correlates the ~39.5-Ma unconformity and Yegua deposition in the Burgos Basin with the last pulse of the Laramide deformation in the Sabinas Basin (Figure 2).

Finally, in the Burgos Basin, the base of the Frio sequence boundary at ~30 Ma (cross section in Figures 4, 5) corresponds to the emplacement of the Magmatic Belt of Eastern Mexico (MBEM) (Robin and Tournon, 1978; Robin, 1982; Aranda-Gómez et al., 2005). This unconformity may confuse the determination of the end of the Laramide deformation because in the Burgos Basin, this unconformity is dated at about 30 Ma and is coeval with deposition of the Frio sequence (Galloway et al., 2000), both being consequences of the MBEM uplift. This uplift eroded parts of the Mesozoic and Paleogene section previously deposited on the Salado-Tamaulipas Arch, which separated the La Popa and Burgos basins. This uplift also imposed a complex facies pattern (lowstand and prograding wedge) on the Frio sequences because of subaerial exposure at ~30 Ma linked to a relative sea level change (Eguiluz, 2007).

On seismic sections (Figure 13), it is also possible to observe that both the Mesozoic (Upper Jurassic–Upper Cretaceous) and lower Tertiary rocks (pre-Vicksburg?) are deformed by gliding of detached blocks because of tilting of the basin, with structural gaps filled by young rocks (Vicksburg ?). The cause of basement tilting toward the Gulf of Mexico that led to the folding and detached sedimentary cover may be related to the Laramide-age and post-Laramide uplifts. Increasing tilt with time changed the stress field, with modification from a detachment with shortening to a detachment with extension.

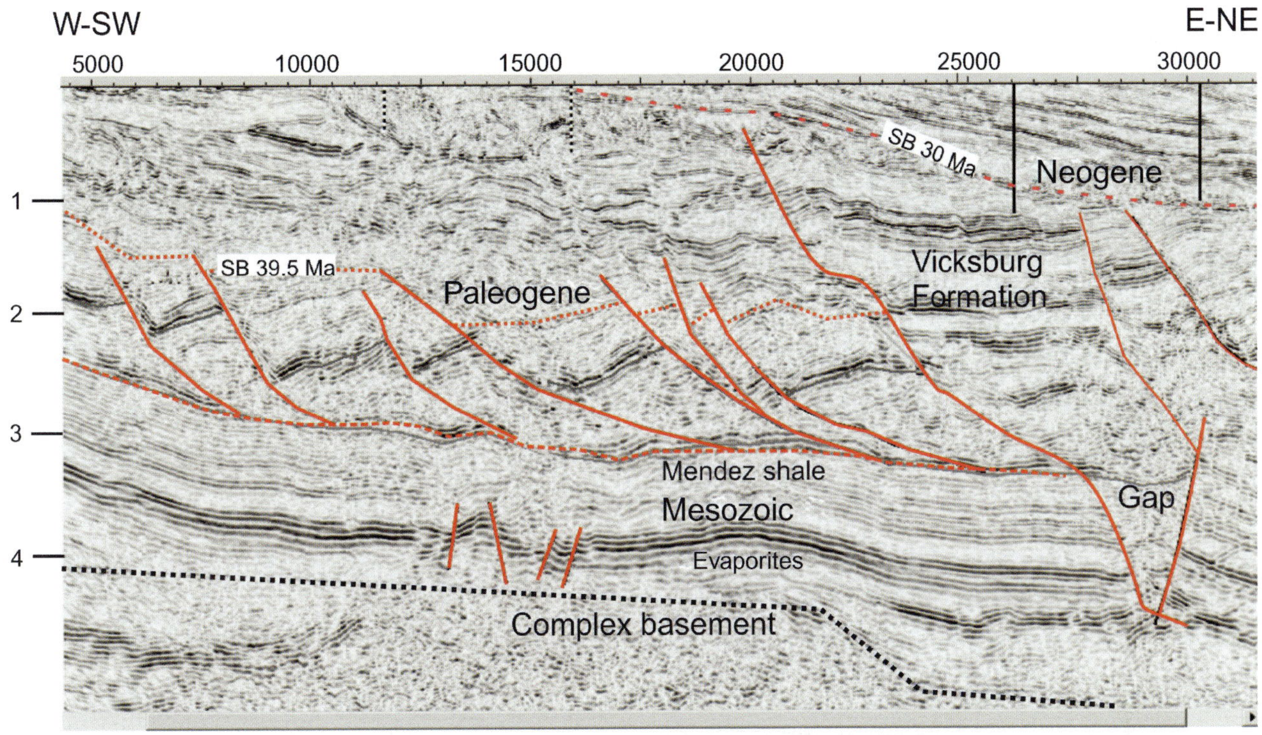

FIGURE 13. Seismic section showing the basement complex tilted toward east-northeast. Mesozoic rocks folded and detached on the Upper Jurassic evaporitic level. Paleogene rocks detached at the top of the Méndez shale. Sequence boundary ~39.5 Ma and lower Paleogene rocks in rotted blocks by listric faults. Extensional blocks with gaps are filled by younger sediments (Vicksburg Formation?). Frio and Neogene rocks underlying the sequence boundary placed at ~30 Ma. The location of this section is next to section IX-IX′ in Figure 3. SB = sequence boundaries.

GENERATING SYSTEM

The mature source rocks of the Burgos Basin are arranged in parallel trends (Figure 1) that are younger to the east and range from shallow-water environments in the west to deep-water environments in the east (Echánove, 1986). The most important source rocks charging the Yegua plays are the Paleocene Midway, Paleocene lower Wilcox, and Eocene upper Wilcox (PEMEX, 1996).

Hydrocarbons in the Eocene and Oligocene trends consist of thermogenic gas and associated condensate. In these trends, no thermogenic gases have been found associated with oil. The highest wet gas content occurs in the central part of the Burgos Basin, related to two factors: (1) stacking of several source rock intervals within the wet gas generating window and (2) anomalously high geothermal gradients of 50–70°C/km.

The Burgos Basin geothermal anomaly is elevated between the Eocene and Oligocene trends where most of the wet gas occurs (PEMEX, 1996). Thermally cool zones produce mainly dry gas to the east (Miocene trend) and west (Paleocene trend) of the Eocene–Oligocene area (Fuentes et al., 2001). This gas may be derived from Jurassic and Neogene source rocks, respectively (Carrillo, 2001; Roman and Holguin, 2001).

Eguiluz (2007) determined that the Eocene–Oligocene contact at the surface overlies deeply buried Jurassic–Cretaceous sedimentary layers that are decoupled and have been displaced, possibly above Callovian evaporites (Figure 13). Therefore, crustal heat flow when the MBEM was emplaced may have been anomalously high leading to increased maturation of the overlying Tertiary and older source rocks. Source intervals in the whole basin are identified by the isotopic signature of the gases, organic matter type, total organic carbon (TOC) content, and the degree of thermal alteration represented by vitrinite reflectance (R_o) (Table 1).

Mesozoic Generating Subsystem

The R_o values and time-temperature model (Figure 14) in the Burgos Basin for the Méndez Formation indicate that this rock is overmature. We assume that it only charged accumulations within the Paleocene trend. The gases in fields of this trend have methane-based carbon isotopes and hydrogen isotopes (Fuentes et al., 2001) indicating that these dry gases are nonassociated and thermally generated from rocks with sapropelic organic matter (kerogen 1 and 2) and maturities of 2–3% R_o

TABLE 1. Properties of the gas generating subsystems in the Burgos Basin.*

Generating Subsystem	*TOC** (%)*	*R_o** (%)*	*Type of Organic Matter*	*Isotopes $\delta^{-13}C/C_2$+ (‰)*	*Environment (West to East)* ⟵⟶	
Upper Jurassic	1 to > 2	2 to >4	2, 3	−25, −30/0		
Cretaceous Méndez		2 to 3	1, 2	−25/0	OS	B
Paleocene Midway	0.5 to > 2	0.5 to 1.2	3, 4	−35, −40/<10	MS	B
Paleocene Wilcox	0.5 to 2	<0.5 to >0.5	3, 4	−35, −45/>5, > 15	Br	OS, B
Reklaw	<0.5 to 1	<0.5 to >4	3, 4		Br	MS
Cook Mountain-Yegua	<0.5 to 1	<0.5 to >4	3, 4		IS	MS

**Values vary from west to east, as do the environments of deposition: IS = inner shelf; MS = middle shelf; OS = outer shelf; B = bathyal; Br = brackish (PEMEX, 1996).*
***TOC = total organic carbon; R_o = vitrinite reflectance; (%) = percent; (‰) = parts per mil.*

(PEMEX, 1996). This high maturity level leads to the conclusion that the hydrocarbons are derived from Mesozoic strata (Carrillo, 2001). West of the Burgos Basin, the source rocks of the Méndez Formation have lower maturity than those of the Lampazos, Anáhuac, and Cadena fields (González and Holguin, 2001), which are sourced from Kimmeridgian-Tithonian beds. All of these source rocks contain humic kerogens with maturities that range, respectively, from 2, 3, and greater than 4% R_o. Therefore, the gas accumulations in the Paleocene trend cannot be derived from Jurassic rocks.

Midway Generating Subsystem

The dry gas reservoirs of the Paleocene trend could not have been generated from the Paleocene Midway

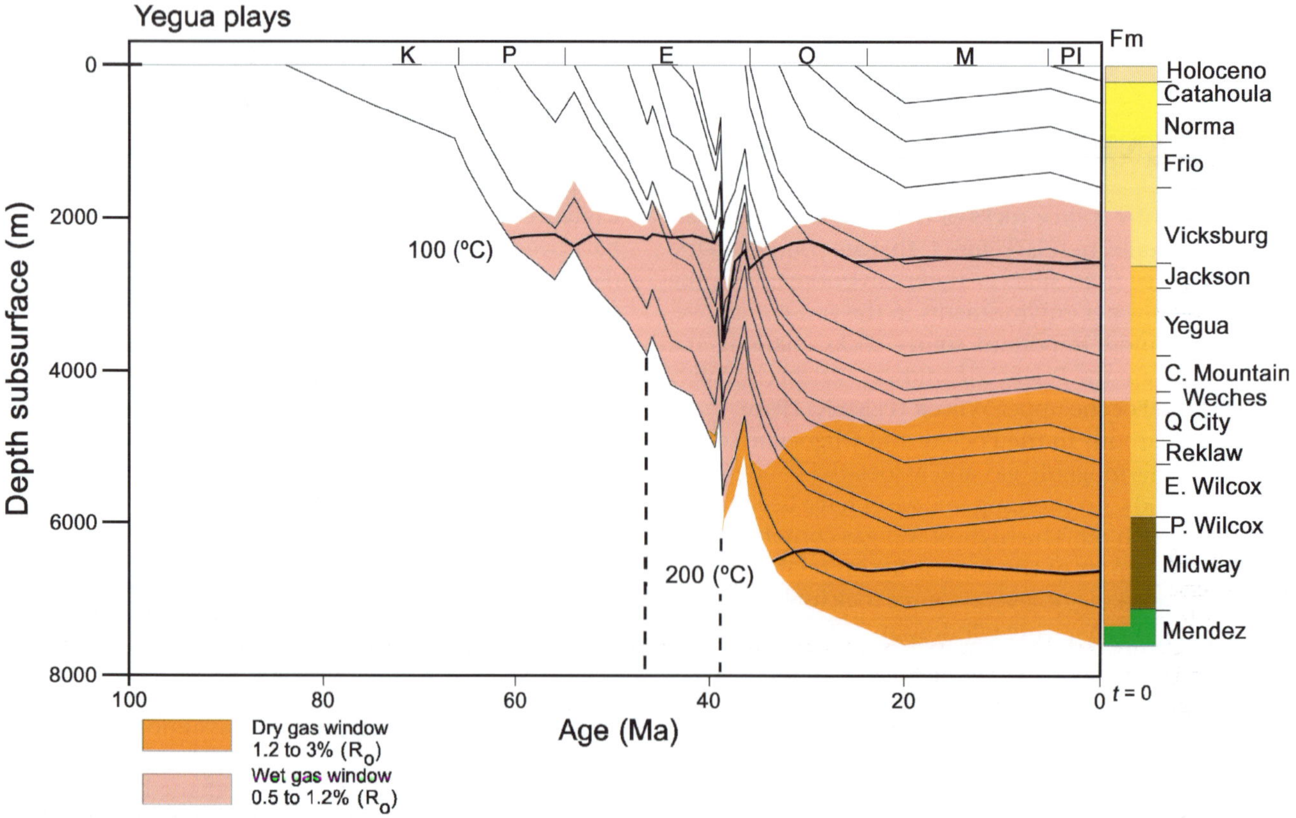

FIGURE 14. Burial history plot of the study area. From Midway to top Queen City overmature rocks in dry gas window; from Weches to Vicksburg rocks in wet gas window. Significant sea level change and not deposits estimated at around 48.5 and 39.5 Ma. K = Cretaceous; P = Paleocene; E = Eocene; O = Oligocene; M = Miocene; Pl = Pleistocene and Pliocene.

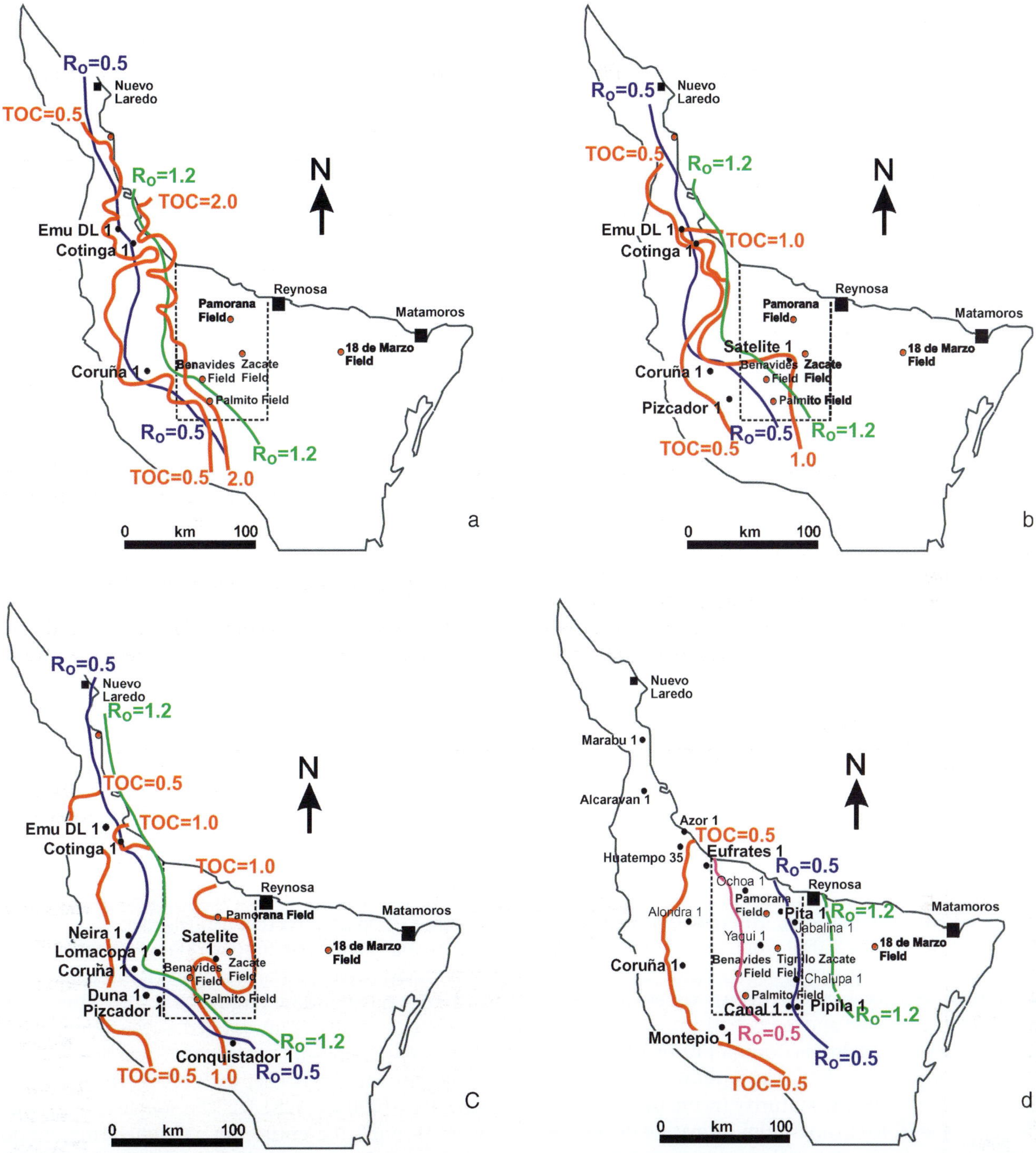

FIGURE 15. Variations in total organic carbon (TOC) in (a) the Midway Formation, (b) the Paleocene Wilcox, (c) the top Eocene Wilcox, and (d) the Cook Mountain-Yegua subsystems. The TOC values (%) ranged from 0.5 to 2. The areas with overlapping TOC offer the best possibility of generation and charging the Yegua plays. Wells are black dots and field gas are red dots. The TOC values are indicated in the text. Kerogen maturation of generating subsystems indicated by the blue line for vitrinite reflectance (R_o) = 0.5 in the wet gas window and the green line for R_o = 1.2 in the dry gas window. The magenta line indicates R_o = 0.5 in the wet gas window of the Cook Mountain; the expansion faults Y and J (Figure 3), are the border to the dry gas window of the Cook Mountain and the Yegua plays.

Formation because this unit has very low thermal maturity (from 0.5 to <1.2% R_o) indicating that it could have only generated wet gas. The carbon isotope ratios of methane versus ethane indicate that the hydrocarbons of the Eocene and Oligocene trends are a mixture of biogenic and thermogenic gases, whereas the hydrocarbons in the Paleocene trend are of thermal origin (PEMEX, 1996; Carrillo, 2001; Fuentes et al., 2001).

The Midway Formation has variable generating potential ranging from poor in the west to very good in the east. The Midway consists of shelf deposits that deepen eastward to upper bathyal environments (Table 1; Figure 15). This easterly change in the Burgos Basin controls variations in hydrocarbon generating capacity. Specifically, the generating capacity is poor to the west, with TOC less than 0.5% (Figure 15), associated with middle shelf environments. Farther east, fair generating capacity (TOC from 0.5 to 1.0%) is associated with outer shelf settings as confirmed by well data (Emu DL 1, 1.0%; Coruña 1, 0.8%; and Cotinga 1, 1.8%). Still farther east, an area of good generating potential (TOC >1.0–2.0%) is associated with a shelf margin depositional environment. To the east of this is an area of very good generating potential (TOC >2%) in upper bathyal environments with possible excellent conditions for the preservation of organic carbon caused by anoxia, low energy, and relatively low sedimentation rates (PEMEX, 1996). The productive Yegua plays occur in this area.

The upper Midway Formation is immature in the western part of the basin, although a tendency towards increasing maturity is observed in the east from the wet gas window (R_o >0.5–1.2%) to the dry gas window (R_o >1.2–4.0%). This is interpreted to result from expansion and deeper burial across the regional faults. Shales of the Midway Formation contain type 3 and 4 kerogen and can only produce dry gas and minor condensate.

Wilcox Generating Subsystem

The Wilcox Formation varies from fluvial environments to the west to bathyal in the east. Its hydrocarbon generating capacity is poor (TOC <0.5%) when associated with the brackish, inner shelf environment and fair (TOC of 0.5–1.0%) when associated with the middle shelf environment (Table 1; Figure 15). The best generating capacity occurs in deeper water facies associated with outer shelf and upper bathyal environments with TOCs of 1.0–2.0 in the Coruña 1 (0.7%), Pizcador 1 (0.7%), Conquistador 1 (1.2%), Satélite 1 (0.8%), and Emu DL 1 (1.0%) wells. Maturity in the upper part of the Paleocene Wilcox varies from immature (R_o <0.5%) to mature (R_o >0.5%) within the wet gas window, where most condensate generation occurs (Figure 15). Maturity also increases towards the east, associated with expansion faults.

In the northern part of the basin, brackish environments of the Paleocene Wilcox have good generating potential (TOC of 1.0–2.0%), possibly corresponding to swampy depositional zones. The Eocene Wilcox in the western part of the area is immature (R_o <0.5%) and does not generate hydrocarbons. Towards the east, however, the basin deepens with maturity within the wet gas window. Well data indicate that source quality increases from fair, associated with the mid-shelf environment, to very good (TOC >2.0%), associated with the bathyal environment (Figure 15). These wells are Pizcador 1 (0.6%), Conquistador 1 (1.1%), Satélite 1 (0.9%), Neira 1 (0.8%), Duna 1 (0.6%), Coruña 1 (0.6%), Lomacopa (0.7%), Cotinga 1 (1.1%), Emu DL 1 (0.8%), and Malinche 1 (visually good). The organic material in these rocks is type 3 and 4, generating mostly methane and scarce wet gas. The best Yegua plays occur in this area.

Reklaw Generating Subsystem

Some Reklaw Formation alluvial environments in the northern Burgos Basin have no generating capacity (TOC <0.5%), although other wells in brackish water environments have some generating potential (Canal 1, 0.8%; Satélite, 0.7%; and Cotinga 1, 1.1%), and perhaps some swampy environment strata have good potential (TOC of 1.0–2.0%). A fair generating potential east of the delta-front and prodelta environments of the Reklaw is observed (Table 1). The maturity of the upper Reklaw varies from immature in the western area to mature in the east (R_o >0.5%). Shales in the Reklaw Formation contain type 3 and 4 kerogen and are therefore capable of generating only dry gas and scarce condensate.

Cook Mountain-Yegua Generating Subsystem

The generating capacity of this unit varies from poor (TOC <0.5%) in the west to fair (TOC between 0.5 and 1.0%) in the east, as well data suggest (Eufrates 1, 0.7%; Montepio 1, 0.8%; Pita 1, 0.6%; Canal 1, 0.9%; and Conquistador 1, 0.9%), where it was deposited at the inner to outer shelf (Figure 15).

The maturity of the upper Yegua Formation varies from immature (R_o <0.5%) to mature (R_o >1.2–4%) within the dry gas window (Figures 14, 15). This abrupt change in maturity is interpreted as the result of a regional expansion fault system that was active during the upper early Oligocene, with blocks down dropped more than 1000 m (3281 ft). The source rocks (shales) passed through the entire wet gas generating window and entered the dry gas window. The Cook Mountain-Yegua source rocks contain type 3 and 4 kerogen and are capable of generating mainly dry gas with scarce condensate (Table 1).

Gas Content

The C_2+ hydrocarbon content in natural gas (C_2+ [% ethane + propane + butane + pentane]) is generally low in the Burgos Basin (<5.0%), especially in Cretaceous reservoir rocks of the Cadena field and in Paleocene rocks of the Pandura 1; Jaribú 1; Jaujal 1; Velero 1; Emú DL 1, DL 2, and 22; Azor 1; Presa 1; Duna 1; and

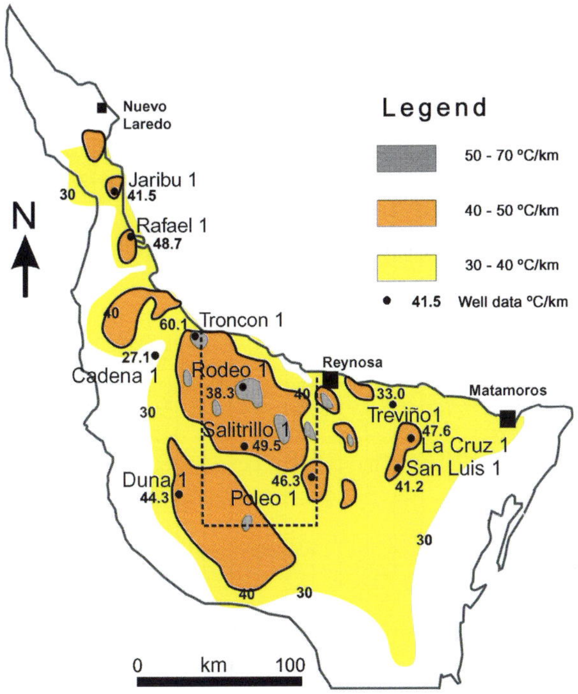

FIGURE 16. Average geothermal gradients in the Burgos Basin measured from reservoir temperatures (PEMEX, 1996). The central part of the basin has foci of higher temperatures that, in the subsurface (Rodeo-Poleo wells), coincide with the decoupling of the Mesozoic cover above a salt (Callovian?) detachment (Figure 13). The Magmatic Belt of Eastern Mexico could cause increases of the higher heat flow emanating from the crust during the Vicksburg and Frio deposition.

Pizcador 1 wells (Figure 15), indicating their derivation from an overmature source.

The greatest concentration of wet gases (C_2+) of 15–20% in the Burgos Basin is produced from Eocene reservoirs, including those within the present study area (Yaqui 1, Zacate 101, and Fronterizo 1 wells, Figure 15). This phenomenon is caused by two factors:

1. The stacking of six Tertiary generating centers (Figure 15): Paleocene Wilcox, Eocene Wilcox, Reklaw, Cook Mountain-Yegua and Jackson, and Oligocene Vicksburg, all within the principal wet gas generating window (R_o >0.5%).
2. A geothermal anomaly near the gas fields (Figure 16), with the highest average geothermal gradient in the Burgos Basin, ranging from 40°C/km to less than 70°C/km. This is located in the subsurface near the gap-detached Mesozoic rocks (Figure 13).

Wet gas content diminishes again towards the east (<3%) in late Oligocene reservoirs of the Treviño 101, G. Prieto 1, 18 de Marzo 9 (Figure 15), and Corzos 2 wells. This is explained by increased burial of the Jackson and Vicksburg generating subsystems across the expansion faults placing them in the dry gas window.

Geothermal Gradient

Reservoir temperatures determined from wells in the Burgos Basin allowed the interpretation of subsurface temperatures between the depths of 600 and 3500 m (1968 and 11,483 ft) (PEMEX, 1996). These temperatures range from 46 to 150°C along different trends, as shown in Figure 16. The Eocene temperature trend is the highest (i.e., the hottest temperatures at the shallowest depth), followed in importance by the Oligocene and Paleocene trends.

These temperature trends, when transformed into geothermal gradients, indicate that the highest values (38–60°C/km) occur in the Eocene trend, diminishing in the Oligocene trend to 30–50°C/km, and in the Paleocene trend to less than 25–53°C/km. Finally, the Miocene trend is the coolest with a gradient of 33°C/km (Table 2).

A regional geothermal view of the Burgos Basin (Figure 16) shows an elevated geothermal anomaly in the central part of the basin (PEMEX, 1996) where the Eocene and Oligocene trends have very high geothermal gradients of 40°C/km to less than 70°C/km. These are bordered to the west (Paleocene trend) and to the east (Miocene trend) by thermally cold zones (gradients of 30°C/km) and both of these trends produce dry gas. The geothermal anomalies are regionally aligned (as are the gas fields) along structural expansion systems that migrated with time towards the east as sedimentation prograded in that direction.

Integration and analysis of data on the generating potential, maturity, generation kitchens, types of organic matter, wet gas contents, regional geothermal gradients, and gas isotopes permit the construction of a composite map that overlays the TOC trends and generating systems charging the Yegua plays (Figure 15), as well as a composite map of the R_o trends (Figure 15).

TABLE 2. Range of depth and temperature from reservoir age and range of geothermal gradients in the Burgos Basin.

Reservoir	*Depth (mbsl)**	*Reservoir Temperature (°C)*	*Geothermal Gradient (°C/km)*
Jurassic	1500–3000	80–150	30–55
Cretaceous	2000–3500	72–150	26–36
Paleocene	1000–3250	60–130	25–53
Eocene	600–2700	46–135	38–60
Oligocene	7–3000	50–140	30–50

**mbsl = meter below sea level.*

Table 3. Production data in the Yegua reservoirs.

FIELD/ WELL	*DISCOVERY/ START PROD*	*WELL* TOT	*WELL* IMPROD	*WELL* PRODS.	*UNIT/ BLOCK*	*DAY PROD MMCF*	*ACUM PROD MMPC*	*NET PAY THICK (m)*	*PRODUCTION INTERVAL (m)*	*POR %*	*Sw %*	*Perm md*	*TEST PROD MMPC*	*TEST PRESSURE kg/cm²* Open	*TEST PRESSURE kg/cm²* Close	*TRAT.**
CABEZA FIELD	1954	15	10													
CABEZA	1954			1	EY	0.6460	836.960	6	1773-1779	16	62	1.3-.2	1.136	65	165	F
CABEZA	1982			3	EY	0.1290	1849.919	10	1725-1735	SD	SD	SD	3.602	168	189	A
CABEZA	1968			3	EY4	0.0000	3235.466	5	1834-1839	13	58	SD	0.890	42	182	A
CABEZA	1968			5	EY4	0.0000	3310.569	10	1950-1960	17	50	SD	2.189	190	266	F, A
CABEZA	1983			5	EY	0.0640	1566.1000	4 + 3 + 9 + 4	1791-95,02-05, 27-36,39-43	20	46	SD	3.308	154	231	F
CABEZA	1968			6	EY	0.0000	2.242	4	1710-1714	19	31	SD	0.700	32	94.5	A
CABEZA	1969			12	EY	0.0640	4004.103	7	1835-1842	SD	SD	SD	0.648	30	175	A
		15	10	5		0.9030	14,805.359									
CLAVEL FIELD	1970	2	1													
CLAVEL	1970			1	EY-m	0.0000	36.938	4.5	1255-1259.5	22	68	SD	0.600	28	105	A
		2	1	1		0.0000	36.938									
COMITAS FIELD	1966	30	29													
COMITAS	1986			14-D	EY	0.0000	0.168	20	2910-2930	9	55	SD	0.127	5	85	A
OROZCO		6	6													
		36	35	1		0.0000	0.168									
CUERVITO FIELD		29	25													
CUERVITO	1992			3	YE-M	0.5400	1910.850	20	1630-1650	SD	SD	SD	2.643	120	SD	F
CUERVITO	SD			10	YE-M	0.5400	271.230	17	1668-1685	SD	SD	SD	2.643	SD	SD	SD
CUERVITO	1969			11	YE-M	0.0000	81.320	4	1611-1615	SD	SD	SD	2.643	90	164	F
CUERVITO	1969			22	YE-1	0.0939	389.619	8	872-880	23	46	SD	1.748	81	88	D
CUERVITO	SD			23	YE-3	0.6230	1992.584	10	1035-1045	23	35	SD	SD	SD	SD	SD
CUERVITO	1970			25-I	YE-3	0.0000	24.141	4	1050-1054	22	45	SD	0.211	9	94	D
CUERVITO	1970			25-S	YE-1	0.0000	322.856	6	491-496	22	41	SD	0.918	42	42	D
CUERVITO	1970			26	YE-3	0.0939	5457.532	5	1027-1032	22	39	SD	2.171	101	115	D
		29	22	7		1.8908	10,450.132									
DIVISADERO FIELD	1966	3	2													
DIVISADERO	1966			1	EY	0.0000	192.733	3	1856-1859	22	49		4.940	219	252	F
TIGRILLO		4	4													
CATARRIN		1	1	0												
		8	7	1		0.000	192.733									
PALMITO FIELD	1963	41	22													
PALMITO	1963			1-I	EY1-BIII N	0.0258	1785.627	3	1205-1208	SD	SD	SD	0.918	4	105	D
PALMITO	1963			1-S	EY-BIII N	0.0387	1368.598	3	1165-1168	15	44	0.8	0.769	37	62	D
PALMITO	1982			2	EY1-BIII N	0.0258	366.799	6	1160-1166	SD	SD	SD	0.918	42	127	F
PALMITO	1964			2-I	EY-BIII N	0.0000	401.590	6	1131-1137	SD	SD	SD	0.847	39	70	A
PALMITO	1964			2-S	EY-BIII I	0.0000	51.016	4	1099-1103	SD	SD	SD	1.175	56	72	A
PALMITO	1982			3	EY1-BIII N	0.0387	349.255	3 + 4 + 4 + 5 + 4	1214-21,28-31, 34-38,44-49, 53-57	SD	SD	SD	SD	87	SD	F

PALMITO	1964			3-I	EY2-BIII N	0.0000	389.290	6	1267-1273	SD	SD	SD	0.829	38	112	A
PALMITO	1964			3-S	EY-BIII N	0.0000	3.073	4	1187-1191	SD	SD	SD	1.052	49	107	A
PALMITO	1982			6-S	EY1-BIII N	0.0387	379.258	7	1281-1288	SD	SD	SD	SD	74	98	F
PALMITO	1965			6-I	EY1-BIII N	0.0000	72.943	7	1281-1288	SD	SD	SD	0.685	31	105	A
PALMITO	1965			6-S	EY-BIII N	0.0000	18.073	5	1245-1250	SD	SD	SD	1.052	49	140	A
PALMITO	1965			8-I	EY2-BIII N	0.0648	1289.109	5	1227-1232	SD	SD	SD	0.762	35	115	F
PALMITO	1965			8-S	EY-BIII I	0.0250	327.530	8	1102-1110	SD	SD	SD	1.426	63	101	F
PALMITO	1984			11	EY-BII	0.0387	382.831	5 + 2	1092-97,1109-11	SD	SD	SD	1.889	87	122	F
PALMITO	1965			11-I	EY5-BII N	0.0000	176.981	2	1348-1350	SD	SD	SD	0.695	37	133	F
PALMITO	1965			11-S	EY3-BII	0.0000	0.611	5	1271-1276	SD	SD	SD	0.493	23	91	F
PALMITO	1966			12	EY-BII	0.0387	482.644	7	1160-1167	14	70	0.1	0.702	35	126	F
PALMITO	1967			13	EY-BII S	0.0387	487.030	3 + 3	1248-51,60-63	9	SD	0.1	1.419	124	144	F
PALMITO	1967			14-I	EY4-BII	0.0258	179.657	4	1480-1484	14	33	SD	1.531	75	194	A
PALMITO	1967			14-S	EY5-BII S	0.0000	167.584	4	1404-1408	17	41	SD	1.531	75	194	A
PALMITO	1967			15	EY5-BII S	0.0000	413.281	4 + 5	1482-86,89-94	SD	SD	SD	0.660	30	185	A
PALMITO	1968			16	EY2-BII	0.0000	516.756	9	1383-1392	SD	SD	SD	0.544	25	108	A
PALMITO	1984			19	EY1-BIII S	0.0000	121.528	8	1380-1388	SD	SD	SD	0.847	38	75	F
PALMITO	1968			19-I	EY1-BIII S	0.0000	5.668	5	1393-1398	SD	SD	SD	0.308	14	164	D
PALMITO	1984			19-S	EY-BII S	0.0000	20.863	6	1313-1319	SD	SD	SD	SD	SD	SD	F
PALMITO	1984			19-S	EY-BII S	0.0000	132.482	10	1340-1350	SD	SD	SD	SD	SD	SD	F
PALMITO	1984			19-S	EY-BIII N	0.0000	84.782	2	1353-1355	SD	SD	SD	SD	51	SD	F
PALMITO	1968			21	EY-BII	0.0000	666.922	6 + 4	1146-1152,54-58	18	43	SD	-0.644	17	133	A
PALMITO	1982			22	EY1-BII S	0.0000	79.798	8	1240-1248	13	47	SD	SD	126	140	F
PALMITO	1982			22	EY3-BII S	0.0000	200.872	11	1265-1276	15	50	SD	SD	126	140	F
PALMITO	1969			22	EY4-BII	0.0000	508.520	10	1437-1447	13	35	SD	0.768	35	87	A
PALMITO	1970			23	EY1-BI	0.0000	426.084	6 + 5	1126-1132, 37-42	18	40	SD	0.298	12	88	A
PALMITO	1979			24	EY2-BI	0.0000	418.161	5	1257-1262	SD	SD	SD	0.170	7	126	A
PALMITO	1983			24-I	EY-BI	0.0000	342.715	11	1081-1092	SD	SD	SD				
PALMITO	1983			24-S	EY-BI	0.0000	342.397	5	1024-1029	SD	SD	SD				
PALMITO	1970			28	EY2-BIII S	0.0000	234.262	4	1508-1512	15	58	SD	0.469	21	115	D
PALMITO	1971			32	EY-BII S	0.0000	1126.110	6	1369-1375	19	35	SD	2.878	133	182	SD
PALMITO	1982			34	EY1-BIII S	0.0000	136.845	4	1493-1497	15	57	SD	1.665	77	175	F
PALMITO	1982			34	EY2-BIII S	0.0000	140.604	1	1514-1515	16	39	SD	1.665	77	175	F
LLANURA	1970	2	1													
LLANURA	1970			1	EY	0.0640	1822.083	7	1581-1588	15	44	SD	1.218	130	56	A
PIAMONTE	1970	1														
PIAMONTE	1970			1	EY3A	0.0387	151.663	4	1496-1500	18	50	SD	0.612	42	161	SD
		44	23	21		0.5021	16,571.895									
PAMORANA FIELD	1951	10	6													
PAMORANA	1963			2	EY10-BB	0.0000	194.993	4	2160-2164	SD	SD	SD	0.565	45	252	HCL
PAMORANA	1999			3	EY5	0.1065	625.354	7	1575-1582	?	?	?	3.117	210	SD	SD
PAMORANA	1964			3	EY6-BA	0.0000	3430.429	4	1616-1620	18	56	SD	3.878	182	203	D
PAMORANA	1982			6	EY7	0.1290	846.710	2 + 2 + 7 + 2	1822-24,26-28, 38-45,63-65	16	42	SD	0.806	115.5	161	HCL
PAMORANA	1985			6-S	EY6-BB	0.0000	558.966	4.5	1704.5-1709				1.936	91	190	A
PAMORANA	1982			6-I	EY9A-BB	0.0000	323.935	2	1936.5-1938.5	11	42	SD	0.610	28	231	SMA
		10	6	3		0.2355	5980.387									

Table 3. Production data in the Yegua reservoirs (cont.).

FIELD/ WELL	DISCOVERY/ START PROD	WELL TOT	IMPROD	PRODS.	UNIT/ BLOCK	DAY PROD MMCF	ACUM PROD MMPC	NET PAY THICK (m)	PRODUCTION INTERVAL (m)	POR %	Sw %	Perm md	TEST PROD MMPC	TEST PRESSURE kg/cm² Open	Close	TRAT.*
PIAMONTE	1969	1	0	WELL INCLUDED IN PALMITO FIELD												
PIEDRAS FIELD	1963	5	3													
PIEDRAS	1965			1	EY-3	0.0000	857.782	4	1265-1269	30	50	SD	2.210	105	193	SMA
PIEDRAS				20	EY	0.0000	684.403	4	1255-1259	22	50	SD	2.242	105	183	SD
SAN PEDRO FIELD	1967	2	2	0												
		5	3	2		0.0000	1542.185									
PITA FIELD	1989	1	0													
PITA	1989			1	EY	0.0856	452.488	15	2473-2488	20	56	SD	0.320	14	308	MCA
		1	0	1		0.0856	452.488									
PRIMAVERA FIELD	1964	23	19													
PRIMAVERA	1969			11	EY	0.1180	64.414	3	659-662	25	40	SD	0.561	25	56	A
PRIMAVERA	1969			13	EY	0.0230	793.368	10	970-980	13	39	SD	0.724	33	110	A
PRIMAVERA	1969			18	EY	0.0000	14.491	7	695-702	17	55	SD	0.812	17	35	A
PRIMAVERA	1969			19	EY	0.3890	236.187	6	584-590	18	52	SD	1.207	56	68	A
4 MILPAS	1964	27	26													
4 MILPAS	1986			25	EY	0.0000	78.700	6	605-631	SD	SD	SD	0.320	14	42	A
BENAVIDES	1960	41	36													
BENAVIDES	1983			4	EY	0.0400	212.590	5	775-780	22	54	SD	SD	21	32	A
BENAVIDES	1968			12	EY	0.0000	477.163	4.5	527.5-532	19	59	SD	0.829	38	49	A
BENAVIDES	1989			15	EY	0.0000	64.624	6	498-504	19	40	SD	0.317	14	35	A
BENAVIDES	1988			24	EY	0.0000	157.340	8	496-504	29	43	SD	0.949	43	45	D
BENAVIDES	1982			27	EY	0.0000	2.450	2 + 5	525-527, 41-46	19	43	SD	SD	14	35	A
Sn BERNARDO	1965	10	8													
Sn BERNARDO	1968			5	EY	0.0260	2388.250	5	974-979	SD	SD	SD		51	101	A
Sn BERNARDO	1968			6	EY	0.0410	41.034	3	908-911	SD	SD	SD		53	70	A
		101	89	12		0.5860	4530.611									
RANCHERIAS FIELD	1935	9	1	8												
RANCHERIAS	1935			1-s	EY	0.0000	SD	SD	SD	SD	SD	SD	SD	SD	SD	SD
RANCHERIAS	1935			1-i	EY	0.0000	SD	SD	SD	SD	SD	SD	1.228	23	86	SD
RANCHERIAS	1935			3-s	EY	0.0000	SD	7	454-467	SD	SD	SD	SD	SD	60	SD
RANCHERIAS	1935			3-i	EY	0.0000	SD	SD	SD	SD	SD	SD	SD	SD	SD	SD
RANCHERIAS	SD			4	EY	0.0000	SD	SD	SD	SD	SD	SD	SD	SD	SD	SD
RANCHERIAS	SD			5	EY	0.0000	SD	SD	SD	SD	SD	SD	SD	SD	SD	SD
RANCHERIAS	SD			6	EY	0.0000	SD	SD	SD	SD	SD	SD	SD	SD	SD	SD
RANCHERIAS	SD			7	EY	0.0000	SD	SD	SD	SD	SD	SD	SD	23	SD	SD
RANCHERIAS	1942			8	EY	0.0000	SD	29	429-458	SD	SD	SD	SD	23	SD	SD
RANCHERIAS	1942			11	EY	0.0000	SD	SD	SD	SD	SD	SD	SD	SD	SD	SD

		9	1	8		0.0000	15000.000									
ZACATE FIELD	1964	8	7													
ZACATE	1967			112	EY-10	0.4470	2803.962	3	1941-1944	18	60	1	1.341	60	238	D
		8	7	1		0.4470	2803.962									
TOTAL	13 FIELD	268	205	63		4.6500	72336.838									
23.5% Success					Average 1148.6 MMPC/well											

Data May 2000.
**Wells in the play. A = acid; F = hydrofracturing; SD = non data; D = shot; HCL = hydrochloric acid; MCA = acetic acid.*

Figure 15 shows that the best hydrocarbon-producing areas overlay two or more generating systems with the highest TOC values. Figure 16 shows that the study area contains a series of parallel trends where rocks that charge the Yegua plays are in the wet gas window, as indicated by the isotopes of the produced gases. From the foregoing, exploration risk can be evaluated for the generating potential of source rocks in the study area. The Wilcox system is the principal source of hydrocarbons for the Yegua plays with the Cook Mountain as another possible source rock is not documented. The Palmito and Cabeza fields are the most prolific in the Yegua play, and they are located where that subsystem exists (Figure 15).

PRODUCTION HISTORY

In 1936, the Burgos Basin established gas production from the fluviodeltaic Yegua play (Rancherías field), but it was not until the 1960s that the principal gas fields in this play were discovered by exploring structural traps. Combination traps (structural and stratigraphic) were commonly found, or in some cases, reservoirs were discovered while looking for Queen City targets (Echánove, 1986). Study of the Yegua and Jackson plays (PEMEX, 2000) identified the stratigraphic traps as basin-floor fans and deltaic deposits, as well as expansion fault zones with prograding wedges, and transgressive and highstand systems. In 2000, the Yegua Formation had a cumulative hydrocarbon production of 72.3 bcf (2.04 billion m^3) of gas from 61 wells in 12 fields (Table 3), and in that year the play produced 3.8 million ft^3 (107,604 m^3) per day from 30 wells in 8 fields (PEMEX, 2000). Before 2000, the following fields had been discovered: Comitas-Orozco, Palmito-Llanura-Piamonte, Cabeza, Cuervito, Divisadero-Tigrillo, Pamorana, Zacate, Piedras-San Pedro, Pita, Clavel, Benavides-San Bernardo-Cuatro Milpas-Primavera, and Rancherías (Figures 3, 15). Several fields have combined production from the Yegua and other formations. The Yegua and Jackson plays had little economic importance compared to other Tertiary reservoirs.

The present study improved exploration success and decreased risk in the Yegua plays, leading to new discoveries and improved development of the existing fields. As a result of renewed activity in the Burgos Basin since 1996, new gas reserves have been added by exploration in the Yegua (Jazmín 1A, Chalupa 1, Figure 15) and additional development wells (Pamorana, Palmito, and Cabeza fields, among others, Figure 3). In 2006, Yegua production was about 40 million ft^3 (1.13 million m^3) per day. No condensate production has been considered.

The largest Yegua fields of Palmito-Llanura and Piamonte (Figure 3) are considered to be stratigraphic traps that may be compartmentalized. Comitas-Orozco

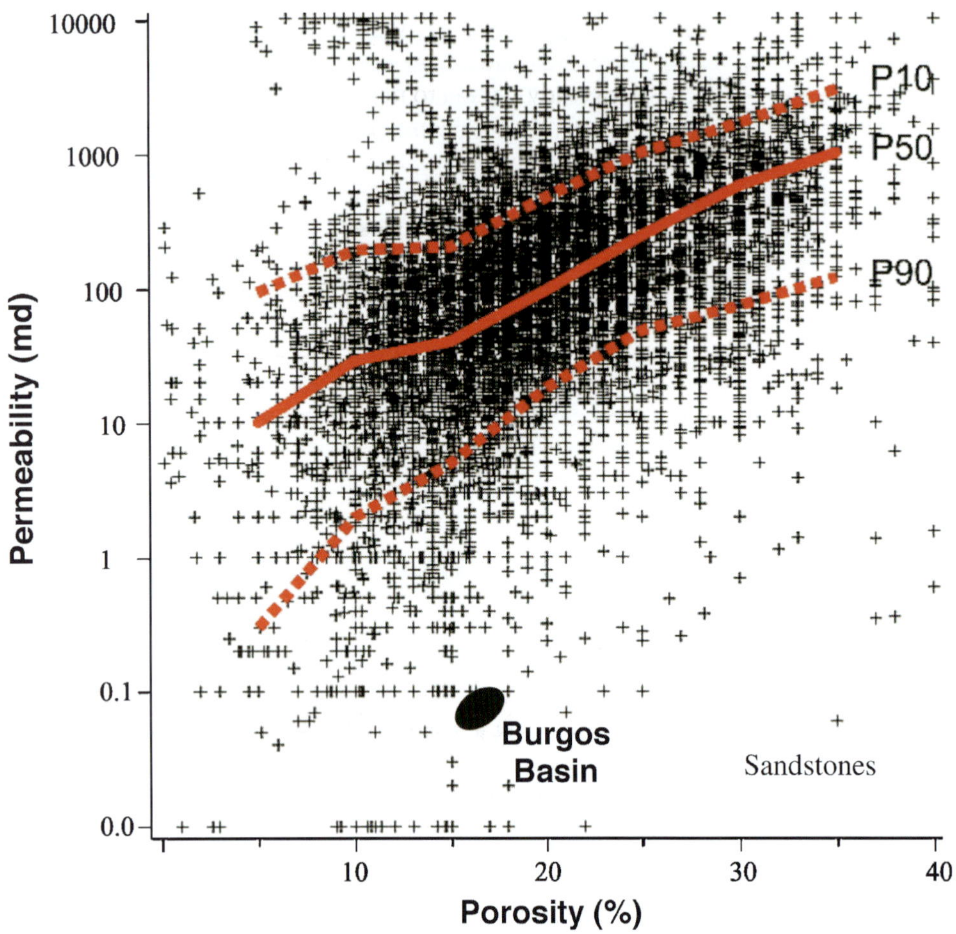

FIGURE 17. Average porosity-permeability relationships of core analysis in sandstones of the Burgos Basin compared to other reservoirs of the world (Ehrenberg and Nadeau, 2005; modified by Galicia-Barrios and Alvarado, 2005).

and Piedras-San Pedro fields as well as Cuervito, Pamorana, and Pita fields in the Yegua play are combined structural-stratigraphic traps associated with growth faults, or small stratigraphic traps in basin-floor fans, channels, and unconformity wedges. Clavel and Cabeza fields are stratigraphic traps, whereas Zacate and Rancherías fields are structural traps. The Divisadero-Tigrillo field is a fold trap against a fault. Within the Primavera, Benevides, San Bernardo, and Cuitlahuac fields, only sporadic wells produce from the Yegua, and it is difficult to unravel the geology in more detail (Table 3).

The depth range of the reservoirs ranges from a minimum of 429 m (1407 ft) to a maximum of 3522 m (11,555 ft). The most common depths for the Yegua are between 1100 and 1800 m (3609 and 5905 ft) (Table 3). The unexpanded plays (mainly stratigraphic traps) are shallower than the downdip plays controlled by growth faults, as in the Pita, Chalupa, and Comitas wells.

The productive sand bodies (net pay) in the Yegua play range from 1 to 29 m (3 to 95 ft) in thickness but most frequently fall between 2 and 8 meters (Table 3). These numbers represent only the hydrocarbon-saturated sands. Porosities for the productive sands range from a minimum of 9% to a maximum of 30% and are most commonly between 14 and 23%. Initial well production is from 1 to 2 million ft^3/day (28,316.84 to 56,633.69 m^3/day), although with modern completion practices, the wells may produce from 10 to 15 million ft^3/day (283,168.46 to 424,752.69 m^3/day) (Calabaza 15 and 12 wells, respectively). Depending on the field, minimum water saturations are 31–39%, whereas maximum values are 60–70%. The most frequent values lie between 42 and 50% (Table 3).

Static reservoir pressures in the Yegua plays range from 32 to 535 kg/cm^2 (3138 to 52,470 kPa), with the most frequent values between 130 and 250 kg/cm^2 (12,750 to 24,520 kPa). Borehole gas in-place (BGI) values range from 0.0030 to 0.022, most frequently ranging from 0.004 to 0.006 for the studied Yegua plays.

Core permeabilities in the sands at atmospheric pressure range from 1 to 10 md, with average porosities from 14 to 23%. An average porosity of 18% is a standard value for the Burgos Basin plays, including the Yegua. A divergence in permeability when compared to worldwide averages for siliciclastic reservoirs is seen (Ehrenberg and Nadeau, 2005). Galicia-Barrios and Alvarado (2005) pointed out that with these porosities Burgos Basin permeabilities should be 10 times higher (Figure 17). This leads to the consideration of other factors, such as natural fractures (seen in Yegua Formation well cores and outcrops), that augment reservoir permeabilities in the Burgos Basin, or the presence of very fine-grained sandstones.

The Yegua in Texas established hydrocarbon production around 1930 from deltaic plays in the updip part of the basin. In the 1970s, expansion fault plays were discovered downdip in the United States Gulf Coast Basin. In 2006, in Texas and Louisiana, the downdip plays had achieved a cumulative production of more than 2.3 tcf (65.1 billion m^3) gas from 221 fields, and the updip trend had produced 1 tcf (28.3 billion m^3) gas from 113 fields (Ewing, 2007). In 1999, the remaining reserves were estimated to be 1.6–2 tcf (45.3–56.6 billion m^3) at depths exceeding 2000 m (6562 ft) (Ewing, 2007). The Yegua plays had accumulated production of more than 18 tcf (509 billion m^3) in the Unites States Gulf Coast Basin (Berman, 2007, personal communication).

Resources estimated by Monte Carlo analysis suggest that the Burgos Basin Yegua plays could have an additional 995 bcf (28.1 billion m^3) recoverable gas in stratigraphic traps. Condensate again was not considered.

CONCLUSIONS

This investigation analyzes the stratigraphy of the Yegua Formation in the Burgos Basin and proposes the division of this formation into three parts. The lower Yegua corresponds to an HST with its top truncated by a regional unconformity corresponding to a sequence boundary at about 39.5 Ma. The middle Yegua is a complex stratigraphic unit with internal unconformities, channels, slumped sediments, and marine basin-floor fans; this interval is identified as a lowstand and complex prograding wedge system, overlying the ~39.5-Ma sequence boundary. The upper Yegua is recognized as a coarsening-upward sequence, possibly being an HST underlying the Jackson Formation. Its contact with the middle Yegua is not always possible to identify.

Based on sequence analysis and with the support of core and well-log, outcrop, and thickness data, a sedimentary model has been established for the Yegua interval. The model generally defines coastal plain, shelf, delta-fan, and estuarine environments, with shallower facies to the west and deeper facies to the east. Based on this model, several hydrocarbon plays have been identified. Expansion faults thicken and deepen the middle and upper Yegua while accumulating sand downdip in the basin; this play is an economically important gas objective.

The unconformity at ~39.5 Ma and the Yegua deposition described in this chapter may be related to the final phase of the Laramide deformation.

The petroleum system (source rock, reservoir, seal, trap, and timing) works efficiently in the Yegua plays of the Burgos Basin. Geological complexity in the Yegua has made hydrocarbon exploration difficult; however, this study has led to new discoveries and has increased gas production by nearly 100% from this stratigraphic interval.

ACKNOWLEDGMENTS

Thanks to Claudio Bartolini for the opportunity to publish this work and opening the door for geoscientific interaction. The manuscript has been substantially improved thanks to constructive comments made by Thomas Ewing, Arthur Berman, Marc Edwards, and Josh Rosenfeld. Special thanks to Natalia Amezcua for her valuable assistance in the preparation and constructive reviews of the manuscript. The author is grateful to Petróleos Mexicanos for allowing the release of the data herein.

REFERENCES CITED

Aranda-Gómez, J. J., J. F. Luhr, T. B. Housh, G. Valdez-Moreno, and G. Chávez-Cabello, 2005, El volcanismo tipo intraplaca del Cenozoico tardío en el centro y norte de México: Una revisión, Boletín de la Sociedad Geológica Mexicana, Volumen conmemorativo del centenario: Temas Selectos de la Geología Mexicana, v. 57, no. 3, p. 187–225.

Berggren, W. A., D. V. Kent, C. C. Swisher III, and M. P. Aubry, 1995, A revised Cenozoic geochronology and chronostratigraphy, *in* W. A. Berggren, D. V. Kent, M. P. Aubry, and J. Hardenbol, eds., Geochronology, time scales and global stratigraphic correlation: Society for Sedimentary Geology Special Publication 54, p. 129–212.

Carrillo, B. J., 1980, Paleocañones terciarios de la Planicie Costera del Golfo de México: Asociación Mexicana de Geólogos Petroleros, v. 32, no. 1, p. 27–55.

Carrillo, H. T. de E., 2001, Estudio geoquímico de gases y condensados de la Cuenca de Burgos: Asociación Mexicana de Geólogos Petroleros, v. 49, no. 1–2, p. 140–143.

Chávez-Cabello, G., 2005, Deformación y magmatismo Cenozoico en el sur de la Cuenca de Sabinas, Coahuila, México, Juriquilla, Querétaro, Universidad Nacional Autónoma de México, Centro de Geociencias: Tesis doctoral en Geología Estructural y Tectónica, 266 p.

Deussen, A., 1914, Geology and underground waters of the southeastern part of the Texas coastal plain: United States Geological Survey Paper 335, 365 p.

Dumble, E. T., 1892, Report of the brown coal and lignite of Texas; character formation, occurence, and fuel uses: Texas Geological Society, 243 p.

Dumble, E. T., 1894, The Cenozoic deposits of Texas: Journal of Geology 2, p. 549–567.

Eargle, D. H., 1968, Nomenclature of formations of Claiborne Group, middle Eocene coastal plain of Texas:

Washington, United State Geological Survey Bulletin 1251-D, 25 p.

Echánove, E. O., 1986, Geología petrolera de la Cuenca de Burgos: Boletín de la Asociación Mexicana de Geólogos Petroleros, v. 38, no. 1, p. 1–74.

Edwards, M. B., 1990, Stratigraphic analysis and reservoir prediction in the Eocene Yegua and Cook Mountain formations of Texas and Louisiana, Gulf Coast: Gulf Coast Section SEPM, 11th Annual Research Conference Program and Abstract, p. 151–164.

Eguiluz, de A. S., 2004, Tectónica Laramide en la Cuenca de Burgos, México: Revista GEOS, v. 24, no. 2, p. 307–308.

Eguiluz, de A. S., 2007, Laramide deformation in the Burgos Basin, northeastern, Mexico: The Paleogene of the Gulf of Mexico and Carribean basins: 57th Annual Convention of the Gulf Coast Association of Geological Societies and Gulf Coast Section SEPM: Processes, events, and petroleum systems, p. 689–702.

Eguiluz, de A. S., G. M. Aranda, and R. Marrett, 2000, Tectónica de la Sierra Madre Oriental, México: Boletín de la Sociedad Geológica Mexicana, v. 53, p. 1–26.

Ehrenberg, N. S., and M. P. Nadeau, 2005, Sandstone vs. carbonate petroleum reservoirs: A global perspective on porosity-depth and porosity-permeability relationships: AAPG Bulletin, v. 89, no. 4, p. 435–445.

Ewing, T. E., 2007, Fairways in the downdip and middip Yegua trend—A review of 25 years of exploration: Gulf Coast Association of Geological Societies, p. 227–248.

Ewing, T. E., and G. Fergeson, 1991, Stratigraphic framework, structural styles and seismic signatures of downdip Yegua gas-condensate fields, Central Wharton County, Texas, Gulf Coast: Transactions-Gulf Coast Association of Geological Societies, v. 41, p. 255–275.

Ewing, T. E., and F. S. Vincent, 1997a, Foundered shelf edges— Examples from the Yegua and Frio, Texas and Louisiana: Gulf Coast Association of Geological Societies Transactions, v. 47, p. 149–157.

Ewing, T. E., and F. S. Vincent, 1997b, Correlation of Yegua/Cockfield genetic cycles, Texas and Louisiana: Gulf Coast Association of Geological Societies Transactions, v. 47, p. 631–633.

Fuentes, P. F., et al., 2001, Análisis isotópicos: Una nueva alternativa en el estudio de las cuencas gasíferas de México: Asociación Mexicana de Geólogos Petroleros, v. 49, no. 1–2, p. 47–67.

Galicia-Barrios, J. G., and P. D. Alvarado, 2005, Yacimientos de gas natural en clásticos terciarios de México, características estáticas y dinámicas: Asociación Mexicana de Geólogos Petroleros, Simposium Reynosa, Tamaulipas, Mexico.

Galloway, E. W., E. P. Ganey-Curry, L. Xiang, and T. R. Buffler, 2000, Cenozoic depositional history of the Gulf of Mexico Basin: AAPG Bulletin, v. 84, no. 11, p. 1743–1774.

Giles, A. K., F. T. Lawton, and J. F. Vega-Vera, 1999, Salt tectonics of Cretaceous-Paleogene La Popa Basin, Nuevo León, México: Asociación Mexicana de Geólogos Petroleros/AAPG, 3rd Joint International Conference Field Trip, sponsored by Dallas Geological Society, 109 p.

Gonzáles, G. R., and Q. N. Holguin, 2001, Las rocas generadoras de México: Asociación Mexicana de Geólogos Petroleros, v. 49, no. 1–2, p. 16–30.

Gradstein, F. M., et al., 2004, A geologic time scale: New York, Cambridge University Press, 589 p.

Gray, G. G., R. J. Pottorf, D. A. Yurewicz, K. I. Mahon, D. R. Pevear, and R. J. Chuchla, 2001, Thermal and chronological record of syn- to post-Laramide burial and exhumation, Sierra Madre Oriental, Mexico, *in* C. Bartolini, R. T. Buffler, and A. Cantu-Chapa, eds., The western Gulf of Mexico Basin, tectonics, sedimentary basins, and petroleum systems: AAPG Memoir 75, p. 159–181.

Haq, B. U., J. Hardenbol, and P. R. Vail, 1988, Mesozoic and Cenozoic chronostratigraphy and cycles of sea level change, *in* C. K. Wilgus, C. G. St. C. Kendall, H. W. Posamentier, C. A. Ross, and J. C. Van Wagoner, eds., Sea level changes— An integrated approach: SEPM Special Publication 42, p. 71–108.

Holroyd, P. A., 2002, New record of Anthracotheriidae (Artiodactyla, Mammalia) from the middle Eocene Yegua Formation (Claiborne Group), Houston County, Texas: The Texas Journal of Science, v. 54, no. 4, p. 301–308.

Hull, A. R., 1995, Shelf-slope break sedimentation and erosion during deposition of the Claiborne Group in Colorado and Wharton counties, Texas: Gulf Coast Association of Geological Societies Transactions, v. 45, p. 267–274.

Jackson, M. P. A., and W. E. Galloway, 1984, Structural and depositional styles of Gulf Coast Tertiary continental margins: Application to hydrocarbon exploration: AAPG Course Notes 25, 226 p.

Jin-wook, K., R. Berg, J. Watkins, and T. Tieh, 2003, Trapping capacity of faults in the Eocene Yegua Formation, East Sour Lake field, southeast Texas: AAPG Bulletin, v. 87 no. 3, p. 415–425.

Kennedy, W., 1892, A section from Terrell, Kaufman County to Sabine Pass on the Gulf of Mexico: Texas Geologicas Society, Annual Report 3, p. 41–125.

Kennedy, W., 1896, The Eocene Tertiary of Texas east of the Brazos River: Academy of Natural Sciences, Philadelphia Press, p. 89–160.

Lagoe, M. B., and T. B. Layman, 1994, Comprehensive, quantitative micropaleontological analysis as a tool for paleoenvironmental interpretation and sequence stratigraphy, with an example from the Yegua Formation, southeast Texas, *in* Gulf Coast Association of Geological Societies and Gulf Coast Section of SEPM meeting (AAPG Gulf Coast Section), Abstracts: AAPG Bulletin, v. 78, no. 9, p 1464.

Luneau, B. A., T. B. Eschner, E. Gomez, and A. J. Scott, 1995, Sequence-stratigraphic framework of the Yegua Formation in the Houston salt basin, Texas Gulf Coast: AAPG Bulletin, v. 79, no. 13, p. 58.

Meckel, L. D., and W. E. Galloway, 1996, Formation of high-frequency sequences and their bounding surfaces: Case study of the Eocene Yegua Formation, Texas, Gulf Coast, U.S.A.: Elsevier Science Sedimentary Geology, v. 102, p. 155–186.

Penrose Jr., R. A. F., 1890, A preliminary report on the

geology of the Gulf Tertiary of Texas from Red River to the Rio Grande, Texas: Geological Society, Annual Report 1, p. 3–101.

Petróleos Mexicanos (PEMEX), 1996, Evaluación de las rocas generadoras en la Cuenca de Burgos: Informe final del Proyecto Integral Cuenca de Burgos (internal file).

Petróleos Mexicanos (PEMEX), 2000, Estudio de los plays Jackson y Yegua, *in* Etapa 1, S. de A. Eguiluz, L. D. Meckel, S. Solano, M. J. Pola, S. Abel, A. Tristán, and G. Ramos, eds., Gerencia de Exploración Reynosa, Tamaulipas, NE-M 2935 (private file), 51 p.

Plummer, F. B., 1932, Cenozoic systems in Texas, *in* E. H. Sellards, W. S. Adkins, and F. B. Plummer, eds., The geology of Texas, v. 1, Stratigraphy: The University of Texas Bulletin, no. 3232, p. 519–996.

Robin, C., 1982, Relations volcanologie-magnetologie-geodinamique: Application au passage entre volcanisms alcalin et andesitique dans le sud mexicain (Axe Trans-mexican et Privince Alcaline Orientale): Annual Sciences De l Universitat Clermont-Ferrand II, v. 503, p. 70.

Robin, C., and J. Tournon, 1978, Spatial relations of andesitic and alkaline provinces in Mexico and Central America: Canadian Journal of Earth Sciences, v. 15, p. 1633–1641.

Rodríguez Lozano, D. E. 1999, Bioestratigrafía de la franja Paleoceno–Eoceno de la Cuenca de Burgos: Asociación Mexicana de Geólogos Petroleros, 50 Aniversario, Publicación Especial 2, La Cuenca de Burgos, p. 227–239.

Román Ramos, J. R., and N. Holguin-Quiñones, 2001, Subsistemas generadores de la Región Norte de México: Asociación Mexicana de Geólogos Petroleros, v. 49, no. 1–2, p. 68–84.

Rosen, N. R., E. B. Bowen, and J. K. Thies, 1994, Subsurface planktonic zonation of the Paleogene of Texas and Louisiana Gulf Coast and its relationship to relative changes of coastal onlap: Transactions of the Gulf Coastal Association of Geological Societies, v. 44, p. 631–639.

Swenson, R. D., 1997, Basal Yegua shelf margin failures along the Texas Gulf Coast: Gulf Coast Association of Geological Societies Transactions, v. 47, p. 571–577.

Tardy, M., 1980, Contribution a l'étude geologique de la Sierra Madre Oriental du Mexique: Tesis doctoral, Université Pierre et Marie Curie de Paris, 445 p.

Tardy, M., J. Sigal, and G. Glacon, 1974, Bosquejo sobre la estratigrafía y la paleogeografía de los flysch cretácicos del Sector Transversal de Parras, Sierra Madre Oriental, México: Universidad Autónoma de México, Instituto de Geología, Serie Divulgación, no. 2, p. 12–28.

Vaugham, T. W., 1896, The stratigraphy of northwestern Louisiana: American Geology, v. 15, p. 205–229.

Vega-Vera, F. J., and M. C. Perrilliat, 1989, On a new species of *Venecardia* from the lower Eocene in northeastern, México (Difunta Group): Tulane Studies in Geology and Paleontology, v. 22, p. 101–106.

Yurewicz, D. A., R. J. Chuchla, M. Richardson, R. J. Pottorf, G. G. Gray, M. G. Kozar, and W. M. Fitchen, 1997, Hydrocarbon generation and migration in the Tampico-Misantla Basin and Sierra Madre Oriental, east central Mexico: Evidence from an exhumed oil field in the Sierra de el Abra, *in* Sedimentation and diagenesis of middle Cretaceous platform margins, east central Mexico: AAPG Annual Meeting Field Trip Guidebook, Dallas Geological Society and SEPM, 24 p.

4

Cantú-Chapa, A., 2009, Upper Jurassic stratigraphy (Oxfordian and Kimmeridgian) in petroleum wells of Campeche Shelf, Gulf of Mexico, *in* C. Bartolini and J. R. Román Ramos, eds., Petroleum systems in the southern Gulf of Mexico: AAPG Memoir 90, p. 79–91.

Upper Jurassic Stratigraphy (Oxfordian and Kimmeridgian) in Petroleum Wells of Campeche Shelf, Gulf of Mexico

Abelardo Cantú-Chapa
Instituto Politécnico Nacional, México, D. F., Mexico

ABSTRACT

The original definitions of the Upper Jurassic Akimpech Formation and Ek-Balam Group of Campeche Shelf (southern Gulf of Mexico) as proposed by Angeles-Aquino and Cantú-Chapa did not incorporate the gamma-ray log responses of these units. Studies of new material from petroleum wells from the southern Gulf of Mexico permit their redefinition as important stratigraphic units for petroleum exploration and production. Additionally, this chapter defines the middle Oxfordian Bacab Formation and its upper member, the Tson anhydrite member.

The middle Oxfordian ammonites *Ochetoceras rioslopezi* n. sp., *Glochiceras* sp., and Perisphinctidae ind. recovered from the oil wells of the southern region of the Gulf of Mexico are described.

INTRODUCTION

Ammonite specimens recovered from petroleum wells in the Campeche Shelf area of the southern Gulf of Mexico were studied and compared with those from other localities to clarify the position of the stratigraphic units in which they occur. Only the Tithonian ammonites in the Edzna Formation had previously been correlated to the gamma-ray logs of wells in which they occur (Cantú-Chapa, 1977, 1982, 2001; Cantú-Chapa and Ortuño-Maldonado, 2003) (Figure 1).

Angeles-Aquino and Cantú-Chapa (2001) formally proposed the names of the Ek-Balam Group, Abkatun Formation, and Edzna Formation for Upper Jurassic (Oxfordian–Tithonian) strata penetrated in wells of this region. In the originally proposed stratigraphic scheme, these units were not defined on the basis of their gamma-ray responses and were only represented on lithofacies maps. Based on new well material, the Akimpech Formation and the Ek-Balam Group are herein redefined based on their lithologies and gamma-ray responses. Their proposed ages are based on ammonite specimens found in cores and compared to previously described examples of these fossils (Cantú-Chapa, 1984).

STRATIGRAPHY

The Oxfordian Ek-Balam Group, with its proposed type locality as the Caan 1 well in the southern Gulf of Mexico, was originally divided into three members by Angeles-Aquino and Cantú-Chapa (2001) (Figure 1). Its upper contact was placed at 4985 m (16,355 ft), and its lower contact was not defined. These members, without

DOI:10.1306/13191079M902804

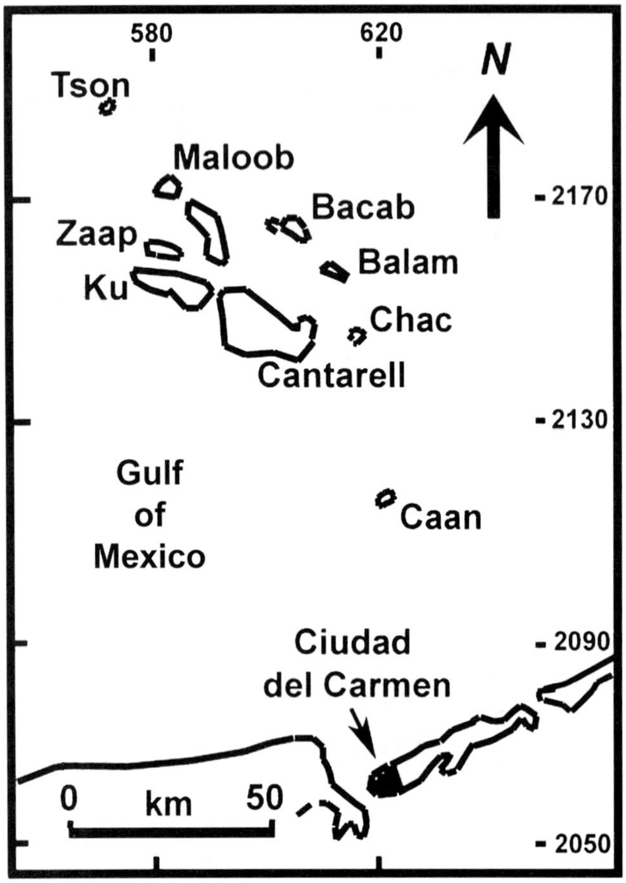

FIGURE 1. Location of the wells studied in Campeche Shelf, southern Gulf of Mexico.

indication of their depths in the well, were described as follows: (1) a lower member consisting of clayey and sandy wackestone to packstone with quartz cement and intercalated evaporites; (2) a middle member consisting of rhythmic alternations of calcareous sandstones, mudstones, and bentonitic shales; and (3) an upper member is characterized by sandy limestones grading into calcareous sandstone and anhydrite.

The age of the Ek-Balam Group was considered to be upper Oxfordian because of the presence of the ammonite *Ochetoceras* sp. in the Balam 101 well (core 3, 4693–4702 m [15,397–15,426 ft]) (Angeles-Aquino and Cantú-Chapa, 2001; Cantú-Chapa, 2001).

The Akimpech Formation was defined as a shale with the Caan 1 well proposed as its type locality. The upper and lower contacts of this unit were placed at 4400 and 4985 m (14,436 and 16,355 ft), respectively. This formation was subdivided into four members (B–E), but none of these members were tied to the gamma-ray log in this well, and the study emphasized the geographic distribution of the different lithofacies represented by these members based on their petrographic characteristics (Angeles-Aquino and Cantú-Chapa, 2001).

The four members were characterized as follows: (1) a lower terrigenous member (B) consisting of mudstone and bentonitic sandy shales; (2) a lower calcareous member (C) consisting of microcrystalline to mesocrystalline dolomites, packstone with incipient dolomitization, and isolated mudstone interbeds; (3) an upper terrigenous member (D) consisting of claystones, mudstones, and sandy interbeds among carbonates; and (4) an upper Calcareous member (E) consisting of mesocrystalline and microcrystalline dolomites.

According to the stratigraphic correlations presented here, only one specimen of *Nebrodites* sp. was recovered from member C in the Chac 1 well (core 15, 4650–4659 m [15,256–15,285 ft]) (Cantú-Chapa, 1977). This ammonite is Kimmeridgian in age according to a biostratigraphic comparison with similar material in north-central Mexico (Burckhardt, 1912, 1930; Imlay, 1980).

The Edzna Formation was subdivided into three members by Angeles-Aquino and Cantú-Chapa (2001) as follows: (1) lower (F) (carbonates), (2) middle (G) (claystone), and (3) upper (H) (bentonitic shale). The Caan 1 well was also proposed as the type locality for this formation (Angeles-Aquino and Cantú-Chapa, 2001).

The sharp upper contact of the Edzna Formation is at 4303 m (14,117 ft) where it underlies dolomitized limestone of the Lower Cretaceous Akal Formation. The gamma-ray curve deflects to the right upon entering the Edzna Formation and remains there except for short intervals with minor inflections to the left indicating the presence of thin bodies of compact, argillaceous limestone. The lower contact of the Edzna Formation with the Akimpech Formation dolomite is also sharp with the gamma-ray curve deflected to the left below the contact. The thickness of the Edzna Formation ranges from 355 to 62 m (1165 to 203 ft) in the Campeche Shelf wells (Cantú-Chapa and Ortuño-Maldonado, 2003).

Tithonian ammonites belonging to several genera were sampled from different wells. The upper Tithonian ammonites include *Durangites*, *Parodontoceras*, *Salinites*, *Protancyloceras*, and *Suarites*, and the lower Tithonian is characterized by *Mazapilites* (Cantú-Chapa, 1977, 1982; Cantú-Chapa and Ortuño-Maldonado, 2003). These genera define biostratigraphic units in several localities of north-central and eastern Mexico (Imlay, 1980; Cantú-Chapa, 2001).

STRATIGRAPHIC COMPARISONS

Geological studies in the Campeche area have proposed three different definitions for the Akimpech Formation and the Ek-Balam Group. These three proposals are compared using the following criteria: (1) regional well data, (2) the original lithologic descriptions at the Caan 1 well type locality (Angeles-Aquino and Cantú-Chapa, 2001), (3) comparisons of definitions for the same

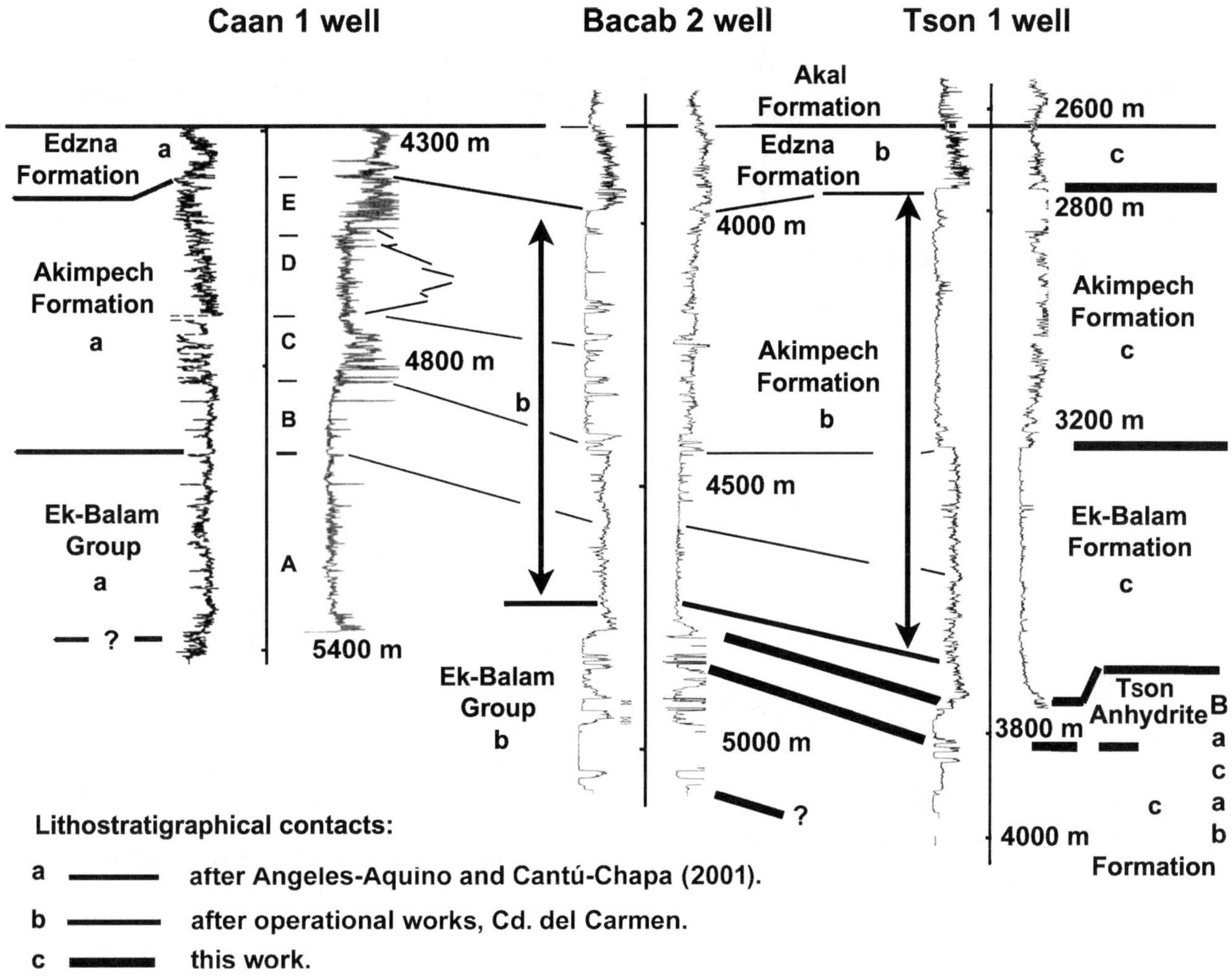

FIGURE 2. Stratigraphic section of the Upper Jurassic formations in three petroleum wells of Campeche Shelf, Gulf of Mexico, according to gamma-ray logs, and comparing different stratigraphic schemes. The correlation horizon is at the top of the Edzna Formation (Tithonian). Gamma ray is on the left side of the wells and neutron log is on the right.

units in unpublished geological reports, and (4) characterizations in this chapter based on the gamma-ray log of the Caan 1 well.

This comparative stratigraphic procedure is applied because the prior formational contacts were not tied to the gamma-ray log of the Caan 1 well (Angeles-Aquino and Cantú-Chapa, 2001). This procedure resulted in either the revival of the original definition based on the descriptions of these formations, when merited, or provided new definitions for these Upper Jurassic units. In the latter case, it is possible to identify lateral facies changes or structural problems manifested in the studied wells (Figure 2).

The Ek-Balam Group

The lithology of this unit is reflected in the gamma-ray curve of the Caan 1 well by strong displacement toward the right beginning at 4830 m (15,846 ft) coinciding with a sharp contact between the overlying dolomitic Akimpech Formation and the underlying Ek-Balam Group shales and sands (Figure 2). The contact between these two formations is more obvious here than where originally described at 4985 m (16,355 ft) within a continuous shale sequence (Angeles-Aquino and Cantú-Chapa, 2001) (Figure 2).

The lower contact of the Ek-Balam Group was placed at 5385 m (17,667 ft) in the Caan 1 well without referring to the gamma-ray log (Angeles-Aquino and Cantú-Chapa, 2001). Unfortunately, this contact is atypical in the Caan 1 well; it is here considered that it was poorly defined in this well and is revised below. This contact is normally sharp and quite evident in other Campeche Shelf wells (Figure 2).

The Akimpech Formation

This formation is not lithologically homogeneous in the Caan 1 well and was subdivided into the aforementioned B, C, D, and E members (Angeles-Aquino and

Cantú-Chapa, 2001). The lower contact of the Akimpech Formation is revised herein. This contact occurs at 4830 m (15,846 ft) in the Caan 1 well where it is characterized by a strong lithological change, in contrast to the originally proposed contact (Figure 2). Upon entering the Akimpech Formation at 4510 m (14,796 ft), the gamma-ray curve deflects to the right and remains there until 4690 m (15,387 ft). This interval consists of claystones and mudstones corresponding to the originally defined D member. Note that the different lithologies of the C and D members are not represented on the gamma-ray log of the Caan 1 well (Angeles-Aquino and Cantú-Chapa, 2001) and are, therefore, undefined (Figure 2).

The Edzna Formation

Only this formation has been defined previously from the gamma-ray log, and it has also been dated by ammonites from several of the Campeche Shelf wells. The stratigraphy of this formation is extrapolated to Caan 1 in this chapter because this well is proposed as the type locality (Cantú-Chapa, 1982; Angeles-Aquino and Cantú-Chapa, 2001; Cantú-Chapa and Ortuño-Maldonado, 2003) (Figure 2).

ANALYSES OF THE UPPER JURASSIC STRATIGRAPHY OF CAMPECHE SHELF

According to Angeles-Aquino and Cantú-Chapa (2001), the nomenclature for the Upper Jurassic stratigraphy in Campeche Shelf has evolved informally from unpublished geological reports. The Ek-Balam Group and the Akimpech Formation do not display the necessary objective physical characteristics on gamma-ray logs for this type of operational work. To document this analysis, the Caan 1 well was integrated with the Bacab 2 and Tson 1 wells in a stratigraphic section (Figure 2) where it is shown that the Ek-Balam Group and the Akimpech Formation have similar, well-defined characteristics that allow them to be easily recognized on the gamma-ray logs of the latter two wells. It is observed in these wells that the stratigraphic sequences were not affected by diagenetic or structural processes, although the Upper Jurassic formations are structurally offset by more than 1000 m (3281 ft) in the Bacab 2 and Tson 1 wells (Figure 2).

The respective logs in these two wells do not show structural deformation or lateral facies changes, although some intervening wells do show facies changes as well as the effects of structural processes, as shown below. For stratigraphic purposes, we note that the Tson 1 well penetrated a more complete Upper Jurassic sequence than the Bacab 2 well. Therefore, this well is the most representative for this part of Campeche (Figure 2).

The Akimpech Formation consists of two thick contrasting lithologic units that are clearly distinguished on the gamma-ray logs of the Tson 1 and Bacab 2 wells. This formation should be represented by a body of dolomite overlying a body of shale, according to the operational work of PEMEX (Figure 2).

The gamma-ray curve is deflected to the left through the upper dolomite and moves to the right in the lower shale. Both units were included in the Akimpech Formation in the unpublished PEMEX exploration reports. The shale in the deeper part is continuous with shales in the underlying unit that were inappropriately included in the Ek-Balam Group. This lower body, in part, was also originally considered part of the Ek-Balam Group in the Caan 1 well by Angeles-Aquino and Cantú-Chapa (2001) (Figure 2).

This lithological evidence used to define the Akimpech Formation in the Bacab 2 and Tson 1 wells does not conform to the established lithostratigraphic concepts required for characterizing and naming a formation (North American Stratigraphic Code). Specifically, the lower contact of the proposed Akimpech Formation is poorly defined on the gamma-ray log in both wells and is, therefore, an arbitrary contact that does not represent significant lithological change (Figure 2) and requires correction.

STRATIGRAPHIC PROPOSALS AND REDEFINITIONS

The unpublished, inconsistent, and incorrect descriptions of the Upper Jurassic formations in the subsurface of Campeche Shelf are herein redefined. Several stratigraphic proposals are made according to the recommendations of the North American Stratigraphic Nomenclature Code (Article 18), consisting of (1) change in lithic designation and (2) inappropriate original lithic designation. These proposals are congruent with the gamma-ray log signatures of these units and will permit their use in petroleum exploration operations.

The Bacab Formation

This name is formally proposed herein for a unit consisting of four members at the base of the Upper Jurassic sedimentary sequence. The name comes from the Mayan language meaning "blow gun." Its members from top to base are (1) Tson anhydrite (100 m [328 ft] thick), (2) translucent quartz sand (100 m [328 ft] thick), (3) white salt (30 m [98 ft] thick), and (4) calcareous-bentonitic shale (240 m [787 ft] thick). The Bacab Formation probably represents the basal sequence of the major Upper Jurassic transgression in southeastern Mexico, as discussed below.

In the absence of index fossils, and based on their stratigraphic position, this formation is considered as possibly being of lower Oxfordian age. The type locality is proposed to be the Bacab 2 well with the following coordinates: X = 590,101; Y = 2,163,942. An alternate reference locality for the Bacab Formation is proposed to be the Tson 1 well. Both wells are shown in Figures 1 and 2.

The gamma-ray log of the Tson 1 well defines the three upper members of the Bacab Formation, but the basal calcareous shale under the white salt is not delineated. The gamma-ray log has three alternating behaviors in this well that reflect the different members of the Bacab Formation: a deflection to the left in the Tson anhydrite at 3750 m (12,303 ft); a deflection to the right in the translucent quartz sand under the anhydrite at 3820 m (12,532 ft); and another deflection to the left upon penetrating the salt at 3950 m (12,959 ft). The basal calcareous shale was not penetrated in Bacab 2 nor was it correctly identified on the gamma-ray log of Tson 1 because this well only reached the calcareous shale member (Figure 2). The top of this formation corresponds to the top of the Tson anhydrite at 3750 m (12,303 ft) in the Tson 1 well and at 4780 m (15,682 ft) in the Bacab 2 well.

The Tson Anhydrite Member

This topmost unit of the Bacab Formation consists of anhydrite alternating with thin shale beds toward its base. The gamma-ray log deflects toward the left upon penetrating this member where it persists until encountering the shale interbeds.

The author formally proposes that this member be named the Tson anhydrite because it is very characteristic, both lithologically and in its gamma-ray log response in Campeche Shelf wells. The Tson anhydrite member has sharp contacts with the overlying shales of the Ek-Balam Formation shales and underlying the translucent quartz sand. These contacts occur, respectively, at 3750 and 3820 m (12,303 and 12,532 ft) in the Tson 1 well. The thickness of the Tson anhydrite member is 100 m (328 ft) in both the Tson 1 and Bacab 2 wells. Because of the absence of fossils, the lower Oxfordian age of the Tson anhydrite member has been inferred from its stratigraphic position.

The Ek-Balam Formation

This formation is redefined in the study area as a shale body approximately 340–500 m (1115–1640 ft) thick, exceptionally exceeding 630 m (2067 ft) in apparent thickness because of structural effects, as in the Maloob 406 well. This formation behaves consistently on the gamma-ray log with deflection to the right throughout its length, contrasting strongly on the gamma-ray log with both the underlying Tson anhydrite member and the overlying Akimpech Formation (Figure 2). The lower contact of the Ek-Balam Formation with the Tson anhydrite member is sharp, as is its upper contact with the dolomite of the Akimpech Formation. In both of these units, the gamma-ray curve is deflected strongly to the left (Figures 2, 3). This lithological consistency has been induced to consider this thick unit to be a formation instead of a group, as originally proposed by Angeles-Aquino and Cantú-Chapa (2001).

The upper contact of the Ek-Balam Formation in the Caan 1 well, based on the gamma-ray log, is placed at the depth of 4830 m (15,846 ft). The lower contact in this well is not well defined on the gamma-ray log but is provisionally placed at a depth of 5360 m (17,585 ft). The Bacab 2 and Tson 1 wells are used to support this proposed stratigraphic definition of the Ek-Balam Formation because both these wells show the typical gamma-ray response for this formation (Figure 2).

The Akimpech Formation

This formation consists of dolomite with thickness ranging from 400 to 490 m (1312 to 1608 ft) in wells of the study area. The gamma-ray curve in this formation has a strong deflection to the left from its sharp upper contact with the Edzna Formation to its sharp lower contact where the Ek-Balam Formation shale is penetrated (Figure 2). The lower contact of the Akimpech Formation proposed in geological operations work is discarded because of the previous attempts to arbitrarily subdivide the thick shale into two units (Figure 2). The Akimpech Formation is here redefined as consisting of a single unit composed predominantly of dolomite in Campeche Shelf.

The Tson 1 well is proposed as the type locality of the Akimpech Formation because it is more typical here than it is in the Caan 1 well. The coordinates of the Tson 1 well are X = 579,044 and Y = 2,175,966. A reference section for the Akimpech Formation is also proposed in the Bacab 2 well where this unit occupies a similar stratigraphic position as in the Caan 1 well. The Akimpech Formation is in contact with both the Edzna and Ek-Balam formations in these wells (Figure 2).

The Akimpech Formation displays lateral facies changes in certain parts of the Campeche Shelf as manifested by variations in the gamma-ray log response. Some of the most significant changes are observed in the Caan 1 well where an upper Terrigenous member D consisting of claystones, mudstones, and sandstones is interbedded with carbonates (Angeles-Aquino and Cantú-Chapa, 2001). This member, with an approximate thickness of 170 m (558 ft), lies between two dolomite bodies in this well (Figure 2). This same lateral facies

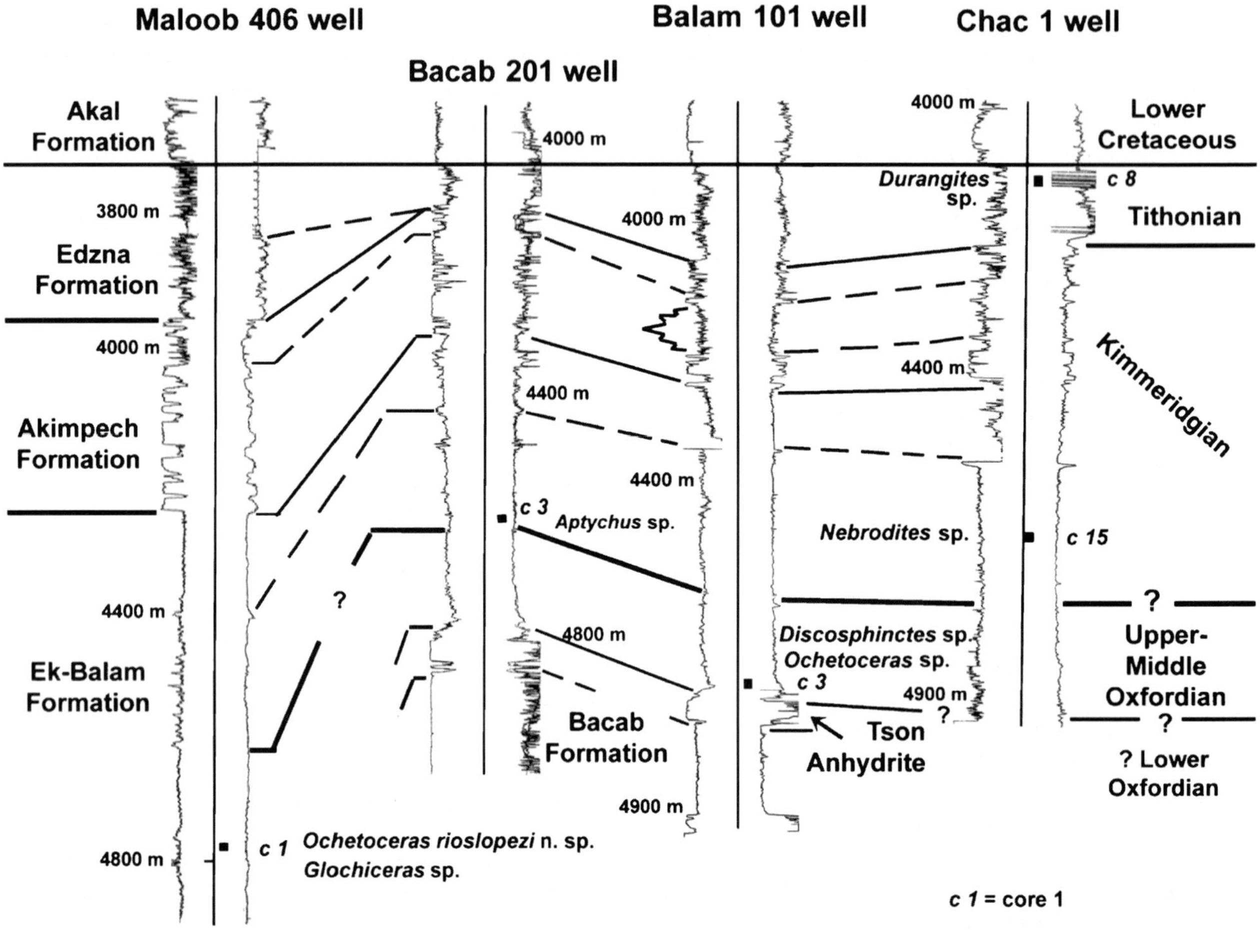

FIGURE 3. Stratigraphic section of the Upper Jurassic formations in petroleum wells of Campeche Shelf, Gulf of Mexico, according to gamma-ray logs and ammonites. The correlating horizon is top of the Edzna Formation (Tithonian).

change is observed in the Balam 101 and Chac 1 wells (Figure 3).

The Akimpech Formation also displays a lateral facies change in the Bacab 2 well where the lower half of the dolomite contains thin shale interbeds below the depth of 4150 m (13,615 ft) (Figure 2). Similarly, the Akimpech Formation contains several shale bodies interbedded with the dolomite in the Maloob 406 well (Figure 3).

AGE

The Upper Jurassic formations in Campeche Shelf wells have yielded scarce but significant ammonite fragments for time correlation. These specimens have been compared with those in the only two Mexican localities where upper Oxfordian and lower Kimmeridgian stratigraphic studies have been conducted. These are at San Pedro del Gallo and Tamán, respectively, in north-central and eastern Mexico (Burckhardt, 1912, 1930; Cantú-Chapa, 1984).

The Bacab Formation (Probably Lower Oxfordian)

This formation has not yielded ammonites. The salt, sandstone, and anhydrite that comprise the three upper members are not carriers of these fossils. Therefore, the age of this formation may only be inferred from its stratigraphic position and could be pre-middle Oxfordian, as previously indicated (Figure 3).

The Ek-Balam Formation (Middle Oxfordian–Lower Kimmeridgian)

The Balam 101 well (core 3, 4693–4702 m [15,397–15,426 ft]) provided ammonite fragments at the base of the Ek-Balam Formation, including a specimen of *Ochetoceras* sp. (Cantú-Chapa, 2001). The Maloob 406 well (core 1, 4754–4763 m [15,597–15,627 ft]) yielded specimens of *Ochetoceras rioslopezi* n. sp., described below, *Glochiceras* sp., and fragments of Ochetoceratinae and Perisphinctidae. The occurrence of these fossils contributed to the age determination of the lower part of the

Ek-Balam Formation (approximately 160 m [525 ft] thick) in these two wells. The age proposed here is based on a stratigraphic correlation with similar materials studied in Europe that were assigned a middle Oxfordian age (zone with transversarium and canaliculatum; Maubeuge, 1975; Ziegler, 1958). *Ochetoceras* sp. from the Balam 101 well is considered fundamental biostratigraphic evidence to establish the indicated age of this formation (Cantú-Chapa, 2001).

Species of this genus have also been studied based on abundant materials from the La Jagua Formation in Cuba (Wierzbowski, 1976) and in the La Casita and Santiago formations of Mexico (Burckhardt, 1912; Cantú-Chapa, 1984). The specimens of this genus from Cuba have not been precisely dated neither by Judoley and Furrazola-Bermúdez (1968), nor by Wierzbowski (1976), nor Myczynski et al. (1998). Callomon (1992) did not provide adequate information when referring to the Cuban material in his theoretical proposal on the Oxfordian biostratigraphy of Mexico. The proposal of this author also did not include the biostratigraphic subdivisions from Tamán in eastern Mexico (Cantú-Chapa, 1984).

The absence of *Gregoryceras* in Mexico and Cuba prevented Myczynski et al. (1998) from correlating the abundant fauna of these countries with those of the Oxfordian of Europe. In their stratigraphic proposals, they inexplicably ignored time correlation from *Ochetoceras* and *Cubaochetoceras* of Mexico and Cuba, whereas Judoley and Furrazola-Bermúdez (1968) found *Ochetoceras canaliculatum* and species of *Cubaochetoceras* in Cuba associated with *Dichotomosphinctes* and *Discosphinctes*.

O. canaliculatum characterizes the Middle Oxfordian (plicatilis-canaliculatum zones) in Europe (Maubeuge, 1975; Cariou et al., 1997) and was studied in San Pedro del Gallo, north-central Mexico, where it was considered a reliable indicator of the middle Oxfordian Transversarius zone of Europe (Burckhardt, 1912).

Representatives of *Ochetoceras* and *Glochiceras* described above represent necessary biostratigraphic elements for picking the base of the middle Oxfordian in the Campeche wells of this study. The biostratigraphic data provided by these fossils range from the base of the Ek-Balam Formation in the Balam 101 well to the lower middle part of the same formation in the Maloob 406 well (Figure 3). Therefore, a possible lower Oxfordian age is inferred for the unfossiliferous Bacab Formation underlying the fossiliferous Ek-Balam Formation, with the proposed age for the Bacab Formation based on its stratigraphic position.

The Lower Kimmeridgian

A specimen of *Nebrodites* from the Chac 1 well (core 15, 4650–4659 m [15,256–15,285 ft]) was studied by Cantú-Chapa (1977). This genus is characteristic of the lower Kimmeridgian in north-central Mexico (Burckhardt, 1912; Imlay, 1980) (Figure 3) and is the only biostratigraphic data point in the Campeche wells. Small aptychi specimens were also obtained from core 3 of the Bacab 201 well at 4588–4598 m (15,052–15,085 ft), but these fossils are not as age diagnostic as ammonites. They do, however, indicate a basinal environment in this well. For these reasons, the Ek-Balam Formation is considered to be of upper Oxfordian–lower Kimmeridgian age in wells of the offshore Campeche region, although it is not yet possible to mark the Oxfordian–Kimmeridgian boundary because of the rarity and wide separation between ammonite specimens within the thick interval in this area's wells (Figure 3).

The Akimpech Formation

The dolomite comprising this formation is not conducive to the preservation of ammonite remains that would permit an age determination. From its stratigraphic position, however, it is inferred to be upper Kimmeridgian because it underlies the Tithonian Edzna Formation (age based on ammonites) and overlies the Ek-Balam Formation of lower Kimmeridgian age.

The Edzna Formation

This formation has yielded ammonite material from several wells in the Campeche marine area that date it as Tithonian. These ammonites have been discussed above (Cantú-Chapa, 1982; Cantú-Chapa and Ortuño-Maldonado, 2003).

STRUCTURAL PROBLEMS

Upper Jurassic rocks in the area between the Tson 1 and Bacab 2 wells present structural problems, as manifested by the absence of the Akimpech Formation in the Bacab 21 well, probably because of reverse faulting. The Ek-Balam Formation is in unconformable contact with a thin body of the Edzna Formation in Bacab 21 at 4020 m (13,189 ft) (Figure 4).

The Bacab Formation consists of salt approximately 270 m (886 ft) thick that first appears in the Bacab 21 well at a depth of 4720 m (15,485 ft) and persists until its total depth of 5000 m (16,404 ft) (Figure 4). The salt underlies the Tson anhydrite member, which appears at 4390 m (14,403 ft) and contains shale interbeds in its lower half. The anhydrite underlies the Ek-Balam Formation. About 300 m (984 ft) of the lower Ek-Balam Formation was cut in the Bacab 21 well (Figure 4). This well lies between the Bacab 2 and 201 wells; the last two wells contain complete Upper Jurassic sedimentary sections, in contrast with the Bacab 21 well (Figure 4). The presence of a fault in Bacab 21 was not reported during petroleum exploration work.

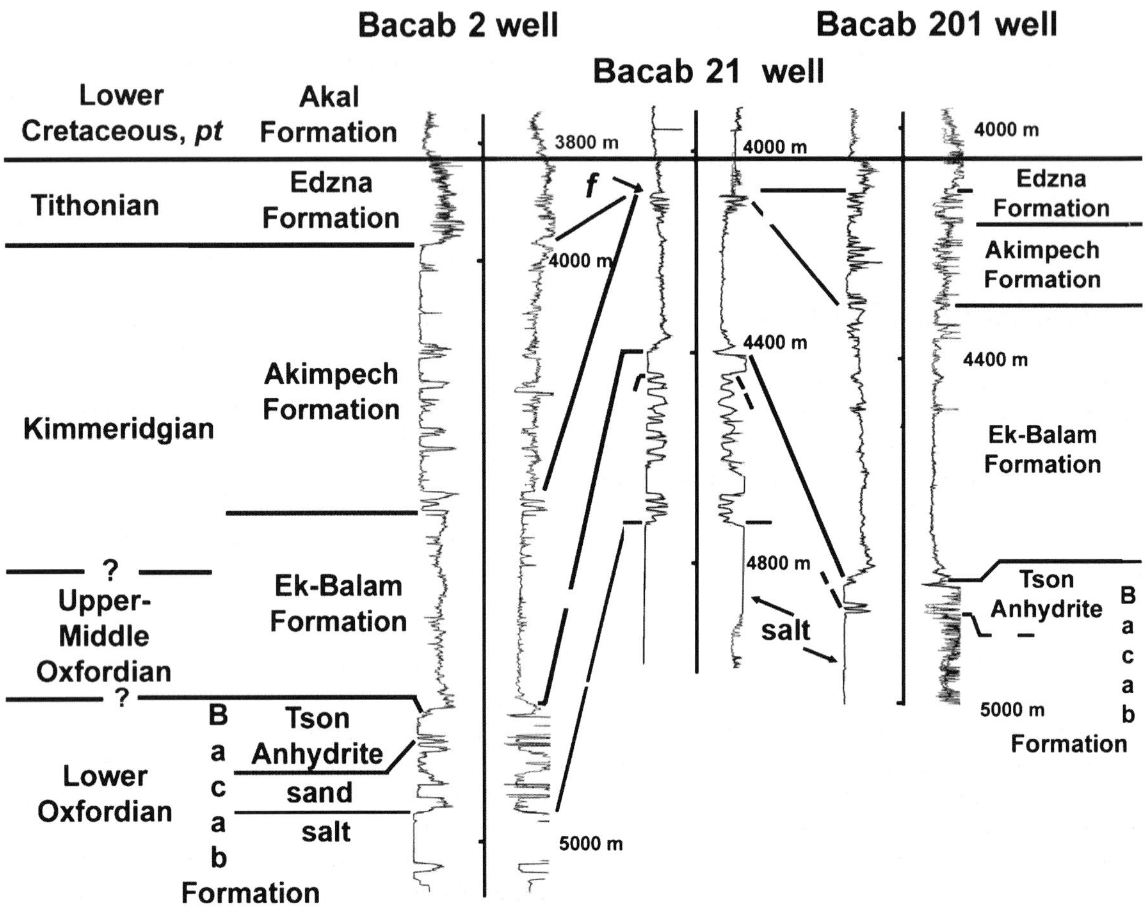

FIGURE 4. Stratigraphic and structural section of the Upper Jurassic in wells of Bacab field in Campeche Shelf. The correlating horizon is top of the Edzna Formation (Tithonian).

SEDIMENTOLOGICAL EVOLUTION OF THE CAMPECHE MARINE JURASSIC SECTION

The Middle to Late Jurassic transgression that flooded the Mexican territory was derived from the Pacific Ocean through the Oaxaca region of southern Mexico starting in the late Bajocian. The transgression bifurcated toward the southeast (Campeche region) during the early Oxfordian and to the east (Poza Rica and Sierra Madre Oriental region) in the late Bathonian–middle Callovian. These two flooded areas joined in the late Oxfordian transgression that covered the Gulf of Mexico also reaching western Cuba and the southeastern United States (Wierzbowski, 1976; Imlay and Hermann, 1984; Cantú-Chapa, 2001).

Oxfordian Transgressive Stages

Two transgressive stages occurred during the Oxfordian in the Campeche region.

The first stage is represented by 240 m (787 ft) of calcareous and bentonitic shale of the Bacab Formation comprising the basal transgressive unit, probably early Oxfordian in age. This was followed by a regression characterized by two strong evaporative phases: a lower phase represented by translucent salt and an upper phase represented by the Tson anhydrite, both within the Bacab Formation. Between these two evaporite units is a body of translucent sand that represents a regressive episode. This was followed by a full marine transgression represented by the ammonite-bearing shale of the Ek-Balam Formation. The lithology and fossils in this unit indicate that marine flooding was related to the first stage of subsidence in the southern Gulf of Mexico during the middle Oxfordian–early Kimmeridgian.

CONCLUSIONS

This study contributes to the identification and dating of middle Oxfordian ammonites collected from deeper depths in PEMEX exploration wells in the southern Gulf of Mexico.

The analyses of gamma-ray logs from several wells were correlated with scarce ammonite remains to update the definition and age determinations for several Oxfordian and Tithonian units in the petroleum-producing area of the Campeche Shelf.

The existing lithostratigraphic nomenclature for the Jurassic and Cretaceous has been extensively revised in detail, and a new lithostratigraphic nomenclature is proposed.

ACKNOWLEDGMENTS

The author is grateful to Claudio Bartolini for guidance through the publication process of this volume. Thanks to Horacio Parent and Blanca Buitrón for their detailed and asserted review of the manuscript. The author was generously assisted by Jaime Rios-López and Rosalba Lima-Velasquez, who provided the material of this research.

SYSTEMATIC PALEONTOLOGY

Fossil preservation: All shells studied here have been crushed and flattened during compaction of the sediments; their suture lines are unknown.

Specimens are housed in the Instituto Politécnico Nacional (IPN), Mexico City.

Superfamily Haploceratoidea von Zittel (1884)

Family Oppeliidae Douvillé (1890)

Subfamily Ochetoceratinae Spath (1928)

Discussion

The Jurassic Ammonite classification proposed by Donovan et al. (1981) is based on subjective dimorphic concepts at the subfamily taxonomic level. This strange way to establish the differences between genera of the subfamily Ochetoceratinae from the Oxfordian–Kimmeridgian is impractical because it hinders the demonstration of their morphological differences. Representatives of this subfamily are characterized by involute shell and falcate ribs that became swollen at the mid-lateral spiral groove. Ammonites of this subfamily have a simple or serrated mid-ventral keel that may be the boundary of two ventrolateral edges, producing a tricarinate aspect.

Representatives of the Glochiceratinae Hyatt (1900) differ from Ochetoceratinae of the same age by their small, evolute, and smooth shell, which is sometimes ornamented with striae at the external part of the flanks; the former also differ by their subrounded venter with no median keel.

The genera *Ochetoceras* Haug (1885) and *Cubaochetoceras* Sánchez-Roig (1951) (see Arkell, 1957) are included into the subfamily Ochetoceratinae. Both genera are apparently present in the Oxfordian sequences of Mexico and Cuba, but their morphological differences have not been clearly established. They were considered to be different genera by Christ (1961), and their systematic positions and age were also discussed.

Genus *Ochetoceras* Haug (1885)

(*Cubaochetoceras* Sánchez-Roig, 1951, in Arkell, 1956)

Type Species

Ammonites canaliculatus von Buch (1831).

Discussion

The subsequent diagnosis of *Cubaochetoceras* is based on a vague comparison with another genus proposed by Arkell (1957, L278) as a Large *Cardioceras*-like forms with indistinct ribbing. Judoley and Furrazola-Bermúdez (1968) did not differentiate *Ochetoceras* from *Cubaochetoceras* in the Cuban specimens studied by them. In the same way, Wierzbowski (1976) differentiated these genera from subjective elements; by the presence or absence of ventrolateral edges and by the style of lateral ribbing. Both genera have involute, flattish, and discoidal shells with a mid-lateral groove. Representatives of these genera have a serrated keel with two ventrolateral edges; this last morphological difference proposed by Wierzbowski (1976) is imperceptible and highly dependant of the preservation of the test.

Separating them into two different genera is impossible, and their respective morphological characters are not well established by this author. *Cubaochetoceras* cannot be considered as a different genus because of the unclear definition furnished by Wierzbowski (1976). As in other Oppelidae, the ornamentation of representatives of *Cubaochetoceras* changes when the shells became larger in size, which produces confusion when attempting to differentiate it from *Ochetoceras*. In the former, the coarse ribs change to striae or very weak ribs on the outer part of the adoral half of the shell (cf. Cantú-Chapa, this volume).

Another important aspect concerning age of these Cuban genera is that they are associated with Perisphinctidae and Aspidoceratidae ammonites, all of which are under discussion. These ammonite assemblages were found in the La Jagua and San Francisco formations of Cuba by Wierzbowski (1976), who pointed out that they range through the middle–upper Oxfordian boundary.

The age of these Cuban ammonites was also discussed by Myczynski et al. (1998), and their proposed

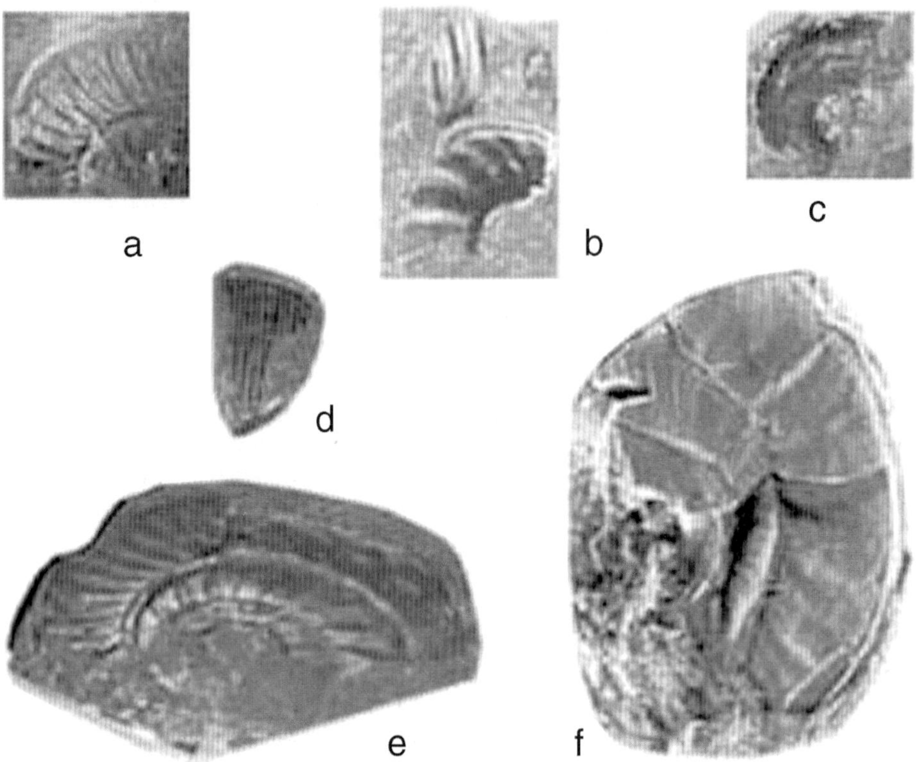

FIGURE 5. (a) Ochetoceratinae ind., (c) *Glochiceras* sp., (d) Perisphinctidae sp. ind., (e) *Ochetoceras rioslopezi* n. sp., and (f) *Ochetoceras* sp. Locality and age: All of these specimens are from the lower part of the Ek-Balam Formation (middle Oxfordian) in the Maloob 406 well (core 1, 4754–4763 m [15,597–15,627 ft]) (Figures 1–5) and the Balam 201 well (core 3, 4693–4702 m [15,397–15,426 ft]), Campeche Shelf, southern Gulf of Mexico. All specimens are shown in their natural size (×1).

time correlation does not vary from that of Wierzbowski (1976). Similarly, the arguments of Oloriz and Villaseñor (in Myczynski et al., 1998) for dating Oxfordian ammonites from Mexico are theoretical and unsupported by the studied material.

Ochetoceras rioslopezi n. sp. (Figure 5e)

Material Studied

A fragmentary specimen showing a flattish flank.

Diagnosis

Shell with a mid-lateral, narrow, deep, spiral groove placed below its mid-height. Two types of lateral ribs; the internal ribs arising from the umbilical wall are strong, prorsiradiate, separated by larger spaces, and swollen at the groove. The outer ribs are uniform, strongly rursiradiate, weakly concave, fine, single, and separated by spaces as large as they are.

The numbers of inner and outer ribs equal about 3–6 per occupied space, respectively. Outer ribs became very distant from one another on the adoral part of the shell. A fragmented fine and serrated ventral keel with a ventrolateral edge is preserved.

Discussion

This specimen shows the typical features of *Ochetoceras* with falcate ribs that are swollen at a mid-lateral groove and by a ventral and serrated keel. Its strong and separated internal ribbing differs from that of *O.* aff. *pedroanum* Burckhardt (Cantú-Chapa, 1984, pl. 1, their figure 3) from Taman, eastern Mexico, whose ribs are separated by spaces as large as they are in this latter specimen. The proposed *O. rioslopezi* n. sp. differs from *O. pedroanum* Burckhardt (1912) (p 9, pl. 1, their figures 13, 14, 15, 16, 17), from north-central Mexico by its internal ribbing; the intercostal spaces are larger than the ribs. The internal ribs are separated by spaces as large as the ribs in *O. pedroanum*. *O. rioslopezi* n. sp. differs from *O. canaliculatum* von Buch (in Maubeuge, 1975, p. 84, their figure G 2427) of the middle Oxfordian from Switzerland by having internal ribs separated by larger intercostal spaces, in contrast to the European specimen whose internal ribs are close together.

The subsurface Mexican specimen studied here is comparable with *O. vignalense* Sánchez-Roig (1951) from Cuba (in Wierzbowski, 1976, p. 237, pl. 7, their figure 12) by having rursiradiate outer ribs, but the internal ribs are close together in the Cuban specimen.

Material Examined

IPN-1112 (Holotype, Figure 5e).

Occurrence

The Maloob 406 well, core 1 (4754–4763 m [15,597–15,627 ft]), southeastern Mexico, middle Oxfordian, Ek-Balam Formation.

Etymology

In honor of Jaime Rios-López, who provided the material studied here.

Other Specimens (Figure 5a, b)

Some fragmented specimens, probably Ochetoceratinae, were found in the same core 1 of the Maloob 406 well with only their outer fine, gradually concave ribs preserved; these are separated by larger spaces and start from the lateral groove, ending at the ventrolateral edge.

Material Examined

IPN-1113, 1114.

Occurrence

The Maloob 406 well, core 1 (4754–4763 m [15,597–15,627 ft]), southeastern Mexico, middle Oxfordian, Ek-Balam Formation.

Ochetoceras sp. (Figure 5f)

Description

A large, fragmented, and crushed specimen shows probably half of the last whorl. This specimen is preserved with a ventral keel and a ventrolateral edge. This specimen seems to be involute with a plane flank and is ornamented with fine and irregularly falcate striae; a lateral groove and suture line are shown.

Material Examined

IPN-1115.

Occurrence

The Balam 201 well, core 3 (4693–4702 m, 15,397–15,426 ft), southeastern Mexico, middle Oxfordian, Ek-Balam Formation.

Discussion

This specimen was illustrated and assigned as *Ochetoceras* sp. by Cantú-Chapa (2001) (their figures 21, 5). This specimen is considered here to be comparable with *Cubaochetoceras imlayi* (Sánchez-Roig, 1951) in Wierzbowski (1976) (p. 240, pl. 7, their figure 3) from Cuba, by its probable large body chamber, although ornamentation varies in that species. The former has weakly falcoidal striae at the flank, and the latter shows larger internal ribs.

Genus *Glochiceras* Hyatt (1990)

Type Species: *Ammonites Nimbatus* Oppel (1863).

Glochiceras sp. (Figure 5c)

Description

A fragmentary specimen, preserved only half of the last whorl. This specimen is subevolute with the flank gently subrounded and smooth with a mid-lateral, narrow, and shallow groove. A lappet is preserved.

Material Examined

IPN-1116.

Occurrence

The Maloob 406 well, core 1 (4754–4763 m [15,597–15,627 ft]), southeastern Mexico, middle Oxfordian, Ek-Balam Formation.

Discussion

This specimen could be assigned to *Glochiceras* (*Glochiceras*) *subclausum* (Oppel) in Ziegler (1958) (p. 107, pl. 10, their figures 3–5) of the Transversarium zone (Middle Oxfordian) from Europe because of its smooth and subrounded flank and its mid-lateral and shallow groove; however, the Mexican specimen is represented by a fragmentary specimen, not enough for definitive specific determination.

G. (*G.*) *amplicanaliculatum* Wierzbowski (1976) (p. 248, pl. 8, their figures 7, 8), of the La Jagua Formation from Cuba was assigned as a new species and differentiated from the mentioned *G.* (*G.*) *subclausum* (Oppel) *in* Ziegler. The differences established by Wierzbowski (1979) for separating them into two different species are based in the subjective position of the lateral groove, which is not enough evidence.

Family Perisphinctidae Steinmann (1890)

Perisphinctidae genus and species indeterminate (Figure 5d)

Description

A small, fragmentary specimen with subrounded flank is ornamented with fine, closely spaced, and gradually sigmoidal simple ribs; these become bifurcate at the middle part of the flank and alternate with simple ribs. No traces of the suture line are observed.

Discussion

This specimen might be assigned to the Perisphinctidae because of its fine ribbing that is bifurcated at the outer part of the flank. This specimen is associated with *Ochetoceras* from the same well and the same core; its ribbing is similar to *Discosphinctes virgulatus* Burckhardt (1912) from the middle Oxfordian of north-central and eastern Mexico (Burckhardt, 1912; Cantú-Chapa, 1984).

Material Examined

IPN-1116.

Occurrence

The Maloob 406 well, core 1 (4754–4763 m [15,597–15,627 ft]), southeastern Mexico, middle Oxfordian, Ek-Balam Formation.

Aptychi

The species and groups of aptychi are not satisfactorily known as to their exact stratigraphic range and classification. Some fragmental and small aptychi were obtained from the Bacab 201 (core 3, 4588–4596 m [15,052–15,079 ft]) and Maloob 406 wells (core 1, 4754–4763 m [15,597–15,627 ft]). They are considered here as marine fossil parts intercalated within the Oxfordian–Kimmeridgian sequences of the wells drilled on the Campeche Shelf. The aptychi obtained here are distributed through sedimentary sequences about 160 m (525 ft) long consisting of shales and sands.

REFERENCES CITED

Angeles-Aquino, F., and A. Cantú-Chapa, 2001, Subsurface Upper Jurassic stratigraphy in the Campeche Shelf, Gulf of Mexico, *in* C. Bartolini, R. T. Buffler, and A. Cantú-Chapa, eds., The western Gulf of Mexico Basin: Tectonics, sedimentary basins, and petroleum systems: AAPG Memoir 75, p. 343–352.

Arkell, W. J., 1957, Mesozoic Ammonoidea, *in* R. C. Moore, ed., Treatise on invertebrate paleontology: Part L (Mollusca 4, Cephalopoda, Ammmonoidea): Lawrence, p. L80–L437.

Burckhardt, C., 1912, Faunes Jurassiques et Cretaciques de San Pedro del Gallo: Boletin Instituto de Geologia de México, v. 29, 264 p.

Burckhardt, C., 1930, Etude synthetique sur le Mesozoïque Mexicain: Mémoire de la Societé Paleontologique Suisse, v. 49–50, 280 p.

Callomon, J. H., *in* A. Hillebrandt, P. Von Smith, G. E. G. Westermann, and J. H. Callomon, 1992, Ammonite zones of the circum-Pacific region, *in* G. E. G. Westermann, ed., The Jurassic of circum-Pacific: New York, Cambridge University Press, p. 247–272.

Cantú-Chapa, A., 1977, Las amonitas del Jurásico Superior del pozo Chac 1, Norte de Campeche (Golfo de México): Revista del Instituto Mexicano del Petróleo, v. 9, no. 2, p. 8–39.

Cantú-Chapa, A., 1982, The Jurassic–Cretaceous boundary in the subsurface of Eastern Mexico: Journal of Petroleum Geology, v. 4, no. 3, p. 311–318.

Cantú-Chapa, A., 1984, El Jurásico Superior de Taman, San Luis Potosi, Este de México, *in* C. Perrilliat, ed., Memoria III Congreso Latinoamericano de Paleontología, México, p. 207–215.

Cantú-Chapa, A., 2001, Mexico as the western margin of Pangea based on biogeographic evidence from the Permian to the Lower Jurassic, *in* C. Bartolini, R. T. Bubbler, and A. Cantú-Chapa, eds., The western Gulf of Mexico Basin: Tectonics, sedimentary basins, and petroleum systems: AAPG Memoir 75, p. 1–27.

Cantú-Chapa, A., 2009, Ammonites of the Cretaceous Taraises and lower Tamaulipas formations in eastern Mexico, *in* C. Bartolini and J. R. Román Ramos, eds., Petroleum systems in the southern Gulf of Mexico: AAPG Memoir 90, p. 191–216.

Cantú-Chapa, A., and E. Ortuño-Maldonado, 2003, The Tithonian (Upper Jurassic) Edzna Formation, an important hydrocarbon reservoir on the Bay of Campeche continental shelf, Gulf of Mexico, *in* C. Bartolini, R. T. Buffler, and J. Blickwede, eds., The Circum-Gulf of Mexico and the Caribbean: Hydrocarbon habitats, basin formation, and plate tectonics: AAPG Memoir 79, p. 305–311.

Cariou, E., R. Enay, F. Atrops, P. Hantzpergue, D. Marchand, and M. Rioult, 1997, Biozonation (Ammonites, Oxfordien), *in* E. Cariou and P. Hantzpergue, eds., Biostratigraphie du Jurassique ouest-européen et méditerranéen: Zonations paralléles et distribution des invertébres et microfossiles: Bulletin du Centre de Recherche Elf-Exploration, v. 17, p. 79–86.

Christ, H. A., 1961, Ueber *Campylites* und *Trimarginites* (Ammonoidea, Jura): Neues Jahrbuch für Geologie und Paläontologie Abhandlungen, v. 3, no. 3, p. 274–325.

Donovan, D. T., J. H. Callomon, and M. K. Howarth, 1981, Classification of the Jurassic ammonitina, *in* M. R. House and J. R. Senior, eds., Systematics association special: The Ammonoidea: London, Academic Press, v. 18, p. 101–155.

Douvillé, H., 1890, Sur la classification des Cératites de la Craie: Bulletin de la Société Géologique de France, v. 3, no. 18, p. 275–292.

Haug, E., 1885, Beiträge zu einer Monographie der Ammonitengattung *Harpoceras*: Neues Jahrbuch für Geologie und Paläontologie Abhandlungen, v. 3, p. 585–722.

Hyatt, A., 1900, Cephalopoda, *in* K. A. Zittel, ed., Textbook of paleontology, 1st English ed. (translated by C. R. Eastman): p. 502–292.

Imlay, R. W., 1980, Jurassic paleobiogeography of the

conterminous United States in its continental setting: U.S. Geological Survey Professional Paper, v. 102, 134 p.
Imlay, R. W., and G. Hermann, 1984, Upper Jurassic ammonites from the subsurface of Texas, Louisiana and Mississippi: Gulf Coast Section, Society for Sedimentary Geology (SEPM) Foundation, 3rd Annual Research Conference Proceedings, p. 149–170.
Judoley, C. M., and G. Furrazola-Bermúdez, 1968, Estratigrafía y fauna del Jurásico de Cuba: Publicación Especial, Instituto Cubano de Recursos Minerales, Departamento Científico Geológico, Academia de Ciencias Cuba, Departamento Geológico, 126 p.
Maubeuge, P. L., 1975, Catalogue des Ammonites du Jurassique supérieur (Callovien à Kimmeridigen) du Musée cantonal de Bâle-Campagne: Tätigkeitsberichte der Naturforschenden Gesellschaft Baselland, v. 29, p. 1971–1974, Liestal, Buchdruckerei Lüdin AG, p. 40–338.
Myczynski, R., F. Oloriz, and A. B. Villaseñor, 1998, Revised biostratigraphy and correlations of the middle–upper Oxfordian in the Americas (southern U.S.A., Mexico, Cuba, and northern Chile): Neues Jahrbuch für Geologie und Paläontologie Abhandlungen, v. 207, no. 2, p. 185–206.
North American Commission on Stratigraphic Nomenclature, 1983, North American Stratigraphic Code.
Oppel, A., 1863, Über die jurassische Cephalopoden: Paläontologische Mitteilungen aus dem Museum des Koeninglich-Bayerischen Stattes, p. 127–266.
Sánchez-Roig, M., 1951, La fauna jurásica de Vinales: Anales de la Academia de Ciencias Médicas, Físicas y Naturales: La Habana, v. 89, no. 2, p. 47–94.
Spath, L. F., 1928, Revision of the Jurassic Cephalopod fauna of Kachh (Cutch): Memoirs of the Geological Survey, India, Paleontological Indica 9, no. 2, p. 73–278.
Steinmann, G., 1890, Cephalopoda, *in* G. Steinmann and L. Döderlein, eds., Elemente der Paläontologie: Leipzig, W. Engelmann, p. 344–475.
von Buch, L., 1831, Recueil de planches de Pétrifications remarquables, *in* A. Oppel, Palaeontologische Mittheilungen, Stuttgart.
von Zittel, K. A., 1884, Handbuch der Palaeontologie: v. 1, no. 2, 893 p.
Wierzbowski, A., 1976, Oxfordian ammonites of the Pinar del Rio province (western Cuba); their revision and stratigraphical significance: Acta Geologica Polonica, v. 26, no. 2, p. 137–260.
Ziegler, B., 1958, Monographie der Ammonitengattung *Glochiceras* im Epikontinentalen Weissjura Mitteleuropas: Palaeontographica, v. 110A, p. 93–164.

Salomón-Mora, L. E., M. Aranda-García, and J. R. Román-Ramos, 2009, Contractional growth faulting in the Mexican Ridges, Gulf of Mexico, *in* C. Bartolini and J. R. Román Ramos, eds., Petroleum systems in the southern Gulf of Mexico: AAPG Memoir 90, p. 93–115.

Contractional Growth Faulting in the Mexican Ridges, Gulf of Mexico

Luis Enrique Salomón-Mora, Mario Aranda-García, and Juan Rogelio Román-Ramos

Petróleos Mexicanos, Exploración y Producción, Edificio Proyectos Marinos, Interior Campo Petróleos Mexicanos (PEMEX), Poza Rica, Veracruz, Mexico

ABSTRACT

The Mexican Ridges fold belt is a Neogene contractional system located on the western continental slope of the Gulf of Mexico Basin. This contractional system formed in response to gravitational slide processes affecting continental shelf deposits along 500 km (311 mi) of the Quetzalcoatl extensional system. Numerous north-south–oriented contractional growth faults are developed mainly above a thick detachment zone consisting of Paleogene argillaceous strata. The main structural style of these fold geometries corresponds to classical detachment folds and detachment folds with late-stage reverse faults. The detachment folds are generally symmetrical, although some folds are asymmetric with eastward vergence in the dominant shortening direction of the fold belt. The contractional folds range in length from 20 to 120 km (12.4 to 74 mi) with amplitudes between 6 and 15 km (3.7 and 9.3 mi) and are important hydrocarbon exploration prospects.

Some examples from the Mexican Ridges indicate that contractional folding belt initiated during the middle Miocene and that the most recent folds are forming in the central and eastern part of the fold belt. Generally, the youngest and highest amplitude folds are found toward the center of the basin.

The growth folds are analyzed on the basis of their stratigraphic relationship to syntectonic sedimentation, which displays different geometries as a consequence of fault movement. Converging seismic reflection geometries are common toward the fold crests and may appear as thinning sets or onlaping above an unconformable surface and are characteristics of fault kinematics and fault activity. In this study, we use growth strata to determinate (1) structural styles of the folds based and their kinematics and (2) the age of contractional deformation. These geological characteristics allow an evaluation of elements such as timing, possible migration pathways, and geological risks of exploration prospects.

DOI:10.1306/13191080M903332

INTRODUCTION

The passive continental margin of the western Gulf of Mexico is characterized by a system of normal growth faults known as the Quetzalcoatl extensional system (Román-Ramos et al., 2001, 2004). This system is associated with the deformation of Tertiary continental shelf and upper slope sediments by gravity sliding above a regional detachment zone located within the Paleogene. Basinward, along the continental slope approximately parallel to the coastline, one of the most extensive submarine features in the world formed in response to the extensional deformation (Figure 1). This system was named the Mexican Ridges fold belt by Bryant et al. (1968).

The structural and stratigraphic characteristics for the occurrence of hydrocarbons, such as trap types and seal and reservoir strata, vary widely along the extensional system on the continental shelf of this part of the Gulf of Mexico as reported by Román-Ramos et al. (2001, 2004), Jennette et al. (2003), Wawrzyniec et al. (2003), and Ambrose et al. (2005), among others.

The Mexican Ridges is a contractional system located in water depths exceeding 1000 m (3281 ft) (Guzmán and Márquez-Dominguez, 2001). The Mexican Ridges comprises numerous folds with dimensions that typically range from 30 to 120 km (18.6 to 74 mi) in length and 8 km (4.9 mi) in width within the area of highest structural relief. The folds and reverse faults demonstrate a pattern of subparallel axial traces aligned approximately north–south. Shortening of the multiple folds occurs mainly along two stratigraphic levels: a basal Paleogene zone of overpressured argillaceous rocks and a Miocene–Oligocene level. These numerous folds constitute attractive hydrocarbon exploration targets.

This chapter consists of three parts: (1) the structural styles of folding are described for the Quetzalcoatl extensional system and the Mexican Ridges fold belt; (2) models are proposed for the detachment folds and faulted detachment folds found in the fold belt; and (3) synsedimentary growth patterns are identified in the folds and used to infer their deformation age (Jáuregui et al., 1997), progressive kinematic (Poblet et al., 1997), and dominant deformational mechanism.

The structural analysis is based on the interpretation of more than 7000 linear kilometers of high-quality two-dimensional (2-D) seismic reflection data over an area of more than 65,000 km^2 (25,097 mi^2). Stratigraphic information obtained from nine exploratory wells located principally on the continental shelf was used to control the structural analysis (Figure 2).

GEOLOGIC SETTING

The passive continental margin of the southwestern Gulf of Mexico comprises two separate tectonic provinces: the Quetzalcoatl extensional system on the shelf and upper continental slope and the Mexican Ridges fold belt covering a large part of the deep-water continental slope (Figure 1).

The Mexican Ridges is bordered on the north by the Neogene Perdido fold belt, whose stratigraphy, structural styles, and hydrocarbon potential have been described and interpreted by Blickwede and Queffelec (1988) and Trudgill et al. (1999) among others. The classic structures of the Perdido fold belt include a thick Jurassic, Cretaceous, and Tertiary stratigraphic section that forms wide, high-amplitude fold traps over a salt detachment. The folds also display Oligocene to Miocene salt diapirism affecting the prekinematic strata. Growth strata of the fold belt indicate that folding is of Miocene–Pliocene age.

Patiño-Ruiz et al. (2004), in their analysis of structural relationships between the Perdido fold belt and the Mexican Ridges, clearly showed the variations in structural style, dominant deformation mechanisms, detachment levels, ages of deformation (using syntectonic sedimentation), and structural trap types within both folded tectonic provinces and in the intervening transition zone.

Toward the south, the fold domain of the Mexican Ridges was separated from the Neogene fold belt of southern Mexico by Aranda-García (1999) across the Teziutlán-Santa Ana Massif on the coastal plain and continental shelf of southern Veracruz state. The southeastern part of the Neogene fold belt of southern Mexico is exposed in the Sierra de Chiapas.

PREVIOUS STUDIES

The Mexican Ridges province was first recognized in 1967 by Jones et al. (De Cserna, 1984), who related its origin to salt tectonic. Later, Bryant et al. (1968) divided this fold belt into four subprovinces characterized by their style of folding. In subprovince 4 (Figure 3), Bryant et al. (1968) suggested contraction associated with gravity sliding over a detachment surface, as well as the vertical movement of salt or shale caused by sediment loading, as the possible causes for compression and shortening of the folded sedimentary rocks. Garrison and Martin (1973) also considered that the Mexican Ridges originated by gravity sliding over a detachment surface consisting of a low-density material that could be salt or shale. In a study of this area, Buffler et al. (1979) showed a fold belt characterized by numerous reverse faults flattening at depth into a zone of low seismic reflectivity inferred to be shaly turbidites (Figure 3). In 1982, Pew interpreted two structural subprovinces in the southern Mexican Ridges (Figure 3) characterized by a detachment level within a sequence of Upper Cretaceous or Paleogene pelagic

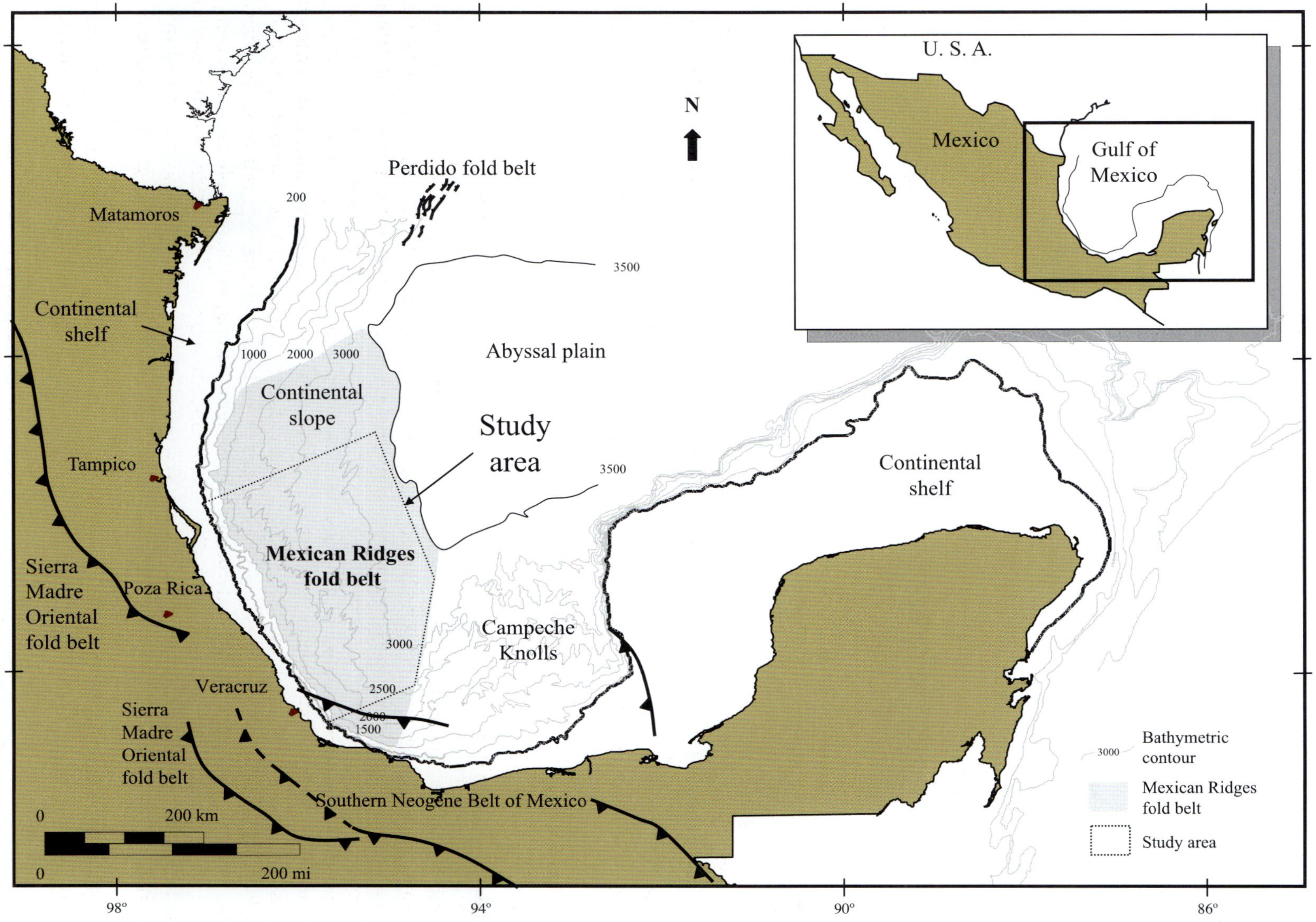

FIGURE 1. Map showing the main physiographic provinces and contractional systems of the southeastern Gulf of Mexico Basin. The Mexican Ridges fold belt is located on the western continental slope of the basin, bordered by the continental shelf to the west, the abyssal plain to the east, the Perdido fold belt to the north, and the Campeche salt province to the south. The study area is shown in the south-central sector of the Mexican Ridges.

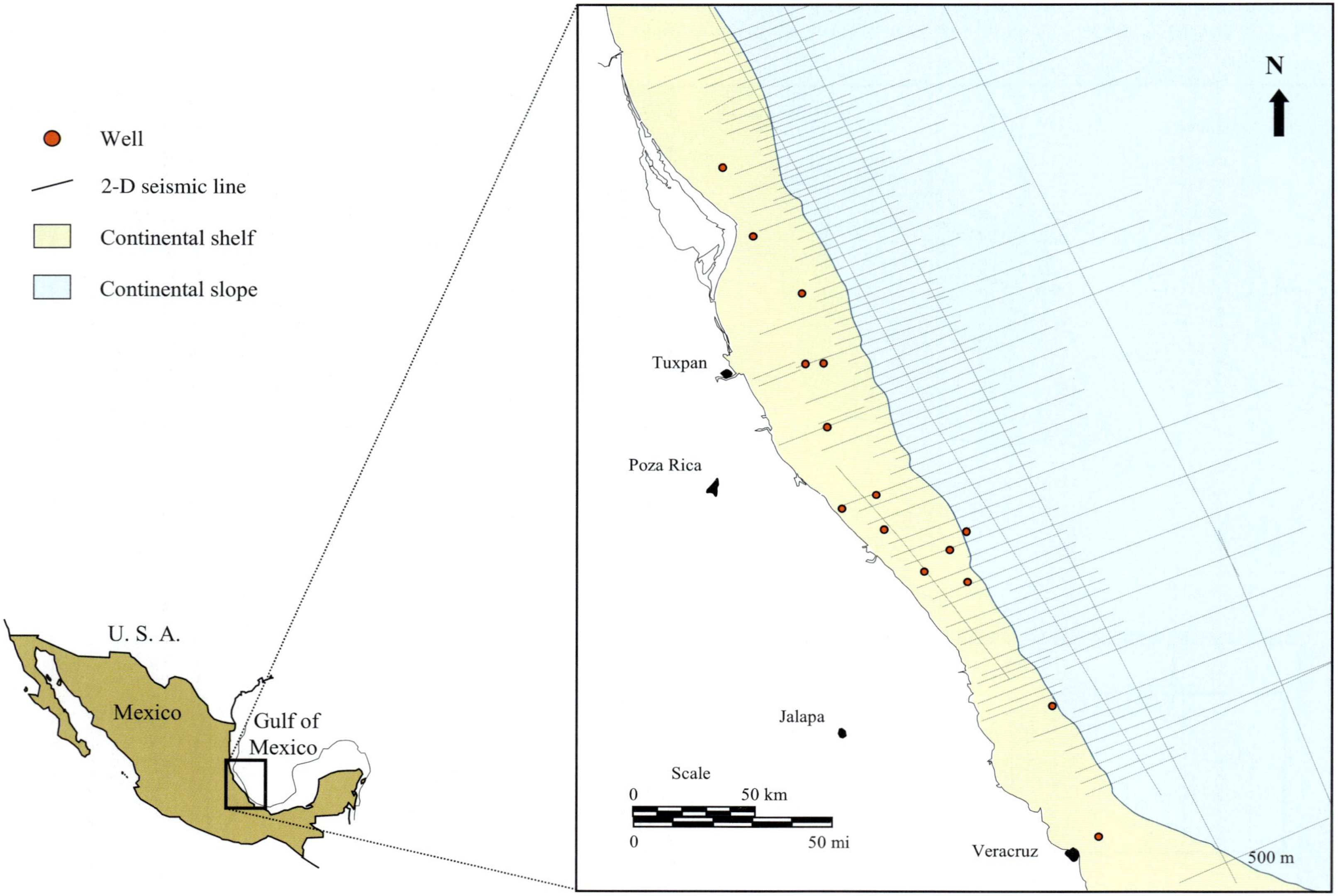

FIGURE 2. The study area in the southern part of the Mexican Ridges is shown. The main wells located on the continental shelf used for stratigraphic ties are indicated as is the southwest-northeast–oriented two-dimensional seismic coverage with 5-km (3.1-mi) spacing.

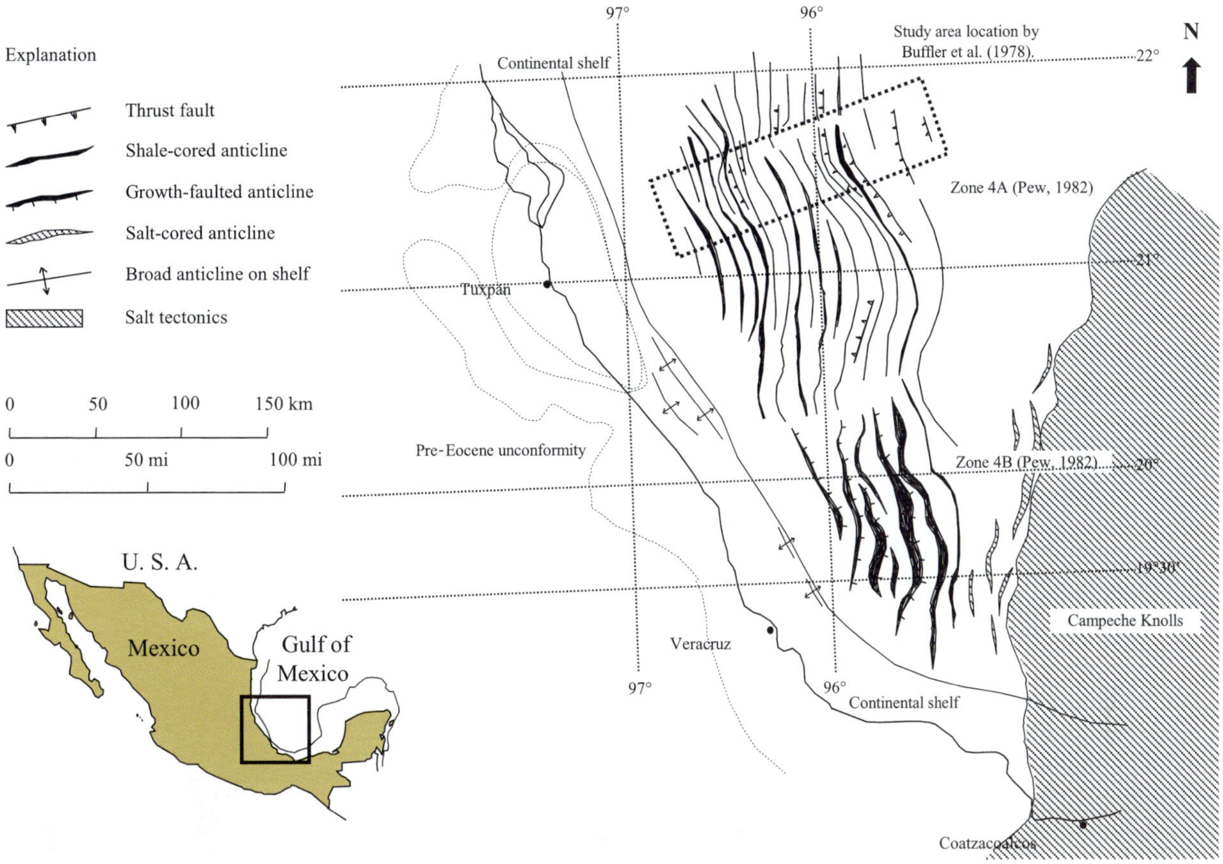

FIGURE 3. The tectonic map of the southern Mexican Ridges corresponding to subprovince 4 of Bryant et al. (1968). Pew (1982) described two zones distinguished by structural and geometric differences in their contractional folds: zone 4A and zone 4B. The study area of Buffler et al. (1979) is shown (modified from Pew, 1982).

shales. According to Pew (1982), the northern zone (4A) is formed by 12 generally symmetric folds associated with reverse faults, with amplitudes increasing and structural relief decreasing toward the basin. The southern zone (4B) is formed by seven asymmetric folds verging toward the basin whose average amplitude is 10–12 km (6.2–7.4 mi), with their structural relief increasing toward the basin.

In a recent work, Román-Ramos et al. (2001) and Salomón-Mora et al. (2004) identified the predominant structural styles of the shelf and western slope of the southern Gulf of Mexico by establishing a deformational mechanism that ties the Quetzalcoatl extensional system and the Mexican Ridges fold belt together through a common detachment at the upper Eocene–Oligocene level. Later, Wawrzyniec et al. (2003) estimated the magnitude and age of the extensional system between the cities of Veracruz and Tampico and recognized that the differential increase in extension toward the north, ranging from 9 to 22 km (5.5 to 13.6 mi), caused an equivalent increase in shortening of the Mexican Ridges fold belt in the deep-water region of the Gulf of Mexico.

SEISMIC STRATIGRAPHY

Because of the difficulty of correlating the stratigraphy from the extensional system located in the shelf to the contractional system, we have supported our interpretations with information from nine wells located mostly on the continental shelf and the seismic character, continuity, and internal geometries of horizons studied on the Gulf of Mexico's abyssal plain.

Seismostratigraphic units named Challenger, Campeche, Mexican Ridges, Cinco de Mayo, and Sigsbee were initially described by Buffler et al. (1979), then discussed in detail by Shaub et al. (1984) in the deeper part of the Gulf of Mexico Basin, and then redefined by Feng and Buffler (1994). Feng (1995) differentiated deep-water seismic facies ranging from A (Upper Cretaceous) to L (lower Pliocene).

These seismostratigraphic units were extrapolated and updated in this study of the Mexican Ridges with recent stratigraphic data from nine Petróleos Mexicanos (PEMEX) wells and information from several 2-D seismic studies (Figure 4). The principal seismic characteristics

Chronostratigraphy: System		Series		Buffler et al., 1978 Shaub et al., 1984	Seismic character	Probable lithology
Tertiary	Neogene	Pleis	U	Sigsbee	Discontinuous reflections	Pelagic-hemipelagic sediments
		Plio	L	Cinco de Mayo	Parallel discontinuous	Turbidite sands and shales
		M	U		Parallel continuous-discontinuous reflections	Turbidite sands and shales. Submarine fans
			M	Upper Mexican Ridges	Parallel-continuous reflections. Variable amplitude	Turbidite sands and shales. Channel/lobe submarine fans
			L			
	Paleogene	O	U	30 ma	Discontinuous reflections. Variable amplitude with chaotic facies	Turbiditic hemipelagic sediments. Lobe systems in submarine fans
			L	Lower Mexican Ridges		
		E	U		Discontinuous reflections. Transparent zone	Turbiditic hemipelagic sediments, shales. Basal detachment zone
			M			
			L			
		P	U	Campeche	Parallel continuous-discontinuous reflections	Fine pelagic hemipelagic sediments. Distal part of submarine fans
			L			
Cretaceous		Upper		94 ma	Continuous reflections, the upper part is a strong reflection	Deep-water carbonates
		Lower				
Jurassic		Upper		Challenger	Parallel-continuous reflections	Transgresive-deep-water carbonates
		Middle				
		Lower			Chaotic reflections	Transitional and oceanic crust

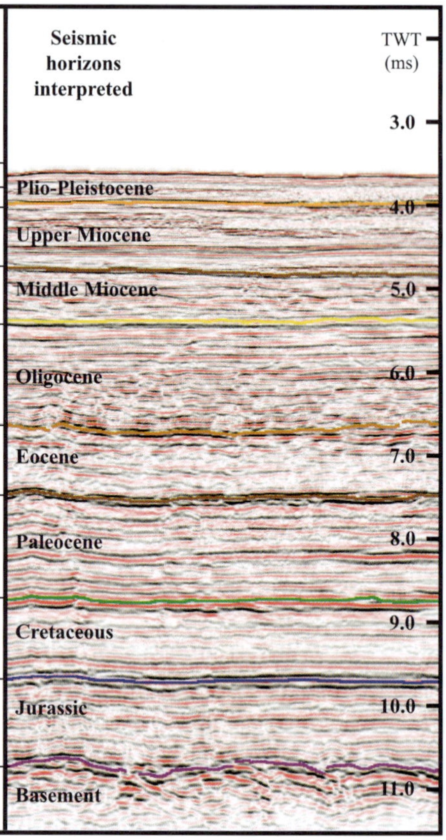

FIGURE 4. Seismic stratigraphy of the Mexican Ridges foreland (modified from Salomón-Mora et al., 2004). TWT = two-way traveltime.

and the probable lithologies of the interpreted seismic horizons are detailed in Figure 4. These horizons tentatively correlate to the following tops: basement, Jurassic, Upper Cretaceous, Paleocene, lower Eocene, upper Eocene, Oligocene, middle Miocene, upper Miocene, and lower Pliocene.

STRUCTURAL STYLES

Major structures of the Quetzalcoatl extensional system and the Mexican Ridges of the western Gulf of Mexico are shown on a regional structural map (Figure 5) and a regional southwest–northeast seismic transect (Figure 6), the transition from the continental shelf to the abyssal plain through the extensional and contractional systems. The regional transect illustrates the morphology of the basement, the Mesozoic and Tertiary sedimentary cover, and the main elements of the extensional and contractional fault systems. Both systems are controlled by subsidence of the prograding Neogene sedimentary load through a series of listric faults connected to a regional detachment zone in the upper Eocene. We divide the Quetzalcoatl system and the Mexican Ridges into three subprovinces, each based on their seismic expression in cross section and mapped structural style.

QUETZALCOATL EXTENSIONAL SYSTEM

The Neogene Quetzalcoatl extensional system is the product of gravity sliding toward the central Gulf of Mexico (Román-Ramos et al., 2001, 2004). This system is a group of east-dipping listric faults with different ages of activation. The normal faults gave rise to the following structural styles (Figure 7):

1. Domino fault system, with a limited number of planar normal growth faults, some of which are slightly listric. The domino system appears to detach above a shallow Upper Cretaceous–Paleogene horizon on a high made up by Mesozoic basement high and Tuxpan carbonate platform.
2. Graben system, characterized by a group of synthetic-antithetic listric faults that form grabens with numerous secondary fault keystones and minor faulted blocks in the structure. The graben structures comprise upper Miocene and Pliocene–Pleistocene depocenters with growth strata on both the synthetic and antithetic faults.
3. Listric growth fault system, with some of these listric faults flatten at depth on a diffuse detachment,

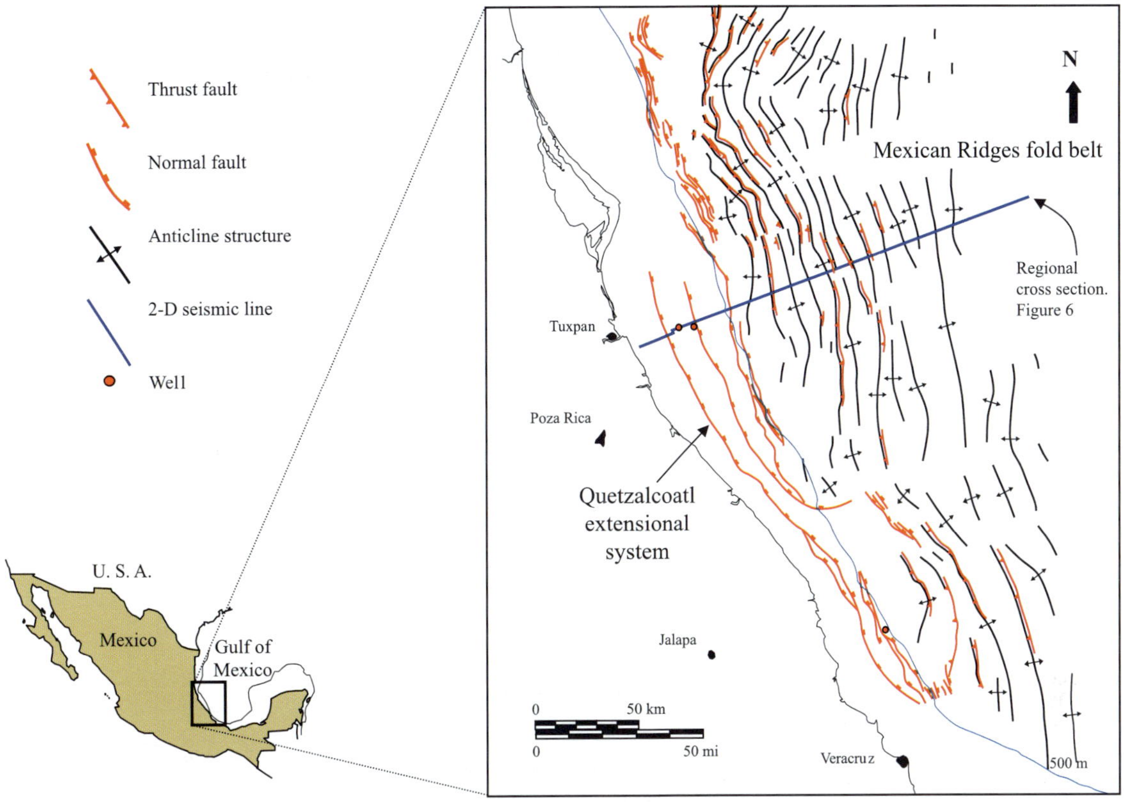

Figure 5. Regional tectonic map of the Quetzalcoatl extensional system and the Mexican Ridges in the southwestern Gulf of Mexico. The principal structural trends were identified and mapped using 2-D seismic surveys (modified from Salomón-Mora et al., 2004).

whereas others display a deep listric geometry with a detachment level clearly associated with the regional detachment of the contractional system. This listric fault system develops the typical half graben with a rollover structure and growth strata in the hanging wall, thickening toward the fault.

The structural interpretation and the seismic correlation across the Quetzalcoatl extensional system indicate that the structural evolution of the late Miocene growth fault systems is synchronous with the contraction of the Mexican Ridge fold belt.

MEXICAN RIDGES FOLD BELT

Folding Styles

Common structural elements created by contractional deformation in the Mexican Ridges fold belt are detachment folds, the detachment zone, and syndepositional rotational wedges. Generally, few contractional structures possess an interpretable record of their evolution from the undeformed state through their final geometry (Aranda-García, 1999). Most contractional belts have been analyzed based on incomplete recognition of their syntectonic geometry and kinetics and by considering only their pre-kinetic stratigraphy (Wallace and Homza, 2004). The principal models of folds associated with faults are based on the importance and magnitude of shortening as follows: detachment folds (Jamison, 1987), fault propagation folds (Suppe and Medwedeff, 1990), and fault-bend folds (Suppe, 1983).

The detachment fold concept was introduced by Jamison (1987) for folding above a detachment surface or bedding parallel thrusting (Figure 8), and the thrusting may have a fixed tip point outside the limits of the fold (Wallace and Homza, 2004). The detachment must consist of a ductile material, such as salt or overpressured shale, that can fill the space created in the nucleus of the fold.

Most of the interpreted folds in the Mexican Ridges fold belt correspond to detachment folds and detachment folds associated with reverse faulting (Figure 9). The principal shortening occurs above a deep detachment zone located in argillaceous Paleogene (upper Eocene) strata, with the principal shortening direction oriented east–west. Most of the back-thrust faults associated with folding are detached along a secondary Oligocene–Miocene detachment level. This shallow detachment zone extends transversally across several

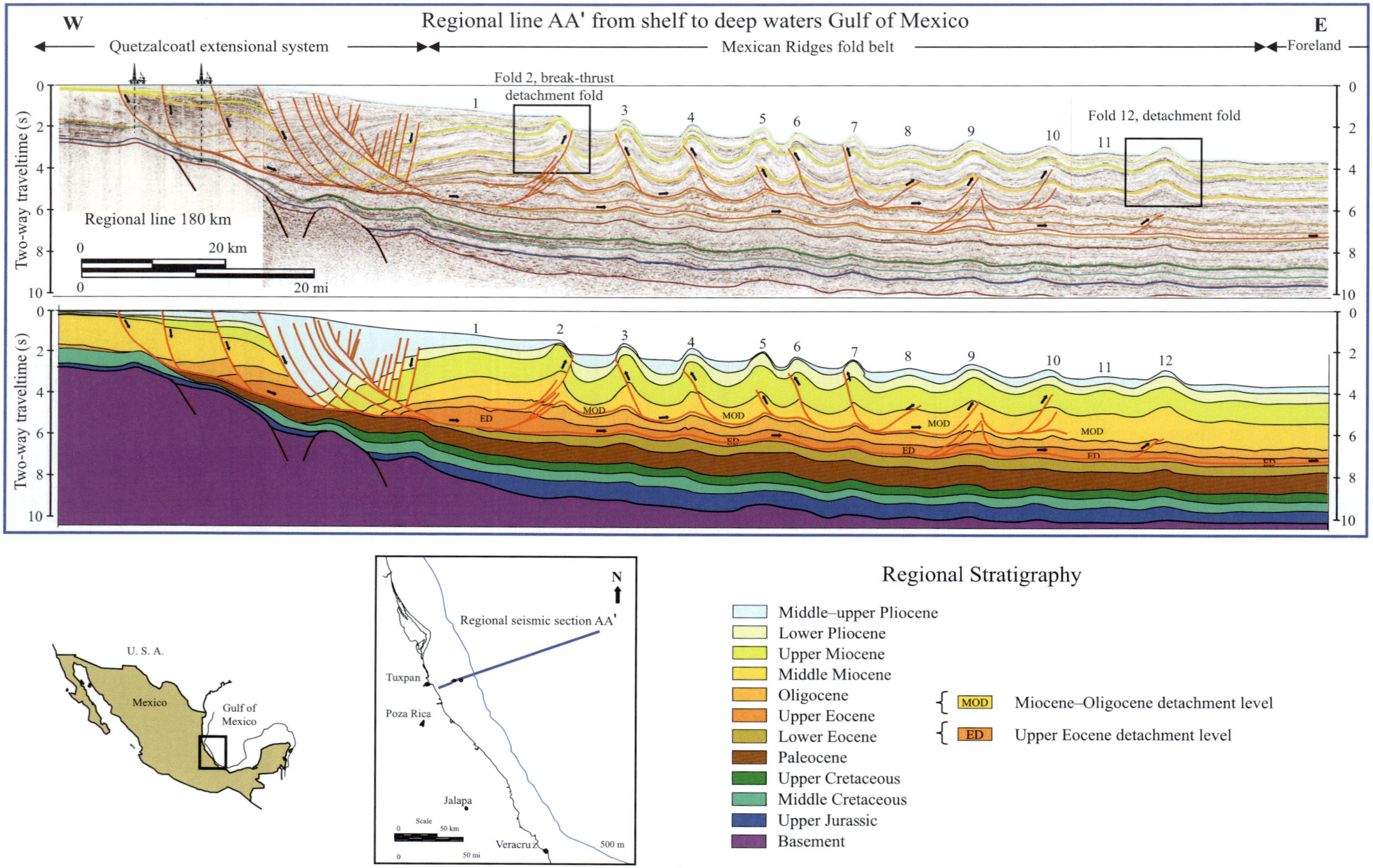

FIGURE 6. Southwest-northeast–oriented seismic section through the Quetzalcoatl extensional system from the continental shelf to the Mexican Ridges in deep water. The interpretation shows the structural deformation of the shelf characterized by numerous listric growth faults, as well as the contractional folded structural style of the Mexican Ridges fold belt.

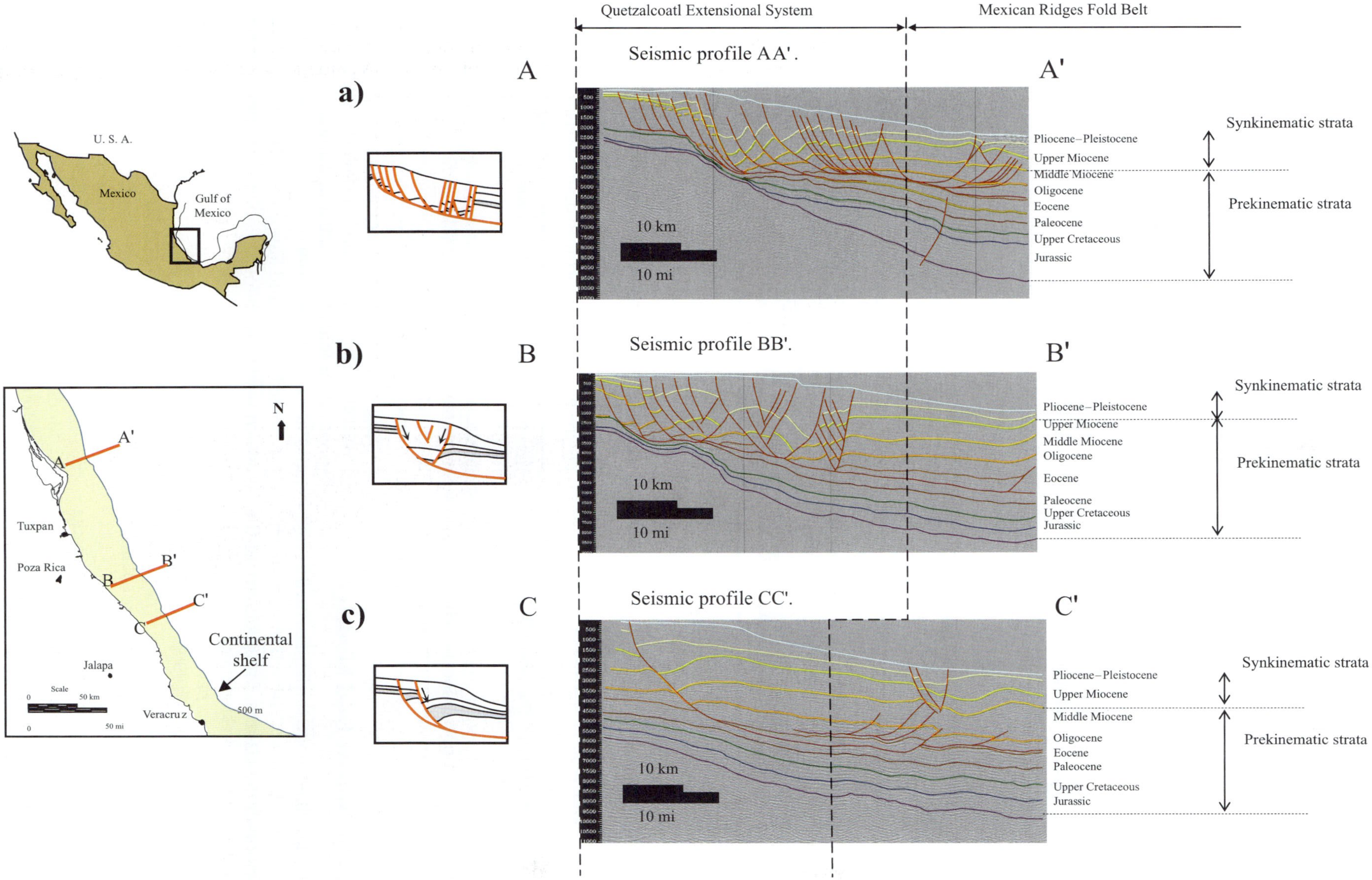

FIGURE 7. Seismic sections and models of the structural styles characteristic of the Quetzalcoatl extensional system on the continental shelf and upper continental slope of the Gulf of Mexico. (a) Line AA′ shows a wide, complex Miocene extensional system associated with the Tuxpan Platform; (b) section BB′ in the central part of the area shows listric growth faults and associated antithetic faults bounding well-developed grabens, mainly in the Pliocene–Pleistocene section; (c) section CC′ shows a half graben with listric faulting and rollover structuring in the Miocene (modified from Román-Ramos et al., 2001, 2004).

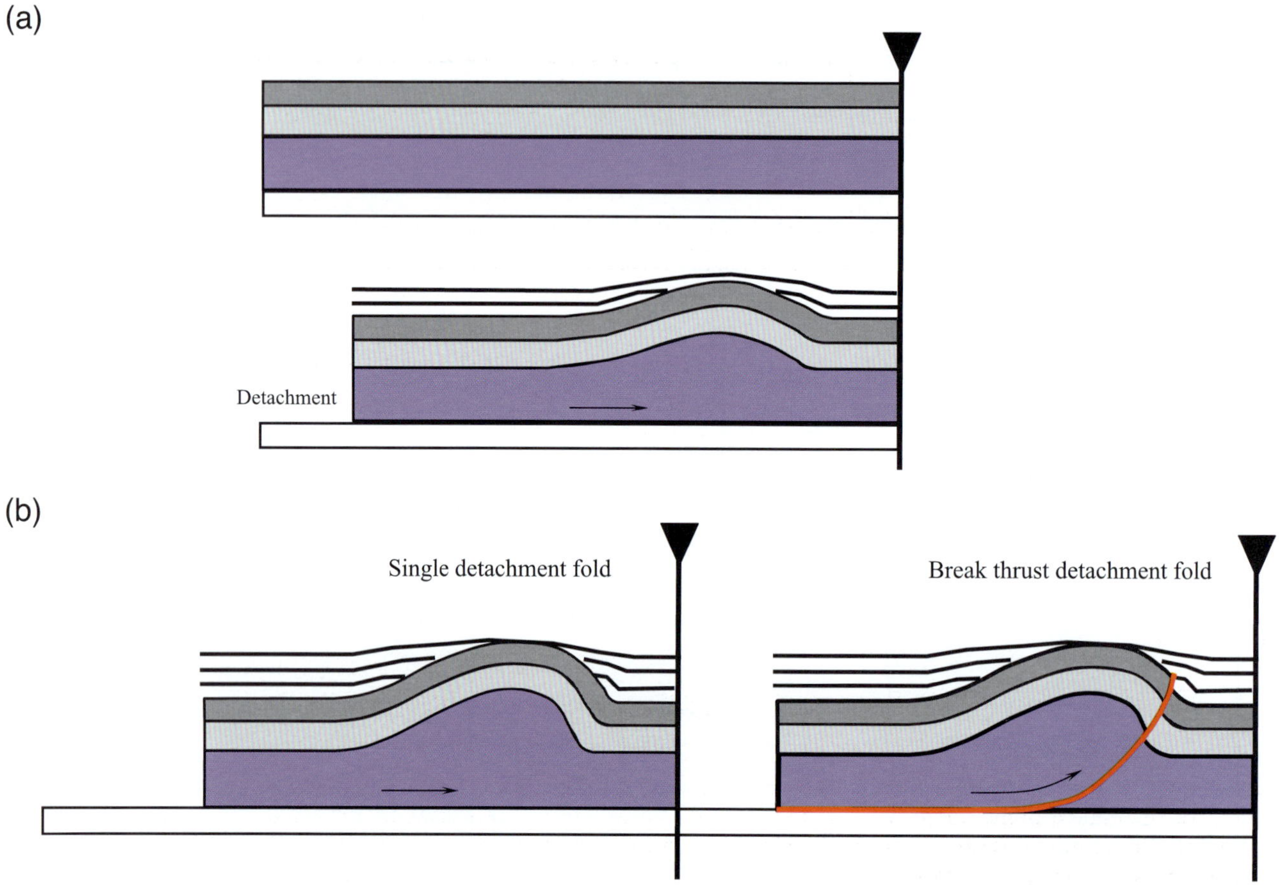

FIGURE 8. Detachment fold (a) free of faulting and (b) associated with late-stage faulting, developed by shortening over a ductile detachment horizon, generally an evaporite or shale sequence.

structural trends and may be activated as a detachment relief zone associated with the deep detachment (Patiño-Ruiz et al., 2004) with the possible presence of late-stage thrust faults (Figure 9).

The growth units that define the ages of the Mexican Ridges folds are the upper Miocene, Pliocene, and Holocene, with migration of the deformation from west to east.

Growth Detachment Fold Models

Generally, described models are represented by folding of prekinematic strata, i.e., those that were present before the deformation occurred. The geometry of the fold models based only on prekinematic units is inadequate and can create confusion between the different types of folding (Aranda-García, 1999). The examples analyzed in the Mexican Ridges preserve synkinematic strata that document the Neogene contractional deformation exceptionally well (Figure 10).

The strata deposited after deformation (post-kinematic) together with the prekinematic units surround the growth strata deposited during deformation. The latter are the macrokinematic deformation indicators that record the progressive evolution of the folding. At seismic scale, they display changes in the thickness of strata deposited on the crest or limbs of the detachment folds with different internal reflection geometries representing three main groups of sequence boundaries: (1) onlapping, (2) unconformable, and (3) conformable permitting the structural interpretation of evolving deformation (Figure 10).

Models identifying the mechanisms for detachment folding in growth structures have been proposed by Poblet et al. (1997) and Storti and Poblet (1997), for fault propagation folds by Zapata and Allmendinger (1996) and Storti and Poblet (1997), and for fault-bend folds by Medwedeff (1989) and Suppe et al. (1992). Models for detachment growth folds (Poblet et al., 1997) are based on three kinematic mechanisms (Figure 11):

1. Model 1. Limb rotation. In this model, folding and shortening occur by limb rotation; this is the variable dips and constant length of the limbs; the axial surfaces rotate to maintain constant stratigraphic thickness while the amplitude of the fold increases.
2. Model 2. Kink band migration. The folds grow through migration of fixed axial surfaces and constant dips, which induces variations in the length of

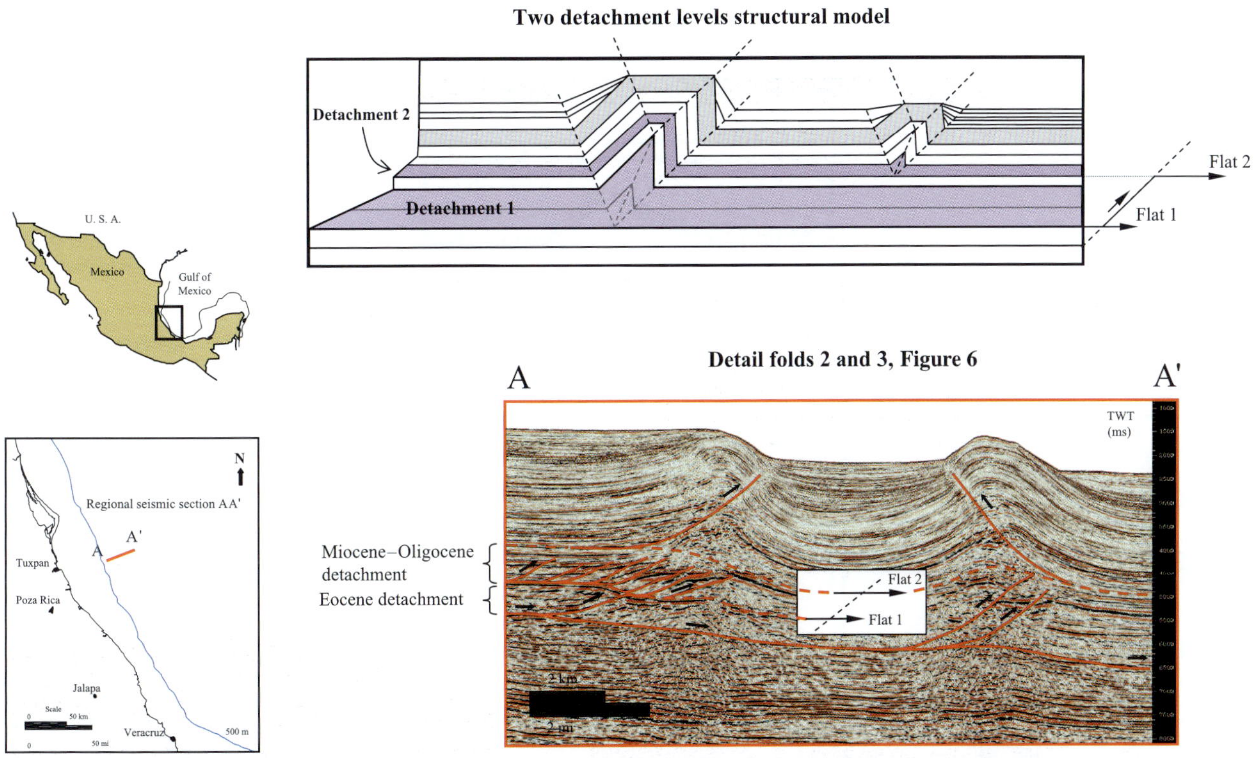

FIGURE 9. Structural model and seismic line AA′ of detachment folds with two levels (shallow and deep) indicating the transfer of the detachment; the growth strata indicate a contemporaneous detachment age for both folds. A similar multilevel folding model is present in the Mexican Ridges fold belt, mainly toward the transition with the Quetzalcoatl extensional system (modified from Patiño-Ruiz et al., 2004). TWT = two-way traveltime.

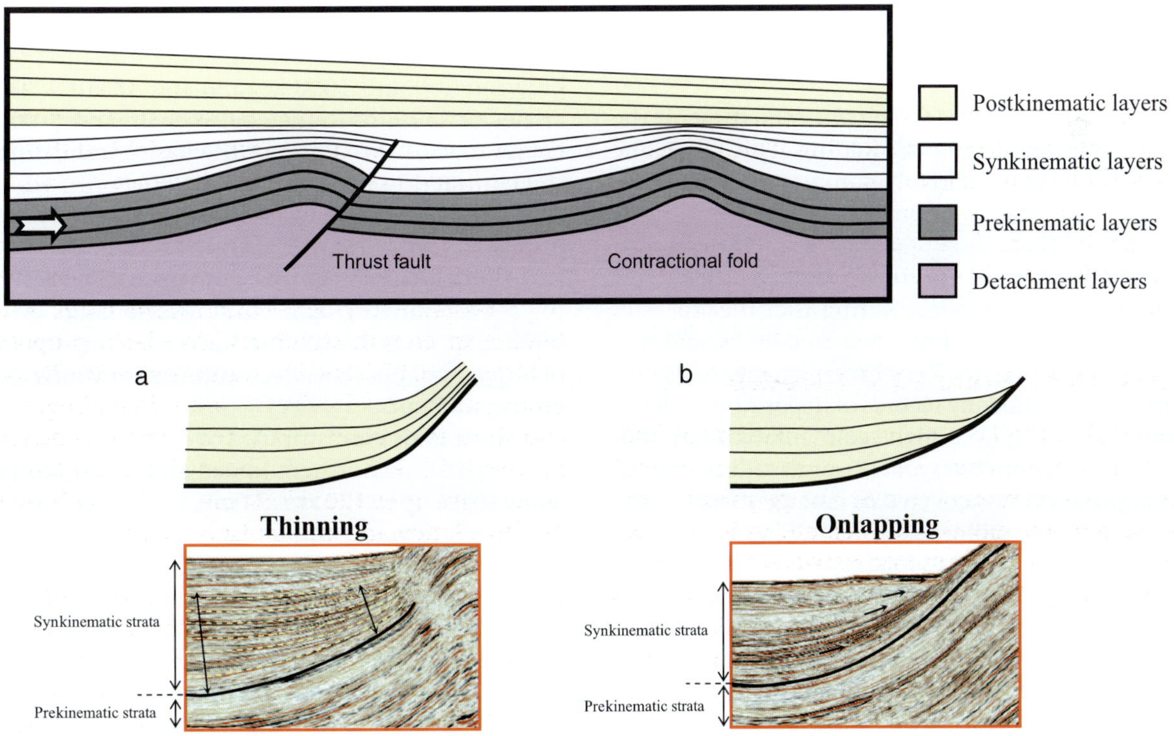

FIGURE 10. Contractional folds with pre-, syn-, and postkinematic units. The synkinematic units can show convergence toward the crest of the growth structure by (a) conformable thinning of the reflectors or (b) onlapping onto an unconformable surface.

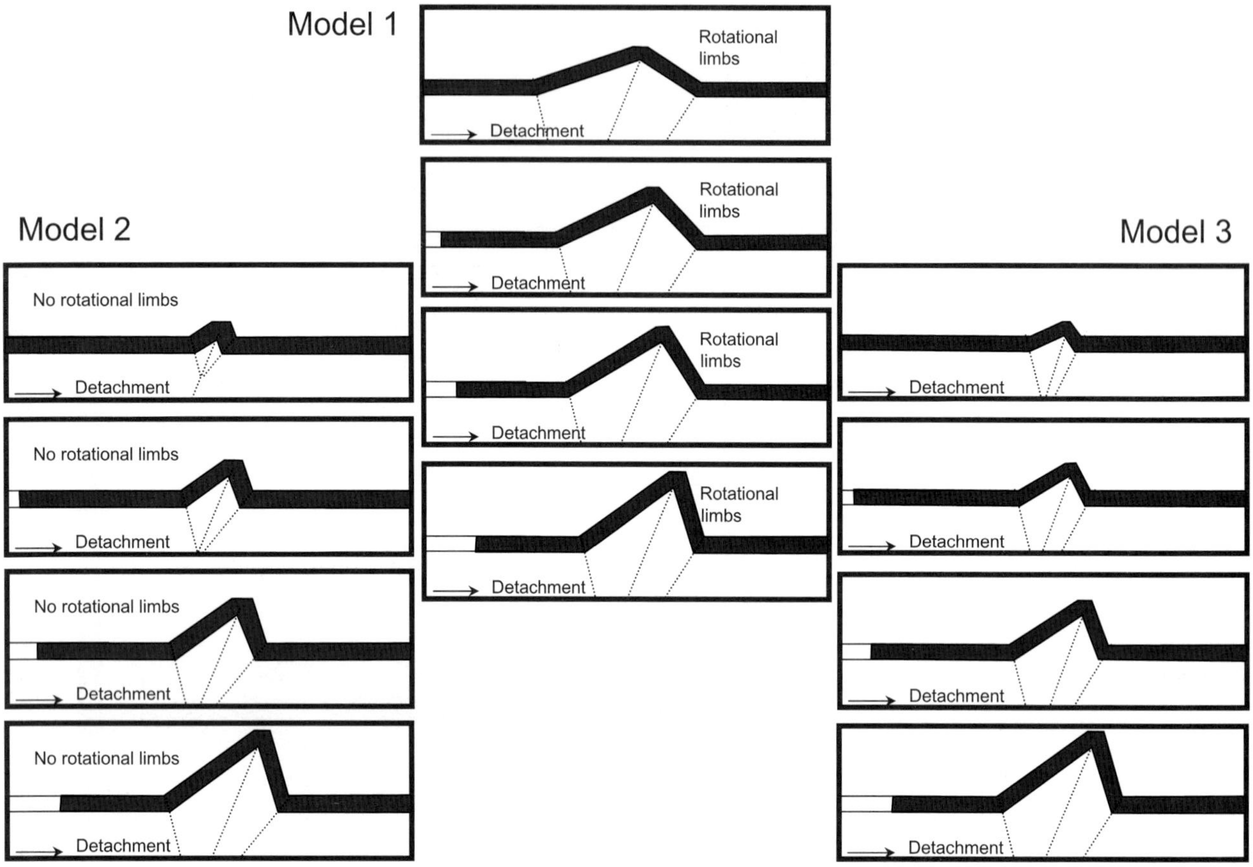

FIGURE 11. Kinematic models of detachment folds. Model 1, by limb rotation; model 2, by migration of the axial surface; model 3, by rotation of the axial surface and limb rotation.

the fold limbs. The amplitude increases as shortening progresses.

3. Model 3. Hybrid model. The dip and length of the limbs increase during the folding. This kinematic folding mechanism involves migration of the axial surface and limb rotation.

The architecture of the growth strata is different for each of the described models. Setting aside the processes of compaction, erosion of the crests and limbs, and syntectonic redeposition, the detachment growth folds may also display the following records of deposition or deformation (Figure 12): (1) a high sedimentation rate that produces buried anticlines and (2) a low sedimentation rate that produces noncovered or emergent anticlines.

In general, the sedimentation rates in the Mexican Ridges are low compared to folding velocities, resulting in clear convergence geometries on the limbs of the anticlines and emerged fold crests compared to the regional sea-floor level. The folds, faults, and syntectonic geometries of the Mexican Ridges correspond to the structural models of detachment growth faults and detachment growth faults with late-stage reverse faults. The detachment folds without associated faults are generally symmetric and oriented north–south. They have an average wavelength of 12 km (7.4 mi) and a structural relief of between 100 and 1000 m (328 and 3281 ft). The dip angles of the limbs range between 1° and 5° and decrease toward the center of the basin, indicating that the contraction decreases in this direction, which includes the central and external part of the study, marked zone A in Figure 13.

The detachment growth faults associated with faulting are asymmetric with blind reverse faults (McClay, 1992) that do not reach the surface and emergent out-of-sequence faults that affect sediments as young as those on the sea floor.

These folds are distributed over two areas (Figure 13) in zone B of the northern folded area. The folds extend along strike up to 120 km (74 mi), with an average wavelength of 6 km (3.7 mi) and a structural relief of 250–2000 m (820–6562 ft). The dip angle of the limbs ranges from 2° to 10°. These folds commonly contain back-thrust faults, i.e., reverse faults that verge in the opposite direction to tectonic transport toward the center of the basin. Seven folds in the southern area B with dip angles of up to 10° are observed, with higher dip angles in the basinward limbs. These structures reach 70 km (43.4 mi) in length with an average wavelength of 10 km (6.2 mi) and a buried structural relief of up to 2500 m (8202 ft).

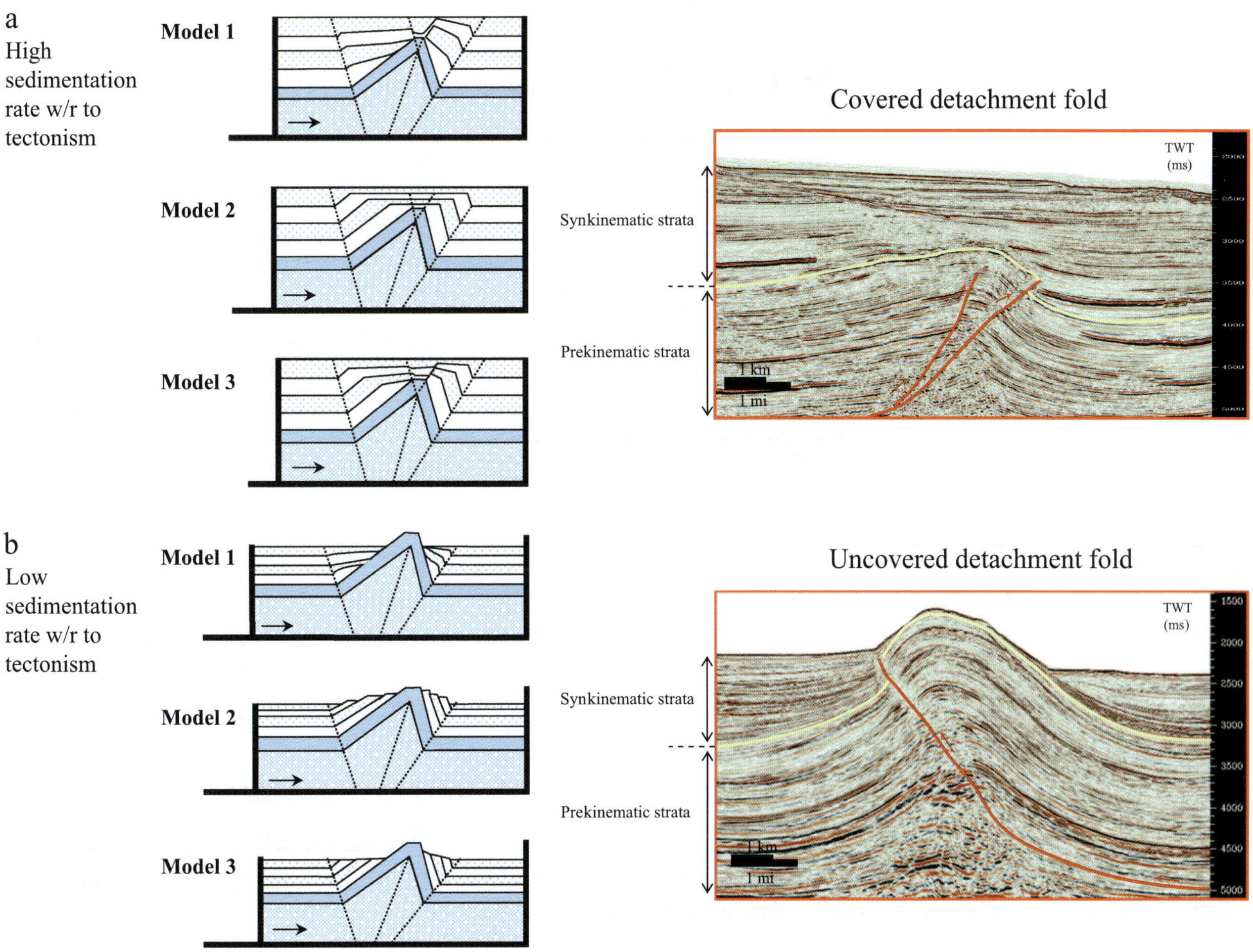

FIGURE 12. Detachment folds developed by different kinematic models. (a) Buried detachment anticlines and (b) emerging noncovered detachment folds. The seismic shows a covered fold crest and a noncovered fold crest, characterized by thinning, onlapping, or exposure of syntectonic strata (modified from Poblet et al., 1997). TWT = two-way traveltime.

One characteristic common to many thrust belts detached on argillaceous material is the dominant presence of asymmetric structures that verge toward the basin; however, according to Rowan et al. (2004), when the fold belts display symmetrical structures, as in the Mexican Ridges, this suggests the possibility of an overpressured detachment zone that facilitates sliding by minimizing friction and inhibiting vergence in a preferred direction.

The analysis of available seismic data for this study supports the presence of overpressure caused by the low velocities of the package comprising the main upper Eocene detachment zone, which indicates high pore pressure as well as the presence of pseudostructures below the detachment zone on seismic time sections caused by velocity effects.

Most of the structures in the fold belt are symmetrical growth folds, although many asymmetrical structures are associated with late reverse faulting on the frontal limb. These structures are interpreted as single detachment folds and then evolved as contraction increases and became asymmetric structures with late-stage reverse faults.

This process of truncation and evolution of detachment folds into detachment folds with reverse faults has been documented by some authors (e.g., Mitra, 2002; Wallace and Homza, 2004) and may be grouped under different concepts, such as transported folds (Jamison, 1987), breakthrough folds (Suppe and Medwedeff, 1990), break-thrust folds (current accepted term) (Fischer et al., 1992), or faulted detachment folds (Mitra, 2002).

Growth Folds in the Mexican Ridges Fold Belt

The facies nomenclature for the sedimentary units of the Mexican Ridges was based on the classification scheme developed by Prather et al. (1998), which recognized three principal classes of seismic facies in the western Gulf of Mexico: chaotic, parallel, and convergent.

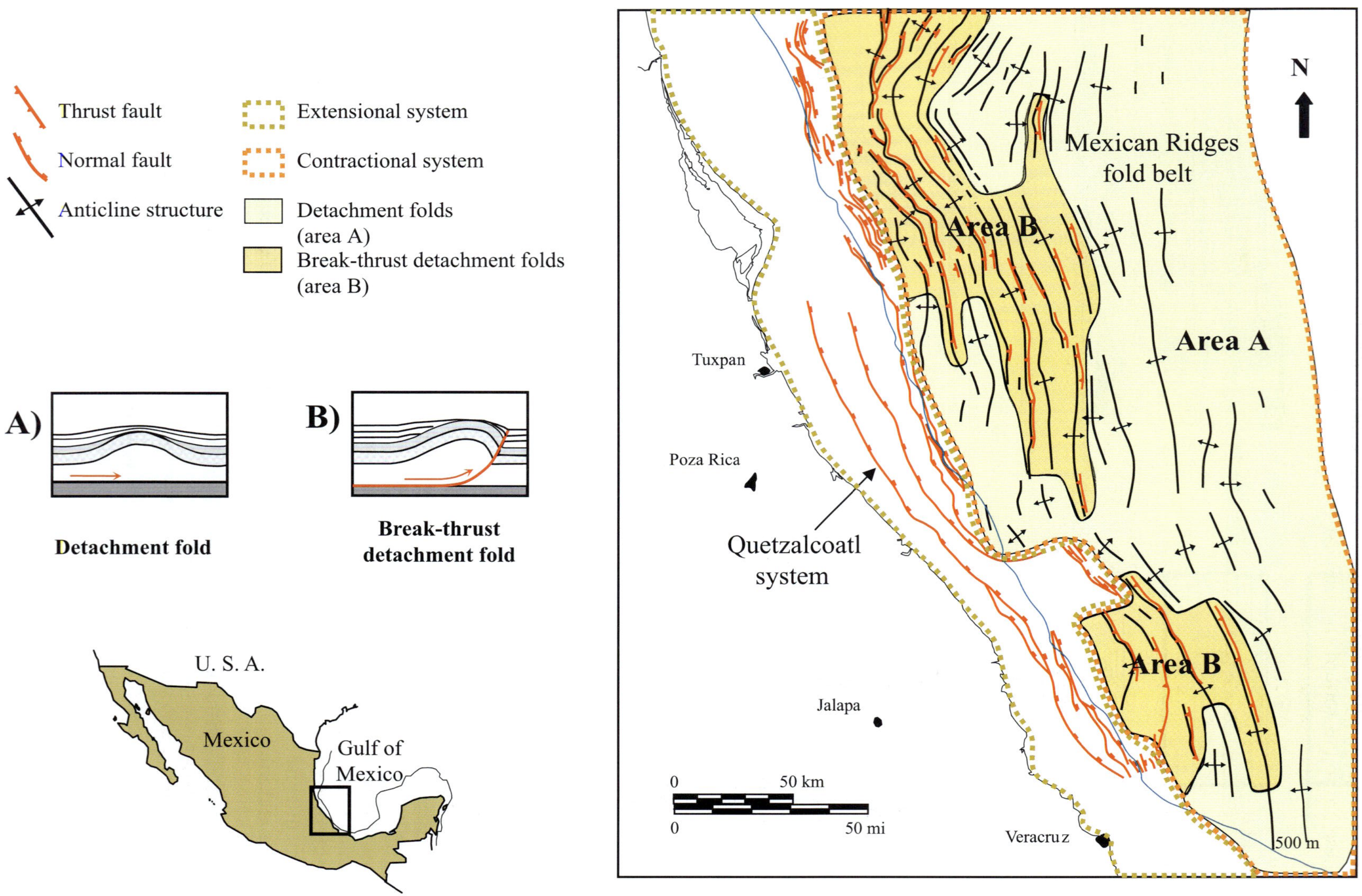

Figure 13. Regional tectonic map with the distribution of detachment fold structural styles (area A) and detachment folds associated with reverse faults (area B) of the Mexican Ridges province. The main contraction in both styles of deformation occurs in the predominantly argillaceous upper Eocene unit, although both detachment levels are identified in younger units.

The chaotic facies comprises seismic units of discontinuous or undulating reflectors with a mounded internal arrangement, commonly bounded at the base by an erosion surface. These are found toward the western margin of the basin and result from the filling of valleys and canyons that cut the shelf and associated continental slope associated with the platform margin of the prograding system that dominated the area during the Tertiary.

The parallel facies is a grouping of regionally continuous reflectors with a uniform thickness that cover local highs in a parallel fashion. This facies is interpreted toward the east in the external part of the study area, comprising the distal part of the unconfined deep-water fan complexes.

The convergent facies is the geometry where the reflectors converge toward the margin of the basin. This facies is common in the syntectonic strata of typical intraslope basins analyzed by Prather et al. (1998) or in the minibasins where structural deformation controls the syntectonic deposition.

The syntectonic sedimentation associated with the contractional deformation that progressively affects the Mexican Ridges displays different geometries and seismic reflector terminations depending on the degree of structural relief. An analysis of the geometry and seismic facies was performed by Román-Ramos et al. (2004) who reported the convergence of the growth strata toward the crests that may be caused by the concordant thinning of the packages or onlap onto an unconformity surface.

In the Mexican Ridges structures, these Neogene synkinematic strata are deposited mostly in the synclines and minibasins where the rapid structural deformation generally confines the sedimentation because of the relatively low sedimentation rate that characterizes this area. A geometric analysis of these growth deposits allows the identification of convergent seismic units toward the axes of the developing anticlines.

The base of the synkinematic deposits is distinguished by their contact with prekinematic strata by seismic reflection termination and stratigraphic thickness. Based on the regional correlation of horizons middle Miocene, upper Miocene, and lower Pliocene from the Quetzalcoatl extensional system to the Mexican Ridges fold belt, we interpret the main phases of contractional deformation of the Mexican Ridges structures. The following describes three contractional fold geometries representing these stages.

Middle Miocene Folds

The oldest detachment folds are in the southern part of the Mexican Ridges fold belt with a middle Miocene inception of growth. The regional seismic section in Figure 14 shows the geometry of three middle Miocene detachment folds with pretectonic strata deposited before the end of the middle Miocene, and the late Miocene to present-day strata are syntectonic. Recognition of the growth limit establishes the age of deformation at the end of the middle Miocene to the present day and is evident from the rotation of sediments on the sea floor and the underlying reflectors. The regional dip toward the center of the basin is only interrupted over the uplifted crest of the structures. The relief of the uplifted crest, the converging syntectonic strata on the limbs of the structure, the regional seismic correlation with the Quetzalcoatl extensional system, and the presence of faulting at the sea floor, in both the extensional system and in the fold belt, indicate the present-day contractional tectonic activity, as well as a low sedimentation rate in relation to the rate of deformation. In this case, no interpreted posttectonic strata are observed.

The deformation of the middle Miocene detachment folds only occurs above the main upper Eocene detachment, which varies in thickness because of increased volume in the nucleus of the structures from which associated reverse faults emerge, principally on the frontal limbs.

From the interpretation of the seismic sections, we know that the detachment folds, which formed since the end of the middle Miocene, may be symmetric or asymmetric, with the latter having reverse faults and a dominant vergence toward the central basin. As seen on the detailed seismic of the central detachment fold in Figure 14, the syntectonic deposits show a progressive rotational onlap. The initial syntectonic units converge by onlapping the back limb while only thinning over the forward limb. As shortening increases, the younger deposits progressively onlap the prior growth strata and are absent over the crest because of the constant structural uplift and slow syndeformational sedimentation rates. The last syntectonic units onlap both fold limbs with an increasing degree of progressive onlap forming depositional systems confined to minibasins between the developing structural trends.

For most of its history, the fold is exposed at the sea floor and its crest is eroded. The sedimentation rate is slow compared to the velocity of deformation, such that the crest of the fold is emergent compared to the minibasins, which have a different fill level. The western minibasin, closest to the coastal sediment source and shelf, captures most of the sediment. Both minibasins are characterized by chaotic seismic facies with discontinuous, chaotic reflectors that onlap erosional surfaces, i.e., the middle Miocene, and can be interpreted as valley and submarine canyon fill controlled by the developing folds.

The structural style is typical of a detachment fold associated with late-stage reverse faulting that can be classified as a model 1 anticline according to Poblet et al. (1997), as shown in Figure 12b.

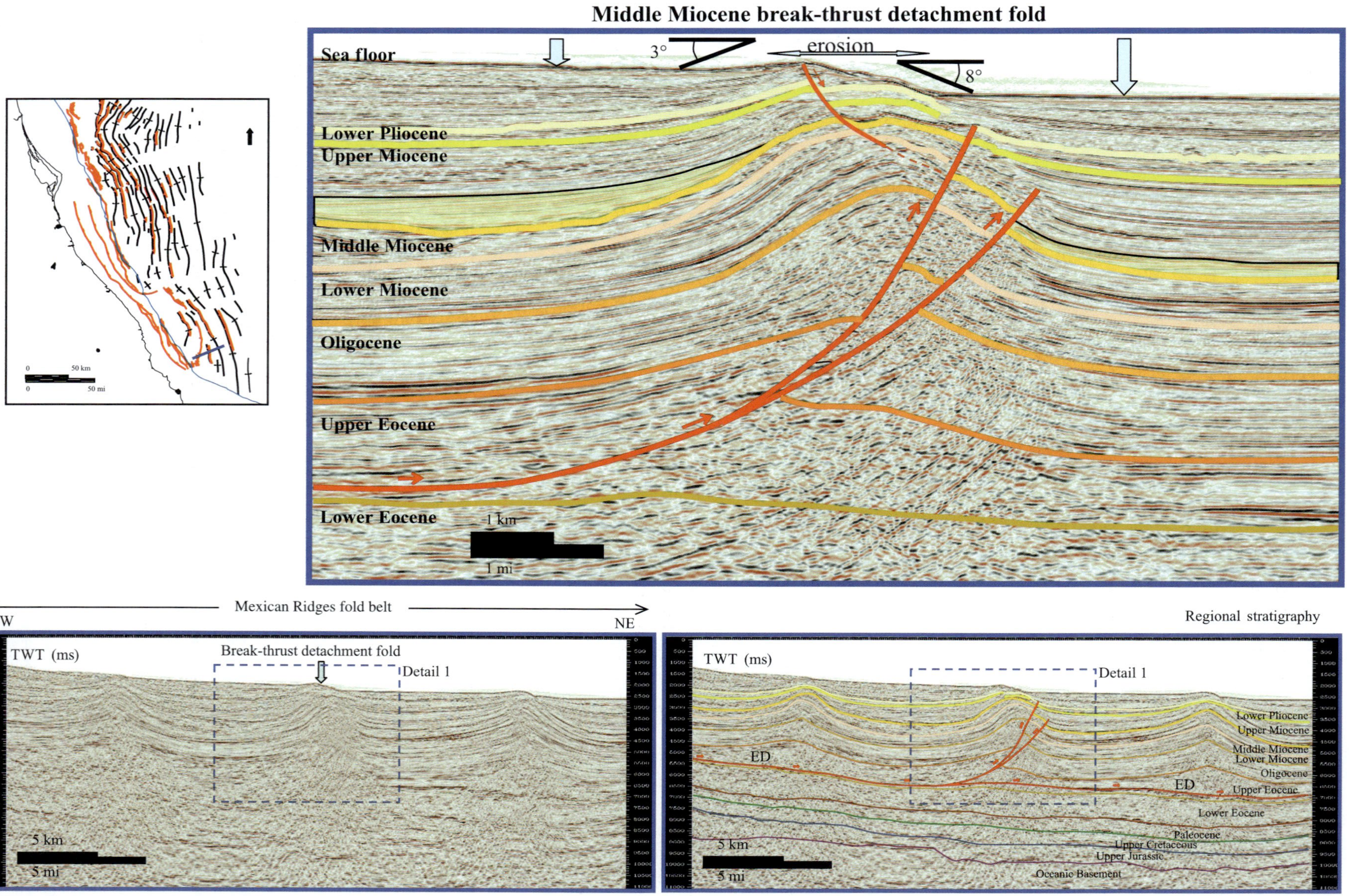

Figure 14. Interpreted and uninterpreted southwest-northeast–oriented seismic lines. The detailed seismic section shows a noncovered detachment fold associated with reverse faulting, with a middle Miocene lower limit to the growth strata. This fold is at a water depth of 1470 m (4823 ft) with a maximum wavelength of 9 km (5.5 mi). Upper Miocene, lower Pliocene, and Holocene growth strata show progressive onlapping. TWT = two-way traveltime; ED = Eocene detachment.

Late Miocene Folds

The boundary between the pretectonic and syntectonic units comprises the upper Miocene horizon and defines the initial contractional deformation that first appears at the end of the late Miocene. This type of structure is one of the growth detachment folds distributed in the western part of the Mexican Ridges. The pretectonic strata with constant thickness occur below the upper Miocene growth limit, and the syntectonic strata, starting with the upper Miocene, reach the sea floor. No posttectonic stage strata are observed.

The seismic section in Figure 15 shows that the detachment folds that began to form at the end of the late Miocene are generally symmetric and free of reverse faults, although some asymmetric structures and associated late-stage reverse faults with minor propagation are present. These folds display a single, upper Eocene detachment surface.

The deformation age is established on the basis of converging geometries of the syntectonic strata toward the crest of the structures. A detailed seismic analysis of the type detachment fold in the eastern sector reveals that the initial syntectonic deposits mainly show thinning and, to a lesser degree, onlap over both limbs of the fold during the first increments of deformation. With continued structuring, growth packages progressively onlap older growth strata on both limbs. The sedimentation rate is low compared to the deformation velocity. A different level of fill in the minibasins is observed, given their position relative to the sediment supply systems and their greater water depth compared to the crest of the fold. For these reasons, this is considered to be an unburied detachment fold. The structural model corresponds to an unburied fold that initially developed under either model 1 or 3. This may also agree with incipient reverse faulting, although the presence of some recent synkinematic units onlapping older strata over the uneroded crest of the structure points to the fold being of model 1 type (Poblet et al., 1997) wherein the onlap is caused by a slow sedimentation rate compared to the deformation rate of the pretectonic units.

Early Pliocene Folds

The youngest structures in the Mexican Ridges are detachment folds with growth beginning in the Lower Pliocene as determined from the contact with pretectonic units. As in the case of the older structures, no posttectonic units are observed, thereby indicating that the activity of structures has been from the end of the early Pliocene until the present day.

According to the available seismic data, many folds of this type are found in the east-central part of the Mexican Ridges. The structures are wide with lower relief when compared to the previously described middle and late Miocene folds, and they may display two detachment zones: the main upper Eocene zone and a shallower zone interpreted to lie between the Oligocene and Miocene units.

From the seismic interpretation of the early Pliocene folding, structures are generally symmetric and unfaulted although some are asymmetric with reverse faults. Some reverse faults show opposing vergence as shown in the detailed seismic section of the western fold in Figure 16.

In this example of an early Pliocene detachment fold, the first syntectonic units display convergence by thinning on both fold limbs followed by a progressive onlap of the older growth strata. More syntectonic units thin over the crest of the structure. The syntectonic sedimentation rate in these pelagic strata is slow in relation to the uplift of the fold principally because of the distance of the structure from the sediment source area on the shelf. This structure is an early Pliocene detachment fold with related late-stage reverse faulting with its main detachment surface at the upper Eocene level, as well as a back thrust detached in the shallower Oligocene–Miocene level. This corresponds, in part, to a noncovered detachment fold of the model 1 type of Poblet et al. (1997) distinguished by the progressive onlap of the syntectonic strata indicating the rotation of the limbs as the structure developed.

DISCUSSION

The Mexican Ridges fold belt is a Neogene contractional system located in the deep-water part of the southwestern Gulf of Mexico. Through the interpretation of diverse 2-D seismic reflection surveys, two structural styles were defined in the southern part of the system: growth detachment folds and growth detachment folds with late-stage reverse faulting.

Nonfaulted folding with symmetric detachments may represent the initial stage of deformation that, as deformation continues, evolves into asymmetric folding with reverse faults resulting in lower amplitude folds with greater structural closure and late-stage fault deformation. This mechanism is evident laterally toward the end of some structural trends where the reverse fault displacement decreases until disappearing, thereby reflecting the temporal and spatial variation in style from symmetric detachment folds toward the end of the fold to asymmetric detachment folds in the central part of the structure where contraction is greater, thus favoring rupture with the presence of reverse faults in the folds.

In fault propagation and fault-bend fold models (Suppe et al., 1992), the folding and reverse faulting occur at the same time; however, in detachment folds associated with faults in the Mexican Ridges, this synchronous behavior does not exist and the folding precedes, and is unrelated to the faulting. In other words, the folding begins without any faulting and is then modified

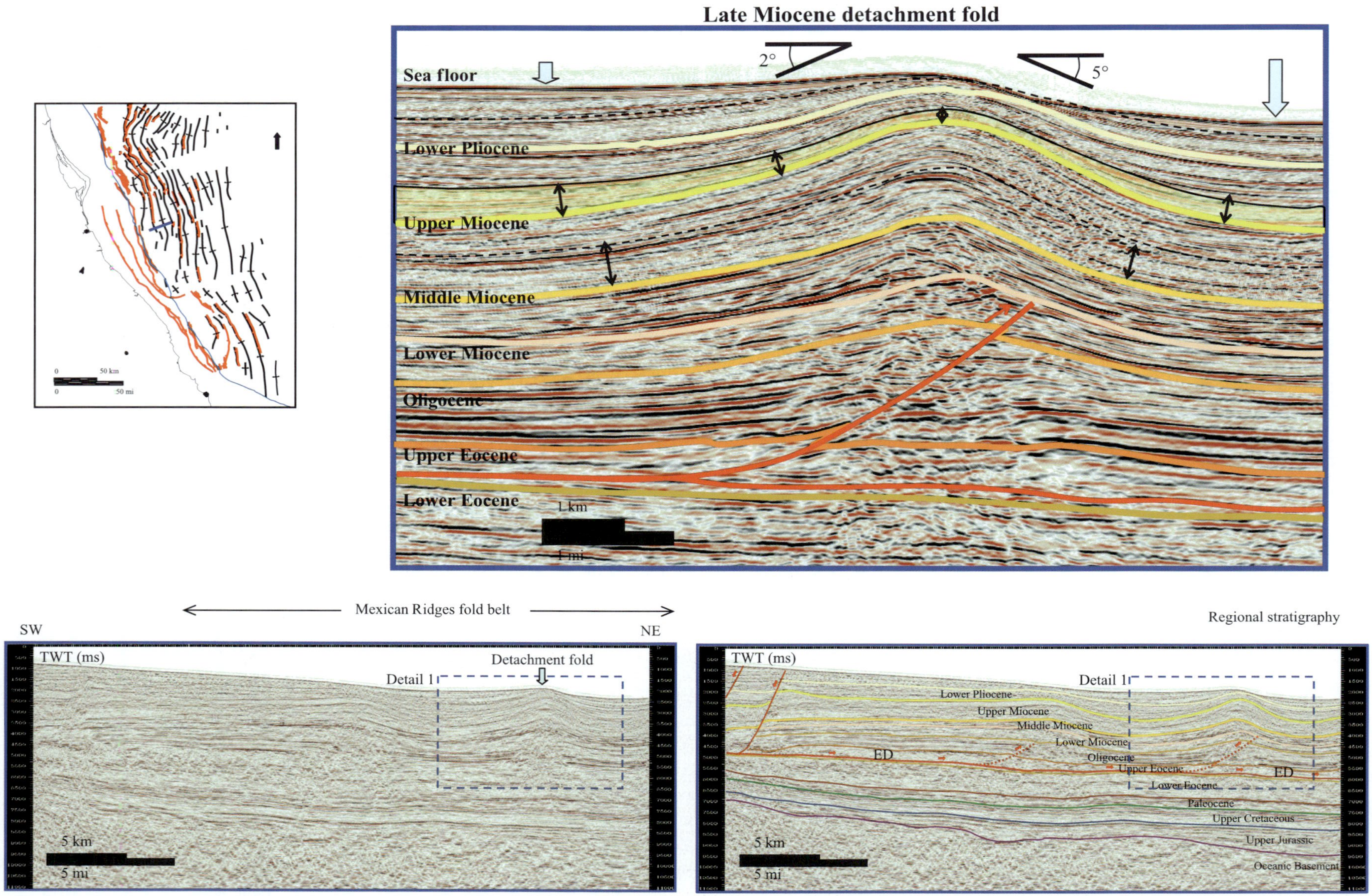

FIGURE 15. Southwest-northeast–oriented seismic section of a symmetric detachment fold with an upper Miocene lower growth limit. The hidden reverse fault does not affect the Pliocene to Holocene syntectonic strata, which thin toward the crest of the structure, although some sequences are onlapping the previous units. This fold is at a water depth of 1380 m (4527 ft) with an average wavelength of 11 km (6.8 mi). TWT = two-way traveltime; ED = Eocene detachment.

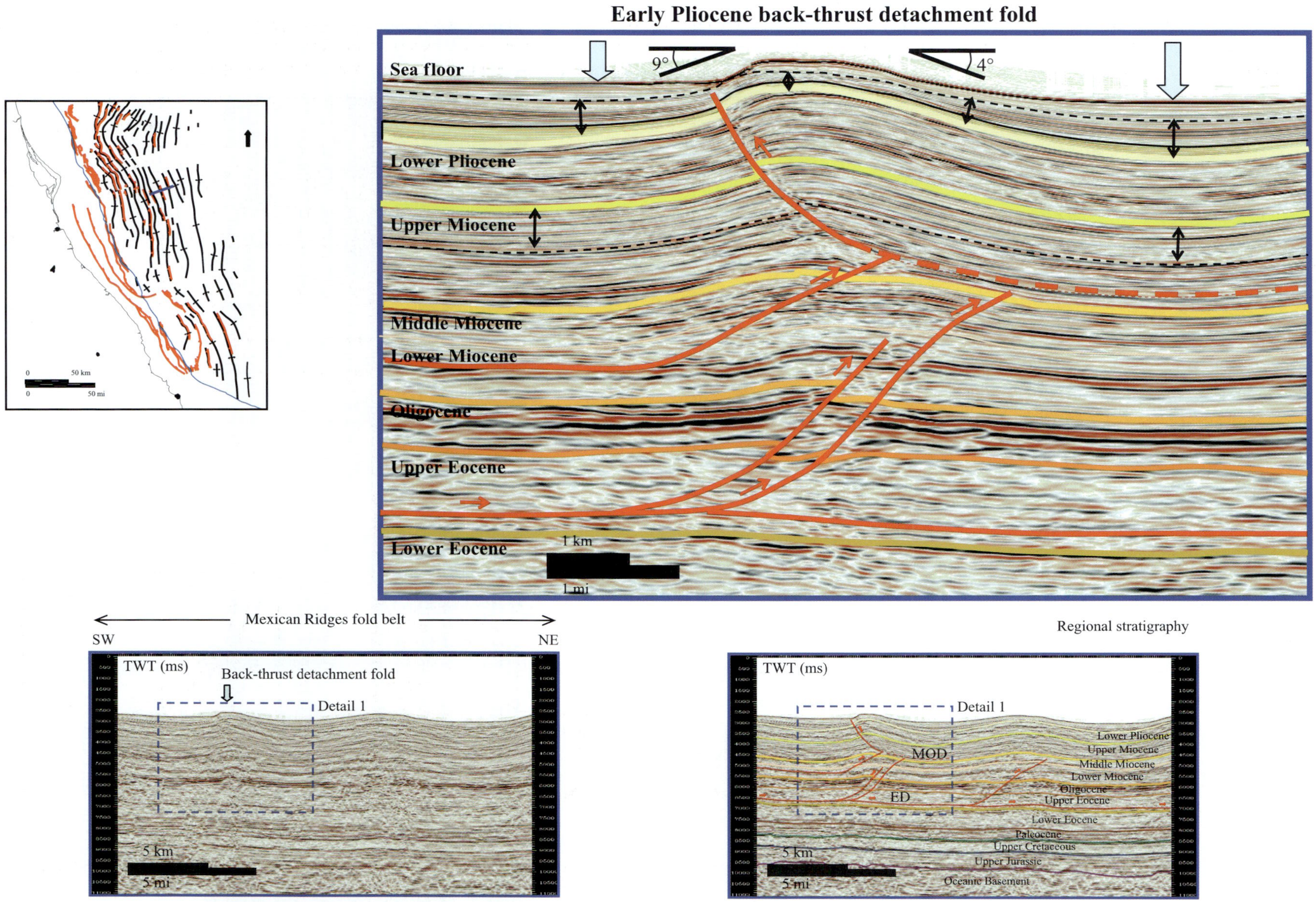

FIGURE 16. Southwest-northeast–oriented seismic section of a moderately asymmetric detachment fold associated with a late-stage reverse fault with opposing vergence. The lower growth limit is lower Pliocene; the syntectonic deposits converge by onlapping in both limbs toward the crest of the structure. This fold is at a water depth of 1930 m (6332 ft) with an average wavelength of 8 km (4.9 mi). TWT = two-way traveltime; MOD = Miocence-Oligocene detachment; ED = Eocene detachment.

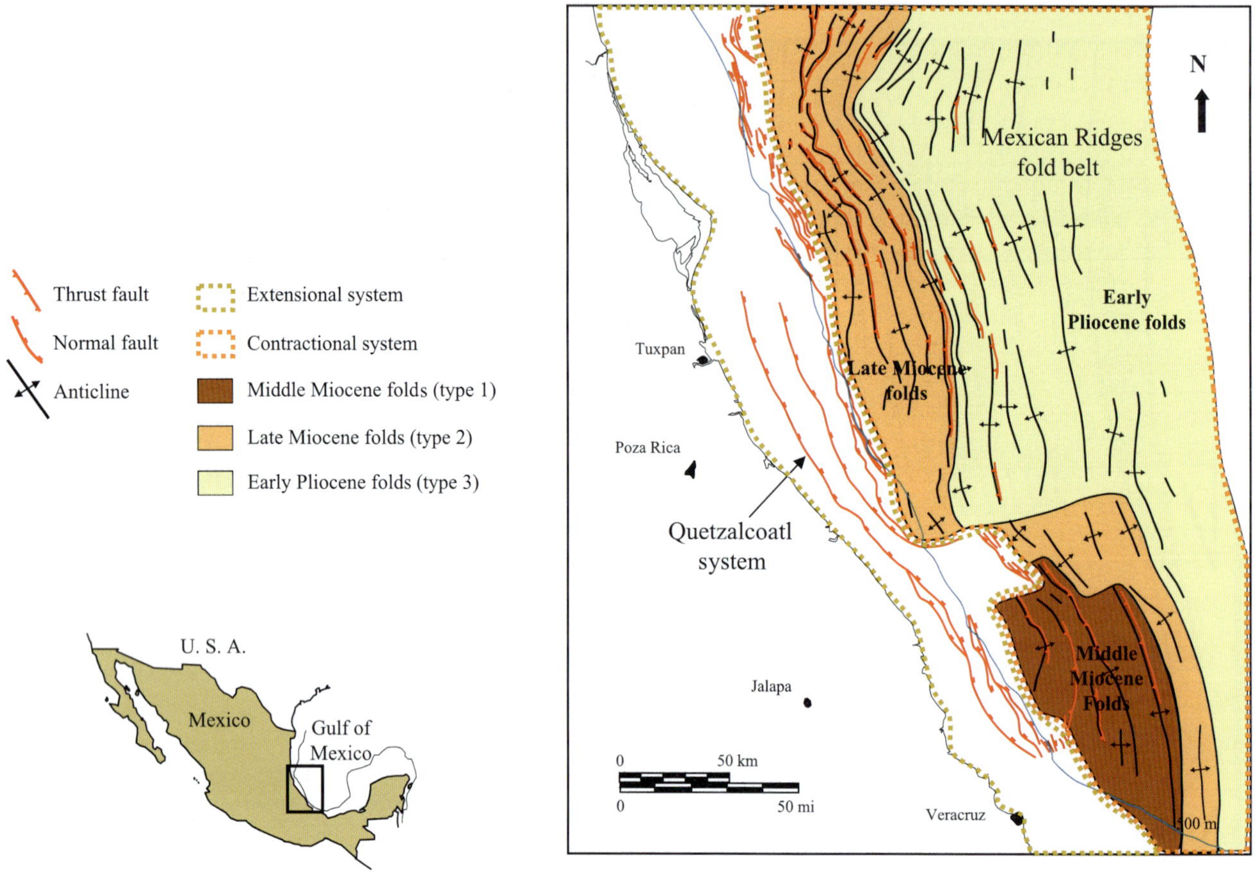

FIGURE 17. Tectonic map showing the age distribution of contractional fold deformation in the Mexican Ridges. Middle Miocene growth folds (type 1) cluster south of the fold belt; upper Miocene growth folds (type 2) are the structures close the Quetzalcoatl extensional system; lower Pliocene folds (type 3) are located in the external part of the study area far from the extensional system.

by a late-stage, out-of-sequence rupture on a reverse fault ramp from the detachment zone with propagation deformation of the rupture.

Three recognized initiation ages for the contractional deformation for the detachment folds in the Mexican Ridges exist: middle Miocene, late Miocene, and early Pliocene, as defined by the growth limits under the synkinematic strata.

Folds with middle Miocene growth are grouped toward the south of the fold belt (Figure 17) and display advanced structural development characterized by reverse faulting, basinward vergence, and erosion of the fold crest. Folds with late Miocene growth show minor structural deformation, are occasionally associated with reverse faults, and may be symmetric or asymmetric with no erosion of their crests. In the northern part, these folds form five or six structures near the extensional shelf system (Figure 17), and the two westernmost fold trends close to the Quetzalcoatl extensional system were completely buried during their development.

Detachment folds with a lower Pliocene growth are found in the eastern external zone of the study area (Figure 17) where they have suffered less compression than the folds closer to the extensional system. These folds are generally symmetric or slightly asymmetric and have greater amplitude with minor structural relief, with most lacking late-stage reverse faults.

The table in Figure 18 lists the main Tertiary tectonic events related to the Quetzalcoatl extensional system and the Mexican Ridges, as well as the principal characteristics of their structural styles, resulting hydrocarbon trap types, and depositional systems present in each province. Also shown are the corresponding deformation stages, growth units, and detachment zones in both tectonic provinces.

The syntectonic stage for the extensional province has been identified from growth strata ranging from the middle Miocene to present, which deform Paleogene and younger units, above a regional detachment. In this zone, the main structural styles are characterized by synthetic and antithetic faults that form grabens and half grabens and fault-bounded extensional blocks. These styles respectively constitute anticlinal rollover traps and normal fault traps.

The regional detachment in the fold belt is well defined as being at the upper Eocene level and below deformed Oligocene to middle Miocene pretectonic units

FIGURE 18. Diagram of the principal Tertiary tectonic events of the southwestern Gulf of Mexico.

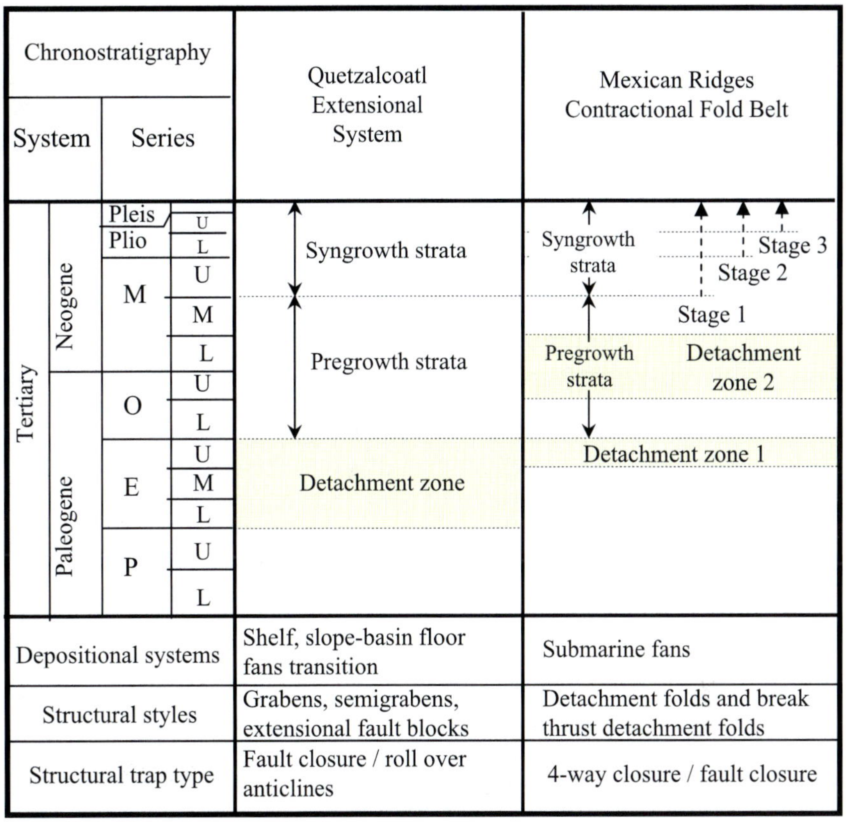

based on seismic correlations. The stages of initial deformation are recognized continuing to the present: middle Miocene, late Miocene, and early Pliocene. In the Mexican Ridges, the main structural styles are detachment folds, with or without associated reverse faults that result in potential hydrocarbon traps with four-way closures or three-way closures against faults. This tectonic province is dominated by turbiditic environments with the development of submarine fans with channelized to lobate facies fed by canyon complexes from the prograding shelf system.

CONCLUSIONS

The Mexican Ridges fold belt in the deep-water Gulf and the Quetzalcoatl extensional system on the shelf and upper continental slope are two tectonic provinces of the eastern Gulf of Mexico that developed simultaneously because of gravity sliding from the shelf. The sedimentary column beneath the Mexican Ridges is composed of Mesozoic rocks that are probably argillaceous carbonates and Tertiary siliciclastics, which together with the remaining Tertiary column reach thicknesses of up to 15 km (9.3 mi).

The Quetzalcoatl extensional system comprises a group of listric normal faults active from the middle Miocene to present, gliding mainly on a regional Paleogene detachment surface. Displacement above the growth faults develops diverse structural styles, including domino type, graben systems, and listric growth fault systems that generate rollover anticlines.

The Mexican Ridges consists of a gravitational system of elongated folds trending approximately north–south associated with a detachment zone in the upper Eocene shales. The Mexican Ridges correspond to growth folds. The geometry of the folds indicates two main structural styles: detachment folds and detachment folds with reverse faulting. The detachment zone is identified as upper Eocene shales. Syntectonic sediments deposited in the folds provide important information for the identification of the mechanisms, kinematic evolution, and ages of deformation to determine the timing and migration of structural traps.

The detachment folds are generally symmetric, although some structures are asymmetric, inclined to the east, and affected by late-stage reverse faults present in sediments up to the sea floor. The importance of the reverse faulting is that they potentially constitute migration pathways toward reservoir units in the anticlinal structures. The dominant deformational mechanism in the Mexican Ridges is detachment folding with variably dipping limbs and constant limb length because of the progressive onlap of growth strata on generally nonburied or emergent fold limbs.

The secondary mechanism affecting the folded structures is the late-stage reverse faulting. The structural evolution of the Mexican Ridges folds is the transition from detachment folding to folding affected by reverse faults because of the increment of contraction provoked by the evolution from symmetric to asymmetric structures with lower amplitude and greater structural relief.

The deformation age of the studied group of folds varies from the end of the middle Miocene to present, with three established contractional deformation stages or periods (middle Miocene, late Miocene, and early Pliocene continuing to the present day) with the youngest and lowest amplitude folds toward the center of the basin.

The petroleum potential of the Mexican Ridges is represented by the large number of identified contractional structures that comprise important exploratory prospects with the potential to contain gas and light oil. The interpreted sedimentary facies and geological history, as well as the data obtained from wells drilled on the continental shelf, support the existence of Tertiary turbidite sand packages as the principal exploration objectives.

REFERENCES CITED

Ambrose, W. A., et al., 2005, Neogene tectonic, stratigraphic, and play framework of the southern Laguna Madre-Tuxpan continental shelf Gulf of Mexico: AAPG Bulletin, v. 89, no. 6, p. 725–751.

Aranda-García, M., 1999, Evolution of Neogene contractional growth structures, southern Gulf of Mexico: Master thesis, University of Texas at Austin, 169 p.

Blickwede, J. F., and T. A. Queffelec, 1988, Perdido fold belt: A new deep-water frontier in western Gulf of Mexico (abs.): AAPG Bulletin, v. 72, p. 163.

Bryant, W. R., J. W. Antoine, M. Ewing, and B. Jones, 1968, Structure of Mexican continental shelf and slope, Gulf of Mexico: AAPG Bulletin, v. 52, p. 1204–1228.

Buffler, R. T., F. J. Shaub, J. S. Watkins, and J. L. Worzel, 1979, Anatomy of the Mexican Ridges, southwestern Gulf of Mexico, *in* J. S. Watkins, L. Montadert, and P. W. Dickerson, eds., Geological and geophysical investigations of continental margin: AAPG Memoir 29, p. 319–327.

De Cserna, Z., 1984, Margen continental de colisión activo en la parte suroccidental del Golfo de México: Revista del Instituto de Geología, Universidad Nacional Autónoma de México, v. 5, no. 2, p. 255–260.

Feng, J., 1995, Post Mid-Cretaceous seismic stratigraphy and depositional history, deep Gulf of Mexico: Ph.D. dissertation, University of Texas at Austin, 253 p.

Feng, J., and R. T. Buffler, 1994, Turbidite systems and their temporal and spatial distribution, deep Gulf of Mexico Basin: Gulf Coast Section SEPM Foundation 15th Annual Research Conference Submarine Fans and Turbidite Systems, December 4–7, p. 115–125.

Fischer, M. P., N. B. Woodward, and M. M. Mitchell, 1992, The kinematics of break-thrust folds: Journal of Structural Geology, v. 14, p. 451–460.

Garrison, L. E., and R. G. Martin, 1973, Geologic structures in the Gulf of Mexico Basin: U.S. Geologic Survey, Professional Paper 773, 85 p.

Guzmán, A. E., and B. Márquez-Dominguez, 2001, The Gulf of Mexico Basin south of the border, *in* M. W. Downey, ed., Petroleum provinces of the 21st Century: AAPG Memoir 74, p. 337–351.

Jamison, W. R., 1987, Geometric analysis of fold development in overthrust terranes: Journal of Structural Geology, v. 9, p. 207–219.

Jáuregui, R. H., R. A. Téllez, and J. J. Valencia, 1997, La sedimentación sintectónica. herramienta de caracterización de evolución de una cuenca petrolera: Gaceta de la Asociación Mexicana de Geólogos Petroleros, v. 4, no. 3, p. 1–3.

Jennette, D., T. Wawrzyniec, K. Fouad, D. B. Dunlap, J. Meneses-Rocha, F. Grimaldo, R. Muñoz, D. Barreda, C. T. Williams-Rojas, and A. Escamilla-Herrera, 2003, Traps and turbidite reservoir characteristics from a complex and evolving tectonic setting, Veracruz Basin, southeastern Mexico: AAPG Bulletin, v. 87, no. 10, p. 1599–1622.

Medwedeff, D. W., 1989, Growth fault-bend folding at southeast Lost Hills, San Joaquin Valley, California: AAPG Bulletin, v. 73, p. 54–67.

McClay, K. R., 1992, Glossary of thrust tectonics terms, *in* K. R. McClay, ed., Thrust tectonics: London, Chapman & Hall, 447 p.

Mitra, S., 2002, Structural models of faulted detachment folds: AAPG Bulletin, v. 86, no. 9, p. 1673–1694.

Patiño-Ruiz, J., L. E. Salomón-Mora, and M. Aranda-García, 2004, Estilos estructurales de los cinturones plegados perdido y Cordilleras Mexicanas (abs.): AAPG Search and Discovery, http://www.searchanddiscovery.net/documents/abstracts/2004intl_cancun/index.htm#qr (accessed August 27, 2009).

Pew, E., 1982, Seismic structural analysis of deformation in the southern Mexican Ridges: Master thesis, University of Texas at Austin, 102 p.

Poblet, J., K. R. McClay, F. Storti, and J. A. Muñoz, 1997, Geometries of syntectonic sediments associated with single-layer detachment folds: Journal of Structural Geology, v. 19, no. 3–4, p. 369–381.

Prather, B. E., J. R. Booth, G. S. Steffens, and P. A. Craig, 1998, Classification, lithologic calibration and stratigraphic succession of seismic facies of intraslope basins, deep-water Gulf of Mexico: AAPG Bulletin, v. 82, no. 5A, p. 701–728.

Román-Ramos, J. R., L. E. Salomón-Mora, M. Aranda-Garcia, C. Rosas-Lara, and T. Cárdenas-Dominguez, 2001, Structural styles and gasiferous potential at the western slope of the Gulf of Mexico (abs.): Cuarta Conferencia Internacional Conjunta Asociación Mexicana de Geólogos Petroleros and AAPG, Veracruz, México, 6 p.

Román-Ramos, J. R., L. E. Salomón-Mora, C. Rosas-Lara, M. Aranda-Garcia, J. Morales-Marin, J. C. Flores-Zamora, E. Sandoval-Solís, and K. Rosas-Lara, 2004, Depositional systems and traps of shallow and deep-water central Gulf of Mexico (abs.): AAPG Search and Discovery, http://www.searchanddiscovery.net/documents/abstracts/2004intl_cancun/index.htm#qr (accessed August 27, 2009).

Rowan, M. G., F. J. Peel, and B. C. Vendeville, 2004, Gravity-driven fold belts on passive margins, *in* K. R. McClay, ed., Thrust tectonics and hydrocarbon systems: AAPG Memoir 82, p. 157–182.

Salomón-Mora, L. E., M. Aranda-García, and J. R. Román-Ramos, 2004, Plegamiento contraccional y sedimentación sintectónica en las Cordilleras Mexicanas: Boletín de la Asociación Mexicana de Geólogos Petroleros, v. 51, no. 2, p. 5–21.

Shaub, F. J., R. T. Buffler, and J. G. Parson, 1984, Seismic stratigraphic framework of the deep central Gulf of Mexico: AAPG Bulletin, v. 68, no. 11, p. 1790–1802.

Storti, F., and J. Poblet, 1997, Growth stratal architectures associated to decollement folds and fault-propagation folds: Inferences on fold kinematics: Tectonophysics, v. 282, p. 353–373.

Suppe, J., 1983, Geometry and kinematics of fault-bend folding: American Journal of Science, v. 283, p. 684–721.

Suppe, J., and D. A. Medwedeff, 1990, Geometry and kinematics of fault-propagation folding: Eclogae Geologicae Helvitae, v. 83, p. 409–454.

Suppe, J., S. T. Chou, and S. C. Hook, 1992, Rates of folding and faulting determined from growth strata, *in* K. R. McClay, ed., Thrust tectonics: London, Chapman & Hall, 447 p.

Trudgill, B. D., M. G. Rowan, J. C. Fiduk, P. Weimer, P. E. Gale, B. E. Korn, R. L. Phair, W. T. Gafford, G. R. Roberts, and S. W. Dobbs, 1999, The Perdido fold belt, northwestern deep Gulf of Mexico: Part 1. Structural geometry, evolution and regional implications: AAPG Bulletin, v. 83, no. 1 and 4, p. 88–113 and 578–612.

Wallace, W. K., and T. X. Homza, 2004, Detachment folds versus fault-propagation folds, and their truncation by thrust faults, *in* K. R. McClay, ed., Thrust tectonics and hydrocarbon systems: AAPG Memoir 82, p. 324–355.

Wawrzyniec, T., et al., 2003, Cenozoic deformational styles of the Laguna-Madre-Tuxpan shelf and Mexican Ridges fold belt, Mexico: Gulf Coast Association of Geological Societies/Gulf Coast Section SEPM Transactions, v. 53, p. 846–858.

Zapata, T. R., and R. W. Allmendinger, 1996, Growth stratal records of instantaneous and progressive limb rotation in the Precordillera thrust belt and Bermejo Basin, Argentina: Tectonics, no. 15, p. 1065–1083.

Ortuño, A. S., E. Soriano M., C. Romero M., J. Rosales R., E. Bahena R., S. Hernández G., and A. Cárdenas, 2009, Two-dimensional and three-dimensional numerical simulation of petroleum systems approaching the deep-water gulf of Mexico (Kayab Area, Campeche Sound, Mexico): Definition of thermally mature and prospective areas, *in* C. Bartolini and J. R. Román Ramos, eds., Petroleum systems in the southern Gulf of Mexico: AAPG Memoir 90, p. 117–136.

6

Two-Dimensional and Three-Dimensional Numerical Simulation of Petroleum Systems Approaching the Deep-Water Gulf of Mexico (Kayab Area, Campeche Sound, Mexico): Definition of Thermally Mature and Prospective Areas

Salvador Ortuño Arzate, Esteban Soriano M., Cuauhtémoc Romero M., Joel Rosales R., and Edgar Bahena R.
Instituto Mexicano del Petróleo, Col. San Bartolo Atepehuacan, Mexico D.F., Mexico

Sergio Hernández G. and Alejandro Cárdenas
Petróleos Mexicano (PEMEX), Exploración y Producción, Región Marina Noreste, Ciudad del Carmen, Campeche, Mexico

ABSTRACT

By integrating geological, geochemical, and geophysical data and using Temispack software, a two-dimensional (2-D) and three-dimensional (3-D) geological-numerical model is being developed for the sedimentary basins and petroleum systems in the Kayab offshore area.

The elements and processes of this petroleum system are intimately tied to its geodynamic history. The Oxfordian, Kimmeridgian, and Tithonian series contain source rocks buried to depths of between 4000 and 7500 m (13,123 and 24,606 ft). These sedimentary units are confined between autochthonous salt horizons, and allochthonous walls and intrusive bodies of salt. Salt is therefore a critical factor in the maturation of organic material (kerogen) in these strata. Salt-bounded subbasins are identified in which hydrocarbons were generated. The proximity to basement and the high thermal conductivity of the salt have generated liquid hydrocarbons from these source rocks. The numerical simulation of the

DOI:10.1306/13191081M903333

shallower subbasins indicates that thermally immature conditions prevail, as in the case of the large Tunich structure.

Emphasis is placed on the importance of compression and halokinesis in focusing hydrocarbon migration toward Cretaceous breccia and limestone reservoirs within compressional stage anticlines and against salt wall structures.

The 2-D and 3-D numerical simulation permits the definition of drainage areas and petroleum accumulations in the Kayab area. In a preliminary manner, it has been possible to define the prospective areas and initially quantify the amount of trapped hydrocarbons. A hierarchy of economically viable prospects has been derived from the numerical simulation in this frontier region. These studies are fundamental in understanding this part of the marine region, presently characterized by heavy oil (Cantarell and Ku-Maloob-Zaap), and in predicting the existence of an important new heavy oil province in the deep-water Gulf of Mexico.

INTRODUCTION

Recent exploration strategies in the Gulf of Mexico have directed studies and activities toward the continental slope and deeper water areas in the eastern part of Campeche Sound. These exploratory activities require regional studies, including basin modeling and petroleum systems, to evaluate exploration risk and identify exploration opportunities in this area. As the first step, exploration activities have begun in the Kayab area and adjacent deeper waters (200–1000 m, 656–3281 ft).

A basic geological model has been developed for the Kayab area, followed by numerical modeling of the evolution of the basin and its petroleum systems, including its thermal history, subsidence history, and kerogen maturation. The goal of this process is to understand the active petroleum systems and define the economic opportunities and petroleum prospects of the Kayab area. Numerical models employed the Temispack two-dimensional (2-D) and three-dimensional (3-D) modeling and simulation program (Ungerer et al., 1990; Rudkiewicz et al., 2000; Schneider et al., 2000), developed by the French Petroleum Institute and distributed by Beicip.

The basis for constructing the numerical models was the 3-D seismic data and wells drilled in the Kayab area. Information from seismic lines and wells adjacent to the Kayab area was also used, along with regional magnetometry, reports, and publications covering Campeche Sound. Among the most important studies conducted in this region is that of Ángeles (1982, 1988, 1993, 1996a, b), as well as numerous unpublished internal reports that have established the basic stratigraphic nomenclature for Campeche Sound. We also refer to studies by Ángeles et al. (1992) and Ángeles and Ortuño (1994) covering the stratigraphy and petroleum geology of this region, and the work of Araujo et al. (1986), Bello (1991), Ornelas et al. (1991), and Ornelas et al. (1994). Important work on the structural geology and tectonics of the area is that of Meneses (1980), Ramírez et al. (1996), and Salomón (1999). More recently, the structural aspect of Campeche Sound has been addressed by Mitra et al. (2005, 2006) on Cantarell-Sihil and Ku-Maloob-Zaap, respectively. Finally, in relation to petroleum geology and geochemistry of petroleum systems in Campeche Sound, we mention the fundamental work by Holguín and Romero (1982), Bañuelos (1993), García (1993), Hernández (1993), Medrano and Romero (1993), Ortega et al. (1995), Medrano et al. (1996), Romero et al. (1996, 2001, 2004), Guzmán and Mello (1999), Santamaría (2000), Santamaría and Horsfield (2000), and Guzmán et al. (2001) and the internal compilations conducted by the Instituto Mexicano del Petróleo (IMP) (2000), "Atlas de las cuencas sedimentarias terciarias y mesozoicas de México," and by the Región Marina Noreste (RMNE) (2002), "Sistemas petroleros en el área Tunich-Chilam." Most recently, Holguín et al. (2005) have presented a review of hydrocarbon seeps in the southern Gulf of Mexico.

GENERAL GEOLOGIC CONTEXT

From the geological perspective, the Campeche marine region, as part of the Maya terrane, originated during the rifting and spreading phase of the Gulf of Mexico. Its basement consists of thinned transitional crust consisting of schists and granitoids intruded by hypabyssal calcalkaline igneous bodies as determined from potential field and geochemical and petrogenetic studies by the Mexican Petroleum Institute (2000). This was an area of internal and external shelf deposition during the Cretaceous and finally became the locus of Cenozoic siliciclastic deposition. This regional framework controlled the Mesozoic and Cenozoic geodynamic evolution of the Campeche Marine area, including its structural and tectonosedimentary characteristics.

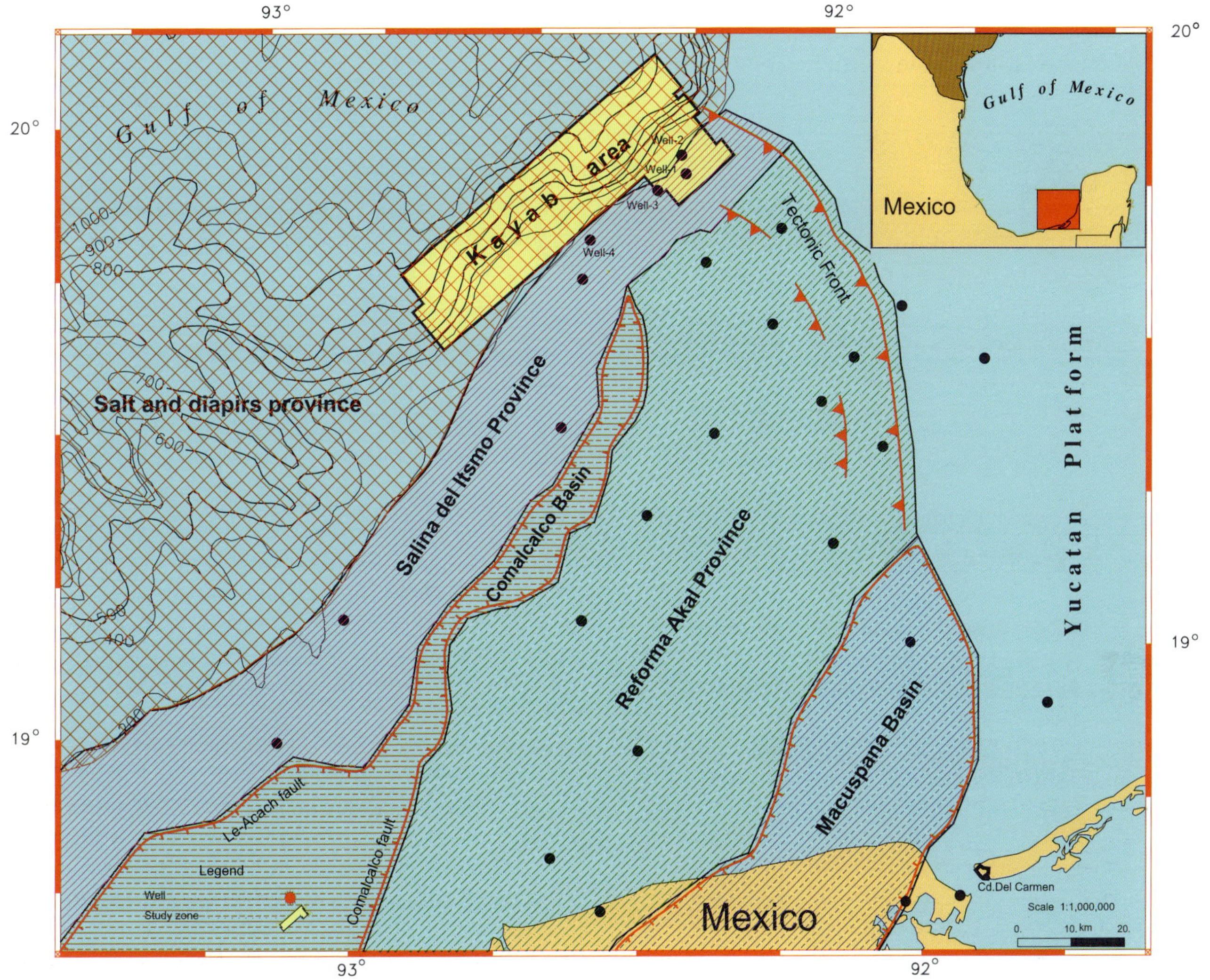

FIGURE 1. Location of the Kayab area in the Campeche Sound.

Figure 1 shows the location of the study area within the context of the geological provinces and tectonic entities of the Gulf of Mexico.

Sedimentary units overlying this basement were deposited throughout the entire Mesozoic and Cenozoic history of the Gulf of Mexico, as seen in the stratigraphic column in Figure 2. The oldest unit, which has not been penetrated by wells in the Marine Region, is inferred to be a siliciclastic sequence of continental origin (red beds) deposited during the initial stages of rifting. This continental sedimentary sequence overlies the basement and is probably of Middle Jurassic age with an average thickness of about 1000 m. This unit is overlain by an evaporite sequence (principally salt) of Callovian age whose thickness ranges from 500 to 4000 m (1640 to 13,123 ft) because of tectonic deformation. The Oxfordian series also varies greatly in thickness, generally between 500 and more than 1500 m (1640 and more than 4921 ft). Some wells have only reached the upper part of this sequence, a facies with source rock characteristics. In contrast, the Kimmeridgian and Tithonian series are thinner, each with average thicknesses of 200–300 m (656–984 ft). Of these, the Tithonian comprises the area's most important hydrocarbon-generating unit. The Cretaceous strata consist essentially of carbonates with thicknesses ranging from 600 to 1000 m (1968 to 3281 ft). Finally, siliciclastic Paleogene and Neogene units vary in thickness depending on their tectonosedimentary evolution during compression and salt tectonics. The average thickness of the Paleogene is about 700 m (2296 ft), and about 4500 m (14,764 ft) for the Neogene (Ángeles, 1996a; Ortuño et al., 2004). A representative Mesozoic–Cenozoic sedimentary column for the study area and Campeche Sound is shown in Figure 2.

Respecting structural styles and tectonic evolution, it is possible to distinguish between the deformation because of compression and that related to halokinesis, the latter consisting of enormous salt intrusions cutting across the Mesozoic and Cenozoic sedimentary sequences.

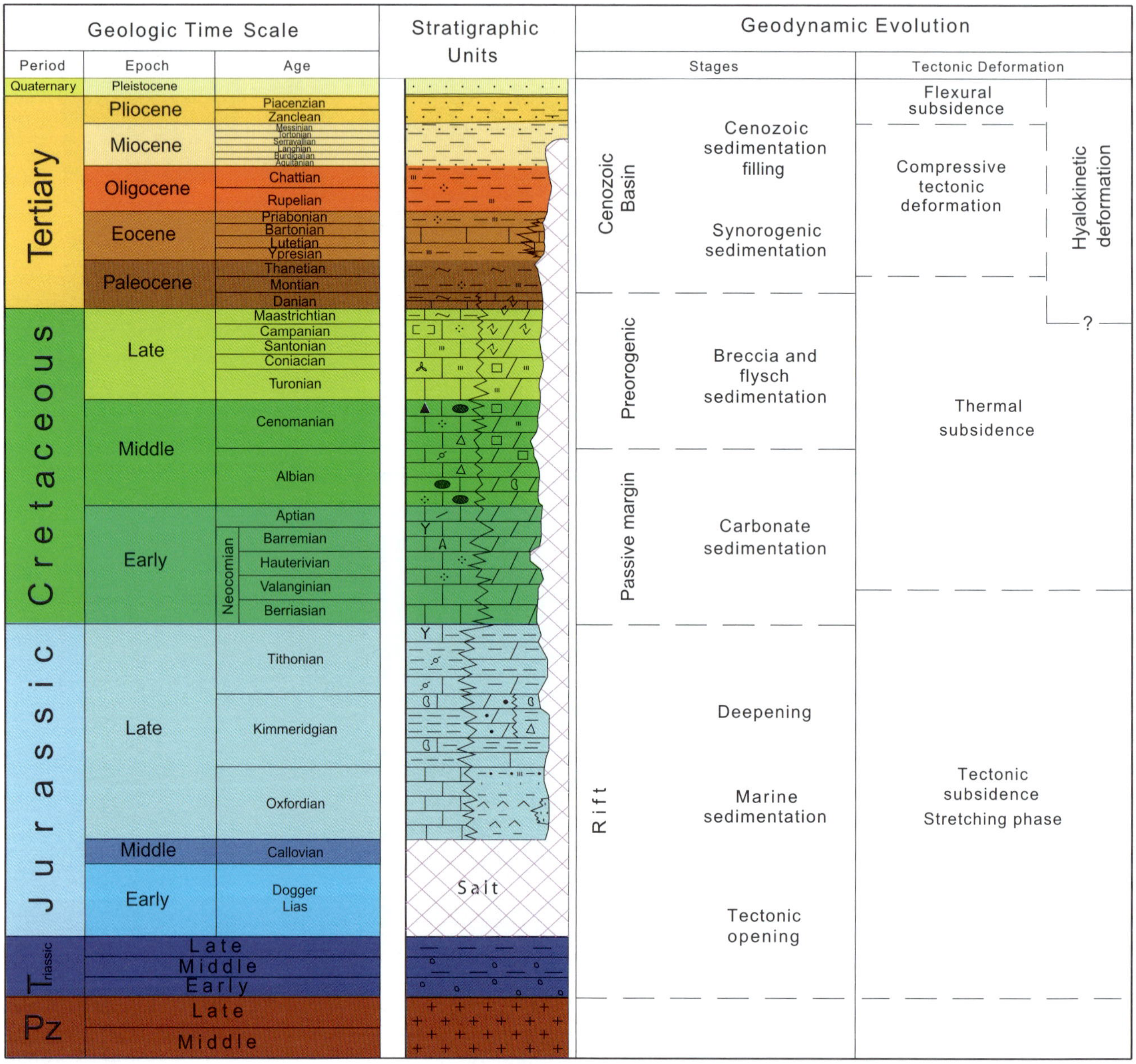

FIGURE 2. Generalized stratigraphic column for the Campeche marine region as it relates to the geodynamic evolution of the Kayab area.

The estimated age of the compressional tectonics is end Paleogene–lower Neogene comprising fault-propagation folds and reverse faults verging toward the northeast. Tectonic shortening is in the order of 10–14% depending on the location within the study area. The compressional tectonic style in the Kayab study area is very similar to that found within the tectonic front of Cantarell, Sihil, and Ku-Maloob-Zaap by Mexican specialists and later described by Mitra et al. (2005, 2006). It is evident that the geometries of the basement and the thick overlying Mesozoic–Cenozoic sedimentary series controlled both the dynamics and final geometry of the halokinesis. Moreover, depositional subbasins that formed during the Neogene (mainly in the Miocene) were a response to halokinetic inversion dynamics (Ortuño et al., 2004). Figure 3 is a schematic representation of the prevailing structural style in the study area along a southwest-northeast–oriented geological section based on seismic data.

The main objective of this study was to build a numerical model of hydrocarbon generation and migration in the Kayab area, as well as identifying the elements and processes of the existing petroleum systems employing the Temispack 2-D and 3-D program.

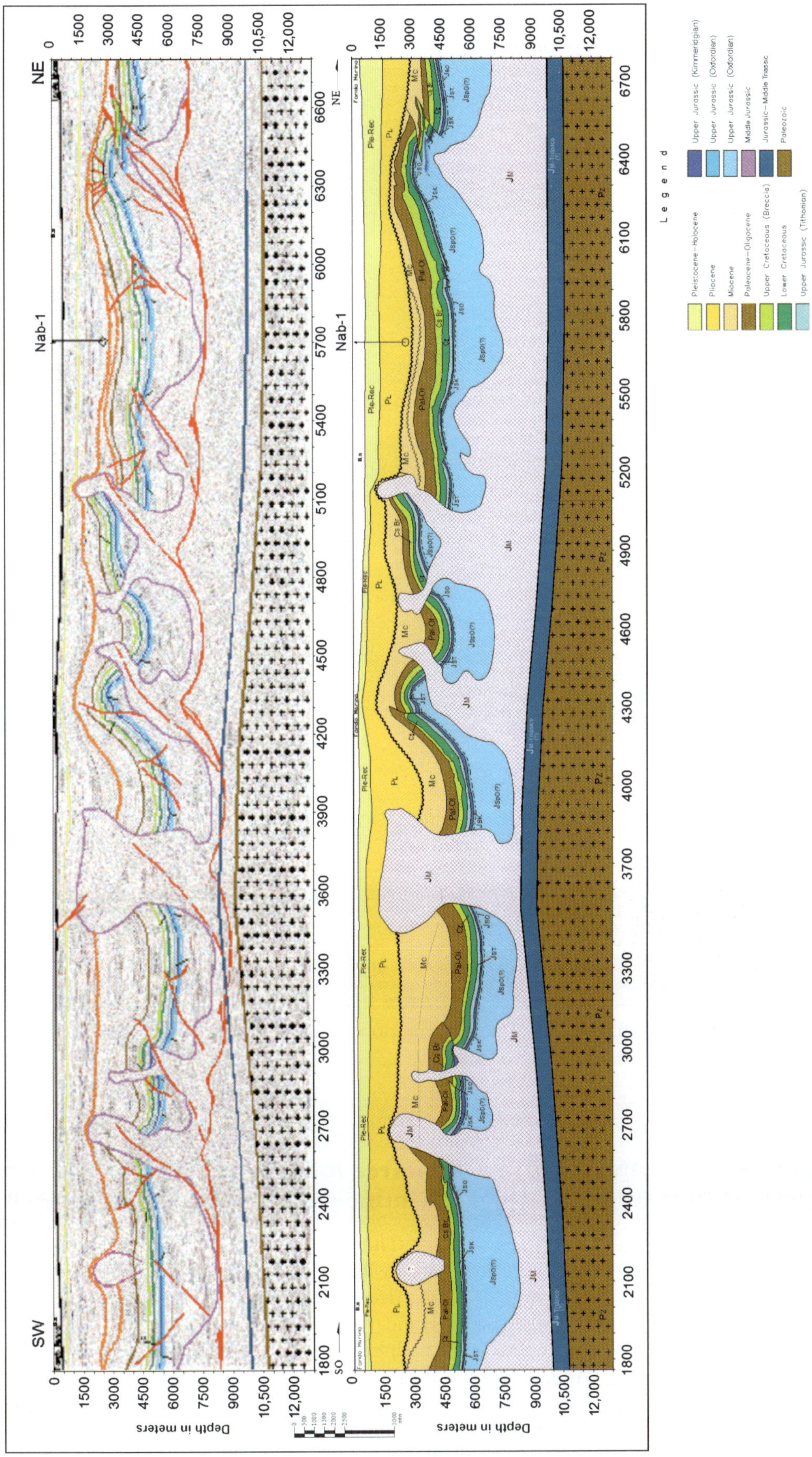

FIGURE 3. Representative geological section for the Kayab area. The geological section shows the complexity of the structure and the general structural style in the study area. Ple-Rec = Pleistocene-Recent; PL = Pliocene; Mc = Miocene; Pal-Ol = Paleocene-Oligocene; Cs Br = Upper Cretaceous (Breccia); Cs = Upper Cretaceous; Ci = Lower Cretaceous; Jst = Upper Jurassic Tithonian; Jsk = Upper Jurassic Kimmeridgian; Jso = Upper Jurassic Oxfordian; JspO (?) = Upper Jurassic (probably Oxfordian); JM = Middle Jurassic; Pz = Paleozoic.

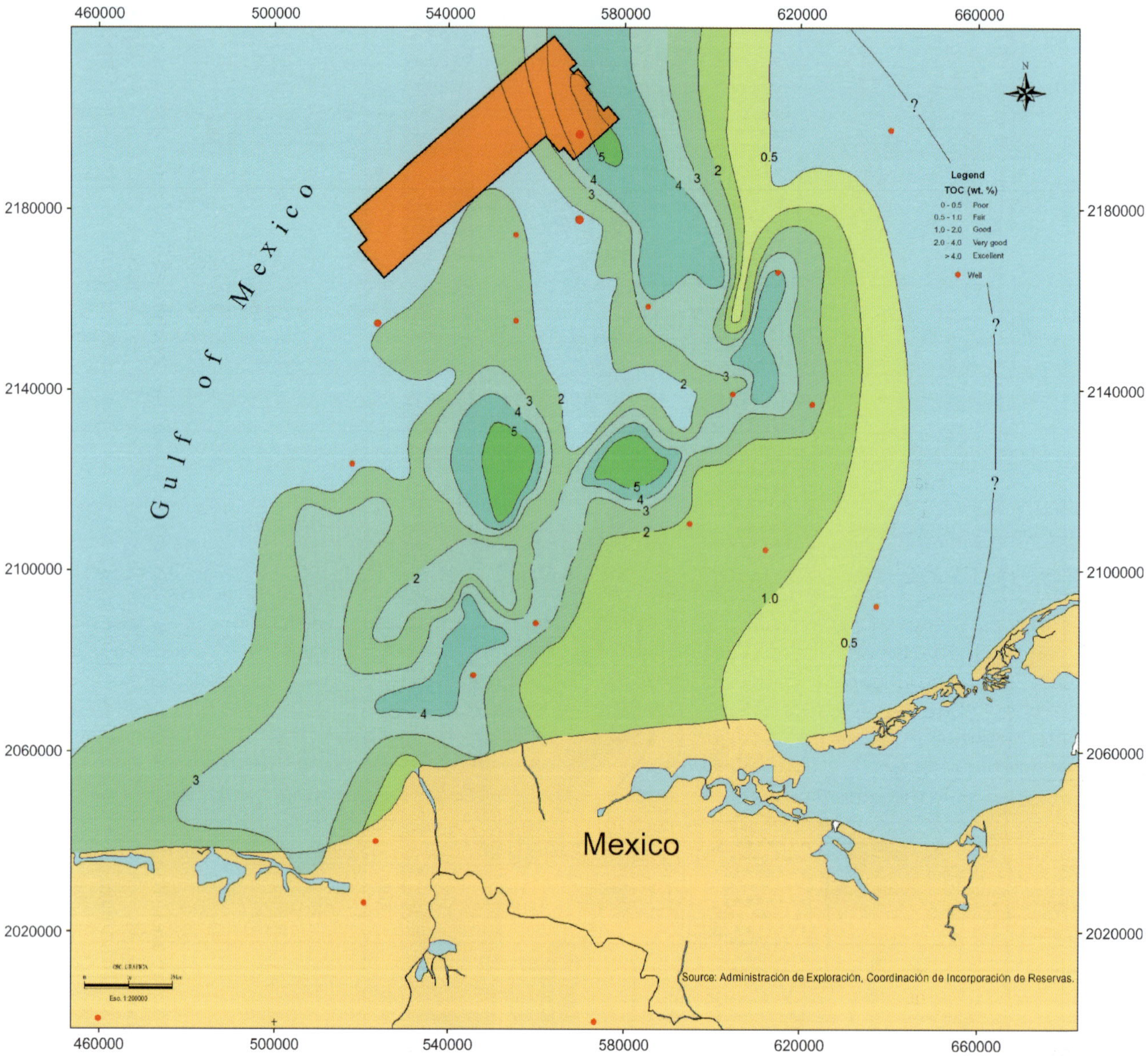

FIGURE 4. Map of Tithonian total organic carbon (TOC) distribution in the Campeche marine region and the study area. Source of data: Activo región Marina, PEMEX.

GENERALIZED ECONOMIC GEOLOGY OF PETROLEUM IN CAMPECHE SOUND

The following discussion is a general panorama of the petroleum factors and processes affecting Campeche Sound and describes the source and reservoir rocks. These updated data integrations were provided by the Reserve Incorporation Coordinating Office of the Northeast Marine Region and served as the basis for numerical modeling in Temispack.

Source Rocks (Tithonian, Kimmeridgian, and Oxfordian)

The predominant lithologies in the Tithonian series are argillaceous limestones and calcareous shales deposited in a deep-water anoxic environment. The Tithonian facies is widely and continuously distributed in the Campeche marine region. This is the principal component of the petroleum system and generated nearly all the hydrocarbons produced from several stratigraphic levels in the Campeche Sound. The Kimmeridgian series, with source rock characteristics in its shaly sections, covers a smaller area than the

Tithonian. Finally, some intervals in the Oxfordian series also have source potential. Tithonian, Kimmeridgian, and Oxfordian source rocks are shown in the stratigraphic chart in Figure 2.

The Oxfordian and Kimmeridgian units in the classic Campeche Sound sections each have an average thickness of about 600 m (1968 ft), whereas the Tithonian may average 200 m (656 ft). It is probable that the Tithonian, Kimmeridgian, and Oxfordian are thinner (~200 m [~656 ft] each) condensed sections in the Kayab area deeper water environment, although their total organic carbon (TOC) values may be higher because of better preservation of their organic content.

The TOC of the Tithonian series across Campeche Sound is quite variable with values ranging from 0.5 to 5.0%. The kerogen is type 1–2 (oil prone) to 2 (oil and gas prone), and occasionally type 2–3. The TOC values in the Kayab area generally range between 4 and 6% (Romero et al., 2001). Organic material within the Kimmeridgian is of good quality derived from algae (kerogen type 2) with TOC values from 0.5 to 3.0% (Ángeles, 1996b). The Oxfordian series in the marine region has TOC contents ranging from very low to very high within a relatively small area; e.g., they are quite variable and spatially discontinuous (Romero et al., 2001). Figure 4 shows the general TOC distribution of the Tithonian series in the study area.

The maturity of the kerogen, or its degree of transformation into hydrocarbon, is defined on the basis of maximum pyrolysis temperature (T_{max}) and vitrinite reflectance (R_o). The T_{max} values in the Tithonian of the marine region are mainly between 435 and 460°C, corresponding to the oil generation window. Over most of the marine region, the Tithonian source rocks are in the oil and gas generation window, becoming overmature only in the central and shoreward parts of the area. The Tithonian is thermally immature in the northeastern part of Campeche Sound.

The R_o values for the Tithonian and Oxfordian series exceed 0.4%, having reached thermal maturity for the generation of liquid hydrocarbons. These rocks have not reached the R_o threshold value of 1.2% required for thermal gas generation. Only the Tithonian has measured values approaching 1.1% R_o (Ángeles, 1996a, b; Romero et al., 2001). Figure 5 shows the R_o value distribution for Campeche Sound, including the study area. The areas of greater thermal maturity (0.5–1.3% R_o) lie in the southern part of the Gulf of Mexico (Figure 5).

With the application of these geochemical parameters to the model, generation-expulsion in Campeche Sound occurred during the Miocene–Pliocene.

Reservoir Rocks (Kimmeridgian, Lower and Upper Cretaceous, and some Cenozoic Intervals)

Several reservoir units for hydrocarbons generated from the Tithonian system are observed, including the Kimmeridgian facies, Lower Cretaceous limestones, Upper Cretaceous breccias, and sandstones at different stratigraphic levels in the Cenozoic (Ortuño et al., 2004). Of these rocks, the most prolific producers are the Upper Cretaceous breccias. These rocks are shown in the stratigraphic column of Figure 2.

Sealing units correspond to shales, and trapping structures are folds related to the compressional and halokinetic episodes. The sealing beds occur at several stratigraphic levels within the Upper Jurassic, Paleogene, and Neogene.

THE MODEL AND NUMERICAL SIMULATION IN TEMISPACK 2-D AND 3-D

The integrated numerical simulation of petroleum systems in the Kayab area consists of two parts: a 2-D simulation evaluated on two-dimensional geological sections and a 3-D simulation wherein several maturity parameters are calculated and visualized in three dimensions.

The particular methodology used in the modeling and numerical simulation consists of the following fundamental steps: (1) construction of the 3-D geological model; (2) the integration of thermal and geochemical parameters; (3) 1-D modeling; (4) 3-D modeling; (5) an estimation of generation processes; and (6) the definition of drainage areas, volumetric calculations, and prioritization of prospects.

For the 2-D simulation, interpreted geological sections were interpreted from seismic data (Figure 6). The specific 3-D Temispack model was constructed from stratigraphic horizons identified in the Kayab 3-D seismic volume. Thirteen stratigraphic horizons were identified and used to construct the 3-D volume in Temispack.

As for the construction of the numerical model, it was assumed that the tectonic deformation only created small folds and a tectonic ramp in the Tunich anticlinal structure, which was slightly displaced. Thus, it is also assumed that the tectonic horizontal displacement did not influence the thermal maturity of the source rocks. Considering this factor, and being aware that horizontal displacements are not handled by Temispack, the geologic model only included the thicknesses of the sedimentary columns, their position at depth, the degree of subsidence, compaction, and the vertical displacements of salt intrusions. Vertical movements related to subsidence and a high rate of Cenozoic sedimentation are the main factors that have placed the source rocks in the oil window.

Construction and Calibration of the Numerical Model

The 3-D model was constructed from depth-converted seismic horizons. After this step, the lithologic and petrophysical properties for each stratigraphic horizon were

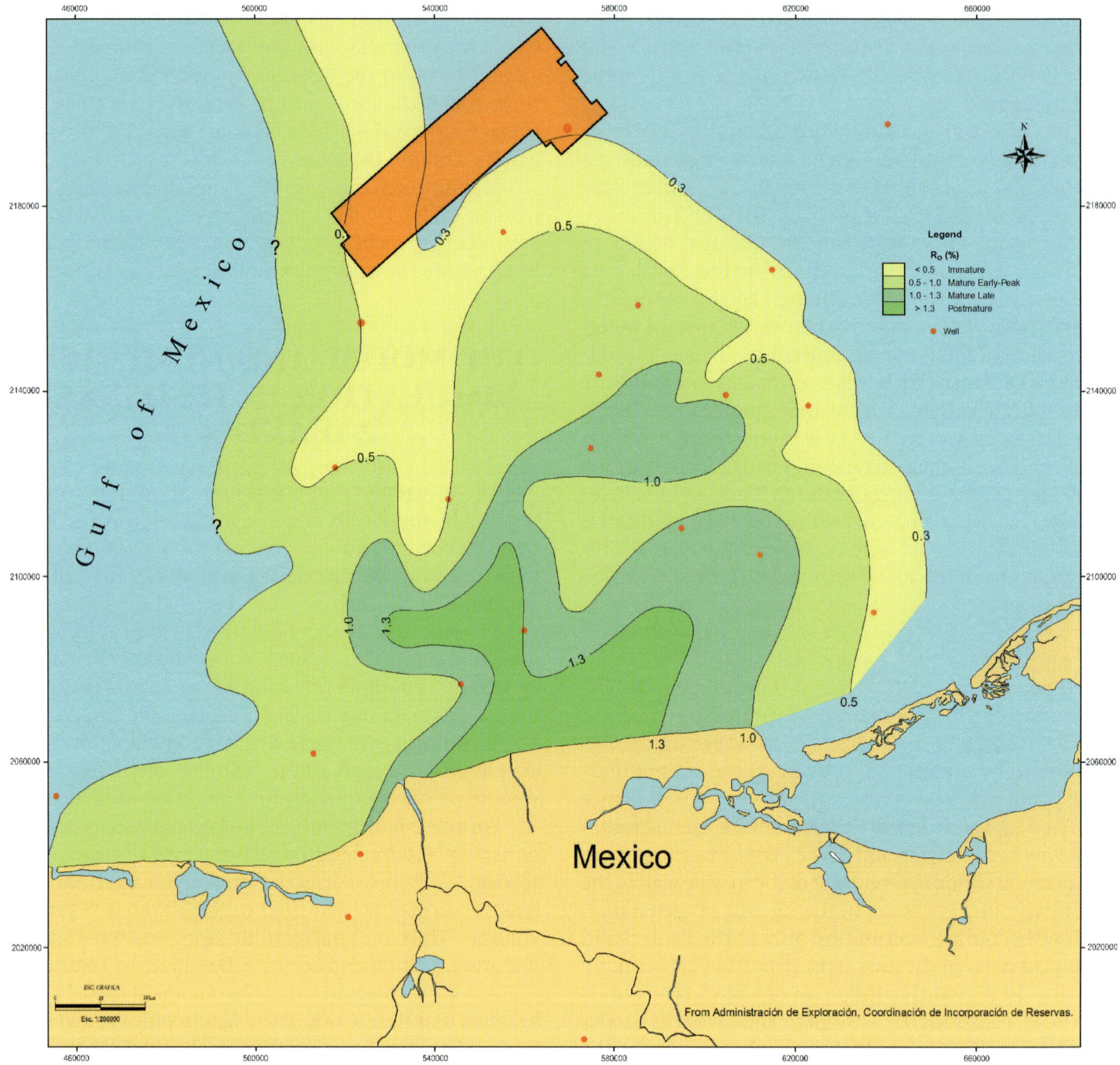

FIGURE 5. Map of Tithonian vitrinite reflectance (R_o) distribution in Campeche Sound and the study area. Source of data: Activo Región Marina, PEMEX.

defined and attributed by integrating all existing data. The 1-D numerical model was derived by simulating the thermal maturity of kerogen in the lithologic column while characterizing the lithofacies and geochemical attributes of all generating levels with their temperature distribution, and other important parameters.

The top and geometry of the basement proceed from the magnetometric and gravimetric model, as well as the petroleum genesis studies conducted by the Mexican Petroleum Institute (2000).

At this stage, calibrations were made of the wells and pseudowells in the study, consisting principally of their temperatures, geothermal gradients, heat flows, R_o values, T_{max} values, etc. The principal calibrations concerned temperature determined from bottom-hole measurements. These calibrations were done for the different models that were constructed. In all cases, the analytical data were adjusted to the modeled results. Generally, the most sensitive and variable of the parameters in the modeling was the temperature. Disparities were noted between the analytical and modeled curves, but these intervals were, in reality, minor. The simulations were run under two scenarios by placing the crust-mantle interface at 1300°C as the heat source with pre-established heat flows. The heat flow range that most closely matched the analytical data was between

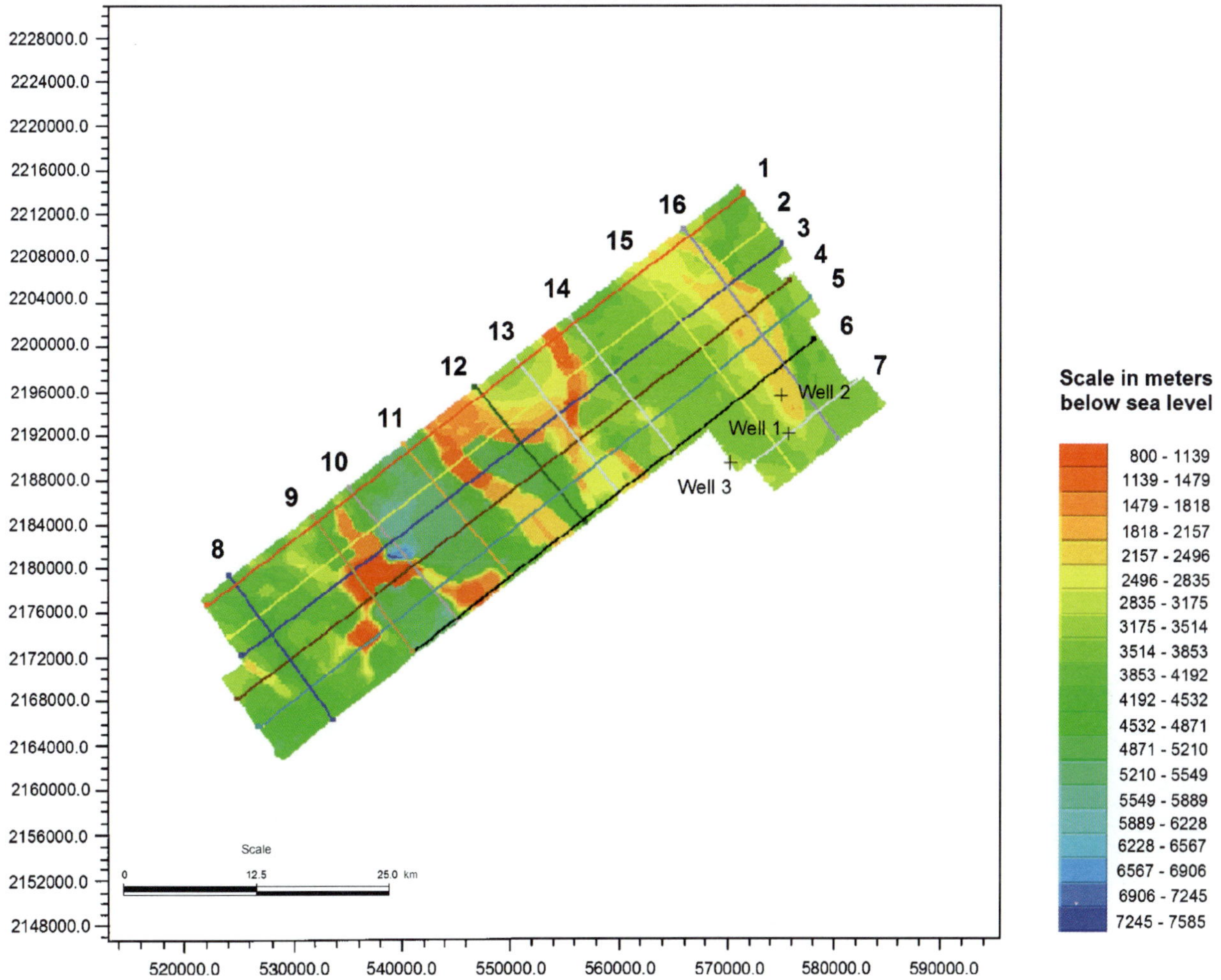

FIGURE 6. Location map for geological sections modeled in Temispack 2-D. The background is the Breccia stratigraphic horizon. The numbers correspond to the modeled geological sections, only a few of which are included in this chapter (the numbers of the 2-D sections are very difficult to see).

20 mW/m^2 and 30 MW/m^2 and, in some cases, by modifying the magnitude of heat flow through the geological history of the basin. The Kayab area is situated on the continental crust in the vicinity of the Yucatan Platform, a region of low to medium heat flow.

Figure 7 shows two of the applied adjustments, i.e., temperature and T_{max}, for the different runs that were conducted. The heat flows obtained through the model were congruent with the limited data available in the study area. Besides the mentioned wells, other nearby wells were used for a more precise calibration. These calibrations were adequate compared to the known bottom-hole temperature, gradient, and heat flow data in the area of Campeche Sound.

The presence of intrusive salt bodies within the Mesozoic and Cenozoic section required a detailed kinematic-structural analysis of the deformation and diapiric emplacement into the sedimentary column using the seismic data and geological cross sections instead of the modeling software. Once the kinematic and tectonic evolution of the intrusive bodies was defined, adjustments were made and thermal history scenarios constructed to adequately model the influence (anomalous conductivities) of the salt bodies on the thermal evolution of kerogen in the source rocks. Comparison of the model results with actual data in the study area was a fundamental aspect in calibrating the simulation.

Two-Dimensional Numerical Simulation on Geological Sections (Generation and Expulsion)

Among the most important parameters generated by the simulation and plotted on the geological sections whose locations are shown in Figure 6, are T_{max}, transformation ratio (TR), fluid saturations in the source rocks, R_o, and temperature distribution in the sedimentary column. Some of these parameters are discussed below and are referred to on the geological sections shown in the figures.

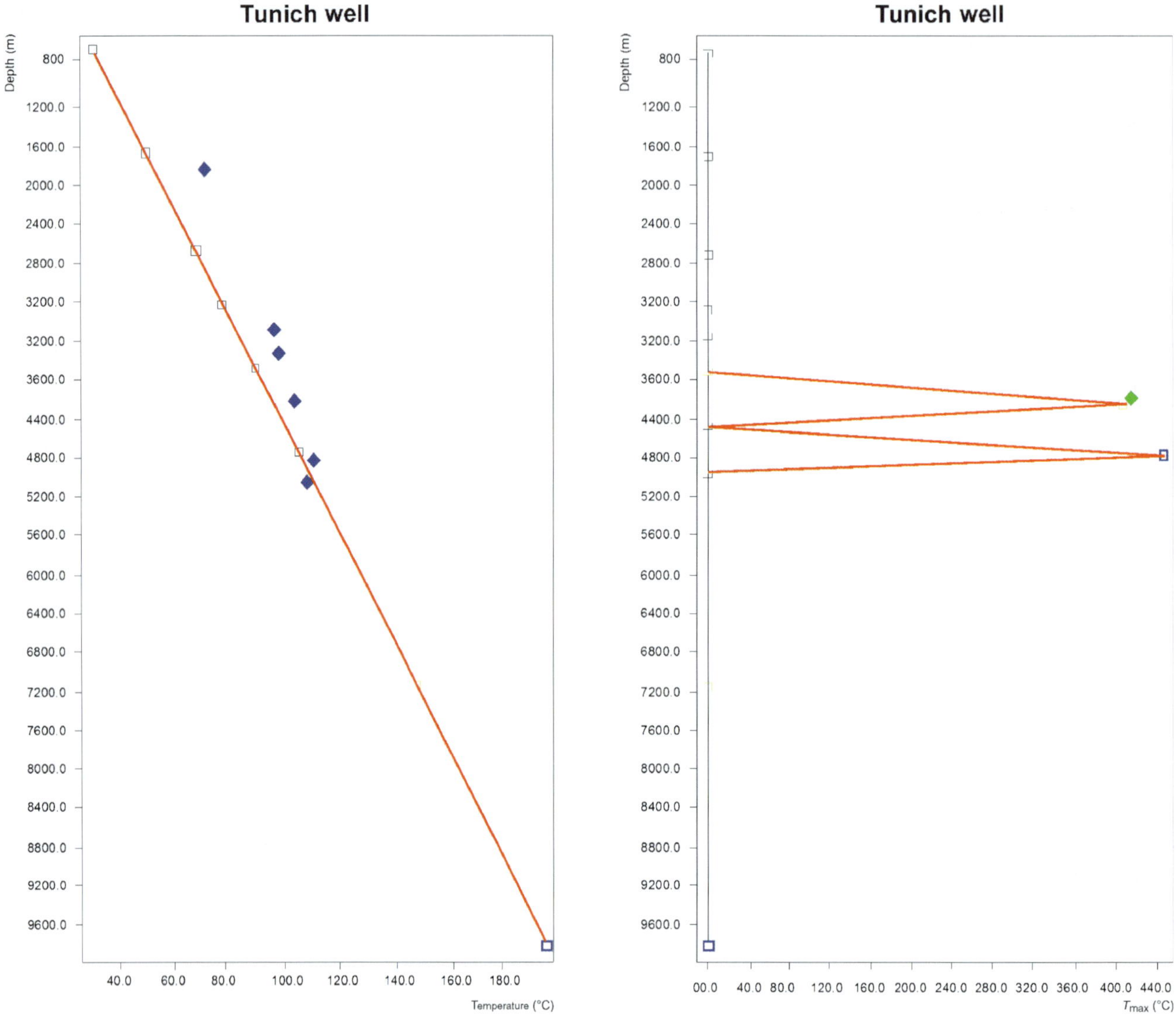

FIGURE 7. Calibrations between temperature and T_{max} data and results of modeling of the Tunich well.

Distribution of Maximum Pyrolysis Temperature

The values of this parameter range from less than 410 to about 490°C for Tithonian to Oxfordian source rocks. The highest values correspond to the deepest subbasins and the lowest in the shallowest subbasins, as in the case of the Tunich structural subbasin.

Transformation Ratio

The evolution of this parameter is shown on geological sections 3 and 11 (Figures 8, 9). The TR values in the subbasins of the southwestern Kayab volume, located between salt intrusions, range from 0.3 to 0.7, reaching perhaps 0.8 or 0.9. Therefore the Tithonian and Oxfordian rocks, as well as the Kimmeridgian, are considered to be in the hydrocarbon-generating window. The TR values in the Tunich well are very low and do not indicate sufficient maturity to have expelled hydrocarbons.

Vitrinite Reflectance

The R_o values range, in general, from 0.5 to 1.0% in Tithonian and Oxfordian source rocks across the study area. According to these results, the deeper subbasins are in the initial stages of maturity (in the case of Tunich area), whereas only the deepest subbasins (areas of the greatest subsidence) would be in the very mature generation stage (block 3-D in Figure 10).

Fluid Migration and Generated Hydrocarbons

From the 2-D numerical simulation of the Kayab sections, it is possible to visualize a nominal scheme

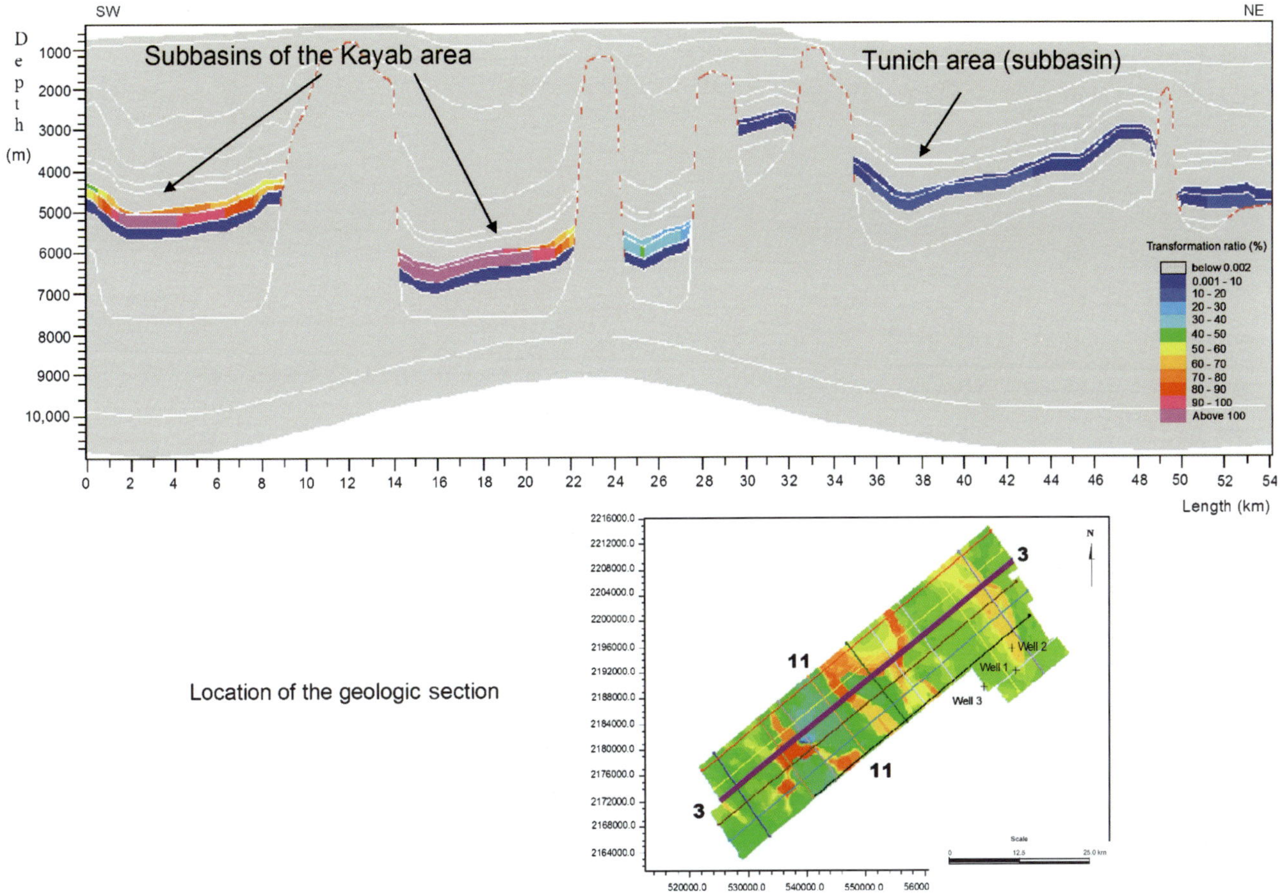

FIGURE 8. Distribution of transformation ratio (TR) magnitudes for the Tithonian, Kimmeridgian, and Oxfordian series along geological section 3. The TR values for these series in the zones of maximum subsidence are 10%, whereas for the Tunich area, the respective values are very low.

for fluid migration pathways through the several different lithofacies that are involved. Among fluids migrating through the rocks is water from the compaction of diverse lithologies, as well as fluids produced during different phases of hydrocarbon generation. Some of these fluid (hydrocarbon) migration pathways are depicted on section 11 (Figure 11). Among these, the most important are vertical migration routes that ascend into Cretaceous strata or toward anticlinal structures that close toward salt intrusions. These migration pathways tend to ascend through the Cretaceous and Cenozoic series in anticlinal structures and along the salt walls.

Thermal History

The numerical simulation provided estimates of some of the thermal parameters that controlled the maturation of the organic matter, including heat flow, temperatures, and geothermal gradients.

Heat flow values of 25 mW/m^2 to more than 40 mW/m^2 were obtained for the Upper Jurassic units in the different subbasins. The higher values correspond to the subbasin in section 11. Temperature values for the source rocks are estimated to range from 100 to more than 160°C, and temperature gradients were critical for generation from the source rocks.

The salt acts as a catalyst for maturation of the kerogen; however, it may also retard maturation (Stover et al., 2001). This is dependent on the existing geometry of the salt and the rocks around it. From the analysis of these thermal parameters, it may be deduced that the subbasins southwest of Kayab are at different stages of hydrocarbon generation. Consistent with the preceding parameters, the Tunich structure is shown to have the lowest heat flow, temperature, and gradient values. In general, the temperature at Tunich structure did not exceed 120°C (Jurassic and Cretaceous at depths of 3500–4500 m [11,483–14,764 ft]), which corresponds to a zone with hypothermal (prematuration) characteristics.

Three-Dimensional Numerical Simulation

This simulation mode incorporates the complete 3-D volume constructed in Temispack, and its visualization

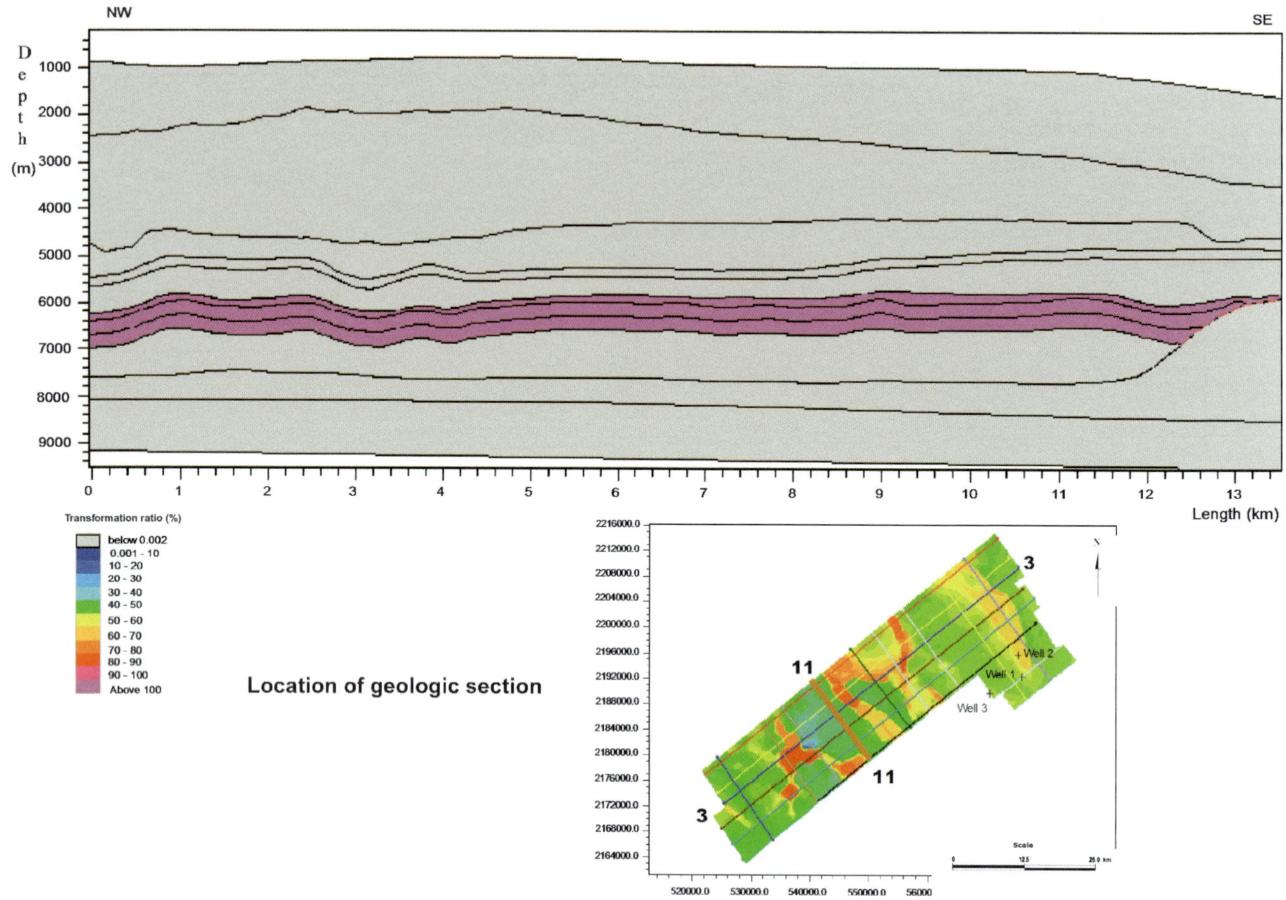

FIGURE 9. Estimate of the transformation ratio (TR) magnitudes for the Tithonian, Kimmeridgian, and Oxfordian series (red-pink) along geological section 11. The corresponding values exceed 60%.

permits the observation of the actual values and evolution of variables that indicate the thermal maturity of the kerogen. Figures 10 and 12 illustrate the results of 3-D modeling for two parameters within the stratigraphic section of the basin, i.e., R_O and temperatures.

The R_O distribution is shown in Figure 10, where values are between 0.5 and 1.0 in the deepest, most subsided subbasins. The values in the 3-D models agree with modeled estimates from the 2-D geological sections. The 2-D sections were extracted from the 3-D model, but an independent 2-D model was run for each of these sections using the same geochemical parameters.

Figure 13 shows the evolution of TR through the late Oligocene, late Miocene, Pliocene, and Pleistocene to Holocene for the Tithonian series. This evolution shows a gradual increase in maturity with the resulting generation-expulsion of hydrocarbons as the subsidence and temperature of the source rocks increased, mostly because of abundant Miocene to Pleistocene sedimentary overburden. The TR, therefore, would have been less than 0.1 during the Paleocene, whereas in the Miocene, TR values varied from 0.1 to 0.4 indicating potential incipient generation at that time. In the Pliocene, however, TR values reached levels of 0.4 to 0.8 in parts of the deepest subbasins, and hydrocarbons were clearly generated from the Tithonian strata. Finally, during the Holocene, TR values range from 0.4 to more than 0.8. Therefore, Tithonian source rocks have generated and expelled hydrocarbons since the Pliocene, and the petroleum system is currently still active. The temperature estimates would be about 140°C. In this way, maturity parameters for the Oxfordian rocks can be estimated. The evolution of this system through geological time is the basis for constructing timing diagrams that show the effects of critical elements and processes on the petroleum system.

Note that the Tithonian and Oxfordian source rocks are thermally immature in the area of the large fault-propagation fold in the northeastern part of the represented area penetrated by the Tunich 1 and 101 and Baksha wells.

A preliminary conclusion from the 3-D model of kerogen maturity is that two active petroleum systems currently generating and expelling hydrocarbons (Tithonian and Oxfordian) in the subbasins of the southwestern Kayab area exist. Of these, the most prolific would be the Tithonian. The proof of this will depend on new exploratory drilling.

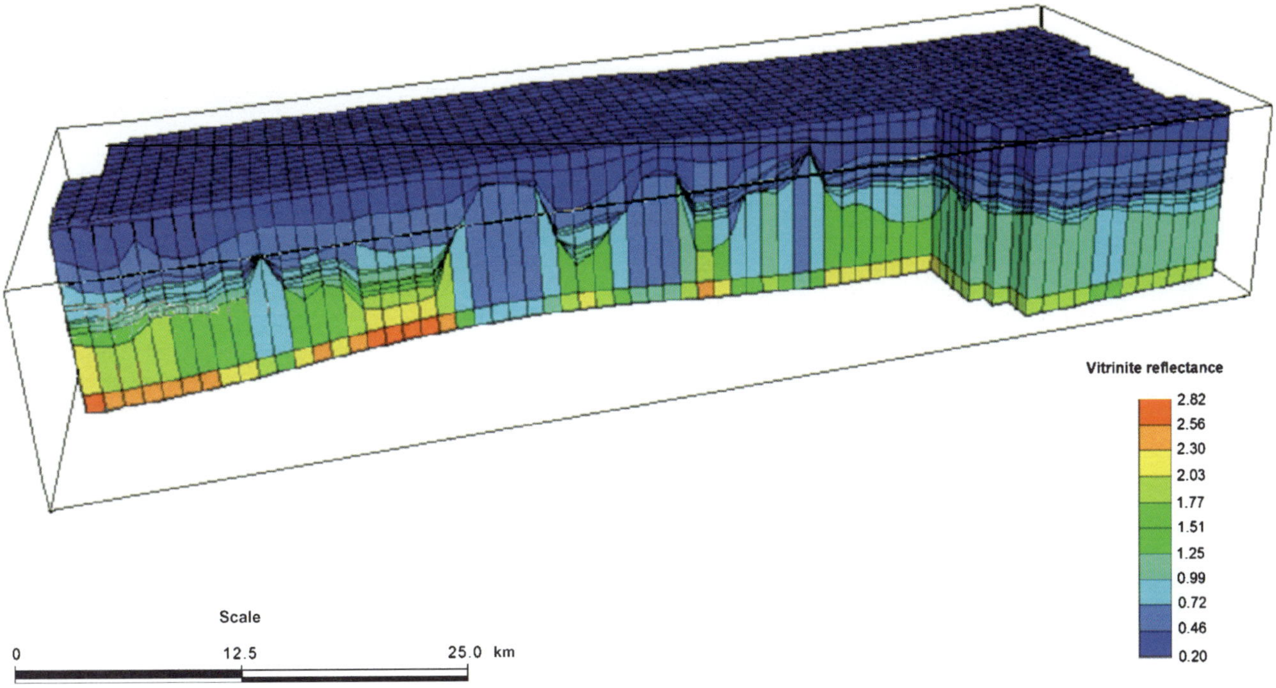

FIGURE 10. Distribution of nominal vitrinite reflectance (R_o) values for the sedimentary sequence across the 3-D area. The nominal R_o values for Tithonian and Oxfordian source rocks range from 0.5 to more than 1.0 in the areas of greatest subsidence. These values are lower in the Tunich area, indicating the immaturity of the kerogens.

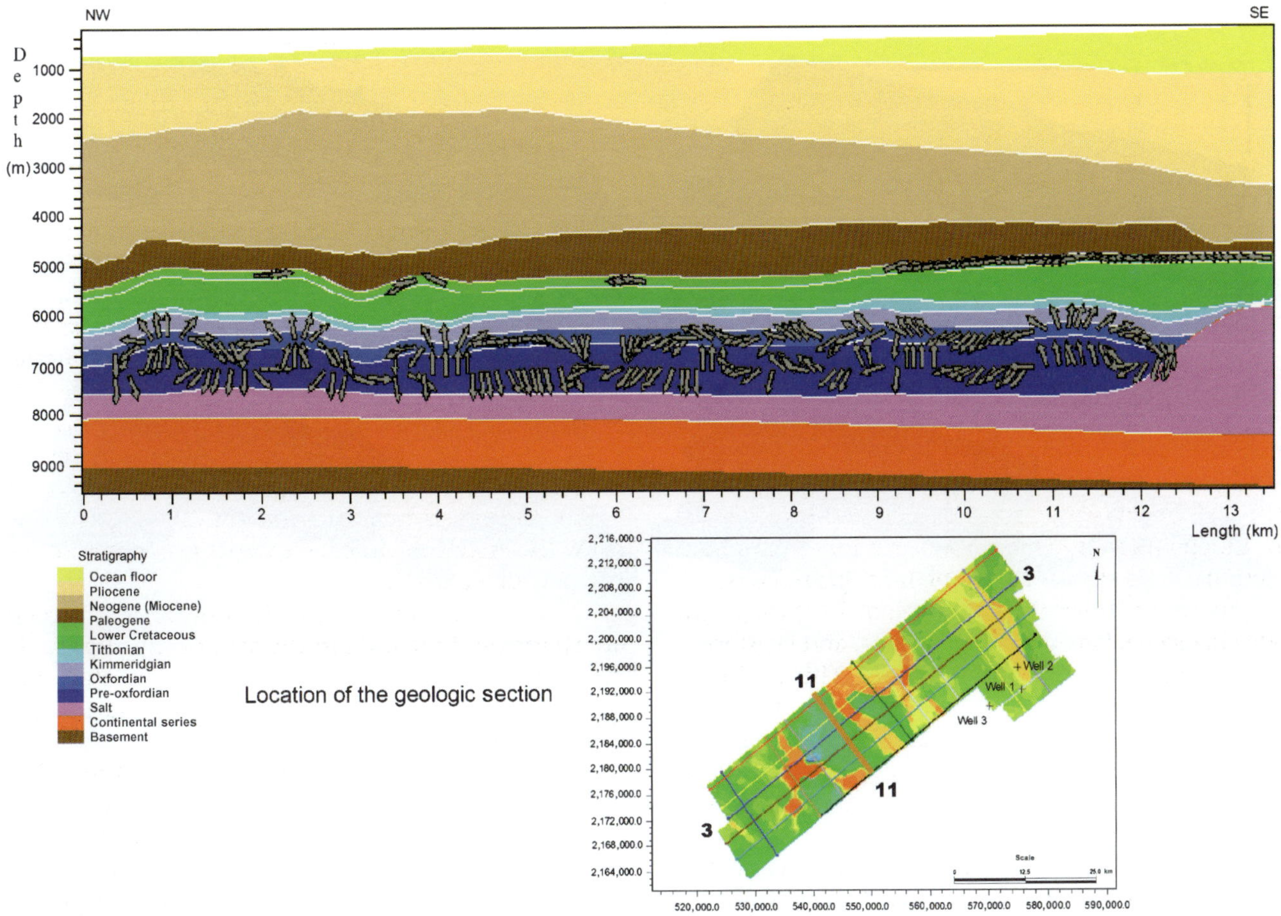

FIGURE 11. Estimate of hydrocarbon migration pathways from the Tithonian, Kimmeridgian, and Oxfordian source rocks along geological section 11.

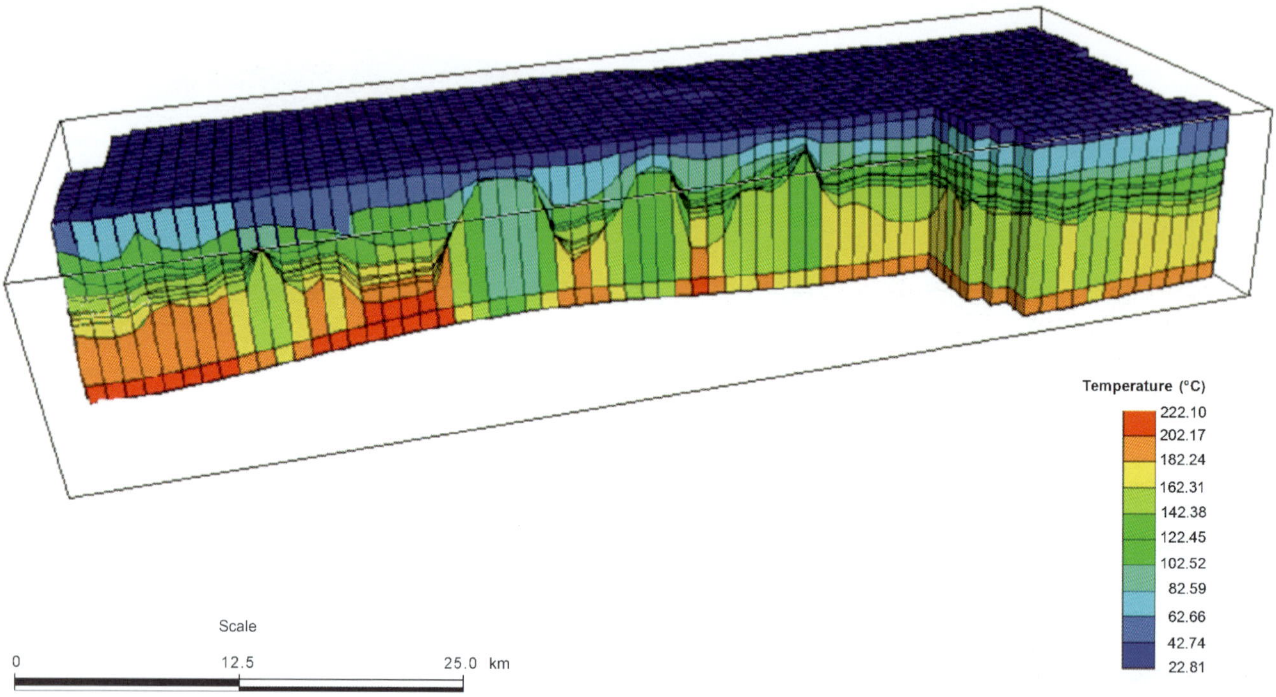

FIGURE 12. Distribution of temperature values across the Kayab 3-D area. The temperatures in the deepest subbasins (at the Upper Jurassic level) oscillate around 135°C (intervals range between 120 and 165°C), whereas the temperature in the Tunich structure is approximately 90 or 100°C.

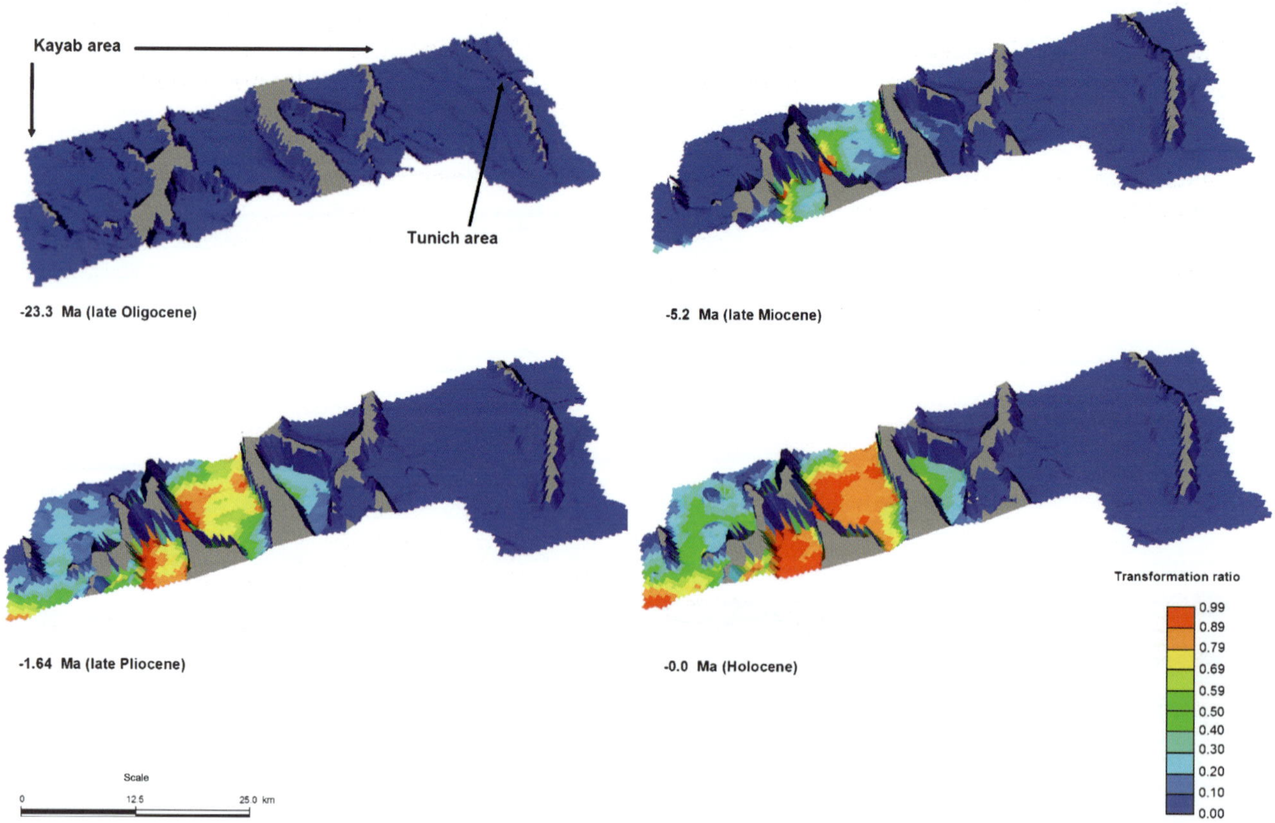

FIGURE 13. Evolution of transformation ratio (TR) magnitudes in the Tithonian rocks during (a) Paleocene, (b) Miocene, (c) Pliocene, and (d) Pleistocene–Holocene. These values continuously increase through time until achieving maxima of around 1.0, with a Pleistocene–Holocene low of 0.3 in the southwestern part of Kayab. The Tunich area shows consistently low TR values (0.1 or less), indicating an immature thermal state.

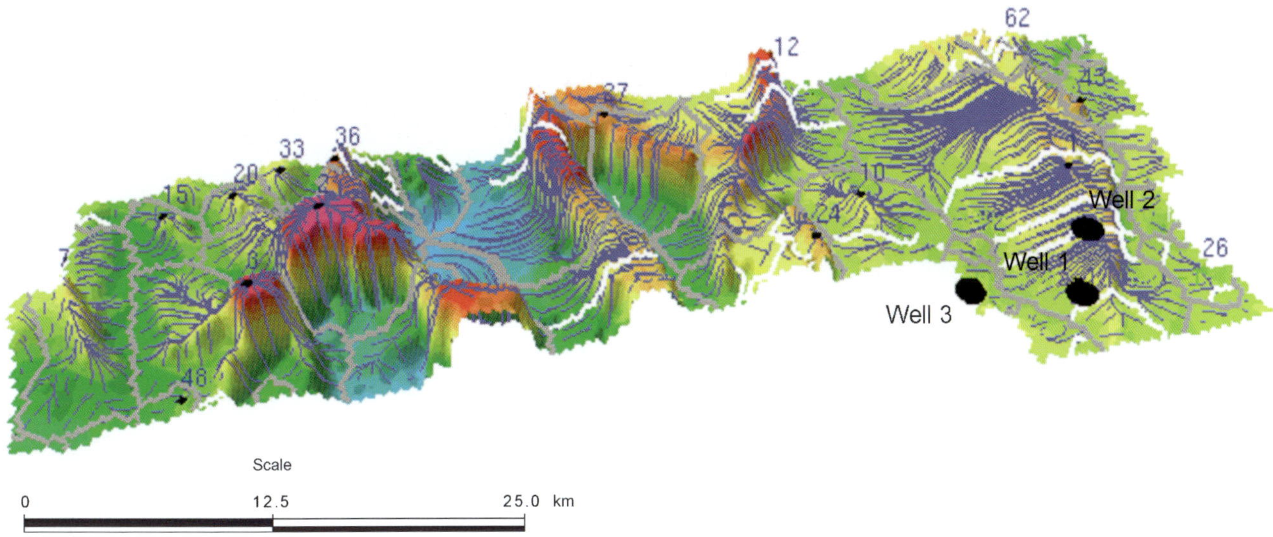

Figure 14. The 3-D configuration of drainage areas showing hydrocarbon flow susceptible to trapping at the Upper Cretaceous Breccia level. Blue lines show the most likely path of gravity migration in the reservoir rocks toward structural highs (or traps). The crests in this depiction are continuous, although in most cases, the Breccia level is interrupted by intrusions of Callovian salt. This estimate is based on the geothermal gradients. The Tunich area is very risky for the breccias level because the source rocks are immature in this area, but the migration possibilities are good.

Identification and Prioritization of Prospects (Drainage Areas)

The evaluation of drainage areas presented in this section corresponds to an analysis of reservoir rocks, sensu stricto; specifically, an estimation of the volume of migrated hydrocarbons trapped in Upper Cretaceous Breccia reservoirs (Figure 14).

The numerical simulation of the Upper Cretaceous Breccia reservoir permits an evaluation of the structural configuration at the top of the unit, as well as quantification of its hydraulic capacity to trap hydrocarbons, and finally, the quantification of theoretical volumetric capacity in these drainage areas. Figures 14 and 15 show the subdivision of drainage areas for the Breccia and its 3-D configuration. The blue lines trace the ascending migration pathways of the hydrocarbon fluids toward the crests of the structures.

An analysis of these volumetric estimates makes it possible to develop a preliminary prioritization of prospects, drainage areas, and hydrocarbon trapping areas. The prioritization is based on the capacity or the volume of stored hydrocarbons for each structure. Figure 15 depicts this prioritization by red zones numbered 1 to 5.

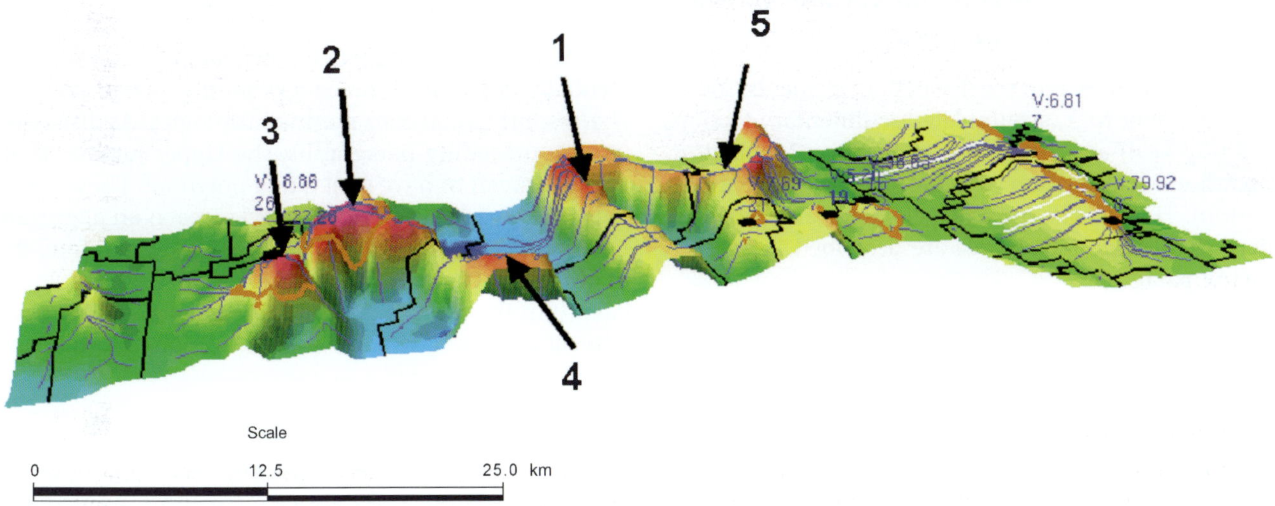

Figure 15. Structural configuration at the top of the Breccia delimiting hydrocarbon drainage areas. The highs are zones of trap development, of which the most important are numbered. Brown lines represent possible structural traps.

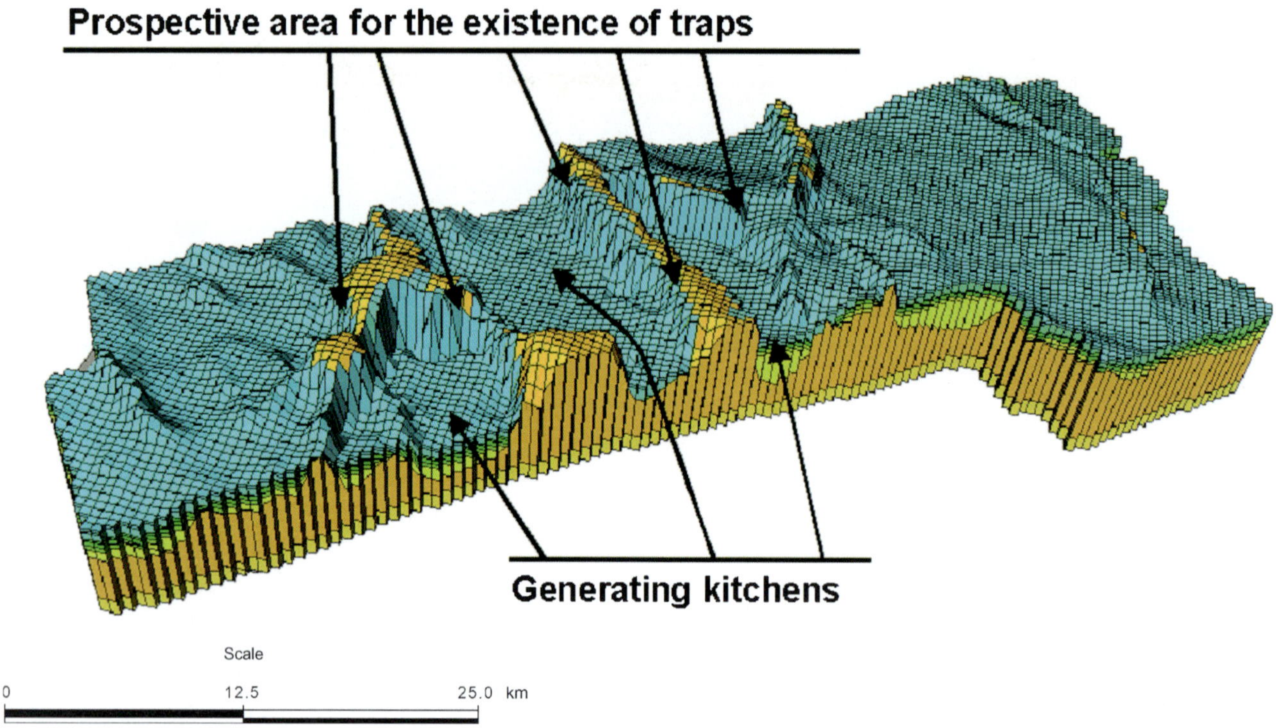

FIGURE 16. The 3-D configuration of the Breccia stratigraphic horizon (in blue) in relation to the intrusions of Callovian salt (golden yellow). The geometries of these two units are fundamentally deterministic in the formation of hydrocarbon traps. The arrangement of generative kitchens in related subbasins and the geometry of the reservoir horizons allow the determination of the most important prospective locations for exploratory wells.

These selected storage or trap areas share the characteristic of being adjacent to the most favorable basins (generating kitchens). The natural drainage areas are shown in Figure 14.

Timing of Petroleum System Elements and Processes - General Integration

Elements of the Petroleum Systems of the Kayab Area

A petroleum system consists of both elements (source rock, reservoir rock, accumulated sedimentary section, and trap seal) and processes (generation, expulsion, migration-accumulation of hydrocarbons, and trap formation). The petroleum system elements and processes identified in the Kayab area are described in the following paragraphs.

Source Rocks

The recognized source rocks in the Kayab area are of Tithonian (average TOC of 5%) and Oxfordian age. Based on lithologic characteristics in neighboring areas, the Kimmeridgian may also generate hydrocarbons. Therefore, the Tithonian, Kimmeridgian, and Oxfordian strata are considered to be the source rocks for the Kayab study area.

The numerical simulation has indicated adequate TR values for the Upper Jurassic strata to have reached maturity in the deepest subbasins of the southwestern Kayab seismic volume. Along with the geodynamic and thermal history of the area, this indicates both generation and expulsion of hydrocarbons. The salt bodies bounding the generating subbasins guided the migrating hydrocarbons (see geological sections 3, 11).

The Buried Sedimentary Sequence

Units in this sequence considered to contain potentially important source rocks are the Jurassic and perhaps some Cretaceous marine lithologies. As discussed in the preceding paragraphs, the Upper Jurassic is already proven to have generating potential. These units, plus the overlying Cenozoic rocks, reach an aggregate thickness of nearly 8000 m (26,247 ft). Therefore, the Tithonian–Kimmeridgian–Oxfordian is at sufficient depth in the southwestern Kayab area to have reached the oil-generating window (see 3-D sections and models).

Reservoir Rocks and Trap Seals

This study considers that the reservoir rocks are contained within the Cretaceous section, particularly the Upper Cretaceous Breccia. In consideration of the calcareous nature of the Lower Cretaceous rocks and

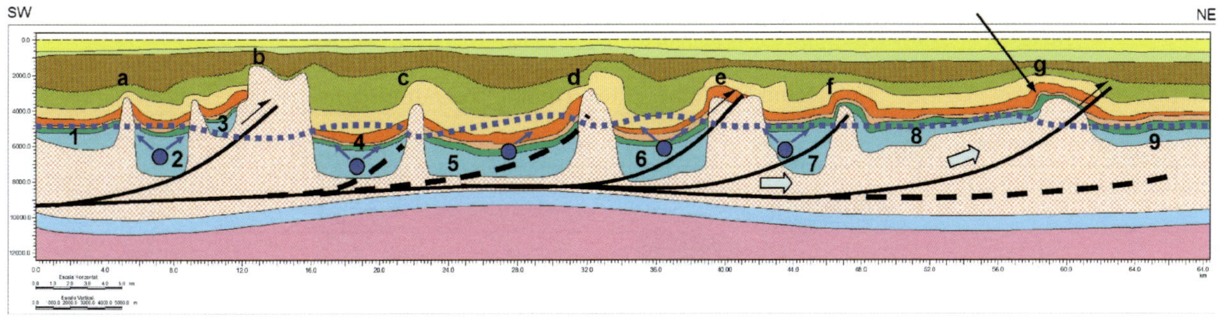

FIGURE 17. Representative geological section for the Kayab area. The areas of greatest thermal maturity correspond to the deepest subbasins with the thickest buried sedimentary accumulations. The blue line in the structural section represents the top oil window. The blue points are the pods of active source rocks in subbasins.

the fracturing they have undoubtedly undergone because of tectonic deformation, these are also considered to be potential reservoirs.

Cretaceous Breccia and limestones are regionally extensive and would be present throughout the study area according to seismic data and sparse subsurface control. The structural configuration of the Breccia horizon, resulting from both compression and salt movement, has favorable hydrocarbon migration and trapping characteristics. Migration pathways are provided by the inherent fracture systems. Seals over the anticlinal highs consist of Cenozoic rocks or salt intrusions (Figures 16, 17). An analysis of this information yields preliminary drilling locations for structures that may contain hydrocarbons, particularly those that are adjacent to the deepest generating subbasins in Kayab.

Regionally distributed Paleocene argillaceous strata, along with diapiric salt, comprise the most important seal rocks for traps in the Kayab area.

Processes in the Petroleum Systems of the Kayab Area

Generation and Expulsion of Hydrocarbons

According to the geological model in Temispack, it is probable that organic facies in the Tithonian, Oxfordian, and possibly Kimmeridgian in the southwestern Kayab area have undergone adequate heating for hydrocarbon maturation. This may be deduced because the entire sedimentary column, including the mentioned section, has a thickness exceeding 8000 m (26,247 ft). Therefore, the generating kitchens would lie between 4500 and 8000 m (14,764 and 26,247 ft) deep, as seen on structural sections of the area.

The numerical simulation of maturation provides important information about the thermal evolution of the oil window in several of the deepest or most subsided subbasins (Figures 16, 17). In this structural context, the positions of the autochthonous and allochthonous salt bodies as they relate to the adjacent sedimentary series are also important because of their influence on kerogen maturation. The salt has high thermal conductivity (6.4 W/m/°C) relative to the other lithologies of the geological model and transmits heat into the overlying sedimentary accumulations, although the presence of salt appears to retard maturation laterally.

The second most important area is in northeastern Kayab (e.g., Tunich) because it is smaller and has undergone less subsidence (Figure 17). This leads us to propose that the area of greater hydrocarbon generation with lower trapping risk is in the southwestern half of the Kayab seismic volume, whereas the part correponding to the large Tunich structure is less prospective.

Migration and Trapping of Hydrocarbons

The hydrocarbon migration and trapping processes may be tied to structuring in the generating subbasins between the salt intrusions and anticlines. The upward flow along migration pathways may move through the flanks of anticlines near salt walls, with the anticlinal crests and upper flanks trapping the migrated hydrocarbons. Salt walls bordering the sedimentary sequences could guide the migrating hydrocarbons toward traps in the upper parts of these structures (Figure 17).

These migration pathways are visualized through simulations of the drainage areas. As already mentioned, these illustrations show the most probable zones of accumulation based on the geometries of the horizons, including permeability barriers and porous levels based on lithofacies distribution (Figures 14, 15).

Trap Formation

Trap formation is tied directly to the sedimentary and tectonic evolution of the entire basin, i.e., textures and sedimentary structures that produce adequate porosity and permeability, and geological structuring, including fracturing and faulting.

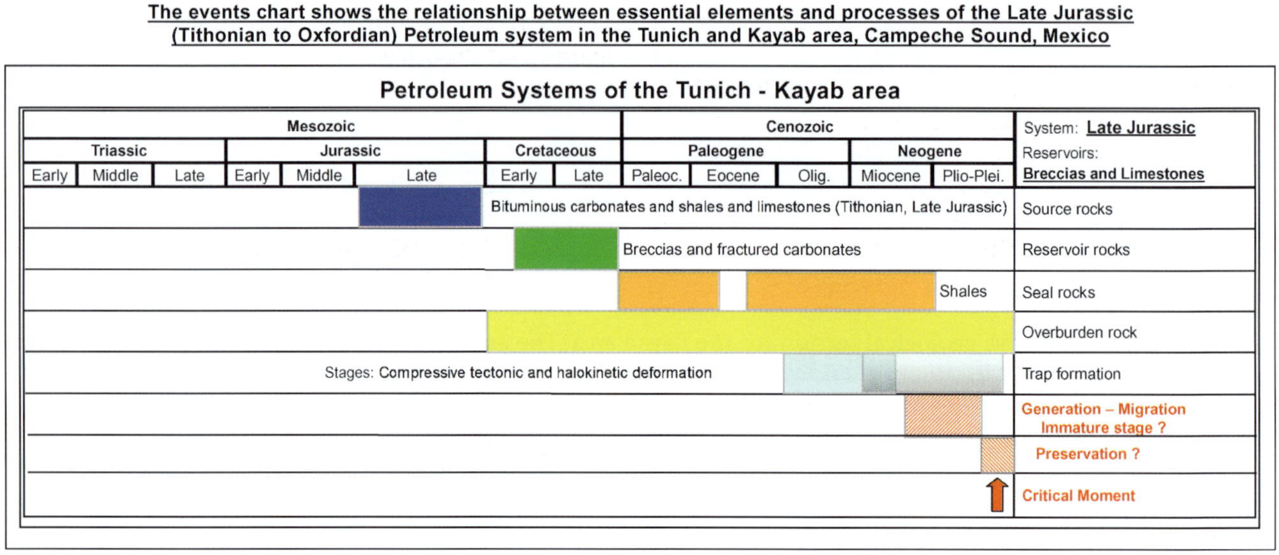

FIGURE 18. Timing diagram of petroleum system elements and processes for the Upper Jurassic sequence (Tithonian, Kimmeridgian, and Oxfordian) in the southwestern Kayab area.

Structural controls on traps in the study area are considered to result mainly from the compression that affected the Jurassic through the Paleogene section, followed by halokinesis and extensional deformation culminating with the Neogene filling of the basin (see geological sections in Figures 16, 17).

Consideration of the data obtained from the geological interpretation and the numerical simulation is of fundamental importance because this defines structural trapping styles and allows the selection of the best locations for exploratory wells having objectives in the anticlinal culminations (fault-propagation folds) or on fold flanks where salt has occupied or broken through the upper part of the structures.

In general terms, the tectonic vector of the compressional phase was oriented southwest–northeast, and the folded structures are generally oriented northwest–southeast. The economic significance of these trends is their parallel alignment to systems in the Ku-Maloob-Zaap and Cantarell province. These aligned structures have now been related to source rocks, petroleum charge systems, migration pathways, and trapping structures for the entire Campeche Sound petroleum province. This preliminary working hypothesis implies that a large part of the petroleum charge of the known fields in Campeche Sound has resulted from long-distance lateral migration originating in generating basins located toward the southwest (i.e., mature zones in Zazil-Ha, southern Kayab, the Coatzacoalcos marine province, etc.), where kerogen is more thermally mature. The migration scenario for Kayab would, therefore, be primarily vertical, with a second phase of migration by lateral flow from deeper reservoirs across spill points toward shallower structures, i.e., the Ku-Maloob-Zaap and Cantarell complexes. Only in this manner can the amount of hydrocarbons produced in Campeche Sound be explained because the local source rocks and vertical migration cannot account for this vast petroleum charge. This hypothesis may be corroborated by additional integrated studies or through detailed syntheses of the existing, active petroleum systems within a regional context.

The Timing of Elements and Processes in the Kayab Area Petroleum System

The following integrated picture has been established for the elements and processes of the Tithonian, Oxfordian, and Kimmeridgian petroleum systems in the Kayab area (Figure 18). During the Paleogene compression, anticlinal structures formed that may have served as hydrocarbon traps for the Cretaceous carbonate reservoir sequence. This was followed by the halokinetic deformation of the initial traps.

Because of the great thickness of siliciclastic sediments accumulated during the Neogene, the Mesozoic sedimentary section was buried sufficiently to initiate generation and expulsion of hydrocarbons from the source rocks, which migrated toward the reservoirs and traps. Generation and expulsion occurred toward the end of the Neogene, especially the Pliocene to Holocene. Tithonian, and perhaps Oxfordian, petroleum systems are still active. Figure 18 shows the timing of the petroleum elements and processes in the southwestern and northeastern parts of the Kayab area.

In summary, geological petroleum risks in the southwestern Kayab area appear to be low, whereas these same risks are higher for the Tunich area in the northeastern part of Kayab.

CONCLUSIONS

Tectonic evolution roughly corresponds with a phase of compressional deformation, a phase of halokinesis, and a phase of deposition and subsequent basin fill during the Paleogene and Neogene.

The 3-D model of the petroleum systems was done in Temispack, highlighting the subbasins with petroleum economic interest, hydrocarbon-generating sedimentary sequences, and possible relationships to trapping structures.

A preliminary visualization of the Kayab area suggests that this region has a very good chance of becoming an important heavy petroleum province in the near future because it contains two active petroleum systems (Tithonian and Oxfordian). The prediction of heavy oil derives from the fact that source rocks are in the early stage of thermal maturity. These petroleum systems charge the structures in southwestern Kayab by vertical migration and also charge the actual productive Cantarell and Ku-Maloob-Zaap area by long-distance lateral migration. Exploration studies scheduled for the near future may more precisely document these conclusions.

According to the sparse kinetic data and numerical modeling of the Tithonian source rocks, petroleum generation began at an R_o of 0.45% (slightly more than 70°C). At R_o greater than 1.0%, gas will be generated. During the early stages of kerogen maturity, heavy, high-sulfur oil may have been generated from the Tithonian. For R_o between 0.57 and 0.81–0.9%, paraffinic, naphthenic, and normal aromatic oils were generated from the type 2 kerogen.

Hydrocarbon expulsion would have begun from the source rocks at a TR of 0.3. When related to temperature, the expulsion of heavy, sulfur-rich oils began at 85°C.

Numerical models and simulations of petroleum system processes and elements can generate more precise concepts and permit the visualization of the parameters and dynamics that culminated in the existence of petroleum fields. This knowledge then allows the development of strategies for petroleum exploration, field development, and new reserve additions because of a better understanding of the geological and physical-chemical parameters, thereby reducing technical risk.

ACKNOWLEDGMENTS

We are grateful to Petróleos Mexicano (PEMEX) for allowing the publication of our manuscript. We are indebted to Francois Roure of the French petroleum Institute and John Bacheller at ExxonMobil for their very constructive reviews of the manuscript. We also thank Claudio Bartolini for the invitation to participate in this AAPG volume and to Salvador Villalbazo for figures and drawings.

REFERENCES CITED

Ángeles, A. F. J., 1982, Estudio sedimentario y diagenético del Campo Abkatún: Sociedad Geofísica Mexicana, VI Convención Nacional, Resúmenes, no. 6, p. 108.

Ángeles, A. F. J., 1988, Estudio estratigráfico-sedimentológico del Jurásico Superior en la Sonda de Campeche: Ingeniería Petrolera, v. 28, no. 1, p. 45–55.

Ángeles, A. F. J., 1993, Breve historia de la Sonda de Campeche: Gaceta de la Asociación Mexicana de Geólogos Petroleros, v. 2, p. 3–4.

Ángeles, A. F. J., 1996a, Estratigrafía del Jurásico Superior del Subsuelo en la Sonda de Campeche (Golfo de México): Tesis de Maestría, Escuela Superior de Ingeniería y Arquitectura (ESIA), Instituto Politécnico Nacional, 93 p.

Ángeles, A. F. J., 1996b, Unidades generadoras del Jurásico Superior en la Sonda de Campeche: Memorias, V Congreso Latinoamericano de Geoquímica Orgánica (Cancún), p. 144–146.

Ángeles, A. F. J., and M. E. Ortuño, 1994, Estudio estratigráfico sedimentológico del Cretácico en la Sonda de Campeche: Sociedad Geológica Mexicana (SGM), XII Convención Nacional, Resúmenes, no. 12, p. 8–9.

Ángeles, A. F. J, J. R. Núñez, and M. J. M. Quezada, 1992, Evolución tectónica de la Sonda de Campeche, estilos estructurales resultantes y su implicación en la generación y acumulación de Hidrocarburos: II Simposio de la Asociación Mexicana de Geólogos Petroleros e Instituto Mexicano del Petróleo, p. 12–13.

Araujo, M. J., L. M. A. Basáñez, and M. A. D. Márquez, 1986, Estudio estratigráfico-sedimentológico de las rocas del Jurásico en el prospecto Área Marina de Campeche (C-5008) inédito: Instituto Mexicano del Petróleo, 92 p.

Bañuelos, M. O., 1993, Roca almacén del Mioceno y Plio-Pleistoceno en la Sonda de Campeche: I Simposio de Geología de Subsuelo, Asociación Mexicana de Geólogos Petroleros, p. 84–91.

Bello, M. R., 1991, Estudio estratigráfico-sedimentológico del Cretácico Superior en la Sonda de Campeche: Instituto Mexicano del Petróleo, Reporte Interno Inédito, Proyecto CAO-3512, 110 p.

García, H. J., 1993, Roca almacenadora Jurásico Superior Oxfordiano: I Simposio de Geología de Subsuelo, Asociación Mexicana de Geólogos Petroleros, p. 49–50.

Guzmán, V. M. A., and M. R. Mello, 1999, Origin of oil in the Sureste Basin, Mexico: AAPG Bulletin, v. 83, no. 7, p. 1068–1095.

Guzmán, V. M. A., L. O. Castro, J. R. R. Román, M. L. M. Medrano, and L. C. Valdez, 2001, El origen del petróleo en las subprovincias Mexicanas del Golfo de México: Boletín de la Asociacion Mexicana de Geologos Petroleros, v. 49, no. 1 and 2, p. 31–46.

Hernández, G. S., 1993, Subsistema almacenador en los horizontes Cretácico Superior y Eoceno Medio de la

Región Marina: Primer Simposio de Geología de Subsuelo, Asociación Mexicana de Geólogos Petroleros, p. 79–83.

Holguín, Q. N., and I. M. A. Romero, 1982, Estudio geoquímico de pozos de exploración en la Sonda de Campeche-Plataforma de Yucatán: Instituto Mexicano del Petróleo, Reporte Interno Inédito, Proyecto C-1142, 60 p.

Holguín, Q. N., et al., 2005, Estudio regional de manifestaciones superficiales de aceite y gas en el sur del Golfo de México, su origen e implicaciones exploratorias: Boletín de la Asociación Mexicana de Geólogos Petroleros, v. 52, no. 1, p. 20–41.

Instituto Mexicano del Petróleo (IMP), 2000, Atlas de las cuencas sedimentarias terciarias y mesozoicas de México: Internal Report, 320 p.

Medrano, M. L. M., and I. M. A. Romero, 1993, Caracterización geoquímica de las rocas generadoras de los sistemas petrolíferos del Jurásico Superior en el área Marina de Campeche: I Simposio de Geología de Subsuelo, Asociación Mexicana de Geólogos Petroleros (AMGP), p. 5–16.

Medrano, M. M. L., I. M. A. Romero, and V. R. Maldonado, 1996, Los subsistemas generadores de la Sonda de Campeche, México: Memorias, V Congreso Latinoamericano de Geoquímica Orgánica (Cancún), p. 85–90.

Meneses, G. J., 1980, Geología de la Sonda de Campeche: Boletín de la Asociación Mexicana de Geólogos Petroleros, v. 32, no. 1, p. 1–26.

Mitra, S., G. C. Figueroa, J. H. García, and A. M. Alvarado, 2005, Three-dimensional structural model of the Cantarell and Sihil structures, Campeche Bay province, Mexico: AAPG Bulletin, v. 89, p. 1–26.

Mitra, S., J. A. D. González, J. G. Hernández, S. H. García, and S. Banerjee, 2006, Structural geometry and evolution of the Ku, Zaap and Maloob structures, Campeche Bay, Mexico: AAPG Bulletin, v. 90, p. 1565–1584.

Ornelas, M., S. Franco, N. Aguilera, H. Alzaga, H. A. Escamilla, and M. Granados, 1994, (res) Secuencias bioestratigráficas del Kimmeridgiano-Tithoniano en pozos de la zona marina de Campeche Sureste de México: Sociedad Geofísica Mexicana, XII Convención Nacional, Resúmenes, no. 12, p. 197.

Ornelas, S. M., F. N. Aguilera, N. S. Franco, M. M. Granado, and M. R. Bello, 1991, Determinación de los límites Oxfordiano-Kimmeridgiano-Tithoniano en pozos de la Zona Marina de Campeche, Plataforma Continental de México: Instituto Mexicano del Petróleo, Reporte Interno Inédito, Proyecto CAO-551, 125 p.

Ortega, L. S., M. L. Medrano, and V. L. Clara, 1995, Definición de los límites reales de generación de las rocas del Tithoniano de las cuencas del Sureste, Sonda Marina de Campeche: Instituto Mexicano del Petróleo, Reporte Interno Inédito, Proyecto CAB-0501, 128 p.

Ortuño, A. S., C. Romero M., J. Rosales R., and E. Bahena R., 2004, Modelado de prospectos en el área de Kayab: Reporte Interno Inédito, Instituto Mexicano del Petróleo, 160 p.

Ramírez, C. L., A. J. Mendoza, G. R. del Valle, and R. P. Gómez, 1996, Mapping salt structure and subsalt imaging in Campeche Bay Mexico, using seismic inversion modeling and prestack attributes: Society of Exploration Geophyscists Expanded Abstracts 15, 588 p.

Región Marina Noreste (RMNE), 2002, Sistemas petroleros terciarios en el área de Tunich-Chilam: PEMEX, Exploración y Producción (PEP), Reporte Interno Inédito, Petróleos Mexicanos (PEMEX), 120 p.

Romero, M. A., J. Hegre, and R. Maldonado, 1996, Modelado 1D Genex en el área marina de Campeche: Memorias, V Congreso Latinoamericano de Geoquímica Orgánica, Cancún, Quintana Roo, México, p. 55–56.

Romero, M. A., L. Medrano, and R. Maldonado, 2001, Subsistemas Generadores de la Región Marina de Campeche: Boletín de la Asociación Mexicana de Geólogos Petroleros, v. 49, nos. 1–2, p. 105–115.

Romero, I. M., C. Cassaigneau, M. Flores, A. Vassayre, R. Navarro, and J. Gaulier, 2004, Future exploration strategy of a frontier offshore Campeche area provided by basin modeling: AAPG International Conference, Cancún, Quintana Roo, México, p. 112–113.

Rudkiewicz, J. L., M. Vandenbroucke, H. L. B. Penteado, F. Brigaud, A. Vear, J. Wendebourg, and S. Düppenbecker, 2000, Integrated basin modeling helps to decipher petroleum systems, *in* M. R. Mello and B. J. Katz, eds., Petroleum systems of south Atlantic margins: AAPG Memoir 73, p. 27–40.

Salomón, M. L., 1999, Modelado estructural de las fallas y su importancia en la exploración de hidrocarburos en el Golfo de México: Biblioteca Instituto Mexicano del Petróleo, TN 872, S35 1999, 85 p.

Santamaría, O. D., 2000, Organic geochemistry of Tithonian source rocks and associated oils from the Sonda de Campeche, México: Ph. D. dissertation, University of Aachen, Germany, 190 p.

Santamaría, O. D., and B. Horsfield, 2000, La Evolución Térmica de las Rocas Generadoras del Tithoniano en el Área Marina de Campeche, México: Boletín Asociación Mexicana de Geólogos Petroleros, v. 49, nos. 1–2, p. 116–136.

Schneider, F., S. Wolf, I. Faille, and D. Pot, 2000, A 3D-basin model for hydrocarbon potential evaluation: Application to Congo offshore: Oil and Gas Science and Technology, Revue de l'Institut Français du Pétrole, v. 55, no. 1, p. 3–13.

Stover, S. C., S. Ge, P. Weimer, and B. C. McBride, 2001, The effects of salt evolution, structural development, and fault propagation on late Mesozoic–Cenozoic oil migration: A two dimensional fluid-flow study along a megaregional profile in the northern Gulf of Mexico Basin: AAPG Bulletin, v. 85, no. 11, p. 1945–1966.

Ungerer, P., J. Burrus, B. Doligez, P. Y. Chenet, and F. Bessis, 1990, Basin evaluation by integrated two-dimensional modeling of heat transfer, fluid flow, hydrocarbon generation and migration: AAPG Bulletin, v. 74, no. 3, p. 309–335.

7

Rueda-Gaxiola, J., 2009, The palynostratigraphy of red-bed and salt units of the Mexican petroleum sub-basins of the Gulf of Mexico, *in* C. Bartolini and J. R. Román Ramos, eds., Petroleum systems in the southern Gulf of Mexico: AAPG Memoir 90, p. 137–154.

The Palynostratigraphy of Red-Bed and Salt Units of the Mexican Petroleum Sub-Basins of the Gulf of Mexico

Jaime Rueda-Gaxiola

Escuela Superior de Ingeniería y Arquitectura-Instituto Politécnico Nacional, Unidad de Ciencias de la Tierra, Calzada Ticomán, Mexico D.F., Mexico

ABSTRACT

Since 1975 Mexican Mesozoic red-bed and salt sequences were dated using the palynostratigraphical method developed at the Mexican Petroleum Institute by Rueda-Gaxiola. They represent the beginning of the Mesozoic marine transgression, which eventually formed the petroleum systems of the region.

The palynostratigraphical method is based on a macro- and microscopic analysis of all the components in palynological residues and of the glycerinated alcohol in which they are preserved. This analysis allows the identification of litho-, bio-, and stratigraphic data, which can be related to tectonosedimentary and diagenetic processes.

The Cahuasas and the Rosario formations (in the Tampico-Misantla subbasin), the Todos Santos Formation (in the southeastern subbasins and in the western Veracruz subbasin), the La Boca Alloformation (in the Huayacocotla-El Alamar Basin), Consuelo Group (in the Tlaxiaco Basin), and some salt units (in the southeastern subbasins) have been dated (J. Rueda-Gaxiola, 1972; J. Rueda-Gaxiola, 1976; L. Gutiérrez-Galicia, 1984; Herrera et al., 1990; J. Rueda-Gaxiola et al., 1993–1996; J. Rueda-Gaxiola; and J. Jiménez-Rentería, 2002).

These data were integrated with those obtained by Kirkland and Gerhard from the cap rock of the Challenger Knoll in the center of the Gulf of Mexico. They have helped understand the sedimentary and tectonic evolutions of the early Gulf of Mexico.

These data indicate the existence of an unconformity between the basement and the Middle Jurassic red beds (La Joya and Todos Santos formations) in the Sabinas and southeastern subbasins of the Gulf of Mexico. This unconformity is also found between the Liassic rocks (La Boca Alloformation and Huayacocotla Formation) and the Middle Jurassic red beds (La Joya and Cahuasas formations) in the Huayacocotla-El Alamar Basin. However, in the marginal Sabinas, Tampico-Misantla, and southwestern subbasins,

DOI:10.1306/13191082M902955

this unconformity is overlain by Middle Jurassic red beds, which is slightly older than the evaporitic rocks deposited in the center of the Gulf of Mexico.

However, some localities of the Tlaxiaco Basin and in the northern part of the Tampico-Misantla Basin, the Liassic Rosario Formation red beds gradually change to the Cualac and Cahuasas formations, which are the basis of the transgressive marine Upper Jurassic–Middle Cretaceous sequence.

INTRODUCTION

The red-bed and salt sequences of Mexico have been generally dated based on their stratigraphic position. Some of them constitute the economic basement of the Mexican Gulf of Mexico petroleum subbasins (Figure 1) and represent the beginning of the Mesozoic marine transgression, which formed the petroleum systems.

Some red-bed formations (Figures 1, 2) have recently been dated using the palynostratigraphic method developed at the Mexican Petroleum Institute: (Rueda-Gaxiola, 1999) i.e., the Cahuasas and the Rosario formations (in the Tampico-Misantla subbasin), the Todos Santos Formation (in the southeastern subbasins and in the western Veracruz subbasin), the La Boca Alloformation (in the Huayacocotla-El Alamar Basin), the Consuelo Group (in the Tlaxiaco Basin), and some salt units in the southeastern subbasins (Figure 3). These data were integrated with those obtained by Kirkland and Gerhard (1971) from the cap rock of the Challenger Knoll in the center of the Gulf of Mexico. They have helped in understanding sedimentary and tectonic evolutions related to the beginning of the early Gulf of Mexico.

Figure 2 shows the age and distribution of the most important Mesozoic red-bed sequences in Mexico (Del Valle-Reyes, 1997). The red-bed and salt sequences are distributed in or around the Gulf of Mexico with ages ranging from Late Triassic to Middle Jurassic. Most are Middle Jurassic and almost always overlain by evaporites that mark the beginning of the characteristic Middle Jurassic transgressive marine sequence (Iturralde-Vinent, 1988).

OBJECTIVES

The objective of this chapter is to show the data obtained from palynostratigraphical studies of the red-bed and salt sequences in the most important Mexican oil basins around the Gulf of Mexico. Data from red-bed sequences can be very useful in determining the tectonic and sedimentary evolution of the Gulf of Mexico and its petroleum subbasins. This information is important because most models proposed to explain the history of the Gulf of Mexico lack reliable and detailed constraints on the initial breakup of the crust during the rifting and drifting stages. The precise age of rifting is difficult to establish because of the lack of fossils and also because the igneous rocks for radiometric and paleomagnetic methods are very deep in the Gulf of Mexico. The only way to determine the age of rifting is by the studying continental fossils in lacustrine and fluvial red beds deposited in grabens above sea level. The marine transgression occurred later during the drift period was blocks subsided under sea level. The subsidence led to the formation of salt above the red beds. These evaporates, in turn, were covered by transgressive calcareous marine rocks, finally forming the Evans Triade, including red beds from epicontinental paralic environments (continental regime) at the base, transitional evaporites from restricted seas (intermediate environment) in the middle, and marine marls (marine environment) at the top. This succession is characteristic of the continental margins around the Atlantic Ocean (Evans, 1978). The only rocks whose age can be determined with macro- and microfossils are those marine units deposited during the drift period. Rocks deposited in transitional environments generally do not contain many fossils to determine geochronologic and paleogeographic evolution. In this situation, palynostratigraphy has proven its value and importance as the best way to date and correlate continental, transitional, and marine rocks, including halite and gypsum, which generally contain few fossils.

Analysis of all the mineral and organic components of the red-bed and salt palynological residues, and multidisciplinary interpretation of these data led to paleoenvironmental and paleogeographic conclusions for placing the red-bed sequence into a plate tectonic framework. The study of red-bed sequence components was done using the palynostratigraphic method, supported by an architectonic stratigraphic analysis (Miall, 1985) of outcrop sequences (Rueda-Gaxiola et al., 1999).

METHODOLOGY

The palynostratigraphic method is a practical way to interpret the organic and inorganic composition of the acid insoluble residue of sedimentary rocks. This method relied on the interpretation of data obtained from the macro- and microscopic analyses of thousands of palynological residues obtained from Mesozoic and

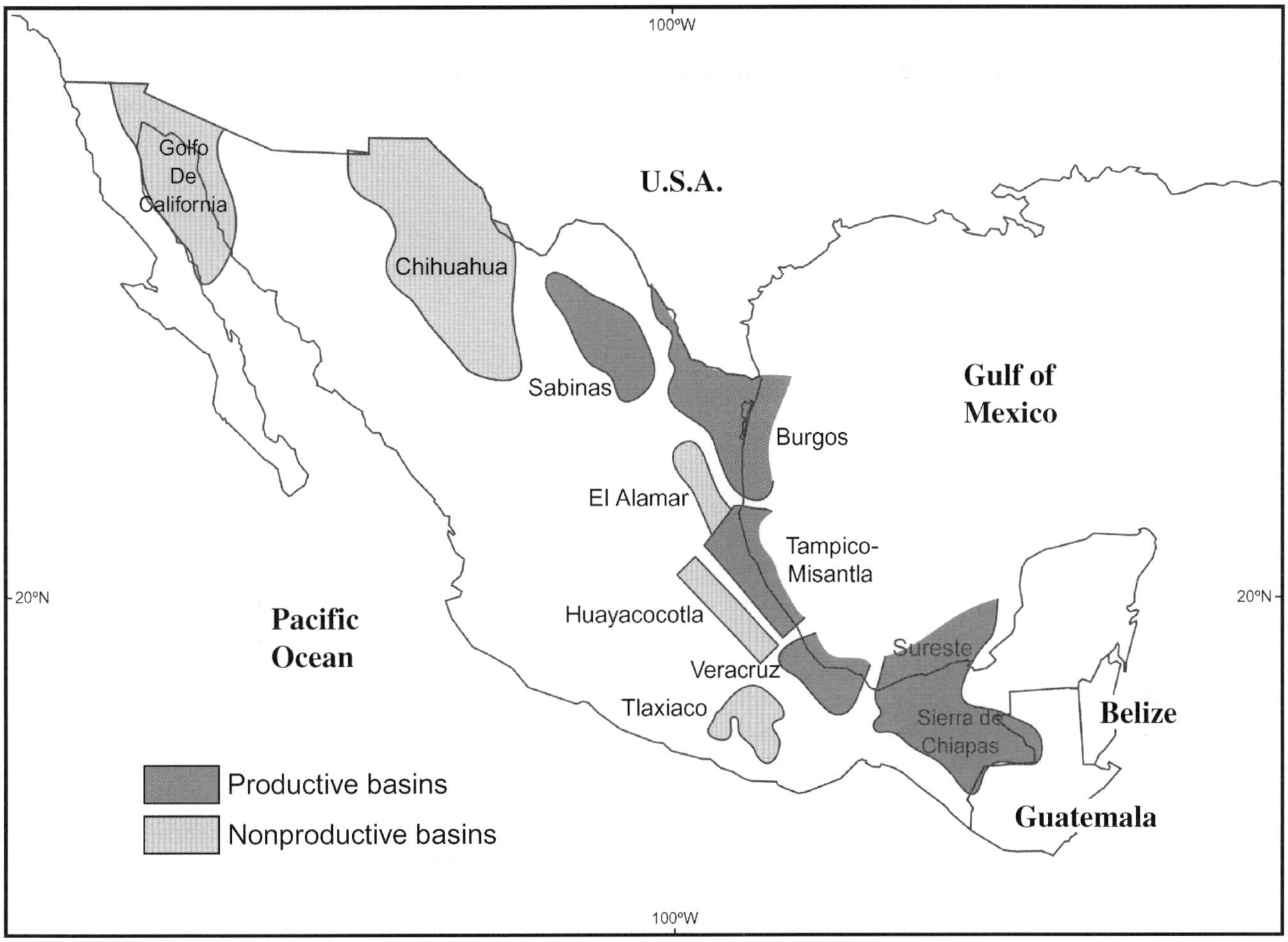

FIGURE 1. Productive and nonproductive Mexican oil basins. All productive basins are directly related to the origin of the Gulf of Mexico Based on González-García and Holguín-Quiñonez (1992).

Cenozoic red-bed and lacustrine rocks of the Mexican petroleum basins (Rueda-Gaxiola, 1999).

Macroscopic observation of palynological residues allows the determination of color and abundance of their components and the color of the glycerinated alcohol in which they are preserved. The color and abundance of palynological residues depend on the type and relative quantity of mineral and organic (acid insoluble) constituents of the rocks. These characteristics and parameters are useful for determining the lithological units in a sedimentary sequence, particularly in red-bed sequences. Additionally, the color of the glycerinated alcohol shows the occurrence of hydrocarbons in the residue.

A microscopic analysis of the palynological residue offers several stratigraphic applications because it contains allocthonous (e.g., heavy minerals) and autochthonous (e.g., pyrite) mineral remains from the original rocks, and the minerals formed during the acid attack. Therefore, this technique helps in the identification of source rocks and their sedimentary conditions. The organic constituents of the residue comprise cells, tissues, and organs of aquatic and terrestrial plants and animals. Their presence also helps to identify the corresponding sedimentary environments. The organic constituents form different types of residual organic matter and fossils. The organic residue is the raw material for coal, oil, and gas formed during diagenesis of sediments and rocks; whereas the fossils determine the age of the rocks and sediments, as well as their paleoecologic and paleogeographic conditions. Additionally, the color of the residual organic particles is related to the thermal evolution of a basin.

In general, the stratigraphic assignment of red beds in Mexico has depended on the age of the overlying and underlying units. Occasionally, the red beds include marine fossiliferous units, but in general, they have not been dated because they are largely unfossiliferous. Therefore, most red-bed sequences in eastern Mexico have been assigned in different chronostratigraphic positions by different authors depending on the methods used and on the region under study. For example, the red-bed sequence exposed in the Huizachal-Peregrina anticlinorium (Figure 4) and at the southern Huasteca

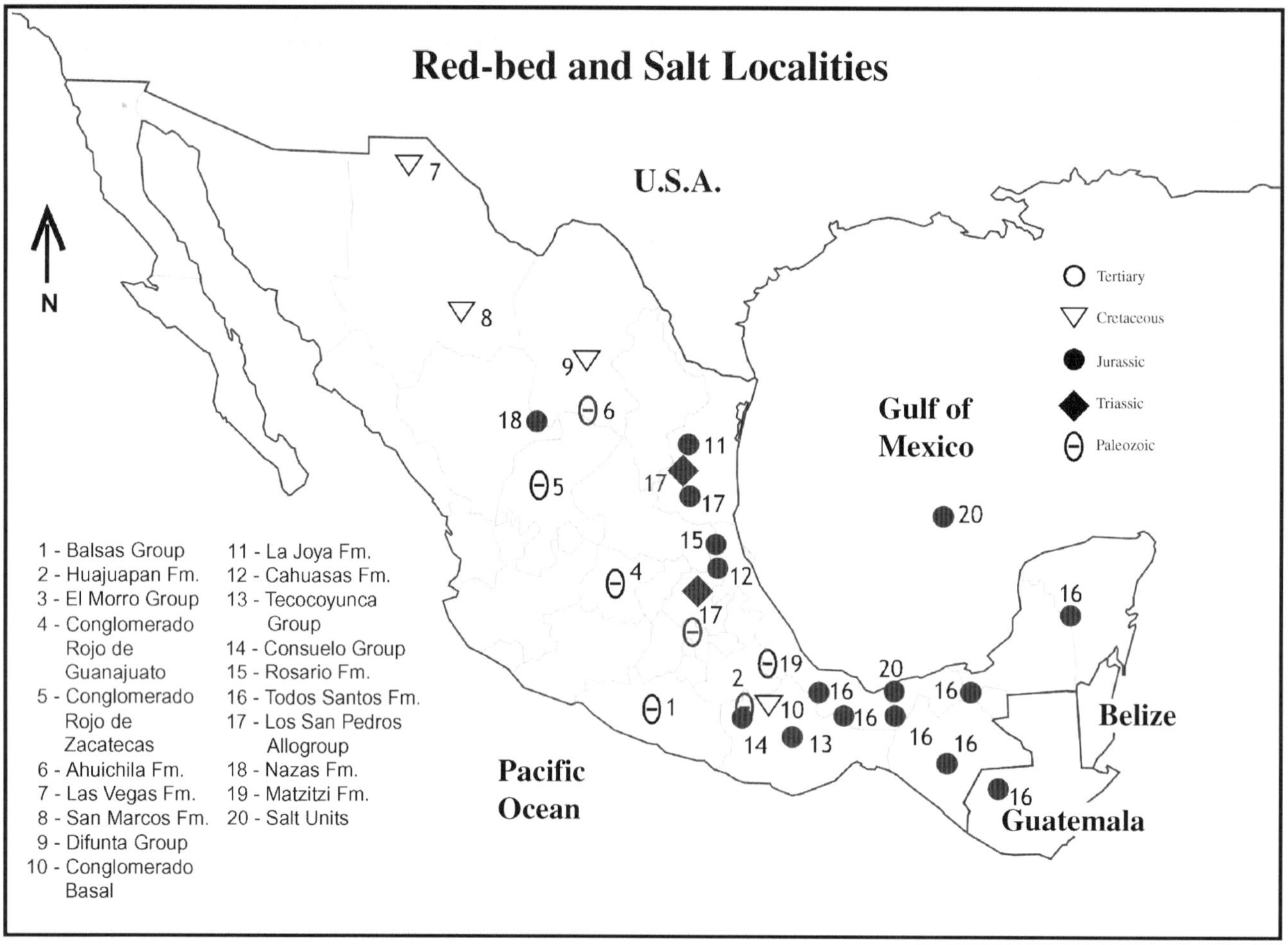

FIGURE 2. Age and distribution of the most important red-bed and salt sequences in Mexico. They range from Miocene to Paleozoic. Red-bed localities are more abundant in eastern Mexico. Jurassic red-bed and salt formations are located in or around the Gulf of Mexico (Del Valle-Reyes, 1997).

region of the Huayacocotla anticlinorium (Figure 4) has been interpreted by different authors as being anywhere from Middle Jurassic to Upper Triassic (Figure 5) (Rueda-Gaxiola et al., 1991a).

MEXICAN RED-BED AND SALT FORMATIONS DATED BY PALYNOSTRATIGRAPHIC ANALYSES

The following red-bed formations in Mexico have been dated by palynologic analysis thus far (Figures 3, 6b–f).

La Boca Alloformation

Since the 19th century, red beds exposed in the Huizachal-Peregrina anticlinorium (Figures 6b, c) between the Zuloaga Formation (Oxfordian) and the Guacamaya Formation (Permian) have been studied in order to establish their chronostratigraphic position (Figure 5). They have been assigned to the Huizachal Group (Callovian–early Oxfordian age) by Imlay et al. (1948), the Huizachal Group composed of the La Joya (Liassic age) and the La Boca (Triassic age) formations by Mixon et al. (1959), and the La Joya (Callovian–early Oxfordian age) and Huizachal (Triassic age) formations by Carrillo-Bravo (1961) and Imlay (1980).

Palynostratigraphic studies (Rueda-Gaxiola et al., 1989), of samples from several red-bed surface sequences in the Huizachal-Peregrina anticlinorium (Figure 4) support formally proposing (Rueda-Gaxiola et al., 1993) the Los San Pedros Allogroup, including the Huizachal and La Boca alloformations (Figure 6b). The organic and inorganic constituents of the residues in these red beds permit the definition of six palynozones and the division of the oldest red beds from the Huizachal-Peregrina anticlinorium correspond to two different sequences. These were deposited during two different periods in two different types of basins: Palynozones A to F form

Formation	Year	Basin	Age	Authors
Cahuasas	1969	Tampico-Misantla Subbasin	Middle Jurassic Bathonian-Callovian?	Rueda-Gaxiola, 1972
Rosario	1975	Tampico-Misantla Subbasin	Liassic-Middle Jurassic Toarcian-Bajocian?	Rueda-Gaxiola (1976, 1977)*
Todos Santos	1982	Western Chiapas-Tabasco-Campeche Subbasin	Upper Jurassic Kimmeridgian	Gutiérrez-Galicia (1984)*
Todos Santos	1986	Eastern Chiapas-Tabasco-Campeche Subbasin	Middle Jurassic Bajocian-Bathonian	Rueda-Gaxiola and Dueñas (1990)*
Todos Santos	1987	Western edge of the Veracruz Subbasin	Middle to Upper Jurassic Bathonian-Oxfordian	Dueñas (1990), in (Herrera et al., 1990)*
La Boca Alloformation. (Los San Pedros Allogroup)	1990	Huayacocotla-El Alamar Basin	Liassic Sinemurian-Pliensbachian	Rueda-Gaxiola, et al. (1993)*
Rosario (Grupo Consuelo)	1996	Tlaxiaco Basin	Liassic	Rueda-Gaxiola and Jiménez-Rentería (2002)

FIGURE 3. Red-bed formations in Mexico, dated by palynostratigraphical analyses.

the Los San Pedros Allogroup, composed of the Huizachal (Late Triassic–Hettangian?) and the La Boca (Sinemurian–Pliensbachian?) alloformations. These units were deposited in a graben (Figure 1) with fluvial-deltaic and interfingered marine sediments at its base, as indicated by the presence of continental and marine palynomorphs (*Dapcodinium priscum*, cf. *Liasidium variabile*, cf. *Nannoceratopsis gracilis*, cf. *Nannoceratopsis ambones*, cf. *Sphaeripollenites* sp., cf. *Inaperturopollenites* sp., *Patinasporites densus*, *Ovalipollis breviformis*) and glauconite. These units were correlated (Rueda-Gaxiola, 1999) with the Huizachal Formation and the Huayacocotla Group (Schmidt-Effing, 1980) that are exposed in the Huayacocotla anticlinorium (Figure 4) about 300 km (186 mi) to the south. These alloformations were found to change to marine facies toward the south where they form the Huayacocotla Group. It was proposed that these continental and marine sediments were deposited in a graben named the Huayacocotla-El Alamar Basin (Rueda-Gaxiola et al., 1993).

The correlation between both anticlinoria (Rueda-Gaxiola et al., 1991, 1993) was confirmed by studies of samples from oil wells southeast of the Huizachal-Peregrina anticlinorium. The palynostratigraphical results allowed differentiation of the palynozones from the Los San Pedros Allogroup and also establishment of the facies change between the La Boca Alloformation (dominantly continental facies) and the Huayacocotla Group (dominantly marine facies). Also, this correlation showed that both anticlinoria were in fact two segments (Figure 7c) of one anticlinorium dextrally shifted by a very important megashear located where the Pánuco River presently flows (Rueda-Gaxiola et al., 1999). As described below, this displacement occurred at the end of the Liassic and produced two tectonic blocks: one to the north of the fault containing the Huizachal-Peregrina

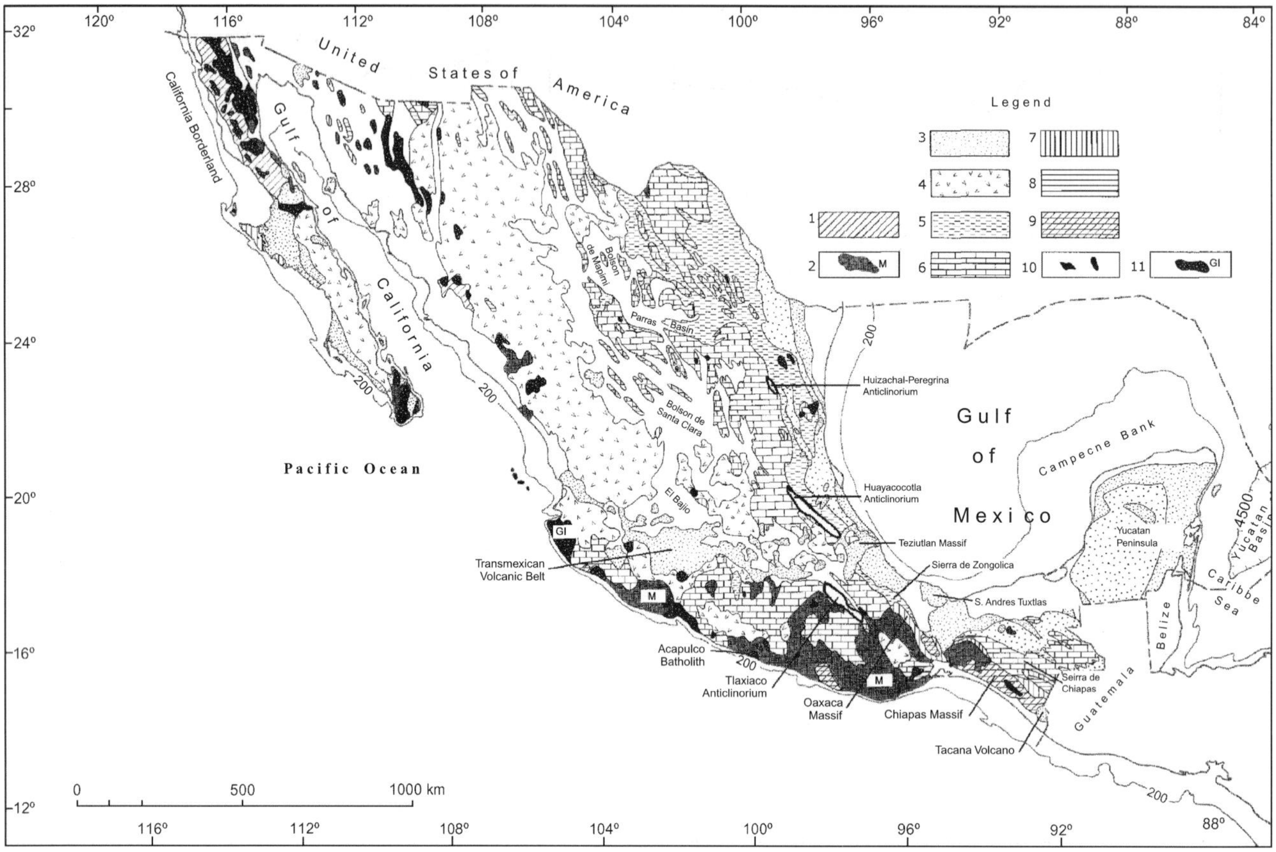

FIGURE 4. Simplified geologic map of Mexico showing principal cities and location of several geologic structures cited in the text. Based on Santiago-Acevedo et al. (1984).

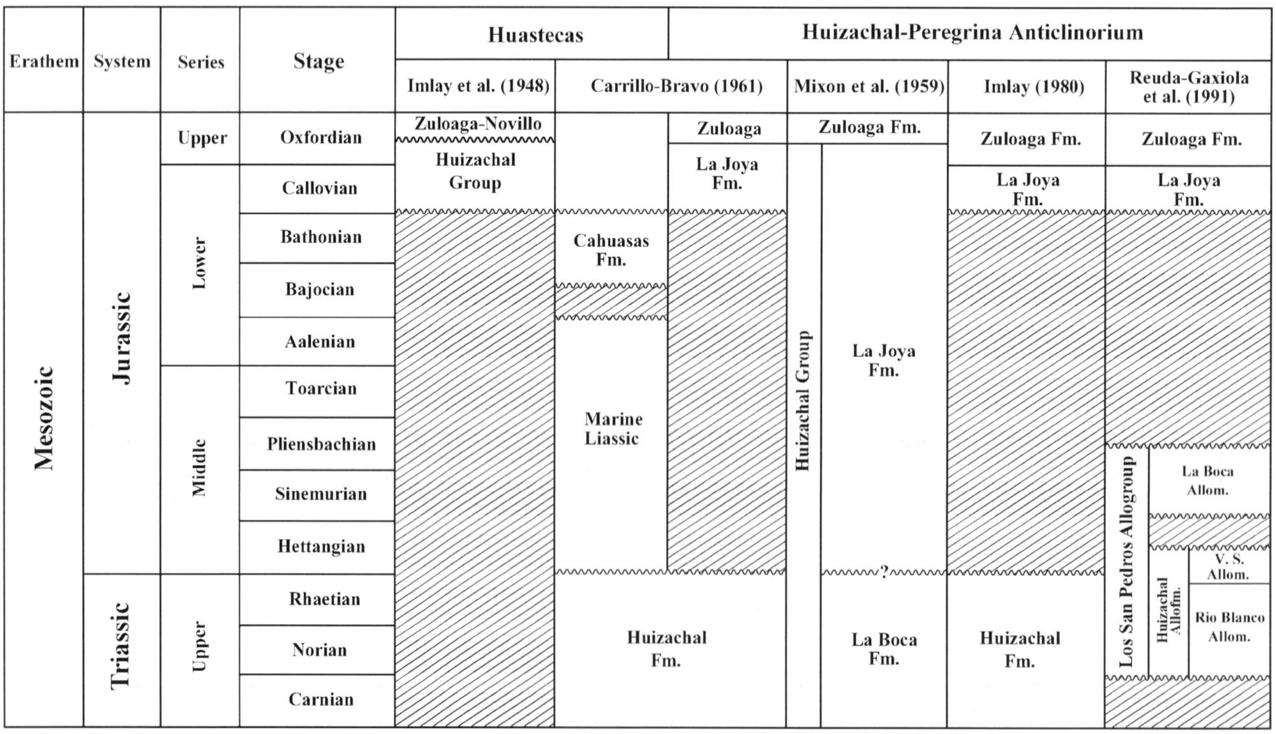

FIGURE 5. Evolution of the chronostratigraphic position of the so-called Huizachal Group from 1948 to 1991.

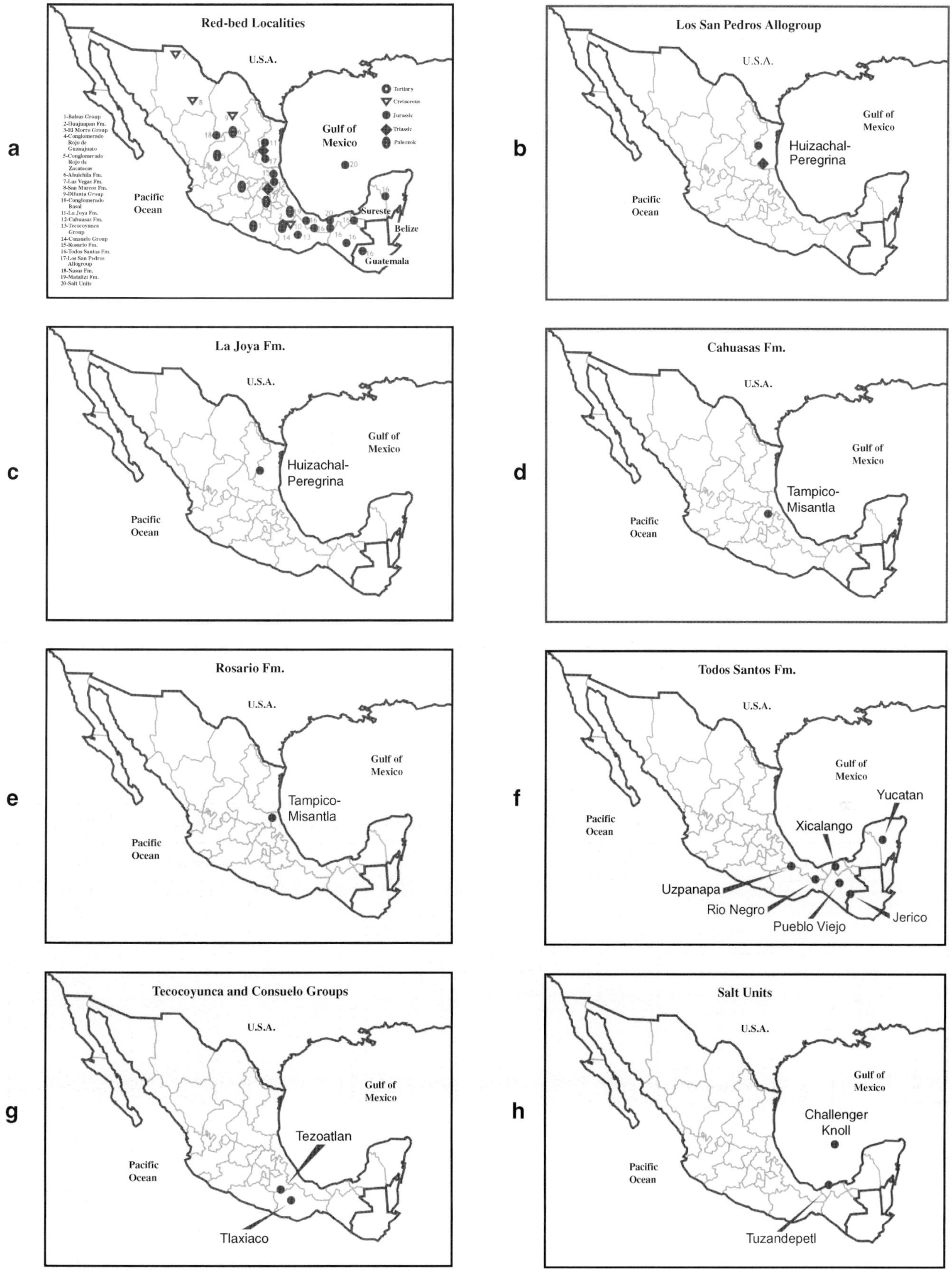

Figure 6. (a) Mexican red-bed and salt localities. (b–h) Red-bed and salt localities cited in Figure 3, dated by palynological analysis.

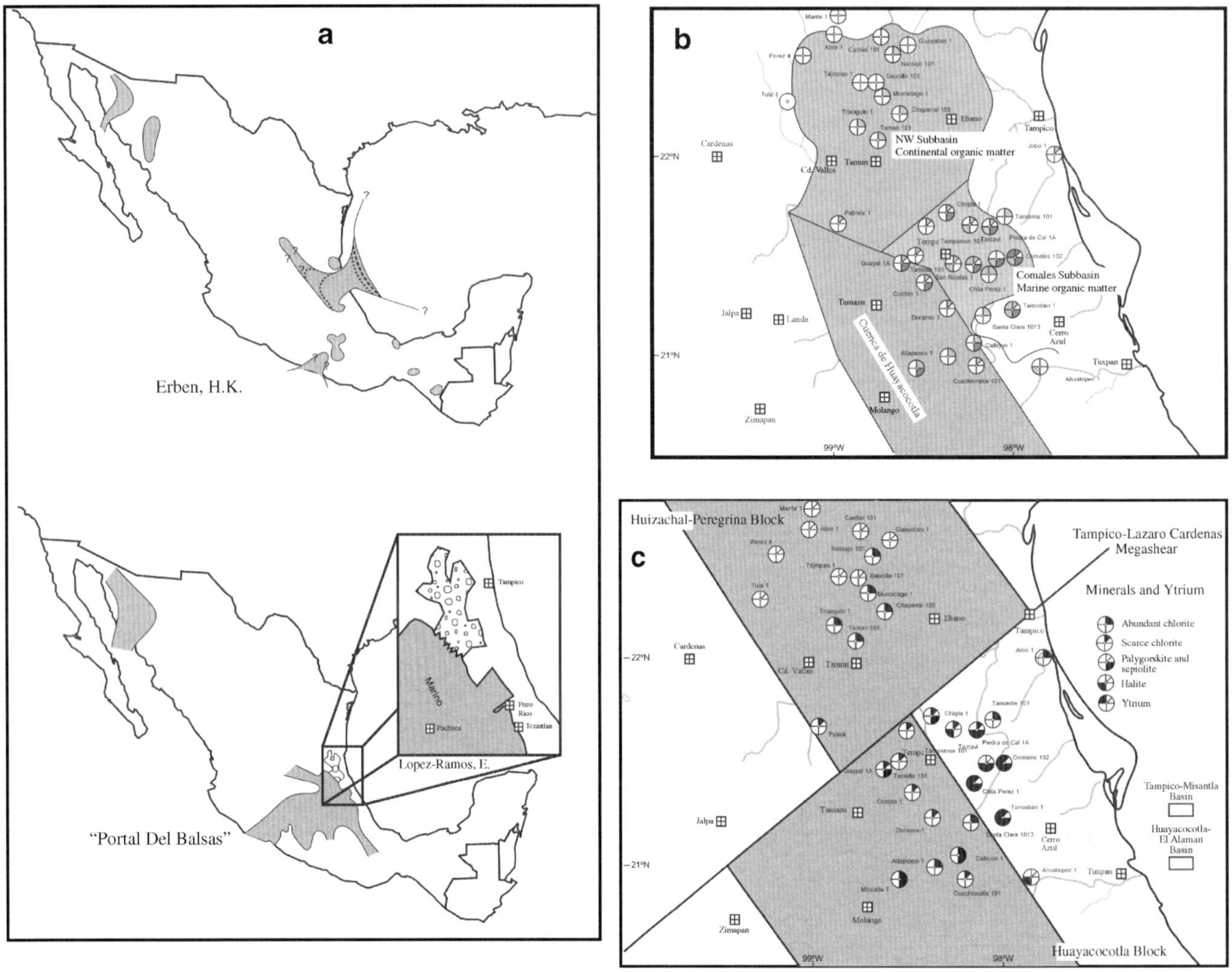

FIGURE 7. (a) Initially (Erben, 1956), the Liassic rocks from eastern Mexico were related to the origin of the Gulf of Mexico. Later (López-Ramos, 1974), these rocks were divided into the marine Huayacocotla Formation and the continental Rosario Formation. This author considered them of the same age but representing a lateral facies change in the same basin open toward the Pacific Ocean. (b) It was later established (Rueda-Gaxiola , 1975; Rueda-Gaxiola et al., 1982) that these formations were deposited in two different basins and at different times. Also, the Rosario Formation Basin was composed of the northwestern and Comales subbasins. (c) Subsequent studies allowed to recognize (Rueda-Gaxiola et al., 1993) that the northwestern subbasin and the Huayacocotla subbasin were two drifted parts of the same Liassic Huayacocotla-El Alamar Basin where La Boca Alloformation and Huayacocotla Formation are deposited. Also, it was clear that the Comales subbasin is the northern part of the Tampico-Misantla Oil Basin. In the later basin, the younger Rosario Formation was deposited above the Huizachal Formation. Drifting followed the Tampico-Lázaro Cárdenas megashear and produced the Huizachal-Peregrina (north) and Huayacocotla (south) blocks during late Liassic.

anticlinorium and the other to the south containing the Huayacocotla anticlinorium (Figures 4, 7c). Both anticlinoria were initially formed by compression of the Triassic–Liassic rocks deposited in the Huayacocotla-El Alamar Basin. This deformation also subsided the northern and eastern parts of the Huayacocotla Block, forming the petroleum Tampico-Misantla subbasin (Figures 1, 7c), whose northern part was initially filled by lacustrine red beds of the Rosario Formation, facing the Liassic rocks of the La Boca Alloformation. This clarifies the origin of what were previously called the Comales and northwestern subbasins (Figure 7b) of the Rosario Formation Basin (Rueda-Gaxiola, 1975), and also demonstrates that two red-bed sequences are present in the Tampico-Misantla Basin, (the Rosario and Cahuasas formations) that represent two different parts of this basin's history: the Rosario Formation as initial sedimentation during the formation of the basin in the late Liassic, and the Cahuasas Formation (Figures 2 , 6d, 8) as the product of the Huayacocotla anticlinorium's first folding and erosion during the Middle Jurassic.

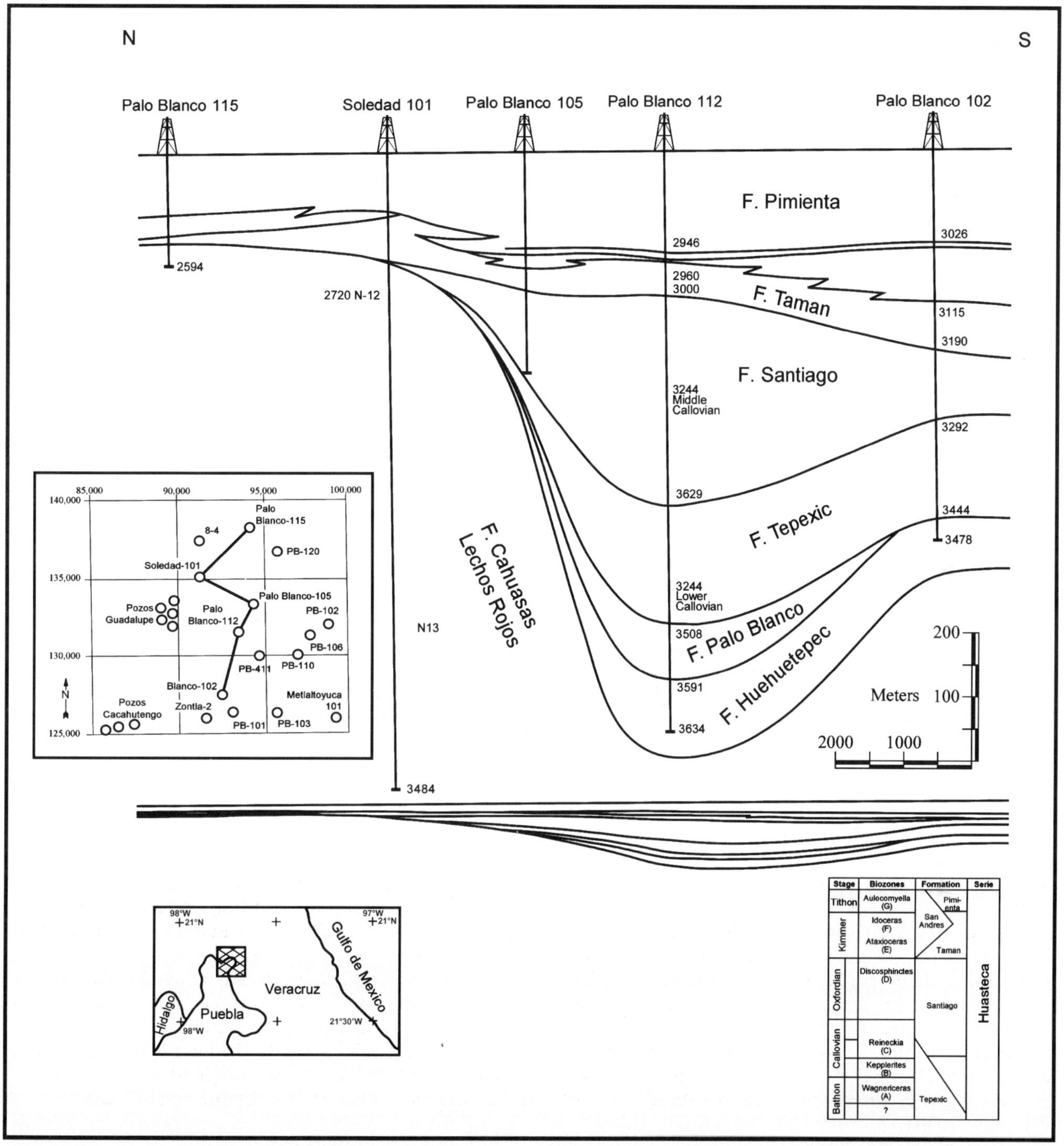

FIGURE 8. The red-bed Cahuasas Formation crops out in the Huayacocotla anticlinorium and found in the subsurface of the Tampico-Misantla Basin. The stratigraphic position of core 13 in the Soledad 101 well and the north–south correlation between the red-bed Cahuasas Formation and the marine rocks of the Huasteca series is based on biostratigraphy (Cantú-Chapa, 1971; Rueda-Gaxiola, 1972).

La Joya Formation

Palynozones I and II of the red-bed sequence analyzed by Rueda-Gaxiola et al. (1993), from the Huizachal-Peregrina anticlinorium, correspond to the younger red-bed La Joya Formation (Figure 6c). This formation is the basal part of the transgressive sequence of the Zuloaga Group (Götte, 1990) and includes lacustrine-fluvial-marginal deposits of a Late Middle Jurassic–early Kimmeridgian transgressive ramp in the Sabinas subbasin (including the Chihuahua, Sabinas, and Burgos basins in Figure 1). This unit unconformably overlaps the Los

San Pedros Allogroup. Because of the lack of palynomorphs or other fossils, this has been dated by stratigraphic position as late Callovian?–early Oxfordian. The La Joya Formation was gradually overlain by the early Oxfordian sulfate-evaporitic Minas Viejas Formation, the middle Oxfordian salt-evaporitic Novillo Formation, or the middle–late Oxfordian calcareous Zuloaga Formation. All of these units were dated from ammonites and microfossils (Bracken, 1982; Michalzik, 1988).

Cahuasas Formation

The Cahuasas Formation has been dated by palynology (Rueda-Gaxiola, 1972). This unit is well exposed in the Huayacocotla Anticlinoriom on the eastern flank of the Sierra Madre Oriental (Figure 6d). This formation is also found in the subsurface in the central region of the Tampico-Misantla subbasin and was identified and dated from core 13 from the Soledad 101 well (Figure 8). Core 13 contained a very thin interval (8 mm, 0.31 in.) of green shale within the very thick (2679–3464 m, 8789–11,365 ft) reddish Cahuasas unit. This interval included a large number of palynomorphs (*Todisporites major*, *Dictyophyllidites mortoni*, *Vitreisporites* sp., *Gliscopollis tarsus*, *Cycadopites* sp., *Exesipollenites tumulus*, *Sphaeripollenites scabratus*, *Eucommiidites* sp., *Classopollis* sp.) of Middle Jurassic age (Bajocian–Bathonian–Callovian?). This unit correlates with the Bajocian–Bathonian red-bed sediments drilled in 1967 by the Glomar Challenger Oceanographic Vessel in the Challenger Knoll (Figure 6h) situated in the central region of the Gulf of Mexico (Kirkland and Gerhard, 1971).

Rosario Formation

The palynostratigraphic method was also used in core samples from the Huayacocotla Formation (9 wells) and the Rosario Formation (27 wells). These units were drilled in the northern part (Figure 6e) of the so-called Tampico-Misantla Basin (Rueda-Gaxiola, 1975). The Huayacocotla Formation was considered to be a marine unit and the Rosario Formation a continental unit of the same age (Figure 7a), thus representing lateral facies in the same basin (López-Ramos, 1974). Palynological data (Rueda-Gaxiola, 1975) prove that each formation represents a separate stratigraphic sequence deposited in two different basins, one in the so-called Huayacocotla Basin and the other in the Rosario Formation Basin (Figures 6e, 7b); the latter as part of the larger Tampico-Misantla Basin. The data also prove that these units have different ages; the former originated during the early Liassic (Sinemurian–Pliensbachian) and the latter at the end of the Liassic–Early Middle Jurassic (Toarcian–Aalenian).

Based on several ammonites at the Pilcuatla 1, Cordón 1, Iztazoquico 1, and Santa Cruz 1A wells (Flores-López, 1974), the Huayacocotla Formation is Sinemurian. However, the Rosario Formation is probably Middle Jurassic as indicated by macroscopic plant remains in cores from the Comales 102 and Piedra de Cal 8 wells (Flores-López, 1974). These plant remains from the Comales 102 well had previously been used by Bonet and Humphrey to correlate this formation to the "Capas con plantas unit," established by Burckhardt in 1930 (in Erben, 1956b). This unit includes the upper part of the Rosario Formation in the southwestern region of Tezoatlán, Oaxaca, in the Tlaxiaco Basin (Figures 1, 4, 6g). The degree of metamorphism in the Huayacocotla Formation was found to be greater than that of the Rosario Formation indicating different tectonic evolution of these formations in two separate basins. From the palynological results, it was also possible to divide the so-called Rosario Formation Basin into two separate subbasins (Figure 7b); one located toward the southeast, called the Comales subbasin, with continuous sedimentation of continental origin at the base and marine origin at the top, and another northwestern subbasin, with a greater continental influence and some erosional events. The northern and eastern regions of the then so-called Rosario Formation Basin were characterized by deltaic environments that were more conspicuous to the east as a product of the erosion of batholiths located in the region now occupied by the Gulf of Mexico. However, tectonically speaking, it was possible to determine that during the late Liassic, the Huayacocotla Basin was affected, by compression events that folded and eroded the Liassic rocks.

These characteristics were still considered as established in 1981 until the Rosario Formation was restudied for a petroleum-economic evaluation by means of palynostratigraphy and inorganic (x-ray) and organic (rock-eval) geochemistry (Rueda-Gaxiola et al., 1982). This study led to the following new important conclusions:

1. In the Comales subbasin, the Rosario Formation contains basal micaceous hematized sandstones, siltstones, and shales that change gradually upward to calcareous subgraywackes and subarkoses, and even to shaly limestones, changing gradually to the younger Cahuasas Formation red beds.
2. The Cahuasas Formation is transitionally or unconformably covered by the Callovian Tepéxic Formation (Figure 4), which represents the beginning of the gulf marine transgressive sedimentation during the Late Middle–Late Jurassic.
3. It was established that the sedimentary sequences in these two basins were different; upward-coarsening for the Huayacocotla Formation and upward-fining for the Rosario Formation.
4. Inorganic geochemical data showed that the Rosario Formation was deposited under alkaline and reducting chemical conditions in the so-called Comales

subbasin, whereas palynostratigraphical data show algaceous organic matter and pyrite formed oxidizing conditions toward the northwestern part of the Rosario Formation Basin.

5. Algaceous organic matter was dominant in the Comales subbasin; it reached the Catagenic zone, according to the thermal alteration index (TAI) (+2) and vitrinite reflectance index (VRI) (1.54) values. The yellow of the palynological residue showed the presence of soluble hydrocarbons in the sequence; these data correlated with the free hydrocarbons shown by rock-eval S1 values.
6. However, in the so-called northwestern subbasin, woody and coaly organic matter was dominant, with higher TAI (3) and VRI (1.61) values. The glycerinated alcohol was uncolored.
7. Palynological residues are darker from the northwestern subbasin (gray and dark gray) than from the Comales subbasin (sepia brown).
8. Abundance of organic matter in the Comales subbasin is higher than in the northwestern region.
9. The presence of white cubic mineral crystals in the palynological residue suggest that the Rosario Formation sedimentary sequence was first deposited under lacustrine evaporitic conditions and then in marine environments.

All these data distinguish two very different regions in the Rosario Formation Basin: the Comales subbasin and the northwestern subbasin. Therefore, to find oil accumulations, drilling the former, based on its more important oil generating characteristics was proposed. In 1989, the Silozuchil 1 well confirmed that this subbasin is productive. In 1993, as previously mentioned, it was understood that the northwestern subbasin is, in fact, a drifted part of the Liassic Huayacocotla Formation deposited in the Huayacocotla-El Alamar Basin, and that the Comales subbasin is the northern part of the Tampico-Misantla Oil Basin where the younger Rosario Formation was deposited above the red-bed Huizachal Formation (Figures 1, 7c, 13).

Todos Santos Formation

This formation was first described near the town of Todos Santos in Guatemala by Sapper (1894). This lithostratigraphic unit is widely distributed (Figures 1, 6, 9, 10), forming a belt along the southern border of the Veracruz Basin. From San Pedro Tentila to Guichicovi, Sierra de Zongolica (Figure 4), it overlies Late Triassic–Liassic andesites or early Paleozoic metamorphic rocks. In the southeastern subbasin, from Matías Romero to Agua Prieta in the Chiapas massif (Figure 4), it overlies late Triassic–Liassic andesites or late Paleozoic igneous intrusive rocks. This red-bed formation has been placed at different levels in the geochronologic scale. Because of the lack of macrofossils, the age of the Todos Santos Formation was established by stratigraphic position as ranging from Triassic to Jurassic. The first known biostratigraphic age was given as Late Jurassic–earliest Cretaceous based on palynomorphs from a surface sequence outcropping on the border of the Chiapas massif in central Chiapas (Blair, 1982, 1986).

Palynostratigraphical data were obtained from the sections cropping out at the Pueblo Viejo and Jericó localities (Figure 6f), northwest and southeast of Tuxtla Gutiérrez, Chiapas, respectively (Gutiérrez-Galicia, 1984). Based on these results, the Todos Santos Formation is interpreted as the red-bed basal unit of the Jurassic marine transgressive sequence. This unit overlaps Middle Jurassic (early Bajocian) andesites that in turn cover the upper Paleozoic granitic rocks of the Chiapas massif (Figure 4). It gradually changes upward to the marine San Ricardo Formation, which contains a wide variety of continental and marine palynomorphs of Callovian to Tithonian age. Therefore, the Todos Santos Formation in this locality has been defined as middle Bajocian–Bathonian.

In 1987 and 1988 (Estavillo-García and Herrera-Soto, 1988; Herrera-Soto and Estavillo-García, 1988), and 1989 (Herrera-Soto et al., 1990), Dueñas performed the most recent palynostratigraphic work on the Todos Santos Formation (data integrated by Herrera-Soto and Villaseñor-Rojas, 1991), from the Mexican states of Oaxaca, Veracruz, and Chiapas in the northeast region of the Sierra de Zongolica and the Chiapas massif (Figure 4). These are the limits of the Veracruz subbasin and southeastern subbasin (Figure 1).

Continental and marine palynomorphs were found, generally, in thin intervals of green and greenish fine-grained strata intercalated within the red-bed sequences. The northernmost locality is in the Uzpanapa region (Figure 6f) where the Bathonian–Callovian red beds represent the basal units of the Upper Jurassic marine transgressive sequence, and unconformably overlap andesitic rocks. These data indicate that the Todos Santos Formation becomes younger toward the southeast (Figure 9).

In the subsurface of the southeastern subbasin, Bajocian?–Bathonian to Oxfordian red beds of the Todos Santos Formation were drilled by the Xicalango 1 well (Figures 6f, 10) from 5240 to 5597 m (17,192 to 18,363 ft) (Rueda-Gaxiola and Dueñas, 1990). This is the oldest biostratigraphic age established for this formation based on palynological information (pollen and spores and dinoflagellate cysts from the following taxa: *Eucommiidites troedssoni*, *Concavissimisporites nutabilis*, *Lygodiumsporites adriennis*, *Perinotriletes perinatus*, *Verreticulisporis giganteus*, *Todisporites major*, *T. minor*, *Concavisporites laticransus*, *Dictyophillidites* sp., *Ischyosporites* sp., *Applanopsis segmentatus*, *Cedripites canadensis*, *Perinopollenites elatoides*, *Callialasporites dampieri*, *Zonalapollenites lucidus*, *Arauariacidites australis*, *Exesipollenites scabratus*, *Valensiella ovula*,

Locality	Age
Northwest	
Uzpanapa, Río Oaxaca, Oaxaca.	Bathonian-Callovian
Matías Romero, Oaxaca	Bathonian-Callovian
Valle Nacional, Río Montenegro, Oaxaca	Late Oxfordian-early Kimmeridgian
Southeast	
Pueblo Viejo, Chiapas.	Middle Bajocian-Bathonian
Pueblo Viejo, Chiapas	Early Middle Kimmeridgian
Jericó, Chiapas	Late Jurassic-Early Cretaceous

FIGURE 9. Older to younger northwest to southeast palynological variation of the red-bed Todos Santos Formation along the southern border of the Veracruz subbasin.

Pareodina prolongata, Tenua capitata, Pareodina ceratophora, Pluriarvalium osmingtonense, Tenua pilosa, Pareodinia spinossisima, Chytrocisphaeridia chytrocides, Scrinodinium sp., *Tenua verrucosa, Tenua villersense, Ellipsoidictyum* sp., *Mendicodinium groenlandicum*). In this sequence, red beds are composed of slightly calcareous bentonitic

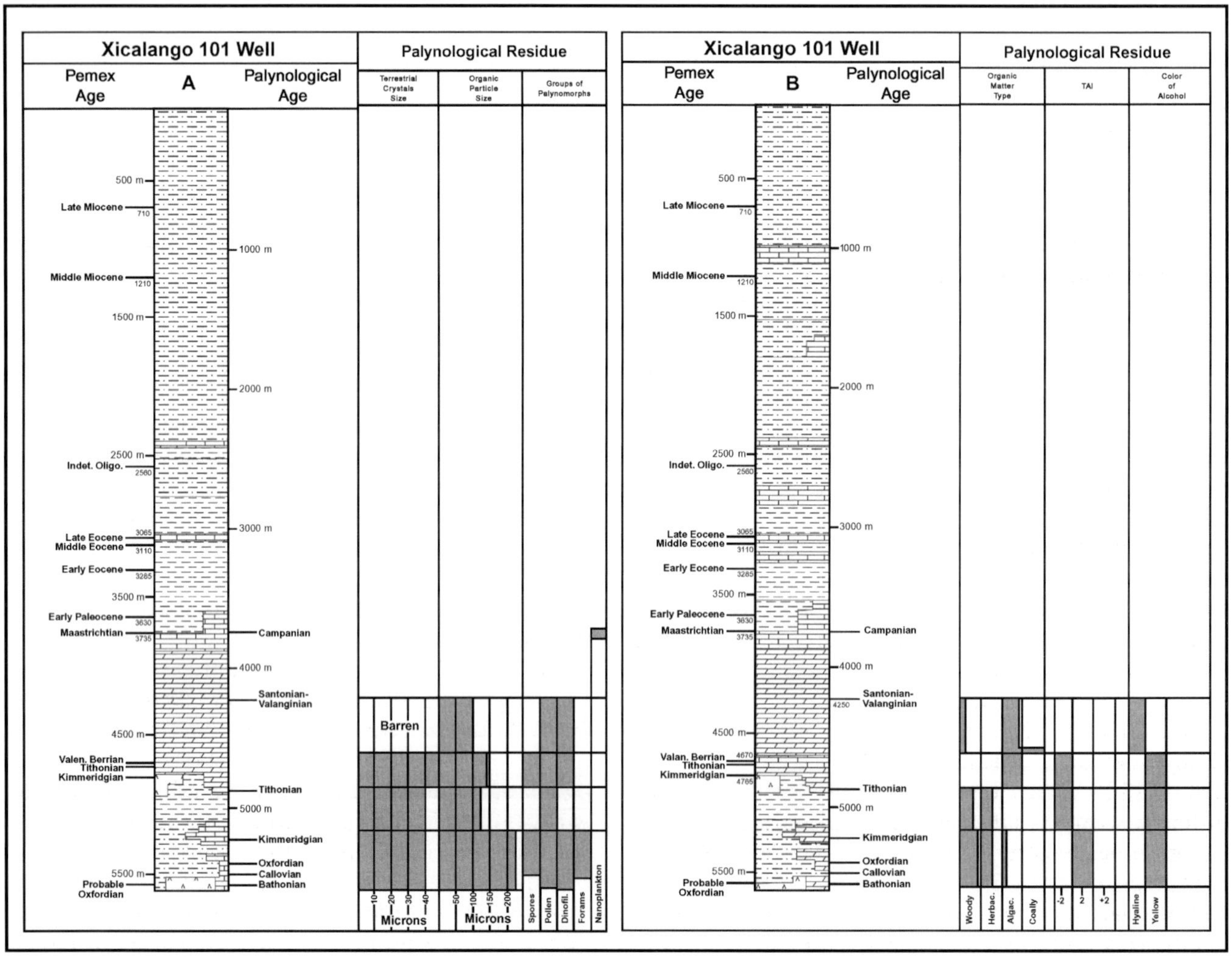

FIGURE 10. (A) Type of palynomorphs and size of organic and inorganic particles in the palynological residue of Bathonian to Campanian rocks from the Xicalango 1 well. (B) Organic matter type, thermal alteration index (TAI), and color of the glycerinated alcohol. Bathonian rocks, some meters above the bottom of the well, correspond to the red-bed Todos Santos Formation; the assemblage of palynomorphs shows that the age beneath is probably Bajocian (Rueda-Gaxiola and Dueñas, 1990).

reddish siltstones and greenish shales with mica and anhydrite. They constitute the base of the Middle–Upper Jurassic marine transgressive sequence that gradually changes from terrestrial to calcareous rocks in the Gulf of Mexico.

Farther south, in the subsurface of the Sierra de Chiapas (Figure 4) and Guatemala, the red beds underlay the evaporitic facies of the Lower Cretaceous Coban Formation. In fact, these rocks belong to the Todos Santos Formation, which is the regional sedimentary basement (Bajocian to Lower Cretaceous by stratigraphic position) differing in age depending on the position of the transgressive sequence above the upper Paleozoic igneous and metamorphic basement. In the subsurface of the Yucatán Peninsula (Figure 4), the oldest sedimentary rocks deposited on the igneous and metamorphic basement are also Lower Cretaceous red beds. These have been also considered as the Todos Santos Formation and form the base of the marine transgressive sequence over the peninsula (López-Ramos, 1979).

Consuelo and Tecocoyunca Groups

During 1995, two preliminary exploration fieldtrips to the northwestern region of the Tlaxiaco Basin, near Tezoatlán, Oaxaca (Figures 1, 2, 6g), allowed us to collect samples from the Consuelo Group (composed of the Rosario, Conglomerado Prieto, and Conglomerado Cualac formations), and from the base of the Tecocoyunca Group (composed by the Zorrillo, Taberna, Simón, Otatera, and Yucuñuti formations), initially described by Erben (1956), in the Tlaxiaco anticlinorium (Figure 4).

From field work and macroscopic and preliminary palynological analysis of outcropping samples (Rueda-Gaxiola and Jiménez-Rentería, 1996; Jiménez-Rentería, 2004), it was possible to identify liquid hydrocarbons in these rocks and also to establish that the Conglomerado Cualac is not the upper unit of the Consuelo Group but the basal unit of the Tecocoyunca Group. In this region, this unit is a breccia not a conglomerate and in some places it is a quarzite, as initially described by Guzmán (1950). This information is very important because the sequence seems continuous, toward the east of the basin but in some places to the northwest, between the Conglomerado Cualac and the Conglomerado Prieto, it is possible to see a disconformity that separates two different sequences (Meneses-Rocha et al., 1994). The lower sequence comprises the Liassic red-bed Rosario Formation, dated by Erben based on macroflora, which changes laterally and upward to the Conglomerado Prieto. This unit has the following palynoflora: *Monosulcites minimus*, *Granulatisporites granifer*, *Lycopodiumsporites trambanensis*, *Baculatisporites comaumensis*, *Cycadopites reticulates*, and *Cyathidites minor*). The upper sequence comprises the Conglomerado Cualac at the base of the late Liassic–Late Jurassic marine transgressive sequence made up of the Zorrillo to Yucuñuti formations, which have been dated based on macroflora and ammonites. The Cualac Formation contains the following taxa: *Cycadopites andrewi*, *Leiotriletes brevis*, *Ischiosporites crateris*, *Cycadopites* cf. *durbamensis*, *Lophotriletes* sp., *Baltisphaeridium micropunctatum*, *Duplicisporites granulates*, and *Cyathidites australis*. These preliminary data suggest that these sequences are sedimentologically and chronologically nearly equivalent to the previously described sequences in the Rosario, Cahuasas, and Tepéxic formations drilled in the Rosario Formation Basin in the northern part of the Tampico-Misantla Basin, and are similar to the Liassic upper part of the Los San Pedros Allogroup (Rueda-Gaxiola et al., 1993), to the Middle Jurassic basal part of the Zuloaga Group (Götte, 1990) described in the Huizachal-Peregrina anticlinorium, and to the Liassic Huayacocotla Group (Schmidt-Effing, 1980) covered by the Middle Jurassic Cahuasas Formation in the Huayacocotla anticlinorium where both groups are separated by a very clear unconformity.

This Tlaxiaco Basin is also a northwest–southeast oriented graben, as proposed by Morán-Zenteno et al. (1993) and Meneses-Rocha et al. (1994), and reported by Williams-Rojas et al. (1997). The latter authors also showed that southwest of the Tlaxiaco Basin (Figure 6g) there is another graben named the Huamuxtitlán Basin, where the Conglomerado Cualac and the Tecocoyunca Group overlap the rhyolitic and andesitic rocks of the Las Lluvias Formation.

Palynological Ages of Salt and Cap Rock in the Southeastern Subbasin

Four age assignments are based on palynological data for the salt structures in the Chiapas-Tabasco-Campeche region also known as the southeastern subbasin (Figures 1, 6h) are observed. These are the oldest drilled sedimentary rocks west of the southeastern subbasin (Figure 11). The center of the Gulf of Mexico is made up of salt, which is considered to have been deposited above red beds (Figure 6h).

On the Challenger Knoll (Figure 6h), the Glomar Challenger Oceanographic Vessel (Kirkland and Gerhard, 1971) drilled cap rock composed of anhydrite, gypsum, calcite, and sulfur. This well established the existence of light and sulfur-rich petroleum of marine origin (Burk et al., 1969). This salt intrusion is one of a series of salt structures that continue southwestward to the Isthmus of Tehuantepec. Some palynomorphs were obtained from an oily gray calcite considered to have migrated from the salt intrusion. These were only carbonized pollenospores indicating continental origin of Middle Jurassic to pre-Tithonian age. The most abundant species were *Sphaeripollenites subgranulatus*, *Classopollis itunensis*, and *C. torosus*; common species were *E. tumulus*, *Classopollis pflugii*, *C. minor*, *Ginkgocycadopites nitidus*, *M.*

Material	Year	Site	Age	Author
Cap rock	1967	Challenger Knoll Center of the Gulf of Mexico	Bajocian-Kimmeridgian	Kirkland and Gerhard (1971)*
Salt	1970	Trinitaria well Southeastern subbasin	Middle Jurassic	Rueda-Gaxiola (unpublished)
Salt	1970	Solidad 102 well Southeastern subbasin	Middle Jurassic	Rueda-Gaxiola (Viniegra-Osorio, 1971)*
Salt	1993	Tuzandépetl Salt Dome Southeastern subbasin	Middle Jurassic	Dueñas (unpublished)

FIGURE 11. Palynological ages of salt and cap rock in the southeastern subbasin.

minimus, and *Perinopollenites* sp. The scarce forms were *Deltoidispora hallii*, *Granulatisporites* sp., *T. minor*, *C. dampieri*, *Vitreisporites pallidus*, *Alisporites thomassii*, *Equisetosporites* sp., and *Eucommiidites* sp. This is almost the same palynomorph assemblage found in core 13 from the Soledad 101 well (Figure 8) in the red-bed Cahuasas Formation, where the presence of the spore *T. major* and the age of the overlapping Tepéxic and Huehuetepec formations led us to limit its age to Middle Jurassic (Bajocian–Bathonian).

A palynological analysis of the salt determined a Middle Jurassic age for this rock. Contreras and Castillón (1968) report three salt samples from the Soledad 102 well, in the so-called Salina del Istmo Basin near the Tuzandépetl wells (Figure 6h). One sample was dated as Cretaceous and the other two as Jurassic–Cretaceous, with abundant pollen of the *Classopollis* and *Exesipollenites* genera and also from Cycadophites. In 1970, Rueda-Gaxiola palynologically dated these same salt rocks (Viniegra-Osorio, 1971) and also evaporites drilled by the Trinitaria 2 well, near the border with Guatemala, as Middle Jurassic (unpublished). Recently, Dueñas (1993, unpublished) analyzed samples from gray and dark gray siltstone and shale fragments in salt from one of the Tuzandépetl wells (Figure 6h) drilled in the Salina del Istmo Basin, near Coatzacoalcos, and found pollenospores and dinoflagellate cysts, which allowed him to determine a Bathonian–Callovian age (unpublished): *Todisporites major*, *Araucariapollenites* cf. *laffitte*, *Ischyopollenites granulatus*, *Trilites* sp., and *Pareodinea prolongata*.

Results of all these palynological analyses suggest that the age of the salt in these localities is Middle Jurassic. Evidences of evaporites over Bathonian red beds were found in the previously mentioned Xicalango 101 well, suggesting a possible post-Bajocian age. However, this age could become younger toward the southeast because the marine transgression could have progressed in this direction. This salt was correlated with the Werner–Louann salt deposits in the northern Gulf of Mexico, of post-Liassic to pre-late Oxfordian age (Salvador, 1991), and probably with the evaporitic rocks of the Minas Viejas Formation in the Sabinas subbasin, of late Callovian?–early Oxfordian age, because in some places they are overlapped by the Zuloaga Formation. The Zuloaga Group (Götte, 1990) comprises the La Joya (continental-transitional), Minas Viejas (evaporitic), Novillo (evaporitic-calcareous), and Zuloaga (calcareous) formations in the Monterrey and Aramberri localities in the Sabinas subbasin, north of the Huizachal-Peregrina anticlinorium. The Zuloaga Group was also correlated with the evaporitic Huehuetepec Formation (Bathonian–Callovian) deposited over the Cahuasas Formation (Figure 8) in the Tampico-Misantla subbasin.

STRATIGRAPHIC RELATIONSHIPS AMONG MESOZOIC RED BEDS AND EVAPORITIC ROCKS

Establishing the relationships among all these dated stratigraphic units, based on palynostratigraphic analyses, is important because they form the sedimentary basement of the oil subbasins in the Mexican part of the Gulf of Mexico.

1. Figure 12 shows an unconformity separating the basement rocks from the Middle Jurassic rocks (La Joya and Todos Santos formations) in the Sabinas and southeastern subbasins (Figure 1) of the Gulf of Mexico and the Liassic rocks (La Boca Alloformation and Huayacocotla Formation) from the Middle Jurassic red beds (La Joya and Cahuasas formations) in the Huayacocotla-El Alamar Basin. Nevertheless, as previously stated, in the Tlaxiaco Basin and in

Age	Sabinas Subbasin	Huizachal-Peregrina Anticlin orium	Huayoco cotla Anticlin orium	Tlaxiaco Anticlin orium	Tampico Mislantla Subbasin	South eastern Subbasin
Middle Jurassic	Minas Viejas La Joya	La Joya *	Cahuasas *	Cualac	Huehue-tepec Cahuas as	Salt Todos Santos *
Lower Jurassic	Basement	La Boca	Huayac-ocotla	Rosario	Rosario	Basement
	Huayacocotla-El Alamar Basin?					

Figure 12. Stratigraphic relationships of red beds and evaporitic rocks.

the northern part of the Tampico-Misantla Basin, in some localities, the Rosario Formation red beds gradually change to the Cualac and Cahuasas formations. Above this unconformity, the red beds are of Middle Jurassic age and slightly older than the evaporitic rocks deposited in the Gulf of Mexico marginal subbasins (Sabinas, Tampico-Misantla, and southeastern subbasins). Therefore, these subbasins are slightly younger than the rest of the Gulf of Mexico. Their red beds correlate with the marine transgression that occurred during the drifting stage, as indicated by the presence of marine palynomorphs present in the salt and marginal red beds. This evidence indicates that the age of the rifting stage is older than the age of the red beds and is possibly related to the Tampico-Lázaro Cárdenas (Figures 8c, 13) and Teziutlán-Acapulco megashears (Toarcian–Aalenian). These megashears allowed the Huayacocotla (Figure 13) and Tlaxiaco blocks to shift southwestward, originating from the Tampico-Misantla and the Tlaxiaco basins.

2. In Figure 12, clearly differentiating the ages of origin of the Huayacocotla-El Alamar Basin and the Gulf of Mexico with its Mexican petroleum sub-basins is possible because (1) in the first case, the oldest red beds belong to the Late Triassic Huizachal Alloformation and, (2) in the second case, the red-bed Todos

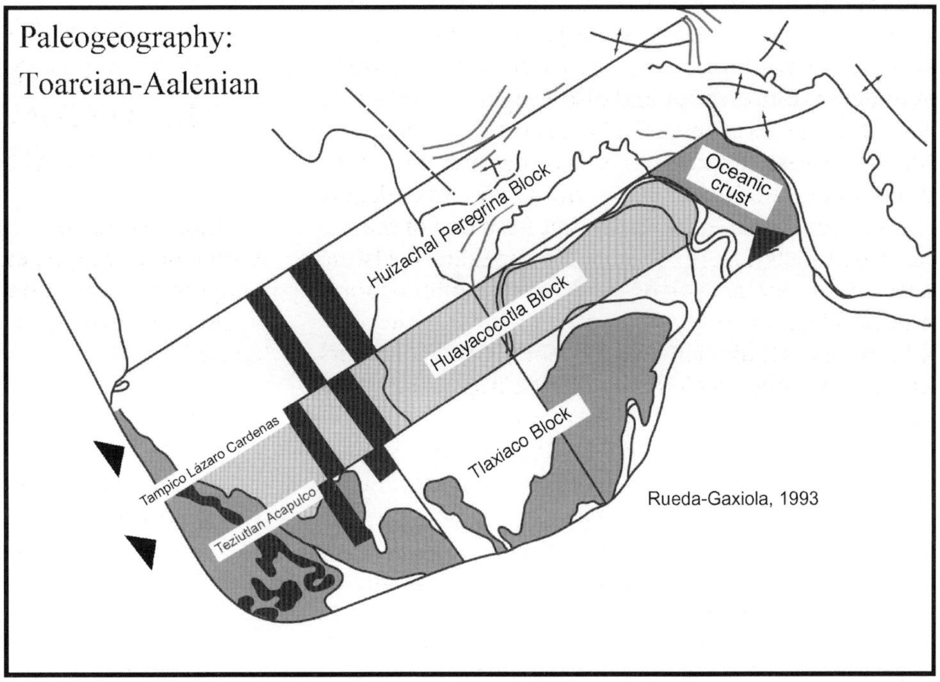

Figure 13. A very careful analysis of the palynological data and geographic, geologic, and tectonic maps of Mexico let us deduce the southwestward shift, during Toarcian–Aalenian ages, of the Huayacocotla and Tlaxiaco blocks through the Tampico-Lazaro Cárdenas and Teziutlán-Acapulco megashears by considering the present geographical positions of the Huizachal-Peregrina, Huayacocotla and Tlaxiaco anticlinoria. This shifting movement divided the Huayacocotla-El Alamar and the Sierra de Catorce-Tlaxiaco basins into three fragments.

Santos Formation could be pre-Bajocian, related to the dextral displacement of the Huayacocotla and Tlaxiaco blocks through the Tampico-Lázaro Cárdenas and Teziutlán-Acapulco megashears (Figure 13). Based on this displacement, it is possible to consider that the Huayacocotla-El Alamar Basin was divided into three rather than two blocks.

3. Based on the last two points, it is possible to understand that the Huayacocotla-El Alamar Basin is, in time and space, a very different basin from the Petroleum subbasins in the Mexican part of the Gulf of Mexico. Careful analysis of the Geologic and Tectonic maps of Mexico suggest a southwestward shift of the Huayacocotla and Tlaxiaco blocks by considering the geographical positions of the Huizachal-Peregrina, Huayacocotla, and Tlaxiaco anticlinoria (Figures 4, 13).

All of this information, obtained from the basal redbed and evaporitic units of the sedimentary sequences from the Huayacocotla-El Alamar Basin and the Gulf of Mexico petroleum subbasins in Mexico, supports a simple tectonic model of the origin and evolution of this basin and these subbasins (Rueda-Gaxiola, 2003).

ACKNOWLEDGMENTS

The author wishes to express his appreciation to Henrik Nohr-Hansen, organizer and chairman of the Mesozoic Palynology Session G6 from the 11th International Palynological Congress, for accepting to be a speaker for my theme in the 2004 IPC in Granada, Spain; Len Hills from the University of Calgary and Javier Helenes from CICESE, Ensenada, Mexico, for pertinent comments, corrections, and suggestions made; and to Joshua Rosenfeld and Claudio Bartolini for the improvements to the final manuscript. I am grateful to the Comisión de Fomento de Actividades Académicas (COFAA) del Instituto Politécnico Nacional, México, for the economic support provided to this project.

REFERENCES CITED

Blair, T. C., 1982, Late Jurassic–earliest Cretaceous alluvial fan deposits of the Todos Santos Formation, central Chiapas, Mexico: Department of Geological Sciences, University of Colorado, p. 1–46.

Blair, T. C., 1986, Paleoenvironments, tectonic and eustatic controls of sedimentation, regional stratigraphic correlation and plate tectonic significance of the Jurassic–lowermost Cretaceous Todos Santos and San Ricardo formations, Chiapas, Mexico: Ph.D. thesis, Arlington, University of Texas, p. 251.

Bracken, B. R., 1982, Environment of deposition and diagenesis of sandstones, La Joya Formation, Huizachal Group, northeastern Mexico: M.A. thesis, University of Texas at Austin, 150 p.

Burk, C. A., M. Ewing, J. L. Worzel, A. O. Beall Jr., W. A. Bergren, D. Bukry, D. Fisher, and E. A. Pessagno Jr., 1969, Deep-sea drilling into the Challenger Knoll, central Gulf of Mexico: AAPG Bulletin, v. 53, no. 7, p. 1338–1347.

Cantú-Chapa, A., 1971, La Serie Huateca (Jurásico Medio-Superior) del centro este de México: Revista del Instituto Mexicano del Petróleo, v. 3, no. 2, p. 17–40.

Carrillo-Bravo, J., 1961, Geología del Anticlinorio de Huizachal-Peregrina al Noreste de Ciudad Victoria, Tamaulipas: Asociación Mexicana de Geólogos Petroleros Boletín, v. 13, no. 1–2, p. 1–99.

Contreras, V. H., and B. M. Castillón, 1968, Morphology and origin of salt domes of Isthmus of Tehuantepec, *in* J. Braunstein and G. D. O'Brien, eds., Diapirism and diapirs: AAPG Memoir 8, p. 244–260.

Del Valle-Reyes, A., 1997, The most important Mesozoic redbeds in Mexico, *in* Mesozoic redbeds of Mexico and related Mesozoic strata, México: Abstracts of the South-Central/Rocky Mountain Section of the Geological Society of America, Session 7, 7 p.

Erben, H. K., 1956, El Jurásico Inferior de México y sus amonitas: 20th Congreso Geológico Internacional, 393 p.

Estavillo-García, C. F., and M. Herrera-Soto, 1988, Estudio estratigráfico-sedimentológico de los lechos rojos de la Formación Todos Santos, del proyecto Pueblo Viejo, Chiapas: Proyecto C-3035, Subdirección de Tecnología de Exploración del Instituto Mexicano del Petróleo, 125 p.

Evans, R., 1978, Origin and significance of evaporites in basins around Atlantic: AAPG Bulletin, v. 57, p. 5–27.

Flores-López, R., 1974, Datos sobre la Bioestratigrafía del Jurásico Inferior y Medio del Subsuelo de la Región de Tampico, Tamaulipas: Revista del Instituto Mexicano del Petróleo, v. 6, no. 3, p. 6–14.

González-García, R., and N. Holguín-Quiñonez, 1992, Las rocas generadoras de México: Asociación Mexicana de Geólogos Petroleros Boletín, v. 42, no. 1, p. 9–23.

Götte, M., 1990, Halotektonische Deformationprozese in Sulfatgestein der Minas Viejas-Formation (Ober-Jura) in der Sierra Madre Oriental, Nordöst-Mexiko: Inaugural-Dissert. dem Fachber. Geowissensch. und Geogr. der Technischen Hochschule Darmstadt, 270 p.

Gutiérrez-Galicia, M. de L., 1984, El Método Palinoestratigráfico aplicado a muestras superficiales mesozoicas de las secciones Pueblo Viejo y Jericó del Estado de Chiapas, México: Tesis de Licenciatura, Escuela Superior de Ingeniería y Arquitectura-Instituto Politécnico Nacional, México, 170 p.

Guzmán, E. J., 1950, Geología del noreste de Guerrero: Asociación Mexicana de Geólogos Petroleros Boletín, v. 2, p. 95–156.

Herrera-Soto, M., and C. F. Estavillo-García, 1988, Estudio estratigráfico de los lechos rojos de la Formación Todos Santos en el área de Matías Romero, Oaxaca: Proyecto C-3034, Subdirección de Tecnología de Exploración del Instituto Mexicano del Petróleo, 50 p.

Herrera-Soto, M., and E. Villaseñor-Rojas, 1991, Integración de los lechos rojos del Mesozoico en el Sureste de México: Proyecto CAO-3517, Subdirección de Tecnología de Exploración del Instituto Mexicano del Petróleo, 112 p.

Herrera-Soto, M., C. F. Estavillo García, and M. Brito, 1990, Estudio estratigráfico-sedimentológico de los lechos rojos en el área de Valle Nacional, Oaxaca: Proyecto C-3048, Subdirección de Tecnología de Exploración del Instituto Mexicano del Petróleo.

Imlay, R. W., 1980, Jurassic paleobiogeography of the conterminous United States and its continental setting: U.S. Geological Survey, Professional Paper 1062, 124 p.

Imlay, R. W., E. Cepeda, M. J. Alvarez, and T. Diaz, 1943, Jurassic formations of the Gulf Region: AAPG Bulletin, v. 27, p. 1407–1533.

Iturralde-Vinent, M. A., 1988, Naturaleza Geológica de Cuba: Cuba, Editorial Científico-Técnica, 146 p.

Jiménez-Rentería, J., 2004, Estudio palinoestratigráfico de los grupos Consuelo y Tecocoyunca (Triásico–Jurásico) en la Barranca Rosario Nuevo, región de Tezoatlán, al noroeste del Estado de Oaxaca: Tesis de Licenciatura, Facultad de Ingeniería-Universidad Nacional Autónoma de México, México, 376 p.

Kirkland, W. D., and E. J. Gerhard, 1971, Jurassic salt, central Gulf of Mexico, and its temporal relation to Circum-Gulf evaporites: AAPG Bulletin, v. 55, no. 5, p. 680–686.

López-Ramos, E., 1974, Geología General y de México: México, D.F., Litográfica Universo, S.A., 509 p.

López-Ramos, E., 1979, Geología de México: México, D.F., Tesis Reséndiz S.A., v. 3, 456 p.

Meneses-Rocha, J. J., M. E. Monroy-Audelo, and J. C. Gómez-Chavarria, 1994, Bosquejo Paleogeográfico y Tectónico del Sur de México durante el Mesozoico: Asociación Mexicana de Geólogos Petroleros Boletín, v. 44, no. 2, p. 18–45.

Miall, A. D., 1985, Architectural-element analysis: A new method of facies analysis applied to fluvial deposits, *in* R. M. Flores, F. G. Ethridge, A. D. Miall, W. E. Galloway, and T. D. Fouch, eds., Recognition of fluvial depositional systems and their resource potential: Society of Economic Paleotologists and Mineralogists, Course No. 19, chapter 3, p. 33–82.

Michalzik, D., 1988, Trias bis Tiefste Unter-Kreide der Nordöstlichen Sierra Madre Oriental, Mexiko-Fazielle Entwicklung eines passiven Kontinentalrandes: Inaugural-Disseert. dem Fachber., Geowissensch. und Geogr. der Technischen Hochschule Darmstadt, 248 p.

Mixon, R. B., G. E. Murray, and G. T. Díaz, 1959, Age and correlation of Huizachal Group (Mesozoic), State of Tamaulipas, Mexico: AAPG Bulletin, v. 43, p. 757–771.

Morán-Zenteno, D. J., C. I. Caballero-Miranda, G. Silva-Romo, B. Ortega-Guerrero, and E. González-Torres, 1993, Jurassic–Cretaceous paleogeography evolution of the northern Mixteca Terrane, southern Mexico: Boletín de Geofísica Internacional, v. 32, no. 3, p. 453–473.

Rueda-Gaxiola, J., 1972, La edad de los lechos rojos del núcleo 13 del Pozo Soledad 101, zona de Poza Rica, Veracruz, México, *in* Resúmenes del I Congreso Latinoamericano de Botánica, México, D. F., p. 9–10.

Rueda-Gaxiola, J., 1975, El estudio de los constituyentes orgánicos e inorgánicos de las formaciones Huayacocotla (Liásico) y Rosario (Jurásico Medio) del E de México y su relación con la tectónica, generación y entrampamiento de hidrocarburos: Proyecto C-3019, Subdirección de Tecnología de Exploración del Instituto: Mexicano del Petróleo, 215 p.

Rueda-Gaxiola, J., 1999, The palynological method: Its applications to understanding the origin, evolution and distribution of redbeds, *in* C. Batolini, J. Wilson, and T. Lawton, eds., Mesozoic sedimentary and tectonic history of north-central Mexico: Geological Society of America Special Paper 340, p. 339–346.

Rueda-Gaxiola, J., 2003, The origin of the Gulf of Mexico Basin and its petroleum subbasins in Mexico, based on red bed and salt palynostratigraphy, *in* C. Bartolini, R. T. Buffer, and J. Blickwede, eds., The Circum-Gulf of Mexico and the Caribbean: Hydrocarbon habitats, basin formations, and plate tectonics: AAPG Memoir 79, p. 246–282.

Rueda-Gaxiola, J., and M. A. Dueñas, 1990, La Palinoestratigrafía de la secuencia cortada por el Pozo Xicalango 101 es básica para el conocimiento de la evolución de la Cuenca del Golfo de México: Investigaciones Recientes en Paleobotánica y Palinología Instituto Nacional de Antropología e Historia, México, no. 42, p. 11–51.

Rueda-Gaxiola, J., and J. Jiménez-Rentería, 1996, Fue la Cuenca de Tlaxiaco una continuidad hacia el sureste de la Fosa triásico-liásica de Huayacocotla-El Alamar?: Memorias del 9th Coloquio de Paleobotánica y Palinología, México, p. 46–47.

Rueda-Gaxiola, J., J. Morales de la Garza, M. A. Dueñas, E. Salas-Gómez, J. L. Rodríguez-Benítez, L. Gutiérrez-Galicia, E. López-Ocampo, R. Trinidad-Reyes, C. León-García, and M. Zepada-Barajas, 1982, Estudio palinológico de las rocas del Jurásico Inferior-Medio del área norte de la Cuenca de Huayacocotla, Ver.: Proyecto C-1115, Subdirección de Tecnología de Exploración del Instituto Mexicano del Petróleo, 282 p.

Rueda-Gaxiola, J., E. López-Ocampo, M. A. Dueñas, and J. L. Rodríguez-Benítez, 1989, Investigación palinoestratigráfica en la región de Huizachal-Peregrina, Tamaulipas: Proyecto C-3503, Subdirección de Tecnología de Exploración del Instituto, Mexicano del Petróleo, 255 p.

Rueda-Gaxiola, J., E. López Ocampo, M. A. Dueñas, M. Minero, G. Uribe, and M. Guerrero-Muñoz, 1991, Investigación palinoestratigráfica de los Lechos Rojos Triásico-Jurásicos del Anticlinorio de Huizachal-Peregrina y su relación tectono-sedimentaria con la secuencia Triásico-Jurásica del Anticlinorio de Huayacocotla: Proyecto C-3507, Subdirección de Tecnología de Exploración del Instituto Mexicano del Petróleo, 142 p.

Rueda-Gaxiola, J., E. López-Ocampo, M. A. Dueñas, and J. L. Rodríguez-Benítez, 1993, Los anticlinorios de Huizachal-Peregrina y de Huayacocotla: Dos parte de la Fosa de Huayacocotla-El Alamar: I. El Alogrupo Los San Pedros: Asociación Mexicana de Geólogos Petroleros Boletín, v. 43, no. 1, p. 1–33.

Rueda-Gaxiola, J., E. López-Ocampo, M. A. Dueñas, J. L. Rodríguez, and A. Torres-Rivero, 1999, The palynological method: Basis for defining stratigraphy and age of the Los San Pedros Allogroup, Huizachal-Peregrina anticlinorium, Mexico, *in* C. Bartolini, J. Wilson, and T. Lawton, eds., Mesozoic sedimentary and tectonic history of north-central Mexico: Geological Society of America Special Paper 340, p. 229–269.

Salvador, A., 1991, Triassic–Jurassic, *in* A. Salvador, ed., The geology of North America: The Gulf of Mexico Basin: The Geological Society of America, v. J., chapter 8, p. 131–180.

Santiago-Acevedo, J., J. Carrillo-Bravo, and B. Martell-Andrade, 1984, Geología Petrolera de México, *in* D. Marmissolle-Daguerre et al., eds., Evaluación de formaciones en México: México, Schlumberger WEC, p. I.1–I.36.

Sapper, C., 1894, Informe de la Geología de los estados de Chiapas y Tabasco: Boletín de Agricultura, Minería e Instr., v. 1, p. 187–211.

Schmidt-Effing, R., 1980, The Huayacocotla Aulacogen in Mexico (Lower Jurassic) and the origin of the Gulf of Mexico, *in* R. H. Pilger, ed., The origin of the Gulf of Mexico and the early opening of central North Atlantic Ocean: Proceedings from Symposium, Louisiana State University, no. 1, p. 79–86.

Viniegra-Osorio, F., 1971, Age and evolution of salt basin of southeastern Mexico: AAPG Bulletin, v. 55, no. 3, p. 478–494.

Williams-Rojas, C. T., J. V. Ortega-González, G. Téllez-Ramírez, R. Sánchez-Granados, A. Toledo-Jiménez, A. Ajactle-Hernández, and S. Sánchez-Martín, 1997, La Cuenca de Huamuxtitlán: Área estratégica del Proyecto de Inversión Tlaxiaco "B", *in* Memorias de Trabajos Técnicos del 35th Congreso Nacional: Asociación de Ingenieros Petroleros de México, v. 2, no. 9, p. 18–21.

8

Chavez Valois, V. M., M. d. L. C. Valdés, J. I. Juárez Placencia, I. A. Ortiz, M. M. Jurado, R. V. Yánez, M. G. Tristán, and S.Ghosh, 2009, A new multidisciplinary focus in the study of the tertiary plays in the Sureste Basin, Mexico, *in* C. Bartolini and J. R. Román Ramos, eds., Petroleum systems in the southern Gulf of Mexico: AAPG Memoir 90, p. 155–190.

A New Multidisciplinary Focus in the Study of the Tertiary Plays in the Sureste Basin, Mexico

Victor M. Chavez Valois, Ma. de Lourdes Clara Valdés, Juan I. Juárez Placencia, Iván Alor Ortiz, Marta Mata Jurado, Ricardo Villagrán Yánez, and Mercedes Guerrero Tristán

Petróleos Mexicano (PEMEX) Exploración y Producción, Villahermosa, Tabasco, Mexico

Santosh Ghosh

Geo-Strat Consulting, Ontario, Canada

ABSTRACT

Our recent multidisciplinary study resulted in a better biostratigraphic, sedimentologic, tectonic-structural, and geochemical understanding of the Tertiary section in Reforma-Comalcalco area as a part of the Sureste Basin of Mexico. Our objectives were to identify, define, and delimit the proven and potential plays.

During the Paleocene to middle Miocene, deep-water organic-rich shales were deposited, whereas debris flows and breccias were deposited on a slope derived from exhumed Cretaceous platforms. These sedimentary facies resulted from a compressive Laramide phase, which generated a broad regional fold belt. The Miocene–Pliocene Chiapaneca orogenic phase redeformed the former structures generating large volumes of terrigenous detritus and consequent rapid progradation of the clastic shelf facies. The paleoenvironments became aggradational-deltaic in nature from the early Pliocene onward; major gravity sliding processes triggered the separation of the Comalcalco and Macuspana subbasins.

Rapid burial during the Tertiary favored hydrocarbon generation and expulsion from the Tithonian source rock. The main generation phase occurred in the late Oligocene. The vertical migration of hydrocarbons along normal faults occurred in the Pliocene–Pleistocene charging the Neogene traps. Diagenetic processes of compaction, cementation, dissolution, and fracturing modified the sandstone porosity and permeability, thus influencing hydrocarbon migration and accumulation in the reservoirs.

DOI:10.1306/13191083M903334

Through our integrated analysis, we have identified nine play fairway complexes with hydrocarbon reserves and future economic potential. Each of them could have multiple and important individual plays.

INTRODUCTION

The main objective of this study was to appraise and evaluate the hydrocarbon potential of the Tertiary siliciclastic strata in the subsurface of the Sureste Basin, especially in the Reforma-Comalcalco region (Figure 1).

The Sureste Basin, considering both, offshore and onshore areas, is the major petroleum province in Mexico and has an effective petroleum system. This basin has highly prolific hydrocarbon source rocks, abundant reservoir facies widely developed in almost all the sedimentary column, and efficient and effective seals. In this southern basin, oil fields that range from super giant, medium, to small have been discovered (Petróleos Mexicanos [PEMEX] Exploración y Producción, 1993, 1999) (Figure 2).

Exploratory activities began in the basin in the early 1900s. The first commercial discoveries occurred in the 1950s, associated with Tertiary successions (Gutiérrez, 1950; Calderón, 1951; Salas and López, 1951; Castillo, 1955). The discovery in 1972 of economically important hydrocarbon accumulations from Upper and Middle Cretaceous reservoirs in the Cactus and Sitio Grande fields, respectively, provided incentives to explore almost exclusively in Mesozoic carbonates (Reyes, 1981; Santiago et al., 1984). High well productivities of the Mesozoic carbonate fields encouraged exploration and production investment in these deep plays. From that time on, the Tertiary sequence remained unexplored for almost 30 years. This explains the lack of discoveries in Tertiary rocks since the 1970s.

This publication is the outcome of a multidisciplinary regional study conducted by PEMEX geoscientists who developed a particular methodology both to understand the proven plays in the discovered Tertiary fields and to delimit potential plays in the unexplored Tertiary columns. This team effort resulted in the identification of nine potential and three conceptual plays. These include some already producing fields, some partially proven whereas others are yet to be proven. The potential volume of estimated reserves in the multiple play fairways is considered significant.

METHODOLOGY AND SCOPE OF THE STUDY

In this study, we subdivided the team's activities based on two main aspects: the risk element of plays and the disciplines involved. Figure 3 broadly outlines the workflows of activities in the study.

The reservoir element was mainly studied by the specialists involved with well logs, stratigraphy, sedimentology, and biostratigraphy. The elements of trap and seal were the responsibility of stratigraphers, structural geologists, and well log experts. Geochemists characterized the generative system and the dynamic process of play, with the support of stratigraphy, seismic interpretation, and structural geology. The seismic interpretation was done by geophysicists, structural geologists, and stratigraphers.

An integrated database was created and developed by the PEMEX multidisciplinary team. A total of 13 regional transects included both 2500 km (1553 mi) of two-dimensional and three-dimensional seismic lines. A stratigraphic correlation, including more than 200 stratigraphic well columns and well logs, was used. The study involved a biostratigraphic correlation, as well as a compilation and synthesis of production history data of Tertiary fields. Additionally, more than 160 conventional cores and hundreds of thin sections were studied under a microscope.

This study is based primarily on the establishment of a sequence-stratigraphic framework for the Tertiary succession, with the identification and correlation of seven third-order sequence boundaries, as well as the determination of sedimentary facies and environments from the Paleocene to the Holocene. Two principal structural styles are represented by the sequence: the fold and thrust belt with two periods of deformation (Paleogene and Miocene–Pliocene) and an extensional phase formed by gravity sliding (Pliocene–Holocene) responsible for the origin of the Comalcalco and Macuspana subbasins.

This study confirmed the importance of the Upper Jurassic Tithonian sequences as the main source rock for hydrocarbons in the region. Geochemical modeling led to the definition of the principal oil kitchens and also established that Mesozoic fault systems acted as important upward migration routes toward the Tertiary reservoirs.

The main tectonic events and their associated sequences were identified and led to a better understanding of the geological evolution of the region. Furthermore, nine play fairways were identified in this study, including several important individual plays, both proven and potential.

BIOSTRATIGRAPHIC ANALYSIS

A microfaunal analysis of conventional core and cutting samples from 142 wells that penetrated the Tertiary sequence was pivotal in the establishment of a

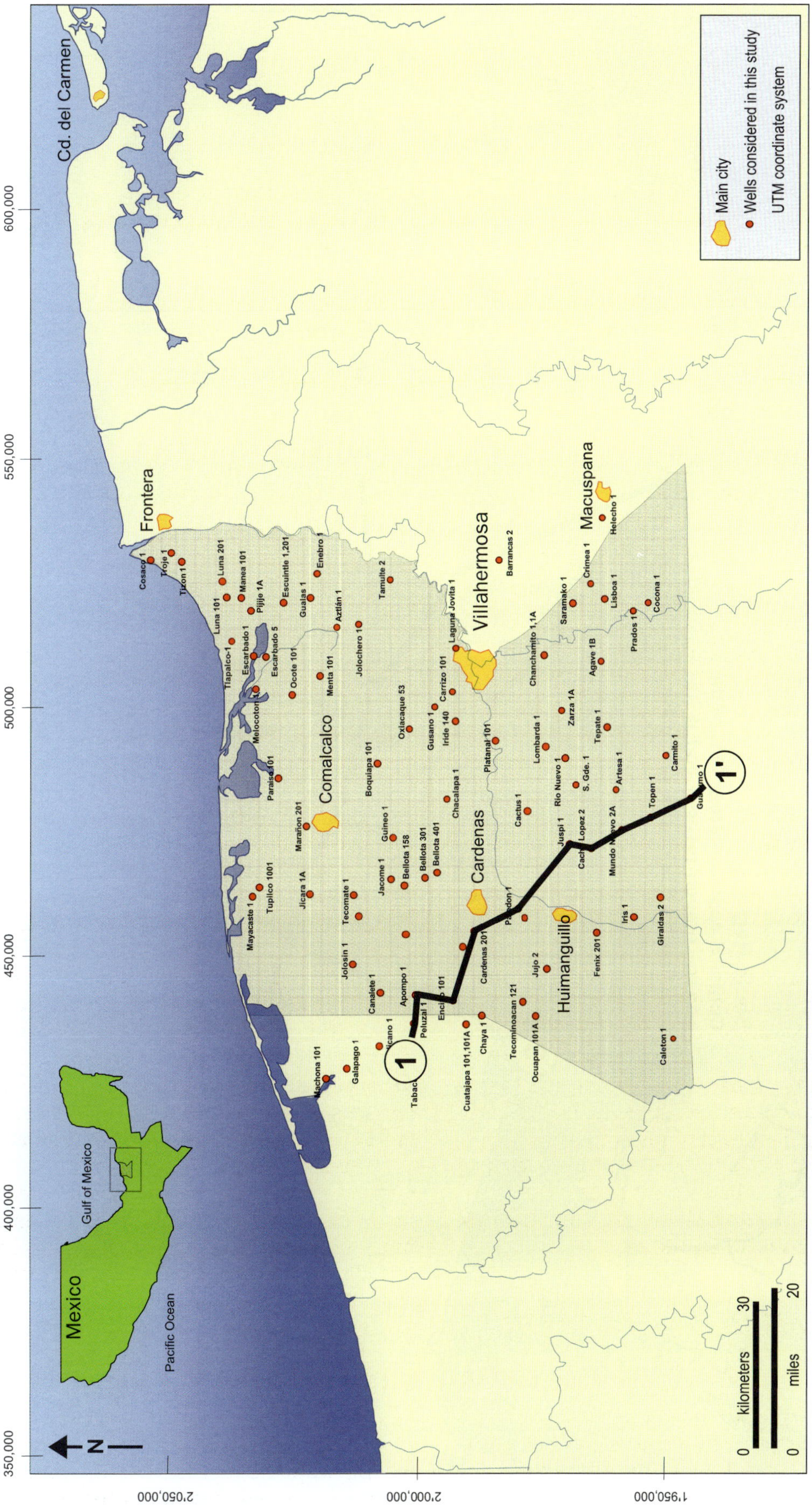

FIGURE 1. Map showing the Gulf of México coastal plain in Tabasco, México. The Reforma-Comalcalco study area is in light-gray shading. UTM = Universal Transverse Mercator.

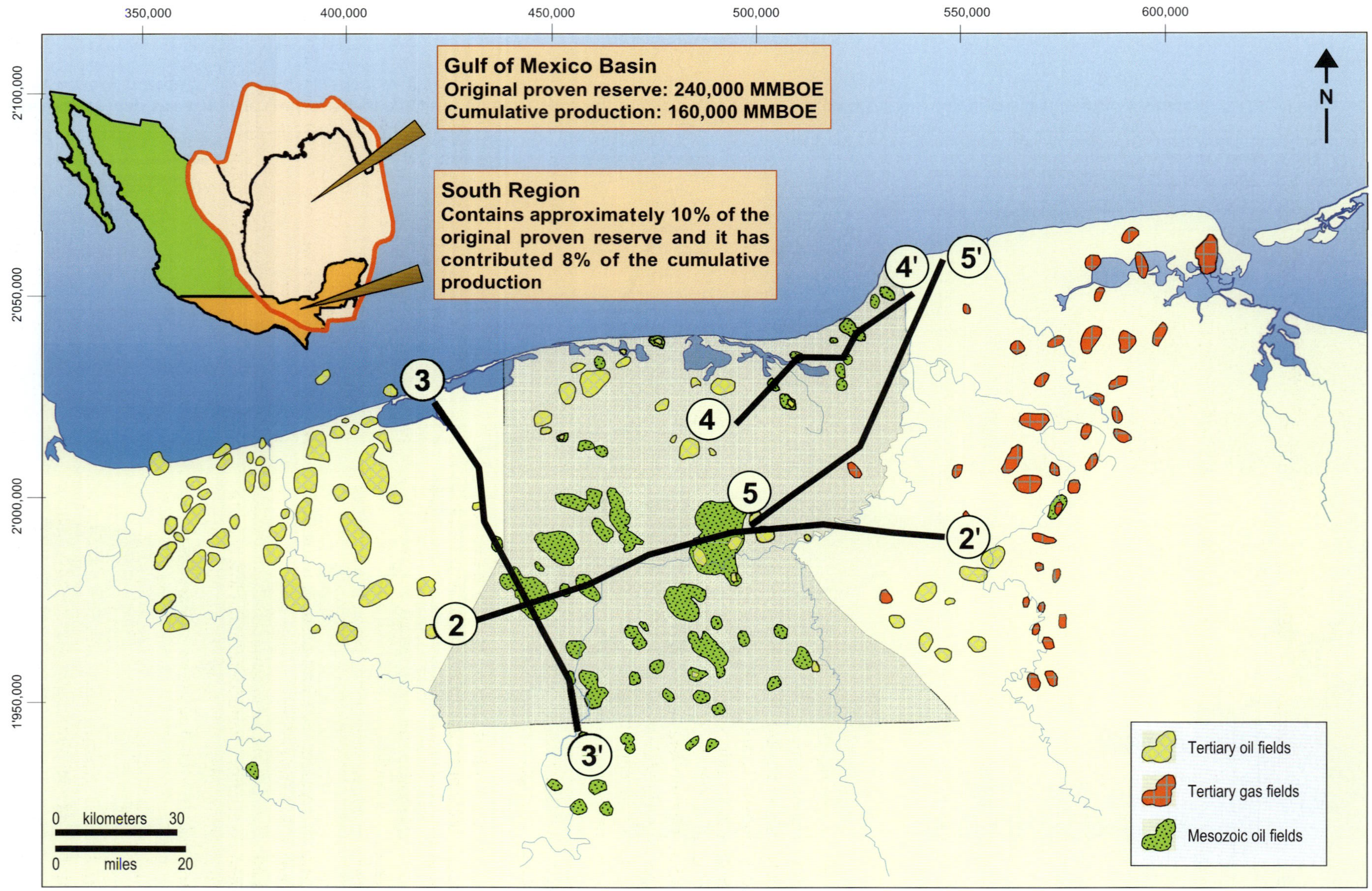

FIGURE 2. The Región Sur contains approximately 30% of the original proven national reserves in reservoirs of oil, humid gas and dry gas in different levels of the sedimentary section from Jurassic to Pliocene. MMBOE = million barrels of oil equivalent.

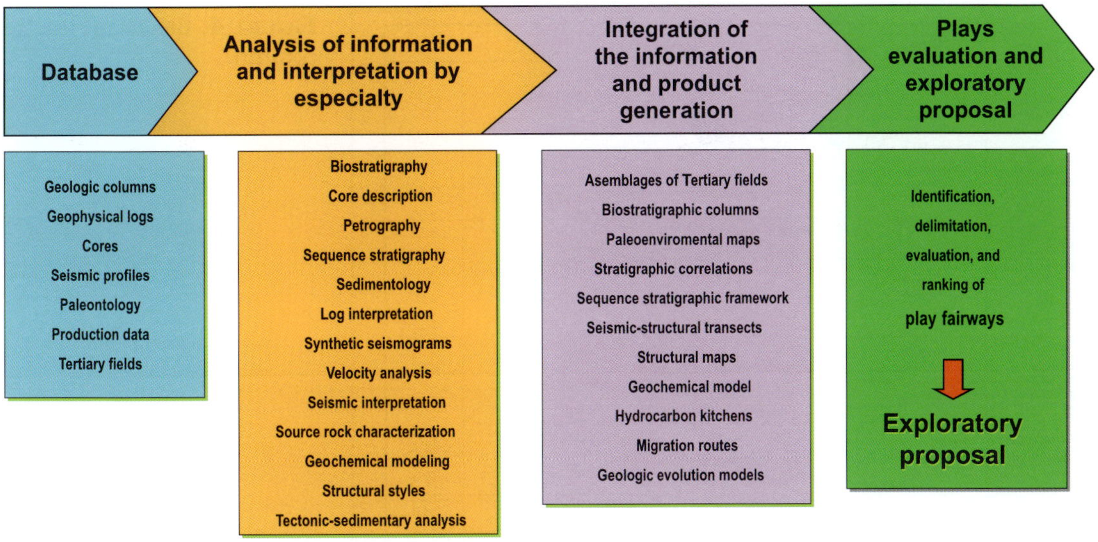

FIGURE 3. Workflow, activities, and products achieved during the course of the study.

regional chronostratigraphic and paleobathymetric framework (Cárdenas and Marín, 2004).

Age determination and nomenclature were based on the biochronostratigraphic charts of several previous authors (Bolli and Saunders, 1985; Berggren et al., 1995; Hardenbol et al., 1998). Forty-five bioevents were identified based on foraminifera and nanoplankton of which 26 correspond to the Paleogene and 18 are within the Miocene–lower Pliocene succession (Figure 4).

All these bioevents coincide with the last occurrences of marker planktonic foraminifera in the area, which may be related to global mass extinction, possibly caused by drastic ecologic changes, or redeposited fauna in younger strata.

Because of the limited density of information and because many bioevents identified in the stratigraphic column of wells were not sufficiently notable, it was not possible to correlate all of them regionally (Figures 4, 5). Only seven events could be reliably used as references for defining a similar number of sequence boundaries (Table 1).

The paleobathymetry was interpreted on the basis of Tipsword's (1962) model, with the appropriate modifications adopted for the area. Seven distinct benthic foraminifera associations were identified, closely correlating with seven bathymetric zones: transitional, inner, middle, and outer neritic, and upper, middle, and lower bathyal (Table 2).

An analysis of samples from wells (cores and cuttings) proved to be an essential element in the statistical control of the existing fauna and the subsequent construction of faunal abundance and density charts. This helped in deciding on locations of condensed stratigraphy and candidates for sequence boundaries.

Standardized chronostratigraphic columns in each well for both the Paleogene and Neogene intervals were used to decipher the biostratigraphic evolution based on foraminifera species taking into consideration the first downhole appearance and the paleobathymetric variations. The observed biostratigraphic distribution reflected local variation, which could have been caused by paleoecologic, evolutionary, structural, or sedimentary changes. A biostratigraphic distribution chart was constructed, which highlighted the difference in five geological subprovinces for the Tertiary column (Table 3). Several important conclusions about some index species arose from an analysis of this chart.

STRATIGRAPHIC-SEDIMENTOLOGIC FRAMEWORK

Sequence analysis (Mitchum et al., 1977; Van Wagoner et al., 1990) was the basis for the understanding of the stratigraphy and sedimentology of the area (Mata et al., 2003). A total of 13 regional transects of stratigraphic correlation were constructed, supported by the identification of seven sequence boundaries (Figure 6; Table 1).

The Tertiary paleogeographic-paleoenvironmental analysis reveals the morphology of a deep basin inherited from the Upper Cretaceous. This phase lasted from the Paleogene up to the early–middle Miocene (Figure 7).

Bathyal slope and basinal facies characterize the Paleogene sediments; dark, slightly calcareous shales

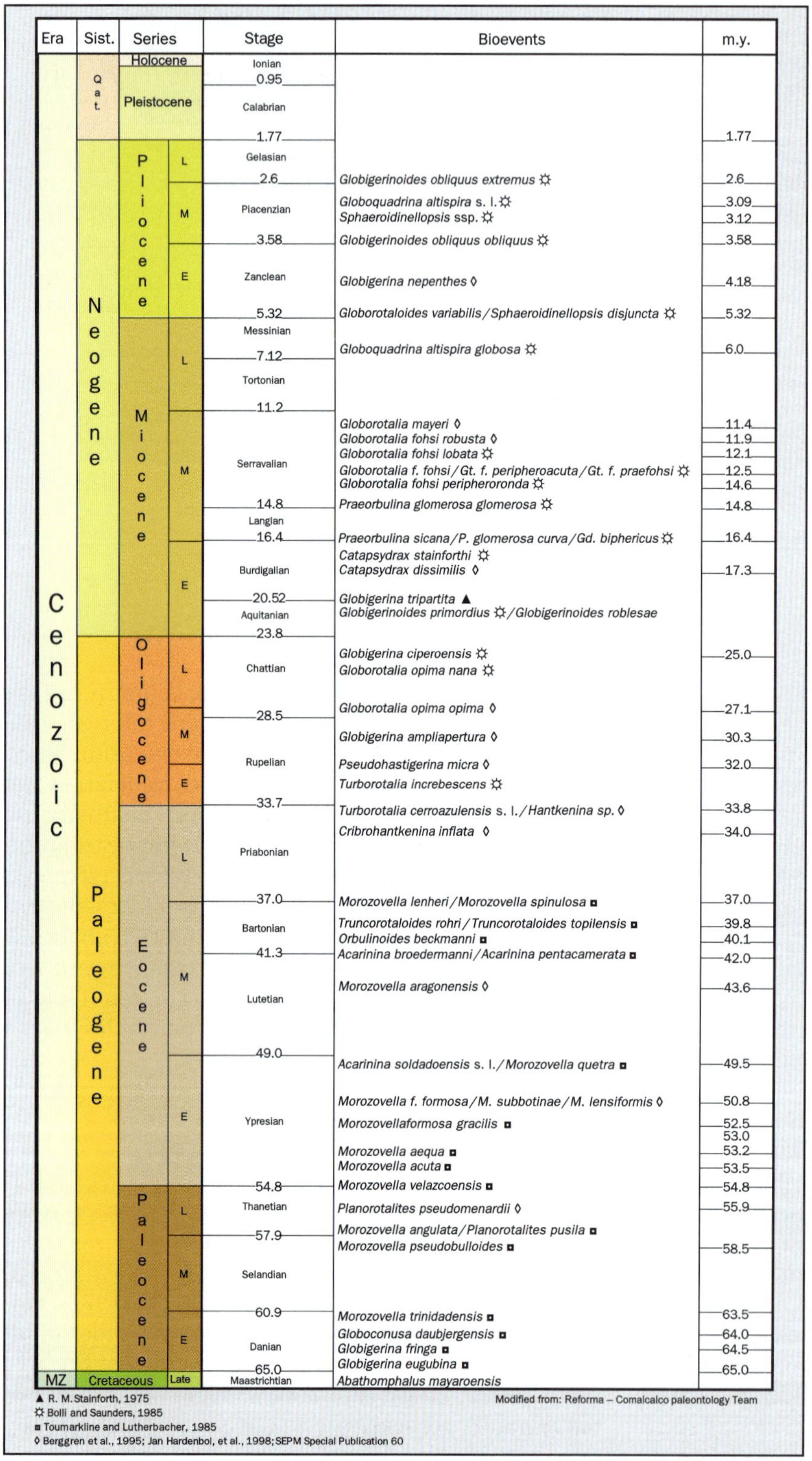

FIGURE 4. Biochronostratigraphic column.

mainly form the Paleocene strata. During the Eocene (Figure 7a), organic-rich, calcareous shales and marly mudstones with abundant planktonic fauna were deposited, whereas debris-flow deposits and calcareous breccias from preexisting platforms were deposited in the southeast part of the area (Figure 7b). The Oligocene was marked by the deposition of shales, which became sandy in the younger beds (Figure 7b).

Neogene strata record a range of depositional facies ranging from outer and inner platform to transitional-paralic facies (Figure 7c–g). Shales with sandy intercalations were deposited during the middle Miocene,

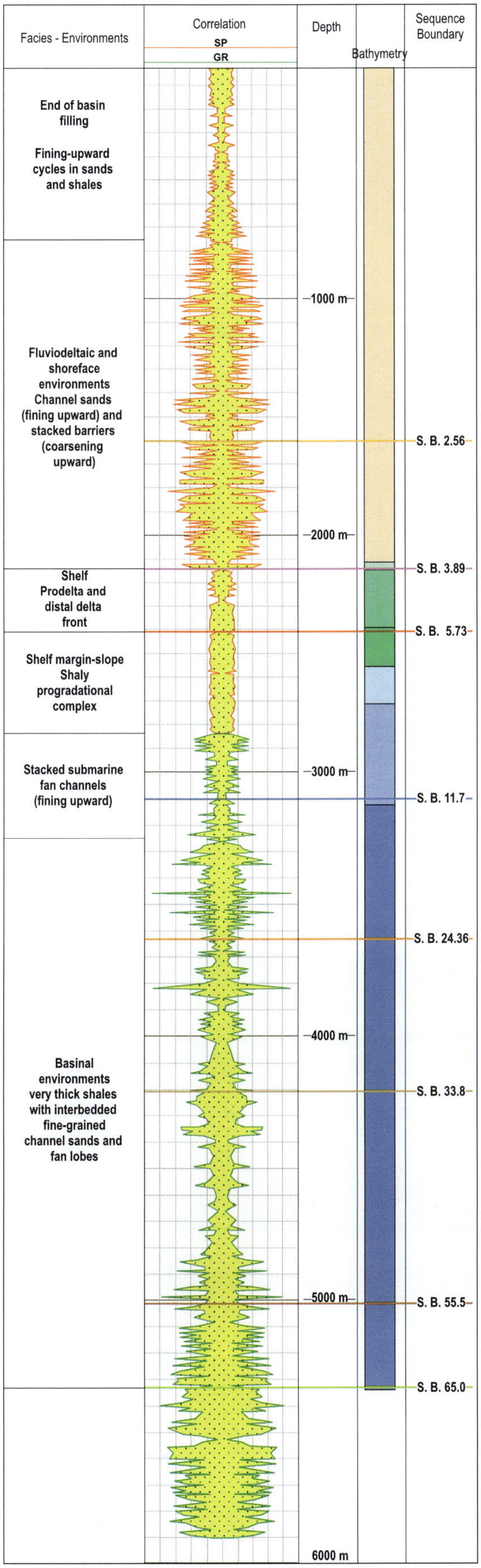

FIGURE 5. Bellota 301 well composite log showing sequence boundaries, depositional environments, and bathymetry (bioevents related to SB's in Figure 4; SP log in red and GR log in green, well location in Figure 1; bathymetry colors in Table 2).

TABLE 1. Tertiary sequences boundaries recognized in the studied area (bioevents related to sequence boundaries in Figure 4).

Sequence Boundary	*Series*	*Stage*
2.56	Late Pliocene	Gelasian
3.89	Early Pliocene	Zanclean
5.73	Late Miocene	Messinian
11.7	Middle Miocene	Serravalian
24.36	Late Oligocene	Chattian
33.8	Late Eocene	Priabonian
55.5	Late Paleocene	Thanetian
65.0*	Late Cretaceous	Maastrichtian

**Considered as the bottom of the Tertiary section.*

whereas sands predominated during the upper Miocene and Pliocene. Sandstones are dominant in the Pleistocene, with subordinate siltstone and mudstone. The sandstones vary from very coarse grained to pebbly in proximal areas, up to very fine grained characteristic of deep-water marine turbidites.

Sedimentologic aspects, one of the main controlling factors of reservoir facies, are related to three main depositional systems: (1) Neogene siliclastic turbidites deposited in a large submarine fan system and turbidites of minibasin origin, (2) upper Miocene–lower Pliocene expanded deltaic deposits, and (3) Eocene platform carbonate breccias derived from the platform and deposited in deep-water environments.

DIAGENESIS AND PROVENANCE OF RESERVOIR ROCK

Petrographic and fluorescence analyses of core and cuttings revealed the detrital composition of the Tertiary sediments. The main constituents are quartz, feldspars, and rock fragments, with minor accessory minerals. The sandstones are classified as arkoses and subarkoses (Krynine, 1935) with a lesser proportion litharenites and lithic arkoses (Figure 8).

Quartz, the main constituent grain, ranges between 40 and 90%, and feldspar between 5 and 30%. Rock fragments are mainly volcanic and metamorphic in origin, with subordinate carbonate grains. Volcanic fragments are primarily microcrystalline tuffs, which are altered to illite, smectite, and sericite, and regionally, these rocks form the most important seals in the region (Robertson Research Internacional Limited, 2004). Accessory minerals include biotite, muscovite, chlorite, and tourmaline, which are indicative of metamorphic and volcanic provenance. Based on the detrital composition of the

Table 2. Paleobathymetric levels, faunistic associations, and proportion of planktonic faunas.

Conti-nental	Transitional	Inner Neritic	Middle Neritic	Outer Neritic	Upper Bathyal	Middle Bathyal	Lower Bathyal
	Brackish water, bay, swamp, lagoon, estuarine	Inner platform	Middle platform	Outer platform	Upper slope	Middle slope	Lower slope
	High Tide	Low Tide					Sea level (0 m)
Without Marine fauna	***Ammonia beccarii, Quinqueloculina lamarckiana***, *Elphidium* spp. *Chara oogonia*, ostrácods, Molluscs.	***Amphistegina lessonii,*** *Eponides antillarum, Florilus* spp. *Hanzawaia* spp. *Lenticulina americana Textularia* spp. *Uvigerina* sp.	***Astacolus vaughani,*** *Astacolus cristi, Brizalina* spp. *Bulimina* spp. *Eponides* spp. *Uvigerina* spp. *Höeglundina elegans*	***Marginulinopsis messinae,*** *Marginulinopsis* spp. ***Praeglobobulimina ovata,*** *Bolivina* spp. *Chilostomella* spp. *Cibicidoides* spp. *Cyclammina cancellata, Gyroidinoides* sp. *Islandiella californica, Plectofrondicularia, Rectuvigerina* spp. *Spiroplectammina*	***Melonis affinis,*** *Melonis* sp. ***Siphouvigerina auberiana,*** *Eggerella bradyi, Heterolepa dutemplei, Karreriella* spp. *Pullenia* spp. *Martinottiella* spp. *Sphaeroidina* spp.	***Gyroidinoides broeckhiana, Planulina wüellerstorfi,*** *Anomalina* spp. *Anomalinoides* spp. *Cibicidoides robertsonianus, Discorbinella cushmani, Osangularia* spp. *Vulvulina* spp.	***Usbekistania charoides,*** *Rzehakina epigona, Trochamminoides* spp.
	Without Planktonics	**Rare Planktonics 0–10%**	**Planktonics 20–30%**	**Planktonics 50%**	**Planktonics 60%**	**Planktonics 80%**	**Planktonics 90–100%**

Depth scale (m): 0 m, 200, 500, 1000, 1500, 2000, 2500

Based on **Tipsword, 1962**; By: Reforma – Comalcalco Paleontology Team.

TABLE 3. Behavior of the Tertiary subbasins based on their bathymetric and biostratigraphic distribution (environments and faunistic associations in Table 2; well locations in Figure 1).

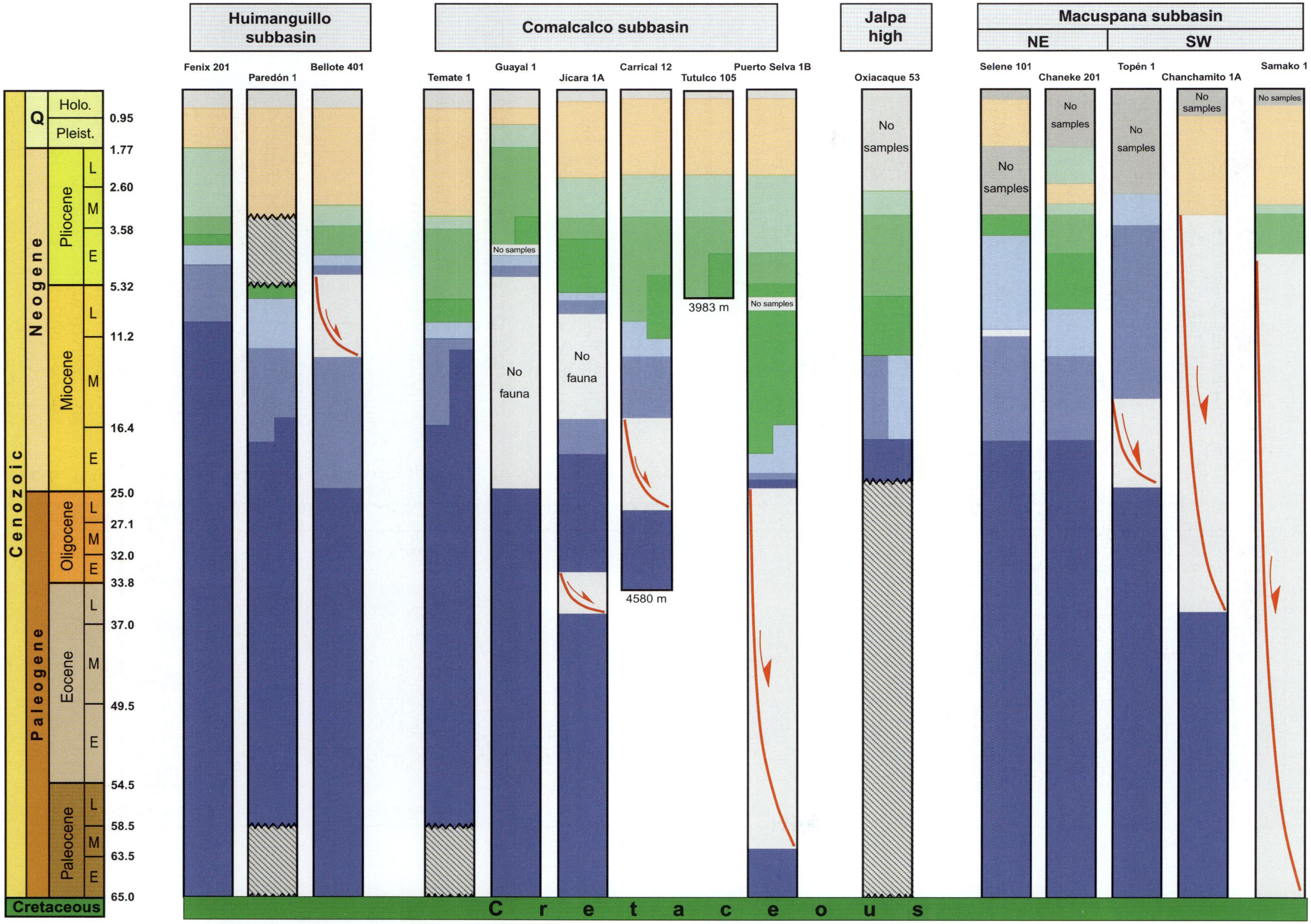

Modified from: Reforma – Comalcalco Paleontology Team.

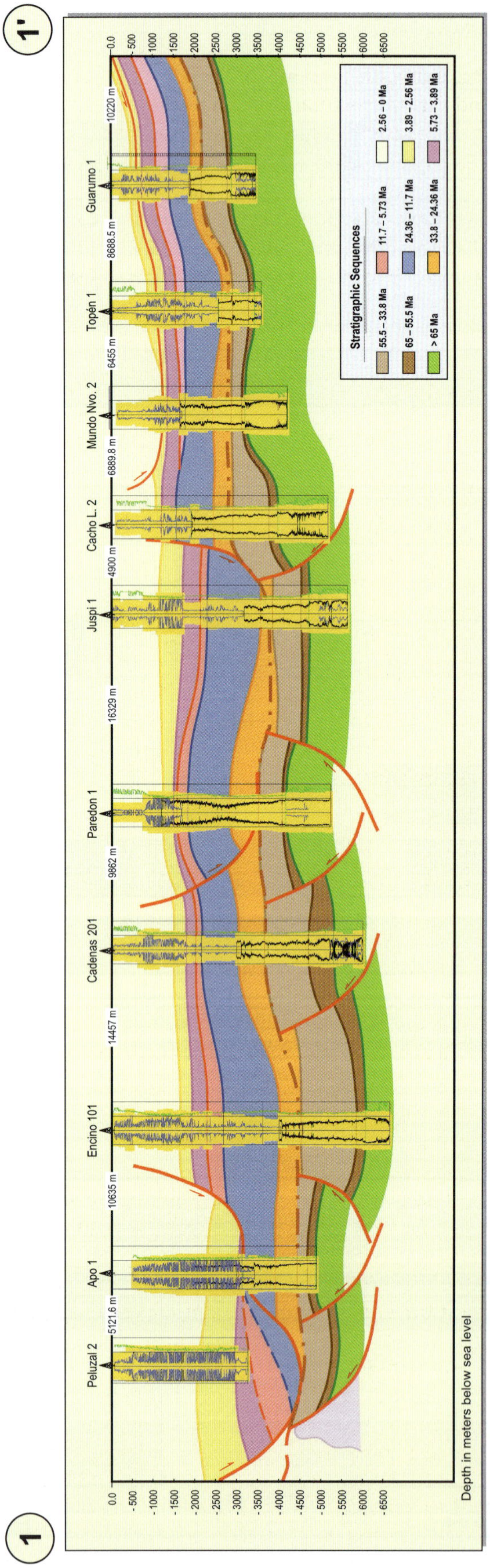

FIGURE 6. Peluzal-Guarumo traverse section corresponding to stratigraphic correlation number 2. See Figure 1 for the location.

Tertiary sequence, the provenance areas have been established in the granitic massif and the Sierra de Chiapas (Odom et al., 1976).

The identification of the principal diagenetic events that control reservoir heterogeneity, porosity, and permeability led to a model for diagenetic evolution and its function on reservoir development (Figure 9).

The more important diagenetic processes affecting reservoir rocks are (1) compaction caused by lithostatic load, which progressively reduces primary porosity; (2) cementation by calcite, siderite, dolomite, silica, authigenic clays, and pyrite; (3) dissolution affecting partially or totally plagioclase or K feldspars, producing secondary porosity; and (4) fracturing associated with tectonic activity and dissolution, which also enhance secondary porosity.

Compaction and cementation are critical factors that control the reservoir quality because they cause progressive loss of porosity in older strata. In contrast, dissolution creates new porosity because it selectively affects rocks with abundant feldspars and lithic fragments.

Porosity types include intergranular, intragranular, moldic, and fracture (Figure 10). Reservoir quality depends on depositional facies and the depth of burial. Observed porosity ranges from 5% for Paleogene sandstone turbidites to 30% for upper Miocene–Pleistocene littoral or fluviodeltaic strata. Hydrocarbon migration and trapping occur late in the diagenesis cycle, where the reservoir facies shows optimum conditions of sandstone reservoir porosity and permeability, as supported from studies of the El Golpe 78, Tupilco 129, Tupilco 132, Tecomate 1, and Luna 201 wells.

STRUCTURAL GEOLOGY

In the Sureste Basin, two main regional structural domains have been recognized: an older and deep-seated event characterized by compressive structures and a younger and shallower episode characterized by extensional structures (Chávez, 1999).

The older structural domain contains a fold and thrust belt above a Jurassic detachment surface within evaporites and limestones (Figure 11a) and affecting Mesozoic and Paleogene rocks (Chávez and Cerrillo, 1999). Most Mesozoic oil fields are encountered in this structural domain. Predominant structural styles are thrust-fold relationships (Figure 11a), and in the subsurface, structural styles are strongly linked to lithologic variations and their location with respect to the main tectonic elements in the basin. Similar structural styles are observed in the Sierra de Chiapas (Sánchez, 1979; Meneses, 1987; Meneses et al., 1987). Regionally, the onset of compressive deformation was at the end of the Late Cretaceous, to be reactivated in the late middle Miocene and forming the present structural styles in the basin.

The younger structural domain is extensional because of gravity sliding in the Pliocene–Pleistocene. Important fault systems represent boundaries of Tertiary subbasins (Figure 11b). Predominant structural styles include listric faults with associated rollover anticlines and growth faults linked to salt or shale diapirs. An important regional detachment surface separates the extensional from the compressive system. This detachment surface, which extends all over the study area, is normally developed within the Eocene–Oligocene argillaceous sedimentary package.

Salt tectonics is an active factor in the whole area. Large volumes of pre-Oxfordian salt were evacuated and emplaced in upper parts of the sedimentary intervals (Figure 12a). Salt has been encountered in many wells within the Mesozoic fold and thrust belt, also affecting the Tertiary sequence in the Salina del Istmo, Comalcalco, Huimanguillo, and Macuspana North subbasins.

In the south-central part of the Macuspana subbasin, shale diapirism, and evacuation phenomena are evident in Oligocene–Miocene strata, which strongly influenced the geometry of Pliocene–Pleistocene rocks (Figure 12b).

Both salt and shale tectonics are important factors in the creation of depocenters and localized depressions, which received a considerable volume of Pliocene–Holocene sediment. The volumes of mobilized salt or shale indirectly affect the dimension of depocenters and their sedimentary fill.

Based on the structural, sedimentary, and biostratigraphic information obtained from the Tertiary successions, four distinct geologic provinces were identified: Comalcalco, Macuspana, Huimanguillo, and the Jalpa high (Figure 13).

Comalcalco Subbasin

This subbasin is a depocenter filled with thick delta front sands that were deposited in the vicinity of a late Miocene–Pliocene platform. This subbasin is associated with a system of normal faults dipping north-northwest and counter-regional faults related to salt evacuation.

Macuspana Subbasin

This late upper Miocene subbasin owes its origin to the interaction of three important events: (1) a large volume of sediment influx, (2) gravity sliding, and (3) deformation and mobilization of shale masses (commonly forming shale ridges). From southwest to northeast, the sedimentary facies vary from fluviodeltaic to transitional marine and turbidite deposition. Along the western and southwestern edges of this subbasin, normal faults cut through the Paleogene and Cretaceous column. The central part of the Subbasin develops shale ridges, salt diapirs, and argillaceous domes.

Huimanguillo Subbasin

This subbasin is noteworthy in that it retains the fold and thrust structural styles of the compressive Chiapaneca regime, with virtually no extensional faults, and the Tertiary stratigraphic column is essentially completely preserved (without unconformities). The subbasin contains a sedimentary column represented by thick turbidites deposited in submarine fans during the middle Miocene and Pliocene–Pleistocene sands deposited in fluviodeltaic environments. All available evidence supports a stable tectonic regime for this area.

Jalpa High

This feature forms some kind of a pivotal point around which the three subbasins developed and where coeval compression, extension, and erosion interacted. The following facts substantiate the above assertion: (1) large faults of Pliocene–Pleistocene age affect the north and east margins of the high, (2) the Mesozoic sequence is at its shallowest level in the region, (3) the thickness of the Tertiary column is reduced because of the absence of the Paleogene, and (4) the Pliocene–Pleistocene successions are represented by fluvial sand deposits. The Jalpa high may be a remnant of Laramide deformation and merits a more detailed tectonic-sedimentary analysis.

TECTONOSEDIMENTARY EVOLUTION

A schematic model is presented to explain the Jurassic to Holocene tectonosedimentary evolution, highlighting three main tectonic events: two compressive phases followed by an extensional episode (Figure 14).

The first compressive phase was associated with the Laramide orogeny or its equivalent coeval regime, which began in the Late Cretaceous and ended in the late Oligocene. The second compressive phase is called the Chiapaneca, and it occurred from the late middle Miocene to the late Pliocene. The Early–middle Miocene is interpreted as an intercompressive stage that is poorly known.

Since about 5.73 Ma, a large-scale gravitational sliding event on a regional detachment surface triggered the separation of the two main tectonic domains. Gravity sliding was responsible for the compartamentalization of the region, generating the two principal depocenters for clastic sediment: the Comalcalco subbasin, which opens up to the Gulf of Mexico, and the Macuspana subbasin, which lies to the east, adjacent to the Yucatan Platform.

In the Comalcalco subbasin, salt masses of Jurassic age are very important factors in its genesis and evolution. Salt evacuation toward the upper part of the sedimentary

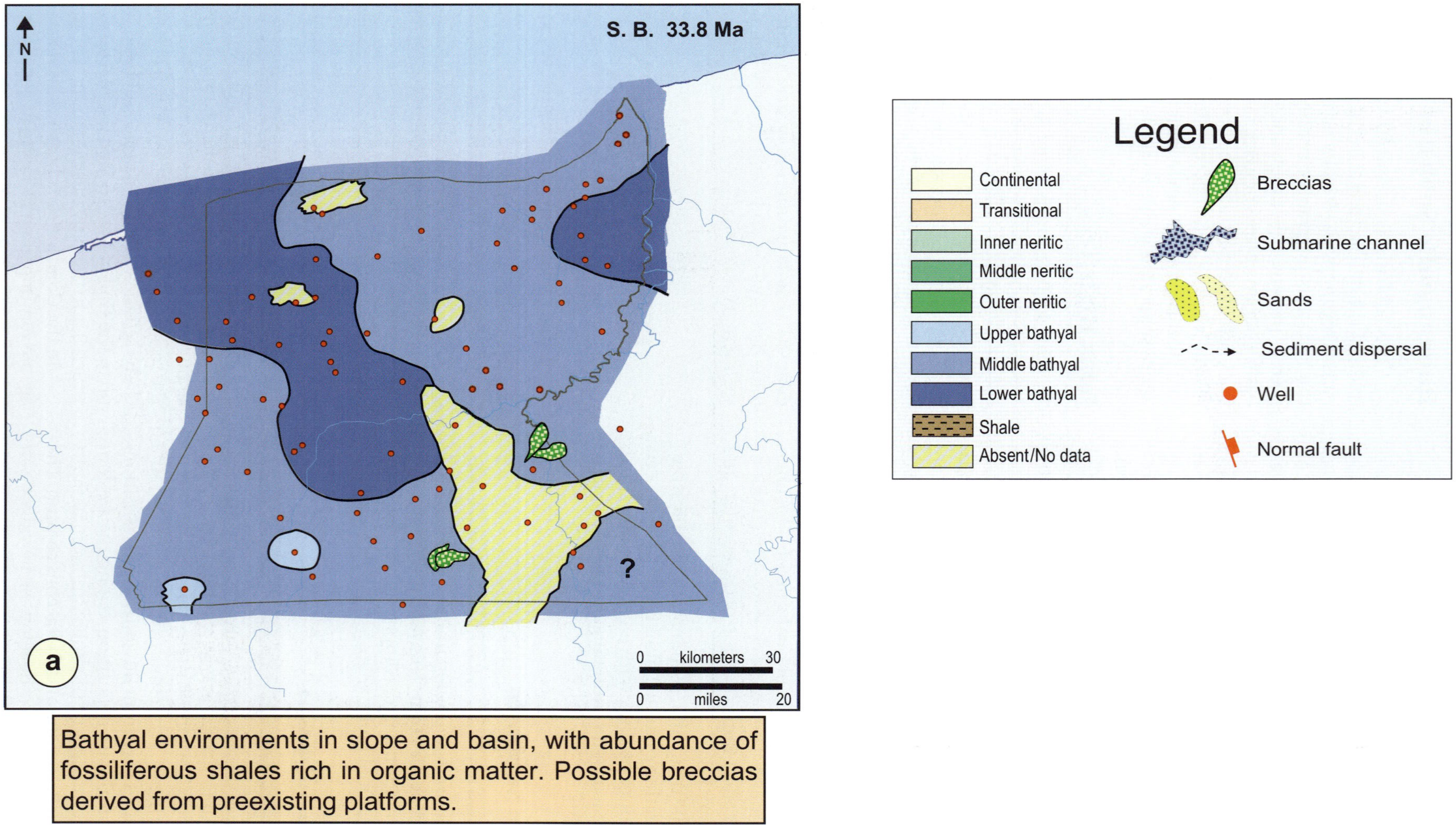

FIGURE 7. Tertiary paleogeographic and paleoenvironmental evolution of the area. S. B. = sequence boundary.

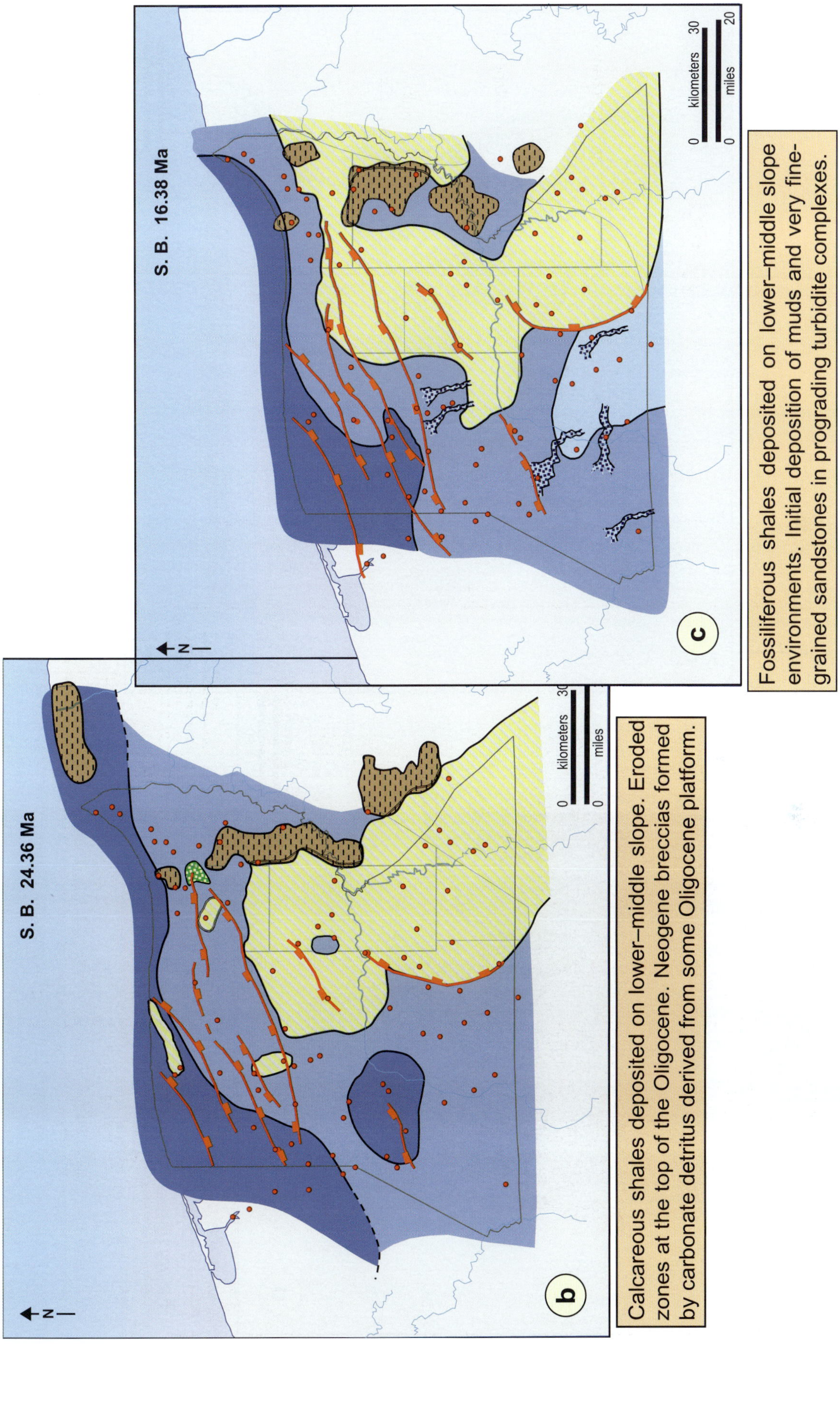

FIGURE 7. (cont.).

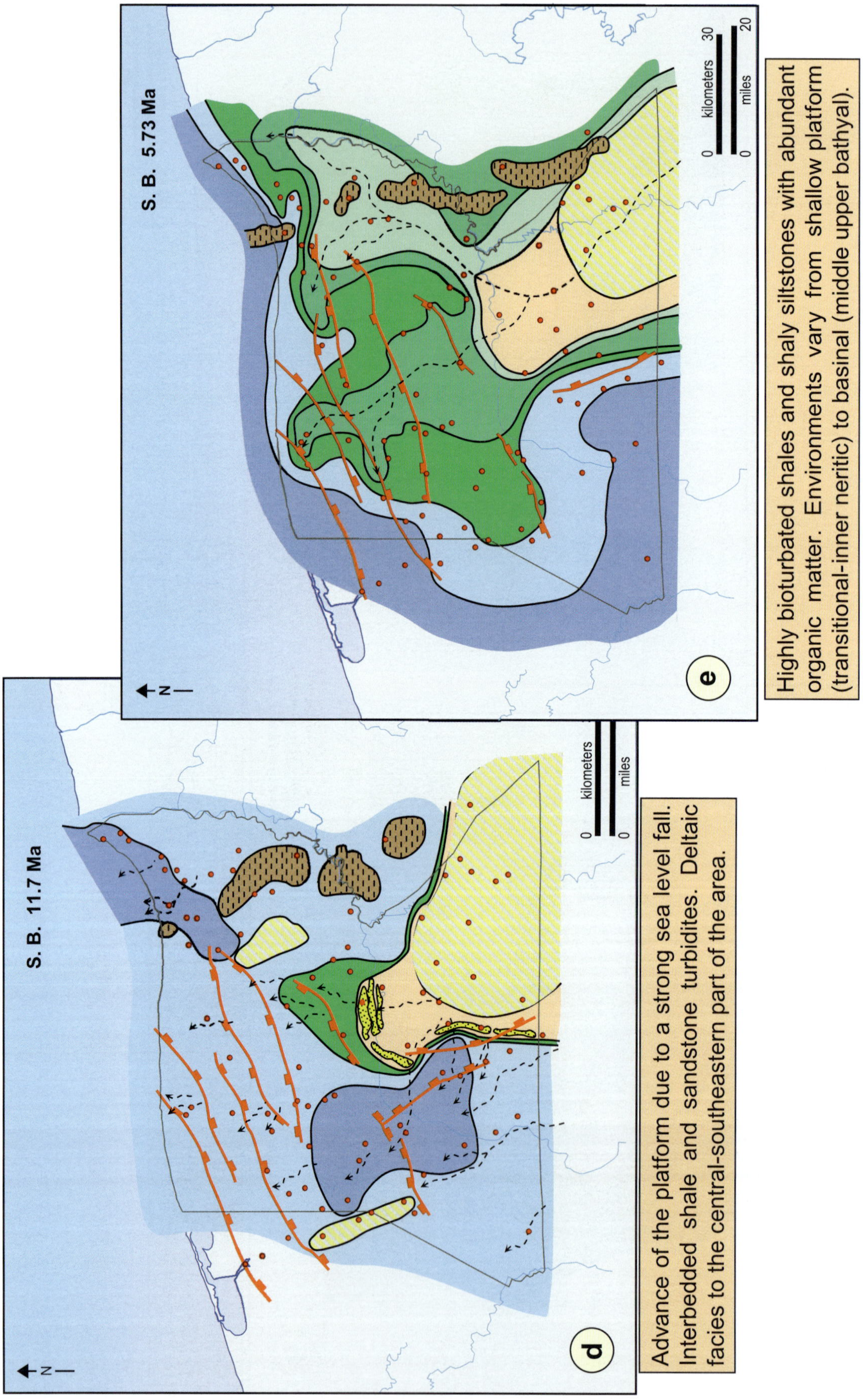

FIGURE 7. (cont.).

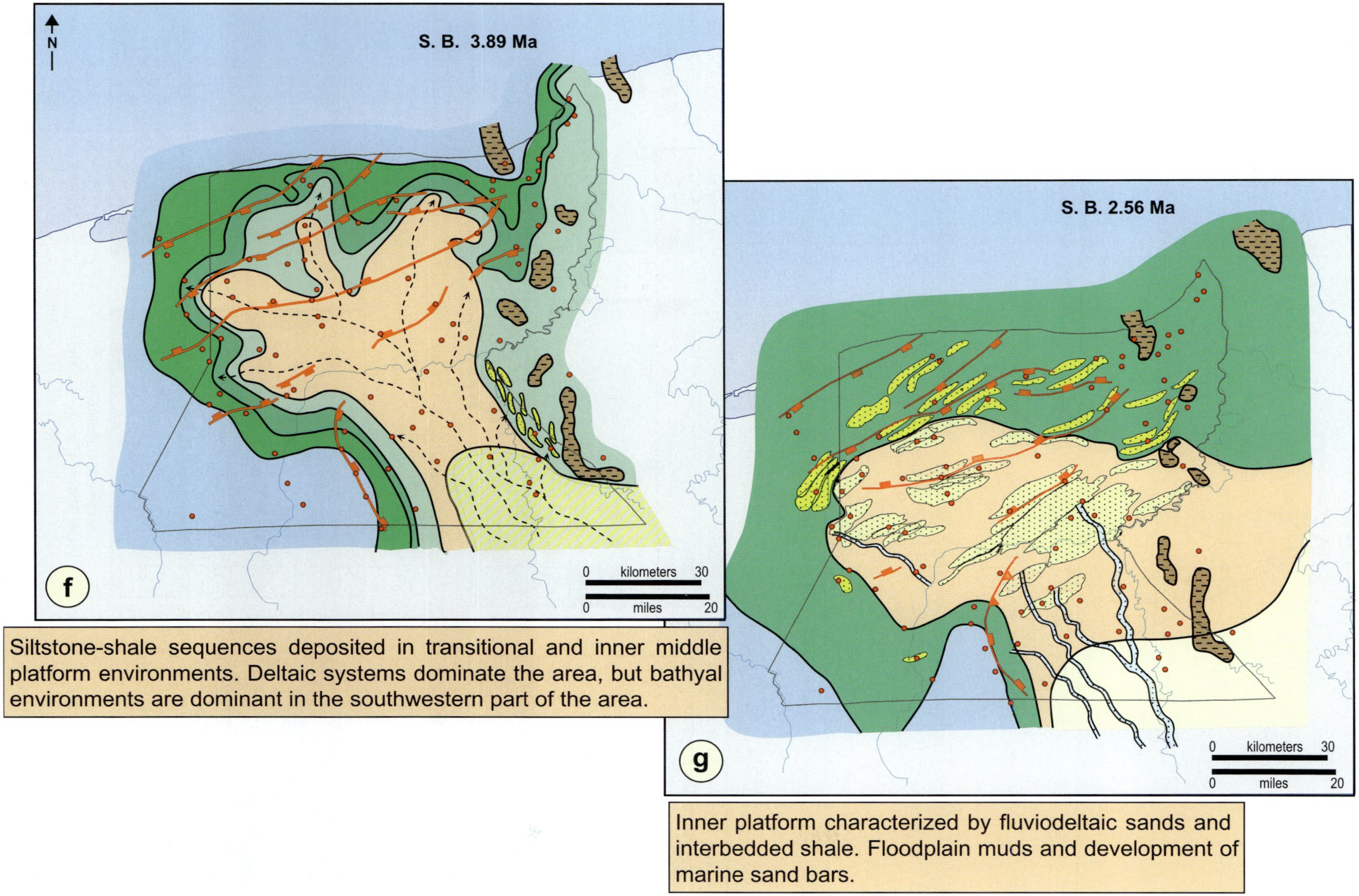

FIGURE 7. (cont.).

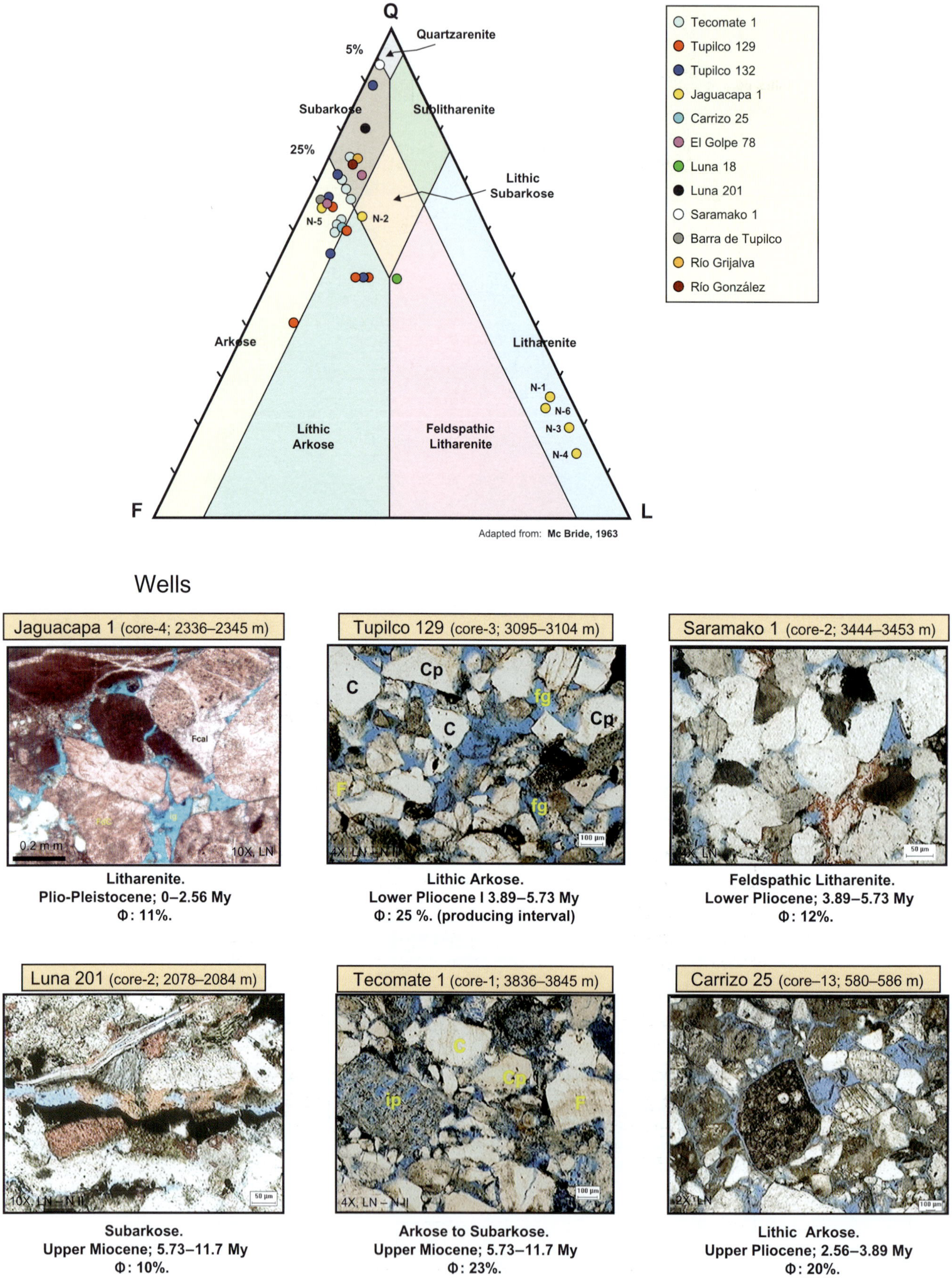

FIGURE 8. Detrital composition and classification of Tertiary sandstones according to Krynine (1935). C = quartz; Cp = polycrystalline quartz; fg = rock fragments; F = feldspars.

FIGURE 9. Diagenetic processes that affect the porosity of the Tertiary rocks, as suggested by cement paragenesis from the El Gole 78 well.

sequence created the necessary accommodation space for the accumulation of great volumes of Pliocene sediments coming from the south (Figure 15).

In the Macuspana subbasin, thick argillaceous sequences of Oligocene and Miocene age underwent plastic deformation. The emplacement and development of growth faults, and the accumulation of high volumes of southerly sourced sediments, triggered the mobilization of shale, which formed diapiric crests and broad domal structures (Figure 16).

Thus, two tectonic compressive events together gave rise to a large fold and thrust belt where the main Mesozoic structural traps containing the largest hydrocarbon reserves so far discovered in the country are located. However, gravity sliding allowed the formation of large growth faults whose association with stratigraphic and combined traps in sandy Tertiary strata also made the accumulation of important hydrocarbon deposits possible.

PETROLEUM SYSTEM

The Sureste Basin of Mexico hosts important oil and gas reservoirs in both carbonate and siliciclastic sequences of Mesozoic and Tertiary age, respectively (PEMEX Exploración y Producción, 1999). Three generative systems have been identified in the basin (Figure 17): (1) shaly limestones of Tithonian age, (2) Cretaceous limestones and evaporites, and (3) Miocene argillaceous terrigenous deposits (Holguin, 1988; PEMEX-Chevron, 1993; Guzmán et al., 1995). The molecular characterizations of oil and gas in the Tertiary intervals indicate the same source rock for all of the hydrocarbons: organic rich, algal, amorphous, dark to black, argillaceous mudstone of Tithonian age, a type 2 kerogen, highly reactive and hydrogen rich, and thermally mature (0.4 to 0.8 R_o) representing early generation of hydrocarbons.

Oil density (50–17° API) reveals thermal maturation processes and secondary alteration in some cases. Evidence of a segregation process and/or evaporative fractionation has been observed in the lightest crudes in Eocene breccias and in gas in rocks of Miocene age.

Geochemical modeling of hydrocarbons suggests that the oil kitchen became active since the late Oligocene (approximately 29 Ma) and is still active at present time. Hydrocarbon migration toward the traps mainly occurred from the late Miocene onward. A moderate geothermal gradient and rapid burial provided favorable conditions for the generation and expulsion of hydrocarbons toward the traps (Figure 18).

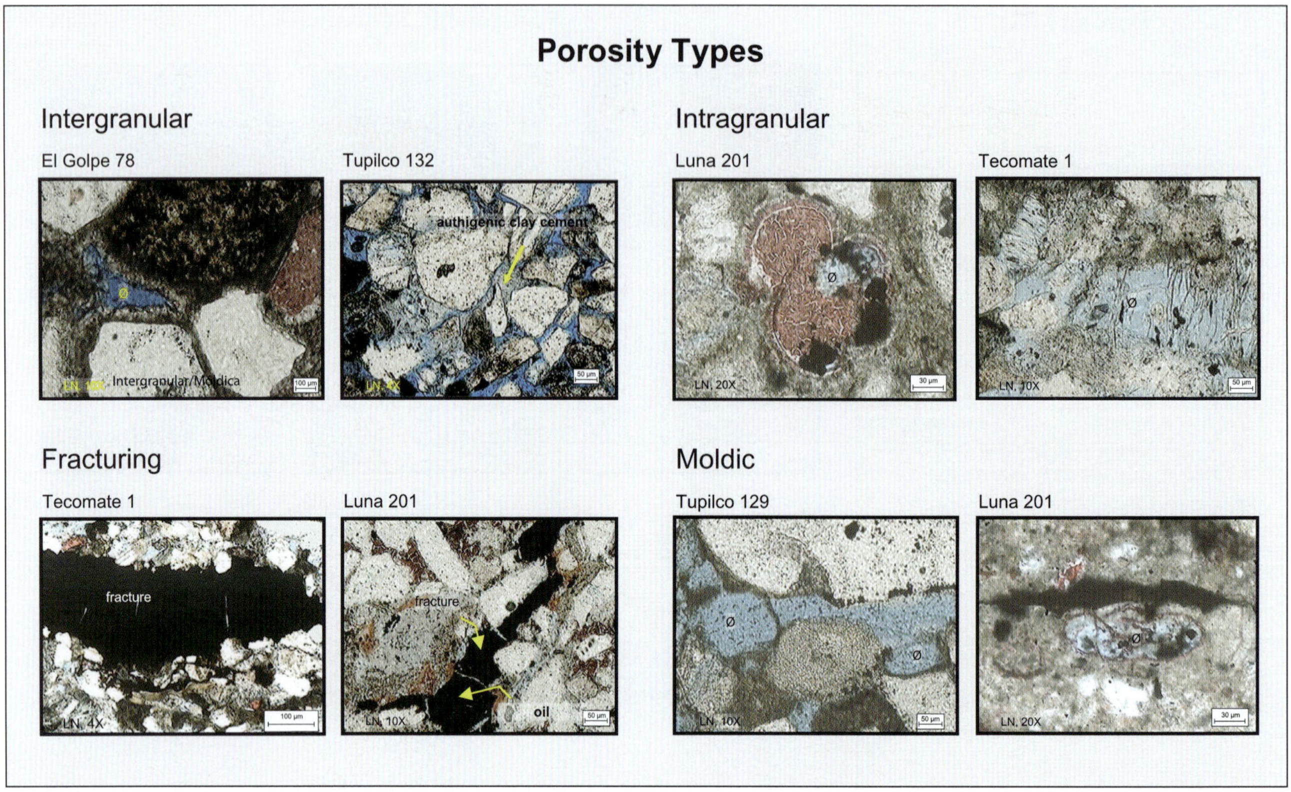

FIGURE 10. Types of porosity in the Tertiary sandstones: intergranular, intragranular, moldic, and by fracturing.

Basin structuring caused by the Laramide deformation phase triggered sediment influx and considerable subsidence, with attendant development of oil kitchens and incipient migration of hydrocarbons. During the Chiapaneca deformation phase, the relief of the granitic massif and the Sierra de Chiapas became more pronounced, initiating the erosion of the higher parts and shedding a great supply of sediment. Associated gravitational extension facilitated transport along normal and inverse faults, causing the upward movement of hydrocarbons to charge the Miocene–Pleistocene reservoir sands (Figure 19).

PETROLEUM POTENTIAL OF TERTIARY PLAYS

Identification of numerous plays, an outcome of this study, should be considered as the first step in their systematic analysis. Nine large play fairways (Figures 20, 21) have been identified in the four structural-stratigraphic provinces established in this study. We define play fairway as a favorable area for hydrocarbon exploration in one or more established stratigraphic sequences characterized by structural and/or stratigraphic factors, where similar geologic elements such as trap, reservoir, seal, and charge form part of the same petroleum system.

All the identified play fairways are related to three main mapped and defined depositional systems.

1) The siliciclastic turbidite system, which consists of (1) an important submarine fan system covering a large sector of the western part of the study area and (2) smaller turbidite systems confined to minibasins whose origin is related to the deformation and displacement of important volumes of shale masses in the northeast part of the study area.
2) Expanded deltaic systems reaching into the shelf margin, associated with less important and secondary fluvial and shallow shelf facies complexes.
3) Debris-flow deposits derived from the erosion of a carbonate platform and deposited in a deep-water bathyal environment.

Inner-Middle Fan Turbidite Play Fairway

This play fairway contains a large volume of sediment in the Huimanguillo and southern part of the Salina del Istmo subbasins (Figure 20; Table 4). These facies are virtually unexplored, although the Ayopreto field is thought to be producing from this sedimentary package.

Sedimentologically, the facies represents the inner and middle part of a submarine fan complex of middle

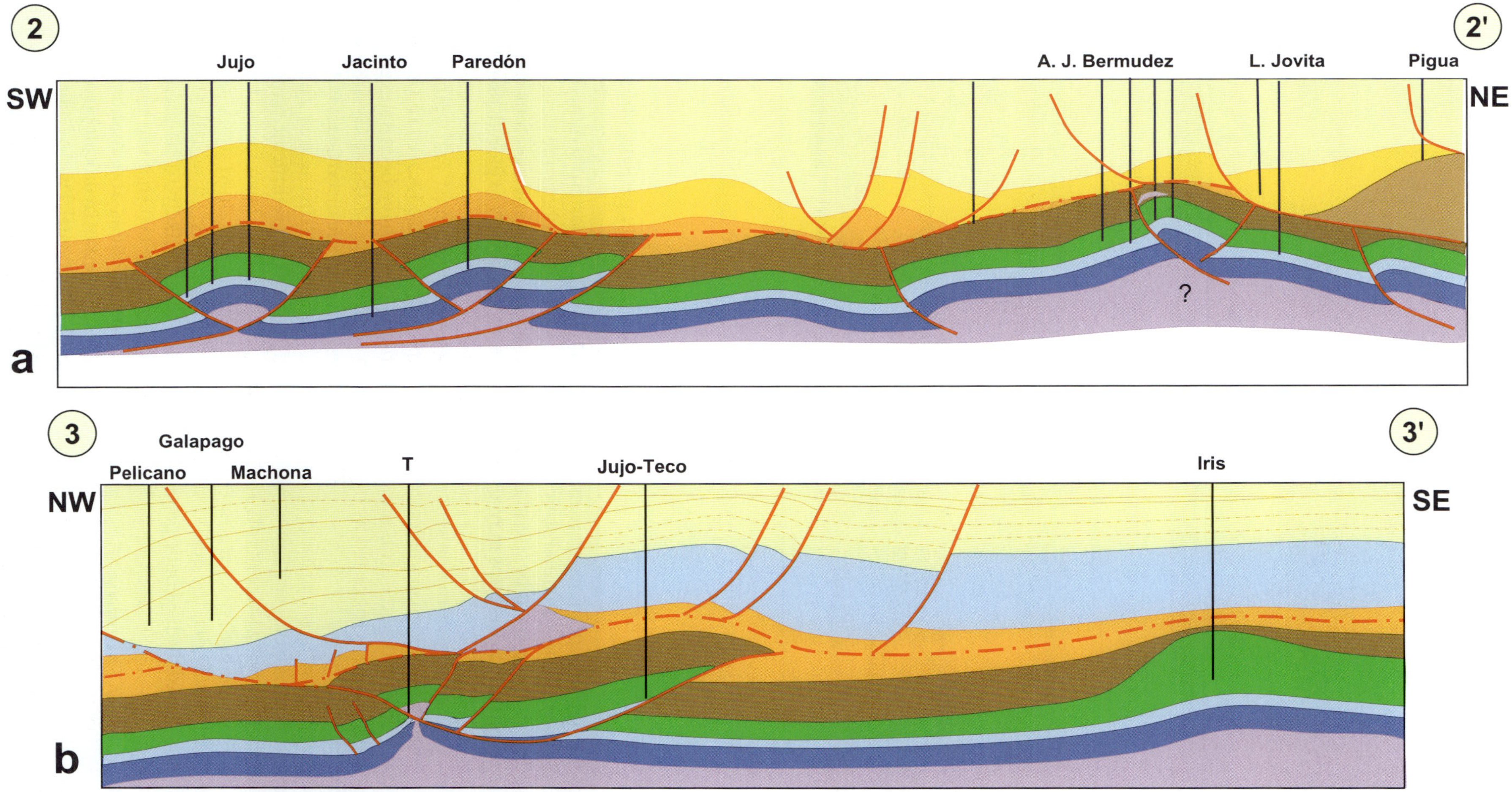

FIGURE 11. (a) Late middle Miocene compressive tectonic event forming the fold and thrust belt. (b) Pliocene–Pleistocene extensional tectonic episode producing gravity sliding. Both tectonic domains are separated by a regional basal detachment. See Figure 2 for the location.

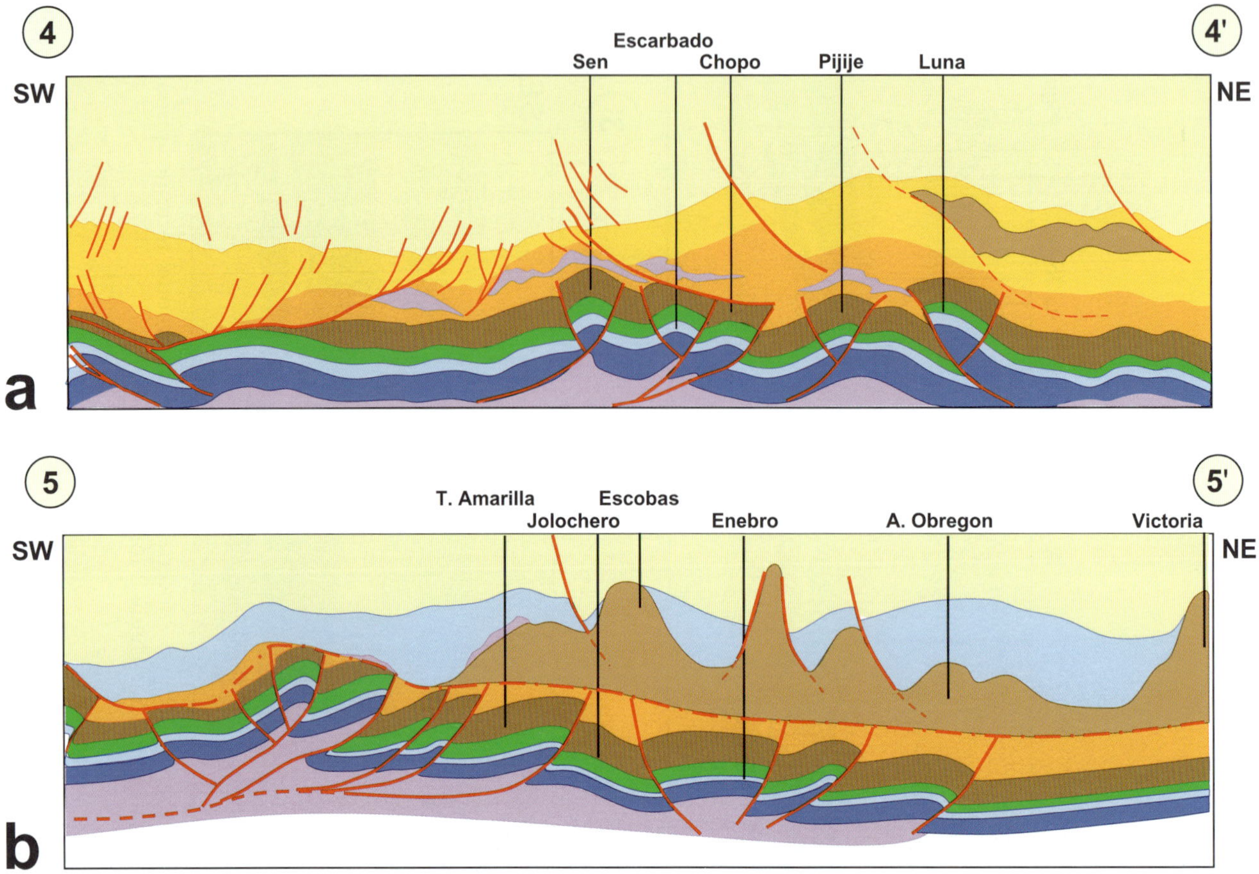

FIGURE 12. (a) Great salt masses have been mobilized by tectonics throughout the basin fill. (b) Shale evacuation is very important in the evolution of the Macuspana subbasin, where it forms domes and diapirs. See Figure 2 for the location.

Miocene to early Pliocene age. The facies thins progressively toward the north.

Stratigraphic, structural, and combined traps may be present in the play fairway. Some anticlinal structures have been identified, which appear to be the most attractive. However, large stratigraphic traps may also be promising.

Geochemical modeling suggests the presence of a large generative kitchen, and although migration routes have not been precisely delimited, some areas may well be relevant migration pathways, as suggested by hydrocarbon shows in the Caletón 1 well.

This is a very attractive play fairway that opens the possibility of discovering giant to super-giant accumulations. However, the limited knowledge of this opportunity renders it a moderate to high risk because of the uncertainty of charge and, secondly, the quality of the reservoir rock, if it was affected by diagenetic processes.

Distal Turbidite Play Fairway

This facies assemblage is widely distributed in the Comalcalco and Salina del Istmo subbasins, extending well into the marine areas to the north (Figure 20; Table 4). The thick turbidite package underlies the play fairway of deltaic sands.

This is an important sedimentary system recording distal facies of a submarine fan system of middle Miocene to early Pliocene age. Similar systems, that have not been sufficiently explored, preceded this succession in the Paleogene.

The southernmost and younger parts of this fan complex system are located in the Huimanguillo subbasin and could have been sourced from the same region to the south of the study area, whereas the provenance area for the distal facies is different. Both appear to be amalgamated in the area, and thus, distinguishing one from another is difficult. Sand bodies up to 100 m (328 ft) thick have been found to be intercalated with the bathyal fine-grained strata in the distal area.

In this fairway, structural, stratigraphic, and combined traps are likely to be present, and their importance lies in the fact that the fairway is associated with important kitchens and numerous migration routes. The effectiveness of hydrocarbon charge is amply demonstrated in the Yagual, Puerto Ceiba, and Castarrical fields, as well as in the Guayo 2, Tres Palmas 1, and Tintal 401 wells, where hydrocarbon shows are very strong.

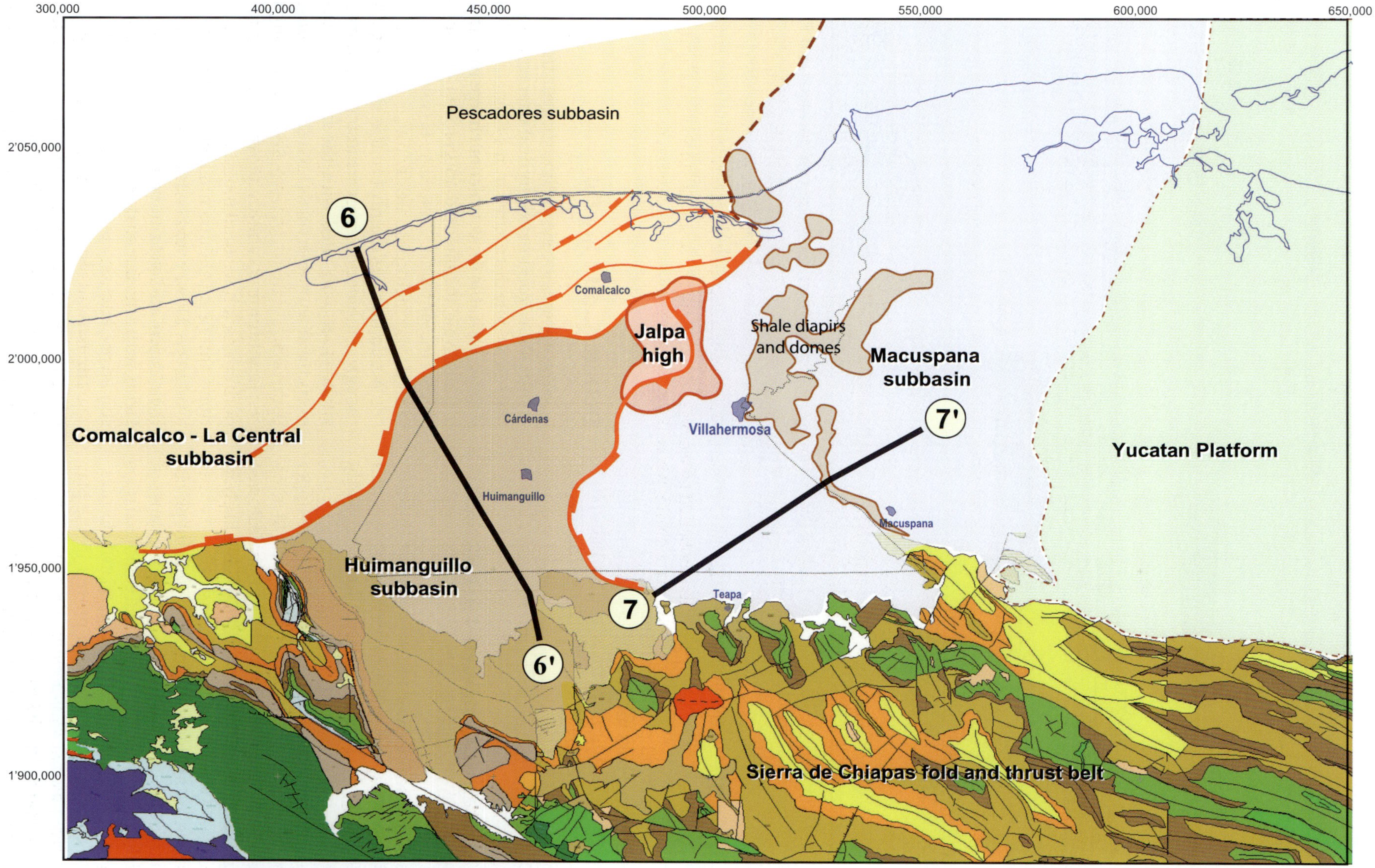

FIGURE 13. Location and distribution of the subbasins in the Sureste Basin and their relation with respect to the fold and thrust belt and the Yucatan Platform.

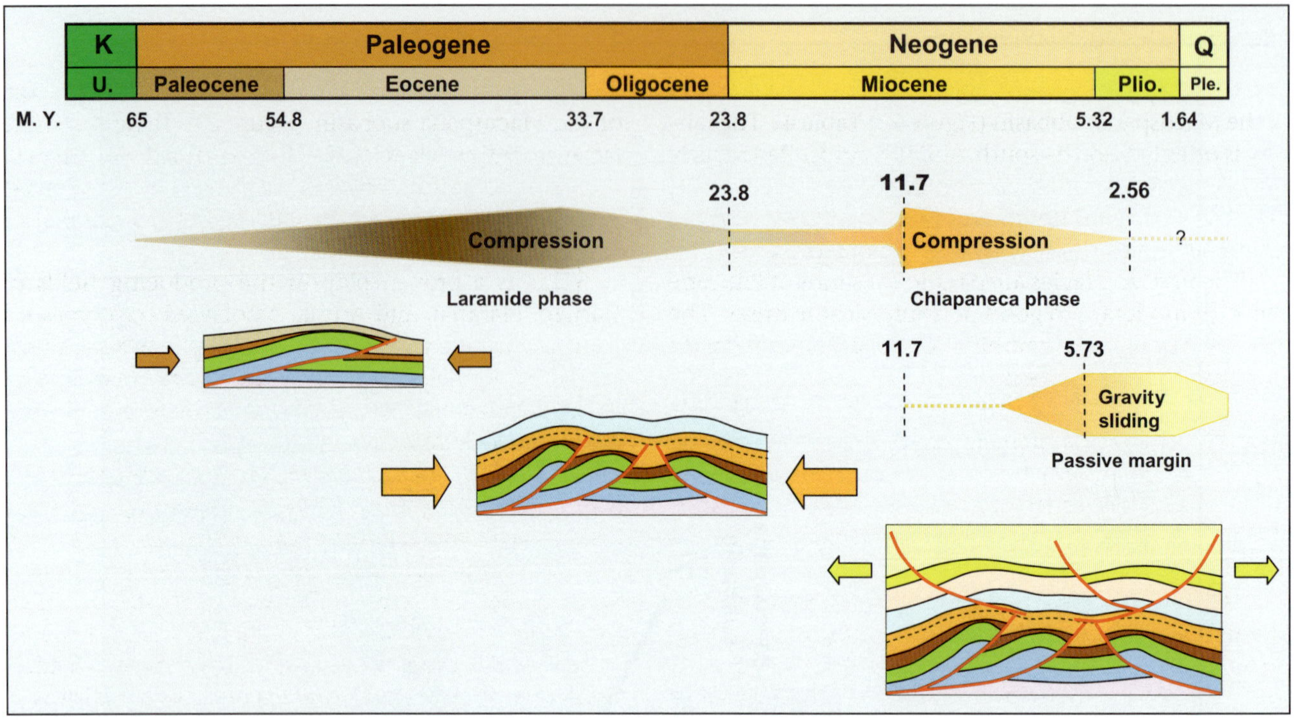

FIGURE 14. Main tectonic phases of the Sureste Basin: Paleogene compression, Neogene compression, and Pliocene–Pleistocene extensional gravitational sliding.

This fairway is economically attractive because some of its exploration targets may be combined with those associated with the deltaic sand fairway, coupled with the well-developed oil production infrastructure in the area. The principal risks include the depth of the objectives, the seismic definition of traps, and the petrophysical characteristics of the sands, which might be affected by diagenesis.

Play Fairway of Confined Turbidites in Minibasins

This play type is located in the northwest onshore part of the Macuspana subbasin, extending seaward (Figure 20; Table 4). This play concept has been partly substantiated in the Luna 301 well.

The reservoir facies is made up of middle to upper Miocene deep-water sands deposited in a system of minibasins. Under progressive lithostatic load, shale flowed into diapirs and crests. The thickness of sands is likely to increase toward the axes of the minibasins.

The traps are mainly stratigraphic because of pinchouts and truncations of sand bodies against the flanks of the mud diapirs; however, structural and combined traps are associated with anticlines and fault-related closures.

Hydrocarbon charge has been proven in the play fairway by the presence of oil accumulation in the Luna 3 well. Although the precise migration routes from the generative kitchen to the reservoir have not yet been established, nevertheless, faults limiting the diapiric crests might have acted as migration routes charging the reservoirs.

Deltaic Sand Play Fairway

The upper Miocene to Pliocene delta front sand reservoirs show good petrophysical characteristics, and they are widely distributed in the Comalcalco subbasin (Figure 21; Table 4). Petroleum charge in this fairway is quite well documented and proven in the Tertiary Mecoacán, Santuario, Caracolillo, El Golpe, Tupilco, Castarrical, and Melocotón fields.

Accumulations mainly occur in structural traps, although combination traps are also present. Important kitchens are associated with the fairway, which connects them with reservoirs through effective migration pathways.

In addition to already proven potential, new opportunities for accumulations exist along the proven trend and in subtle structural and stratigraphic multilevel traps. These could be highly attractive because of their depth and proximity to existing infrastructure. Fields with the size of those already discovered could still be present.

The fairway represents an area of low risk, which still adds moderate reserves and important production supplement in a short period of time, offsetting the natural decline of reserves in existing fields.

Paralic Sand Play Fairway

This play fairway is located in the southwestern part of the Macuspana subbasin (Figure 21; Table 4). The fairway is oriented north–south, and it is controlled mainly by large masses of mud crests, domes, and diapirs. The play has been partially proven by the Morales and Tamulté fields and recently by the Saramako 1 well.

The reservoir facies are paralic bar sands of Pliocene age with moderate to good petrophysical features. The identified traps are of combination type, associated with truncations and pinch-outs related to mud diapirism.

Hydrocarbon charge of the traps has been already documented in the aforementioned fields. However, the migration routes may be complex and unpredictable because of the presence of mobile shale masses. Faults affecting the Pliocene sequence will indeed facilitate hydrocarbon migration.

This play has good potential for new discoveries, in as much as the facies remains mostly unexplored. Based on our study, coupled with the success in the Saramako 1 well, it is possible to delimit more precisely the area of extension and to design a more systematic exploratory program for this play fairway.

We expect accumulations of moderate magnitude. One important and attractive aspect of this fairway is the repetitive cyclic nature of the reservoir facies in time and space. Moreover, the proximity to infrastructure will ensure that any accumulation discovered is cost effective. Two critical risks associated with this fairway, which will control the trap volume, are the seismic determination of trap configuration and the presence of an effective regional seal.

Fluvial and Bar Sand Play Fairway

This play fairway, limited to the southern part of the Comalcalco subbasin (Figure 21; Table 4), is intimately related to the expanded sedimentary sequence formed as a result of the effect of the Comalcalco system of faults, limiting its southern boundary.

The reservoir facies here are of middle Pliocene to Pleistocene age. The facies comprises primarily fluvial channel bar sands and associated facies, as shown by wireline logs. These sequences have been successfully drilled in the Ayapa and Tintal fields, as well as in the Alameda 1 and Sen 32 wells.

The more common trap types will be structural, particularly rollover anticlines associated with low-angle listric faults, although stratigraphic and combination traps are also expected.

The reservoir facies, likely to be present between 900- and 2500-m (2953- and 8202-ft) depths, will be an attractive factor for drilling. Although the reservoir facies may not pose any risk, seal may be the main risk, being ineffective or even absent.

Fluvial-Deltaic Sand Play Fairway

This play fairway is found in the southernmost part of the Macuspana subbasin (Figure 21; Table 4). These facies are intimately related to the expanded sequences of the so-called Topen-Nispero fault. Reservoir rocks consist of Pliocene–Pleistocene paralic to shallow marine deltaic sands and silts.

This is a proven play in the producing fields of Carrizo, Platanal, and Artesa. Associated trap types include growth successions, stratigraphic pinch-outs against faults (Artesa field), and inversion-triggered structural highs (Carrizo field). Similar traps are likely to be found in future exploratory efforts.

The limited size of the traps may be a negative factor in prospecting this play fairway, although the traps may prove to be attractive in case of multiple reservoirs. In addition, subtle stratigraphic traps may be an important element in this play. The average depths at which traps may be encountered are shallow, whereas the oil to be found is expected to be of a heavy type. An additional, relevant economic factor of this play fairway is the existence of numerous Mesozoic producing fields beneath, thus, exploration could be focused in both Mesozoic and Tertiary successions. Furthermore, oil production infrastructure is extensive and access to the area is easy. All these factors will allow the addition of new reserves and their timely production.

Fluvial Sand Play Fairway

The fluvial sand play fairway is located in the central part of the study area, between the Comalcalco system of faults to the north and the Topen-Nispero system to the south (Figure 21; Table 4). The Tertiary column is the thinnest here among all the play fairways, resting on the Jalpa high, where the Neogene sequence is almost entirely absent.

The reservoir facies consists of Pliocene–Pleistocene fluvial sand bodies in the form of channel and/or stacked or amalgamated bars. This play fairway produces in the Samaria, Girasol, and Gusano fields, with a spectrum of hydrocarbons ranging from gas to very heavy oil.

The known reservoirs occur at a depth range between 800 and 2000 m (2625 and 6562 ft), making them an attractive exploration target. Nevertheless, heavy oils in Gusano field will negatively affect the economic aspects of prospects developed in this play fairway. Lithofacies and depositional environments indicate the presence of multiple traps, particularly stratigraphic and combined, associated with an important Mesozoic structural high. Important fault systems might have served as migration routes from the nearby oil kitchens to the prospective younger traps.

The principal risk will be the presence and distribution of seals, which are expected to be thin and of

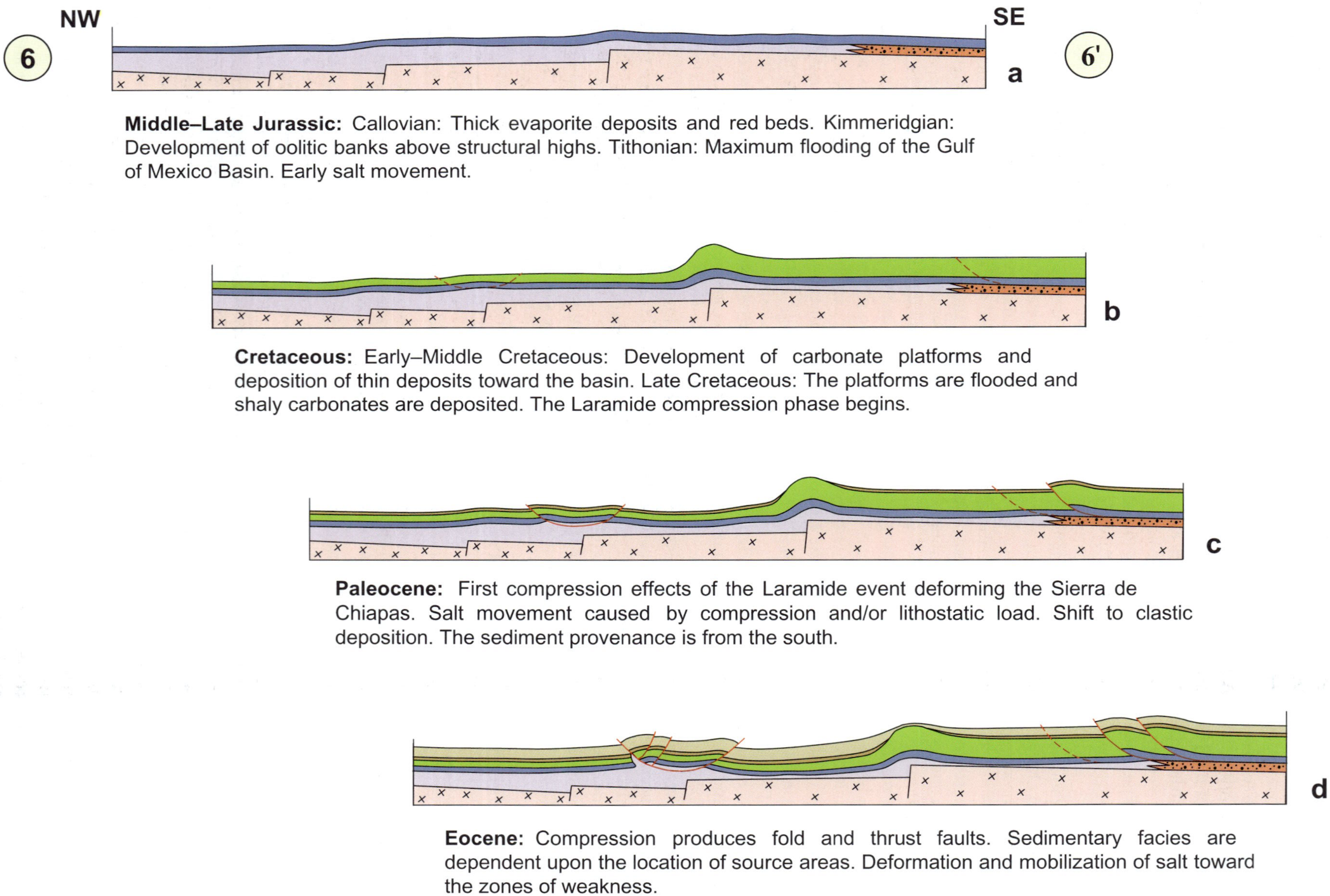

FIGURE 15. Schematic evolutionary model for the Comalcalco subbasin. See Figure 13 for the location.

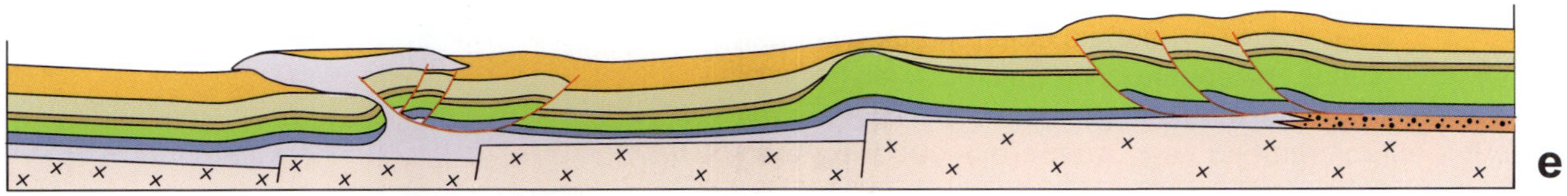

Oligocene: Compressive deformation continues. Salt intrusion occurs in different parts of the basin, and diapirs sometimes reach the sea floor.

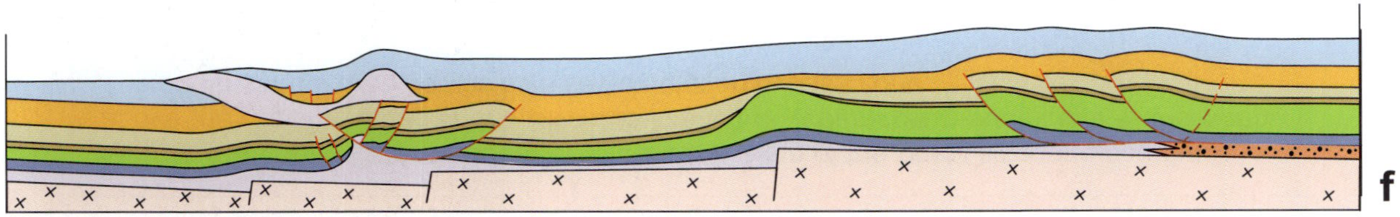

Early–Middle Miocene: Tectonic stability stage. Deformation seems to be associated mainly with salt movement. Sedimentary sequences tend to thin from south to north and the stratification continues and is tabular.

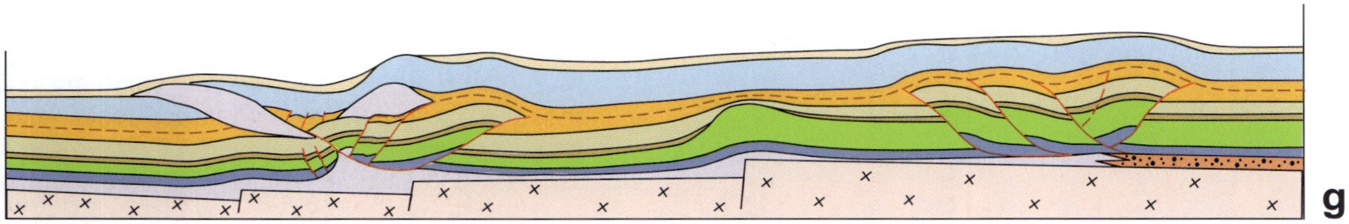

Late Miocene: Peak of compression of the Chiapaneca event, which redeforms previously formed compressive structures. Development of a regional detachment that separates the older fold and thrust belt from the younger extensional domain. Salt continues to mobilize. Sedimentary environments reflect shallow water facies.

FIGURE 15. (cont.).

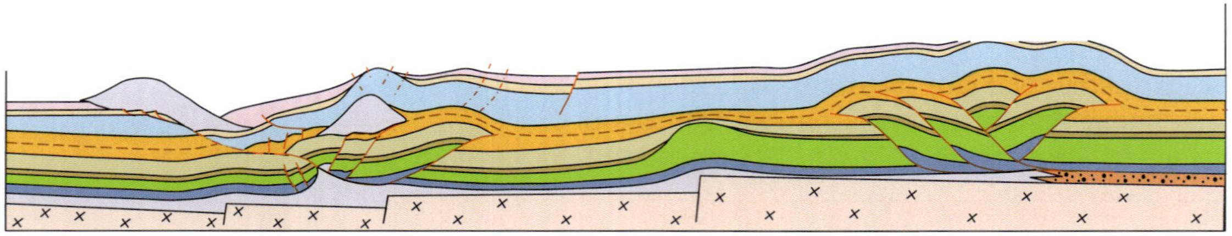

h

Early Pliocene: Rapid uplift of the Sierra de Chiapas generated great volumes of sediment and gravity sliding. Salt evacuation forms minibasins that are subsequently filled with sediments. Development of canopies and welds.

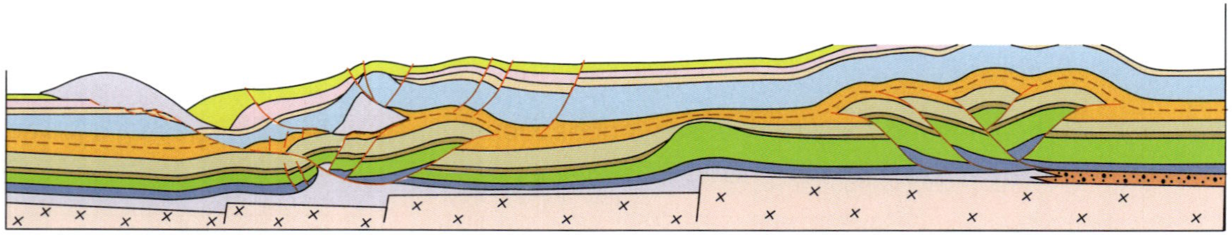

i

Middle Pliocene: Extension results in the separation of the post-Oligocene sequence above the regional detachment. Siliciclastic shelf environments predominate to the north. The south-central area serves as a sedimentation bypass zone.

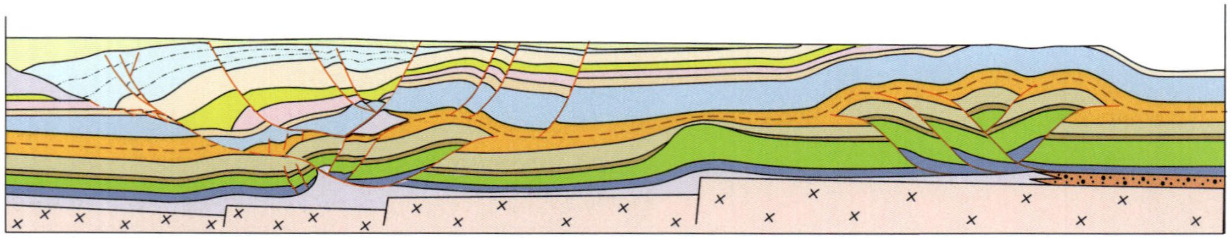

j

Late Pliocene–Holocene: Salt migrates to the north, leaving isolated salt bodies and welds behind. The accommodation space is compensated through the formation of counterregional faults. The remaining accommodation space is rapidly filled with sediment through various various sedimentation stages.

FIGURE 15. (cont.)

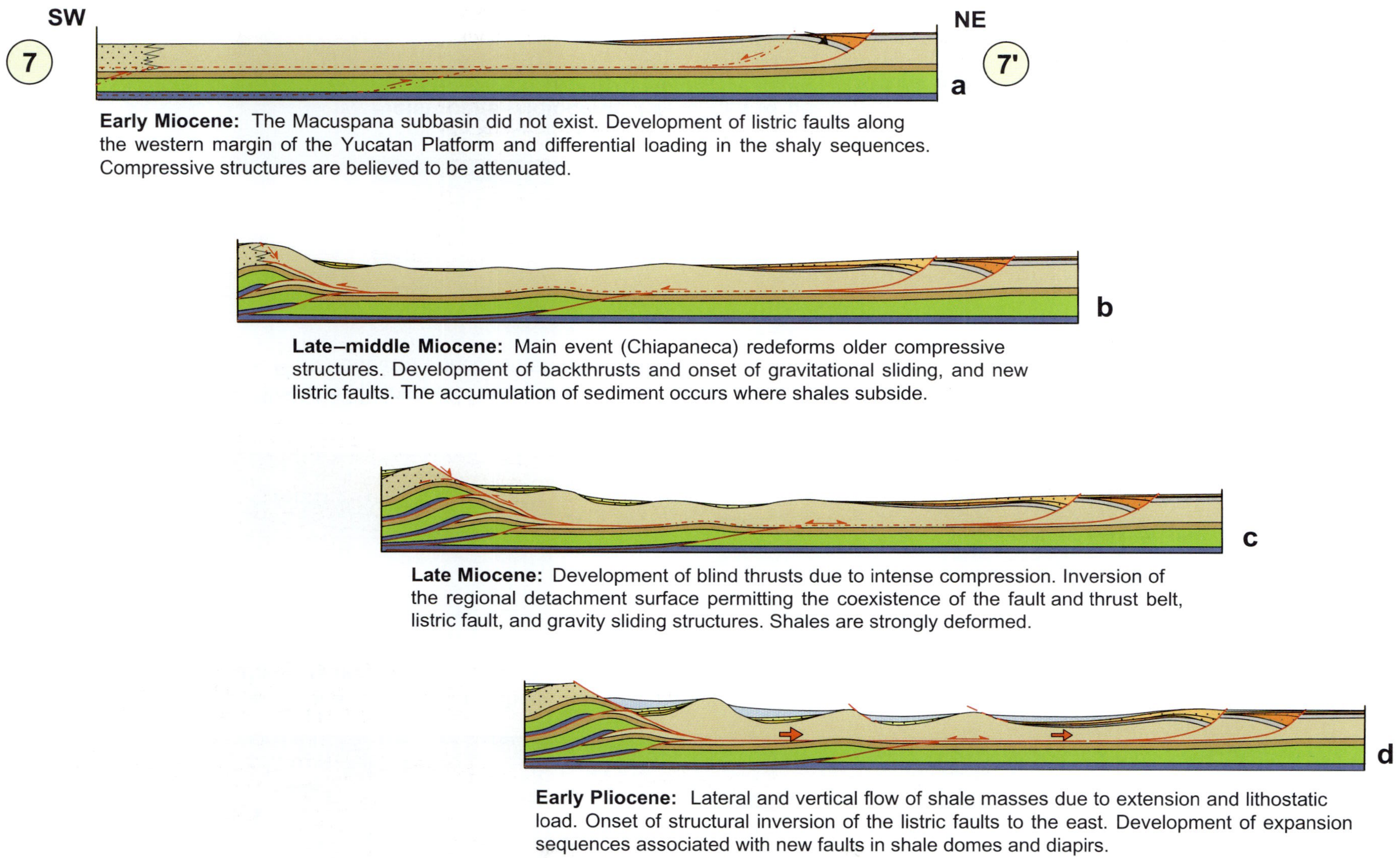

FIGURE 16. Schematic evolutionary model for the Macuspana subbasin. See Figure 13 for the location.

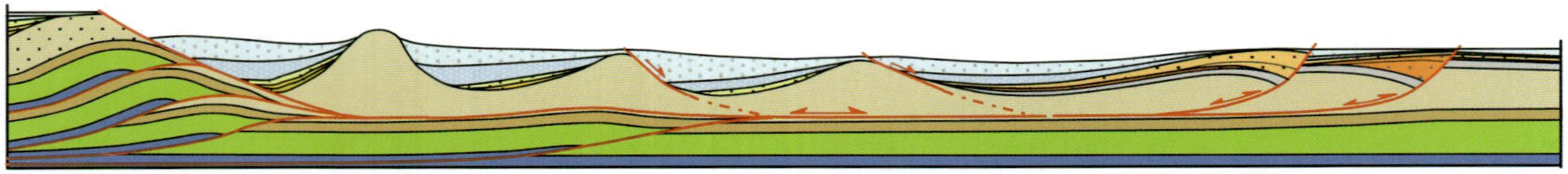

Middle Pliocene: Increase of sedimentation. Shale diapirism continues and depressions start to isolate. Structural inversion is totally extended with deformation of more stable blocks.

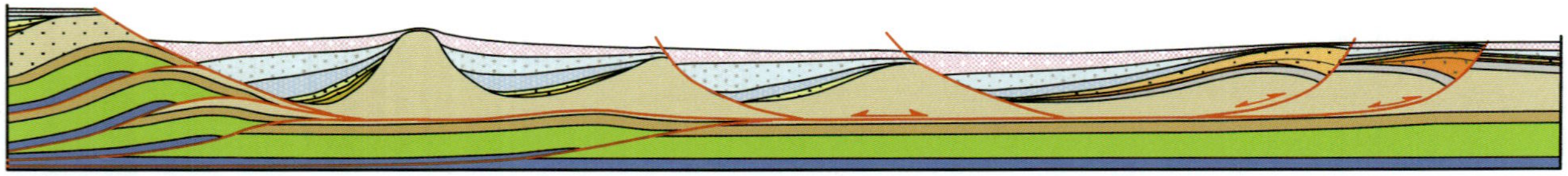

Late Pliocene: Increase of sedimentation rate and development of growth faults along shale domes. Pinch-out and truncation of the strata caused by shale diapirs.

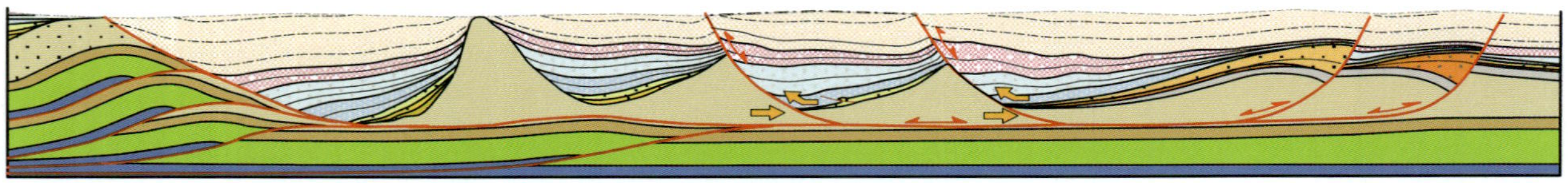

Pleistocene–Holocene: Filling of the basin. Minibasins, diapirs, and domes are fully developed. Late structural inversion related to the later movement of diapirs.

FIGURE 16. (cont.)

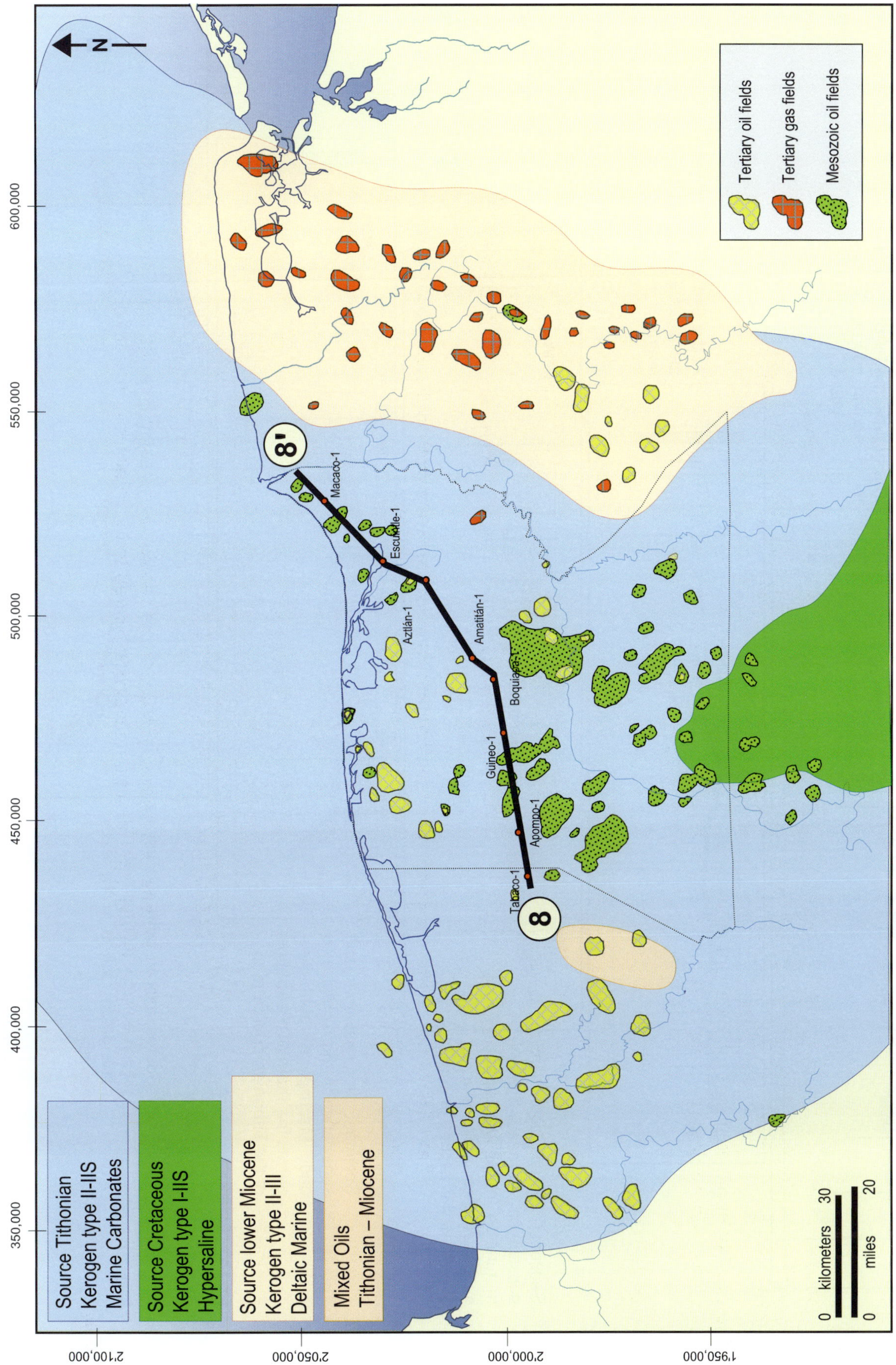

FIGURE 17. Distribution of the petroleum systems in the Sureste Basin. The Tithonian marine source rock is the best known because of its wide distribution and richness.

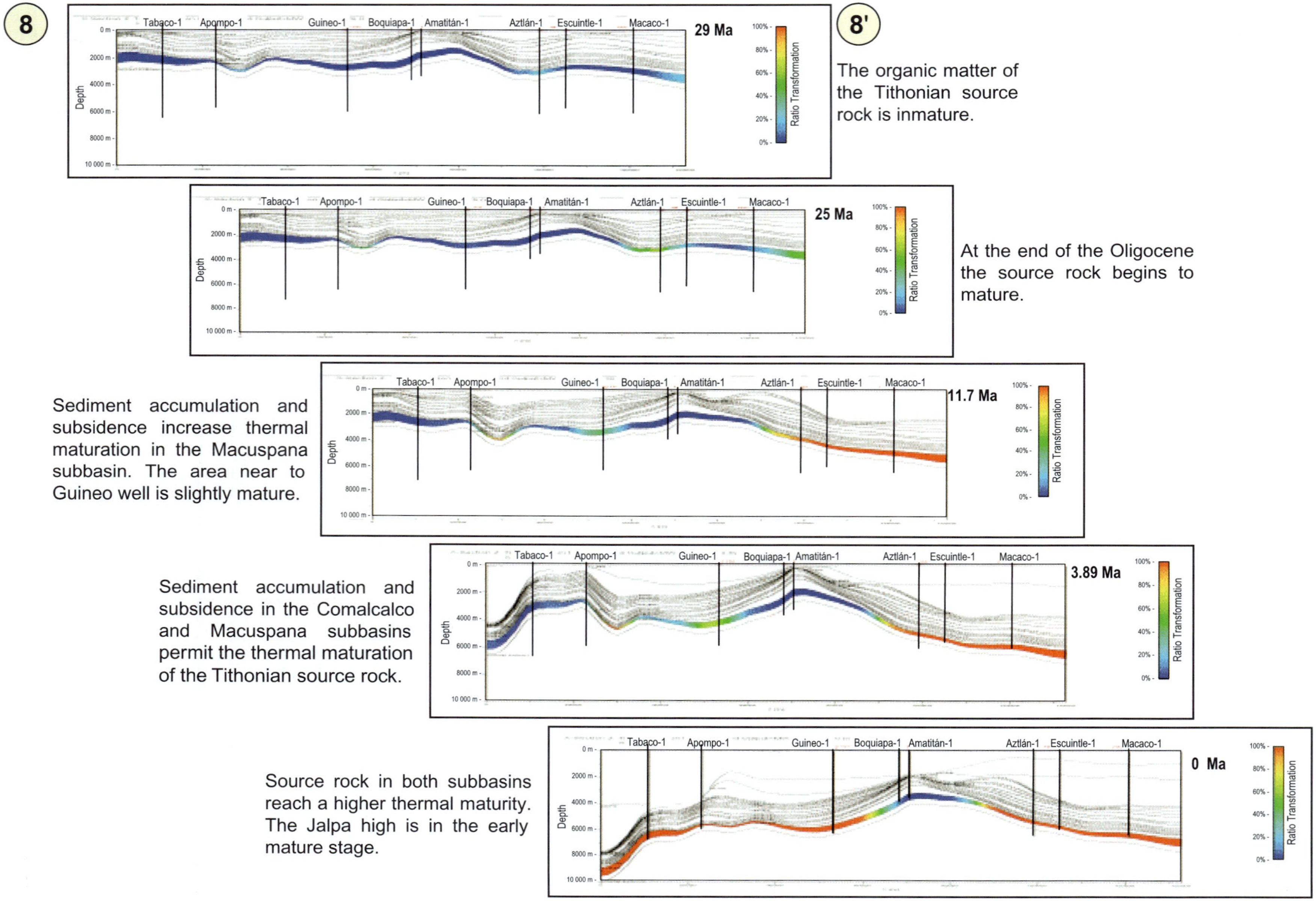

FIGURE 18. Petroleum generative systems in the Sureste Basin.

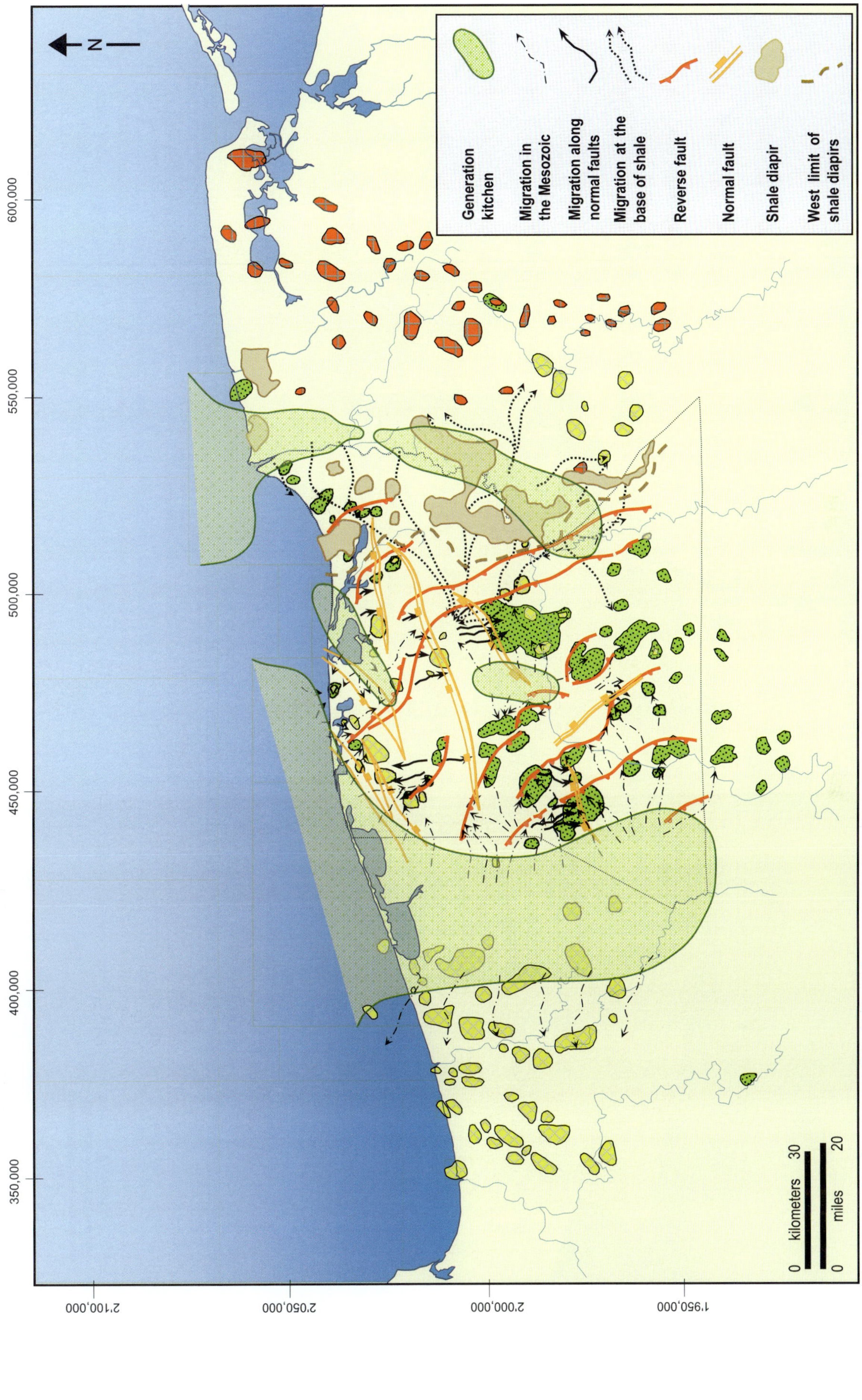

Figure 19. Kitchens and migration pathways. Mesozoic traps were filled along reverse faults, but when these are crosscut by normal faults, hydrocarbons migrated upward into the Tertiary sequences. Lateral migration is interpreted to charge the farther reservoirs.

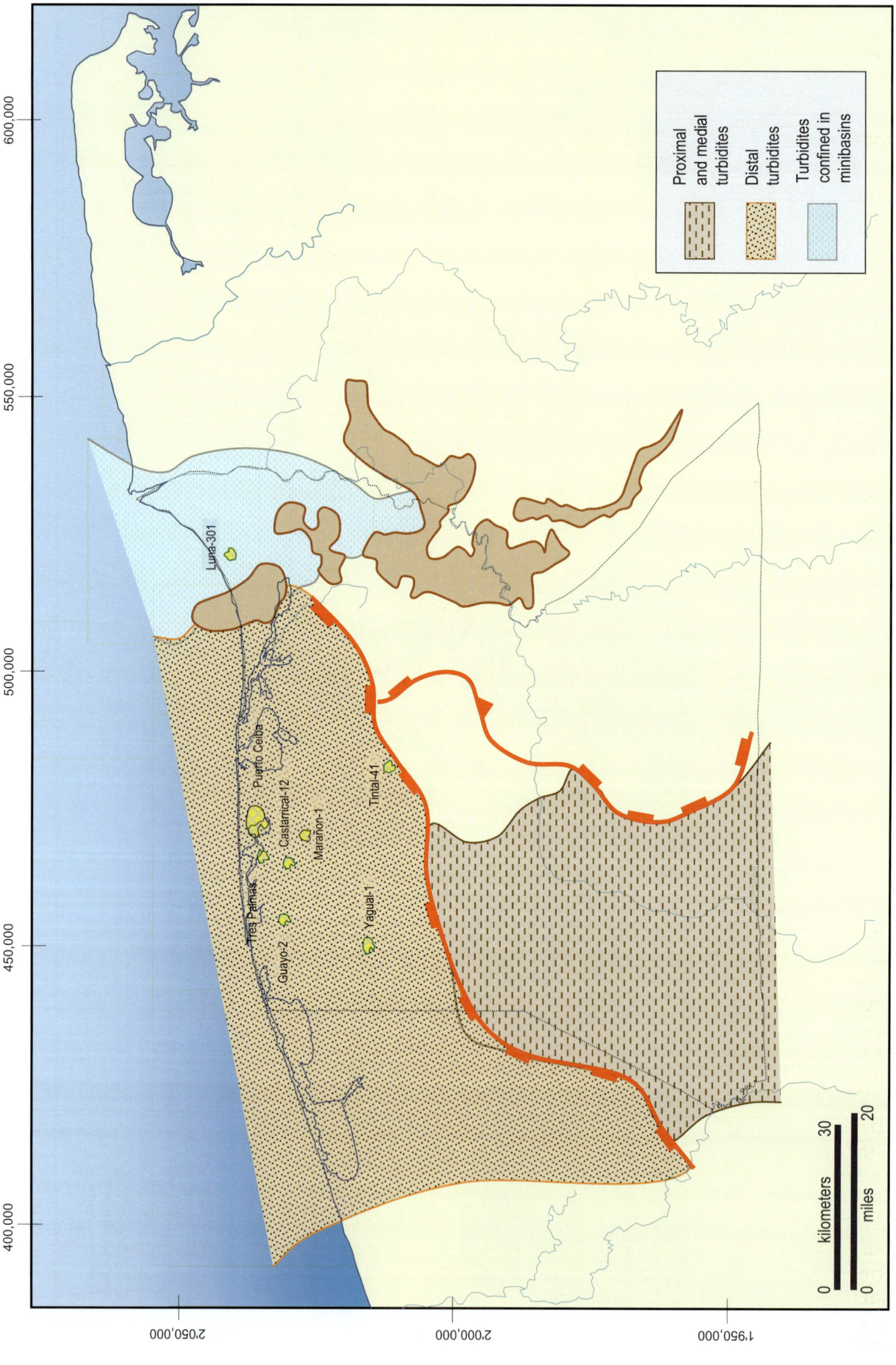

FIGURE 20. Turbidite play fairways.

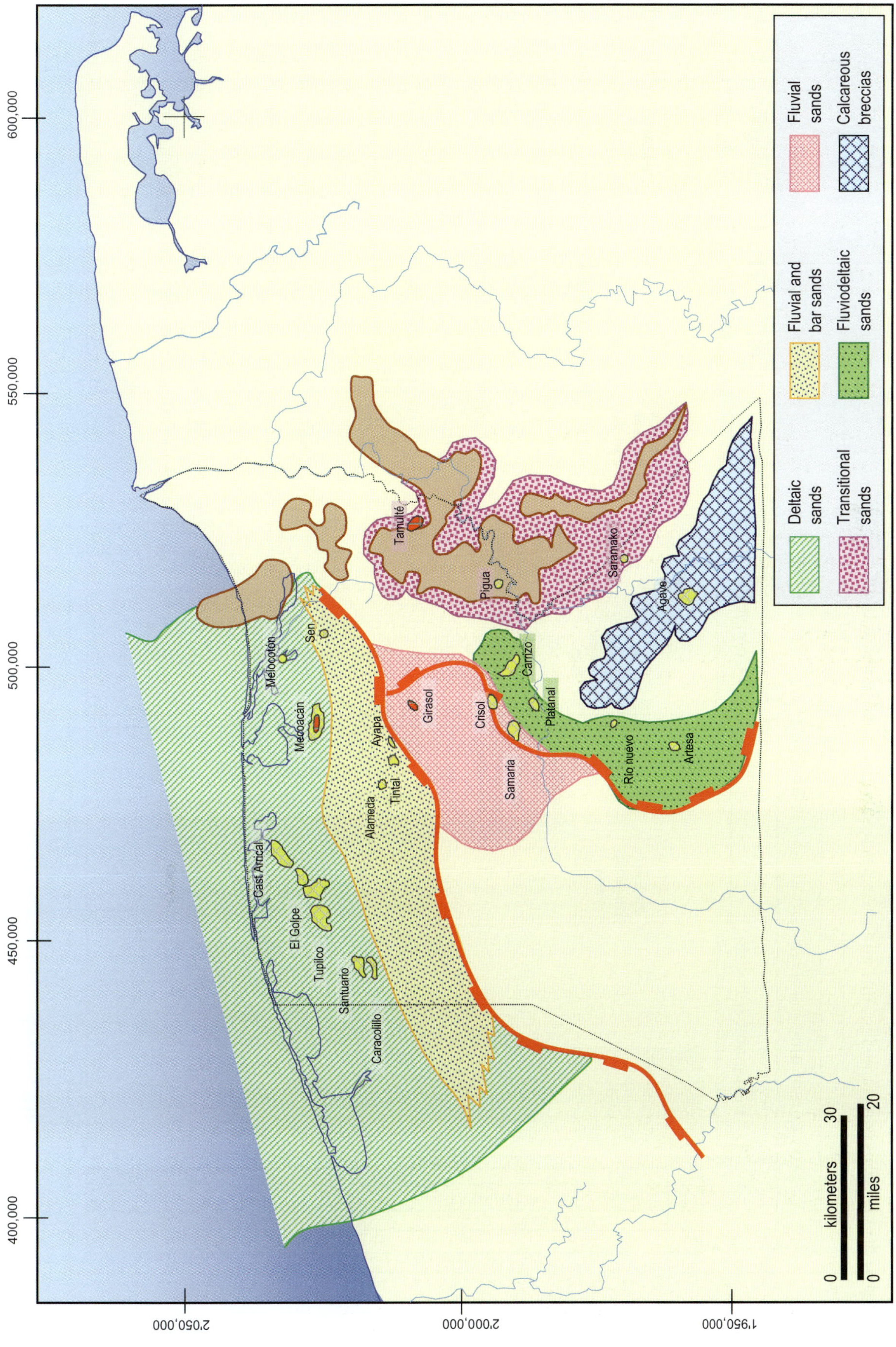

FIGURE 21. Deltaic and debris-flow play fairways.

TABLE 4. Play fairways constrained in time and their position within the studied subbasins.

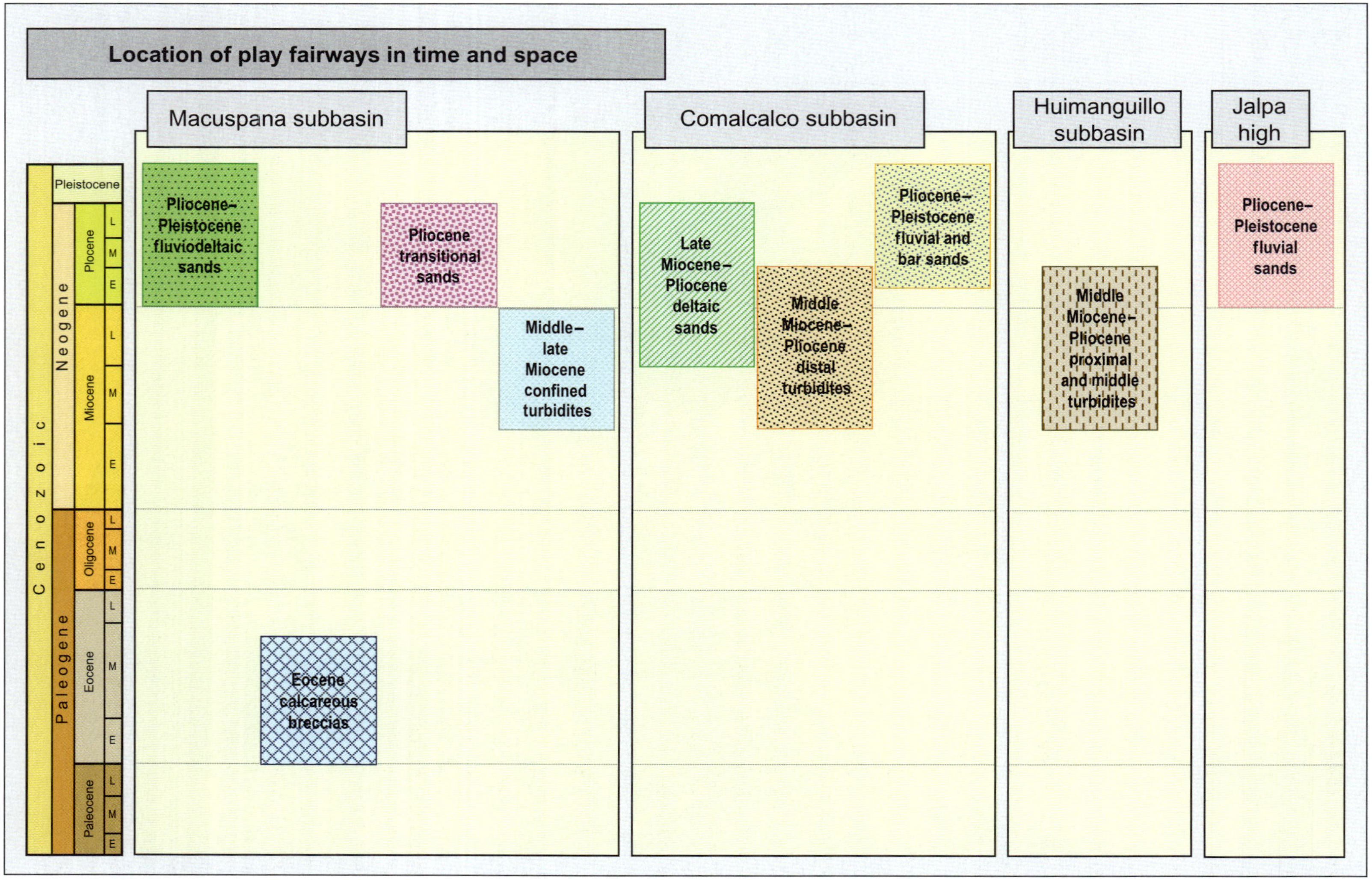

local nature. Economic viability of these plays is bolstered by the production infrastructure in place.

Calcareous Breccia Play Fairway

This play fairway occurs in the southwestern part of the Macuspana subbasin, being the only known Paleogene play type (Figure 21; Table 4). Its distribution and characteristics are not yet well understood.

It consists of Eocene debris flows made up of calcareous breccias deposited in a deep-water environment from the erosion of a preexisting, emergent Cretaceous platform lying to the south. This play fairway has been tested successfully in the Agave field. Similar lithofacies encountered in the Zarza and Lombarda wells are yet to yield positive results.

The calcareous breccia play fairway holds great promise because it has not yet been sufficiently explored despite the success in the Agave field. Good possibilities exist for similar plays in other areas and in different stratigraphic levels. The expected, common trap types are stratigraphic and combined stratigraphic-structural.

The principal risks involve the reservoir facies because of the erratic distribution of debris-flow deposits and diagenetic changes, such as compaction, cementation, fracturing, and dissolution. Also, the presence and efficiency of seals will be a factor because unconformities and large faults decapitating the structures are common in the area.

CONCLUSIONS

The sequence-stratigraphic framework as well as the sedimentological characteristics of the Tertiary column for this important sector of the Sureste Basin were established in this study by a PEMEX multidisciplinary team. The potential for best reservoir facies lies in the Neogene turbidite and fluviodeltaic systems.

Diagenetic history in representative Tertiary wells was interpreted; diagenetic processes most favoring the reservoir facies are dissolution and fracturing, whereas compaction and cementation are the main negative factors affecting the reservoir facies.

The petroleum system study established the distribution of the generative kitchens, timing of onset of hydrocarbon generation, different pulses of expulsion, and the potential migration routes.

Four large distinctive Tertiary subprovinces have been identified and delimited, each showing particular stratigraphic characteristics and structural histories. The progression of the principal tectonic phases affected the evolution of the basin and associated subbasins.

By understanding the tectonic and sedimentary history of the basin, coupled with the petroleum system, we were able to identify nine potential play fairways and three conceptual ones. We evaluated and ranked them according to the criteria of immediate economic viability and lesser exploratory risks, which would allow us in the short term to reverse the tendency toward dwindling reserves in known fields and to add new reserves. We showed that the Tertiary column in the Sureste Basin holds great hydrocarbon potential, which has not yet been fully explored or exploited.

REFERENCES CITED

Cárdenas, L. C., and S. C. Marín, 2004, Biostratigraphy as a fundamental tool in the analysis of the Sureste Basin of Mexico: AAPG International Conference and Exhibition, Cancun, October 24–27, 2004, Technical Program Abstracts, p. A41.

Berggren, W. A., D. V. Kent, C. C. Swisher III, and M. P. Aubry, 1995, A revised Cenozoic Geochronology and Chronostratigraphy, *in* W. A. Berggren, D. V. Kent, M. P. Aubry, and J. J. Hardenbol, eds., Geochronology time scales and global stratigraphic correlation: SEPM Special Publication 54, p. 129–212.

Bolli, H. M., and J. B. Saunders, 1985, Oligocene to Holocene low latitude planktic foraminifera, *in* H. M. Bolli, J. B Saunders, and K. Perch-NielSan, eds., Plankton stratigraphy volume 1: Great Britain, Cambridge University Press, p. 155–262.

Cabrera, C. R., and R. J. E. Lugo, 1988, Estratigrafía-sedimentología de las Cuencas Terciarias del Golfo de México: Boletín de la Asociación Mexicana de Geólogos Petroleros, v. 3, no. 7–8, p. 229–257.

Calderón, G. A., 1951, Condiciones estratigráficas de las formaciones miocénicas de la Cuenca Salina del Istmo de Tehuantepec: Boletín de la Asociación Mexicana de Geólogos Petroleros, v. 36, no. 1, p. 3–55.

Castillo, T. C., 1955, Bosquejo estratigráfico de la Cuenca Salina del Istmo de Tehuantepec: Boletín de la Asociación Mexicana de Geólogos Petroleros, v. 7, no. 5–6, p. 173–212.

Chávez, V. V., 1999, El sistema petrolífero como herramienta para la revitalización de la Cuenca del Sureste, México: Documento Interno (inédito), PEMEX Internal Publication, 12 p.

Chávez, V. V., and C. J. Cerrillo, 1999, Marco Estructural de la Cima del Mesozoico, Cuenca del Sureste: Mapa de configuración estructural de la cima del Cretácico Superior entre las coordenadas UTM 435,000–550,000 y 1,940,000–2,050,000 (inédito), 16 p.

García, R. A., S. M. Herrera, and H. A. Escamilla, 1989, Estudio sedimentológico del Terciario en el prospecto Cuenca de Comalcalco, Tabasco: Instituto Mexicano del Petróleo: Proyecto, C-3042, 24 figuras, 18 anexos, Informe Final (inédito), 76 p.

Gutiérrez, G. R., 1950, Yacimientos petroleros en la región de Macuspana, Tabasco: Boletín de la Asociación Mexicana de Geólogos Petroleros, v. 2, no. 8, p. 499–510.

Guzmán, V. M., V. M. Clara, and M. L. Medrano, 1995, El Hábitat del Petróleo de las Cuencas del Sureste, Un

Enfoque con Biomarcadores: Proyecto IMP-Petróleos Mexicano (PEMEX), CAB 0501/95, 44 figuras, 11 anexos, Informe Final (inédito), 57 p.

Hardenbol, J., J. Thierry, M. B. Farley, T. Jacquin, P. Ch. De Graciansky, and P. R. Vail, 1998, Mesozoic and Cenozoic sequence chronostratigraphic framework of European basins: SEPM Special Publication 60, 764 p.

Holguin, Q. N., 1988, Evaluación Geoquímica del Sureste de México: Boletín de la Asociación Mexicana de Geólogos Petroleros, v. 37, no. 1, p. 3–48.

Krynine, P. D., 1935, Arkose deposits in the humid tropics, a study of sedimentation in southern Mexico: American Journal of Sciences, v. 29, p. 353–363.

Mata, J. M. E., S. M. Varela, P. J. I. Juárez, Y. J. R. Villagrán, L. C. Y. Cárdenas, and S. C. Marín, 2003, Aplicación de la bioestratigrafía en el análisis secuencial de las subcuencas terciarias del Sureste de México: II Simposio de Bioestratigrafía en la exploración petrolera de México y exposición. Reynos, Tamaulipas, México, p. 16.

McBride, E. F., 1963, A classification of common sandstones: Journal of Sedimentary Petrology, v. 33, no. 3, p. 664–669.

Meneses, R. J., 1987, Marco tectónico y paleogeográfico del Triasico Tardío-Jurásico en el Sureste de México: Boletín de la Asociación Mexicana de Geólogos Petroleros, v. 39, no. 2, p. 3–69.

Meneses, R. J., M. I. Gutierrez, and Ch. J. Gómez, 1987, Evaluación geológica regional de la Sierra de Chiapas: Informe Geológico No. 1031, Z. SE: Petróleos Mexicanos (inédito), PEMEX Internal Publication, 44 figs., 17 tablas, 69 anexos, 395 p.

Mitchum, R. M., P. R. Vail, and S. Thompson, 1977, Seismic stratigraphy and global changes of sea level: Part 2. The depositional sequence as a basic unit for stratigraphic analysis in seismic stratigraphy— Applications to hydrocarbon exploration: AAPG Memoir 26, p. 53–62.

Odom, I. E., T. W. Doe, and R. H. Dott Jr., 1976, Nature of feldspar grain size relations in some quartz rich sandstones: Journal of Sedimentary Petrology, v. 46, no. 4, p. 862–870.

Petróleos Mexicano (PEMEX)-Chevron, 1993, Proyecto de la Cuenca del Sureste: Informe Final, Pemex Internal Publication, tomo I, texto y figuras, 92 figuras (inédito), 110 p.

Petróleos Mexicano (PEMEX) Exploración y Producción (Gerencia de Producción, Región Sur), 1993, Censo de Yacimientos Petroleros, Campos Región Sur: Vilahermosa, Tabasco, México, PEMEX, 15 tomos (inédito), no. 3, 556 p.

Petróleos Mexicano (PEMEX) Exploración y Producción, 1999, Las reservas de hidrocarburos de México: Volumen I Evaluación al 1 de Enero de 1999, Volumen II Los principales campos de petróleo y gas de México: PEMEX Exploración y Producción, México, D. F., 398 p.

Reyes, N. J., 1981, Características estructurales del área mesozoica Chiapas-Tabasco: Boletín de la Asociación Mexicana de Geofísicos de Exploración, v. 22, no. 1–2, p. 1–22.

Ricoy, S. U., and M. I. A. Rodríguez, 1981, Interpretación de ambientes sedimentarios a través de secciones sísmicas: Revista de Ingeniería Petrolera, v. 21, no. 11, p. 5–16.

Robertson Research Internacional Limited, 2004, Estudio de distribución y calidad de sellos, Activo Reforma-Comalcalco, México: Informe No. 8793/Id, Proyecto No. Id/GK377, 3 tablas, 6 figuras, 6 apéndices, 8 montajes, 8 anexos (inédito), 34 p.

Salas, P. G., and R. E. López, 1951, Geología y Tectónica de la región de Macuspana Tabasco y parte norte de Chiapas: Boletín de la Asociación Mexicana de Geólogos Petroleros, v. 3, no. 1–2, p. 3–56.

Sánchez, M. de O. R., 1979, Geología petrolera de la Sierra de Chiapas: Boletín de la Asociación Mexicana de Geólogos Petroleros, v. 31, no. 1–2, p. 67–97.

Santiago, A. J., B. J. Carrillo, and A. B. Martell, 1984, Geología petrolera de México: Evaluación de formaciones en México: Paris, France, Schlumberger Offshore Services, 32 figs., p. 1–36.

Tipsword, H. L., 1962, Tertiary foraminifera in Gulf Coast petroleum exploration and development, *in* E. H. Rainwater and R. P. Zingula, eds., Geology of the Gulf Coast and central Texas: Houston Geological Society Guidebook, Houston Geological Society, p. 16–57.

Van Wagoner, J. C., R. M. Mitchum, K. M. Campion, and V. D. Rahmanian, 1990, Siliciclastic sequence stratigraphy in well logs, cores and outcrops: Concepts for high resolution correlation of time and facies: AAPG Methods in Exploration Series 7, 55 p.

Cantú-Chapa, A., 2009, Ammonites of the Cretaceous Taraises and lower Tamaulipas formations in eastern Mexico, *in* C. Bartolini and J. R. Román Ramos, eds., Petroleum systems in the southern Gulf of Mexico: AAPG Memoir 90, p. 191–216.

Ammonites of the Cretaceous Taraises and lower Tamaulipas formations in eastern Mexico

Abelardo Cantú-Chapa

Instituto Politecnico Nacional, Mexico, D.F., Mexico

ABSTRACT

The Taraises and lower Tamaulipas formations of the Early Cretaceous (Berriasian–Valanginian) are basinal sedimentary sequences in northeastern and eastern Mexico, respectively. The first unit was penetrated in the San Javier 1 and 2 wells at depths between 1800 and 3200 m (5905 and 10,499 ft), and the second was drilled in the Bejuco 6 and the La Laja 8 wells between 2000 and 2500 m (6562 and 8202 ft). In the Bejuco 6 well, the lower Tamaulipas Formation represents a condensed section. Geophysical logs and ammonites allow the characterization and age dating of these units.

Two important ammonite groups (Ancyloceratoidea and Perisphinctoidea) are described from cores in the four wells and from surface occurrences. Four new genera are proposed, *Misantlites*, *Wiedmannites*, *Parvaites*, and *Parrasites*; new species are *M. reyesi*, *Aegocrioceras eguiluzi*, *Capeloites lajense*, *C. peyroullesense*, *Thurmanniceras kleini*, *Neohoploceras bartolinii*, *Parvaites aguirrei*, *P. leanzai*; and one new subfamily is proposed, Capeloitinae.

INTRODUCTION

Geological Setting and Location

The Taraises and lower Tamaulipas formations were found in petroleum wells in northeastern and eastern Mexico. The core samples with ammonites were correlated with the corresponding well logs to determine the precise depths from which they were obtained. These depths range from 1800 to 3300 m (5905 to 10,827 ft) below sea level (Figure 1). The ammonites from the wells were crushed and fragmented, which made complete systematic determinations difficult.

However, many fragmented specimens recovered from wells are very valuable for stratigraphic purposes and in no case should they be discarded because this paleontologic material is obtained from depth at high cost. A study of this type of material allows a determination of the chronostratigraphic position of the rock in which the fossils are found, particularly with support of specimens from neighboring wells and integration of geophysical logs.

The find in eastern Mexico La Laja 8 well (core 9; 2537–2546 m [8323–8353 ft]) of a specimen with a fine ribbing of olcostephanid type, considered to be *Olcostephanus* sp., is significant, suggesting that the rock from

DOI:10.1306/13191084M902804

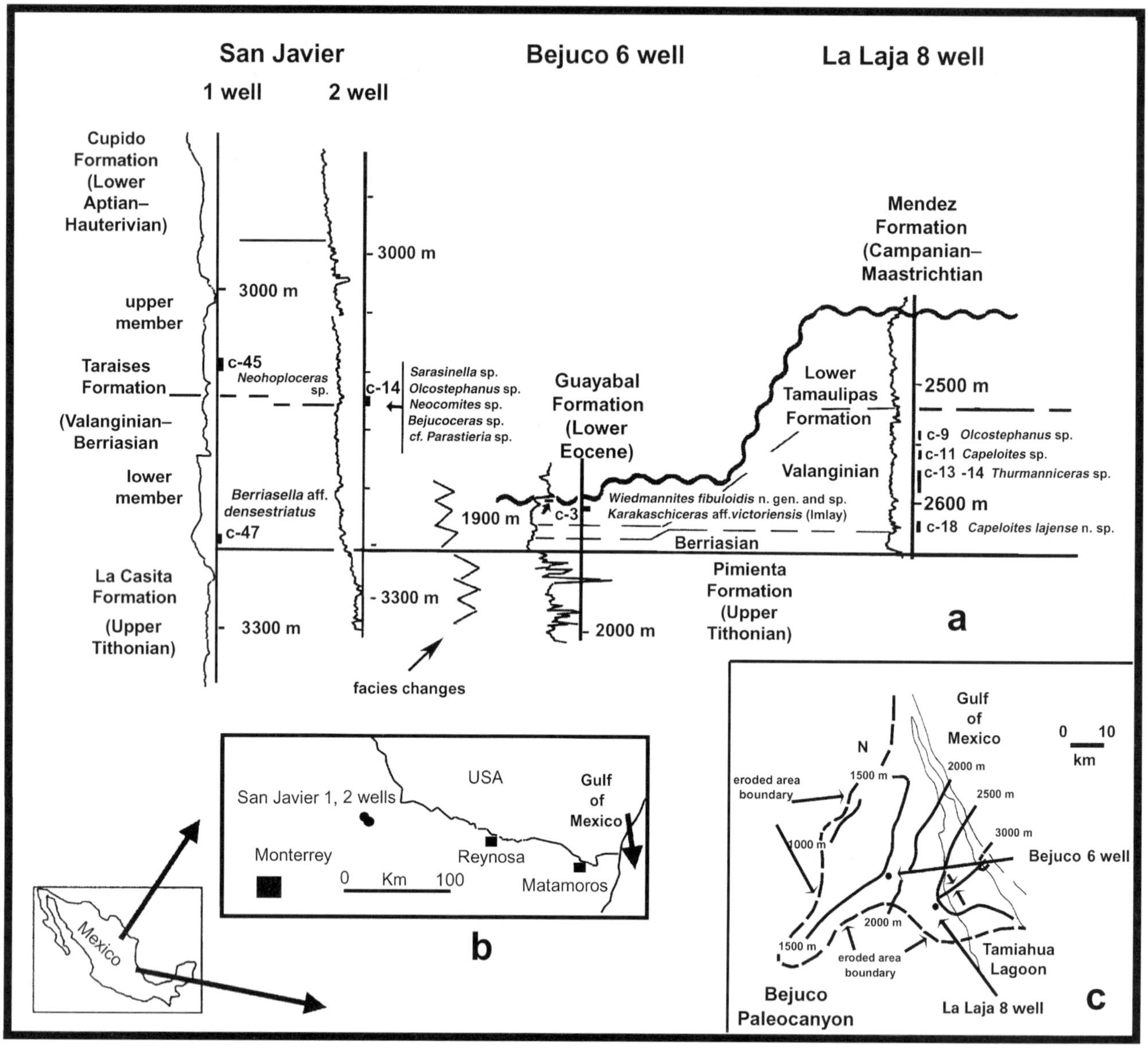

FIGURE 1. Cross-section of the Taraises and lower Tamaulipas formations (Lower Cretaceous) in oil wells in northeastern and eastern Mexico based on gamma-ray logs and ammonites. The stratigraphic datum is top of Tithonian.

which it was recovered is the Valanginian lower Tamaulipas Formation (Figure 2).

THE TARAISES FORMATION (BERRIASIAN–VALANGINIAN) IN THE SAN JAVIER 1 AND 2 WELLS IN NORTHEASTERN MEXICO

These two wells were drilled in the Paras fault system (Cantú-Chapa, 1989), and three ammonite specimens were studied. Some ammonites from the San Javier 1 well were determined by Imlay (1980).

The Taraises Formation consists of alternating beds of shaly limestone and shale subdivided into lower and upper members. This sedimentary sequence is quite recognizable on the gamma-ray log. The Taraises Formation overlies the Tithonian La Casita Formation and underlies the Hauterivian–lower Aptian Cupido Formation in this part of eastern Mexico (Cantú-Chapa, 2001) (Figure 1).

The lower member consists of a dense limestone approximately 140 m (459 ft) thick. A specimen of the ammonite *Berriasella* was recovered from the base of this unit in the San Javier 1 well (core 47; 3199–3207 m [10,495–10,522 ft]). A study of this specimen allowed the characterization of beds containing *Spiticeras* of Berriasian age in this well, as studied in north-central Mexico (Burckhardt, 1930).

FIGURE 2. A fragmented specimen was cut by the borehole in the lower Tamaulipas Formation from the La Laja 8 well (core 9, 2537–2546 m [8323–8353 ft]), eastern Mexico. The specimen shows only a dense ribbing and probably corresponds to *Olcostephanus* sp. from the Valanginian age. Scale = × 1.

The upper member consists of shaly limestone about 130 m (426 ft) thick. This member includes two ammonite associations; the lower containing *Bejucoceras-Sarasinella* and the upper with *Karakaschiceras-Olcostephanus*. These are found at the base and top of this member, respectively (Cantú-Chapa, 2001).

Two fragmented new ammonites were also obtained within the basal upper member of the Taraises Formation in the San Javier 2 well (core 14). These specimens were identified as cf. *Parastieria* sp. and *Sarasinella* sp., suggesting an early Valanginian age because they are associated with previously studied specimens of this age. For chronostratigraphic purposes, these are comparable to similar materials from Europe (Kemper et al., 1981; Company, 1987; Cantú-Chapa, 2001).

A new species of *Neohoploceras* is also described from the upper member of the Taraises Formation in the San Javier 1 well (core 45; 3056–3068 m [10,026–10,066 ft]). This genus characterizes the Valanginian in Argentina, Spain, and France (Company, 1987; Reboulet, 1996; Aguirre-Urreta, 1998).

The lower member of the Taraisas Formation is dated as Berriasian–basal early Valanginian, whereas the upper member is early Valanginian (pertransiens zone) to basal late Valanginian (verrucosum zone). These ages were established in northeastern Mexico from the stratigraphic position of the studied material and by correlation with chronostratigraphic units recognized in Europe (Cantú-Chapa, 2001) (Figure 1).

THE LOWER TAMAULIPAS FORMATION (EARLY CRETACEOUS) IN THE BEJUCO 6 AND LA LAJA 8 WELLS IN EASTERN MEXICO

Both wells were drilled in the Bejuco paleocanyon, which represents an important area of subaqueous erosion that caused the Neocomian–Eocene unconformity. Depths to the unconformity in the studied wells were plotted to show the Bejuco paleocanyon trending from the southwest toward the east. This paleogeographic element trends perpendicular to the modern coast of the Gulf of Mexico and occurs at depths between 1200 and 3000 m (3937 and 9842 ft); the latter depth is toward the Gulf (Cantú-Chapa, 1987) (Figure 1).

Two uncoiled ammonites were obtained from core 3 of the Bejuco 6 well (1907–1919 m, 6256–6296 ft) herein denominated *Wiedmannites* gen. n. from its looped ribbing, which differs from that of similar genera. These fossils are associated with specimens of *Karakaschiceras* in this well and characterize the base of the upper Valanginian according to established comparisons with similar material from France, Spain, Germany, and Poland (Kemper et al., 1981; Company, 1987; Kutek et al., 1989; Reboulet, 1996).

The La Laja 8 well (core 11) provided a specimen of *Capeloites* sp. in the micritic limestone of the lower Tamaulipas Formation of early Valanginian age (Cantú-Chapa, 2001). New specimens from this well are reported herein derived from cores 9 (2537–2546 m, 8323–8353 ft), 11 (2555–2564 m, 8382–8412 ft), 13 (2573–2582 m, 8442–8471 ft), 14 (2582–2591 m, 8471–8501 ft), and 18 (2618–2627 m, 8589–8619 ft). The gamma-ray log curve of this interval is offset to the left and shows little variation representing the massive limestone of this formation (Figure 1).

The lower Tamaulipas Formation in the La Laja 8 well is covered unconformably by marl of the Méndez Formation. The latter formation in this well contains Globotruncanidae foraminifera of Campanian–Maestrichtian age. In a prior study, this unit was considered as part of the Guayabal Formation (middle Eocene), but this has been corrected based on a study of the foraminifera (Cantú-Chapa, 2001).

The La Laja 8 well provides new taxa of the genera *Capeloites* Lissón (1937) and *Thurmanniceras* Cossman (1901). The first genus includes small forms without precise stratigraphic distribution. A Valanginian–early Hauterivian age was considered for similar material in France, Spain, and Peru (Wright et al., 1996).

Some specimens of *Capeloites* in the material from northeastern Mexico were dated as late Valanginian because of their association with *Garcites* Cantú-Chapa

France, and Spain, Company (1987), Reboulet (1996)	Estern Mexico, Bejuco-6 well Cantú-Chapa (1976)	Estern Mexico, La Laja-8 well (this work)	Formation
Lower Valanginian		- - -	
Neocomites campylotoxus (= *Neolissoceras salinarium*)	*Valanginites* sp.	*Olcostephanus* sp. (core 9, 2,539-2,547 m) *Capeloites* sp. (core 11, 2,556-2,562 m)	lower
Thuramanniceras pertransiens	*Thurmanniceras* sp.	*Thurmanniceras kleini* sp. nov. - *Olcostephanus* sp. (core 13, 2,575-2,584 m)	Tamaulipas
Thurmanniceras otopeta		cf. *Bejucoceras* sp. (core 14, 2,583-2,592 m)	*pt.*
		Capeloites lajense (core 18, 2,619-2,628 m)	

FIGURE 3. Ammonite distribution in the lower Tamaulipas Formation (Lower Cretaceous), in two oil wells from eastern Mexico, and correlation with zonal schemes from France and Spain.

(2001) and *Karakaschiceras* Thieuloy (1969) (Cantú-Chapa, 2001).

Capeloites lajense n. sp. is described herein from core 18, corresponding to a layer below that containing *Capeloites* sp.; the latter specimen is obtained from core 11 in the La Laja 8 well (Cantú-Chapa, 2001). Therefore, *C. lajense* has an older stratigraphic position than *Capeloites* sp. Both specimens are from the lower Tamaulipas Formation and are separated by approximately 65 m (213 ft) in this well.

C. lajense is older than the *Thurmanniceras kleini* n. sp.-*Olcostephanus* sp. association established from core 13 of the La Laja 8 well (Figures 1, 3). The 72-m (236-ft) thickness of the lower Tamaulipas Formation is herein dated as early Valanginian because of the predominance of *Thurmanniceras*. This thickness is represented in cores 11 to 18 of the La Laja 8 well. In contrast, the age-equivalent unit in the Bejuco 6 well from the same area is represented by a calcareous condensed section only 22 m (72 ft) thick (Figure 1). Some species of *Thurmanniceras* were dated as early Valanginian by Imlay (1937) in outcrops of northeastern Mexico; however, there are doubts concerning this age (Kemper et al., 1981).

Ammonites from Outcrops

Three interesting heteromorphic ammonite specimens of the Family Crioceratitidae were collected at three outcrops of the lower Tamaulipas Formation in northeastern Mexico and are described herein (Figure 4). These are *Misantlites reyesi* n. gen. and sp., *Aegocrioceras eguiluzi* n. sp., and *Crioceratites* sp. The first is from Misantla, Veracruz, eastern Mexico; *Aegocrioceras* was collected in the Sierra de San Francisco, Coahuila; and *Crioceratites* is from Miquihuana, Tamaulipas; both in northeastern Mexico (Figure 4).

Aegocrioceras and *Crioceratites*, described here, represent the first known specimens from Mexico. *Misantlites* is proposed as a new genus distinct from the crioceratids of the rest of the world.

Systematic Considerations

Three Tithonian–early Cretaceous systematic ammonite groups corresponding to the families Olcostephanidae, Berriasellidae, and Neocomitidae are studied here because each includes genera with very contrasting morphologies and insufficient systematic studies as defined in the *Treatise*, part L, by Wright et al. (1996).

CONCLUSIONS

These studies have made four important achievements possible:

1. The stratigraphic position of fragmented and crushed ammonites coming from great depths in Petróleos Mexicanos (PEMEX) wells along the coastal plain of Gulf of Mexico was established.
2. A Berriasian–Valanginian age was determined for several beds in two PEMEX wells.
3. A condensed Valanginian age sequence in the lower Tamaulipas Formation was recognized through the systematic study of ammonites recovered from the PEMEX Bejuco 6 and La Laja 8 wells.
4. A revision of the systematic classifications of the Lower Cretaceous ammonite groups was proposed (crioceraitids, olcostephanids, and berriasellids).

ACKNOWLEDGMENTS

The author was generously assisted by N. Monroy Ayala, M. A. Gómez Galindo, J. Morales Marin, and R. Orozco Charles in the research of the Tampico oil

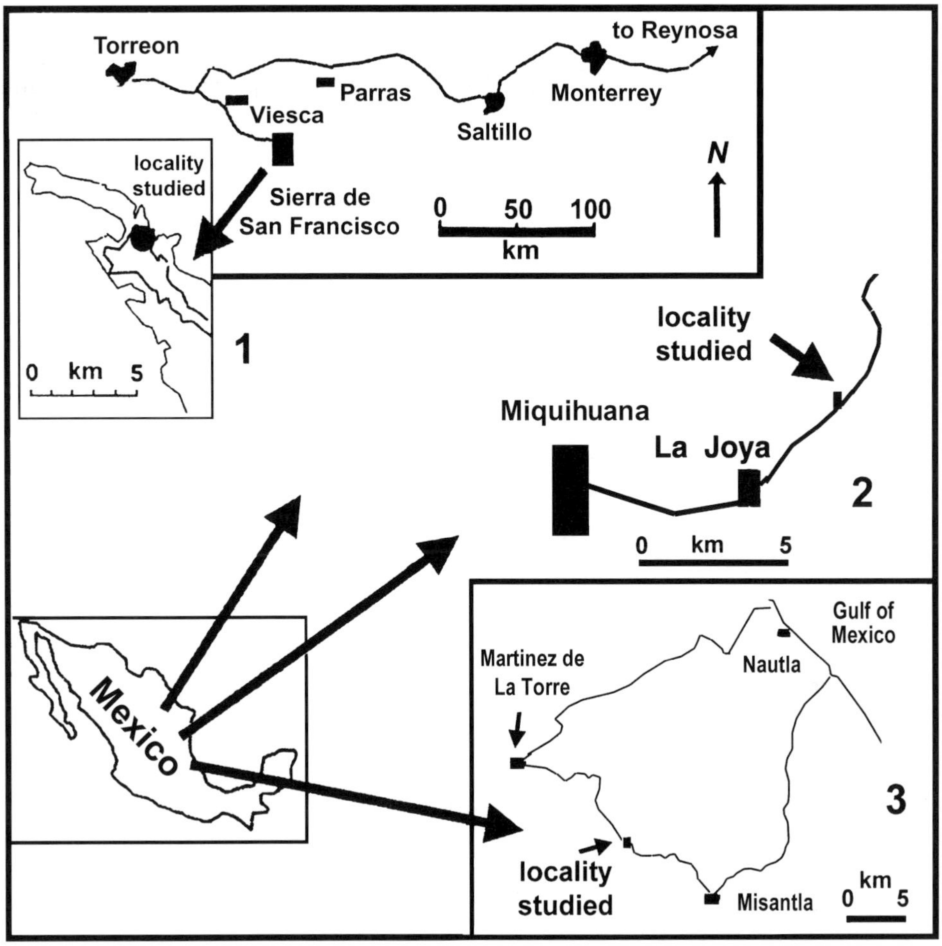

FIGURE 4. Map showing the ammonite localities of the Taraises and lower Tamaulipas formations (Lower Cretaceous). (a) Sierra de San Francisco, Coahuila; (b) Miquihuana, Tamaulipas; (c) Misantla, Veracruz, northeastern and eastern Mexico.

well regions. The author thanks H. Amezcua for the assistance with photography. I express my gratitude to J. Klein and Robert Scott for providing constructive reviews of the manuscript. My special thanks to J. Klein for helping with the bibliographic research. I am grateful to Claudio Bartolini for the translation of the manuscript.

SYSTEMATIC PALEONTOLOGY

The morphological terminology follows Wright et al. (1996). All dimensions are given in millimeters. Described specimens are housed in the Geological Department at the Instituto Politécnico Nacional (IPN), Mexico D.F. 07738, Mexico (Table 1).

TABLE 1. Shell measurements (in millimeters) and proportions of Valanginian–Hauterivian (Cretaceous) ammonites from San Francisco, Coahuila, and San Javier 1 and 2 wells, northeastern Mexico; Misantla, Veracruz; and La Laja 8 and Bejuco 6 wells, eastern Mexico.*

Specimen	*D*	*U*	*(U/D)*	*H*	*W*
Misantlites reyesi n. gen. and sp. IPN-1098	147	97	(0.66)	31	26
Aegocrioceras eguiluzi n. sp. IPN-1097	21	10	(0.47)	6.5	–
Capeloites lajense n. sp. IPN-1102	20	8	(0.40)	8	–
Neohoploceras bartolinii n. sp. IPN-1110	42	11.5	(0.27)	19	–
Thurmanniceras kleini n. sp. IPN-1107	68	23	(0.34)	26	–
Sarasinella sp. A IPN-1112	47	14	(0.30)	19	–

**D* = shell diameter; *H* = whorl height measured from umbilical seam to venter; *W* = whorl width; and *U* = umbilical diameter (U/D).*

Order Ammonoidea von Zittel (1884)

Suborder Ancyloceratina Wiedmann (1966)

Superfamily Ancyloceratacea Gill (1871)

Family Protancyloceratidae Breistroffer (1947)

Discussion

Dimitrova (1970) characterized the families Protancyloceratidae and Bochianitidae Spath (1922) by the form of the shells and suture lines. The Protancyloceratidae show an uncoiled shell, simple ribs, and eventually ventral tubercles; the suture line is characterized by an S1 with a large base, irregularly bifid at the top; it is from the Tithonian to Hauterivian (Wright et al., 1996).

The latter has a straight shell; the suture line with S1 and S2 is rectangular, wide, symmetrically divided at the top, with narrow bases; it is known from the Tithonian to lower Aptian.

Wiedmann (1973) proposed the same systematic classification as was pointed out by Dimitrova (1970) but at a subfamily level.

In contrast, Wright et al. (1996) did not analyze the morphological differences between the two heteromorph ammonoides taxa. Without giving any explanation, Wright et al. (1996) included the Subfamily Protancyloceratinae in the Family Bochianitidae.

The Family Protancyloceratidae is here accepted as it was recognized by Dimitrova (1970) and Wiedmann (1973). This family is a useful group that differs from the Family Bochianitidae by its loosely coiled whorls.

Genus *Wiedmannites* new genus

Type Species

Wiedmannites fibuloidis n. gen. and sp. Upper Valanginian, Bejuco 6 well, Veracruz state, eastern Mexico, lower Tamaulipas Formation (Figure 1c).

Including Species

Wiedmannites aff. *fibuloidis* n. sp. from Cuba.

Upper Valanginian, Bejuco 6 well, Veracruz state, eastern Mexico, lower Tamaulipas Formation (Figures 1, 3).

Diagnosis

Small uncoiled whorls. Fine, flexuous looped ribs joined by small, fine ventrolateral spines.

Etymology

In memory of Jost Wiedmann from Germany, who studied the Ancyloceratoidea.

Description

The small shell with flat flanks is ornamented with slightly flexuous looped ribs; they are branched from the internal region of the flank and joined by pair in a small, radial ventrolateral spine; the looped ribs are separated by larger interspaces, but the secondary ribs are separated by narrower spaces. The ventral region is not observed.

Discussion

The proponed genus *Wiedmannites* is distinguished from *Leptoceras* Uhlig (1883) and *Parapedioceras* Collignon (1962) by its looped ribs, which are joined by ventrolateral fine spines. The last two genera have simple, rectiradiate, or rursiradiate ribs; *Wiedmannites* without tubercles and the *Leptoceras* and *Parapedioceras* with feeble inner and ventrolateral tubercles. The upper Valanginian *Juddiceras* Spath (1924) differs from *Wiedmannites* by its irregularly alternating weak and strong ribs and by its inconstant ventral tubercles.

Two small heteromorph specimens with uncoiled shell were described as *Paracrioceras* Spath (1924) from Cuba by Myczynski (1977). They possess similar looped ribbing as *Wiedmannites*, but they do not have other conclusive characteristics.

Wiedmannites fibuloidis new species (Figure 5b, c)

Cf. *Protancyloceras* sp. Cantú-Chapa (1976, p. 65, pl. 6, figure 4, 4a).

Diagnosis

As the genus.

Description

The small specimens show only a flat flank; the whorl height increases moderately. The ventral region and suture line are not observed.

Etymology

Fibula, Latin fibula, plus *oide*, Greek form, alluding to the ribbing.

Type

Holotype, IPN-1095 (Figure 5b); paratype IPN-1096 (Figure 5c).

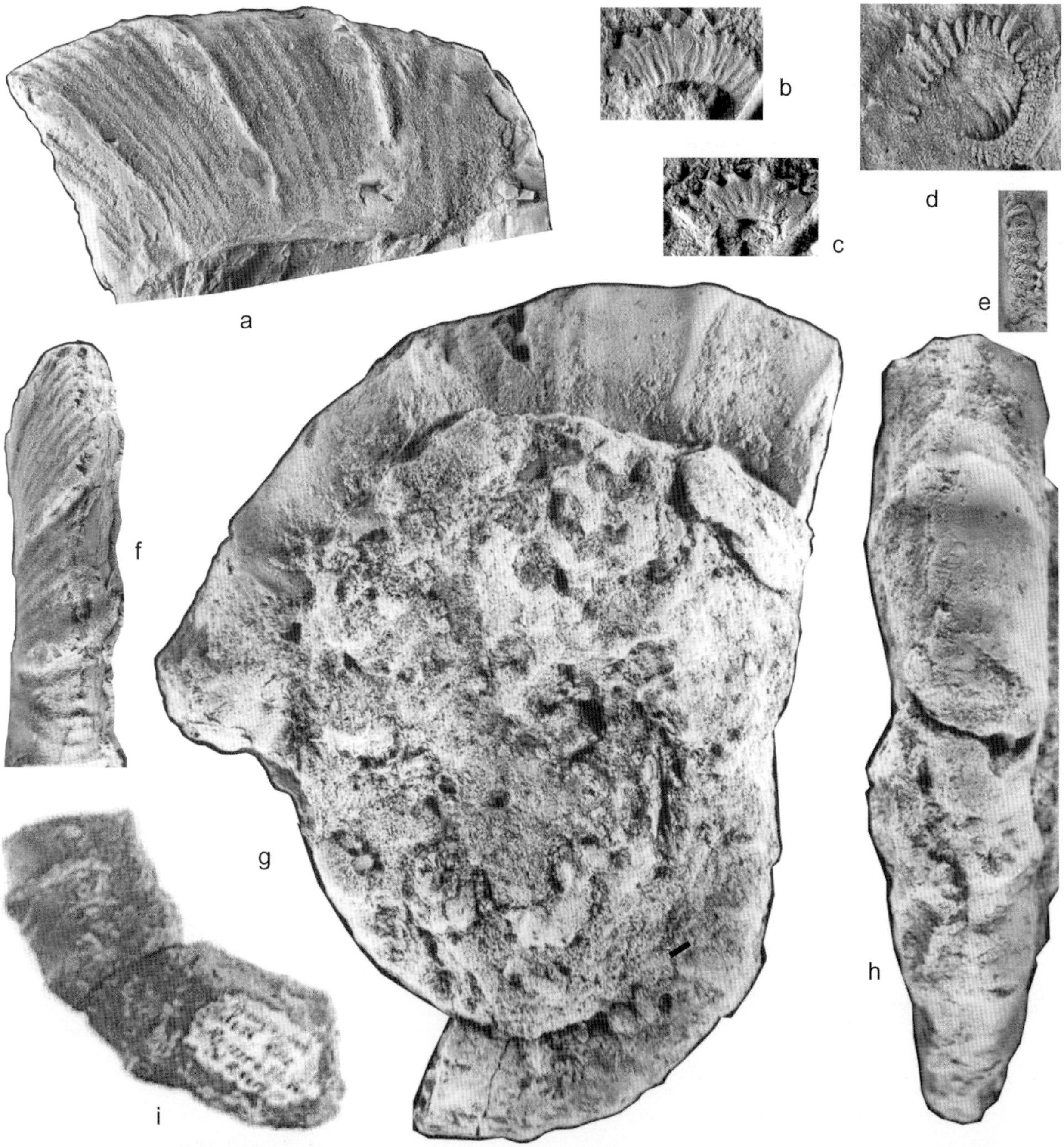

FIGURE 5. (a, f) *Crioceratites* n. sp., lateral and ventral views, IPN-1099. (b, c) *Wiedmannites fibuloides* n. gen. and sp., lateral views: (b) holotype, IPN-1096; (c) paratype, IPN 1095. (d, e) *Aegocrioceras eguiluzi* n. sp., lateral and ventral views, holotype, IPN-1097. (g–i) *Misantlites reyesi* n. gen. and sp., lateral and ventral views, holotype, IPN-1098. Specimens a and f are from Miquihuana; b and c are from the Bejuco 6 well, northeastern and eastern Mexico; d and e are from Sierra de San Francisco; and g–i are from Misantla. Scale = (all specimens except b, c) ×1; (b, c) ×1.5. Specimens are coated with ammonium chloride.

Occurrence

Bejuco 6 well, core 3 (1904–1910 m, 6247–6266 ft), Veracruz state, eastern Mexico; lower Tamaulipas Formation, upper Valanginian (Figure 1).

Discussion

These small, fragmental specimens were previously assigned to cf. *Protancyloceras* sp. (Cantú-Chapa, 1976, p. 65, pl. 6a, figure 4, 4a) on the basis of the resemblance to the loosely coiled whorls of this genus. It was then noted that the looped ribbing differs from all previously described genera of the Protancyloceratidae.

In addition, two small specimens with uncoiled whorls were assigned to *Paracrioceras* cf. *elegans* (Koenen, 1902) by Myczynski (1977, p. 158, pl. 5, figure 9; pl. 6, figure 1); they are from the Polier Formation in western Cuba. These specimens are similar to the proposed *W. fibuloidis* by their looped ribs and uncoiled whorls. No other ammonites were found with the Cuban specimens; they were, however, dated as Barremian by Myczynski (1977).

Family Ancyloceratidae Gill (1871)

Subfamily Crioceratitinae Gill (1871)

Discussion

The genera of this subfamily are grouped here by the uncoiled whorls (1–5), the straightened shaft (6), and the slightly involute shell (7).

The ornamentation also characterizes them as follows:

1. simple, radial, and nontuberculate ribs cross normally the ventral region such as *Aegocrioceras* Spath (1924), *Parancyloceras* Spath (1924), *Simancyloceras* Kemper (1973), and *Bejucoceras* Cantú-Chapa (1976);
2. simple, radial, and trituberculate ribs such as *Spinocrioceras* Kemper (1973);
3. trituberculate, radial, or slowly flexuous major ribs that are separated by some minor and finer ribs such as *Menuthiocrioceras* Collignon (1949), *Eocrioceratites* Wiedemann (1973), *Crioceratites* Léveillé (1837), and *Acantholytoceras* Spath (1923a);
4. flexuous ribs with ventrolateral tubercles such as *Shasticrioceras* Anderson (1938);
5. radial, simple, or irregularly bifurcate ribs with inconstant inner and ventrolateral tubercles as *Pedioceras* Gerhardt (1897) and *Hemicrioceras* Spath (1924);
6. straightened long shaft with dense, fine, and nontuberculate ribs alternating with constrictions such as *Megacrioceras* Delanoy et al. (1987); and
7. coiled ammonite shells with fine ribs branching from umbilical tubercles such as *Pseudothurmannia* Spath (1923a).

Wright et al. (1996) considered it unnecessary to distinguish *Toxoceras* d'Orbygny (1842) and *Himantoceras* Thieuloy (1965) as different genera of Crioceratitinae because of their very loosely coiled shells; both genera were assimilated by these authors into *C.* (*Crioceratites*). However, this argument is not convincing to invalidate these genera; they must persist as different taxa because of their distinctive uncoiled shells and their stratigraphical distribution.

Genus *Aegocrioceras* Spath (1924)

Type Species

Hamites capricornu Roemer (1841) by original designation.

Discussion

Wright et al. (1996) indicated that *Aegocrioceras* was found in Mexico because they synonymized the Mexican genus *Bejucoceras* with it. The differences between these genera have been already pointed out (Cantú-Chapa, 1976, 2001).

Aegocrioceras eguiluzi new species (Figure 5d, e)

Diagnosis

Uncoiled shell; single, rectiradiate, sharp ribs, separated by narrow interspaces, crossing the venter transversely without interruption.

Description

The whorl shell is moderately compressed, with flat flanks and gently subrounded venter. The whorl height increases moderately in relation to diameter. Twenty ribs are in the adoral half whorl.

Etymology

Named in honor of Samuel Eguiluz Antuñano, engineer for Petróleos Mexicanos, who collected the holotype.

Type

Holotype, IPN-1097.

Occurrence

Sierra de San Francisco, Coahuila state, northeastern Mexico (Figure 4a).

Discussion

This specimen is assigned to *Aegocrioceras* by its uncoiled shell and by its rectiradiate, simple ribs that cross the ventral region transversely.

Rawson (1975) proposed several characters to distinguish *Aegocrioceras* species; they are recognized by the number, size, and lateral direction of the ribbing, as well as by the relative separation of the whorls.

The proposed *A. eguiluzi* n. sp. resembles *A. spathi* Rawson (1975, p. 145, pl. 4, figures 1, 2, 5) and *A. raricostatum* (Rawson, 1975, p. 143, pl. 4, figures 6–10; pl. 5, figures 3, 4, 7) from England by its crioceratitid coiling; it differs from them by its rectiradiate and sharp ribs that are separated by narrower spaces; ribs are curved gently backward over the venter in the two last species.

Genus *Misantlites* new genus

Type Species

Misantlites reyesi n. gen. and sp. by monotype. Probably Hauterivian, lower Tamaulipas Formation; Misantla, Veracruz (Figure 4c).

Diagnosis

Uncoiled in open spiral shell; coarse, single, straight major ribs arise from umbilical bullae, weaken at the flanks, are enlarged at the venter from ventrolateral tubercles, and are separated by minor ventrolateral ribs.

Etymology

After the Misantla village, Veracruz state, eastern Mexico (Figure 4c).

Description

Uncoiled shell in a regular plane spire. The whorl section is subrounded to subquadrate; the whorl height increases moderately, and the flanks and venter are planar to subrounded.

The principal ribs arise from the subrounded edge; they become weakened at the flanks, enlarged across the venter, and are separated by some minor ventrolateral ribs.

Discussion

Menuthiocrioceras, *Eocrioceratites*, *Crioceratites*, and *Acantholytoceras* have trituberculate major ribs alternating with some minor ribs on the flanks and all cross the ventral region normally; although *Misantlites* n. gen. possesses major ribs arising from umbilical bullae, ribs become weakened at the flanks and enlarged across the venter from the ventrolateral tubercles.

The upper Valanginian *Himantoceras* Thieuloy (1965) from France and Morocco differs from *Misantlites* by having major trituberculated ribs that weaken at the ventral region and by its open great uncoiled shell (Thieuloy, 1965; Ettachfini, 2004).

The Hauterivian *Shasticrioceras* from California have flexuous ribs at the flanks and ventrolateral tubercles, and it differs from *Misantlites* in lacking enlarged major ribs at the venter. This last genus is here assigned to the second group of the Crioceratitinae mentioned above.

Age

Its stratigraphic position is unknown. *Misantlites* could be dated as Hauterivian by morphological comparison with the genera of the subfamily Crioceratinae.

Misantlites reyesi new species (Figure 5g–i)

Diagnosis

As the genus.

Description

As the genus.

Etymology

In memory of Eliseo Reyes, engineer for Petróleos Mexicanos, who collected the fossil described.

Type

Holotype, IPN-1098.

Occurrence

Misantla, Veracruz; lower Tamaulipas Formation.

Discussion

As the genus. One incomplete specimen was assigned to *Menuthiocrioceras*? sp. by Cantú-Chapa (1963). Later, the uncoiled specimen IPN-1098 was found in the IMP collections, and it allows a more complete description and reassignment of the fragmented specimen.

Age

Probably Hauterivian.

Genus *Crioceratites* Léveillé (1837)

Type Species

Crioceratites duvalii Léveillé, 1837. By subsequent designation of Diener (1925).

Discussion

The systematic classification of *Crioceratites* was established based on its uncoiled or coiled whorls, its ornamentation, and its suture line (Sarkar, 1955; Thomel, 1964; Wiedmann, 1973; Immel, 1978).

This genus is divided into subgenera by Wright et al. (1996) after the following morphological characteristics:

1. uncoiled whorls with major trituberculated ribs alternating with some minor ribs in *C.* (*Crioceratites*);
2. scarce minor ribs in *C.* (*Paracrioceras*) Spath (1924);
3. major bituberculated ribs on early whorls that became divided at the umbilical edge and are nontuberculate on the outer whorls in *C.* (*Spathicrioceras*) Sarkar (1955); and
4. coiled ammonitic whorls with trituberculated ribs in *C.* (*Sornayites*) Wiedmann (1962). This subgenus is based on an old illustration by Parona (1898), which is not well known; its systematic position is discussed by Thomel (1965).

The *Crioceratites* subgenera are herein considered as genera following Reboulet (1996) and Ettachfini (2004).

Crioceratites sp. (Figure 5a, f)

Description

A fragmental specimen with uncoiled shell, compressed whorl section, flat flanks, subrounded umbilical edge, narrow venter, and dorsal region gradually depressed, and sharp, prorsiradiate, and trituberculate major ribs; the internal and lateral tubercles are radially elongated. The ventral tubercles are inclined forward. One spine is preserved from an internal tubercle.

Major ribs are separated by six or seven fine, prorsiradial, simple ribs; all ribs arise from the subrounded internal shoulder, crossing the ventral region without interruption where they are projected forward.

Material Examined

IPN-1099.

Occurrence

Near Miquihuana, state of Tamaulipas, northeast Mexico; lower Tamaulipas Formation, Hauterivian (Figure 4b).

Discussion

This fragmental specimen shows the distinctive characters from a heteromorphic ammonite with an uncoiled shell that shows its dorsal region; it is assigned to *Crioceratites* sp. by its sharp and trituberculated major ribs alternating regularly with some fine, minor ribs. *Paracrioceras* Spath (1924) differs from *Crioceratites* by having stronger ribs with spines and a smaller number or even an absence of intermediate fine ribs (Kakabadze and Hoedemacker, 2004).

This is the first time that a specimen of *Crioceratites* is described and illustrated from Mexico. Representatives of this genus were mentioned by Burckhardt (1930) from the lower Hauterivian in north-central Mexico; its presence in this country has been successively mentioned by Imlay (1944), Arkell et al. (1957), Wright et al. (1996), and Ettachfini (2004).

Crioceratites sp. indet. from western United States (Imlay, 1960, p. 196, pl. 26, figures 1, 5) resembles our specimen by having trituberculated major ribs alternating with 8 to 9 simple, flexuous minor ribs, but some ribs bifurcate on the outer part of the flanks.

The specimens assigned to *Crioceratites* in the Lower Cretaceous from Cuba by Myczynski (1977, p. 158, pl. 5, figures 7, 8) have constant, dense, fine ribs; they differ from the Mexican specimen herein described by their finer ribbing. The Cuban specimens resemble the upper Hauterivian *C. duvalii* Léveillé from Germany (Kemper, 1992, pl. 57, figures 1, 7) by the fine, flexuous ribs.

This Mexican specimen resembles *C. andinus* (Gerth, 1925, in Aguirre-Urreta, 1993, figure 4.1, 4.2) of the upper Hauterivian from Argentina and Chile by its major ribs. The South American specimens, however, have fewer minor ribs alternating between pairs of major ribs than the Mexican specimen. The outer lateral tubercles of *C. andinus* are not well known so they cannot be fully compared with the Mexican specimen.

The Hauterivian *C. duvalii* Léveillé and *C. nolani* (Kilian, 1907, in Thomel, 1965a) from France have trituberculate major ribs such as the Mexican specimen; but the first differs from the Mexican species by its slightly flexuous minor ribs and the latter by its flexuous and irregularly bifurcated minor ribs.

Suborder Ammonoidea Hyatt (1889)

Superfamily Perisphinctaceae Steinmann (1890)

Family Olcostephanidae Haug (1910)

Subfamily Capeloitinae new subfamily

Type Genus

Capeloites Lissón, 1937.

FIGURE 6. (a, g) *Thurmanniceras kleini* sp. nov.; (a) lateral view, holotype, IPN-1107. The arrow shows a small specimen of cf. *Bejucoceras* sp.; (g) paratype, IPN-1108. (b) *Sarasinella* sp. A, lateral view, IPN-1110. (c) *Olcostephanus* sp., lateral views, IPN-1104. (d) *Berriasella* aff. *densestriatus* (Burckhardt, 1912), ventrolateral view, IPN-1106. (e) *Thurmanniceras* sp. A, lateral view, IPN-1109. (f) cf. *Parastieria* sp., lateral view, IPN-1103. (h) *Capeloites lajense* n. sp., lateral views, holotype, IPN-1102. (i) *Neohoploceras bartolinii* n. sp., lateral view, holotype, IPN-1110. Specimens a, c, e, g, and h are from the La Laja 8 well; b and f are from the San Javier 2 well; and d and i are from the San Javier 1 well, northeastern and eastern Mexico. Scale = (all specimens) ×1; (f, h) ×1.5. Specimens are coated with ammonium chloride.

Included Genus

Parastieria Spath, 1923b.

Age

Upper Valanginian–Lower Hauterivian.

Locality

France, Spain, England, Peru, and Mexico.

Diagnosis

Small shell; moderately evolute, compressed whorl section; subrounded venter.

Dimorphic ornamentation in two different positions at the last whorl:

1. Sharp, rectiradiate, or looped and sigmoidal primary ribs with umbilical bullae, ventrolateral tubercles, and inconstant constrictions in the adapical half; fine, dense, bifurcated, or bundled sigmoidal

ribs; inconstant midventral tubercles in the adoral part.
2. Fine, dense, slightly sinuous, closely spaced ribs in the adapical half, and stronger, slightly convex forward, cuneiform, distant ribs in the adoral part; ribs cross normally the ventral region; suture line with long and parallel S1, S2, and L.

Discussion

Ammonites with dimorphic ornamentation on the last whorl are scarce in the Family Olcostephanidae; this type of ornamentation is present in *Capeloites* and *Parastieria* from the Valanginian to the lower Hauterivian.

Both genera were assigned to this family by Kemper et al. (1981) and Wright et al. (1996). Nevertheless, the strong change in the ornamentation on the last whorl in these genera was pointed out by Thieuloy (1969), Kemper et al. (1981), Autran (1989), Bulot (1990), and Cantú-Chapa (2001).

Representatives of the Olcostephanidae possess uniform ornamentation in the last whorl, including genera of the Subfamily Spiticeratinae Spath (1924), which differ from the two genera mentioned above by this type of ornamentation.

The suture line of Capeloitinae genera is only known after *Olcostephanus* (*Subastieria*) *hispanicus* (Mallada, 1882) (= ? *Parastieria*? *hispanica* in Klein, 2005). The suture line was illustrated by Tzankov (1943, in Cooper, 1981, figure 20H, J); it possesses long, narrow, and parallel S1 and S2.

Capeloites and *Parastieria* are herein grouped in the Capeloitinae new subfamily to characterize their dimorphic ornamentation on the last whorl. The former has a widespread geographic distribution from Peru, Mexico, France, and Spain; the latter is only known from England. Both genera are Valanginian and lower Hauterivian in age (Wright et al., 1996).

The position of the dimorphic ornamentation on the last whorl differs in each one of these genera; it generates confusion and produces different opinions because of the position of the dimorphic ornamentation on the last whorl of the shell; this will be discussed farther on.

Genus *Capeloites* Lissón (1937)

Type Species

Capeloites larozai Lissón, 1937. By original designation.

Discussion

Capeloites from Peru was proposed by Lissón (1937); it was later recognized in France and Mexico by Thieuloy (1969), Autran (1989), and Cantú-Chapa (2001). An old illustration of *Ammonite perelegans* (Matheron, 1878) from France was assigned to this genus by Thieuloy (1969).

Capeloites lajense new species (Figure 6h)

Diagnosis

Small, evolute, and dimorphic shell; slightly subrounded flanks and venter. Sharp, rectiradiate, distant primary ribs arise from periumbilical bullae, alternating with one or two fine ribs; two prorsiradial constrictions in the adapical part. Fine, dense, closely spaced, radial to sigmoidal ribs, irregularly bifurcated in the adoral part of the shell.

Description

This small, evolute specimen with dimorphic ornamentation has a subrounded flank. Seven sharp, rectiradiate, and distant primary ribs arise from periumbilical bullae; they are separated by one or two secondary fine ribs. Two prorsiradial constrictions are intercalated in the adapical part of the shell.

The adoral part possesses fine, dense, sigmoidal, closely spaced primary ribs; some ribs are bifurcate, alternating irregularly with simple ribs, which arise from the middle part of the shell. Sixteen primary ribs and 49 secondary ribs exist in the adoral part against 7 primary ribs in the adapical part of the shell.

Etymology

The name refers to the La Laja 8 well.

Type

Holotype, IPN-1102 (Figure 6h).

Occurrence

La Laja 8 well, core 18 (2619–2628 m, 8592–8622 ft); lower Valanginian, lower Tamaulipas Formation; Tampico, eastern Mexico (Figure 1a, c).

Discussion

The proponed *Capeloites lajense* is preserved as a complete shell; with the exception of the ammonite illustrated by Matheron (1878), all other specimens of this genus from France and Peru are preserved as a half whorl (Lissón, 1937; Thieuloy, 1969; Autran, 1989).

C. lajense differs from all species described as *Capeloites* by having two constrictions and rectiradiate, distant primary ribs in the adapical half of the shell; by having fine, dense, sigmoidal, closely spaced, and bifurcated ribs in the adoral part; and by the absence of

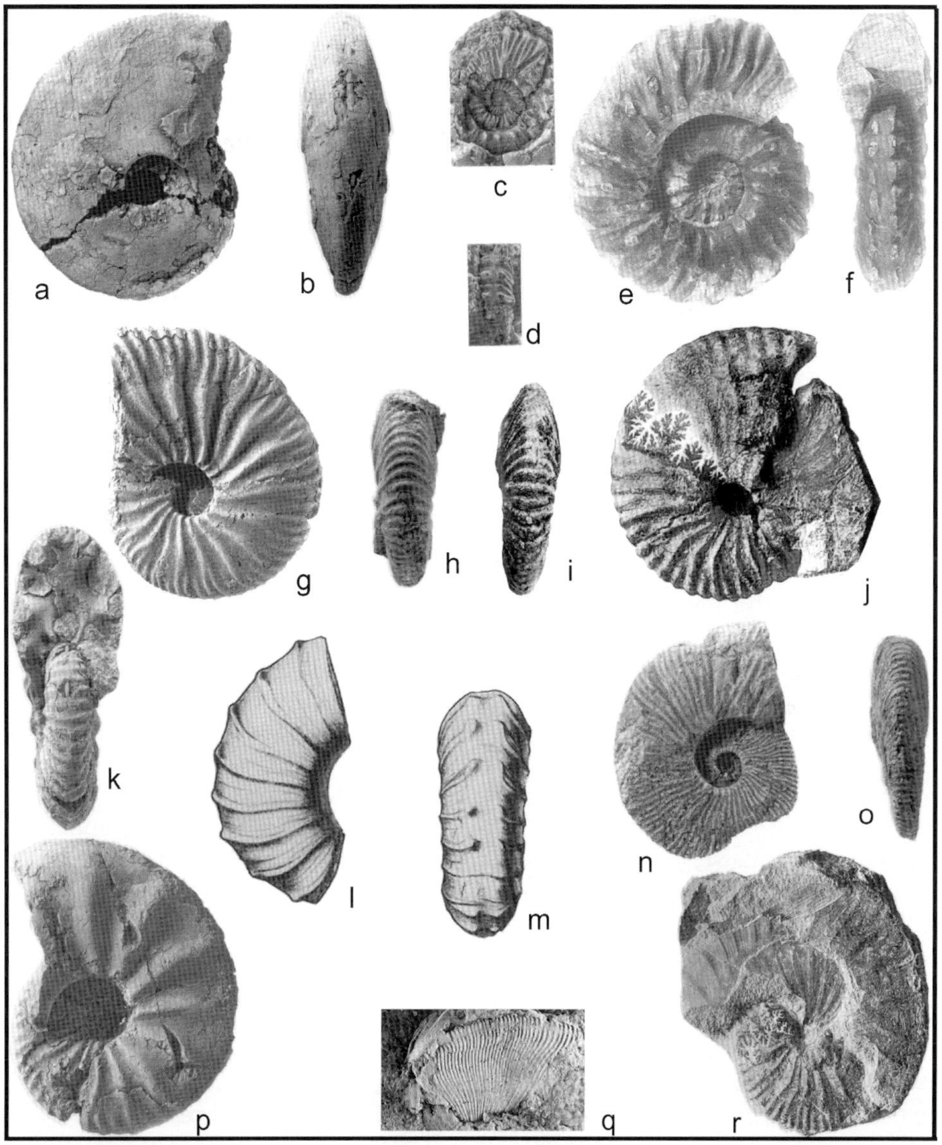

FIGURE 7. Genera and species of the Family Berriasellidae with different ornamentations and form of shells. (a–d) *Banikoceras involutum* Howarth (1992, p. 650, pl. 12, figures 1–4): (a, b) lateral and ventral views; (c, d) internal whorls, lateral and ventral views. (e, f) *Neocosmoceras* aff. *Sayni* Mazenot (1939, p. 183, pl. 29, figure 2a, b), lateral and ventral views. (g, h) *Chigaroceras wetzeli* Howarth (1992, p. 643, pl. 10, figures 4, 5), lateral and ventral views. (i, j, r) *Mazapilites symonensis* Burckhardt (1919–1921, p. 4, pl. 1, figures 2, 3, 9), ventral and lateral views. (k, p) *Chigaroceras banikense* Howarth (1992, p. 643, pl. 10, figures 1, 2), apertural and lateral views. (l, m) *Lythohoplites burckhardti* (Burckhardt 1900, p. 17, pl. 26, figures 1, 2), lateral and ventral views. (n, o) *Tirnovella alpillensis* (Mazenot, 1939) in Howarth (1992, p. 648, pl. 9, figures 3, 4), lateral and ventral views. (q) *Berriasella* aff. *densestriatus* (Burckhardt, 1912), ventrolateral view; this work, IPN-1106. All specimens are without scale.

ventral tubercles at some ribs in the adapical part of the last whorl.

C. neoleonense Cantú-Chapa (2001) from northeastern Mexico differs by the absence of secondary ribs intercalated between the coarse and rectiradiate primary ribs in the adapical half of the shell as *C. lajense* has.

C. perelegans (Matheron, 1878) from France and *C. lajense* are comparable in having a similar type of ribbing on the adapical part of the shell, but *C. perelegans* differs by the absence of constrictions.

C. perelegans (Matheron) differs from *C. perelegans larozai* (Lissón in Thieuloy, 1969, p. 256, figures B1–B3), also from France, by its coarse, radial principal ribs alternating with fine secondaries on the adapical part of the shell.

This later subspecies differs from the former by its convex forward ribbing that is sharp and divided from periumbilical bullae, producing ventral tubercles on the adapical part of the shell; ribs are very dense, fine, and divided on the internal part of the flanks in the adoral half of the shell.

The specimen described by Thieuloy (1969) as *C. perelegans larozai* (Lissón in Thieuloy, 1969, p. 256, figures B1–B3) is here proposed as the holotype of *C. peyroullense* n. sp. It differs from *C. lajense* by its convex ribs at the flanks and from *C. larozai* by the absence of looped ribs on the adapical part of the shell.

C. larozai Lissón (1937) from Peru differs from *C. lajense* by its sigmoidal ribbing; its sharp, looped ribs arising from periumbilical bullae in the adapical part change to very fine, dense, and bifurcated ribs on the adoral part of the shell.

All *Capeloites* species from France, Peru, and Mexico differ from *C. lajense* by the lower ratio of U/D, which varies from evolute (0.40–0.42) in *C. lajense* and

Capeloites sp., respectively, from Mexico to relative involute (0.26–0.36) in *C. peyroullense* and *C. hispanicus* (Mallada in Autran, 1989) from France and *C. neoleonense* (0.25) from Mexico (Cantú-Chapa, 2001).

Genus *Parastieria* Spath (1923b)

Type Species

Acanthoceras? *peltoceroides* Pavlow (1892). By original designation.

Discussion

The dimorphic ornamentation in the last whorl of the shell produces confusion between *Parastieria* and *Capeloites*.

Kemper et al. (1981) pointed out the presence of looped ribs in the adoral part of the shell in one specimen from England; it was not specified which one of these two genera have these structures. The coarse ribbing occurs in the adoral part in *Parastieria wrightae* Rawson from England or in the adapical half of the shell in *Capeloites larozai* Lissón (1937) from Peru.

The specimen mentioned above was assigned to *Parastieria* sp. nov. by Kemper et al. (1981, p. 273, pl. 36, figures 5, 6); its coarse, looped ribs are separated by larger spaces on the adoral part, and it contrasts with the fine ribbing on the adapical half of the shell. Klein (2005) rescues and renames it as *Parastieria wrightae* Rawson (1966) because it was originally proposed by P. F. Rawson (1966).

The position of the dimorphic ornamentation with coarse ribs in the adoral part of the shell characterizes *Parastieria*, but the presence of looped ribs in this part of the shell in *Parastieria* sp. nov. (Kemper et al., 1981) differs from all species assigned previously to *Parastieria*.

This type of ribbing differs from *Parastieria peltoceroides* (Pavlow, 1892, pl. 18 (11), figures 21a, b, in Kemper et al., 1981, p. 273, text, figures 6e–g); its principal ribs became larger and cuneiform on the outer part of the flanks in the adoral half of the whorl.

Autran (1989) differentiated *Parastieria* from *Olcostephanus* by the absence of umbilical tubercles in the former; however, Wright et al. (1996) recognized the presence of slight umbilical bullae in *Parastieria* without specifying in which part of the last whorl they are located.

Cf. *Parastieria* sp. (Figure 6f)

Description

Small and evolute shell, dimorphic ornamentation in the last whorl; inclined and subrounded umbilical wall, slightly subrounded flank.

First whorls are smooth, but the last whorl is ornamented by primary, fine, simple, prorsiradial to slightly flexuous, and closely spaced ribs. They arise freely at the umbilical margin in the adapical part and become strong, slightly convex forward, and distant in the adoral part of the shell.

Material Examined

IPN-1103, negative ammonite mold.

Occurrence

San Javier 2 well, core 14 (3120–3129 m, 10,236–10,266 ft), northeastern Mexico; Taraises Formation, base of the upper member, lower Valanginian (Figure 1a, c).

Discussion

This specimen is assigned to cf. *Parastieria* sp. because of its dimorphic ornamentation in the last whorl; its fine, dense ribbing in the adapical part changes to stronger ribs that are separated by spaces with the same size in the adoral part of the shell.

However, its fragmented condition does not allow a conclusive generic determination. It differs from *P. peltoceroides* (Pavlow, 1892, p. 510, pl. 18 (11), figure 21a–c) in its simple and fine ribs that arise at the umbilical margin in the adapical part and in the presence of coarse, distant ribs in the adoral part of the shell.

Two specimens from Spain were assigned to *Capeloites* by Wilke (1988, pl. 2, figures 3, 4); the dimorphic ornamentation changes from fine ribbing in the adapical part to large, widely spaced ribs in the adoral half of the last whorl; they are herein considered as *Parastieria*.

The specimen from the San Javier 2 well described above resembles the Spanish *P.* aff. *peltoceratoides* Pavlow (1892) in Wilke (1988, pl. 2, figure 4a, b) (= *Capeloites perelegans larozoi*, in Klein, 2005) by its evolute shell; its great number of fine, closely spaced secondary ribs in the adapical part; and its coarse, separated ribs in the adoral part of the shell.

Another Spanish specimen was illustrated by Wilke (1988, pl. 2, figure 3a, b) as *Capeloites perelegans larozoi*; it shows the typical dimorphic ornamentation of *Parastieria*, consisting of fine ribs in the adapical part and coarse ribbing in the adoral part of the shell; this specimen is rather involute, and it is herein considered to be a *Parastieria* sp.

Subfamily Olcostephaninae Haug (1910)

Genus *Olcostephanus* Neumayr (1875)

Type Species

Ammonites astierianus d'Orbigny (1842). By original designation.

Olcostephanus sp. (Figure 6c)

Description

Small shell, moderately evolute with a convex flank and a lappet at the peristome. The fine, gently convex forward ribs are separated by spaces as large as them. It shows small tubercles near the peristomal area, and very fine striae arise from small tubercles at the adapical part of the shell, corresponding to the eroded rib margins. The umbilical area is partially eroded.

Material Examined

IPN-1104.

Occurrence

La Laja 8 well, core 9 (2539–2547 m, 8330–8356 ft), Tampico, eastern Mexico; lower Valanginian, lower Tamaulipas Formation (Figure 1a, c).

Discussion

This specimen has the distinctive characteristics of *Olcostephanus*, consisting of a gently convex flank with fine ribbing. The most characteristic ornamentation is the small umbilical tubercles and the fine striae, which represent the rest of the eroded ribs at the adapical part of the shell; its whorl section is unknown.

Family Berriasellidae Spath (1922)

Discussion

This family is incorrectly treated as a subfamily of the Family Neocomitidae Salfeld in the *Treatise*. Genera with differently coiled shells and ornamentation are included together; they vary from

1. involute, smooth, and discoidal shells such as *Banikoceras* Howarth (1992) (Figure 7a, b);
2. evolute, with alternating fine and coarse ribs such as *Lytohoplites* Spath (1925) (Figure 7l, m), or with lateral and ventrolateral tubercles, leaving a smooth ventral band such as *Neocosmoceras* Blanchet (1922) (Figure 7e, f);
3. subevolute with fine, closely spaced ribs such as *Tirnovella* Nikolov (1966) (Figure 7n, o);
4. *Banikoceras* possesses umbilical tubercles, looped ribs, and constrictions on the internal whorls (Figure 7c, d).

The diverse set of genera in the Family Neocomitidae (Subfamily Berriasellinae) cannot be retained as having phylogenetic unity because of the great heterogeneity of shells and ornamentation (Figure 7).

For instance, the upper Tithonian *Chigaroceras* Howarth (1992) from Iraq was placed in the Subfamily Berriasellidae in the *Treatise*. Nothing shows that this genus is similar to the *Dalmasiceras* Djanelidzé (1922) to be assigned to this family (Howarth, 1992).

In contrast, *Chigaroceras* must be assigned to the Subfamily Mazapilitinae Spath (1928) according to its involute and compressed shell and its coarse, foldlike, branched ribbing that crosses normally the narrow and subrounded venter.

Chigaroceras is very close to the lower Tithonian *Mazapilites* Burckhardt (1919) from Mexico; in the former, any tubercles are really conspicuous at the umbilical marge and at the midlateral shell, as was pointed out by Howarth (1992). Similar involute and compressed shell and ribbing are present in the both genera, but they occur at different stratigraphic levels through the Tithonian. Compare Figure 7g, h, k, and p (*Chigaroceras*) with Figure 7i, j, and r (*Mazapilites*).

The upper Tithonian–Berriasian (?) *Submazapilites* Cantú-Chapa (1963) from Mexico has an involute shell with an oxycone whorl section. The principal ribs bifurcate from a lateral tubercle and they alternate with two secondaries from the midlateral area; all ribs have ventrolateral tubercles. This genus differs from *Chigaroceras* by its rows of two tubercles.

Chigaroceras and *Mazapilites* are similar in ornamentation and even in their dimorphic ornamentation stages, which are observed in *Ch. banikense* Howarth (1992, p. 643, pl. 10.1-3) and *M. symonensis* Burckhardt (1919, p. 4, pl. 1.1). Both genera have the typical mazapilitinae ribbing in the adapical part of the shell; the former changes to very large, coarse, separated, and rectiradial ribs; the latter varies to very weakened ribs in the adoral half of the shell (Figure 7p, q).

Involute shells with subrounded venter and coarse, foldlike, branched ribbing characterize the Subfamily Mazapilitinae Spath (1928) (*Mazapilites*, *Submazapilites*, and *Chigaroceras*) instead of the Subfamily Berriasellidae where the last genus was included in the *Treatise*.

These genera are here integrated into the Subfamily Mazapilitinae instead in the Subfamily Taramelliceratinae Spath (1928). The presence of one *Mazapilites* specimen in the lower Tithonian beds from Germany (Berckheimer and Hölder, 1959) are not enough to reassign it in the Taramelliceratinae as it was proposed by Donovan et al. (1980).

These subfamilies have different morphological characters and suture lines that allow to separate them.

Another genus with a strange form of shell is the upper Tithonian *Banikoceras* from Iraq (Howarth, 1992); it is included in the Subfamily Berriasellinae in the *Treatise*. The internal whorls are subevolute, the ventral area is smooth, and the looped ribs have umbilical and ventrolateral tubercles; this ornamentation changes to a smooth, involute, and oxyconic shell in the outer whorl (Figure 7a–d).

This specimen is herein assigned to the Subfamily Suaritinae Cantú-Chapa (1998), from the Family Neocomitidae Salfeld (1921) because of its subevolute shell and its looped ribs in the internal whorls.

Family Berriasellidae Spath (1922)

Discussion

This family included genera with fine primary simple ribs that bifurcate at the midflanks; the ribs cross perpendicularly the subrounded ventral region or are interrupted at a smooth or grooved midventral band; exceptionally, the ribs arise from umbilical tubercles as in *Elenaella* Nikolov (1966).

The absence of umbilical tubercles characterize genera such as *Berriasella* Uhlig (1905), *Substeueroceras* Spath (1923), *Parodontoceras* Spath (1939), *Riasanites* Spath (1939), *Andiceras* Krantz (1928), and *Parandiceras* Spath (1939).

The Subfamily Richterellinae Sapunov (1977) differs from the Berriasellinae by its ribbing projected forward at the ventral region and by its suture line (Cantú-Chapa, 1999, 2006).

Genus *Berriasella* Uhlig (1905)

Type Species

Ammonites privasensis Pictet (1867); SD by Roman (1938, p. 324).

Discussion

The systematic concepts of this genus here does not follow the definition of *Berriasella* (*Berriasella*) of Wright et al. (1996). It differs from *B.* (*Elenaella*) by the absence of umbilical tubercles. Both genera were integrated into the former by Wright et al. (1996) but later were separated by Klein (2005). His proposition is followed herein.

Berriasella aff. *densestriatus* (Burckhardt, 1912) (Figure 6d)

1912 *Neocomites densestriatus* Burckhardt, p. 190, figure 45.1–45.15, 45.19.

1964 *Neocomites densestriatus* Burckhardt, in Peña-Muñoz, p. 25, figure 7.6, 7.7.

2005 *Tirnovella? densestriatus* (Burckhardt, 1912) in Klein, p. 255.

Material Examined

One fragmented specimen shows the lateral and ventral ribbing. IPN-1106.

Locality

San Javier 1 well, core 47 (3199–3207 m, 10,495–10,522 ft); Taraises Formation, lower Valanginian; northeastern Mexico (Figure 1a, b).

Description

Shell with a planar to slightly convex flank. Narrow subrounded ventral region, partially with a middle smooth band. Very fine, closely spaced, and slightly flexuous ribs, some of them are bifurcate at different levels from the midflank; secondary ribs are intercalated on the outer half of the flank; all ribs are separated by very fine spaces and weaken at the venter.

Discussion

This specimen is herein assigned to Berriasellinae by its fine ribs arising free from the umbilical shoulder, branching at the middle part of the flanks, and crossing perpendicularly the venter where they weaken.

It resembles *Neocomites densestriatus* Burckhardt (1912, p. 190, pl. 45, figures 1–15, 19) by its slightly planar flank; by its subrounded and narrow venter with a smooth band; and by its very fine, dense, closely spaced ribs that arise free at the umbilical shoulder that bifurcate at the midflank; some secondary ribs are intercalated.

Peña-Muñoz (1964) described a specimen from northern Mexico and assigned it to *Berriasella densestriatus*, but unfortunately the illustration is very bad. This Mexican species is assigned with doubt to *Tirnovella* Nikolov (1966) in the Family Neocomitidae by Klein (2005); this last genus differs from *Berriasella* by its fine and flexuous ribs that born from umbilical tubercles (Figure 7n, o).

B. densestriatus is important to date the Taraises Formation base because it represents the Berriasian beds (zone with *Spiticeras*) from north-central Mexico (Burckhardt, 1912; Imlay, 1944).

Family Neocomitidae Salfeld (1921)

Genus *Thurmanniceras* Cossman (1901)

Type Species

Ammonites thurmanni Pictet and Campiche (1860). By original designation.

Discussion

Some Neocomitidae genera of this family differ from *Thurmanniceras* by the presence of umbilical tubercles such as *Fauriella* Nikolov (1966), *Jabronella* Nikolov

(1966), and *Busnardoites* Nikolov (1966); in contrast, *Thurmanniceras* has periumbilical bullae. *Tirnovella* Nikolov (1966) and *Neocomites* Uhlig (1905) differ from *Thurmanniceras* by their subinvolute shells; *Fauriella* differs also from *Thurmanniceras* by its straight, dense, and fasciculate fine ribbing, contrary to the flexuous angular ribs of the latter genus.

Thurmanniceras kleini new species (Figure 6a, g)

Diagnosis

Evolute shell with slightly convex flank, oblique, low umbilical wall; sharp ribs arising from umbilical bullae, becoming strongly prorsiradiate on the inner flank, producing small swellings at midflank, and dividing irregularly into two or three ribs; all ribs curved forward at the mid-outer part of the flank. Shallow, flexuous, and forward-inclined constrictions irregularly intercalated with the ribs. The ventral region is unknown.

Etymology

Named for Jaap Klein, in recognition of his studies on the Lower Cretaceous ammonites.

Material Examined

Holotype, IPN-1107 (Figure 6a), paratype, IPN-1108 (Figure 6g).

Occurrence

La Laja 8 well, core 13 (2575–2584 m, 8448–8478 ft), Tampico District, eastern Mexico; lower Tamaulipas Formation, lower Valanginian (Figure 1a, c).

Discussion

This specimen is assigned to *Thurmanniceras* Cossman (1901) based on its evolute coiling, slightly convex flank, and oblique umbilical wall; it is ornamented with sharp, falcoid, and furcated ribs arising from umbilical bullae and with forwardly projected constrictions.

Some *Thurmanniceras* species described by Imlay (1937) from northeastern Mexico differ from *T. kleini* by the rectiradiate ribs in *T. angusticostatus* Imlay (1937) and *T. miquihuanense* Imlay (1937) and by the umbilical tubercles in *T. novihispanicus* Imlay (1937); by these structures, this last species is herein assigned to *Busnardoites* Nikolov (1966).

T. jenkinsi Anderson (1938) from western United States was assigned to *Thurmanniceras* instead of *Neocomites* because it lacks umbilical tubercles (Imlay, 1960), but it differs from *T. kleini* by its very fine and closely spaced ribbing.

T. kleini differs from the species of this genus previously described by Böse (1923), Imlay (1937, 1938), and Cantú-Chapa (1963) from Mexico by its ribbing, which is projected strongly forward on the internal part of the flank that becomes irregularly bifurcated or trifurcated at the midflank, and by its falcoid style of curving on the mid-outer part of the flank.

Another specimen from the same core is preserved in a half of whorl (Figure 6g). The specimen is subevolute with gently convex flank; the flexuous primary ribs arising singly or irregularly bifurcated from the inclined umbilical margin are prorsiradiate at the midflank; they are irregularly bifurcated or trifurcated from small swellings at midflank and curve forward on outer flank. All ribs are separated by fine interspaces. This specimen is assigned to *T. kleini* n. sp. The type of its lateral ribbing described above and its subevolute shell allow it to be assimilated to this genus and species.

Thurmanniceras sp. A (Figure 6e)

Material Examined

IPN-1109.

Occurrence

La Laja 8 well, core 13 (2575–2584 m, 8448–8478 ft), Tampico District, eastern Mexico; lower Tamaulipas Formation, lower Valanginian (Figure 1a, c).

Description

A fragment of an evolute shell shows a half of the outer whorl; early whorls are eroded. Flexuous ribs arise free or bifurcate from the umbilical shoulder; they are irregularly bifurcated at the midflank and are separated by narrower spaces on the outer flank.

This specimen differs from *T. kleini* by its slightly flexuous ribbing in contrast to the falcoid ribs from the former. This specimen was previously assigned to *Thurmanniceras* sp. on the basis of its ribbing arising from periumbilical bullae.

Genus *Neohoploceras* Spath (1939)

Type Species

Ammonites sub-Martinii (Mallada, 1882).

Neotype

Kilianella (Neohoploceras) submartini (Mallada) Wiedmann (1966).

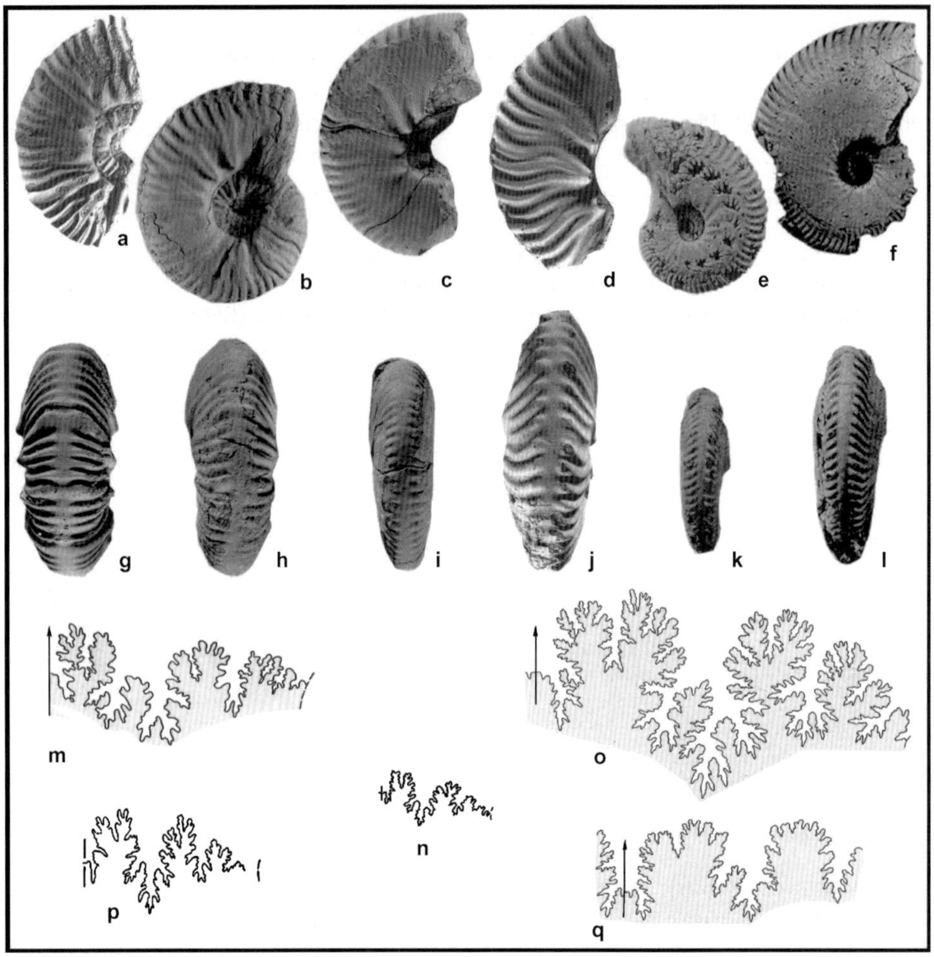

FIGURE 8. Four Neocomitidae genera with different ventral ribbing. (a, g) *Neohoploceras arnoldi* (Pictet and Campiche, 1860) in Aguirre-Urreta (1998, p. 51, figure 12.13, 12.14), lateral and ventral views. (b, h) *Parvaites aguirrei* n. gen. and sp. in Aguirre-Urreta (1998, p. 51, figure 12.3, 12.4), lateral and ventral views. (c, i) *Parvaites leanzai* n. gen. and sp. in Aguirre-Urreta (1998, p. 44, figure 9.3, 9.4), lateral and ventral views. (d, j) *Karakaschiceras neumayri* (Behrendsen, 1892) in Aguirre-Urreta (1998, p. 44, figure 12.15, 12.16), lateral and ventral views. (e, k) *Parrasites flexuosa* (Imlay, 1938, p. 582, figure 11.5, 11.6), lateral and ventral views. (f, l) *Parrasites victoriensis* (Imlay, 1938, p. 581, figure 12.1, 12.2), lateral and ventral views. (m) *Karakaschiceras attenuatus* (Behrendsen, 1892) in Aguirre-Urreta (1998, figure 5.1), suture line. (n) *Parrasites victoriensis* (Imlay, 1938, p. 581, figure 4.11), suture line. (o) *Karakaschiceras pronecostatum* (Felix, 1891) Wiedmann in Kutek et al. (1989, text, figure 5A), suture line. (p) *Parrasites flexuosa* (Imlay, 1938, p. 582, figure 4.10), suture line. (q) *Karakaschiceras subgibbosum* Wiedmann in Kutek et al.(1989, text, figure 5D), suture line. Without scale.

Discussion

Spath (1939, p. 105) applied the name *Neohoploceras* to a group of species assigned to *Leopoldia (Hoplitides)* by Sayn (1907). Spath described this group as Neocomitids with "...rib bundles, starting from umbilical tubercles and bearing lateral tubercles higher up, also single ribs without tubercles between..."

This important morphological character proposed by Spath is not mentioned in subsequent discussions of this genus; its systematic position is based on the dimorphic shell, which is characterized by two or three growth stages according to different authors.

Reboulet (1996) distinguished *Sarasinella* from *Neohoploceras* by its trituberculated ribbing in the inner whorls and by the absence of constrictions. He also pointed out that the latter genus and *Kilianella* have lateral tubercules but without umbilical tubercles in the last one.

Representatives of *Neohoploceras* have been described from fragmented specimens with inconstant bituberculated ribs in half of the last preserved whorl (Koenen, 1902; Gerth, 1925; Arkell et al., 1957; Kemper et al., 1981; Company, 1987; Wright et al., 1996; Aguirre-Urreta, 1998).

Other forms possess inconstant lateral tubercles grouped in a small part of the last whorl (Company, 1987; Reboulet, 1996), and some specimens without lateral tubercles were assigned to this genus by Aguirre-Urreta (1998). This latter distinctive ribbing will be discussed further on.

The dimorphic stages and the position of the bituberculate ribs vary in the last whorl from the adapical half of the shell (*Hoplites* spec., Baumberger, 1906, p. 67, figure 7.5a, b = *Neohoploceras arnoldi* [Pictet and Campiche] in Klein, 2005, p. 333) to the adoral part (*N. submartini* [Mallada, 1987] in Company, 1987, p. 155, figure 13.4; Reboulet, 1996, p. 66, figure 2.8).

The umbilical tubercles are inconsistent in a part of the last whorl and produce single or bifurcate ribs; in the former, they become bifurcate or trifurcate from a lateral tubercle that alternates irregularly with single ribs. This ornamentation is observed in some French, Spanish, and Argentine species (Company, 1987; Reboulet, 1996; Aguirre-Urreta, 1998).

Finally it is recognized that *Neohoploceras* has a widespread geographic distribution and occurs from early Valanginian to early late Valanginian (Reboulet, 1996).

Neohoploceras bartolinii new species (Figure 6i)

Hoplites spec., Baumberger (1906, p. 67, figure 7.5a, b).

Diagnosis

Compressed and evolute shell with shallow umbilicus; inclined and low umbilical wall and flat flank; bifurcated or single, fine, falcoid ribs irregularly arising from umbilical nodes, dividing or resting singly with fine, inconstant nodes at the midflank, curving forward at the outer flank, and ending at ventrolateral small nodes. Shallow and flexuous constrictions. The suture line is unknown.

Etymology

Named in honor of Claudio Bartolini in recognition of his Mexican geological studies.

Material Examined

Holotype, IPN-1110.

Occurrence

San Javier 1 well, core 45 (3056–3068 m, 10,026–10,066 ft), northeastern Mexico; Taraises Formation, lower Valanginian (Figure 1a, b).

Discussion

This specimen is assigned to *Neohoploceras* by its evolute shell with umbilical nodes, from which falcoid and bifurcated or single ribs originate; by its inconsistent midlateral nodes on the middle part of the last whorl; and by its small ventrolateral nodes.

N. imlayi Baraboshkin and Mikhalova (1994) (= *Acanthodiscus bernii* Imlay, 1938, p. 575, pl. 4, figures 7, 9) from northern Mexico is readily distinguished from the proposed *N. bartolinii* by its greater U/D radio (0.46 in the former vs. 0.26 in the latter) and by its radial and strong ribs alternating irregularly with fine ribs; some ribs bifurcate or trifurcate from strong tubercles.

N. submartini (Reboulet, 1996, p. 66, figure 2.8) from France and *N. bartolinii* possess falcoid and fine ribs; the former species differs by its principal, sharp, simple ribs arising from umbilical tubercles, becoming trifurcate from a lateral tubercle; ribs alternate with simple ribs at the adoral part of the preserved shell; the latter species possess inconsistent alternating bifurcate and simple ribs arising from umbilical nodes, some of which have inconsistent lateral nodes on the middle part of the last whorl.

N. bartolinii resembles *Hoplites* spec. (Baumberger, 1906, p. 67, figure 7.5a, b) by its fine ribs, some of which are bituberculate, and by the similar U/D ratio (0.27 in the former, 0.25 in the latter). They differ by the falcoid ribbing style in the Mexican specimen and are rectiradial and slightly inclined forward in the European specimen. However, considering that *Hoplites* spec. (Baumberger) was illustrated by a drawing, these differences are minor; it is here assigned to *N. bartolinii* n. sp.

Klein (2005) assigned *Hoplites* spec. (Baumberger) to *N. arnoldi*, but this species differs by having sharp flexuous ribs and inconsistent umbilical and lateral tubercles that originate as bituberculated ribs in specimens from Spain and Argentina (Aguirre-Urreta, 1998).

Neohoploceras sp. (Kemper et al., 1981, p. 288, figure 41.14, 41.15) (= *N. jacobi* in Klein, 2005 with doubt) from Germany differs from *N. bartolinii* by its lateral tubercles that occur in consecutive ribs; the same tubercles are inconsistent in the Mexican specimen.

N. arnoldi from Spain differs from *Hoplites* spec. (Baumberger) by its stronger lateral tubercles and flexuous ribs (Company, 1987). In addition, *N. arnoldi* from Spain differs from species in Argentina (Company, 1987, figure 13.1, 13.3; Aguirre-Urreta, 1998, p. 51, figure 12.3, 12.4); the former has inconsistent umbilical tubercles and stronger lateral tubercles, the later has inconspicuous lateral tubercles.

The *Neohoploceras* specimens without lateral tubercles that were described by Aguirre-Urreta (1998) from Argentina characterize the following new genus.

Genus *Parvaites* new genus

Type Species

Parvaites aguirrei n. sp. (= *Neohoploceras arnoldi* [Pictet and Campiche, 1860] in Aguirre-Urreta, 1998, p. 51, figure 12.1–4).

Included Species

Parvaites leanzai n. sp. (= *Karakaschiceras neumayri* [Behrendsen, 1892] in Aguirre-Urreta, 1998, p. 42, figure 9.3, 9.4).

Diagnosis

Subevolute shell; whorl section is much higher than wide; subrounded or tabulated venter with a smooth midband; flattened or subrounded flanks; vertical or slightly convex umbilical wall; umbilical tubercles give rise to falcoid, fasciculate, and sharp ribs alternating with simple, bifurcated, or trifurcated ribs at the midflanks, ending perpendicularly at tiny nodes in front of the ventral band or in long and radial tubercles corresponding to the constrictions. The suture line is unknown.

Etymology

After the Cerro La Parva, Neuquen Basin, Argentina (Aguirre-Urreta, 1998).

Occurrence

Idem.

Discussion

The proposed *Parvaites* n. gen. is represented by two shell forms, with subrounded flanks and venter and with constrictions or with planar flanks and venter and without constrictions.

Both forms of *Parvaites* differ from *Neohoploceras* by the absence of lateral tubercles and from the Valanginian *Karakaschiceras* Thieuloy (1971) by its secondary ribs ending perpendicularly at the subrounded or planar venter in front of the midventral band; the ventrolateral tubercles are only on the ribs corresponding to the constrictions (Figure 8b, c, h, i).

In contrast, the *Karakaschiceras* ribbing forms a marked forward flexure that terminates in radial tubercles at the ventrolateral shoulder (Figure 8d, j). Representatives of this genus with this type of ventral tubercles are *K. heteroptychum* (Pavlow, 1892), *K. quadristrangulatum* (Sayn, 1907), and *K. subgibbosum* Kutek et al. (1989, figure 1.1B–1.3B), all from Poland.

Some specimens from Argentina were assigned to *Neohoploceras* by Aguirre-Urreta (1998, figure 12.15, 12.16, 12.22, 12.23); they are here assigned to *Karakaschiceras* because of the forward flexure of the ventrolateral ribbing and the absence of lateral tubercles (Figure 8d, j).

The ventral regions of some representatives of *Karakaschiceras* from Spain, France, and Mexico were not illustrated, so it is not possible to consider them as appertaining to this genus (Kilian, 1905; Cantú-Chapa, 1976, 2001; Contreras, 1977; Company, 1987; Young, 1988; Reboulet, 1996).

In addition, Imlay (1938) described *Leopoldia victoriensis, L. flexuosa, L. crassicostata*, and *L. bakeri* from Mexico; they were later assigned to *Karakaschiceras* by Young (1987) and Cantú-Chapa (2001). These species do not show this type of ribbing with a ventrolateral flexure that characterizes *Karakaschiceras* (Figure 8e, f, k, l).

This type of ribbing was described by Imlay (1938, p. 582) as "...the ventral terminations are inclined strongly forward at all stages, became swollen on the outer whorls..."; it is present in the species mentioned above and occurs also in *L. victoriensis* (Peña-Muñoz, 1964, p. 28, figure 9.1.2) and cf. *Karakaschiceras* sp. (Cantú-Chapa, 2001, p. 366, figure 8.4, 8.4a). In contrast *L. bakeri* shows this type of ventral terminations only in the specimen illustrated by Imlay (1938, figure 12.6, 12.7).

This last ventral ribbing style characterizes *Parrasites* n. gen., which will be described further; however, it is not observed in representatives of *Karakaschiceras* from South America and Europe (Kutek et al., 1989; Aguirre-Urreta, 1998).

Parvaites aguirrei **new genus and species (Figure 8b, h)**

Neohoploceras arnoldi (Pictet and Campiche, 1860) in Aguirre-Urreta (1998, p. 51, figure 12.1–4).

Holotype

Specimen studied by Aguirre-Urreta (1998, p. 51, figure 12.1–4) from the lower Agrio Formation (Valanginian), Cerro La Parva, Neuquen Basin, Argentina.

Diagnosis

Subevolute shell with subrounded flanks and venter; subovate whorl section; vertical umbilical wall with subrounded shoulder; principal ribs arising from radial umbilical tubercles, bifurcated or not at the midflanks, alternating irregularly with secondaries, ending perpendicularly at the venter in front of a smooth midband; flexuous and irregular constrictions, some of which produce radial gently ventrolateral tubercles.

Etymology

Named in honor of Beatriz Aguirre-Urreta, in recognition of her studies on Cretaceous ammonites from Argentina.

Occurrence

Idem.

Discussion

The specimen described above does not have the ventrolateral flexure termination of the ribbing that characterizes *Karakaschiceras*; instead, its ribs end perpendicularly at the venter in front of a smooth midband; it is herein assigned to *Parvaites aguirrei* n. sp. (= *Neohoploceras arnoldi* [Pictet and Campiche, 1860] in Aguirre-Urreta, 1998, p. 51, figure 12.1–4).

Other specimens from the same Argentinian locality are fragmented, so it is impossible to assign them to this genus (Aguirre-Urreta, 1998, figure 12.7–10).

Parvaites leanzai new genus and species (Figure 8c, i)

Karakaschiceras neumayri (Behrendsen, 1892) in Aguirre-Urreta (1998, p. 42, figure 9.3, 9.4).

Holotype

Specimen figured above (Aguirre-Urreta, 1998); paratype figures 9.5, 9.6, after Aguirre-Urreta (1998). All are from the lower Agrio Formation, Valanginian, Arroyo Truquico, Neuquen Basin, Argentina.

Diagnosis

Subevolute shell; plane flanks and venter; subtrapezoidal whorl section is higher than wide; nearly flat flanks; broad, smooth, and slightly convex venter; vertical umbilical wall with a subrounded shoulder. Principal ribs are flexuous arising furcated from umbilical tubercles, irregularly bifurcated, alternating with secondaries at the midflanks, and ending perpendicularly at the venter in radial nodes. The suture line is unknown.

Etymology

Named in honor of Héctor Leanza in recognition of his studies on Lower Cretaceous ammonites from Argentina.

Occurrence

Idem.

Discussion

The proposed *Parvaites leanzai* differs from *P. aguirrai* by its planar flanks and venter and by its homogenous ribs ending perpendicularly at the venter in a small, radial node.

Genus *Parrasites* new genus

Type Species

Leopoldia victoriensis Imlay (1938, p. 581, figure 12.1–4).

Included Species

Leopoldia victoriensis Imlay (1938), *L. flexuosa* Imlay (1938), *L. crassicostata* Imlay (1938), and *L. bakeri* Imlay (1938, p. 583, figure 12.6, 12.7) from Mexico, and with doubt, *K. pruszowskii* Kutek et al. (1989, p. 73, pl. 2, figure 1A–C) from Poland.

Diagnosis

Subevolute shell, vertical umbilical wall, flat flanks; flattened and smooth venter; whorl section is much higher than wide. Ribs arise singly, rarely doubly, from blunt thickenings at the umbilical shoulder; simple ribs intercalate and bifurcate at the midflank, strongly falciform with ventral and weak terminations inclined diagonally forward. Faint constrictions. Suture line with S1 subrectangular, irregularly and superficial bifurcated; S2 long and irregularly bifurcated; L asymmetrically subdivided.

Etymology

After the Sierra de Parras, Coahuila, Northern Mexico.

Occurrence

Idem.

Discussion

Some specimens assigned to *Leopoldia* by Imlay (1938) from northern Mexico were later considered to be *Karakaschiceras* by Young (1987), Cantú-Chapa (2001), and Klein (2005). The above mentioned species described by Imlay (1938) are here grouped in *Parrasites* n. gen. because of their particularly ventrolateral termination of the ribbing.

The new genus differs from *Karakaschiceras* by its flat flanks, by its slightly flat smooth venter, and by its ventral ribbing with diagonally forward-inclined terminations without swellings (Figure 8e, f, k, l).

The suture line of *Karakaschiceras* has been illustrated from some European and Argentinian species; unfortunately, some of these illustrations do not correspond to the respective ventral views from the same specimens to establish the exact comparison (Riccardi et al., 1971; Kutek et al., 1989; Aguirre-Urreta, 1998). Also, it is difficult to compare both elements with those of *Parrasites*.

However, the suture line of *Karkaschiceras* varies from S1 irregularly and deeply bifurcate, with a narrow base in the Argentinian specimens, to strongly rectangular with superficial divisions at the top and a large base in the Poland specimens (Figure 8o, q); both groups have L irregularly divided with a strong structure at its external base (Kutek et al., 1989; Aguirre-Urreta, 1998).

This type of S1 differs from that of *Parrasites*, which is superficially and irregularly divided at the top. The L is trifurcated with small denticulations (Figure 8m, n, q).

Karakaschiceras pruszowskii Kutek et al. (1989, p. 73, pl. 2, figure 1A–C, text, figure 5C) from central Poland has the same type of ribbing with ventral terminations in its internal whorl, as proposed for *Parrasites* from Mexico.

The suture line of this species, illustrated by these authors, shows that S1 is apparently small, strongly inclined to L, and irregularly divided, and that it is smaller than S2; both are profusely denticulated with narrow bases. The Polish species mentioned above differs from *P. pruszkowskii* by S1, which is rectangular and large.

This latter Polish species has a planar and smooth venter in the internal part of the shell, but it differs from the Mexican *Parrasites* species by its rounded venter and by its flanks that are ornamented with large and swollen primary ribs in the outer whorl. However, it is here doubtfully assigned to *Parrasites*.

Age

Parrasites victoriensis, *P. flexuosa*, *P. crassicostata*, and *P. bakeri* were collected from the Taraises Formation, lower part of the upper member, at locality 58, Sierra de Parras, Coahuila, northern Mexico (Imlay, 1938). They were associated with *Olcostephanus*, *Mexicanoceras*, *Neocomites*, and *Acanthodiscus*; the age of this association was not well constrained to the Valanginian or the lower Hauterivian (Imlay, 1938).

However, *P. pruszkowskii* was dated by Kutek et al. (1989) from the *Saynoceras verrucosum* zone (upper Valanginian) in Poland.

Genus *Sarasinella* Uhlig (1905)

Type Species

Hoplites ambiguus Uhlig (1902). Subsequent designation by Lemoine (1906).

Discussion

Company (1987) pointed out two types of ornamentation on representatives of this genus that are based on the ribbing patterns and in the number and position of the lateral tubercles. The first whorls have rectiradiate ribs, some of them bifurcate from lateral tubercles; the last whorl possesses flexuous and bifurcate ribs arising from umbilical tubercles. Reboulet (1996) separated *Sarasinella* from *Neohoploceras* by its trituberculate ribbing and by the absence of constrictions.

Sarasinella sp. A (Figure 6b)

Material Examined

IPN-1110.

Locality

San Javier 2 well, core 14 (3120–3129 m, 10,236–10,266 ft); Taraises Formation, lower Valanginian; northeastern Mexico (Figure 1a, b).

Description

The fragmented specimen shows the lateral ornamentation, probably of the last whorl; it is evolute with a slightly convex flank.

The ornamentation consists of rectiradiate, sharp, and bifurcate ribs arising from umbilical tubercles; some ribs are bifurcated or trifurcated from a lateral tubercle; these ribs alternate irregularly with one simple and fine rib; all ribs are inclined forward on the external part of the flank and are separated by narrower spaces. The ventral region is not preserved.

Discussion

This fragmented specimen is assigned to *Sarasinella* by this type of ornamentation, consisting of two types of ribs, which bifurcate or trifurcate from lateral tubercles, the ribs alternate irregularly with simple ribs. This specimen corresponds to the species grouped with umbilical and lateral tubercles as pointed out by Company (1987).

This specimen does not show its ventral tubercles because of its incomplete preservation state. It resembles *Sarasinella* aff. *subspinosa* (Uhlig, 1910, in Imlay, 1960, p. 219, pl. 42, figure 16; Imlay and Jones, 1970, p. 50, figure 12.13–15) from California by the same type of irregular ribbing.

Sarasinella eucyrta (Sayn, 1907) described by Company (1987, p. 141, pl. 8, figures 13–15) from the lower Valanginian of Spain is less evolute than the Mexican ammonite described here. It differs also by its principal and trituberculate ribs, which are weaker. Our specimen is considered as *Sarasinella* sp.

REFERENCES CITED

Aguirre-Urreta, M. B., 1993, Neocomian ammonite biostratigraphy of the Andean basins of Argentina and

Chile: Revista Española de Paleontología, v. 8, no. 1, p. 57–74.
Aguirre-Urreta, M. B., 1998, The ammonites *Karakaschiceras* and *Neohoploceras* (Valanginian Neocomitidae) from the Neuquén Basin: Journal of Paleontology, v. 72, no. 1, p. 39–59.
Anderson, F. M., 1938, Lower Cretaceous deposits in California and Oregon: Geological Society of America Special Paper 18, 340 p.
Arkell, W. J., B. Kummel, and C. W. Wright, 1957, Mesozoic Ammonoidea, *in* R. C. Moore, ed., Treatise on invertebrate paleontology: Part I. Mollusca 4 (Cephalopoda, Ammonoidea): Lawrence, Geological Society of America and University of Kansas Press, p. L80–L437.
Autran, G., 1989, L'évolution de la marge nord-est provençale (Arc de Castelane) du Valanginien Moyen à l'Hauterivien à travers l'analyse biostratigraphique des séries de la région de Peyroules: Séries condensées, discontinuités et indices d'une tectgenèse distensive. Paléobiologie: Thése Doctorat, Faculté des Sciences et Techniques de l'Université de Nice, 233 p.
Baraboshkin, E., and I. A. Mikhailova, 1994, New and poorly known Valanginian ammonites from south–west Crimena: Bulletin de l'Institut Royal des Sciences Naturelles de Belgique: Sciences de la Terre, v. 70, p. 89–120.
Baumberger, E., 1906, Fauna der untern Kreide im westscherzeischen Jura II. Teil. Die Ammonitiden der unteren Kreide im westschweizerische Jura: Abhandlungen der Scheizerischen Paläontologischen Gesellschaft, v. 32, no. 3, 80 p.
Behrendsen, O., 1892, Zur Geologie des Ostabhanges der argentinishen Cordillere, Theil 2: Zeitschrift der Deutschen Geologischen Gesellschaft, v. 44, 42 p.
Berckheimer, F., and H. Hölder, 1959, Ammoniten aus dem Oberen Weißen Jura Süddeutschlands: Beihefte zum Geologischen Jahrbuch, v. 35, 135 p.
Blanchet, F., 1922, Sur un groupe d'ammonites éocrétacées dérivées des *Cosmoceras*: Compte Rendus Sommaire des Séances de la Société Géologique de France, v. 13, p. 158–160.
Böse, E., 1923, Algunas faunas cretácicas de Zacatecas, Durango y Guerrero: Boletin Instituto Geológico de México, v. 42, 219 p.
Breistroffer, M., 1947, Notes de nomenclature paléozoologique: Proces-verbaux Mensuels de la Société Scientifique du Dauphiné, v. 195, 5 p.
Bulot, L., 1990, Révision des types et figurés de la collection Matheron 2. *Olcostephanus* (*Olcostephanus*) *perinflatus* (Matheron, 1878) et *Olcostephanus* (*Olcostephanus*) *mittreanus* (Matheron, non d'Orbigny, 1850): Mésogée, v. 50, p. 3–8.
Burckhardt, C., 1900, Profils géologiques transveraux de la Cordillére Argentino-Chilienne: Museo La Plata, Sección Geológica y Mineralógica, Anales 11, 136 p.
Burckhardt, C., 1912, Faunes jurásiques et crétaciques de San Pedro del Gallo (Etat de Durango, México): Instituto Geológico de México, v. 29, 260 p.
Burckhardt, C., 1919–1921, Faunas jurásicas de Symon (Zacatecas) y faunas cretácicas de Zumpango del Rio (Guerrero): Instituto Geológico de México, v. 33, 135 p.
Burckhardt, C., 1930, Etude synthétique sur le Mesozoïque mexicain: Mémoire de la Société Paléontologique Suisse 49–50, 280 p.
Cantú-Chapa, A., 1963, Étude biostratigraphique des ammonites du centre et de l'Est du Mexique (Jurassique supérieur et Crétacé): Mémoire de la Société Géologique de France 99, no. 5, 103 p.
Cantú-Chapa, A., 1976, El contacto Jurásico–Cretácico, la Estratigrafía del Neocomiano, el Hiato Hauteriviano Superior–Eoceno Inferior y las amonitas del pozo Bejuco 6 (Centro-Este de México): Boletín de la Sociedad Geológica Mexicana, v. 37, p. 60–83.
Cantú-Chapa, A., 1987, The Bejuco Paleocanyon (Cretaceous–Paleocene) in the Tampico district, Mexico: Journal of Petroleum Geology, v. 10, no. 2, p. 207–218.
Cantú-Chapa, A., 1989, La Peña Formation (Aptian): A condensed limestone-shale sequence form the subsurface of NE Mexico: Journal of Petroleum Geology, v. 12, no. 1, p. 69–84.
Cantú-Chapa, A., 1998, Suaritinae new subfamily and *Galeanites garcense* new genus and new species of the Upper Jurassic Perisphinctaceae ammonites: Revista del Instituto de Geología, Universidad Nacional Autómata de México, v. 15, no. 1, p. 21–24.
Cantú-Chapa, A., 1999, Confrontation of stratigraphic methods to define the Jurassic–Cretaceous boundary in eastern Mexico subsurface, *in* C. Bartolini, J. L. Wilson, and T. F. Lawton, eds., Mesozoic sedimentary and tectonic history of north-central Mexico: Geological Society of America Special Paper 340, p. 93–103.
Cantú-Chapa, A., 2001, The Taraises Formation (Berriasian–Valanginian) in northeastern Mexico: Subsurface and outcrop studies, *in* C. Bartolini, R. Buffler, and A. Cantú-Chapa, eds., The western Gulf of Mexico Basin: Tectonics, sedimentary basins, and petroleum systems: AAPG Memoir 75, p. 353–370.
Cantú-Chapa, A., 2006, New Upper Tithonian (Jurassic) ammonites from the Chinameca Formation in southern Veracruz, Eastern Mexico: Journal of Paleontology, v. 80, no. 2, p. 294–308.
Collignon, M., 1949, Faune néocomienne des couches à Criocèras de Belohasifaka (Cercle de Sitampiky) (Madagascar): Annales Geologiques du Service de Mines, Madagascar, v. 15, p. 53–86.
Collignon, M., 1962, Atlas des fossiles caractéristiques de Madagascar (Ammonites), Fascicule 8 (Berriasien, Valanginien, Hauterivien, Barremien): Service Géologique, Tananarive, 96 p.
Company, M., 1987, Los ammonites del Valanginense del sector oriental de las Cordilleras Béticas (SE de España): Tesis, Universidad de Granada, Spain, 295 p.
Contreras, M. B., 1977, Bioestratigrafía de las formaciones Taraises y La Peña (Cretácico Inferior) de la Goleta, Coahuila y Minillas, Nuevo León: Revista del Instituto Mexicano del Petróleo, v. 9, no. 1, p. 8–29.
Cooper, M. R., 1981, Revision of the late Valanginian cephalopos from the Sundays River Formation of South Africa, with special reference to the genus *Olcostephanus*: Annals of the South African Museum, v. 83, no. 7, 366 p.

Cossman, M., 1901, Rectifications de nomenclature: Revue Critique de Paleozoologie, v. 11, 64 p.

Delanoy, G., G. Autran, and G. Thomel, 1987, Proposition d'un nouveau genre d'Ammonoidea de l'Hauterivien supérieur: *Megacrioceras*, après révision d'une espèce méconnue de la littérature paléontologique: *Ancylocerasdoublieri* Jaubert et comparison avec les formes affines: Comptes Rendus de l'Academie des Sciences, v. 2, no. 305, p. 311–315.

Diener, C., 1925, Ammonoides neocretacea, *in* C. Diener, ed., Fossilium catalogus: I. Animalia: Berlin, W. Junk, v. 29, 244 p.

Dimitrova, N., 1970, Phylogénèse des Ammonites heteromorphes du Crètacé inférieur: Bulletin of the Geological Institute, Sofia (Series Paleontology), v. 19, p. 71–110.

Djanelidzé, A., 1922, Les *Spiticeras* du sud-est de la France: Mémoire pour Servir à 'Explication de la Carte géologique détaillée de la France 6, 255 p.

Donovan, D. T., J. H. Callomon, and M. K. Howarth, 1980, Classification of the Jurassic Ammonitina, *in* M. R. House and J. R. Senior, eds., The Ammmonoidea: Systematics Association Special: London, Academic Press, v. 18, p. 101–155.

d'Orbigny, A., 1842, Paléontologie française: Térrains crétacés: I. Céphalopodes: Paris, Masson, 662 p.

Ettachfini, M., 2004, Les ammonites Néocomiennes dans l'Atlas atlantique (Maroc). Biostratigraphie, Paléontologie, Paléobiogéographie et Paléoécologie: Actes du Laboratoire de Géologie Sédimentaire et Paléontologie de l'Université Paul-Sabatier Toulouse, Strata, v. 43, no. 2, 225 p.

Felix, J., 1891, Versteinerngen aus der mexicanischen Jura- und Kreide-Formation: Palaeontographica 37, p. 1–74.

Gerhardt, K., 1897, Beitrag zur Kenntniss der Kreideformation in Columbien, *in* G. Steinmann, ed., Beiträge zur Geologie und Palaeontologie von Südamerika: Neues Jahrbuch fuer Mineralogie, Geologie und Palaeontologie, v. 11, p. 118–208.

Gerth, H., 1925, La fauna neocomiana de la Cordillera Argentina en la parte meridional de la Provincia de Mendoza: Actas de la Academia Nacional de Ciencias, Argentina, v. 9, p. 57–132.

Gill, T., 1871, Arrangement of the families of mollusks: Smithonian Miscellaneous Collections, v. 227, 49 p.

Haug, E., 1910, Période Crétacé, *in* E. Haug, ed., Traité de géologie: 2. Les Périodes Géologiques: Paris, Colin, v. 2, p. 1153–1398.

Howarth, M. K., 1992, Tithonian and Berriasian ammonites from the Chia Gara Formation in northern Iraq: Paleontology, v. 35, no. 3, p. 597–655.

Hyatt, A., 1889, Genesis of the Arietitidae: Smithsonian contributions to knowledge: Washington, D.C., 67:XI+238 p.

Imlay, R. W., 1937, Lower Neocomian fossils from the Miquihuana region, Mexico: Journal of Paleontology, v. 11, no. 7, p. 552–574.

Imlay, R. W., 1938, Ammonites of the Taraises Formation of northern Mexico: Geological Society of America Bulletin, v. 49, p. 539–602.

Imlay, R. W., 1944, Cretaceous formations of central America and Mexico: AAPG Bulletin, v. 29, p. 1416–1469.

Imlay, R. W., 1960, Ammonites of Early Cretaceous age (Valanginian and Hauterivian) from the Pacific Coast states: U.S. Geological Survey Professional Paper 334-F, p. 167–228.

Imlay, R. W., 1980, Jurassic paleobiogeography of the conterminous United States in its continental settings: U. S. Geological Survey Professional Paper 1062, 134 p.

Imlay, R. W., and D. L. Jones, 1970, Ammonite from the *Buchia* zones in northwestern California and southwestern Oregon: U.S. Geological Survey Professional Paper 647-B, 59 p.

Immel, H., 1978, Die Crioceratiten (Ancyloceratina, Ammonoidea) des mediterranen und boreales Hauterive-Barrem (Unterkreide): Paleontographica, v. 167, 85 p.

Kakabadze, M. V., and Ph. J. Hoedemacker, 2004, Heteromorphic ammonites from the Barremian and Aptian strata of Colombia, *in* S. K. Donovan, ed., Early Cretaceous ammonites from Colombia, compiled by Ph. J. Hoedemacker: Scripta Geologica, v. 128, p. 39,182.

Kemper, E., 1973, Die Unerkreide im Untergrund der Gehrdener Berge und in der Deister-Mulde: Bericht der Naturhistorischen Gesellschaf zu Hannover, v. 117, p. 29–54.

Kemper, E., P. F. Rawson, and J.-P. Thieuloy, 1981, Ammonites of Tethyan ancestry in the early Lower Cretaceous of northwest Europe: Paleontology, v. 24, p. 252–311.

Kilian, M. W., 1905, Sur Quelques Fossiles remarquables de l'Hauterivien de la Région d'Escragnolles (A-M): Extrait du Bulletin de la Société Géologique de France, v. 4, no. 2, p. 864–867.

Kilian, M. W., 1907, Erste Abteilung: Unterkreide (Palaeocretacicum): Lieferung 1: Allgemeines über Palaeocretacicum; Unterkreide im südostlichen Frankreich, Einleitung, *in* Fritz Frech, Lethaea Geognostica: II. Das Mesozoicum (Kreide): Stutgart, Schweizerbart, v. 3, p. 1–168.

Klein, J., 2005, Fossilium catalogus, *in* W. Riegraf, ed., I Animalia, Pars 139: Lower Cretaceous ammonites I: Leiden, Backhuys Publishers, 484 p.

Koenen, A. Von, 1902, Die Ammonitiden des Norddeutschen Neocom (Valanginien, Hauterivien, Barrêmien und Aptien): Abhandlungen der Koeniglich Preussischen Geologischen Landesantalt und Bergakademie, Berlin, v. 24, 449 p.

Krantz, F., 1928, La fauna del Titono superior y medio de la Cordillera Argentina en la parte meridional de la provincia de Mendoza: Actas de la Academia Nacional de Ciencias, v. 10, p. 5–37.

Kutek, J., R. Marcinowski, and J. Wiedmann, 1989, The Wawal section, central Poland—An important link between Boreal and Tethyan Valanginian, *in* J. Wiedmann, ed., Cretaceous of the western Tethys: Proceedings of the 3rd International Cretaceous Symposium, Tübingen 1987: Stuttgart, E. Schweizerbart'sche Verlagsbuchhandlung, p. 717–754.

Lemoine, P., 1906, Études géologiques dans le nord de Madagascar. Contribution à l'histoire géologique de l'Ocean Indien: Annales Hebert, Paris 3, 520 p.

Léveillé, C., 1837, Description de quelques nouvelles coquilles fossiles du département des Basses-Alpes: Mémoires de la Société Geologique de France 2, p. 313–315.

Lissón, C. I., 1937, Dos amonitas del Peru: Boletín de Minas, v. 3, 9 p.

Mallada, L., 1882, Sinopsis paleontológica de España. Cretáceo: Boletín de la Comisión del Mapa Geológico de España, v. 9, 88 p.

Mallada, L., 1987, Sinopsis de las especies fósiles que se han encontrado en España, Tomo III. Terreno Mesozoico (Cretácico inferior): Boletín de la Comisión del Mapa Geológico de España 14, 171 p.

Matheron, P., 1878–1880, Recherches paléontologiques dans le Midi de la France: Marseille, 12 p.

Mazenot, G., 1939, Les Paleohoplitidae tithoniques et berriasiens du sud-est de la France: Mémoires de la Société Géologique de France, v. 18, no. 41, 303 p.

Myczynski, R., 1977, Lower Cretaceous ammonites from Sierra del Rosario (western Cuba): Acta Palaeontologica Polonica, v. 22, no. 2, p. 139–173.

Neumayr, M., 1875, Die Ammoniten der Kreide und die Systematik der Ammonitiden: Zeitschrift der Deutschen Geologischen Gesellschaft, v. 27, p. 854–942.

Nikolov, T. G., 1966, New genera and subgenera of ammonites of Family Berriasellidae: Doklady Bolgarskoi Akademii Nauk, v. 19, p. 839–841.

Parona, F., 1898, Descrizione di alcune Ammoniti del Neocomiano Veneto: Paleontographia Italica, v. 3, p. 137–144.

Pavlow, A. P., 1892, Ammonites de Speeton et leurs rapports avec les ammonites des autres pays, *in* A. P. Pavlow and G. W. Lamplugh, Argiles de Speeton et leurs équivalents: Bulletin de la Société Imperiale des Naturalistes de Moscou 8 (N.S.), v. 5, no. 4, p. 455–570.

Peña-Muñoz, M. J., 1964, Amonitas del Jurásico Superior y del Cretácico Inferior del extremo oriental del Estado de Durango, México: Paleontología Mexicana, v. 20, 33 p.

Pictet, F.-J., 1867, Etudes paléontologiques sur la Faune à *Terebratula diphyoides* de Berrias (Ardèche): Melanges Paléontologiques, Bâle-Geneve, v. 4, p. 207–312.

Pictet, F.-J., and G. Campiche, 1860, Description des fossiles du terrain Crétacé des environs de Sainte-Crox: Part I: Materiaux pour la Paleontologie Suisse, v. 2, 380 p.

Rawson, P. F., 1966, Hauterivian ammonites from the northeast England, with notes on some Valanginian forms: Thesis, University of Hull, 364 p. (manuscript not consulted).

Rawson, P. F., 1975, Lower Cretaceous ammonites from north-east England: The Hauterivian heteromorph *Aegocrioceras*: Bulletin of the British Museum (Natural History), Geology, v. 26, p. 129–159.

Reboulet, S., 1996, L'évolution des ammonites du Valanginien–Hauterivien inférieur du Bassin Vocontien et de la plate-forme provençale (Sud-est de la France): Rélations avec la stratigraphie séquentielle et implications biostratigraphiques: Documents des Laboratoires de Géologie Lyon, Centre des Sciences de la Terre, Université Claude-Bernard Lyon I, v. 137, 294 p.

Riccardi, A., G. E. G. Westermann, and R. Levy, 1971, The Lower Cretaceous Ammonitina *Olcostephanus*, *Leopoldia* and *Favrella* from west-central Argentina: Palaeontographica, v. 136A, p. 83–121.

Roemer, F. H., 1841, Die Versteinerungen des Norddeustsches Kreidegebirges, Hannover, Hahn, v. 2, p. 49–145.

Roman, F., 1938, Les ammonites jurassiques et crétacées. Essai de genera: Paris, Masson, 554 p.

Salfeld, H., 1921, Kiel-und Furchenbildung auf der Schalenaussenseite der Ammonoideen in ihrer Bedeutung für die Systematik und Festlegung von Biozonen: Zentralblatt fuer Mineralogie, Geologie und Paläontologie, p. 342–347.

Sapunov, I., 1977, Ammonite stratigraphy of the Upper Jurassic in Bulgaria: IV. Tithonian: Substages, zones and subzones: Geologica Balcanica, v. 7, no. 2, p. 43–64.

Sarkar, S. S., 1955, Révision des ammonites déroulées du Crétacé inférieur du Sud-est de la France: Mémoires de la Société Géologique de France (Série 5) 34, no. 72, 176 p.

Sayn, G., 1907, Les ammonites pyriteuses des Marnes valanginiennes du sud-est de la France: Part 2: Mémoires de la Société Géologique de France (Sayn 1901–1907), Paléontologie 15, Mémoires 23, p. 29–68.

Spath, L. F., 1922, On Cretaceous Ammonoidea from Angola, collected by Professor J. W. Gregory, D.Sc., F.R.S: Transactions of the Royal Society of South Africa, v. 10, p. 113–148.

Spath, L. F., 1923a, A monograph of the Ammonoidea of the Gault: Part I: Palaeontographical Society Monographs, London, 72 p.

Spath, L. F., 1923b, On the ammonite horizons of the Gault and the contiguous deposits. Summary of progress of the Geological Survey for 1922: Memoirs of the Geological Survey of Great Britain, p. 139–149.

Spath, L. F., 1924, On the ammonites of the Speeton Clay and the subdivisions of the Neocomian: Geological Magazine, v. 61, p. 73–89.

Spath, L. F., 1925, VII. Ammonites and Aptychi, *in* J. W. Gregory, ed., The collection of fossils and rocks from Somaliland made by Messrs, Wyllie and Smelie: Monographs of the Geological Department of the Hunterian Museum, Glasgow University, v. 1, p. 111–164.

Spath, L. F., 1928, Revision of the Jurassic cephalopod faunas of Kachh (Cutch): Memoirs of the Geological Survey of India, Palaeontologia Indica 9, no. 2, 945 p.

Spath, L. F., 1939, The Cephalopoda of the Neocomian Belemnites Beds of the Salt Range: Memoirs of the Geological Survey of India (new series), Paleontologia Indica 25, no. 1, 154 p.

Steinmann, G., 1890, Cephalopoda, *in* G. Steinmann and L. Döderlein, eds., Elemente der Paläontologie: Leipzig, W. Engelmann, p. 344–475.

Thieuloy, J.-P., 1965, Un Céphalopode remarquable de l'Hauterivien basal de la Drôme: *Himantoceras* nov. gen.: Bulletin de la Societé geologique de France, v. 7, no. 6, p. 205–213.

Thieuloy, J.-P., 1969, Sur la présence du genre *Capeloites* Lissón (Ammonoidea) dans le Néocomian des Basses-Alpes et la signification des espèces migratrices transatlantiques: Compte Rendus Sommmaire des Séances de la Société géologique de France, v. 1969, p. 256–258.

Thieuloy, J.-P., 1971, Réflexions sur le genre *Lyticoceras* Hyatt, 1900 (Ammonoidea): Comptes Rendus des Seances de l'Académie des Sciences, v. 272, p. 2297–3000.

Thomel, G., 1964, Contribution à la connaissance des céphalopodes crétacés du sud-est de la France. Notes sur les ammonites déroulées du Crétacé inférieur vocontien: Mémoires de la Société Géologique de France 43, no. 101, 80 p.

Thomel, G., 1965, Contribution à la connaissance des Céphalopodes crétacés du Sud-est de la France. Sur deux espèces du Crétacé Inférieur des Basses-Alpes: Bulletin de la Société Géologique de France, v. 7, no. 7, p. 417–420.

Tzankov, V., 1943, Contribution à l'étude du genre *Holcostephanus* Neumayr, 1875: Spisanie na Bulgarskoto Geologischesko Druzhestvo, v. 14, no. 3, p. 167–206.

Uhlig, V., 1883, Die Cephalopodenfauna der Wernsdorfer Schichten: Denkschriften der Kaiserlichen Akademie der Wissenschaften, Wien, Mathematisch-Naturwissenshaftliche Klasse, v. 46, p. 127–290.

Uhlig, V., 1902, Über die Cephalopodenfauna der Teschener und Grodischer Schichten: Denkschriften der Kaiserlichen Akademie der Wissenschaften, Wien, Mathematisch-Naturwissenschaftliche Klasse, v. 72, 88 p.

Uhlig, V., 1905, Einige Bemerkungen über die Ammonitengattung *Hoplites* Neumayr: Sitzungsberichte der Kaiserlichen Akademie der Wissenschaften in Wien, Mathematisch-Naturwissenshaftliche Klasse, v. 114, p. 591–636.

Uhlig, V., 1910, The fauna of the Spiti shales (1903–1910): Memoirs of the Geological Survey of India (series 15, Himalayan fossils), Paleontologia Indica 4, no. 3, p. 133–306.

von Zittel, K. A., 1884, Cephalopoda, *in* K. A. von Zittel, ed., Handbuch der Palaeontologie, Oldenbourg: Muniche & Leipzing, v. 1, no. 2, 3, p. 329–522.

Wiedmann, J., 1962, Unterkreide-Ammoniten von Mallorca: I. Lieferung: Lytoceratina, Aptychi: Abhandlungen der Mathemathisch-Naturwissenschaftlichen Klass, Akademic Der issenschaften und der Literatur in Mainz 1962, 148 p.

Wiedmann, J., 1966, Stammesgeschichte und System der Posttriadischen Ammonoideen, ein Überblick: Neues Jahrbuch fuer Geologie and Paleontologie, Abhandlungen, v. 127, p. 13–81.

Wiedmann, J., 1973, Ancylocertina (Ammonoidea) at the Jurassic/Cretaceous boundary, *in* A. Hallam, ed., Atlas of Paleobiogeography: Amsterdam, Elsevier, p. 309–316.

Wilke, H.-G., 1988, Stratigraphie und Sedimentologie der Kreide im Nordwesten der Provinz Alicante (SE-Spanien): Berliner Geowissenschaftliche Abhandlungen Reihe, v. A95, 72 p.

Wright, C. W., J. H. Calloman, and M. K. Howarth, 1996, Cretaceous Ammonoidea, *in* R. L. Kaesler, ed., Treatise on invertebrate paleontology: Part L. Mollusca 4: The Geological Society of America and The University of Kansas, 362 p.

Young, K., 1988, *Karakaschiceras* and the late Valanginian of northern Mexico and Texas, *in* J. Wiedmann and J. Kullmann, eds., Cephalopods—Present and past: Schweizerbart'sche Verlagsbuchhandlung, Stuttgart, p. 621–632.

10

Martínez-Medrano, M., R. Vega-Escobar, F. Flores-Cruz, D. Ángeles-Marín, and C. López-Martínez, 2009, Integrated seismic and petrographic analyses of the sandstone reservoirs of the Tertiary Veracruz Basin, Mexico, *in* C. Bartolini and J. R. Román Ramos, eds., Petroleum systems in the southern Gulf of Mexico: AAPG Memoir 90, p. 217–235.

Integrated Seismic and Petrographic Analyses of the Sandstone Reservoirs of the Tertiary Veracruz Basin, Mexico

Martín Martínez-Medrano, Ricardo Vega-Escobar, Fernando Flores-Cruz, Daniela Ángeles-Marín, and Cesar López-Martínez

Petróleos Mexicanos, Exploración and Producción, Veracruz Integrated Asset Team, Veracruz, Mexico

ABSTRACT

The study area within the Tertiary Veracruz Basin has undergone at least five regional geological events associated with voluminous deep-water terrigenous sediments represented by the following six sequences: MS-PL-5.7, MS-6.9, MS-9.2, MM-11.7, MM-16.3, and MI-17.3.

Metamorphic provenance together with evidence from seismic data suggest that during the early Miocene sequence (MI-17.3), deep-water turbidites fed the basin via canyons from the west. A change in depositional pathways and provenance are evident during the middle Miocene sequence (MM-11.7). Seismic stratigraphy and isochron patterns suggest a southwest-derived sediment fill as demonstrated by increasing igneous and metamorphic fragment content and calcareous constituents in the sedarenite reservoir rocks. Sandstones from this interval prove to be very productive hydrocarbon reservoirs. The upper Miocene sequence (MS-6.9) marks a drastic change in fill history and source for the Tertiary Veracruz Basin. Seismic evidence for northwest- to southeast-trending turbidite pathways together with sandstone compositions dominated by felsic to mafic volcanic provenance forms the basis for this interpretation. Prograding deltas begin to dominate the depositional environment beginning in the late Miocene to Pliocene (MS-PL-5.7).

From a relative comparison of the core data, the reservoirs from these key sequences can be ranked in decreasing order of rock quality: first, the MS-PL-5.73 sequence, characterized by maximum porosity approaching 34% and permeability near 1 d; second, the MS-6.9 sequence with maximum porosity of 32% and permeability as high as 795 md; third, the MM-11.7 sequence with 29% maximum porosity and 943-md maximum permeability; and finally, the MI-17.3 sequence with maximum porosity of 23% and 212-md permeability.

The integration of reservoir quality data with seismic character mapping has resulted in innovative concepts and the definition of new exploration potential in recent

DOI:10.1306/13191085M903335

years. Prior to 2001, only three fields had been discovered in the Tertiary Veracruz Basin, whereas now 16 fields exist. This exploratory success is caused by major acquisition of three-dimensional seismic data and multidisciplinary interpretation work, such as that presented here, conducted by Petróleos Mexicano Exploration and Production personnel in Veracruz City.

INTRODUCTION

The Tertiary Veracruz Basin is the second most important basin in Petróleos Mexicano's (PEMEX) northern region. To date, cumulative gas production is approximately 1.9 tcf (579.12 billion m^3), 3P reserves are 1.2 tcf (365.76 billion m^3), and daily production is roughly 1 bcf (304.8 million m^3) per day. The conservative evaluation of the basin's potential assigns it a P50 volume of 11.4 tcf (3.47 trillion m^3), leaving at least 8 tcf (2.43 trillion m^3) of gas yet to be discovered. Despite these numbers, exploration in the basin was dwindling in the late 1990s. Seismic acquisition beginning in 1997 resulted in a database consisting of thousands of kilometers of two-dimensional (2-D) seismic data and two localized three-dimensional (3-D) volumes. Integration and interpretation of these data did not support drilling the PEMEX prospect inventory carried at that time. Thus, to remedy this situation, 3-D seismic coverage of the northern Tertiary Veracruz Basin has been conducted since 2001, which has successfully guided exploration into areas outside of those with earlier known success.

Thus, since 2001, the process for generating drill locations has been successfully transformed, with a net impact of increasing the number of wells drilled from 1 in 1998 to 21 wells in 2005. One of the key factors in this profound change was the use of 3-D seismic information and the systematic integration of seismic attribute analysis, structural geology, sequence stratigraphy, and the application of PEMEX's exploration evaluation process and risk analysis. The results of these integrated studies permitted the identification, definition, and characterization of the key clastic reservoir intervals in the Tertiary Veracruz Basin.

Gas production in the basin comes from five analyzed sequences: LS-MS-5.7, LS-MS-6.9, LS-MS-9.2, LS-MM-11.7, and LS-MI-17.3. Reservoirs in these sequences were discovered essentially from the analysis of amplitude anomalies. Most of the areas with high seismic amplitude anomalies have been drilled or are in the process of being drilled. This is significant because it generated high-quality geological information but there is an urgent need to identify and characterize reservoirs that have no amplitude response. To approach this problem, the five producing sequences discussed here have been correlated, representing regional events relating all of the discovered reservoirs. Because every well drilled outside of these sequences has been unproductive, understanding these stratigraphic sequences with great detail is critical for improving exploration and production in the Tertiary Veracruz Basin.

GEOLOGIC SETTING

The Veracruz Basin is located in southeastern Mexico between the Tampico Misantla and Isthmus Saline basins (Figure 1). This basin consists of the Cretaceous Córdoba Platform in the west and the Tertiary Veracruz Basin in the east. The western limit of the Tertiary Basin is the frontal thrust fault of the Sierra Madre Oriental fold and thrust belt (Figure 1). The eastern limit has not been defined and extends into the Gulf of Mexico.

The Tertiary Veracruz Basin is a foreland basin formed during the Paleocene–Oligocene (PEMEX, 1991, 1995, 1996; PEMEX et al., 1995; Rojas and McClay, 1999). The evolution of the foreland basin began in the Paleocene with the folding and exposure of the Córdoba Platform area (Figure 2). Paleocene–Oligocene sedimentation occurred under conditions of tectonic instability because of the structural stacking of thrust sheets along the Sierra Madre Oriental fold and thrust belt. Highly unstable slopes, developed in response to structural uplift, generated intermittent, carbonate-rich clast-dominated debris flows on the continental slope and base of slope (PEMEX, 1995, 1996; Rojas and McClay, 1999). During the Miocene, erosion along canyons that cut through the fold and thrust belt shed the sediments that filled the basin.

The gradual inversion of the basin began during the early Miocene and reached its maximum development at the end of the middle Miocene when the ancient foreland basin was divided into at least three depressions separated by two structural trends. The structural inversion is related mainly to two regional faults, one with west vergence and a second one that is east verging (Figure 2). In the Miocene, the sediment provenance areas for the basin began to change from a pre-Miocene primarily carbonate source, to a carbonate and siliciclastic source area in the early Miocene, to a solely siliciclastic source from the middle Miocene to Pliocene. Uplift ceased in the late Miocene followed by the deposition of a prograding shelf sequence.

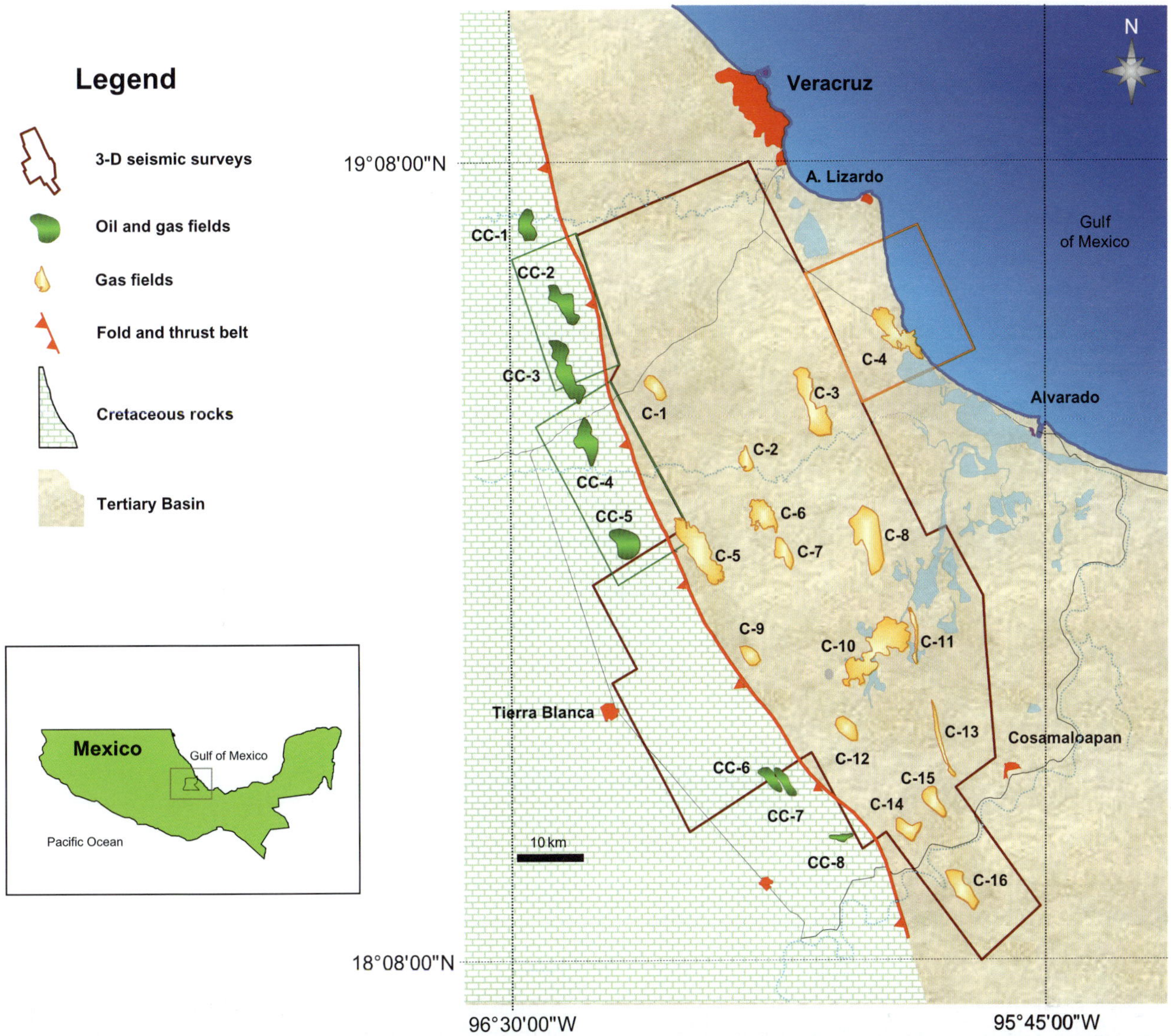

FIGURE 1. Location map of the study area. The Tertiary Veracruz Basin is shown in light brown and the Cretaceous carbonates within the fold and thrust belt are in light green.

Provenance studies of sandstones record source areas in igneous (Santa Ana high), metamorphic (the Mixtequita, Sierra Juárez, and Chiapas massif), and carbonate terrane (Córdoba Platform) (Olivas, 1953; De La Fuente-Navarro, 1959; Mena-Rojas, 1962; Flores, 1970; Menes, 1977; Jacobo and Murillo, 1989; Delgado-Argote et al., 1992; Molina-Garza et al., 1992). The thickness of the sedimentary package exceeds 9000 m (29,527 ft) (Cruz-Helú, 1970), consisting of terrigenous shelf and basinal deposits (Viniegra, 1950; Meneses De Gyves, 1953).

The basin reached its maximum extent in the early Miocene as a depression between the structurally high elements of Los Tuxtlas (?), Santa Ana, and the Córdoba Platform thrust front. This period marks the beginning of uplift in the basin. Clast compositions vary from carbonates to metamorphics in the southwest and igneous to calcareous in the northwest.

LITHOFACIES DESCRIPTION

The Miocene–Pliocene lithology of the Tertiary Veracruz Basin consists of at least two well defined lithologic types: (1) sandstone deposited by turbidity currents (Figure 3) and (2) debrite deposits of debris flow origin (Figure 4).

Turbidites

As indicated by studied cores (Figure 3), submarine fans are composed of four facies associations: channels,

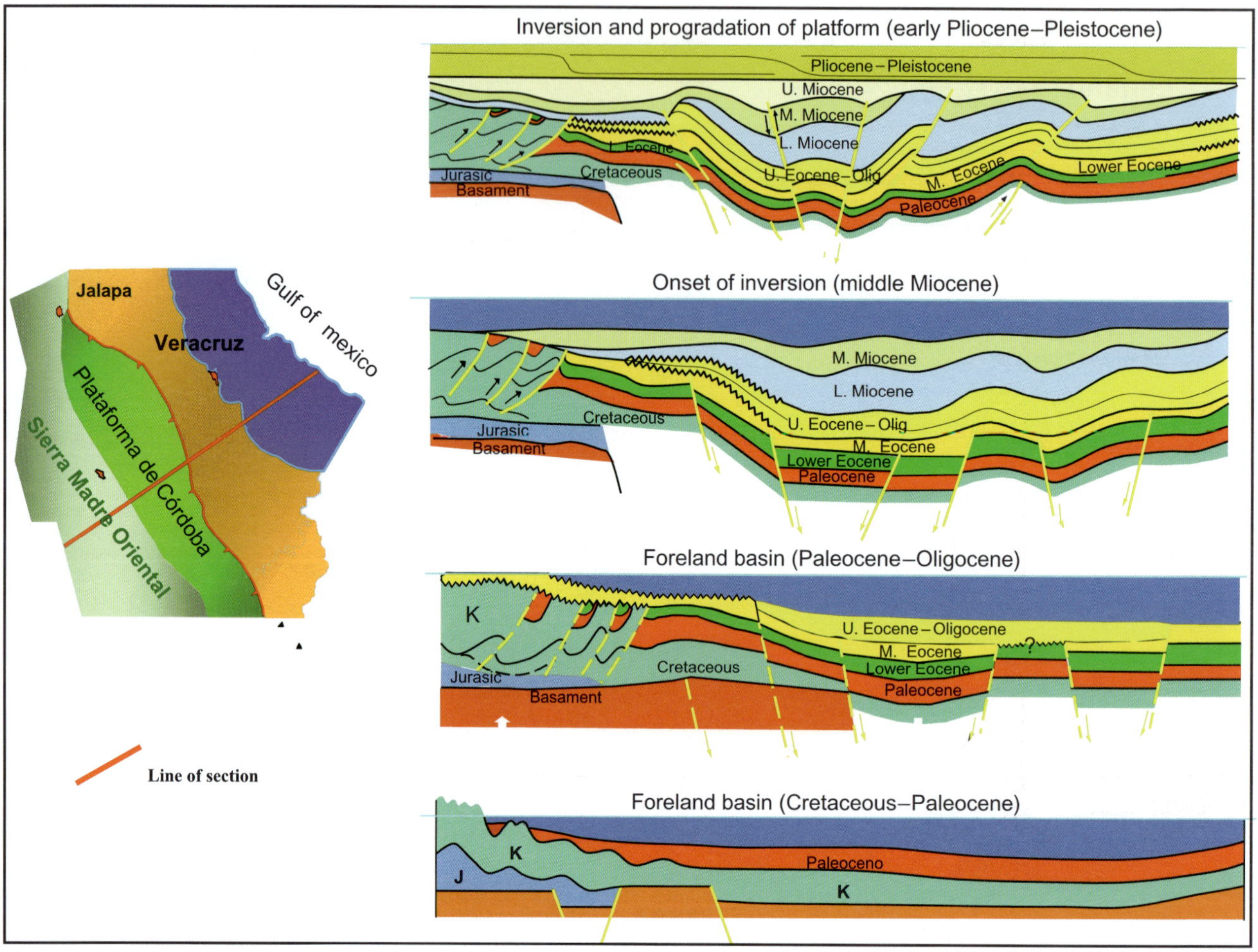

FIGURE 2. Tectonic evolution of the Veracruz Basin.

proximal overbank, distal overbank deposits, and lobe or sheet sands. Channel deposits occur with undulating, rarely flat erosional bases and predominantly massive sands (Bouma A and B facies types). In most cases, these sands are composed of medium- to fine-grained litharenites with argillaceous clasts 2 mm–10 cm (0.078–3.9 in.) floating in the matrix. The clasts are subangular to subrounded, with good to fair sorting. The thickness of these deposits ranges from 20 cm to 1.5 m (7.8 in. to 4.9 ft). Proximal overbank deposits dominantly occur as levees of various sizes with secondary channels. These deposits are characterized by interbedded shales and sandstones from 2 to 20 cm (0.78 to 7.8 in.) thick. The shales are normally bioturbated (burrowed). The sandstone bedding types are parallel, cross-bedded, rippled, and normally graded. Grain size normally varies from fine to very fine. Sand dikes are occasionally seen. Distal overbank deposits contain sandy lenses with cross-bedding, migrating ripples, and flaser structures. Some of these deposits may be considered to be splays from the main channel. Lobes and/or sheet sands have been detected in a few cases.

In the Tertiary Veracruz Basin, these turbidite-derived sandstones are characterized by cyclic, fining-upward deposition. Normally, sands have a basal medium- to fine-grained sandstone that grades upward to shale. These deposits generally contain clay clasts and sedimentary structures such as normal grading, parallel and cross-bedding, ripple marks, and bioturbation. Sand-bed thickness ranges from a few centimeters to up to 1.0–1.5 m (3.2–4.9 ft). These sandstones are normally massive within submarine fan channels or meandering channel systems and occur as sandy lenses within overbank deposits.

Debris Flows and Debrites

Rocks in debris flows (Figure 4) were deposited toward the center and western side of the basin at the base of slope near canyon outlets. They are common in lower and middle Miocene strata. These sediments occur in a wide range of fabrics, from clast-supported conglomerates to shales with floating rock fragments. The clasts are limestone and metamorphic quartz. Soft sediment structures, such as slumps and convoluted bedding, are common. Debris flows are present in lithologies ranging from mud-rich debrites (shales with scarce

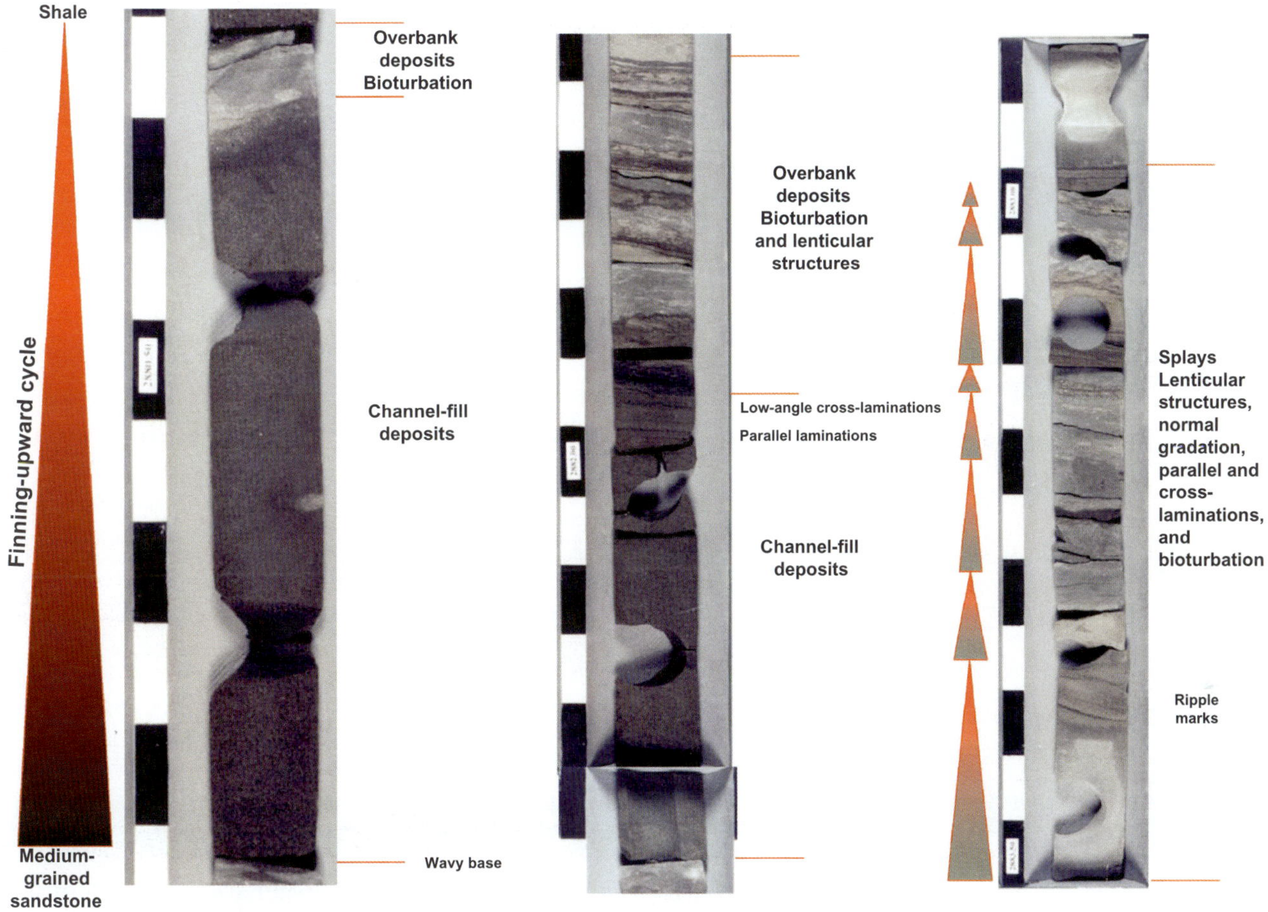

FIGURE 3. Stratigraphic sequences studied in conventional and sidewall cores.

FIGURE 4. Sequences studied in conventional and sidewall cores.

Age	Sequence Boundary	Unit Name	Depositional Environment	Producing Fields
Pliocene	MSI_PI_04_20	PI-R-4		
	MSI_PI_05_00	PI-5		
	LS_MS_05_73	MS-PI-5.7	Meandering channels	Cocuite and Lizamba
Upper Miocene	MSI_MS_06_00	MS-6		
	LS_MS_06_98	MS-6.9	Meandering channels and basin plain submarine fans	Playuela, Vistoso Apertura, Cocuite
	MSI_MS_07_08	MS-7		
	LS_MS_09_26	MS-9	Basin plain submarine fans Multiple sources	Cocuite = Breña, Arena G. Lizamba = Arquimia, Kepler.
	MSI_MM_11_20	MS-11.2		
Middle Miocene	LS_MM_11_70	MM-11.7	Basin plain submarine fans Multiple sources	Cocuite, Apertura, Madera, Uloa, Novillero
	LS_MI_16_38	MM-16		
Lower Miocene	LS_MI_17_34	MI-17	Continental rise submarine fans and debris flows	Cocuite Apertura, Lizamba Novillero
	LS_OL_24_36	MI-24		
Oligocene				
Eocene			Continental slope and rise	Perdiz

FIGURE 5. Twelve Neogene depositional sequences were detected in the basin, five of them have an unconformable base (LS) with tops represented by maximum flooding surfaces. These are the only productive sequences in the entire basin: MS-PL-5.7, MS-6.9, MS-9.2, MM-11.7, and MI-17.3. An erosional base has only been detected in two of them (MI-24 and MM-16), neither of which has established production.

rock fragments floating in a matrix) to clast-supported gravelly debrites. These basinal deposits have no preferred grain arrangement; i.e., they are poorly sorted with rounded to subrounded grains. Debrites are commonly found near the paleoslopes, predominantly in the lower Miocene. Their clasts vary from rock fragments (calcareous, igneous, metamorphic, sandstones, and even mudstones) to gravel-size quartz and feldspar clasts.

MIOCENE–PLIOCENE SEQUENCES

Stratigraphic analysis has defined 12 Miocene–Pliocene sequences (Figures 5, 6), five of which produce gas. In the following, the five producing units are described here in terms of their age, lithology, and depositional environment, described from younger to older. All of the sequences were identified on electric logs using paleontologic data derived from cuttings and cores. They were then identified and mapped on the seismic data (Figure 7).

Sequence UM_PL_5.7

The age range established for this sequence is from 5 to 5.7 m.y. (Figure 6). This period was characterized by deposition of submarine fans dominated by meandering channels (Figure 8).

Lithology

Sandstones of this sequence are classified as litharenites with three dominant rock fragment types (Figure 9): (1) sedimentary rock fragments (limestones and dolomites) subclassified as sedarenites, (2) intermediate volcanic rock fragments subclassified as volcano-arenites, and (3) metamorphic rock fragments; although samples with abundant clasts of this type were scarce. These sandstones were subclassified as phyllarenites. The grain size of these sandstones ranges from very fine to medium; they are fair to well sorted and subangular to subrounded. The Quartz (Q), Feldspars (F), and Lithics (L) diagram shows that the sandstones are composed of rock fragments with a lesser proportion of quartz and feldspar corresponding to litharenites. The LS-LV-LM litharenite triangle has

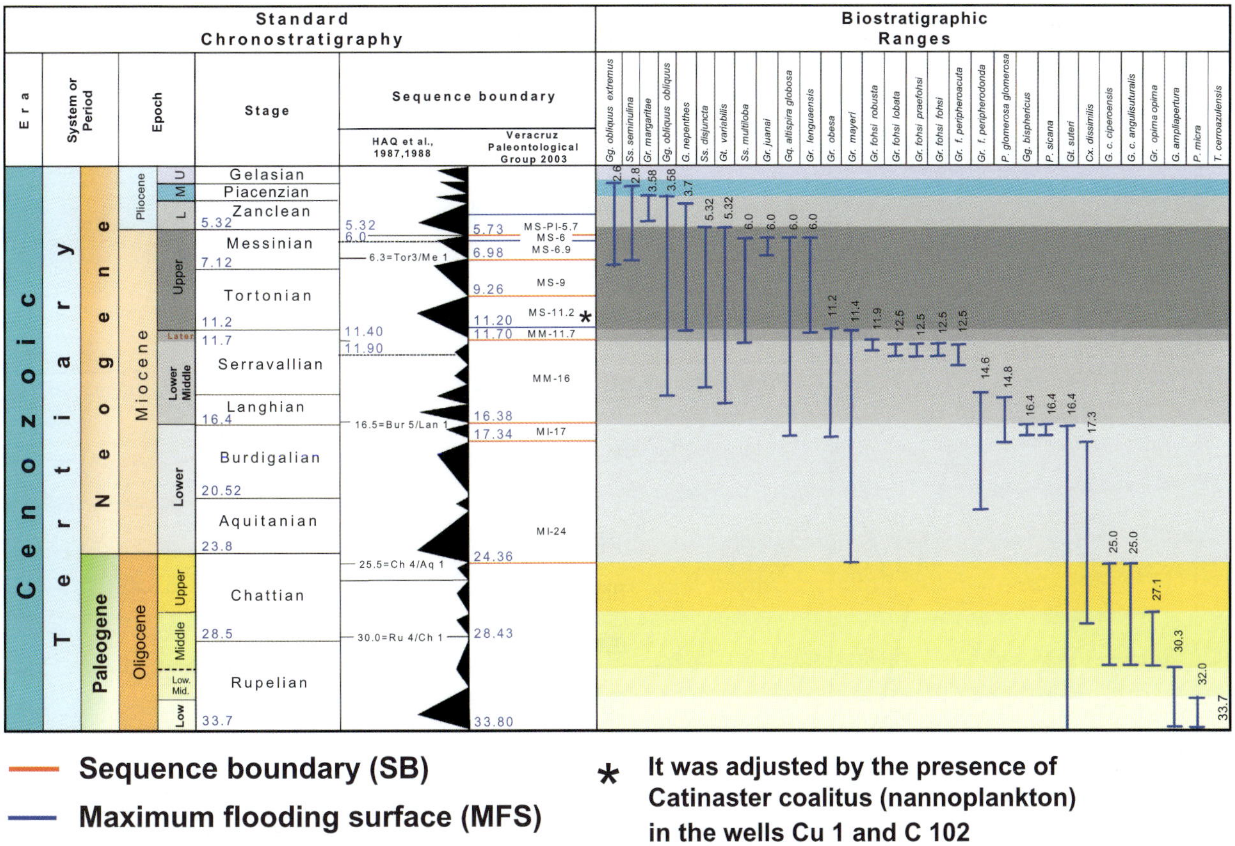

FIGURE 6. Biochronostratigraphic chart based on paleontological information from wells throughout the basin and geological events described by Haq and Hardenbol (1987). Seven sequence boundaries and five maximum flooding surfaces were established. The biostratigraphic ranges of the index fossils are based on Bolli et al. (1989).

two fields: one dominated by volcanic rock fragments and one by sedimentary rock fragments. The source of these fragments was to the northwest, where mafic to felsic extrusive rocks, as well as carbonates exist.

Depositional Environment

By this time, the basin was tectonically stable with steep slopes to the north and a deep central depression. Under these conditions, a submarine fan was deposited with straight channels that became braided and meandering toward the central basin (Figure 10). Lobes of argillaceous material were deposited at the mouths of the channels. Wells that penetrate this prograding system commonly find a basal section of channel fills and an upper section consisting of levees and lobes.

Sequence UM_6.9

The age range established for this sequence is 6 to 6.9 m.y. (Figure 6). This sequence consists of deposits representing a transition from basin-floor fans with wide lateral development to channel-dominated fans. It consists of interfingering, meandering channels and basin-floor fan deposits (Figure 11). The last deformational pulses occurred at this time, although they were not intense. The middle Miocene deformation produced paleobathymetric relief that was eventually buried. The central part of the area suffered the most deformation producing two northwest-southeast–oriented depocenters, one to the west and the other to the east (Figure 11). The southern part of the western depocenter (Estanzuela 1, Chúcaro 1, and Zarin 1 wells) acted as a closed basin bordered by the thrust front (west) and the Novillero-Veinte structural high (east).

Lithology

Samples from this sequence are classified as litharenites (Figure 12) consisting of sedimentary clasts (limestone) and volcanic rock fragments in a 1:1 proportion. Thus, they are further classified as volcano-arenites and sedarenites. Additionally, they include a group of sandstones that are a mixture of rock fragments and feldspars representing a transition from pure litharenites to lithic arkoses. Very few samples have been classified as lithic arkose. Sandstones vary from fine to medium grained and are well sorted, and grains are subangular to subrounded.

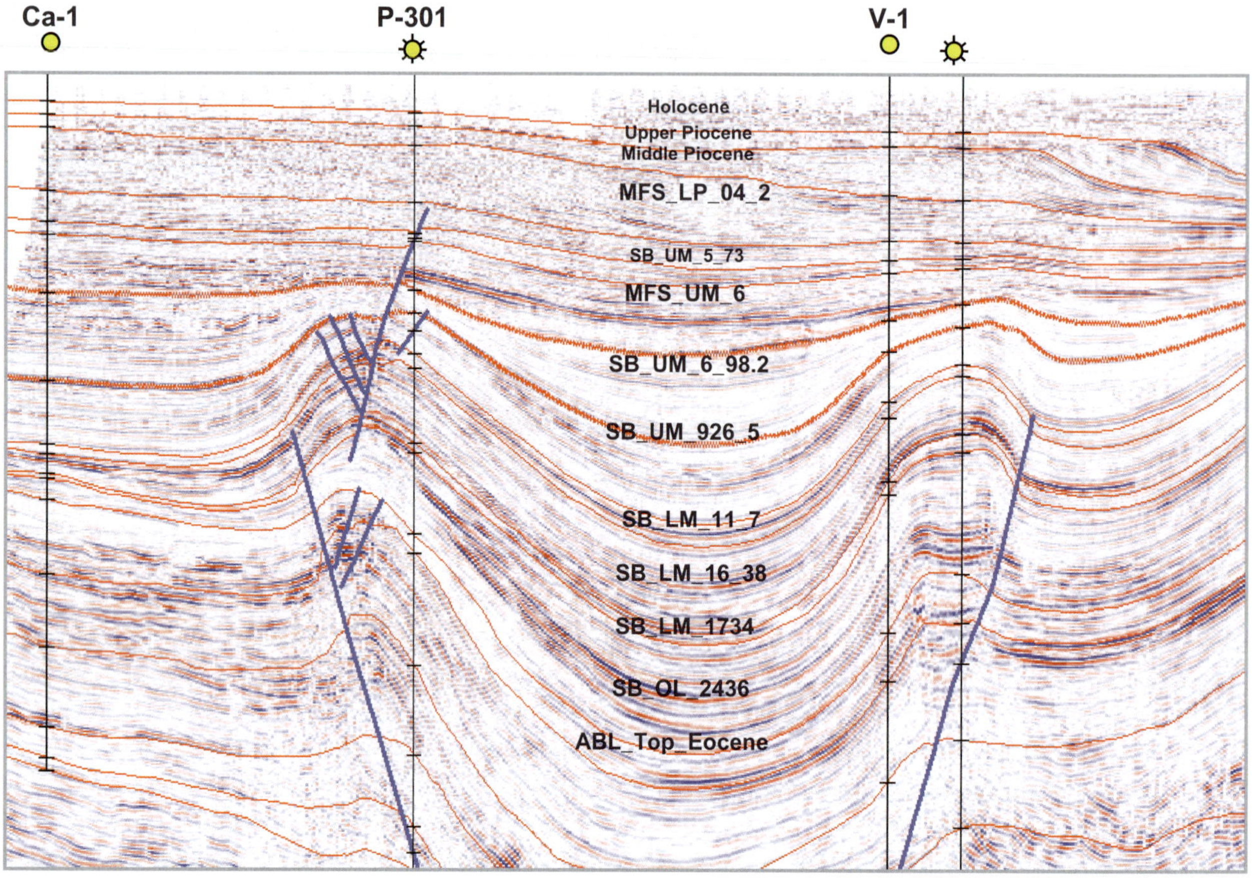

FIGURE 7. East–west representative seismic section between the Playuela and Vistoso fields showing the distribution of the different sequences and the ties between respective wells.

Lithologic data are derived from cores in Playuela, Vistoso, Cocuite, and Apertura-Madera fields (Figure 12). Figure 12a shows the QFL triangle and Figure 12b shows the LS-LV-LM triangle. The QFL diagram indicates that the sandstones of this sequence are mainly lithic with a lesser proportion of quartz and feldspar. Clearly, the Playuela field has considerably higher feldspar and quartz content. On the LS-LV-LM diagram, two main compositional fields are observed: a field of sedimentary rock fragments (Vistoso) and another of volcanic rock fragments (Playuela and Apertura). Cocuite field contains a mixture of sedimentary and volcanic rock fragments. The sandstones of this sequence are classified as litharenites (calcarenite and sedarenite). The source for these sediments is possibly located to the northwest and consists of mafic to felsic volcanic rocks, as well as platform carbonates.

Depositional Environment

At this time, the basin had remnant paleobathymetry derived from the earlier deformation. A slope to the north with northwest-southeast–oriented re-entrants exists (Figure 13). To the west was a depression bordered farther west by slopes and to the east by the Novillero-Vibora-Chachalacas (NVC) structural trend, which was being eroded along its crest (Figure 11). To the east was another depression bordered by the NVC trend and an indeterminate boundary on the east.

These basins were gradually filled from the northwest by submarine fans. Deposition was controlled by paleobathymetric conditions; therefore, fans filling both depressions were deposited parallel to the structural trends (Figure 13). When the western depression was full, deposition overflowed to the east. These fans are progradational and consist of a series of lobes or secondary fans and meandering channels. At least three groups of lobes and stacked channels can be distinguished: (1) eastern lobes formed by stacking of channels, levees, and lobes; (2) meandering channels and central lobes with the basal section dominated by braided channels and an upper fan section of meandering channels and lobes; and (3) western fans with basal meandering channels and lobes.

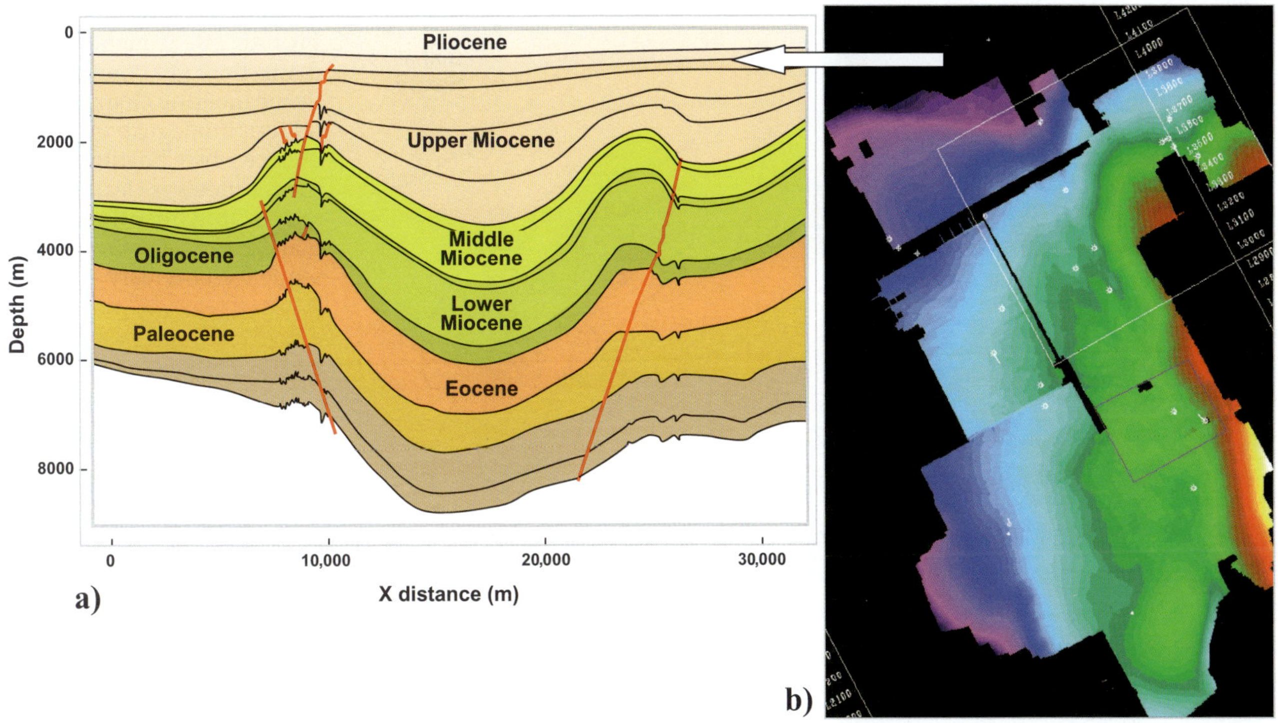

FIGURE 8. Distribution of sequence MS-PL-5.7 in the northern Tertiary Veracruz Basin. At this time, deformation was incipient with a large northwest-southeast–oriented depocenter limited on the north by the progradational slope of this sequence and to the west by slopes produced by the deformation. Sedimentation occurred as debris flows.

FIGURE 9. Sandstone composition in sequence MS.5.7. SRF = sedimentary rock fragments; VRF = volcanic rock fragments; MRF = metamorphic rock fragments.

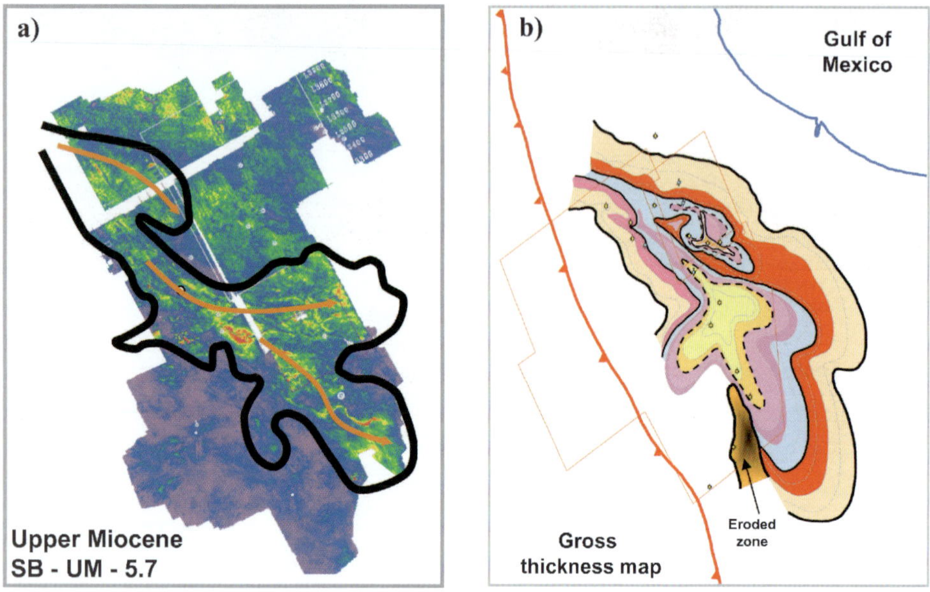

FIGURE 10. Sequence MS-5.7 (basin-floor fan dominated by meandering channels). (a) Amplitude map of this sequence where brightening of the meanders is clearly observed. (b) Gross thickness map of this sequence. SB = sequence boundary.

Sequence UM_9.2

The age range established for this sequence is 7 to 9.2 m.y. (Figure 6). This represents a second regional event with widely developed basin-floor fans formed by sediment derived from several point sources or zones (Figure 14).

Lithology

Sandstones deposited during this time show a relative abundance of feldspar and quartz with diminishing rock fragments. They are transitional between litharenites and lithic arkoses, with a greater inclination toward the former (Figure 15). The litharenites were

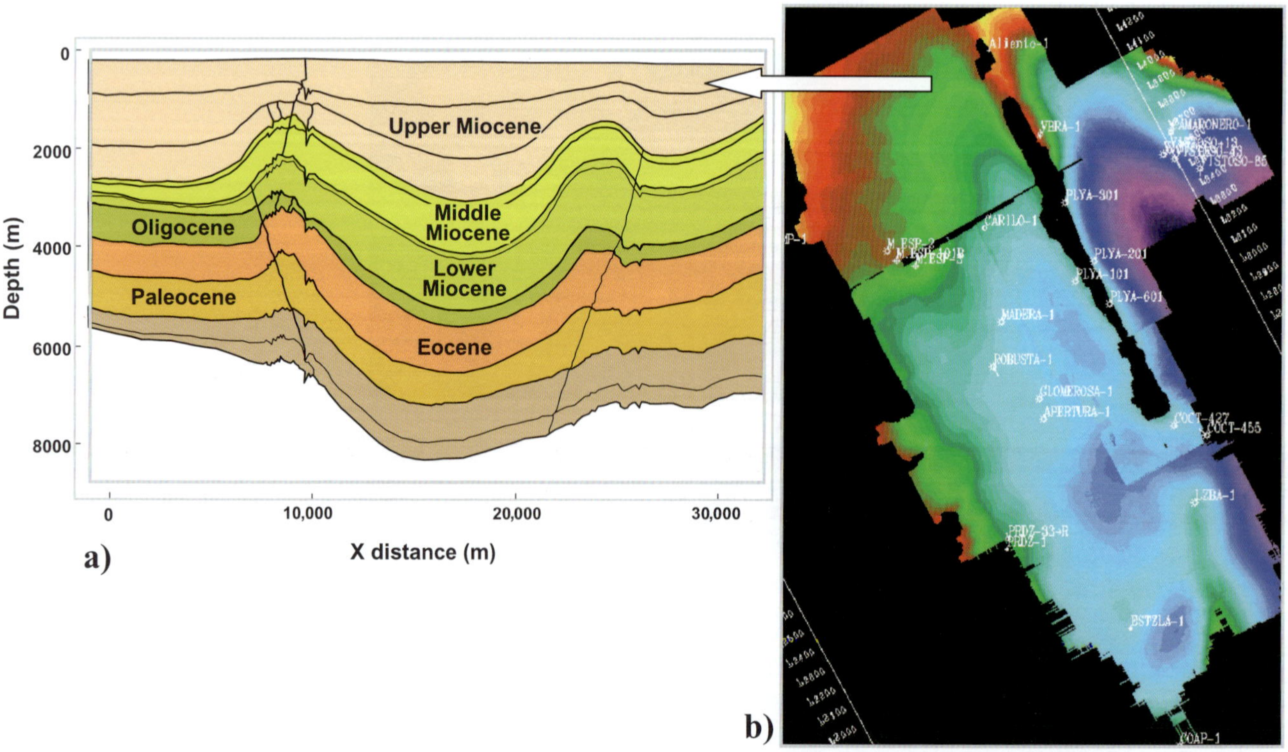

FIGURE 11. Distribution of sequence MS-6.9.

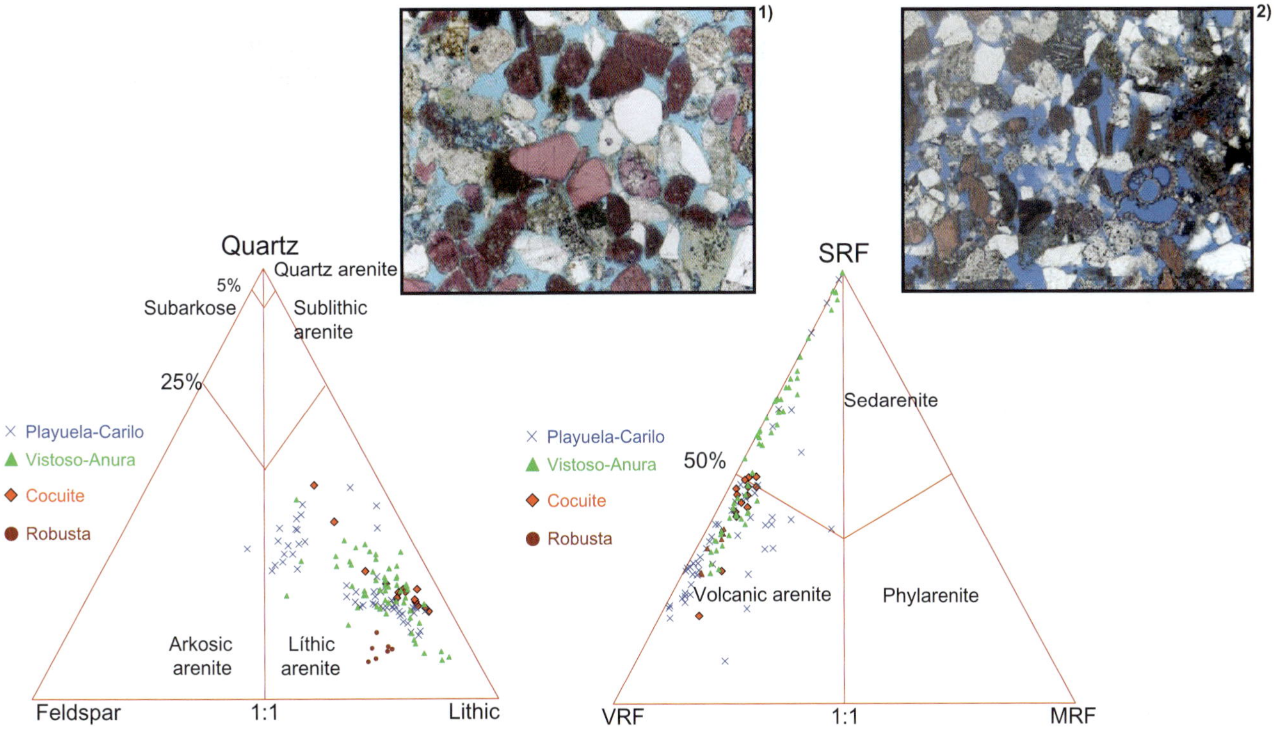

FIGURE 12. Sandstone composition of sequence MS-6.9. SRF = sedimentary rock fragments; VRF = volcanic rock fragments; MRF = metamorphic rock fragments.

subdivided into two main groups: volcano-arenites and phyllarenites, but some sedarenites are also present. Sandstones are medium to coarse grained, moderately sorted, and with subangular grains.

Data shown in Figure 15 are from cores in Cocuite, Vistoso, Lizamba, and Estanzuela fields. Figure 15a shows the QFL triangle and Figure 15b shows the LS-LV-LM triangle. The QFL diagram shows that these sandstones consist mainly of rock fragments; however, in contrast with sequences MS-6.9 and MS-PL-5.7, quartz, feldspar, and metamorphic rock fragments have a considerable contribution. The LS-LV-LM diagram shows that this

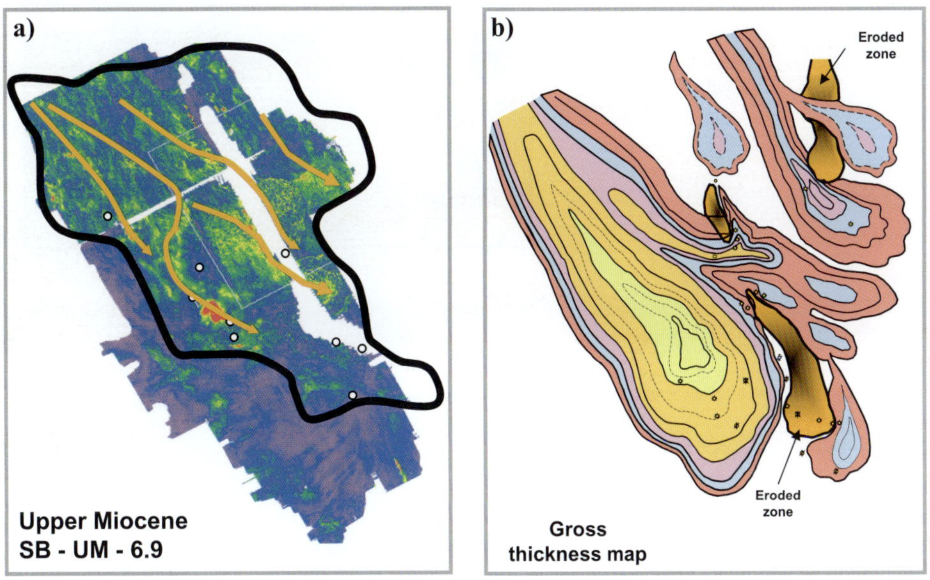

FIGURE 13. Aggrading basin-floor fan dominated by lobes and channels. (a) Amplitude map for sequence MS-6.9 that demonstrates the wide lateral extent of the fan. On the amplitude and gross thickness maps, two depocenters are observed to be separated by a paleohigh. The major depocenter is to the west and is filled by an aggrading submarine fan penetrated by wells of the Apertura-Madera producing field (b). The eastern depocenter is filled by a group of coalesced submarine fans in Playuela, Vistoso, and Cocuite fields. Perdiz, Lizamba, Estanzuela 1, Chúcaro 1, Zarin 1, and Coapa 1 wells penetrated shales of this age. SB = sequence boundary.

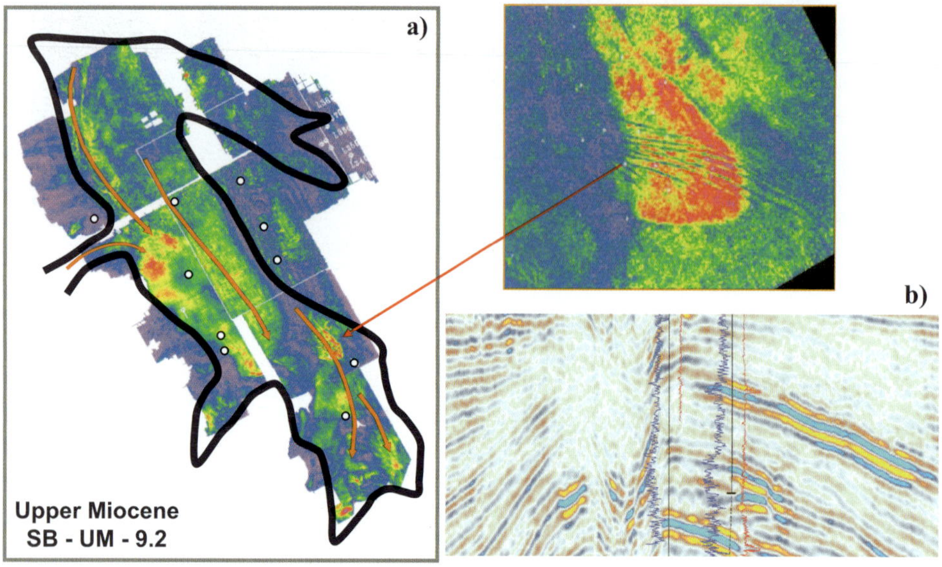

Figure 14. Basin-floor fans with a wide lateral distribution of sequence MS-9.2. (a) Amplitude map for this sequence showing two areas with high amplitudes. (b) Map of gross thickness for this sequence. From these two maps and well analyses, two depocenters are known to exist: one to the north (Perdiz and Apertura-Madera fields) filled by an argillaceous fan and one to the south (Cocuite, Lizamba, Estanzuela, and Mirador fields) filled by widely distributed basin-floor fans. SB = sequence boundary.

sequence received sediments from multiple sources, i.e., volcanic rocks from the northwest, metamorphic rocks from the southwest, and calcareous rocks from the platform to the west. Therefore, these sandstones are classified as litharenites and arkosic arenites. The litharenites are subdivided into volcano-arenites, phyllarenites, and sedarenites.

Depositional Environment

This sequence represents a transition in the direction of sediment supply to the basin. The western sources were being lost, and sediment contribution increased from the northwest (Figure 16). Furthermore, the possibility exists that the basin had no significant

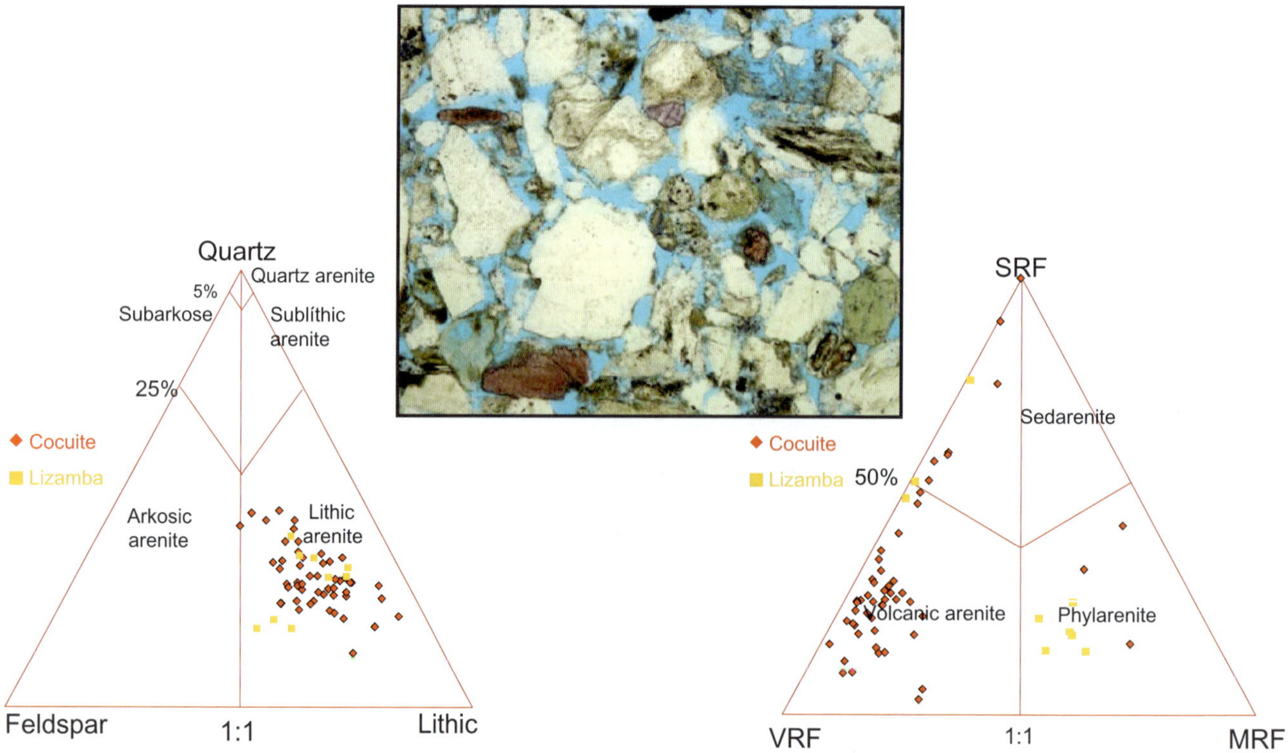

Figure 15. Sandstone composition of sequence MS-9.2. SRF = sedimentary rock fragments; VRF = volcanic rock fragments; MRF = metamorphic rock fragments.

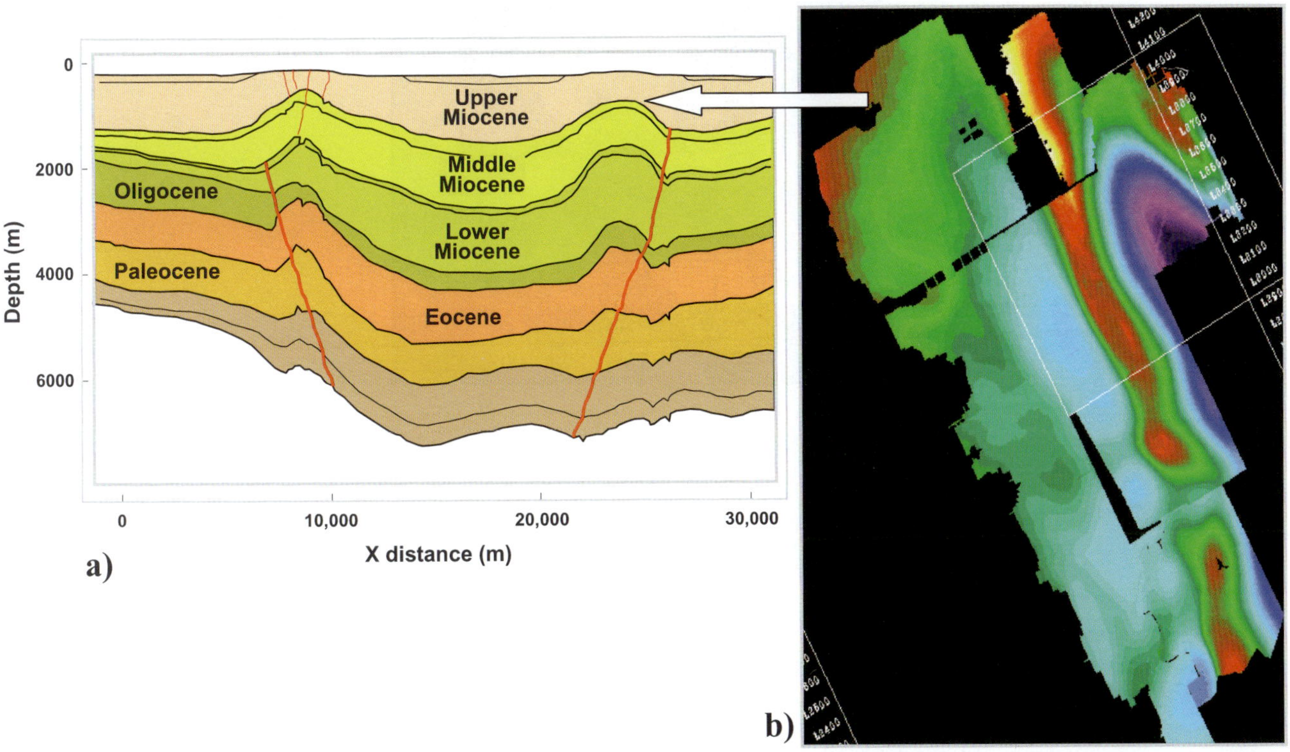

FIGURE 16. Distribution of the MS-9.2 sequence in the northern part of the Tertiary Veracruz Basin. The final stage of deformation of the basin is possible to occur at this time. The development of large slopes along the entire western and northern parts of the basin produced a northwest-southeast–oriented depocenter.

bathymetric barriers to impede lateral sediment distribution resulting in basin-floor fans of wide lateral extent. Wells that penetrate this sequence find basin-floor fans that consist of channels and proximal and distal overbank deposits coalescing into a larger fan.

Sequence MM_11.7

The age range established for this sequence is 11.2 to 11.7 m.y. (Figure 6). This sequence represents the first instance of a regional basin-floor fan fed from

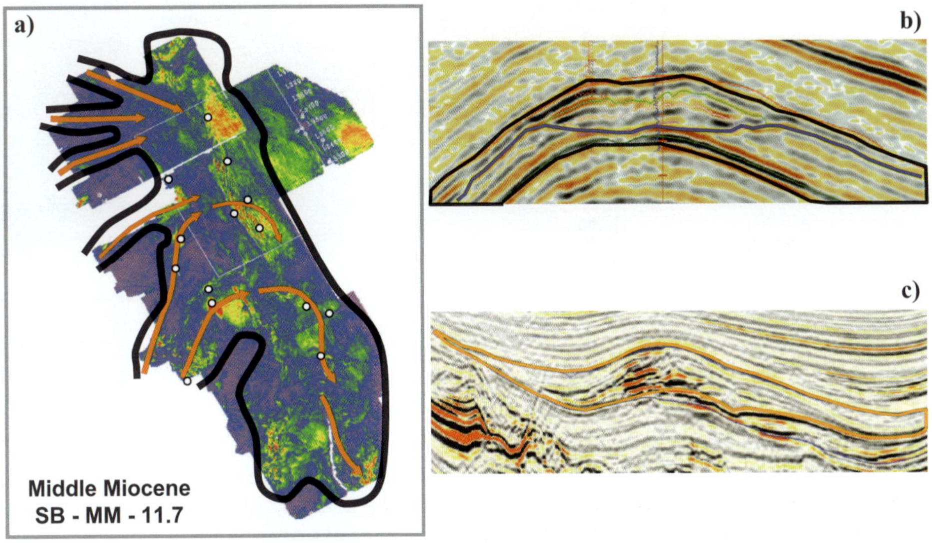

FIGURE 17. Basin-floor and slope fan dominated by lobes with multiple sediment sources. (a) Amplitude map for sequence MM-11.7. SB = sequence boundary.

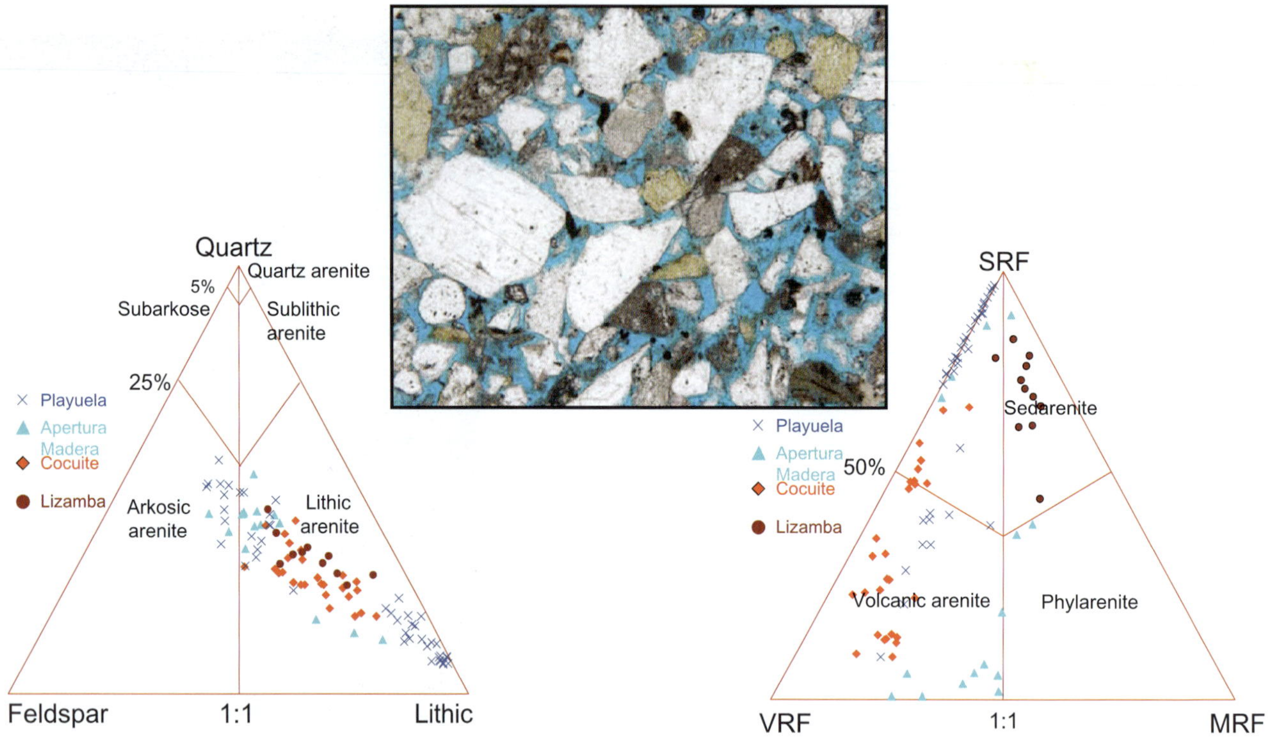

FIGURE 18. Sandstone composition of sequence MM-11.7. SRF = sedimentary rock fragments; VRF = volcanic rock fragments; MRF = metamorphic rock fragments.

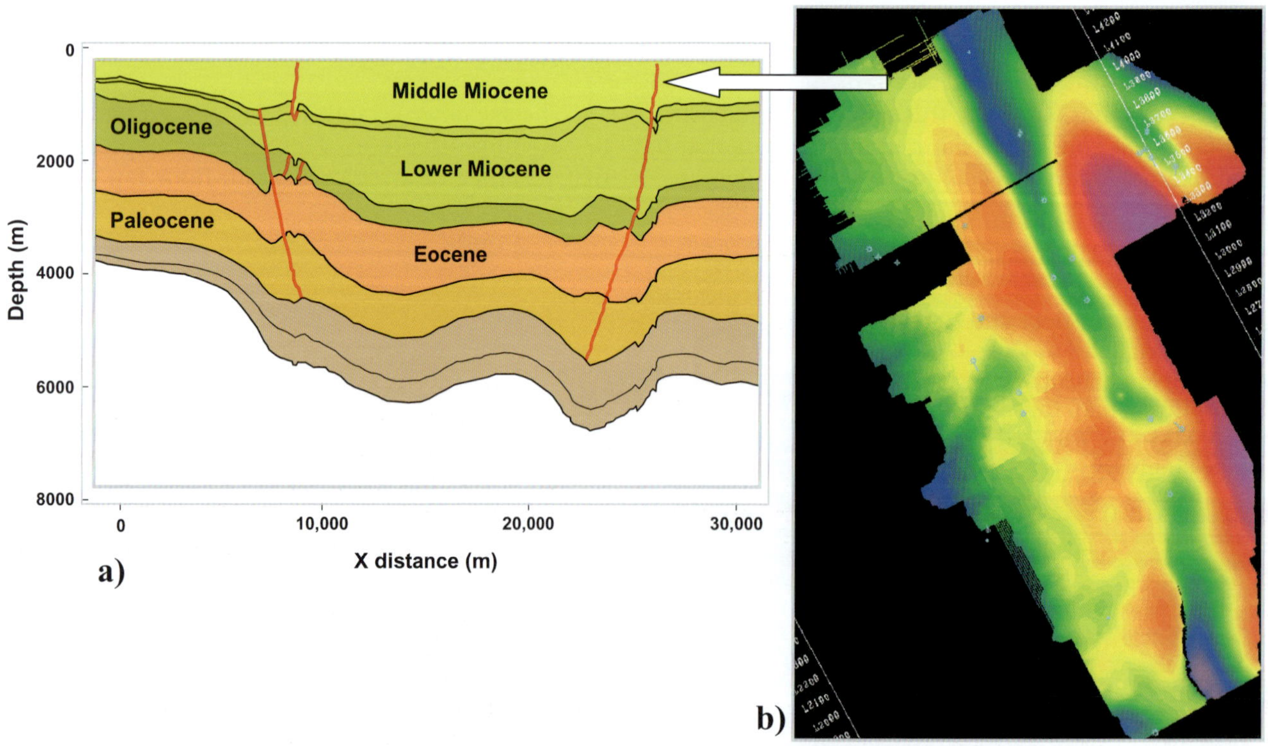

FIGURE 19. Distribution of sequence MM-11.7 in the northern part of the Tertiary Veracruz Basin. This sequence may represent the final period of major subsidence, as well as the beginning of basin inversion. At this time, slopes existed west and north of the basin from which sediment moved toward a large northwest-southeast–oriented depocenter.

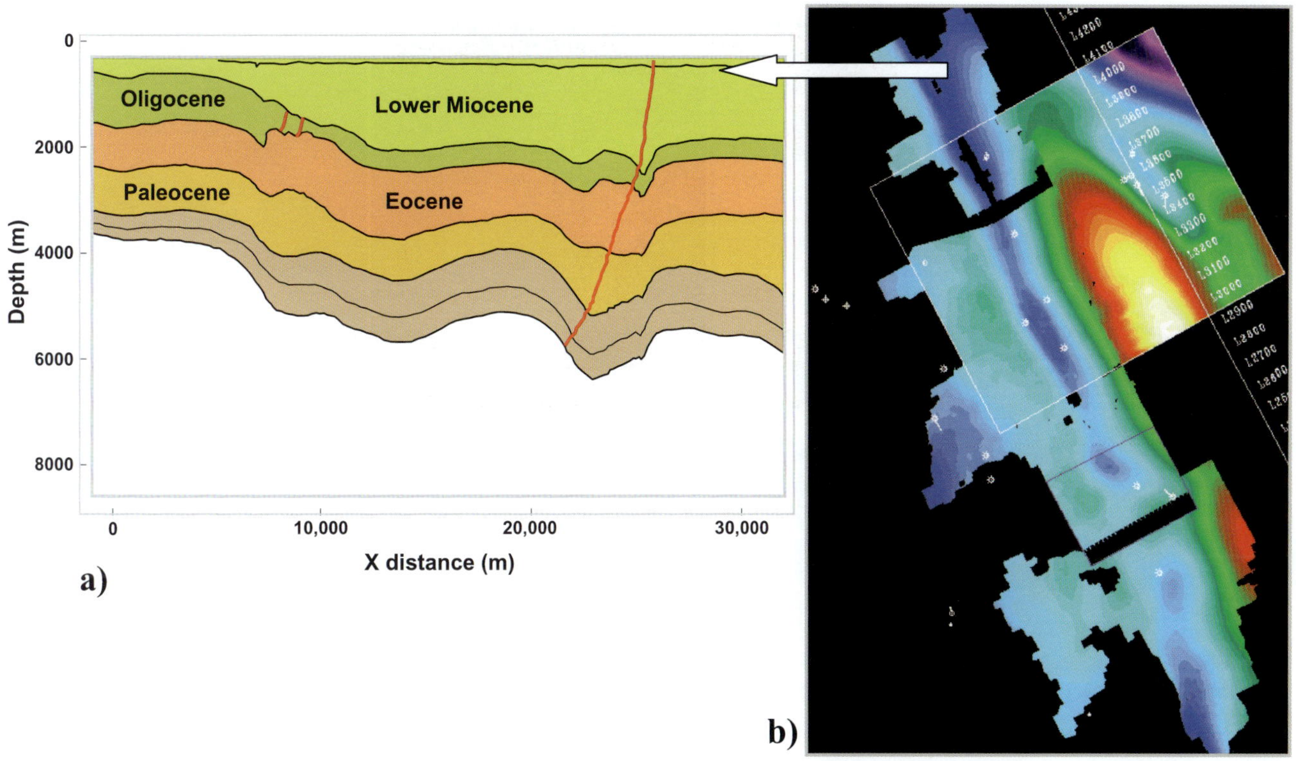

FIGURE 20. This figure shows the distribution of sequence MI-17.3. At this time, the basin had a large depression in the east bordered to the west by slopes cut by canyons through which sediments entered the basin. The predominant sedimentation at this time was base-of-slope fans and submarine canyon fill consisting of debris flows.

multiple sediment source areas, as well as the first sedimentary event after infill of the remnant early Miocene paleocanyons (Figure 17a). This map shows sediment supply from the west and north of the basin deposited in the east-west–oriented base-of-slope fans (Apertura, Madera, and Estanzuela fields). Part of this sediment was carried into the depocenter as basin-floor fans with wide lateral distribution and northwest-southeast depositional orientation (Cocuite, Uloa, and Novillero fields). Figure 17b and c show the principal seismic patterns of this sequence.

Lithology

This was the period of maximum quartz and feldspar deposition, with minor rock fragments being deposited, producing two groups of sandstones: arkosic arenites and litharenites (Figure 18). This variation in supply is geographic, with arkosic arenites dominating in Playuela and Apertura-Madera fields and a transition to litharenites in Cocuite and Lizamba fields. The litharenites are of three different types: (1) volcano-arenites and phyllarenites, (2) volcano-arenites and sedarenites, and (3) sedarenites and phyllarenites.

Data in Figure 18 are from cores in Cocuite, Apertura-Madera, Lizamba, and Playuela fields. Sediment composition indicates that during that time, the area continued to receive quartz and feldspar from the west (Apertura-Madera and Playuela fields) whereas the depocenter (Lizamba and Cocuite fields) only received lithic fragments. The latter sandstones are subclassified as volcanic arenites, sedarenites, and phyllarenites (Figure 18).

Depositional Environment

During the deposition of this sequence, the basin was a depression bounded by slopes to the north and west and by a relatively low positive structure on the east (Figure 19). The depression was oriented northwest–southeast. Sediments arrived from multiple sources to the west and northwest (Figure 17a).

The western slopes retained remnants of the older canyons that continued to serve as sediment conduits. Upon arriving in the basin, the east and southeast sediment flow changed to a more southerly direction. Three types of deposits occur: (1) base-of-slope fans near the outlets of the older canyons with west to east and northwest to southeast depositional directions, (2) basin-floor fans within the depression with a northwest to southeast depositional direction, and (3) northwest-southeast–oriented meandering channel systems.

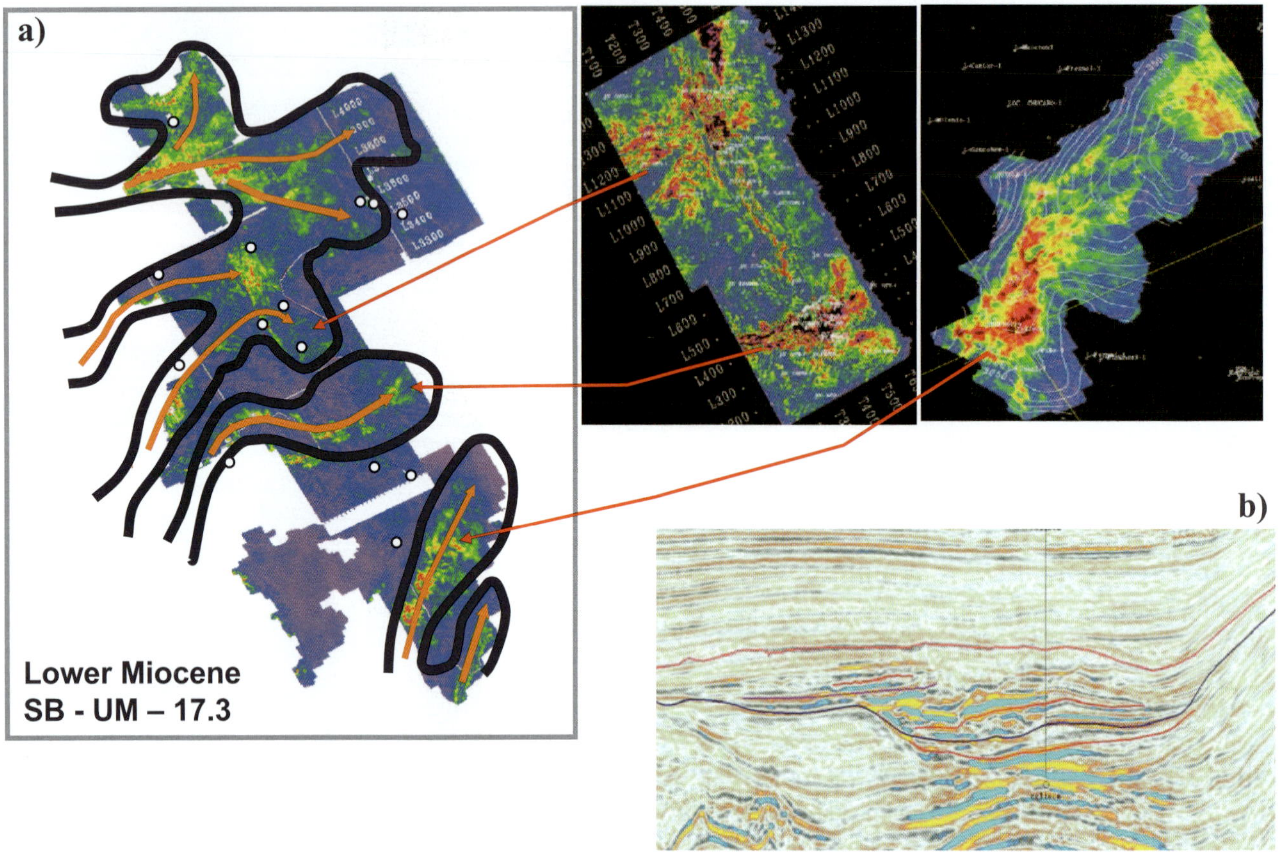

FIGURE 21. Canyon fill and base-of-slope fans. (a) Map of anomalies in sequence MI-17.3. The map shows how the sediment entered the basin from the west through at least six submarine canyons: two in the south (Novillero-Veinte-Mirador fields); one in the center, west of Cocuite field; and three to the north, forming the large fan northwest of Playuela field. (b, c) Amplitude maps indicating the lobate shape of these fans. (d) Seismic line showing the erosional nature of these canyons. SB = sequence boundary.

Sequence MM_16

The age range established for this sequence is 11.7 to 16.38 m.y. (Figure 6). This sequence seems to represent the last infill stage of the early Miocene canyons, nevertheless, it has not been studied in detail, therefore no illustrations were included to document its depositional framework and sandstone compositional characteristics.

Lithology

This sequence consists of interbedded conglomerates and sandstones. The conglomerates contain abundant metamorphic rock and quartz fragments and are of two types: (1) proximal conglomerates with clay matrix and (2) distal conglomerates with well-developed calcite and silica cements. Clasts are well rounded, and sorting is fair to poor. The sandstones of this sequence are classified as litharenites and subclassified as phyllarenites. Clearly, sediment provenance at this time was from metamorphic rocks.

Depositional Environment

This possibly represents the last period when canyons were truly active as major sediment conduits into the basin. Base-of-slope fans occur at the canyon mouths. Some canyons in the central part of the basin (Cocuite and Playuela fields) continued into the basin with large fans at their mouths and channel systems in the canyon fill.

Sequences LM_17.3 and LM_24.3

The age range established for sequence LM_17.3 is from 16.38 to 17.3 m.y. (Figure 6). Sequence LM_24 was deposited from 17.3 to 24.36 m.y. (Figure 5). The distribution of sequence MI-17.3 is shown in Figure 20, where these sequences can be distinguished in the distal part of the basin; however, they converge and join to the west. Together, these sequences are characterized by the presence of canyons along the western edge of the basin through which sediments were transported down to form large, conglomerate-dominated fans at the base of the slope, as well as on the basin floor (Figure 21).

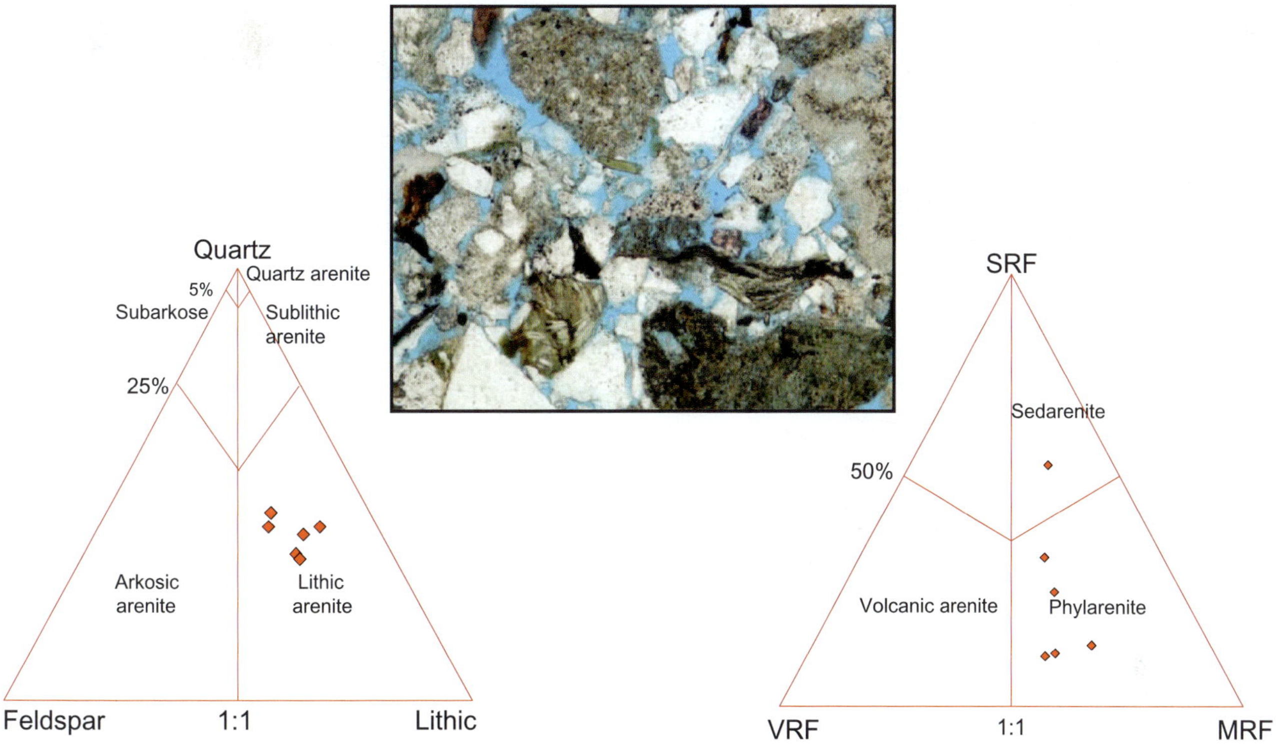

FIGURE 22. Sandstone composition of sequence MI-17.3. SRF = sedimentary rock fragments; VRF = volcanic rock fragments; MRF = metamorphic rock fragments.

Lithology

These sequences contain conglomerates interbedded with sandstones. The conglomerates have abundant calcareous clasts. Sandstones in this sequence are classified as litharenites, subclassified as sedarenites, with predominantly carbonate fragments (Figure 22). This suggests the possibility that the initial early Miocene sediment supply was derived from the erosion of Cretaceous carbonates in the Córdoba Platform west of the basin.

The data depicted in Figure 22 are from cores in Cocuite, Lizamba, Perdiz, and Apertura fields and the Cardúmenes and Chalpa wells. Figure 22 shows the QFL triangle and the LS-LV-LM triangle. Quartz and feldspar supply to the basin began during deposition of this sequence (Apertura, Chalpa, and Guinea wells), resulting in arkosic and sublithic arenites. The rocks from which these sediments are derived were volcanic, calcareous, and metamorphic.

Depositional Environment

The submarine canyons were fully developed at this time; i.e., they served as both sediment conduits and sediment sources. The fill of these canyons is composed of debris flows and base-of-slope fans.

Relative Reservoir Ranking

Although all intervals discussed here support gas production, the core analyses also allow a ranking of these intervals in terms of their porosity and permeability characteristics. Sands with the best reservoir quality occur in the LS_MS_5.7 sequence, where maxima in porosity and permeability are 34% and 2.2 d, respectively. The next best sequence is LS_MS_6.9 with maxima of 32% porosity and 795-md permeability. The best sequences that follow are sequence LS_MM_11.7, maxima of 29% porosity and 943-md permeability, and LS_MI_16.38, maxima of 23% porosity and 212-md permeability (Figure 23).

CONCLUSIONS

Gas production in the Tertiary Veracruz Basin is restricted to the five sequences analyzed here: LS-MS-5.7, LS-MS-6.9, LS-MS-9, LS-MM-11.7, and LS-MI-17.3. These sequences represent regional events relating all of the discovered reservoirs. Understanding their character and distribution is fundamental for exploration and production in the Tertiary Veracruz Basin.

Through Tertiary filling of the basin, a classic transition occurs from submarine fans on the continental

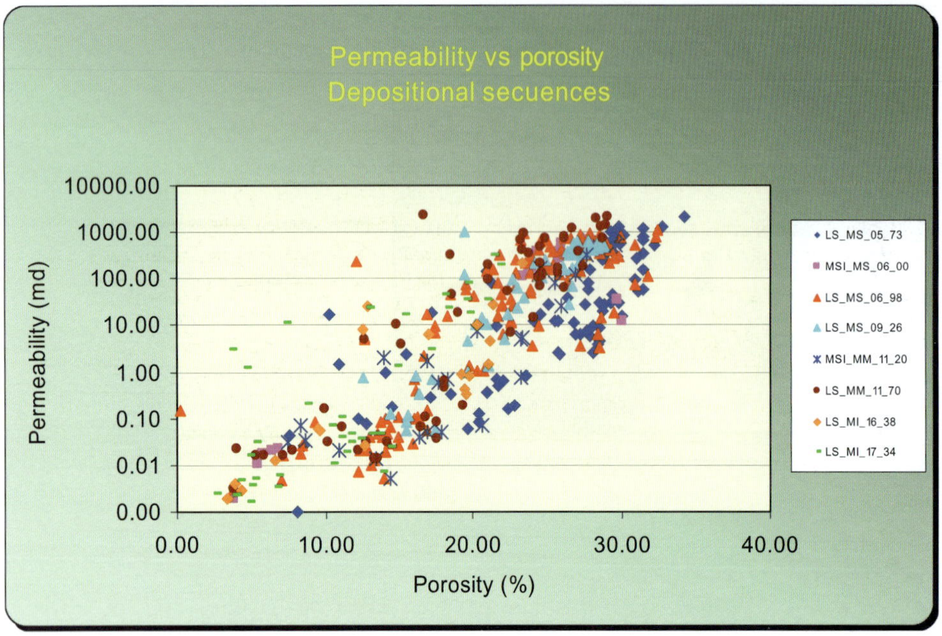

FIGURE 23. Graph of porosity vs. permeability for all the cores cut in the studied sequences showing that the shallowest cores have the best porosity and permeability. This indicate that one of the main controls on porosity in this part of the Tertiary Veracruz Basin is lithostatic load.

rise in the lower Miocene (sequence LS-MI-17.3) to multiply-sourced basin plain submarine fans through the middle to upper Miocene (sequences LS-MM-11.7, LS-MS-9, LS-MS-6.9) to meandering channel systems in the upper Miocene to Pliocene (sequences LS-MS-6.9 to LS-MS-PL-5.7).

The petrography presented here gives insight into the character and origin of the sediments within the five key stratigraphic intervals. Sandstones interbedded with conglomerates in LS-MI-17.3 represent the onset of quartz and feldspar delivery to the basin. Sediment sources for these sequences include volcanics, calcareous, and metamorphic rocks. Sediments from LS-MM-16 occur as interbedded conglomerates and litharenites and phyllarenites. Rock fragments in this sequence are metamorphic and quartz fragments suggesting metamorphic provenance. LS-MM-11.7 represents the time of maximum quartz and feldspar delivery, and subsequent lithologies are dominantly arkosic arenites and litharenites. Reservoir intervals within the LS-MS-9 sequence represent a transition between lithic arkoses and litharenites, because quartz and feldspar contents are higher relative to rock fragments contained in subsequent sequences. LS-MS-6.9 contains litharenites with near equal volcanic and limestone rock fragment content. LS-MS-5.7 contains litharenites with three dominant rock fragment types: limestones or dolomites and intermediate volcanic rocks. Carbonates and basaltic, intermediate, and acidic extrusives to the northwest represent the source for these fragments.

Detailed petrographical analyses of cores and thin sections permit the ranking of the best sandstone reservoirs in the basin. Although all of the intervals support production, the LS_MS_5.7 sequence contains sands with the best quality: maximum porosity of 34% and greater than 2-d permeability.

ACKNOWLEDGMENTS

We thank PEMEX Exploration and Production for allowing the publication of this chapter. We are indebted to Dr. Thomas Buerkert whose highly detailed review, constructive criticism, and re-organization of the original manuscript made possible this final version. We are grateful to Claudio Bartolini for his efforts in putting together this valuable volume.

REFERENCES CITED

Bolli, H. M., J. B. Saunders, and K. Perch-Nielsen, 1989, Plankton stratigraphy: Great Britain, Cambridge University Press, v. 1., 599 p.

Cruz-Helú, P., 1970, Estudio del Campo Mata Pionche, Formación Orizaba: Reporte Interno de Petróleos Mexicano (PEMEX), Inédito, 41 p.

De La Fuente-Navarro, J. M., 1959, Paleogeografía y distribución de los sedimentos Terciarios en la Cuenca de Veracruz: Boletín de la Asociación Mexicana de Geólogos Petroleros, v. 11, p. 21–50.

Delgado-Argote, L. A., M. López-Martínez, D. Cork, and C. M. Hall, 1992, Geologic framework and geochronology of ultramaphic complexes of southern Mexico: Canadian Journal of Earth Sciences, v. 29, p. 1590–1604.

Flores, R. J. V., 1970, Estudio estratigráfico del Jurásico Superior en la Sierra Madre Oriental, entre Teotitlán,

Oaxaca y Zongolica, Veracruz: Reporte Interno de Petróleos Mexicano (PEMEX), Inédito, 70 p.

Haq, B. U., J. Hardenbol, and P. R. Vail, 1987, Chronology of fluctuating sea levels since the Triassic (250 million years ago to present), Science, v. 235, no. 4793, p. 1156–1167.

Jacobo, A. J., and M. G. Murillo, 1989, Integración e interpretación petrogenética del basamento de México: Instiuto Mexicano del Petroleo, Informe Interno, 95 p.

Mena-Rojas, E., 1962, Geología y posibilidades petroleras del Jurásico Marino de la Región de Córdoba: Boletín de la Asociación Mexicana de Geólogos Petroleros, v. 14, p. 77–84.

Menes, L. S., 1977, Estudio sedimentológico del Jurásico Superior, área: Fortín-Jalapa-Zongolica: Reporte Interno de Petróleos Mexicano (PEMEX), Inédito, 26 p.

Menses De Gyves, J., 1953, Condiciones Estratigráficas de los sedimentos Terciarios en la Cuenca de Veracruz: Boletín de la Asociación Mexicana de Geólogos Petroleros, v. 5, p. 105–112.

Molina-Garza, R. S., R. Van der Voo, and J. Urrutia-Fucugauchi, 1992, Paleomagnetism of the Chiapas massif, southern México: Evidence for rotation of the Maya block and implications for the opening of the Gulf of Mexico: Geological Society of America Bulletin, v. 104, p. 1156–1168.

Olivas, M. R., 1953, Influencia del Macizo de Teziutlán sobre los sedimentos Mesozoicos y Terciarios, y posibilidades petroleras de estos: Reporte Interno de Boletín de la Asociación Mexicana de Geólogos Petroleros, 35 p.

Petróleos Mexicano (PEMEX), 1991, Evaluación Geológica regional de los sedimentos de la Cuenca Terciaria de Veracruz: Reporte Interno de PEMEX, Inédito, 43 p.

Petróleos Mexicano (PEMEX), 1995, Tectónica del cinturón plegado y cabalgado de Zongolica y de la Cuenca Terciaria de Veracruz: Reporte Interno de PEMEX, Inédito, 179 p.

Petróleos Mexicano (PEMEX), 1996, Excursión geológica al Cinturón Plegado y Cabalgado de Zongolica: Reporte Interno de PEMEX, Inédito, 42 p.

Petróleos Mexicano (PEMEX), Amoco, and Instituto Mexicano del Petróleo, 1995, Tertiary stratigraphy, basin evolution and economic potencial of the Veracruz Basin, east-central Mexico: Reporte Interno de PEMEX, Inédito, 78 p.

Rojas, L., and K. R. McClay, 1999, Tectonic evolution of the Zongolica fold-thrust belt and the Veracruz Basin, eastern Mexico: Master thesis, University of London, Royal Holloway, 75 p.

Viniegra, O. F., 1950, Breve Análisis Geológico de la llamada Cuenca de Veracruz: Boletín de la Asociación Mexicana de Geólogos Petroleros, v. 2, p. 281–290.

Gómez-Cabrera, P. T., and M. P. A. Jackson, 2009, Neogene stratigraphy and salt tectonics of the Santa Ana area, offshore Salina del Istmo Basin, southeastern Mexico, *in* C. Bartolini and J. R. Román Ramos, eds., Petroleum systems in the southern Gulf of Mexico: AAPG Memoir 90, p. 237–255.

Neogene Stratigraphy and Salt Tectonics of the Santa Ana Area, Offshore Salina del Istmo Basin, Southeastern Mexico

Pedro T. Gómez-Cabrera
Petróleos Mexicanos (PEMEX), Activo Integral Bellota-Jujo, Comalcalco, Tabasco, Mexico

Martin P. A. Jackson
Bureau of Economic Geology, Jackson School of Geosciences, University of Texas at Austin, Austin, Texas, U.S.A.

ABSTRACT

The Santa Ana area in the Bay of Campeche features a complex Neogene interaction of sedimentation, salt-related deformation, and regional tectonics. We synthesized this interaction using three-dimensional seismic data and logs and paleontological data from six wells. These provide a tectonostratigraphic framework for seven Neogene sequence-stratigraphic boundaries between 12.5 and 1.4 Ma. We analyzed these data using seven isochore maps, two restored cross sections, and a geohistory plot. The Neogene history begins with the Agua Dulce fold belt, which formed between 8.0 and 6 Ma. As the Agua Dulce folding waned, at about 6 Ma, extrusive salt blanketed much of the late Miocene fold belt. Sediment loading began to redistribute this allochthonous salt immediately after the discontinuous canopy was emplaced. Thin overburden sags may have originated as synclines during the waning stages of regional buckling, which also gently deformed the base of the salt canopy. After 3.8 Ma, the mean sedimentation rate more than doubled to 0.7 mm/yr (0.03 in./yr) as sediments became trapped against extensional, mostly counterregional, faults. Between 2.6 and 2.4 Ma, mean aggradation rates increased as much as sixfold to 4 mm/yr (0.16 in./yr) as major extension continued. A north block remained partly starved of sediment whereas depocenters formed proximally because of underlying salt evacuation. After 2.4 Ma, mean aggradation rates fell to 1 mm/yr (0.04 in./yr), probably because the shelf break crossed the study area. The boundary between thick distal sediment and thin proximal sediment

DOI:10.1306/13191086M9037

shifted northward as the shelf edge prograded. Some depocenters merged in depressions in counterregional hanging walls. After 1.4 Ma, the north and west blocks acted as one, tilting seaward and accommodating the thickest strata.

INTRODUCTION

The Salina del Istmo Basin in the Bay of Campeche in the southwestern Gulf of Mexico (Figure 1) is the southern equivalent of the northern Gulf of Mexico. However, much less has been published on salt tectonics in the southern gulf, especially in English. Most publications on the Salina del Istmo Basin focused onshore (e.g., Contreras and Castillon, 1960, 1968; Viniegra, 1971; Wilson, 1993). Even fewer publications describe the offshore part of the basin. García-Molina (1994) used onshore and offshore seismic data to study structural evolution in a broad area in southeastern Mexico and the adjoining Bay of Campeche. A Petróleos Mexicanos (PEMEX)-British Petroleum (1994) team was the first to study the Salina del Istmo Basin using modern concepts of salt tectonics, but this study has not been published, apart from a thesis (Oviedo-Perez, 1996).

The Salina del Istmo Basin has a complex tectonic history. The study area comprises (1) Pliocene–Pleistocene diapirs and minibasins, which underwent gravity-driven extension overlying (2) partly evacuated end-Miocene allochthonous salt sheets that extruded and partly coalesced into a canopy, which in turn overlie (3) orogenically driven Miocene fold belts.

Hydrocarbon exploration has great potential in the offshore Salina del Istmo Basin. On the shelf, oil plays are present in suprasalt settings, such as sand bodies pinching out landward in rollover anticlines in the hanging wall of counterregional faults or sand bodies pinching out near salt diapirs, and subsalt traps in the Miocene fold belts sealed by salt sheets.

The tectonic setting and regional salt tectonics, especially the Neogene salt-related structures, are described and interpreted in a companion chapter based on two-dimensional (2-D) seismic data (Gomez-Cabrera and Jackson, this volume). That chapter also reviews previous work.

The present chapter focuses on the Santa Ana area in the continental shelf of the Salina del Istmo Basin (Figure 1). Together with the companion chapter (Gomez-Cabrera and Jackson, this volume), this is to our knowledge the first detailed description of allochthonous salt tectonics in the southern Gulf of Mexico. In the present chapter, time-migrated three-dimensional (3-D) seismic data combined with logs and paleontological data from six wells were used to map seven horizons, create isochore maps, conduct structural restorations, and determine sedimentation rates. This allowed us to track deformation associated with salt evacuation. No wells in the study area reached salt, so we identified salt bodies by their 3-D seismic character using criteria listed by Gomez-Cabrera and Jackson (this volume).

SEQUENCE BOUNDARIES AND BIOSTRATIGRAPHY

Stratigraphic signatures and stacking patterns in strata result from the interaction of tectonics, eustasy, and sediment supply. Several models explain the stratal stacking patterns on seismic profiles. We followed the Exxon approach (Vail et al., 1977), placing sequence boundaries along unconformities created by fall of relative sea level. To analyze the salt-tectonic history, we based the chronostratigraphy on the 3-D seismic survey, then extrapolated this framework using 2-D seismic profiles to the larger study area described in the companion chapter. We interpreted and mapped seven Neogene sequence-stratigraphic boundaries in the 3-D seismic survey (Figure 2). The five oldest (12.5–2.4 Ma) were defined by biostratigraphic and seismic data. The five biostratigraphic horizons are *Globorotalia foshi lobata* (12.5 Ma), *Globorotalia juanai* (5.5 Ma), *Globigerina nepenthes* (3.8 Ma), *Globoquadrina altispira* (2.6 Ma), and *Globorotalia miocenica* (2.4 Ma). Planktonic foraminiferal markers supplied by PEMEX provided absolute ages, defined by the last appearance datum (LAD). The biostratigraphic horizons and the two shallower horizons were age adjusted to the chronostratigraphic chart of Wornardt and Vail (1993). Because the onshore Salina del Istmo Basin lacks reliably dated upper Pliocene and Pleistocene strata, we correlated the two shallowest sequence boundaries by seismic and well-log character.

PROGRADATION AND CHANGING SEDIMENTATION RATES

Figure 3 shows how shelf margins shifted seaward by about 100 km (60 mi) during the last 8 Ma (PEMEX-British Petroleum, 1994; Oviedo-Perez, 1996). Progradation quickened in the Pleistocene, and the shelf margin is now about 30 km (20 mi) seaward of the study area. The position of the shelf margin over time was a major control on the amount of shelf-derived sediment redeposited by slope processes. Because the shelf margin is a

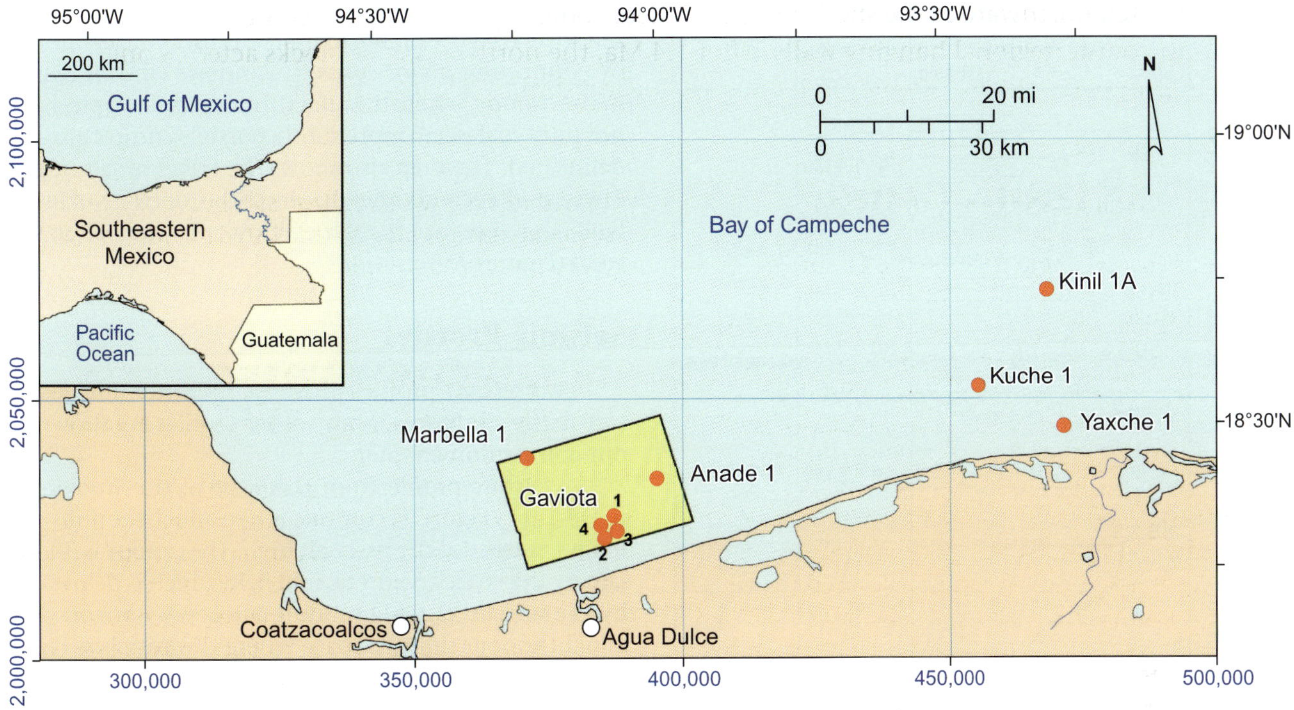

FIGURE 1. Location map of the 3-D seismic survey defining the study area (yellow) on the continental shelf of the southern Gulf of Mexico (inset map). Red dots show the well locations. Universal Transverse Mercator (UTM) projection.

locus of active sediment accumulation (Woodbury et al., 1973), increasing volumes of shelf-derived sediments are likely to have reached the study area as the shelf margin neared.

Table 1 shows the maximum, minimum, and mean interval thickness between each of the following horizons: 12.5, 5.5, 3.8, 2.6, 2.4, 1.9, and 1.4 Ma. Calculated aggradation rates for each time span are not corrected for compaction, so they are lower than the actual rates. Aggradation rates varied greatly, reflecting changing proximity to sediment sources, shifting depositional processes, and variations in accommodation created by extension or salt evacuation. The slower aggradation was above salt uplifts.

The mean accumulation rates in Figure 4 show that Pliocene and Pleistocene sedimentation was vigorous and on the order of 1 mm/yr or 1 km/Ma (0.04 in./yr or 0.62 mi/Ma). The deposition rate shows a spike in the late Pliocene, lasting from about 3.7 to 2.4 Ma. Because of this short duration, small inaccuracies in interpreting seismic horizons could lead to large errors in aggradation rates. Thus, the spike could be smaller or larger than estimated. If the spike is actually smaller than shown, then the aggradation rate in the bracketing time intervals would be higher, making a lower and broader spike. Even so, some kind of surge is indicated. This coincided with the early evacuation of the Sal Somera salt canopy (Gomez-Cabrera and Jackson, this volume). In the Geohistory section, we speculate on the causes for this surge in sedimentation rate.

STRUCTURAL GEOMETRY

Precanopy Fold Belt

Our companion chapter describes four fold belts in the offshore Salina del Istmo Basin. All have separate names, but we argue there that because of their similarities, these fold belts could be merely better imaged patches of one giant Chiapas-Campeche fold belt. The Agua Dulce fold belt underlies the present study area (Figure 5). Parallel reflectors suggest that strata older than about 8 Ma predated the rise of these particular anticlines because older reflectors within them are more or less parallel, suggesting deposition before folding began. The top of this prekinematic interval marks the onset of shortening. After that, sedimentary thicks infilled subsiding synclines, whereas the rising anticlines had thin synkinematic drape (left anticline in Figure 5) or were locally eroded (right anticline). The top of the synkinematic interval is slightly below the sequence boundary 5.5 Ma. Thus, the age of the shortening here is roughly 8–6 Ma. The overlying unit is postkinematic with respect to the shortening but includes thickness changes related to Pliocene–Pleistocene salt tectonics, as described below.

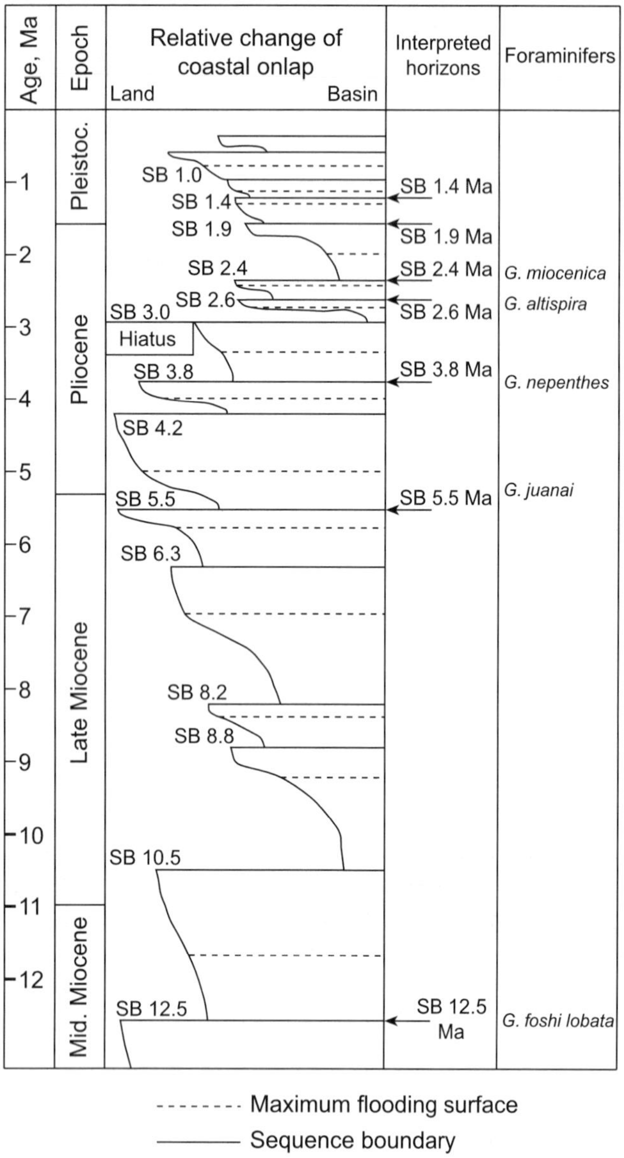

Figure 2. Chronostratigraphic chart showing onlap cycles, sequence boundaries (SB), maximum flooding surfaces, interpreted horizons, and paleontologic markers used in this study (modified from Wornardt and Vail, 1993).

The trend of these folds cannot be reliably estimated, as illustrated below in the section on isopach maps.

Top-Salt Structure

A depth map of the top of salt (Figure 6) shows the geometry of allochthonous salt, or equivalent salt weld, in this part of the Sal Somera canopy. The salt top ranges in depth from 1900 to 4900 m (6200 to 16,000 ft) below sea level. The canopy comprises at least three and possibly as many as six components, which range in major-axis length from 5 to 24 km (3 to 15 mi). Salt feeders presumably supplied each component of the canopy, but no feeders are imaged. Salt sutures join these individual salt sheets (Figure 6). We mapped sutures conservatively by connecting major cusps in windows (shown blank) in the canopy, where the allochthonous salt diapirs have not fully coalesced; more subtle possible sutures are also delineated. The map also shows the traces of salt ridges. These mark second-order diapirs in the footwalls of listric faults that sole into the salt canopy or its equivalent salt weld (Figures 7, 8).

Seismic Profiles

Two seismic profiles illustrate the structural geometry in the study area. Many other profiles are shown in our companion chapter.

A seismic profile from the west of the study area (CDP-1460, Figure 7) comprises two allochthonous salt bodies, sutured at the base-salt high. The canopy is underlain by the eroded Miocene fold belt, which is delineated by the subsalt 12.5-Ma (middle Miocene) horizon. The allochthonous salt fills space where the anticline crests would normally be. This preferential location is most likely a result of buckling folds tending to nucleate in the weak, thin, bulging stratal roofs of preexisting diapirs. Early diapirism cannot be documented within the study area because of poor seismic quality below salt, but precanopy passive diapirs are common in the surrounding area (companion chapter). Extension above the salt diapir created a complex fault pattern above the suture. We interpret a salt weld in the south, but this is not necessary to explain the geometry; instead, this southern area could comprise a single primary basin, resting on autochthonous salt.

Seismic profile CDP-2710 is near the Anade 1 well in the east of the study area (Figure 8). The seismic profile comprises three allochthonous salt bodies and their overburden, aged lowest Pliocene (5.5 Ma) to Holocene. The allochthonous salt occupies what would normally have been the crest of anticlines in the Miocene fold belt. Thus, the diapiric feeders for these sheets are possibly situated in the anticlines, as in the other profile. Two landward-dipping faults sole into the allochthonous salt bodies in the south. We interpret a salt weld connecting two salt sheets, but this is not necessary to explain the geometry, instead this central area could comprise a single primary basin, resting on autochthonous salt.

STRUCTURAL EVOLUTION

Restorations

The dip-oriented seismic sections (Figures 7, 8) were depth converted and structurally restored using Midland Valley's software 2DMove (Figures 9–11). We describe the results forward in time to match how the

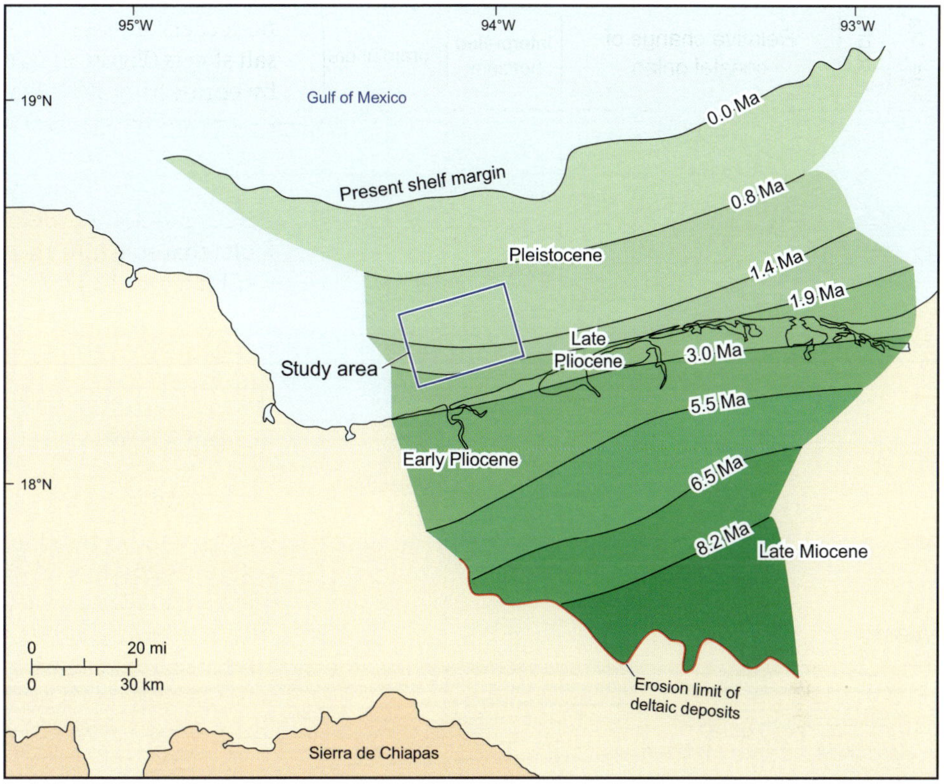

FIGURE 3. Schematic map of the prograding shelf margin since the late Miocene (after Oviedo-Perez, 1996). The shelf margin prograded through the study area between 2 and 1 Ma and is now about 30 km (19 mi) seaward of there. Universal Transverse Mercator (UTM) projection.

structures evolved. Restorations assumed negligible deformation of subsalt strata during post–salt time. This simplification was adopted because the subsalt structure is so poorly imaged. Another assumption is the bathymetry of the paleoslope. The present-day slope is so gentle (~0.5° dip) that one could assume that the paleoslope was horizontal. However, even small variations in slope greatly alter the inferred welding history of thin salt sheets. So we assumed both dipping and horizontal paleoslopes. The restorations suggest that the total salt area in the cross sections decreased through time by 52 and 67%, respectively. Such a decrease could be explained by salt dissolution before burial and by export of salt either out of plane or down regional slope after burial (as has been assumed in restorations from the northern Gulf of Mexico) (e.g., Worrall and Snelson, 1989; Diegel and Cook, 1990; Seni, 1992; McBride et al., 1998).

For the western profile (CDP-1460), the structural restoration at 5.5 Ma (Figure 9a) shows thin wedges of overburden that partly covered a canopy of coalesced salt glaciers above the Miocene fold belt. The extrusive salt may have also been covered by a veneer of sediment below seismic resolution. Thickness variations in the uppermost Miocene sediments (Figure 9a) and their onlap against the salt (Figure 7) show that the allochthonous salt was mobile and had bathymetric relief immediately

TABLE 1. Sediment accumulation rates for the Neogene sediments in the Santa Ana area in southeastern Mexico.

Interval Age (Ma)	*Time Span (m.y.)*	*Sediment Isopach (m)*			*Accumulation Rate (mm/yr or km/Ma)*		
		Maximum	*Minimum*	*Mean*	*Maximum*	*Minimum*	*Mean*
1.4–0	1.4	2200	200	1200	1.57	0.14	0.86
1.9–1.4	0.5	700	50	375	1.52	0.11	0.82
2.4–1.9	0.5	1050	150	600	1.94	0.28	1.11
2.6–2.4	0.2	1500	300	900	7.50	1.50	4.50
3.8–2.6	1.2	1400	350	875	1.17	0.29	0.73
5.5–3.8	1.7	1050	50	550	0.62	0.03	0.32
12.5–5.5	7.0	1800	500	1150	0.26	0.07	0.16

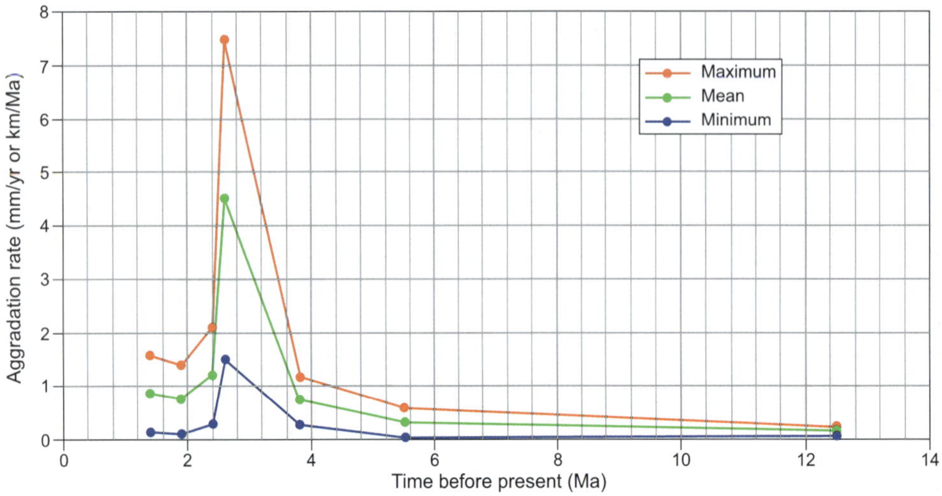

FIGURE 4. Sediment accumulation rates since the middle Miocene based on isochore mapping of 3-D seismic data. Note the massive surge in sediment aggradation between 2.4 and 2.6 Ma (see Table 1).

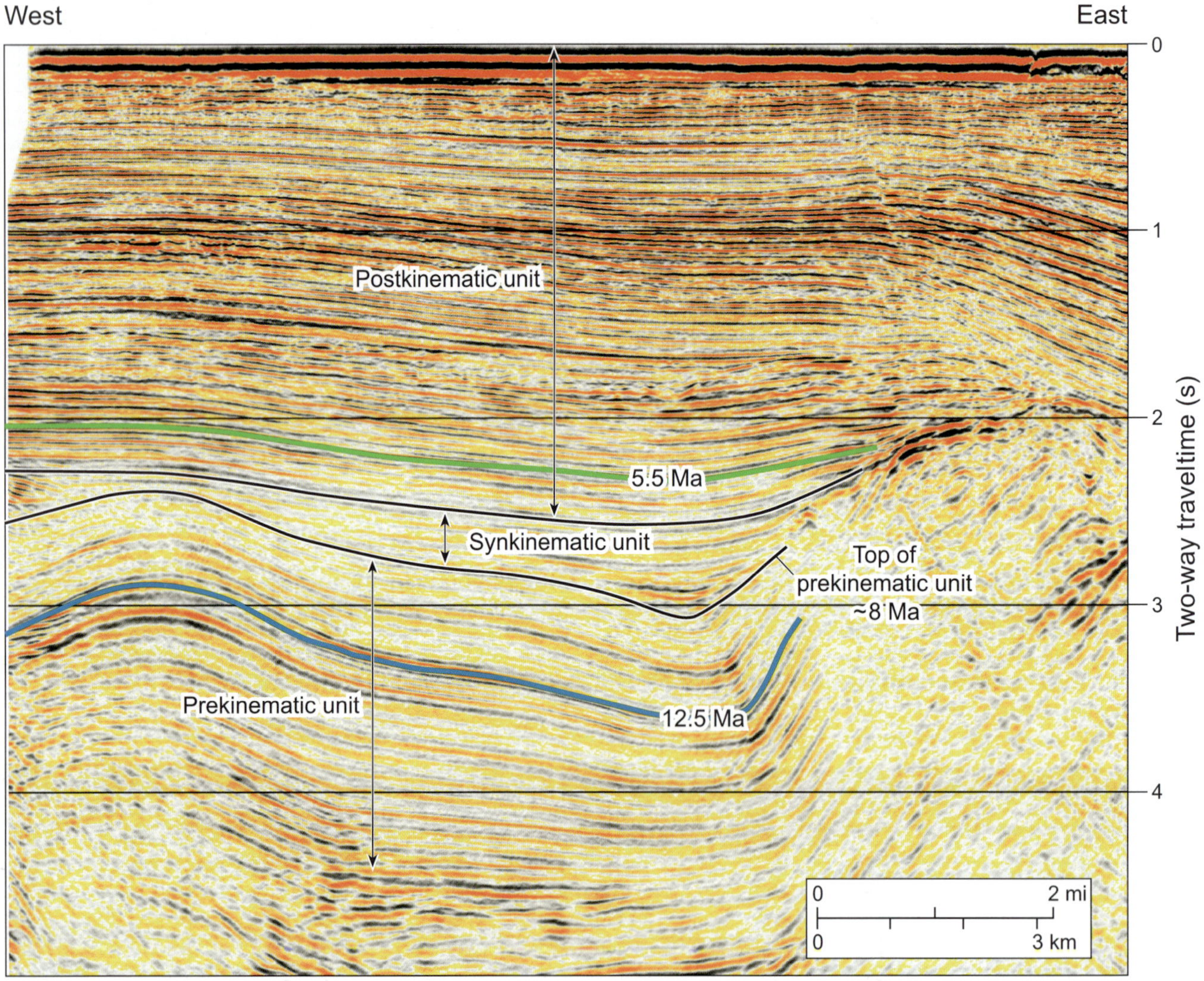

FIGURE 5. Two-dimensional time-migrated seismic profile showing the synkinematic unit recording the time of folding in the Agua Dulce fold belt between 8 and 6 Ma. This unit is underlain by a prekinematic unit having parallel reflectors. The lower part of the postkinematic unit onlapped the eroded crest of the eastern anticline. The location of the profile is shown in Figure 12.

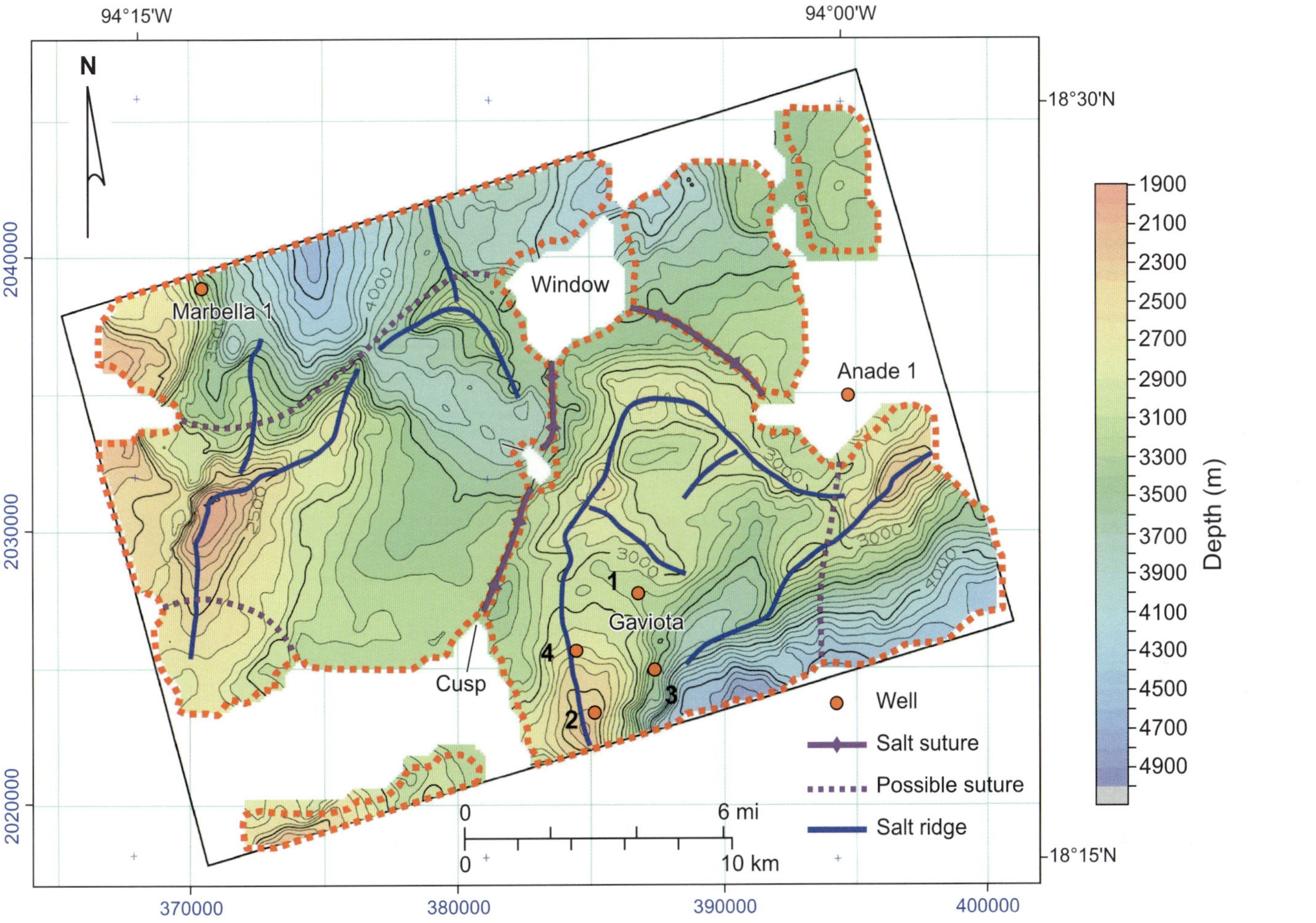

FIGURE 6. Structural depth map of the top of the allochthonous salt canopy, or equivalent weld, in the Santa Ana area. The relief of the salt, cusps in its margin, and salt-free windows were used to define salt sutures where salt stocks coalesced to form the canopy. Blank areas are mostly windows in the original canopy, but some may be completely welded parts of the canopy. The depth contour interval is 100 m (330 ft). Because of the structural complexity and dense faulting, small faults with a throw less than 100 ms were disregarded. Universal Transverse Mercator (UTM) projection.

after the canopy was emplaced and before 5.5 Ma. Differential sedimentary loading would have displaced underlying salt. Because these unconsolidated sediments were likely less dense than salt, the origin of these suprasalt minibasins is problematic. One possibility is that these sags were synclines formed by gentle buckling during the waning stages of Agua Dulce folding, which also folded at least the base of the salt canopy. As the sags formed, salt would have flowed from beneath the growing minibasins and supplied the intervening passive diapirs, which formed structural highs near the sediment-water boundary (Figure 9a–c).

At 3.8 Ma (Figure 9b), sediments continued to prograde across and onlap against the displaced salt. By 2.6 Ma, a relic of downbuilding salt remained exposed at the sea floor. A small depocenter near this central diapir formed by expulsion of salt as a rollover syncline (sensu Ge et al., 1997). Major subsidence at the landward (south) side of the cross section at 2.6 Ma (Figure 9c) may have begun forming a salt weld. This speculative weld is not clear in the seismic profile (Figure 7). The former existence of the salt sheet here is compatible with the restoration, which assumed that paleobathymetry was flat. If, instead, the sea floor sagged in the south at 5.5 Ma above a fold-belt syncline, the restoration would not need salt here, so no salt weld would have formed there. Between 2.6 and 2.4 Ma, the restoration shows the base of salt in the south subsiding abruptly by 1 km (0.62 ft). This subsidence is constrained by the simplistic assumption of horizontal paleobathymetry. Reconstructing the correct paleobathymetry is especially needed if the timing of this weld is important for inferring the hydrocarbon charge history above the canopy.

At 2.4 Ma (Figure 9d), a large influx of sediment rapidly buried the emergent salt, and a crestal graben formed above the central diapir. The normal faults converge downward to the tip of the diapir, and the overlying strata are not arched, indicating that the buried

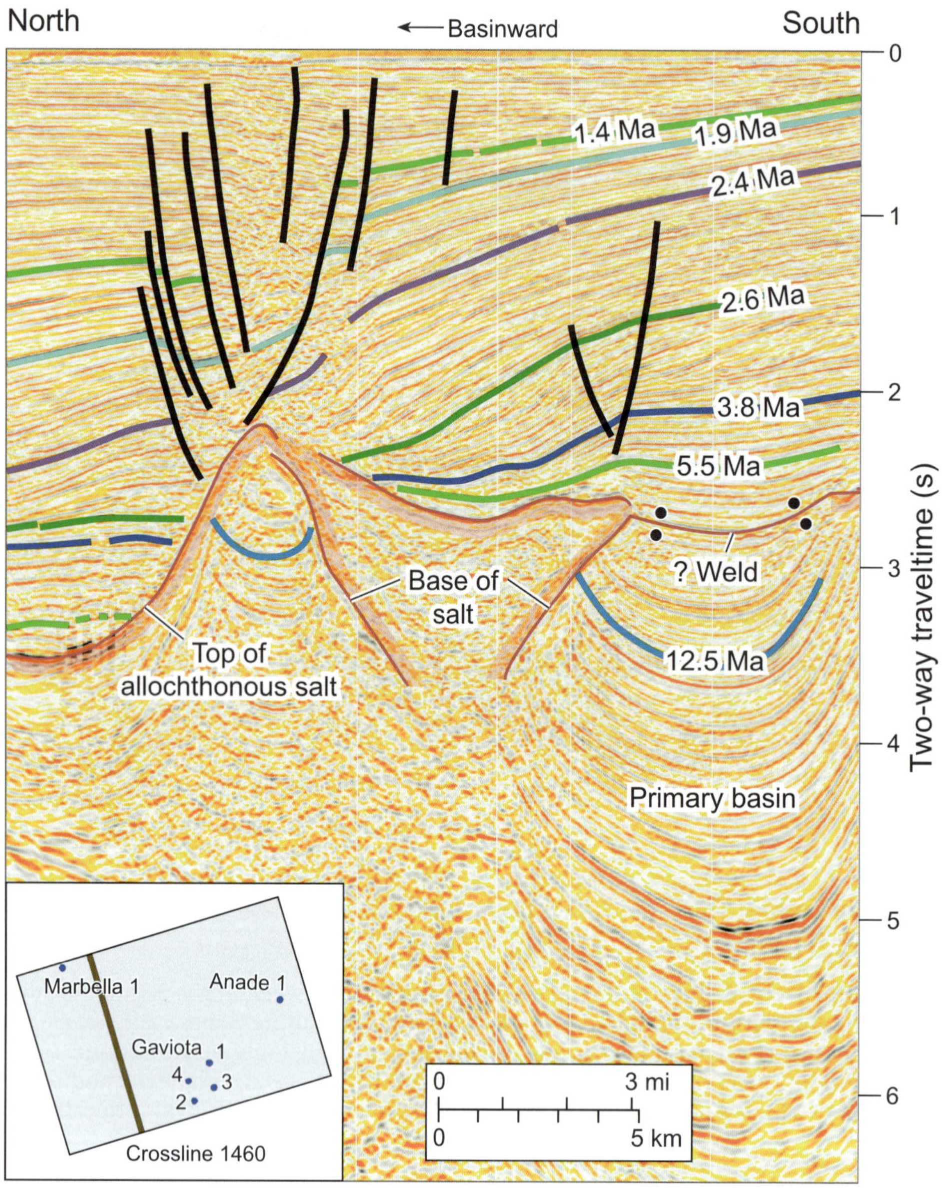

FIGURE 7. Three-dimensional seismic profile, common depth point (CDP) 1460, restored in Figure 9. Based on suprasalt sediment velocities, vertical exaggeration is about 3:1.

passive diapir became a reactive diapir (Vendeville and Jackson, 1992a) during a regional extension on the upper slope.

By 1.9 Ma (Figure 9e), continued regional extension and salt evacuation converted the reactive diapir into a falling diapir (Vendeville and Jackson, 1992b). The landward-dipping fault above its crest dominated the next stage at 1.4 Ma (Figure 9f).

Between 1.4 Ma and the present (Figure 9g), basinward-dipping faults above the falling diapir dominated as the half graben evolved into a graben. As the underlying salt evacuated, the basinward minibasin continued to subside markedly as it slid northward. The regional extension continued into the late Pleistocene (Figure 7), although the shelf margin had prograded seaward of the study area.

We restored the eastern profile in two ways to demonstrate the effect of different assumptions for the paleobathymetry. The first restoration (Figure 10) assumed a seaward slope of 3°. To honor the observed local thickness variations in the overburden, we distorted subsalt strata slightly between restoration stages, thus simulating continued deformation of the subsalt strata during burial by younger overburden.

The first stage in this restoration (Figure 10a) shows an extensive allochthonous salt canopy capped by a sedimentary veneer equivalent in age to the Miocene–Pliocene boundary (5.5 Ma). Small sags formed above the salt canopy, probably separated by condensed sections. Differential sedimentary loading by these sags could have started salt flow (the unconsolidated sediments are likely to have been less dense than salt, thus

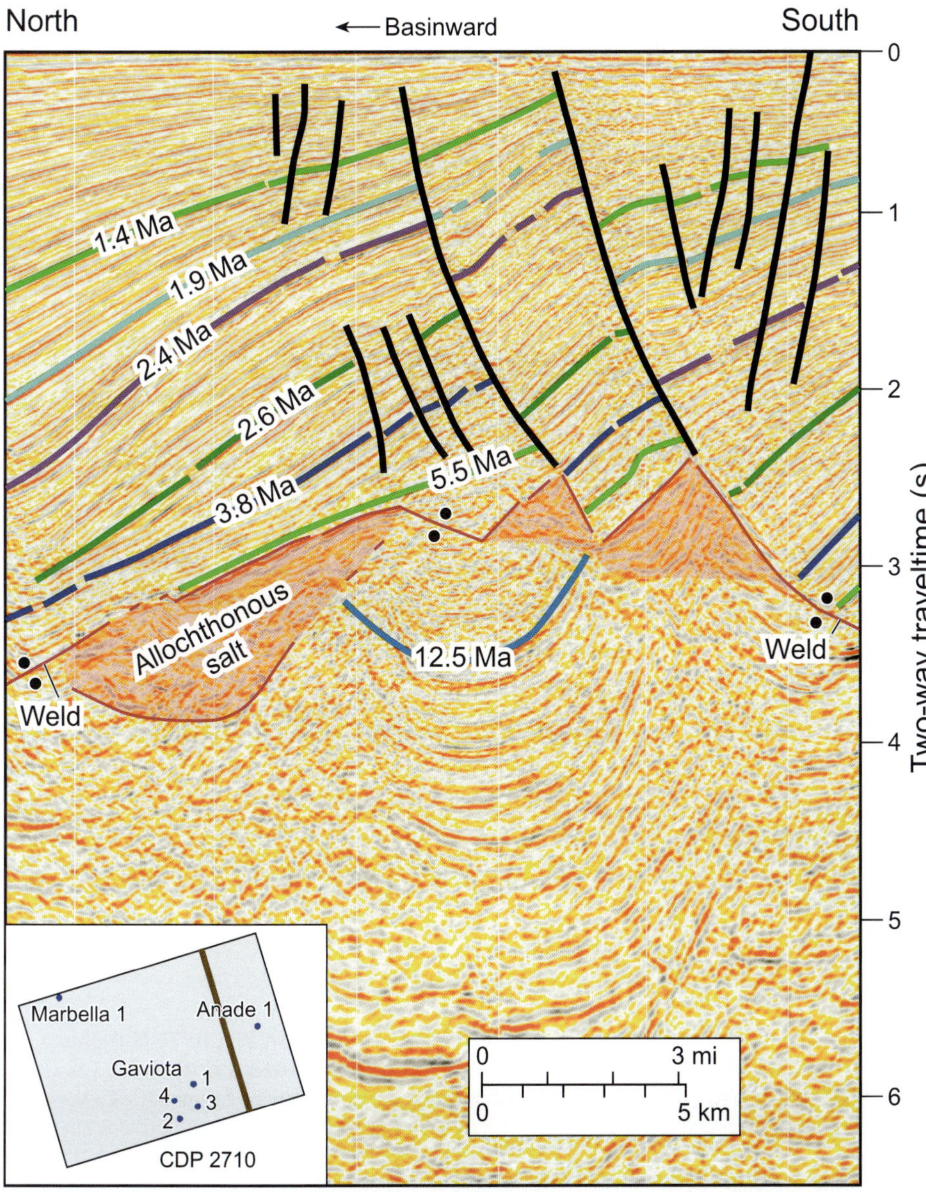

FIGURE 8. Three-dimensional seismic profile, common depth point (CDP) 2710, used for alternative structural restorations in Figures 10 and 11. Based on suprasalt sediment velocities, vertical exaggeration is about 3:1.

eliminating buoyancy as a trigger). These sags may have formed by gentle buckling of the canopy veneer during the waning stages of the Agua Dulce folding, as in the other restored section (Figure 9). If so, their dominant wavelength could be smaller than that of the underlying Agua Dulce fold belt, influenced by the thickness and strength of three mechanical units being buckled: a strong subsalt clastic package, the weaker salt canopy, and the mostly unconsolidated suprasalt veneer.

From 5.5 to 3.8 Ma (Figure 10b), sediments of uniform thickness blanketed the allochthonous salt. At 2.6 Ma (Figure 10c), major extension began as two landward-dipping faults detached on the allochthonous salt sheet. As restored, these are normal faults, not welds, and so must involve extension. Growth faulting continued at 2.4 Ma (Figure 10d) while parts of the canopy began to weld.

Between 1.9 and 1.4 Ma (Figure 10e, f), units thickened northward as the area transitioned from the upper continental slope to the shelf. By the present day (Figure 10g), the salt bodies stopped affecting sedimentation patterns, although faulting continued.

An alternative restoration assumed a flat sea floor and rigid subsalt (Figure 11). With this assumption, problems arise at the seaward end. Salt was gradually expelled from the sheet, originally up to 3 km (1.8 mi) thick as opposed to 2 km (1.2 mi) thick in the previous restoration of the same profile. This salt evacuation accords with the rest of the cross section until after 1.4 Ma when the 1-km (0.62-mi) thickness of salt was expelled in the basinward end (Figure 11f, g). This style of salt evacuation seems implausible because we would expect that salt evacuation this large would be recorded by local thickness changes above the junction of the salt weld

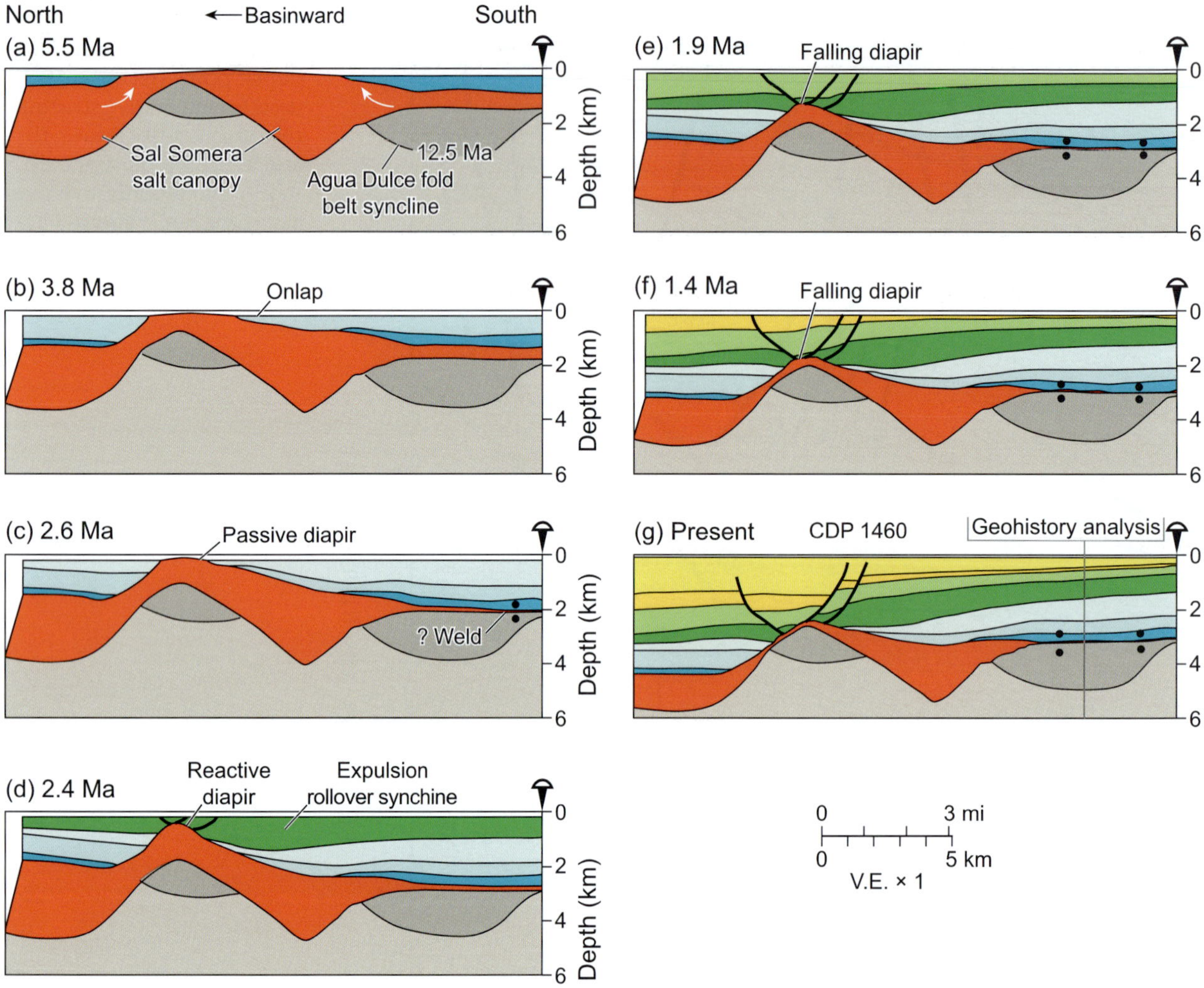

FIGURE 9. Sequential restoration of a depth section (g) calculated from the seismic profile in Figure 7. The oldest restoration is the base of the Pliocene (a). For simplicity, each stage of the restoration assumed a horizontal sea floor and rigid subsalt strata. No vertical exaggeration is present. The geohistory analysis is shown in Figure 19. CDP = common depth point; V.E. = vertical exaggeration.

and canopy, instead, local thickness changes ended before 2.4 Ma.

Isochore Maps

Because isochore maps record paleohighs and paleolows, they reveal much about tectonics during the mapped time interval. Because the upper and lower bounding horizons for each isochore differ in shape, each isochore map covers less area than either of its two bounding horizons. The isochore mapping was not corrected for the effects of compaction. Letters mark the main depocenters to identify them from map to map.

12.5- to 5.5-Ma Isochore

This upper Miocene interval below the salt canopy is the oldest mapped (Figure 12). During this time, regional shortening began and ended in most of the study area. Because the mapped area is small, the regional trend of the Agua Dulce fold belt is ambiguous and could be northeast, north, or northwest.

5.5- to 3.8-Ma Isochore

This lower Pliocene interval is the oldest above the allochthonous salt (Figure 13). The interval ranges between 50 and 1050 m (160 and 3400 ft) in thickness and averages about 550 m (1800 ft). These thicknesses give minimum, maximum, and mean aggradation rates of 0.03, 0.6, and 0.3 mm/yr (0.001, 0.02, and 0.01 in./yr), respectively (Table 1). The map also shows an arcuate array of en echelon faults, labeled as the northern, central, and southern faults, which divide the study area into three main blocks. All three main faults dip landward. At this stage, the northern fault was probably emergent salt (Figure 9, restoration), which evolved into a fault

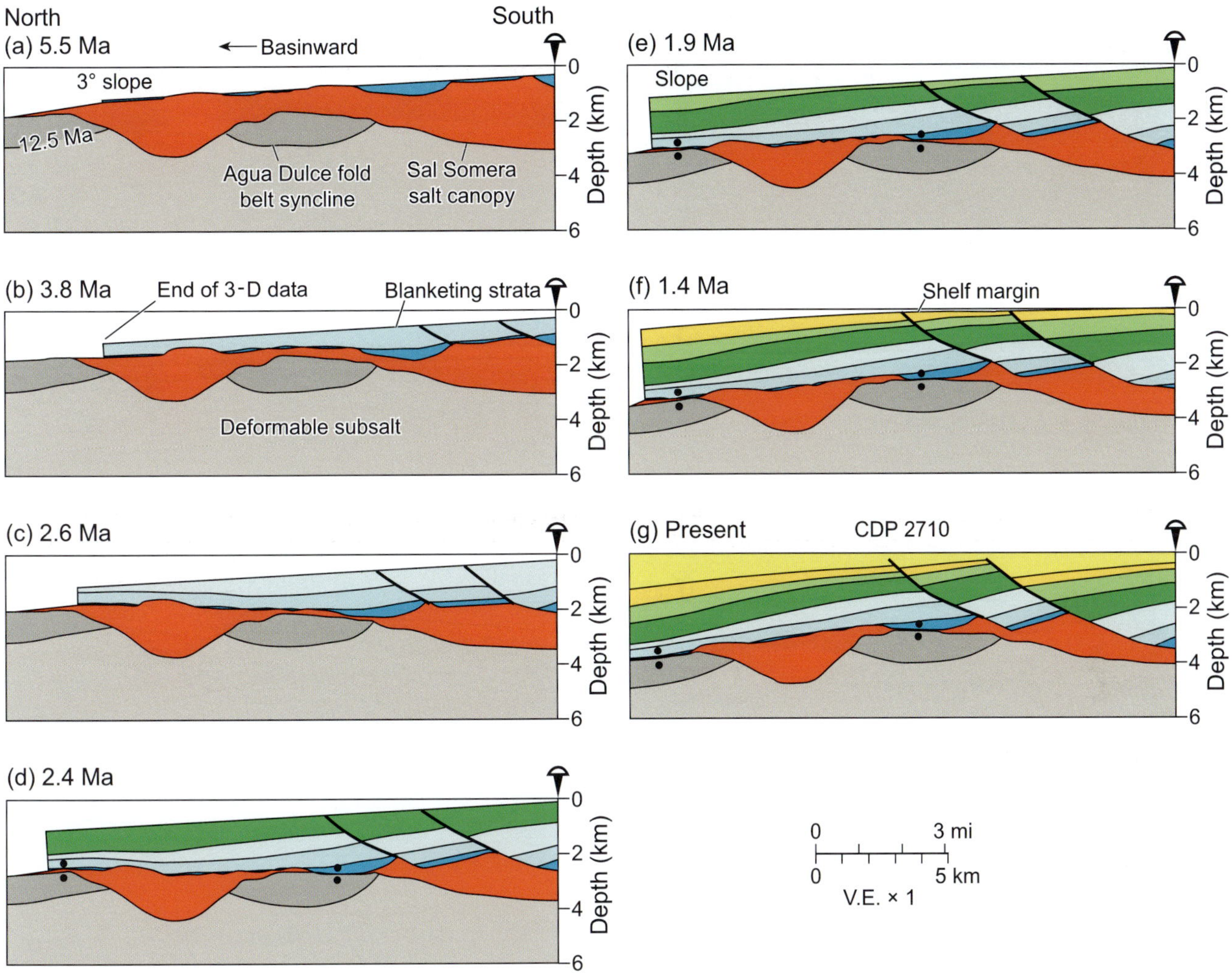

FIGURE 10. Sequential restoration of a depth section (g) calculated from the seismic profile in Figure 8. The oldest stage of the restoration is the base of the Pliocene (a). The restoration assumes a gently dipping sea floor on the former continental slope. Maintaining compatibility between salt thickness and overlying stratal thickness variations required slight deformation of subsalt strata. Compare with the restoration in Figure 11. No vertical exaggeration is present. CDP = common depth point; V.E. = vertical exaggeration.

only after the canopy was buried. Data gaps in each fault result from either inadequate seismic data or later extension. No evidence of these extensional growth faults starting to slip during the early Pliocene is observed, but this early onset is possible because (1) these strata rest unstably on mobile allochthonous salt and (2) the study area was on the upper continental slope, whose seaward dip provided a gravitational instability for regional extension.

Early evacuation of allochthonous salt seems to have created all the depocenters. The main lower Pliocene depocenter, in the southeast near the Anade 1 well, has a maximum thickness of 1050 m (3400 ft) and is underlain by allochthonous salt. Younger faulting divides it into two (C and D). The other depocenters (A, B, E, F, and G) are thinner. Depocenter A may be a primary basin, as discussed in Figures 7 and 9. The Gaviota 4 well area may be the result of nondeposition over a paleohigh.

3.8- to 2.6-Ma Isochore

Mean sedimentation rates more than doubled. Minimum, maximum, and mean sediment accumulation rates are 0.3, 1.2, and 0.7 mm/yr (0.01, 0.05, and 0.03 in./yr), respectively (Table 1). Only five depocenters (B, C, E, F, and G) are evident (Figure 14). Depocenters A and D disappeared, suggesting that the underlying salt had been completely expelled.

Extensional growth faulting, which may have started earlier, cut the study area into the three blocks: north, west, and south. The main sediment catchment was the west block, comprising depocenters B and C, trapped in the landward hanging walls of the northern and central faults. The north block in the lee of the three faults received only thin sediments (depocenters E and F). Depocenter G occupies the hanging wall of the landward-dipping southern fault.

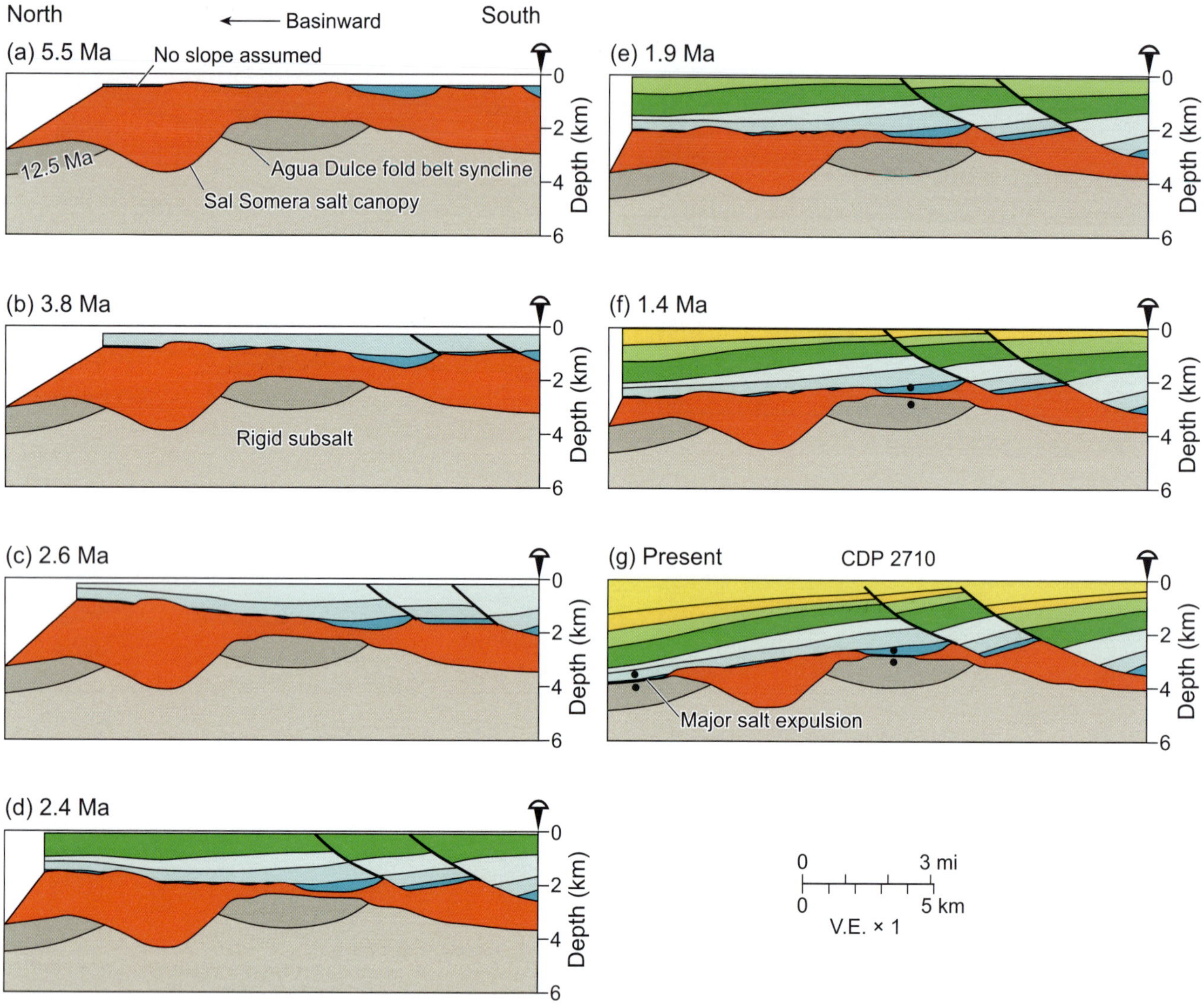

FIGURE 11. Sequential restoration of a depth section (g) calculated from the seismic profile in Figure 8. The oldest stage of the restoration is the base of the Pliocene (a). Each stage of the restoration assumed a horizontal sea floor and rigid subsalt strata. Compare with the restoration in Figure 10, which assumed a gently dipping paleoslope. No vertical exaggeration is present. CDP = common depth point; V.E. = vertical exaggeration.

Depocenters were also divided by allochthonous salt. In the west, horizons 3.8 and 2.6 Ma onlap a formerly extrusive salt body that separated the north and west blocks. Cutoffs of adjoining subsalt strata suggest that this salt extruded both northward and southward.

2.6- to 2.4-Ma Isochore

Mean aggradation rates increased to perhaps six times the previous isochore intervals. Estimated rates were 1.5, 7.5, and 4.5 mm/yr (0.06, 0.3, and 0.2 in./yr) for minimum, maximum, and mean values, respectively (Table 1). Despite accelerated sedimentation, the pattern of the 2.6–2.4-Ma isochore interval is similar to that of the underlying interval (compare Figures 14 and 15). The north block (depocenters E and F) remained sediment starved because of competition from depocenters B and C in the west block. These formed by salt evacuation in the hanging wall of counterregional faults and the landward flank of emergent allochthonous diapirs.

2.4- to 1.9-Ma Isochore

Mean aggradation rates fell to one quarter or 1.1 mm/yr (0.04 in./yr) (Figure 16). A dashed red line shows the strike-parallel boundary between areas of thin proximal sediment and thick distal sediment. Thin sediments overlie older isochore thicks on the 2.6- to 2.4-Ma isochore map, especially former depocenter C (Figure 15). We attribute this thinning to the fourfold drop in sedimentation rate as the shelf break crossed the southern boundary of the study area (Figure 3).

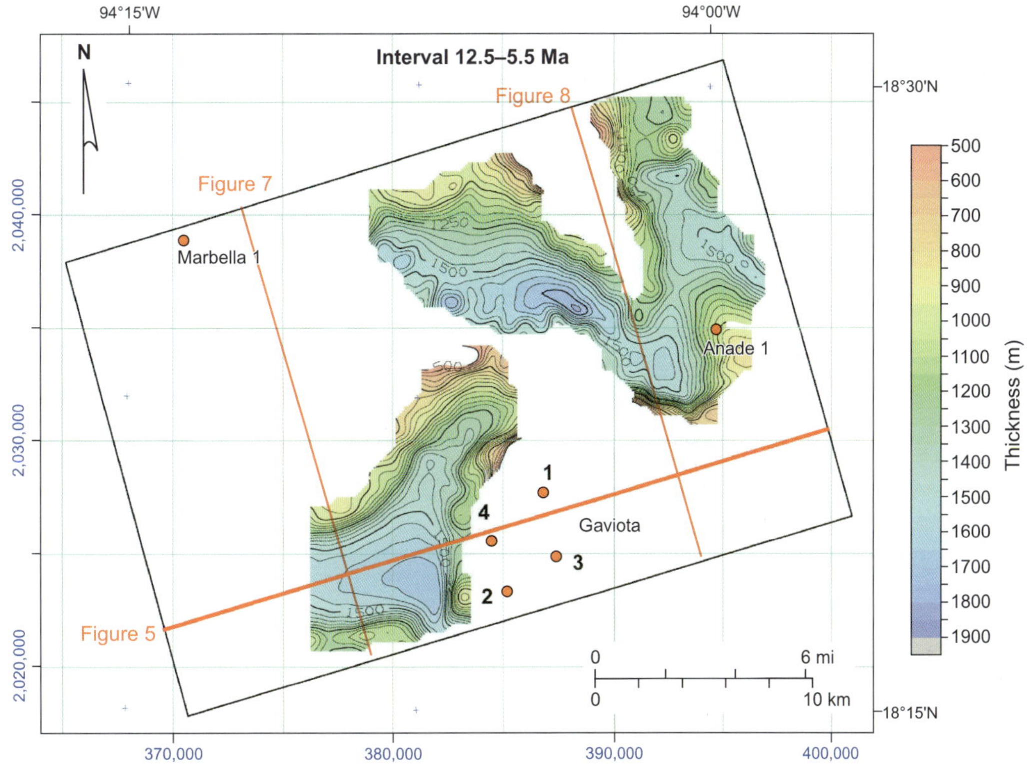

FIGURE 12. Isochore map for the middle–upper Miocene 12.5–5.5-Ma interval. The limited coverage of the map is because of poor coincidence between the map areas of upper and lower bounding surfaces. The contour interval is 50 m (160 ft). Universal Transverse Mercator (UTM) projection.

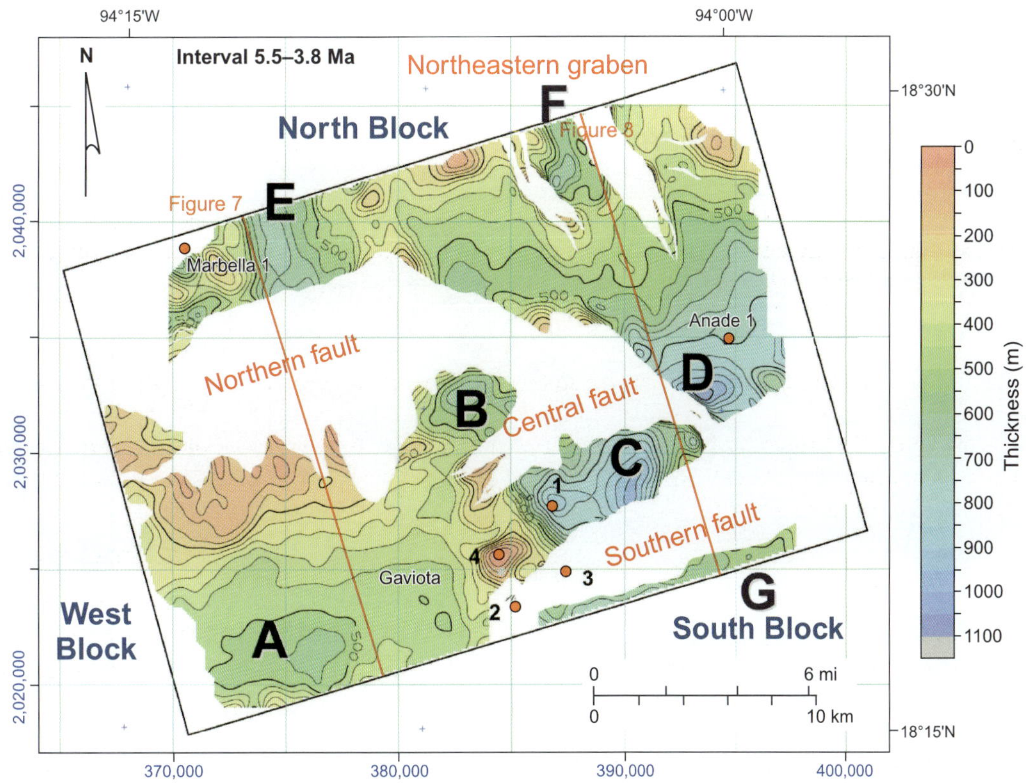

FIGURE 13. Isochore map of the lower Pliocene 5.5–3.8 Ma-interval. Younger faulting separated depocenters C and D. The contour interval is 50 m (160 ft). Letters A–G represent depocenters or minibasins. Universal Transverse Mercator (UTM) projection.

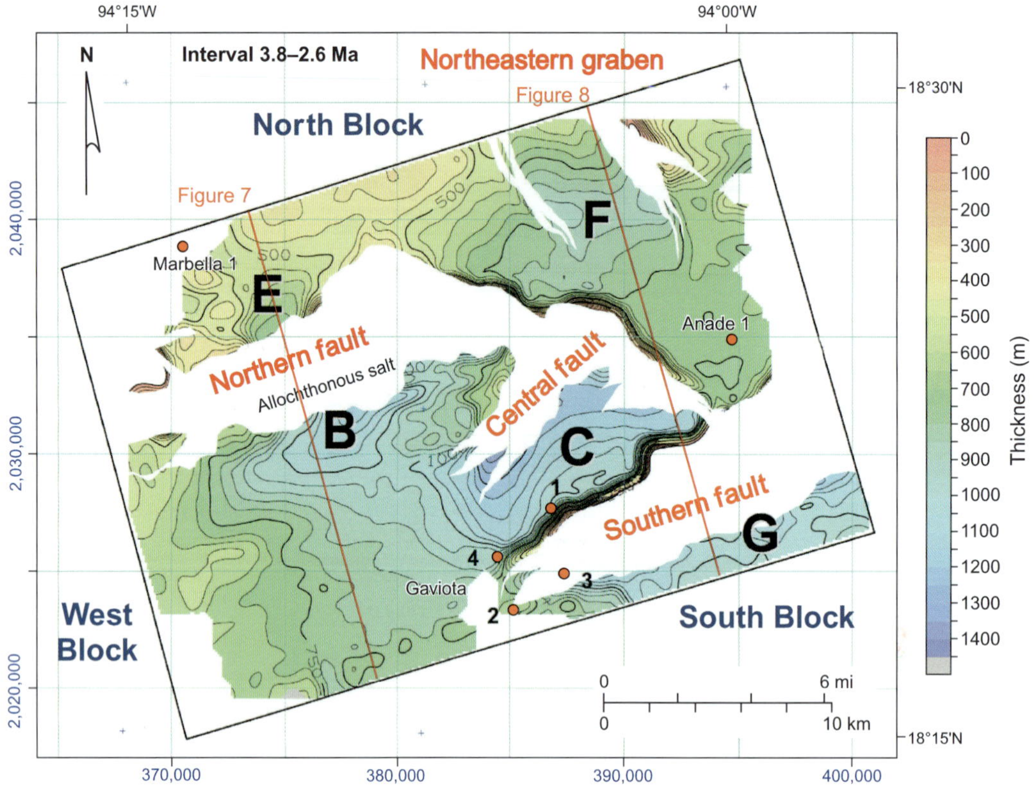

FIGURE 14. Isochore map of the upper Pliocene 3.8–2.6-Ma interval. The contour interval is 50 m (160 ft). Letters B, C, E, F, and G represent depocenters. Universal Transverse Mercator (UTM) projection.

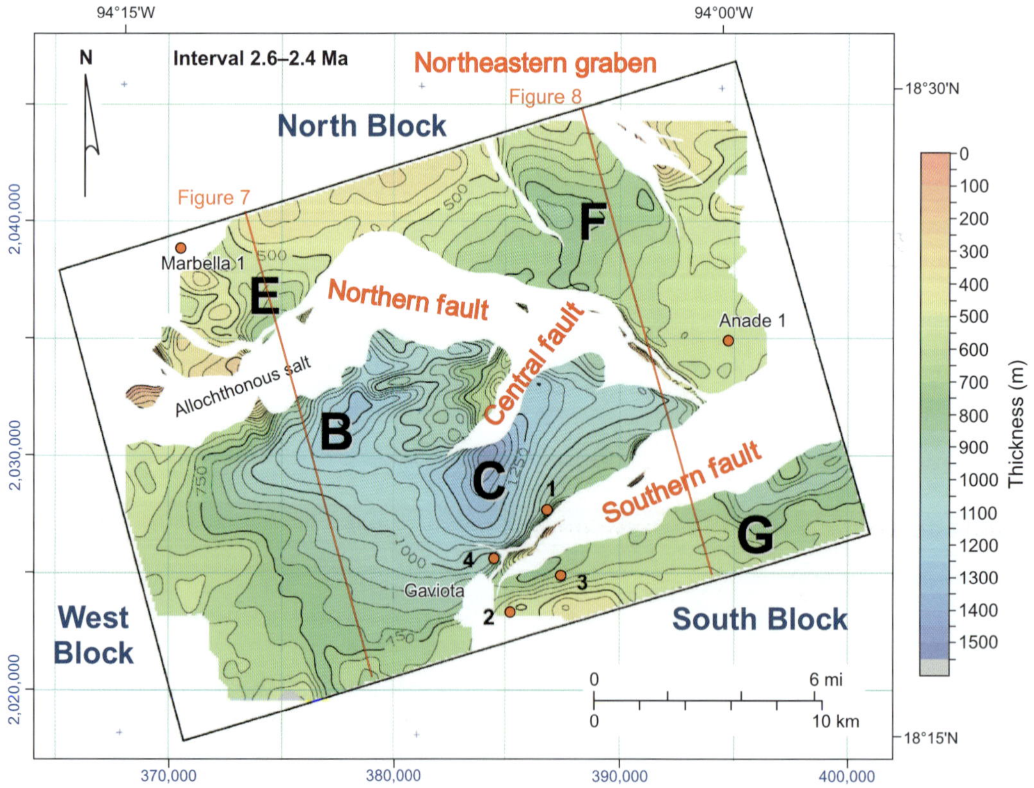

FIGURE 15. Isochore map of the upper Pliocene 2.6–2.4-Ma interval. The contour interval is 50 m (160 ft). Letters B, C, E, F, and G represent depocenters. Universal Transverse Mercator (UTM) projection.

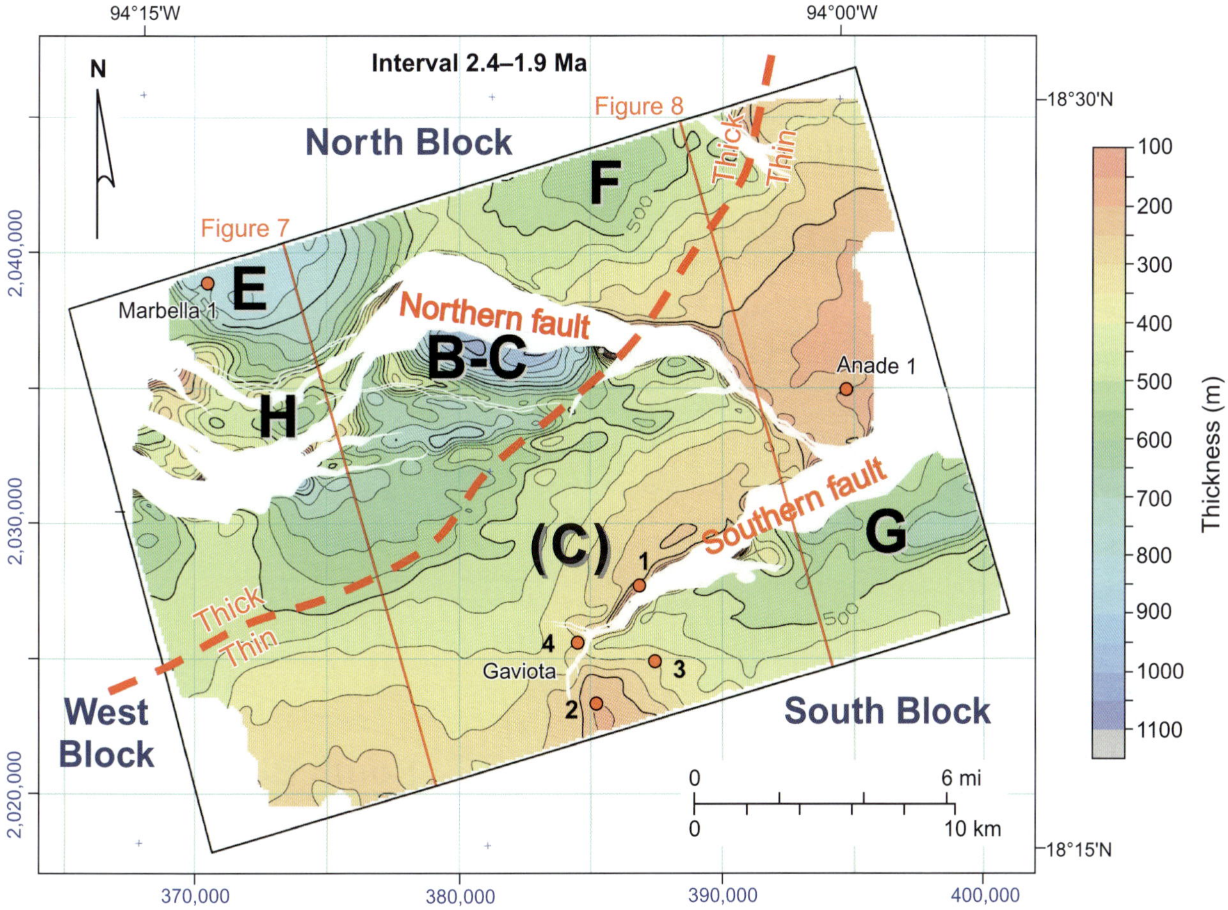

FIGURE 16. Isochore map of the upper Pliocene 2.4–1.9-Ma interval. The contour interval is 50 m (160 ft). Letters B, C, E, F, G, and H represent depocenters. The position of former depocenter C is shown in parentheses. Universal Transverse Mercator (UTM) projection.

Depocenters B and C crowded together in a depression in the hanging wall of the northern fault. The west end of the northern fault splayed as a new depocenter, H, formed in the crestal graben above the salt glacier that formerly separated the north and west blocks. Depocenter E, around the Marbella 1 well, was thicker and wider than previously as salt evacuation accommodated prograding sediments. The south block maintained thicknesses and widened because of the landward fault dip.

1.9- to 1.4-Ma Isochore

Aggradation rates continued to fall slightly during the early Pleistocene: rates were 0.1, 1.5, and 0.8 mm/yr (0.004, 0.06, and 0.03 in./yr) for minimum, maximum, and mean rates, respectively. Sedimentation was concentrated in depocenters B and C in a bend in the hanging wall of the northern counterregional fault (Figure 17). Depocenter H merged with depocenters B and C as the northern fault continued to splay. Most depocenters did not shift laterally since the previous time interval.

The central part of the northern fault was the most active, as shown by large isochore differences on each side of the fault. However, the overall thicknesses on each side of the fault are less than those in the underlying interval. The south block remained a distinct depocenter, G, because the bounding fault scarp remained as a bathymetric barrier and because underlying allochthonous salt could still be expelled to accommodate overburden.

In the center and west, the boundary between areas of thick and thin sediment shifted northward as the shelf edge prograded through the study area.

1.-4 to 0-Ma Isochore

Aggradation rates remained high during the late Pleistocene, with rates of 0.1, 1.6, and 0.9 mm/yr (0.004, 0.06, and 0.04 in./yr) for minimum, maximum, and mean values, respectively.

The isochore map (Figure 18) shows that the north and west blocks acted mostly as one, as faulting diminished except in the west. The thickest sediment accumulated in the north, where greater water depths on the upper slope accommodated aggradation. In contrast, the shelf in the south received only a veneer of sediment. Perturbing this uniform regional pattern, local depocenter

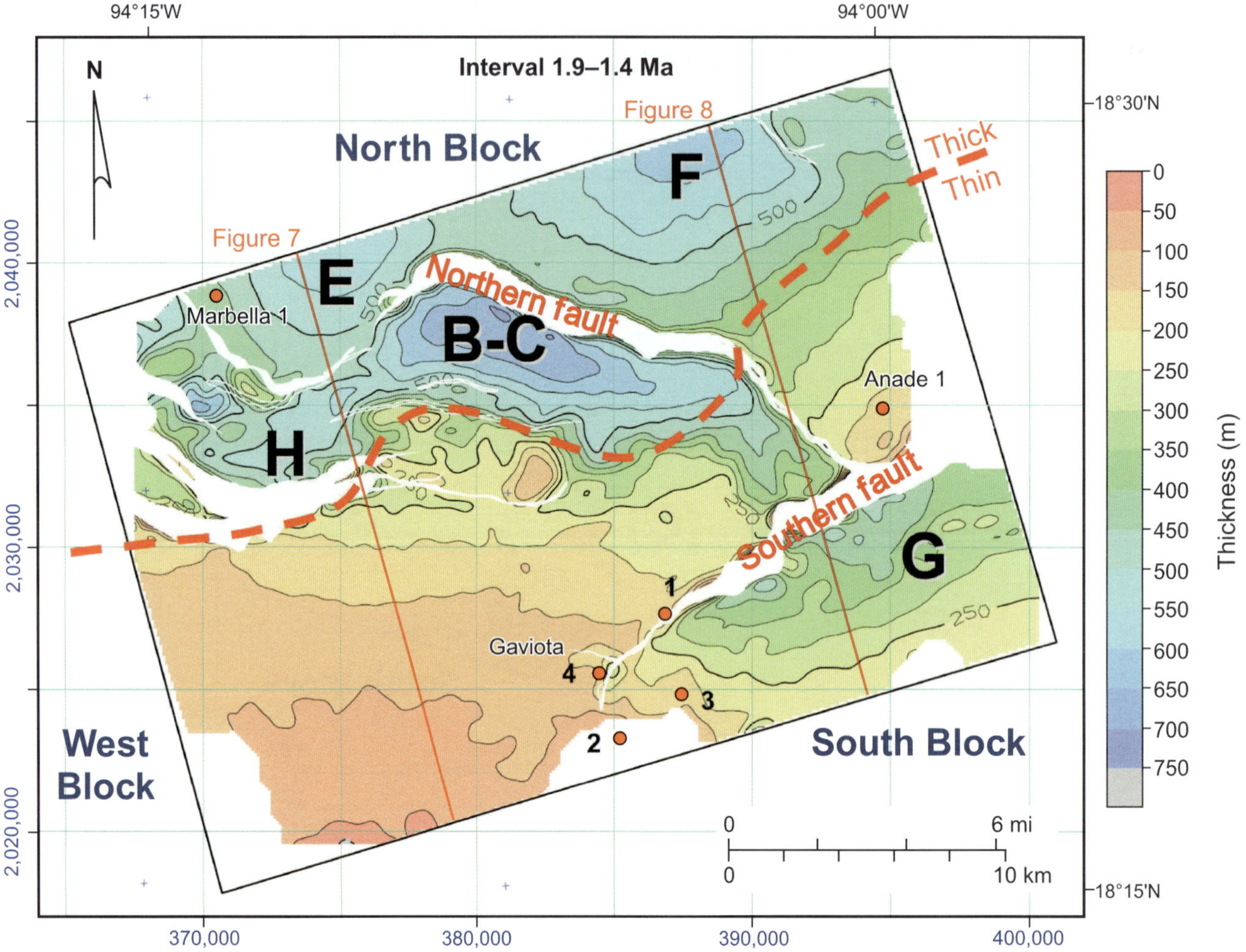

FIGURE 17. Isochore map of the lower Pleistocene 1.9–1.4-Ma interval. The contour interval is 50 m (160 ft). Letters B, C, E, F, G, and H represent different depocenters. Universal Transverse Mercator (UTM) projection.

G remained in the hanging wall of the counterregional southern fault.

Geohistory Analysis

Figure 19 shows a geohistory plot (Van Hinte, 1978) created from the restoration of line CDP 1460. The location marked in Figure 9 was chosen for geohistory analysis because it lacks evidence of irregular salt evacuation and should record regional subsidence. The geohistory analysis includes the effects of decompaction but is only semiquantitative because (1) the paleobathymetry is approximate and (2) changes in absolute sea level, which are minor on the scale of the plot, were not accounted for. Even so, the analysis is accurate enough to document major changes in burial history. Since 5.5 Ma, the curve shows a continuous linear subsidence interrupted by a spike of rapid subsidence and sedimentation from 2.6 to 2.4 Ma (Table 1; Figure 4). This surge in subsidence could be because of several causes, which are not mutually exclusive.

First, the prodigious sediments shed from the Sierra de Chiapas rising onshore would rapidly load and displace the underlying allochthonous salt, a form of local accommodation.

Second, regional subsidence could also have increased because the base of allochthonous salt subsided, although slightly less than did the suprasalt strata (Figure 19). This subsalt subsidence could result from two processes. Deeper salt, probably at the Jurassic level, could be expelled despite the long history of previous autochthonous evacuation required to form the salt canopy. Alternatively, the cause might be deeper in the pre-Jurassic basement, lower crust, and mantle, for example by plate tectonic movement of the Chortis block or increased sedimentary isostatic loading.

Third, a pulse of regional extension could accelerate subsidence. This extension would have to affect suprasalt strata, allochthonous salt, and subsalt strata because the base of the salt canopy also subsided. Both restorations provide evidence of supracanopy regional extension since the late Pliocene. In one of the restored cross

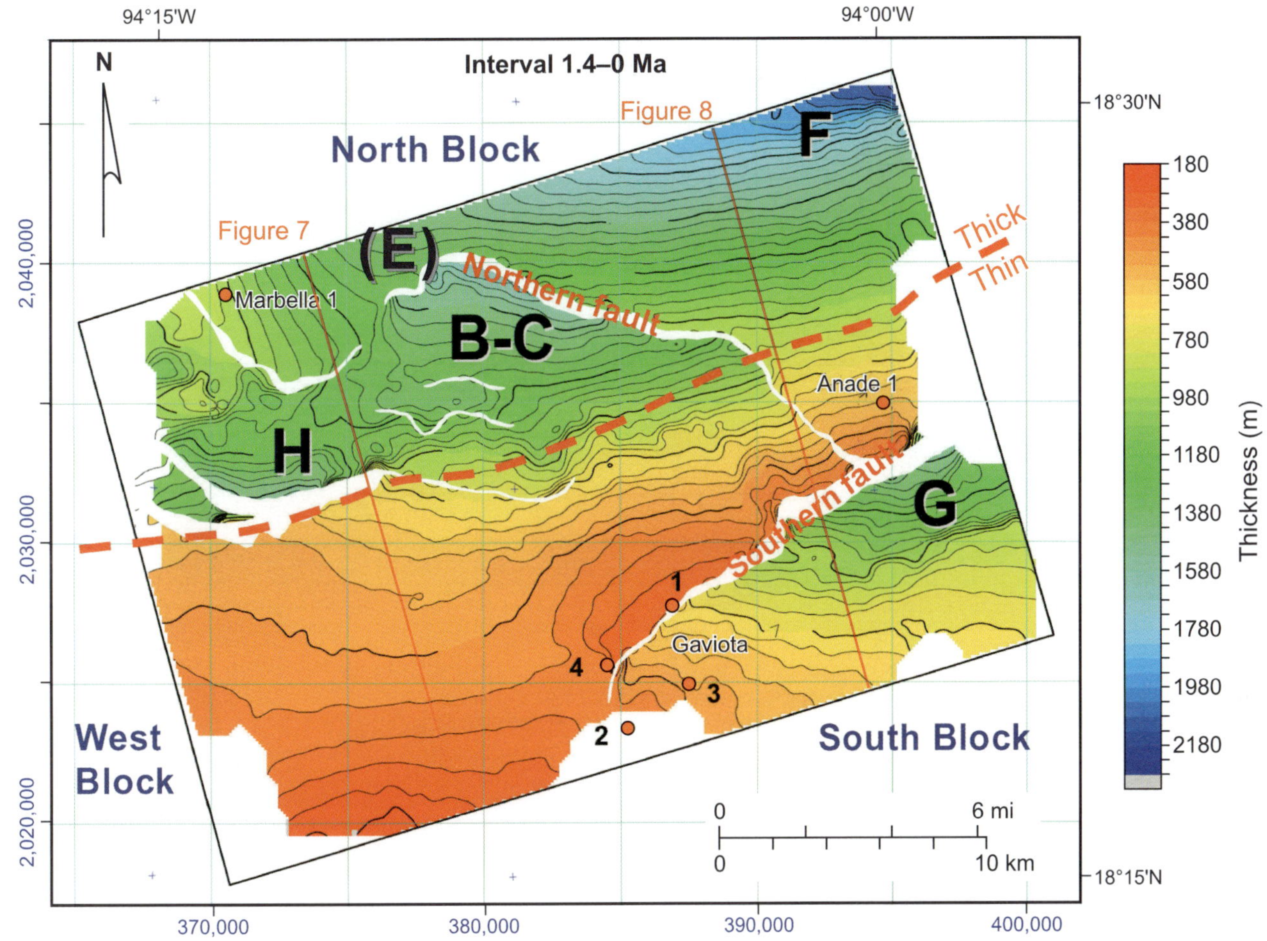

FIGURE 18. Isochore map of the upper Pleistocene 1.4–0-Ma interval. The contour interval is 50 m (160 ft). Letters B, C, F, G, and H represent depocenters. The position of former depocenter E is shown in parentheses. Universal Transverse Mercator (UTM) projection.

sections (Figure 17), regional extension was active while subsidence and aggradation surged from 2.6 to 2.4 Ma. However, seismic quality is too poor to show a direct sign of any subsalt extension, so this mechanism for increased subsidence is particularly speculative.

Besides subsidence, the geohistory plot may also record salt welding, which was incorporated in the restoration although evidence for welding is equivocal. In the graph, the top and base of the allochthonous salt converged through time as the salt thinned. Thinning continued until about 1.8 Ma when the salt weld formed and the top and base of salt became locked together in a common subsidence history.

SUMMARY AND CONCLUSIONS

Precanopy Shortening (8–5.5 Ma)

The Agua Dulce fold belt is one of several Neogene fold belts in the offshore Salina del Istmo Basin (Gomez-Cabrera and Jackson, this volume). Seismic onlap and thickness variations suggest that this fold belt formed between 8 and 6 Ma in the study area. The isochore map (Figure 12) allows several possible trends for the fold belt. The preferential location of diapirs in anticlinal crests most likely reflects the tendency of buckle folds and thrusts to nucleate in the weak, thin, bulging stratal roofs of preexisting diapirs. These precursor diapirs may have been growing since the Jurassic.

Salt-Canopy Emplacement (Ending ~6 Ma)

As the Agua Dulce folding waned in the late Miocene, extrusive salt blanketed much of the fold belt. Salt extrusion was enhanced by tectonic squeezing of diapirs, which expelled salt from canopy feeders, and by shortening-related uplift.

Evacuation of the Salt Canopy (5.5–2.6 Ma)

The oldest strata above the salt canopy are of uppermost Miocene age. Thin sags of overburden partly buried

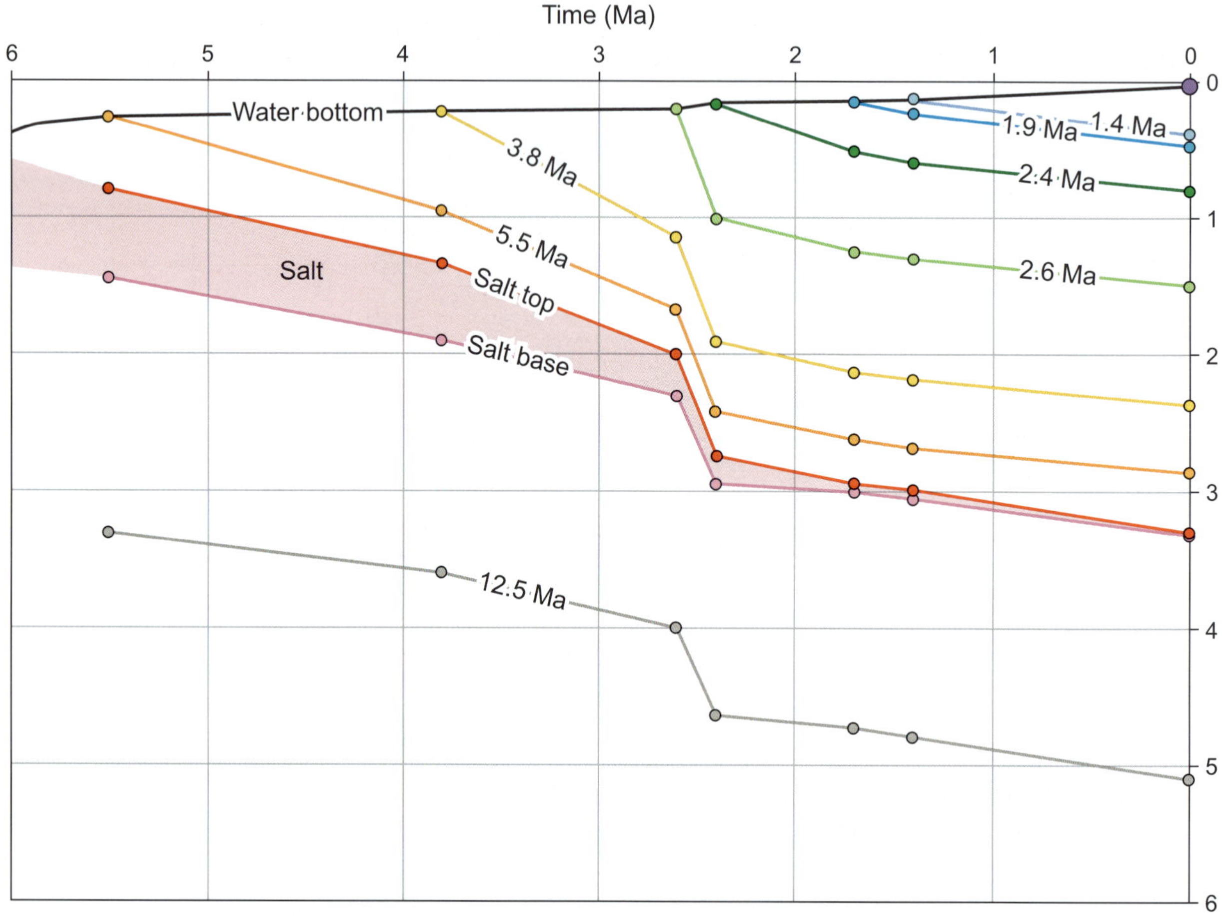

FIGURE 19. Geohistory diagram showing qualitative burial trends typical of a mature passive margin having continuous subsidence. Rapid subsidence between 2.6 and 2.4 Ma interrupted this trend. Both the top and base of the allochthonous salt subsided abruptly, with the top salt subsiding faster. Curves for the top and base of salt gradually converged during inferred salt welding, after which strata above and below subsided in tandem. Water-bottom curve shows shoaling to the present-day shelf setting. Figure 9 shows the location of this one-dimensional analysis.

the allochthonous canopy. Thickness variations and onlap against the top of the allochthonous salt show that the salt was redistributed immediately after the canopy was emplaced. These overburden sags may have originated as synclines formed by gentle buckling during the waning stages of Agua Dulce folding, which may also have folded the base of the salt canopy.

After 3.8 Ma, mean sedimentation rates more than doubled to 0.7 mm/yr (0.03 in./yr), and sediment became trapped against extensional faults. An arcuate array of landward-dipping, en echelon faults, called the northern, central, and southern faults, divided the study area into north, west, and south blocks. The main sediment catchment was the west block, where sediments filled the landward hanging walls of the northern-central fault system, and the south block, which fringes the study area. The north block was sediment starved because of competition from the west block on its landward flank, which drove slip on the northern fault. In the west, an allochthonous diapir separated depocenters in the north and west blocks.

From 2.6 to 2.4 Ma, mean aggradation rates increased to as much as sixfold, to 4.5 mm/yr (0.18 in./yr). Loading by this prodigious quantity of sediment enhanced salt evacuation as major extension continued. The north block remained partly starved of sediment, whereas depocenters formed in the west block because of salt evacuation in the hanging wall of counterregional faults and in the landward flank of emergent allochthonous diapirs. The canopy continued to thin until salt became locally welded.

Shelf Subsidence (2.4–0 Ma)

Mean aggradation rates fell to one quarter or 1.1 mm/yr (0.04 in./yr). Slowing sedimentation is attributed to decreased sediment supply as the shelf break crossed the study area. In the center and west, the boundary between areas of thick and thin sediment shifted northward as the shelf edge prograded. Some depocenters crowded together in depressions in counterregional hanging walls. Buried passive diapirs became reactive or falling diapirs as regional extension continued and salt welding progressed. After 1.4 Ma, the north and west blocks acted as a single structural block, except in the far west, tilting seaward and receiving the thickest sediment. In contrast, the shelf in the south received only a veneer of sediment, except for a depocenter in the hanging wall of the southern fault. By the present day, the salt bodies mostly stopped affecting sedimentation patterns, although faulting continued.

ACKNOWLEDGMENTS

We thank the following for their contributions: Petróleos Mexicanos provided the data and permitted publication. William L. Fisher cosupervised the Ph.D. dissertation (Gómez-Cabrera, 2003) on which this article was based and provided valuable stratigraphic insights. Peer reviewers Mark Rowan and Lou Liro provided a wealth of cogent suggestions for improving the article. Contributions by the second author (M.P.A.J.) were funded by the Applied Geodynamics Laboratory consortium of oil companies and by the Jackson School of Geosciences. Midland Valley supplied 2DMove restoration software, and Halliburton provided Landmark's SeisWorks interpretation software. Nancy Cottington drew the illustrations, and Lana Dieterich edited the manuscript. The chapter is published by permission of the director, Bureau of Economic Geology, Jackson School of Geosciences, the University of Texas at Austin.

REFERENCES CITED

Contreras V., and B. Castillon, 1960, Morfología y origin de los domos salinos del Istmo de Tehuantepec: Boletín de la Asociación Mexicana de Geólogos Petroleros, v. 12, p. 221–242.

Contreras, V., and B. Castillon, 1968, Morphology and origin of salt domes of Isthmus of Tehuantepec: AAPG Memoir 8, p. 244–260.

Diegel, F. A., and R. W. Cook, 1990, Palinspastic reconstruction of salt-withdrawal growth-fault systems, northern Gulf of Mexico (abs.): Geological Society of America Abstracts with Programs, v. 72, no. 7, p. A48.

García-Molina, G., 1994, Structural evolution of the SE Mexico (Chiapas-Tabasco-Campeche) offshore and onshore : Ph.D. dissertation, Rice University, 161 p.

Ge, H., M. P. A. Jackson, and B. C. Vendeville, 1997, Kinematics and dynamics of salt tectonics driven by progradation: AAPG Bulletin, v. 81, p. 398–423.

Gómez-Cabrera, P. T., 2003, Stratigraphic and structural analysis of the Neogene sediments of the offshore part of the Salina del Istmo Basin, southeastern Mexico: Ph.D. dissertation, the University of Texas at Austin, 274 p.

Gómez-Cabrera, P. T., and M. P. A. Jackson, 2009, Regional Neogene salt tectonics in the offshore Salina del Istmo Basin, southeastern Mexico, *in* C. Bartolini and J. R. Román Ramos, eds., Petroleum systems in the southern Gulf of Mexico: AAPG Memoir 90, p. 1–28.

McBride, B. C., M. G. Rowan, and P. Weimer, 1998a, The evolution of allochthonous salt systems, northern Green Canyon and Ewing Bank (offshore Louisiana), northern Gulf of Mexico: AAPG Bulletin, v. 82, no. 5B, p. 1013–1036.

Oviedo-Perez, A. E., 1996, Evolución neógena de las Cuencas Terciarias del Sureste: Tesis de Maestría, División de Estudios de Posgrado, Facultad de Ingeniería, Universidad Nacional Autónoma de México, 91 p.

Petróleos Mexicanos (PEMEX)-British Petroleum, 1994, Proyecto cuencas terciarias del sureste y área Marbella: Unpublished Internal Report, 297 p.

Seni, S. J., 1992, Evolution of salt structures during burial of salt sheets on the slope, northern Gulf of Mexico: Marine and Petroleum Geology, v. 9, no. 4, p. 452–468.

Vail, P. R., R. G. Todd, and J. B. Sangree, 1977, Seismic stratigraphy and global changes of sea level: V. Chronostratigraphic significance of seismic reflections, *in* C. E. Payton, ed., Seismic stratigraphy—Applications to hydrocarbon exploration: AAPG Memoir 26, p. 99–116.

Van Hinte, J. E., 1978, Geohistory analysis—Application of micropaleontology in exploration geology: AAPG Bulletin, v. 62, p. 201–222.

Vendeville, B. C., and M. P. A. Jackson, 1992a, The rise of diapirs during thin-skinned extension: Marine and Petroleum Geology, v. 9, p. 331–353.

Vendeville, B. C., and M. P. A. Jackson, 1992b, The fall of diapirs during thin-skinned extension: Marine and Petroleum Geology, v. 9, p. 354–371.

Viniegra, O. F., 1971, Age and evolution of salt basins of southeastern Mexico: AAPG Bulletin, v. 55, p. 478–494.

Wilson, H. H., 1993, The age of salt in the Gulf of Mexico basin: Journal of Petroleum Geology, v. 16, no. 2, p. 125–152.

Woodbury, H. O., I. B. Murray Jr., P. J. Pickford, and W. H. Akers, 1973, Pliocene and Pleistocene depocenters, outer continental shelf Louisiana and Texas: AAPG Bulletin, v. 54, p. 2428–2439.

Wornardt, W. W., and P. R. Vail, 1993, Third order depositional sequences and systems tract: Variations in thickness and lithofacies with paleobathymetry, *in* J. M. Armentrout, R. Bloch, H. C. Olson, and B. F. Perkins, eds., Rates of geological processes: Gulf Coast Section SEPM Foundation, 14th Annual Research Conference, p. 277–285.

Worrall, D. M., and S. Snelson, 1989, Evolution of the northern Gulf of Mexico, with emphasis on Cenozoic growth faulting and the role of salt, *in* A. W. Bally and A. R. Palmer, eds., The geology of north America: An overview: Geological Society of America, v. A, p. 97–138.

12

Ríos-López, J. J., and A. Cantú-Chapa, 2009, Stratigraphy and sedimentology of Middle Eocene Kumaza Calcarenites Member in the Ku, Maloob, and Zaap oil fields, offshore Campeche, Mexico, *in* C. Bartolini and J. R. Román Ramos, eds., Petroleum systems in the southern Gulf of Mexico: AAPG Memoir 90, p. 257–277.

Stratigraphy and Sedimentology of Middle Eocene Kumaza Calcarenites "Member" in the Ku, Maloob, and Zaap Oil Fields, Offshore Campeche, Mexico

Jaime J. Ríos-López
Petróleos Mexicanos (PEMEX) Exploration and Production, Northeast Marine Region, Campeche, Mexico

Abelardo Cantú-Chapa
Instituto Politécnico Nacional, Mexico City, Mexico

ABSTRACT

The middle Eocene Ku, Maloob, and Zaap (Kumaza) Calcarenites "Member" is formally proposed to describe a calcareous unit represented by one to three, calcite-cemented packstone to grainstone calcarenites in a thick undescribed sequence of shales that occur in the subsurface of Campeche Shelf, southeastern Gulf of Mexico. These calcarenite bodies are here called units 1-A, 1-B, and 1-C; they are separated by one to two shale layers denominated as units Lu-1 and Lu-2.

The type section is in the Maloob 97 well. The Kumaza Member has been studied using seismic data, core, chip samples, and log information from 57 wells drilled in the Kumaza fields in the Campeche Shelf. The Kumaza Member contains the most important reservoir rocks in this region and is mappable more than approximately 100 km^2 (39 mi^2). These calcarenite banks are here inferred to have originated on a shallow water platform and were later redeposited in deeper waters from a restricted area.

The age of the rock units in the Kumaza Member is established from foraminifera, including *Morozovella lenheri, Morozovella spinulosa, Truncorotaloides rohri, Orbulinoides beckmanni*, and *Lepidocyclina (Polylepidina) antillea.*

DOI:10.1306/13191087M903336

INTRODUCTION

The Ku, Maloob, and Zaap (Kumaza) fields are located offshore on the Campeche Shelf in the southern part of the Gulf of Mexico (Figure 1). The Kumaza fields are located within the Akal-Reforma horst domain. The fields are bounded to the northeast by the Comalcalco fault and to the east by the Frontera fault.

These fields are subdivided into blocks whose reserves and production come from reservoirs ranging in age from Jurassic Kimmeridgian (Ángeles-Aquino and Cantú-Chapa, 2001) to middle Eocene (Figure 2).

The first core and cutting samples obtained from calcarenites in this area were recovered from the Ku 1 well in 1979 (drilled in the same location as Ku 22). The first oil production from calcarenites in the Campeche offshore came from the Ku 10 well, drilled in 1986.

Middle Eocene calcarenites have a predominant packstone-grainstone texture. Minor amounts of wackestones and mudstone textures are also present, as well as argillaceous minerals such as illite. These are the main reservoir units within the Kumaza fields.

Paleontologic sampling was not systematic or continuous in the wells drilled in middle Eocene calcarenites. Therefore, the proposed correlation is lithostratigraphic and therefore possibly diachronous over large distances beyond the field scale.

The calcarenites are assigned a middle Eocene age based on the fossils *Morosovella lenheri*, *Morosovella spinulosa*, *Truncorotaloides rohri*, *Truncorotaloides topilensis*, *Porticulasphaera mexicana*, *Lepidocyclina (Polylepidina) antillea*, and *Rotalia cushmani* (Table 1).

In this chapter, stratigraphic correlations were based on well data in the three fields calibrated to seismic data. In particular, the physical criteria of lithologic correlation were applied, considering the continuity of the different sedimentary rocks inferred between the wells from the Kumaza fields. The geographical and stratigraphic limits of the calcarenite are also inferred, defining the extent of the oil reservoir.

Calcarenites of presumed Paleocene and early Eocene ages occur at several localities on the Campeche Shelf, but their ages have not been constrained paleontologically, as in the Cantarell field (Gutiérrez et al., 1998; Ortuño et al., 2001). Previous interpretations of the age and depositional facies of calcarenites are somewhat ambiguous (Hernández-García, 1992; Gardner et al., 1997; Juárez-Aguilar and Ríos-López, 1999). Our studies provide revised ages and a new sedimentary model for the origin of these sequences.

The Kumaza Calcarenites "Member" is formally proposed here to distinguish the calcarenite rocks of middle Eocene age from units that are lithologically similar but of different age and which occur in other fields in the Campeche Shelf.

This work reviews a stratigraphic and sedimentological study of the middle Eocene calcarenites of the Kumaza fields and provides a guide to interpret and forecast reservoir fluid flow and production characteristics. Results from this study will subsequently be used for reservoir management strategies in these mature fields.

LITHOSTRATIGRAPHY AND MICROFOSSILS IN THE KUMAZA FIELDS

Lower Eocene

The appearance and presence of *Morozovella aragonensis*, *Morozovella formosa*, *M. spinulosa*, *Morozovella subbotinae (=Globorotalia rex)*, and *Hantkenina aragonensis* mark the top of the lower Eocene as identified in the Maloob 101 and 103, Ku 101, Zaap 1-A, and Kutz 1 wells. The lithology of the lower Eocene is described as calcareous, benthonic shale with interbedded mudstone-wackestone units that have a distinctive gamma-ray response.

From the top of the lower Eocene to the lower middle Eocene and at the base of the lower Kumaza Member, a gradual decrease in the gamma-ray readings (from the high at the top of the lower Eocene) is observed. This decrease occurs across a 10–40-m-thick (33–131-ft-thick) interval (Figure 3).

Middle Eocene

The fossil assemblages recognized in middle Eocene rocks in the Kumaza fields (Table 2) are outlined below.

The top of the middle Eocene was identified in the Maloob 101, 103, and 97; KU 401 and 101; Zaap 1-A and 106; and Kutz 1 wells at the first appearance and presence of *Amphistegina parvula*, *Morozovella lenheri*, *M. spinulosa*, *Truncorotaloides topilensis*, *Orbulinoides beckmanni*, *Lepidocyclina (Polylepidina) antillea*, *Lepidocyclina pustulosa*, *Dictyocunus floridanus*, *Heterodictyoconus* sp., and *Lepidocyclina* sp.

T. topilensis, *M. lenheri*, and *M. spinulosa* occur in the Maloob 103 (Murillo and Perez, 1992), Maloob 101 (García-Genis and Pérez, 1992), and Kutz 1 (Díaz and Landeros, 1979) wells in a slightly calcareous, grayish-green shale.

Pararotalia roblesi, *Heterodictyoconus* sp., *M. spinulosa*, *M. lenheri*, *H. aragonensis*, and *Smoutina* sp. occur in shale interbedded with calcareous shales (fragment 32, core 1) in the Maloob 97 well (Romero-Peñaloza and Noriega, 1998; Torres-Rodríguez, 1998). The same core yielded *Amphistegina parvula* and *Lepidocyclina pustulosa* from calcarenitic and brecciated sediments.

Lepidocyclina (Polylepidina) antillea, *Amphistegina parvula*, *Quinqueloculina* sp., and *Spiroloculina* sp. were identified in cores 1 and 2 from the Zaap 106 well (Cabrera

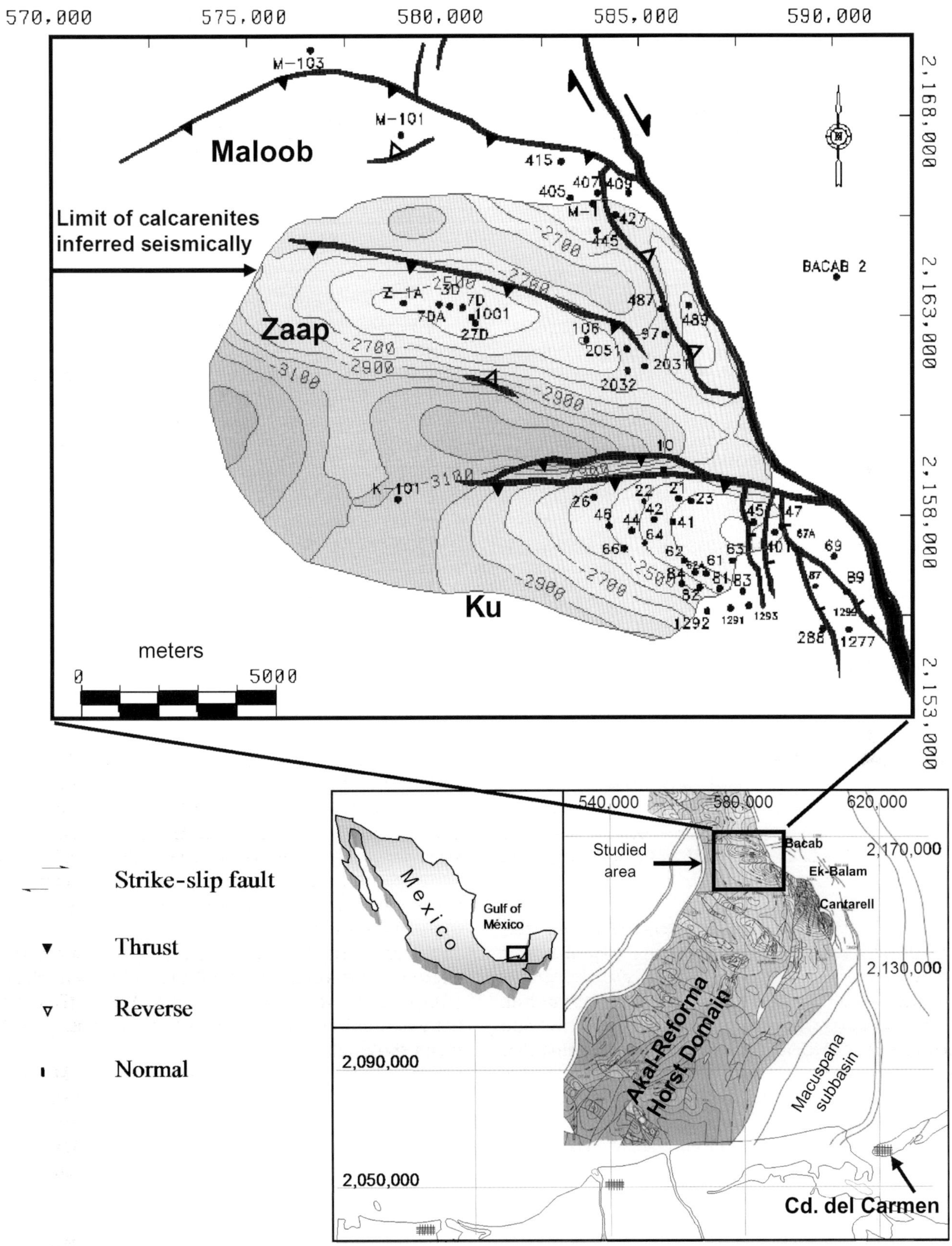

FIGURE 1. Location of the study area and simplified structural map of the top middle Eocene calcarenites of the Ku, Maloob, and Zaap oil fields.

Age	Period / Epoch		Series / Stage		Lithost. Units
Cenozoic	Paleogene	Eocene	Upper		
			Middle		Kumaza "member"
			Lower		
		Paleocene	Upper		
			Lower		Abkatun
Mesozoic	Cretaceous		Upper	Maastrichtian	Cantarell
				Campanian	
				Santonian	Chac
				Coniacian	
				Turonian	
				Cenomanian	
			Lower	Albian	Akal
				Aptian	
				Barremian	
				Hauterivian	
				Valanginian	
				Berriasian	
	Jurassic		Upper	Tithonian	Edzna
				Kimmeridgian	Akimpech
				Oxfordian	Ek-Balam Group
			Middle	Callovian	Isthmian Salt

FIGURE 2. Simplified stratigraphic column for the subsurface Campeche Shelf, from the Callovian to Eocene. From Ángeles-Aquino and Cantú-Chapa (2001), and Cantú-Chapa and Landeros-Flores (2001).

et al., 1997) and *Lituonella* sp., *A. parvula*, *Rotalia cushmani*, and *Lepidocyclina (Polylepidina) antillea* were identified in core 1 from the Ku 401 well (Gardner et al., 1997).

The top of the middle Eocene is recognized on gamma-ray logs where an inflection toward the right indicates a zone where radioactivity values are high. The well-log signature of the middle Eocene was used to correlate between exploratory and development wells in the Kumaza fields because they lacked samples from these horizons or contacts (upper to middle Eocene).

Gamma-ray logs of calcarenites within the middle Eocene are characterized by descending values related to an upward-coarsening log signature (Figure 3b). This abrupt change or truncation in the logs is interpreted to mark a transgressive surface because a strong contrast between the calcarenites and the marine sediments of lower energy (deep waters) characterized by calcareous shales is observed.

Additionally, the top of the middle Eocene could be related to a maximum flooding surface associated with transgression, as on the gamma ray, the surface where the maximum inflection point is observed is typical of zones with high contents of argillaceous minerals and/or mudstone.

Middle Eocene Kumaza Member

The Kumaza Calcarenites "Member" is proposed for the three calcarenite bodies predominantly made up of packstones and grainstones, minor wackestones, and mudstones with occasional illite. The type section is in the Maloob 97 well, from 2634 to 2826 m (8642 to 9272 ft)

TABLE 1. Fossil content recognized in drill cuttings* and cores within the Middle Eocene.**

Well (of reference)	Assemblage Fossil†	Paleogene: Pal. E	Pal. L	Eocene E	Eocene M	Eocene L	Olig. E	Olig. M	Olig. L	Lithology
Z - 1A*	*Truncorotaloides rohri*				■					Wackestone-packstone
M-103*, M-101*, Kutz-1*	*Truncorotaloides topilensis*				■					Dark shale
ZAAP 106, K-101*, Z-27D	*Spiroloculina* sp.	▒	▒	▒	▒	▒	▒	▒	▒	Packstone-grainstone
Z-1001, Z-106, K-401, M-97	*Amphistegina parvula*				■	■				Packstone-grainstone
Z-1001, Z-27D	*Amphistegina* sp.				▒ ¿?	▒				Packstone-grainstone
Z-1001, M-97	*Dictyoconus* sp.			■	■	■				Wackestone-packstone
Z-1001, Z-7DA	*Dictyoconus floridanus*				■	■				Packstone-grainstone
K-401, K-101*	*Lituonella* sp.			■	■	■				Packstone-grainstone
K-401, K-101*	*Lepidocyclina (Polylepidina) antillea*				■					Packstone-grainstone
M-97	*Pararotalia roblesae*			■	■	■	■	■	■	Packstone-grainstone
K-101*, Z-27D	*Pararotalia* sp.			■	■	■				Packstone-grainstone
M-97	*Triloculina* sp.	■	■	■	■	■	■	■	■	Packstone-grainstone
M-97	*Sigmoillopsis* sp.	■	■	■	■	■	■	■	■	Packstone-grainstone
M-97	*Smoutina* sp.			■	■	■				Packstone-grainstone
M-97, K-101*, M-97, Z-106, Z-27D	*Quinqueloculina* sp.	■	■	■	■	■	■	■	■	Packstone-grainstone
M-97	*Heterodictyoconus* sp.				■	■				Packstone-grainstone
M-97	*Hantkenina aragonensis*			■						Shale
M-97	*Lepidocyclina (Neolepidina) pustulosa*				■	■				Wackestone-packstone
K-101*, K-401, Z-106, M-97	*Lepidocyclina (Polylepidina) antillea*				■					Packstone
K-101*, Z-27D	*Lepidocyclina* sp.				■	■	■	■	■	Packstone-grainstone
K-101*	*Orbulinoides beckmanni*				■					Packstone-grainstone
K-101*, K-401	*Rotalia cushmani*			▒	▒	▒				Packstone-grainstone
M-103*, M-101*, Kutz-1*, Z-1*, K-101*, M-97	*Globorotalia lenheri*				■					Calcareous shales
M-103*, M-101*, M-97	*Globorotalia spinulosa*			■	■					Shale

Fossil content recognized in drill cuttings and cores within the middle Eocene.

**Most of the benthonic fossils identified in the middle Eocene calcarenites are widely distributed in shallow or subtidal environments (including all platform subenvironments) and transitional and lagoonal environments. Some organisms found in wells outside of the calcarenitic sediment area (for example: M-103, M-101, Kutz-1) correspond generally with fossil organisms deposited in deep water environments. All of the fossils are reported in PEMEX Biostratigraphy unpublished notes for the Marine Region.*

†The benthonic organisms before mentioned in the table generally have other organism fragments (without detailing its classification) with predominant amounts: amphisteginids, discocyclinids, echinoderms, miliolids, nummulitids, rotalids, textularids, mollusks, pelecypods, gastropods, bryozoans, coralline algae: blue-green, brown, and red.

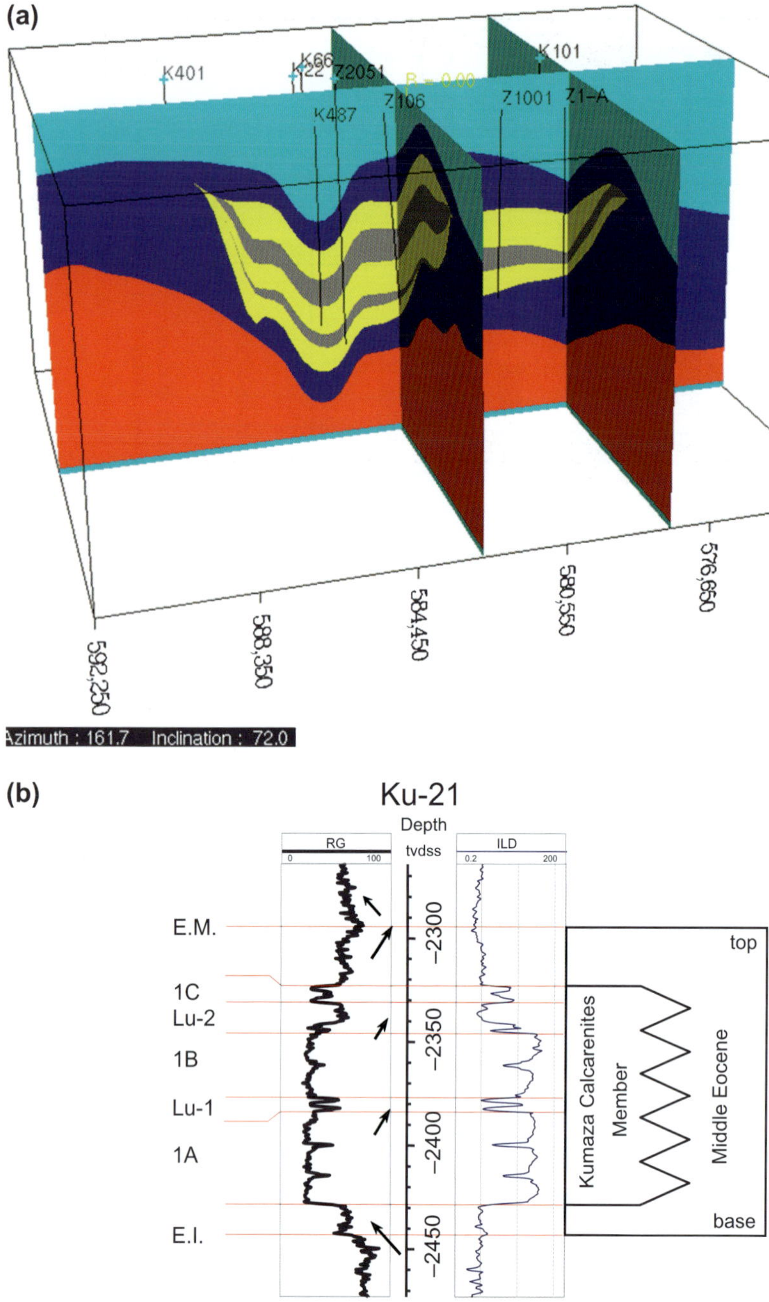

FIGURE 3. (a) Block diagram showing lithostratigraphic correlation in the middle Eocene Kumaza Member proposed in this study. (b) Gamma-ray (RG) and induction logs (ILD) showing qualitative grain size influences such as fining (upward right arrows) or coarsening successions (upward left arrows) within the Kumaza Calcarenites "Member" (including units 1-A, Lu-1, 1-B, Lu-2, and 1-C). Abrupt interruptions (cutoffs) are observed in most of the successions. TVDss = true vertical depth subsea; E.M. = middle Eocene; E.I. = lower Eocene.

true vertical depth subsea (TVDss). The Kumaza Member is present in the Kumaza oil fields, as well as in some wells of the Cantarell complex.

The calcarenites of the Kumaza Member are identified as units 1-A, 1-B, and 1-C. The calcarenite bodies are separated by two thin layers of slightly calcareous shales and very argillaceous greenish mudstones, referred to as units Lu-1 and Lu-2, respectively. Locally, these rocks are brecciated, fractured, dolomitized, and silicified (e.g., Maloob 97 well).

The thickness of the Kumaza Member ranges from 190 m (623 ft) in the northern part of the area (Maloob 97; Figure 4A) to about 5 m (16 ft) in the south and central parts of the field (e.g., Ku 81) where it pinches out (Figure 1).

The upper contact of the Kumaza Member is at an abrupt inflection of the gamma ray toward the left. The contact is inferred to be abrupt because of lithologic changes caused by the presence of slightly calcareous shales overlying the Kumaza Member. Commonly, the gamma-ray curve for the Kumaza Member displays 1 or 2 short inflexions to the right when some argillaceous layers are intercalated within the calcarenite units.

The contact between the base of the Kumaza Member and the underlying lower middle Eocene rocks shows an abrupt change on the gamma-ray log. The curve is

TABLE 2. Semiquantitative petrographic description of the Kumaza calcarenites.*

Allochemical and Orthochemical Components

Well	Core / Sample / Depth	Blocky calcite cement / sparite	Elongate calcite / Microesparite	Isopachous bladed calcite	Thin dogtooth spar	Magnesian calcite	Calcita sintaxial	Fibrous aragonite	Microcrystalline dolomite	Pyrite	Chert / Quartz grains	Pellets / Peloids	Ooids	Intraclasts
KU-1	C-1(2430-30.8)													
KU-46	C-1 (2838-47)													
KU - 66	C-1 (2950-59 md, 2570-2572 mv)	23	20		X	X	X	X				45		
KU - 66	F32-M1-H	16	18							0.1		40		
M-97	C-1 (2840-48)									X	X			
M-97						X	X		X	X			X	
KU-401	2214	4	1	1			X			0.1	0.1	15		5
KU-401	2214.43	5	1							0.1	0.1	34.9		3
KU-401	2215.1	5	1							0.1		25		3
Z-7D	C-1 (2555-59)	X	X				X					X	X	X
Z-7DA	2616.24	10.2				X	X	X	0.8	0.4	1.6	8		X
Z-7DA	2619.65	8.4								0.8	0.8	8	Tr	
Z-7DA	2622.23	7.6								0.8	1.6	10	0.4	
Z-27D	(2740-49 md),F5-MA-V	5	20			X	X	X				35		X
Z-27D	2744.4, F5-M2-H	5	20									35		
Z-27D	2741.15, F19-M3-V	4	24							0.1		40		
Z-106	C-2 (3605-14), F6													X
Z-1001	C-1, 2582.76	4								0.1		20		3
Z-1001	2585.36	3	0.1							0.1		35		4
Z-1001	2587.54	20							1.0	2		18		
Z-1001	C-2, 2610.71	2	1							0.1		36		6
Z-1001	2614.56	5								0.1		22.9		4

Biogenics — Foraminifera (not detailed) (*Bentonic & Planktonic); Algae

Well	Depth	Foraminifera (not detailed)	Bentonics (not detailed)	Planktonic (not detailed)	*Miliolids	*Amphisteginids	*Rotalids	*Nummulitids	*Discocyclinids	*Textularids	Algae fragments	Coralline algae	Brown algae	Green algae	Red algae	Blue green algae	Molusc shell fragments	Echinoderms	Bryzoans	Undifferentiated fossil fragments
KU-1	C-1(2430-30.8)																			
KU-46	C-1 (2838-47)																			
KU - 66	C-1 (2950-59 md, 2570-2572 mv)	12			X	X							X					X		X
KU - 66	F32-M1-H	16.0																		
M-97	C-1 (2840-48)	X	X	X														X		
M-97					X	X	X							X			X	X		
KU-401	2214	47.8			X		X	X	X		6	5			X	X	2	2	2	5
KU-401	2214.43	37									5	3					2	1.0	3	2
KU-401	2215.1	39.9									5	6					2	3	2	2
Z-7D	C-1 (2555-59)			X	X		X					X					X	X		
Z-7DA	2616.24	14.8			X	X				X		24.8	X				2	3.2	X	15.2
Z-7DA	2619.65	19.2										35.6					1.6	2.4		7.2
Z-7DA	2622.23	25.6										29.2					3.2	0.8		4.8
Z-27D	(2740-49 md),F5-MA-V	25	X		X					X	X						X	X		
Z-27D	2744.4, F5-M2-H	25															X	X	X	
Z-27D	2741.15, F19-M3-V	20															X			
Z-106	C-2 (3605-14), F6																			
Z-1001	C-1, 2582.76	31			X		X	X	X		8	13			X	X	0.1		1	
Z-1001	2585.36	34.9									7	4					2	2	1	
Z-1001	2587.54	15									8	3						1		4.0
Z-1001	C-2, 2610.71	35									5	4					2	5	2	
Z-1001	2614.56	25									8	9					0.1	4	2	

Allochemical and Orthochemical Components (continued; Clays) and Mineralogical Contents

Well	Depth	Micrite	Bentonite	Phosphate	Mica	Clay (not detailed)	Illite	Smectite	Organic material	Bitumen	Calcite	Dolomite	Quartz	Clay (V_{clay})	Pyrite	Organic	Titanium oxide
KU-1	C-1(2430-30.8)																
KU-46	C-1 (2838-47)																
KU - 66	C-1 (2950-59 md, 2570-2572 mv)	X									X						
KU - 66	F32-M1-H	10									100		0.05		0.05		
M-97	C-1 (2840-48)					X					X		X		X		
M-97		X							X			X					
KU-401	2214	4		0.05					0.05		100		0.05		0.05	0.05	
KU-401	2214.43	3								0.1	100		0.1		0.1	0.1	
KU-401	2215.1	6								0.1	100				0.1		
Z-7D	C-1 (2555-59)						X						X		X		X
Z-7DA	2616.24	12.4									98		2				
Z-7DA	2619.65	4.4									99		1.0				
Z-7DA	2622.23	2.8									98		2				
Z-27D	(2740-49 md),F5-MA-V	X	X			X											
Z-27D	2744.4, F5-M2-H	15															
Z-27D	2741.15, F19-M3-V	1 2															
Z-106	C-2 (3605-14), F6	X	X														
Z-1001	C-1, 2582.76	14.9				3	X	X	2		95			3	0.1	2	
Z-1001	2585.36	7									99		1				
Z-1001	2587.54	22				6					92			6	2		
Z-1001	C-2, 2610.71	2									100				0.1		
Z-1001	2614.56	20									100				0.1		

Petrographic porosity determ.

Well	Depth	Vug	Fracture	Intercrystalline	Interparticle	Intraparticle / intrafossil	Moldic - dissolution	Micro-porosity (cement)	Shelter / Moldic	Total porosity
KU-1	C-1(2430-30.8)									
KU-46	C-1 (2838-47)									
KU - 66	C-1 (2950-59 md, 2570-2572 mv)				X		X		X	
KU - 66	F32-M1-H				X				X	
M-97	C-1 (2840-48)				X					
M-97					X		X			
KU-401	2214	tr		2	21	1			3	27
KU-401	2214.43	1		1	19	1			5	27
KU-401	2215.1			2	17	1			4	24
Z-7D	C-1 (2555-59)									
Z-7DA	2616.24				3.6	3.2	1.2	5.6		13.6
Z-7DA	2619.65		0.4		8	4.8	0.4	4.4		18
Z-7DA	2622.23				13.6	4	0.4	3.6		21.6
Z-27D	(2740-49 md),F5-MA-V				X					
Z-27D	2744.4, F5-M2-H				X					
Z-27D	2741.15, F19-M3-V				X					
Z-106	C-2 (3605-14), F6	X			X		X			
Z-1001	C-1, 2582.76	0.1		4	13	2			1	20.1
Z-1001	2585.36			5	11	2			2	20
Z-1001	2587.54	0.1		2	2					4.1
Z-1001	C-2, 2610.71	0.1		2	18	1			2	23.1
Z-1001	2614.56			5	12	1			1	19

*After Gardner et al. (1997).

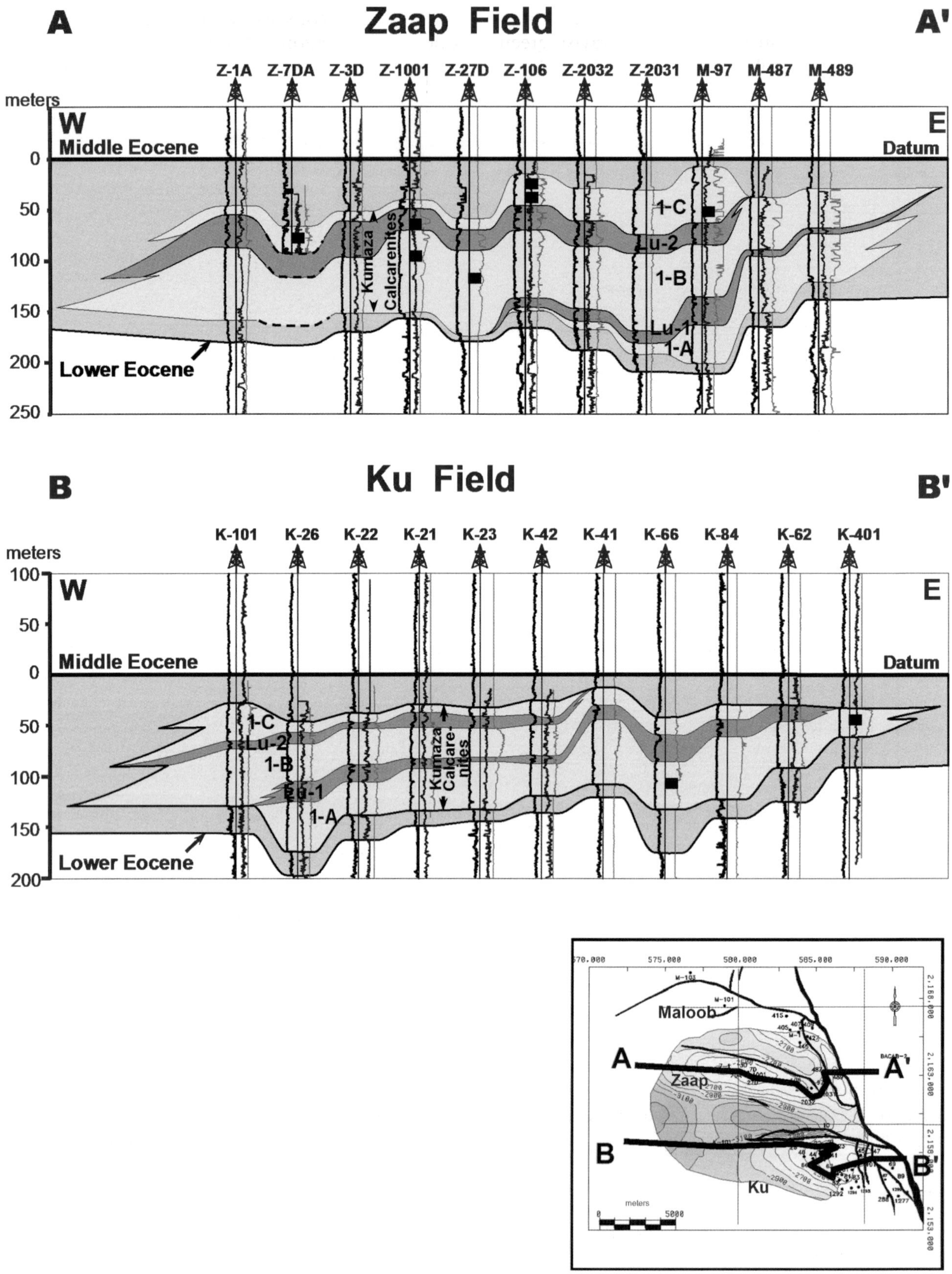

FIGURE 4. Stratigraphic sections showing the lithostratigraphic unit distribution referred to in this study as units 1-A, Lu-1, 1-B, Lu-2, and 1-C in the Zaap (AA′) and Ku (BB′) fields. Correlations were built using gamma-ray, induction, and porosity (from left to right) log curves in most of the wells. Black squares correspond to cores. The sections, which in this case use vertically oriented wells with constant spacing, are shown on the map.

displaced to the right as a result of the lithologic contrast between calcarenite sediments and grayish-green, slightly calcareous middle Eocene shales, which display relatively high gamma-ray values. The truncation in the basal part, seen in the lower section of the middle Eocene calcarenites, is interpreted to be probably related to a regressive erosional surface of possible lowstand origin.

DESCRIPTION OF CALCARENITE UNITS IN KUMAZA FIELDS

A detailed lithostratigraphic correlation of the middle Eocene Kumaza calcarenites has been conducted in the Kumaza fields to determine the distribution of the units. This correlation highlights the predominant lithologic and petrophysical characteristics of the middle Eocene Kumaza Member (Ríos-López, 2002) and permits the definition of the five distinctive lithologic units described below (Figure 4).

Unit 1-A

Unit 1-A is the lowermost depositional unit and overlies a calcareous, grayish-green shale and slightly argillaceous unit within the lower part of the middle Eocene. The geographic distribution of this unit in Kumaza fields is shown in the isopach map (Figure 5).

Three core samples are recovered from this unit (Ku 46 [C-1: 2838–2847 m, 9311–9340 ft], Ku 66 [C-1: 2950–2959 m, 9678–9708 ft], and Ku 401 [C-1: 2214–2223 m, 7264–7293 ft]); all measured depths consist of packstone to grainstone, with rounded foraminifera and pellet fragments cemented by sparite. Bedding is generally greater than 20 cm (8 in.) and occasionally massive with rare laminations. Fragments of *Lepidocyclina* sp., *Amphistegina*, *Lituonella*, and *Anomalina* fossils were identified in the Ku 46 and Ku 401 wells.

The maximum thickness of unit 1-A is 79 m (259 ft). Unit 1-A is missing in the central and western Ku and Zaap fields (Figure 5). This unit is the only one of the three recently named units in the Ku field (Ku 401, 45, 61, 81). The upper contact is with middle Eocene grayish-green, slightly calcareous shale.

Unit Lu-1

Unit Lu-1 overlies unit 1-A. In most of the wells where the Lu-1 has been identified, the upper contact of this unit is with rocks of unit 1-B (Figure 4). Where unit 1-B is absent (Ku 62, 62A, and 84 wells in the Ku field), unit Lu-1 is in direct contact with middle Eocene sediments.

Unit Lu-1 consists of stratified calcarenitic and calcareous-shale zones. No core samples from this unit have been recovered, but the interpretation of logs (gamma-ray spectrometry, microresistivity, and magnetic resonance) provides this interpretation. Also, the lithologic and petrophysical characteristics of Lu-1 are similar to Lu-2 (Ríos-López, 2002). Because of the separation of radioactive components (thorium and potassium) recorded by the gamma-ray tools, the inferred lithology is assumed to contain the clay minerals illite and smectite because these have been identified in cores obtained from Lu-2.

The maximum thickness of the Lu-1 is 49 m (161 ft). The isopach map of Lu-1 (Figure 5) illustrates its geographic distribution within the Kumaza fields and the lateral thinning in the study area (Figure 4), controlled by the morphology of previous units.

Unit 1-B

Unit 1-B commonly overlies Lu-1. In the absence of units Lu-1 and 1-A, particularly in the central area of the Zaap field and the west end of the Ku field, unit 1-B is at the base of the Kumaza Member. Where units 1-C and Lu-2 are absent, unit 1-B represents the top of the calcarenitic deposits in the middle Eocene, as in the Ku 41, 66, and 84 and Maloob 487 and 489 (Figure 4) wells.

Unit 1-B is composed of packstone to grainstone and rounded biogenic fragments, mostly benthonic foraminifera and algae cemented by fine sparite. Evidence of bioturbation, scarce fracturing, and clay (illite) is seen. Three cores from wells in the Zaap field were recovered from this unit (Table 3). Benthonic foraminifera, including *Pararotalia* sp., *Lepidocyclina* sp., *A. parvula*, *Quinqueloculina* sp., *Spiroloculina* sp., and *Dictyoconus floridanus* mixed with abundant fragments of mollusks, bryozoans, echinoderms, calcareous algae, brown algae, and pellets, characterize the fossil assemblage in this unit.

The maximum thickness of unit 1-B is 86 m (282 ft) in the Zaap field (Z-27D well). The geographic distribution of unit 1-B, within the Kumaza fields, mainly based on well logs, is illustrated in Figure 5.

This unit shows very few changes on the basis of gamma-ray and density logs. The argillaceous mudstones or calcareous-shale layers are also rare or nonexistent (Figure 4).

The common gamma-ray response for unit 1-B is a constant displacement toward lower values; it eventually shows inflection points toward the right, indicating the presence of thin argillaceous lenses.

Unit Lu-2

Unit Lu-2 overlies unit 1-B and underlies unit 1-C. Unit Lu-2 consists of packstones with interbeds of green smectite and illite-rich highly calcareous shale and argillaceous

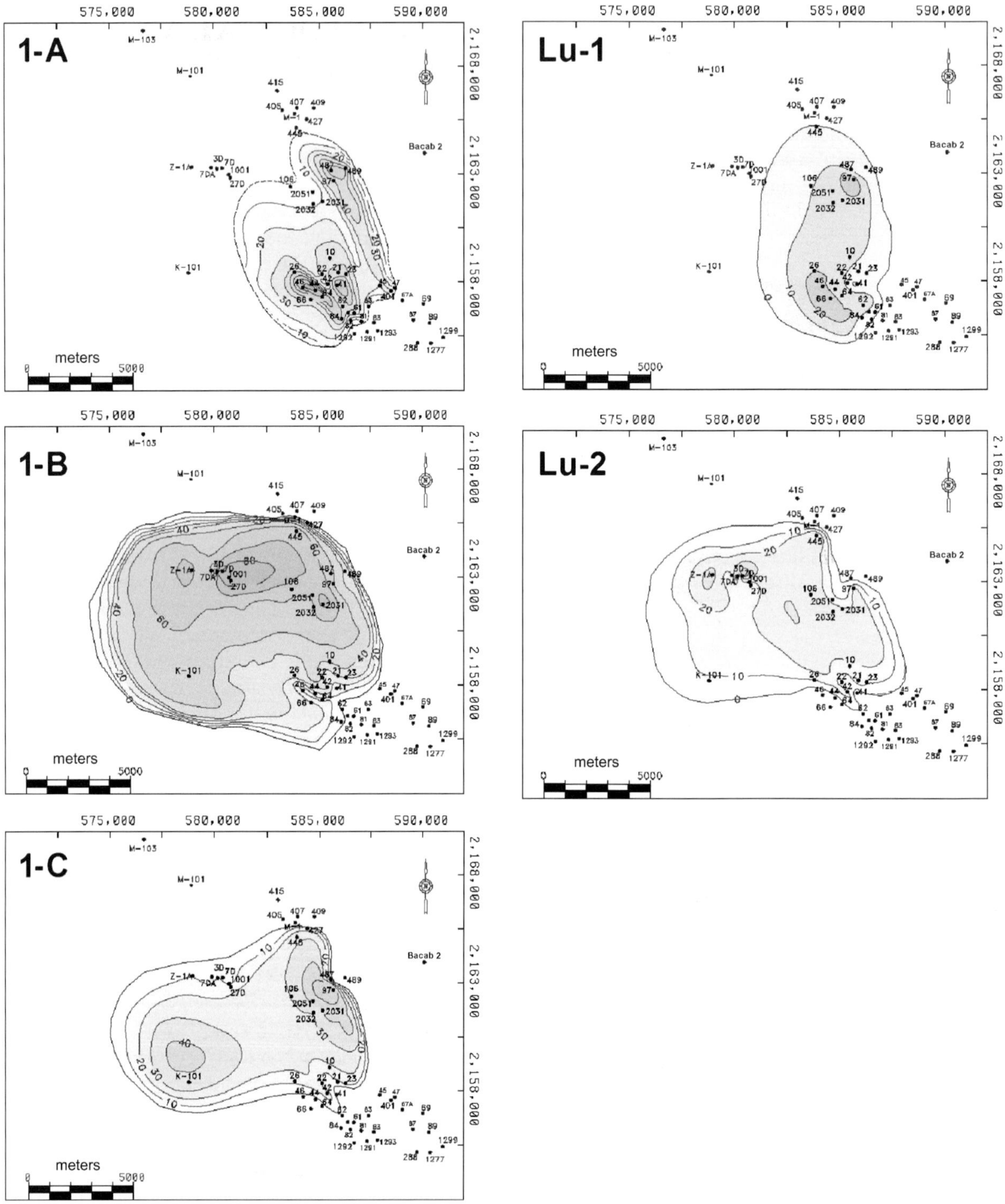

FIGURE 5. Isopach maps by unit in the Kumaza Member. Maps show the total thickness and southeast–northwest transport direction inferred from areas of thickest incised valley fill. See the text for details.

mudstone. Evidence of burrowing is seen. Dominant components include benthonic foraminifera fragments: miliolids, nummulitids, and rotalids with minor amounts of discocyclinids; coralline and blue-green algae; and secondary amounts of mollusks and bryozoans. A core (C-1, Z-1001) from the Zaap field was recovered from this interval.

The maximum thickness of unit Lu-2 is 38 m (125 ft). The distribution of the unit Lu-2 was mapped mostly from well logs (Figure 5). Unit Lu-2 contains calcarenites with a series of calcareous-shale interbeds that become more abundant toward the top. In most wells where the unit has been identified, a 5–15-m-thick (16–49-ft-thick) shale and calcareous-shale section caps the unit. The

Lu-2 lithologic breaks have a log response similar to that of unit Lu-1.

Unit 1-C

Unit 1-C is identified within the Kumaza Member and overlies unit Lu-2. As with the other lithologic units, the extent of unit 1-C is limited. Based on three cores collected from this interval in Zaap field (C-1, Z-7DA; C-1; and C-2, Z-106) and one core (C-1, M-97) from the Maloob field, unit 1-C is composed of packstones with rounded to subrounded foraminifera fragments (benthonic > planktonic), algae, pellets, and intraclasts. Mudstone interbeds and clay mixed with unmodified micrite and organic matter are the dominant rock types. Locally, the unit is silicified, fractured, brecciated, and dolomitized, as in the Maloob field. The maximum thickness of unit 1-C is 54 m (177 ft) (Figure 5).

The fossil assemblage identified in this interval include *A. parvula*, *H. aragonensis*, *Lepidocyclina pustulosa*, *Lepidocyclina (Polylepidina) antillea*, *Pararotalia roblesae*, *Heterodictyoconus* sp., *Smoutina* sp., miliolids (*Triloculina* sp., *Quinqueloculina* sp., *Sigmoilopsis* sp.), discocyclinids, echinoderm fragments, and coralline algae.

Upper Eocene

The upper Eocene contact in these fields is not precisely defined. The upper Eocene consists of approximately 60 m (197 ft) of slightly calcareous, gray shales with *Turborotalia cerroazulensis*, *Hantkenina alabamensis*, *Globigerinatheka semiinvoluta*, and *Globigerinateka mexicana barri*.

The gamma-ray signature for the upper Eocene displays constant and high readings compared to that of the middle Eocene (Figure 3b).

Analysis of Isopach Maps

Isopach maps show that the depocenter for the Kumaza Member (1-A, Lu-1, 1-B, Lu-2, 1-C) is oriented south-southeast to north–northwest, a trend inferred to correspond to the sediment transport direction. This interpretation is supported by a morphological analysis of lithologic units 1-A, Lu-1, 1-B, Lu-2 and 1-C, which suggests a submarine fan-type depositional environment. Accumulations that have been mapped with elongated shapes are interpreted to represent parallel channels.

The isopach map for each unit shows that relatively thin underlying sections are compensated for by a thicker overlying package and vice versa, an evidence of reciprocal sedimentation. Areas of greater topographic relief in underlying units are inferred to have been subject to major erosional effects, constraining the updip limit of local depocenters.

LITHOLOGICAL ANALYSIS OF THE MIDDLE EOCENE KUMAZA MEMBER

The limestone classification used here is that by Dunham (1962). Samples are grainstones and packstones with very little or no matrix between grains. Calcareous-shale facies (unit Lu-2 in the Zaap 1001 well) consist of interbeds of green, smectite-rich, highly calcareous shale and argillaceous mudstone and packstone. Evidence of burrowing, with some of the underlying grainstone bed burrows (without identify what organism type) filled with green clay, is observed. Sorting is good to very good and most grains appear to be abraded. A breccia (core 1, Maloob 97) consists of biomicritic limestone fragments and abundant fractures. Grainstones and packstones develop an incipient planar, gently dipping lamination resulting from the alignment of tabular grains.

Petrographic Analysis

A petrographic summary is included on Table 2. Reliable component determinations can be observed in some samples, and such values are representative of corresponding samples. Where it was not possible to make specific determinations of components, the letter "X" was marked.

Orthochemical Constituents

Matrix

Micrite is the predominant matrix in samples from the Zaap 1001 well from 2586 to 2587 m (8484 to 8487 ft). The reduction in both the amount and size of allochems, along with a greater percentage of micritic mud, indicates a restricted, low-energy environment. Most of these rocks are classified as packstones and wackestones.

Other matrix constituents include organic material and clay. Organic material is present throughout the matrix and is concentrated within some stylolites. Clay laminations are most prevalent within the Zaap 1001 samples. These intervals contain between 3 and 6% clay, of which the predominant constituents are smectite and illite.

Cement

The most common secondary cement is blocky, low-magnesium calcite (LMC) that precipitated in interparticle pore spaces and in the sheltered pores of foraminifera. This cement occurs in minor amounts (3 to 5%). Another cement type (Figures 6, 7) is elongated, isopachous, bladed calcite surrounding allochemical grains (commonly less than 1%).

Table 3. Summary of the diagenetic characteristics observed and interpreted in the Middle Eocene Kumaza Member.

Unit/Well	*Geologic Observations*	*Diagenetic Processes*	*Diagenetic Evidences*	*Diagenetic Environments*	*Porosity Evolution*
1-A/Ku 46, 66, 401	High content of pellets and peloids, biogenic fragments with a very good interparticle porosity preservation. Micrite presence.	Shallow marine waters environments		Sedimentation in shallow marine waters.	
	Fine calcite type dogtooth spar surrounding some grains (Ku 66)	Cementation	Calcite type dogtooth spar could be a recrystallization product by magnesian calcite or from original aragonite but it is possible to find it in meteoric influence zones.	Phreatic marine active zone with transition to saturated fresh-water influence	Good primary porosity preservation
	Moldic porosity in foraminifera and unstable particles	Micritic matrix and grain dissolution		Meteoric environment and probably with meteoric vadose influence	Moldic porosity development
	Intergranular porosity eventually sealed by calcite. Blocky low magnesium-calcite cement that infills the interparticle pore spaces and the sheltered pores of the foraminifera (Ku 401).	Calcite spar precipitation on porous cavities	Minimal exposure in the solution zones because of moldic and vugs porosities are not found in significant amounts.	Meteoric influence, probably in a solution zone	
	Calcite syntaxial overgrowth and calcite spar precipitation in porous cavities.	Magnesian calcite cement and fibrous aragonite between intergranular porosity	A significant exposure in an active meteoric-phreatic environment has promoted abundant amounts of calcite cement (20% in Ku 66).	Phreatic meteoric environment	Primary porosity decrease
1-B/Z-7D, 27D, 1001 (C-2)	Bioclasts constituted by benthonic organisms, algae, peloids and ooids.	Shallow platform marine waters		Sedimentation in shallow marine waters	
	Micritization is observed on fragments' edges.	Bioturbation-micritization	Clasts and matrix micritization.	Marine diagenetic environment	
	Magnesian calcite cement and fibrous aragonite.	Cementation	Cementation of calcite spar over fibrous aragonite, particularly around porous cavities.	Marine environments and phreatic active marine zones	Primary porosity lightly altered

	Calcite spar is infilling and covering the pore walls. Syntaxial overgrowth on echinoderm fragments.	Precipitation	Calcite spar around porous cavities and grain's edges cemented previously. Calcite syntaxial overgrowth on echinoderm plates.	Transition from phreatic active marine zones to fresh-water meteoric zones	Pore size decrease
	Cement and micritic limestone mud dissolution, generating intraparticle, interparticle, or intergranular porosity.	Dissolution	Different leaching degrees are present in many grains. Selective dissolution of skeletal components.	Vadose zone and probably exposure meteoric conditions	Moldic, vugular, and intergranular porosity
	Micritic matrix recrystallized to fine pseudospar calcite.	Recrystallization			
	Mechanic compaction	Compaction		Deep burial	
	Few fractures are present, probably by fluid expansions.	Fracturing			Decrease of porosity by compaction
	Stylolite presence in clasts and grain's edges (until 5 micras)	Pressure-solution			
1-C/Zaap 106, M-97	Bioclasts and ooids. Allochemical components are highly packaged. Intergranular porosity is occupied by spar calcite (Z-106).			Internal platform environments	
	Brecciated limestone fragments mainly constituted by foraminifera biomicrite. The micritic matrix has argillaceous minerals. Foraminifera are not well preserved, and they were partially replaced by microsparite. Feldspars and quartz (200 micras) and pyrite (50–100 micras) have been observed in M-97.	Deposition and sedimentation	Syntectonic appearance of matrix and fragments because the matrix has an alignment with fragments' edges. This could be associated gravitational displacements of lithified (fragmented) rocks.	External platform debris (associated to a self-brecciating process), which is mixed with basin or talus deposits	Primary porosity as a consequence micritic matrix
	Moldic porosity associated with the internal structure of the foraminifera, and vuggy porosity may be present in clasts.	Early dissolution and cementation (stable calcite)	HMC and aragonite are preferentially dissolved by undersaturated calcite fluids, consequently, vuggy and moldic porosity may be present. Spar calcite cementation between fragments and matrix discontinuities. Intergranular and syntaxial overgrowth on echinoderm plates	Probably meteoric influence	
	Bioclasts flow and/or self-brecciating process.	Self-brecciating	Some fragments are with limited dissolution and braked before to be fractured.	Shallow burial	

Table 3. Summary of the diagenetic characteristics observed and interpreted in the Middle Eocene Kumaza Member (cont.).

Unit/Well	*Geologic Observations*	*Diagenetic Processes*	*Diagenetic Evidences*	*Diagenetic Environments*	*Porosity Evolution*
		Mechanic compaction	Fossils and bioclasts have been microfractured or they are broken.	Deep burial	
	Microfractures	Microfracturing	Microfractures are cutting to grains and cements. Microfractures are cemented by microcrystalline calcite.		
	Stylolites	Chemical and mechanical compactation	Stylolites are present in some fragments by pressure-solution effects.		
		Recrystallization	Micrite by microsparite replacement in packstone fragments.		
	Microfractures	Microfracturing	Fracturing without calcite cementation (cutting partially cemented fractures), and eventually fractures are partially cemented by dolomite. This fracturing is also cutting the stylolites.		Permeability increasing
	Bioclasts, clays, and siliceous trace presence have been affected by dolomitization (preferentially microcrystalline) and silication.	Replacement	Incipient matrix dolomitization with isolated crystals. The replacement process is explained by silicic and magnesian fluid presence. The increase in siliceous amounts could be possible because of the primary separation of magnesium to create dolomite.		Decrease of porosity when the clays and micritic matrix are increasing
		Hydrocarbon emplacement	The hydrocarbon emplacement was probably an obstruction to micrite development and dolomitic cement precipitation.		
Lu-2/Zaap 1001	Allochemical grain constituents similar to those found in units 1-A, 1-B, and 1-C but almost replaced by micrite. Matrix is an argillaceous mudstone with interbeddings of illite-smectite highly calcareous shales.	Biological activity as evidence of burrowing is observed		Marine environment	Decrease of porosity when the clays and micritic matrix are increasing

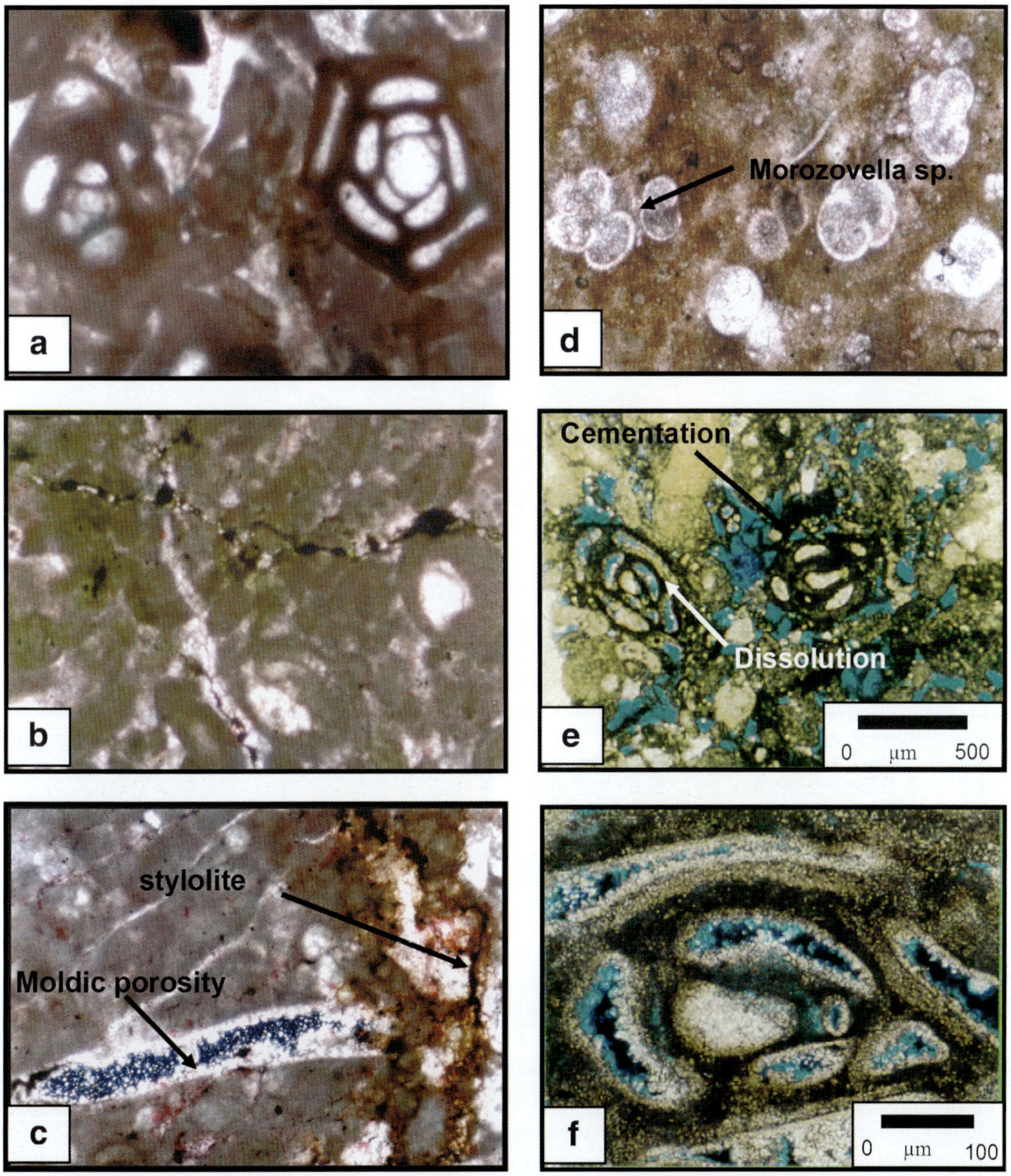

FIGURE 6. Thin-section photomicrographs from calcarenite cores. (a) Fossiliferous wackestone-packstone formed by the abundance of ooids and benthonics (miliolids). Well M-97, C-1, F-26. (b) Fracture details. Well M-97, core C-1, photo F-32. (c) Selective dissolution of some fossils, stable calcite cementation, presence of microfractures cutting cements and grains, and mechanical and chemical compactation features. Well M-97, C-1, F-27. (d) Planktonic foraminifera in micritic matrix with clay and organic matter. Well M-97, C-1, F-44. (e) Fossiliferous packstone with intergranular porosity, dissolution of microfossils, and spar calcite cementation. Well Z-7DA, C-1, F-4. (f) Fibrous calcite cement inside microfossil cavities, calcite spar cement saturating cavities, and dissolution. Well Z-7DA, C-1, F-4.

Allochemical Constituents

Micritic peloids are the major allochemical constituent and occur as rounded grains ranging from 0.1 to 0.2 mm (0.003 to 0.007 in.) (core from the Zaap field). These pellets are primarily present as homogenous aggregates of microcrystalline calcite. Sorting of the grains is good to very good. Peloids range from 10 to 25% in samples from units 1-C (Maloob 97 and Zaap 106) and Lu-2 (core 1 from Zaap 001) and from to 20 to 45% in

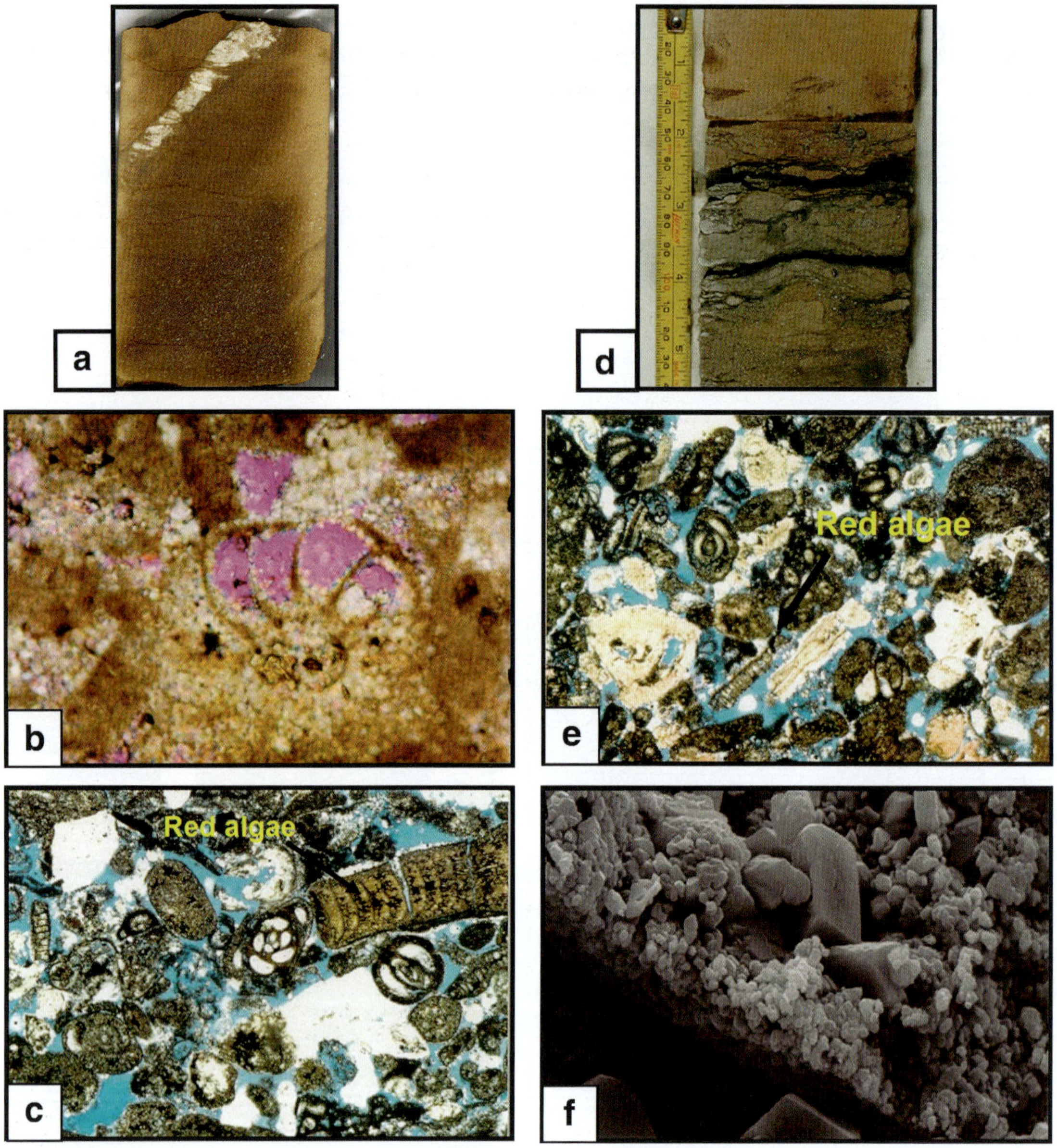

Figure 7. Core photographs and thin-section photomicrographs from calcarenite rocks in the Ku and Zaap fields. (a) Packstone bio- and intraclasts with sporadic fractures sealed by calcite and stylolites. Well Z-106, C-1, F-36. (b) Intragranular and intrafossil porosities. Well Z-106, N-2 F-26. (c) Grainstone consisting of skeletal grains, foraminifera, algae, pellets and blocky calcite cement, late cements, and shelter porosity. Well K-401, C-1, 2214 m (7264 ft). (d) Packstone-wackestone containing argillaceous intercalations and mudstone with evidence of burrowing and sediment churning. Well Z-1001, C-1, 2588 m (8491 ft). (e) Packstone to grainstone containing biogenic fragments and red algae. Some syntaxial overgrowth on echinoderm fragments is observed. Well Z-1001, C-1, 2582 m (8471 ft). (f) Electron microscope image showing intraparticle porosity, calcite spar crystals, microsparite, and micrite, as well as micritization around the skeleton. Well Z-7D, C-1.

samples from units 1-A (Ku 46, 66, 401) and 1-B (Zaap 7D, 7DA, 27D, 1001).

Foraminifera are most abundant in the calcarenite samples, ranging from 15 to 45%. Foraminifera percentages are inversely proportional to pellet content. Foraminifera, dominated by miliolids, with secondary amounts of nummulitids and rotalids, contribute significantly to the overall limestone framework. Minor amounts of discocyclinids are also present. Some of the biomicrite samples from brecciated rocks within unit 1-C in Maloob 97 present a higher proportion of planktonic foraminifera with respect to benthonic foraminifera.

Algae are also present in significant amounts (5–35%), with coralline algae being prevalent within unit 1-B samples (core 2 of Zaap 1001 and 7DA). Blue-green, red, and minor brown algae are also present (Figure 7c, e).

Other fossil fragments present in lesser amounts include mollusk shell fragments, pelecypods, and gastropods (<2%); bryozoans (1–3%); and echinoderm fragments and plates (1–6%). Intraclastic grains occur in amounts ranging between 3 and 6%.

Porosity Types

The prevalent porosity type in Kumaza calcarenites (Table 2) is primary interparticle porosity, followed by intercrystalline and shelter porosities. Secondary porosity, such as vuggy, moldic, and fractures, are subordinate. Figures 6 and 7 illustrate the main types of porosity identified from core samples from the different units of the Kumaza Calcarenites "Member."

Porosities that may or may not depend on the calcarenite structure include the breccias (Maloob 97) and bioturbation (Zaap 1001, core 1).

ORIGIN AND DEPOSITIONAL ENVIRONMENT OF KUMAZA CALCARENITES

Depositional Environments of Units 1-A, 1-B, and 1-C

Unit 1-A

The depositional environment corresponding to pellets and biogenic fragments (foraminifera, echinoderms, and algae) with micrite mud, identified on samples obtained from Ku 66 (core 1), is shallow marine, probably lagoonal or proximal lagoonal. In addition, core samples from the Ku 401 well (C-1, F-2214, Figure 7c) contain a large amount of benthonic foraminifera and abundant miliolids and rotalids with secondary amounts of nummulitids, indicative of a marine lagoon, shallow platform, or shallow-water carbonate-shoal environment. The presence of coralline algae, together with these features, supports an interpretation of shallow-water, platform deposition (Figure 8).

Unit 1-B

Bioclasts consisting of algae fragments and benthonic foraminifera coexisted with bryozoans, mollusks, and echinoderms. Evidence of bioturbation, which indicates that these deposits may have been formed in a shallow-marine or lagoonal setting, is also observed. The abundance of blue-green and red algae in most of the core samples studied (C-1, Zaap 7D), in association to miliolids and other benthonic foraminifera, indicates shallow-shelf deposition, probably intertidal-subtidal.

Unit 1-C

Samples from 1-C (core 1 from Zaap 7DA, cores 1 and 2 from Zaap 106, and core 1 from Maloob 97) contain a similar proportion of components to those observed in units 1-A and 1-B. However, in the Maloob 97 well, brecciation and predominance of planktonic foraminifera over benthonic foraminifera indicate that the calcarenitic sediments, originally deposited in a platform setting, were ultimately transported to slope or platform-edge settings. These allochthonous calcarenite deposits were then quickly covered to preserve the original, syndepositional porosity and diagenetic fabric.

Unit 1-C in Maloob 97, core 1 (2846–2847 m, 9337–9340 ft), is composed of brecciated fossiliferous limestone, foraminiferal biomicrite, and a higher proportion of planktonic over benthonic forams, observed on photomicrographs (Figure 6d). This could be associated with a self-brecciating process resulting from gravitational displacements of lithified sediments from platform edges. These would have been swept away with foreslope debris and mixed suspension sediments from shallow waters. Gravity-induced deposits, similar to "fore reef slope deposits" (Enos and Moore, 1983), could also be associated with the fragmentation of lithified rocks produced by slope instability.

Depositional Environments of Units Lu-1 and Lu-2

The presence of miliolids, nummulitids, rotalids, bryozoans, echinoderm fragments, red algae, and green-blue algae, along with burrowing, indicates shallow-water, platform environments. Illite and smectite in shale and mudstone may be associated with low-energy facies and are mixed with calcareous argillaceous sediments from internal platform subtidal environments.

The calcareous argillaceous sediments in core 1 from the Zaap 1001 well (unit Lu-2) comprise argillaceous mudstones with interbeds of green smectite-rich, highly calcareous shales. In addition, two types of illite are present: authigenic and allogenic, both rich in potassium. Analogous potassium-rich shales in the Cisco and lower Wolfcampian detrital in the southeastern part of Andrews, Texas (Saller et al., 1999), have been interpreted as deltaic or fluvial in origin.

Authigenic illite or low-swelling smectite-illite is associated with sporadic micaceous grains (Baker et al., 2000). Illite is the predominant argillaceous mineral in the shales and mudstones. According to Serra (1990), the origin of the illites, as allogenetic minerals, is strongly related to the diagenetic alteration of feldspars and micas (minerals rich in Al and K). The optimum conditions for their formation are linked to environments with high pH.

Clay is confined to low-energy, restricted environments. According to the constituents and type of argillaceous materials in the calcarenite units (unit Lu-2), no

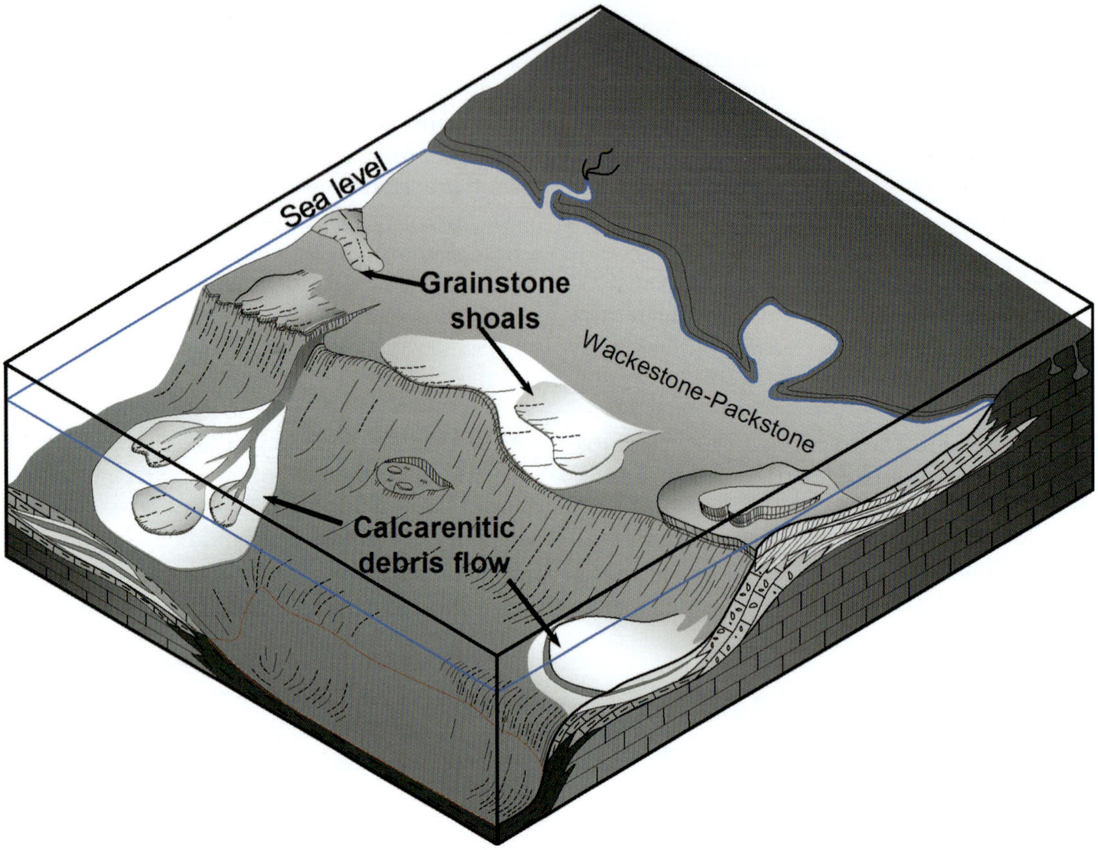

FIGURE 8. Schematic block diagram showing the paleogeography during deposition of the various units of the middle Eocene Kumaza Member.

evidence of sandstone, indicating high-depositional energy, or significant subaerial exposure is observed. Illite and smectite contained in units Lu-1 and Lu-2 are believed to have been transported, probably by submarine currents, to their present position.

Diagenesis of the Calcarenites in the Kumaza Area

Grainstones and packstone-grainstone samples from calcarenitic units 1-A, 1-B, and 1-C contain syntaxial cements surrounding echinoderm plates. This cement is scarce in rocks with porosities greater than 25%. Both grains and matrix are uniformly micritized. Many grains are wholly or partially leached.

No evidence of compaction either at the grain-to-grain level or in the form of stylolites is observed. Rare fractures exist, some of which have been solution enlarged and other fractures partially cemented by calcite. A few irregular vugs occur and attest to dissolution during an eogenetic stage that was a major factor in developing the high-porosity characteristics of these facies.

Syntaxial calcite cement, present in small amounts, but prevalent over other types of cements identified within the calcarenites, indicates a very early diagenetic phase (Cander, 1995), probably contemporaneous with blocky calcite cement.

Most clasts from samples obtained from core 1 (Zaap 7D) and core 2 (Zaap 1001) indicate micritization by bacterial action. Micrite is only observed along grain edges, as most rocks are cemented by sparite. Grain micritization indicates alteration during an early diagenetic stage produced by organic acids (Williams and Barceló-Duarte, 1996), probably by algae, because these occur in greater amounts on samples analyzed for the 1-A, 1-B, and 1-C units.

Cement type and abundance were examined in all calcarenite samples. Cements obtained from the Ku 46, 66, and 401 and Zaap 7D, 27D, Zaap 1001 wells (core 2) in units 1-A and 1-B, together with the development of echinoderms, indicate a minor influence of an active diagenetic-meteoric-phreatic environment that would have encouraged the precipitation of abundant calcite cement. Had the reservoir rock experienced significant exposure to this diagenetic environment, reservoir quality would have been destroyed.

The presence of magnesian calcite cement and fibrous aragonite was observed in intergranular porosity in the form of remnants around biogenic fragments (Zaap

7DA, 27D, and 66 wells). Calcite spar is observed above and/or around these cements, particularly in porous cavities. Early marine diagenesis is assumed because the aragonite and high-magnesium calcite (HMC) particles are stable in marine waters. Consequently, initial marine cementation typically consists of acicular aragonite and micritic calcite with high magnesium content. When interparticle pore spaces are present, the cement forms a uniform thickness jacket (isopach cement) with crystals that if elongated and perpendicularly oriented to the pore wall are known as radial fibrous crystals (Lloyd, 1987).

The HMC and aragonite are preferentially dissolved by meteoric waters until the percolation of saturated solutions begins the precipitation of LMC, whereupon most soluble constituents, such as HMC and aragonite, are replaced by less soluble constituents such as LMC.

Cementation of calcite spar over magnesian calcite and fibrous aragonite (Zaap 7DA and 27D and Ku 66) suggests the effect of near-surface diagenesis. In phreatic environments influenced by saturated fresh water, having skeletal aragonites replaced by mosaics of pseudospar calcite is common, preserving some relic microstructure (Folk, 1964; Beach, 1995).

Petrographic observations of and diagenetic evidence from calcarenite samples were grouped by calcarenitic units and are summarized in Table 3. Description of alteration processes, interpretation of environments, and final porosity effects in the Kumaza calcarenites are major headings in this table.

In the solution zone, circulating waters are still undersaturated with respect to $CaCO_3$, and consequently, vuggy and moldic porosity may be present. Moldic and vuggy porosities do not occur in significant amounts within the Kumaza Calcarenites "Member." Therefore, this zone experienced minimal exposure.

Theoretically, a more detailed zoning of diagenetic environments can be observed in continental facies. Nevertheless, the same characteristics could also be present in the barrier zone of the external shelf, which in this case is the area with meteoric influence. This explains why some of the samples from the Ku 66 and Zaap 1001 well contain up to 20% calcite spar cement with additional moldic and vuggy porosities. This also explains the presence of cementing materials produced by the influence of fresh water and those resulting from marine diagenesis.

A dissolution event becomes noticeable with the emerging of a new (intrafossil) porosity type, possibly under the influence of fresh water derived from rock exposure to meteoric conditions. The minor amount of vadose dissolution indicates that most of the Kumaza Member is not significantly affected by meteoric vadose influences. Nevertheless, sample Zaap 1001 (2587 m, 8487 ft) contains an abundance of blocky LMC (up to 20%). This sample represents shallow, restricted conditions. For a short period, these rocks were exposed to subaerial processes with fresh-water influence.

Undersaturated water eventually became oversaturated at depth, where the most soluble constituents, such as HMC and aragonite, were replaced by less-soluble LMC constituents (Purdy, 1964). Either caused by sediment accumulation or sea level regression, such shallow-water facies could have been initially exposed to meteoric conditions and thus contain a greater percentage of blocky calcite.

Calcarenites deposited in shallow marine waters typically have good interparticle porosity preservation and minimum cement. The most prevalent porosity within the Kumaza Member is primary interparticle porosity followed by intrafossil secondary porosity. The high reservoir quality is because of the preservation of primary porosity, which is 20% in calcarenitic units 1-A, 1-B, and 1-C.

Paragenesis

The proposed paragenetic sequence (Table 3) for the Kumaza Member is (1) shallow tidal deposition, (2) bioturbation-micritization, (3) redeposition (shelf edge), (4) early dissolution and cementation (stable calcite), (5) dissolution of grains and micritic matrix, (6) calcite spar precipitation in pores, (7) intergranular cementation caused by magnesian calcite and fibrous aragonite, (8) recrystallization, (9) chemical and mechanical compaction, (10) fracturing, (11) pressure-solution, and (12) hydrocarbon emplacement.

Variations in depositional environments and diagenetic processes are explained by the sedimentary model, which assumes that the initial deposition of calcarenitic sediments from units 1-A, 1-B, and 1-C was positionally inverted to their final deposition. This means that unit 1-C was located on the base of the sequence (older sediments) and unit 1-A (most recent sediments) was located toward the top.

Consequently, the diagenetic processes affecting the calcarenites could also have had an effect on the overlying units, depending on the length of exposure for each unit and their geographic location.

Additional processes identified in unit 1-C are (1) deposition and sedimentation (shelf edge), (2) self-brecciation, (3) microfracturing (at least two stages), and (4) replacement.

Although the different diagenetic processes and environments are useful for defining the origin and final disposition of the Kumaza Calcarenites "Member," the effect of diagenesis on porosity was considered to be less relevant for reservoir quality. This is because of the good preservation of porosity during the mesogenetic stage, especially for units 1-A, 1-B, and 1-C. These are the permeable units of most interest, which also have increased hydrocarbon storage capacity.

CONCLUSIONS

1. The well-log correlation was performed using only lithostratigraphic concepts because different beds of the Kumaza calcarenite are characterized by them. The correlation type proposed here has been confirmed with pressure measurements and predicted production rates, which are according to the detailed internal reservoir architecture.
2. The lithostratigraphic correlation was based on the recognition of upward-fining or upward-coarsening successions based on detailed observations of well logs and impact of lateral facies changes.
3. The upward fining or upward coarsening were interpreted as cycles or patterns of log behavior and were useful in the identification of regressive and transgressive successions related to changes in sea level.
4. Reservoir fluid characteristics such as water/gas ratios and production-decline rate are elucidated by the new stratigraphic and sedimentological model proposed here.
5. The middle Eocene Kumaza sediments are inferred here to have been originally deposited in shallow-water environments and later were subsequently transported to deeper water environments.
6. Diagenetic evidence in the calcarenites indicates that they were exposed to diagenetic processes beginning at the time of deposition. These processes occurred in shelf environments, with the initial alteration occurring in marine environments and phreatic active marine zones, and progressing through mixing or transitional environments with meteoric influence.

ACKNOWLEDGMENTS

We thank Petróleos Mexicanos (PEMEX) Exploration and Production for releasing the contents of the material for external publication. This chapter is published with the permission of Gustavo Hernández-García, Luis Ramos-Martínez, and Adan E. Oviedo-Pérez of PEMEX. We especially thank Pedro Salmeron-Ureña, Francisco Angeles-Aquino, and Cesar Cabrera-Cuervo for their constructive comments and assistance and many stimulating discussions. We thank Monica Betancourt-Marchant for translating the manuscript to English. The manuscript has benefited from critical reviews of Alula Damte, Tomaz L. Jerzykiewicz, Claudio Bartolini, Robert Gardner, and William A. Ambrose.

REFERENCES CITED

Ángeles-Aquino, F., and A. Cantú-Chapa, 2001, Subsurface Upper Jurassic stratigraphy in the Campeche Shelf, Gulf of Mexico, *in* C. Bartolini, R. T. Buffler, and A. Cantú-Chapa, eds., The western Gulf of Mexico Basin: Tectonics, sedimentary basins, and petroleum systems: AAPG Memoir 75, p. 343–352.

Baker, J. C., P. J. Havord, K. R. Martin, and A. R. Ghori, 2000, Diagenesis and petrophysics of the Early Permian Moogooloo Sandstone, southern Carnavon Basin, western Australia: AAPG Bulletin, v. 84, no. 2, p. 250–265.

Beach, K. D., 1995, Controls and effects of subaerial exposure on cementation and development of secondary porosity in the subsurface of Great Bahama Bank, *in* D. A. Budd, A. H. Saller, and P. M. Harris, eds., Unconformities and porosity in carbonate strata: AAPG Memoir 63, p. 1–33.

Cabrera, T. C., R. Romero, and C. Medina, 1997, Descripcion integrada de nucleos 1 y 2 del pozo Zaap-106-D: Petróleos Mexicanos Unpublished Technical Report, 6 p.

Cander, H., 1995, Interplay of water-rock interaction efficiency, unconformities, and fluid flow in carbonate aquifer: Floridan aquifer system, *in* D. A. Budd and P. M. Harris, eds., Unconformities and porosity in carbonate strata: AAPG Memoir 63, p. 103–124.

Cantú-Chapa, A., and R. Landeros-Flores, 2001, The Cretaceous–Paleocene boundary in the subsurface Campeche Shelf, southern Gulf of Mexico, *in* C. Bartolini, R. T. Buffler, and A. Cantú-Chapa, eds., The western Gulf of Mexico Basin: Tectonics, sedimentary basins, and petroleum systems: AAPG Memoir 75, p. 389–395.

Díaz, A., and R. Landeros, 1979, Descripción de muestras de canal del pozo Kutz 1: Petróleos Mexicanos Unpublished Technical Report, 19 p.

Dunham, R. J., 1962, Classification of carbonate rocks according to depositional texture, *in* W. E. Ham, ed., Classification of carbonate rocks: AAPG Memoir 1, p. 108–121.

Enos, P., and C. H. Moore, 1983, Fore-reef slope environment, *in* P. A. Scholle, D. G. Bebout, and C. H. Moore, eds., Carbonate depositional environments: AAPG Memoir 33, p. 507–537.

Folk, R. L., 1964, Some aspects of recrystallization in ancient limestones, *in* L. C. Pray, ed., Dolomitization and limestone diagenesis, a symposium: SEPM Special Publication 13, p. 169.

García-Genis, J., and L. Pérez, 1992, Informe final del pozo Maloob-101: Petróleos Mexicanos Unpublished Technical Report, 100 p.

Gardner, R. I., E. Klovan, and B. L. Anderson, 1997, Ku-Maloob-Zaap integrated reservoir study (Sección Petrofísica y Geología): Petróleos Mexicanos Internal Technical Report, 85 p.

Gutiérrez, M. I., et al., 1998, Evaluación de reservas de las calcarenitas del Campo Cantarell: Petróleos Mexicanos (PEMEX) Unpublished Technical Report, 162 p.

Hernández-García, S., 1992, Localizaciones para desarrollar el Campo Ku (Terciario): Petróleos Mexicanos Unpublished Technical Report, 18 p.

Juárez-Aguilar, L., and J. Ríos-López, 1999, Análisis petrográfico y modelo sedimentario de las calcarenitas del Eoceno medio en las estructuras Ku-Zaap-Maloob: Petróleos Mexicanos-Instituto Mexicano del Petróleo Unpublished Technical Report, 21 p.

Lloyd, M. R., 1987, Porosity evolution in carbonate rocks: GL507 Petroleum Geology International Human Resources Development Corporation Video Library for Exploration and Production Specialists, 112 p.

Murillo, A., and L. Perez, 1992, Informe final del pozo Maloob-103: Petróleos Mexicanos Unpublished Technical Report, 73 p.

Ortuño M. E., G. del Ángel, J. Gorian, and L. Juárez, 2001, Evaluación petrofísica de las calcarenitas del Eoceno Paleoceno del Complejo Cantarell: Manuscrito presentado en el 39th Congreso Nacional AIPM, Villahermosa Tab., 20 p.

Purdy, E. G., 1964, Diagenesis of recent marine carbonate sediments (abs.): Toronto Symposium on Diagenesis of Carbonate Rocks, *in* L. C. Pray, ed., Dolomitization and limestone diagenesis a Symposium: SEPM Special Publication 13, p. 169.

Ríos-López, J. J., 2002, Indicadores petrofísicos para la determinación de unidades permeables y sus propiedades en rocas carbonatadas de textura calcarenítica: Memorias del Tercer Congreso Nacional de Registros Geofísicos de Pozos, Reynosa, México, 12 p.

Romero-Peñaloza, R., and R. Noriega, 1998, Descripción integrada del nucleo 1 pozo Ku 97: Petróleos Mexicanos Unpublished Technical Report, 7 p.

Saller, A. H., J. A. D. Dickson, and F. Matsuda, 1999, Evolution and distribution of porosity associated with subaerial exposure in upper Paleozoic platform limestones, west Texas: AAPG Bulletin, v. 83, no. 11, p. 1835–1854.

Serra, O., 1990, Clay, silt, sand, shales: Houston, Texas, Schlumberger, 211 p.

Torres-Rodríguez, V., 1998, Estudios geológicos y petrográficos de los núcleos 1 y 2 del pozo Maloob-97: Petróleos Mexicanos-Instituto de Investigaciones Eléctricas Unpublished Report, 74 p.

Williams, R. C., and J. Barceló-Duarte, 1996, Estudio diagenético en el Campo Catedral, Estado de Chiapas: Revista Ingeniería Petrolera, v. 36, no. 12, p. 44–53.

13

Cantú-Chapa, A., 2009, Validation of the stratigraphic method to build Tithonian (Jurassic) subsurface structural maps, Campeche Shelf, Southern Gulf of Mexico, *in* C. Bartolini and J. R. Román Ramos, eds., Petroleum systems in the southern Gulf of Mexico: AAPG Memoir 90, p. 279–284.

Validation of the Stratigraphic Method to Build Tithonian (Jurassic) Subsurface Structural Maps, Campeche Shelf, Southern Gulf of Mexico

Abelardo Cantú-Chapa

Instituto Politecnico Nacional, Mexico, D.F., Mexico

ABSTRACT

Stratigraphic interpretation of gamma-ray logs and identification of ammonites from the Kix 1A and Pich 1 wells confirm the existence of the Sinan-Mison syncline in the Campeche Shelf, southern Gulf of Mexico. The structural configuration of the syncline was determined once the contact between the Edzna-Akal formations was established at the Jurassic–Cretaceous boundary on the basis of depth data from wells in the region.

Subsurface data (1) yield new insights into the stratigraphic and structural position of the contact between the Edzna and Akal formations, which have great petroleum potential and (2) validate the certainty of the methodology applied.

The systematic study of specimens *Aspidostephanus* sp., *Cuyaniceras* sp., and cf. *Corongoceras* sp. from the Kix 1A and Pich 1 wells are characteristic of the Tithonian and the Berriasian. This is the first known occurrence of Berriasian ammonites in the subsurface of southern Mexico.

INTRODUCTION

Ammonites are the most important but scarce fossils used in dating Upper Jurassic strata from wells drilled in southeastern Mexico. The precise depth and age of the sequences from where the conventional cores containing the ammonites were collected are established through (1) the systematic study of those fossils, (2) their restricted stratigraphic distribution, and (3) gamma-ray logs.

Deformed and fragmented ammonites coming from deeper depths in the Kix 1A and Pich 1 wells are hard to identify. These specimens should not be thrown away because they yield ages of older beds that can be used in chronostratigraphic studies.

GEOLOGICAL BACKGROUND

This work represents a number of projects conducted by the author in several regions of southeastern Mexico over the years. These investigations began with the study of upper Tithonian ammonites from the Chac 1 and 2 and Tunich 1 wells (Cantú-Chapa, 1977, 1982).

DOI:10.1306/13191088M902804

The genera *Salinites*, *Durangites*, *Parodontoceras*, and *Protancyloceras* are characteristic of the late Tithonian in Mexico, Cuba, and Louisiana, and they represent the same paleogeographic conditions at the top of the Jurassic (Cantú-Chapa, 1982, 1989; Imlay, 1980; Imlay and Hermann 1984; Myczynski, 1999).

The identification of these fossils permitted the dating of the formations and contacts in the southeastern region of Mexico. In the same way, the combination of gamma-ray logs and ammonites from oil wells has been used in stratigraphic and structural studies in several regions of Mexico (Cantú-Chapa, 1992).

In this study, the stratigraphic data were digitized to construct two structural maps at the top and base of the Tithonian in the Campeche Shelf. The objective was to prove the presence of the same structures in both maps (Cantú-Chapa and Ortuño-Maldonado, 2003).

In the southern Gulf of Mexico, the top of the Edzna Formation ranges from 1850 to 5500 m (6069 to 18,045 ft) below sea level. The Tithonian age of this formation was established with ammonites from several offshore oil wells. The structural maps of the Tithonian Edzna Formation were built using paleontologic and biostratigraphic data, as well as the gamma-ray log. The latter curve was used to determine the precise depth at which cores containing the specimens were collected.

Tithonian Stratigraphy of Campeche

Strata of this age were penetrated by all wells in the southern Gulf of Mexico. The Tithonian Edzna Formation generally consists of mudstone and calcareous, bentonitic shale, which is underlain and overlain by dolomitized limestone units. This formation had previously been divided in three members (Angeles-Aguino and Cantú-Chapa, 2001).

The transitional contact between the Akal (Lower Cretaceous) and Edzna (Tithonian) formations is characterized by a displacement to the right of the gamma-ray curve caused by the calcareous and bentonitic content of the shale.

The contact between the Edzna and Akimpech (Kimmeridgian) formations records a displacement to the right by left of the gamma-ray curve caused by a change in lithology of the Akimpech, which consists of microcrystalline dolomites (Angeles-Aguino and Cantú-Chapa, 2001; Cantú-Chapa and Ortuño-Maldonado, 2003). Commonly, this contact reflects abrupt changes of lithologies.

Structural Maps of the Tithonian Edzna Formation in the Campeche Shelf

Two structural maps depict the most well known regional structures affecting this formation: (1) the Ceeh-Cantarell anticlinorium in the northeast, (2) the Sinan-Mison syncline in the center, and (3) the May-Okan anticline in the south (Figure 1a).

The Sinan-Mison syncline is 70 km (43 mi) long and strikes west–east. In this structure, the top of the Edzna ranges from 3283 to 5594 m (10,771 to 18,353 ft) (Cantú-Chapa and Ortuño-Maldonado, 2003).

Studied ammonites come from conventional cores collected from the Pich 1 and Kix 1A wells, which are located along the flanks of the syncline (Figure 1b). These specimens define the Jurassic–Cretaceous boundary (Edzna and Akal contact). The specimens are identified as *Aspidostephanus* sp., *Cuyaniceras* sp., and *Corongoceras* sp. The first two genera correspond to the late Berriasian, and the last one is of late Tithonian age.

The stratigraphic data yielded by the ammonites are comparable to those previously used to construct two Tithonian maps in the same region of Mexico (Cantú-Chapa and Ortuño-Maldonado, 2003).

Comparison of Berriasian and Tithonian Subsurface Stratigraphic and Structural Data

The Pich 1 and Kix 1A wells were drilled in the Sinan-Mison syncline. Both wells penetrated the transitional contact between the Akal and Edzna formations at different depths. The contact in both wells is characterized by a displacement of the gamma-ray curve toward the left (Figure 1b).

Aspidostephanus sp. and *Cuyaniceras* sp. were obtained from core 1 at the interval 5415–5424 m (17,766–17,795 ft) in the Kix 1A well. Both genera of the upper Berriasian had only been recognized in Argentina (Steuer, 1897; Leanza, 1945; Wright et al., 1996). The presence of these specimens in Mexico will yield new insights into the biostratigraphic and paleogeographic relationships between the two regions.

One specimen assigned as cf. *Corongoceras* sp. was obtained from core 3 at the interval 4520–4529 m (14,829–14,859 ft) in the Pich 1 well. This genus has been reported from the upper Tithonian of Argentina, Austria, Cuba, Spain, Madagascar, Mexico, and Rumania (Klein, 2005).

The specimens of *Aspidostephanus* sp. and *Cuyaniceras* sp. are 175 m (574 ft) above the contact between the Edzna and Akal formations in the Kix 1A well (core 1), whereas the specimen of *Corongoceras* was obtained at the contact between the same formations in the Pich-1 well (Figure 1b).

Structural Observations

The new stratigraphic data, along with the structural data, were used to build the top of the Tithonian map in the Campeche Shelf (Cantú-Chapa and Ortuño-Maldonado,

FIGURE 1. (a) Upper Tithonian structural map in the Campeche Shelf. (b) Gamma-ray logs showing the Akal-Edzna contact and core depths. (c) Sinan-Mison syncline and location of the Pich 1 and Kix 1A wells. Dotted lines indicate the new structural value along the axis of the structure.

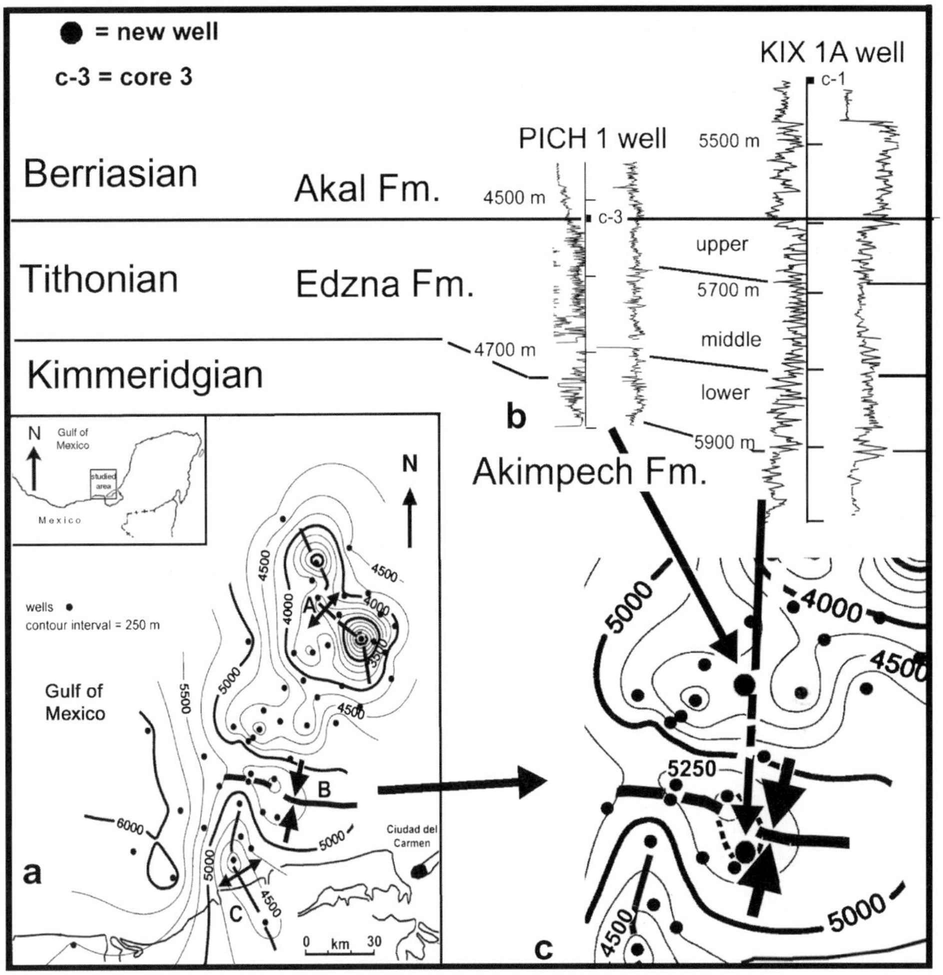

2003). In addition, biostratigraphic data and radioactive well logs (gamma ray) support the mapping elements.

The exact structural position of the new stratigraphic data in both wells is confirmed in the original map, confirming thus, the validity of the methodology applied to build the top Tithonian map (Cantú-Chapa and Ortuño-Maldonado, 2003). The new subsurface data from both wells are located in the flanks of the Sinan-Mison syncline.

The Pich 1 well was drilled in the north flank of the Sinan-Mison syncline in a local structure perpendicular to the syncline axis. The structural position of the ammonite obtained (4520–4529 m, 14,829–14,859 ft) from the Tithonian–Cretaceous contact (Akal-Edzna formations) corresponds to the north flank of the Sinan-Mison syncline. At the well site, the new structural value changes from 4500 to 4530 m (14,764 to 14,862 ft), which is meaningless at the regional scale at which the map was built.

However, the contact between the Akal-Edzna formations changes from 5250 to 5500 m (17,224 to 18,045 ft) in the Kix 1A well. Berriasian ammonites obtained from the interval 5415–5424 m (17,766–17,795 ft) show that these fossils are 170 m (558 ft) above the Tithonian–Cretaceous contact, and thus, their position does not alter the originally proposed structural configuration (Cantú-Chapa and Ortuño, 2003). The precision of the new structural data documented for both wells coincides or is very close to the data used to build the original map. Thus, the Berriasian ammonites from the Kix 1A well are the deepest occurrences of these specimens in Mexico.

CONCLUSIONS

The results of this investigation demonstrate that an appropriate use of paleontology, stratigraphy, and well-log data are adequate to build subsurface structural maps.

New stratigraphic and structural information from two wells in the Campeche Self validates the mapping results previously achieved for this region, confirming the correctness of the methodology applied in earlier studies (Cantú-Chapa, 1982; Cantú-Chapa and Ortuño, 2003).

SYSTEMATIC PALEONTOLOGY

Family Aspidoceratidae Zittel (1895)

Subfamily Simoceratinae Spath (1924)

Genus *Aspidostephanus* Spath (1925)

Type species

Holcostephanus (*Astieria*) *depressus* Steuer (1897).

Discussion

Aspidostephanus is known from the Berriasian of Argentina after an old description and illustration (Steuer, 1897; see Wright et al., 1996). *Aspidostephanus* is assigned here to the family and subfamily mentioned above by its evolute shell, which is ornamented with simple, large, and widely spaced ribs, and by its suture line with S1 rectangular, large base, and superficial subdivisions.

The latter structure characterizes the Subfamily Simoceratinae instead of the Subfamily Spiticeratinae, which was previously assigned by Wright et al. (1996) and Klein (2005).

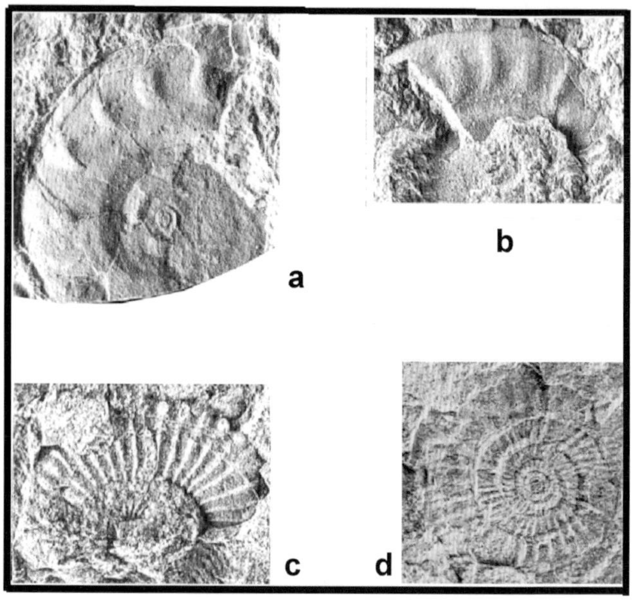

FIGURE 2. (a, b) *Aspidostephanus* sp.; (c) *Cuyaniceras* sp.; (d) cf. *Corongoceras* sp. All specimens with lateral view. (a–c) From the Kix 1A well (core); (d) from the Pich 1 well, Campeche Shelf, southern Gulf of Mexico. The scale for all specimens is ×2. Specimens are coated with ammonium chloride. (a, b) IPN-1113-1114; (c) IPN-1115; (d) IPN-1116.

Aspidostephanus sp. (Figure 2a, b)

Description

Small, evolute, and planulate shell with plain flanks; shallow umbilicus with inclined and low wall; and a row of long, large, slightly lateral nodes, inclined forward born freely at the internal part of the flank and separated by space as large as them. The ventral region, whorl cross section, and suture line are unknown.

Discussion

The two small, evolute, and fragmented specimens have well-preserved lateral sides. The specimens resemble the Argentine *Aspidostephanus depressus* (Steuer, 1897) by their evolute shell, shallow umbilicus, and simple, large, and separated lateral and long nodes, but the Mexican specimens differ by the lateral nodes that occur in the middle part of the flank instead of the ribs that characterize the Argentinian form.

Material

Two small and fragmented specimens. Kix 1A well, core 1 (5415–5424 m, 17,766–17,795 ft). IPN-1113-1114, respectively.

Age

Upper Berriasian: lower part of the Akal Formation.

Family Neocomitidae Salfeld (1921)

Genus *Cuyaniceras* A. F. Leanza (1945)

Type species

Odontoceras transgredians Steuer (1897); SD by Arkell (1952).

Cuyaniceras sp. (Figure 2c)

Description

Flat and evolute shell with fine, simple, and slighty flexuous ribs that are separated by larger interspaces, ending in small ventrolateral tubercles.

Discussion

This genus is from Argentina group species with evolute shell and fine and simple ribs, some of them branching from the umbilical margin. Our Mexican specimens resemble *Corongoceras raripartitum* (Steuer) *in* Leanza (1945, pl. 13, figure 3, p. 56) by its evolute shell and by its simple ribbing ending in ventrolateral tubercles. By the poor state of preservation, these specimens are assigned as genus level.

Material

Two fragmented specimens. Kix 1A well, core 1 (5415–5424 m, 17,766–17,795 ft). IPN-1115.

Age

Cuyaniceras characterizes the upper zone with *Spiticeras damesi* of the Berriasian from Argentina (Leanza, 1945).

Family Himalayitidae Spath (1925)

Genus *Corongoceras* Spath (1925)

Type species

Corongoceras lotenoense Spath (1925)

cf. *Corongoceras* sp. (Figure 2d)

Description

Very evolute and planulate shell, with a shallow umbilicus; it is ornamented with fine, simple, rectiradiate, evenly spaced primary ribs born freely at the umbilical edge. Ribs at unevenly spaced intervals and bifurcate alternate with simple ribs in the internal whorls. The ventral region is unknown.

Discussion

This crushed specimen with very evolute coiling and simple, radial ribs bears considerable resemblance to *Corongoceras cordobai* Verma and Westermann (1973, p. 248, pl. 52, figures 4, 5) from the upper Tithonian of Real de Catorce, central Mexico.

A piece of the preserved flank of this specimen is unidentifiable.

Material

One fragmented specimen. Pich 1 well, core 3 (4520–4529 m, 14,829–14,859 ft). IPN-1116.

ACKNOWLEDGMENTS

The author would like to thank Claudio Bartolini for his assistance with the English manuscript. Special thanks to Blanca Buitrón for critical review of this manuscript.

REFERENCES CITED

Angeles-Aguino, F., and A. Cantu-Chapa, 2001, Subsurface Upper Jurassic stratigraphy in the Campeche Shelf, Gulf of Mexico: AAPG Memoir 75, p. 343–352.

Arkell, W. J., 1952, Jurassic Ammonoidea: Journal of Paleontology, v. 26, p. 860–861.

Cantú-Chapa, A., 1977, Las amonitas del Jurásico Superior del pozo Chac 1, Norte de Campeche (Golfo de México): Revista del Instituto Mexicano del Petróleo, v. 9, no. 2, p. 8–39.

Cantú-Chapa, A., 1982, The Jurassic–Cretaceous boundary in the subsurface of Eastern Mexico: Journal of Petroleum Geology, v. 4, no. 3, p. 311–318.

Cantú-Chapa, A., 1989, Precisiones sobre el limite Jurásico-Cretácico en el subsuelo del Este de México: Revista de la Sociedad Mexicana de Paleontologia, v. 2, no. 1, p. 26–69.

Cantú-Chapa, A., 1992, The Jurassic Huasteca series in the subsurface of Poza Rica, eastern Mexico: Journal of Petroleum Geology, v. 15, no. 3, p. 259–282.

Cantú-Chapa, A., and E. Ortuño-Maldonado, 2003, The Tithonian (Upper Jurassic) Edzna Formation, an important hydrocarbon reservoir on the Campeche Shelf, Gulf of Mexico, *in* C. Bartolini, R. T. Buffler, and J. Blickwede, eds., The Circum-Gulf of Mexico and the Caribbean: Hydrocarbon habitats, basin formations, and plate tectonics: AAPG Memoir 79, p. 305–311.

Imlay, R. W., 1980, Jurassic paleobiogeography of the conterminous United States in its continental setting: U.S. Geological Survey Professional Paper, v. 102, 134 p.

Imlay, R. W., and G. Hermann, 1984, Upper Jurassic ammonites from the subsurface of Texas, Louisiana, and Mississippi, *in* W. S. Ventress, D. G. Bebout, B. F. Perkins, and C. H. Moore, eds., SEPM, Gulf Coast Section Foundation, 3rd Annual Research Conference Proceedings, Baton Rouge, Louisiana, 1982, p. 149–170.

Klein, J., 2005, Lower Cretaceous ammonites I, *in* W. Riegraf, ed., Fossilium Catalogus, I Animalia, Pars 139: Leiden, Backhuys Publishers, 484 p.

Leanza, A. F., 1945, Ammonites del Jurásico superior y del Cretácico inferior de la Sierra Azul, en la parte meridional de la provincia de Mendoza: Anales del Museo de La Plata, Paleontología (Nueva Serie), v. 1, 99 p.

Myczynski, R., 1999, Some ammonite genera from the Tithonian of western Cuba and their paleobiogeographic importance: Studia Geologica Polonica, v. 111, p. 99–112.

Salfeld, H., 1921, Kiel-und Furehenbildung auf der Schalenaussenseite der Ammonoideen in ihrer Bedeutung für die Systematik und Festlegung von Biozonen: Zentralbland für Mineralogie, Geologie und Paläontologie, p. 343–347.

Spath, L. F., 1924, On the ammonites of the Speeton Clay and the subdivisions of the Neocomian: Geological Magazine, v. 61, p. 73–89.

Spath, L. F., 1925, VII, Ammonites and Aptychi, *in* J. W. Gregory, ed., The collection of fossils and rocks from Somaliland made by Messrs: Wyllie and Smellie, Monographs of the Geological Department of the Hunterian Museum, Glasgow University, v. 1, p. 111–164.

Steuer, A., 1897, Argentinische Jura-Ablagerungen: Ein Beitrag zur Kenntniss der Geologie und Palaeontologie der argentinischen Anden: Palaeontologische Abhandlungen, Jena (new series), v. 7, pt. 3: 127–222, p. 192 (66).

Verma, H. M., and G. E. G. Westermann, 1973, The Tithonian (Jurassic) ammonite fauna of Sierra Catorce, San Luis Potosi, Mexico: Bulletins of American Paleontology, v. 63, no. 277, p. 107–320.

Wright, C. W., J. H. Callomon, and M. K. Howarth, 1996, Cretaceous Ammonoidea: Treatise on invertebrate paleontology: Part L. Mollusca 4 revised: Colorado and Lawrence, Kansas, The Geological Society of America and The University of Kansas Press, 362 p.

Zittel, K. A. Von, 1895, Grundzüge der Paleontolgie (Palaeozoologie): München, Oldenbourg, 971 p.

14

Ortuño, A. S., M. Flores F., C. Romero M., and M. Moreno L., 2009, Basin modeling of the Upper Jurassic petroleum systems (Tithonian and Oxfordian) in the Akalan-Chilam area of the Campeche Sound in the Southern Gulf of Mexico, *in* C. Bartolini and J. R. Román Ramos, eds., Petroleum systems in the southern Gulf of Mexico: AAPG Memoir 90, p. 285–313.

Basin Modeling of the Upper Jurassic Petroleum Systems (Tithonian and Oxfordian) in the Akalan-Chilam Area of the Campeche Sound in the Southern Gulf of Mexico

Salvador Ortuño Arzate, Cuauhtémoc Romero Moguel, and Marcelino Moreno López

Instituto Mexicano del Petróleo, San Bartolo Atepehuacán, Mexico D.F., Mexico

Marco Flores Flores

Activo Regional de Exploración Marina, Petróleos Mexicanos (PEMEX), Exploración y Producción, Ciudad del Carmen, Campeche, Mexico

ABSTRACT

Integration of geological, geochemical, and geophysical data and information has commenced in the southern part of the Campeche Sound in the Gulf of Mexico. In the Akalan-Chilam study area, two-dimensional (2-D) and three-dimensional (3-D) seismic data were integrated with information from exploration wells to reveal the region's oil potential and estimate its economic value.

The initial study approach was to integrate all of the available data to construct numerical models of the petroleum systems to ascertain their thermal evolution and the region's petroleum potential, with an emphasis on hydrocarbon volumes available to charge the reservoirs. This required the identification and generation potential of source rocks, an appraisal of expulsion efficiency, definition of migration pathways, and estimation of trapped volumes in plays and prospects of the study area. The ultimate goal was to define exploration prospects in the study area.

DOI:10.1306/13191089M903333

Three-dimensional numerical models were constructed by integrating regional 2-D and 3-D seismic data in the study area. Numerical modeling and simulations of thermal maturity and fluids were accomplished with the Temispack program for 2-D and 3-D petroleum systems.

INTRODUCTION

Exploration of the Campeche Sound in the southern Gulf of Mexico requires accurate documentation to identify plays, evaluate their petroleum potential, and support recommendations for proposed well locations. Constructing models encompassing the geodynamics of this sedimentary basin and its petroleum systems requires the integration of all available information for Campeche Sound, particularly in the Akalan-Chilam study area.

Campeche Sound, with its vast petroleum resources, has been the subject of many geological, geophysical, and geochemical studies. Among the most important publications in this region are those of Ángeles (1982, 1988, 1993, 1996a, 1996b), which, along with numerous unpublished reports, established the basic stratigraphic nomenclature of the Campeche Sound. Furthermore, important contributions were made by Araujo et al. (1986), Bello (1991), Ornelas et al. (1991), Ángeles et al. (1992), Ángeles and Ortuño (1994), and Ornelas et al. (1994) concerning the petroleum stratigraphy and geology of this region. The structural geology and general tectonics of the area were addressed by Meneses (1980), Ramírez et al. (1996), and Salomón (1999). More recently, structural work was done by Mitra et al. (2005) on the Cantarell-Sihil field and Mitra et al. (2006) on the Ku-Maloob-Zaap field. The following authors have discussed the petroleum geology and geochemistry of Campeche Sound's petroleum systems: Holguín and Romero (1982), Bañuelos (1993), García (1993), Hernández (1993), Medrano and Romero (1993), Ortega et al. (1995), Medrano et al. (1996), Romero et al. (1996), Guzmán and Mello (1999), Santamaría (2000), Guzmán et al. (2001), Romero et al. (2001), Santamaría and Horsfield (2001), and Romero et al. (2004). Integrated studies were documented by an internal report of the Instituto Mexicano del Petróleo (IMP) (2000), entitled Atlas de las Cuencas Sedimentarias Terciarias y Mesozoicas de México and the Northeast Marine Region (RMNE) (2002) of Petróleos Mexicanos (PEMEX) entitled Sistemas Petroleros en el Area Tunich-Chilam. Recently, Holguín et al. (2005) reviewed the hydrocarbon seeps of the southern Gulf of Mexico. Finally, petroleum system modeling by the marine region is discussed in a PEMEX internal report (RMNE, 1999) and reports by Beicip-Franlab (2004), Ortuño et al. (2004), and Ortuño et al. (2006).

Several general exploration studies as well as specific geophysical, geological, and geochemical work have been conducted in the Akalan-Chilam area. Seismic, gravimetric, and some geochemical data have been gathered, and three exploration wells were drilled in this area. Several planned well locations have also been documented.

In this Akalan-Chilam area study, petroleum systems were modeled with the Temispack software program based on the theoretical and technological work of the French Petroleum Institute and the Beicip-Franlab company (Ungerer et al., 1990; Rudkiewicz et al., 2000; Schneider et al., 2000).

REGIONAL GEOLOGICAL AND GEODYNAMIC FRAMEWORK

Stratigraphic and Sedimentological Synthesis of Campeche Sound

The study area is located in southwestern Campeche Sound near the city of Ciudad del Carmen in the southernmost Gulf of Mexico (Figure 1). This area is bordered to the east by the Yucatan Platform, to the west by the 200-m (656-ft) isobath, and to the south by the coasts of Tabasco and Campeche states. The basement is considered to be transitional, or thinned continental crust, composed of a metamorphic complex with Paleozoic and Mesozoic intrusions (IMP, 2000) called the Maya block. Diabase dikes intrude the metamorphic basement, which consists of mylonitic and amphibolitic gneisses with radiometric ages of about 500 m.y. (Ángeles, 2006).

Through most of its history, the Yucatan Platform has been an area of predominantly carbonate sedimentation (Aguayo, 1966). Figure 2 displays the general stratigraphic table for Campeche Sound, including the Akalan-Chilam study area. From a geodynamic standpoint, this area evolved as a passive margin after the Jurassic rifting opened the Gulf of Mexico. During the Cretaceous, inner and outer carbonate shelf sedimentation typified this area, followed by Tertiary siliclastic sedimentation.

The sedimentary sequence is inferred to have started with Late Triassic to Middle Jurassic continental red beds more than 1000 m (3281 ft) thick deposited during rifting. Their nearest onshore equivalents are the Cahuasas and Todos Santos formations. This continental sequence is overlain by Middle Jurassic (Callovian) evaporites, consisting mostly of salt, which has intruded the overlying Upper Jurassic, Cretaceous, and Cenozoic strata.

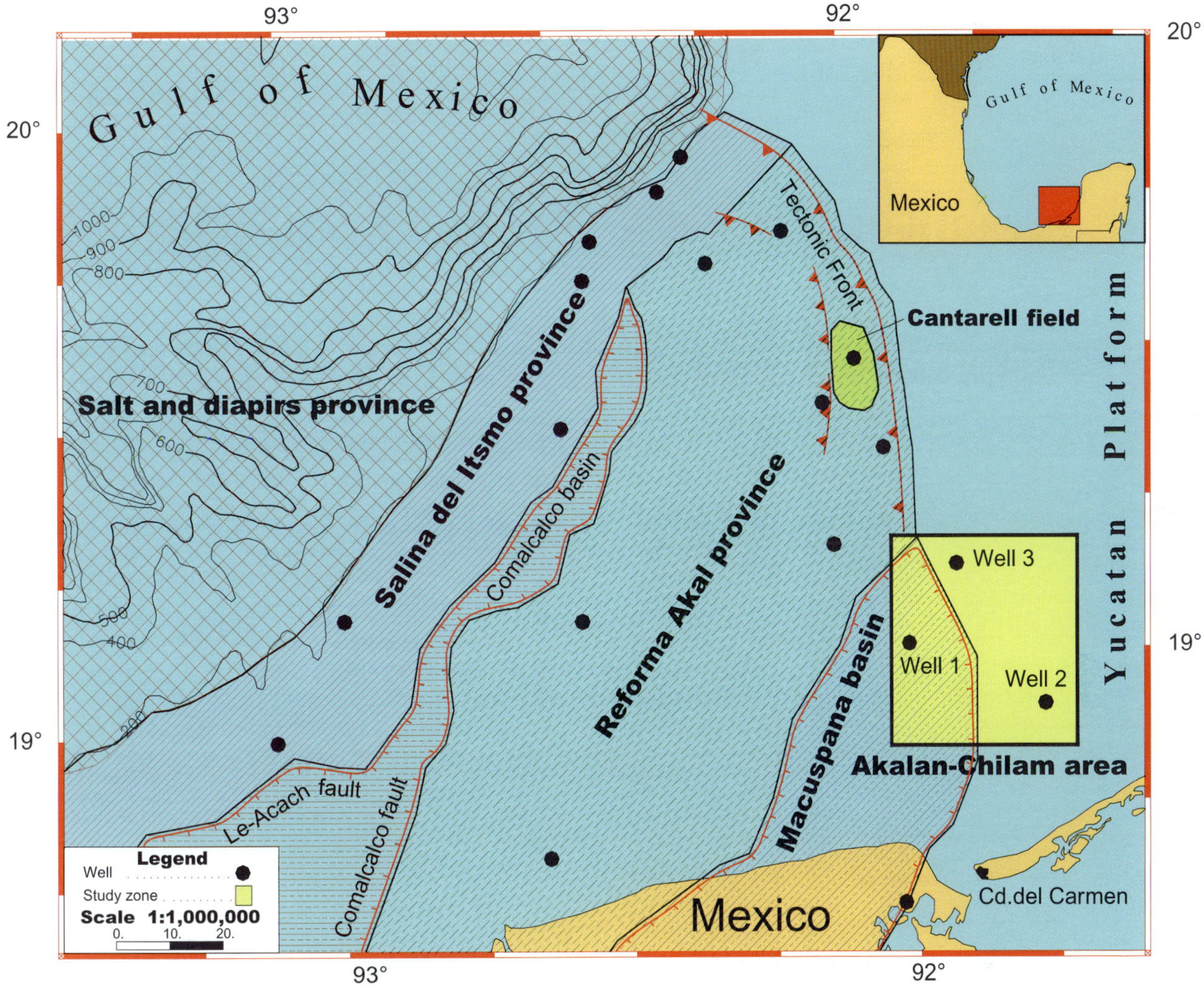

FIGURE 1. Location map of the Akalan-Chilam area within the Campeche Sound.

The Campeche salt averages about 2000 m (6562 ft) thick, thins to less than 500 m (1640 ft) to the east in the Akalan-Chilam area, and is generally absent on the Yucatan Platform.

Open-marine deposition began at the end of the Callovian and became widespread in the Oxfordian. The Yucatan Platform remained emergent and was the source area for clastic marine sedimentation (Ángeles, 1996a). During the Oxfordian, therefore, sandstones and carbonates were being deposited, whereas oolitic carbonate deposition characterized the Kimmeridgian. This was followed in the Tithonian by organic-rich argillaceous carbonate deposition, and these high total organic carbon (TOC) facies constitute the main source rocks in Campeche Sound.

Ángeles (1996a) provided a detailed stratigraphy for the Oxfordian, Kimmeridgian, and Tithonian by subdividing this sequence into eight lithofacies from oldest to youngest named A, B, C, D, E, F, G, and H, respectively. Lithofacies A encompasses the entire Oxfordian, whereas lithofacies B to E correspond to the Kimmeridgian. Lithofacies F, G, and H make up the Tithonian. These lithofacies are grouped into lithostratigraphic units named the Ek-Balam Group (Oxfordian), Akimpech Formation (Kimmeridgian), and Edzná Formation (Tithonian). The Cretaceous is represented by carbonates with facies that vary from inner to outer platforms in the western Campeche Sound to shallow water environments in the Akalan-Chilam area. The uppermost Cretaceous is represented by carbonate breccias. Figure 2 shows the Mesozoic stratigraphy of this area in general terms. During the Cenozoic, extensive clay and sand-rich sedimentation occurred in inner to outer neritic environments dominated by a regression that started in the Paleocene and continued through most of the Neogene.

Post-Laramide compression affected the Mesozoic–Paleogene sequence, accompanied by halokinetic deformation. This tectonic front is significant in the Taratunich, Zaap, Ku, Batab, Abkatun, Bacab, Balam, and Cantarell structures (Ángeles, 1996a). The structure of

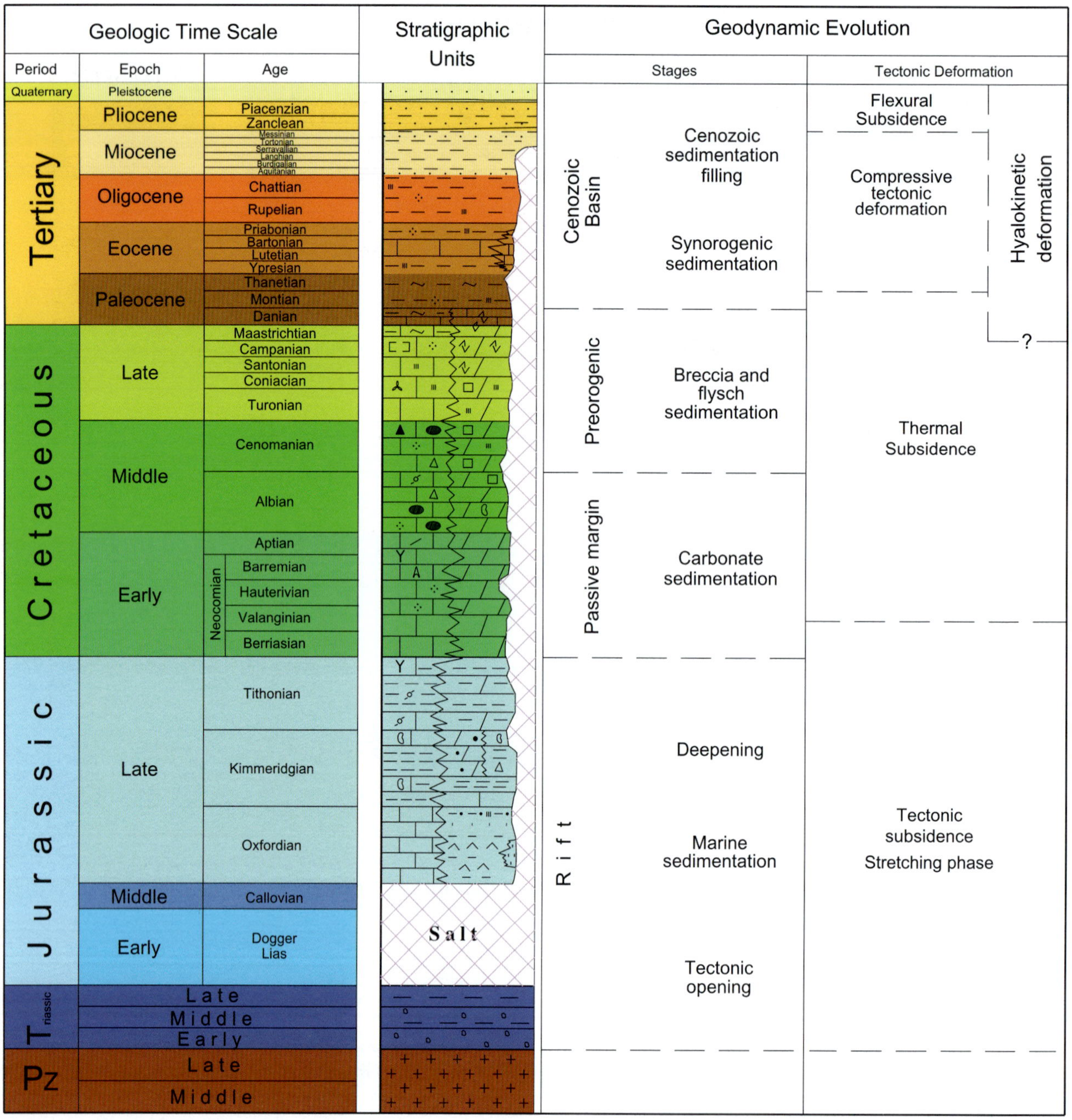

FIGURE 2. General stratigraphic chart of the Campeche Sound and the Akalan-Chilam area (Ortuño et al., 2006).

the Akalan-Chilam study area is generally homoclinal with little deformation. A subsequent Miocene–Pliocene extensional phase created the Macuspana and Comalcalco basins that are dominated by regional-scale listric faults, whereas the Akalan Trench formed in the Akalan-Chilam area (Figure 3).

This tectonic history produced six morphostructural provinces in Campeche Sound: a salt dome area, the Comalcalco Basin, the Akal horst, the Macuspana Basin, a continental slope area, and the Yucatan Platform (Ángeles, 1996a) (Figure 1).

DATA PREPARATION AND INTEGRATION FOR NUMERICAL PETROLEUM SYSTEM MODELING

Seismic Interpretation and Results

Conversion of seismic data from time to depth required applicable velocity functions to create a velocity model for each seismic-stratigraphic sequence. Figures 4

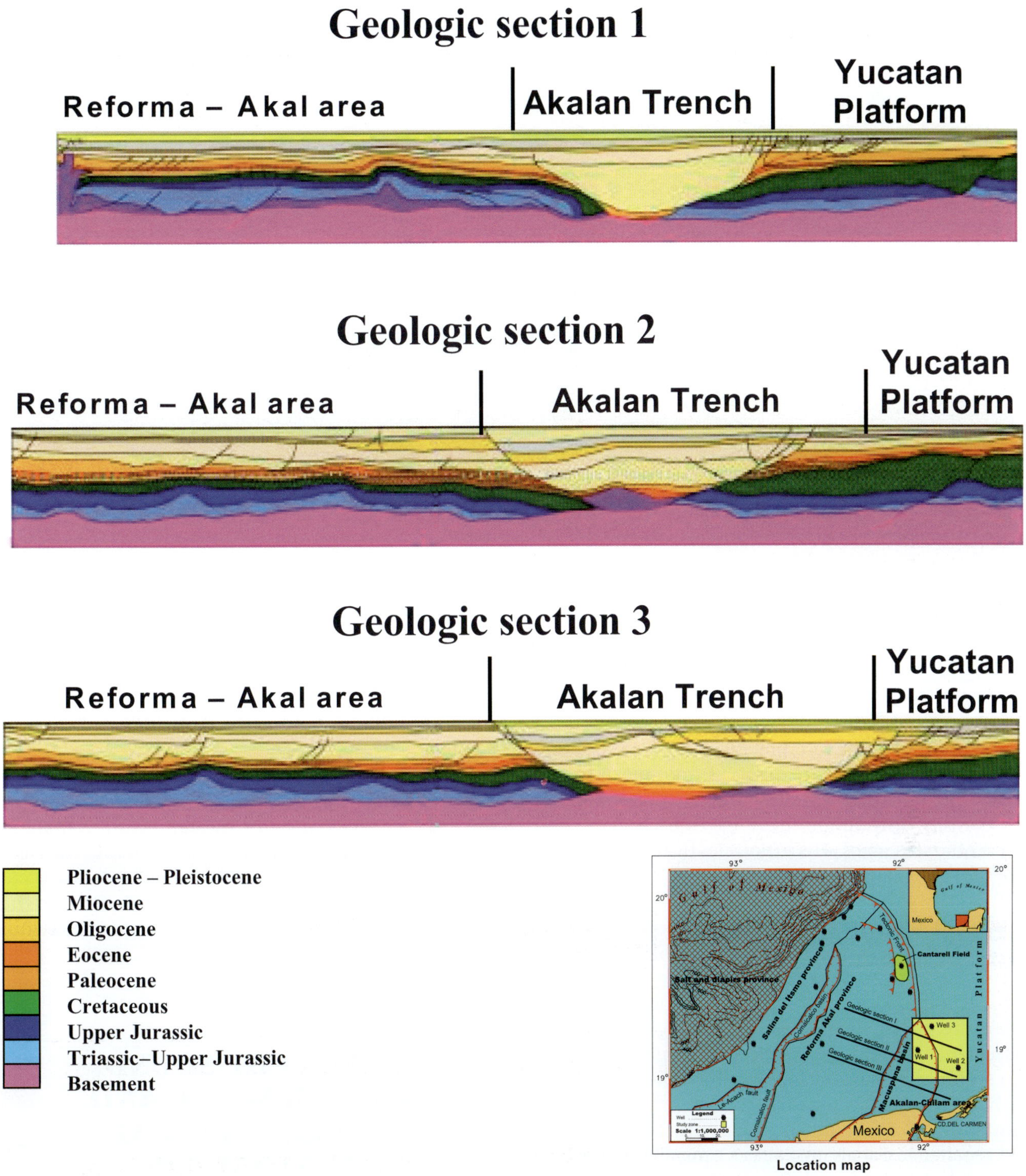

FIGURE 3. Structural sections (1, 2, and 3) of the Akalan-Chilam region from Villaseñor and España in Ortuño et al., (2006).

and 5 show the velocity functions and three-dimensional (3-D) velocity panels for all of the depth-converted horizons in each sequence. These seismic horizons were selected as stratigraphic horizons and designated by their corresponding geological ages. Once the depth conversion of the horizons was accomplished, several seismic depth profiles were constructed. Some of these are shown in Figure 6. These seismic-stratigraphic horizons formed the basis for constructing the 3-D numerical models used to evaluate the petroleum systems.

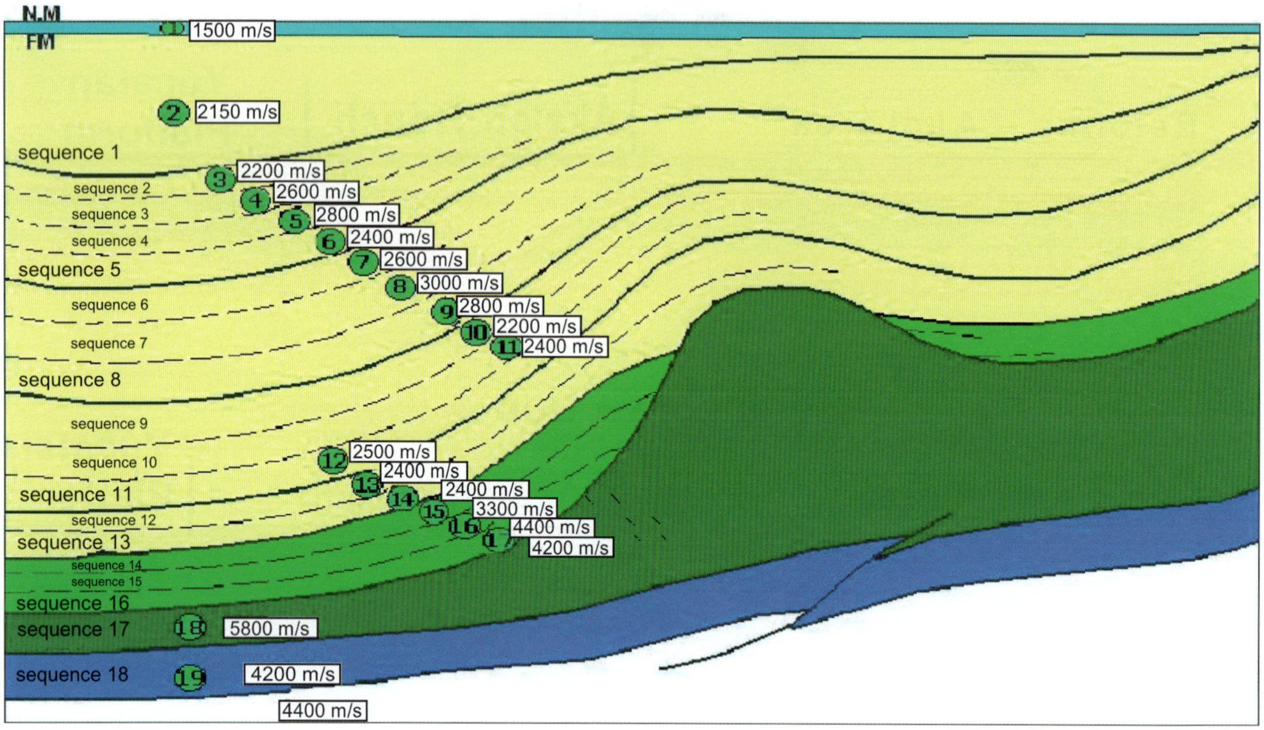

FIGURE 4. Seismic velocity models for each of the interpreted seismic-stratigraphic horizons.

Additionally, the following maps were generated based on the seismic interpretation: (1) time structure maps, (2) depth structure maps, (3) root-mean-square (RMS) attribute maps, (4) lithology distribution maps, (5) porosity distribution maps, (6) isopach maps, and (7) geohazards maps. The lithofacies distribution maps are shown in Figure 7.

Geological and Geochemical Interpretations

All of the existing geological and geochemical data were interpreted, including the following: (1) Magnetic basement map; (2) 18 seismic-stratigraphic horizons; (3) sedimentologic and stratigraphic lithofacies distribution maps for the Jurassic, Cretaceous, and Cenozoic horizons; (4) local and regional structure sections; (5) geochemical analyses, i.e., TOC, maximum pyrolysis temperatures (T_{max}), vitrinite reflectance (R_o), hydrogen indices (HI), organic facies, pyrolysis, etc. The offshore Jurassic and Cretaceous lithofacies comprise extensive areas of shallow-water carbonate sedimentation, from shoreline to normal-wave base (inner ramp), normal-wave base to storm-wave base (middle ramp), and finally, between storm-wave base and the pycnocline (suboxic stratified external ramp) (Burchette and Wright, 1992). Cenozoic paleoenvironments depict neritic areas (epicontinental seas) and pelagic areas (bathyal and abyssal regions) as illustrated for the Akalan-Chilam area in Figure 8.

Jurassic and Cretaceous sedimentary rocks in the Akalan-Chilam area occur in two structural domains: the Akalan Trench and the Yucatan Platform margin. The Cenozoic sedimentary rocks thicken into the Neogene Akalan Trench. Therefore, the sedimentary accumulation forms a homocline from the platform to the Akalan Trench, as shown in Figure 3 (from Villaseñor and España in Ortuño et al., 2006).

CAMPECHE SOUND: GEOCHEMICAL AND PETROCHEMICAL FRAMEWORK

The most important source rock in the offshore Campeche Sound area, including the study area, is the Tithonian series. The second most important is the Oxfordian series and, to an even lesser degree, the Kimmeridgian and lower to middle Miocene strata. Hydrocarbons generated in Campeche Sound Jurassic source rocks are produced from Upper Jurassic (Kimmeridgian and Oxfordian), Cretaceous, and Eocene reservoirs.

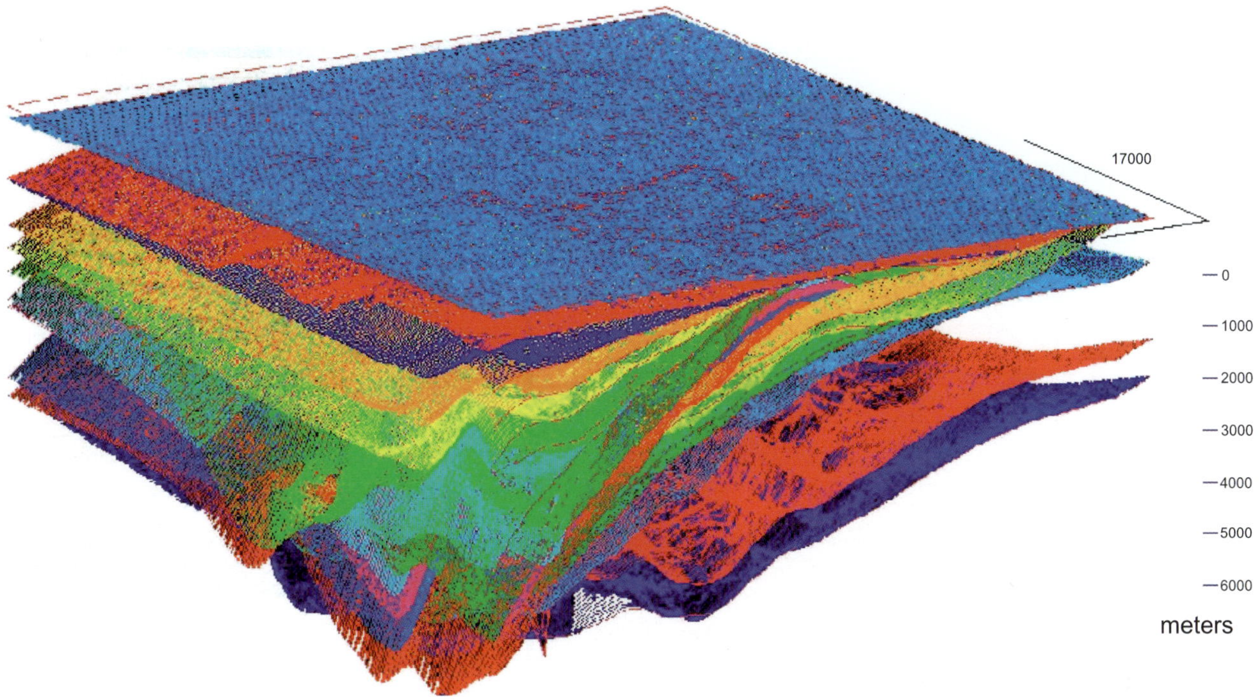

FIGURE 5. Three-dimensional construction of the velocity panels for each seismic sequence identified.

The main geochemical parameters used to characterize the Tithonian and Oxfordian source rocks are thicknesses, TOC contents, organic facies, generating potential, maturity measured by R_o, T_{max}, and HI.

Characterization of Tithonian and Oxfordian Source Rocks

Lithofacies and Depositional Environments

Tithonian source rocks are extensively distributed in the Campeche Sound (Figure 9) and consist of marls and calcareous shales deposited in a relatively deep-water anoxic environment. The Oxfordian source rocks are inner shelf marls with a geographic distribution similar to that of the Tithonian sequence. Both the Tithonian and Oxfordian rocks in the Akalan-Chilam area have average thicknesses of 100 to 150 m (192 ft), whereas in other parts of the Campeche Sound they may be as much as 350 m (1148 ft) thick.

Organic Richness

The TOC content of the Tithonian rocks in the Campeche Sound ranges from less than 0.5% (low) to greater than 4% (excellent) (Romero et al., 2001), with 1–4% (high to very high) TOC predominating. Tithonian TOC in the Akalan-Chilam area ranges from 0.5 to 2.0% (Figure 9a), and Oxfordian source beds have similar TOC values.

Kerogen Quality and Type

The Tithonian and Oxfordian rocks contain similar kerogen types, varying from type I-II (oil precursor) to type II (oil and gas precursor). The latter is more common, but considerable differences in the quality of organic matter between the Tithonian and Oxfordian source rocks are observed; see for example the hydrogen index in Table 1.

Kerogen Maturity

Available laboratory information allows a good understanding of Tithonian and Oxfordian source rock maturity in the Campeche Sound. These data include T_{max}, R_o, downhole temperature, geothermal gradient, and the optical analysis of organic material.

The T_{max} values for the Tithonian rocks range from 410 to nearly 480°C, indicating the presence of both mature and immature areas. The oil- and gas-generating areas occur toward the southwestern and central parts of the Campeche Sound, whereas less mature areas are toward the northeast near the tectonic front (Tunich, Ku-Maloob-Zaap, Cantarell, and Chilam), as observed by several authors, including Santamaria and Horsfield (2001), Romero et al. (2001), internal reports of the PEMEX Marine Region, and the Geochemical Laboratory of the Instituto Mexicano del Petróleo. Figure 9b shows the distribution of Tithonian source rock maturity in the Akalan-Chilam area with R_o values ranging from 0.4 to 1.3. Less abundant maturity data are available

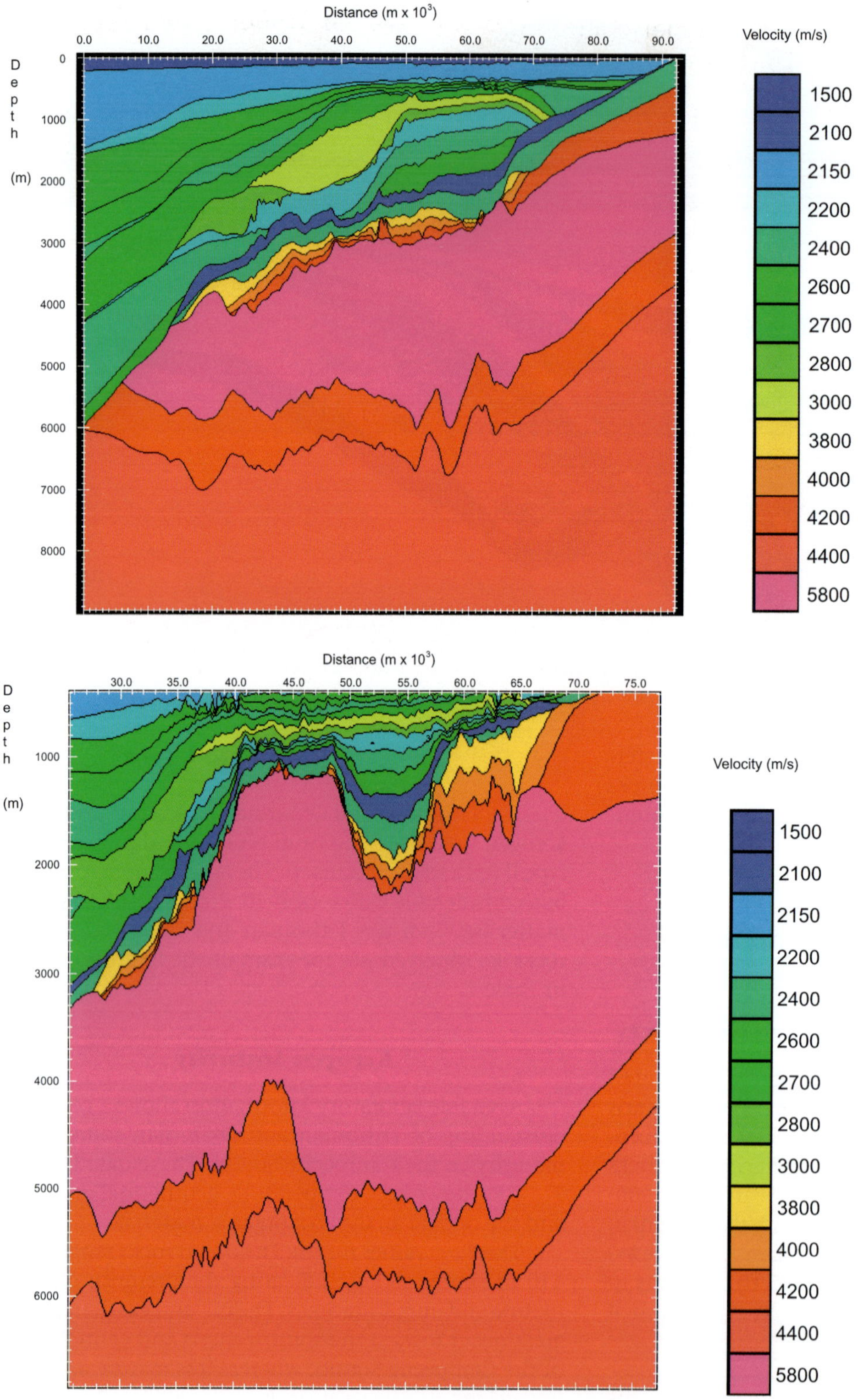

FIGURE 6. Profiles or seismic sections in depth through the interpreted area.

for the Oxfordian source rocks; however, this parameter in the Akalan-Chilam area is similar to the Tithonian rocks.

By integrating various geochemical parameters for the Tithonian series, such as TOC, T_{max}, R_o, HI, and S2, the expulsion threshold was determined to be at an average burial depth of between 4300 and 4500 m (14,108 and 14,764 ft) in the Campeche Sound. This would also apply to the Akalan-Chilam area, although thick Cenozoic units in the Macuspana Basin or the Akalan Trench would produce higher thermal maturity levels favoring the generation and expulsion of gaseous hydrocarbons.

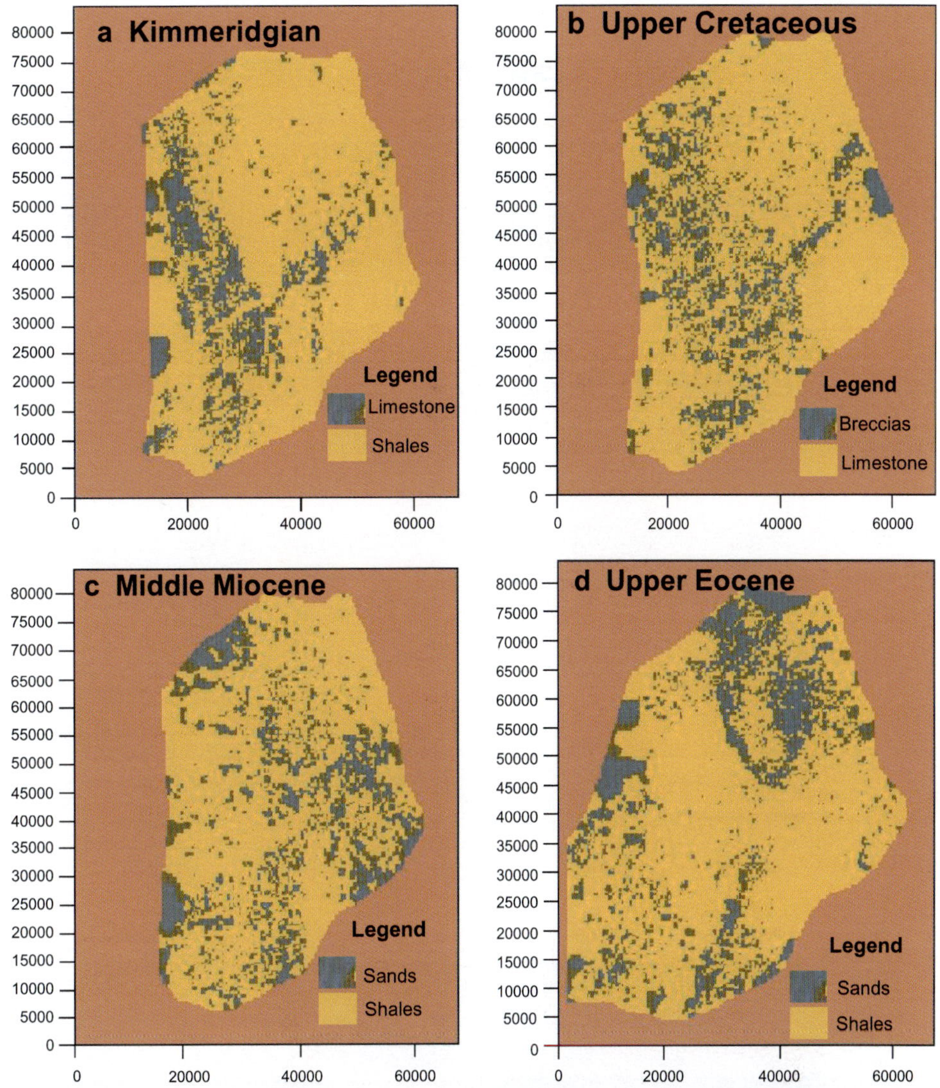

FIGURE 7. Maps of lithofacies from the interpreted sequences based on seismic data: (a) sequence 7 (Kimmeridgian), (b) sequence 6 (Upper Cretaceous), (c) sequence 4_1 (upper Eocene), and (d) sequence 3 (middle Miocene). The dark areas generally correspond to sands and the yellow background to predominantly shaly lithologies.

Recent rock-fluid correlations using Tithonian samples and oils from Kimmeridgian, Cretaceous, Eocene, and Miocene reservoirs suggest that the hydrocarbons in the Campeche Sound are generated mainly from Tithonian source rocks (RMNE, 2002; Guzmán et al., 2001) indicating the predominance of long-distance lateral migration over vertical migration.

Campeche Sound Reservoirs

Multiple horizons contain hydrocarbon accumulations generated mainly from Tithonian source rocks; these include Kimmeridgian facies, fractured Lower to Upper Cretaceous facies, the Cretaceous–Paleocene boundary breccia, and sandstones at several Cenozoic levels. The Upper Cretaceous breccia is particularly important because of its storage capacity and drainage efficiency and widespread distribution of the overlying Paleocene sealing strata.

Shales, and occasionally evaporites, form seals at various stratigraphic levels in the Upper Jurassic, Paleogene, and Neogene, whereas traps are generally characterized by compressional folds.

NUMERICAL SIMULATION METHODOLOGY USED TO MODEL PETROLEUM SYSTEMS

Construction of the Conceptual 3-D Numerical Model

The 3-D and 2-D geological models were prepared from interpreted seismic volumes. Lithofacies distributions were ascertained for 28 stratigraphic horizons, as well as their corresponding paleobathymetries, petrophysical parameters, and basement characteristics. The

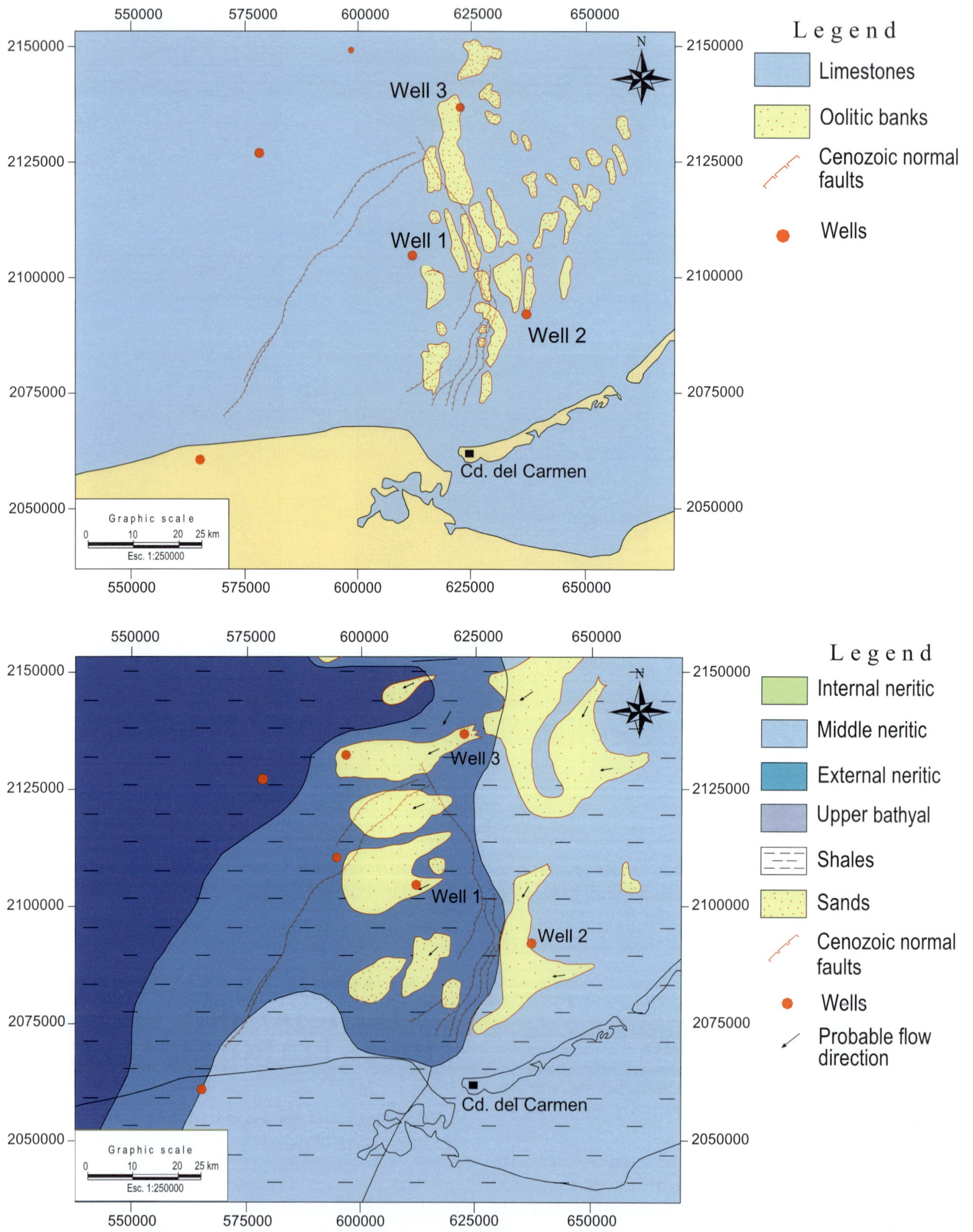

FIGURE 8. Lithofacies distribution charts based on preexisting data and seismic interpretation: (a) Kimmeridgian and (b) middle Miocene. Similar maps were prepared for each stratigraphic horizon.

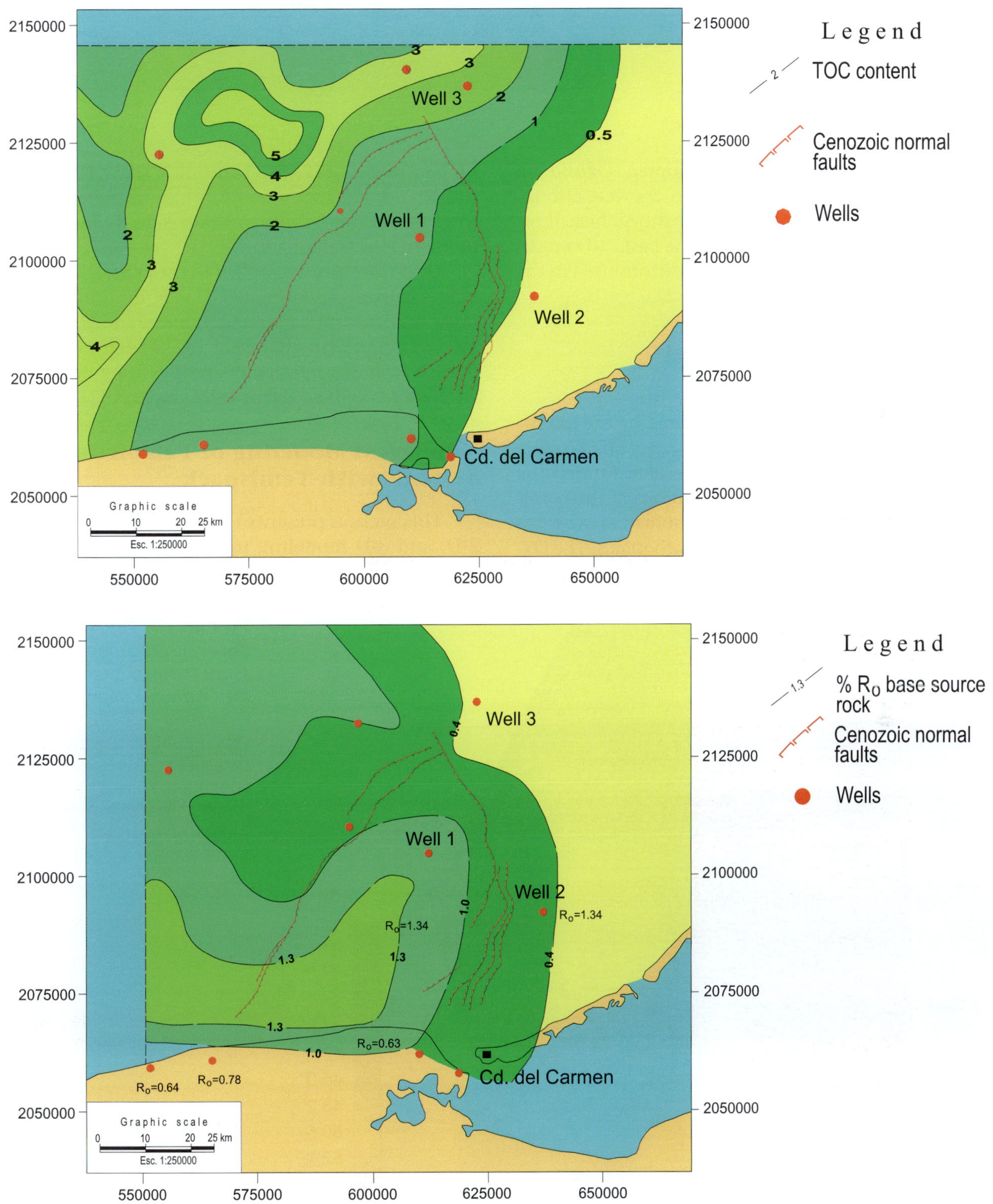

FIGURE 9. Distribution charts of total organic carbon (TOC) and vitrinite reflectance (R_o) for Tithonian rocks at the Holocene epoch. Data are derived from laboratory and subsurface sources. Other geochemical parameter maps were also interpreted.

Table 1. Main geochemical parameters of source rocks in the Akalan-Chilam area.

Source Rock	*Kerogen Type*	*Hydrogen Index (mg Hc/g TOC)*	*Average TOC (wt.% TOC)*	*Characteristics*
Tithonian	I	900–600	3.0–3.5	Excellent
Oxfordian	I-II (II)	600–300	3.0	Medium
Lower–middle Miocene	III	<300	2.0	Probable

geochemical parameters of the Tithonian, Oxfordian, and lower–middle Miocene source rocks were also included, such as maturity, thermal history, heat flow, thermal gradient, and other data. The basic 3-D model is shown in Figure 10. The seismic horizons used in the Temispack model are listed in Table 2.

Integration of Thermal and Geochemical Parameters

The source rocks were characterized by the following geochemical parameters: TOC, T_{max}, HI, kerogen type, and kinetic parameters. These data formed the basis for the simulations and calculations in the petroleum system models (Table 1). The Tithonian source rocks consist of shaly to silty bituminous carbonates. The Oxfordian source rocks are bentonitic shaly bituminous carbonates. The high-quality, oil-prone organic matter (type I-II) of the Tithonian strata is derived from algae (Romero and Maldonado, 1995).

Heat-flow history applied to the generative models ranges between an estimated 43 mW/m^2 during the Jurassic to 23 mW/m^2 at present (Romero and Maldonado, 1995). To complement the kinetic data derived from the Tunich well, the parameters proposed by Behar et al. (1997) for type II kerogens were employed (Table 3).

Numerical Modeling of Petroleum Systems with Temispack

This section presents the final results of the 1-D, 2-D, and 3-D modeling in the Akalan-Chilam area.

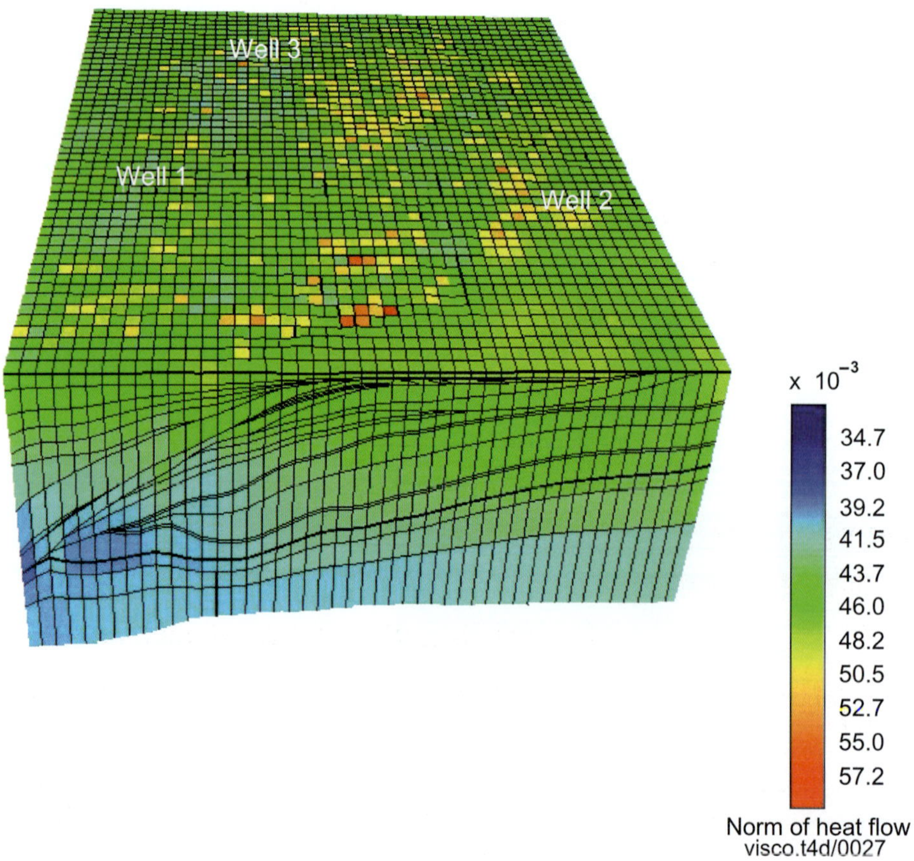

Figure 10. Three-dimensional cube constructed as the basic numerical model for Temispack simulations. Data come from depth conversion of the original seismic data. Several sections were constructed based on this two-dimensional maturity and migration analysis.

Table 2. Stratigraphic horizons in the 3-D Temispack numerical model.

Stratigraphic Horizon (Peak Sequence)	*Age (m.y.)*	*Period*	*General Lithology*
Marine seabed	0	Holocene	Siliclastics
Sequence 1	1.8	Pleistocene	Siliclastics
Sequence 1_1	2.05	Upper Pliocene	Siliclastics
Sequence 1_2	2.26	Upper Pliocene	Siliclastics
Sequence 1_3	2.6	Upper Pliocene	Siliclastics
Sequence 2	2.8	Upper Pliocene	Siliclastics
Sequence 2_1	3.6	Lower Pliocene	Siliclastics
Sequence 2_2	5.3	Upper Miocene	Siliclastics
Sequence 3	11.2	Middle Miocene	Siliclastics
Sequence 3_1	16.4	Lower Miocene	Siliclastics
Sequence 3_1 RG	20.1	Lower middle Miocene	Siliclastics
Sequence 3_2	23.8	Upper Oligocene	Siliclastics
Sequence 4	28.5	Lower Oligocene	Siliclastics
Sequence 4_1	33.7	Upper Eocene	Siliclastics
Sequence 5	54.8	Upper Paleocene	Siliclastics
Sequence 5_1	57.9	Upper middle Paleocene	Siliclastics
Sequence 5_2	61.0	Lower Paleocene	Siliclastics
Sequence 6	65	Upper Cretaceous	Breccias and spars
Sequence 6_1	70	Upper Cretaceous	Spars and caliches
Sequence 6_2	107	Lower Cretaceous	Caliches
Tithonian sequence	144	Tithonian	Lutites and calcareous siltstone with organic matter
Sequence 7	151	Kimmeridgian	Caliches and oolitic banks
Sequence 7 inter.	152	Kimmeridgian	Caliches
Sequence 8	154	Oxfordian	Sandstones and lutites (70% sandstone, 30% lutite)
Sequence 8 RG	155	Oxfordian	Clays and caliches (80% lutite, 20% carbonate)
Oxfordian sequence 2	157	Oxfordian	Sandstones
Callovian sequence	159	Callovian	Salt and evaporites
Middle Jurassic	180	Middle–Lower Jurassic	Continental sandstones (60% sandstone, 40% lutite)
Basement	248	Paleozoic	Igneous and metamorphic

The calibration process is discussed, followed by the 2-D geological sections, and then the 3-D modeling with a synthesis of the results.

Calibration of Geochemical Parameters

Parameters used to calibrate the wells or pseudowells included temperature (derived from downhole readings), geothermal gradient, heat flow, R_o, and T_{max}. In all cases, an effort was made to adjust the analytical data to the model results, with temperature being the most variable parameter. Heat-flow sensitivities from 20 to 40 mW/m^2 were applied and, in several cases, were adjusted through geological time. The calibration also incorporated the densities and compositions of hydrocarbons found in seeps, exploration wells, and developed fields of the region; i.e., gas, light oil, saturated oil, and aromatic and heavy oils. Figure 11a and b show some of the adjustments and calibrations used in the model runs. Calibration wells are within or near the study area (Pech, Chilam, Nix, Pek, Xicalango, etc.).

2-D Modeling of Geological Sections

Five geological section models were constructed: section 1, Akalan Trench to the Yucatan Platform; section 2, Pech to Chilam wells; section 3, Akalan to Nix wells; section 4, Nix well to the Yucatan Platform; and section 5, Yucatan Platform (Figure 12). The modeling parameters are discussed below.

Table 3. Kinetic parameters for Type II kerogen (Behar et al., 1997).

Type II Kerogen	*Activation Energy (kcal/mol)*	*Arrhenius Coefficient (1/s)*	*Partial Potential (mg/g °C)*	*Sum of fractions (%)*	*C6+ (%)*	*C1–C5 (%)*	*Coke (%)*
1	46	1.6E14	1.4	100	92.9	7.1	0E0
2	48	1.6E14	10.7	100	89.7	10.3	0E0
3	50	1.6E14	67.3	100	91.9	8.1	0E0
4	52	1.6E14	185.4	100	92.4	7.6	0E0
5	54	1.6E14	285.9	100	94.7	5.3	0E0
6	56	1.6E14	29.0	100	51.0	49.0	0E0

Stratigraphy and Lithology

The stratigraphic and lithofacies representation of the sections (Figures 13–15) incorporates their petrophysical properties with the lithological variability of the Cenozoic sandstones shown in sections 1 and 2.

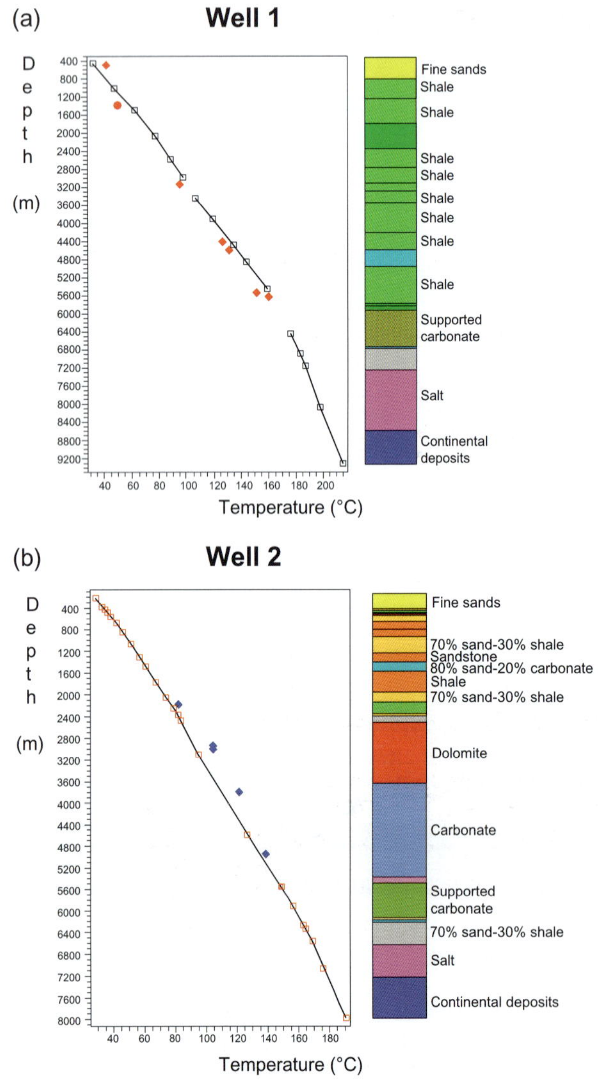

Figure 11. (a, b) Temperature calibration of analytical laboratory data and modeling results for wells: (a) Pech 1 (well 1) and (b) Chilam 1 (well 2).

Thermal Maturity

Kerogen maturity was estimated using transformation relationships (TRs), T_{max}, and R_o magnitudes. For example, in section 1 of Figure 16, TR values of 0.8 to 0.9 were obtained for the Tithonian rocks, whereas the TR values for Oxfordian rocks were close to 1.0. The maturity of the Miocene rocks (values less than 0.1) is insufficient for hydrocarbon generation to have occurred.

The T_{max} values from 440 to more than 510°C were applied for the Tithonian and Oxfordian rocks in section 1 (Figure 17). The higher thermal maturities correlate to greater depths, whereas Tithonian and Oxfordian rocks closer to the Yucatan Platform (right part of the section) display values closer to 430 or 440°C.

Modeled geological section 1 (Figures 12, location map; 16, TR; 17, T_{max}; 18, R_o) shows that the highest maturity indices within the Tithonian and Oxfordian generating kitchen correspond to the Akalan Trench, and that maturation gradually diminishes toward the platform. The thermal maturity of geological section 4 (Nix-Yucatan Platform; Figure 12, location map; Figure 19, hydrocarbon-generated distribution) indicates lower maturity where the source rocks are shallow in this part of the study area, particularly toward the Yucatan Platform. Hydrocarbon generation and expulsion occurred mainly during the Miocene in response to the high rate of subsidence of the Akalan Trench.

Thermal Parameters

Nominal R_o, from modeling, in geological section 1 (Figure 18) is 0.8–1.0% in the source kitchen, i.e., the Akalan Trench, with average heat-flow values of 30 to 40 mW/m^2. The geothermal gradients are congruent

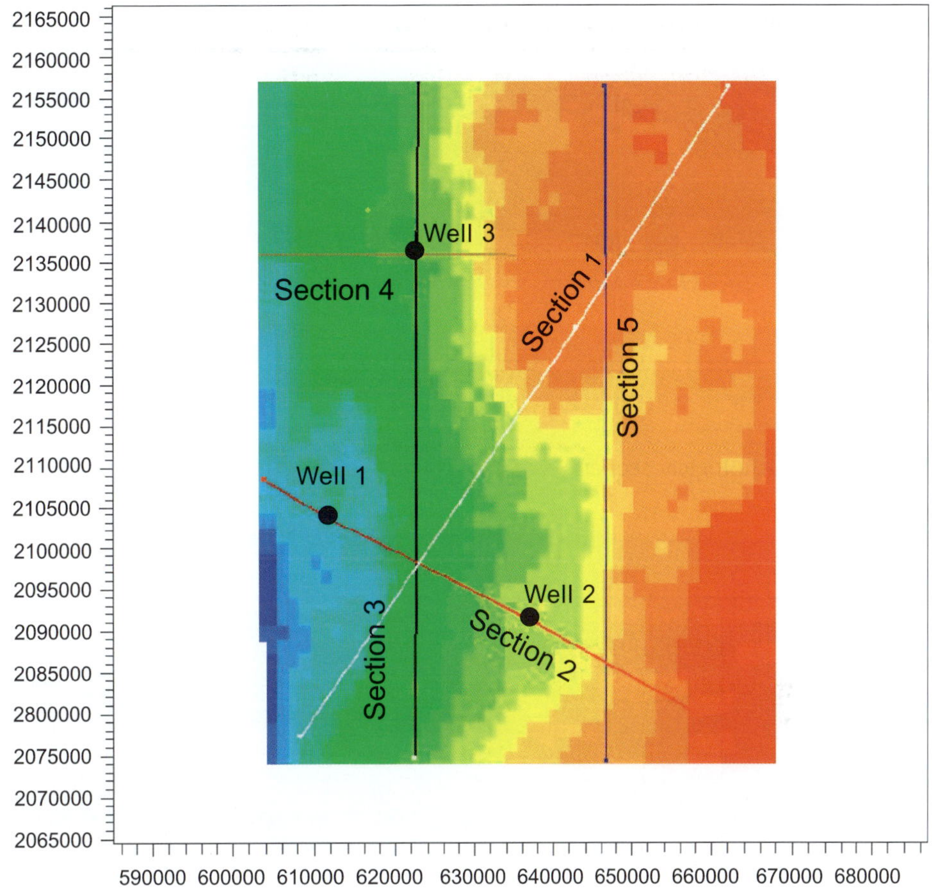

Figure 12. Location map for geological sections modeled in the study area. The structural configuration corresponds to the top of the Upper Cretaceous. The modeled geological sections are section 1, Akalan-Yucatan Platform (white); section 2, Pech-Chilam (red); section 3, Akalan Trench-Nix (black); section 4, Nix-Yucatan Platform (brown); and section 5, Yucatan Platform (blue).

with the heat flow and temperatures in all of the modeled geological sections (i.e., Figure 20, temperature distribution in the geological section). In geological section 5 (Figure 21, R_o) on the edge of the Yucatan Platform, the R_o and other parameters indicate a lower thermal maturity.

Hydrocarbon Generation, Expulsion, and Migration Parameters

Based on the models, it was possible to estimate the total amount of hydrocarbons generated from the Jurassic rocks. Expulsion of the hydrocarbons commences

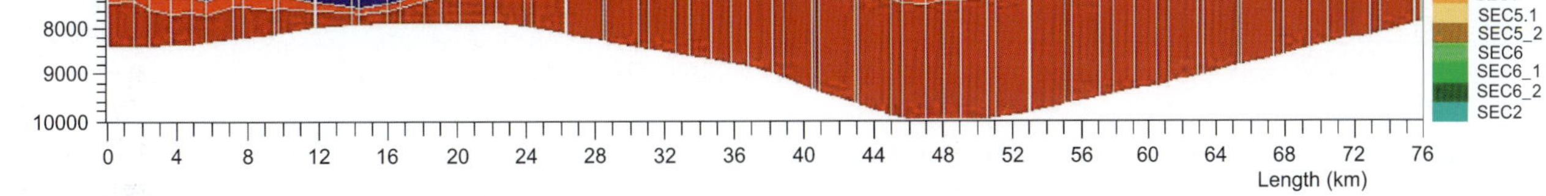

Figure 13. Geological section of the Akalan Trench-Yucatan Platform (section 1), showing the Mesozoic and Cenozoic stratigraphic sequence. Green corresponds to the Cretaceous.

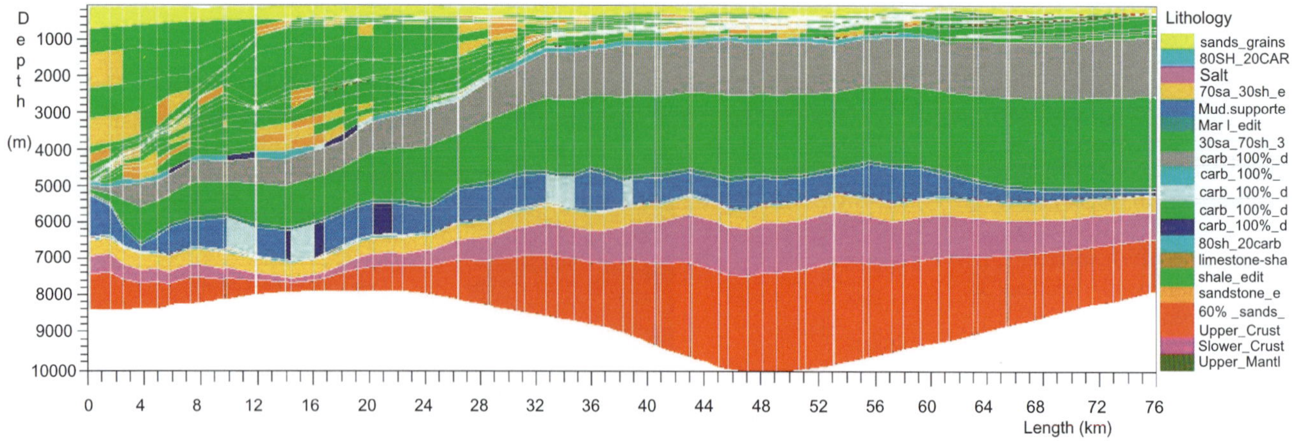

FIGURE 14. Lithological sequences of the Akalan Trench-Yucatan Platform geological section (section 1).

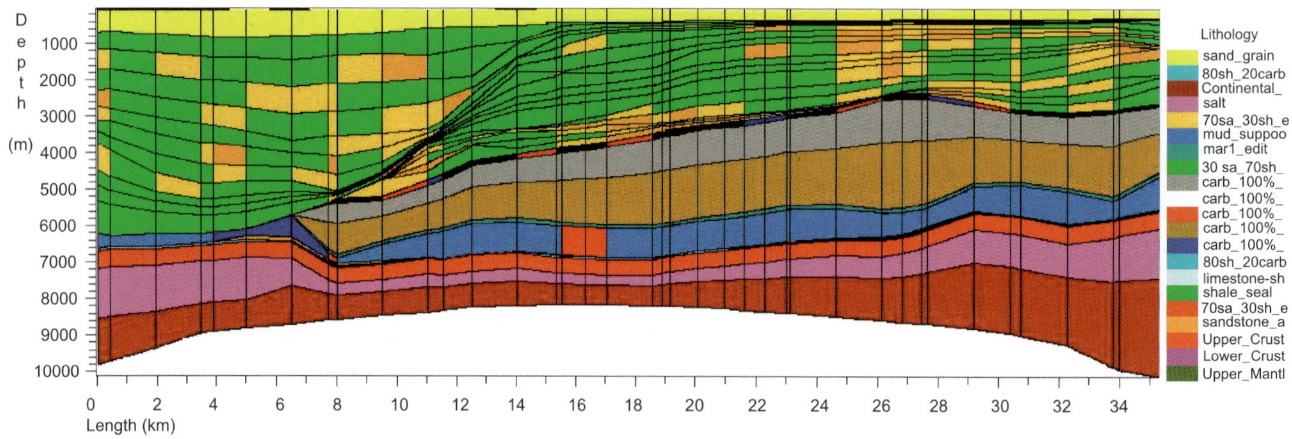

FIGURE 15. Geological section 2, Pech-Chilam, representing several lithofacies in the stratigraphic sequence. Sand distribution is shown in the Cenozoic section. The representation in the squares corresponds to the resolution adopted.

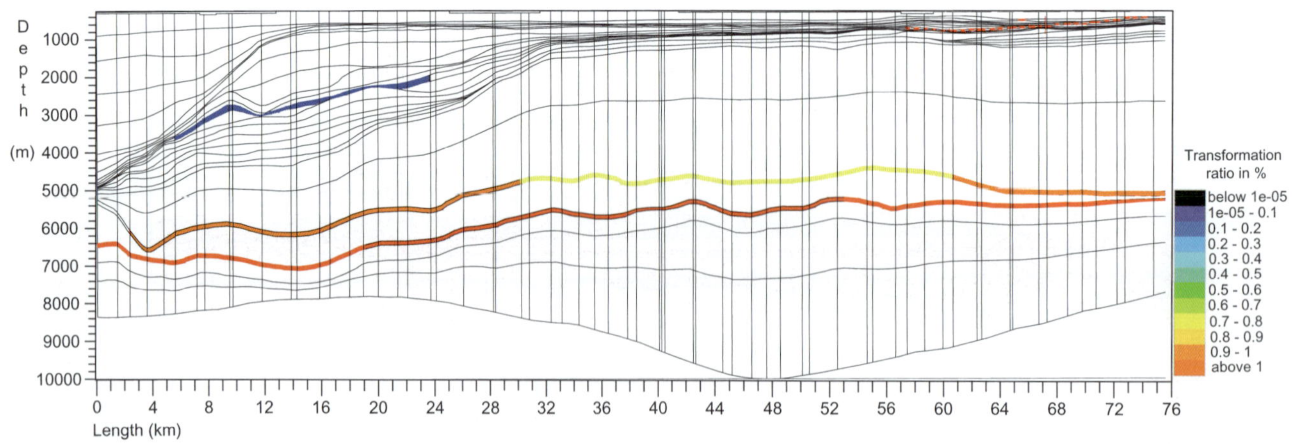

FIGURE 16. Thermal maturity represented by the transformation ratio for source rocks in geological section 1, Akalan Trench-Yucatan Platform. Sufficient maturity achieved for generation from the Tithonian and Oxfordian series (ranging from 0.8 to 1.0) is evident, particularly in the deeper sections (lower left). The Miocene sequence is still thermally immature.

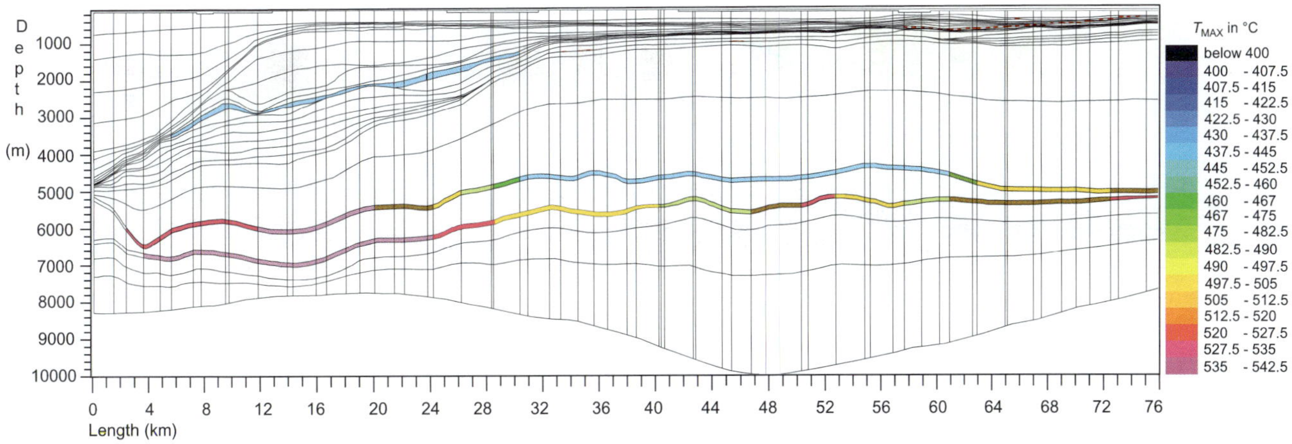

FIGURE 17. Distribution of maximum pyrolysis temperatures (T_{max}) in Tithonian, Oxfordian, and Miocene source rocks. The mature rocks are in the Upper Jurassic, with values of 445°C in the shallower parts to more than 500°C in the deeper parts of section 1, Akalan Trench-Yucatan Platform (lower left of the section).

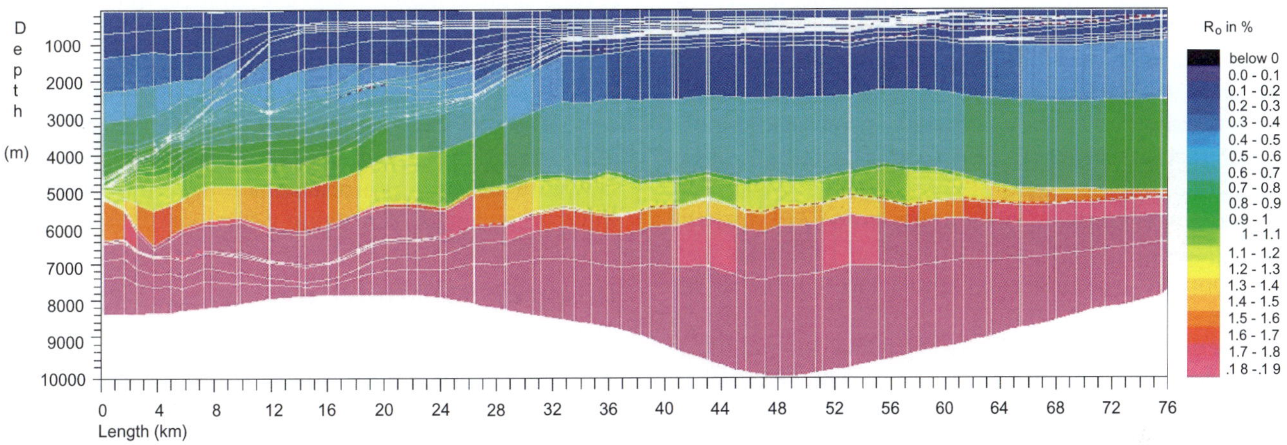

FIGURE 18. Nominal vitrinite reflectance (R_o) in section 1, Akalan Trench-Yucatan Platform. The R_o of the Upper Jurassic source rocks generally exceeds 1.0.

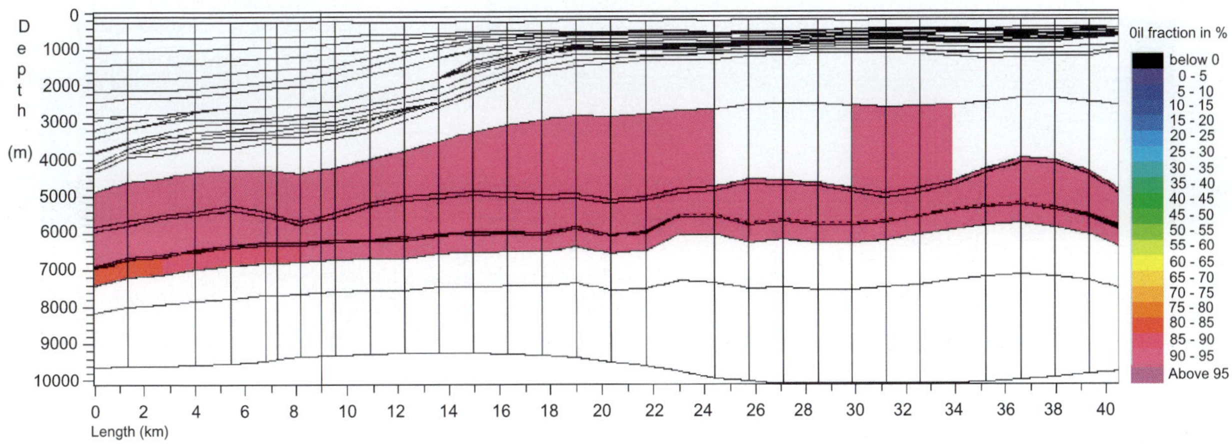

FIGURE 19. Percentage of liquid hydrocarbons expelled and migrated toward neighboring rocks through section 4 (Nix-Platform). The presence of liquid hydrocarbons is solely because of the thermal maturity level, which did not allow the development of lighter hydrocarbons.

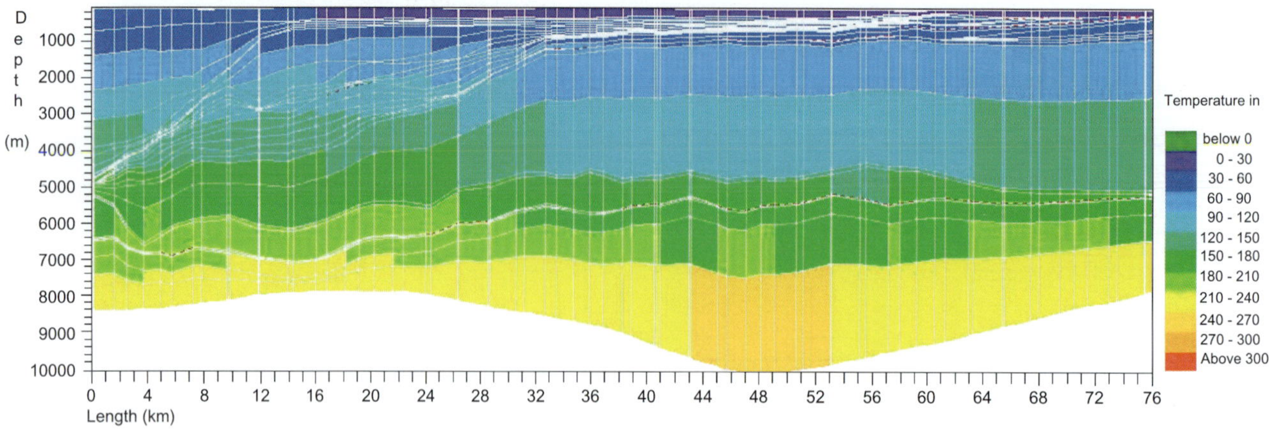

Figure 20. Temperatures throughout the sedimentary sequence in geological section 1, Akalan-Yucatan Platform. Temperatures exceed 150°C in the deeper Tithonian and Oxfordian source rocks.

at the saturation thresholds, as seen in Figure 22. Hydrocarbon volumes were different for each section modeled, expressed as ratios of liquid to gaseous hydrocarbons (Figures 19, 23–26). Predicting the reservoir fluids (water, liquid and gaseous hydrocarbons) as well as compositional fractions, such as C1–C5, C6+, saturated, aromatics (i.e., Figures 27–29), and API gravities, is also possible. The average modeled oil gravity is 30° API and the compositions, phases, and densities of the fluids were corroborated with internal exploration reports for the region.

Defining probable migration pathways through the entire sedimentary section was also possible (Figures 30, 31). The arrows indicate the migration of hydrocarbons through the rocks based on their petrophysical characteristics, with Cretaceous carbonates and Cenozoic sandstones possessing the best porosities and permeabilities, as well as fracture-fault systems. Normal and reverse fault systems in the 2-D models were assigned appropriate permeabilities, according to their associated rock types (Figure 32).

Modeled hydrocarbon compositions in both the 3-D and 2-D geological sections enabled verification and correlation with the existing results of sea-floor geochemical sampling and the hydrocarbons found in exploration wells (for example, see Figure 22).

Integral 3-D Modeling

The assigned source rock parameters for basic 3-D geological modeling were similar to those of the 2-D models, as follows:

- Tithonian source rocks: Kerogen type II; TOC from 0.7 (east) to 4% (west) depending on the identified sedimentary environments; and heat flow of

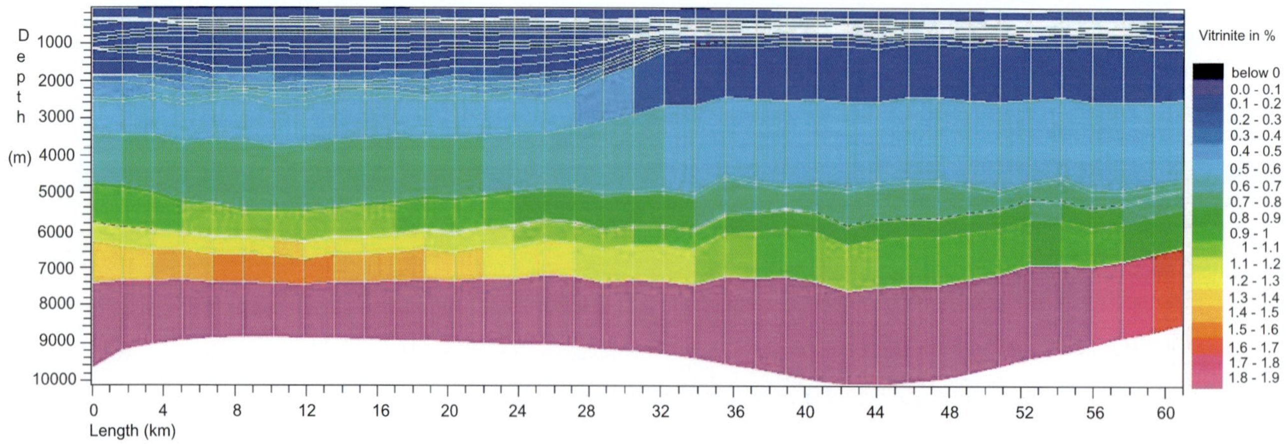

Figure 21. Nominal R_o magnitudes in the sedimentary sequence of geological section 5, on the edge of the Yucatan Platform.

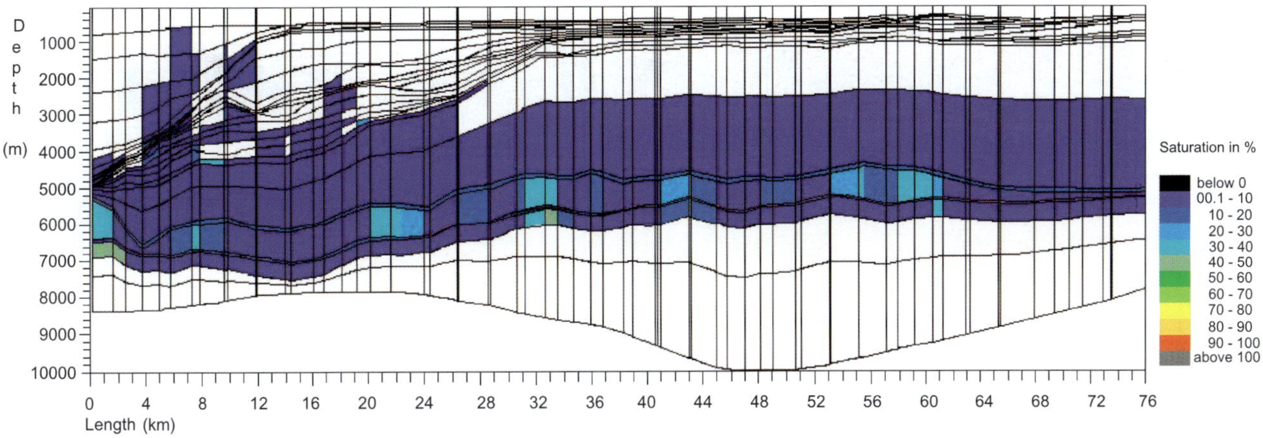

FIGURE 22. Saturation pattern within sedimentary rocks from section 1, Akalan-Yucatan Platform. The presence of hydrocarbons near the current seabed is evident.

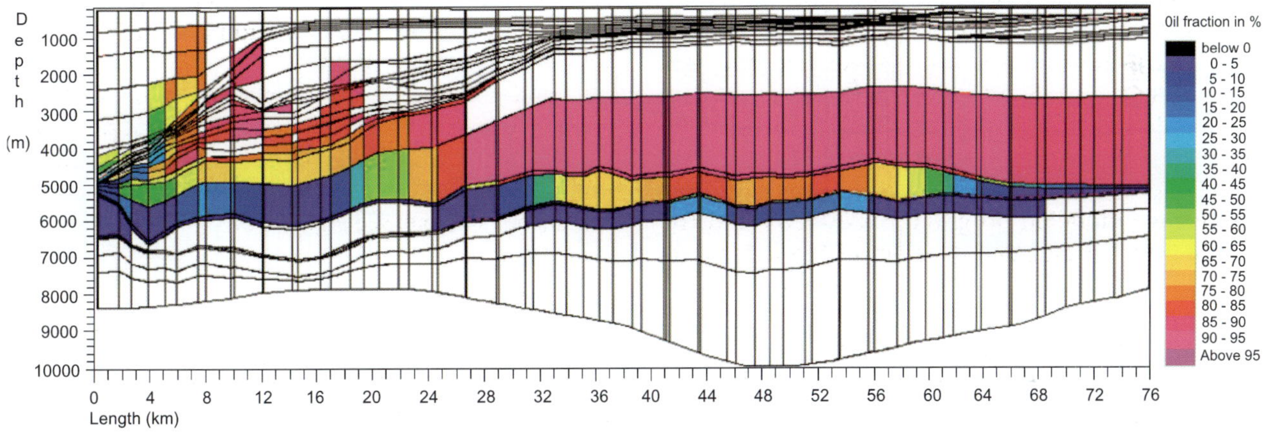

FIGURE 23. Percentage of liquid hydrocarbons in adjacent reservoir rocks. The presence of liquid hydrocarbons diminishes in the deeper zones (Akalan Trench to the left of the cross section 1), whereas liquid hydrocarbons are predominant toward the carbonate platform (right part of the section).

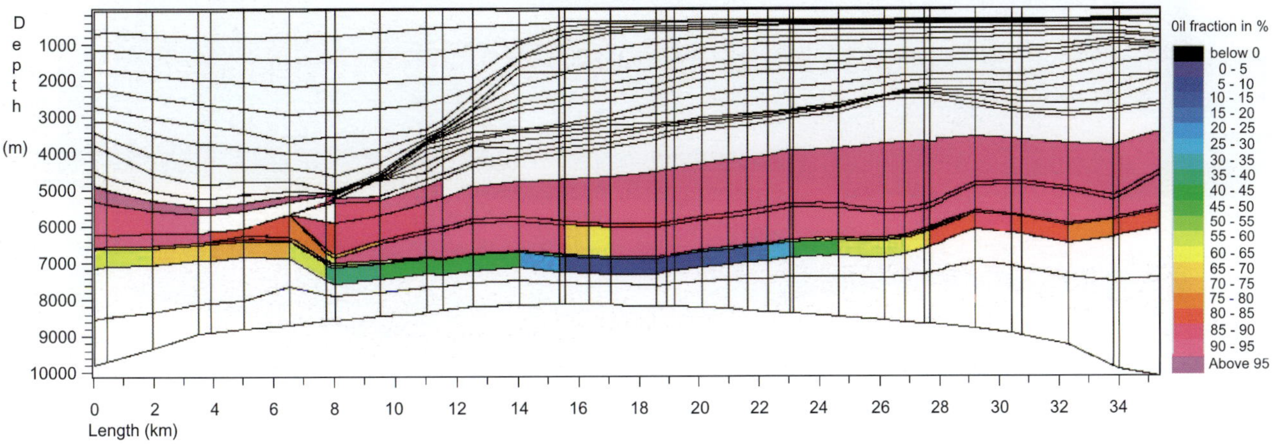

FIGURE 24. Percentage of liquid hydrocarbons in the vicinity of the source rocks. Percentages higher than 70% are reached in Cretaceous rocks. Section 2, Pech-Chilam.

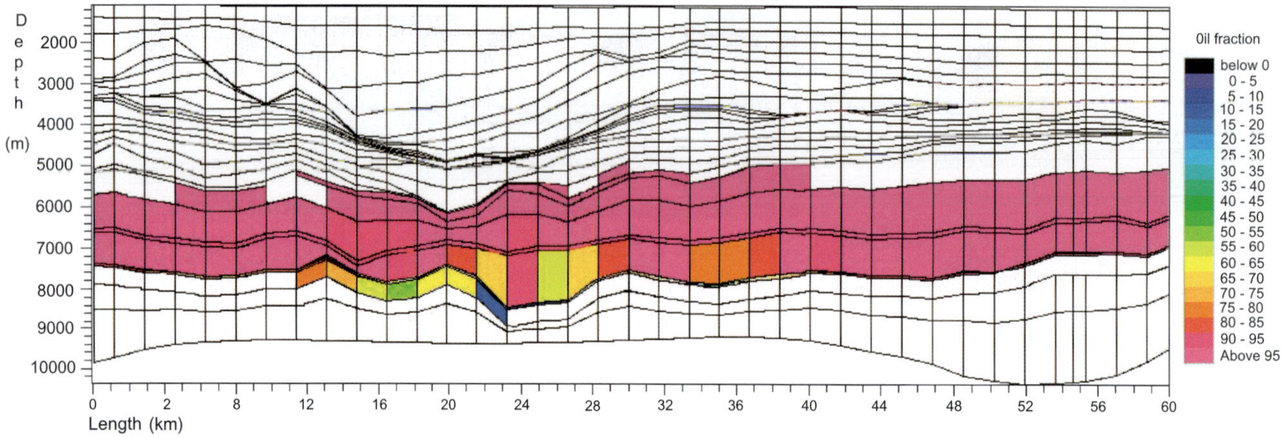

FIGURE 25. Percentage of liquid hydrocarbons in reservoirs of section 3, Akalan-Nix. The predominant percentages exceed 80%.

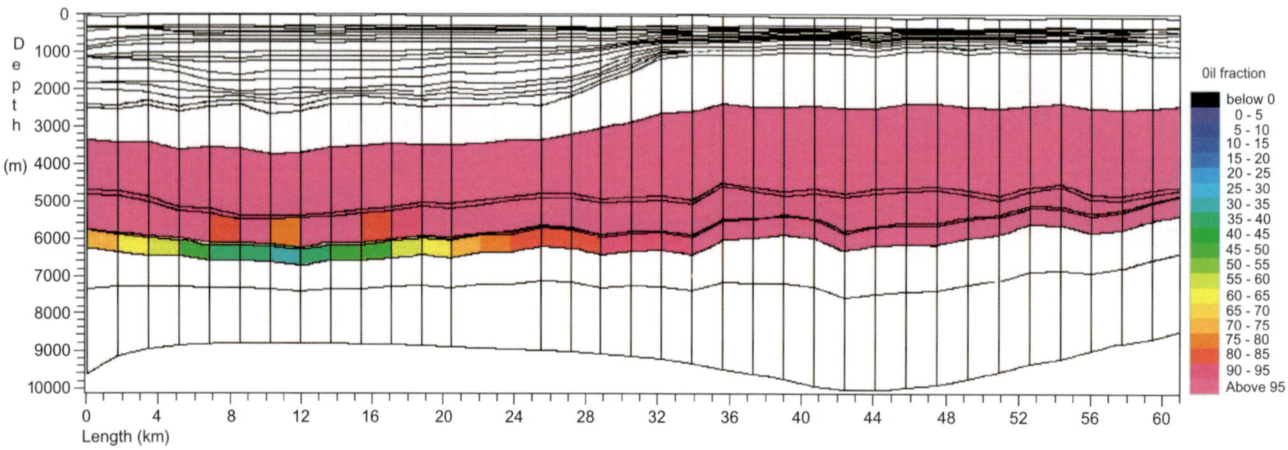

FIGURE 26. Percentage of liquid hydrocarbons in reservoirs of geological section 5 on the carbonate platform. The higher quantity of generated and migrated hydrocarbons corresponds to the liquid fraction. Some gas exists only in the deeper zones.

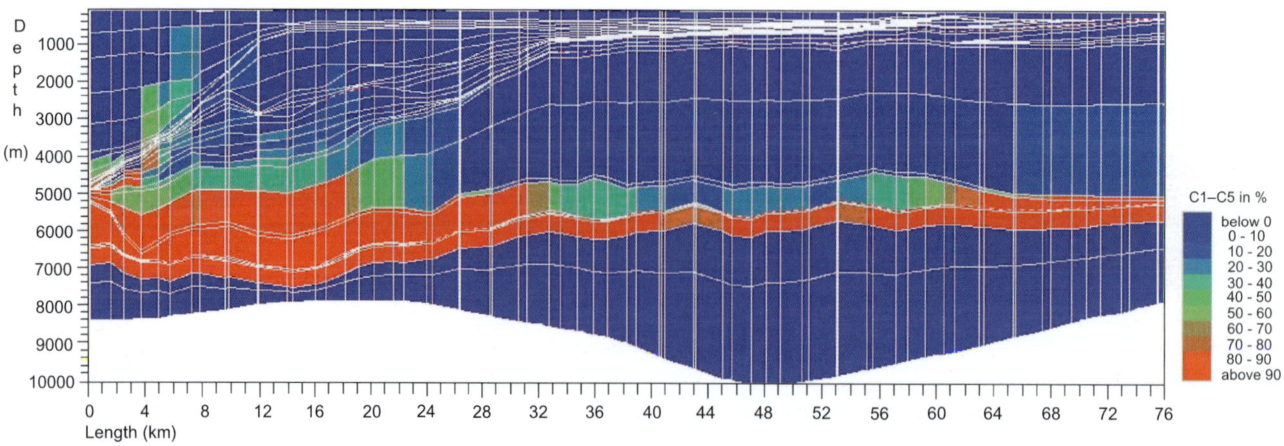

FIGURE 27. Percentage of C1 to C5 hydrocarbons present in the sedimentary sequence of geological section 1. Their predominance can be observed in the deeper reservoir rocks where thermal maturity was higher.

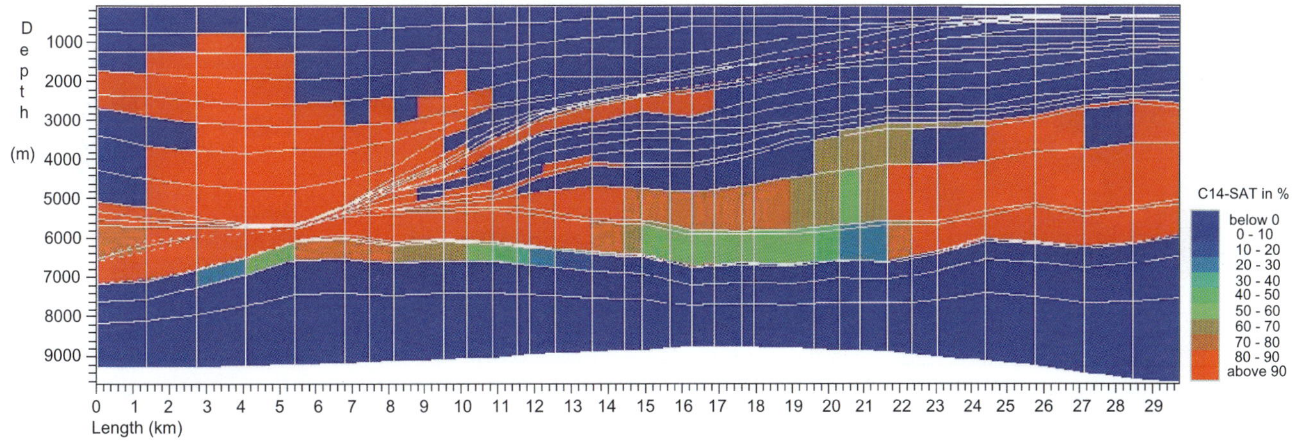

FIGURE 28. Percentage of C14-saturated (SAT) hydrocarbons in geological section 2 (Pech-Chilam). Its abundance is 90% on average.

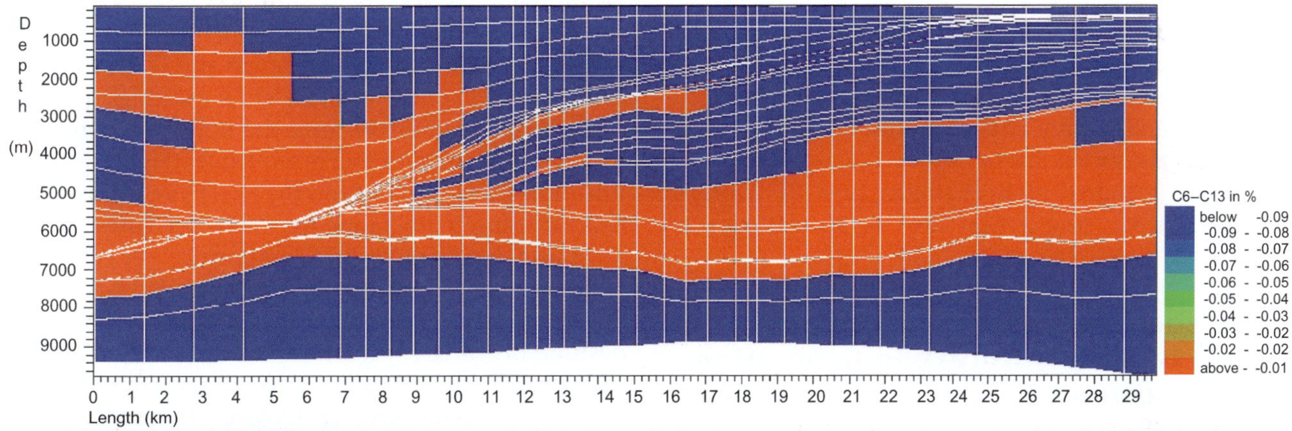

FIGURE 29. Saturation pattern of the C6–C15 fraction in section 2, Pech-Chilam.

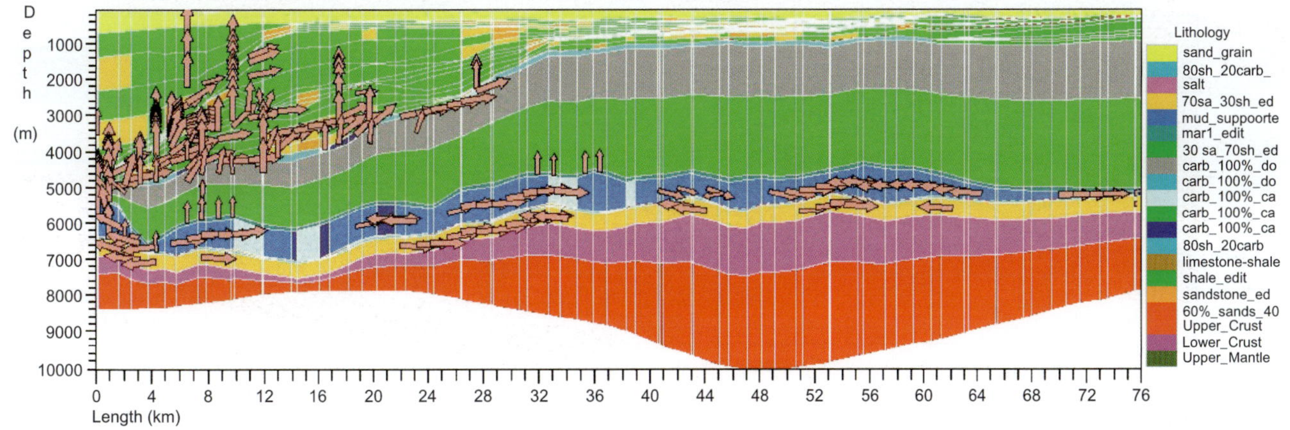

FIGURE 30. Hydrocarbon migration patterns defined by the petrophysical properties of the rocks and by the presence of faults (i.e., geological section 1). The arrival of hydrocarbons to the present-day sea floor is evident.

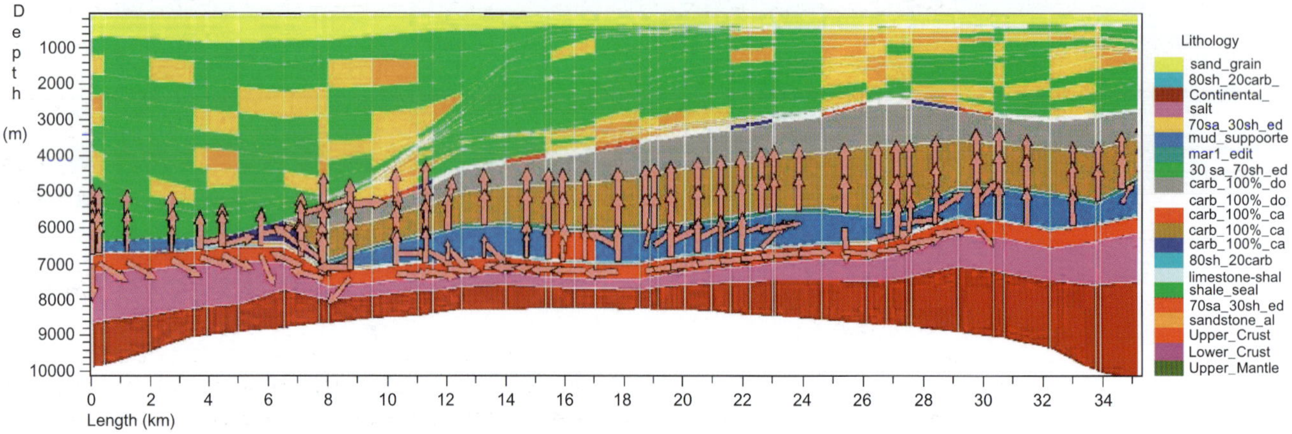

FIGURE 31. Hydrocarbon migration patterns in Cretaceous rocks of section 2, Pech-Chilam. Migration is enabled by the favorable petrophysical characteristics of the carbonate rocks.

30 mW/m^2 based on existing geothermal gradients and time-dependent variations.

- Oxfordian source rocks: Kerogen type I-II; TOC from 1.8 (east) to 4% (west) depending on the identified sedimentary environments; and a heat flow of 30 mW/m^2, as above.
- Miocene source rocks: Kerogen type III; average TOC of 3%; and a heat flow of 30 mW/m^2, as above.

3-D Modeling Results: Source Rock Maturity and Hydrocarbon Generation and Expulsion

The principal 3-D modeling results are briefly described in terms of transformation ratio (TR), R_o, heat flow, temperature, hydrocarbon saturation, and the amount of generated hydrocarbons.

Transformation ratios observed in the 3-D cube indicate the areas of greatest subsidence that constitute generating kitchens. In the study area, the Akalan Trench, with high TR at both the Tithonian and Oxfordian levels, is such a kitchen. The modeling indicates that this kitchen has been active from the upper Miocene into the Holocene for both source sequences (Figures 33, 34). Vitrinite reflectance confirms the TR (Figure 34), as do the heat flow and temperature parameters (Figures 34, 35).

Estimated amounts and compositions of generated hydrocarbons were based on the simulation (Figure 36a, b). The estimated distributions (saturations) of three compositionally different hydrocarbons (C1–C5, C6+, and the heavy fraction) were obtained for the Kimmeridgian oolites, Cretaceous breccias (Figure 37), and Cenozoic sandstones adjacent to the source rocks. Proper evaluation explains the densities (°API) and compositional characteristics of the known hydrocarbons in the area (Ortuño et al., 2006).

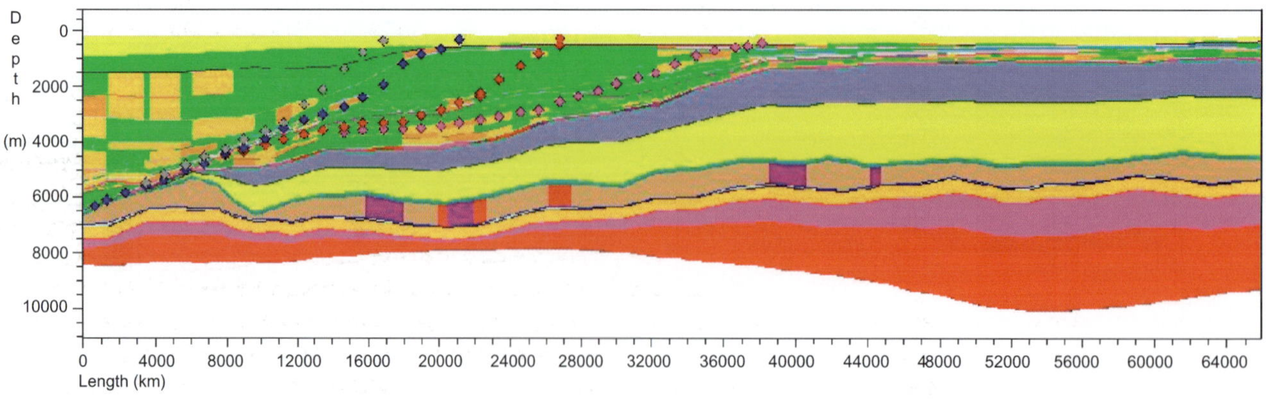

FIGURE 32. Section 1, Akalan-Yucatan Platform, where normal faults affecting the Cenozoic sedimentary series are represented schematically. Their geometry is related to the formation of the Akalan Trench, which was a fundamental factor in hydrocarbon migration and trapping. The fault zones are more permeable than neighboring lithologies.

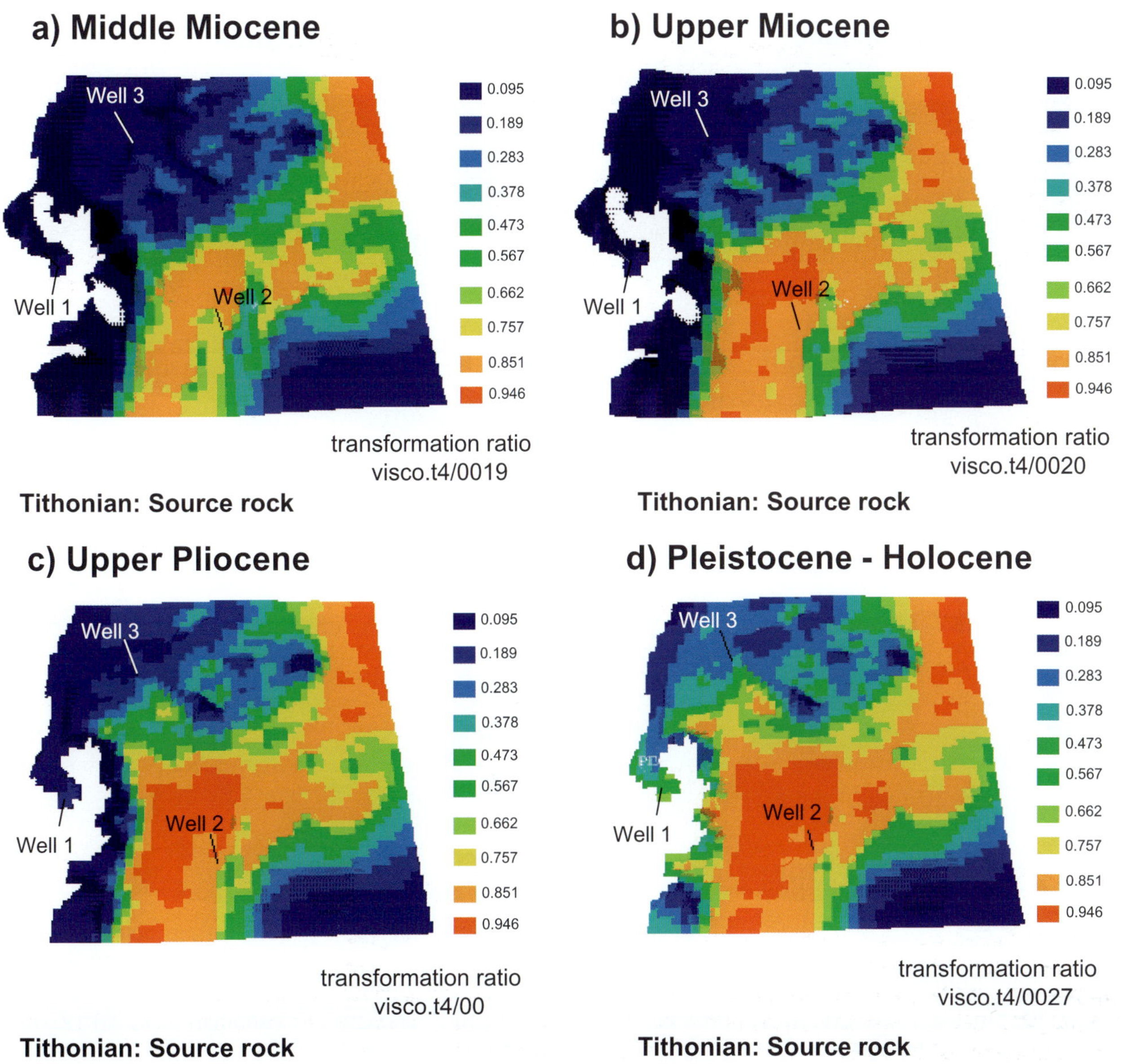

Figure 33. Evolution of thermal maturity (transformation ratio) in Tithonian rocks at the following stages: (a) middle Miocene, (b) upper Miocene, (c) upper Pliocene, and (d) Pliocene–Holocene.

The 3-D modeling gives insights into the generation and expulsion processes for the Tithonian and Oxfordian source rocks. Migration efficiency averages 30%, depending on the lithologies involved and their petrophysical characteristics. Migration would have occurred mainly during the Miocene–Pliocene with maximum kerogen maturity reaching the gas window in the deepest parts of the Akalan Trench. Both vertical and lateral migration occurred, although long-distance migration toward the high-energy facies rocks bordering the Yucatan Platform did not result in trapping because of the lack of good seals. The timing of generation from the two Jurassic source horizons (Oxfordian and Tithonian) is very similar given their physical proximity and shared geological history.

Definition of Drainage Areas, Trapping, and Prospect Ranking

The modeling allowed quantification of the drainage areas and hydrocarbon volumes available for migration into structural and stratigraphic traps, as well as the relative ranking of prospects. The most important reservoir horizons are (1) Kimmeridgian (particularly oolitic horizons), (2) Lower and Upper Cretaceous carbonates (mainly fractured breccias and fractured porous limestones and dolomites), and (3) Paleogene and Neogene deep-water fans. Figure 38 ranks identified Kimmeridgian and Lower and Upper Cretaceous prospects, based on their economic interest and exploration risk. Specific petroleum system elements of the Tithonian

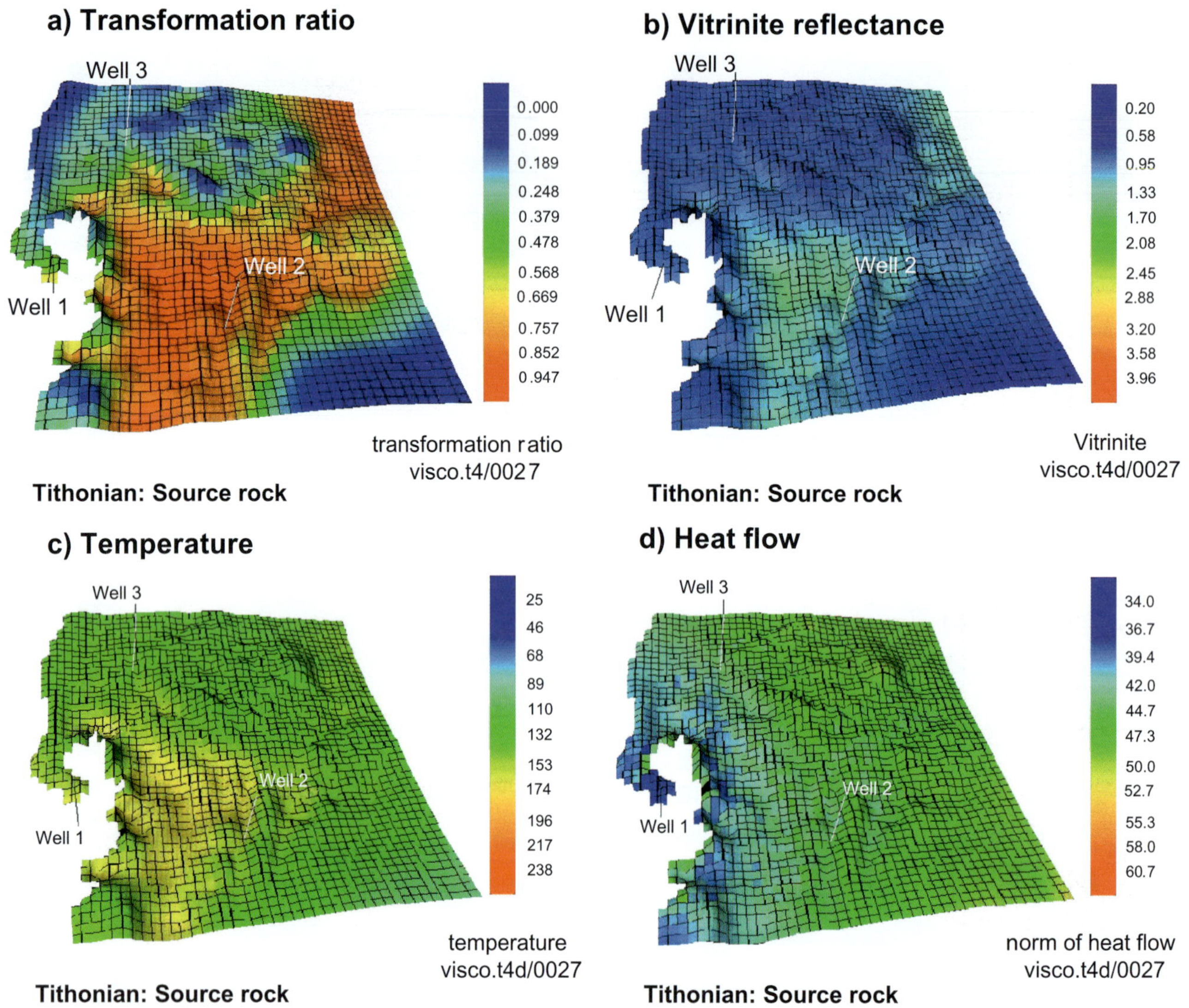

FIGURE 34. Three-dimensional display of the magnitudes of modeling parameters in Tithonian rocks: (a) TR, (b) R_o, (c) temperature, and (d) heat flows. Kerogen maturation is evident in these rocks.

sequence are shown in Figure 39. Based on current knowledge, these elements are also generally valid for the Oxfordian series.

Opportunities and Perspectives

The following reservoirs are considered to represent the best Mesozoic and Tertiary exploration opportunities:

1. Kimmeridgian oolitic horizons. Note the location of the oolitic banks in relation to the migration pattern from the drainage area
2. Upper and Lower Cretaceous fractured carbonates, particularly the breccias at the top of the Upper Cretaceous. The main negative factor for these rocks in the study area is their presence in an area of homoclinal dip (trap risk) and possibly ineffective seals
3. Paleogene (Eocene-Oligocene) calcarenites and Neogene (Miocene-Pliocene) sandstones
4. Areas along the edge of the carbonate platform between the Akalan Trench and the continental slope

CONCLUSIONS

The Akalan Trench mainly formed during the Miocene to Pleistocene. Significant subsidence affected the trench, as well as the Yucatan Platform, allowing source rocks to reach a high level of thermal maturity.

Tithonian source rocks generated liquid hydrocarbons and gas. According to the modeling, the gravity of the generated hydrocarbons was about 30° API; however, actual hydrocarbons in the study area probably originated in several kitchens due the predominantly lateral migration.

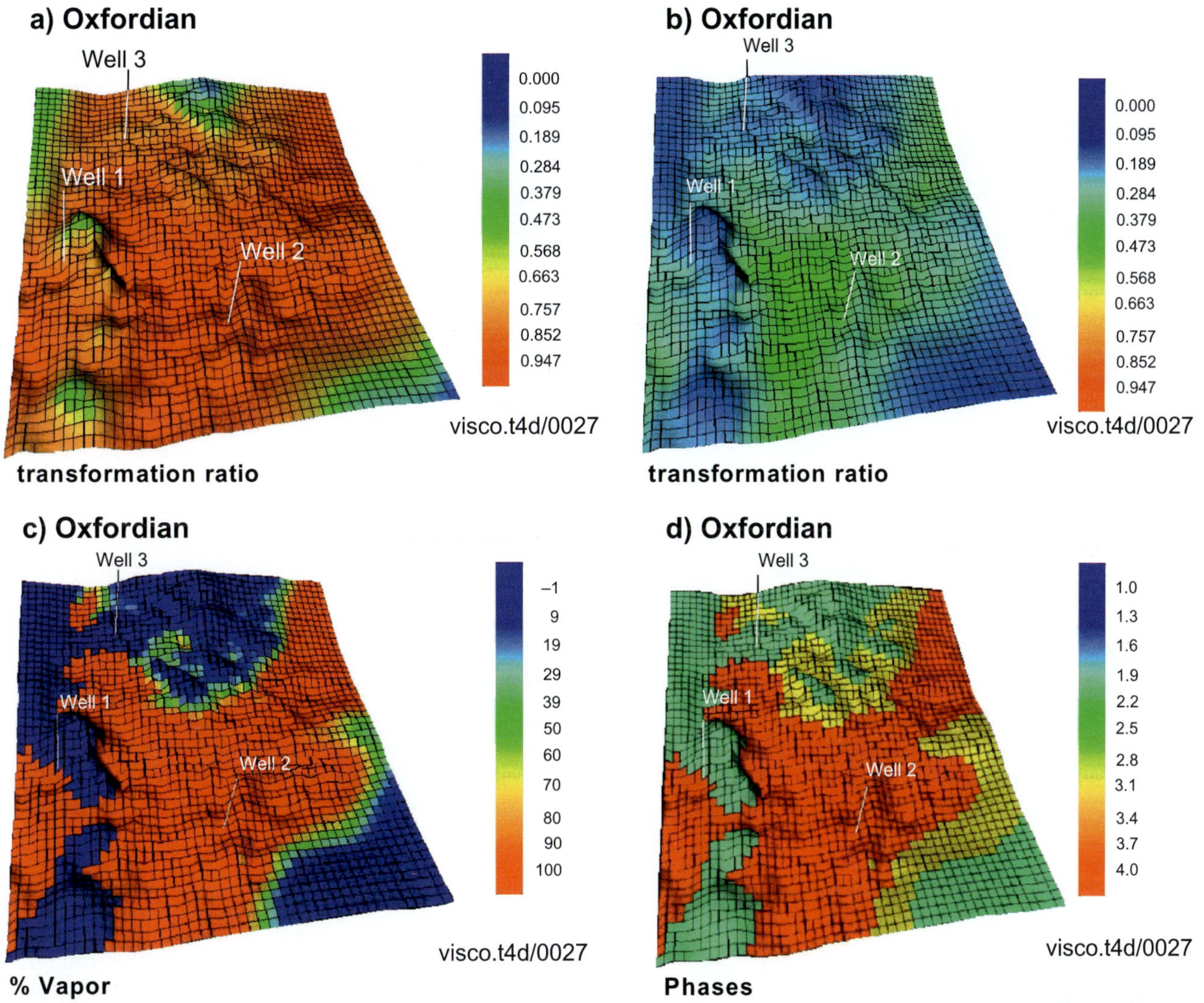

FIGURE 35. (a) Three-dimensional view of TR distribution in the Oxfordian rocks. (b) Three-dimensional view of R_o distribution in the Oxfordian rocks. (c) Percentage distribution of the gas in the Oxfordian rocks. The abundance of gaseous hydrocarbons is because of higher kerogen maturity levels. (d) Fluid phases present in Oxfordian rocks as a result of thermal evolution. Phases: (red) gas, (yellow) oil-gas, and (green) oil.

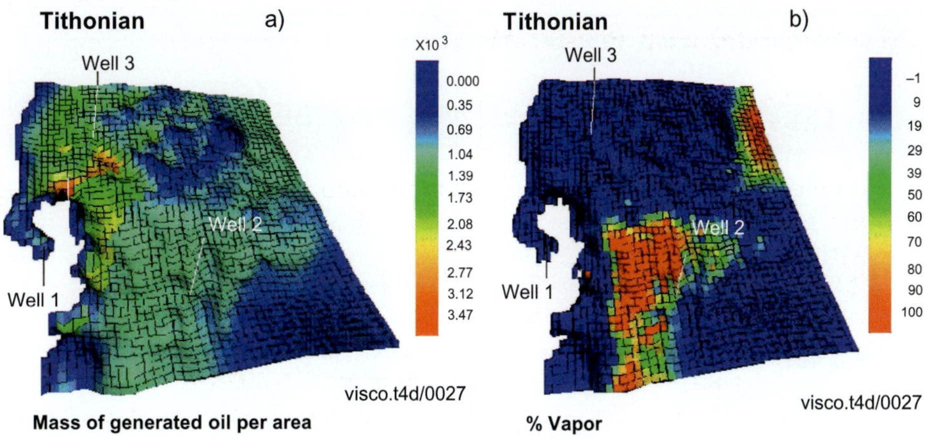

FIGURE 36. (a) Hydrocarbon mass generated per area in the Tithonian rocks; three-dimensional visualization. (b) Percentage of gas generated in Tithonian rocks, compared to liquid hydrocarbons.

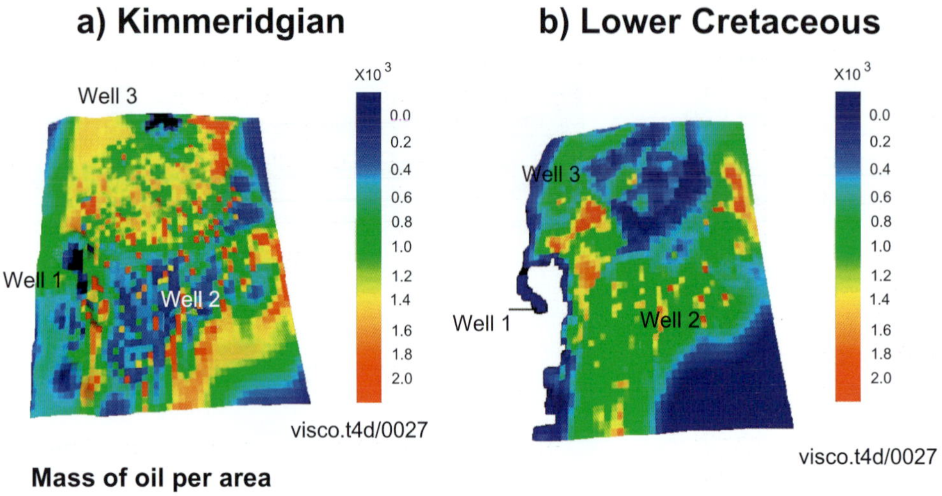

FIGURE 37. (a, b) Distribution of the hydrocarbon mass reservoired in (a) Kimmeridgian rocks and (b) Lower Cretaceous rocks. The greatest accumulations of hydrocarbons occur in relation to Kimmeridgian oolite shoals or porous Cretaceous strata.

The Oxfordian petroleum system had a slightly higher level of thermal maturity than the Tithonian, reaching the gas generation window in the Akalan Trench, and contributed to hydrocarbon migration into the surrounding Cretaceous and Cenozoic rocks toward the Yucatan Platform and in the Chilam area. This kitchen covers both the deep trench and its eastern margin toward the Yucatan Platform, although maturity diminishes eastward into the less deeply buried platform area.

Compositional modeling enabled estimates for three hydrocarbon fractions (C1–C5, C6+, and heavy oil) and is capable of modeling up to 12 components. These models confirmed the presence of methane, as well as the other petroleum fractions actually present in the area. Exploration decision making can capitalize on forecasts of the nature of hydrocarbons yet to be discovered.

In general terms, both of the Jurassic petroleum systems are mature and continue to be active. However, a

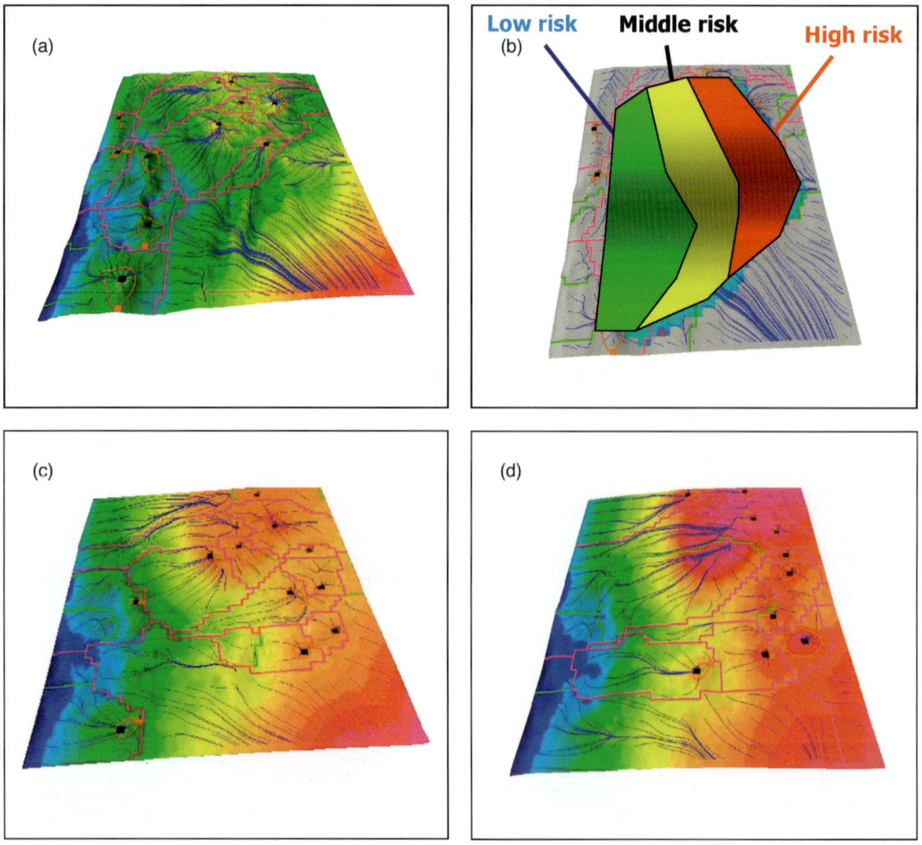

FIGURE 38. (a) Distribution of drainage areas in Kimmeridgian rocks relative to the structural configuration of the horizon of the same age. (b) Distribution of drainage areas in the Kimmeridgian rocks relative to the distribution of oolitic shoal development. A correspondence is noted between the drainage areas (topographic-structural configuration) and the oolitic developments, resulting in hydrocarbon traps and the definition of areas with both high geological risk and high opportunity. (c) Distribution of drainage areas in Lower Cretaceous rocks. The configuration of this stratigraphic horizon is very similar to that of the Kimmeridgian because of a parallelism and nearly homogeneous thickness of stratigraphic units. (d) Distribution of drainage areas in Upper Cretaceous rocks relative to the structural configuration at the top of this stratigraphic horizon.

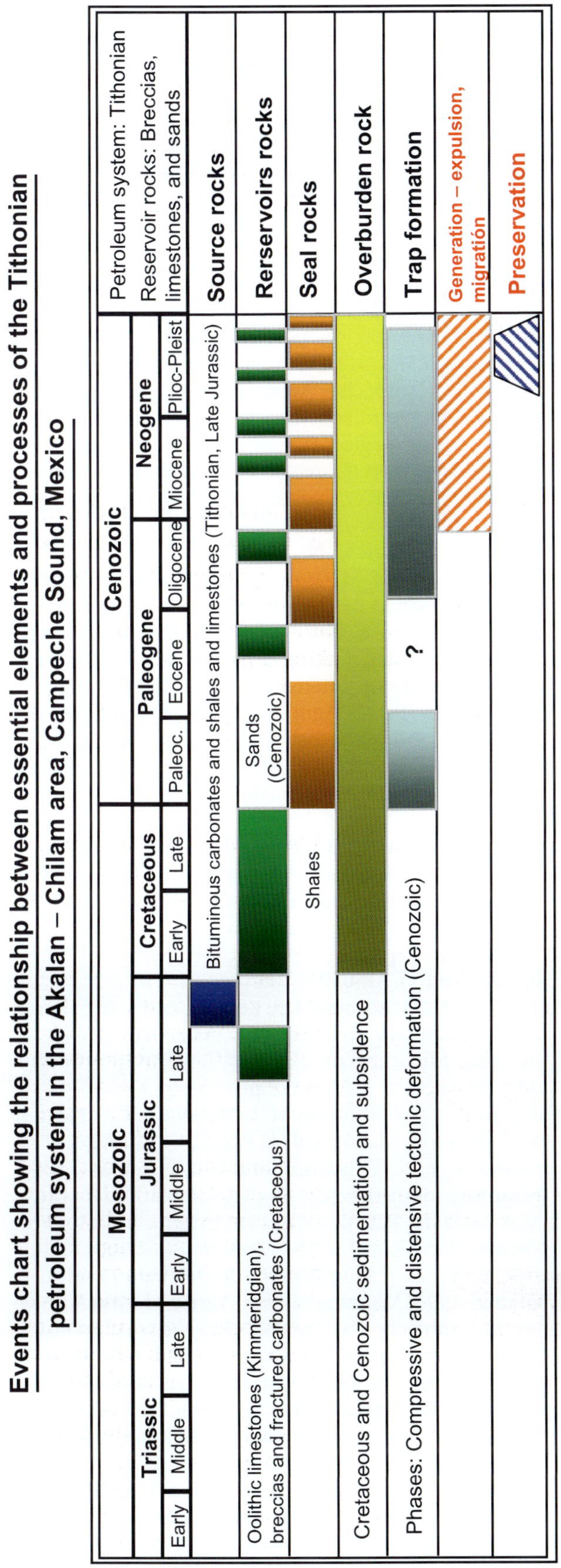

FIGURE 39. Timing diagram of Tithonian petroleum system elements and processes in the Akalan-Chilam area.

negative factor is probably the ineffectiveness of the seals because of the shallow depositional environments that prevailed throughout the Cenozoic.

Evidence from modeling and geological experience in the area indicates that both vertical and horizontal migration occurred, with both short-range and long-range horizontal migration taking place. Faults are an essential factor in the migration.

ACKNOWLEDGMENTS

We are grateful to Marco Flores Flores and Antonio Escalera Alcocer for supporting the publication of this manuscript in an AAPG forum. Finally, we are grateful to Emilio Garciacaro for his constructive review of the manuscript and to Salvador Villalbazo for figures and drawings.

REFERENCES CITED

Aguayo, C. E. J., 1966, Estudio Geológico de la Zona Transicional Comprendida entre la Cuenca de Macuspana, Campeche y la Porción Occidental de la Plataforma de Yucatán: Tesis Profesional, Facultad de Ingeniería, Universidad Nacional Autónoma de México, 118 p.

Ángeles, A. F. J., 1982, Estudio sedimentario y diagenético del Campo Abkatún: Sociedad Geofísica Mexicana, VI Convención Nacional, Resúmenes no. 6, p. 108.

Ángeles, A. F. J., 1988, Estudio estratigráfico-sedimentológico del Jurásico Superior en la Sonda de Campeche: Ingeniería Petrolera, v. 28, no. 1, p. 45–55.

Ángeles, A. F. J., 1993, Breve historia de la Sonda de Campeche: Gaceta de la Asociación Mexicana de Geólogos Petroleros, v. 2, p. 3–4.

Ángeles, A. F. J., 1996a, Estratigrafía del Jurásico Superior del Subsuelo en la Sonda de Campeche (Golfo de México): Tesis de Maestría: ESIA, Instituto Politécnico Nacional, 93 p.

Ángeles, A. F. J., 1996b, Unidades generadoras del Jurásico Superior en la Sonda de Campeche: Memorias, V Congreso Latinoamericano de Geoquímica Orgánica (Cancún), p. 144–146.

Ángeles, A. F. J., 2006, Monografía petrolera de la Zona Marina: Boletín Especial de la Asociación Mexicana de Geólogos Petroleros, volumen especial, p. 77.

Ángeles, A. F. J., and M. E. Ortuño, 1994, Estudio estratigráfico sedimentológico del Cretácico en la Sonda de Campeche: Sociedad Geológica Mexicana, XII Convención Nacional, Resúmenes no. 12, p. 8–9.

Ángeles, A. F. J., J. R. Núñez, and M. J. M. Quezada, 1992, Evolución tectónica de la Sonda de Campeche, estilos estructurales resultantes y su implicación en la generación y acumulación de Hidrocarburos: II Simposio de la Asociación Mexicana de Geólogos Petroleros e Instituto Mexicano del Petróleo, p. 12–13.

Araujo, M. J., L. M. A. Basáñez, and M. A. D. Márquez, 1986, Estudio estratigráfico-sedimentológico de las rocas del Jurásico en el prospecto Área Marina de Campeche: Instituto Mexicano del Petróleo, Reporte Interno Inédito, Proyecto C-5008, 92 p.

Bañuelos, M. O., 1993, Roca almacén del Mioceno y Plio-Pleistoceno en la Sonda de Campeche: I Simposio de Geología de Subsuelo, Asociación Mexicana de Geólogos Petroleros, p. 84–91.

Behar, F., M. Vandenbroucke, Y. Tang, F. Marquis, and J. Espitalié, 1997, Thermal cracking of kerogen in open and closed systems: Determination of kinetic parameters and stoichiometric coefficients for oil and gas generation: Organic Geochemistry, v. 26, p. 321–339.

Beicip-Franlab, 2004, Estudio de modelado de prospectos del bloque Akalan: Reporte Interno Inédito, Petróleos Mexicanos (PEMEX), Región Marina, 93 p.

Bello, M. R., 1991, Estudio estratigráfico-sedimentológico del Cretácico Superior en la Sonda de Campeche: Instituto Mexicano del Petróleo, Reporte Interno Inédito, Proyecto CAO-3512, 110 p.

Burchette, T. P., and V. P. Wright, 1992, Carbonate ramp depositional systems: Sedimentary Geology, v. 79, p. 3–57.

García, H. J., 1993, Roca almacenadora Jurásico Superior Oxfordiano: I Simposio de Geología de Subsuelo: Asociación Mexicana de Geólogos Petroleros, Ciudad del Carmen, Camp, México, p. 49–50.

Guzmán, V. M. A., and M. R. Mello, 1999, Origin of oil in the Sureste Basin, Mexico: AAPG Bulletin, v. 83, no. 7, p. 1068–1095.

Guzmán, V. M. A., L. O. Castro, J. R. R. Román, M. L. M. Medrano, and L. C. Valdez, 2001, El origen del petróleo en las subprovincias Mexicanas del Golfo de México: Boletín de la Asociación Mexicana de Geólogos Petroleros, v. 49, no. 1 and 2, p. 31–46.

Hernández, G. S., 1993, Subsistema almacenador en los horizontes Cretácico Superior y Eoceno Medio de la Región Marina: Primer Simposio de Geología de Subsuelo, Asociación Mexicana de Geólogos Petroleros, p. 79–83.

Holguín, Q. N., and I. M. A. Romero, 1982, Estudio geoquímico de pozos de exploración en la Sonda de Campeche-Plataforma de Yucatán: Instituto Mexicano del Petróleo, Reporte Interno Inédito, Proyecto C-1142, 60 p.

Holguín, Q. N., et al., 2005, Estudio regional de manifestaciones superficiales de aceite y gas en el sur del Golfo de México, su origen e implicaciones exploratorias: Boletín de la Asociación Mexicana de Geólogos Petroleros, v. 52, no. 1, p. 20–41.

Instituto Mexicano del Petróleo (IMP), 2000, Atlas de las Cuencas Petroleras de México: Reporte Interno Inédito, 320 p.

Medrano, M. L. M., and I. M. A. Romero, 1993, Caracterización geoquímica de las rocas generadoras de los sistemas petrolíferos del Jurásico Superior en el área Marina de Campeche: I Simposio de Geología de Subsuelo, Asociación Mexicana de Geólogos Petroleros, p. 5–16.

Medrano, M. M. L., I. M. A. Romero, and V. R. Maldonado, 1996, Los subsistemas generadores de la Sonda de Campeche, México: Memorias, V Congreso Latinoamericano de Geoquímica Orgánica (Cancún), p. 85–90.

Meneses, G. J., 1980, Geología de la Sonda de Campeche: Boletín de la Asociación Mexicana de Geólogos Petroleros, v. 32, no. 1, p. 1–26.

Mitra, S., G. C. Figueroa, J. H. García, and A. M. Alvarado, 2005, Three-dimensional structural model of the Cantarell and Sihil structures, Campeche Bay province, Mexico: AAPG Bulletin, v. 89, p. 1–26.

Mitra, S., J. A. D. González, J. G. Hernández, S. H., García, and S. Banerjee, 2006, Structural geometry and evolution of the Ku, Zaap and Maloob structures, Campeche Bay, Mexico: AAPG Bulletin, v. 90, p. 1565–1584.

Northeast Marine Region (RMNE), 1999, Modelado de Cuencas *TEMIS 3D* del área Zazilha-Tunich, en la Sonda de Campeche, México: Reporte Interno Inédito, RMNE, Petróleos Mexicanos (PEMEX), 115 p.

Northeast Marine Region (RMNE), 2002, Sistemas Petroleros Terciarios en el Área de Tunich-Chilam: Petróleos Mexicanos (PEMEX), Exploración y Producción, Reporte Interno Inédito, RMNE, PEMEX, 110 p.

Ornelas, M., S. Franco, N. Aguilera, H. Alzaga, H. A. Escamilla, and M. M. Granados, 1994, (res) Secuencias bioestratigráficas del Kimmeridgiano–Tithoniano en pozos de la zona marina de Campeche Sureste de México: Sociedad Geofísica Mexicana, XII Convención Nacional, Resúmenes no. 12, p. 197.

Ornelas, S. M., F. N. Aguilera, N. S. Franco, M. M. Granados, and M. R. Bello, 1991, Determinación de los límites Oxfordiano–Kimmeridgiano–Tithoniano en pozos de la Zona Marina de Campeche, Plataforma Continental de México: Instituto Mexicano del Petróleo, Reporte Interno Inédito, Proyecto CAO-5511, 125 p.

Ortega, L. S., M. L. Medrano, and V. L. Clara, 1995, Definición de los límites reales de generación de las rocas del Tithoniano de las cuencas del Sureste, Sonda Marina de Campeche: Instituto Mexicano del Petróleo, Proyecto CAB-0501, 128 p.

Ortuño, A. S., C. M. Romero, J. R. Rosales, and E. R. Bahena, 2004, Modelado de prospectos en el área de Kayab: Instituto Mexicano del Petróleo, Reporte Interno Inédito, Proyecto C-30559, 177 p.

Ortuño, A. S., C. M. Romero, M. F. Flores, G. R. Gutiérrez, M. L. Moreno, M. López de la Paz, C. G. Santiago, D. G. Ibáñez, P. E. R. Villaseñor, and J. A. P. España, 2006, Estudio e identificación de plays terciarios: Instituto Mexicano del Petróleo, Reporte Interno Inédito, Proyecto C-30567, 252 p.

Ramírez, C. L., A. J. Mendoza, G. R. del Valle, and R. P. Gómez, 1996, Mapping salt structure and subsalt imaging in Campeche Bay Mexico, using seismic inversion modeling and prestack attributes: Society of Exploration Geophysicists Expanded Abstracts 15, p. 588–591.

Romero, I. M., C. Cassaigneau, M. Flores, A. Vassayre, R. Navarro, and J. Gaulier, 2004, Future exploration strategy of a frontier offshore Campeche area provided by basin modeling: AAPG International Conference, Cancún, Quintana Roo, Mexico, p. 112–113.

Romero, I. M. A., and R. Maldonado V., 1995, Subsistema Generador Oxfordiano: Reporte Interno Inédito, Región Marina, Petróleos Mexicanos (PEMEX), 147 p.

Romero, M. A., J. Hegre, and R. Maldonado, 1996, Modelado 1-D Genex en el área marina de Campeche: Memorias, V Congreso Latinoamericano de Geoquímica Orgánica, Cancún, Quintana Roo, México, p. 55–56.

Romero, M. A., L. Medrano, and R. Maldonado, 2001, Subsistemas Generadores del Área Marina de Campeche, México: Boletín de la Asociación Mexicana de Geólogos Petroleros, v. 49, no. 1–2, p. 105–115.

Rudkiewicz, J. L., M. Vandenbroucke, H. L. B. Penteado, F. Brigaud, A. Vear, J. Wendebourg, and S. Dûppenbecker, 2000, Integrated basin modeling helps to decipher petroleum systems, *in* M. R. Mello and B. J. Katz, eds., Petroleum systems of South Atlantic margins: AAPG Memoir 73, p. 27–40.

Salomón, M. L., 1999, Modelado estructural de las fallas y su importancia en la exploración de hidrocarburos en el Golfo de México: Instituto Mexicano del Petróleo, Reporte Interno Inédito, TN 872, S35 1999, 85 p.

Santamaría, O. D., 2000, Organic geochemistry of Tithonian source rocks and associated oils from the Sonda de Campeche, Mexico: Ph.D. dissertation, University of Aachen, Germany, 190 p.

Santamaria, O. D., and B. Horsfield, 2001, Tendencias de Evolución Térmica de la Materia Orgánica en el Área Marina de Campeche, México: Boletín Asociación Mexicana de Geólogos Petroleros, v. 49, no. 1–2, p. 116–136.

Schneider, F., S. Wolf, I. Faille, and D. Pot, 2000, A 3-D basin model for hydrocarbon potential evaluation: Application to Congo offshore: Oil and Gas Science and Technology, Revue de l'Institut Français du Pétrole, v. 55, no. 1, p. 3–13.

Ungerer, P., J. Burrus, B. Doligez, P. Y. Chenet, and F. Bessis, 1990, Basin evaluation by integrated two-dimensional modeling of heat transfer, fluid flow, hydrocarbon generation and migration: AAPG Bulletin, v. 74, no. 3, p. 309–335.

15

Grajales-Nishimura, J. M., G. Murillo-Muñetón, C. Rosales-Domínguez, J. C. Bermúdez-Santana, L. G. Velasquillo-Martínez, J. García-Hernández, J. A. Arz, and I. Arenillas, 2009, The Cretaceous–Paleogene boundary Chicxulub Impact: Its effect on carbonate sedimentation on the western margin of the Yucatan Platform and nearby areas, *in* C. Bartolini and J. R. Román- Ramos, eds., Petroleum systems in the southern Gulf of Mexico: AAPG Memoir 90, p. 315–335.

The Cretaceous–Paleogene Boundary Chicxulub Impact: Its Effect on Carbonate Sedimentation on the Western Margin of the Yucatan Platform and Nearby Areas

J. M. Grajales-Nishimura, G. Murillo-Muñetón, C. Rosales-Domínguez, J. C. Bermúdez-Santana, and L. G. Velasquillo-Martínez

Instituto Mexicano del Petróleo, Mexico D.F., Mexico

J. García-Hernández

Petróleos Mexicanos (PEMEX) Exploración y Producción, Región Marina Noreste, Ciudad del Carmen, Campeche, Mexico

J. A. Arz and I. Arenillas

Universidad de Zaragoza, Spain

ABSTRACT

Outcrops and offshore Campeche borehole data clearly document the presence of a carbonate facies succession, including calcareous breccia, on the western Yucatan Platform (Campeche Sound) and the Chiapas-Tabasco Platform. This carbonate sequence is associated with ejecta that contains altered glass, shocked minerals, and accretionary lapilli derived from the Chicxulub impact on the Yucatan Platform. The Cretaceous–Paleogene (K-Pg) boundary sedimentary succession is found at the El Guayal, Bochil, and Chilil outcrops of Tabasco and Chiapas, and offshore Campeche, 300–500 km (186–311 mi) west of the Chicxulub structure center. From base to top, this succession consists of four subunits: (1) carbonate breccia, 40–300 m (131–984 ft) thick, without ejecta; (2) fine- to medium-grained carbonate breccia, 10–20 m (33–66 ft) thick, mixed with sparse ejecta; and (3) siltstone, shale, and carbonate sand facies, 9–30 m (29–98 ft) thick, containing abundant ejecta (altered glass and shocked quartz). This unit culminates in a nearly pure clay layer (~2 cm [~0.7 in.] thick) with the well-known

DOI:10.1306/13191090M902954

iridium anomaly at the top. Unit 4 is a conglomeratic breccia ranging from 10 to 20 m (33 to 66 ft) thick containing ejecta that is interbedded with or overlays subunit 3 (the ejecta layer) in some wells. Subunits 1, 2, and 3 are highly dolomitized in offshore Campeche, and the glass in subunit 3 is altered to clay minerals (smectite). Subunits 1 and 2 constitute hydrocarbon reservoir facies, whereas subunit 3 corresponds to the sealing layer of these reservoirs. Regionally, this sequence displays a gradational structure that represents a large debris flow followed by ballistic and clastic sedimentation with materials reworked by currents. Moreover, well logs, areal distribution, and stratigraphic relationships suggest that the thick K-Pg boundary sedimentary succession is a base-of-slope apron deposit. Based on the stratigraphy, sedimentology, and distribution of impact materials in the carbonate sedimentary succession, the following sequence of events can be inferred: megaseismic shaking that induced the collapse of the platform margin and produced the lower breccia facies (subunits 1 and 2); ballistic emplacement of ejected material (carbonate fragments, shocked minerals and glass) that supplied components to subunit 2 and formed the ejecta layer (subunit 3), the latter acting as the seal for Cantarell and neighboring oil fields; and reworking of the ejecta layer and coarser-grained carbonate fragments by the effect of one or more impact-generated tsunami waves to form a conglomeratic breccia (subunit 4) within subunit 3.

INTRODUCTION

The Chicxulub meteorite impact at the Cretaceous–Paleogene (K-Pg) boundary strongly influenced the sedimentary processes that occurred 65 Ma in the area of the Yucatan Platform and nearby localities such as the Campeche Sound and the present Chiapas-Tabasco range. Furthermore, this event produced a broad range of impact-related materials and formed an ejecta-rich layer present in many K-Pg boundary sections worldwide located both nearby and distant from Chicxulub (Alvarez et al., 1980, 1995; Hildebrand and Boynton, 1990; Smit, 1999; Claeys et al., 2002; Arenillas et al., 2006) (i.e., Figure 1). Petróleos Mexicanos (PEMEX) technical personnel (Cornejo-Toledo and Hernández-Osuna, 1950; Penfield and Camargo, 1981) originally reported a set of nearly circular geophysical anomalies on the Yucatan Peninsula. Later, Kring et al. (1991) and Hildebrand et al. (1992) discovered shocked quartz and altered impact melt in samples from the Yucatan 6 borehole, and Hildebrand et al. (1991b) recognized the Chicxulub structure as a 65-Ma, 180-km (112-mi) diameter impact crater, as confirmed by ^{40}Ar-^{39}Ar analysis (Sharpton et al., 1992; Swisher et al., 1992). A link between the Chicxulub impact and the K-Pg event was also indicated by the thickness of the K-Pg ejecta in the North American region (Hildebrand and Boynton, 1990), which decreases with the distance exactly as expected if it had been ejected from the Chicxulub crater (Hildebrand and Stansberry, 1992). Universidad Nacional Autónoma de México (UNAM) investigators and the International Continental Scientific Drilling Program (Urrutia-Fucugauchi et al., 1996a, b; 2004) recently drilled into the structure.

Prior work interpreted the Yucatan Peninsula as a shallow platform during the Maastrichtian with carbonate sedimentation along its margins and lagoonal and evaporitic sedimentation toward the interior. In the north and northwest of the peninsula, evaporites decrease in relative volume with more abundant carbonate sedimentation forming limestone and marl deposits (López-Ramos, 1979; Santiago-Acevedo et al., 1984). On the northeastern margin of the Yucatan Peninsula Platform, Deep Sea Drilling Project (DSDP) leg 10, site 95, core 19 found pelagic and marly deposits of middle–late Maastrichtian age (Sohl et al., 1991). Similarly, in the southwest, south, and southeastern parts of the Yucatan Platform, evaporite deposition decreased in favor of pelagic limestones (Jolpabuchil Formation) and marls (Méndez Formation). Carbonate breccias were deposited on top of these pelagic marine rocks (Meneses de Gyves, 1980; Quezada-Muñetón, 1990; Smit, 1999; Grajales-Nishimura, 2001; Grajales-Nishimura et al., 2003). At the end of the Maastrichtian, Yucatan represented a large isolated platform separated from the Chiapas-Tabasco Platform by a depression defined by the respective slopes of each platform (Figure 2). Hydrocarbons in the Campeche Sound are mainly produced from four stratigraphic units: (1) Oxfordian sandstones and dolostones, (2) Kimmeridgian oolitic carbonates, (3) Lower to Middle Cretaceous dolomitized limestones, and (4) Upper Cretaceous–Paleocene calcareous breccias. The last unit was the most important hydrocarbon reservoir in Mexico about 10 yr ago (PEMEX Exploración y Producción, 1999).

Many K-Pg boundary sequences include complex tsunami and gravity-flow deposits triggered by the seismicity of the Chicxulub impact (Smit et al., 1992, 1996;

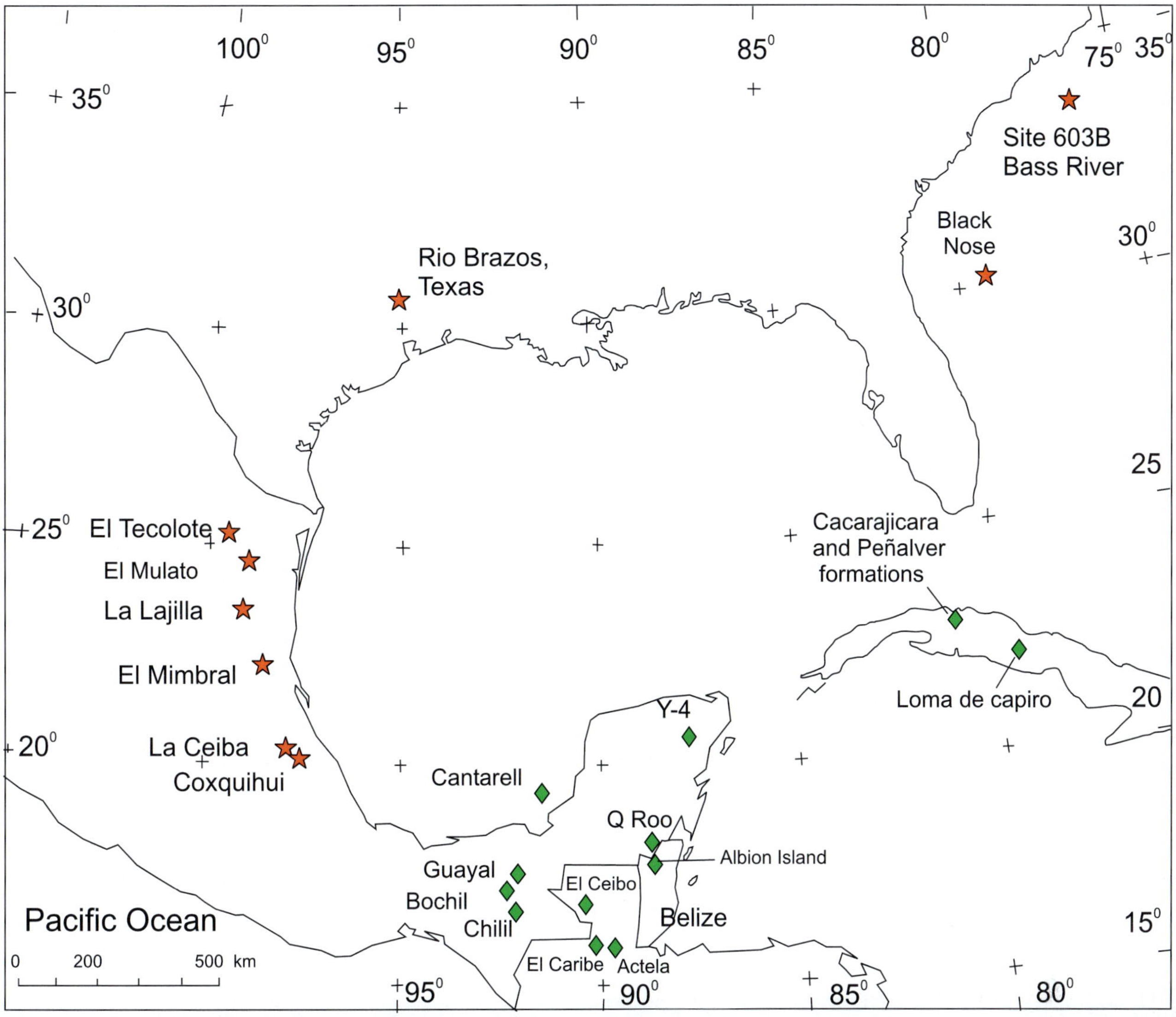

FIGURE 1. The principal localities where K-Pg boundary outcrops have been studied or visited on siliciclastic (star) and carbonate platform environments (rhombus).

Grajales-Nishimura et al., 2000a, b; Takayama et al., 2000; Arz et al., 2001a,b; Tada et al., 2003). The previously mentioned sedimentological process mixed different lithologies and fossils of diverse age that resulted in the carbonate debris sequence known as the "KT boundary cocktail" (Bralower et al., 1998; Díaz-Otero et al., 2000; Arz et al., 2004; Alegret et al., 2005) or complex clastic unit (CCU) (Arenillas et al., 2006). Slumping processes in the latest Maastrichtian chalk and/or pelagic limestone are also linked to seismicity generated by the Chicxulub impact as documented by different workers (Klaus et al., 2000; Soria et al., 2001; Schulte et al., 2003).

In this chapter, we focus on locations close to the impact zone in an offshore region situated to the west of the Chicxulub structure and in the Chiapas-Tabasco range to the southwest. However, we also refer to distal deposits to compare and contrast the sedimentary sequences discussed here. The objectives are to analyze and describe the stratigraphy and petrography of the K-Pg boundary carbonate sequence, including the hydrocarbon producing breccias, to integrate and correlate the litho- and biostratigraphic data, to interpret the implied sedimentologic processes, and to describe new outcrops from the Chiapas range. Its aim is to present the sedimentologic and geological scenario in which the related hydrocarbon reservoir rocks were deposited in the Campeche Sound.

Previous Work

In the Gulf of Mexico region, two main types of high-energy K-Pg boundary sedimentary deposits have been described: a southeastern 50–300-m-thick (164–984-ft-thick) carbonate debris sequence originated by gravity-flow processes and a northwestern high-energy clastic turbidite deposit 1–4-m-thick (3–13-ft-thick).

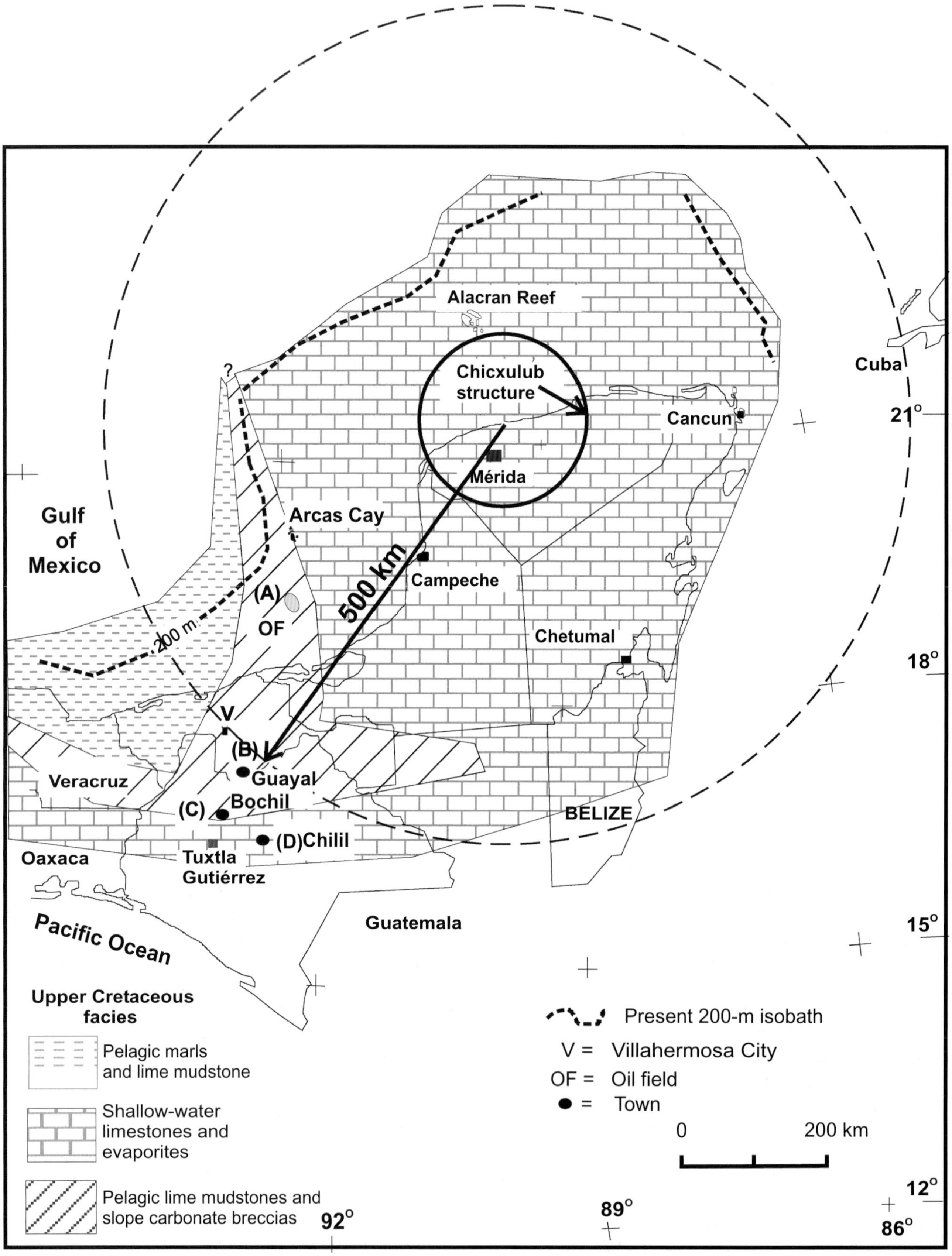

FIGURE 2. Distribution map of the K-Pg boundary facies in southeastern Mexico. The distribution of the pelagic, slope, and shelf environments is also shown.

In Chiapas-Tabasco and Yucatan, the K-Pg boundary is frequently represented by a greater than 100-m-thick (328-ft-thick) carbonate breccia composed of blocks that range from a few centimeters to several meters in diameter. This chaotic breccia has been interpreted as a normal slope deposit (Salas and López-Ramos, 1951; López-Ortíz, 1962; Meneses de Gyves, 1980; Santiago-Acevedo, 1980, 1984; Michaud, 1987; Quezada-Muñetón, 1990; Hernández-García, 1993) and, in more recent investigations, as a base-of-slope breccia deposit of the K-Pg boundary impact event (Hildebrand et al., 1993; Limón-Gonzáles et al., 1994; Montanari et al., 1994; Grajales et al., 1996, Pécheux and Michaud, 1997; Bralower et al., 1998; Murillo-Muñetón et al., 2002). In localities near the ancient platform margin, the sequence is graded with a basal breccia of limestone blocks containing rudist fragments and shallow-water carbonates, covered by finer grained breccia with fragments containing altered glass, culminating in sandstone and claystone beds with abundant shocked minerals displaying planar structures.

In the outcrops of Tabasco and Chiapas, the K-Pg carbonate sequence is observed resting on top of both pelagic and shallow-water marine limestone and is covered by pelagic limestone and marls. Immediately overlying the basal breccia is a 2–12-m-thick sequence of medium-grained sandstone, siltstone, and claystone. In the Bochil section, the very fine-grained siltstone to claystone at the top of the CCU, has an anomalous high Iridium concentration of 1.5 ppb (Montanari et al., 1994). This iridium anomaly was recently found in the El Guayal section of Chiapas (Tagle and Claeys, 2001). A greater than 100-m-thick (328-ft-thick) chaotic carbonate breccia was also reported next to the ancient margin of the Campeche marine platform (Limón-Gonzáles et al., 1994; Grajales et al., 1996) (Figure 1). Another similar K-Pg age breccia up to 450 m (1476 ft) thick is present in Cuba (Takayama et al., 2000; Tada et al., 2003). In the El Caribe locality in Guatemala, Hildebrand et al. (1993) described a 6-m-thick (20-ft-thick) sequence of fining-upward clastic carbonates with fragments as much as 10 cm (4 in.) in diameter in the lower part and calcarenites in the upper part. In this same region, Stinnesbeck et al. (1997) also reported a carbonate sequence with breccia in the lower part and calcarenites associated with glassy material toward the top, and concluded that the breccia was deposited before the K-Pg boundary. At two localities in the El Petén region (El Ceibo and El Caribe), Fourcade et al. (1997) reported a 2.5-m-thick breccia containing rudists, alveolinids, and dasycladacean algae, and attributed its origin to tectonic events. In later work, these same authors reported a 15-m-thick sequence at the Actelá section (Guatemala) with carbonate breccia that gradates to fine-grained material, including glass. They assigned a K-Pg age and related the carbonate sequence to the Chicxulub impact (Fourcade et al., 1998).

In the Burgos Basin of northeastern Mexico, where sedimentation was dominantly siliciclastic, sandstone bodies within shales and/or marls are reported in the Upper Cretaceous (Sohl et al., 1991) and Tertiary (Galloway et al., 1991). These siliciclastic levels suggest that the Late Cretaceous terrigenous sedimentation continued with little change during the Paleocene. In these localities, the base of a sandstone body that is generally thicker than the underlying and overlying sandstones marks the K-Pg boundary. This siliciclastic unit normally contains altered spheroids (tektites), shocked quartz, and the iridium anomaly (Alvarez et al., 1992a; Smit et al., 1992, 1996).

Different specialists have interpreted that the K-Pg siliciclastic unit from northeastern Mexico, Texas, and Alabama is linked to the Chicxulub impact event (Bourgeois et al., 1988; Alvarez et al., 1992b; Smit et al., 1992, 1996) (Figure 1). Originally, the siliciclastic unit was interpreted as the result of erosion during a brief uplift (Muir, 1936). Later, it was reinterpreted as a rapidly deposited sedimentary unit from the passage of tsunamis as a consequence of the Chicxulub impact (Alvarez et al., 1992b; Smit et al., 1992). Some authors suggested that the whole sequence is not K-Pg in age and concluded that it was deposited as a channel fill hundreds of thousands of years prior to the K-Pg boundary (Keller et al., 1993; Stinnesbeck et al., 1993). Others have interpreted the clastic unit as a turbidite and debris deposit triggered by the Chicxulub impact (Bohor and Betterton, 1993, Bohor 1996; Soria et al., 2001).

In a well from Bass River, New Jersey, Olsson et al. (1997) reported a basal spheroid bed (tektites) directly on top of the youngest Maastrichtian marine sediments. This discovery clearly demonstrates that the spherule-bearing unit was deposited precisely at the K-Pg boundary above late Maastrichtian marls. This evidence supports the interpretation that sandstones with tektites that overlie the Méndez Formation and underlie the earliest Paleocene rocks (Velasco Formation) at El Mimbral, La Lajilla, El Tecolote, La Ceiba, Coxquihui, and several other localities in Tamaulipas, Nuevo Leon, and Veracruz were also deposited at the K-Pg boundary (Alvarez et al., 1992b; Longoria and Grajales-Nishimura, 1993; Smit et al., 1996; Arz et al., 2001a, b; Soria et al., 2001).

STRATIGRAPHIC SECTIONS

This chapter presents lithostratigraphic and biostratigraphic data from three K-Pg outcrop sections (El Guayal, Bochil, and Chilil) in Tabasco and Chiapas states and describes a composite K-Pg section from the Cantarell field offshore Campeche state (Figure 2).

El Guayal Section (Tabasco)

The El Guayal locality (17°32′37.73″N, 92°36′07. 61″W) in southeastern Mexico is near the northern boundary between the Sierra de Chiapas subprovince (López-Ramos, 1979), also known as the Chiapas-Tabasco Mesozoic area (Santiago-Acevedo et al., 1984), and the southern boundary of the Tertiary Macuspana Basin (Salas and López-Ramos, 1951). Parts of the following stratigraphic units, from bottom to top, were studied at this locality: Maastrichtian pelagic limestone equivalent to the Jolpabuchil Formation (Quezada-Muñetón, 1990); the Lomas Tristes carbonate breccia (Salas and López-Ramos, 1951; López-Ramos, 1979; Grajales et al., 1996); and an early Paleocene sequence of marl, shale, and thin limestone equivalent to the Soyatal Formation in the northern part of the Sierra de Chiapas (Quezada-Muñetón, 1990). Salas and López-Ramos (1951) were the first to study the breccia outcrops at El Guayal and measured a 10-m-thick (33-ft-thick) section and assigned it a middle Cretaceous age. PEMEX geologists continued to study the carbonate breccia in other localities near El Guayal, such as those at Puxcatán and Pasa Monos, assigning it a Maastrichtian age (Quezada-Muñetón, 1990).

Bochil Section (Chiapas)

The Bochil locality in the state of Chiapas (17°00′34″N and 92°56′46″W) is located in the northern part of the Sierra de Chiapas subprovince. The sedimentary succession at Bochil composed of carbonate debris (coarse-grained breccia, fine-grained breccia, sandstones, and claystones) was originally described in an unpublished work by PEMEX (Medina de la Paz and Duran-Rodriguez, 1991) and later by Montanari et al. (1994) who reported an iridium anomaly at the top of the sequence and assigned it a K-Pg boundary age. Later, Smit et al. (1996), Smit (1999), and Grajales-Nishimura et al. (2000a, b) interpreted this sequence as a debris flow initiated by seismic shaking caused by the impact at Chicxulub. Arenillas et al. (2002) confirmed a basal Danian age for the sequence above the K-Pg boundary sedimentary succession (Figures 1, 2).

Chilil Section (Chiapas)

The small town of Chilil (16°40′49″N and 92°29′46″W) is located about 20 km (12 mi) to the east-southeast of the city of San Cristóbal de las Casas, Chiapas. Quezada-Muñetón (1990) first reported the Chilil K-Pg boundary locality. According to Quezada-Muñetón, the breccia facies corresponds to a slope deposit within the rudist-rich shallow-water marine carbonates of the Maastrichtian Angostura Formation. Sánchez-Montes de Oca (1969, cited by Quezada-Muñetón, 1990) first defined the Angostura Formation. The type locality, at the north area of La Angostura Dam, is located about 55 km (34 mi) to the southwest of Chilil. The breccia marks the contact between the Campanian–Maastrichtian shallow-water limestone of the Angostura Formation and the overlying Paleocene outer platform lime mudstones of the Soyaló Formation.

Offshore Campeche

The K-Pg carbonate sequences were studied in oil fields located about 80 km (48 mi) offshore Campeche, north-northeast of Ciudad del Carmen. We studied K-Pg carbonate sequences from Cantarell and other nearby oil fields, such as Balam and Bacab (Figure 2).

STRATIGRAPHY AND PETROGRAPHY

This key stratigraphic package, from base to top, consists of three marine sedimentary sequences. The subdivison into three sequences is based on their distinctive lithologies, ages, and relative stratigraphic position. The first lithostratigraphic sequence (1) corresponds to Maastrichtian pelagic limestone with 10-, 15-, and 20-cm-thick (4-, 6-, and 8-in.-thick) beds. Above these limestones is (2) a graded detrital K-Pg boundary carbonate sequence consisting of coarse calcareous breccia, fine calcareous breccia, sandstone, and claystone. This, in turn, is overlain by (3) a basal Paleocene package of shaly limestones and marls (see Figure 3).

Maastrichtian Deep- and Shallow-Water Marine Limestone

These units are observed in outcrops at El Guayal, Bochil, and Chilil and in cores from the Campeche Sound oil fields. The pelagic deep marine limestone exposed at both El Guayal and Bochil is dark gray, hard and fine grained, with well-defined beds 10–20 cm (4–8 in.) thick containing diagenetic chert nodules. The Jolpabuchil Formation, defined by González-Alvarado (1963, cited by Quezada-Muñetón, 1990), is the formational equivalent of this sequence. The original description of the Jolpabuchil Formation from an unpublished PEMEX report is "...light gray, dark gray and steel gray limestone, lithographic, cryptocrystalline, dense and compact, conchoidal fracture, well bedded, thin bedded, with chert bands and nodules, of Turonian–Coniacian age" (González-Alvarado, 1963, cited by Quezada-Muñetón, 1990, p. 15). Quezada-Muñetón (1990) refers to the unpublished work by Sánchez-Montes de Oca who extended the age of the Jolpabuchil Formation into the Maastrichtian. In contrast,

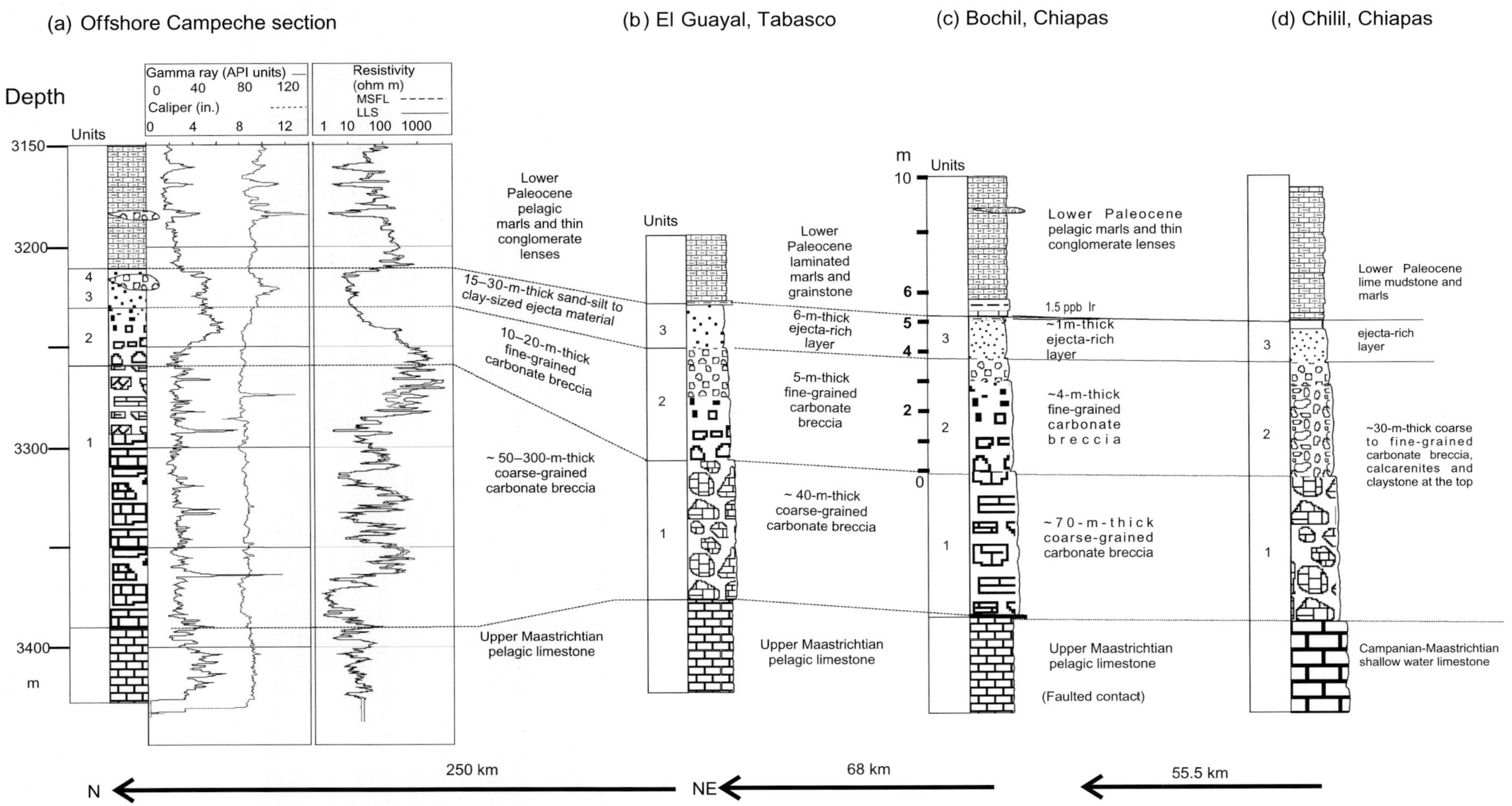

FIGURE 3. Stratigraphic correlation of subsurface sections offshore Campeche Sound and the Cretaceous–Tertiary boundary outcrops in Chiapas and Tabasco. (a) Composite column of hydrocarbon-producing zones (Cantarell, Bacab, and Balam fields). (b, c) The K-Pg boundary-graded carbonate succession at El Guayal, Bochil, and Chilil. MSFL = micro spherical focused log; LLS = shallow lateralog.

the Maastrichtian limestone from Chilil is a shallow-water rudistid-rich facies. This limestone is light gray and well-bedded in 0.20–2-m-thick (0.65–6-ft-thick) strata, with abundant rudistids and benthic foraminifera. This limestone was defined and described as the Angostura Formation in an internal PEMEX report (Sánchez-Montes de Oca, 1969, cited by Quezada-Muñetón, 1990).

Pelagic limestones studied at El Guayal consist of 14 m (46 ft) of dark gray, thin-bedded lime mudstone (5–20-cm [2–8-in.] beds) and thin beds (20–30 cm [8–12 in.] thick) of fine-grained breccia interbedded among the limestones. Abundant diagenetic chert nodules are seen near the central part of the section. Examined microscopically, these limestones are planktonic foraminifer-rich lime mudstones and wackestones. They contain abundant calcite-filled veinlets and microveinlets with intercrystalline spaces frequently containing solidified petroleum. Other carbonate constituents include shallow-water bioclasts such as rudist fragments, echinoderm plates, mollusk fragments, and benthic foraminifera.

Similarly, in Bochil, mudstone-wackestone facies with abundant planktonic foraminifera are present. Additionally, in some wackestone beds toward the top of the sequence, we recognized biogenic structures that characterize the *Zoophycus* ichnofacies. The presence of this ichnogenera associated with abundant planktonic foraminifera confirms a deep-water paleoenvironment for this limestone at Bochil. In cores from offshore Campeche, we observed thin-bedded lime mudstone to wackestone with abundant planktonic foraminifera. Eventually, thin carbonate breccia lenses from centimeter to a few meters thick are interstratified in the lime mudstone to wackestone. However, it is also common that the Upper Cretaceous carbonate from offshore Campeche is commonly 100% dolomitized; in this case, the original texture or the microfossils cannot be determined.

In contrast to the previous described sequences, the Angostura Formation facies at Chilil consist of 0.20–2-m-thick (0.65–6-ft-thick) limestone beds that include: bioclastic grainstone to packstone, packstone with mixed fossil, and rudistid (caprinid) breccia-like rudstone. Benthic foraminifera in the packstone with mixed fossil include: miliolids, *Chubina* sp., and other alveolinids. Scarce planktonic foraminifera have also been reported from the packstone-grainstone beds of the La Angostura Formation at Chilil (Quezada-Muñetón, 1990).

The K-Pg Boundary-Graded Detrital Carbonates (Clastic Complex Unit)

The stratigraphic architecture of the K-Pg boundary-graded carbonate sequence generally suggests a single continuously graded, fining-upward deposit. For descriptive purposes, it may be subdivided, from base to top, into four main subunits: (1) coarse-grained carbonate breccia (40–300 m [131–984 ft] thick); (2) medium- to fine-grained carbonate breccia (10–30 m [33–98 ft] thick); (3) graded calcarenite, siltstone, and claystone; (4) fine-grained calcareous breccia that in some wells is interbedded with or overlies subunit 3.

The coarse-grained K-Pg boundary carbonate breccia (subunit 1) overlies the Maastrichtian pelagic limestones and is composed of a graded carbonate breccia ranging from 40 m (131 ft) thick at El Guayal, more than 70 m (230 ft) thick at Bochil, 30 m (98 ft) thick at Chilil, and up to 300 m (984 ft) thick in Campeche Sound. Near its base, in outcrop, are blocks of pelagic limestones up to 2 m (6 ft) in diameter, and toward the top, clasts decrease continuously to a maximum size of 30 cm (12 in.). The pelagic limestone fragments contain black chert nodules very similar to those in the underlying limestones. Fragment size continues to decrease upward in subunit 2, as well as displaying a notable change in clast composition. Toward the top of subunit 2, the fragments consist of shallow-water limestones with rudists and macroforaminifera, such as *Orbitoides*, *Asterorbis*, *Sulcoperculina*, and *Chubbina*. The shallow-water fragments are mostly alveolinid and miliolid grainstones, whereas others consist of bioclastic limestones with rudist fragments. A scarce, though well preserved matrix around the limestone fragments contains abundant planktonic microfossils. Stylolites are common, indicating pressure solution, and vugular porosity is also observed. Hydrocarbon impregnations are observed in some fractures and vugs.

The 10–30-m-thick (33–98 ft-thick), fine- to medium-grained carbonate breccia (subunit 2) of the K-Pg boundary carbonate sequence lies conformably and gradationally above the finest grained beds of the coarse-grained carbonate breccia (subunit 1). The fine-grained carbonate breccia of subunit 2 also grades upward to a coarse-grained sandstone. The fine-grained breccia contains limestone fragments and abundant bentonite clasts 3 mm to 1 cm (0.11 to 0.39 in.) in diameter. Toward the top of this unit, an 80-cm-thick (31-in.-thick) bed of poorly consolidated, coarse-grained sand containing accretionary lapilli and quartz with planar deformation structures (PDFs) is observed (Stöffler and Langenhorst, 1994; Grieve et al., 1996) (Figure 4).

The coarse-grained sand with accretionary lapilli layer in the upper part of subunit 2 grades upward into a well-sorted, medium- to fine-grained calcarenite (subunit 3) that, in turn, grades upward into siltstone and claystone. The thickness of subunit 3 is variable; i.e., 3 m (10 ft) at Bochil, about 10 m (33 ft) at El Guayal, and more than 15 m (49 ft) in Campeche Sound. The CCU is overlain by a dark clay layer that can be correlated globally with the K-Pg boundary dark clay from different marine settings (e.g., Arenillas et al., 2006).

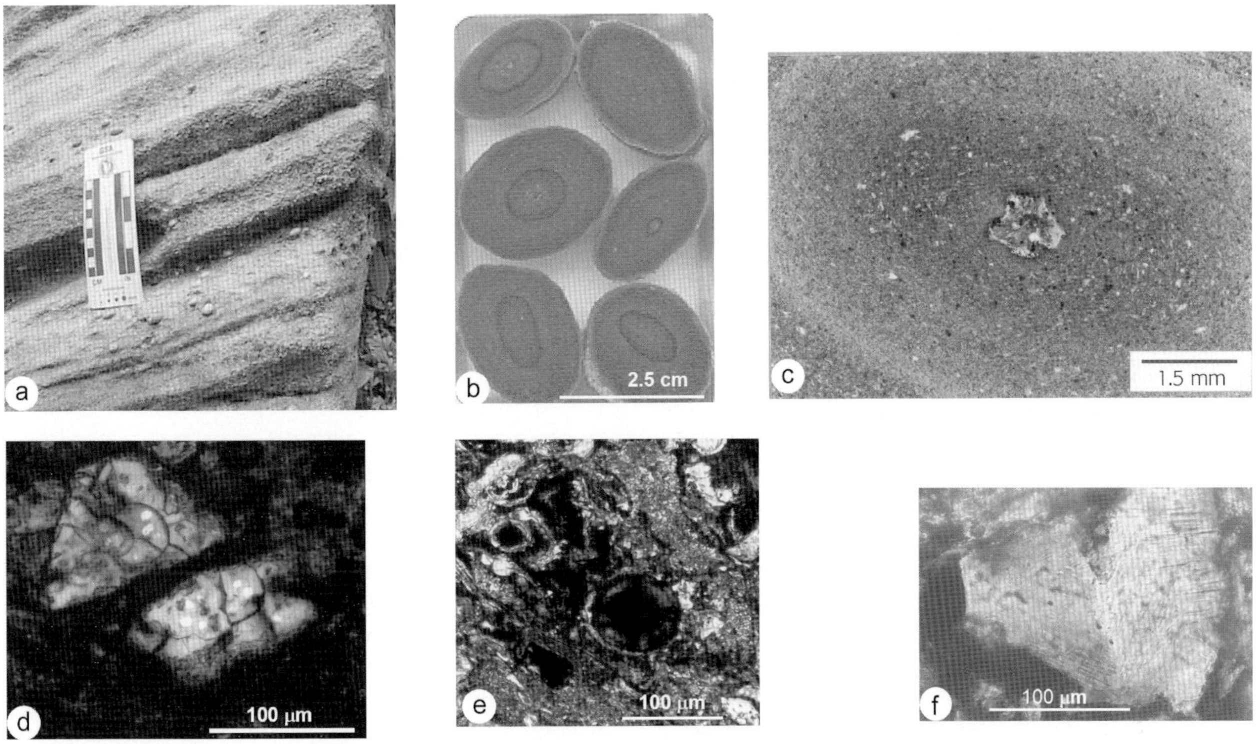

FIGURE 4. Photographs of accretionary lapilli and other impact materials. (a) Layer with accretionary lapilli from an outcrop at El Guayal, Tabasco. (b) Sections of accretionary lapilli. (c) Accretionary lapilli in thin section showing the nucleus and shocked quartz grains. (d) Ballen-type quartz, well C227N1. (e) Glass alteration textures, well C227N1. (f) Shocked quartz in ejecta at Chilil, Chiapas. At least three sets of intersecting planar deformation features are prominent in the photographs. FOD = first occurrence data; LOD = last occurrence data.

The exotic mineralogy found in subunits 2 and 3 is distributed as follows: a 70–80-cm-thick (27–31-in.-thick) interval located in the upper part of subunit 2 contains abundant accretionary lapilli (Figure 4). Within this layer and into the overlying sandy to argillaceous strata (subunit 3), abundant altered glass and quartz fragments with planar deformation features exist, as described below.

Accretionary lapilli may form in the pyroclastic clouds of volcanic eruptions and meteorite impacts. Lapilli of volcanic origin has been documented in Phanerozoic and Precambrian rocks (Moore and Peck, 1962; Reimer, 1983a, b). They consist of a nucleus (generally a lithic fragment) surrounded by concentric layers of fragmented material with a generally pyroclastic aspect (Figure 4). Concentric layers around the nucleus form by the accretion of particles within the eruption clouds, probably from electrostatic attraction under certain conditions of humidity and temperature.

The lapilli at El Guayal consists of dark, hard, calcitized, dolomitized, partially silicified, slightly elongate ellipsoids 1.0–1.8 cm (0.3–0.7 in.) in diameter (Grajales et al., 1996). Their nuclei are partially fused rock fragments within a concentric matrix, or layered structure, formed by the accretion of smaller mineral particles. This concentric matrix is composed of glass splinters that have been replaced by calcite and/or chlorite; quartz, with and without planar deformation features; and feldspar and mica. The general texture of the concentric matrix resembles the pyroclastic texture of tuffs but differs by its content of quartz with planar deformation features. The lapilli at El Guayal are similar to those reported from the Ries Crater in Germany (Graup, 1981; Newson et al., 1990).

Quartz with PDFs was found at the top of the fine-grained breccia (subunit 2) and in the calcareous sandstones of subunit 3. Quartz occurs in both the sandy matrix around the accretionary lapilli and around the nuclei of the lapilli in the internal matrix (Grajales et al., 1996). In thin section, the quartz grains are anhedral with up to four directions of planar deformation. At Bochil, quartz with PDFs occurs in the siltstone and reddish shale layer towards the top of subunit 3. This level also has an anomalously high iridium content of 1.5 ppb (Montanari et al., 1994). Quartz with impact deformation has also been identified in subunits 3 and 4 from Campeche Sound (Figure 4).

Paleocene Pelagic Limestone and Marls

Pelagic limestone, marl, calcareous shale, and argillaceous limestone overlie the upper unit of the K-Pg boundary breccia. This facies association is equivalent to the Soyaló Formation described by Quezada-Muñetón (1990) and Castañón and Medina (1991). The Soyaló Formation was deposited in an open marine environment (Quezada-Muñetón, 1990).

For this study, we only measured a small portion of the basal part of the Paleocene. The detailed micropaleontological studies were conducted from the basal part of the Guayal (1 m, 3 ft), Bochil (2.5 m, 8.2 ft) and Chilil (1 m, 3 ft) sections. Because of the lack of a core that precisely cuts the basal Paleocene, detailed analyses from sequences of this age in offshore Campeche have not been possible. However, the outcrops collected along the lower Paleocene and the upper part of the K-Pg boundary sequence, as well as some cores from near the base of the Paleocene seem to suggest that the same biozones identified in the outcrops are also present in offshore Campeche.

Paleocene facies from the four different locations are very similar. In Bochil and Guayal, they correspond mainly to dark gray lime mudstone or wackestone; scarce clayed mudstone beds are also present. The planktic foraminifera are abundant in Bochil, but scarce in Guayal. In Chilil, the Paleocene facies are light to gray lime mudstone or wackestone with scarce planktic microfossils. This basal section is composed of 5-cm-thick (2-in.-thick) beds of dark to light gray marl, calcareous shale, and argillaceous limestone. These beds overlay the yellowish silty and argillaceous beds in the upper part of subunit 3. A petrographic analysis of the oldest beds in the Soyaló Formation reveals that the basal 8 cm (3 in.) contains very scarce microfossils and many worm borings (bioturbation).

BIOSTRATIGRAPHY AND BIOCHRONOLOGY

The reported sedimentary sequences were dated mainly with planktonic foraminiferal assemblages. In the biostratigraphic study, we used the planktonic foraminiferal zonations and subzonations proposed by Arz and Molina (2002) for Upper Cretaceous and Arenillas et al. (2004) for the lower Danian, which are a slight modification of previous biozonations (Caron, 1985; Berggren et al., 1995; Molina et al., 1996). For the definition of the biozones and subbiozones, they used the last occurrence data (LOD) of *Plummerita hantkeninoides* and *Abathomphalus mayaroensis*, and the first occurrence data (FOD) of *Contusotruncana plicata, Rugoglobigerina rotundata, Rugoglobigerina scotti, Planoglobulina acervulinoides, Racemiguembelina fructicosa, A. mayaroensis, Pseudoguembelina hariaensis, P. hantkeninoides, Parvularugoglobigerina longiapertura, Parvularugoglobigerina eugubina, Eoglobigerina simplicissima, Parasubbotina pseudobulloides, Subbotina triloculinoides*, and *Globanomalina compressa* (see Figure 5). The occurrence ages of the index species and the biozonal boundaries are based on biomagnetostratigraphic calibrations from complete and very expanded worldwide sections (Arz and Molina, 2002; Arenillas et al., 2004). Scanning electron microscope (SEM) images of some Upper Cretaceous and Danian planktonic foraminifer are illustrated in Figures 6 and 7, respectively.

A correlation chart for planktonic foraminiferal zonations and index species is also shown in Figure 5. The biozones recognized, from the oldest to the youngest, are the following:

- *Gansserina gansseri* Zone (upper Campanian–early Maastrichtian): Biostratigraphic interval between the FOD of the nominal taxon and the FOD of *A. mayaroensis*. This biozone was further divided into five subzones by Arz and Molina (2002): *C. plicata* subzone, *R. rotundata* subzone, *R. scotti* subzone, *P. acervulinoides* subzone, and *R. fructicosa* subzone.
- *Abathomphalus mayaroensis* Zone (upper Maastrichtian): Biostratigraphic interval between the FOD of the nominal taxon and the LOD of *A. mayaroensis* and *P. hantkeninoides*. This biozone was further divided into three subzones by Arz and Molina (2002): *Abathomphalus mayaroensis, Pseudoguembelina hariaensis*, and *Plummerita hantkeninoides* Subzones.
- *Guembelitria cretacea* Zone (lower Danian): Biostratigraphic interval between the LOD of *P. hantkeninoides*, precisely at the K-Pg boundary, and the FOD of *P. eugubina*. This biozone was further divided into two subzones by Arenillas et al. (2004): *Hedbergella holmdelensis* subzone and *P. longiapertura* subzone. The FOD of *P. longiapertura* (the base of this biozone) approximately corresponds with the lowest record of Tertiary species, therefore, the *H. holmdelensis* subzone is equivalent to the zone P0 of Berggren et al. (1995).
- *Parvularugoglobigerina eugubina* Zone (lower Danian): Biostratigraphic interval between the FOD of the nominate taxon and the FOD of *P. pseudobulloides*. This biozone was further divided into two subzones by Arenillas et al. (2004): *Parvularugoglobigerina sabina* subzone and *E. simplicissima* subzone.
- *P. pseudobulloides* Zone (lower Danian): Biostratigraphic interval between the FOD of the nominal taxon and FOD of *G. compressa*. This biozone was further divided in two subzones by Arenillas et al. (2004): *E. trivialis* subzone and *S. triloculinoides* subzone.

Upper Cretaceous Units

The biostratigraphic study of the pelagic limestones at El Guayal reveals the presence of the planktonic foraminifer *Guembelitia cretacea, G. trifolia, Heterohelix pulchra, H.*

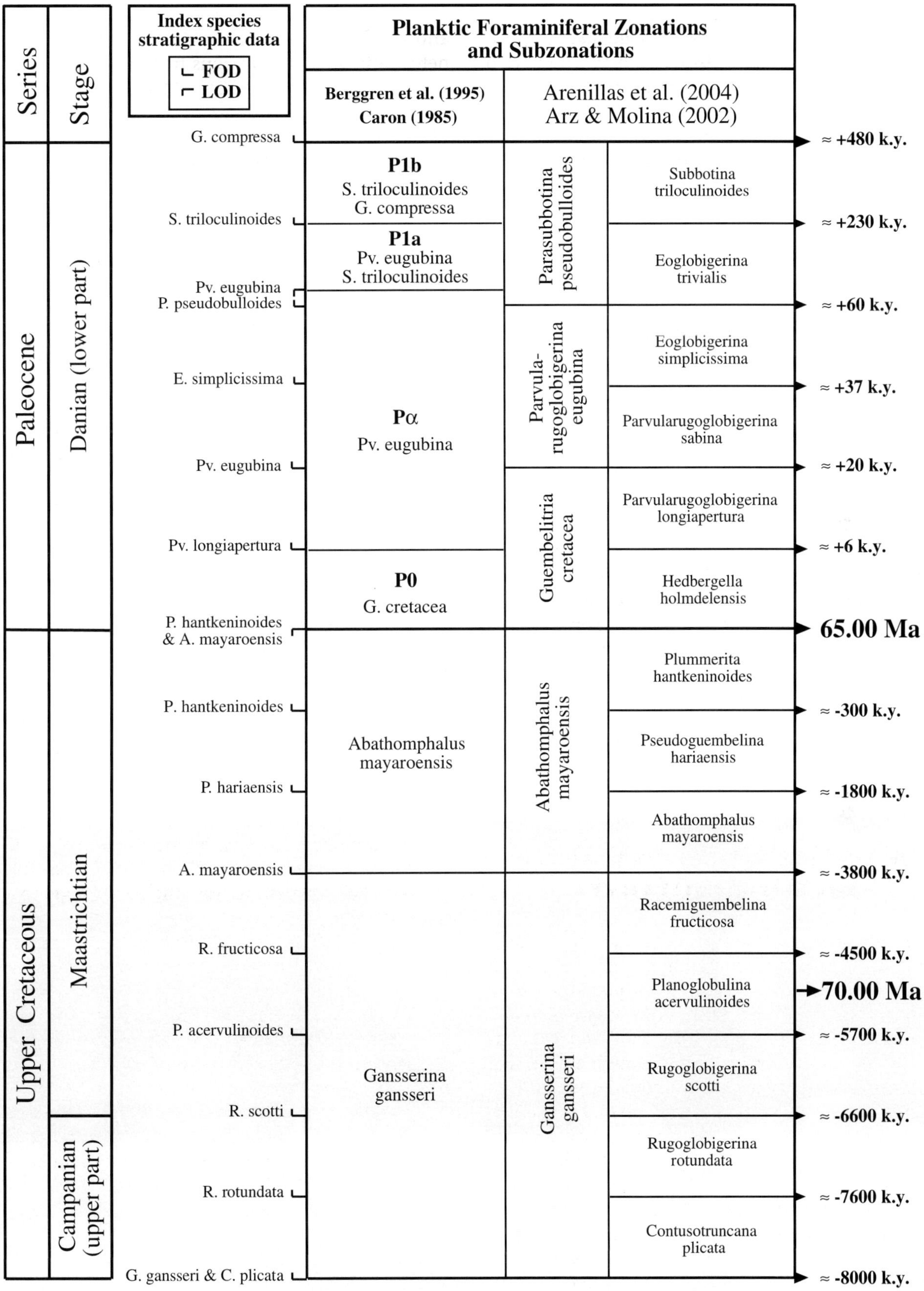

FIGURE 5. Comparison of the planktonic foraminiferal zonations for the Campanian, Maastrichtian (Caron, 1985; Arz and Molina, 2002), and lower Danian (Berggren et al., 1995; Arenillas et al., 2004), estimated ages for the biozonal boundaries and first and last occurrence data of several index species.

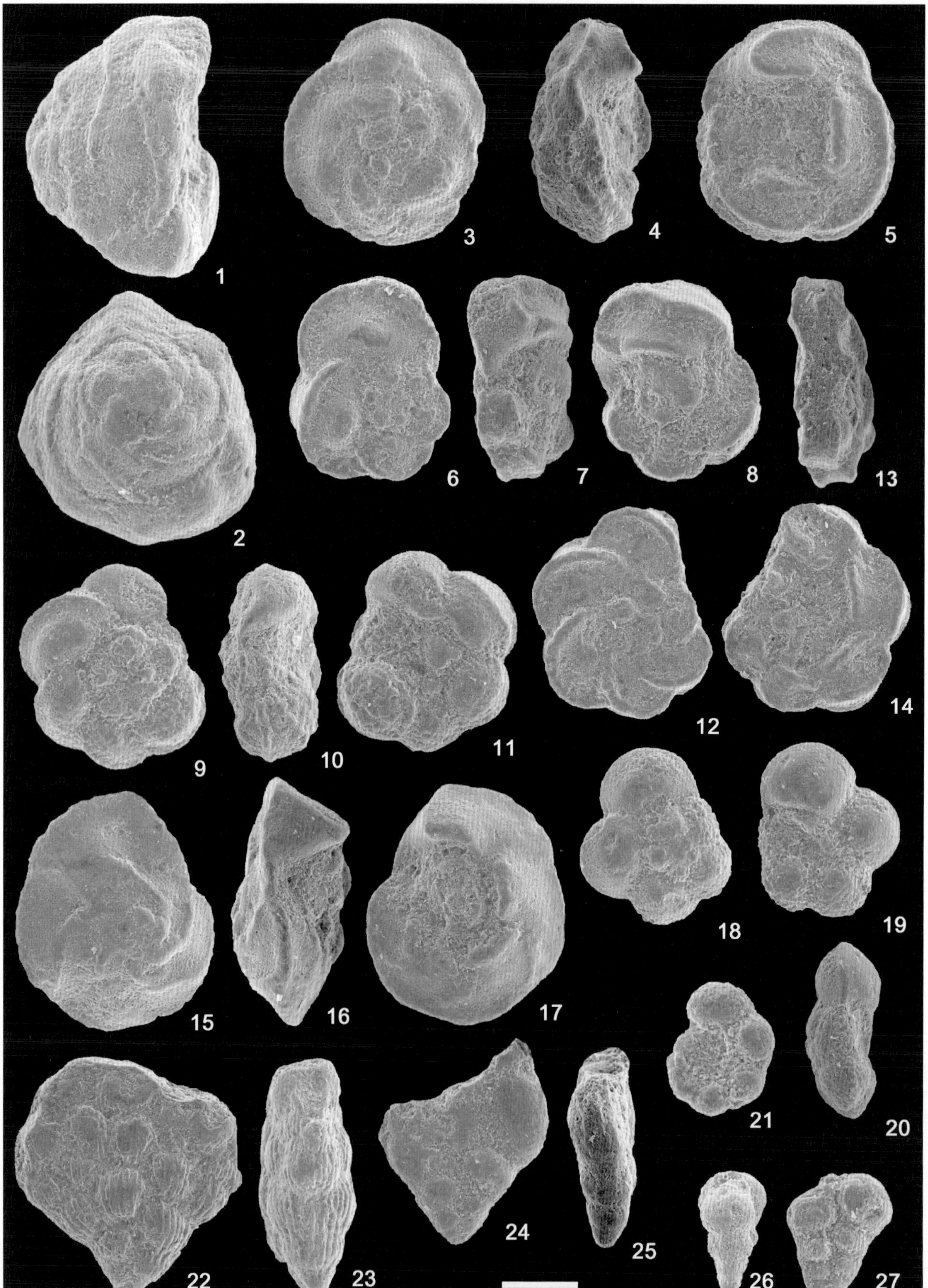
1
2
3
4
5
6
7
8
9
10
11
12
13
14
15
16
17
18
19
20
21
22
23
24
25
26
27

planata, H. globulosa, H. navarroensis, H. punctulata, H. labellosa, H. postsemicostata, Pseudotextularia nuttalli, P. elegans, P. intermedia, Racemiguembelina fructicosa, Pseudoguembelina kempensis, P. costulata, P. excolata, Planoglobulina acervulinoides, P. carseyae, P. multicamerata, Globigerinelloides rosebudensis, G. prairiehillensis, G. volutus, G. yaucoensis, Hedbergella holmdelensis, H. monmouthensis, Globotruncanella havanensis, Archaeoglobigerina cretacea, A. blowi, Rugoglobigerina rugosa, R. hexacamerata, R. reicheli, R. scotti, Globotruncana arca, G. aegyptiaca, G. linneiana, G. bulloides, G. falsostuarti, G. mariei, G. ventricosa, Globotruncanita stuartiformis, G. fareedi, Contusotruncana contusa, C. patelliformis, C. fornicata, and *C. plummerae*, coexisting with calcispherulids. This association is typical of the middle Maastrichtian *R. fructicosa* subzone (*G. gansseri* Zone), between about 4.5 and 3.8 m.y. before the K-Pg boundary (Figure 5).

The planktonic foraminiferal assemblages in the Late Cretaceous lime mudstones and marls from Bochil are quite diverse and include: *Guembelitia cretacea, G. trifolia, Heterohelix pulchra, H. planata, H. globulosa, H. navarroensis, H. punctulata, H. labellosa, Pseudotextularia nuttalli, P. elegans, P. intermedia, Pseudoguembelina kempensis, P. palpebra, P. costulata, P. costellifera, P. excolata, Planoglobulina acervulinoides, P. carseyae, P. riograndensis, P. multicamerata, Gublerina acuta, G. cuvillieri, Globigerinelloides rosebudensis, G. subcarinatus, G. prairiehillensis, G. volutus, G. yaucoensis, Hedbergella holmdelensis, H. monmouthensis, Globotruncanella havanensis, G. petaloidea, G. pschadae, Archaeoglobigerina cretacea, A. blowi, Rugoglobigerina rugosa, R. hexacamerata, R. pennyi, R. scotti, R. macrocephala, Globotruncana arca, G. aegyptiaca, G. linneiana, G. bulloides, G. falsostuarti, G. mariei, G. orientalis, Globotruncanita stuarti, G. stuartiformis, G. insignis, G. fareedi, G. angulata, G. dupeublei, Contusotruncana contusa, C. patelliformis, C. walfischensis, C. fornicata, C. plummerae, Gansserina wiedenmayeri*, and *Abathomphalus intermedius*. These planktonic foraminifera assemblages belong to the *P. acervulinoides* subzone (*G. gansseri* Zone) and indicates a middle Maastrichtian age, between about 5.7 and 4.5 m.y. before the K-Pg boundary (Figure 5).

The paleontological assemblages found at the Angostura Formation facies at Chilil include both benthic and planktonic microfossils as well as macrofossils. The planktonic foraminiferal assemblage is composed of scarce *H. globulosa, G. yaucoensis, H. holmdelensis*, and *H. monmouthensis*. Benthic foraminifera are abundant and include *Vaughanina cubensis, Orbitoides* sp., and Maastrichtian alveolinids such as *Chubbina jamaicensis* as well as miliolids. The macrofossils are mainly rudists and have not been taxonomically studied yet. The paleontological assemblage is indicative of a Late Cretaceous age (Campanian–Maastrichtian) for the Angostura Formation.

Carbonate Breccia Trough Calcareous Sandstone Succession (Clastic Complex Unit)

Although the reworked specimens cannot be used in biostratigraphic studies, they can be used for the biochronological interpretations to constrain the age of the graded detrital carbonates emplacement. In the CCU of the Bochil and Guayal sections, planktonic foraminiferal specimens such as *Racemiguembelina powelli, R. fructicosa, C. contusa, P. intermedia, G. linneiana, C. fornicata*, and *P. hariaensis* were identified. The last three species have biochronological ranges that do not overlap in time, indicating that the planktonic foraminiferal assemblages identified are reworked and mixed. The presence of *P. hariaensis* suggests that part of the eroded carbonate rock is late Maastrichtian in age because the first appearance of *P. hariaensis* was approximately 1.8 m.y. before the K-Pg boundary (Arz and Molina, 2002). Furthermore, reworked specimens of *P. hantkeninoides* have also been identified in the subunit 3 at Bochil (Keller and Stinnesbeck, 1996). The total range of this species spans approximately the latest 300 k.y. of the Maastrichtian, supporting the hypothesis that uppermost Maastrichtian facies were eroded in this area and accumulated in the CCU (Arenillas et al., 2006).

A similar reworked assemblage to that of the Angostura Formation was found in the overlying CCU at the Chilil section. The basal carbonate breccia contains abundant rudist fragments as well as abundant benthic microfossils such as *Chubina* sp., *Orbitoides* sp., and *Lepidorbitoides* sp. We also found reworked planktonic foraminifera in the fine-grained carbonate breccia at the top of the sequence, including *H. globulosa, H. planata, H. labellosa, G. yaucoensis, G. prairiehillensis, H. holmdelensis, H. monmouthensis, R. rugosa*, and *R. scotti*. This planktonic assemblage is indicative of a Maastrichtian age for the source of the sediments.

FIGURE 6. The SEM images of some examined Upper Cretaceous planktonic foraminiferal species from Bochil (BK), Guayal (GK), and Chilil (CHK) sections (scale bar = 100 μm). The numbers of each sample are referred to centimeters below the base of the complex clastic unit (= K-Pg boundary). (1, 2) *Contusotruncana contusa*, Bochil, BK-5-7; (3–5) *Contusotruncana fornicata*, Guayal, GK-0-3; (6–8) *Globotruncana aegyptiaca*, Bochil, BK-0-3; (9–11) *Globotruncana bulloides*, Guayal, GK-0-3; (12–14) *Globotruncana linneiana*, Bochil, BK-0-3; (15–17) *Globotruncanita stuartiformis*, Bochil, BK-0-3; (18–20) *Globotruncanella havanensis*, Bochil, BK-5-7; (21) *Globigerinelloides yaucoensis*, Chilil, CHK-0-3; (22, 23) *Planoglobulina acervulinoides*, Bochil, BK-5-7; (24, 25) *Gublerina acuta*, Bochil, BK-5-7; (26, 27) *Heterohelix globulosa*, Chilil, CHK-0-3.

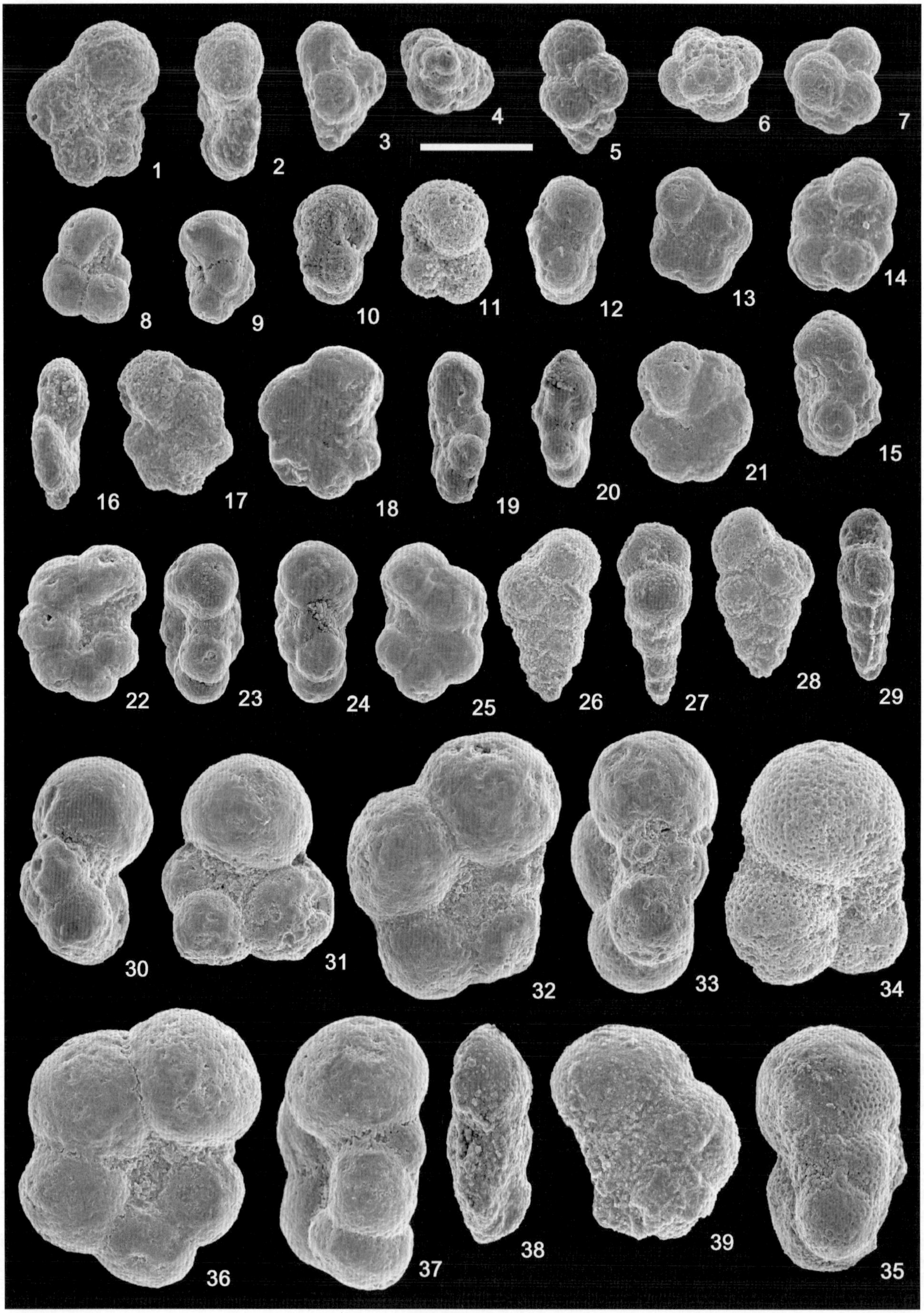
1
2
3
4
5
6
7
8
9
10
11
12
13
14
15
16
17
18
19
20
21
22
23
24
25
26
27
28
29
30
31
32
33
34
35
36
37
38
39

The Paleocene

Previous studies indicate that the basal part of pelagic limestones and marls at Bochil contains *Globigerina* sp. and *P. eugubina* typical of a basal Paleocene age (Montanari et al., 1994; Smit, 1999). The lithology and micropaleontological assemblages in the marls and limestones indicate an outer shelf depositional environment. A 6–8-cm-thick dark clay layer is observed overlying the uppermost part of the CCU at Bochil. The planktonic foraminiferal assemblages identified in this unit correspond to the lower part of the *G. cretacea* Zone (*H. holmdelensis* Subzone) characterized by a sharp increase in the abundance of *Guembelitria*, which is precisely the initial evolutionary event of the planktonic foraminifer after the K-Pg boundary mass extinction (Arenillas et al., 2002, 2006). This *Guembelitria* peak (Figure 5) is globally recognized as the basal Paleocene and occurs at Bochil precisely above subunit 3, which contains the finest grained clastics of the K-Pg boundary succession in the Gulf of Mexico. The part of the basal Paleocene marls described herein corresponds to the Soyaló Formation (Quezada-Muñetón, 1990).

At the Guayal outcrops, a 3–4-cm-thick (0.098–0.131-ft-thick) dark clay bed is observed overlying the finest upper part of the CCU (subunit 3). This clay bed is assigned to the *H. holmdelensis* subzone by Arenillas et al. (2006) or biozone P0 by Berggren et al. (1995). On top of the dark clay bed is a sequence of calcareous shale with laminated lime mudstone beds at its basal part. The Paleocene planktic foraminiferal assemblages of Bochil and Guayal (described by Arenillas et al. [2006]) indicate that the basal Danian *G. cretacea*, *P. eugubina*, and *P. pseudobulloides* Zones (and all their subbiozones) are recorded at both sections (Figure 5).

At the Chilil outcrops, the first 84 cm (33 in.) of clayey mudstone and marls resting on top of the carbonate breccia sequence was sampled in detail for micropaleontological studies. These facies contain scarce, poorly preserved planktonic foraminifera, including *G. cretacea*, *G.* cf. *trifolia*, *Palaeoglobigerina alticonusa*, *Palaeoglobigerina fodina*, *Palaeoglobigerina luterbacheri*, *Palaeoglobigerina extensa*, *Palaeoglobigerina minutula*, *P. longiapertura*, *Parvularugoglobigerina umbrica*, *Parvularugoglobigerina perexigua*, *P. sabina*, *Woodringina claytonensis*, and *Woodringina hornerstownensis*. Approximately 40 cm (116 in.) above the calcareous sandstone, the LOD of *P. eugubina* was identified. The foraminiferal assemblages suggest that the studied Paleocene stratigraphic interval from Chilil includes the *G. cretacea* and *P. eugubina* Zones, which are diagnostic of the lowermost Danian (Figure 5).

SEDIMENTOLOGIC AND GEOLOGIC SCENARIO OFFSHORE CAMPECHE

In a previous work, the Late Cretaceous Jolpabuchil Formation was considered to have been deposited in deep marine environments with poor circulation (Quezada-Muñetón, 1990). The dark gray color of the lime mudstone, the thin to medium stratification, the chert nodules, and the thin beds containing bioclasts of benthic organisms intercalated with some beds of lime mudstone with planktonic foraminifera indicate that the Jolpabuchil Formation facies studied at Guayal and Bochil correspond to outer shelf carbonates. The planktonic/benthic foraminifera ratio from Bochil suggests a depositional paleoenvironment on the outer shelf or the upper continental slope (probably more than 200 m [656 ft] of water depth). Of note is the presence of *Zoophycus*, an ichnofossil commonly ascribed to the bathyal environment, although it can occur in both shallow-water and deep-water deposits (Pemberton et al., 2001). However, the thin beds containing bioclasts and benthic organisms appear to represent turbidite or debrite deposits, suggesting the proximity of a platform margin. From the data obtained to date, we consider that the pelagic limestone was deposited in an outer shelf environment in water depths of tens to hundreds of meters. In contrast, the rudistids and benthic foraminifera-rich facies from Chilil indicate a shallow lagoon environment of deposition for the Angostura Formation.

Concerning the CCU, diverse mechanisms might lead to deposition of fragmented slope carbonates

FIGURE 7. The SEM images of some examined lowermost Danian planktonic foraminiferal species from Bochil (BP), Guayal (GP), and Chilil (CHP) sections (scale bar = 100 μm). The numbers of each sample are referred to centimeters above the top of the complex clastic unit. (1, 2) *Hedbergella holmdelensis*, Bochil, BP+2+4; (3, 4) *Guembelitria cretacea*, Bochil, BP+24+26; (5) *G. cretacea*, Chilil, CHP+48+50; (6, 7) *Palaeoglobigerina alticonusa*, Bochil, BP+45+50; (8, 9) *Palaeoglobigerina fodina*, Bochil, BP+38+40; (10, 11) *Palaeoglobigerina minutula*, Guayal, GP+62+67; (12, 13) *Palaeoglobigerina luterbacheri*, Bochil, BP+24+26; (14, 15) *Parvularugoglobigerina sabina*, Bochil, BP+155+160; (16, 17) *Parvularugoglobigerina longiapertura*, Bochil, BP+38+40; (18, 19) *P. longiapertura*, Chilil, CHP+48+50; (20, 21) *P. longiapertura*, Guayal, GP+62+67; (22, 23) *Parvularugoglobigerina eugubina*, Bochil, BP+192+197; (24, 25) *P. eugubina*, Guayal, GP+62+67; (26, 27) *Woodringina hornerstownensis*, Bochil, BP+192+197; (28, 29) *Chiloguembelina taurica*, Bochil, BP+183+188; (30, 31) *Eoglobigerina simplicissima*, Bochil, BP+218+223; (32, 33) *Parasubbotina pseudobulloides*, Bochil, BP+235+240; (34, 35) *Subbotina triloculinoides*, Guayal, GP+260; (36, 37) *Praemurica inconstans*, Bochil, BP295+300; (38, 39) *Globanomalina compressa*, Guayal, GP+260.

because of the collapse of the shelf. One mechanism is the instability of the platform margin because of a thick sedimentary overburden (Cook, 1983; Read, 1985; Coniglio and Dix, 1992). The overpressuring caused by pore water sealed in permeable and porous facies during relative lowering of the sea level is another suggested process (Spence and Tucker, 1997). The storms and tsunamis acting on shelf-edge sediments prone to failure may also originate debris flows to form sedimentary breccia (Cook, 1983; Spence and Tucker, 1997), and finally, seismicity may trigger rock fall or the collapse of the platform edge (Cook, 1983; Coniglio and Dix, 1992). The large continuity and enormous extension of the carbonate breccia succession suggest that a single geological event, which activated many of the previous enlisted mechanisms, is responsible for this regional-scale deposit.

In the offshore Campeche wells, the carbonate breccias have been encountered in a belt measuring some 200 km (124 mi) along the ancient edge of the platform, and extending for about 40 km (25 mi) basinward, with an average maximum thickness of about 300 m (984 ft) (Grajales-Nishimura et al., 2000b). We interpret that the carbonate debris was deposited as a continuous sheet with variable thickness hundreds of kilometers around the Chicxulub structure (Figure 2). This interpretation is supported by several discoveries of K-Pg boundary sequences with impact materials on and around the Yucatan Platform (Hildebrand et al., 1993; Stinnesbeck et al., 1997; Fourcade et al., 1998; Takayama et al., 2000; Tada et al., 2003; Grajales-Nishimura et al., 2000b).

In a previous work, it was suggested that the origin of the K-Pg boundary calcareous breccia sequence in the petroleum fields of Campeche Sound can be explained by two principal mechanisms: (1) dissolution-collapse caused by subaerial exposure of the basal carbonate substrate and (2) erosion of shallow-water carbonates from the Yucatan Platform and deposition in deep-water locations. However, the stratigraphic and micropaleontological analyses in offshore Campeche indicate instantaneous deposition, contradicting the two previous interpretations that would have required more geological time. The most likely scenario is that deposition of the breccia occurred along the margins of the Yucatan and Chiapas-Tabasco Platforms in outer shelf and base-of-slope environments on a pelagic or shallow-water carbonate substrate (Meneses de Gyves, 1980; Santiago-Acevedo, 1980; Grajales-Nishimura et al., 2000a, b; Grajales-Nishimura, 2001). We have suggested and provided evidence in previous work that the sedimentary succession was originated by the impact at Chicxulub (Grajales-Nishimura et al., 2000a, b; Grajales-Nishimura, 2001; Arenillas et al., 2002, 2006). We contend that this is the only reasonable, instantaneous process that could have deposited a 50–300-m-thick (164–984-ft-thick) carbonate sequence of enormous extent in deep- and shallow-water environments. This explains other phenomena such as the precise K-Pg boundary age, very short time of deposition, minerals diagnostic of bolide impact, and graded sedimentary structures in the carbonate deposit (Figure 3). This carbonate sequence is the result of complex interactions among geologically instantaneous mechanisms triggered by the Chicxulub impact that produced sedimentary deposits by gravity and ballistic processes, followed shortly thereafter by tsunami-induced currents that partially reworked parts of the previously formed sequence. The term "ballistic sedimentation" was coined by Oberbeck (1975) to explain the mechanism for the ejected and sedimented material observed around lunar impact craters. This same author later described the hypothetical lithostratigraphic sequence that would result from an oceanic impact on the Earth, including breccia through argillaceous material with diverse stratigraphy dependent on the distance from the crater (Oberbeck et al., 1993).

The coarser breccia material at the base of the unit likely represents a base-of-slope apron, whereas finer material at the top, containing calcareous sand, silt and clay minerals, represents a distinctive facies denominated the "ejecta layer" consisting of shocked mineral and other material ejected from the impact zone and deposited within the upper part of the carbonate clastic sequence. Most of the coarser lithoclasts were deposited on the outer shelf by vigorous gravity-flows almost simultaneously with deposition of the ejected material. We infer that meteorologic and sedimentologic conditions were unstable immediately following the megaquake caused by the impact. The seismicity and meteorologic perturbations may have originated gigantic waves (tsunamis) that inundated the coastal regions. Vigorous currents formed in this manner would have dragged a huge volume of material in the form of debris flows on their return to the sea. In some places, this debris flow represents a complete sequence from coarse to fine sediments, but in others, they represent only the fine-grained conglomerates of subunit 4 that is found at the top of the sedimentary succession in some offshore Campeche wells (Figure 3a).

In contrast to the deep-water deposits, the processes of sedimentation in the shallow-water environment seem to be different. The sedimentology of the K-Pg boundary carbonate sequence at Chilil has been studied only preliminarily. An outstanding sedimentological feature at this outcrop, also found in deep-water conditions, is the fining-upward structure of the sequence. This sedimentologic array starts at the base with coarse-grained carbonate breccia facies grading upward to coarse- and medium-grained breccia and calcareous sandstone, and ending at the top in calcareous siltstone and claystone facies (Figure 3d). The yellowish-colored, fine-grained horizon corresponds to the ejecta-rich layer. The graded pattern of the sequence suggests a single depositional event. Discrete to clear parallel lamination and very low-angle cross-bedding are distinguished in the medium- to fine-grained calcareous breccia and the finest grained facies of the K-Pg boundary sequence, respectively. This type of sedimentological features (parallel lamination) in sandstones and fine-grained conglomerates and breccias has been reported from different depths in inner shelf and outer shelf settings, and its origin has been related to storms or tsunamis (Bourgeois et al., 1988; Smit et al., 1992). The overall graded structure of the K-Pg boundary deposit is consistent with a storm-related depositional mechanism. Therefore, it is interpreted that a large tsunami caused erosion of the shallow-water sediments, and transported and redeposited them under relatively shallow-water conditions. These hydrodynamic conditions resulted in the redistribution of shallow-water bioclasts in the coarse-grained part of the deposit, mainly in the upper part of the sequence.

The fossil content of the Angostura Formation suggests relatively shallow-water conditions, probably a few meters of depth, during deposition of the K-Pg boundary carbonate clastic sequence. However, if this sequence is approximately 30 m (98 ft) thick, then an accommodation space of at least 30 m (98 ft) was required for the deposition of the impact-related sediments. This relation seems to be inconsistent; however, as mentioned above, the K-Pg boundary carbonate clastic succession is normally overlain by carbonate facies with planktonic foraminifera, which suggests relatively deep-water conditions. During the time of deposition of the K-Pg boundary carbonate succession, this region was likely relatively deep, probably related to the beginning of a transgressive episode, which may have drowned the Angostura carbonate platform. Alternatively, the presence of ancient submarine topographic depressions cannot be ruled out to explain the accommodation of a 30-m-thick (98-ft-thick) section in a shallow-water environment.

The presence of Danian planktonic microfossils in thin-bedded argillaceous-calcareous strata in the Bochil, Guayal, and Chilil sections suggests that the lower Danian sequence was deposited in an outer shelf or basinal environment, confirming previous observations made by Quezada-Muñetón (1990). Furthermore, this study determined that deposition of the basal Soyaló Formation occurred on top of a detrital carbonate sequence (CCU) grading upward from breccia through siltstone and claystone.

CONCLUSIONS

The sedimentologic-mineralogical-stratigraphic evidence indicates that the CCU in southeastern Mexico constitutes a continuous gradational succession. The biostratigraphic data demonstrate that, in many places, this sequence was deposited in a deep marine environment at the K-Pg boundary. The presence of sandy, silty, and clayed facies with abundant minerals displaying planar deformation features typical of high pressure and temperature demonstrates that the graded K-Pg boundary carbonate clastic sequence was deposited as a result of the meteorite impact at Chicxulub, Yucatan, 65 m.y. Recognition of this stratigraphic arrangement has contributed to the construction of a more consistent sedimentary model for understanding the internal architecture of the reservoir in Cantarell field. This catastrophic event created the most important reservoir, as well as the sealing strata, for several offshore petroleum fields in Campeche Sound, including the supergiant Cantarell field.

ACKNOWLEDGMENTS

We thank PEMEX and the Instituto Mexicano del Petróleo (IMP) for permitting the publication of this work. This investigation was supported by projects D.1003 of IMP; CGL2004-00738 and CGL2007-63724 of the Spanish Ministerio de Ciencia y Tecnología; and by the Grupo Consolidado EO5 of the Gobierno de Aragón (Spain). This chapter is dedicated to Rafael Sánchez Montes de Oca, a very good friend and gifted PEMEX geologist, who contributed significantly to the knowledge of the geology of the Sierra de Chiapas and was one of the pioneers in exploring southeastern Mexico. Rafael Sánchez Montes de Oca passed away last year at the age of 69. Thanks to Jon Blickwede for his critical review, which improved the manuscript significantly. We also thank Claudio Bartolini for the invitation to publish this chapter and for his valuable revision of the manuscript.

REFERENCES CITED

Alegret, L., I. Arenillas, J. A. Arz, C. Díaz, J. M. Grajales-Nishimura, A. Meléndez, E. Molina, R. Rojas, and A. R. Soria, 2005, Cretaceous/Paleogene (K/Pg) boundary clastic complex at Loma Capiro, Central Cuba: Evidence for a K/Pg impact origin: Geology, v. 33, p. 721–724.

Alvarez, L. W., W. Alvarez, F. Asaro, and H. V. Michel, 1980, Extraterrestrial cause for the Cretaceous–Tertiary extinction: Science, v. 208, p. 1095–1108.

Alvarez, W., P. Claeys, and S. W. Kieffer, 1995, Emplacement of Cretaceous–Tertiary boundary shocked quartz from Chicxulub Crater: Science, v. 269, no. 5226, p. 930–935.

Alvarez, W., J. Smit, W. Lowrie, F. Asaro, S. V. Margolis, P. Claeys, M. Kastner, and A. R. Hildebrand, 1992a, Proximal impact deposits at the Cretaceous–Tertiary boundary in the Gulf of Mexico: A restudy of DSDP Leg 77 Sites 536 and 540: Geology, v. 20, p. 697–700.

Alvarez, W., J. M. Grajales-Nishimura, S. Martínez, R. Romero, M. P. Ruiz, L. E. Guzmán, A. M. Zambrano,

N. H. M. Swinburne, and S. V. Margolis, 1992b, The Cretaceous–Tertiary boundary impact-tsunami deposit in northeast Mexico: Geological Society of America Annual Meeting, p. A331.

Arenillas, I., L. Alegret, J. A. Arz, C. Liesa, A. Meléndez, E. Molina, A. R. Soria, E. Cedillo-Pardo, J. M. Grajales-Nishimura, and M. C. Rosales-Domínguez, 2002, Cretaceous/Tertiary boundary planktonic foraminifer mass extinction and biochronology at the La Ceiba, Bochil (Mexico) and El Kef (Tunisia) stratotypes: Geological Society of America Special Paper 356, p. 253–263.

Arenillas, I., J. A. Arz, and E. Molina, 2004, A new high-resolution planktonic foraminiferal zonation and subzonation of the lower Danian: Lethaia, v. 37, p. 79–95.

Arenillas, I., J. A. Arz, J. M. Grajales-Nishimura, G. Murillo-Muñetón, W. Alvarez, A. Camargo-Zanoguera, E. Molina, and M. C. Rosales-Domínguez, 2006, Chicxulub impact event is Cretaceous/Paleogene boundary in age: New micropaleontological evidence: Earth and Planetary Science Letters, v. 249, p. 241–257.

Arz, J. A., and E. Molina, 2002, Bioestratigrafía y cronoestratigrafía con foraminíferos planctónicos del Campaniense superior y Maastrichtiense de latitudes subtropicales y templadas (España, Francia y Tunicia): Neues Jarbuch fur Geologie und Paläontologie, Ahhandlungen, v. 224, p. 161–195.

Arz, J. A., I. Arenillas, A. R. Soria , L. Alegret , J. M. Grajales-Nishimura, C. L. Liesa, A. Meléndez, E. Molina, and M. C. Rosales, 2001a, Micropaleontology and sedimentology across the Cretaceous/Tertiary boundary at La Ceiba (Mexico): Impact-generated sediment gravity flows: Journal of South American Earth Sciences, v. 14, p. 505–519.

Arz, J. A., L. Alegret, I. Arenillas, C. Liesa, E. Molina, and A. R. Soria, 2001b, Extinción de foraminíferos del límite Cretácico/Terciario en Coxquihui (México) y su relación con las evidencias de impacto: Revista Española de Micropaleontología, v. 33, no. 2, p. 221–236.

Arz, J. A., L. Alegret, and I. Arenillas, 2004, Foraminiferal biostratigraphy and paleoenvironmental reconstruction at Yaxcopoil-1 drill hole (Chicxulub crater, Yucatan Peninsula): Meteoritics and Planetary Science, v. 39, p. 1099–1111.

Berggren, W. A., D. V. Kent, C. C. Swisher III, and M. P. Aubry, 1995, A revised Cenozoic geochronology and chronostratigraphy: Society of Economic Paleontologists and Mineralogists Special Paper 54, p. 129–131.

Bohor, B. F., 1996, A Sediment gravity flow hypothesis for siliciclastic units at the K/T boundary, northeastern, México: Geological Society of America Special Paper 307, p. 151–186.

Bohor, B. F., and W. J. Betterton, 1993, Arroyo El Mimbral, Mexico, K/T unit: origin as debris-flow/turbidite, not a tsunami deposit (abs.): 24th Lunar and Planetary Science Conference, Houston, Texas, Lunar and Planetary Institute, p. 143–144.

Bourgeois, J., T. A. Hansen, P. L. Wiberg, and E. G. Kauffman, 1988, A tsunami deposit at the Cretaceous–Tertiary boundary in Texas: Science, v. 241, p. 567–570.

Bralower, T., Ch. K. Paull, and R. M. Leckie, 1998, The Cretaceous–Tertiary boundary cocktail: Chicxulub impact triggers margin collapse and extensive sediment gravity flows: Geology, v. 26, no. 4, p. 331–334.

Caron, M., 1985, Cretaceous planktonic foraminifera, *in* H. M. Bolli, J. B. Saunders, and K. Perch-Nielsen, eds., Plankton stratigraphy: Cambridge, United Kingdom, Cambridge University Press, p. 17–86.

Castañón, H. J. L., and F. U. Medina, 1991, Hojas geológicas del Prospecto Azufrito-Potioja: Informe No. 956, Gerencia de Exploración Región Sur, Petróleos Mexicanos (PEMEX), Inédito.

Claeys, P., W. Kiessling, and W. Alvarez, 2002, Distribution of Chicxulub ejecta at the Cretaceous–Tertiary boundary: Geological Society of America Special Paper 356, p. 55–68.

Coniglio, M., and G. R. Dix, 1992, Carbonate slopes, *in* R. G. Walker and N. P. James, eds., Models response to sea level change: Newfoundlands, Geological Association of Canada, p. 349–373.

Cook, H. E., 1983, Ancient carbonate platform margins, slopes, and basins, *in* H. E. Cook and H. T. Mullins, eds., Platform margin and deep water carbonates: SEPM Short Course 12, p. 1–5.

Cornejo-Toledo, A., and A. Hernández-Osuna, 1950, Las anomalías gravimétricas en la Cuenca Salina del Istmo, planicie costera de Tabasco, Campeche y Península de Yucatan: Boletín de la Asociación Mexicana de Geólogos Petroleros, v. 2, p. 453–460.

Díaz-Otero, C., M. Iturralde-Vinent, and D. E. García-Delgado, 2000, The Cretaceous–Tertiary boundary "Cocktail" in western Cuba, Greater Antilles: Lunar and Planetary Institute Contribution 1053, p. 76–77.

Fourcade, E., et al., 1997, The K/T boundary in southwestern Petén (Guatemala): Comptes Rendus de la Academie de Sciences Paris, Sciences de la Terre et des Planètes, v. 325, p. 57–64.

Fourcade, E., R. Rocchia, S. Gardin, J.-P. Bellier, P. Debrabant, E. Masure, E. Robin, and W. Toni Pop, 1998, Age of the Guatemala breccias around the Cretaceous–Tertiary boundary: Relationships with the asteroid impact on the Yucatan: Comptes Rendus de la Academie de Sciences Paris, Sciences de la Terre et des Planètes, v. 327, p. 47–53.

Galloway, W. E., D. G. Bebout, W. L. Fisher, J. B. Dunlap Jr., R. Cabrera-Castro, J. E. Lugo-Rivera, and T. M. Scott, 1991, Cenozoic, *in* A. Salvador, ed., The Gulf of Mexico Basin: Geological Society of America, The Geology of North America, v. J, p. 245–324.

Grajales, J. M., D. J. Morán, P. Padilla, M. A. Sánchez, E. Cedillo, and W. Alvarez, 1996, The Lomas Tristes Breccia: A K–T impact-related breccia from southern Mexico: Geological Society of America Abstracts with Programs, v. 28, no. 7, p. A–183.

Grajales-Nishimura, J. M., 2001, Origen, diagénesis e importancia petrolera de la secuencia carbonatada del límite Cretácico–Terciario en el sureste de México: Tesis de Doctorado, Universidad Nacional Autónoma de México, Instituto de Geología, 230 p.

Grajales-Nishimura, J. M., E. Cedillo-Pardo, and R. Martínez-Ibarra, 2000a, Burial dolomitization of the KT carbonate breccia, subsurface Campeche, southeastern México (abs.): AAPG Annual Convention, v. 9, p. A57.

Grajales-Nishimura, J. M., E. Cedillo-Pardo, M. C. Rosales-Domínguez, D. J. Moran-Zenteno, W. Alvarez, P. Claeys, J. Ruiz-Morales, J. Garcia-Hernandez, P. Padilla-Avila, and A. Sanchez-Rios, 2000b, Chicxulub impact: The origin of reservoir and seal in the southeastern Mexico oil fields: Geology, v. 28, p. 307–310.

Grajales-Nishimura, J. M., G. Murillo-Muñetón, C. Rosales-Domínguez, E. Cedillo-Pardo, and J. García-Hernández, 2003, Heterogeneity of lithoclast composition in the deep-water carbonate breccias of the K/T boundary sedimentary succession, southeastern Mexico and offshore Campeche, *in* C. Bartolini, R. T. Buffler, and J. Blickwede, eds., The Circum-Gulf of Mexico and the Caribbean: Hydrocarbon habitats, basin formation, and plate tectonics: AAPG Memoir 79, p. 312–329.

Graup, G., 1981, Terrestrial chondrules, glass spherules, and accretionary lapilli from the suevite, Ries Crater, Germany: Earth and Planetary Science Letters, v. 55, p. 407–418.

Grieve, R. A. F., F. Langenhorst, and D. Stöffler, 1996, Shock metamorphism of quartz in nature and experiment: II. Significance in geoscience: Meteoritics and Planetary Science, v. 31, p. 6–35.

Hernández-García, S., 1993, Subsistema almacenador en los horizontes Cretácico Superior y Eoceno Medio de la Región Marina: Asociación Mexicana de Geológos Petroleros: I Simposio de Geología del Subsuelo El Sistema Petrolífero y su Evaluación, Ciudad del Carmen, Campeche, p. 79–83.

Hildebrand, A. R., and W. V. Boynton, 1990, Proximal Cretaceous–Tertiary boundary impact deposits in the Caribbean: Science, v. 248, p. 843–847.

Hildebrand, A. R., and J. A. Stansberry, 1992, K/T boundary ejecta distribution predicts size and location of Chicxulub crater: Lunar and Planetary Science, v. 23, p. 537–538.

Hildebrand, A. R., D. A. Kring, and W. V. Boynton, 1991a, A Cretaceous/Tertiary boundary ejecta layer, ~50 cm thick, from Beloc: Haiti: Meteoritics, v. 25, p. 370–371.

Hildebrand, A. R., G. T. Penfield, D. A. Kring, M. Pilkington, Z. A. Camargo, S. B. Jacobsen, and W. V. Boynton, 1991b, Chicxulub crater: A possible Cretaceous/Tertiary boundary impact crater on the Yucatan Peninsula, Mexico: Geology, v. 19, p. 867–871.

Hildebrand, A. R., M. Pilkington, R. A. F. Grieve, P. B. Robertson, and G. T. Penfield, 1992, Recent studies of the Chicxulub crater, Mexico: Lunar and Planetary Science, v. 24, p. 539–540.

Hildebrand, A. R., S. Bonis, J. Smit, and M. Attrep, 1993, Cretaceous/Tertiary boundary deposits in Guatemala: Evidence for impact waves and slumping on a platform scale?: Sociedad Mexicana de Paleontología, A.C., IV Congreso Nacional de Paleontología, August 19, 1993, p. 133–138.

Keller, G., and W. Stinnesbeck, 1996, Near-K/T age of clastic deposits from Texas to Brazil: Impact, volcanism and/or sea level lowstand: Terra Nova, v. 8, p. 277–285.

Keller, G., N. MacLeod, J. B. Lyons, and C. B. Officer, 1993, Is there evidence for Cretaceous–Tertiary boundary-age deep-water deposits in the Caribbean and Gulf of Mexico?: Geology, v. 21, no. 9, p. 776–780.

Klaus, A., R. N. Norris, D. Kroon, and J. Smit, 2000, Impact-induced mass wasting at the K–T boundary: Black Nose, western North Atlantic: Geology, v. 28, p. 319–322.

Kring, D. A., A. R. Hildebrand, and W. V. Boynton, 1991, The petrology of a andesitic melt rock and a polimictic breccia from the interior of the Chicxulub structure, Yucatan, Mexico: Lunar and Planetary Science, v. 22, p. 755–756.

Limón-González, M., E. Cedillo-Pardo, J. M. Quezada-Muñetón, J. M. Grajales-Nishimura, W. Alvarez, A. R. Hildebrand, M. A. Sánchez-Rios, M. C. Rosales-Domínguez, and V. González-Casildo, 1994, Cretaceous–Tertiary boundary sedimentary breccias from southern Mexico: Normal sedimentary deposits or impact related breccias?: AAPG Annual Convention, Denver, Colorado, v. 3, p. 199.

Longoria, J. F., and J. M. Grajales-Nishimura, 1993, Field trip guide to selected K/T boundary localities in Tamaulipas and Nuevo León, northeast Mexico: Revista de la Sociedad Mexicana de Paleontología, v. 6, no. 2, p. 73–85.

López-Ortiz, R., 1962, Geología y posibilidades petroleras de los sedimentos cretácicos en la parte sureste del frente de la sierra madre de Chiapas: Boletín de la Asociación Mexicana de Geólogos Petroleros, v. 14, no. 5–6, p. 135–151.

López-Ramos, E., 1979, Geología de México: Mexico, D.F., Secretaría de Educación Pública, v. 3, 446 p.

Medina de la Paz, J., and V. Duran-Rodriguez, 1991, Sección estratigráfica Bochil-Soyaló-1: Prospecto Bochil, Petróleos Mexicanos (PEMEX), Zona sureste (Villahermosa), Inédito.

Meneses de Gyves, J., 1980, Geología de la Sonda de Campeche: Boletín de la Asociación Mexicana de Geólogos Petroleros, v. 32, no. 1, p. 1–26.

Michaud, F., 1987, Stratigraphie et Palèographie du Mèsozoïque du Chiapas (Sud-Est. Mexique): Thèse de Doctorat, Universitè Pierre et Marie Curie, Paris Mèmoire de Stratigraphie 6, 301 p.

Molina, E., I. Arenillas, and J. A. Arz, 1996, The Cretaceous/Tertiary boundary mass extinction in planktonic foraminifera at Agost (Spain): Revue de Micropaléontologie, v. 39, no. 3, p. 225–243.

Montanari, S., P. Claeys, F. Asaro, J. Bermudez, and J. Smit, 1994, Preliminary stratigraphy and iridium and other geochemical anomalies across the KT boundary in the Bochil section (Chiapas, southeastern Mexico), *in* New developments regarding the KT event and other catastrophes in Earth history: Lunar and Planetary Institute Contribution 825, p. 84–85.

Moore, J. G., and D. L. Peck, 1962, Accretionary lapilli in volcanic rocks of the western continental United States: Journal of Geology, v. 70, p. 182–193.

Muir, J. M., 1936, Geology of the Tampico region, Mexico: Tulsa, Oklahoma, AAPG, 280 p.

Murillo-Muñetón, G., J. M. Grajales-Nishimura, E. Cedillo-Pardo, J. García-Hernández, and S. Hernández-García,

2002, Stratigraphic architecture and sedimentology of the main oil-producing stratigraphic interval at the Cantarell oil field: The K/T Boundary sedimentary succession: Society of Petroleum Engineers, SPE Paper 74431, 5 p.
Newson, H. E., G. Graup, D. A. Iseri, J. W. Geissman, and K. Keil, 1990, The formation of the Ries Crater, west Germany, evidence of atmospheric interactions during a larger cratering event: Geological Society of America Special Paper 247, p. 195–206.
Oberbeck, V. R., 1975, The role of ballistic erosion and sedimentation in lunar stratigraphy: Reviews Geophysics and Space Physics, v. 13, p. 337–362.
Oberbeck, V. R., J. R. Marshall, and H. Aggarwal, 1993, Impacts, tillites, and the breakup of Gondwanaland: The Journal of Geology, v. 101, p. 1–19.
Olsson, R. K., K. G. Miller, J. V. Browning, D. Habib, and P. Sugarman, 1997, Ejecta layer at the Cretaceous–Tertiary boundary, Bass River, New Jersey (Ocean Drilling Program leg 174 AX): Geology, v. 25, p. 759–762.
Pecheux, M., and F. Michaud, 1997, Yucatan subsurface stratigraphy: Implications and constraints for the Chicxulub impact: Comment: Geology, v. 25, p. 92.
Pemberton, S. G., M. Spila, A. J. Pulham, T. Saunders, J. A. MacEachern, D. Robbins, and I. K. Sinclair, 2001, Ichnology and sedimentology of shallow to marginal marine systems: Ben Nevis and Avalon reservoirs, Jeanne D'Arc Basin: Geological Association of Canada Short Course Notes, Memorial University of Newfoundland St. John's, Newfoundland, v. 15, 353 p.
Penfield, G. T., and Z. A. Camargo, 1981, Definition of a major igneous zone in the central Yucatan Platform with aeromagnetics and gravity, *in* Technical Program, abstracts and bibliographies: Society of Exploration Geophysicists, 51st Annual Meeting, Tulsa, Oklahoma, Society of Exploration Geophysicists, p. 37.
Petróleos Mexicanos (PEMEX) Exploración y Producción, 1999, Las reservas de hidrocarburos de México: Los principales campos de petróleo y gas de México: México, D.F., Petróleos Mexicanos, v. 2, 398 p.
Quezada-Muñetón, J. M., 1990, El Cretácico Medio-Superior y el límite Cretácico Superior-Terciario Inferior en la Sierra de Chiapas: Boletín de la Asociación Mexicana de Geólogos Petroleros, v. 39, no. 1, p. 3–98.
Read, J. F., 1985, Carbonate platform models: AAPG Bulletin, v. 69, p. 1–21.
Reimer, T. O., 1983a, Accretionary lapilli in volcanic ash falls: Physical factors governing their formation, *in* T. M. Peryt, ed., Coated grains: Berlin, Springer Verlag, p. 56–68.
Reimer, T. O., 1983b, Accretionary lapilli and other spheroidal rocks from the Archaean Swaziland Supergroup, Barberton Mountain Land, South Africa, *in* T. M. Peryt, ed., Coated grains: Berlin, Springer Verlag, p. 619–634.
Salas, G. P., and E. López-Ramos, 1951, Geología y Tectónica de la región de Macuspana, Tabasco y parte norte de Chiapas: Boletín Asociación Mexicana de Geólogos Petroleros, v. 3, no. 1–2, p. 3–56.
Santiago-Acevedo, J., 1980, Giant fields of the Southern Zone-México, *in* M. T. Halbouty, ed., Giant oil and gas fields of the decade: 1968–1978: AAPG Memoir 30, p. 339–385.
Santiago-Acevedo, J., J. Carrillo-Bravo, and B. Martell-Andrade, 1984, Geología Petrolera de México, *in* D. Marmissolle-Daguerre, ed., Evaluación de Formaciones de Formaciones en México: México, D.F., Schlumberger, p. I-1–I-36.
Schulte, P., W. Stinnesbeck, D. Stüben, U. Kramar, Z. Berner, G. Keller, and T. Adatte, 2003, Fe-rich and K-rich mafic spherules from slumped and channelized Chicxulub ejecta deposits in the northern La Sierrita area, NE Mexico: International Journal of Earth Sciences (Geologische Rundschau), v. 92, p. 114–142.
Sharpton, V. L., G. B. Dalyrymple, L. Marin, G. Ryder, B. C. Schuraytz, and J. Urrutia-Fucugauchi, 1992, New links between the Chicxulub impact structure and the Cretaceous–Tertiary boundary: Nature, v. 359, p. 819–821.
Smit, J., 1999, The global stratigraphy of the Cretaceous–Tertiary boundary impact ejecta: Annual Review Earth Planetary Science, v. 27, p. 75–113.
Smit, J., S. Montanari, N. H. M. Swinburne, W. Alvarez, A. R. Hildebrand, S. V. Margolis, P. Claeys, W. Lowrie, and F. Asaro, 1992, Tektite-bearing deep-water clastic unit at the Cretaceous–Tertiary boundary in northeastern Mexico: Geology, v. 20, p. 99–103.
Smit, J., T. B. Roep, W. Alvarez, S. Montanari, P. Claeys, J. M. Grajales-Nishimura, and J. Bermúdez-Santana, 1996, Coarse-grained, clastic sandstone complex at the K/T boundary around the Gulf of Mexico: Deposition by tsunami waves induced by the Chicxulub impact?: Geological Society of America Special Paper 307, p. 151–186.
Sohl, N. F., E. Martínez, P. Salmerón-Ureña, and F. Soto-Jaramillo, 1991, Upper Cretaceous, *in* A. Salvador, ed., The Gulf of Mexico Basin, Boulder, Colorado: Geological Society of America, The Geology of North America, v. J, p. 205–244.
Soria, A. L., C. Liesa, P. Mata, J. A. Arz, L. Alegret, I. Arenillas, and A. Meléndez, 2001, Slumping and a sandbar deposit at the K/T boundary in the El Tecolote sector (northeastern Mexico): An impact-induced sediment gravity flow: Geology, v. 29, no. 3, p. 231–234.
Spence, G. H., and M. E. Tucker, 1997, Genesis of limestone megabreccias and their significance in carbonate sequence-stratigraphic models: A review: Sedimentary Geology, v. 112, p. 163–193.
Stinnesbeck, W., et al., 1993, Deposition of channel deposits near the Cretaceous–Tertiary boundary in northeastern Mexico: Catastrophic or "normal" sedimentary deposits: Geology, v. 21, p. 797–800.
Stinnesbeck, W., G. Keller, J. de la Cruz, C. de León, N. Mac Leod, and J. E. Whittaker, 1997, The Cretaceous–Tertiary transition in Guatemala: Limestone breccia deposits from the South Petén Basin: International Journal of Earth Sciences (Geologische Rundschau), v. 86, p. 689–709.
Stöffler, D., and F. Langenhorst, 1994, Shock metamorphism of quartz in nature and experiment: I. Basic observation and theory: Meteoritics, v. 29, p. 155–181.

Swisher III, C. C., et al., 1992, Coeval $^{40}Ar/^{39}Ar$ ages of 65 million years ago from Chicxulub crater melt rock and Cretaceous–Tertiary boundary tektites: Science, v. 257, p. 954–958.

Tada, R., et al., 2003, K/T boundary deposits in the paleo-western Caribbean basin, *in* C. Bartolini, R. T. Buffler, and J. Blickwede, eds., The Circum-Gulf of Mexico and the Caribbean: Hydrocarbon habitats, basin formation, and plate tectonics: AAPG Memoir 79, p. 582–604.

Tagle, R., and P. Claeys, 2001, Platinum group element (PGE) distribution in ejecta and impactites from Chicxulub: Abstract Book, 6th European Science Foundation-IMPACT Workshop "Impact Marker in the Stratigraphic Record," Granada, Spain, p. 129.

Takayama, H., et al., 2000, Origin of the Peñalver Formation in northwestern Cuba and its relation to the K/T boundary impact event: Sedimentary Geology, v. 135, p. 295–320.

Urrutia-Fucugauchi, J., L. Marín, and A. Trejo-García, 1996a, Initial results of the Universidad Nacional Autónoma de México scientific drilling program on the Chicxulub impact structure: Rock magnetic properties of UNAM-7 Tekax borehole: Geofísica Internacional, v. 35, p. 125–133.

Urrutia-Fucugauchi, J., L. Marin, and A. Trejo-Garcia, 1996b, Universidad Nacional Autónoma de México scientific drilling program of Chicxulub impact structure—Evidence for a 300-kilometer crater diameter: Geophysical Research Letters, v. 23, no. 13, p. 1565–1568.

Urrutia-Fucugauchi, J., J. Morgan, D. Stöffler, and P. Claeys, 2004, The Chicxulub Scientific Drilling Project (CSDP): Meteoritics & Planetary Science, v. 39, p. 787–790.

16

Clara Valdés, M. d. L., L. Villanueva Rodríguez, and E. Caballero García, 2009, Geochemical integration and interpretation of source rocks, oils, and natural gases in southeastern Mexico, *in* C. Bartolini and J. R. Román Ramos, eds., Petroleum systems in the southern Gulf of Mexico: AAPG Memoir 90, p. 337–368.

Geochemical Integration and Interpretation of Source Rocks, Oils, and Natural Gases in Southeastern Mexico

Ma. de Lourdes Clara Valdés, Laura Villanueva Rodríguez, and Ernesto Caballero García

Petróleos Mexicanos (PEMEX) Exploración y Producción, Activo Regional Exploración Sur, Mexico

ABSTRACT

The identification and characterization of source rocks, represented by their organic richness, quality, distribution, and maturity, are important for evaluating the petroleum systems in a sedimentary basin. The geochemical characterization of the source rocks is accomplished by the analysis and evaluation of their contained organic carbon. Used parameters (total organic carbon, free hydrocarbon (S_1), potential hydrocarbon (S_2), maximum temperature (T_{max}), hydrogen index (HI), oxygen index (OI), production index (PI), etc.) are derived from pyrolysis studies of outcrop, cuttings, and core samples. Other tools are gas chromatography-mass spectrometry (GC-MS), isotopes of specific compounds, and carbon isotopes of surface seep material, oil, and gas that define source rock affinity, depositional environment, and maturity level. The following reional source rock systems have been identified in the southeastern basins of Mexico: Upper Jurassic Oxfordian (!), Upper Jurassic Tithonian (!) argillaceous facies (JS1), Upper Jurassic Tithonian (!) calcareous-argillaceous facies (JS), Lower Cretaceous (!) carbonate-evaporite, and lower Miocene (!) marine deltaic siliciclastic.

INTRODUCTION

The 55,000-km^2 (21,236-mi^2) study area (Figure 1), geologically, is part of the southeastern basins of Mexico (Isthmus Saline Basin, Comalcalco Basin, Macuspana Basin and Reforma-Chiapas Mesozoic area, and the southern part of the Veracruz Basin). The Isthmus Saline Basin produces light oil from middle–upper Miocene and Pliocene reservoirs. The Comalcalco Basin has light oil production in Kimmeridgian, Cretaceous, and middle–upper Miocene reservoirs. The Macuspana Basin produces condensates and wet and dry gases from middle–upper Miocene and Pliocene reservoirs. The Reforma-Chiapas Mesozoic area has production of condensates, volatile oils, and

DOI:10.1306/13191091M903337

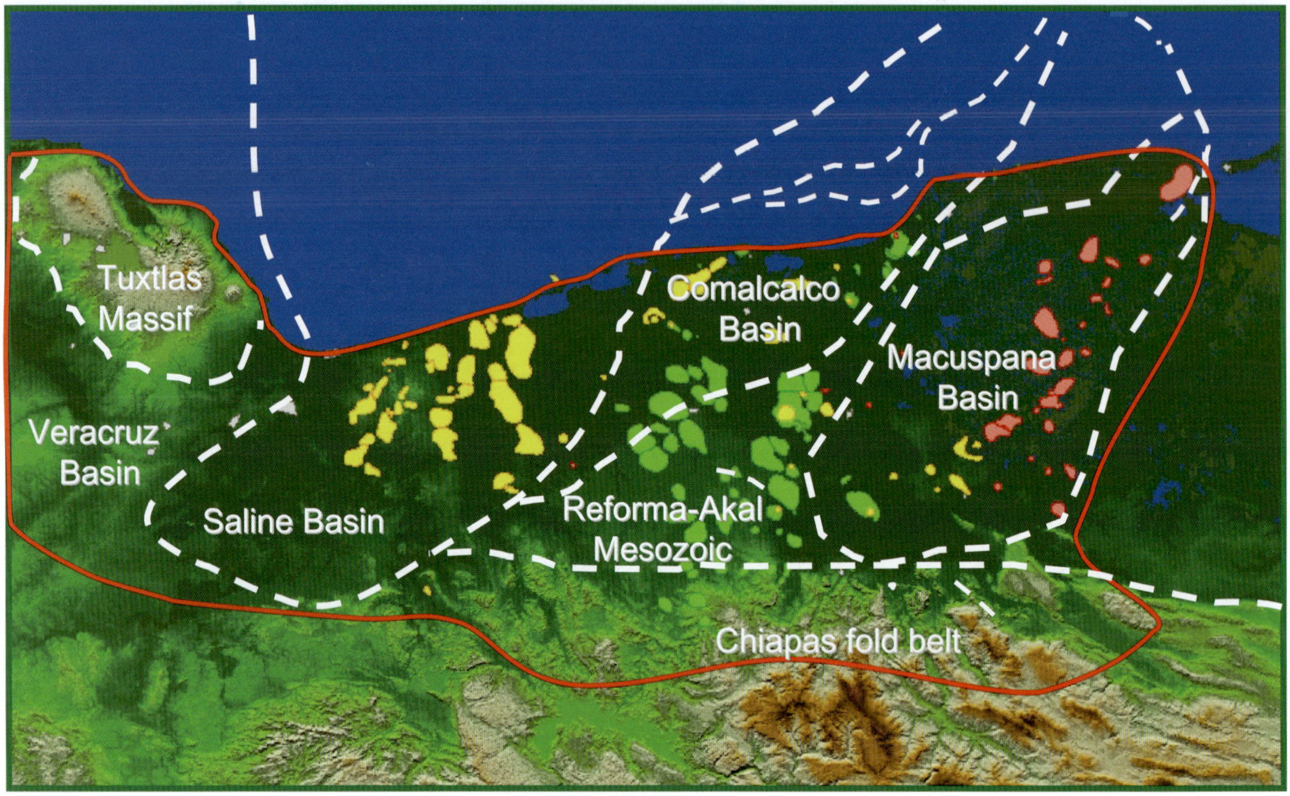

FIGURE 1. Location of the study area.

light and heavy oils in Kimmeridgian, Cretaceous, Eocene, middle–upper Miocene, and Pliocene reservoirs. The Veracruz Basin produces principally wet and dry gases from middle–upper Miocene reservoirs. The total cumulative production in the latter area (onshore) is 9319 MMBO and 26,184.5 cubic feet million miles of gas (MMMPCE). The objective of this study is the integration of geochemical information derived from rocks and fluids to identify and characterize petroleum systems for the oil and gas plays of Petróleos Mexicanos' (PEMEX) southern region.

Geochemical analyses of rocks (total organic carbon [TOC], Rock-eval, extracts, type of organic matter [TMO], vitrinite reflectance [R_o]), oils (isotopes, gas chromatography [GC], gas chromatography-mass spectrometry [GC-MS], isotopes of specific compounds, density [API], sulphur [S]), and gases (isotopes and compositions) conducted by different laboratories (French Petroleum Institute, Mexican Petroleum Institute, Sunmark, Humble, HRT, Biotechnology, and DGSI) were integrated into our database and used to identify and characterize generating source rocks.

Additionally, several oil and gas parameters were plotted to determine their relationships to the identified source rocks in this area; they will be discussed further in the manuscript.

Organic richness data from 170 wells and 20,000 geochemical analysis wells permitted the identification of several stratigraphic levels with source rock properties according to Peters (1986) (TOC > 0.5%, $S_2 > 2$ mg HC/g of rock, $T_{max} > 430°C$, PI < 0.3); the average values show mapping according to paleoenvironment distribution. The effectiveness of these systems is corroborated by the presence of their produced fluids. Geochemical signatures of these source rocks were mapped to understand their distribution and their characteristics, as discussed below.

UPPER JURASSIC (TITHONIAN) GENERATING SYSTEM

The organic richness of the source rock found in exploration wells indicates a poor to medium residual generating potential (Clara, 1985; Holguín, 1985; Sosa and Clara, 2001). Surface samples display originally high generation potential with TOC values of 2–15%, hydrocarbon potentials of 2–74 mg HC/g of rock, HI of 250–800, T_{max} of less than 430°C, mainly amorphous to algal organic matter with little herbaceous or woody contribution (Figure 2), and relative abundance of 10–40%. The maturity of these surface sections, as indicated by R_o, ranges from 0.4 to 0.6, indicating that some of this rock is still immature. Subsurface samples range from 0.6 to 2 R_o, indicating areas that range from immature

Figure 2. Amorphous-algal organic material in Tithonian outcrops as seen under a microscope. (a) transmitted light (b) fluorescent light.

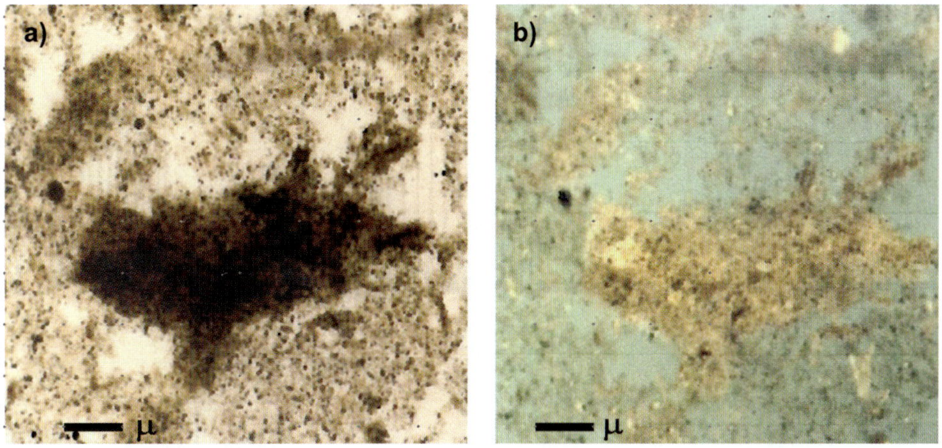

to overmature. These characteristics combine to show that the Tithonian source rocks are poor to excellent as shown in the natural series of the subsurface (Figure 3). The Tithonian geochemical characteristics were mapped (Figure 4) showing the entire range of hydrocarbons found in the southeastern basins. Existing variations depend on the thermal maturity of the source rock, physical-chemical changes during migration, and changes occurring within

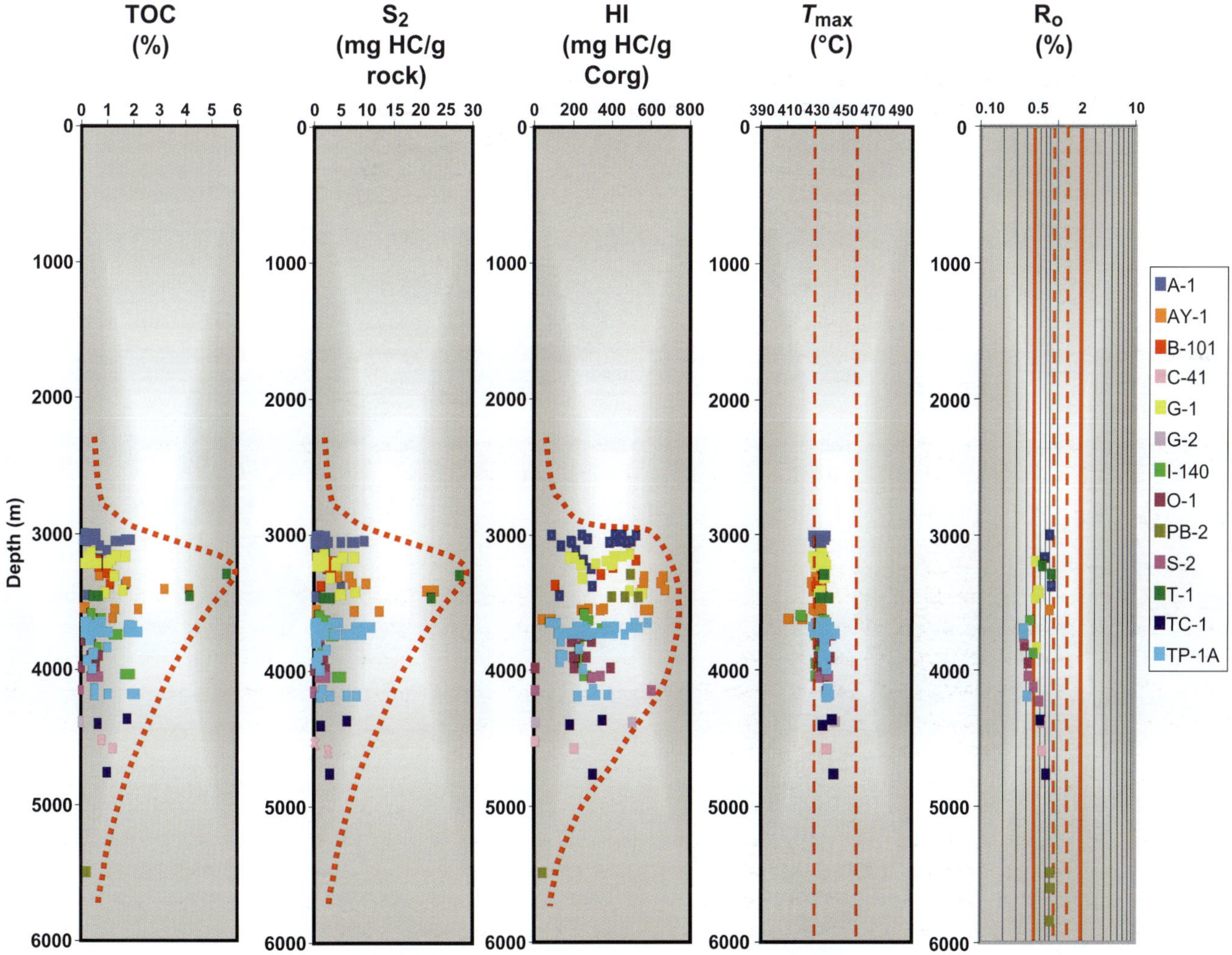

Figure 3. Natural series of Tithonian source rocks in exploratory wells. TOC = total organic carbon; R_o = vitrinite reflectance; S_2 = potential hydrocarbon; HI = hydrogen index; T_{max} = maximum temperature.

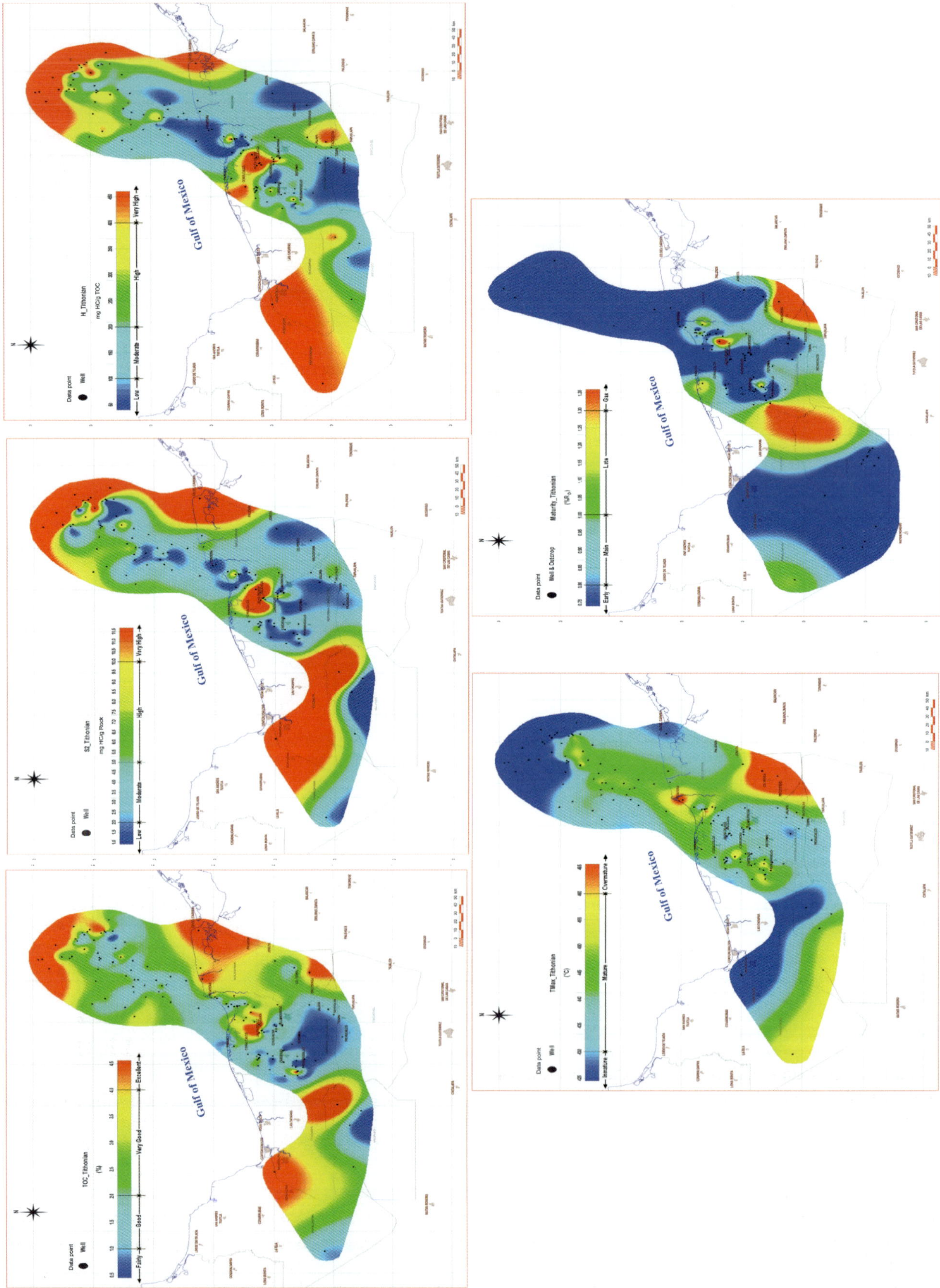

Figure 4. Maps of Upper Jurassic (Tithonian) geochemical parameters.

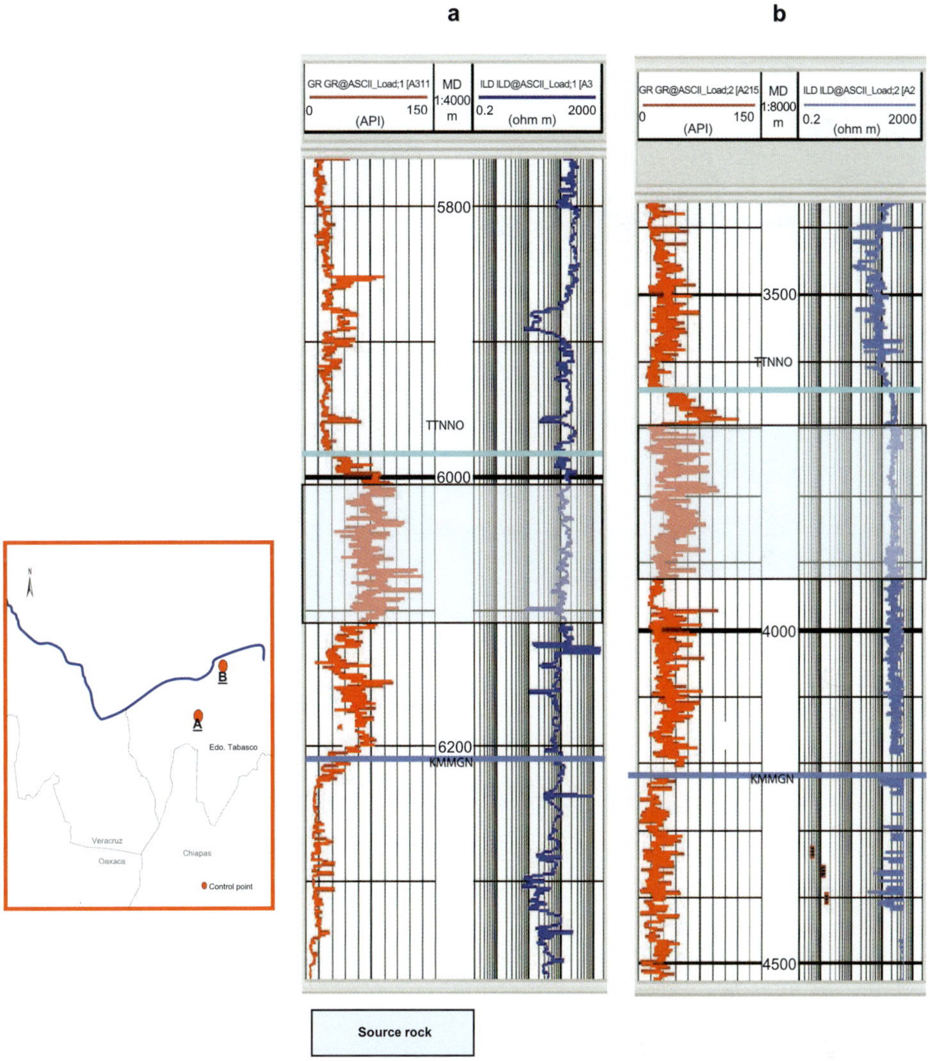

FIGURE 5. An example of a geophysical log illustrating variations in thickness for the exploration wells in which organic richness has been measured.

the reservoirs. Areas with poor values do not necessarily signify the lack of organic richness and may simply indicate that the source has nearly exhausted its generating capacity.

This is the most widely distributed source rock in the area; its distinctive signature on geophysical well logs (Figure 5) combined with its organic richness clearly represents a transgressive event that provided an ideal environment for the production and preservation of organic matter. Net source rock thickness ranges from tens of meters to as much as 460 m (1509 ft), depending on location.

The integration of sedimentologic, stratigraphic, paleontologic, and petrographic studies (Figure 6) indicates the gradual middle to upper Tithonian progression through the inner ramp to the outer ramp to basinal environments. The organic-rich sedimentary strata consist of dark gray to black, highly bituminous shales; dark gray to black argillaceous mudstones; and a marly sequence. The association of these rocks with fish fragments indicates the presence of phosphatic material within the source sequence. The rocks also contain a high concentration of pyrite and pyritized radiolaria, whereas parts of the sequence have been dolomitized.

LOWER CRETACEOUS GENERATING SYSTEM

A few available measured organic richness values from five exploration wells cluster between 0.6 and 1% TOC, with poor to fair generating potential (0.5–6 mg HC/g of rock). Type II kerogen is represented by algal material with a relative abundance of 4–25% cumulos and colloids, along with scarce woody material and migrated bitumens fluorescing yellow, orange, and brown, which is characteristic of hydrocarbons in the oil window (Figure 7) (Sosa, 1994). Identified biomarkers indicate the presence of microbial matter, and the

FIGURE 6. An Upper Jurassic outcrop in the Cerro Pelón anticline consisting of dark gray to black argillaceous mudstone-wackestone with strong hydrocarbon odor and evidence of impregnations. Petrographically (lower part), this is characterized as a radiolarian wackestone with hydrocarbon impregnations and fish bone remnants, indicating phosphatic contribution to the source rock.

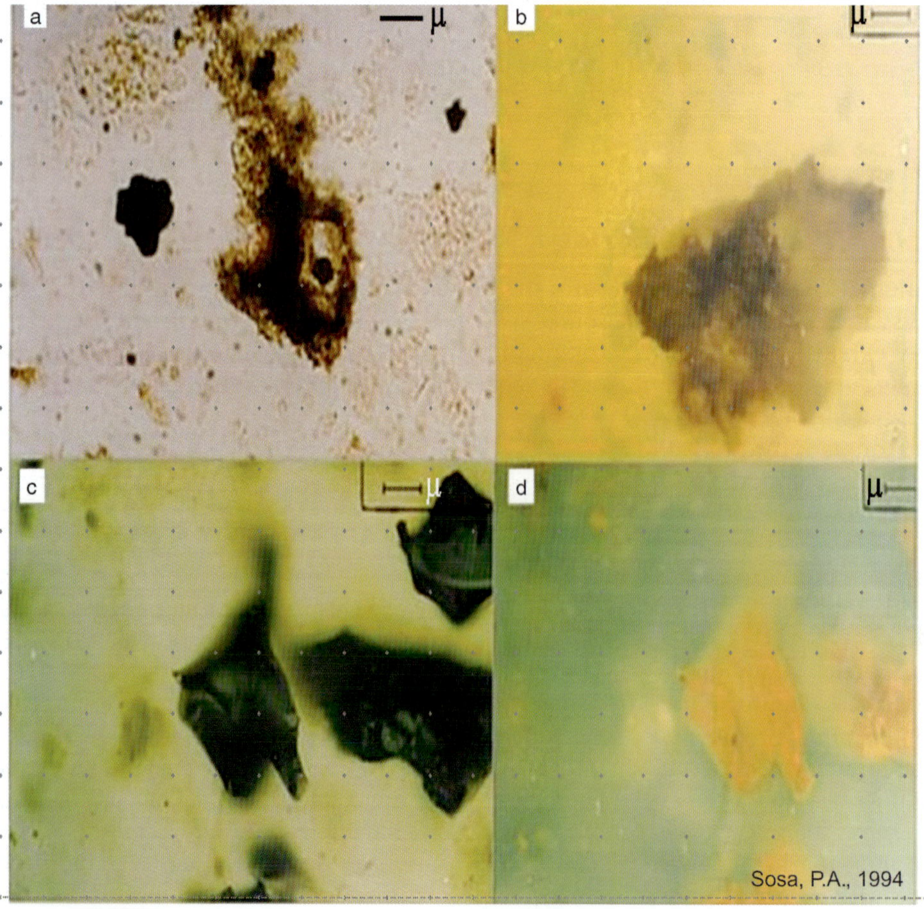

FIGURE 7. Types of organic matter in a Lower Cretaceous well. (a) Cumuliform, fragmental, and disseminated algal organic matter in transmitted light with brown coloration. (b) Organic matter fluorescing high-intensity yellow-orange in the catagenetic stage. (c) Pellet under reflected light showing black bitumens. (d) Pellet under fluorescent light; the bitumens are brown and the organic material is orange, indicating that they are undergoing catagenesis with the potential to produce medium to light oil.

type of organic matter assemblage is considered a precursor to oil and gas. Based on the maximum pyrolysis temperatures of 430–437°C, these rocks are thermally mature.

Proving the organic richness of these rocks has been difficult, however, based on the study of fluids related to hypersaline conditions toward the Artesa-Mundo Nuevo Platform and the Sierra de Chiapas. These oils show characteristics similar to those of El Petén in northern Guatemala, which are associated with a carbonate-evaporite facies (Connan, 1986); i.e., calcareous-evaporite sequences consisting of anhydrite with mudstone-wackestone interbeds and containing benthonic foraminifera, oil-impregnated dolomite, blue-green algal mats, ostracods, etc. This high-salinity anoxic sedimentary environment allows the enhanced production and preservation of sapropelic, microbial, diatomaceous, green algal, and cyanobacterial organic matter.

LOWER MIOCENE GENERATING SYSTEM

Although the lower Miocene is present in all of the Tertiary southeastern basins, it has only been recognized as an effective source rock in the Macuspana Basin where it is a basinal facies deposited under conditions favorable for the preservation of organic matter. Much of the organic matter is derived from higher plants mixed with algal material in an aqueous environment. In contrast, the preservation potential of organic matter in the fluvial and deltaic facies is quite variable because of exposure to the atmosphere and/or shallow-water conditions.

The residual organic richness data (Figure 8) indicate good hydrocarbon potential (1–2% TOC, 2 to >5 mg HC/g of rock, HI values from 300 to 500; T_{max} data indicate that this rock is thermally immature in the northern part of the area [<430°C] and mature to the west and south [430 to >440°C]). The kerogen is principally type II/III (amorphous-woody mixture, Figure 9) corresponding to organic facies B-C to D. This type of organic matter can generate oil and gas, depending on the local maturity level.

FLUID CHARACTERIZATION

The southeastern basins produce heavy oil, light oil, volatile oil, super-light oil, condensate, wet gas, and dry gas (Figure 10) from large segments of the stratigraphic column. Geochemical studies have been applied to determine the physical-chemical processes that affected the fluids before, during, and after their accumulation, as well as to understand the parameters that elucidate the origin, composition, density, maturity, and distribution of the hydrocarbons.

The known plays are:

Kimmeridgian: This sequence was deposited in an inner ramp with oolite shoals and is characterized by ooid packstone-grainstone with intraclasts and bioclasts that change laterally to mudstones-wackestones with peloids and intraclasts. The produced oils have densities of 28–45° API.

Tithonian: This was a period of environmental transition from the inner ramp to the outer ramp to basin conditions. Reservoir rocks consist of argillaceous mudstones-wackestone that have produced oils with densities of 38–44° API.

Lower Cretaceous: This period is represented by dolomitized flow breccias with subrounded cream to light brown clasts in a calcareous-argillaceous matrix. The rocks are fractured with vugular cavities partially filled with white dolomite and residual oil. A second facies consists of dark gray microdolomite (argillaceous mudstone-wackestone) with black chert nodules, abundant fractures, and microfractures. These facies correspond, respectively, to slope and basinal environments. The produced oils have densities of 25–54° API.

Middle Cretaceous: The most important play in this sequence produces from patch reef facies (intraclast and bioclast packstone-grainstone), lagoonal facies (miliolid packstone-grainstone), slope facies (dark gray, dolomitized flow breccias), and basinal facies (argillaceous mudstone). All are dolomitized with abundant fractures. The produced oils have densities of 26–53° API.

Upper Cretaceous: This sequence produces from its basal part; i.e., Turonian–Coniacian–Santonian. Only one field produces from the Campanian–Maastrichtian. Characteristic paleoenvironments are platform (patch reefs), slope (calcareous flow breccias), and basin (highly fractured, argillaceous mudstone-wackestone with chert banding). The produced oils have densities of 30–53° API.

Tertiary: Lower to middle Eocene siliciclastic reservoirs consist of carbonate debris flows surrounded by argillaceous matrix with oil gravities ranging from 43 to 53° API. Other important producing horizons are within the middle Miocene section. Reservoirs in most of these fields have good production from deep-water turbidite sand systems associated with slope and basin-floor fans. Produced oils have densities that vary from 24 to 44° API. The upper Miocene and lower Pliocene show a pattern of shallowing depositional environments. There also exist deltaic and fluviodeltaic reservoirs consisting of isolated or stacked shallow-water bar sands with oil densities of 17–60° API.

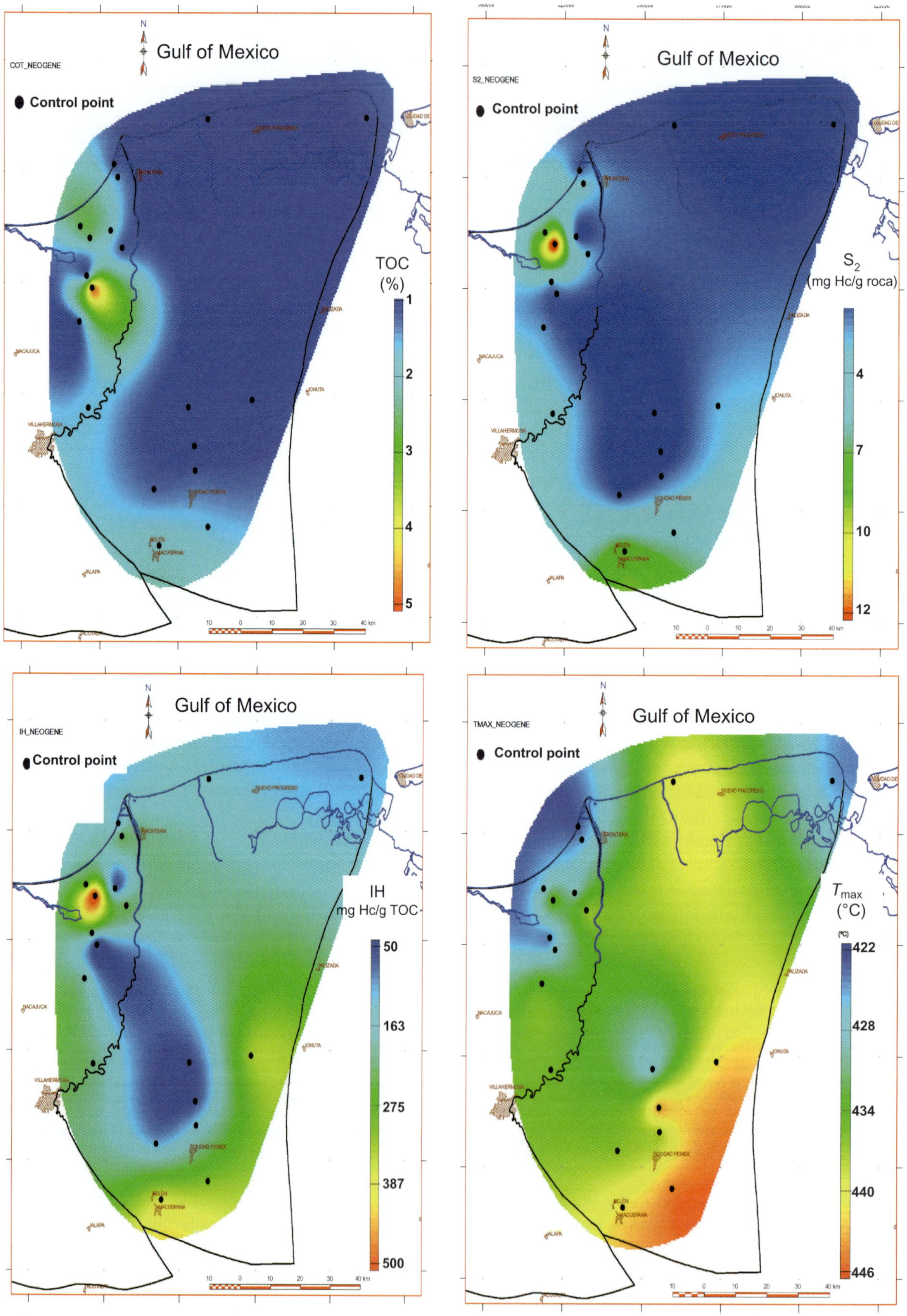

FIGURE 8. Distribution of Lower Miocene geochemical parameters. TOC = total organic carbon; S_2 = potential hydrocarbon; HI = hydrogen index; T_{max} = maximum temperature.

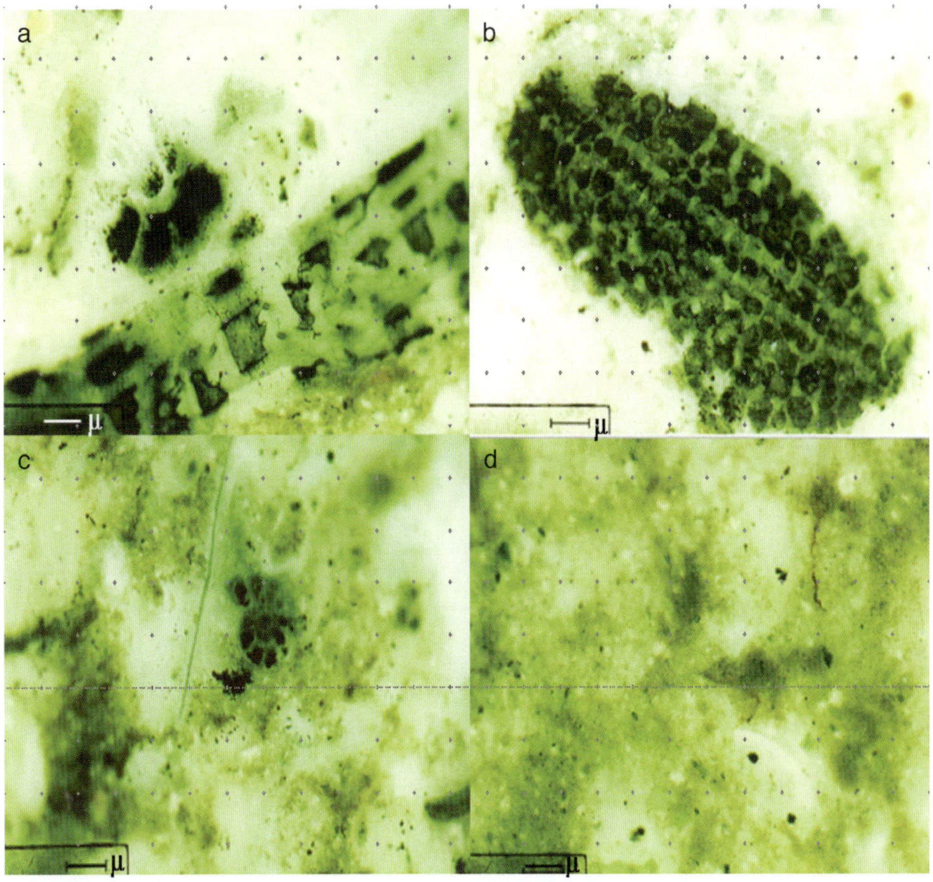

FIGURE 9. Types of Lower Miocene organic matter. (a, b) Structured organic matter with hydrocarbon impregnation. (c, d) Amorphous algal organic matter under fluorescent light with greenish color and hydrocarbon impregnation in animal remains.

Variations in API gravity and sulfur content of oils in the southeastern basins is related to thermal evolutionary processes (Figure 11) starting with the source rocks, i.e., high API gravity and low sulfur content mean high thermal maturity and vice versa. Notably, the actual behavior of the oil may be masked if these properties are altered either during migration toward traps, secondary alteration within the reservoir, or sulfur contamination from proximity to salt domes or canopies.

The results of GC and biomarkers (GC-MS) from oils and surface seeps (Figure 12) were established to clarify factors such as the affinity of the hydrocarbon source, depositional environments, maturities, migration routes, and secondary alteration processes. These analyses are based on physical-chemical properties and the behavior of steranes and hopanes, isotopic compositions, and by crossplotting these parameters. Results from surface seeps, oils, and rock extracts have allowed the identification

FIGURE 10. A variety of fluids reservoired in different plays in the southeastern basins run the gamut through heavy oil, light and volatile oils, superlight oils, and condensates.

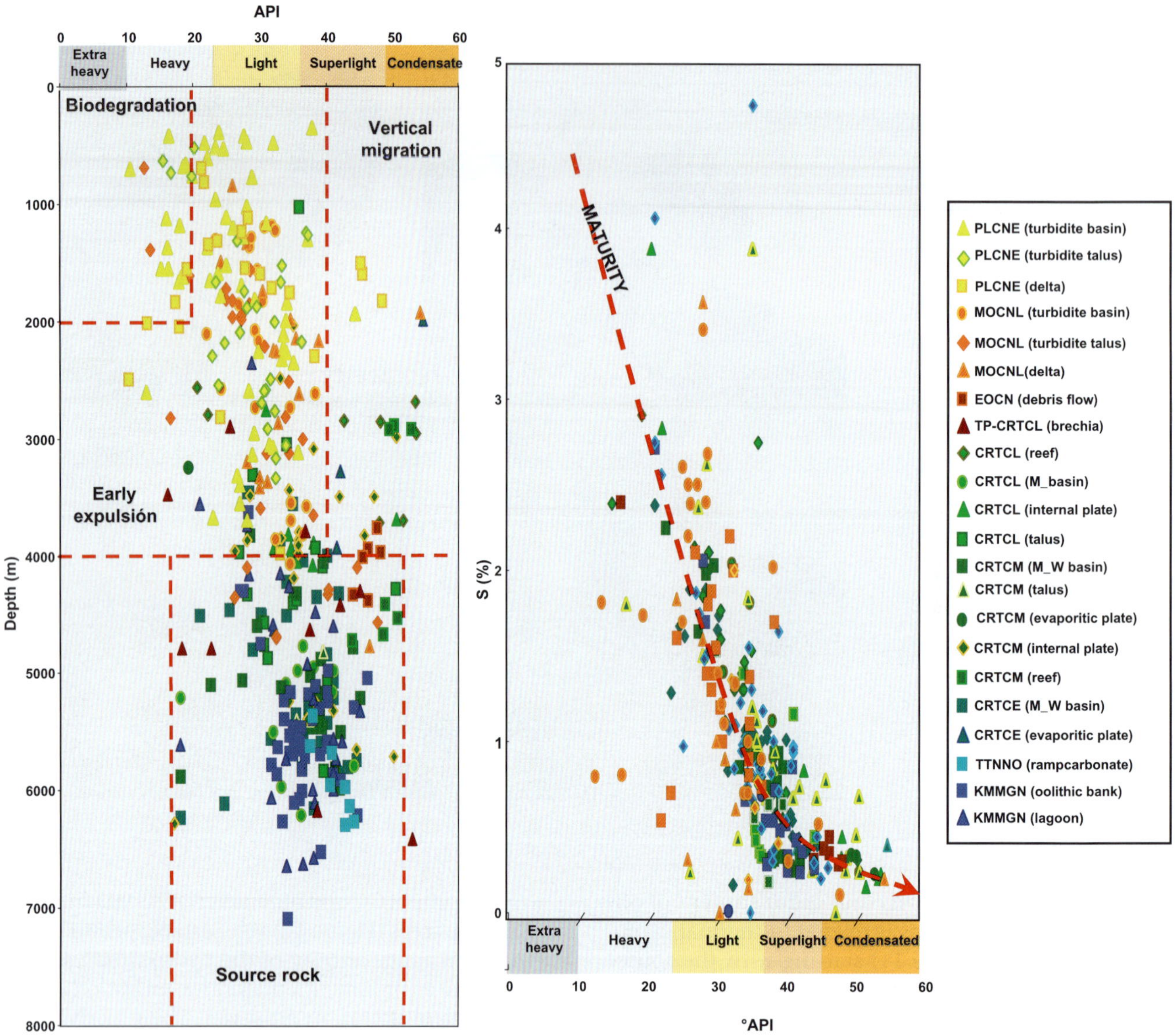

FIGURE 11. Plot of API gravity vs. depth and sulfur content vs. depth by stratigraphic level and play.

of the following petroleum systems in the southeastern basins:

1. The Oxfordian (!)–Oxfordian–Kimmeridgian (Figure 12) petroleum system (Guzman and Mello, 1994; Guzmán et al., 1995, 2001) has been identified in localized areas and reflects geochemical characteristics related to source rock deposition in a hypersaline marine carbonated environment, such as $\delta^{13}C$ values of −25.41 to −25.6, high Tetra 24/26 Tri ratios, high C_{35}/C_{34} ratios (>>1), Ts/Tm < 1, $C_{29}/C_{30} < 1$, $C_{29} > C_{27}$, low disteranes, high dibenzothiophanes, 27 nor > 24 nor, and 27 nor > 21 nor. This system is efficient in charging Oxfordian sandstones in Campeche Sound and onshore Kimmeridgian reservoirs of pellet wackestone-packstone in the Palancares area.

2. The Tithonian (!)–Miocene/Pliocene (Figure 12) petroleum system (Guzman and Mello, 1994; Guzmán et al., 1995, 2001; Mello and Moldowan, 2006) has been identified mainly in the Isthmus Saline Basin with middle to upper Miocene turbidite sands and Pliocene fluviodeltaic sands. The oil affinity is directly related to black, bituminous shales of Tithonian age of the JS1 family with $\delta^{13}C$ of −26.4 to 27.5; high sulfur values; Pr/Ft > 1; $C_{29}/C_{30} < 1$; C_{35}/C_{34} less than or equal to 1; Ts/Tm greater than or equal to 1; the presence of hopanes in extension; the 30-norhopane family represented by the compounds 17α (H)-29,30 bisnorhopane and 17α (H)-30-nor-29 homohopane; high ratios of $C_{29}Ts/C_{29}$; relative abundances of 17α (H)-dihopane compounds; predominance of C_{29} in the steranes; high relative proportions of disteranes; and the presence of C_{30} steranes,

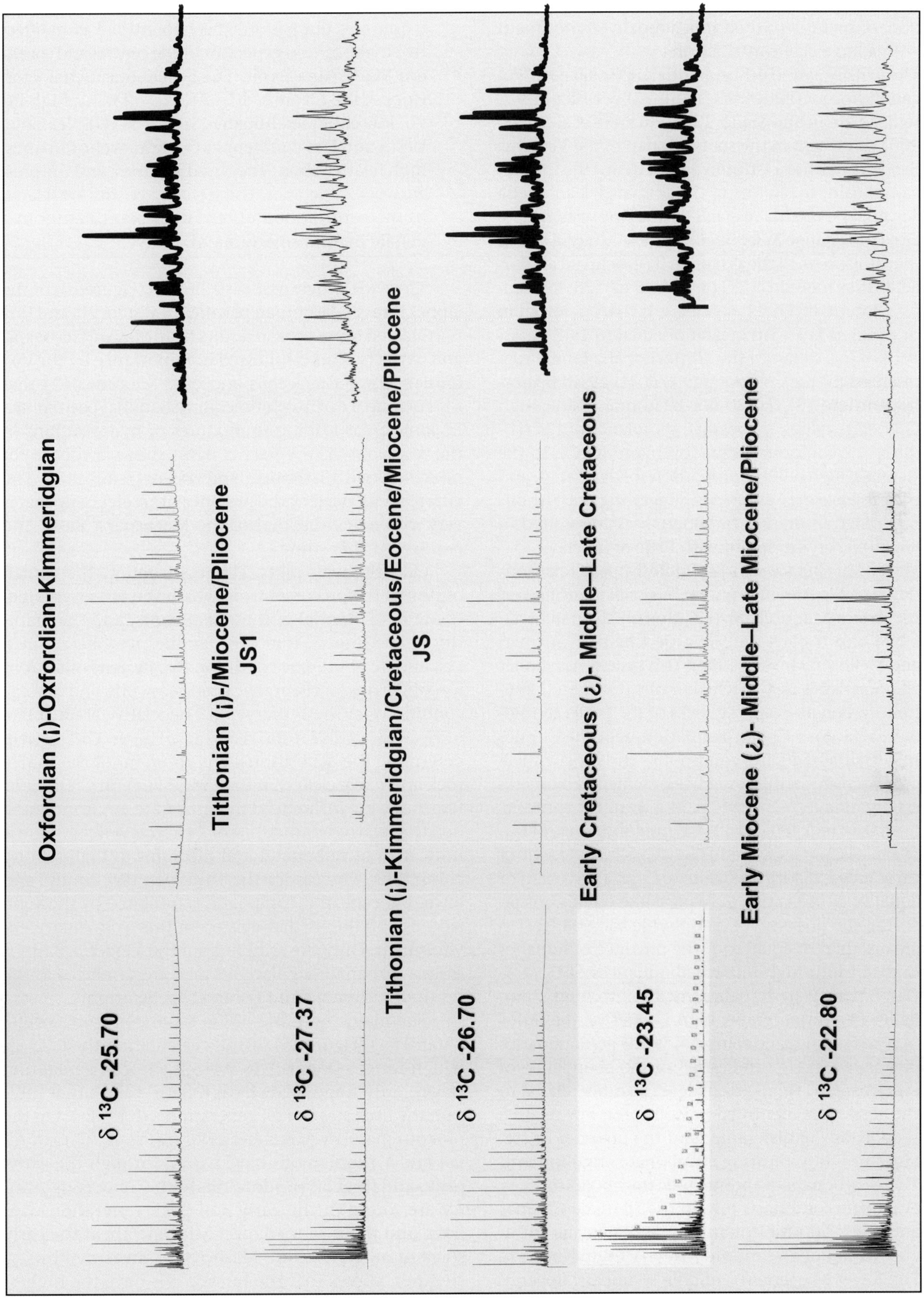

FIGURE 12. Geochemical signature of the petroleum systems.

suggesting that the oils originated in marine carbonate rocks deposited in a suboxic environment with a large clay contribution.

3. The Tithonian (!)–Kimmeridgian/Cretaceous/Eocene/Miocene/Pliocene (Figure 12) petroleum system (Guzman and Mello, 1994; Guzmán et al., 1995, 2001) is present in the southern part of the Veracruz Basin, the eastern Isthmus Saline Basin, the Comalcalco Basin, the Reforma-Chiapas area, Campeche Sound, and the Macuspana Basin. This system is related to a change in sedimentary facies to argillaceous mudstone named the JS family. Its geochemical characteristics include $\delta\ ^{13}C$ of −25.45 to −28; low sulfur content; $Pr/Ph > 1$; $C_{29}/C_{30} > 1$; C_{35}/C_{34} less than or equal to 1; Ts/Tm greater or equal to 1; the presence of C_{30} steranes; the 30-norhopane family represented by the compounds 17α (H)-29,30 bisnorhopane and 17α (H)-30-nor-29 homohopane; high $C_{29}Ts/C_{29}$ values; the relative abundance of 17α (H)-dihopane compounds; predominance of C_{29} in the steranes; high relative proportions of disteranes; and the presence of C_{30} steranes, suggesting that the oils originated in marine carbonate rocks deposited in an anoxic environment with little or no clay.
4. The Lower Cretaceous (.)–Middle/Upper Cretaceous (Figure 12) petroleum system has been identified on the Artesa-Mundo Nuevo Platform (Guzmán et al., 1995) and toward the Sierra de Chiapas (Guzman and Mello, 1994; Sosa, 1994). This system has similar characteristics to Guatemalan oils (Connan, 1986) from carbonate-evaporite rocks of the Lower to middle Cretaceous Coban Formation deposited under hypersaline-reducing conditions. The geochemical characteristics of this system, represented by biomarkers, include $\delta\ ^{13}C$ −23.6 to 24.6%; sulfur contents from 0.13 to 3.18%; $Pr/Ph > 1$; predominance of C_{29} over C_{30}; $C_{29}/C_{30} >> 1$ and $C_{34} > C_{33}$ characteristic of carbonate-evaporite sequences; C_{34}/C_{33} less than or equal to 1, indicative of hypersaline conditions favoring the presence of halophyllic bacteria; Ts/Tm greater than or equal to 1, the presence of hopanes in extension; high relative abundances of C_{24}-C_{27} Des_E tetracyclic terpanes; medium to high abundance of gammacerane; high $C_{29}Ts/C_{29}$; the ratios of 17α (H)-dihopane compounds; the predominance of steranes in C_{27}, indicative of the abundance of cyanobacteria (blue-green algae) possibly related to the algal mats identified in well cores; low relative proportions of disteranes; and the presence of steranes in C_{30}, indicating a carbonate environment. These carbonate-evaporite indicators point to more evaporitic conditions toward the Sierra de Chiapas and less evaporitic conditions with a greater influence of carbonates toward Artesa-Mundo Nuevo.
5. The lower Miocene (!)–middle to upper Miocene/Pliocene (Figure 12) petroleum system (Guzman and Mello, 1994; Guzmán et al., 1995) is found in the whole area but has only been identified as an effective hydrocarbon generator in the central and southern Macuspana Basin. The geochemical characteristics of the oil are $\delta\ ^{13}C$ −21.6 to −23.7‰, high Pr/Ph, low extended hopanes, high 18α (H)-oleanane, Des-A and Des-E present in C_{24} tetracyclic terpanes, high relative abundances of disteranes, and the presence of C_{30} steranes. These characteristics are related to the contribution of terrestrial plant matter in a deltaic marine environment.

The wide geographic distribution (Figure 13) of the Upper Jurassic Tithonian petroleum system (JS and JS1) is indicated by analytical results from oils, surface seeps, and extracts. This is followed in importance by the Oxfordian, Lower Cretaceous, and lower Miocene. Note that the coexistence of two petroleum systems has been shown in some areas leading to mixtures of hydrocarbons in the reservoirs. The Veracruz Basin shows evidence for mixed oils with Tithonian and Eocene(?) affinities. The latter is evidenced by the presence of an oleanane that is very different from that of the Macuspana Basin and requires further study.

Oil parameter plots (Figure 14) reflect the contribution of the precursor from which the oils originated, the redox potential of these sediments, and the maturity of the fluids. From this, families may be defined; i.e., anoxic or suboxic conditions for Jurassic oils, salinity conditions for the anoxic Cretaceous oils, and suboxic conditions for the Tertiary oils. The relative abundances of H_{29}/H_{30} vs. 24Tet/26Tri and H_{29}/H_{30} vs. Ol/H_{30} help determine the depositional environment of the source rock and make clear that the Tithonian and Cretaceous families were influenced by carbonate environments; i.e., the relative abundance of 24-tetracyclic terpane is indicative of carbonates and medium- to high-salinity conditions, whereas for the Tithonian (JS1 family) and lower Miocene, this compound is inhibited by the argillaceous conditions that existed in the sedimentary environment. Only the sample from the Veracruz Basin is from a carbonate-argillaceous mixture suspected as the result of Tithonian and Eocene cogeneration.

The maturity of the oils is estimated from several parameters (Figure 15), first by calculating the $R_{o\ equiv}$, which ranges from 0.7 to 0.9% (i.e., peak generation stage); only some fields have $R_{o\ equiv}$ values that indicate the final stage of oil generation (>1.1% R_o). On the plot of a maturity parameter as $R_{o\ equiv}$ vs. $\delta\ ^{13}C$, Oxfordian and Tithonian oils have passed through the early, peak, and final oil window; oils with Cretaceous affinity are found in the early and peak generation window, and those of the Lower Miocene are at the early stage of oil generation. Maturity, as measured by C_{29} steranes, shows that the Jurassic oils have the highest maturities and are reaching the equilibrium stage, and

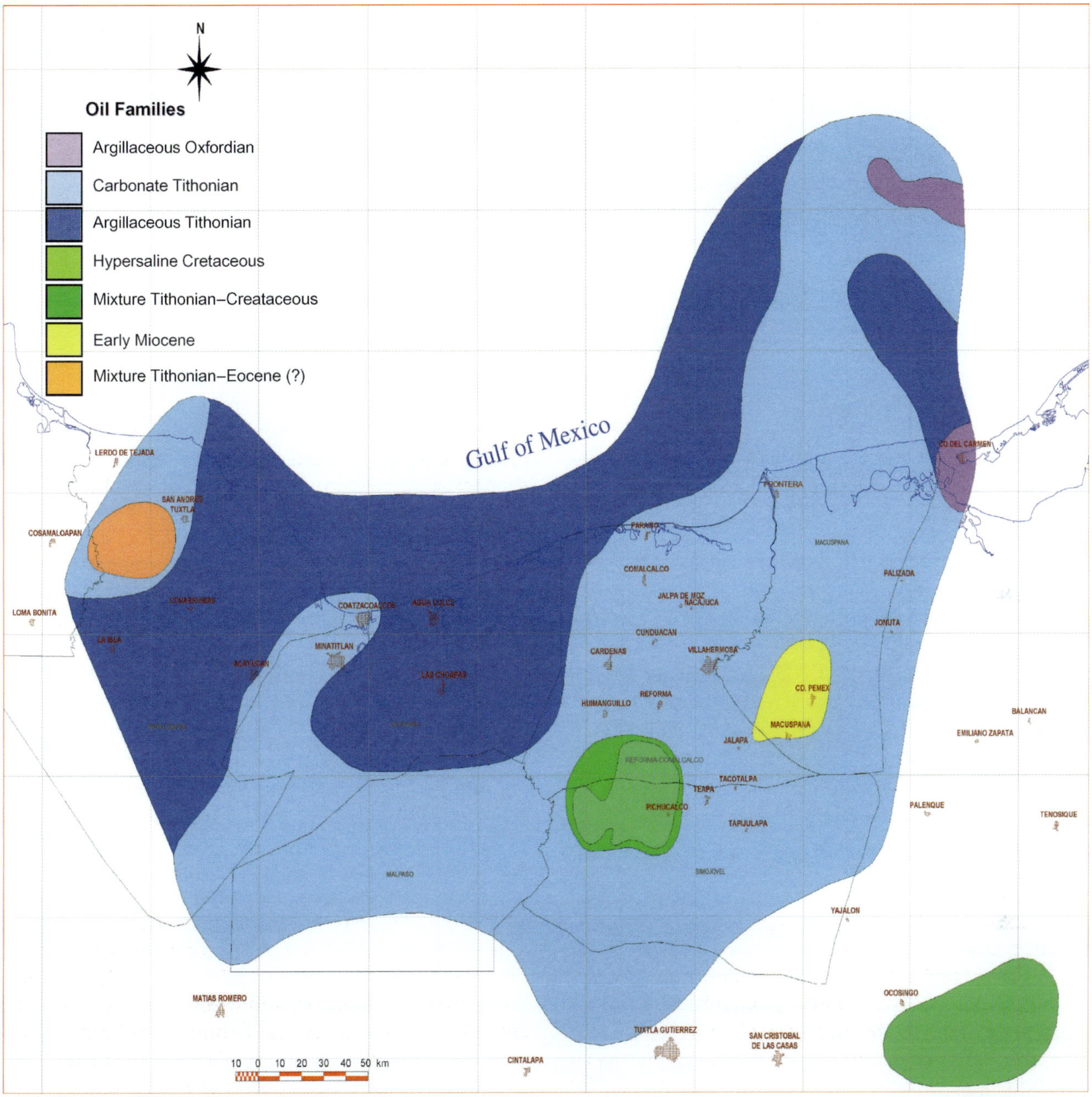

FIGURE 13. Regional distribution of oil families.

the least mature oils are those of the Isthmus Saline Basin. The Cretaceous oils have moderate maturity and are also reaching the point of equilibrium, whereas Lower Miocene oils, as shown on the plot, are in the initial stages of maturity.

The Ts/(Ts + Tm) ratio vs. API gravity (Peters and Moldowan, 1993) also serves as a maturity indicator and relates oils to their source rocks. This relationship increases according the level of maturity and anoxia in the sedimentary environment. Generally, a tendency toward higher maturity in the Jurassic oils, advanced maturity in the Cretaceous oils, and peak generation in the lower Miocene oils exists. Oils that appear on this plot without clearly defined trends are the products of secondary processes such as biodegradation or fractionation during hydrocarbon migration.

Secondary alteration processes were recognized (Figure 16a, b), such as (1) biodegradation caused by aerobic bacteria in the subsurface at low reservoir temperatures (60–80°C) and (2) through recharging by surface waters with the presence of oxygen and/or aerobic bacteria that initiate the degradation of petroleum. In general, the secondary alteration process provokes an increase in sulfur, nitrogen, and oxygen content (nitrogen, sulphur, oxygen compounds); increased asphaltenes; and decreasing API gravity of the oils. The degree of

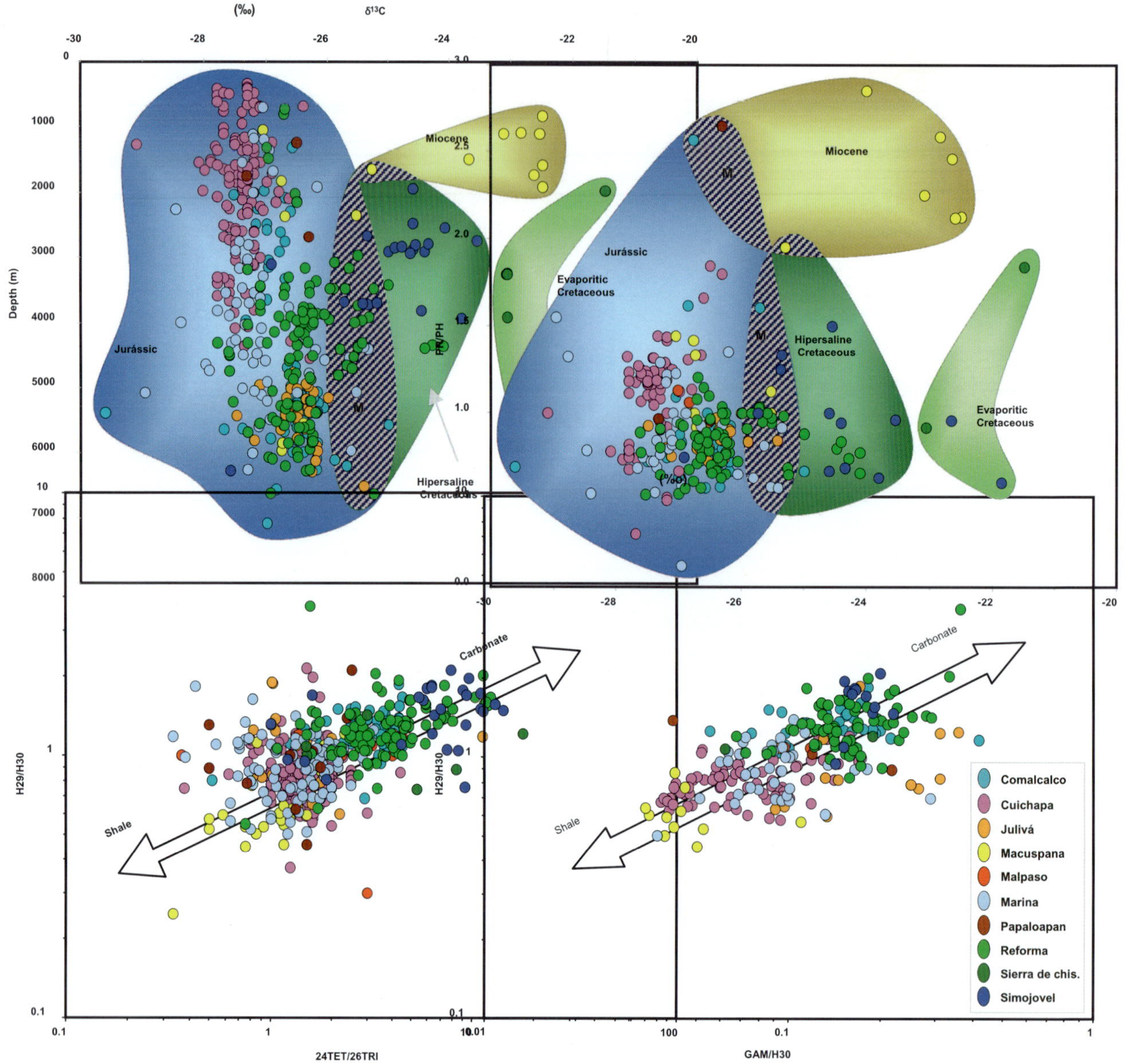

FIGURE 14. Graph of depth vs. carbon isotopes (δ ^{13}C) and Pr/Ph vs. carbon isotopes (δ ^{13}C) showing the separation into oil families; H_{29}/H_{30} vs. 24 Tet/26 Tri and H_{29}/H_{30} vs. Gam/H_{30}, indicating the environment of deposition of the source rock.

biodegradation is estimated by employing the classification of Peters and Moldowan (1993) in the scale of 1 to 10; i.e., light, moderate, heavy, very heavy, and severe according to the affected fraction starting with the n_paraffins, isoprenoids, steranes, hopanes, diasteranes, aromatics, and porphyrins.

Oils reservoired at depths of less than 1500 m (4921 ft) in the upper Miocene and Pliocene plays are lightly to moderately biodegraded. The paraffins are affected, and in some case the isoprenoids (Figure 16a). Additionally, in some apparent normally evolved oils, or in oils with a mixed pattern of heavy and light hydrocarbons, 25-norhopane is present (Figure 16b) indicating paleobiodegradation. This is caused by biodegradation of an earlier accumulation followed by a second charge of oil that diluted the biodegraded oil in the reservoir. This process is recognized from biomarkers resistant to microbial attack, which preserves the affinity or correlation to the hydrocarbon sources.

ISOTOPIC SIGNATURE OF SPECIFIC COMPOUNDS

An analysis of specific biomarker compounds (hopanes and steranes) and diamondoids in the n-alkane fraction was applied to oils of extremely high thermal

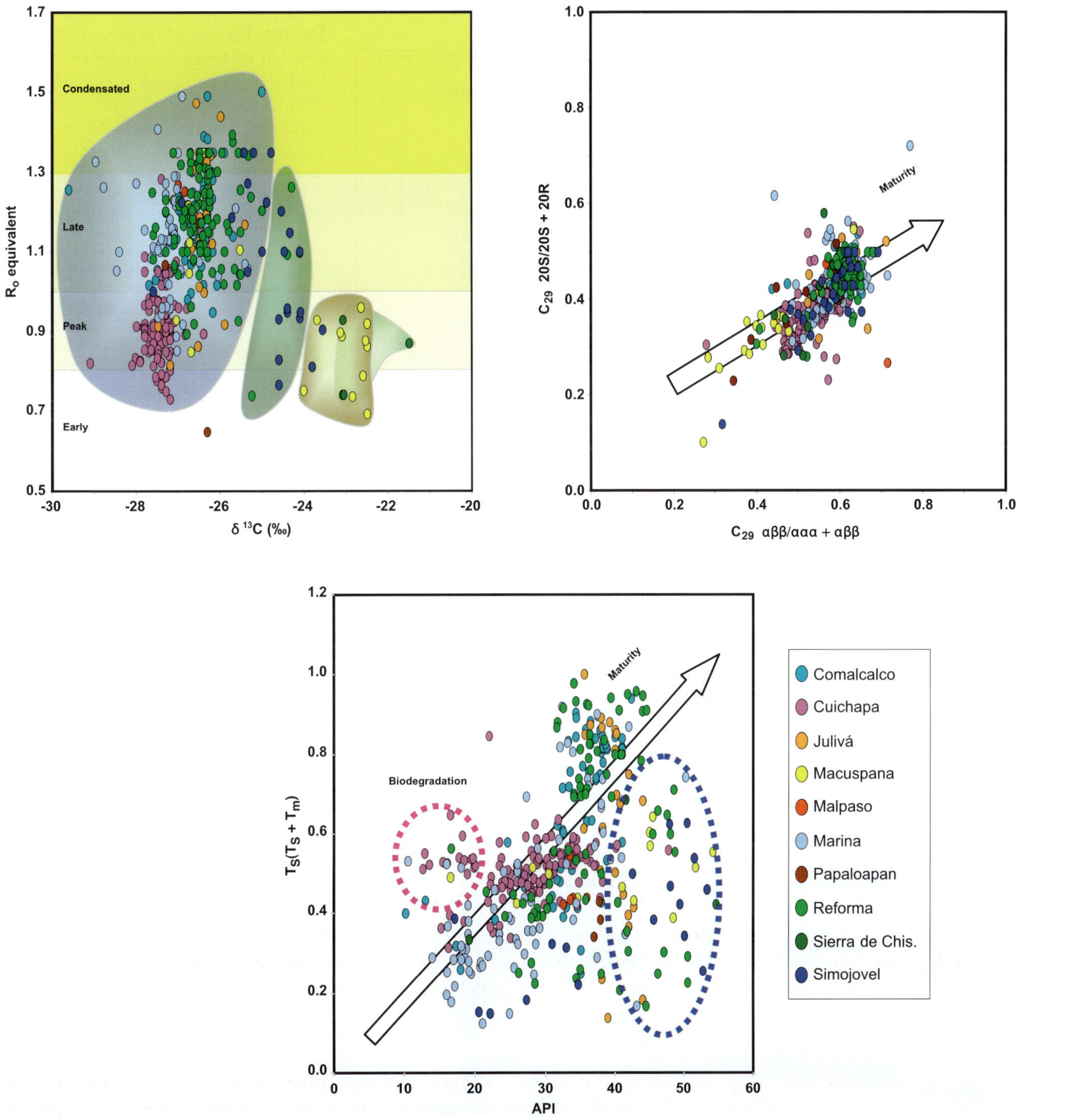

FIGURE 15. Maturity plots (R_o equivalent vs. δ ^{13}C, C_{29} steranes, and Ts/(Ts + Tm) hopanes from oils).

maturity. Generative sources were identified, as well as the presence of mixed oils, either from different sources or caused by multiple migrations from the same source. Four main sources were differentiated: a first group associated with Oxfordian and Tithonian rocks, a second group related to Lower Cretaceous and Miocene source rocks; and mixtures of oil that could be Tithonian + lower Miocene and/or Tithonian + Lower Cretaceous (Figure 17a, b).

Diamondoid studies in each of the basins indicate the following (Figure 18):

1. Many of the oils in the Isthmus Saline Basin were generated in the early oil window and only a few have high thermal maturity with one of the latter being a possible mixture from two sources.
2. Oils from the Comalcalco Basin are also from the early generation window. Some are mixtures caused

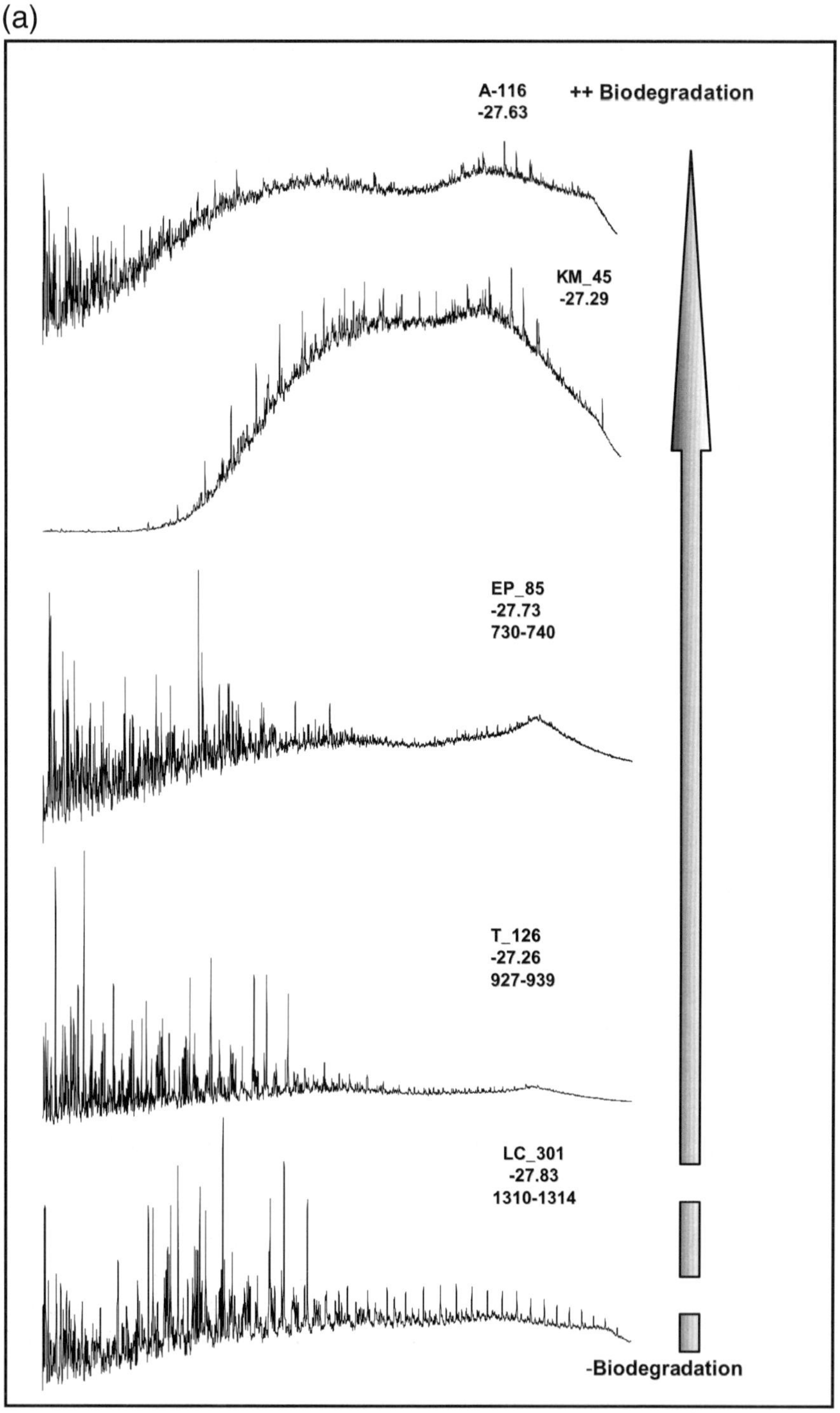

Figure 16. (a) Examples of oils affected by biodegradation showing the loss of n-paraffins. (b) Detail of the identification of 25-norhopane indicative of paleobiodegradation of an oil with apparently normal thermal evolution.

by different migration pulses from one source. Very few show higher maturity and are entering the cracking stage.

3. Toward the Julivá area, one early-stage oil is found, with the remainder in the final oil window and starting the cracking stage.
4. The Reforma-Simojovel area oils may be interpreted as mixtures from one source at different maturity levels or different migration pulses. Some may be mixtures from two generating sources (Tithonian + Cretaceous). Some oils are in the late generation window and are cracking, which indicates their advanced thermal evolution.
5. In the Macuspana Basin, the oils derived from siliciclastic rocks are in the early through late generation stage, with only one interpreted to be a mixture. Oils

(b)

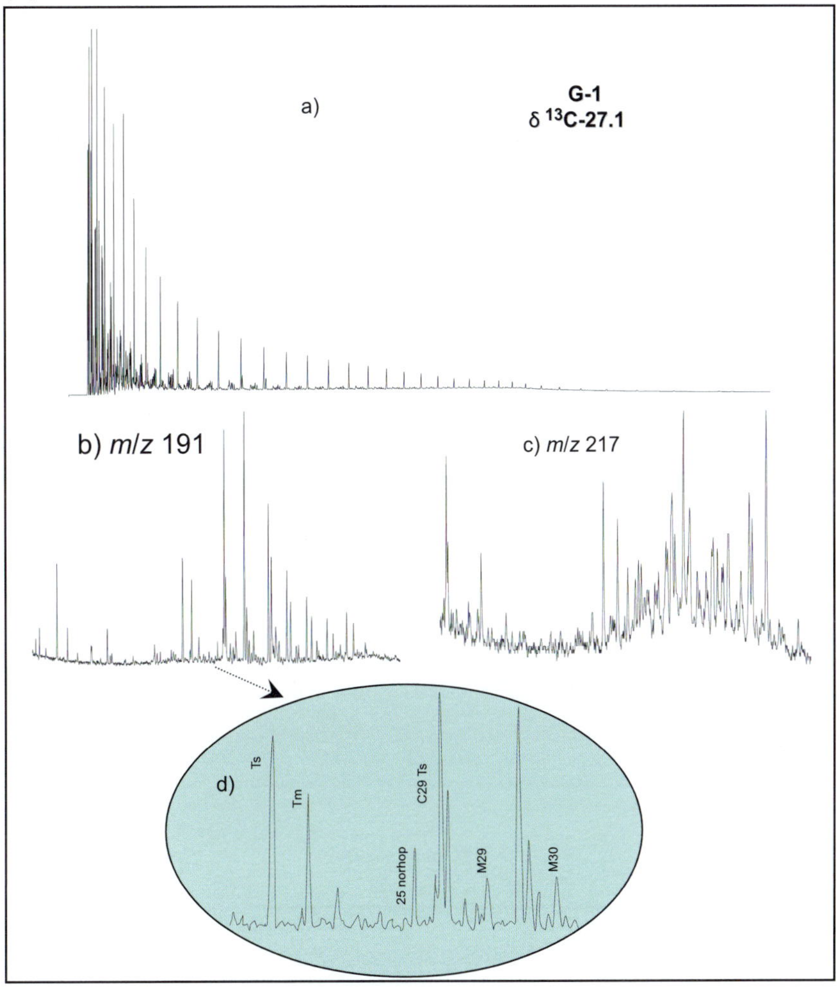

FIGURE 16. (cont.).

in the Mesozoic strata are in the late oil window and perhaps some of these are beginning to crack.

6. Finally, one oil in the Sierra de Chiapas from the early generating window is found, another from the late generating window, and yet another oil that is very highly evolved.

GENETIC CHARACTERIZATION OF GASES

The integration and interpretation of isotopic analysis and molecular composition for 268 gas samples from 80 fields were used to determine the behavior and origin of gases in the southeastern basins. Approximately 80% of the analyzed gases come from Miocene and Pliocene reservoirs; the other 20% are from Mesozoic reservoirs.

This study analyzed and interpreted the molecular composition and isotope data for methane (C1), ethane (C2), propane (C3), butane (nC4), and isobutane (iC4) applying the methodologies and concepts of Bernard, Berner, Clayton, Prinzofer, Chung, Pernaton, and Schoell (1983), Lorant et al. (1998), among others. By combining isotopic ratios and molecular compositions, determining the presence and maturity index of thermogenic gas, biogenic (microbial) gas, mixtures of thermogenic and biogenic gases, isotopic fractionation by diffusion, and possible sources was possible.

Gases in the southern region may be classified into two large groups (Figure 19) (Caballero et al., 1998; Sosa, 2000); the most represented is group 1 with heavier isotopes. Isotopic values for ethane ($\delta\,^{13}C2$) are more consistent and have wider distribution than methane isotopes ($\delta\,^{13}C1$). Group 1 also has a higher C2/C1 ratio and a wetness percentage of 3.5–52%. In contrast, group 2 has isotopically lighter gases; the methane isotopes are more homogeneous than those of ethane, and the C2/C1 ratio is generally low, with wetness less than 1%.

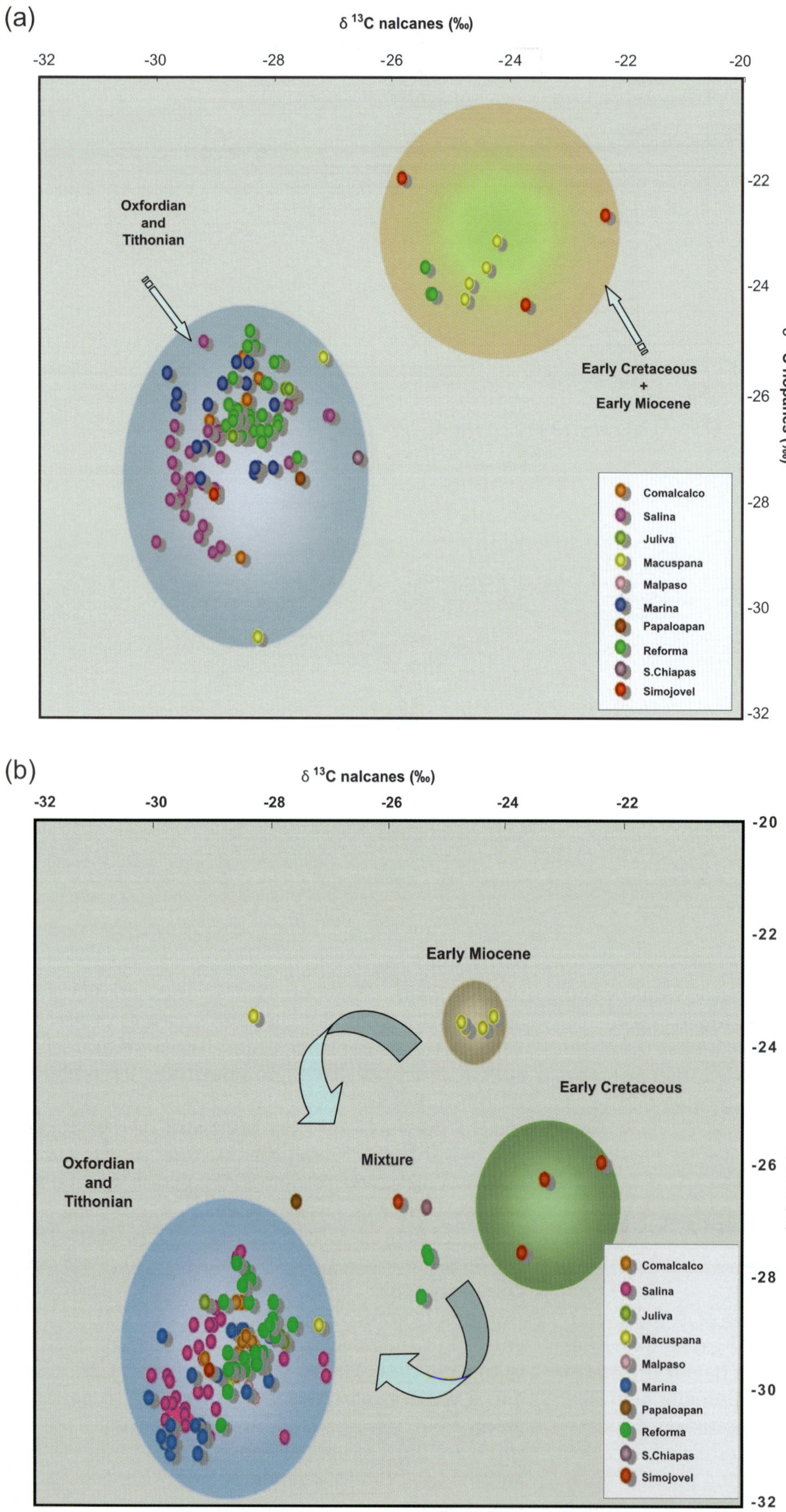

FIGURE 17. (a) Plot of δ ^{13}C n-alkanes vs. δ ^{13}C hopanes and (b) δ ^{13}C n-Alkanes vs. δ ^{13}C steranes differentiating hydrocarbon predecessors in the southeastern basins.

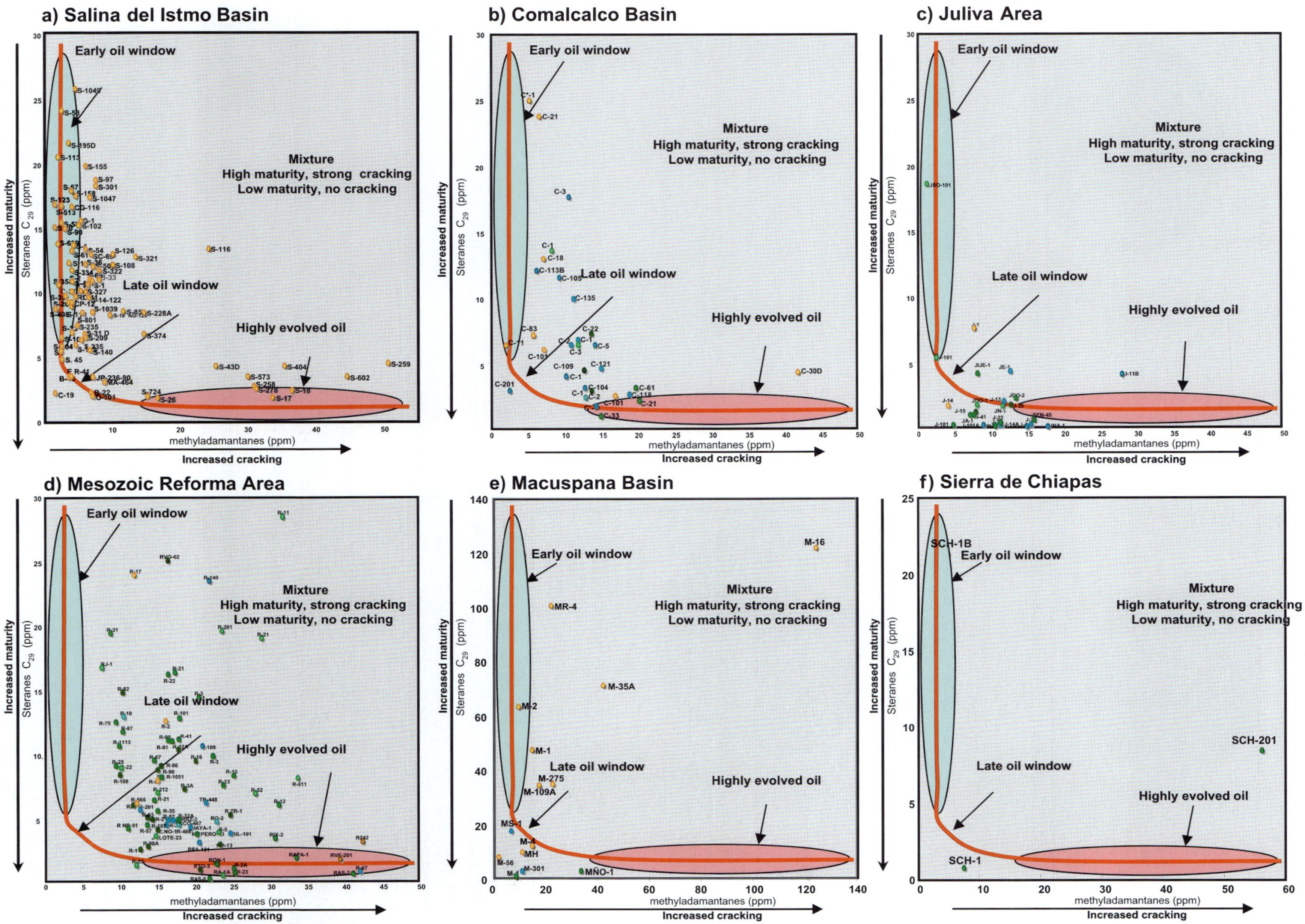

FIGURE 18. Plots of diamondoids by basin applied to highly evolved oils showing the actual stage of evolution at which oil was expelled from the source rock and showing the maturity of these fluids.

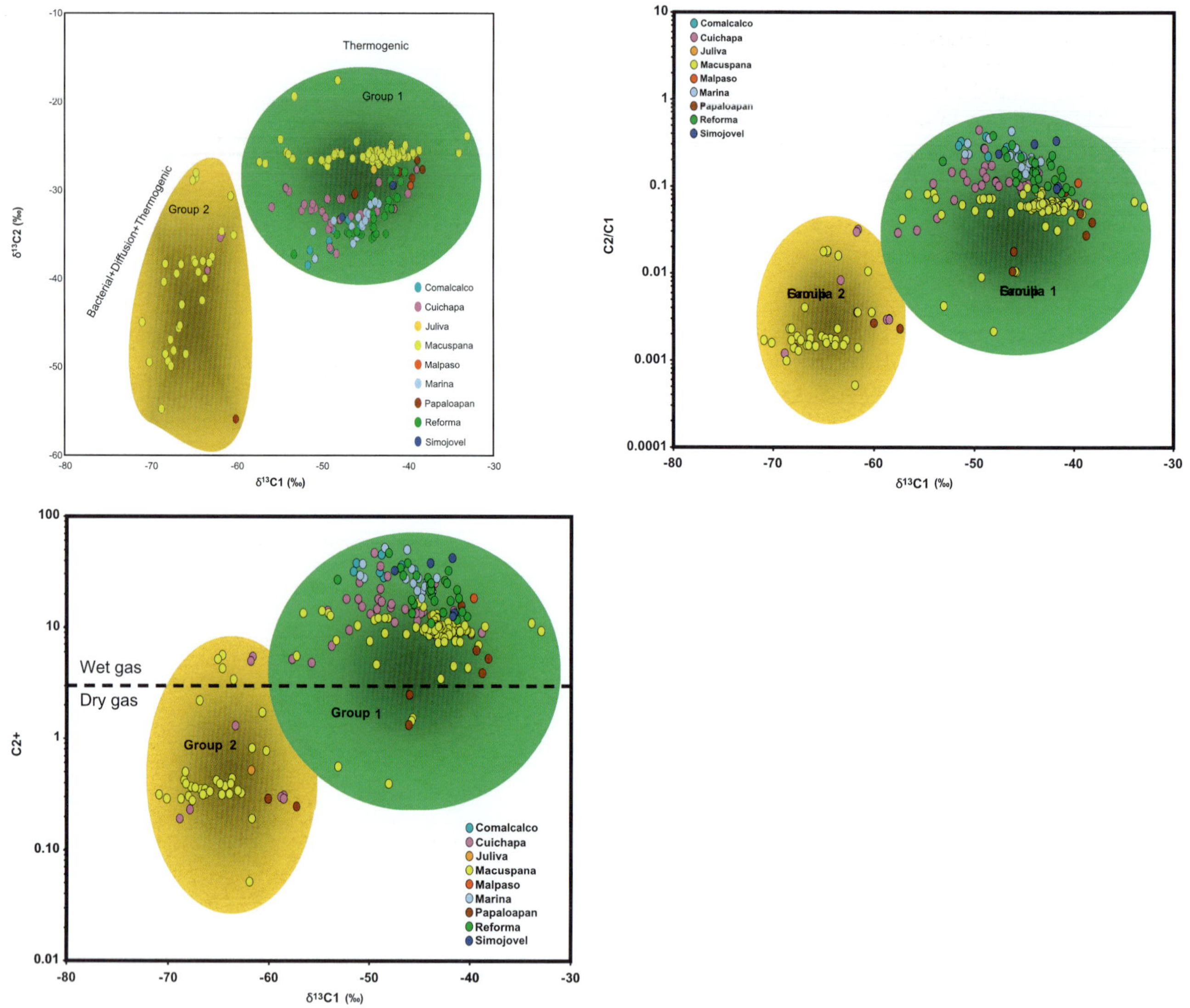

Figure 19. Plot of δ ^{13}C2 vs. δ ^{13}C1 and C2/C1 vs. δ ^{13}C1 differentiating the two gas families; C2+ vs. δ ^{13}C1 shows the two gas families and their relationship to wetness percentages in the southern region.

The C1/(C2 + C3) vs. δ ^{13}C1 plot (Bernard et al., 1978) shows the occurrence of biogenic gas, thermogenic gas, and mixed gas. Most of the gases can be seen in the thermogenic gas field, with a smaller grouping in the microbial gas and mixed gas space (Figure 20).

The presence of mixing and diffusion is distinguished on the C2/C1 vs. δ ^{13}C1 plot (Prinzhofer et al., 1999, 2000). The gases plot along a straight line on a linear scale and along a curved line on a semilogarithmic scale indicating a trend of diffusion during migration. A maturity trend is also observed with gases plotting above a curved line on a linear scale and a straight line on a semilogarithmic scale with methane isotopes increasing to more positive values (Figure 21). This is evidence that only Macuspana, Julivá, Cuichapa, and Papaloapan show significant changes in methane isotope ratios because of a strong isotopic fractionation during migration and because of the presence of microbial gas.

The existence of microbial gases, thermogenic gases with differing maturities, and strongly segregated thermogenic gases allows us to presume that although purely thermogenic and biogenic gases might exist, most gas accumulations in the southern region consist of mixtures.

Group 1 contains mixtures of thermogenic gases of different maturities as well as thermogenic gas > biogenic gas. These are the wettest gases in the southern region. This family also includes dry gas consisting of strongly segregated thermogenic gas mixed with small quantities of microbial gas (Figures 20, 21).

Group 2 is distinguished by biogenic gases mixed with strongly fractionated thermogenic gases that are the dryest in the southern region. To a lesser degree, this

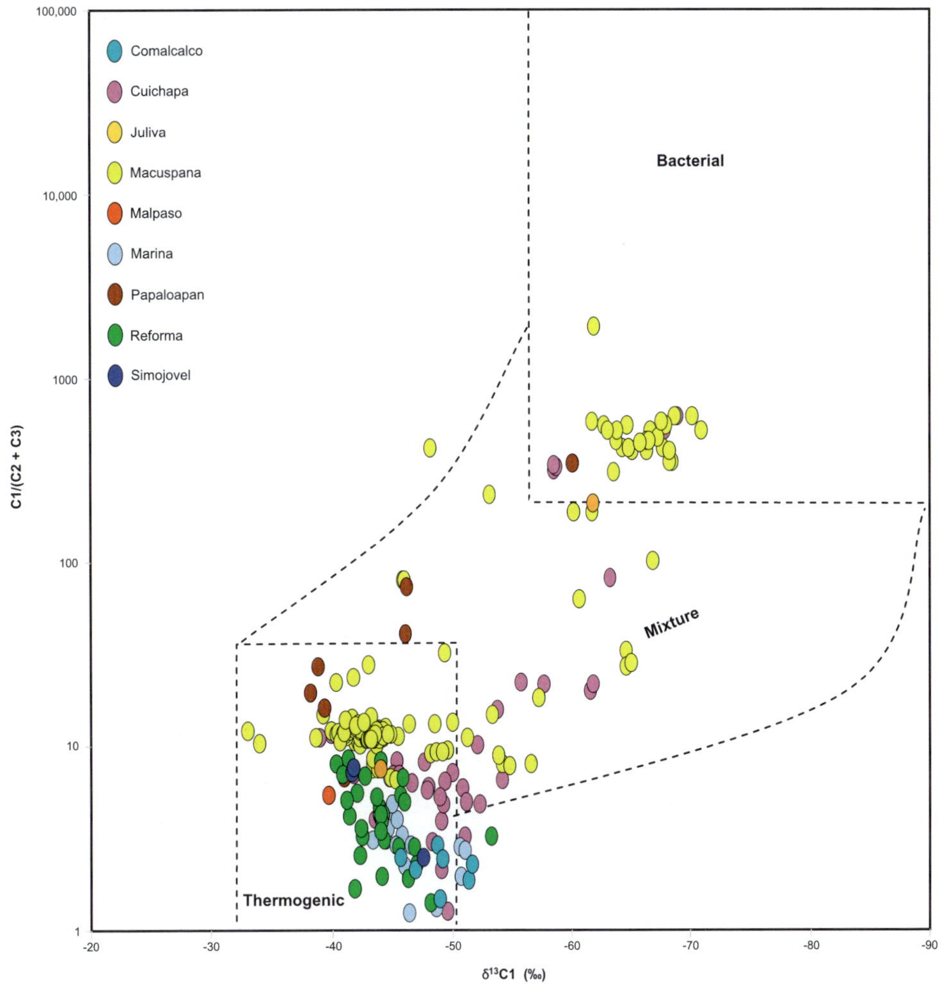

FIGURE 20. Diagram of C1/(C2 + C3) vs. δ^{13} C1 showing the presence of microbial gas, thermogenic gas, and a mixture of the two in the southern region.

family also includes mixtures of biogenic and thermogenic gases similar to those of group 1 (Figures 20, 21).

In calculating the mixture of thermogenic and biogenic gases (Chung et al., 1988) in the Macuspana Basin, group 1 (Figure 22a) low-maturity thermogenic gases are mixed with gas of microbial origin in different proportions; the thermogenic gases are associated with kerogen with an isotopic signature of −21‰ that correlates to oils of Tertiary origin (−22‰). Group 2 (Figure 22b) has two groups of thermogenic gases, one of low maturity closely related to group 1 and another of high maturity with strong isotopic fractionation associated with a deeper, possibly Jurassic, source. Both groups are admixed with a high proportion of microbial gas.

In the Papaloapan area (Figure 23), group 1 comprises two groups of gases, one of low maturity (δ^{13}C values of ethane and butane with a lower gradient) associated with kerogen with an isotopic signature of −25‰ that correlates to Jurassic oil in the area (−26‰); however, this low maturity may also be caused by mixing with gas from a younger source rock (Eocene?) whose isotopic signature is similar to that of the Jurassic as is the case in the Burgos Basin. The other group with high maturity is associated with a deeper, possibly Jurassic, source. A single sample of gas from family 2 showing strong isotopic fractionation is mixed with abundant microbial gas.

The Reforma-Simojovel area (Figure 24) contains gases of moderate to high maturity whose propane and butane δ^{13}C values occasionally cross over, indicating the probability that these gases are derived from two different sources (Jurassic and Cretaceous).

Group 1 gases in the Cuichapa area (Figure 25) are of moderate to low maturity and rarely of high maturity. Their propane and butane δ^{13}C values occasionally cross over implying that the gases associated with Jurassic-derived oils are occasionally mixed with gases from Tertiary-derived oil and comprise the only group of high-maturity gases in the marine region (Figure 26). Their propane and butane δ^{13}C values show two trends that occasionally cross over, indicating that these gases are probably derived from different sources (Tithonian and Oxfordian). This single group of high-maturity gases derived from the Jurassic (Tithonian) is distinguished in the Comalcalco Basin (Figure 27).

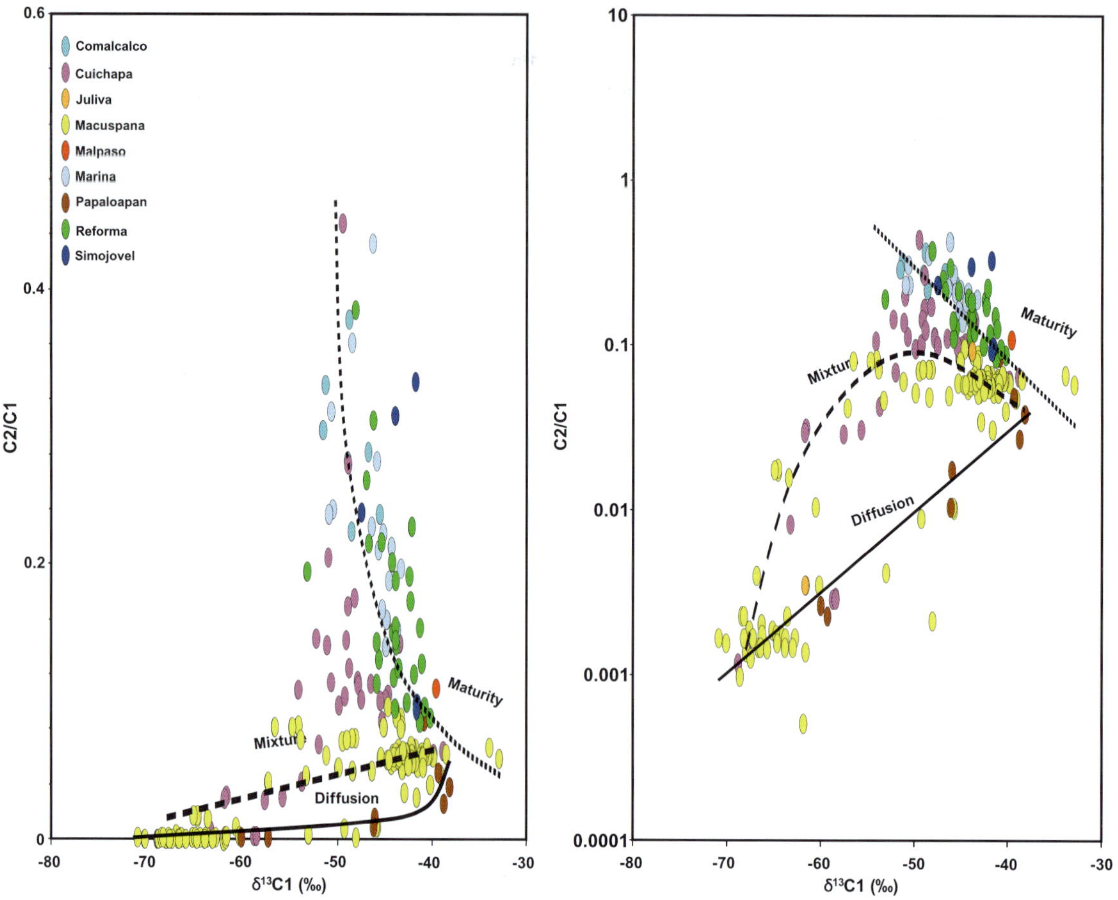

FIGURE 21. Correlation plots relating C2/C1 vs. $\delta^{13}C1$ in linear and logarithmic scales, indicating the mixture and diffusion of gases.

ORIGIN OF BIOGENIC GAS

Biogenic gases of the southern region are mixed with thermogenic gases in different proportions, and finding purely microbial gas is rare. All of the conditions necessary for the generation, migration, accumulation, and preservation of biogenic gas existed in the Neogene, principally in upper Miocene–lower Pliocene shales with sufficient immature type III organic matter (TOC of 0.5–3%) to generate microbial methane. Microbial methane is produced at depths with temperatures of 0–80°C; therefore, generation in Miocene–Pliocene sediments has continued to an approximate depth of 3000 m (9842 ft) depending on the thermal gradients in the study area. The most significant generation and accumulation occur at depths of 500–2500 m (1640–8202 ft) (Figure 28), although the content of microbial gas diminishes gradually as the depth increases. Preservation is mostly related to the degree of communication of the reservoirs with thermogenic gases produced from adjacent source rocks or migrated from greater depths.

ORIGIN AND MATURITY OF THERMOGENIC GASES

The behavior of thermogenic gases, as well as their genesis and relative maturities, may be portrayed on a plot of $\delta^{13}C2$-$\delta^{13}C3$ vs. C2/C3 (Prinzhofer and Pernaton, 1997) and the GASTAR diagram (Prinzhofer et al., 2000), which allows a comparison of the gases based on the average of 11 normalized geochemical parameters.

The $\delta^{13}C2$-$\delta^{13}C3$ vs. C2/C3 plot shows that most thermogenic gases of the southern and marine regions were expelled by primary cracking at a variable range of maturities in the oil window. These gases are considered to be derived from the four identified generating systems in the southern region. Those along the high-maturity trend are highly migrated and isotopically fractionated, possibly from the secondary cracking of oil (Figure 29). The GASTAR plot allows the evaluation of relative maturity indices, migration, mixing, and accumulation capacity for both gas groups of the southern region (Figure 30).

Group 1 gases of the Comalcalco, Reforma-Simojovel, Marina, and Cuichapa areas show very homogeneous

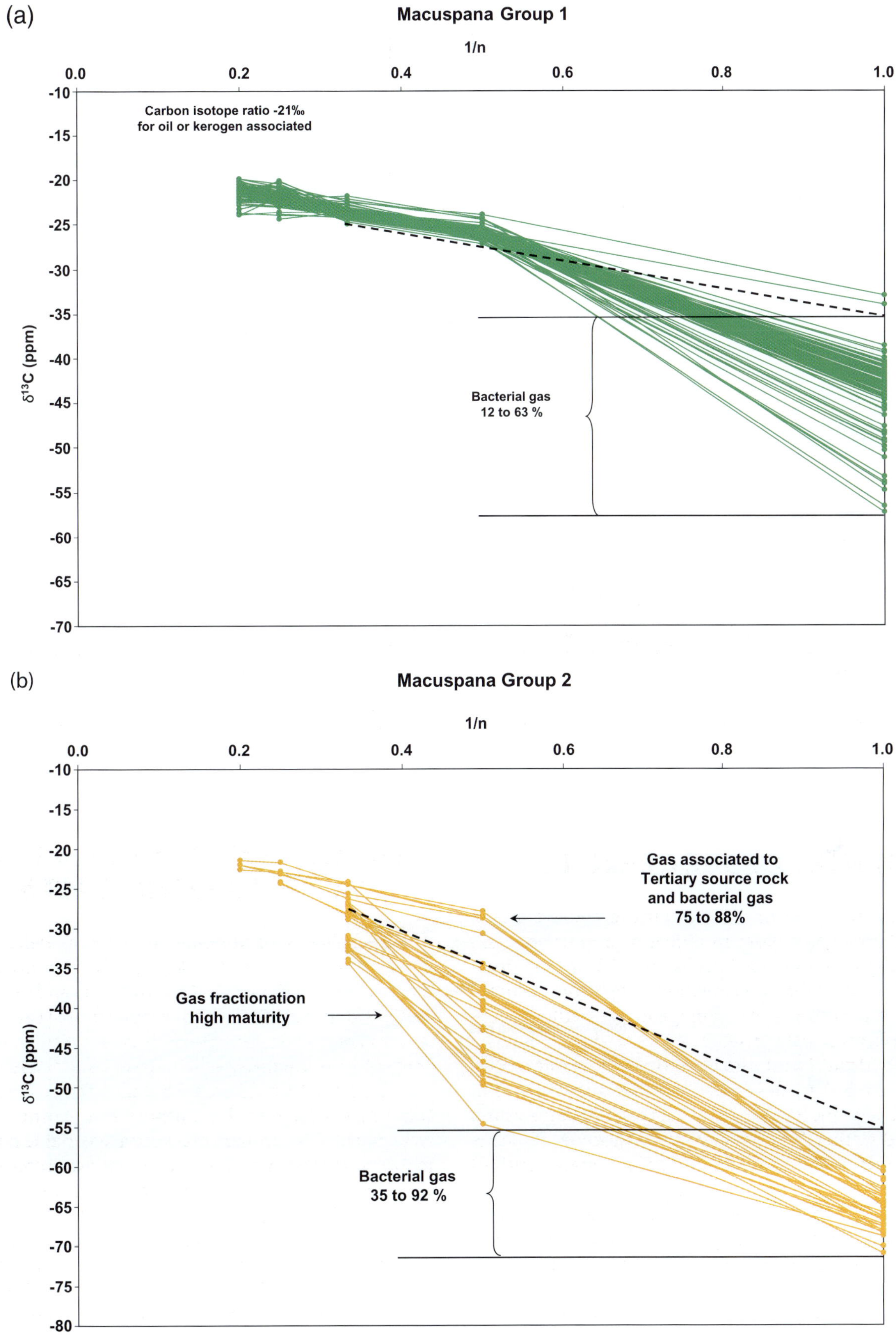

Figure 22. (a) Group 1 thermogenic gases mixed with microbial gases. (b) Thermogenic gases of different maturities showing strong isotopic fractionation.

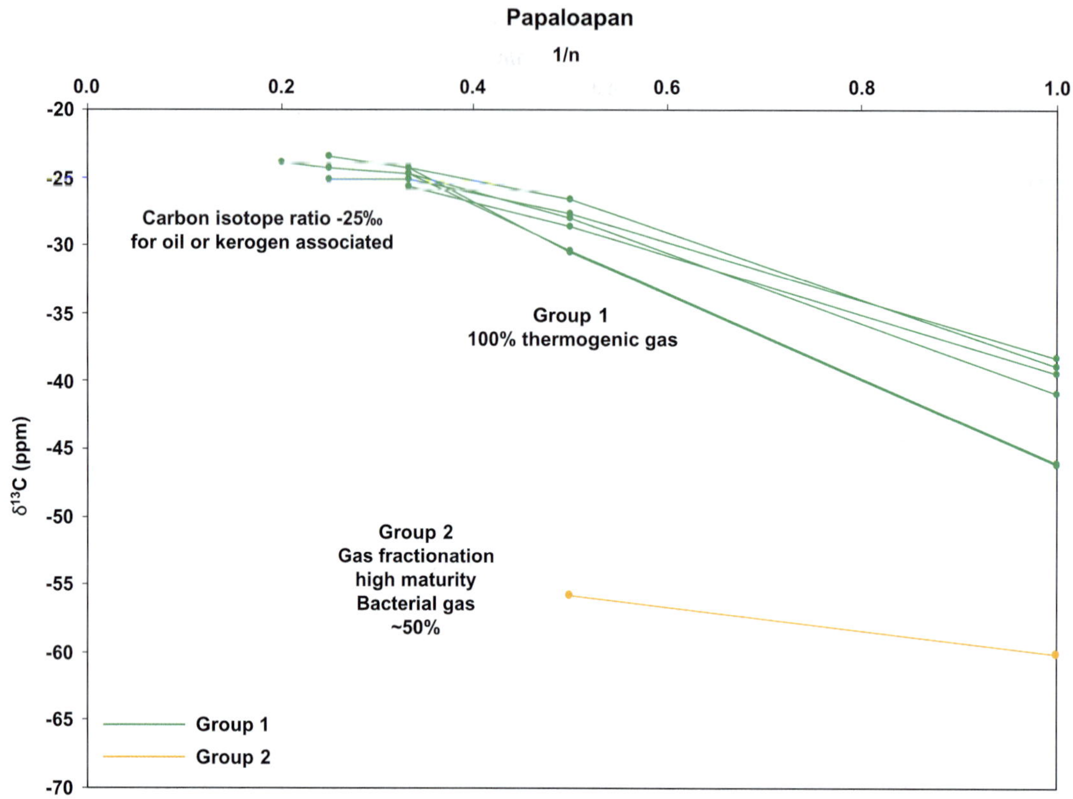

FIGURE 23. Group 2 thermogenic gases showing strong isotopic fractionation caused by high maturity.

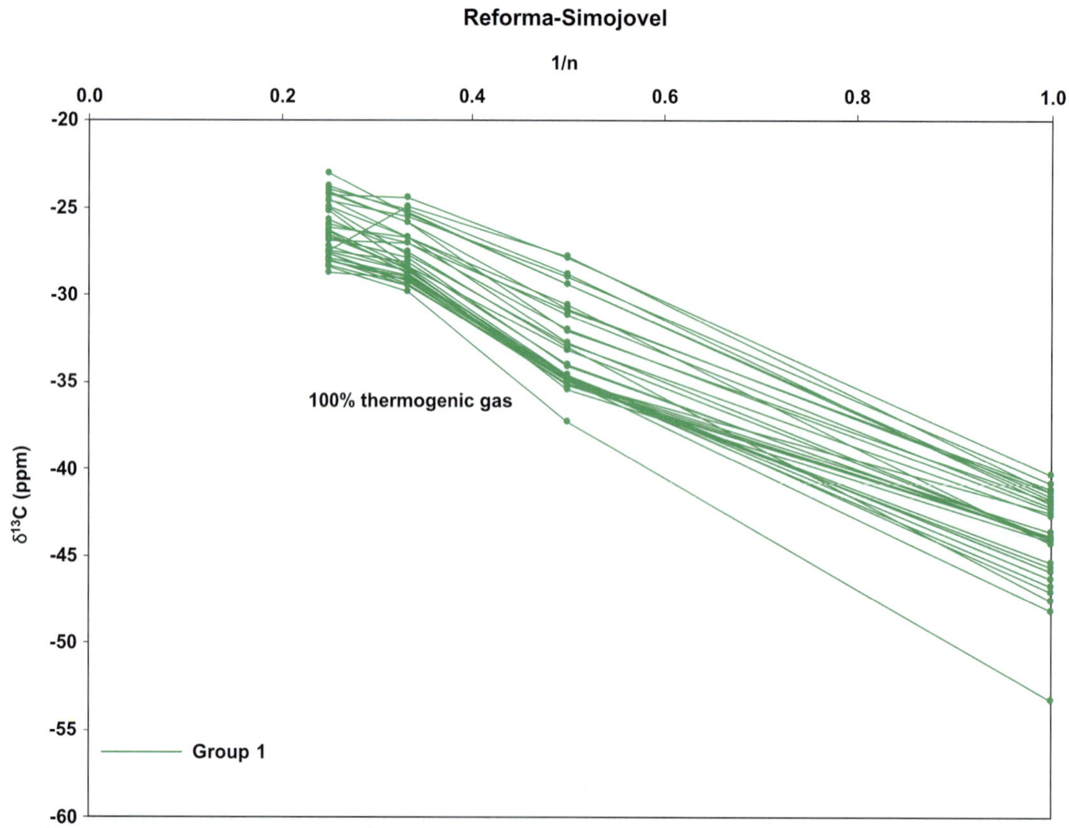

FIGURE 24. Thermogenic gases with a cross over of propane and butane, indicating two generative sources.

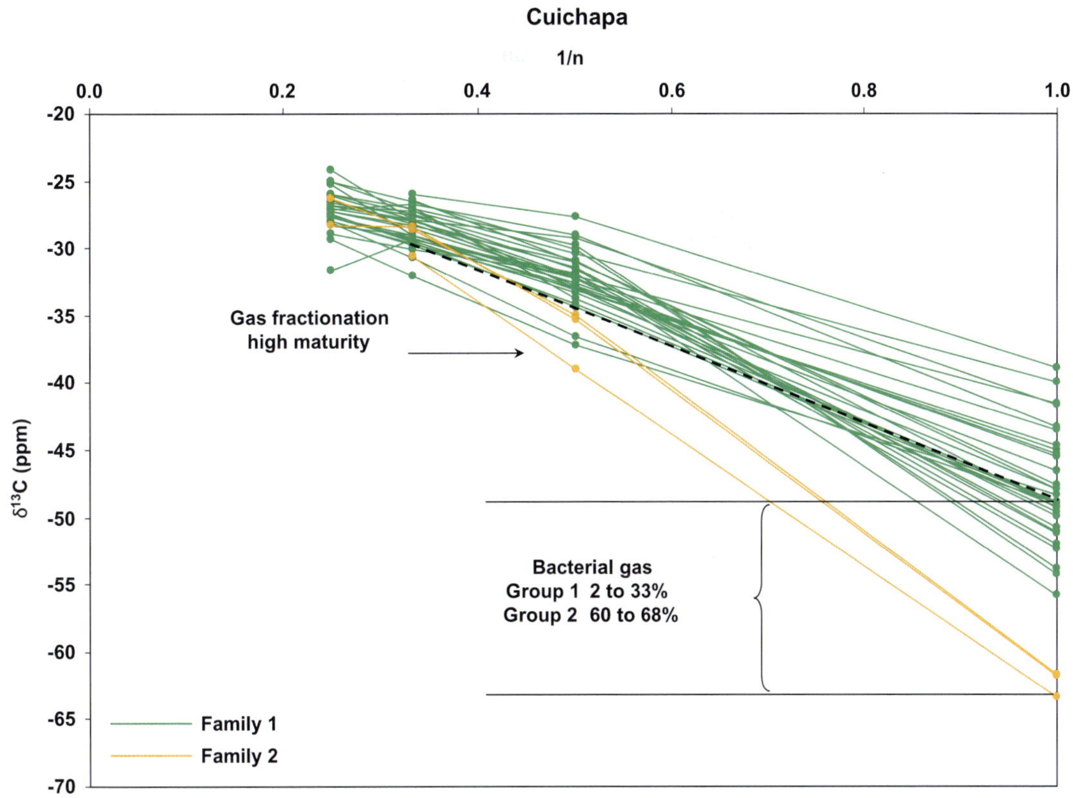

FIGURE 25. Thermogenic gases. Group 1 implies a Jurassic source with a Tertiary contribution; group 2 shows high maturity and isotopic fractionation.

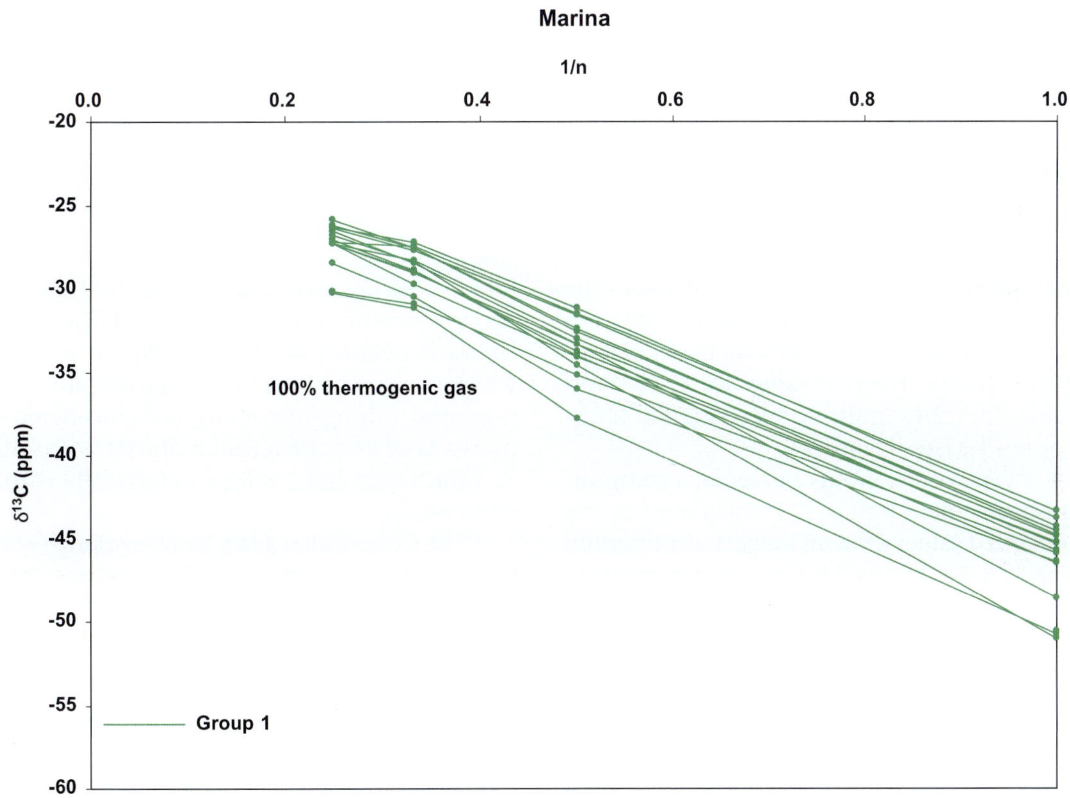

FIGURE 26. Thermogenic gases. The cross over indicates that two deep sources, such as the Tithonian and Oxfordian, contribute to the presence of these gases.

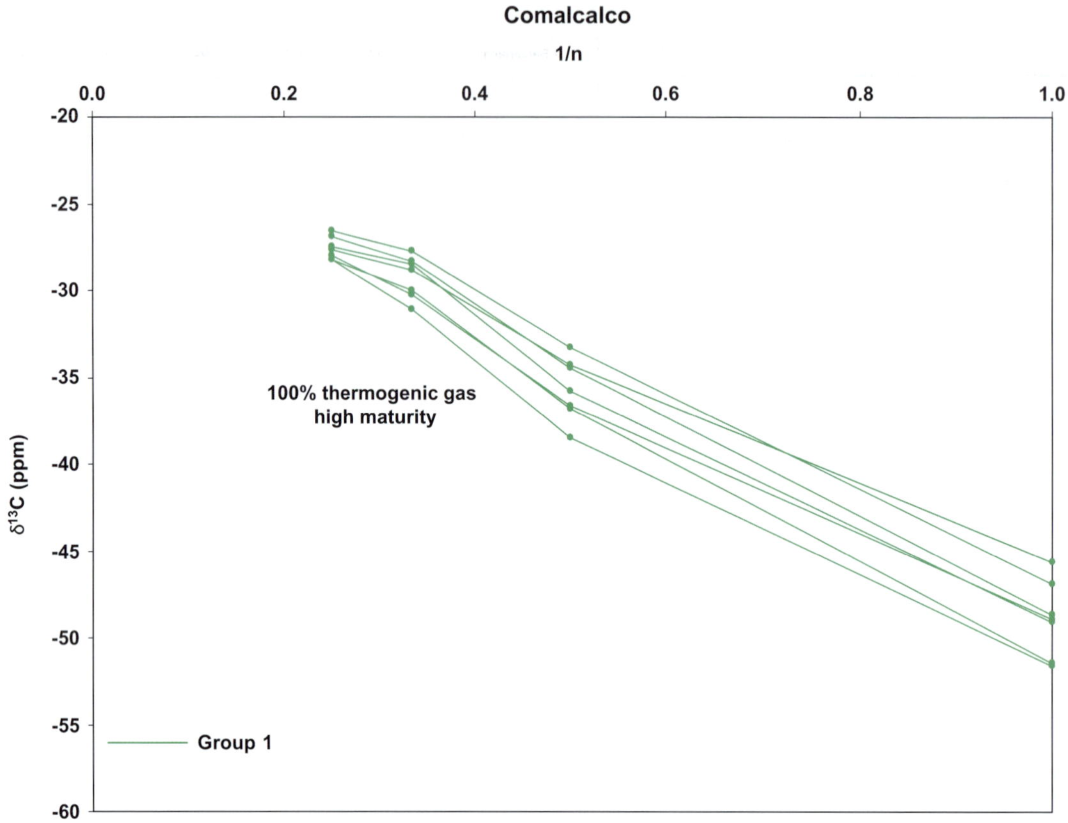

FIGURE 27. Thermogenic gases of Tithonian origin.

values of iC4/nC4 to $\delta\ ^{13}$nC4, indicting a single maturity range that is relatively less mature than Macuspana and Papaloapan gases, which are displaced toward more positive values. The $\delta\ ^{13}$C2-1 and C1/C2 parameters show signs of microbial gas mixing in Macuspana and Cuichapa. The values of $\delta\ ^{13}$n-iC4 and $\delta\ ^{13}$C3-C2 indicate a greater accumulation capability for Comalcalco, Reforma-Simojovel, and Marina gases, which lead to the observation that they may be more closely related to liquid hydrocarbons than the gases of Cuichapa, Macuspana, and Papaloapan. Positive values for $\delta\ ^{13}$C1 and negative values for C2/C3 indicate short migration distances for group 1 gases (Figure 30).

Because several of the group 2 gases lack compositional and isotopic data for C4, C3, and even C2, the average normalized values show an irregular distribution; however, the $\delta\ ^{13}$C2-C1 and C1/C2 parameters point to a mixture with microbial gas, which is notably more abundant in Macuspana than in Cuichapa and Papaloapan. Furthermore, strong tendencies toward negative values of $\delta\ ^{13}$C1 and positive values of C2/C3 reveal that all of the thermogenic gases of group 2 have migrated for long distances and therefore show no relationship to liquid hydrocarbons.

Maturities of the thermogenic gases (R_o equivalent) were obtained for each part of the study area from $\delta\ ^{13}$C3 vs. $\delta\ ^{13}$C2 plots (Berner et al., 1995) supported by the isotopic distributions of C1, C2, C3, and C4 obtained from the relationship $\delta\ ^{13}$C gas-$\delta\ ^{13}$C oil (Clayton, 1991). Trends for cogenetic pairs were constructed for type II kerogen (Figure 31) using isotopic values of −27 and −25 for gases of Jurassic origin, −23 for Cretaceous gases, and −21 for Tertiary gases (isotopic data for total oil from Guzman and Mello, 1999) observing that the low-maturity Tertiary gases of Macuspana are in the early peak oil window with R_o values of 0.7–1.0%, whereas gases of Jurassic origin, despite isotopic fractionation during migration, are considered to be the products of very advanced maturity with interpolated R_o values reaching 1.5%; i.e., practically within the gas window.

The Comalcalco gases show evidence of migration because they are displaced −25 from the kerogen trend; however, by interpolating the data in the opposite direction from migration, they fall within the high maturity range at approximately 1.0–1.1% R_o, precisely at the end of oil generation.

The Reforma-Simojovel gases are slightly displaced from the corresponding trends; however, by interpolating in the opposite direction from migration, one can estimate that gases derived from the Tithonian are highly mature, with an approximate R_o of 1.1–1.2%, whereas

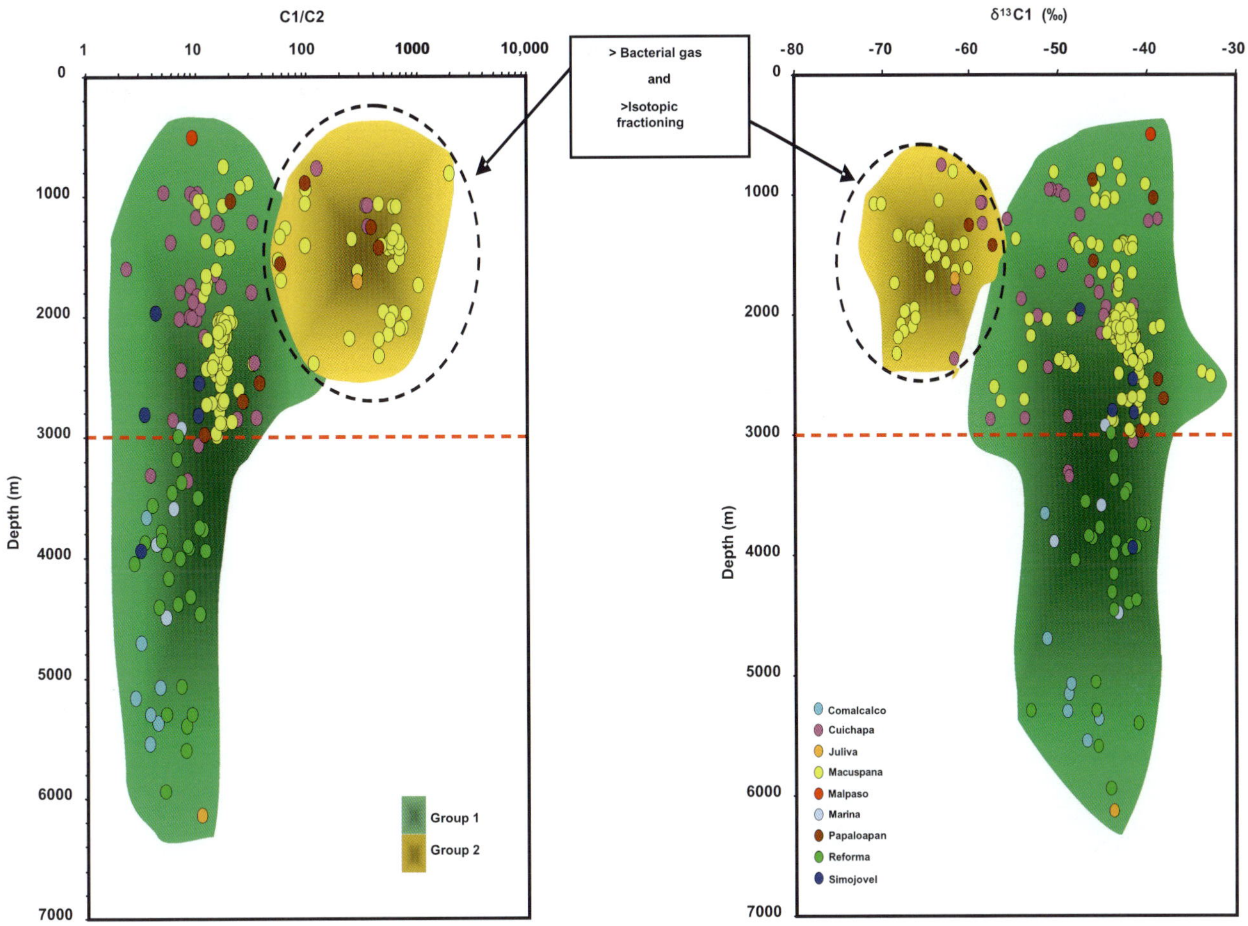

FIGURE 28. Plot of the parameters C1/C2 and δ ^{13}C 1 vs. depth. The two gas families are indicated, as well as the limit of microbial gas.

those of the Cretaceous are in the peak oil generation window with R_o from 0.9 to 1.0%.

Gases of both groups 1 and 2 in the Saline Basin (Cuichapa) correlate to oils with Jurassic affinity (δ ^{13}C − 25), including one gas sample from Malpaso; therefore, they are derived from a Tithonian source rock. In Figure 31, some gases from group 1 are observed to have moderate maturity with R_o of 0.9–1.0% (peak oil generation stage) and kerogen at −23. Others are slightly displaced and, upon interpolation in the opposite direction from migration, are at the end of the oil-generating window with approximate R_o values of 1.1–1.2%. Segregated gases of group 2 were possibly expelled within the gas window ($R_o > 1.3\%$).

Geographically (Figure 32), group 1 thermogenic gases occur throughout the southeastern basins. Group 2 dry gases, although more restricted in their distribution, occupy the largest area in the Macuspana Basin starting from a narrow belt in the southeast that widens considerably to the north. To a lesser degree, group 2 gases are also found north and southwest of Cuichapa, north of Julivá, and northwest of Papaloapan.

Ethane (δ ^{13}C2) and propane (δ ^{13}C3) isotope distributions indicate migrating thermogenic gases. Highly segregated gases related to deeper source rocks are mixed with large quantities of microbial gas, mainly in the Macuspana and Papaloapan areas.

CONCLUSIONS

Identification of the source rocks of this sedimentary basin is based on organic richness, quality, distribution, and maturity obtained from pyrolysis studies (TOC, S_1, S_2, T_{max}, HI, OI, PI, etc.) and from optical studies of outcrop, cuttings, and core samples. Additionally, physical-chemical properties of fluids were incorporated in this investigation.

The Upper Jurassic (Oxfordian) is identified principally in Campeche Sound and recently in the onshore Palancares area. Source rock to oil correlation indicates that these are laminated carbonates from a restricted nearshore marine environment. These source rocks charge

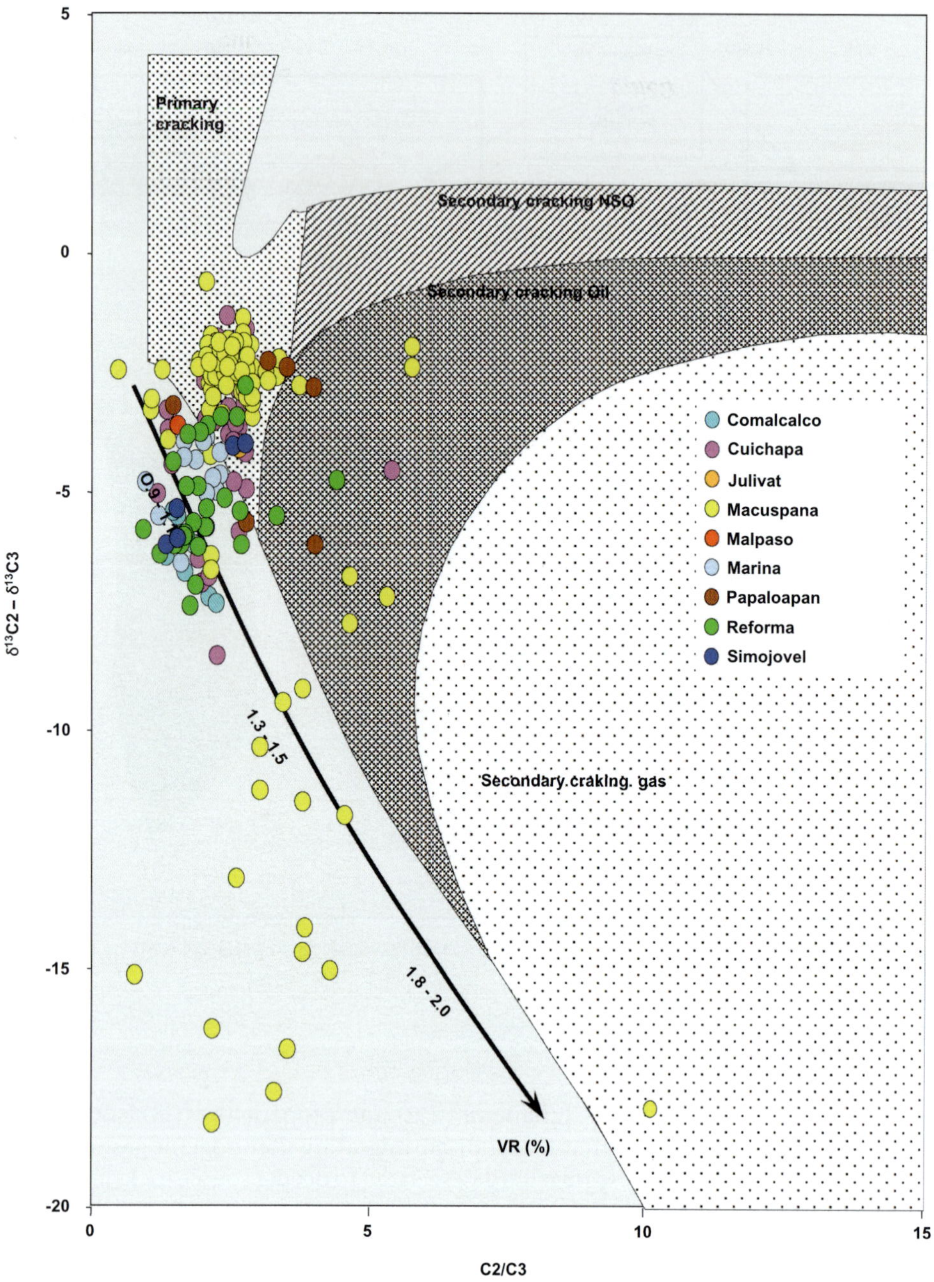

FIGURE 29. Plot of $\delta^{13}C2$-$\delta^{13}C3$ vs. C2/C3. Gases with primary cracking and others in the southern region may be distinguished along a trend of very high maturity. VR = vitrinite reflectance.

the Oxfordian and Kimmeridgian reservoirs with 20–35° API oils.

The Upper Jurassic (Tithonian) has the greatest regional importance. The Upper Jurassic (Tithonian) is rich in type II organic matter with TOC values of 4–15%, where immature, and with a residual richness of 1–3% TOC. The average net generating thickness is 250 m (820 ft). Two large subfamilies are present: the JS1 family related to shale and the JS family related to an argillaceous mudstone. This system has the capacity to charge plays in the Kimmeridgian, Tithonian, Cretaceous, Eocene, Miocene, and Pliocene with fluids ranging in density from 10° to 58° API.

The Lower Cretaceous carbonate-evaporite sequence has poor to fair generating potential. The presence of type II kerogen prone to oil and gas generation is related to bacterial and algal-amorphous organic matter. The differences in salinity suggest the presence of highly evaporitic sabkha or restricted lagoon environments, or a possible high-salinity incipient platform.

The Lower Miocene has provided very few organic richness data points; however, those that exist indicate good organic content with fair to very good generating potential. The presence of oleanane indicates an important contribution from higher plants mixed with algal-amorphous matter in shales of this age. These shales

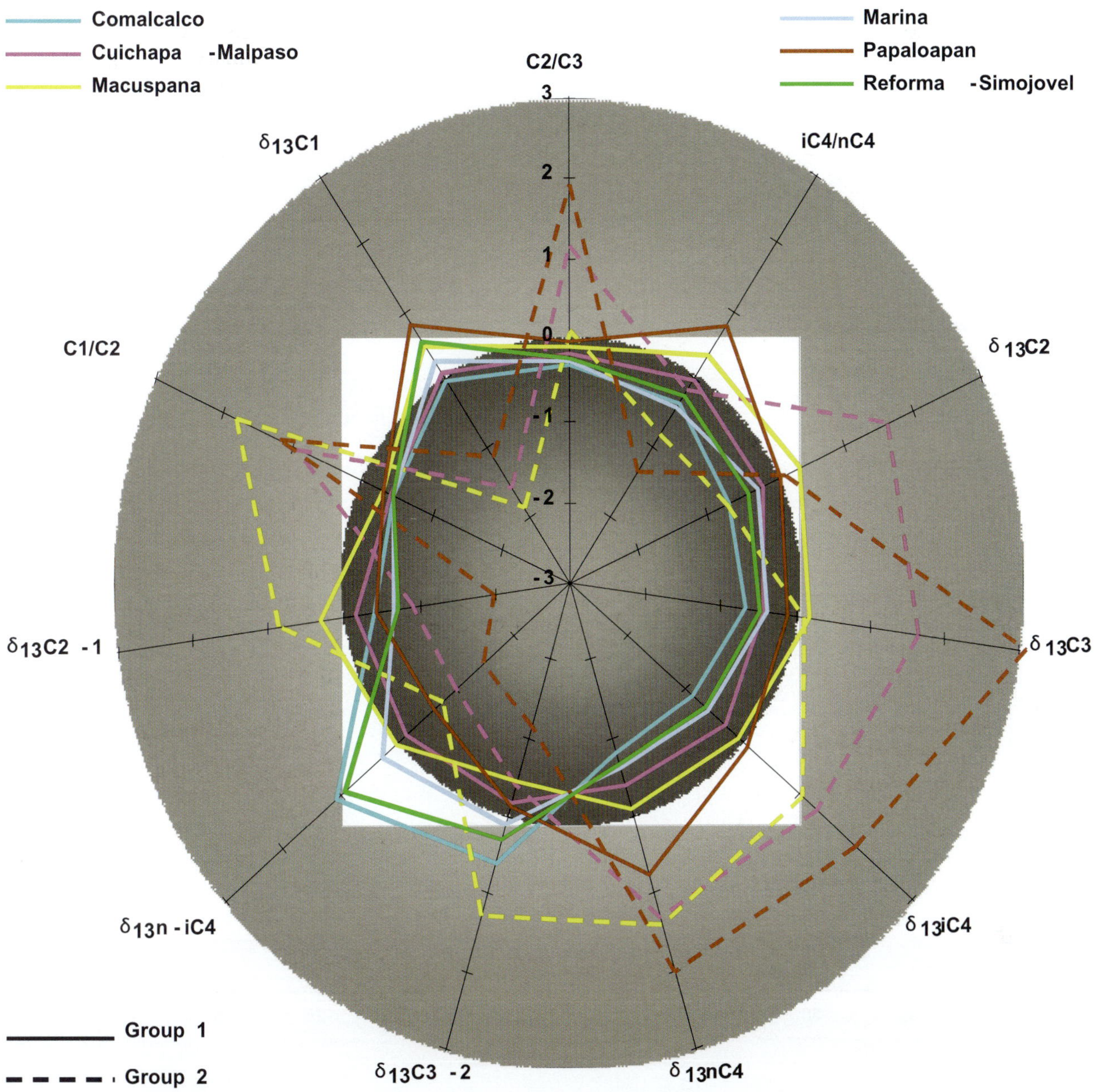

FIGURE 30. The GASTAR diagram showing the average of 11 normalized parameters for gases in the southern region.

principally produce condensates, wet gas, and dry gas from siliciclastic reservoirs of middle to upper Miocene and Pliocene age.

Analytical results of diamondoids and isotopes of specific compounds prove that many of the oils are hydrocarbon mixtures, either from one source with different migration pulses (Tithonian + Tithonian); Jurassic mixtures, e.g., Oxfordian + Tithonian (probably in the marine area); mixtures from two very different sources, e.g., Tithonian + Lower Cretaceous, for some of the oils of the Artesa-Mundo Nuevo Platform; oil mixtures from Tithonian + lower Miocene sources, in the case of the Macuspana Basin; or Tithonian + Eocene mixtures toward the Veracruz Basin in the Otapan area.

Gases of the southern and marine regions are divided into two large groups; each with characteristic isotopic signatures and molecular compounds. Group 1 basically contains wet gas of thermogenic origin with heavier isotopes and low percentages of methane. Group 1 is distributed throughout the study area. In contrast, group 2 mainly includes dry gas of microbial origin, lighter isotopes, and high concentrations of methane. Group 2 is present only in the Macuspana, Cuichapa, and Papaloapan areas. Mixed gases were also identified

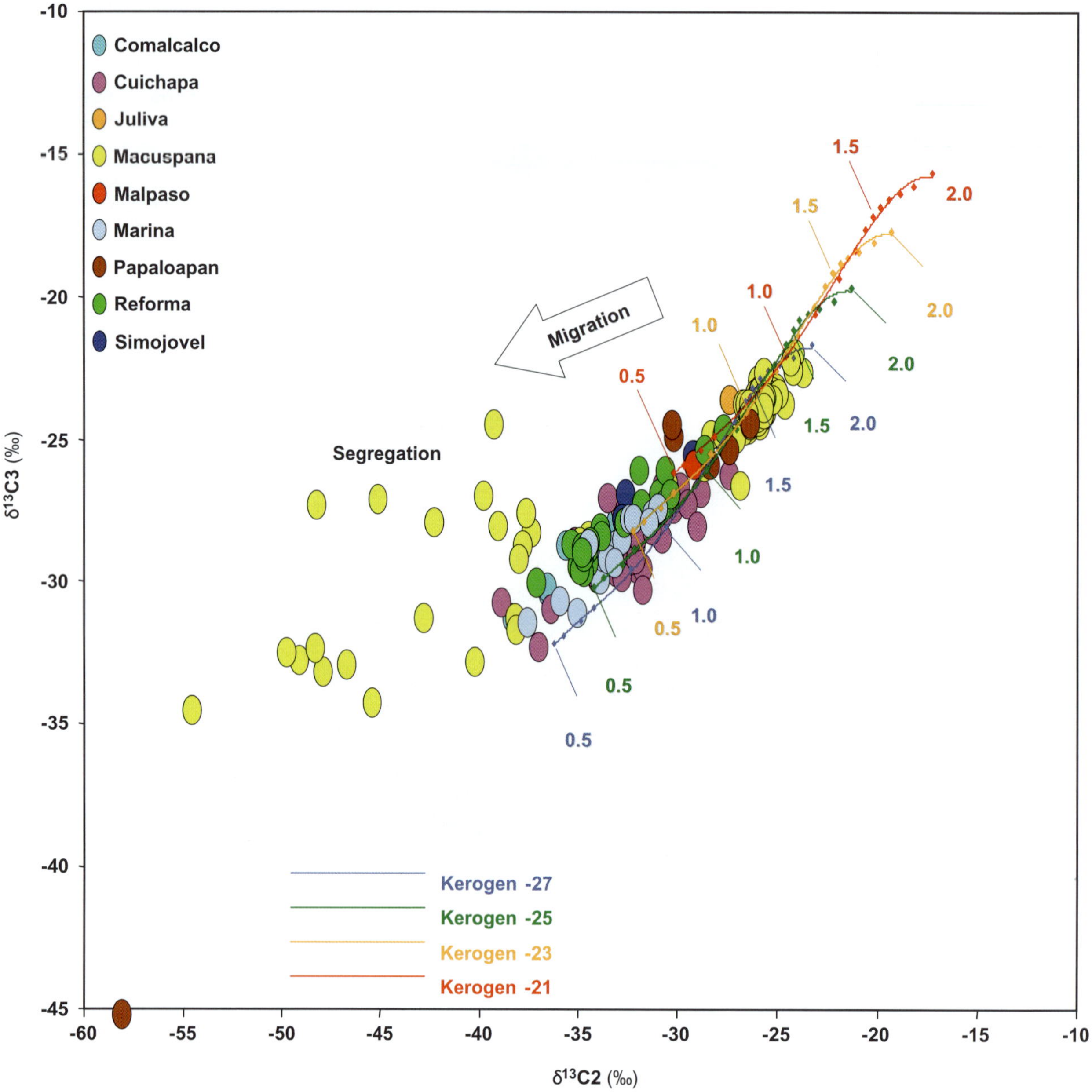

FIGURE 31. Plot of δ ^{13}C3 vs. δ ^{13}C2 showing trends for cogenetic pairs to estimate the maturity (R_o equivalent) of gases in the Macuspana Basin.

from both groups, indicating isotopic fractionation or very fractionated.

Although purely thermogenic or biogenic gases may exist, most accumulations in the southern region consist of mixtures. Group 1 contains mixtures of thermogenic gases with variable maturity ranges that are associated with minor proportions of microbial gas and thermogenic gas with isotopic fractionation in the Macuspana, Cuichapa, and Papaloapan areas. Group 2 is characterized by a high percentage of biogenic gas mixed with highly fractionated thermogenic gas.

Conditions that favor the generation, migration, accumulation, and preservation of biogenic gas exist principally in sediments of upper Miocene to lower Pliocene age at shallow depths ranging from 500 to 2500 m (1640 to 8202 ft) at temperatures no higher than 80°C.

Most of the thermogenic gases were expelled by primary cracking at variable maturities within the oil window.

Group 2 thermogenic gases of the Macuspana Basin are derived from a Tertiary source rocks in the initial phase of oil generation (between 0.7 and 1.0% R_o) and are mixed with a low percentage of microbial gas. Thermogenic gases of group 1 are generated from a highly mature Tithonian source rock, practically in the gas

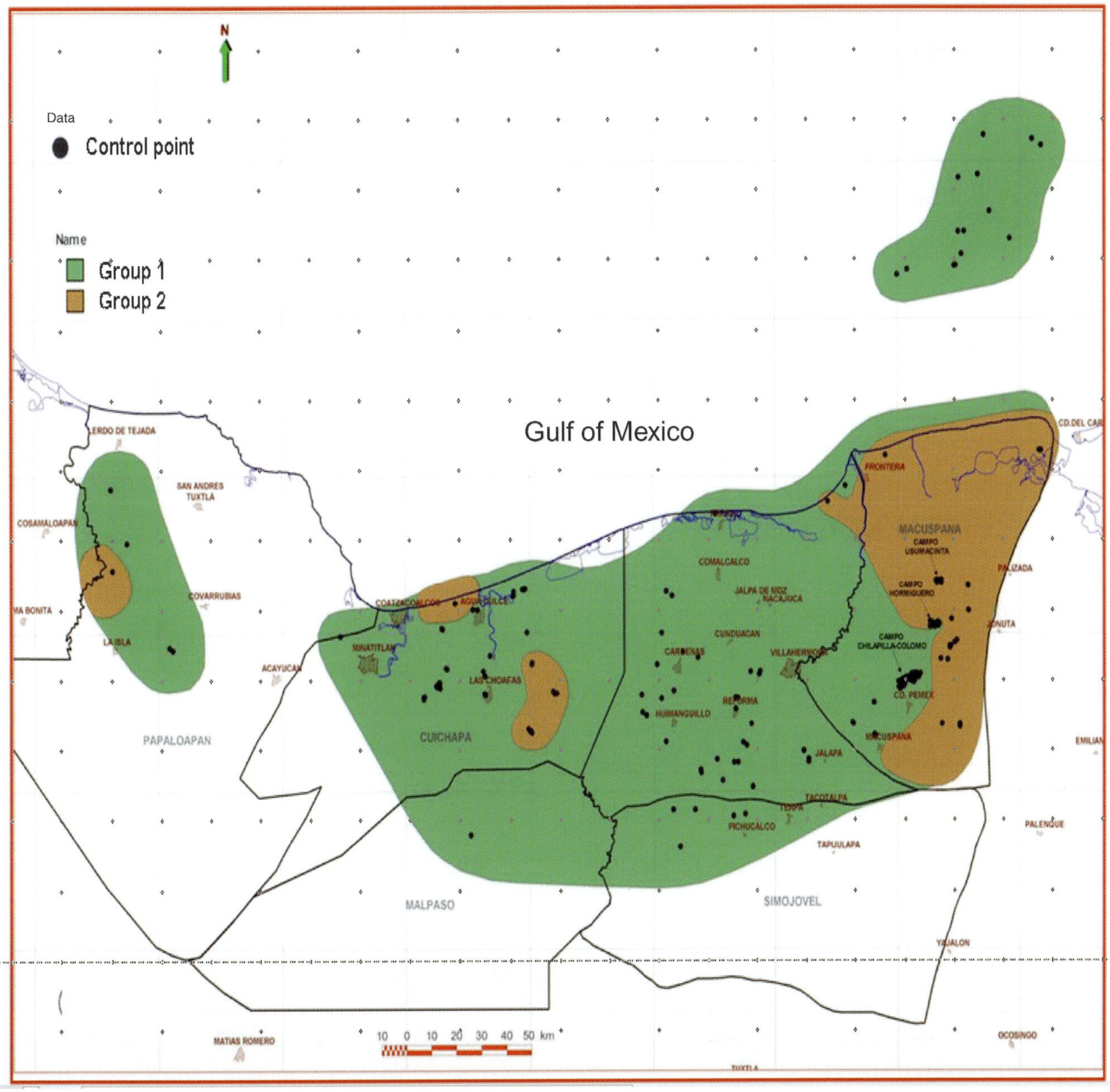

FIGURE 32. Location map of gas families in the southern region.

window (>1.3% R_o), showing strong isotopic fractionation and mixed with high percentages of microbial gas.

Thermogenic gases of the Comalcalco Basin are associated with oil and correspond to group 1. They were generated from a Tithonian source rock at the end of the oil generation (1.0–1.1% R_o).

The thermogenic gases of Reforma-Simojovel belong to group 1 and are derived from different source rocks; one of Tithonian age with maturity at the end of oil generation (1.1–1.2% R_o) and the other of Lower Cretaceous age at peak oil generation (0.9–1.0% R_o); both are associated with oils.

All of the thermogenic gases at Cuichapa-Malpaso (groups 1 and 2) are of Tithonian origin and were generated between the peak oil and final oil generation stages (0.9–1.1% R_o).

Thermogenic gases of the marine region belong to group 1. They are associated with oils of Tithonian and, very possibly, Oxfordian age. They are of low maturity having been generated during the initial stage of the oil window (0.5–0.8% R_o).

Group 1 in the Papaloapan area contains moderately fractionated thermogenic gases related to a Lower Cretaceous source with maturities ranging from peak oil to the end of oil generation (0.9–1.2% R_o). The thermogenic gases of group 2 were expelled in the gas window and are strongly fractionated with high percentages of microbial gas.

The thermogenic gases of group 1 have varied migration histories. Those of the central and southwestern Macuspana Basin and central and southern Cuichapa have little migration; those of the marine region have little to moderate migration; and those of the Comalcalco Basin, the northern and northwestern parts of Cuichapa, and the Reforma and Simojovel areas have had moderate to extensive migration. All the gases in group 2 show evidence of having migrated extensively.

ACKNOWLEDGMENTS

We are grateful to Robert Katz for the time spent reviewing this manuscript. Katz's suggestions and comments were very constructive and helpful.

REFERENCES CITED

Bernard, B. B., J. M. Brooks, and W. M. Sackett, 1978, Light hydrocarbons in recent Texas continental shelf and slope sediments: Journal of Geophysical Research, v. 83, p. 4053–4061.

Berner, U., E. Faber, G. Scheeder, and D. Panten, 1995, Primary cracking of algal and landplant kerogens: Kinetic models of isotope variations in methane, ethane and propane: Chemical Geology, v. 126, no. 3 and 4, p. 233–246.

Caballero, E., A. F. Callejon, N. Holguin, and L. Elrod, 1998, Geochemical correlation of natural gases from Macuspana Basin, Mexico: Petróleos Mexicanos (PEMEX)-Houston Advanced Research Center report, 40 p.

Chung, H. M., J. R. Gormly, and R. M. Squires, 1988, Origin of gaseous hydrocarbons in subsurface environments: Theoretical considerations of carbon isotope distribution: Chemical Geology, v. 71, p. 97–103.

Clara, V. M. L., 1985, Interpretación geoquímica del Proy, Malpaso: Petróleos Mexicanos (PEMEX), Inedito, 91 p.

Clayton, C., 1991, Carbon isotope fractionation during natural gas generation from kerogen: Marine and Petroleum Geology, v. 8, p. 232–240.

Connan, J. J., 1986, The microbial input in carbonate-anhydrite facies of a sabkha paleoenvironments from Guatemala, a molecular approach: Advances in geochemistry 1985: Part I: Organic Geochemistry, v. 10, p. 29–50.

Guzman, M. A., and M. R. Mello, 1994, The habitat of petroleum of Mexican southeastern basins: A biological marker approach: Unpublished report, Instituto Mexicano del Petróleo y Petróleo de Brasil, 65 p.

Guzman, M. A., and M. R. Mello, 1999, Origin of oil in the Southeast Basin: AAPG Bulletin, v. 83, p. 1068–1095.

Guzmán, V. M., V. L. Clara, and M. L. Medrano, 1995, Estudio de biomarcadores: Parte II. El hábitat de petróleo en las Cuencas del Sureste, un enfoque con biomarcadores: Proy. CAB-501 Instituto Mexicano del Petróleo-Petróleos Mexicanos (PEMEX), Inèdito, 50 p.

Guzmán, V. M., L. Castro, R. J. Roman, V. L. Clara, M. L. Medrano, C. E. Vázquez, and G. Ziga, 2001, El origen del petróleo en las subprovincias mexicanas del Golfo de México: Boletin Asociacion Mexicana de Geologos Petroleros, v. 49, no. 1 and 2, p. 31–36.

Holguín, Q. N., 1985, Evaluación geoquímica del Sureste de México: Boletin Asociacion Mexicana de Geologos Petroleros, v. 37, no. 1, 25 p.

Lorant, F., A. Prinzhofer, F. Behar, and A. Y. Huc, 1998, Carbon isotope and molecular constraints on the formation and the expulsion of thermogenic hydrocarbon gases: Chemical Geology, v. 147, p. 249–264.

Mello, M. R., and J. M. Moldowan, 2006, Caracterización de los Sistemas Petrolíferos de México basado en Análisis Geoquímicos de Alta Resolución High Resolution geochemistry technology de rocas, Aceites, Condensados y Gases: Petróleos Mexicanos (PEMEX)-High Resolution Technology, Informe Final, Inédito, 297 p.

Peters, K., 1986, Guidelines for evaluating petroleum source rock using programmed pyrolysis: AAPG Bulletin, v. 70, p. 318–329.

Peters, K. E., and J. M. Moldowan, 1993, The biomarker guide: Interpreting molecular fossils in petroleum and ancient sediments: Prentice Hall, Englewood Cliffs, New Jersey, 363 p.

Prinzhofer, A., and E. Pernaton, 1997, Isotopically light methane in natural gases: Bacterial imprint or segregative migration?: Chemical Geology, v. 142, p. 193–200.

Prinzhofer, A., M. A. Guzman-Vega, A. Battani, and M. Escudero, 1999, Gas geochemistry of the Macuspana Basin (Mexico): Thermogenic accumulations in sediments impregnated by bacterial gas: Europe Annual Organic Geochemistry, Report, 12 p.

Prinzhofer, A., M. R. Mello, and T. Takaki, 2000, Geochemical characterization of natural gas: A physical multivariable approach and its applications in maturity and migration estimates: AAPG Bulletin, v. 84, p. 1152–1172.

Schoell, M., 1983, Genetic characterization of natural gases: AAPG Bulletin, v. 67, no. 12, p. 2225–2238.

Sosa, A., 2000, Caracterización Genética de los gases de la Cuenca de Macuspana: Petróleos Mexicanos (PEMEX), Inédito, 90 p.

Sosa, P. A., 1994, Caracterización geoquímica de la secuencia calcáreo-evaporitica y aceites del Campo Nazareth Chiapas: México Tesis de maestría, Inédito, 60 p.

Sosa, P. A., and V. L. Clara, 2001, Subsistemas generadores en el sureste de México: Asociacion Mexicana de Geologos Petroleros, v. 49, no. 1 and 2, p. 31–36.

17

Sánchez Rios, M. A., M. L. Ayala Nieto, J. C. González Lara, A. del Valle Reyes, J. Sánchez-Durán, P. Padilla Avila, D. García Urbano, C. Pérez Castillo, B. A. Reyna Téllez, A. Reyes Vázquez, P. A. Fuentes Franco, J. Rico Pérez, L. Aguirre Meza, and G. Quintanilla Rodríguez, 2009, High-resolution biostratigraphy in Miocene–Pleistocene sediments from two wells in the western Gulf of Mexico, *in* C. Bartolini and J. R. Román Ramos, eds., Petroleum systems in the southern Gulf of Mexico: AAPG Memoir 90, p. 369–395.

High-Resolution Biostratigraphy in Miocene–Pleistocene Sediments from Two Wells in the Western Gulf of Mexico

María Antonieta Sánchez Rios, Mónica L. Ayala Nieto, Julio C. González Lara, Aarón del Valle Reyes, Janett Sánchez-Durán, Patricia Padilla Avila, Daniel García Urbano, Cristina Pérez Castillo, Bruno A. Reyna Téllez, Antonio Reyes Vázquez, and Paula A. Fuentes Franco

Instituto Mexicano del Petróleo, Col. San Bartolo Atepehuacan, México,

Juan Rico Pérez, Lidia Aguirre Meza, and Guillermo Quintanilla Rodríguez

Petróleos Mexicanos, Edificio Aguila (Area de Paleontología), Mexico

ABSTRACT

An integrated high-resolution biostratigraphic study was conducted using calcareous nannofossils, planktonic and benthic foraminifers, and palynomorphs from two wells in the western Gulf of Mexico. Sediments range in age from early Pleistocene to middle Miocene.

The early Pleistocene is represented by nannoflora of biozone NN19 and microfauna from the *Globorotalia truncatulinoides truncatulinoides* biozones and by the dinocyst species *Spiniferites delicatus*, *Spiniferites membranaceous*, *Multispinula quanta*, and *Hystrichokolpoma rigaudiae*. These sediments were deposited under inner to middle neritic bathymetric conditions.

The Pliocene was characterized by nannofossil biozones NN18, NN17, NN16, and NN15, and the NN14–NN12 interval; foraminifera biozones of *Globorotalia miocenica*, *Globorotalia tosaensis tosaensis*, *Globorotalia margaritae*, and the dinoflagellate cyst species; and *Polysphaeridium zoharyi*, *Lingulodinium machaerophorum*, *Melitasphaeridium choanophorum*, and *Operculodinium crassum*. Paleobathymetry was inner neritic-upper bathyal.

Late Miocene sediments produced microfossils indicating nannofossil biozones NN11, NN10, NN9, and the *Globorotalia humerosa* and *Globorotalia acostaensis* planktonic foraminiferal biozones. Dinoflagellate cysts found in these sediments were *Spiniferites mirabilis*, *Dapsilidinium pastielsii*, *Sumatradinium hispidum*, and *Achomosphaera andalusiensis*. These sediments were deposited under outer neritic-upper bathyal conditions.

DOI:10.1306/13191092M903338

Middle Miocene strata are represented by calcareous nannofossil biozones NN9, NN8, NN6, and NN5, and by the planktonic foraminiferal biozones *Globorotalia mayeri*, *Globigerinoides rubber*, *Globorotalia fohsi lobata-Globorotalia fohsi robusta*, *Globorotalia fohsi fohsi*, and probably by the upper part of the *Globorotalia fohsiperipheroronda* biozone. These strata were deposited mainly in outer neritic-upper bathyal environments.

Nannofossil diversity and abundance show notable variations throughout the two wells. A strong predominance of calcareous nannofossils and marine palynomorphs is present in outer neritic settings, and a planktonic foraminifer increase is observed mainly in the upper bathyal of well A. Continental palynomorphs are abundant in inner neritic-middle neritic settings, and a reduction is observed in the outer neritic-upper bathyal in well B.

INTRODUCTION

High-resolution biostratigraphy (HRB) is a new methodology applied in the petroleum industry in Mexico. In this study, HRB is defined as the integrated study of different microfossil groups, such as calcareous nannoplankton, planktonic and benthic foraminifera, and marine and continental palynomorphs. Bioevents are determined by species diversity, microfossil abundance, and their highest occurrence (last appearance datum or LAD). Ages and paleobathymetry of the sediments are determined. This study is mainly focused on hydrocarbon exploration.

The HRB has been applied with great success, mainly in Neogene sediments, in which planktonic and benthic foraminifera may be scarce or have a wide biostratigraphic range. For this reason, calcareous nannofossils and palynomorphs are used to complete the Mexican Neogene column. Biostratigraphic data are also calibrated with well logs.

GEOLOGICAL SETTING

Location Area

Two wells (wells A and B) are chosen for this study. They are located in the offshore Burgos Basin, western part of the Gulf of Mexico (Figure 1). In the Burgos Basin, the sedimentary fill contains Upper Jurassic to Holocene sediments. The Mesozoic is characterized by carbonates, clastic rocks, and evaporites, whereas the Tertiary sedimentary column consists of a Paleocene to Holocene predominantly siliciclastic sequence (mainly shale and sandstone) (Echánove, 1986).

Tertiary sedimentary rocks in this region have a general north-northwest–south-southeast orientation. Oldest deposits in the basin are located in the west and younger strata are toward the east, where the section is significantly thicker. The lithologic sequence is represented by alternating shale and sandstone recording mainly regressive deposits, with short transgressive periods, deposited in environments that varied from shallow marine to bathyal (Santiago et al., 1984).

METHOLODOLOGY AND STUDY TECHNIQUES

The methodology used in this HRB study consists of a thorough identification of calcareous nannofossils, planktonic and benthic foraminifera, and palynomorphs in each of the 600 cuttings at 20-m (66-ft) intervals and core samples. In addition, abundance and diversity counts of microfauna and microflora were calculated for subsequent analysis, and the ratios for organic to inorganic material (OM/IM) in the sediments were determined. Biostratigraphic, diversity, and abundance data of the different microfossil groups were calibrated with the lithologic data and gamma-ray logs from each well.

Calcareous nannofossil taxonomy and zonation references can be found in Perch-Nielsen (1985), Aubry (1984–1987), and Bown (1999).

Biostratigraphic ranges of continental and marine palynomorph are taken from Germeraad et al. (1968), Lorente (1986), Kalgutkar and Jansonius (2000), Drugg and Stover (1975), Williams and Bujak (1985), Williams et al. (1993), Wrenn et al. (1986), Matsuoka and Bujak (1988), Powell (1992), and Head and Wrenn (1992). The marine palynologic index (MPI) was calculated in accordance with the criteria proposed by Helenes et al. (1998) and Helenes and Somoza (1999).

The planktonic foraminifera references used in the study are Banner and Blow (1965), Postuma (1971), Stainforth et al. (1975), Bolli and Saunders (1985), Lipps (1979, 1993), González-Lara (1994, 2000), Berggren et al. (1995a), and Berggren et al. (1995b). For the benthic foraminifera, the depth criteria proposed by Morkhoven et al. (1986), Robles (1967), and Sansores and Flores-Covarrubias (1972) were followed. These models have been adjusted to the depositional environments proposed for Mexico.

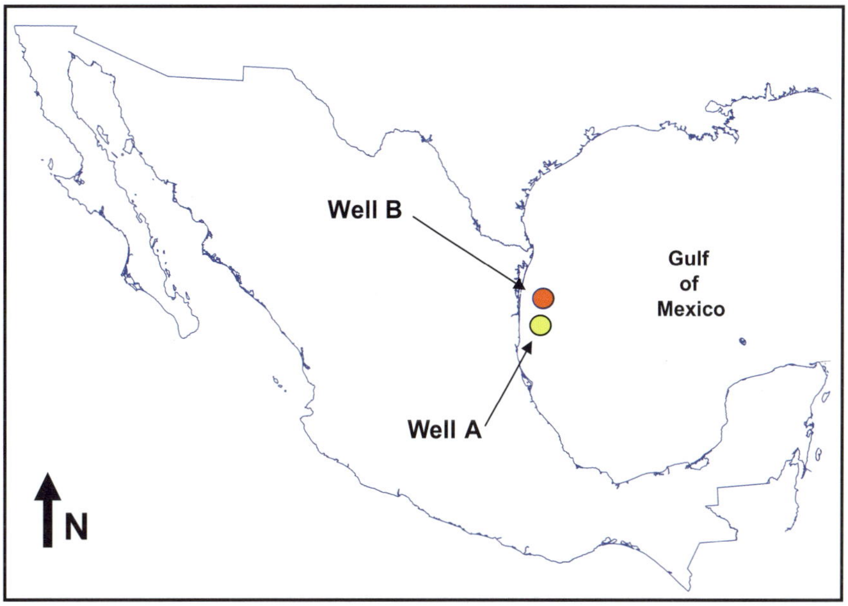

FIGURE 1. Location map.

The following sections are the techniques applied for each of the microfossil groups.

Calcareous Nannofossils

Standard nannofossil smear slide techniques are used to prepare samples. Unprocessed sediment is smeared on a glass coverslip, dried, and then mounted on a glass slide using Norland optical adhesive 61 mounting media. The nannofossil biostratigraphy presented here is based on the examination of each sample; quantitative and qualitative abundance and diversity are recorded from 60 random traverses.

Foraminifera

Foraminifera sample processing is done by the classic technique of washing a 100-g sample of sediment. The washing technique has to be done with abundant flowing water and detergent. After the washing step, the sediment collected is dried. Quantitative and qualitative analysis is made dividing the sample and counting the species diversity and abundance. When the foraminifera are very abundant, the samples are divided in two parts using the Otto fractionator.

Palynomorphs

Chemical processing for palynological material uses 60 g of sediment. Samples are prepared following the standard treatment proposed by Wood et al. (1996), which consists of consecutive acids (HCl and HF) digestion of the samples and flotation of the residue in a heavy liquid (sodium polytungstate).

Quantitative and qualitative palynomorph analysis is made by counting all the species and specimens observed on the smear slide of each sample. The MPI is calculated using the formula:

$$\mathrm{MPI} = \left(\frac{\text{Marine diversity}}{\text{Continental diversity} + 1}\right)100$$

The MPI values allow us to determine the depositional environment of the sediment. Low values indicate shallow environments, and high values indicate marine environments.

The HRB, as mentioned before, consists of the analysis of different microfossil groups; a thorough systematic study; and the integration of taxonomic data, microfossil abundance, and species diversity. In addition, biostratigraphic checklists of microfossil analysis are created, and bioevents are recorded. The HRB studies done in Mexico so far include Sánchez-Rios et al. (2000, 2001, 2002a, b, 2003, 2004, 2005, 2006).

The HRB results from wells A and B are presented as follows.

The results of the present study on HRB are shown on Figures 2 and 3. Identified species are depicted in Plates I–XIII (located at the end of this chapter prior to the references). Plates I–IV show nannofossils, Plates V–VIII show planktonic and benthic foraminifers, and Plates IX–XIII show palynomorphs. Consensus ages of the two wells range from Pleistocene to middle Miocene age (Figures 4, 5). Because the samples are mainly drill cuttings, the age descriptions are shown on measured depths.

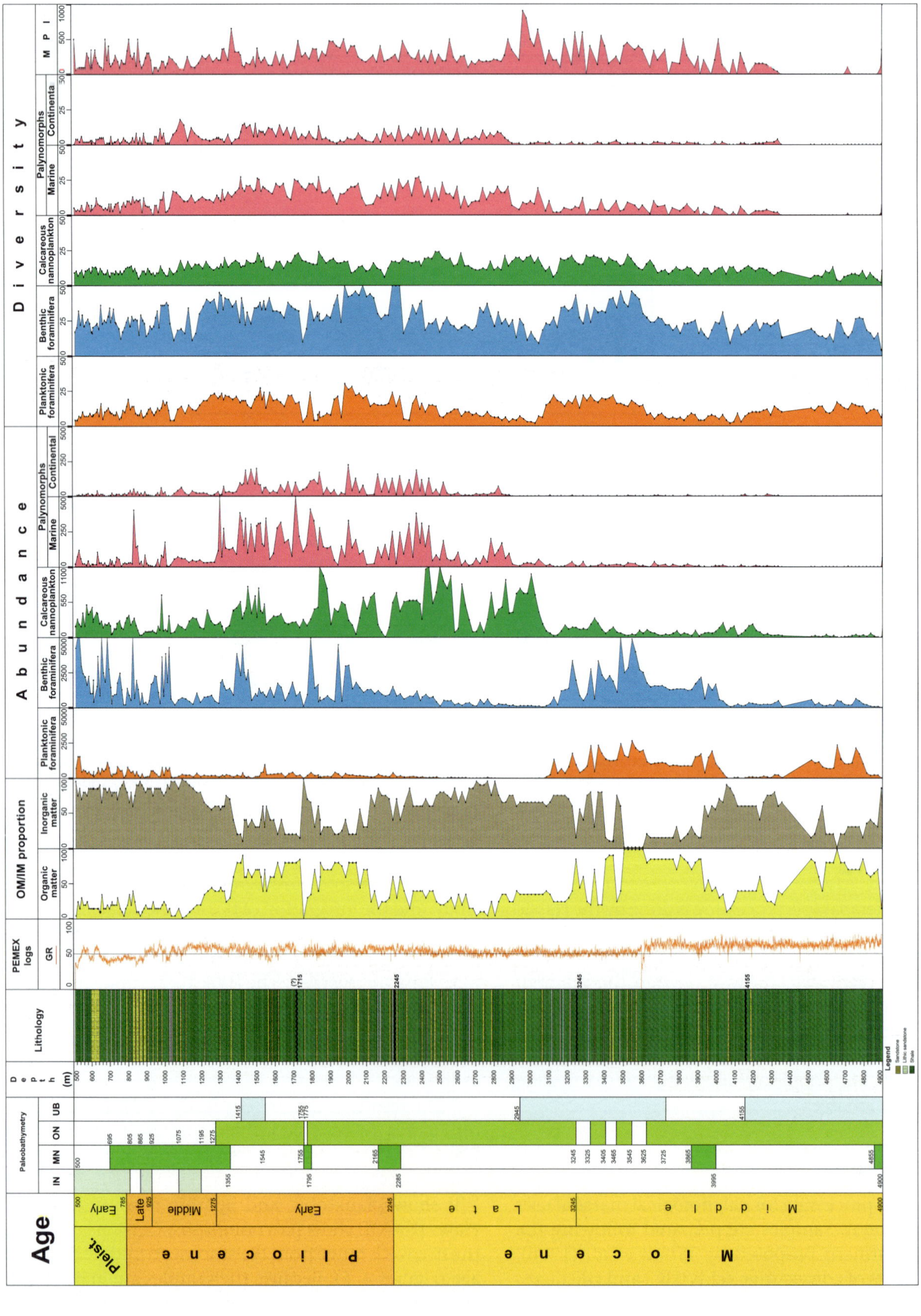

FIGURE 2. Diversity and abundance in well A. The calcareous nannoplankton and dinoflagellate cyst abundance in the outer neritic zone is remarkable; planktonic foraminifera show an increase in the upper bathyal. GR = gamma ray; OM/IM = organic to inorganic material; MPI = marine palynologic index; IN = inner neritic; MN = middle neritic; ON = outer neritic; UB = upper bathyal.

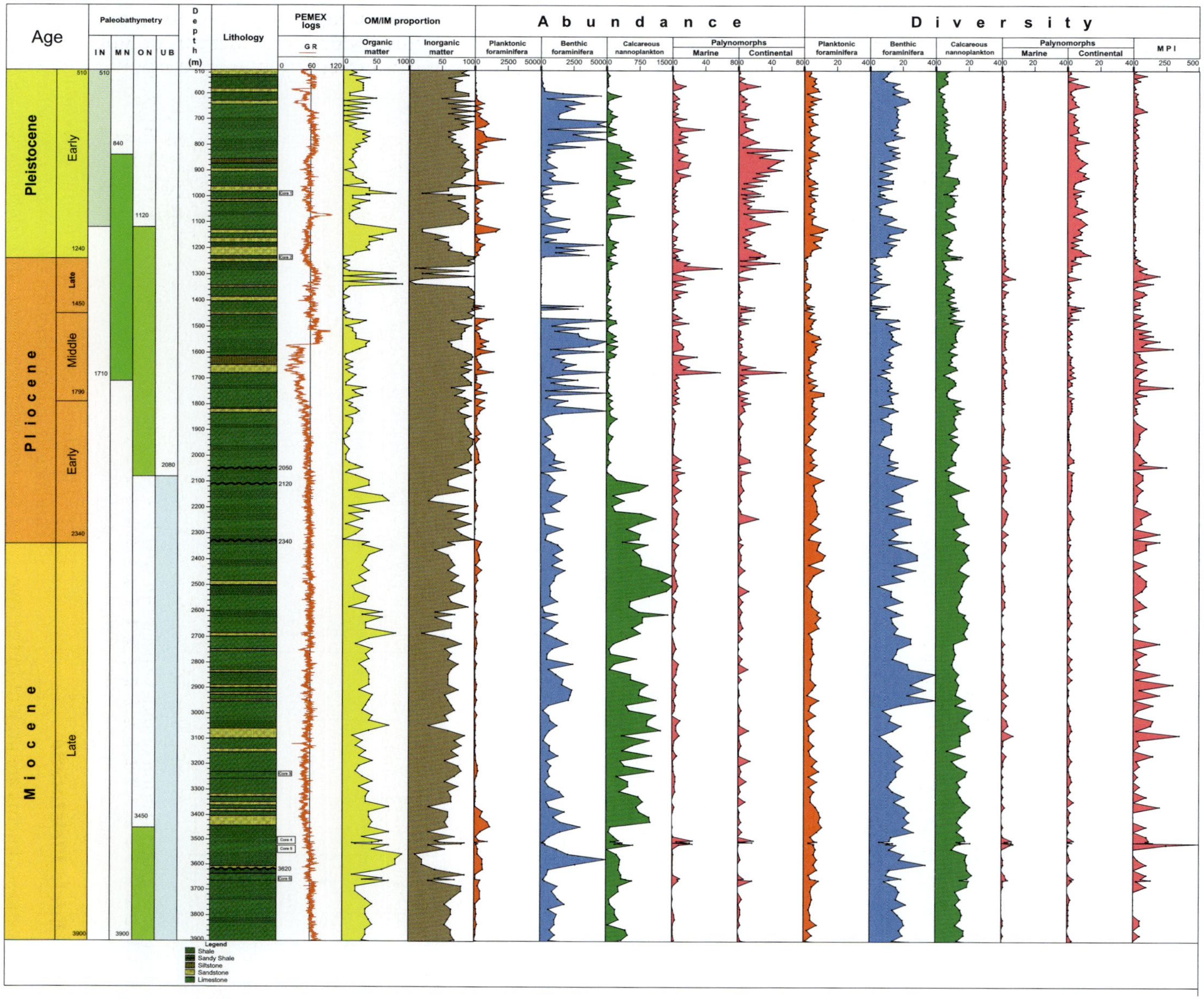

Figure 3. Diversity and abundance of microfossils in well B. The parameters of abundance and diversity show important variations throughout the entire column. The predominance of pollen and spores in shallow environments and their decrease in deep environments are clearly shown. Calcareous nannofossils are abundant in upper bathyal environments. GR = gamma ray; OM/IM = organic to inorganic material; MPI = marine palynologic index; IN = inner neritic; MN = middle neritic; ON = outer neritic; UB = upper bathyal.

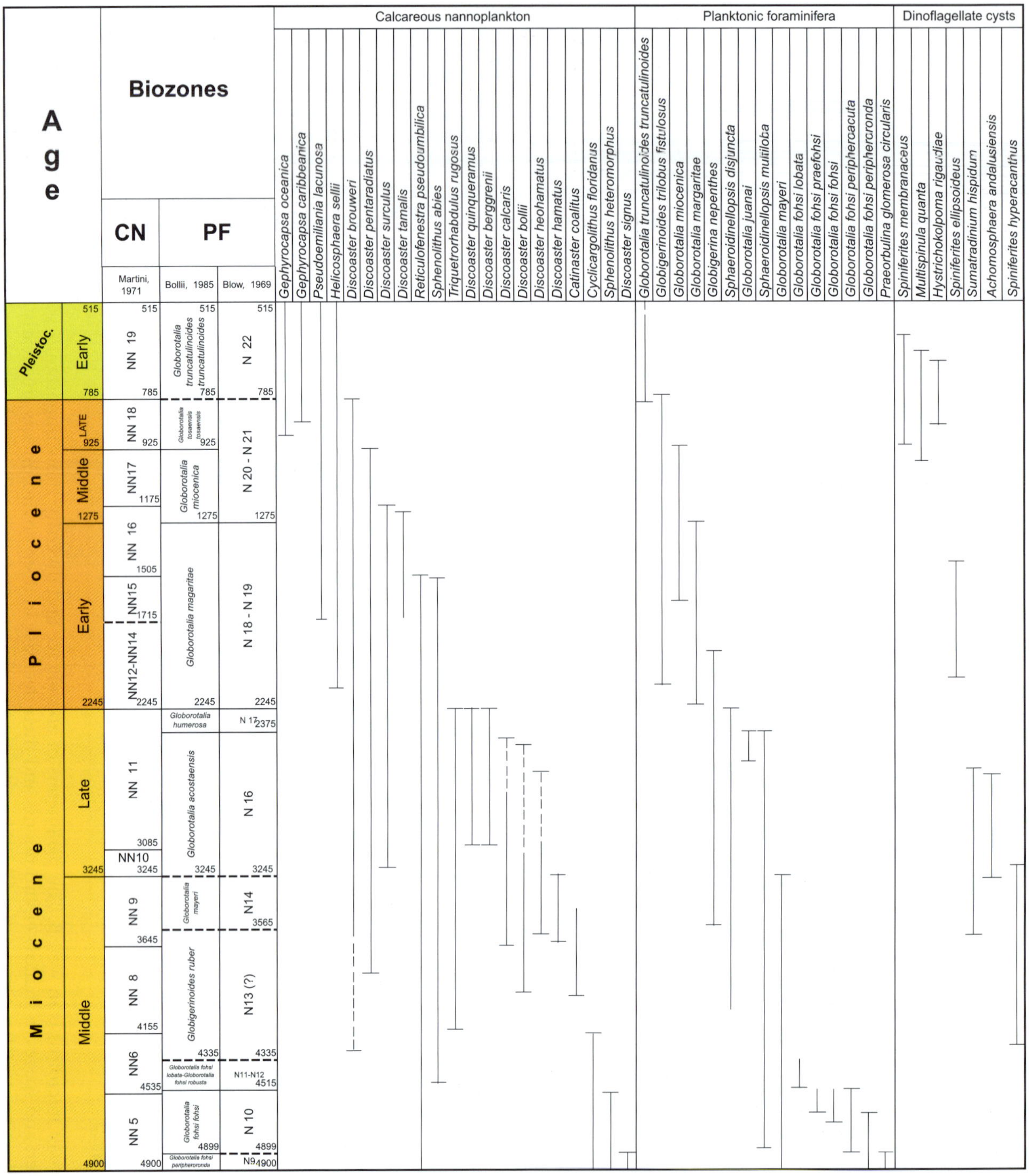

FIGURE 4. Stratigraphic distribution of index species in well A. CN = calcareous nannoplankton; PF = planktonic foraminifera.

Early Pleistocene

Early Pleistocene sediments are present in the two wells. They are characterized by the presence of *Pseudoemiliania lacunosa* without discoasters. The floral association indicates the *P. lacunosa* (NN19) biozone. Other important species present in this biozone are *Gephyrocapsa oceanica, Gephyrocapsa caribbeanica, Calcidiscus macintyrei,* and *Helicosphaera sellii.*

Planktonic foraminifera are mainly *Globorotalia truncatulinoides truncatulinoides, Globorotalia crassaformis viola, Globorotalia tenuitheca,* and *Globorotalia tosaensis s.l.* This association indicates the *G. truncatulinoides truncatulinoides* biozone.

Dinoflagellate cyst species found in these sediments are *Spiniferites delicatus* in well B and *Spiniferites membranaceous, Multispinula quanta,* and *Hystrichokolpoma rigaudiae* in well A.

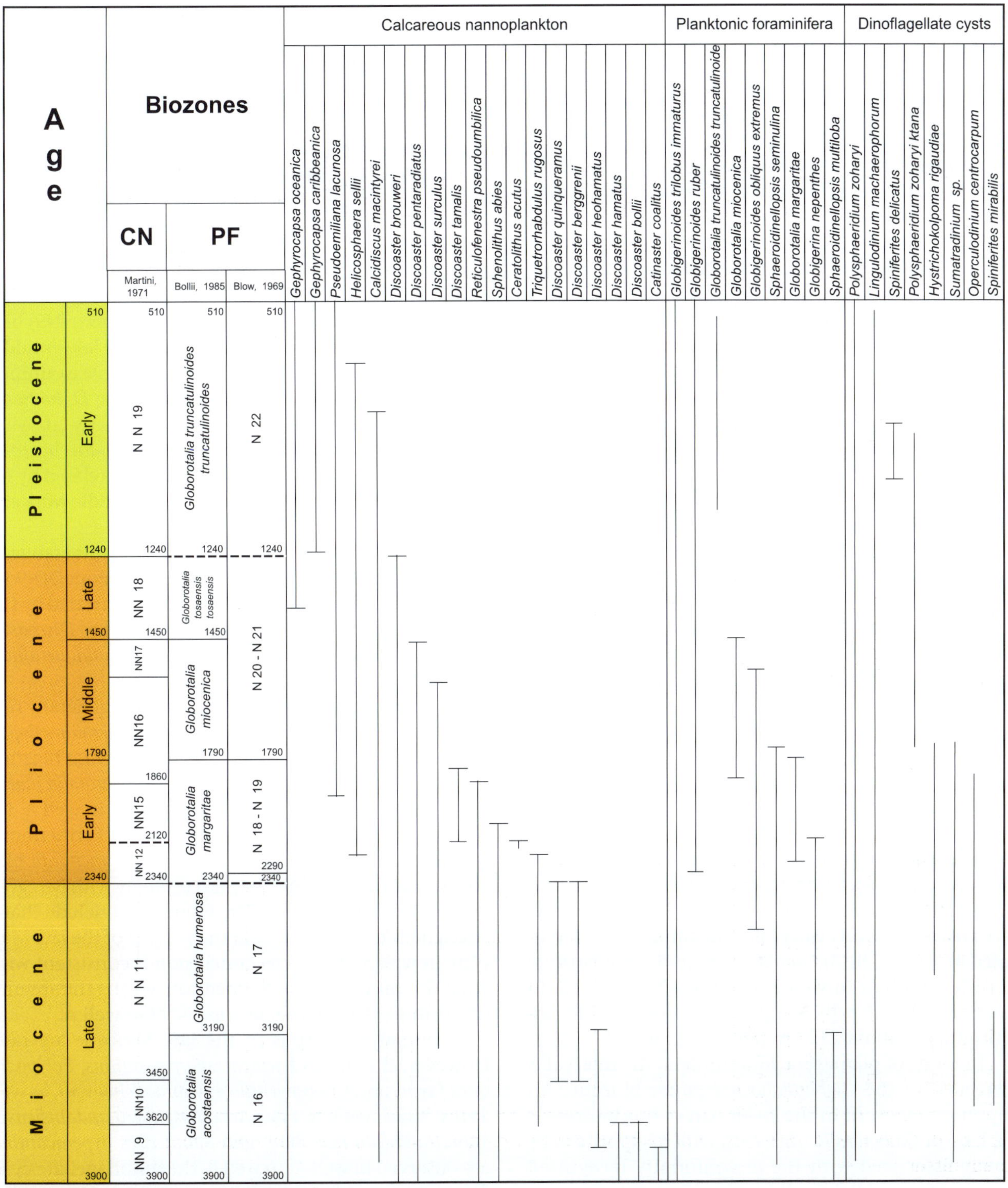

FIGURE 5. Stratigraphic distribution of index species in well B. CN = calcareous nannoplankton; PF = planktonic foraminifera.

Pliocene–Pleistocene Boundary (785 m [2575 ft] in Well A; 1240 m [4068 ft] in Well B)

The Pliocene–Pleistocene boundary is picked at the LAD (i.e., extinction point or first downhole occurrence) of *Discoaster brouweri*, which is very close to the lowest occurrence of *G. truncatulinoides truncatulinoides* and the extinction of *Globigerinoides trilobus fistulosus* in well A. The latter foraminiferal event was used by Berggren (1995a, b) for the recognition of the Pliocene–Pleistocene boundary.

Late Pliocene (785–925 m [2575–3035 ft] in Well A; 1240–1450 m [4068–4757 ft] in Well B)

The late Pliocene is represented by the *D. brouweri* (NN18) biozone, with the nominate species present. In terms of foraminifera, the interval is assigned to the *Globorotalia tosaensis tosaensis* biozone. Also, this biozone is characterized by the presence of *G. trilobus fistulosus* in well A.

The dinoflagellate cyst species representative of this age found in well B are *Polysphaeridium zoharyi*, *Lingulodinium machaerophorum*, and *Melitasphaeridium choanophorum*. In well A, *Operculodinium crassum*, the spore *Laevigatosporites major*, and the pollen grain *Fenestrites spinosus* are also present, which support a late Pliocene age.

Middle Pliocene (925–1275 m [3035–4183 ft] in Well A; 1460–1790 m [4790–5873 ft] in Well B)

Middle Pliocene calcareous nannofossil biozones include *Discoaster pentaradiatus* (NN17) and a part of *Discoaster surculus* (NN16). The *Globorotalia miocenica* biozone characterizes this age; its upper boundary marks the upper limit of the middle Pliocene. Palynomorph species that characterize the middle Pliocene in well B are *M. choanophorum*, *Nematosphaeropsis labyrinthea*, and *H. rigaudiae*.

Early Pliocene (1275–2245 m [4183–7365 ft] in Well A; 1790–2340 m [5873–7677 ft] in Well B)

In both wells, late early Pliocene sediments produced nannofossil assemblages belonging to the lower part of the *D. surculus* (NN16) and the *Reticulofenestra pseudoumbilica* (NN15) biozones. Biozones NN12–NN14 of the lower part of the early early Pliocene cannot be differentiated in well A. The *Amaurolithus tricorniculatus* biozone (NN12) is present in well B, but the NN13 and NN14 biozones are not, which indicates a stratigraphic discontinuity at about 2120 m (6955 ft) in well B.

In terms of planktonic foraminifera, the interval is correlated to the *Globorotalia margaritae* biozone. An important bioevent in the lower part of this biozone is the LAD of *Globorotalia nepenthes*, which happens to be closely associated with the disconformity recognized by nannos in well B. In well A, palynomorphs are scarce, but *Spiniferites ellipsoideus*, which becomes extinct in the early Pliocene, is present.

Miocene–Pliocene Boundary (2245 m [7365 ft] in Well A; 2340 m [7677 ft] in Well B)

This boundary is commonly established by the LAD (or highest occurrence) of the calcareous nannofossils *Discoaster quinqueramus* and/or *Triquetrorhabdulus rugosus*, following Perch-Nielsen (1985). In both wells, this boundary is established instead by the LAD of *D. quinqueramus* together with *Discoaster berggrenii*; thus, a stratigraphic discontinuity is determined at this level.

Late Miocene (2245–3245 m [7365–10,646 ft] in Well A; 2340–3900 m [7677–12,795 ft] in Well B)

The Late Miocene calcareous nannofossil associations recorded in the wells indicate the *D. quinqueramus* (NN11) biozone, the *Discoaster calcaris* (NN10) biozone, and the upper part of the *Discoaster hamatus* (NN9) biozone. The *D. hamatus* biozone straddles the late–middle Miocene boundary. In well B, the last sample examined from 3900 m (12,795 ft) is assigned to the *D. hamatus* biozone correlated by foraminifera to the early late Miocene. However, in well A, the entire *D. hamatus* biozone recognized below 3245 m (10,646 ft) is correlated by foraminifera to the youngest part of the middle Miocene (Figure 6).

Biozone NN11 is characterized by the total range of *D. quinqueramus* and *D. berggrenii* and biozone NN9 by that of *D. hamatus*. Biozone NN10 is characterized by the presence of *D. calcaris*, *Discoaster neohamatus*, *Discoaster bolli*, and others, and in the absence of *D. quinqueramus*, *D. berggrenii*, and *D. hamatus*.

Although planktonic foraminifera are generally rare in the late Miocene, the occurrence of *Sphaeroidinellopsis disjuncta*, indicating the *Globorotalia humerosa* biozone of upper Miocene, and occurrence of *Globorotalia juanai* and *Sphaeroidinellopsis multiloba* stand out in well A, indicating the *Globorotalia acostaensis* biozone. Furthermore, no record was found for the *Globorotalia menardii* biozone of the early late Miocene (accordingly with Berggren et al., 1995a, b) in well A. Therefore, we conclude that a stratigraphic discontinuity at the bottom of the late Miocene in well A exists. The conclusion is consistent with what was previously mentioned concerning the absence of the upper part of the biozone NN9 in well A.

The palynomorphs in the late Miocene are rare. However, the spores *Laevigatosporites nitidus*, *Polypodiidites favus*, and *Crassoretitriletes vanraadshooveni*, as well as the dinoflagellate cysts *Achomosphaera andalusiensis*, *Sumatradinium hispidum*, and *Spiniferites hyperacanthus* are observed in well A. In well B, the dinoflagellate cysts *Spiniferites mirabilis*, *Dapsilidinium pastielsii*, *H. rigaudiae*, *Operculodinium centrocarpum*, and the spore *Cicatricosisporites claricristata* have been identified. All these support a general late Miocene age for the interval.

Late–Middle Miocene Boundary (3245 m [10,646 ft] in Well A)

In well A, the late–middle Miocene boundary is placed at the LAD of *Globorotalia mayeri*, which is very

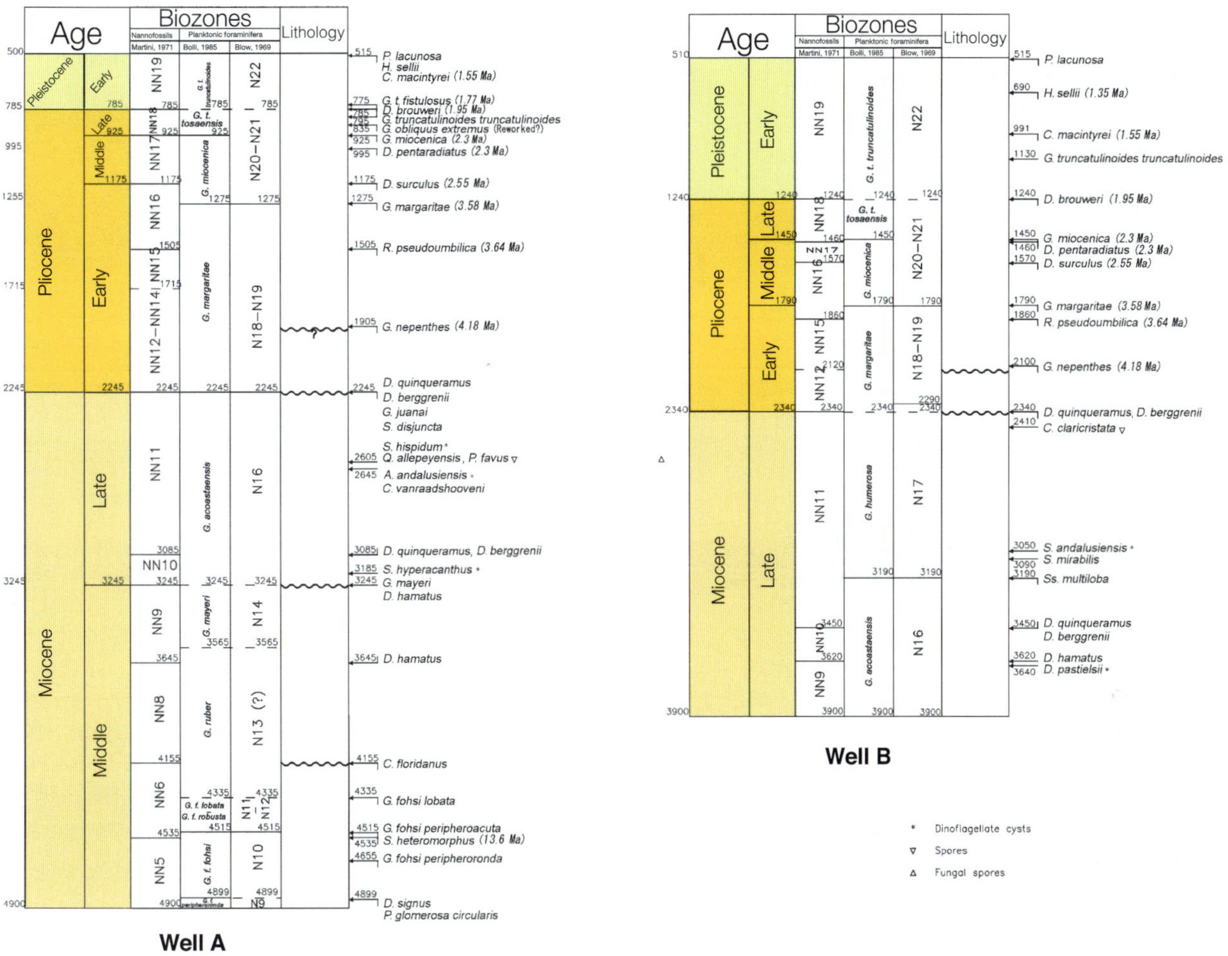

FIGURE 6. Biochronostratigraphic correlation of wells A and B.

close to the top of the dinoflagellate cyst *S. hyperacanthus* in well A.

Middle Miocene (3245–4900 m [10,646–16,076 ft] in Well A)

The middle Miocene sediments in well A are correlated with the lower part of the *D. hamatus* (NN9) biozone, the *Catinaster coalitus* (NN8) biozone, the *Discoaster exilis* (NN6) biozone, and the *Sphenolithus heteromorphus* (NN5) biozone. The first two and the last biozones are characterized by the presence of the nominate species. Biozone NN6 is characterized by the first downhole occurrence (highest occurrence) of *Cyclicargolithus floridanus*. The *Discoaster kugleri* (NN7) biozone was not present in the well. Consequently, we conclude that a stratigraphic discontinuity occurs at 4155 m (13,632 ft) at the top of zone NN6 in well A.

Planktonic foraminifera are sparse in the middle Miocene of well A. The biozones represented are the *G. mayeri* biozone, the *Globigerinoides ruber* biozone, the undifferentiated *Globorotalia fohsi robusta* and *Globorotalia fohsi lobata* biozones, the *Globorotalia fohsi fohsi* biozone, and the upper part of the *Globorotalia fohsi peripheroronda* biozone.

Middle Miocene palynomorphs are very scarce; only *S. hyperacanthus* has been identified in the middle Miocene of well A.

Paleobathymetry

Paleobathymetry is established using mainly benthic foraminifera and dinoflagellate cysts. The planktonic and benthic ratio, diversity and abundance of different groups of microfossils, and the MPI are criteria that allow us to estimate proximity to the coastline. Gamma-ray logs are critical in the identification of sandstone and shale intervals. The paleobathymetry is proposed after integrating biostratigraphic data and well logs (gamma ray and spontaneous potential).

The paleobathymetry for each well and the characteristic features in terms of the parameters mentioned above are presented as follows.

In well A, sediments are mainly shale with alternating thin sandstone layers. The highest sandstone:shale ratio is toward the top of the well. These sediments were

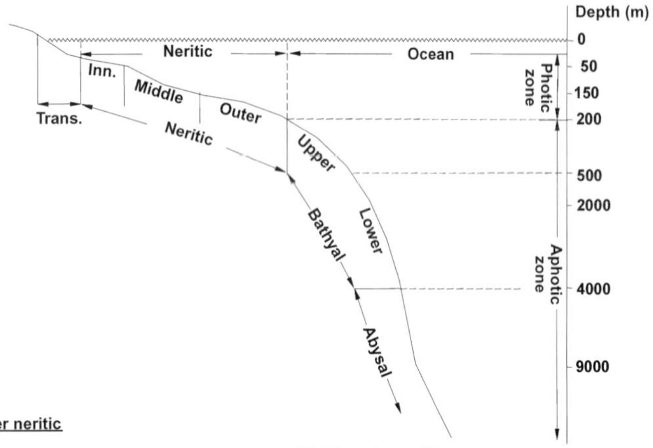

Inner neritic

Ammonia beccarii
Amphistegina lessoni
Bigenerina irregularis
Bigenerina nodosaria
Criboelphidium advenum
Elphidium sp.
Eponides antillarum
Florilus communis
Florilus scaphus
Hanzawaia isidroensis
Lenticulina americana var. *spinulosa*
Nonionella sp.
Quinqueloculina lamarckiana
Reusella sp.
Reusella spinulosa
Sigmoilinita tenuis
Textularia mayori
Tifarina rutila

Middle neritic

Amphistegina lessonii
Brizalina alata
Brizalina subaenariensis
Bulimina sp.
Bulimina striata
Cancris sagra
Cassidulina laevigata
Cassidulina sp.
Cibicidoides sp.
Lenticulina sp.
Nonionella turgida
Uvigerina spp.

Middle-outer neritic

* *Spiniferites bentorii*
* *Spiniferites mirabilis*
* *Spiniferites elongatus*
* *Spiniferites ramosus*
* *Spiniferites membranaceous*
* *Tuberculodinium vancampoae*
* *Polysphaeridium zoharyi*

Outer neritic

Bolivina floridana
Bulimina marginata
Chilostomella czizeki
Cibicidoides pseudoungerianus
Cyclammina sp.
Gyroidinoides sp.
Gyroidinoides altiformis
Heterolepa sp.
Höeglundina elegans
Marginulinopsis goajiraensis
Oridorsalis westi
Plectofrondicularia californica
Planulina ariminensis
Praeglobobulimina sp.
Praeglobobulimina ovata
Pullenia quinqueloba
Rectuvigerina sp.
Rectuvigerina transversa
Sigmoilopsis schlumbergeri
* *Polysphaeridium zoharyi*
* *Operculodinium israelianum*
* *Spiniferites ramosus*
* *Spiniferites elongatus*
* *Lingulodinium machaerophorum*

Outer neritic-upper bathyal

* *Impagidinium* sp.
* *Nematosphaeropsis labyrinthea*
* *Spiniferites mirabilis*
* *Lingulodinium machaerophorum*
* *Operculodinium israelianum*
* *Spiniferites pseudofurcatus*
* *Spiniferites mirabilis*
* *Achomosphaera andalusiensis*

Upper bathyal

Anomalinoides trinitatensis
Bulimina alazanensis
Cyclammina cancellata
Eggerella bradyi
Gyroidinoides broeckhiana
Martinottiella sp.
Melonis affinis
Melonis pompilioides
Planulina mantaensis
Pullenia bulloides
Siphonodosaria sp.
Shipouvigerina auberiana
Usbekistania charoides

* Dinoflagellate cysts

FIGURE 7. Indicative flora and fauna for paleobathymetry in well A.

deposited in inner neritic to upper bathyal water depths. Figure 7 shows the microflora and microfauna used for determining paleobathymetry in well A.

In these inner and middle neritic settings, inorganic matter represented by thick sediments prevails over organic material, benthic foraminifera are more abundant than planktonic foraminifera, and calcareous nannofossils are scarce. Unexpectedly, the relation between marine and continental palynomorphs is opposite to what would be predicted. That is, in shallow waters, there should be a predominance of continental forms (Figure 2), but here, it is the opposite: continental forms are scarce. This is probably because of clastic dilution caused by rapid sedimentation and high-energy conditions, which tend to destroy the microfossils.

In outer neritic settings, the proportion of organic matter increases, and a strong predominance of calcareous nannofossils and dinoflagellate cysts is observed. Both the diversity and abundance of dinoflagellate cysts are greater than pollen and spores. In this environmental setting, planktonic and benthic foraminifera are not very abundant. However, a slight predominance of benthic forms is observed, which may be caused by sediment transport toward the deeper part of the basin. Salmerón-Ureña (2002, p. 260) concluded that "the fossils can be frequently transported towards very different environments from which they are restricted to or characterized."

In the deeper part of well A, sediments consist of semi-indurated, slightly calcareous arenaceous shale, deposited in an outer neritic-upper bathyal paleoenvironment. Planktonic foraminiferal abundance increases. Palynomorph assemblages, which are dominated by continental forms, and calcareous nannofossils decrease in abundance until almost disappearing at the base of the well. This reduction in nannofossils is probably caused by indurated sediments, which did not allow the extraction of small microfossils. The disappearance of pollen and spores toward the bottom of well A is probably caused by the distance from the coastline, indicating deep waters.

The paleobathymetric profile in well B (Figure 8) consists of an inner neritic environment at the top of the well, becoming middle neritic to outer neritic and continuing until the upper bathyal toward the middle- lower part of the well. Toward the bottom of the well, the paleobathymetry is undifferentiated outer neritic to upper bathyal.

Benthic foraminifera and continental palynomorphs are particularly abundant in the inner-middle neritic sediments, whereas marine palynomorphs are less abundant. Low MPI values are characteristic of near coastal environments. Toward middle-outer neritic depths, the change in abundance between continental and marine palynomorphs is evident as the former predominates over the latter. At the bottom of the well, nannofossil diversity increases gradually while MPI values also increase. This indicates a location far from the coastline. In the upper bathyal deposits, a strong predominance of calcareous nannofossils is observed. The MPI values are highest in this part of the well, indicating deep water. Palynomorph abundances decrease slightly at the bottom of the well, but MPI values still clearly indicate a marine environment (Figure 3).

Note the predominance of benthic over planktonic foraminifera throughout well B, probably caused by the transport of these microfossils into the basin (Salmerón-Ureña, 2002). In terms of the OM/IM proportion, an OM dominance is observed in deep water.

Biocronostratigraphic Correlation

Figure 6 shows the biochronostratigraphic correlation between the two studied wells. Absolute ages are in accordance with the criteria of Aubry (1993) and particularly Berggren et al. (1995a, b).

The early Pleistocene in both wells is identified by the *P. lacunosa* (NN19) biozone and the *G. truncatulinoides truncatulinoides* biozone. The highest appearance (LAD) of *H. sellii* (1.35 m.y.) in well B is also considered an important event.

We recognize the Pliocene–Pleistocene boundary at the LAD of *D. brouweri* (1.95 m.y.).

Bolli and Saunders (1985) placed the Pliocene–Pleistocene boundary at the first appearance of *G. truncatulinoides truncatulinoides* and/or the last occurrence of *G. crassaformis s.l.* However, these species are scarce in the studied wells. Berggren et al. (1995b) proposed the boundary at the last appearance of *Globigerinoides fistulosus* (1.77 m.y.), but unfortunately, this species was only found in well A. Consequently, we choose to use the highest occurrence of *D. brouweri* (1.95 m.y.), which occurs in both wells as the Pliocene–Pleistocene boundary.

Pliocene bioevents reliable for subsurface correlation between the two wells are the last appearances of *D. brouweri* (1.95 m.y.), *D. pentaradiatus* (2.3 m.y.), *G. miocenica* (2.3 m.y.), *D. surculus* (2.55 m.y.), *G. margaritae* (3.58 m.y.), *R. pseudoumbilica* (3.64 m.y.), and *Globigerina nepenthes* (4.18 m.y.).

The Miocene–Pliocene boundary is marked at the LAD of *D. quinqueramus* and *D. berggrenii*. The simultaneous disappearance of these two species at the boundary in the two wells implies a stratigraphic discontinuity. Unfortunately, index foraminifera are scarce at this level in the two wells.

In the Miocene, the most important bioevents of calcareous and foraminiferal nanofossils are the LAD of *S. heteromorphus* (13.6 m.y.), *G. fohsi peripheroronda*, and *Preorbulina glomerosa circularis* in well A. *C. floridanus* is found reworked in almost all late Miocene and Pliocene sediments.

CONCLUSIONS

An HRB study has been conducted for the early Pleistocene to middle Miocene interval, using two wells from the western Gulf of Mexico.

The calcareous nannofossils biozones identified are *P. lacunosa* (NN19), *D. brouweri* (NN18), *D. pentaradiatus* (NN17), *D. surculus* (NN16), *R. pseudoumbilica* (NN15), *A. tricorniculatus* (NN12), *D. quinqueramus* (NN11), *D. calcaris* (NN10), *D. hamatus* (NN9), *C. coalitus* (NN8), *D. exilis* (NN6), and *S. heteromorphus* (NN5).

The planktonic foraminifera biozones are *G. truncatulinoides truncatulinoides*, *G. tosaensis tosaensis*, *G. miocenica*, *G. margaritae*, *G. humerosa*, *G. acostaensis*, *G. mayeri*, *G. ruber*, *G. fohsi lobata*, *G. fohsi robusta*, *G. fohsi fohsi*, and probably a part of *G. fohsiperipheroronda*.

The important palynomorph species found in the two wells are *P. zoharyi*, *L. machaerophorum*, *S. membranaceous*, *M. quanta*, *H. rigaudiae*, *S. delicatus*, *M. choanophorum*, *N. labyrinthea*, *O. crassum*, *S. ellipsoideus*, *S. mirabilis*, *D. pastielsii*, and *S. hispidum*.

Nannofloral and microfaunal associations allow the identification of four stratigraphic discontinuities: in the early Pliocene, at the Pliocene–Miocene boundary, at the late middle Miocene boundary, and in the middle Miocene. The first two are found in both wells and the latter two are recorded only in well A.

The proposed paleobathymetry varies from inner neritic to deep upper bathyal. Planktonic foraminifera associations show some predominance in deep water, and the benthic and continental forms are more abundant in shallow water.

PLATES

Plates I–IV: Nannofossils
Plates V–VIII: Foraminifera
Plates IX–XIII: Palynomorphs

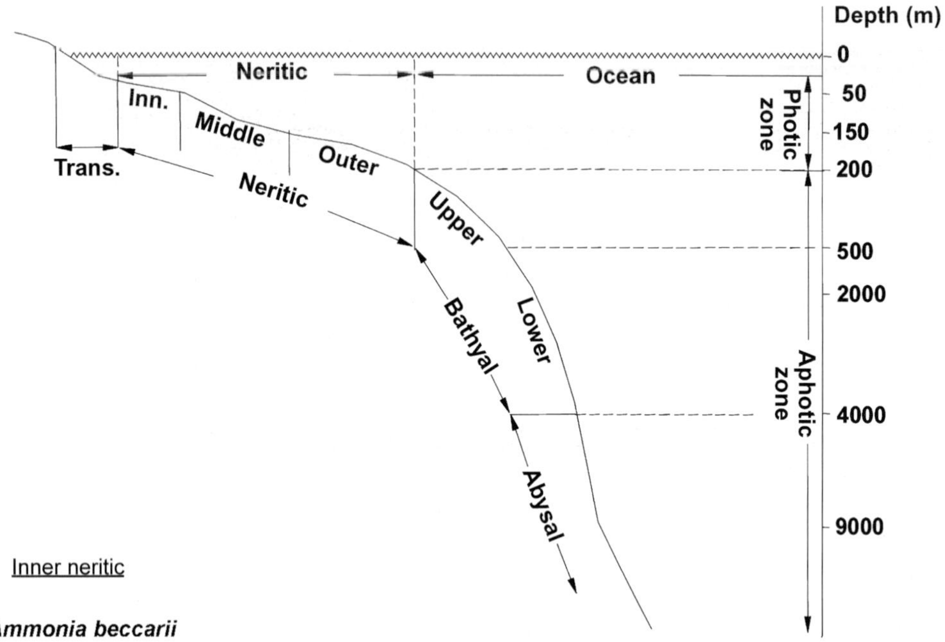

Inner neritic

Ammonia beccarii
Criboelphidium sp.
Eponides sp.
Lenticulina sp.
Lenticulina americana
Nonionella sp.
Quinqueloculina lamarckiana
Reusella sp.
Hanzawaia concentrica
Hanzawaia isidroensis
Hanzawaia sp.
Elphidium sp.
Eponides antillarum
Siphonina sp.
Florilus spp.
Textularia sp.
* *Polysphaeridium zoharyi*
* *Polysphaeridium zoharyi ktana*
* *Spiniferites delicatus*

Middle neritic

Uvigerina spp.
Lenticulina sp.
Brizalina sp.
Brizalina subaenariensis
Cibicidoides sp.
Cassidulina laevigata
Cassidulina sp.
Bulimina sp.
Amphistegina lessonii
Bulimina striata

Outer neritic

Sigmoilopsis schlumbergeri
Plectofrondicularia californica
Cibicidoides pseudoungerianus
Gyroidinoides sp.
Gyroidinoides altiformis
Bolivina sp.
Bolivina floridana
Bulimina marginata
Planulina sp.
Planulina ariminensis
Praeglobobulimina sp.
Praeglobobulimina ovata
Chilostomella czizeki
Höeglundina elegans
Cyclammina cancellata
Oridorsalis sp.
Oridorsalis westi
* *Operculodinium israelianum*
* *Sumatradinium* sp.
* *Spiniferites andalusiensis*
* *Dapsilidinium pastielsii*

Middle-outer neritic

* *Polysphaeridium zoharyi*
* *Polysphaeridium zoharyi ktana*

Upper bathyal

Pullenia bulloides
Melonis affinis
Pullenia quinqueloba
Shipouvigerina auberiana
Bulimina alazanensis
Planulina mantaensis
Melonis pompilioides
* *Impagidinium* sp.
* *Hystrichokolpoma rigaudiae*
* *Nematosphaeropsis labyrinthea*
* *Spiniferites ramosus*

* Dinoflagellate cysts

FIGURE 8. Indicative flora and fauna for paleobathymetry in well B.

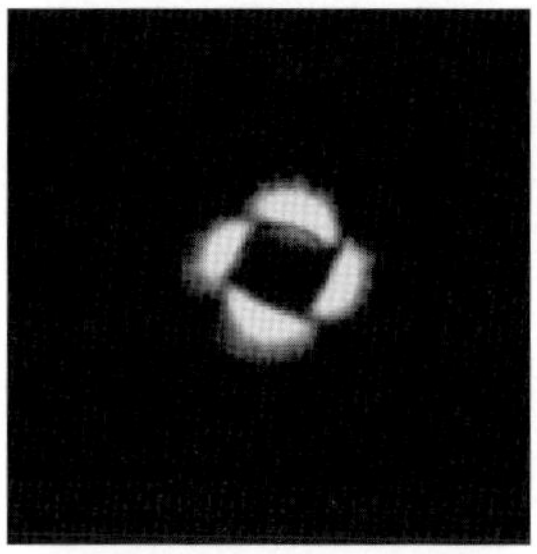

1
Pseudoemilinia lacunosa
[X1600] P.L. depth 515 m
Well A

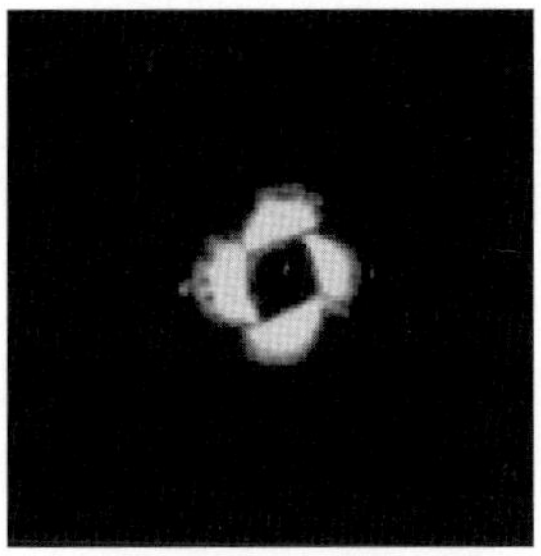

2
Pseudoemilinia lacunosa
[X1600] P.L. depth 610 m
Well A

3
Gephyrocapsa aperta
[X1600] P.L. depth 630 m
Well A

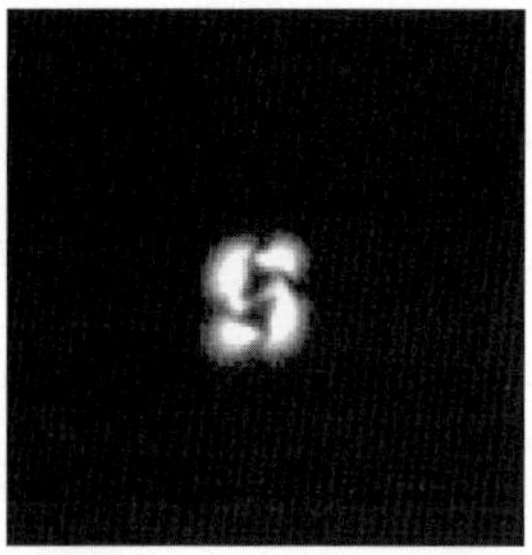

4
Gephyrocapsa caribbeanica
[X1600] P.L. depth 985 m
Well A

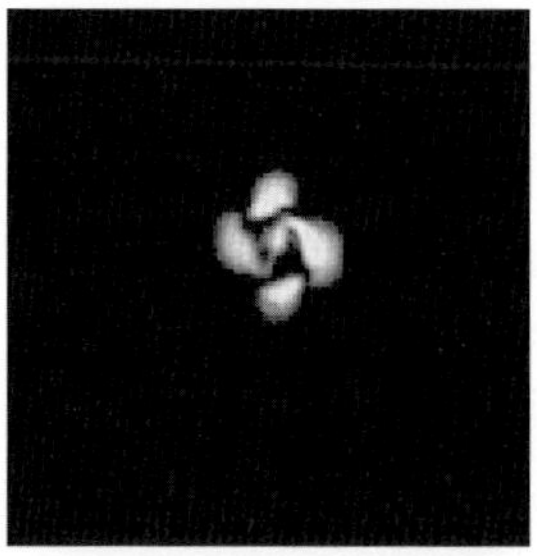

5
Gephyrocapsa oceanica
[X1600] P.L. depth 515 m
Well A

6
Helicosphaera sellii
[X1600] P.L. depth 515 m
Well A

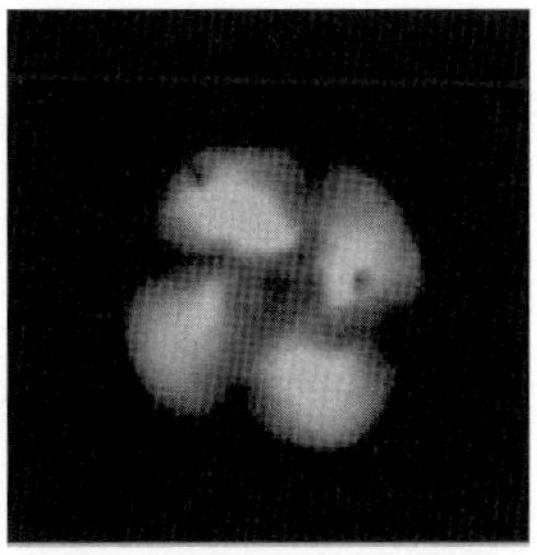

7
Calcidiscus macintyrei
[X1600] P.L. depth 795 m
Well A

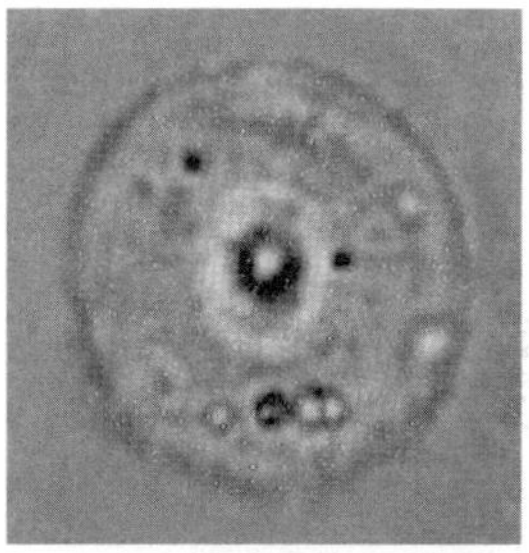

8
Calcidiscus macintyrei
[X1600] N.L. depth 795 m
Well A

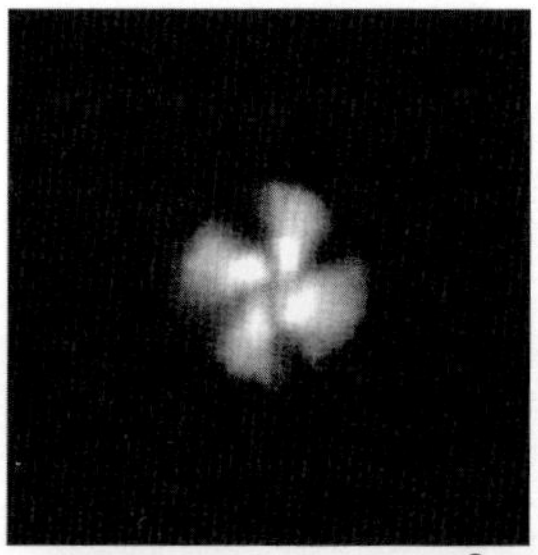

9
Calcidiscus leptoporus
[X1600] P.L. depth 815 m
Well A

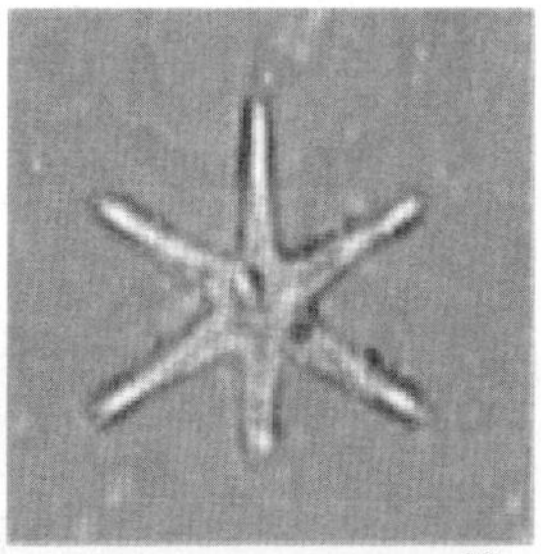

10
Discoaster brouweri
[X1600] N.L. depth 1075 m
Well A

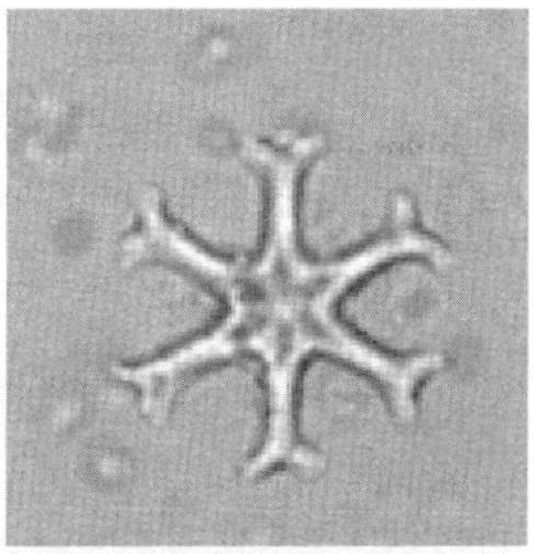

11
Discoaster variabilis
[X1600] N.L. depth 1215 m
Well A

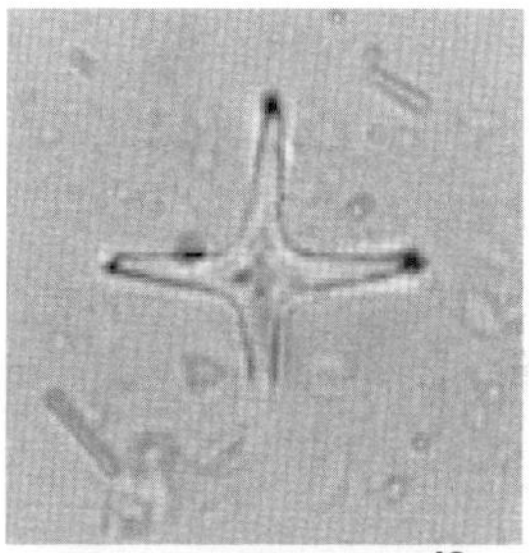

12
Discoater tamalis
[X1600] N.L. depth 1455 m
Well A

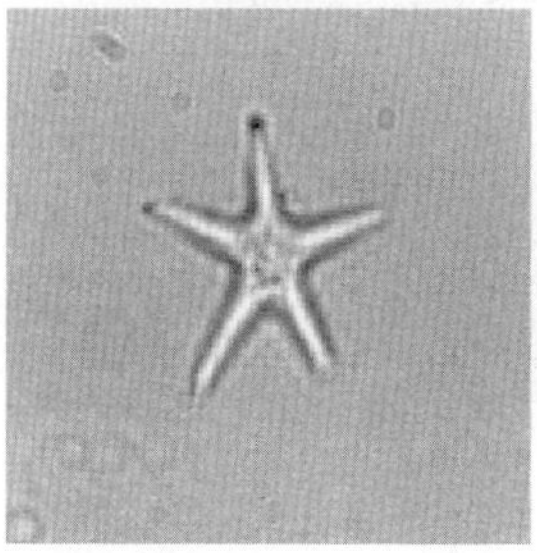

13
Discoaster asymmetricus
[X1600] N.L. depth 1175 m
Well A

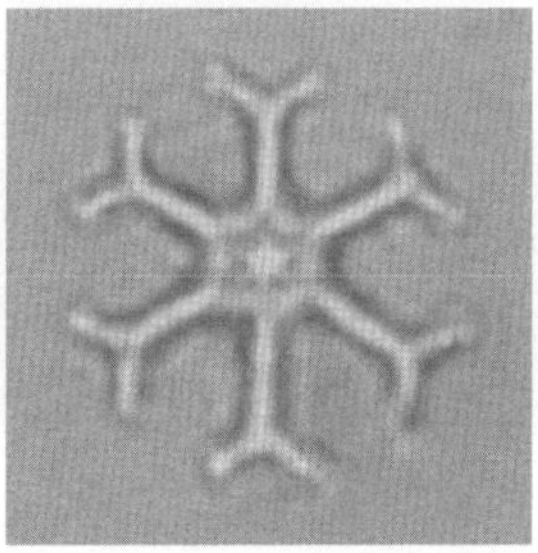

14
Discoaster challengeri
[X1600] N.L. depth 2465 m
Well A

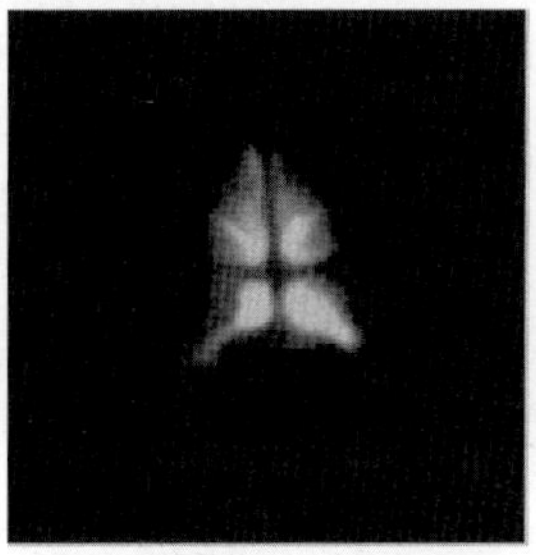

15
Sphenolithus abies
[X1600] P.L. depth 1525 m
Well A

16
Sphenolithus abies
[X1600] P.L. depth 1525 m
Well A

Plate I. Nannofossils.

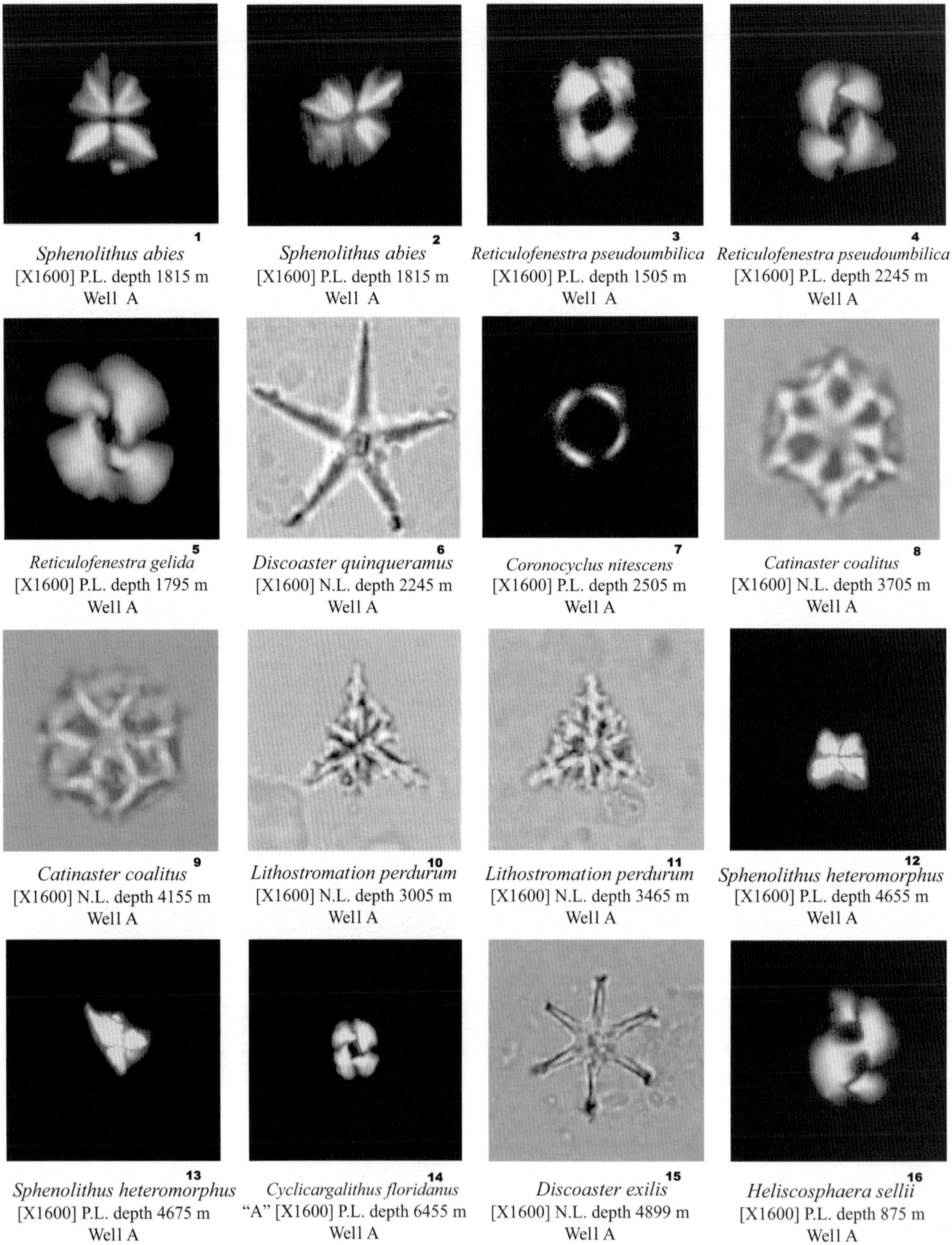

1 *Sphenolithus abies* [X1600] P.L. depth 1815 m Well A

2 *Sphenolithus abies* [X1600] P.L. depth 1815 m Well A

3 *Reticulofenestra pseudoumbilica* [X1600] P.L. depth 1505 m Well A

4 *Reticulofenestra pseudoumbilica* [X1600] P.L. depth 2245 m Well A

5 *Reticulofenestra gelida* [X1600] P.L. depth 1795 m Well A

6 *Discoaster quinqueramus* [X1600] N.L. depth 2245 m Well A

7 *Coronocyclus nitescens* [X1600] P.L. depth 2505 m Well A

8 *Catinaster coalitus* [X1600] N.L. depth 3705 m Well A

9 *Catinaster coalitus* [X1600] N.L. depth 4155 m Well A

10 *Lithostromation perdurum* [X1600] N.L. depth 3005 m Well A

11 *Lithostromation perdurum* [X1600] N.L. depth 3465 m Well A

12 *Sphenolithus heteromorphus* [X1600] P.L. depth 4655 m Well A

13 *Sphenolithus heteromorphus* [X1600] P.L. depth 4675 m Well A

14 *Cyclicargalithus floridanus* "A" [X1600] P.L. depth 6455 m Well A

15 *Discoaster exilis* [X1600] N.L. depth 4899 m Well A

16 *Heliscosphaera sellii* [X1600] P.L. depth 875 m Well A

PLATE II. Nannofossils.

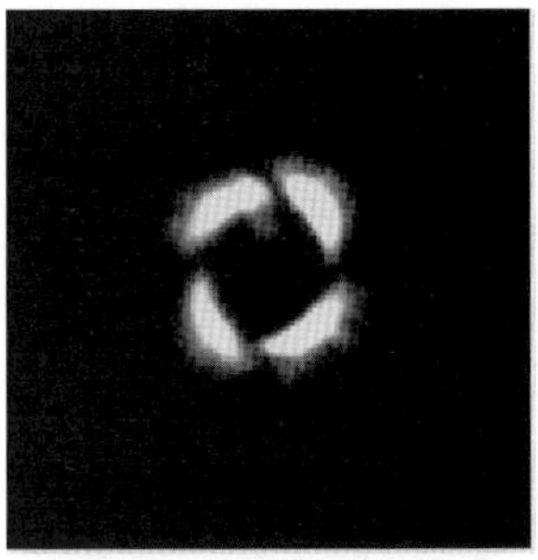

1
Pseudoemilinia lacunosa
[X1600] P.L. depth 510 m
Well B

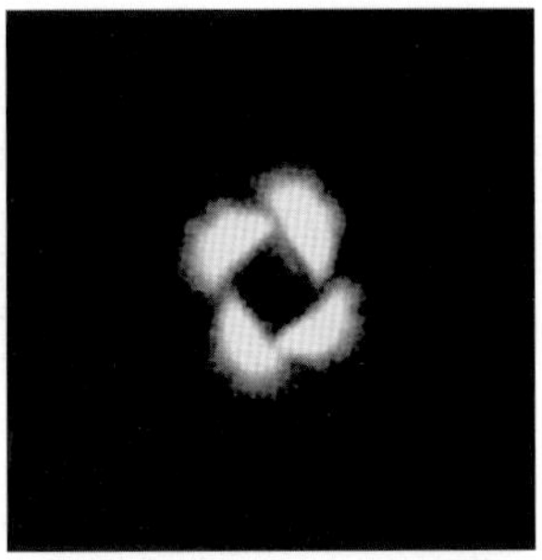

2
Pseudoemilinia lacunosa
[X1600] P.L. depth 1690 m
Well B

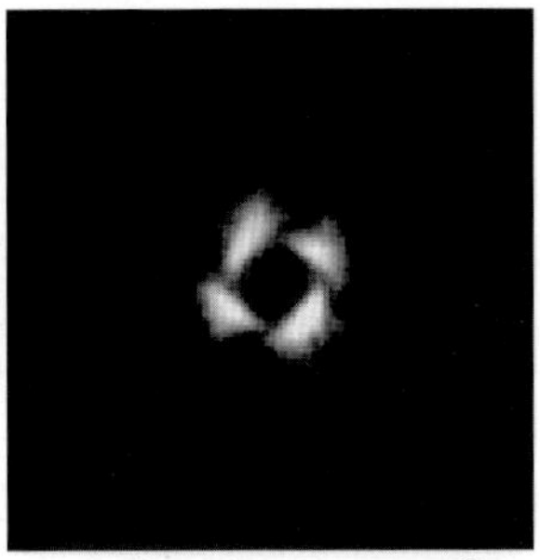

3
Pseudoemilinia lacunosa
[X1600] P.L. depth 1350 m
Well B

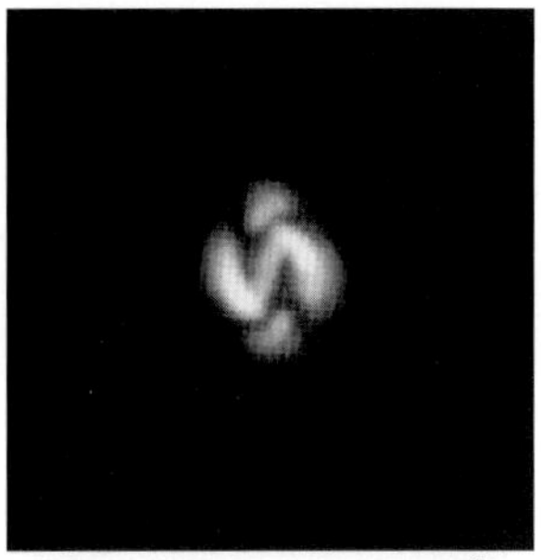

4
Gephyrocapsa caribbeanica
[X1600] P.L. depth 530 m
Well B

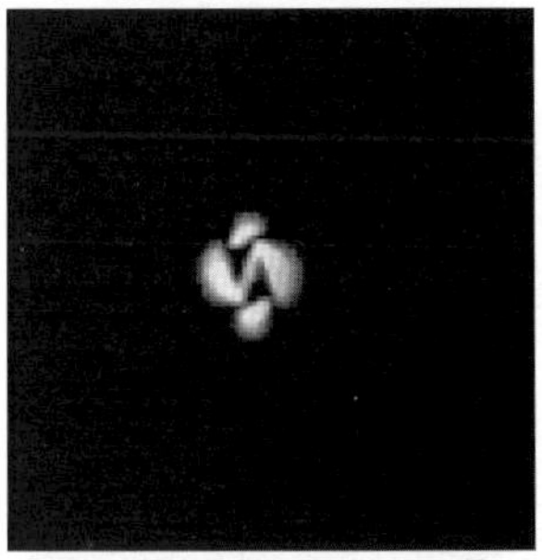

5
Gephyrocapsa oceanica
[X1600] P.L. depth 510 m
Well B

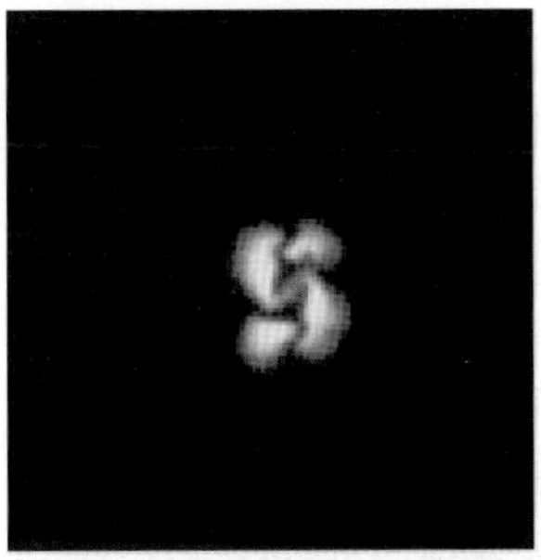

6
Gephyrocapsa oceanica
[X1600] P.L. depth 710 m
Well B

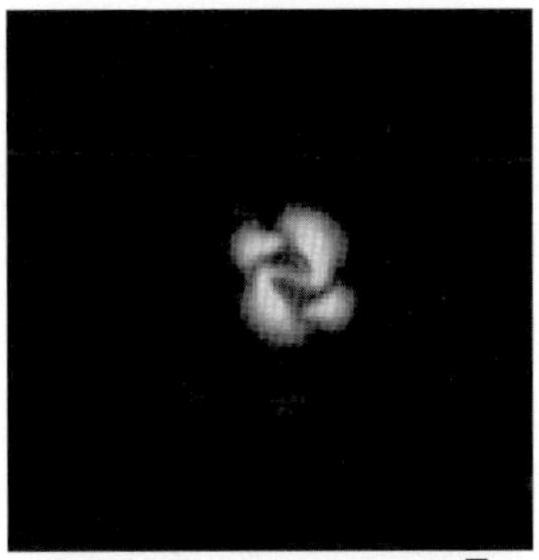

7
Gephyrocapsa oceanica
[X1600] P.L. depth 890 m
Well B

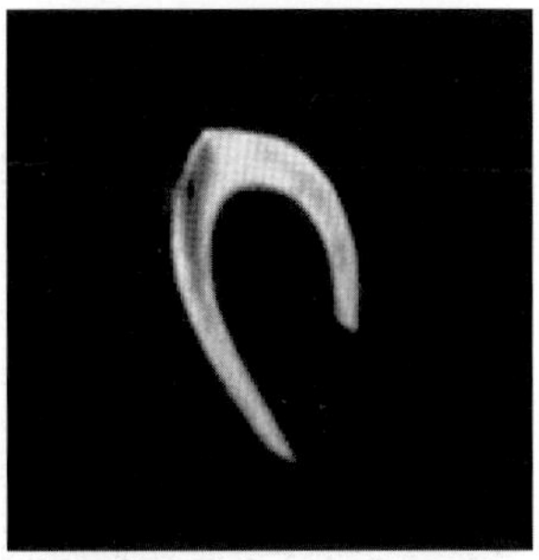

8
Ceratolithus telesmus
[X1600] P.L. depth 1245 m
Well B

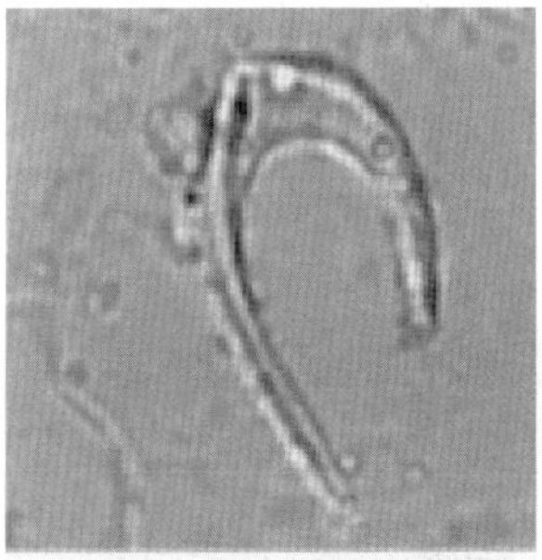

9
Ceratolithus telesmus
[X1600] N.L. depth 1245 m
Well B

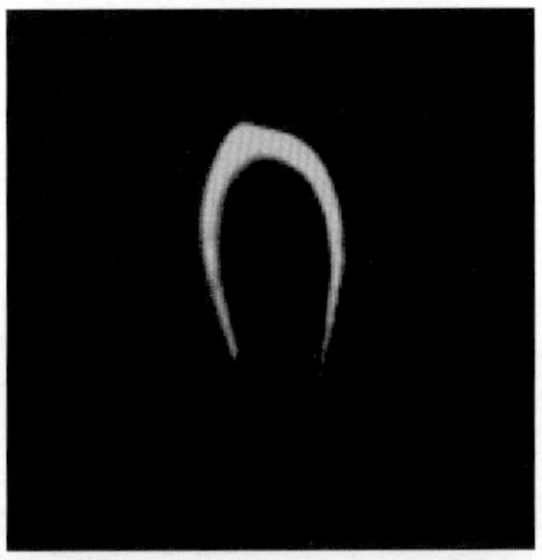

10
Ceratolithus cf telesmus
[X1600] P.L. depth 1350 m
Well B

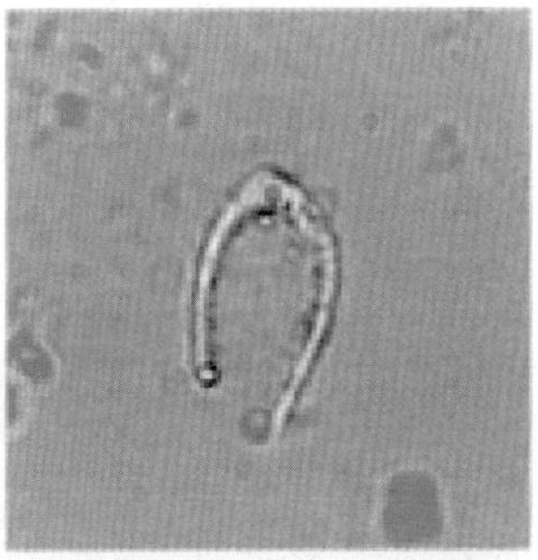

11
Ceratolithus cf telesmus
[X1600] N.L. depth 1350 m
Well B

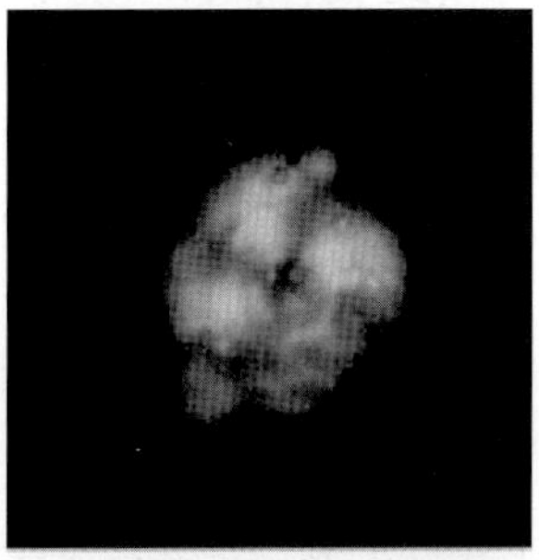

12
Calcidiscus macintyrei
[X1600] P.L. depth 991 m
Well B

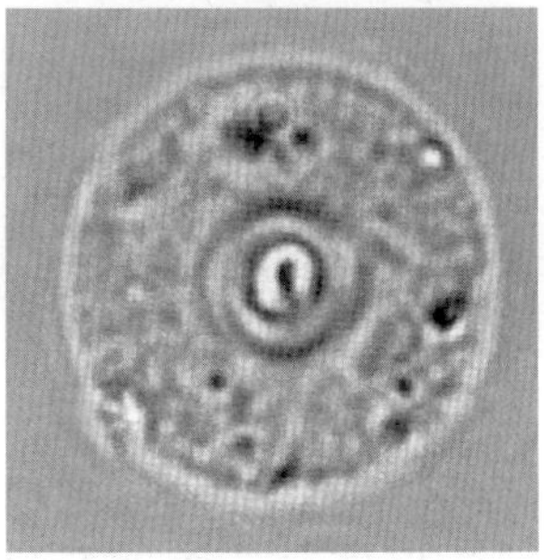

13
Calcidiscus macintyrei
[X1600] N.L. depth 991 m
Well B

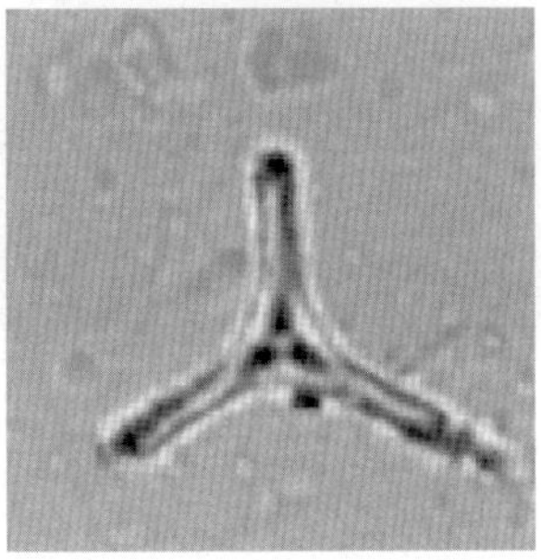

14
Discoaster triradiatus
[X1600] N.L. depth 2270 m
Well B

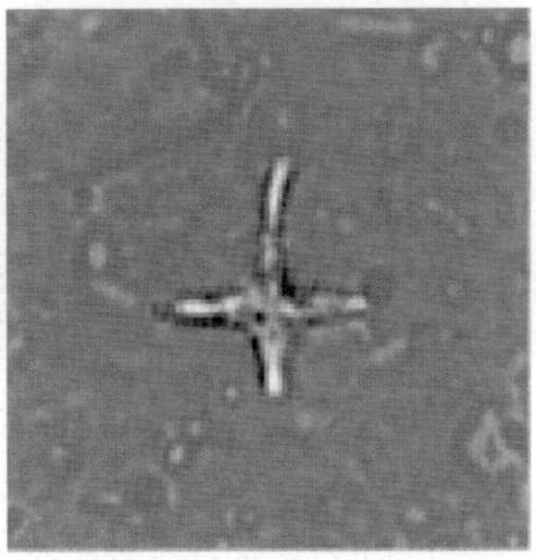

15
Discoaster tamalis
[X1600] N.L. depth 1820 m
Well B

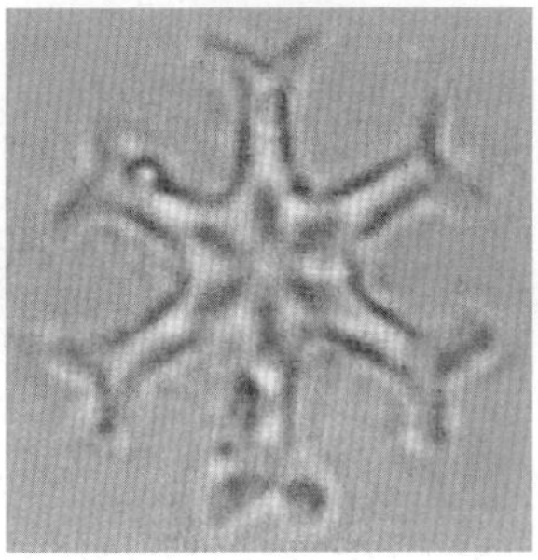

16
Discoaster variabilis
[X1600] N.L. depth 1710 m
Well B

Plate III. Nannofossils.

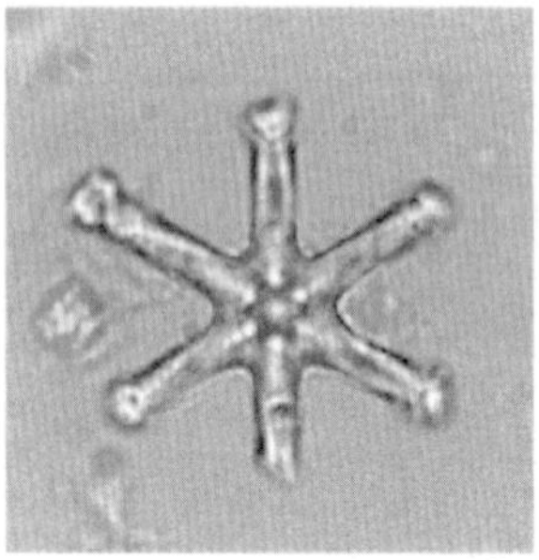

1
Discoaster surculus
[X1600] N.L. depth 1860 m
Well B

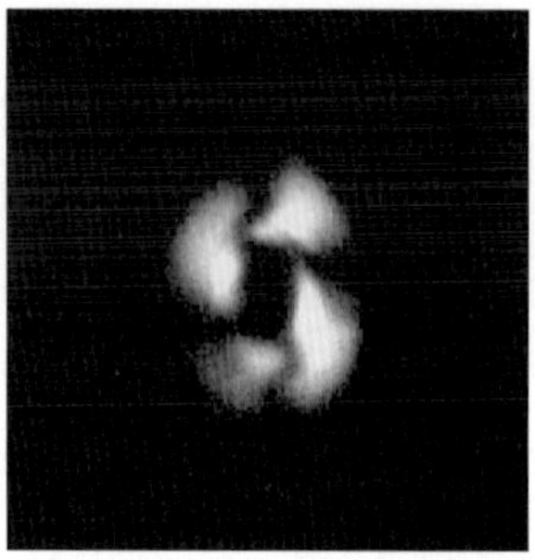

2
Reticulofenestra pseudoumbilica
[X1600] P.L. depth 2040 m
Well B

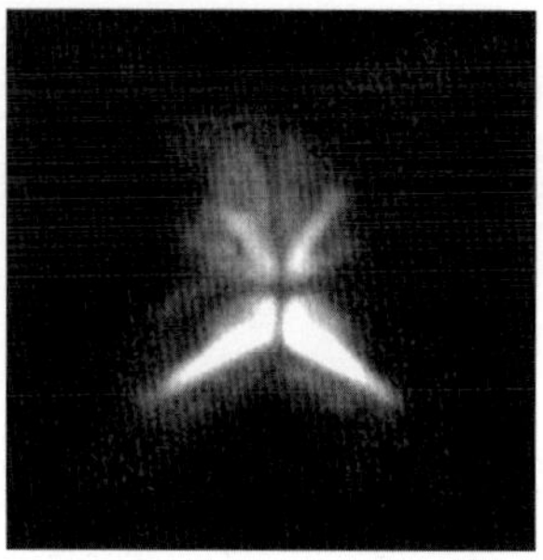

3
Sphenolithus abies
[X1600] P.L. depth 2100 m
Well B

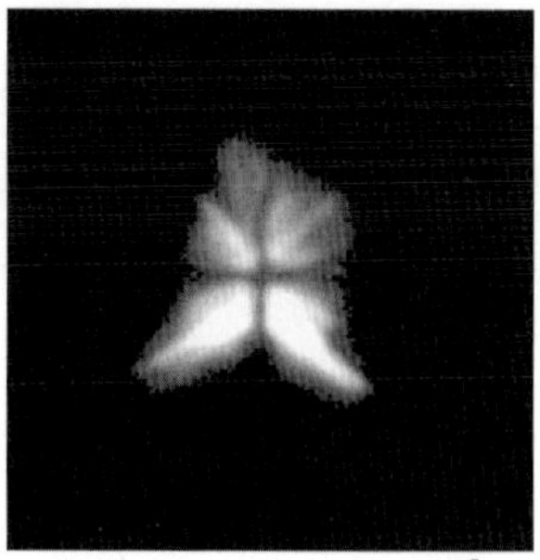

4
Sphenolithus abies
[X1600] P.L. depth 2510 m
Well B

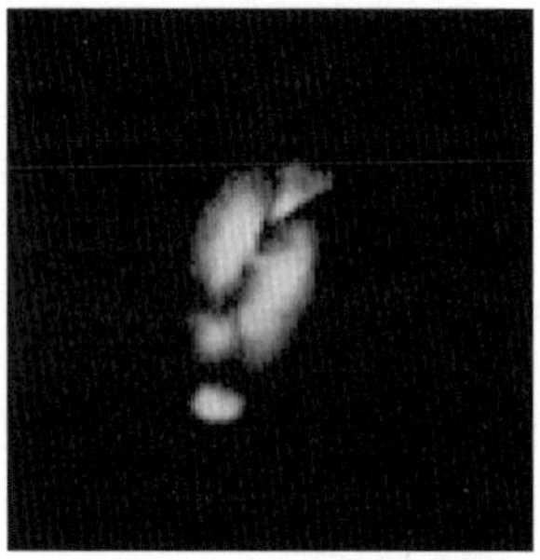

5
Helicosphaera Kamptneri
[X1600] P.L. depth 1380 m
Well B

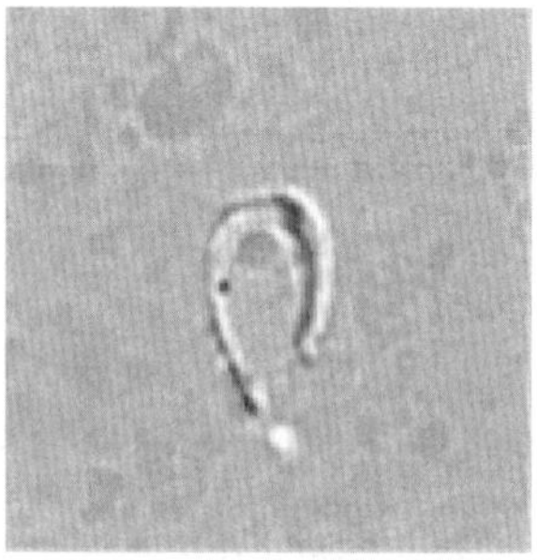

6
Amaurolithus delicatus
[X1600] P.L. depth 2350 m
Well B

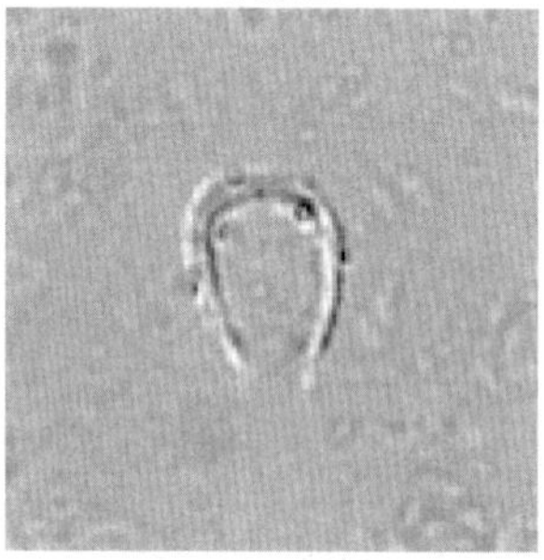

7
Amaurolithus delicatus
[X1600] P.L. depth 2690 m
Well B

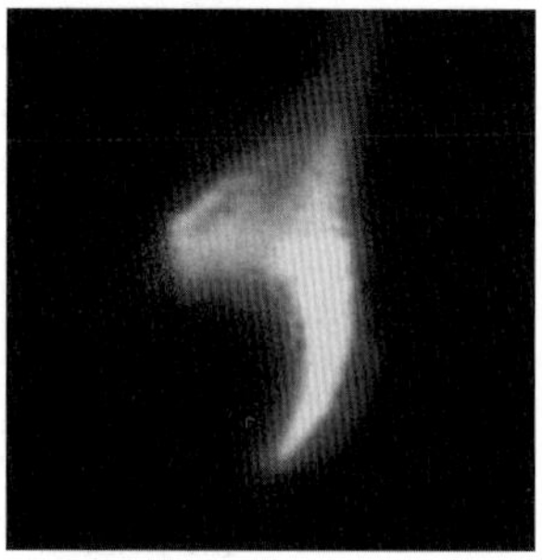

8
Ceratolithus acutus
[X1600] P.L. depth 2120 m
Well B

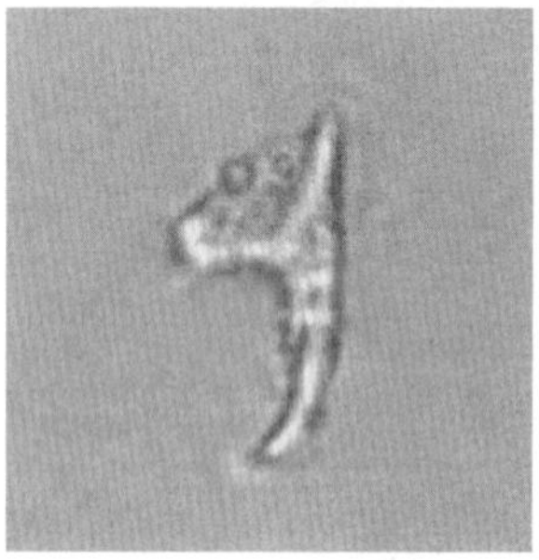

9
Ceratolithus acutus
[X1600] N.L. depth 2120 m
Well B

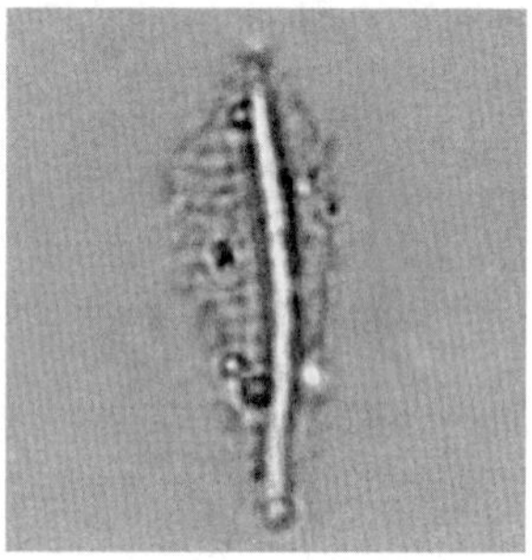

10
Triquetrorhabdulus rugosus
[X1600] N.L. depth 2200 m
Well B

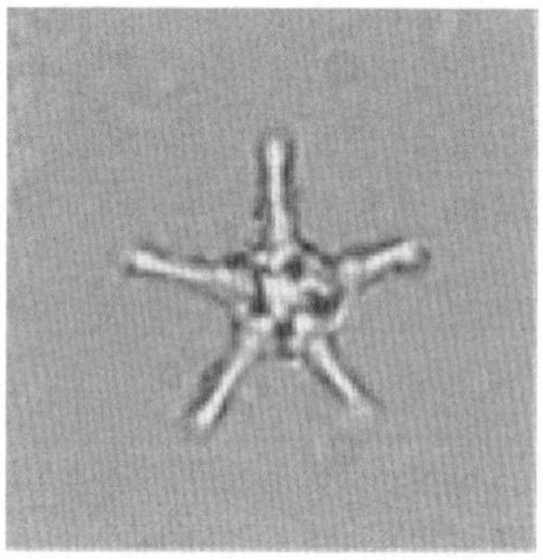

11
Discoaster berggrenii
[X1600 N.L. depth 2340 m
Well B

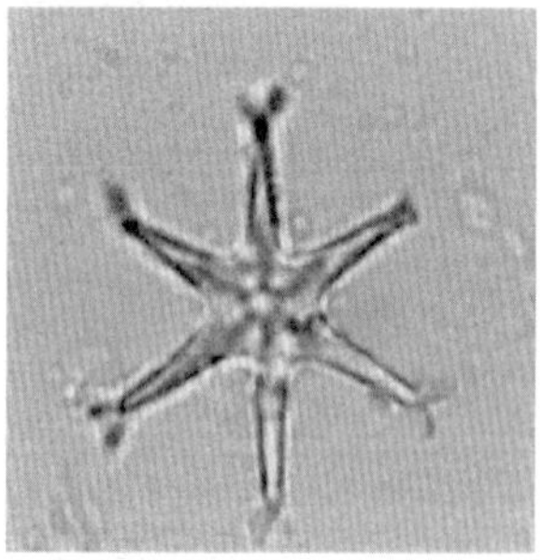

12
Discoaster exilis
[X1600] N.L. depth 3815 m
Well B

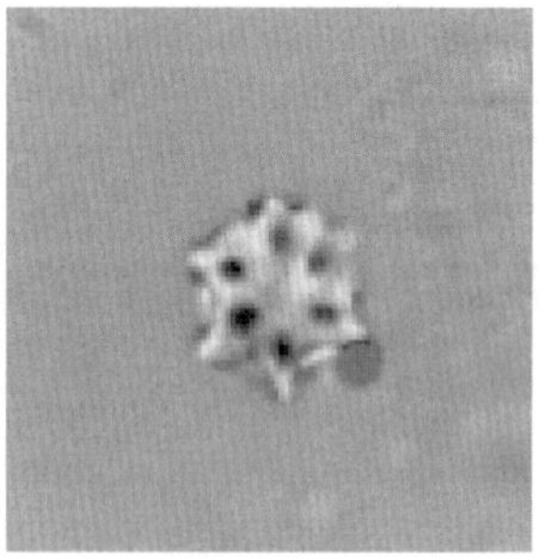

13
Catinaster coalitus
[X1600] N.L. depth 3735 m
Well B

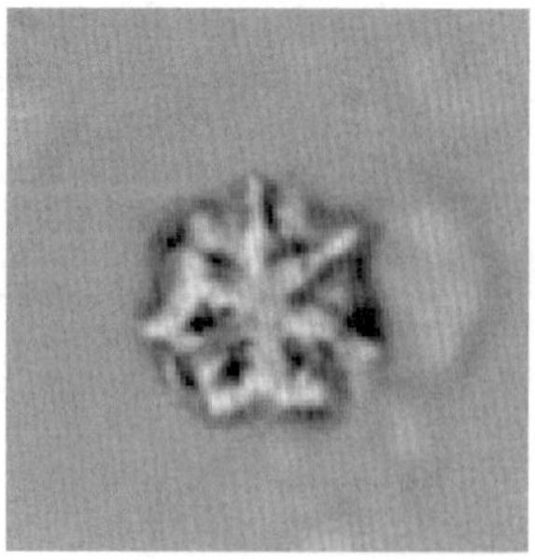

14
Catinaster coalitus
[X1600] N.L. depth 3875 m
Well B

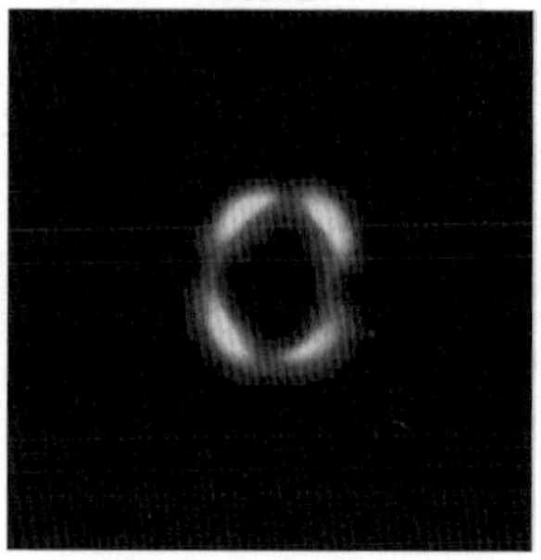

15
Coronocycus nitescens
[X1600] N.L. depth 855 m
Well B

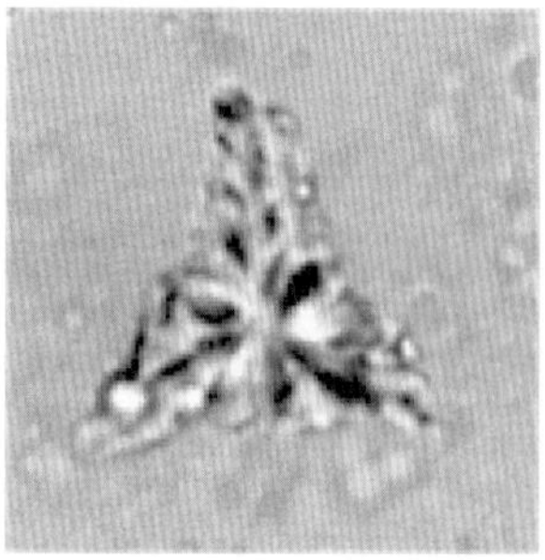

16
Lithostromation perdurum
[X1600] N.L. depth 855 m
Well B

Plate IV. Nannofossils.

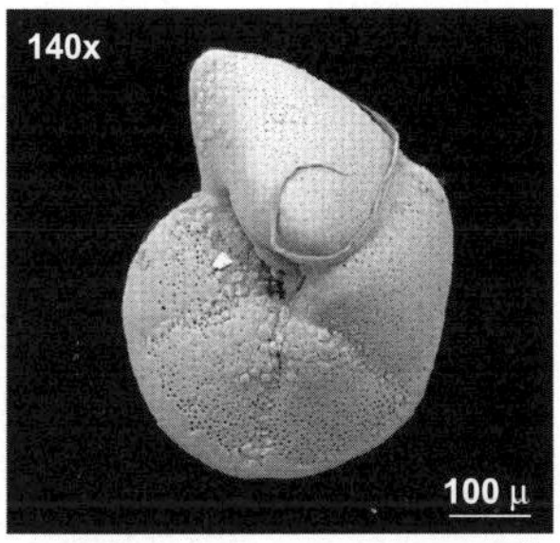

1
Globorotalia truncatulinoides truncatulinoides depth 805 m
Well A

2
Globigerinoides trilobus fistulosus depth 995 m
Well A

3
Globigerinoides obliquus extremus depth 1005 m
Well A

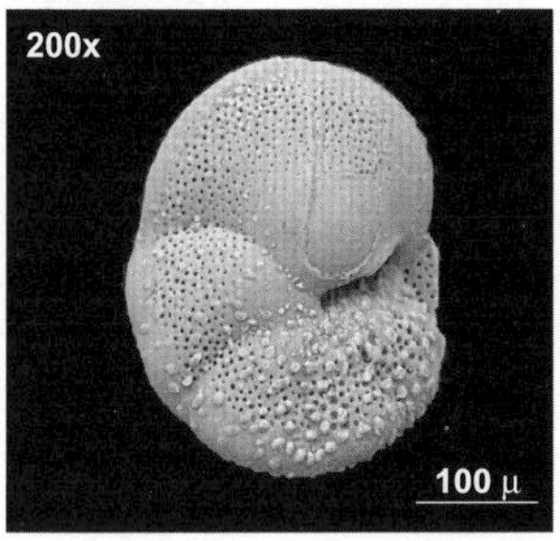

4
Globorotalia margaritae
depth 1675 m
Well A

5
Globigerina nepenthes
depth 2905 m
Well A

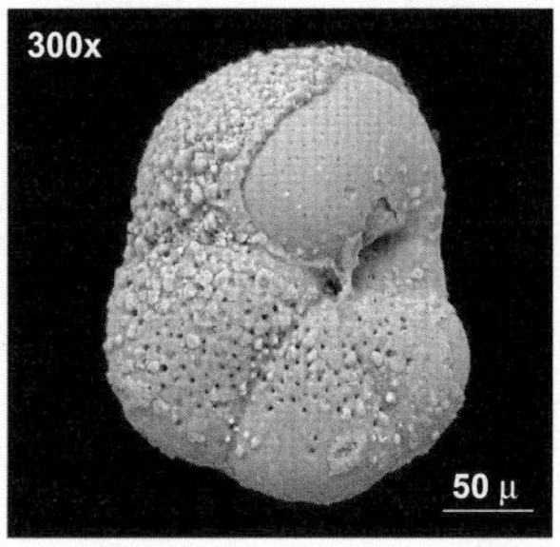

6
Globorotalia juanai
depth 2505 m
Well A

7
Globorotalia mayeri
depth 3385 m
Well A

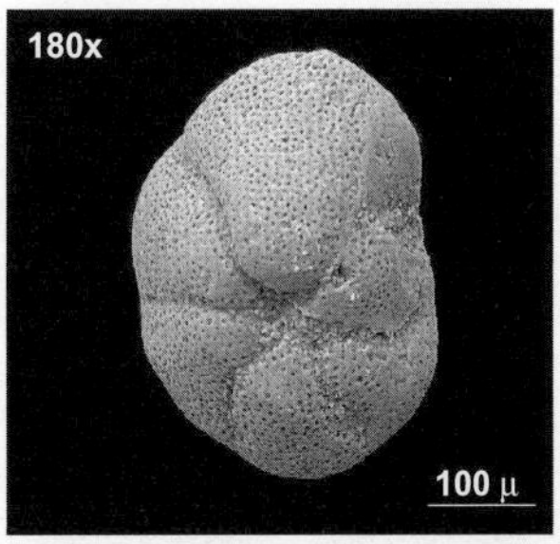

8
Globorotalia fohsi peripheroacuta
depth 4615 m
Well A

9
Globorotalia fohsi peripheroronda
depth 1310 m
Well A

Plate V. Foraminifera.

PLATE VI. Foraminifera.

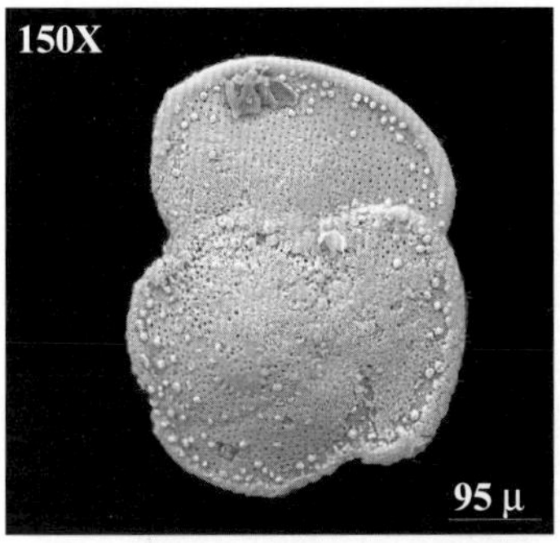

1
Globorotalia crassaformis viola
depth 590 m
Well B

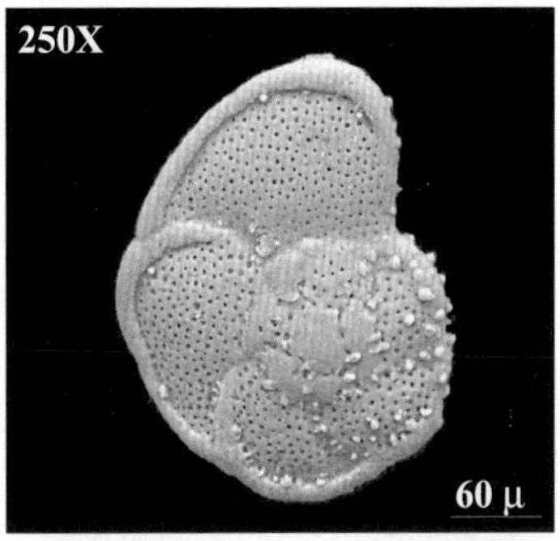

2
Globorotalia tosaensis tenuitheca
depth 600 m
Well B

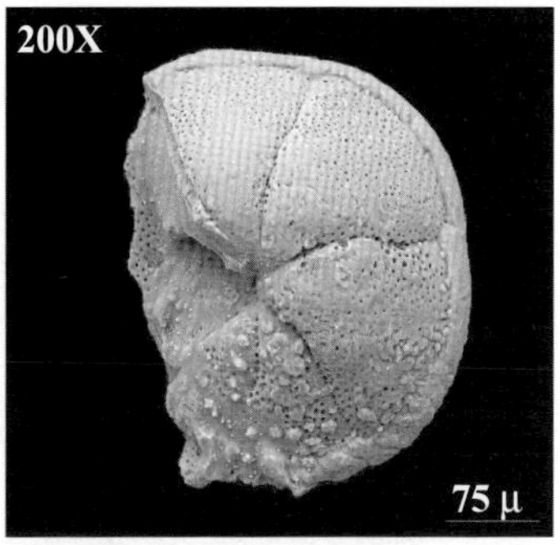

3
Globorotalia miocenica
depth 1450 m
Well B

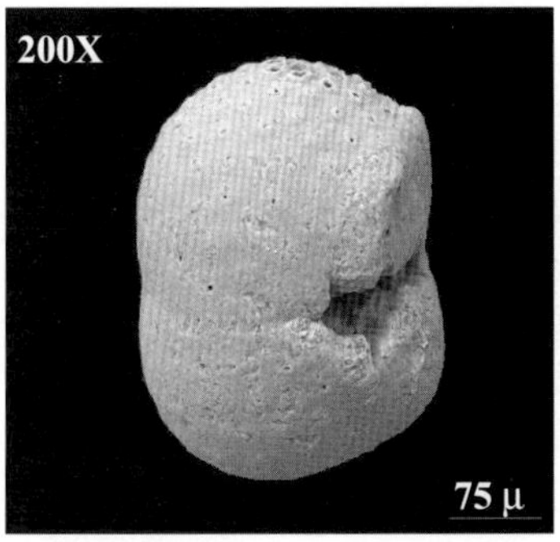

4
Sphaeroidinellopsis seminulina
depth 3600 m
Well B

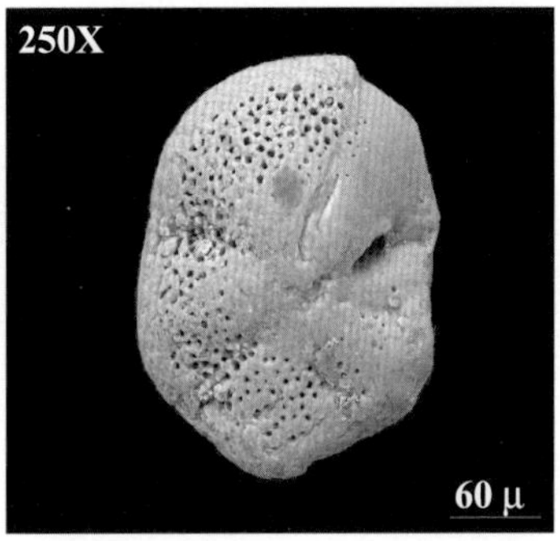

5
Globorotalia margaritae
depth 2230 m
Well B

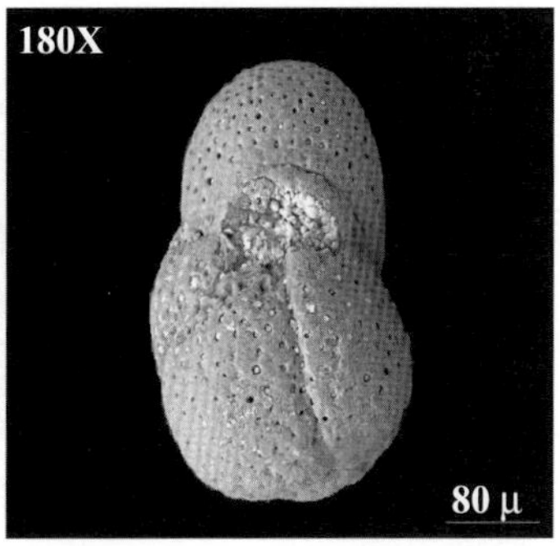

6
Globigerina nepenthes
depth 2100 m
Well B

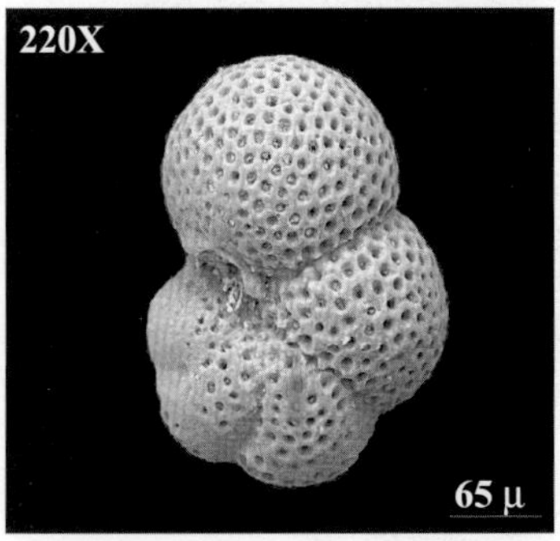

7
Globorotaloides variabilis
depth 3600 m
Well B

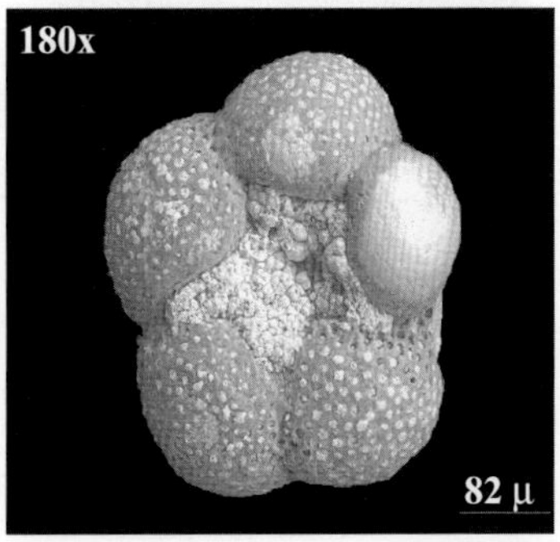

8
Globoquadrina altispira
globosa depth 3895 m
Well B

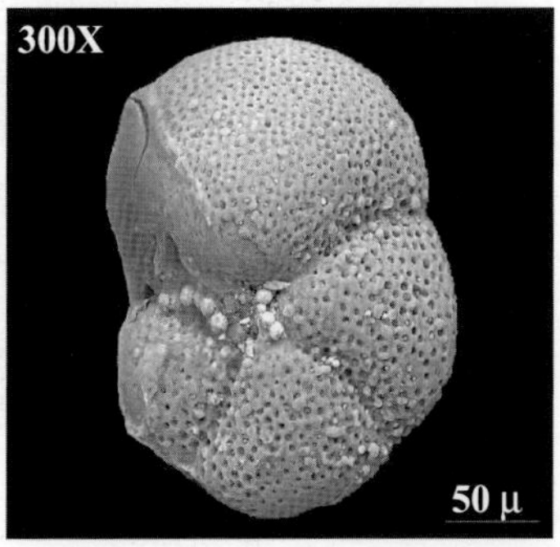

9
Globorotalia cf. juanai
depth 3470 m
Well B

Plate VII. Foraminifera.

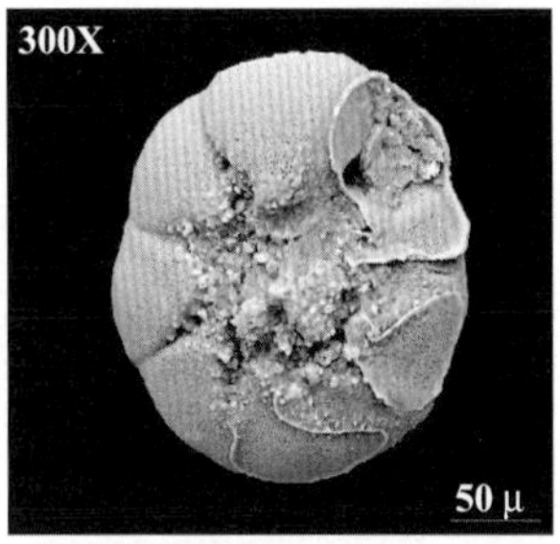

1
Ammonia beccarii
depth 620 m
Well B

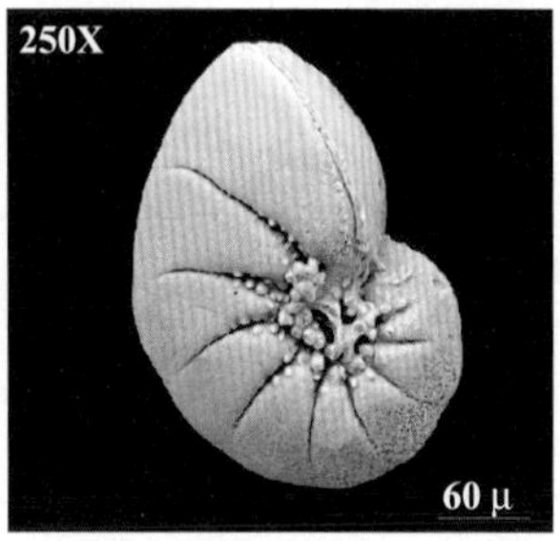

2
Florilus sp.
depth 620 m
Well B

3
Uvigerina hispidocostata
depth 640 m
Well B

4
Uvigerina sp.
depth 640 m
Well B

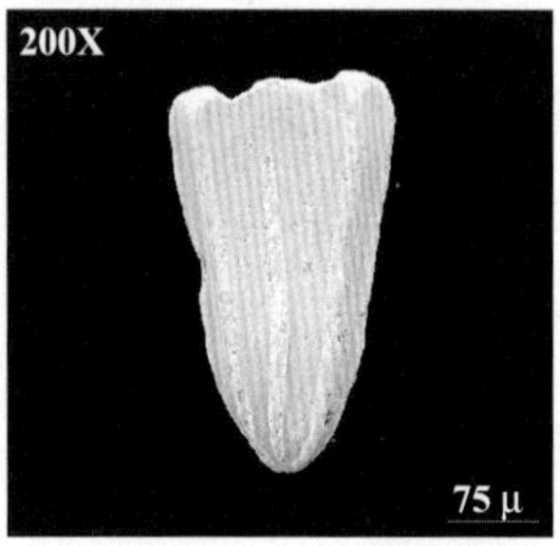

5
Plectofrondicularia californica
depth 2250 m
Well B

6
Sigmoilopsis schlumbergeri
depth 2460 m
Well B

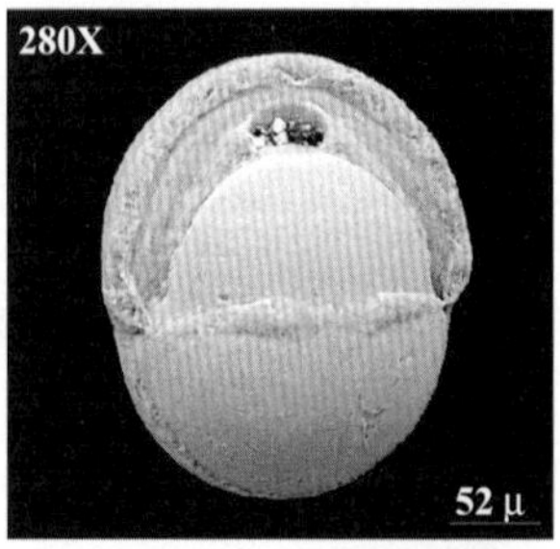

7
Pullenia bulloides
depth 2950 m
Well B

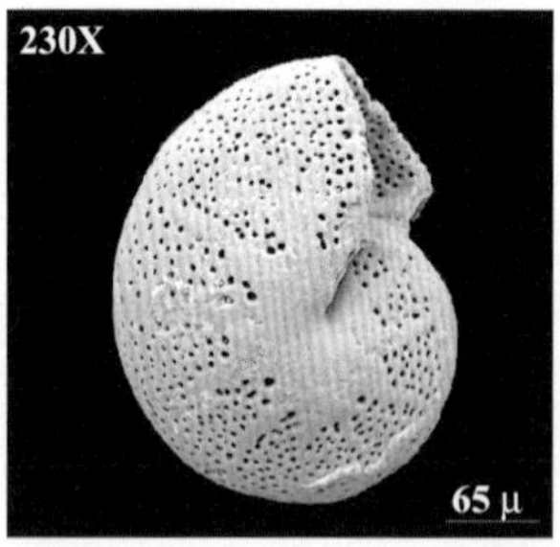

8
Melonis affinis
depth 2220 m
Well B

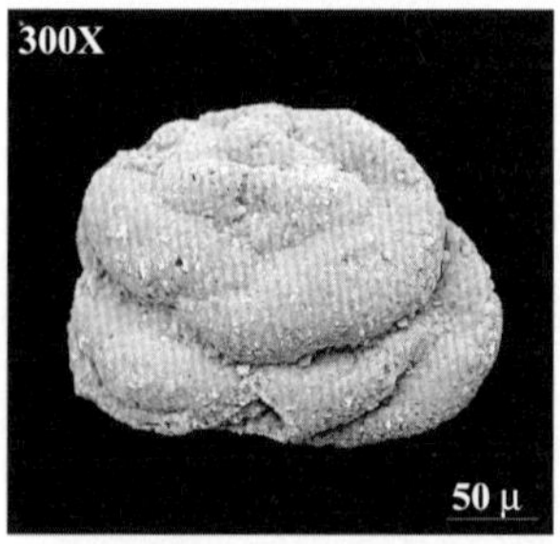

9
Usbekistania charoides
depth 3663 m
Well B

Plate VIII. Foraminifera.

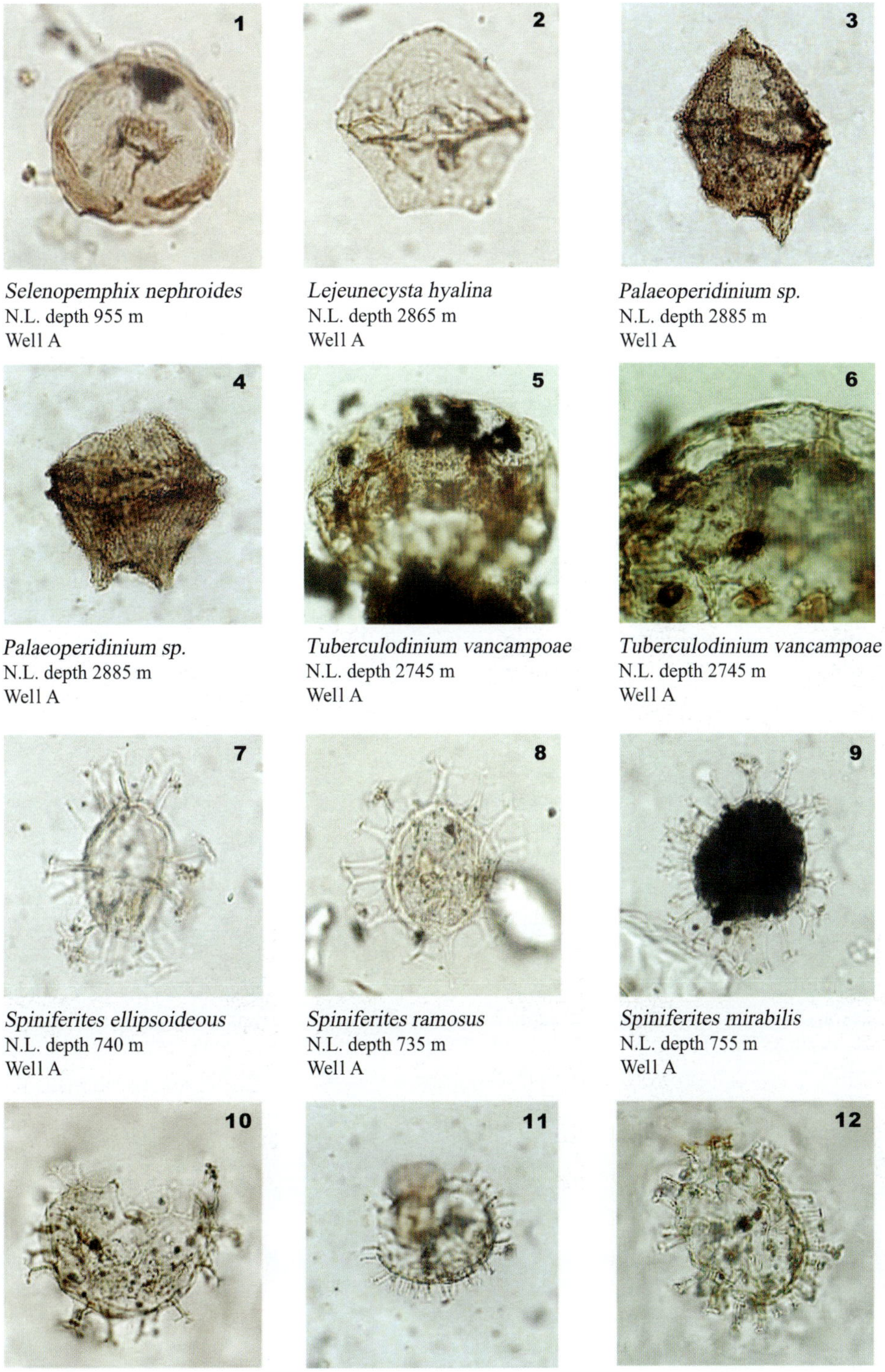

1. *Selenopemphix nephroides*
N.L. depth 955 m
Well A

2. *Lejeunecysta hyalina*
N.L. depth 2865 m
Well A

3. *Palaeoperidinium sp.*
N.L. depth 2885 m
Well A

4. *Palaeoperidinium sp.*
N.L. depth 2885 m
Well A

5. *Tuberculodinium vancampoae*
N.L. depth 2745 m
Well A

6. *Tuberculodinium vancampoae*
N.L. depth 2745 m
Well A

7. *Spiniferites ellipsoideous*
N.L. depth 740 m
Well A

8. *Spiniferites ramosus*
N.L. depth 735 m
Well A

9. *Spiniferites mirabilis*
N.L. depth 755 m
Well A

10. *Sphiniferites hyperacanthus*
N.L. depth 2525 m
Well A

11. *Polysphaeridium zoharyi*
N.L. depth 855 m
Well A

12. *Polysphaeridium zoharyi ktana*
N.L. depth 2585 m
Well A

Plate IX. Palynomorphs.

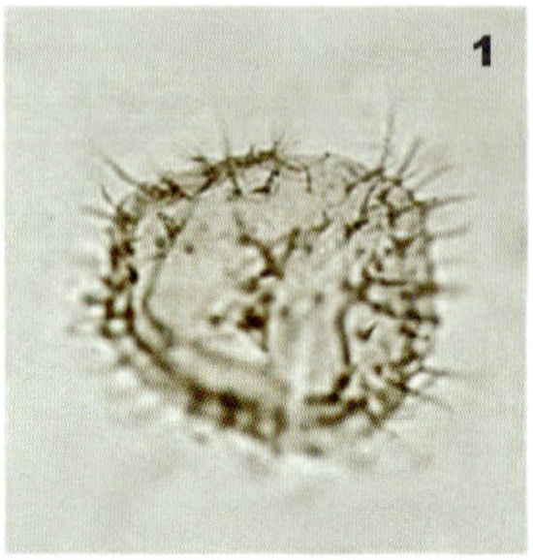

Multispinula quanta
N.L. depth 710 m
Well A

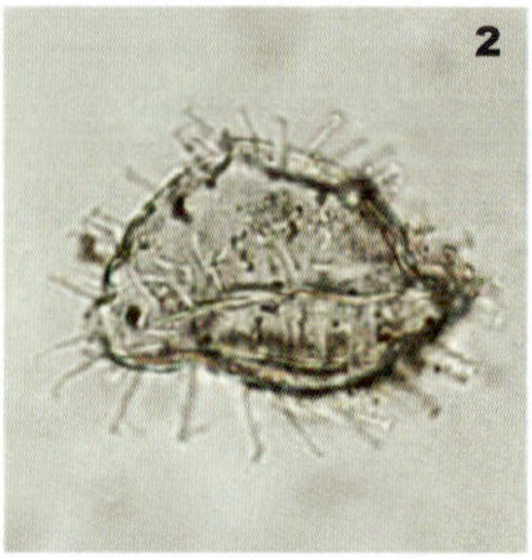

Sumatrodinium hispidum
N.L. depth 775 m
Well A

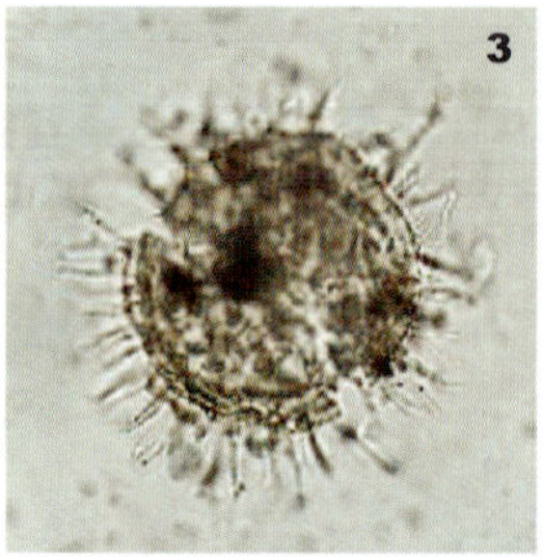

Operculodinium centrocarpum
N.L. depth 895 m
Well A

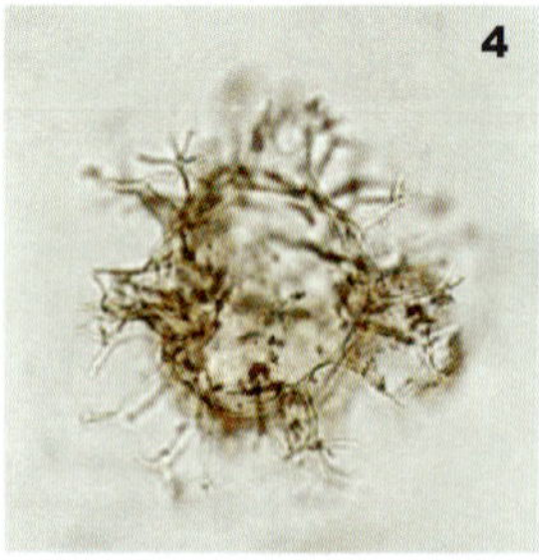

Achomosphaera andalusiensis
N.L. depth 2505 m
Well A

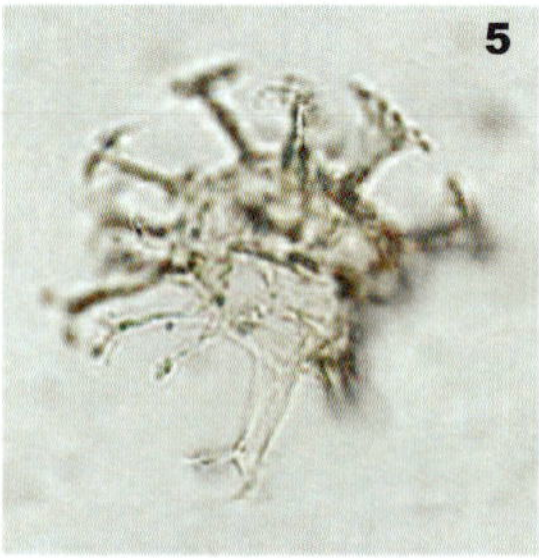

Oligosphaeridium complex
N.L. depth 2745 m
Well A

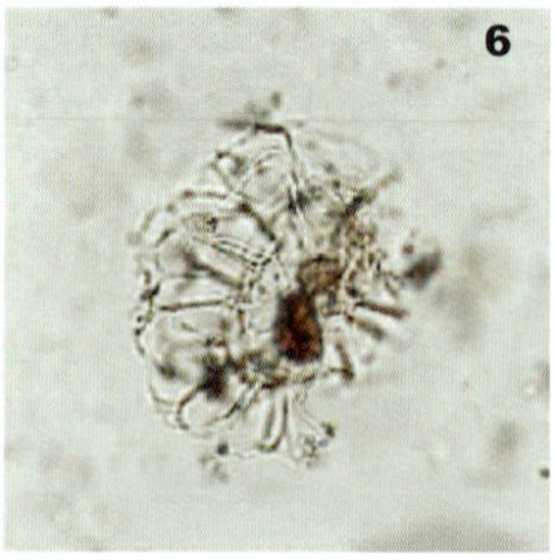

Nematosphaeropsis labyrinthea
N.L. depth 1455 m
Well A

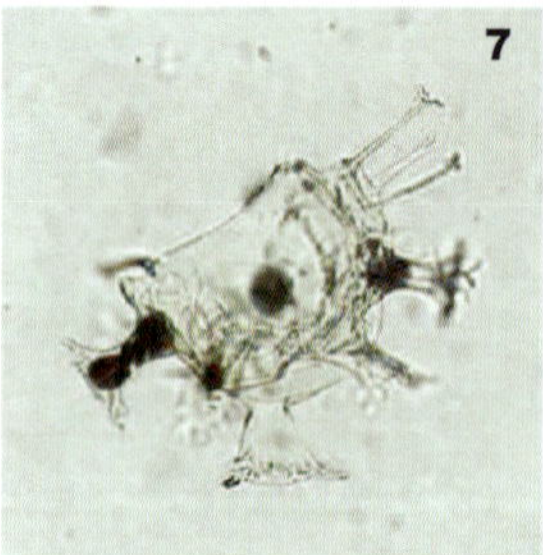

Homotryblium tenuispinosum
N.L. depth 855 m
Well A

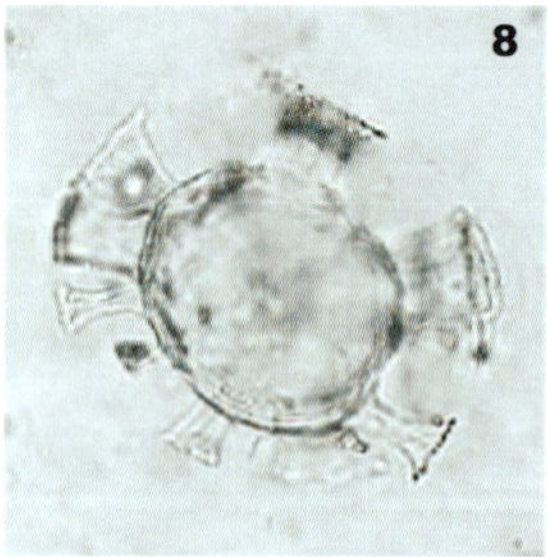

Homotryblium plectilum
N.L. depth 755 m
Well A

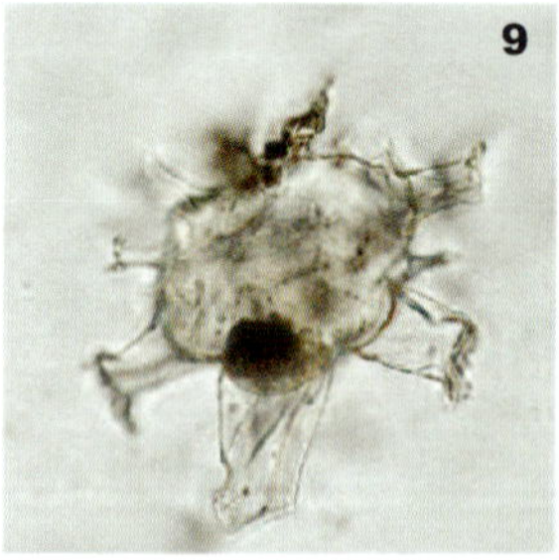

Hystricholcopoma rigaudiae
N.L. depth 755 m
Well A

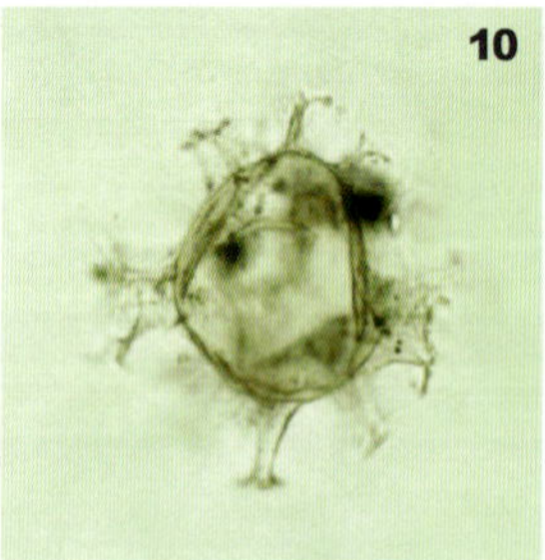

Spiniferites membranaceus
N.L. depth 715 m
Well A

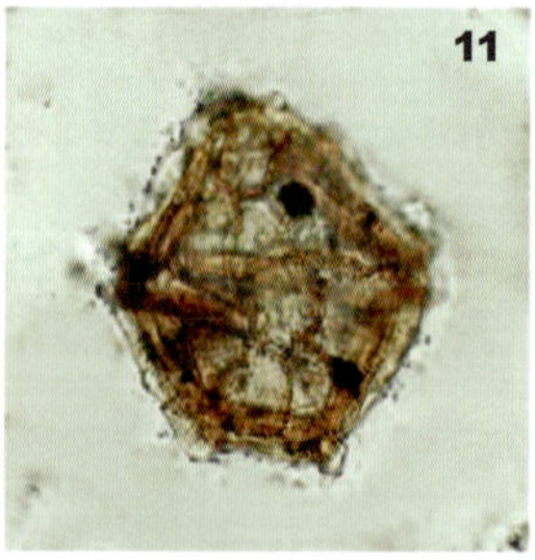

Impagidinium strialatum
N.L. depth 2585 m
Well A

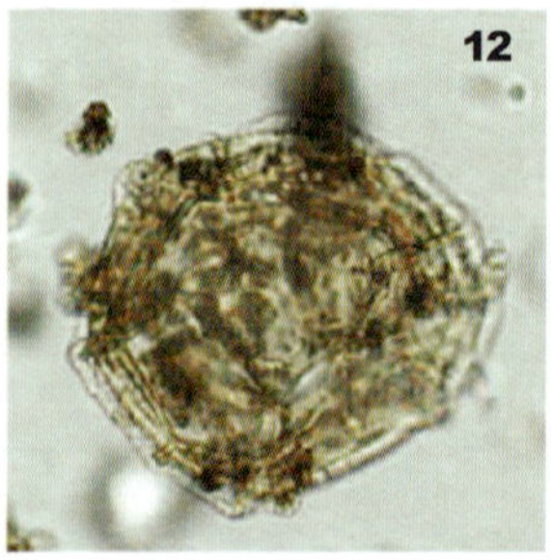

Impagidinium sp.
N.L. depth 3045 m
Well A

Plate X. Palynomorphs.

Quilonia allepeyensis
N.L. depth 875 m
Well A

Phragmothyrites kiandresis
N.L. depth 985 m
Well A

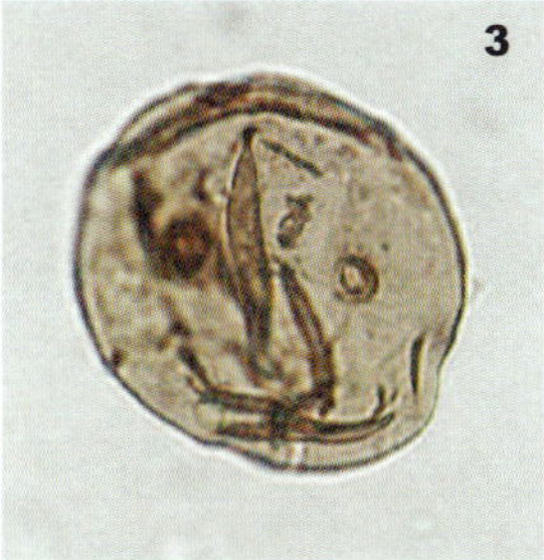

Diporipollis sp.
N.L. depth 755 m
Well A

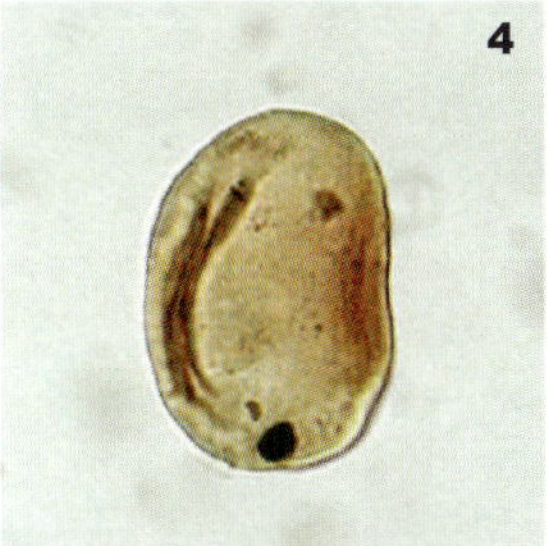

Laevigatosporites nutidus
N.L. depth 2805 m
Well A

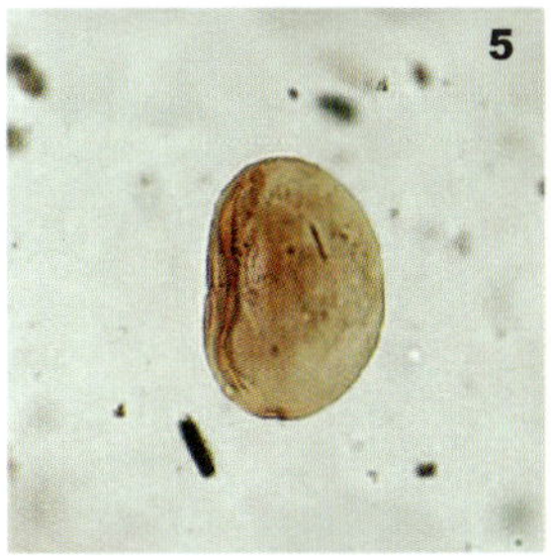

Laevigatosporites major
N.L. depth 985 m
Well A

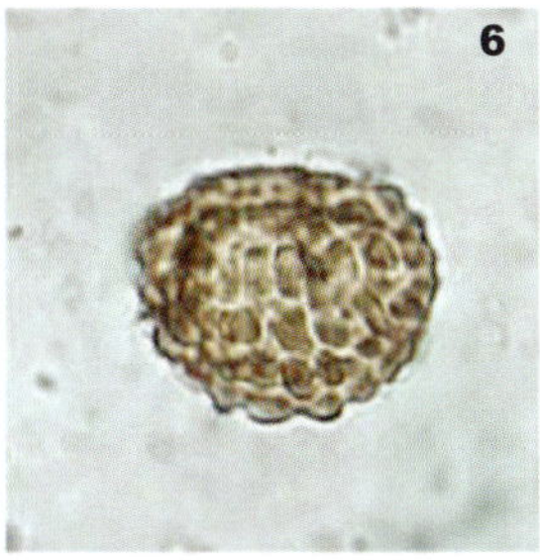

Polypodidiites usmensis
N.L. depth 1095 m
Well A

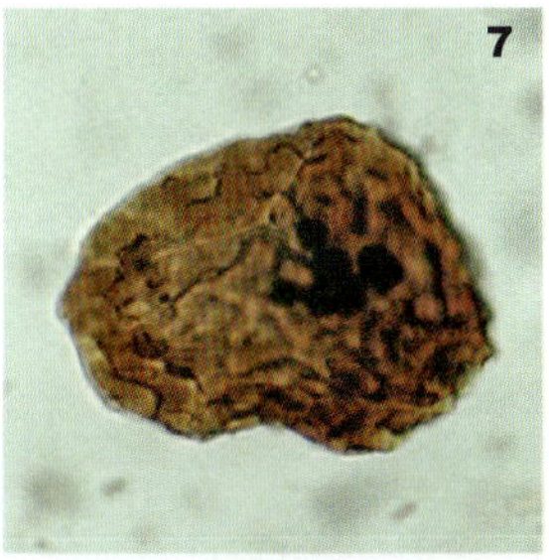

Crassoretitriletes vanraadshooveni
N.L depth 2525 m
Well A

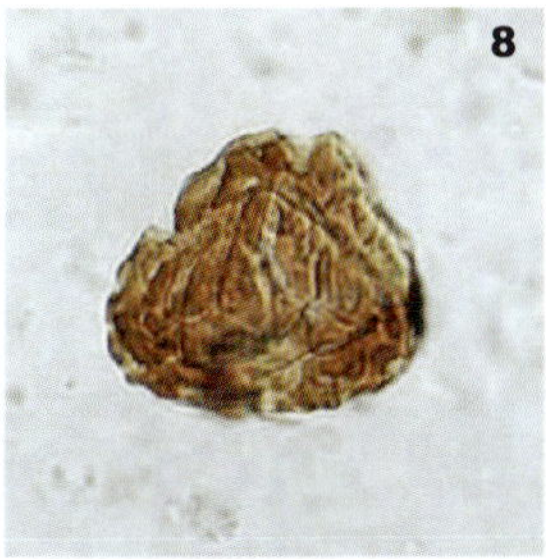

Contignisporites multimuratus
N.L. depth 1455 m
Well A

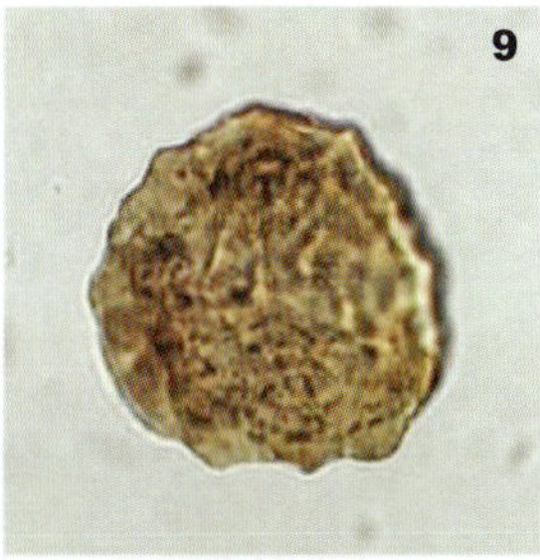

Criptogramasporis sp.
N.L. depth 1095 m
Well A

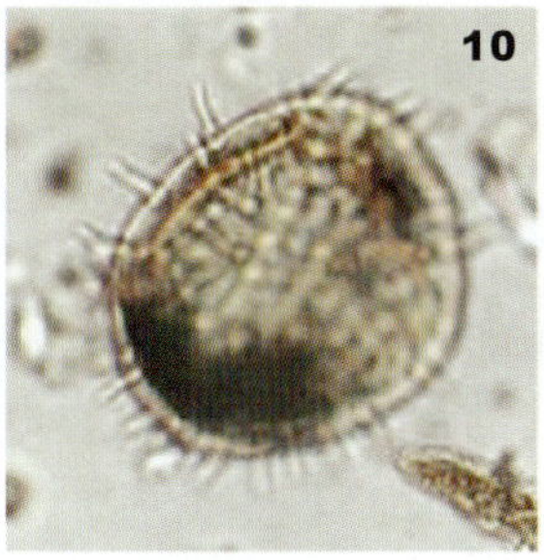

Spinizonocolpites echinatus
N.L. depth 1075 m
Well A

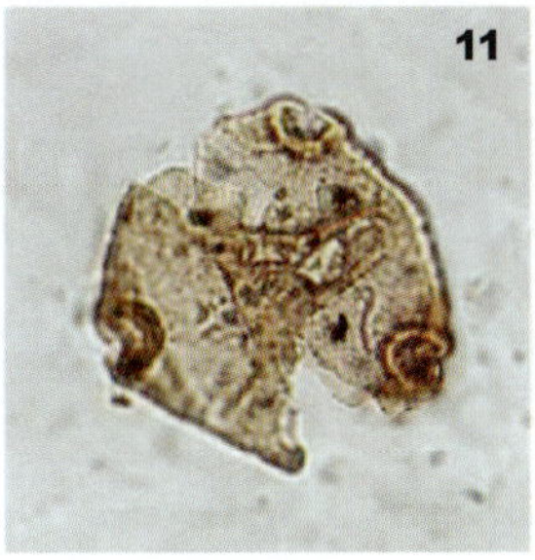

Corsinipollenites sp.
N.L. depth 1075 m
Well A

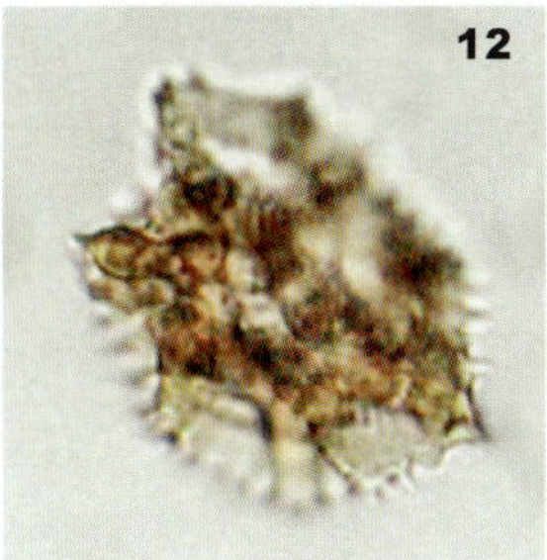

Fenestrites spinosus
N.L. depth 1055 m
Well A

Plate XI. Palynomorphs.

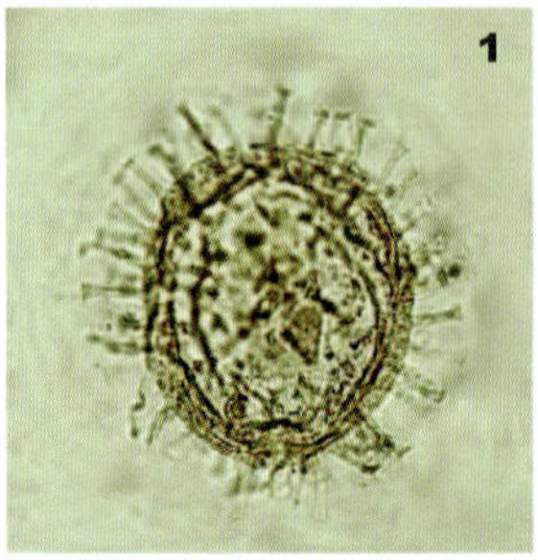

Polysphaeridium zoharyi
N.L. depth 1750 m
Well B

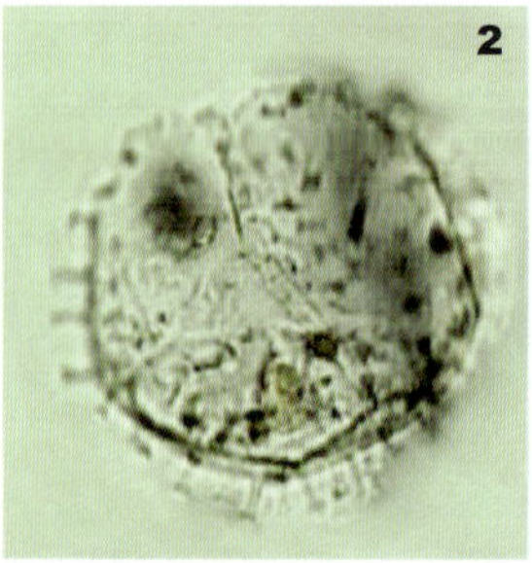

Polysphaeridium zoharyi ktana
N.L. depth 1210 m
Well B

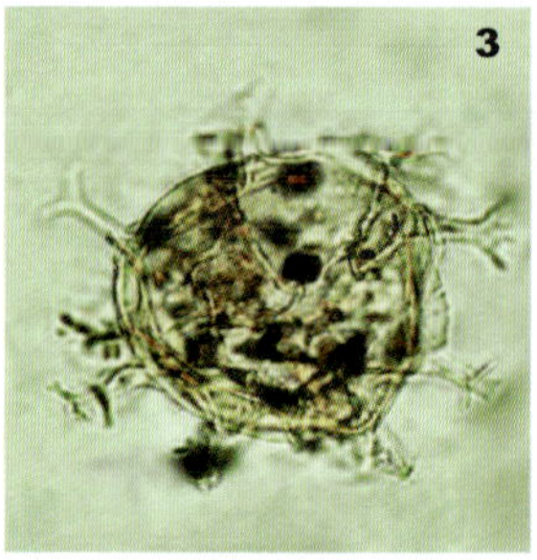

Spiniferitus delicatus
N.L. depth 870 m
Well B

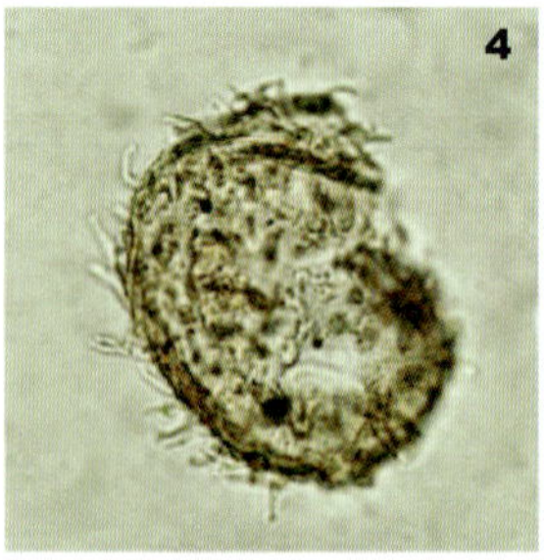

Sumatradinium hispidum
N.L. depth 2620 m
Well B

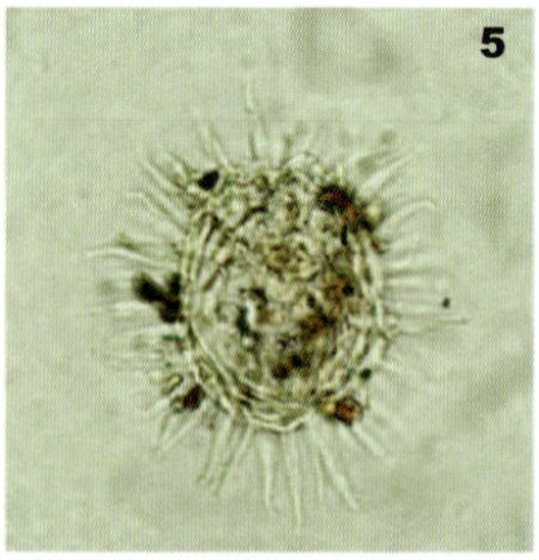

Lingulodinium machaerophorum
N.L. depth 950 m
Well B

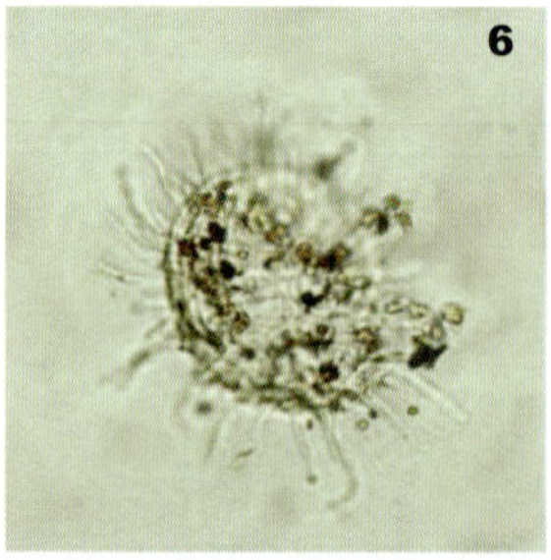

Operculodinium centrocarpum
N.L. depth 950 m
Well B

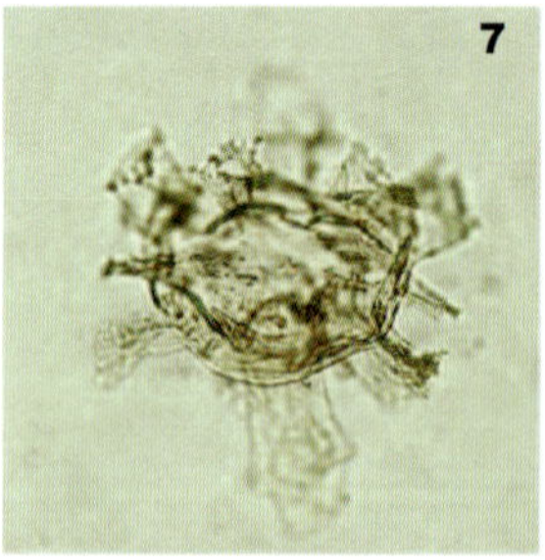

Hystricholcolpoma rigaudiae
N.L. depth 1750 m
Well B

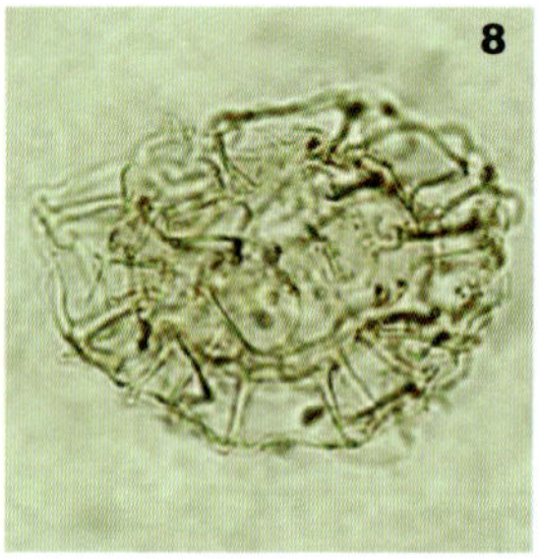

Nematosphaeroysis labyrinthea
N.L. depth 1500 m
Well B

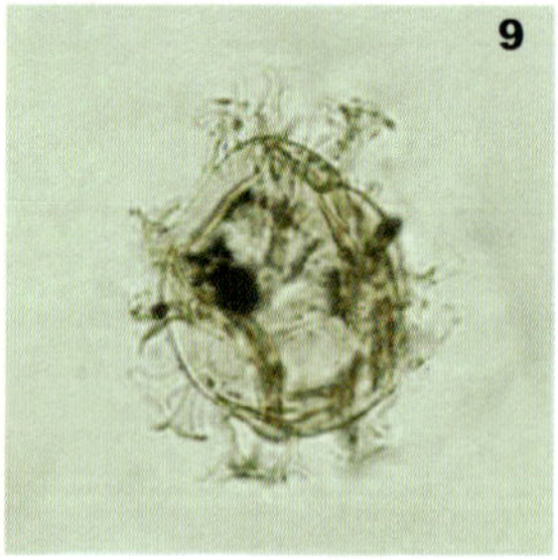

Homotryblium plectilum
N.L. depth 1310 m
Well B

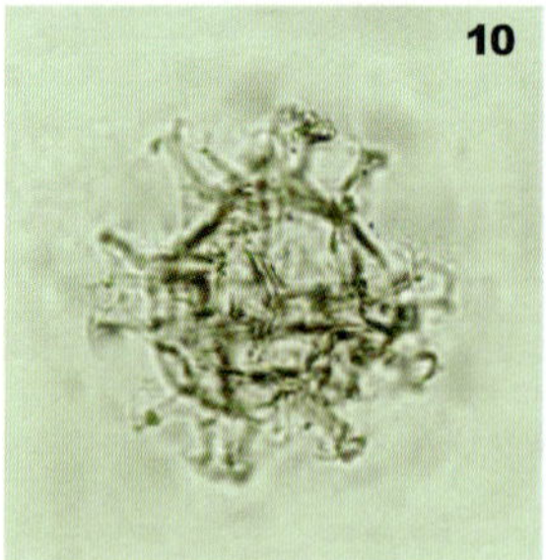

Spiniferites membranaceous
N.L. depth 1960 m
Well B

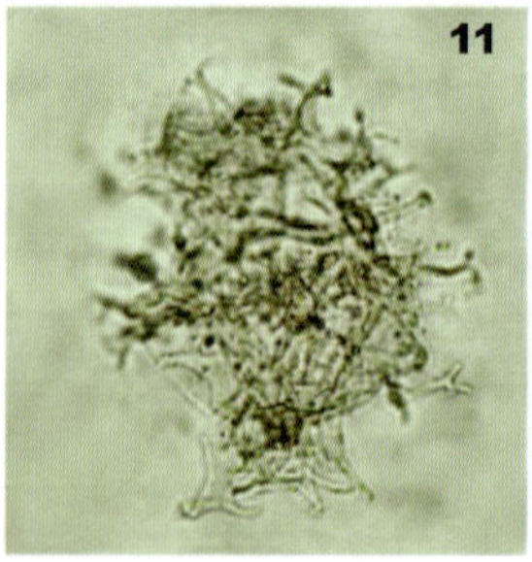

Sphiniferites mirabilis
N.L. depth 3090 m
Well B

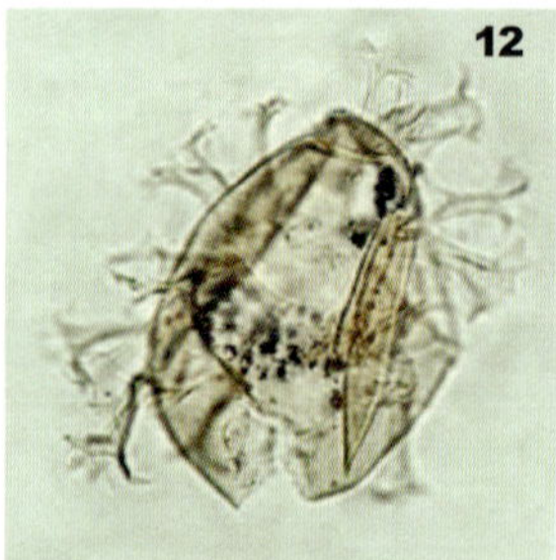

Spiniferites andalusiensis
N.L. depth 1250 m
Well B

Plate XII. Palynomorphs.

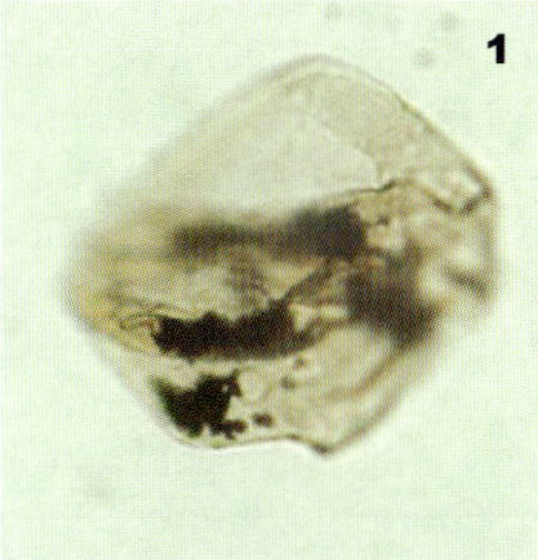

Lejeunecysta diversiforma
N.L. depth 1490 m
Well B

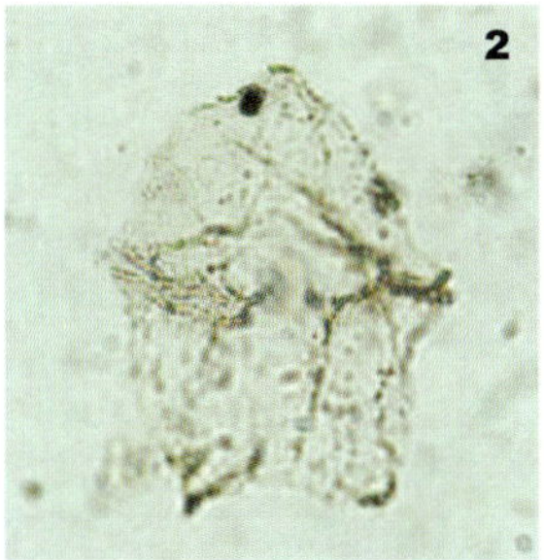

Lejeunecysta tenella
N.L. depth 3506 m
Well B

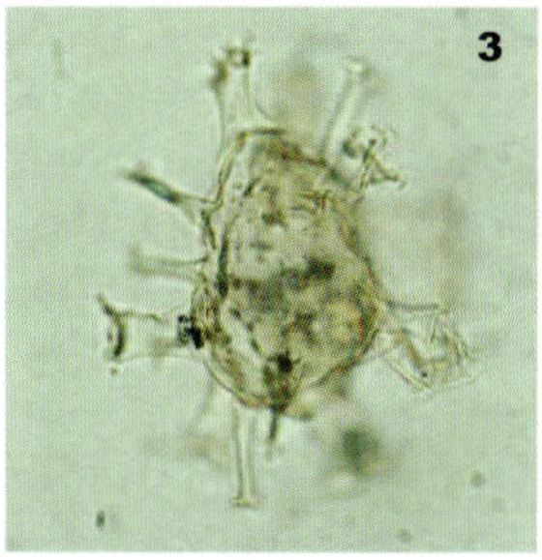

Homotryblium oceanicum
N.L. depth 1620 m
Well B

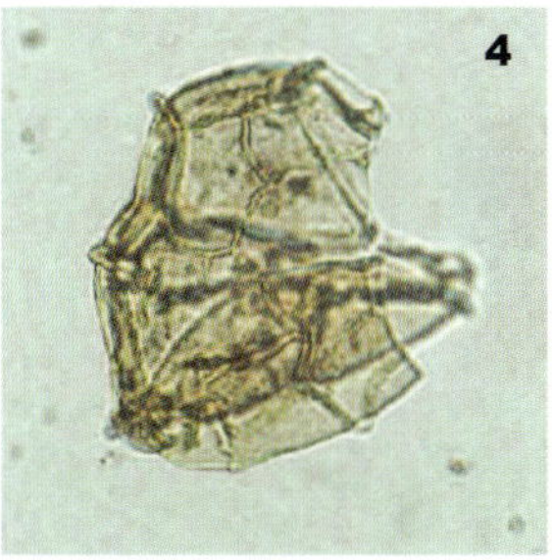

Impagidinium sp.
N.L. depth 2250 m
Well B

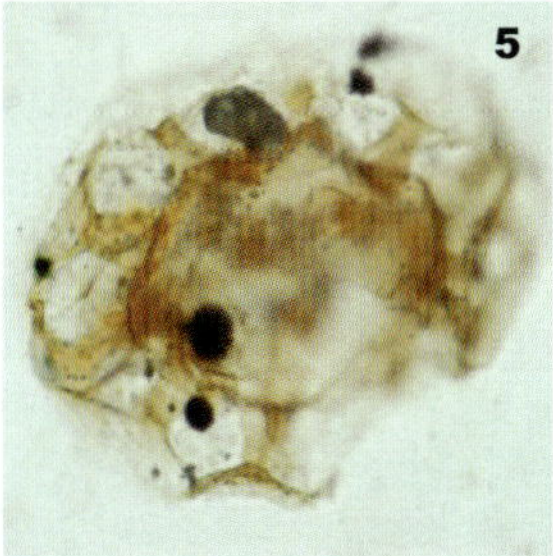

Tuberculodinium vancampoae
N.L. depth 2350 m
Well B

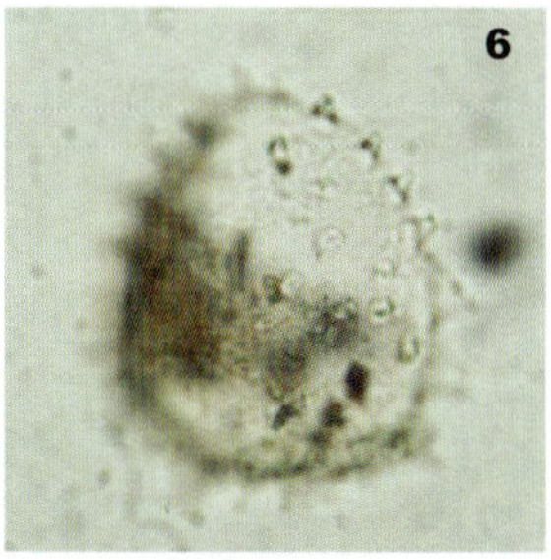

Operculodinium janduchenei.
N.L. depth 830 m
Well B

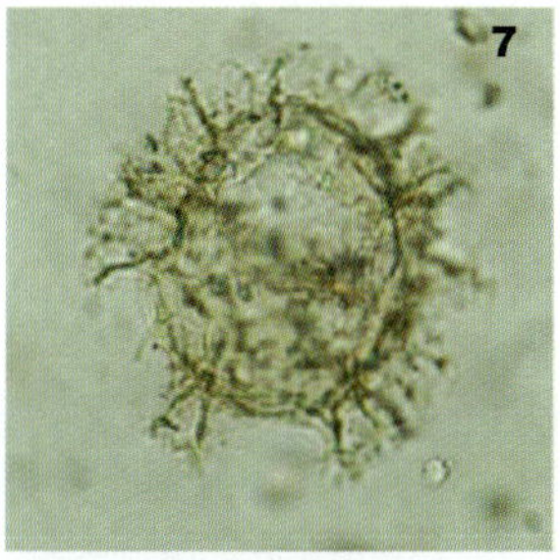

Spiniferites ramosus
N.L. depth 3090 m
Well B

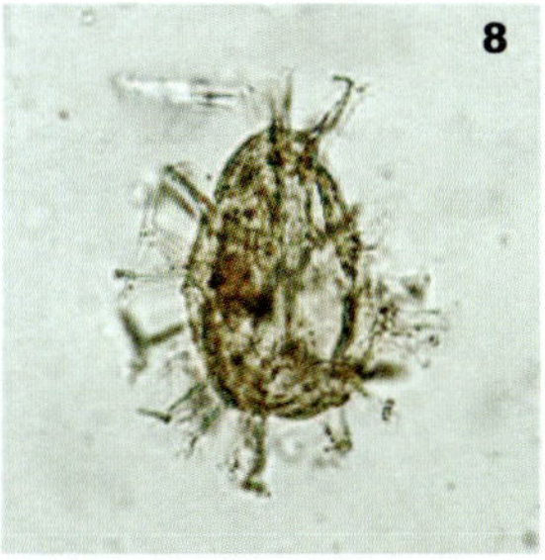

Spiniferites cf. elongatus
N.L. depth 2930 m
Well B

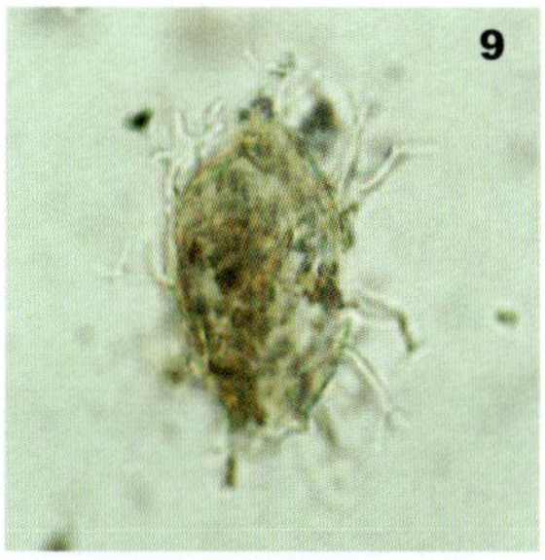

Spiniferites cf. ellipsoideus
N.L. depth 930 m
Well B

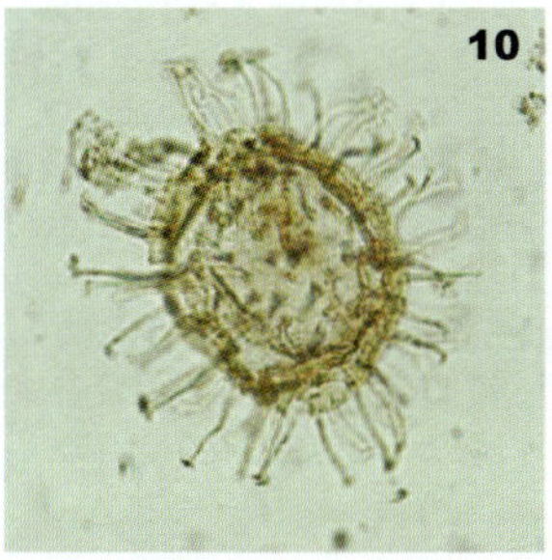

Dapsilidinium pastielsii
N.L. depth 3640 m
Well B

Frasnacritetrus taugourdeaui
N.L. depth 1230 m
Well B

Cicatricosisporites claricristata
N.L. depth 2410 m
Well B

Plate XIII. Palynomorphs.

ACKNOWLEDGMENTS

Special thanks to Adán Oviedo Pérez, former subdirector of the Exploration Technical Coordination for authorizing the publication of this work. Our special gratitude goes to Mario Limón González, former manager of Regional Geologic Models, for coordinating this research project. The same appreciation is extended to Fernando Navarro Vaca and Jorge E. Lugo. We also thank Dionisio Rodríguez Figueroa of the Regional Exploration Activo in Tampico for the amenities provided to develop this project. Our recognition goes to Yolanda Cabañas P., Daniel Pérez Rodríguez, Rafael Espejel Correa, and Miguel Martínez Nevárez for their technical support. We are indebted to Rashel Rosen, Mark Jiang, Norman Rosen, and Christopher Denison for their detailed and constructive review of the manuscript.

REFERENCES CITED

Aubry, M. P., 1984–1987, Handbook of Cenozoic calcareous nannoplankton: Micropaleontology handbook series, New York: Micropaleontology Press, The American Museum of Natural History, v. 1–4, 1205 p.

Aubry, M. P., 1993, Neogene allostratigraphy and depositional history of the De Soto Canyon area, northern Gulf of Mexico: Micropaleontology, v. 39, no. 4, figs. 1–19, and tables 1–5, p. 327–366, plates 1–3.

Banner, F. T., and W. H. Blow, 1965, Progress in the planktonic foraminiferal biostratigraphy of the Neogene: Nature, v. 208, p. 1164–1166.

Berggren, W. A., D. V. Kent, M. P. Aubry, and J. Hardenbol, 1995a, A revised geochronology and chronostratigraphy, geochronology, times scales and global stratigraphic correlation: Society for Sedimentary Geology Special Publication 54, p. 129–212.

Berggren, W. A., F. J. Hilgen, C. G. Langereis, D. V. Kent, J. D. Obradovich, I. Raffi, M. E. Raymo, and N. J. Shackeleton, 1995b, Late Neogene chronology: New perspectives in high-resolution stratigraphy: Geological Society of America Bulletin, v. 107, p. 1272–1287.

Bolli, H. M., and J. B. Saunders, 1985, Oligocene to Holocene low latitude planktonic foraminifera, *in* H. M. Bolli, J. B. Saunders, and K. Perch-Nielsen, eds., Plankton stratigraphy: New York, University Press, p. 155–328.

Bown, P. R., 1999, Calcareous nannofossil biostratigraphy: London, Kluwer Academia Publishers, 313 p.

Drugg, W. S., and L. E. Stover, 1975, Stratigraphic range charts of selected fossil dinoflagellates—Selected Cenozoic dinoflagellates, *in* W. R. Evitt, ed., Proceedings of a Forum on Dinoflagellates: American Association of Stratigraphic Palynologists Contribution Series 4, p. 73–90.

Echánove, E. O., 1986, Geología petrolera de la Cuenca de Burgos: Parte 1. Consideraciones geológico petroleras: Boletín de la Asociación Mexicana de Geólogos Petroleros, v. 38, no. 1, p. 3–74.

Germeraad, J. H., C. A. Hopping, and J. Muller, 1968, Palynology of Tertiary sediments from tropical areas: Review of Paleobotany and Palynology, v. 6, p. 189–348.

González-Lara, J. C., 1994, Estudio bioestratigráfico de una secuencia arcillo-arenosa del Neógeno, basado en foraminíferos planctónicos y bentónicos del sur de Veracruz y noroeste de Chiapas: Tesis de Maestría en Ciencias, Instituto Politécnico Nacional, México, D.F., 120 p.

González-Lara, J. C., 2000, Le Paléocène du Chiapas (SE du Mexique): Biostratigraphie, Sédimentologie et Stratigraphie Sequentielle: Thèse du Doctorat, Université Joseph Fourier, Grenoble, France, 236 p.

Head, J. M., and J. H. Wrenn, eds., 1992, Neogene and Quaternary dinoflagellate cysts and acritarchs: Elsevier, Amsterdam, The Netherlands, American Association of Stratigraphic Palynologists Foundation, 432 p.

Helenes, J., and D. Somoza, 1999, Palynology and sequence stratigraphy from the Cretaceous of eastern Venezuela: Cretaceous Research, v. 20, p. 447–463.

Helenes, J., C. De Guerra, and J. Vasquez, 1998, Palynology and chronostratigraphy of the Upper Cretaceous in the subsurface of the Barinas area, western Venezuela: AAPG Bulletin, v. 82, no. 7, p. 1308–1328.

Kalgutkar, R. M., and J. Jansonius, 2000, Synopsis of fossil fungal spores, mycelia and fructifications: American Association of Stratigraphic Palynologists Foundation Contributions Series 39, 429 p.

Lipps, J. H., 1979, Ecology and paleoecology of planktonic foraminifera: SEPM, no. 6, p. 62–104.

Lipps, J. H., 1993, Fossil prokaryotes and protists: Boston, Blackwell Scientific Publication, 342 p.

Lorente, M. A., 1986, Palynology and palynofacies of the upper Tertiary in Venezuela: Ph.D. dissertation, Dissertationes Botanicae, Universiteit van Amsterdam, Band 99, 222 p.

Matsuoka, K., and J. P. Bujak, 1988, Cenozoic dinoflagellate cysts from the Navarin Basin, Sound and St. George Basin, Bering Sea: Bulletin of the Faculty of Liberal Arts, Nagasaki University, Natural Science, v. 29, no. 1, p. 1–147.

Morkhoven, V., W. A. Berggren, and A. S. Edwards, 1986, Cenozoic cosmopolitan deep-water benthic foraminifera: Bulletin des Centres Recherches Exploration-Production Elf-Aquitaine, Mèmoire 11, Pau, p. 422.

Perch-Nielsen, 1985, Oligocene to Holocene low latitude planktonic foraminifera. Plankton stratigraphy, *in* H. M. Bolli, J. B. Saunders, and K. Perch-Nielsen, eds., Cambridge Earth Science Series 1, p. 155–262.

Postuma, J. A., 1971, Manual of planktonic foraminifera: Amsterdam, New York, Elsevier Publishing Company, 420 p.

Powell, A. J., ed., 1992, A stratigraphic index of dinoflagellate cysts: London, Chapman & Hall, 290 p.

Robles, R. M. L., 1967, Catálogo Ilustrado de la Superfamilia Globigerinacea, Laboratorio de Micropaleontología del Cenozoico: México, D.F., Instituto Mexicano del Petróleo, 98 p.

Salmerón-Ureña, P., 2002, Uso de datos biosedimentológicos para determinar la columna cronoestratigráfica y paleobatimétrica del Terciario, en los pozos Tixán-1 y Chem-1A, de la Sonda de Campeche, México: México, D.F., Universidad Nacional Autónoma de México, p. 251–265.

Sánchez-Rios, M. A., et al., 2000, Bioestratigrafia de alta resolución y su aplicación en la estratigrafia de secuencias en sedimentos del Mioceno-Pleistoceno de la Cuenca de Macuspana, Instituto Mexicano del Petróleo, Informe Inédito, 12 cuadros, 65 p.

Sánchez-Rios, M. A., et al., 2001, Bioestratigrafía de alta resolución y su aplicación en la estratigrafía de secuencias en sedimentos del Mioceno–Plioceno de las cuencas de Burgos, Veracruz y Macuspana: México, D.F., Instituto Mexicano del Petróleo, Informe Inédito, 19 cuadros, 20 láminas, 47 p.

Sánchez-Rios, M. A., et al., 2002a, Bioestratigrafía de alta resolución y su aplicación en la estratigrafía de secuencias en sedimentos del Mioceno–Pleistoceno de la Cuenca de Macuspana: México, D.F., Instituto Mexicano del Petróleo, Informe Inédito, 11 figs., 19 cuadros, 37 p.

Sánchez-Rios, M. A., et al., 2002b, Bioestratigrafía de alta resolución del Terciario en Pozos de la Cuenca de Veracruz y la Sonda de Campeche: México, D.F., Instituto Mexicano del Petróleo, Informe Inédito, 15 figs., 28 cuadros, 65 p.

Sánchez-Rios, M. A., et al., 2003, Bioestratigrafía de alta resolución en sedimentos del Neógeno del Golfo de México: México, D.F., Instituto Mexicano del Petróleo, Informe Inédito, 26 figs., 19 cuadros, 25 láminas, 145 p.

Sánchez-Rios, M. A., et al., 2004, Bioestratigrafía de alta resolución en sedimentos del Neógeno del Golfo de México: México, D.F., Instituto Mexicano del Petróleo, Informe Inédito, 56 figs., 21 cuadros, 48 láminas, 135 p.

Sánchez-Rios, M. A., et al., 2005, Bioestratigrafía de alta resolución en sedimentos del Neógeno del Golfo de México: México, D.F., Instituto Mexicano del Petróleo, Informe Inédito, 33 láminas, 43 figs., 15 cuadros, 182 p.

Sánchez -Rios, M. A., et al., 2006, Bioestratigrafía de alta resolución en sedimentos del Neógeno del Golfo de México: México, D.F., Instituto Mexicano del Petróleo, Informe Inédito, 36 figs., 20 cuadros, 107 p.

Sansores, J. C., and C. Flores-Covarrubias, 1972, Foraminíferos bentónicos del Terciario Superior de la Cuenca Salina del Istmo de Tehuantepec, México: México, D.F., Instituto Mexicano del Petróleo, v. 1 and 2, 535 p.

Santiago, A. J., B. J. Carrillo, and A. Martel, 1984, Geología Petrolera de México. Evaluación de Formaciones en México: México, D.F., Schlumberger, p. 1–36.

Stainforth, R. M., J. L. Lamb, H. Luterbacher, J. H. Beard, and R. S. Jeffords, 1975, Cenozoic planktonic foraminiferal zonation and characteristics of index forms, University of Kansas, Paleontological Contributions, Art. 62, pt. 426.

Williams, G. L., and J. P. Bujak, 1985, Mesozoic and Cenozoic dinoflagellates, *in* H. M. Bolli, J. B. Saunders, and K. Perch-Nielsen, eds., Plankton Stratigraphy, Cambridge Earth Science Serie, p. 847–964.

Williams, G. L., L. E. Stover, and E. J. Kidson, 1993, Morphology and stratigraphic ranges of selected Mesozoic-Cenozoic dinoflagellate taxa in the north hemisphere: Geological Survey of Canada, v. 92, no. 10, p. 1–137.

Wood, G. D., A. M. Gabriel, and J. C. Lawson, 1996, Palynological techniques-processing and microscopy, *in* J. Jansonius and D. C. McGregor, eds., Palynology: Principles and applications, American Association of Stratigraphic Palynologists Foundation, v. 1, p. 29–50.

Wrenn, J. H., S. L. Duffield, and J. A. Stein, eds., 1986, Papers from the first Symposium on Neogene dinoflagellate cyst biostratigraphy: American Association of Stratigraphic Palynology Foundation, Contribution Series 17, 225 p.

18

Gutiérrez Paredes , H. C., M. Martinez Medrano, and H. L. Sessarego, 2009, Provenance for the middle and upper Miocene sandstones of the Veracruz Basin, Mexico, *in* C. Bartolini and J. R. Román Ramos, eds., Petroleum systems in the southern Gulf of Mexico: AAPG Memoir 90, p. 397–407.

Provenance for the Middle and Upper Miocene Sandstones of the Veracruz Basin, Mexico

Hilda Clarisa Gutiérrez Paredes
Petróleos Mexicanos (PEMEX), Region Marina Suroeste, Mexico

Martin Martinez Medrano
Petróleos Mexicanos (PEMEX), Region Norte, Mexico

Horacio Luis Sessarego
Independent Geologist, Calgary Alberta, Canada

ABSTRACT

Middle and upper Miocene sandstones are the best gas-producing reservoirs in the Veracruz Basin. Sandstone source areas for two of the main gas-producing sequences in Cocuite, Vistoso, and Playuela fields were determined by petrographic analysis with plate tectonics considered to be the key control linking the provenance areas to the basin.

Upper Miocene litharenites and feldspathic litharenites in the Vistoso and Playuela fields are derived from magmatic arc sources in the Transmexican volcanic belt and the Anegada high. Middle Miocene lithic felsarenites and feldspathic litharenites in Playuela field are derived mainly from a recycled orogen and in Cocuite field from a dissected arc. Source areas for the middle Miocene sandstones were the Sierra de Juárez and the Transmexican volcanic belt, and carbonate fragments in both sequences were derived from the Sierra Madre Oriental.

INTRODUCTION

Novillero and Cocuite fields are the traditional gas producers in the Veracruz Basin; the former from lower Miocene conglomerates and the latter from lower Pliocene, upper Miocene, and middle Miocene sandstones. In recent years, new gas deposits have been discovered in the upper Miocene sands. The discovery and further development of these fields motivated this clastic provenance study.

Provenance studies are based principally on the relative proportions of different types of detrital grains in

DOI:10.1306/1093M903335

FIGURE 1. Location map of the Veracruz Basin.

the sandstones that serve as guides to identify the original source rocks (Dickinson and Valloni, 1980). The relationship between provenance and the depositional basin is governed mainly by plate tectonics, which controls the distribution of the different rock types. The final composition of the sandstones is also affected by other factors such as the climate and environment of deposition (Dickinson and Valloni, 1980).

The relationship between sandstone provenance and the depositional basin is important for hydrocarbon exploration because variations in detrital composition control their diagenetic response and therefore their tendency to lose porosity.

Objectives

The main objective of this study is to determine the provenance areas and lithologic sources for the productive middle and upper Miocene sands in Cocuite, Vistoso, and Playuela fields.

Location

The Veracruz Basin is located in the south-central part of the state of Veracruz. The basin covers an area of 30,000 km^2 (11,583 mi^2) bounded to the north by the Santa Ana high; to the west by the Sierra Madre Oriental; to the south by the Sierra de Chiapas, Mixtequita, and Sierra de Juárez metamorphic complex; to the southeast by the Isthmian Saline Basin; and to the east by the Anegada high, the Mexican ridges, and the Los Tuxtlas volcanic center (Figure 1).

Methodology

Sandstones in the productive intervals were identified by age within the stratigraphic framework. Upper Miocene sandstones were placed in the LS_MS_6.98 Ma sequence, and middle Miocene sandstones were placed in the LS_MM_11.7 Ma sequence (Petroleos Mexicanos [PEMEX]-Schlumberger, 2003, confidential report).

Petrographic information derived from point count analyses (Table 1) was integrated into a database of normalized values from which quartz, lithics, feldspars (QLF) and monocrystalline quartz, feldspars, total lithics (QmFLt) ternary diagrams (Dickinson and Suczek, 1979) and polycrystalline quartz, metavolcanic lithics, metasedimentary lithics, (QpLvmLsm) and metamorphic lithics, volcanic lithics, sedimentary lithics (LmLvLs) diagrams (Ingersoll and Suczek, 1979) were constructed.

VOLCANISM

Two large magmatic provinces are recognized in the coastal plain of the Gulf of Mexico: the Transmexican volcanic belt and the alkaline province of eastern Mexico (Figure 2). The Transmexican volcanic belt has been active since the early Miocene, and its greatest activity appears to have occurred from 20 to 15 Ma and from 9 to 6 Ma. Activity dated as 3.0 Ma is concentrated in the southern part of this trend, and older calc-alkaline lavas have been recognized at the eastern terminus of the Transmexican volcanic belt in the Palma Sola massif (Cantagrel and Robin, 1979). In contrast with the calc-alkaline (siliceous andesites, rhyolites, and minor microdiorites) of the Transmexican volcanic belt, alkaline to hyper-alkaline magmatism occurred in the alkaline province of eastern Mexico from late Oligocene to Holocene along a north–south trend from Tamaulipas to the Veracruz region.

Volcanic activity in the Transmexican volcanic belt and south to Los Tuxtlas occurred principally during the Pliocene and Quaternary (Jacobo, 1997). The northern Veracruz Basin is covered by this volcanic material consisting of air fall and flow pyroclastics, as well as basaltic and andesitic flows erupted from Pico de Orizaba and Cofre de Perote volcanoes. Magmatic activity in Los Tuxtlas is heterogeneous comprising rocks with both calc-alkaline and alkaline affinities in five differentiated magmatic phases that range in age from 8.3 Ma to the Holocene (Jacobo, 1997).

PETROGRAPHIC ANALYSIS

The upper Miocene sandstones (LS_MS_6.98) have an average grain size of 0.17 mm (0.006 in.) with a minimum of 0.07 mm (0.002 in.). Sorting is moderate to good, and clast shapes are subangular to subrounded. Middle Miocene sandstones have an average grain size measured from core plugs of 0.20 mm (0.7 in.). Coarse-grained, moderately sorted sandstone with subangular clasts has also been observed in hand specimens.

The sandstones of sequence LS_MS_6.98 are litharenites and feldspathic litharenites, and in sequence LS_MM_11.7, they are lithic felsarenite and feldspathic litharenite (Figure 3).

Compositional differences between the upper and middle Miocene sandstones are represented by the relative proportions of the detrital fraction (clasts) and the greater quantity of carbonate cement in the younger unit.

The most common lithic clasts observed in thin sections from wells in Cocuite, Playuela, and Vistoso fields are the following:

1. Quartz: This occurs as three varieties with monocrystalline quartz being the most abundant. Note that the quantity of quartz in the lower unit MM_11.7 is accompanied by a lesser content in the matrix accompanied by a slight increase in porosity and permeability.
2. Plagioclase: The most common type of plagioclase in the MM_11.7 unit is albite; the plagioclase occurs as phenocrysts within igneous fragments with sodic plagioclase predominant over potassic plagioclase.
3. Lithoclasts: Igneous (volcanic and plutonic) rocks predominate over carbonates with a lesser proportion of other sedimentary and metamorphic lithoclasts (Figure 4).

Provenance

Sequence LS_MS_6.98 sandstones in Vistoso and Playuela fields are derived from magmatic arcs, although some variation among them is observed.

Vistoso field sandstones are derived from an eroded arc, whereas Playuela field sandstones are from a transitional arc, with indications of greater stability in the former because of the presence of larger amounts of quartz (Figure 4). The detrital grains in Playuela field indicate derivation from a more mature mixture of magmatic arc volcanic and plutonic clasts with local or intermittent exposure of the plutons (Figure 5).

The QpLvmLsm and LmLvLs plots (Figure 6) define the provenance of sequence LS_MS_6.98 from well-defined tectonic areas with most constituents being of volcanic origin. The sources for LS_MS_6.98 sequence sandstones are the Transmexican volcanic belt and the Anegada volcanic center (Figure 7).

Sandstones of the LS_MM_11.7 sequence were derived mainly from the recycled Sierra de Juárez orogen and the Transmexican volcanic belt magmatic arc (Figure 8).

The sedimentological models for both sequences show that the carbonate clasts are derived from the buried Laramide tectonic front.

CONCLUSIONS

The provenance studies assist with the analyses and selection of mineralogical and petrophysical characteristics of the rocks.

Table 1. Point count analyses, including mean and standard deviation data.

Field	Samples	Age	Q	Qm	Qp	F_{total}	Fk	Fp	L	Lt	Chert	Lign	Lm	Lc	Ls (are)
Playuela	1	MS (6.9)	15.4	10.7	4.0	12.7	4.0	8.7	81.0	86.4	0.7	70.0	1.0	9.3	0.7
	2	MS (6.9)	16.7	11.7	1.7	8.3	3.0	5.3	87.4	95.7	3.3	73.0	0.7	12.7	1.0
	3	MS (6.9)	17.0	10.7	3.3	7.4	2.7	4.7	91.7	101.0	3.0	78.0	0.7	12.3	0.7
	4	MS (6.9)	16.3	11.3	3.0	7.3	2.3	5.0	89.4	96.4	2.0	76.0	0.7	11.7	1.0
	5	MS (6.9)	18.6	13.3	2.3	8.0	2.7	5.3	85.9	94.2	3.0	75.0	1.0	9.6	0.3
	6	MS (6.9)	14.4	10.0	2.7	7.0	2.3	4.7	92.7	98.8	1.7	80.0	0.7	11.0	1.0
	7	MS (6.9)	15.9	12.3	1.3	9.7	3.7	6.0	84.0	89.9	2.3	72.0	1.0	10.3	0.7
	8	MS (6.9)	15.3	11.0	2.3	9.3	3.3	6.0	84.2	90.5	2.0	74.0	1.3	8.6	0.3
	9	MS (6.9)	15.3	14.7	0.3	10.0	3.3	6.7	80.9	81.8	0.3	66.0	0.3	14.6	0.0
	10	MS (6.9)	15.4	13.7	0.7	9.7	2.0	7.7	80.0	82.7	1.0	66.0	0.0	14.0	0.0
	11	MS (6.9)	17.7	13.7	0.7	8.4	1.7	6.7	82.3	89.6	3.3	65.0	0.7	15.6	0.0
	12	MS (6.9)	18.4	14.7	1.0	9.0	2.0	7.0	82.3	88.7	2.7	67.0	0.3	15.0	0.0
	13	MS (6.9)	16.0	10.0	1.3	8.0	1.7	6.3	83.6	94.3	4.7	68.0	0.0	15.6	0.0
	14	MS (6.9)	14.0	9.0	2.0	7.3	1.3	6.0	93.0	101.0	3.0	80.0	0.3	12.7	0.0
	15	MS (6.9)	14.0	8.0	3.0	6.3	1.3	5.0	87.3	96.3	3.0	76.0	0.3	11.0	0.0
	16	MS (6.9)	13.3	9.3	1.7	6.7	2.7	4.0	99.0	105.3	2.3	86.0	1.0	11.0	1.0
	17	MS (6.9)	19.7	10.7	7.0	7.3	1.3	6.0	98.7	109.7	2.0	87.0	1.0	10.0	0.7
	18	MS (6.9)	15.3	13.7	0.3	11.7	3.0	8.7	90.6	93.5	1.3	80.0	1.0	9.3	0.3
	19	MS (6.9)	14.0	11.7	1.3	11.6	4.3	7.3	88.0	91.3	1.0	80.0	0.7	5.6	0.7
	20	MS (6.9)	12.7	6.7	3.3	10.3	1.3	9.0	100.6	109.3	2.7	86.0	1.0	13.3	0.3
	21	MS (6.9)	15.0	10.7	2.3	12.7	3.0	9.7	97.0	103.3	2.0	80.0	0.7	15.3	0.0
	22	MS (6.9)	17.4	10.0	2.7	7.0	1.7	5.3	99.0	111.1	4.7	76.0	0.3	21.7	1.0
	23	MS (6.9)	13.0	5.3	4.7	8.7	2.7	6.0	105.3	116.0	3.0	83.0	1.0	20.6	0.7
	24	MS (6.9)	14.0	7.7	3.3	9.4	1.7	7.7	99.3	108.6	3.0	82.0	1.0	15.3	1.0
	25	MS (6.9)	11.0	5.3	1.7	7.7	2.7	5.0	106.0	115.7	4.0	85.0	0.4	19.6	1.0
	26	MS (6.9)	15.0	9.3	2.0	8.0	2.0	6.0	79.0	88.4	3.7	58.0	0.3	20.0	0.7
Mean			15.4	10.6	2.3	8.8	2.5	6.4	90.3	97.7	2.5	75.7	0.7	13.4	0.5
Standard deviation			2.0	2.6	1.5	1.8	0.9	1.5	8.1	9.8	1.1	7.6	0.4	4.0	0.4
Vistoso	1	MS (6.9)	16.3	11.7	3.3	8.0	0.7	7.3	64.0	69.9	1.3	43.0	3.3	16.7	1.0
	2	MS (6.9)	16.7	14.0	2.0	11.0	1.7	9.3	59.1	62.5	0.7	39.0	2.7	16.7	0.7
	3	MS (6.9)	18.0	16.3	1.0	12.0	1.3	10.7	52.7	55.1	0.7	33.0	2.7	16.0	1.0
	4	MS (6.9)	15.7	13.7	1.0	5.4	0.7	4.7	52.8	55.8	1.0	41.0	1.7	9.4	0.7
	5	MS (6.9)	19.0	14.7	1.0	5.3	1.0	4.3	61.3	68.9	3.3	44.0	1.3	15.0	1.0
	6	MS (6.9)	22.0	20.0	0.3	4.3	0.3	4.0	58.3	62.0	1.7	43.0	1.0	13.0	1.3
	7	MS (6.9)	22.0	20.0	0.7	4.3	0.0	4.3	67.4	70.7	1.3	49.0	1.3	16.4	0.7
Mean			18.5	15.8	1.3	7.2	0.8	6.4	59.4	63.6	1.4	41.7	2.0	14.7	0.9

Standard deviation			2.6	3.2	1.0	3.2	0.6	2.7	5.5	6.5	0.9	4.9	0.9	2.7	0.2
Playuela	1	MM (11.7)	29.7	18.7	10.0	16.0	6.0	10.0	15.3	27.3	1.0	3.3	5.0	7.0	0.0
	2	MM (11.7)	35.5	17.3	15.2	17.4	5.7	11.7	16.1	37.3	3.0	3.7	5.7	6.7	0.0
	3	MM (11.7)	26.8	16.8	8.7	20.7	8.7	12.0	15.9	27.2	1.3	1.0	2.3	12.3	0.3
	4	MM (11.7)	32.1	23.1	6.3	21.7	6.7	15.0	14.7	26.4	2.7	1.7	1.0	10.7	1.3
	5	MM (11.7)	31.7	22.7	8.3	20.0	6.7	13.3	15.4	25.1	0.7	4.0	2.0	8.7	0.7
	6	MM (11.7)	35.8	22.8	11.3	19.6	7.3	12.3	11.0	25.7	1.7	2.3	2.0	6.7	0.0
	7	MM (11.7)	32.4	22.8	9.3	21.7	5.7	16.0	12.4	22.3	0.3	2.7	1.0	7.7	1.0
	8	MM (11.7)	35.4	28.4	6.3	22.3	7.3	15.0	12.6	20.3	0.7	3.3	1.0	8.3	0.0
	9	MM (11.7)	38.5	31.2	6.3	18.7	6.7	12.0	11.6	19.9	1.0	2.0	1.3	8.0	0.3
	10	MM (11.7)	38.6	25.3	12.0	22.3	7.3	15.0	15.0	29.6	1.3	4.0	1.3	9.0	0.7
	11	MM (11.7)	31.4	18.4	11.3	18.6	6.3	12.3	21.7	36.4	1.7	4.7	3.0	14.0	0.0
	12	MM (11.7)	37.6	23.0	11.3	18.3	4.3	14.0	40.6	58.5	3.3	29.2	3.7	6.7	1.0
	13	MM (11.7)	34.1	23.7	8.7	19.3	5.0	14.3	43.5	55.6	1.7	28.8	4.3	8.7	1.7
	14	MM (11.7)	27.0	21.7	5.0	23.0	4.0	19.0	41.1	46.7	0.3	24.8	3.7	12.3	0.3
	15	MM (11.7)	30.7	20.7	8.0	22.0	4.3	17.7	40.5	52.5	2.0	25.2	3.7	10.3	1.3
	16	MM (11.7)	35.7	23.7	10.0	19.6	5.3	14.3	44.9	58.9	2.0	27.2	4.7	10.0	3.0
	17	MM (11.7)	24.7	19.0	5.0	20.7	2.0	18.7	33.4	39.8	0.7	15.4	3.0	14.3	0.7
	18	MM (11.7)	25.3	17.3	6.7	22.7	5.7	17.0	39.9	49.2	1.3	25.2	3.3	10.7	0.7
	19	MM (11.7)	26.6	16.3	9.0	26.4	8.7	17.7	27.3	38.9	1.3	15.6	3.3	7.7	0.7
	20	MM (11.7)	24.3	15.7	7.3	26.0	7.3	18.7	34.7	44.6	1.3	17.4	7.3	9.7	0.3
	21	MM (11.7)	30.0	18.7	10.0	22.4	4.7	17.7	44.3	56.9	1.3	36.6	4.7	2.3	0.7
	22	MM (11.7)	33.3	18.0	13.3	13.4	3.7	9.7	36.4	53.7	2.0	27.4	4.0	4.3	0.7
Mean			31.7	21.2	9.1	20.6	5.9	14.7	26.7	38.8	1.5	13.9	3.2	8.9	0.7
Standard deviation			4.5	4.0	2.7	3.0	1.6	2.8	13.0	13.6	0.8	12.0	1.7	2.9	0.7
Cocuite	1	MM (11.7)	21.0	16.3	4.7	24.0	6.7	17.3	21.8	26.5	0.0	2.0	2.3	17.5	0.0
	2	MM (11.7)	32.1	26.2	5.4	23.5	5.4	18.1	26.3	32.6	0.4	2.3	2.3	21.7	0.0
	3	MM (11.7)	25.7	18.0	6.7	21.3	4.0	11.0	23.0	31.7	1.0	2.0	0.7	20.3	0.0
	4	MM (11.7)	24.0	19.0	2.7	13.6	2.0	8.7	28.0	35.3	2.3	2.0	2.0	24.0	0.0
	5	MM (11.7)	19.3	16.3	2.0	26.3	4.7	12.3	21.3	25.3	1.0	2.3	1.0	18.0	0.0
	6	MM (11.7)	25.6	20.3	5.3	17.4	1.7	9.3	17.4	22.7	0.0	1.7	2.3	13.4	0.0
	7	MM (11.7)	13.0	6.7	6.0	17.4	0.7	10.3	76.0	82.6	0.3	60.1	9.3	6.3	0.3
	8	MM (11.7)	13.3	5.0	7.3	17.4	0.3	7.3	85.9	95.2	1.0	69.9	11.0	4.0	1.0
	9	MM (11.7)	13.7	6.7	6.3	17.4	0.0	9.7	79.2	86.9	0.7	68.5	6.0	4.0	0.7
Mean			20.9	14.9	5.2	19.8	2.8	11.6	42.1	48.8	0.7	23.4	4.1	14.4	0.2
Standard deviation			6.7	7.2	1.8	4.2	2.4	3.8	29.0	30.0	0.7	32.2	3.8	7.8	0.4

Q = quartz; Qm = monocrystalline quartz; Qp = polycristalline quartz; F_{total} = total feldspars; Fk = plagioclases; Fp = potassic feldspars; L = lithics; Lt = total lithics; Chert = pedernal; Lign = igneous lithics; Lm = metamorphic lithics; Lc = carbonate lithics; Ls = sedimentary lithics.

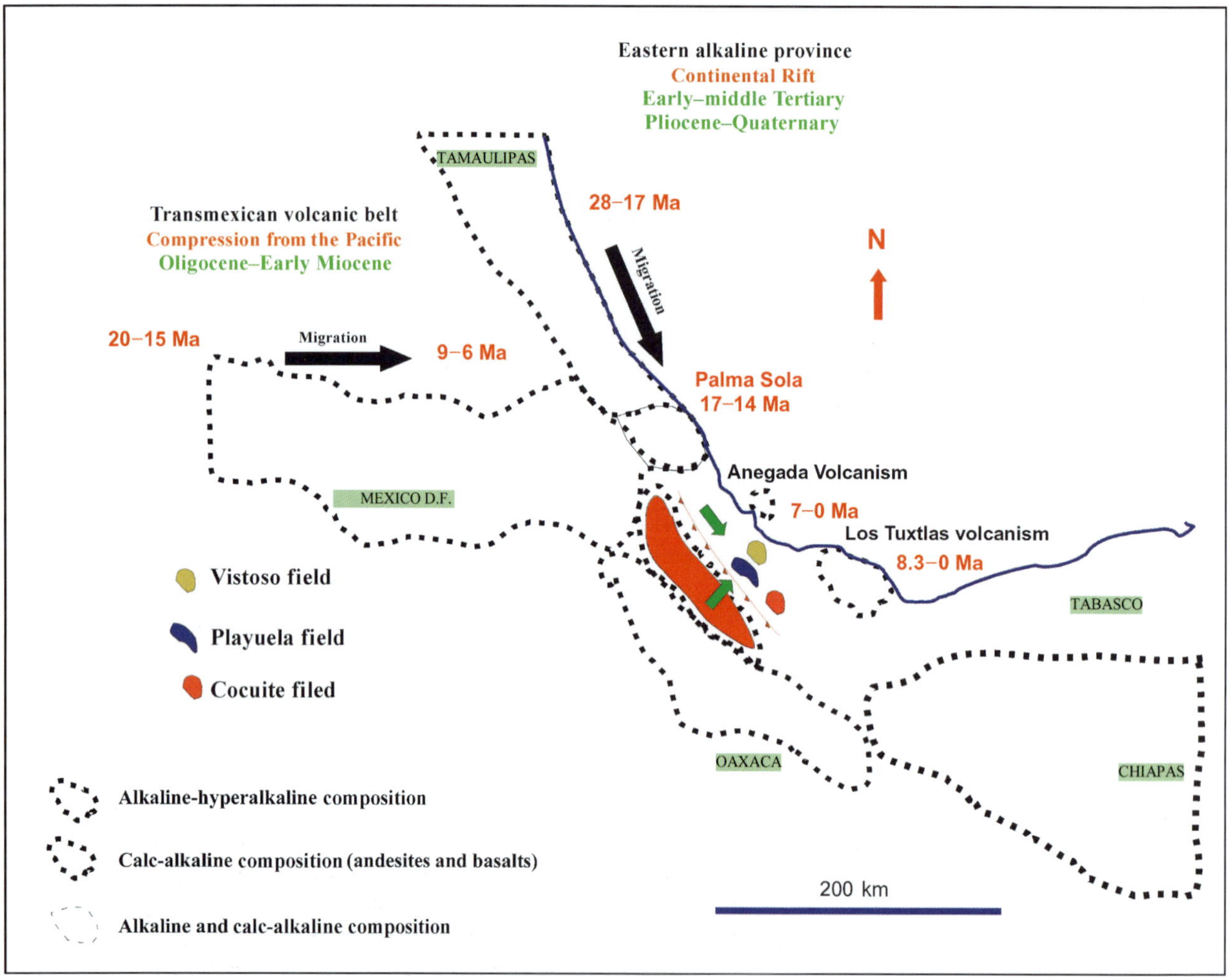

FIGURE 2. Volcanism in the Transmexican volcanic belt.

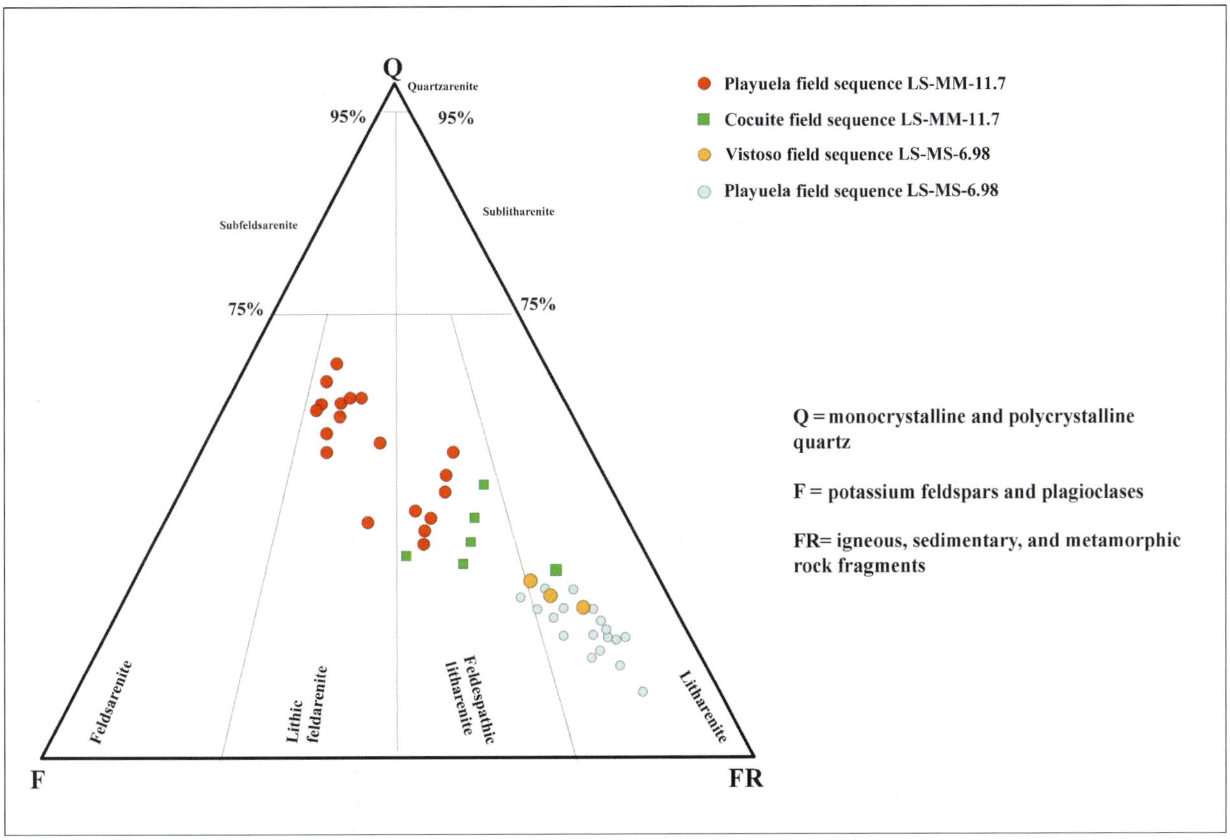

FIGURE 3. Classification of sandstones in sequences LS_MS_6.98 and LS_MM_11.7.

Sandstones of the LS_MS_6.98 sequence in Vistoso and Playuela fields are derived from the magmatic arcs of the Transmexican volcanic belt and the Anegada volcanic center.

Sandstones of the LS_MM_11 sequence are derived from volcanic arcs and recycled orogens with the main source areas being the Sierra de Juárez and the Transmexican volcanic belt.

Carbonate rock fragments are derived from the buried tectonic front of the Sierra Madre Oriental fold belt.

These studies are of great importance in exploration for hydrocarbons because knowledge of the detrital composition of the sandstones can help predict their diagenetic response and trends of porosity reduction with burial depth.

Field	Age	#	Qm	Qp	P	F	Lt	Lv	Lm	Lc	Ls
Playuela	MS–6.9	26	5–14	.3–7	4–9.7	1.7–4.3	37–62	40–55	0.3–1	6.6–21	0.3–1
Vistoso	MS–6.9	7	11–20	0.3–3.3	4–10.7	0–1.7	34–48	40–49	1–3	9–16	0.7–1.3
Playuela	MM–11.7	22	15–31	5–15	9–19	2–8.7	19–45	15–22	1–7.3	2–14	0–1.7
Cocuite	MM–11.7	9	5–26	2–7	7–18	0–6	31–60	6–35	0.7–11	4–54	0–1

FIGURE 4. Table summarizing modal data and point counts.

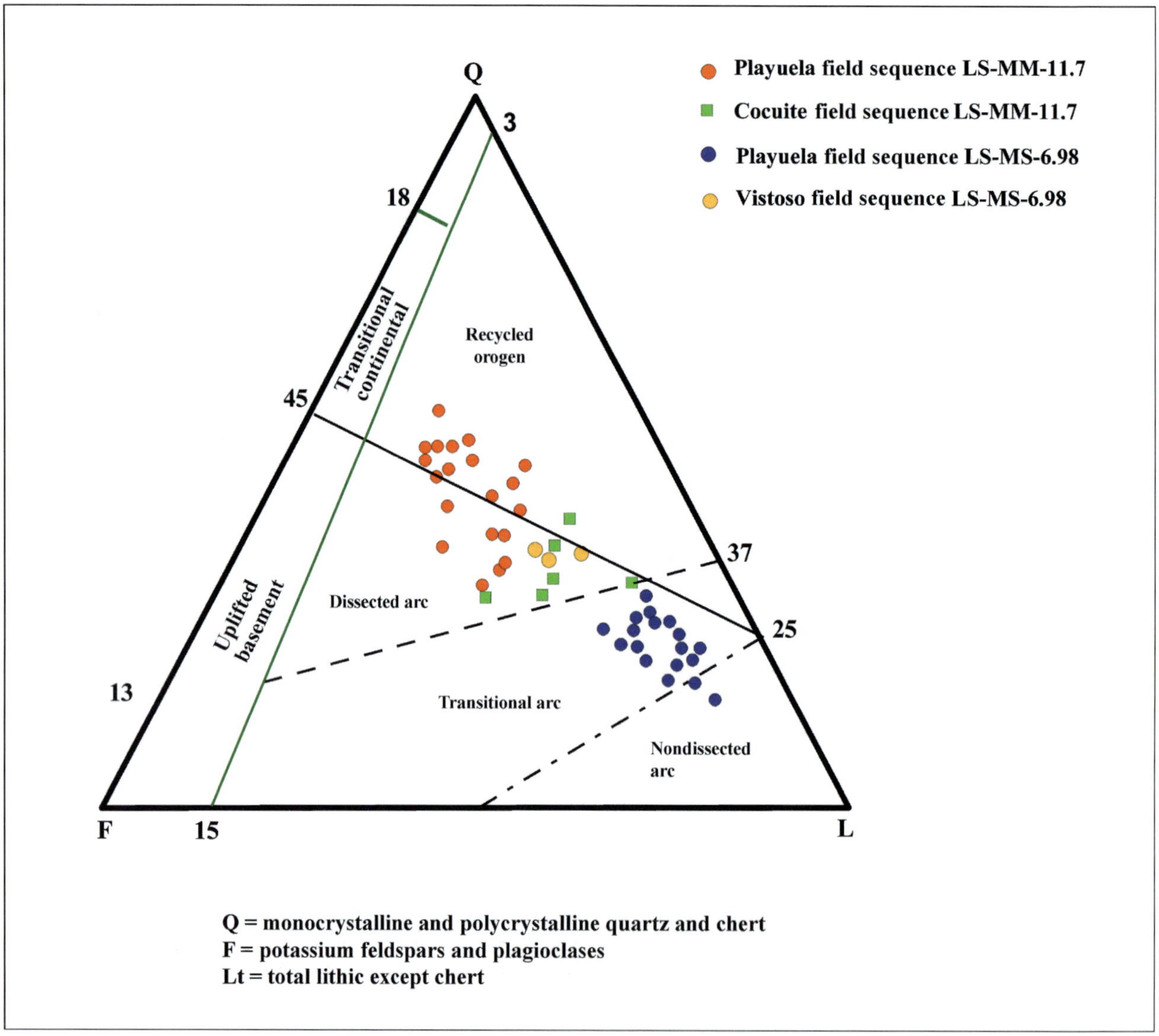

FIGURE 5. Ternary quartz, feldspars, lithics (QFL) diagrams for both sequences.

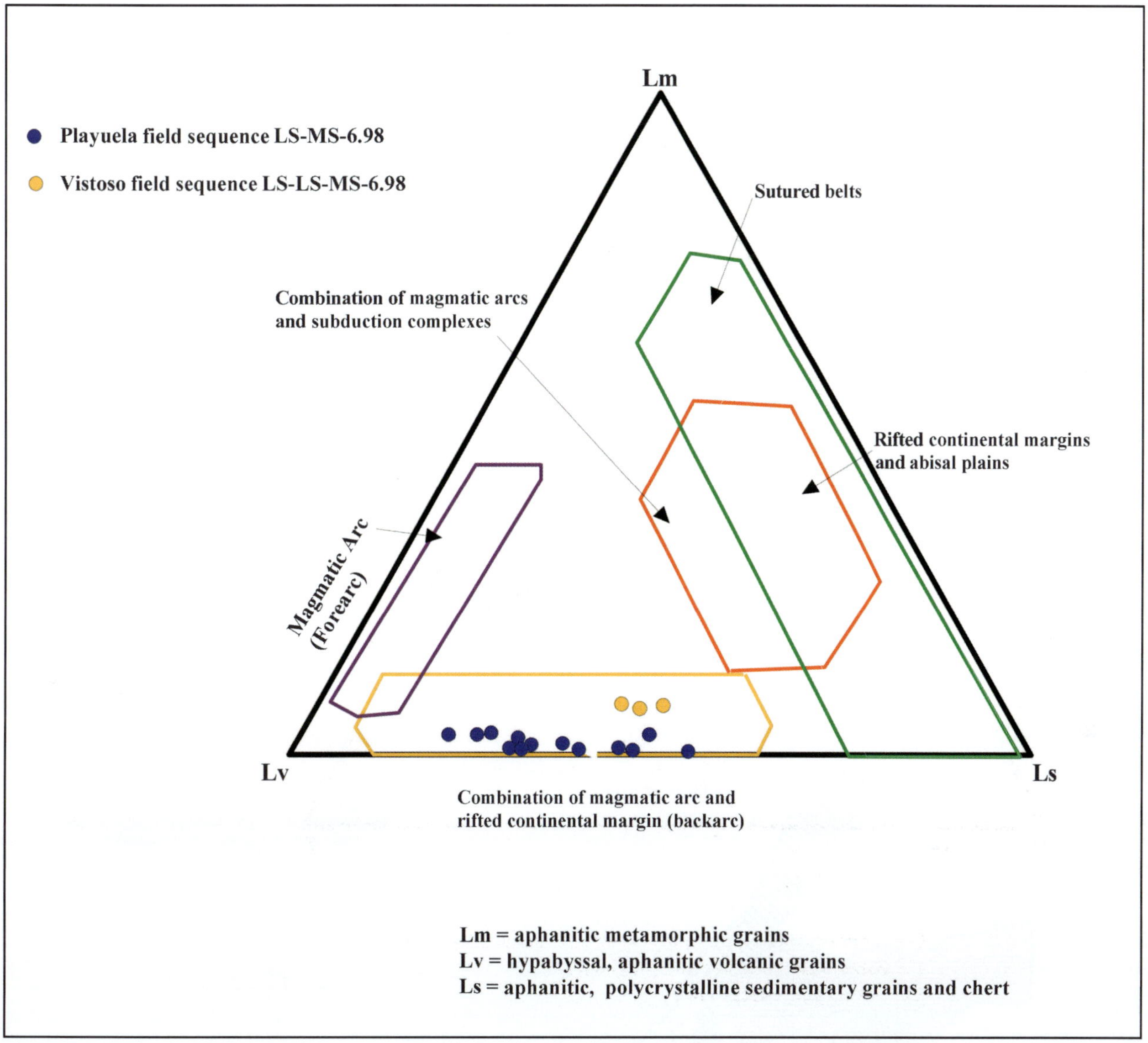

FIGURE 6. Volcanic lithics (Lv), sedimentary lithics (Ls), and metamorphic lithics (Lm) diagram for sequence LS_MS_6.98.

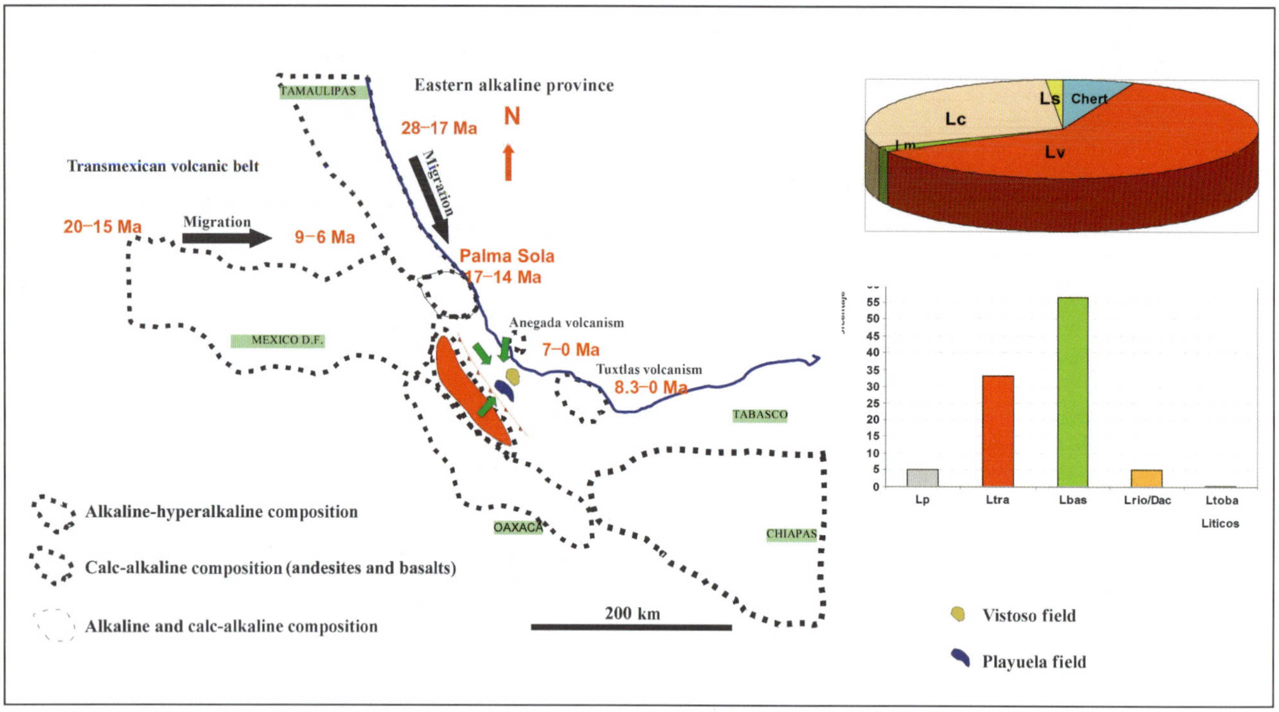

FIGURE 7. Source areas for the upper Miocene sandstones (LS_MS_6.98).

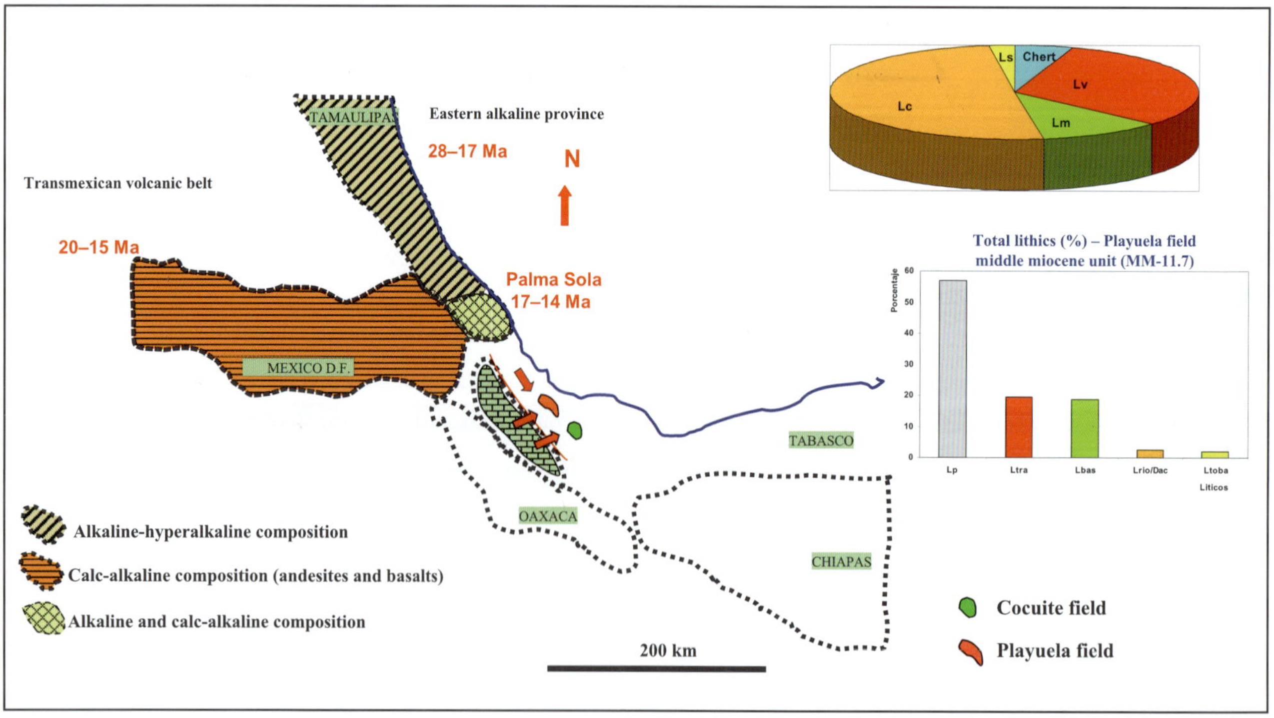

FIGURE 8. Source areas for the middle Miocene sandstones (LS_MM_11.7).

The petrographic information comes from the microscopic point-counting of sandstone samples taken from all the studied cores.

ACKNOWLEDGMENTS

We thank Petroleos Mexicanos (PEMEX) for permitting the publication of our manuscript. We are grateful to Abhijit Basu for the constructive review of the manuscript. We also thank Claudio Bartolini for assistance during the different stages of manuscript preparation.

REFERENCES CITED

Cantagrel, M., and C. Robin, 1979, K-Ar dating on eastern Mexican volcanic rocks— Relations between the andesitic and the alkaline provinces: Journal of Volcanology and Geothermal Research, v. 5, p. 99–114.

Dickinson, W., and C. Suczek, 1979, Plate tectonics and sandstone compositions: AAPG Bulletin, v. 63, no. 12, p. 2164–2182.

Dickinson, W. R., and R. Valloni, 1980, Plate settings and provenance of sands in modern ocean basins: Geology, v. 8, p. 82–86.

Ingersoll, R., and C. Suczek, 1979, Petrology and provenance of Neogene sand from Nicobar and Bengal fans, DSDP sites 211 and 218: SEPM, v. 49, p. 1217–1223.

Jacobo, A. J., 1997, Studio Geologico e Petrologico del Complesso Vulcanico de los Tuxtlas, Stato9 de Veracruz Messico: Ph.D. tesis, Universita degli Studi de Pisa, Italia, 145 p.

Petróleos Mexicanos (PEMEX)-Sclumberger, 2003, Informe de Caracterización Inicial del Campo Playuela, Informe Inédito, 245 p.

19

Román Ramos, J. R., M. A. Cruz Mercado, L. E. Salomón Mora, C. Rosas Lara, F. Sanchez Ferrer, E. Biegert, and E. Bartsch, 2008, Continental-oceanic boundary deep structure in a shear margin: Western main transform, offshore Veracruz, Southern Gulf of Mexico, *in* C. Bartolini and J. R. Román Ramos, eds., Petroleum systems in the southern Gulf of Mexico: AAPG Memoir 90, p. 409–420.

Continental-Oceanic Boundary Deep Structure in a Shear Margin: Western Main Transform, Offshore Veracruz, Southern Gulf of Mexico

Juan R. Román Ramos, Miguel A. Cruz Mercado, Luis E. Salomón Mora, and Carlos Rosas Lara

Petróleos Mexicanos (PEMEX), Exploracion y Producción, Mexico

Fernando Sanchez Ferrer, Ed Biegert, and Erik Bartsch

Shell International E&P, Houston, U.S.A.

ABSTRACT

Gravity, magnetic, and heat-flow data suggest that the western main transform shear margin is associated with a transform deformation zone having considerable lateral extent. Inboard of this zone lays a continental crust, whereas outboard to the east of the large gravity anomaly associated with this structure lays an oceanic crust. The basement complex in the deformation zone is likely to be a heavily modified continental crust intruded with magnetic and higher density crustal material.

INTRODUCTION

The Mexican ridges, in the southwestern Gulf of Mexico, are part of a linear trend of offshore contractional features forming pronounced bathymetric highs parallel to the Mexican coastline (Figure 1). The Mexican ridges are interpreted as the down-dip contractional part of a linked extensional-contractional belt soling onto a lower Tertiary detachment horizon (Figure 2). The deep-water fold belt extends 200 km (124 mi) from onshore Mexico to the abyssal plain (Garrison and Martin, 1973; Worral and Snelson, 1989; Buffler, 1991; Feng et al., 1994). We ascribed the formation of the Mexican ridges to large-scale gravity sliding, triggered by the coastal uplift of the Sierras farther west (5–7 km [3–4 mi] of uplift according to Gray et al., 2001). Our own palinspastic restoration indicates that folding started in the middle Miocene, peaked around the Late Pliocene, and continued to the present day. We estimate the total extension and contraction in the Mexican ridges to be around 11% (Figure 3).

In contrast to the salt-cored Perdido folds farther north or the Campeche folds farther south (Figure 1),

DOI:10.1306/13191094M903416

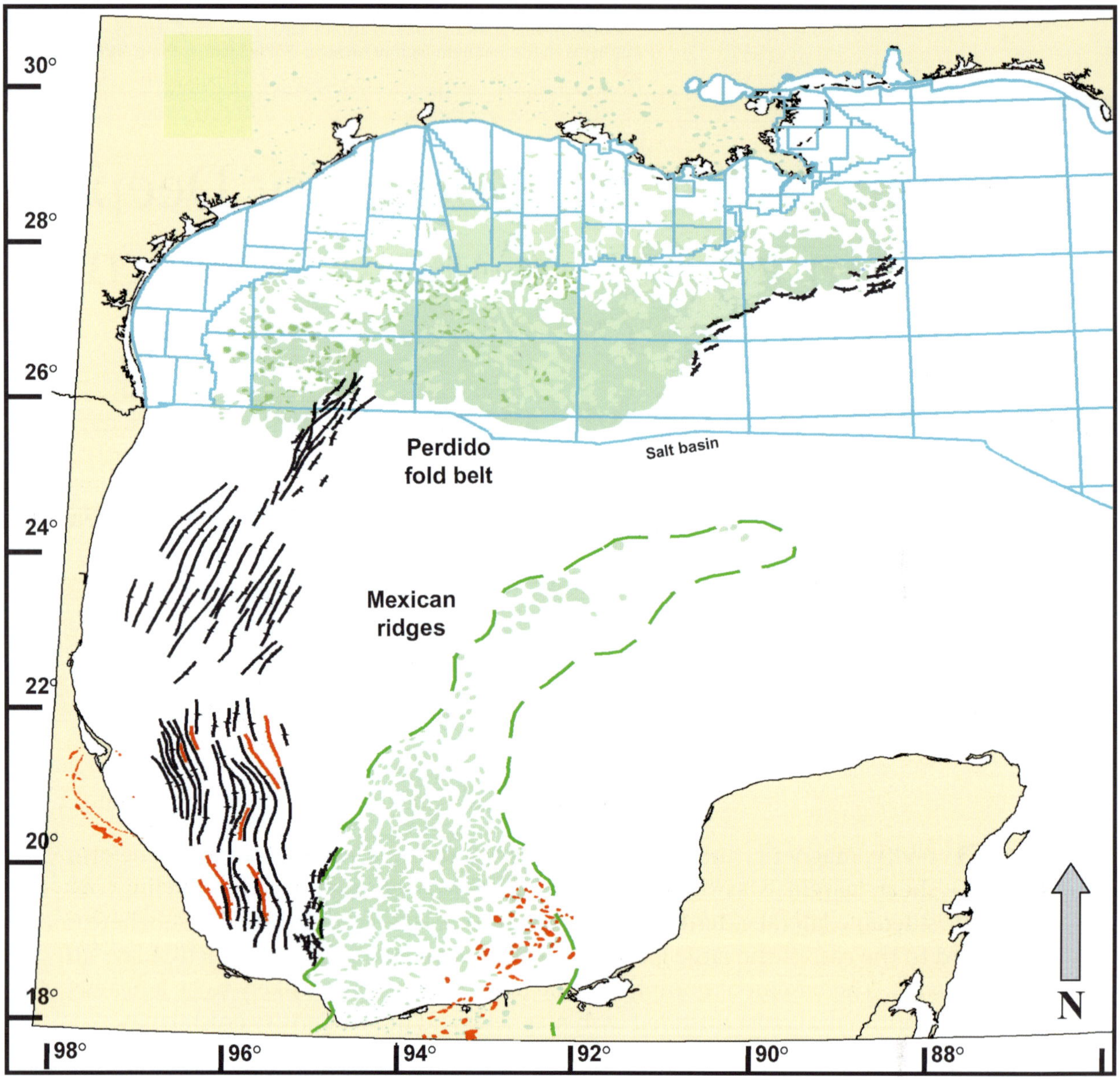

FIGURE 1. Western Gulf of Mexico tectonic elements.

no evidence of the Middle Jurassic Louann Salt is found in the area of the Mexican ridges. The lack of Louann Salt in the Mexican ridges is a result of its tectonic evolution, namely its situation above an oceanic crust generated during the opening of the Gulf of Mexico (Jacques and Clegg, 2003).

The regional framework history of the Gulf of Mexico is generally accepted to involve a simple two-stage model for the opening of the basin (Figure 4).

The initial stage of crustal stretching was the result of the Gondwana plate (specifically South America) pulling away from North America in a relative northwest–southeast sense. The Yucatan block is interpreted to have initiated a counterclockwise rotation during this phase. The shear margin or western main transform discussed by Marton and Buffler (1994), Ross and Scotese (1988), Bird et al. (2005), and others is commonly considered to be the main western transform boundary associated with this Jurassic rifting and the rotation of the Yucatan block.

Toward the end of this initial rifting phase, a great thickness of the Louann Salt was deposited across the entire basin.

The second stage saw the onset of the emplacement of the oceanic crust in the Upper Jurassic, which lasted into the Early Cretaceous. The spreading center was segmented by many northeast-trending oceanic transform faults, which formed during the continued rotation of Yucatan away from the Texas margin. The Yucatan block continued its counterclockwise rotation with a pole near the straights of Florida.

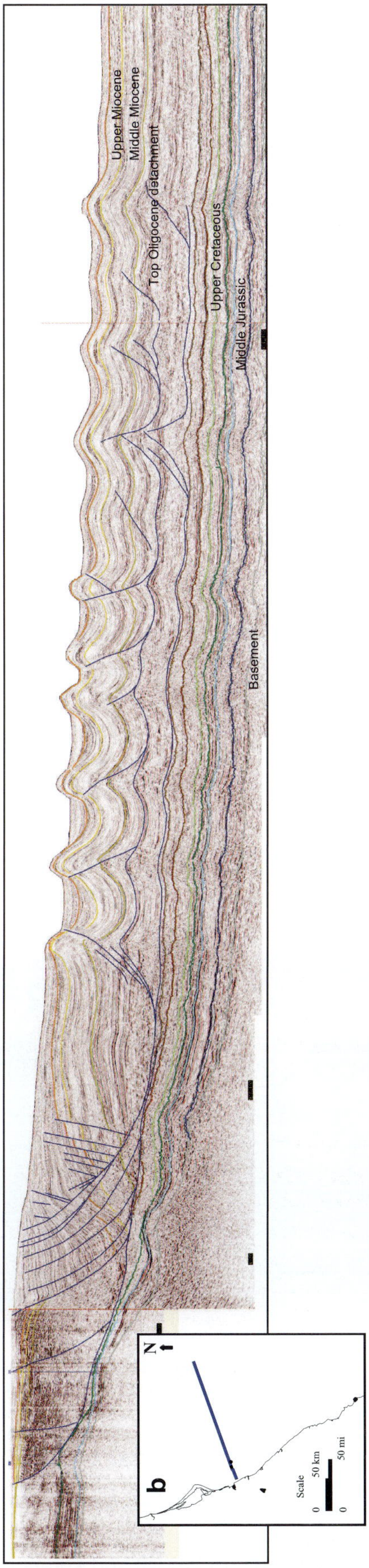

FIGURE 2. (a) Two-dimensional regional dip-oriented seismic line throughout the Mexican ridges. The seismic line stretches 200 km (124 mi) across the Lankahuasa Profundo region, from onshore Mexico to the deep-water Mexican ridges fold belt. (b) Location map for the seismic line used for palinspastic restoration (Figure 3).

Data

The key gravity and magnetic data for this study were acquired and processed between 1993 and 1999. Two-dimensional (2-D) reflexion seismic lines, as well as refraction seismic profiles complemented these data.

The regional reduced to pole magnetic data were used to infer crustal type and estimate the depth to the basement. Linear features and faults interpreted from the truncation of magnetic anomalies shown in gravity-magnetic deep structure (Figure 3) also helped define the basement fabric.

Tectonic Elements and Qualitative Interpretation from Potential Methods

The regional RTP magnetic data of (Figure 5) provide an overview and introduction to the regional tectonic elements. These data are essential for reconstructing the opening of the Gulf of Mexico, inferring crustal type (continental versus oceanic) and identifying volcanic zones and intrusives. As discussed below, these magnetic data were used to define regional basement structures, thus allowing for necessary constrains in the integrated 2-D modeling.

The composite regional Bouguer gravity map (Figure 6), compensated for bathymetry effects, exhibits a strong gradient increasing seaward that indicates a shallowing Mohorovicic discontinuity (Moho) (upper mantle) and the transition to the oceanic crust (see the red areas). The large positive gravity anomaly in the central Gulf of Mexico is associated with the oceanic crust and shows a good correlation with the location of the continental-oceanic boundary from Pindell and Kennan's (2002) kinematic models, as illustrated in Figure 7. Furthermore, the total horizontal gradient (THG) of the regional Bouguer gravity shown in Figure 8 shows an improved definition of contacts and correlates well with crustal elements illustrated in Figure 9 (Winker and Buffler, 1988).

Of particular interest in both the Bouguer and the THG map is the prominent positive gravity anomaly in the western Gulf of Mexico associated with the shear margin and the rotation of the Yucatan. Ross and Scotese (1988) and Marton and Buffler (1994) referred to this zone as the western main transform. As discussed below, we correlate this gravity anomaly with a zone of heavily modified continental crust, marking a wide transition between the continental and oceanic crust.

QUANTITATIVE INTERPRETATION

Overview

We used the topography, bathymetry, seismic horizons, and basement grids to build two 2-D models down

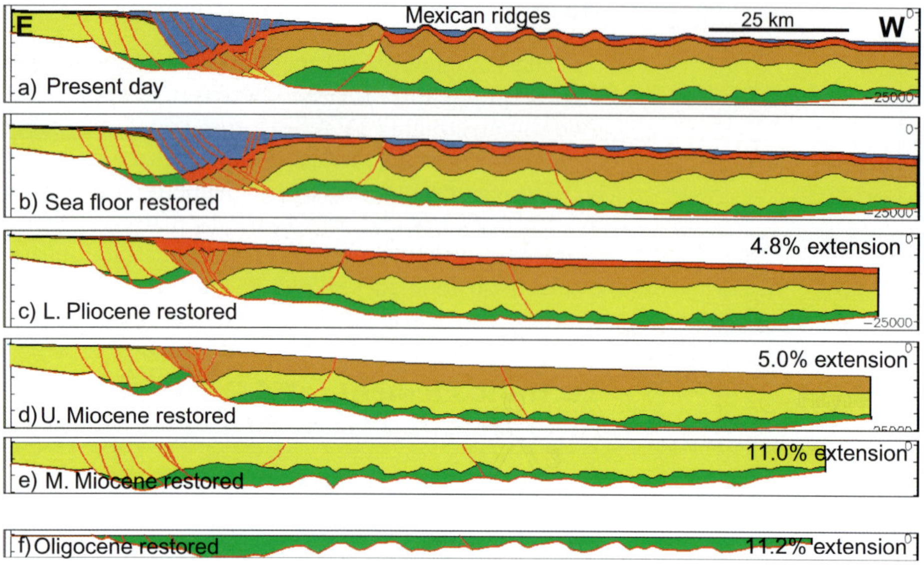

FIGURE 3. Palinspastic restoration of the seismic line in Figure 2. Downdip contraction in Mexican ridges appears to be linked to the updip extension. Extensional strain restored at each increment marked.

to the basement. Modeling the observed magnetic and gravity fields in two dimensions along selected seismic lines gives a good overview about the gravity response of certain geological features interpreted in the seismic data. Determining the sensitivity of the depth structure against densities or susceptibilities used during the modeling process is also extremely useful, especially when the amount of well data is limited. The results of the two

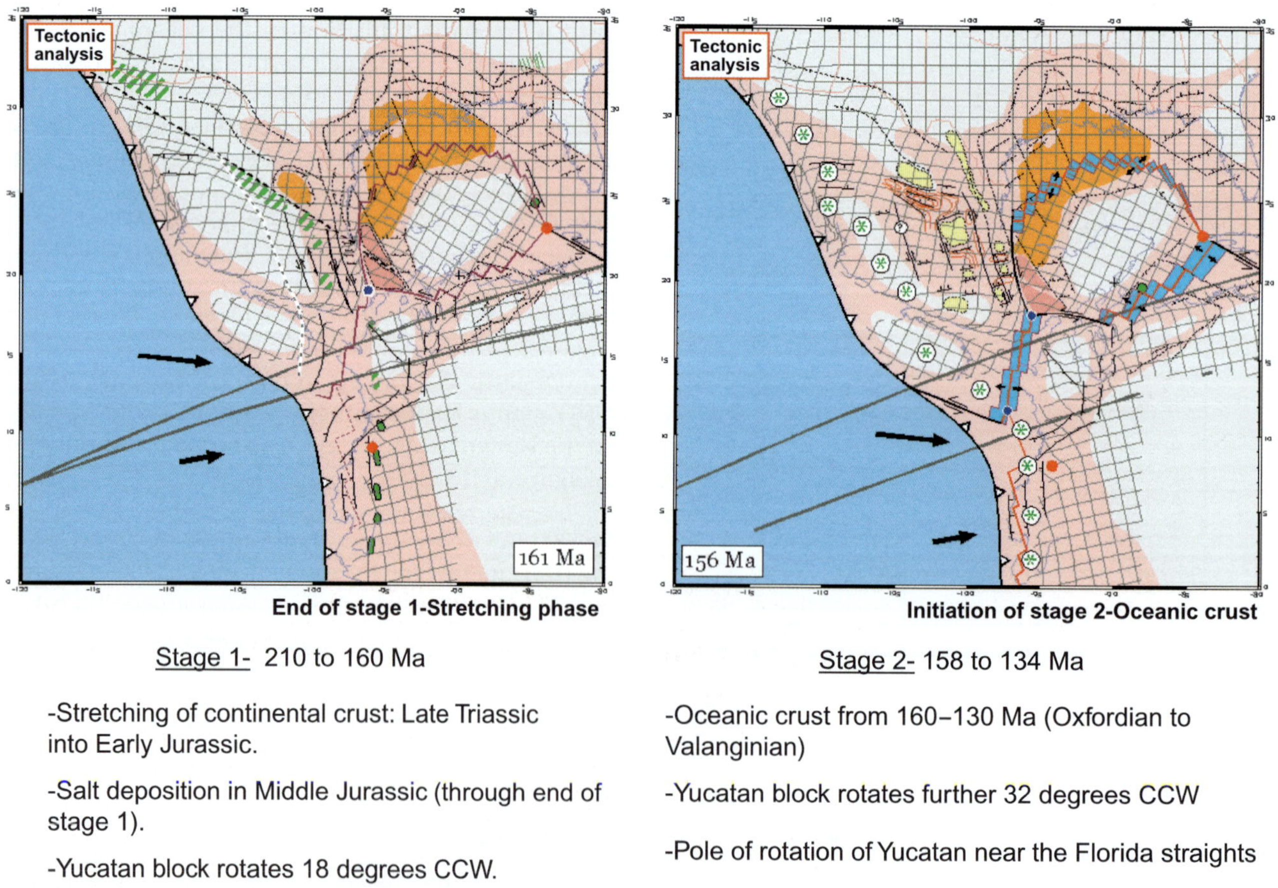

FIGURE 4. Kinematics summary of the two-stage model. CCW = counterclockwise (From Pindell and Kennan, 2002).

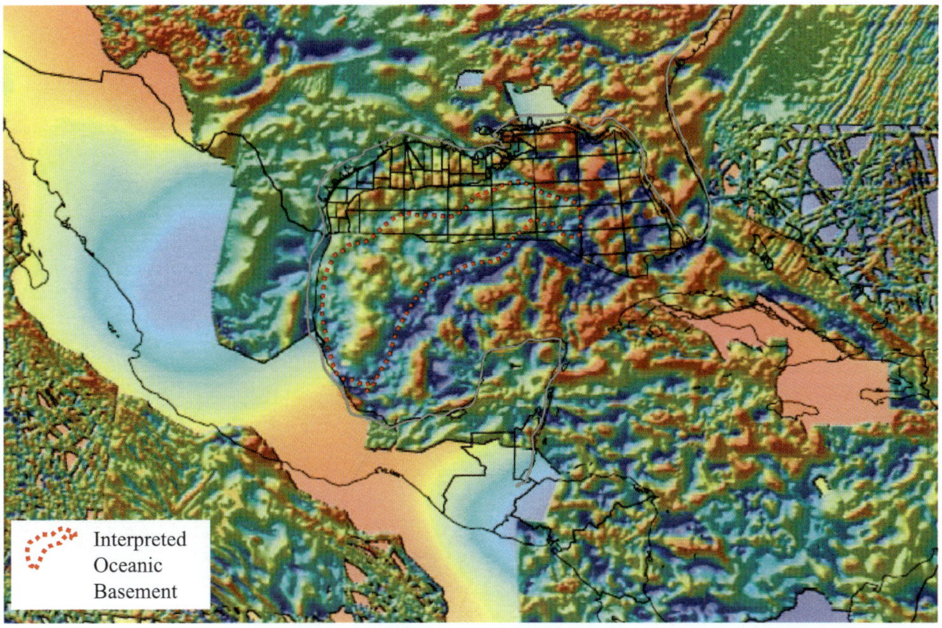

FIGURE 5. Regional magnetic data (unmerged).

dimensional modeling can be used as a calibrated starting point to produce a three-dimensional surface of the metamorphic basement for further studies.

Figure 10 shows the two primary potential field data sets used in the 2-D modeling of lines L1 and L2. The free-air gravity shown in the upper half of the figure is large and positive onshore, reflecting the shallow continental basement. The field diminishes slightly on the continental shelf as the water depth gradually increases, and then the field greatly decreases as the effect of the deep bathymetry dominates. We interpret the continental-oceanic boundary as following the green-blue transition in the gravity map. The map shows the two seismic lines modeled in this study, along with refraction data, which provide the depth estimated to the basement and to the Moho. These data unambiguously tag the oceanic crust in the indicated outboard position.

A prominent near-shore gravity anomaly, trending north–south and seemingly associated with the shear transform, is strikingly evident in the mapped gravity field. We interpret this feature as a zone of denser material representing modified transitional crust in between the continental and oceanic realms.

The onshore and nearshore RTP magnetic field shown in the lower half of Figure 10 reveals short-wavelength and high-amplitude anomalies, indicative of a shallow basement nearshore on the continental crust. These are replaced offshore by low-amplitude long-wavelength magnetic anomalies associated with a deeper oceanic

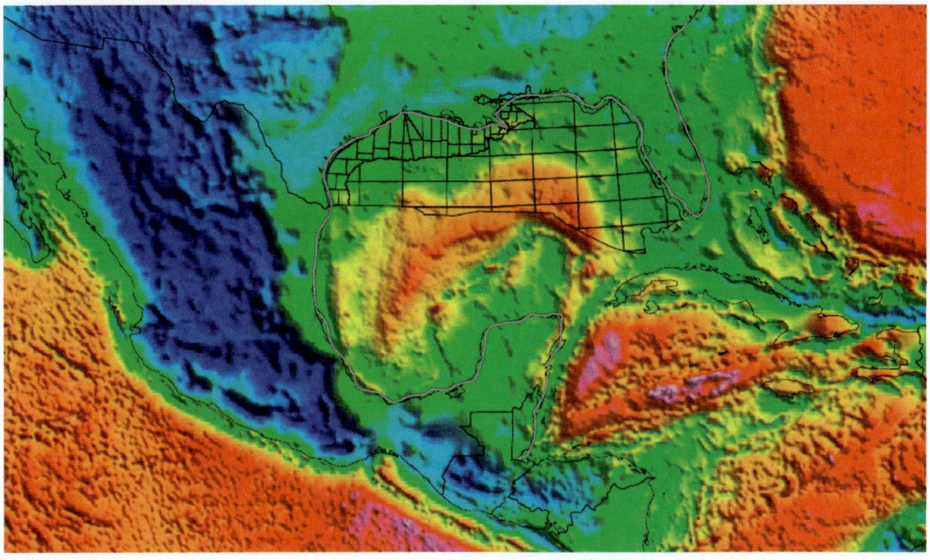

FIGURE 6. Regional Bouguer gravity.

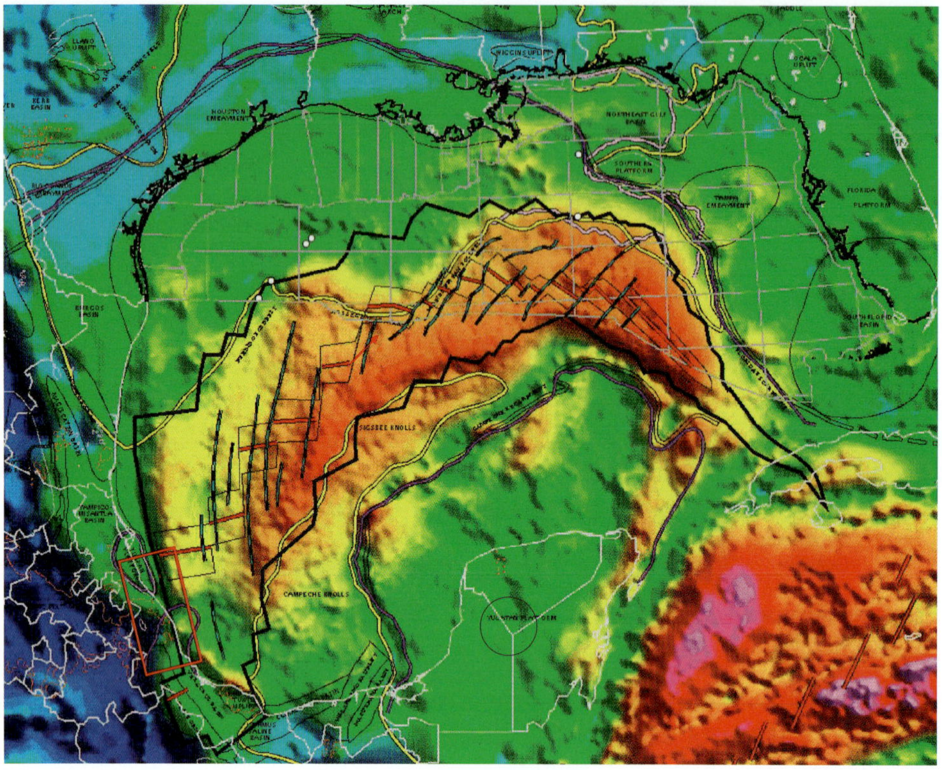

FIGURE 7. Regional Bouguer gravity and main tectonic elements of the Gulf of Mexico. This figure shows good correspondence between potential fields and Pindell and Kennan's (2002) kinematic model. The red outline shows the area of detailed study.

basement. In the intervening area, the shear transform zone is marked by intermediate wavelength anomalies, which we ascribe to magnetic intrusives.

In general, one expects large-amplitude magnetic anomalies to arise from variation in the magnetic susceptibility within the basement, corresponding to intrabasement lithology changes. Similarly, smaller amplitude magnetic anomalies might be expected to develop from structure on the basement surface or suprabasement features.

Our map of merged basement is depicted in Figure 11, which also shows the location of the seismic deep refraction control over the oceanic crust and the two traverses used in the 2-D models. The numbers on the refraction line indicate the depth to the basement and to the Moho in kilometers. These basement depth estimates were used to constrain the gravity modeling, as discussed next.

2-D Modeling and Sensitivity Analysis

To constrain the gravity modeling exercise and infer the deeper crustal structure, a quantitative estimation of the depth to the basement is needed. Such basement depth estimates were deduced from the RTP magnetic data, constrained with onshore basement control from well penetrations, outboard seismic refraction control, and basement picked from the seismic data in the study area. By way of illustration, Figure 12 shows the shallow sediment and basement control along depth-converted seismic line L1, which was used in our gravity model. Sediment densities were taken from local well logs. A similar control was used for line L2.

Using this shallow control as a constraint, we used the free-air gravity to derive a model of the depth to the Moho, the boundary separating the crust from the mantle. In seismic terms, this boundary is typically interpreted as denoting the level where the P-wave velocity changes abruptly from 6.7–7.2 km/s (4.1–4.4 mi/s) in the lower crust to 7.6–8.6 km/s (4.7–5.3 mi/s) in the upper mantle. We modeled both free-air gravity and its first vertical derivative to ensure that we fit not only the deep structure but also the sedimentary section. In Figure 13, the models for line L1 are shown in the upper half of the figure and those for seismic line L2 are shown in the lower half. The location of the shear margin (the western main transfer zone) is indicated by a vertical line. Two different models fitting the gravity response are presented. The computed and observed free-air gravity is shown on the top of each model.

Deep Structure of the Western Main Transform from Potential Methods

Inverting the gravity field for the deep structure in the 2-D models of lines L1 and L2, we found two different structural styles that fit the observed gravity data, as shown in Figure 13. Although both reproduce the gravity field, however, only one style agrees well with other geophysical data.

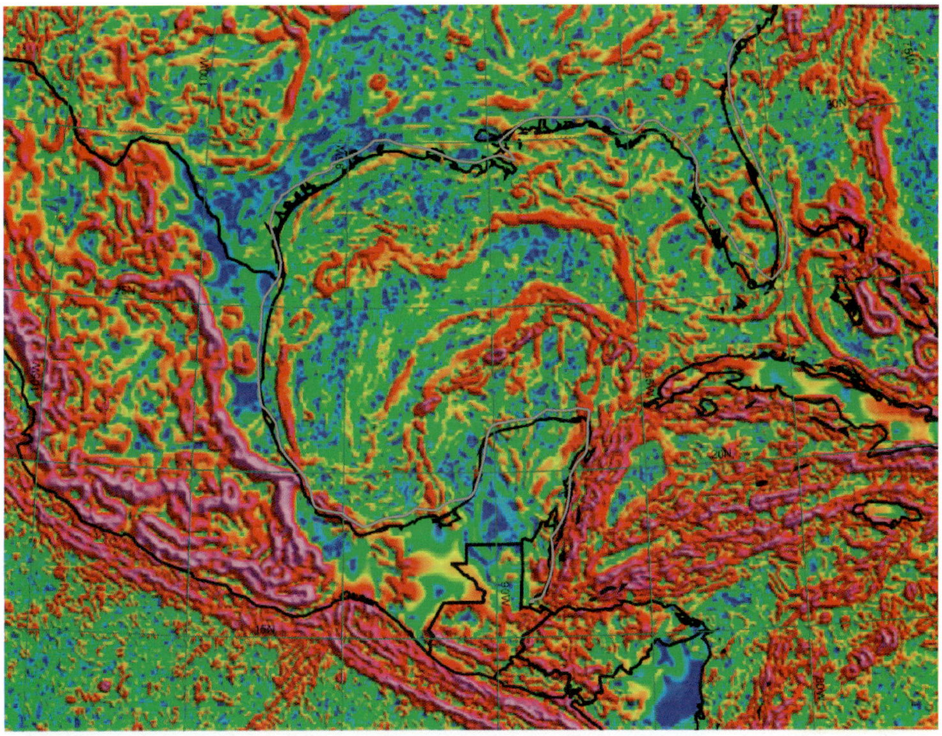

FIGURE 8. Total horizon gradient of Bouguer gravity showing improved definition of contacts between crustal elements (see Figure 9 for the interpretation).

Sharp Transition Zone Model

The left of Figure 13 depicts a structural style that might occur if the shear margin was very abrupt and the oceanic crust was juxtaposed to the continental crust with a very small transition zone; that is, there would be very little extension or thinning of the continental crust. In this structural style, we would place the continental-oceanic boundary near the peak of the large positive gravity anomaly corresponding to the western main transfer zone. Referring back to the free-air gravity map of Figure 10, this would place the continental-oceanic

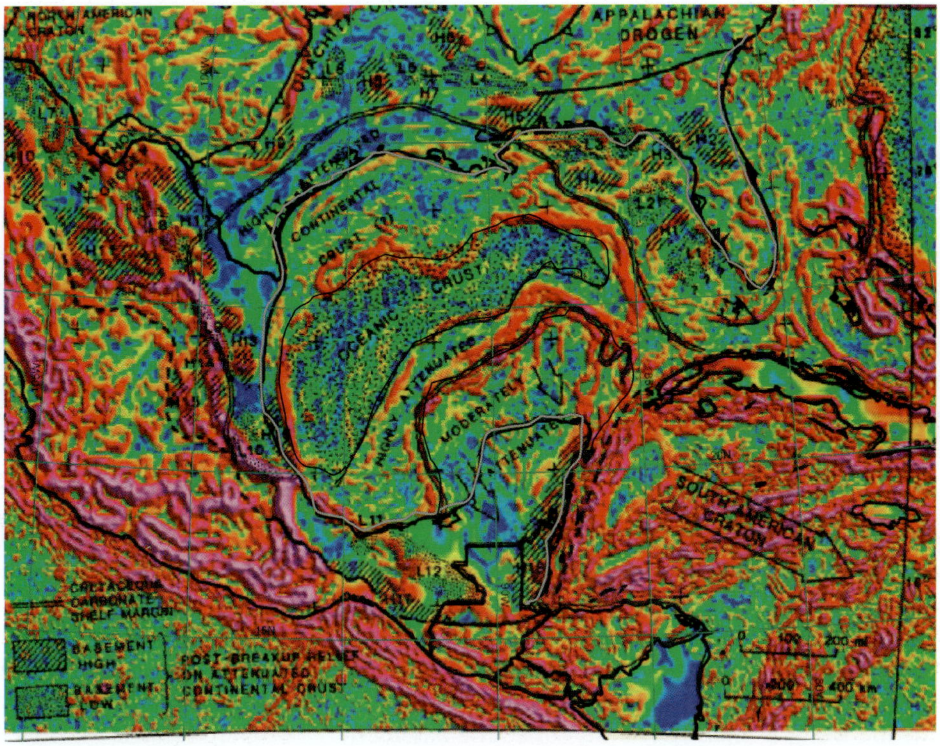

FIGURE 9. Crustal elements in the Gulf of Mexico (Winker and Buffler, 1988) and total horizontal gradient.

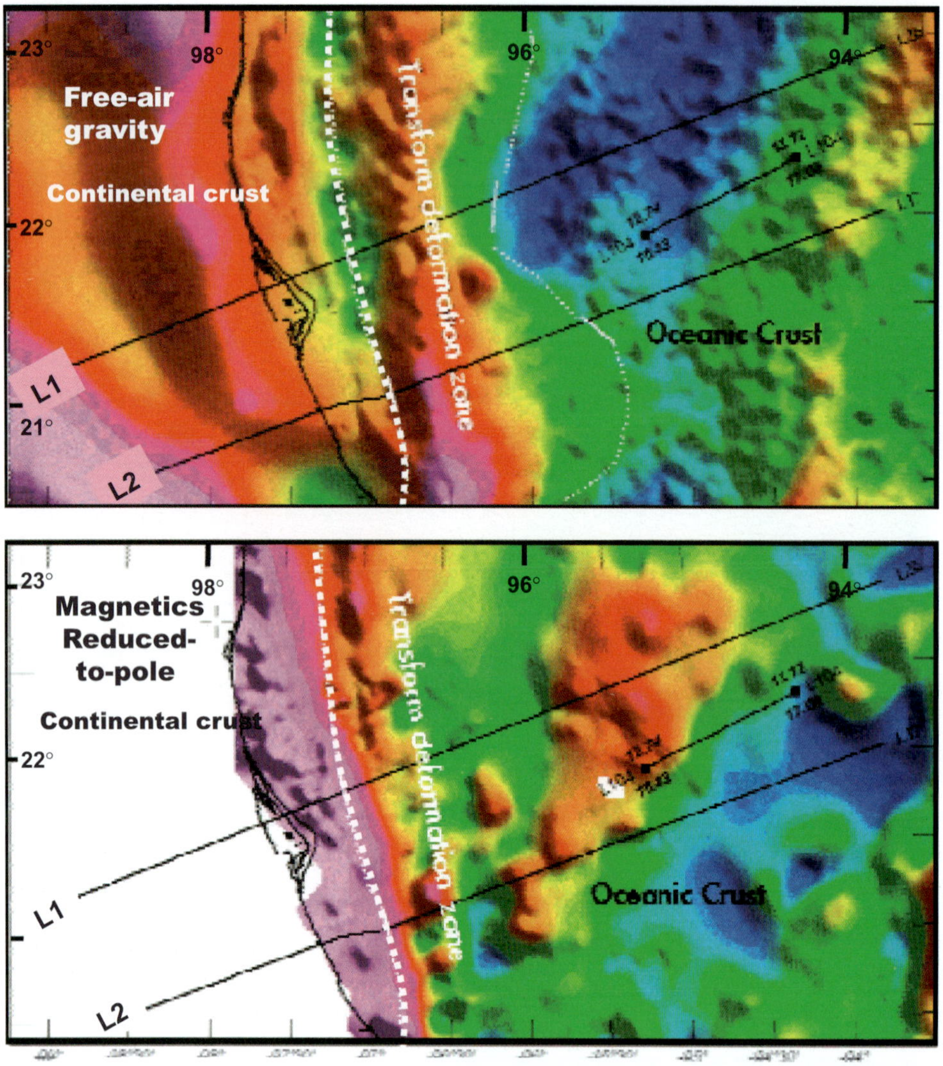

FIGURE 10. Potential field maps used in the interpretation of the study area.

boundary along the gravity ridge about where the text "transform deformation zone" is situated. Adopting this structural style thus places a large part of the study area on the oceanic crust. With such sharp transition zone, we would expect to see a relatively abrupt change in character of the magnetic field data from continental character to oceanic character, which is not observed. Additional contradictory data are that existing heat-flow data do not support this sharp transition zone model, as discussed below.

Extended Deformation Zone Model

Our preferred structural style is shown in the right of Figure 13 (see also Figure 14 for detail). This model not only fits the gravity data but also predicts the proper character of the observed magnetic data and is consistent with crustal heat-flow estimations.

The preferred model thus involves a continental crust inboard of the large gravity anomaly associated with the western main transform zone. The basement complex in the deformation zone is likely to be a heavily modified continental crust intruded with magnetic and higher density crustal material. The heat flow would be expected to be much more like the continental than the oceanic crust, which lies farther outboard. Magnetic anomalies associated with the intrusives in this zone are large and have intermediate wavelengths, consistent with the observed magnetic data. In Figure 10, we place the continental-oceanic boundary on the outboard edge of the large gravity anomaly defining the shear margin at the transition between green and blue on the map; this agrees very well with the heat-flow data.

DISCUSSION

The shear margin, or western main transform discussed by Ross and Scotese (1988), Marton and Buffler (1993, 1994), and others, is generally interpreted to have accommodated the spreading of the Gulf of Mexico as Yucatan rotated counterclockwise from the present-day

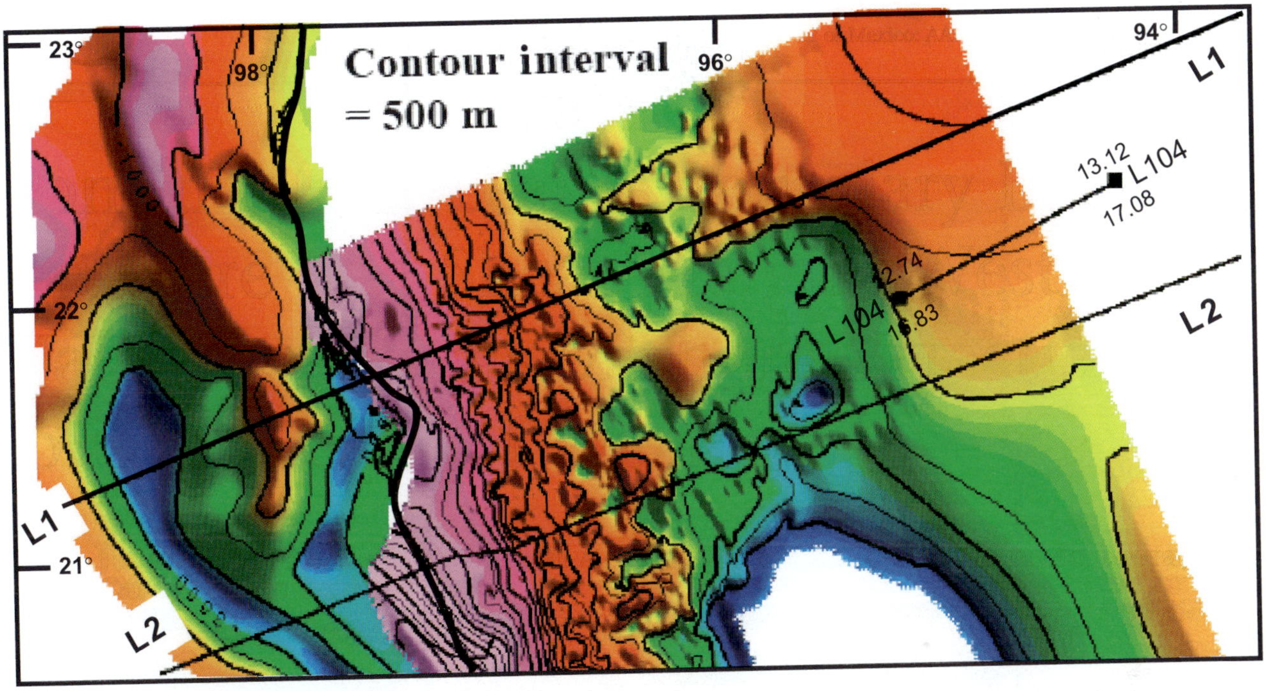

FIGURE 11. Basement control used in gravity modeling.

Texas and Louisiana coast around a pole of rotation near the straits of Florida (Figure 9). The transform zone and spreading center are apparent in the illustrations of gravity and magnetic data presented previously.

Possible thermal effects from this tectonic boundary include (1) unknown radiogenic component of heat flow relative to crustal composition in a shear zone and (2) increased heat-flow flux as a result of magmatic intrusion.

Although some transform boundaries may be characterized as sharp transition zones separating the continental and oceanic crust, our analysis suggests that the western main transform zone is not such a discrete boundary. Instead, our analysis indicates that it is more likely to have considerable lateral extent, as illustrated in Figure 15, which compares the gravity data in our study area to that of a possible analog. The shear margin of the Exmouth Plateau offshore northwest Australia as described by Lorenzo et al. (1991) has a very similar gravity response. A weakened crustal zone along the shear zone would have facilitated widespread intrusive and igneous plutonic activity, producing a highly

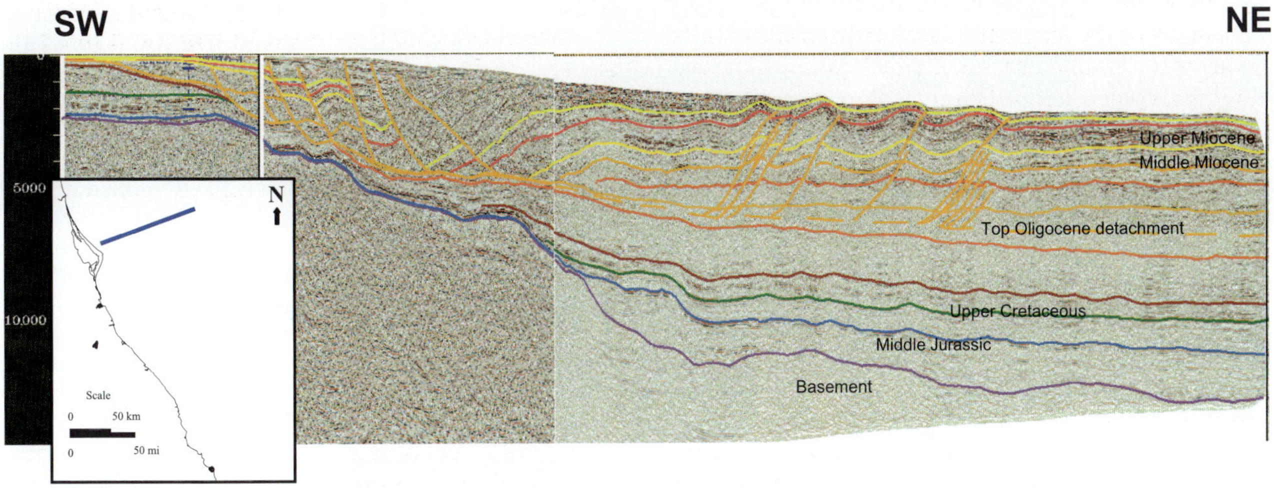

FIGURE 12. Dip-oriented seismic line (line 25) used to constrain the shallow horizons in gravity modeling.

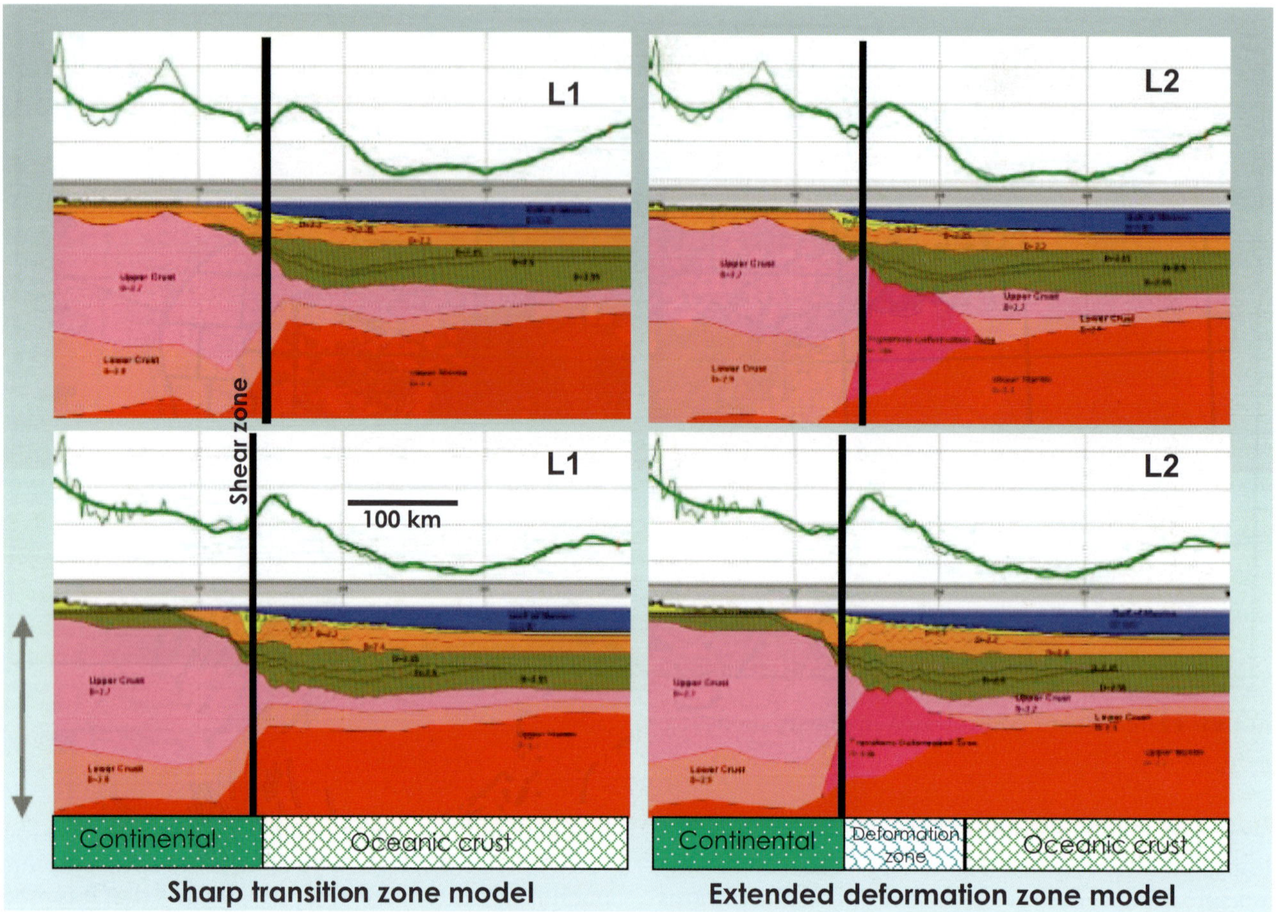

FIGURE 13. Alternative crustal models fitting the gravity response (shown on top of figures). The location of the two dip transects depicted above (lines L1 and L2) is shown in Figure 11.

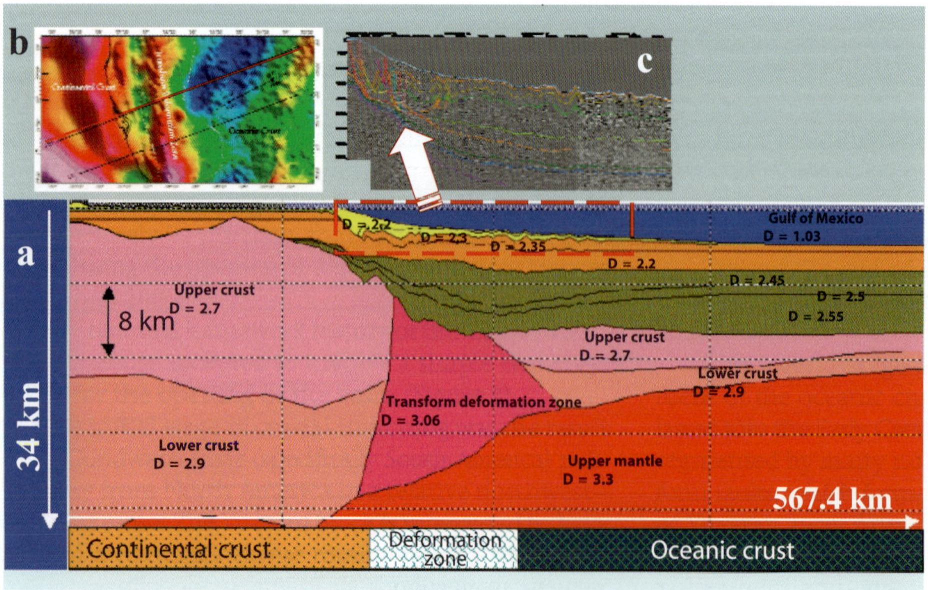

FIGURE 14. (a) Preferred crustal model along the dip transect through the area of interest (see also Figure 13). Both potential methods and inferred crustal thermal gradients suggest that the basement in the deformation zone is likely to be a heavily modified continental crust by intrusion of magnetic and higher density crustal material. (b) Location of the 2-D profile shown on the gravity map. (c) Seismic control used and modeling of shallow gravimetric response.

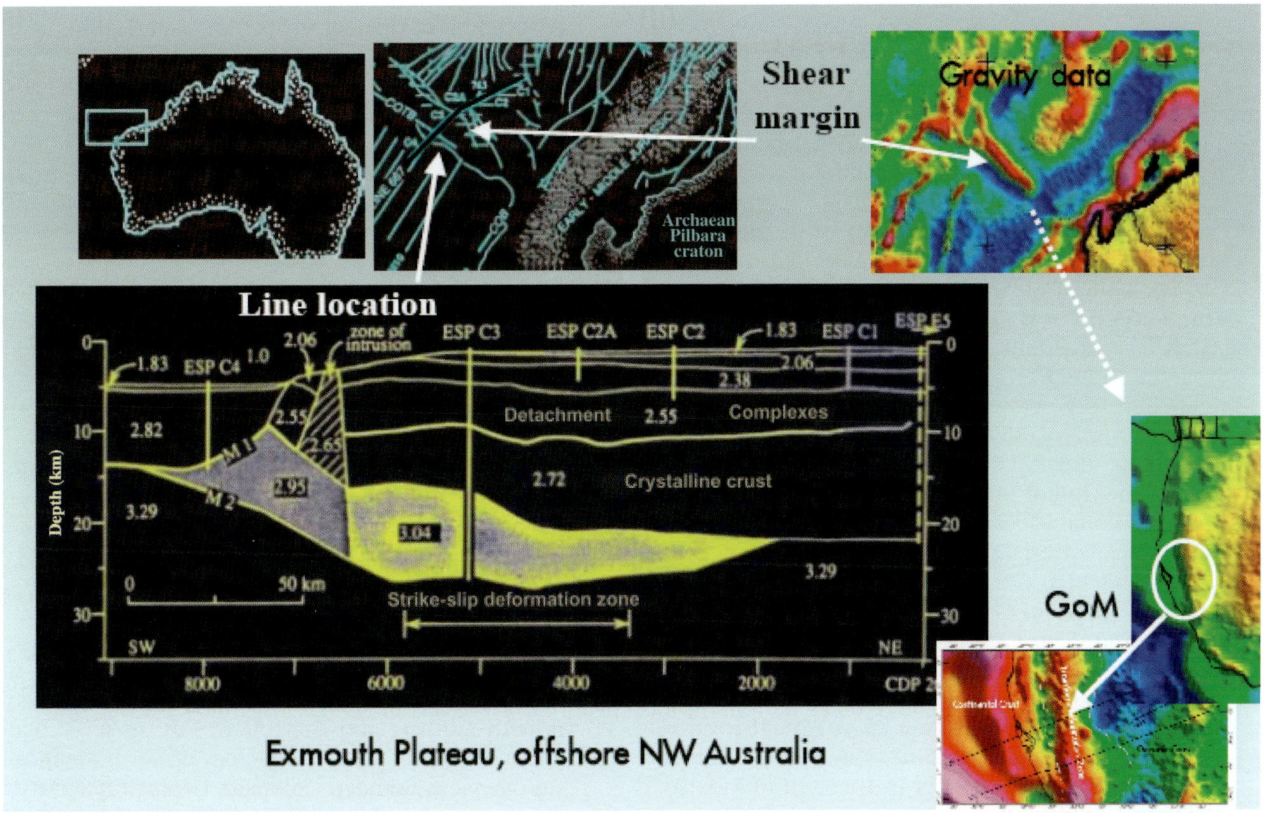

FIGURE 15. Comparison of the western main transform with a possible crustal analog, the Exmouth Plateau, offshore northwest Australia (Lorenzo et al., 1991). See also Figure 14 for gravity modeling in the area of interest. CDP = common depth point. GoM = Gulf of Mexico.

intruded crystalline basement complex of heavily modified and stretched continental crust enriched by higher density material from below.

The 2-D gravity modeling across the northwest Australia anomaly suggests that the optimum crustal density to fit the gravity signature is one that shares a mixed continental (silicic) and oceanic (mafic) constituency (Figure 15). Hypotheses for the origin of this mixed composition vary, but a likely model is that as the shear zone accommodates transtensional motion, dilatation of fractures in the original (continental) crust provides enough weakness for the intrusion of swarms of mafic dykes radiating from the mantle below. The result is that there will likely be fewer radiogenic minerals (potassium, thorium, etc.) in this crust relative to a typical transitional crust that underlies most continental margins.

This line of evidence was explored by deriving present-day heat flow for the study area from crustal history estimations. Modeled heat flow was iterated until it predicted thermal gradients consistent with nearby well temperatures and, to a lesser extent, surface heat-flow measurements. The results of the basin model are consistent with observed data. The above thermal evidence lends additional support to our hypothesis that the western main transform does not constitute a sharp boundary between continental and oceanic basements but instead involves a wide zone of heavily modified continental crust (Figure 14).

CONCLUSIONS

Subsurface data integration reveals new insights on the deep continental-oceanic boundary structure in a shear margin the western main transform of Ross and Scotese (1988) and Marton and Buffler (1994).

Evidence from both potential methods (gravity and magnetics) and basin modeling (inferred crustal thermal gradients) indicates that the basement complex in the western main transform is likely to be a wide belt of heavily modified continental crust intruded with magnetic and higher density crustal material.

In view of its similar gravimetric and thermal responses, the Exmouth Plateau, offshore northwest Australia (Lorenzo et al., 1991), is suggested as a possible tectonic analog for the west main transform.

According to our model, the oceanic crust lies outboard of the western main transform deformation zone, which may have favorable implications on petroleum systems in the area.

REFERENCES CITED

Bird, D. E., S. A. Hall, and J. F. Casey, 2005, Gulf of Mexico tectonic history: Hotspot tracks, crustal boundaries, and early salt distribution: AAPG Bulletin, v. 89, no. 3, p. 311–328.

Buffler, R. T., 1991, Seismic stratigraphy of the deep Gulf of Mexico Basin and adjacent margins, *in* A. Salvador, ed., The Gulf of Mexico Basin: Geological Society of America Centennial Volume, v. J, p. 353–389.

Feng, J., R. T. Buffler, and M. A. Kominz, 1994, Laramide orogenic influence on late Mesozoic–Cenozoic subsidence history, western deep Gulf of Mexico Basin: Geology, v. 22, p. 359–362.

Garrison, L. E., and R. G. Martin, 1973, Geologic structures in the Gulf of Mexico Basin: U.S. Geological Survey Professional Paper 773, 85 p.

Gray, G. G., R. J. Pottorf, D. A. Yurewicz, K. I. Mahon, D. R. Pevear, and R. J. Cuchla, 2001, Thermal and chronological record of syn- to post-Laramide burial and exhumation, Sierra Madre Oriental, *in* C. Bartolini, R. T. Buffler, and A. Cantu-Chapa, eds., The western Gulf of Mexico Basin: Tectonics, sedimentary basins, and petroleum systems: AAPG Memoir 75, p. 159–181.

Jacques, J. M., and H. Clegg, 2002, Late Jurassic source rock distribution and quality in the Gulf of Mexico: Inferences from plate tectonic modeling: Gulf Coast Association of Geological Societies Transactions, v. 52, p. 429–439.

Lorenzo, J., J. C. Mutter, and R. L. Larson, 1991, Development of the continent ocean transform boundary of the southern Exmouth Plateau: Geology, v. 19, p. 843–846.

Marton, G., and R. T. Buffler, 1993, Application of simple-shear model to the evolution of passive continental margins of the Gulf of Mexico Basin: Geology, v. 21, p. 495–498.

Marton, G., and R. T. Buffler, 1994, Jurassic Reconstruction of the Gulf of Mexico Basin: International Geology Review, v. 36, p. 545–586.

Pindell, J., and L. Kennan, 2002, Exploration framework atlas series, *in* Tectonic Analysis Ltd., ed., Mexico and Gulf of Mexico: West Sussex, England, v. 4, 46 p.

Ross, M. I., and C. R. Scotese, 1988, A hierarchical tectonic model of the Gulf of Mexico and Caribbean region: Tectonophysics, v. 155, p. 139–168.

Winker, C. D., and R. T. Buffler, 1988, Paleogeographic evolution of early deep-water Gulf of Mexico and margins, Jurassic to middle Cretaceous (Comanchean): AAPG Bulletin, v. 72, p. 318–345.

Worral, D. M., and S. Snelson, 1989, Evolution of the northern Gulf of Mexico with emphasis on Cenozoic growth faulting and the role of salt, *in* A. W. Bally and A. R. Palmer, eds., The geology of North America—An overview: Boulder, Colorado, Geological Society of America, v. A, p. 97–138.

20

Cobiella-Reguera, J. L., and F. Olóriz, 2009, Oxfordian–Berriasian stratigraphy of the North American paleomargin in western Cuba: Constraints for the geological history of the proto-Caribbean and the early Gulf of Mexico, *in* C. Bartolini and J. R. Román Ramos, eds., Petroleum systems in the southern Gulf of Mexico: AAPG Memoir 90, p. 421–451.

Oxfordian–Berriasian Stratigraphy of the North American Paleomargin in Western Cuba: Constraints for the Geological History of the Proto-Caribbean and the Early Gulf of Mexico

Jorge L. Cobiella-Reguera

Department of Geology, University of Pinar del Rio, Cuba

Federico Olóriz

Department of Stratigraphy and Paleontology, University of Granada, Spain

ABSTRACT

Many models that attempt to interpret the regional geology of the Caribbean-Gulf of Mexico area share the handicap of paying little attention to precise data from Cuba despite its location at the North American-proto-Caribbean paleoboundary. The North American Mesozoic paleomargin crops out along northern Cuba, an area where many wells have encountered Jurassic and Lower Cretaceous rocks. In this chapter, we present a stratigraphic interpretation of the North American passive continental paleomargin recognized in the Guaniguanico mountains (western Cuba) and identify the main geological events recorded in their Oxfordian to Berriasian sections. We correlate these sections with other Mesozoic passive paleomargin sections in Cuba, including the Escambray and Isle of Youth metamorphic terranes. These sections and those in the southeastern Gulf of Mexico record early stages of the disintegration of Pangea in Mesoamerica and the early development of the southeastern Gulf of Mexico and westernmost Tethys (proto-Caribbean). In all, two main sequences can be distinguished: (1) a lower terrigenous one and (2) an upper marine carbonate sequence. The transition from terrigenous to carbonate sedimentation occured during the late middle Oxfordian in western Cuba (and possibly in the Escambray mountains of central Cuba). In Placetas zone of central Cuba, the terrigenous-carbonate transition spans from Kimmeridgian to Berriasian. Transgression occurred close to the Kimmeridgian–Tithonian

DOI:10.1306/13191304M903339

boundary. Deeper depositional conditions started during the Tithonian along the paleomargin in northern Cuba. This event may be correlated with the beginning of the drowning and stepback of the small carbonate platforms developed in the southeastern Gulf. Spreading and rifting ceased in the southeastern Gulf of Mexico when the Yucatan block reached its present position in the earliest Cretaceous. The unconformity that records the change in tectonic setting is called the late Berriasian surface, but our reinterpretation of biostratigraphic data indicates that the age of generalized flooding is middle Berriasian. In middle and late Berriasian times, a belt of deep-water carbonates surrounded the northern margin of the proto-Caribbean (westernmost Tethys). Extensive carbonate banks developed on the Bahamas-Florida platform and surrounded the emergent Yucatan Platform.

INTRODUCTION

Jurassic and Cretaceous sediments have been reported from Cuba since the 19th century, as shown on the map drawn in 1868 by Fernández de Castro (Hayes et al., 1901). Not until the mid-20th century, however, was there significant progress toward modern stratigraphic research regarding the Upper Jurassic and Lower Cretaceous, and this was mostly a result of oil exploration (Rigassi-Studer, 1963; Hatten, 1967; Pardo, 1975; Hatten et al., 1988, among others). Regional studies, such as those by Furrazola-Bermúdez (1964), Khudoley and Meyerhoff (1971), and Pardo (1975), reflect geological knowledge prior to the general acceptance of plate tectonics. Since the 1970s, three main factors have promoted the knowledge of Jurassic and lowermost Cretaceous stratigraphy in Cuba: (1) the introduction of plate tectonics theory; (2) geological surveys at scales of 1:250,000 for the whole country (Pushcharovsky, 1988) as well as of 1:100,000 and 1:50,000 for large regions; and (3) the drilling of many oil wells. Regional studies incorporating updated interpretations have been published since the late 1980s (Hatten et al., 1988; Cobiella-Reguera, 1996a, b, 1998, 2000; Iturralde-Vinent, 1996a, 2003, 2006; Millán Trujillo, 1997a, b, c; Alvarez Castro et al., 1998; Sánchez et al., 1998; Pszczolkowski, 1999; Moretti et al., 2003; Pszczolkowski and Myczyñski, 2003; Iturralde-Vinent and Lidiak, 2006; Pindell et al., 2006, among others). These provide relatively detailed data on the Cuban Jurassic and lowermost Cretaceous in relation to data available for other areas in the Caribbean and the Gulf of Mexico. Nevertheless, many models that attempt to interpret the regional geology (Ladd and Sheridan, 1987; Salvador, 1987; Winker and Buffler, 1988; Lewis and Draper, 1990; Pindell and Barrett, 1990; Stephan et al., 1990; Montgomery et al., 1992; Buffler and Thomas, 1994; Marton and Buffler, 1994; Pindell, 1994, among others) share the handicap of having little precise data from Cuba despite its location at the North American-proto-Caribbean paleoboundary. Therefore, its crucial importance for interpreting the geologic history in Mesoamerica has been overlooked (Cobiella-Reguera, 1996a, 2000; Pszczolkowski, 1999; Iturralde-Vinent, 2003; Pszczolkowski and Myczyñski, 2003; Pindell et al., 2006).

The North American Mesozoic paleomargin crops out across northern Cuba (Figure 1), an area where many wells have drilled into Jurassic and Lower Cretaceous rocks. Recent articles present several interpretations of the extent and composition of these rocks (Pszczolkowski, 1982, 1999; Echevarría-Rodríguez et al., 1991; Draper and Barros, 1994; Iturralde-Vinent, 1996a, 1997; Hutson et al., 1998, Cobiella-Reguera, 2000; Moretti et al., 2003; Pszczolkowski and Myczyñski, 2003, Schaffhauser et al., 2003, among others). This is when the Gulf of Mexico opened and widened, with oceanic lithosphere forming in the central part because of the Maya (Yucatan) block separation, rotation, and final emplacement (e.g., Marton and Buffler, 1994, 1999; Pindell et al., 2006). In this chapter, we present a stratigraphic interpretation of the North American passive continental paleomargin recognized in western Cuba (but see Pessagno et al., 1999; Pszczolkowski, 1999; Iturralde-Vinent, 2003; Pessagno and Martin, 2003; Pszczolkowski and Myczyñski, 2003, for rather different approaches) and identify the main geological events recorded in its Oxfordian to Berriasian sections. In addition, we evaluate correlations with other Mesozoic passive paleomargin sections in Cuba and offer some comments concerning regional stratigraphic correlations, especially involving the southeastern Gulf of Mexico.

STRATIGRAPHY

Upper Jurassic–Berriasian Rocks in Cuba

These fall into four basic groups (Figure 1):

1. The mainly sedimentary passive paleomargin sections of northern Cuba comprise three distinct areas (see below): western Cuba (Guaniguanico Cordillera), northern-central Cuba, and Maisí, in easternmost Cuba.

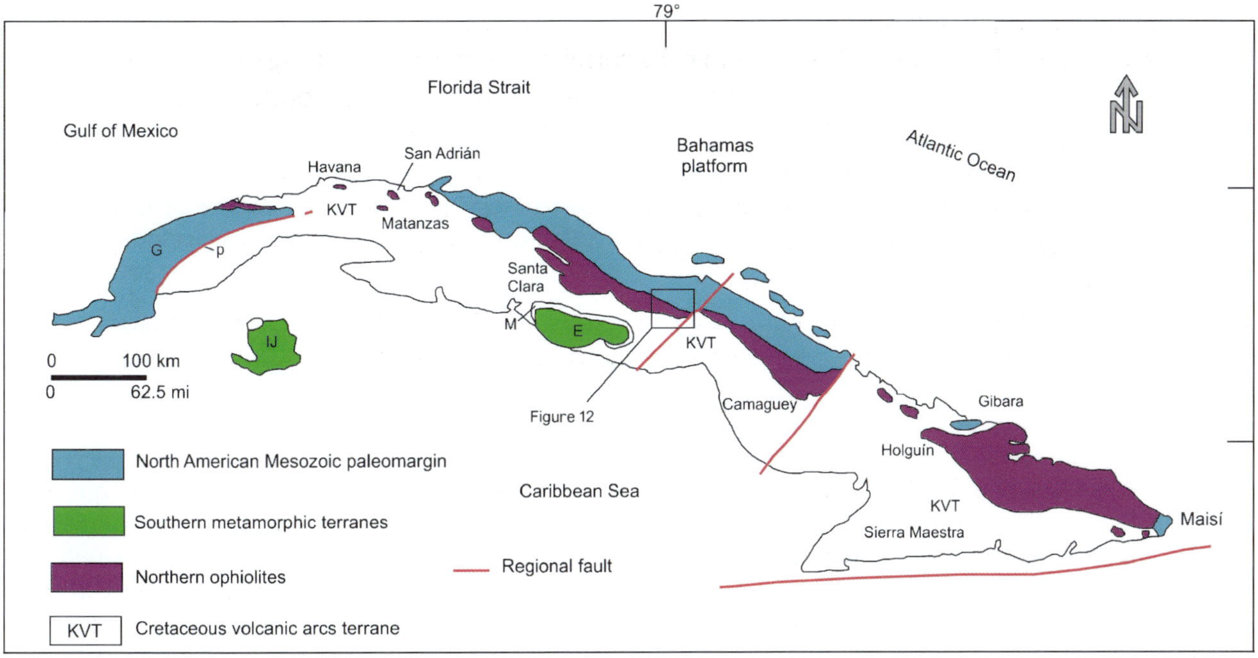

Figure 1. Mesozoic paleotectonic units of Cuba. G = Guaniguanico Cordillera; KVT = Cretaceous volcanic terrane; IJ = Isle of Youth (Isla de la Juventud); E = Escambray (Guamuhaya) massif; p = Pinar fault; violet areas: northern ophiolite belt; M = Mabujina complex. Note the inset for the location of Figure 12.

2. The Ophiolitic belt in northern Cuba, which comprises a huge mélange (Knipper and Cabrera, 1974; Cobiella-Reguera, 2005a; Garcia-Casco et al., 2006), is tectonically superposed on the Mesozoic paleomargin of northern Cuba. In the Ophiolitic belt, Andó et al. (1996), Iturralde-Vinent (1996a, b), and Llanes Castro et al. (1998) identified upper Tithonian to Albian rocks. Oxfordian rocks could be present in the ophiolites, considering their possible link with the fissure magmatism observed in the western Cuban paleomargin (Cobiella-Reguera, 1996a). The intense block tectonics and occurrence of transgressive pulses toward the end of the Middle Jurassic in the southwestern rim of the Gulf of Mexico Basin support the existence of deep waters in the region during the Late Jurassic interpreted by Schlager et al. (1984).
3. Successions are mainly composed of amphibolites and ultramafic rocks, which crop out most notably in the central Cuba Mabujina complex (Figure 1). The Mabujina complex not only is considered the oceanic basement of the Cretaceous volcanic terrane, but also includes some sections of metavolcanic arc rocks (Haydoutov et al., 1989; Millán Trujillo, 1996; Cobiella-Reguera, 2000; Pindell et al., 2006). Geochemical data have allowed some authors to interpret the Mabujina complex as mainly arc-derived metamorphic rocks (Grafe et al., 2001; Blein et al., 2003; Garcia-Casco et al., 2006). In general, this terrane rests tectonically on the ophiolites of northern Cuba. In south-central Cuba, the Mabujina complex and the Cretaceous volcanics rest, with tectonic contact, on the metamorphic Mesozoic successions of the Escambray (Guamuhaya) massif (Iturralde-Vinent, 1996a; Stanek et al., 2006).
4. Rocks of a southern extensional margin, subjected to Cretaceous metamorphism, are exposed on the Isla de la Juventud (Pinos) and the Escambray massif in southern central Cuba (Figure 1). Sedimentary protoliths of these southern metamorphic successions, commonly interpreted as terranes (e.g., Pszczolkowski, 1999; Iturralde-Vinent, 2003; Iturralde-Vinent and Lidiak, 2006; Pindell et al., 2006), are very similar to those of the Guaniguanico Cordillera (Furrazola-Bermúdez et al., 1964; Millán Trujillo, 1997a, b, c; Stanek et al., 2006, among others). Similarities suggest that Jurassic deposits in the three areas share a common origin; therefore, some authors have combined them within a single domain: the southwestern metamorphic terranes (Iturralde-Vinent, 1996a, 1997). Nevertheless, their actual tectonic positions are very different (Cobiella-Reguera, 1996a, 2000, see below) (Figure 1).

Only the passive continental margin type successions 1 and 4 are discussed in this chapter. Succesions 2 and 3 are discussed by Iturralde-Vinent (1996a, b), Cobiella-Reguera (2000, 2005a), Lewis et al. (2006), and Pindell et al. (2006), among others.

Upper Jurassic–Berriasian in the North American Paleomargin of Western Cuba (Guaniguanico Cordillera)

The low mountains of the Guaniguanico Cordillera contain the most extensive continuous outcrops of the North American Jurassic paleomargin in Cuba (Figure 2). To the south, the Guaniguanico successions are bounded by the Pinar fault, which separates them from the Cenozoic Los Palacios Basin. The largest and most impressive outcrops of alpine tectonics in the Greater Antilles are present in these highlands (Rigassi-Studer, 1963; Hatten, 1967; Piotrowska, 1978). To the northeast, the Guaniguanico section is covered by thrusted ophiolites and Cretaceous volcanic terranes (Pszczolkowski, 1994b; Cobiella-Reguera, 2000, 2005a). The Guaniguanico successions rest tectonically on sediments of the Gulf of Mexico (Iturralde-Vinent, 1996a; Gordon et al., 1997; Moretti et al., 2003). A thrust system divides rocks of the Guaniganico Cordillera into various tectonic units. Thrust sheets of the Sierra de los Organos comprise the lowest tectonic units, overridden by those of the Alturas Pizarras del Sur and Sierra del Rosario-Alturas de Pizarras del Norte (Figure 2). Minimum horizontal displacement of several tens of kilometers can be assumed between the lowest and highest units in the nappe pile (Cobiella-Reguera, 2005b). The age of the tectonic event is late Paleocene to earliest Eocene.

Throughout the Guaniguanico Cordillera, the Jurassic sections reveal (1) a lower part, composed almost entirely of slightly metamorphosed siliciclastic beds of the San Cayetano Formation high-pressure metasiliciclastic rocks, outcrops in the Cangre belt (Arroyo Cangre Formation, Figures 2, 3) (Somin and Millán 1981) and (2) mainly carbonate deposits rest on the siliciclastic section and extend to the Cretaceous. The upper San Cayetano Formation contains Oxfordian fossils, and thus the limit between the siliciclastic and carbonate successions is located within the Upper Jurassic. Oxfordian–Berriasian beds in western Cuba consist of sedimentary rocks and, to a lesser extent, igneous rocks that accumulated during the early drifting stage of Pangea and the formation of the proto-Caribbean (Pszczolkowski, 1999; Cobiella-Reguera, 2000, and references therein).

Figure 3 gives the Upper Jurassic–Berriasian stratigraphy in western Cuba, as updated in the present study.

San Cayetano Formation

The San Cayetano Formation is a terrigenous deposit (perhaps several thousands of meters thick in the most complete sections in Alturas de Pizarras) mainly composed of sandstones, siltstones, and claystones, with some conglomeratic beds (Haczewski, 1976; Pszczolkowski, 1978, 1999; Hutson et al., 1998) (Figure 4). Ore bodies with exhalative sedimentary sulfide mineralization (e.g., Valdés-Nodarse, 1998, and references therein) can be found in this unit. Thin discontinuous limestone horizons appear toward the top (Pszczolkowski, 1999) and no erosional discontinuity with the overlying carbonates is observed (Pszczolkowski, 1999, and references therein). At present, offering a detailed characterization of the terrigenous succession is impossible because of its homogeneity, the acute shortage of fossils, and tectonic complexity. The best stratigraphic approach is the facies pattern proposed by Haczewski (1976), who distinguished nine facies and interpreted deposition in the context of a widespread deltaic environment. Sierra de los Organos (SO) and at least part of the Alturas de Pizarras del Sur (APS) sections record fluvial, transitional, and shallow marine paleoenvironments (Figure 4a), whereas deeper water turbidites are present in Sierra del Rosario and Alturas de Pizarras del Norte (Figure 4b) (Haczewski, 1976; Cobiella-Reguera et al., 2000). Therefore, in Figure 2, the southwest corner of Alturas de Pizarras del Norte (APN) highlands is included in the APS tectonic unit (see also Figure 4a). Following Haczewski (1976) and (Cobiella-Reguera et al., 1997; Cobiella-Reguera, 2000) the authors believe that the formation accumulated in a major delta complex on a young passive continental margin during the proto-Caribbean early drift stage. Perhaps some sediment arrived from other sources located along an opposite margin, as suggested by preliminary studies on the mineral composition of San Cayetano sandstones (Pszczolkowski, 1986a; Cobiella-Reguera et al., 1997, 2000; Hutson et al., 1998; Rojas-Agramonte et al., 2008) and paleocurrent data (Haczewski, 1976). Total organic carbon (TOC) averages 1.1% (Moretti et al., 2003). Some authors consider the unit a synrift accumulation (Linares Cala, 1999; Moretti et al., 2003), but as Hutson et al. (1998) stated, neither the predominant quartz-rich sandstone composition (Pszczolkowski, 1986a; Cobiella-Reguera et al., 1997) nor the depositional architecture is suggestive of formation as a synrift unit.

The age of the San Cayetano Formation is controversial. A Triassic age was proposed by Iturralde-Vinent (2003) for the lower San Cayetano Formation, but no definitive Triassic fossils have been found. The oldest layers might be Early Jurassic (Areces-Mallea, 1991; Cobiella-Reguera, 1996b). The youngest beds are middle Oxfordian (Myczyñski and Pszczolkowski, 1976) (Figure 3) and contain *Perisphinctes (Dichotomosphinctes) cayetanensis, Perisphinctes (Dichotomosphinctes?) cf. anconensis,* and *Perisphinctes (Discosphinctes) pichardoi* in the Sierra del Rosario. Myczyñski et al. (1998) suggested that among the *Dichotomosphinctes* identified, forms such as *D. cayetanensis* may appear with a primitive morphology (Vertebrale-Antecedens chron in Europe).

Focusing on the Upper Jurassic, we are only concerned with the age of the upper part of the San Cayetano Formation. According to Myczyñski et al. (1998), the oldest horizons with ammonites contain morphologically

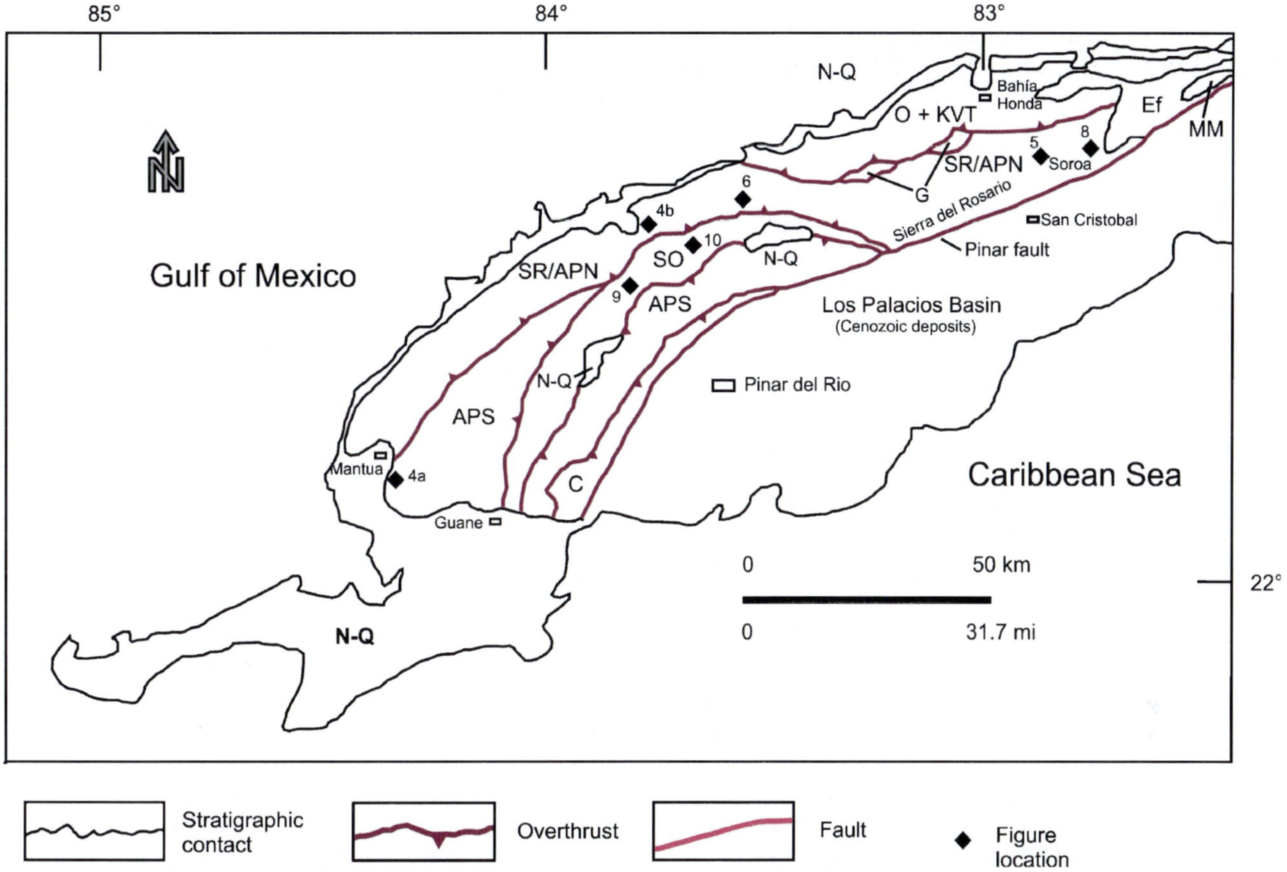

FIGURE 2. Tectonic map of the Guaniguanico Cordillera and its surroundings. N-Q = Neogene and Quaternary sediments; Ef = lower Eocene deposits (mainly turbidites); O + KVT = Cretaceous volcanic terrane + ophiolites; SR/APN = Sierra del Rosario-Alturas de Pizarras del Norte and Esperanza zone tectonic sheets; G = Guajaibón tectonic unit (Albian–Cenomanian carbonate bank); SO = Sierra de los Organos tectonic sheets; APS = Alturas de Pizarras del Sur tectonic sheet; C = Cangre belt; MM = Martín Mesa erosional window (outcrop of Sierra del Rosario sections). The ophiolites, the volcanic terrane, and the Guaniguanico units were emplaced during the Cuban orogeny of late Paleocene–early Eocene age in western Cuba. Los Palacios Basin is a Cenozoic structure. The locations of the outcrops in Figures 4–6, 8–10 are indicated by the corresponding numbers.

primitive *Dichotomosphinctes*, comparable with those from the Vertebrale-Antecedens chron in Europe, whereas the youngest horizons with ammonites contain morphologically innovative *Dichotomosphinctes* and probable *Gemmellarites* that belong to the Antecedens chron (European standard) of the middle Oxfordian. In other words, the ammonite-bearing stratigraphic interval within the Upper Jurassic of the San Cayetano Formation belongs to the lower–middle middle Oxfordian. Unfortunately, no ammonites have been found in the uppermost San Cayetano Formation, and thus, we cannot interpret the upper boundary of this unit at the ammonite zone-subzone level.

Francisco Formation

The Francisco Formation (Figure 3) crops out with a thickness of 13–25 m (43–82 ft) in the Sierra del Rosario and is composed of argillites and well-stratified micritic limestones, with fine intercalations of sandstones (Myczyñski, 1976a). According to Myczyñski and Pszczolkowski (1976), shales with calcareous concretions are present in the type locality, overlain by laminated limestones with thin shale beds (5 m [16 ft] thick), sandstones, and shales (2 m [6 ft]), and topped by shales with thin limestone intercalations. This formation contains the transition between the underlying terrigenous San Cayetano Formation and the overlying, mainly carbonate Artemisa Formation. The absence of the thin Francisco Formation in many sections might be caused by tectonic factors (shearing during thrusting). According to Myczyñski and Pszczolkowski (1976), water depth did not exceed 300 m (984 ft); Schlager et al. (1984) interpreted neritic deposition, and Pszczolkowski (1999) interpreted outer shelf deposits. We interpret that Francisco deposits accumulated at neritic depths under marine-restricted environments.

Two ammonite assemblages were identified in the Francisco Formation; the lower one with *Vinalesphinctes*,

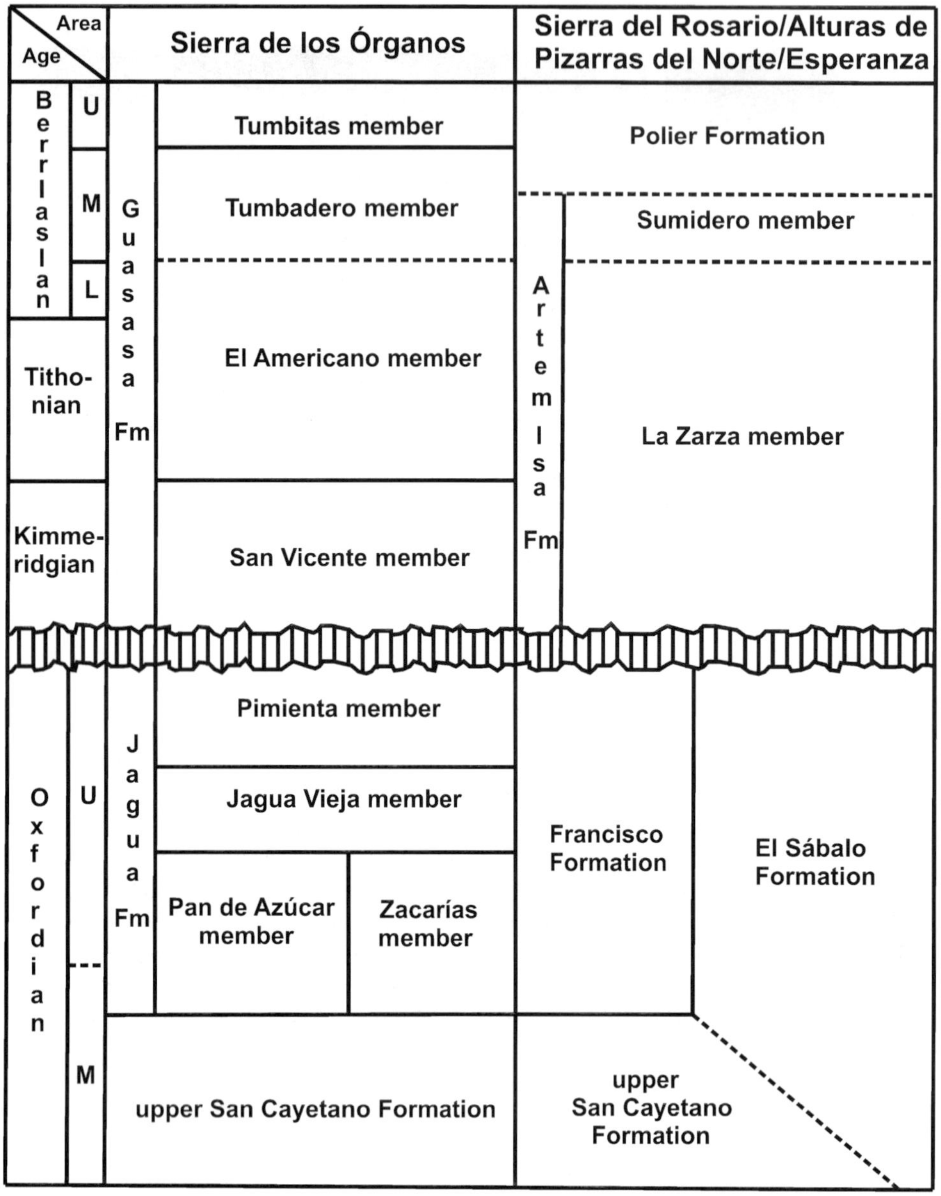

FIGURE 3. Schematic correlation chart of Upper Jurassic–Berriasian deposits in western Cuba. The main features of each lithostratigraphic unit are discussed in the chapter. U = upper; M = middle; L = lower.

Roigites, *Antilloceras*, *Cubaochetoceras*, and *Glochiceras*, and the upper one with *Euaspidoceras*, *Mirosphinctes*, *Cubaspidoceras*, and *Glochiceras*. Kutek et al. (1976) and Myczyñski (1976a, 1994) first interpreted these ammonite assemblages as indicative of late–middle and early–late Oxfordian age. Myczyñski et al. (1998) later reinterpreted the assemblages noting the predominance of *Vinalesphinctes* and *Cubasphinctes*, as well as *"Discosphinctes"* (= *Subdiscosphinctes*), *Ochetoceras*, *Cubaochetoceras*, and *Glochiceras*, and ancillary *Euaspidoceras*. These authors (1998, p. 198) assigned a Bifurcatus age to the assemblage dominated by *Vinalesphinctes* and *Cubasphinctes* in the Guaniguanico Cordillera in view of its correlation with the Chilean and North American findings of *Gregoryceras* and associated fauna (Gygi and Hillebrandt, 1991; Young and Oloriz, 1993). These findings have recently been extended to Mexico by Villaseñor et al. (2004). According to Myczyñski et al. (1998), the ammonite assemblages dominated by *Vinalesphinctes* and *Cubasphinctes* correspond to periods of strong endemism and can be correlated with the lower–upper Oxfordian in Europe. Particularly interesting for biogeography and relative paleogeography is the frequent occurrence of this ammonite assemblage in the Guaniguanico Cordillera (Francisco Formation in the Sierra del Rosario and Jagua Formation in the Sierra de los Organos) and its local, biostratigraphically coherent record in east-central Mexico (López-Palomino et al., 2006). Myczyñski et al. (1998) interpreted the upper assemblages with *Euaspidoceras* and *Mirosphinctes*, and with *Cubaspidoceras* and *Mirosphinctes* as early and middle Bimammatum chron, respectively (middle latest Oxfordian).

FIGURE 4. Outcrops of San Cayetano Formation. (a) Cross-bedded laminated quartzose sandstones at Veinte de Mayo, near the Guane-Mantua road (Figure 2). Cross-bedded sets attain several meters thickness. Some isolated well-rounded clasts 1–2 cm (0.3–0.6 in.) in diameter are present. The outcrop corresponds to Haczewski's (1976) facies A and belongs to Alturas de Pizarras del Sur unit. Coordinates: 22°12′51″N; 84°11′52.7″. (b) Well-bedded distal turbidites at Merceditas, Alturas de Pizarras del Norte, on the Viñales-Puerto Esperanza road (Figure 2). The sandstone interbeds increase to the right and become younger. Facies G in Haczewski's model. Coordinates: 22°41′30.4″ N; 83°44′0.7″.

El Sábalo Formation

Defined by Pszczolkowski (1994a) in the Sierra del Rosario, the El Sábalo Formation comprises successions dominated by mafic rocks (diabases, basalts, and, occasionally, gabbros) with several-meter-thick intercalations of well-stratified laminated siltstones, claystones, and black limestones with high organic content (Figure 5). El Sábalo Formation thickness can reach 400 m (1312 ft) (Cobiella-Reguera, 1996a, Iturralde-Vinent, 1996c, Pszczolkowski, 1999). Contacts are commonly tectonic, the upper one showing breccias locally (Pszczolkowski, 1999). Typically, carbonate intercalations appear only in the upper part of the El Sábalo Formation, whereas terrigenous layers are distributed throughout the succession. The presence of synsedimentary folds and faults involving mafites indicates that El Sábalo deposits were deposited on an inclined and/or episodically unstable bottom (Figure 5) (Cobiella-Reguera, 1996a). Deposition of the El Sábalo Formation on a poorly oxygenated bottom is accepted (Pszczolkowski and de Albear, 1983; Cobiella-Reguera, 1996a; Pszczolkowski, 1999), although the depositional environment was interpreted as continental slope (Cobiella-Reguera, 1996a) or outer shelf (Pszczolkowski, 1999). We interpret that low oxygenation is related to active magmatism and the pattern of marine currents in the area. As marine-restricted environments can be inferred for the coeval Francisco and Jagua formations, low oxygenation most probably was widespread, but it cannot be determined if it was basinwide with the information available.

Scarce microfossils are known from the El Sábalo Formation, although Pszczolkowski, (1994a) reported *Globochete alpina*, *Didemnoides moreti*, *Didemnoides* sp., *Didemnum carpaticum*, *Didemnum minutum*, "*Colomisphaera*" cf. *nagyi*, "*Colomisphaera*" cf. *pieniniensis*, and "*Colomisphaera*" sp. in carbonate horizons. We correlate the upper El Sábalo with the Francisco Formation (lower–upper Oxfordian). Supporting this interpretation

FIGURE 5. El Sábalo Formation outcrop in a road cut at the type locality (Figure 2). Massive dark rocks are diabases and laminated shales (s); the light ones are limestones (L). Note the broken limestone layer and the irregular sediments-mafite contact in some places, suggesting synsedimentary deformation. Hammer for scale, in circle. Coordinates: 22°49′16.4″N; 83°05′00″W.

FIGURE 6. Well-bedded, slightly cherty, micritic dark, commonly laminated limestones of Artemisa Formation (Sumidero member) at Vega Nueva quarry, west of La Palma (Figure 2), Alturas de Pizarras del Norte. Strongly weathered foliated shale interbeds, 1–2 cm (0.3–0.6 in.) thick, constitute nearly 10% of the section. Terrigenous intercalations are more frequent in the Artemisa Formation than in coeval Sierra de los Organos sections. Coordinates: 22°44′02.5″N; 83°37′15.5″W.

are (1) microfossil horizons underlying the Artemisa Formation, the base of which is the upper Oxfordian Bimammatum zone (Kutek et al., 1976), probably Bimammatum subzone (Myczyñski et al., 1998), or younger as envisaged below (p. 11), and (2) the lithological similarity between limestones and claystones in the upper part of the El Sábalo Formation and the Francisco Formation. The lower, terrigenous El Sábalo Formation, without limestone beds, may also be correlative with the uppermost San Cayetano Formation (Cobiella-Reguera, 1996a). Although the upper and lower contacts are tectonic, the stratigraphic interval assumed for correlation should correspond to the Oxfordian interval that includes the upper San Cayetano and the Francisco formations (Figure 3). Yet, it is difficult to be conclusive with the data available. The occurrence of basalts in the Francisco Formation (Pszczolkowski, 1999, and references therein) reinforces this stratigraphic interpretation.

Cobiella-Reguera (1996b) correlated mafites of the El Sábalo Formation with the youngest diabase dikes reported from the Catoche knoll, in the southern Gulf of Mexico (Schlager, Buffler et al., 1984; Schlager et al., 1984). Based on the stratigraphic position assumed by Cobiella-Reguera (1996b), biochronostratigraphy as proposed by Myczyñski et al. (1998), and revisions of recent geochronology (UNESCO and IUGS, 2000 scale), the late intrusive phase recorded at Catoche knoll (160 Ma in Schlager et al., 1984; Schlager, Buffler et al., 1984) occurred earlier than the late Oxfordian magmatism recorded in the El Sábalo Formation. In fact, the youngest magmatic event in Catoche knoll can be placed at the latest Callovian (e.g., Hardenbol et al., 1998), or very close to the Callovian Oxfordian boundary, according to Witrock et al. (2003) (i.e., at 159.4 Ma). Therefore we assume some genetic asssociation of the magmatic phases in the El Sábalo Formation, and its related units in the Guaniguanico Cordillera, with the youngest intrusive rocks recorded in the southeastern Gulf (see below).

Artemisa Formation

This mainly calcareous unit represents the onset of a carbonate succession eastward in the Guaniguanico Cordillera (Sierra del Rosario). The bulk of the unit can be divided into two members: La Zarza and Sumidero. In the lower and middle part, the La Zarza member contains well-bedded and commonly laminated black micritic limestones with occasional intercalations of shales and siltstones. Pszczolkowski (1978) reported local interbeds of siltstones and fine-grained sandstones at the base of the Artemisa Formation and breccias, including shallow-water carbonate debris, from the lowermost horizons. Local occurrences of thick-bedded dolomitic limestones point to the restricted record of shallow-water San Vicente-type carbonates within the Artemisa Formation (Pszczolkowski, 1978). The upper La Zarza member contains bioclastic limestone horizons and some coquinas made up of ammonites and aptychi (Pszczolkowski, 1999). Calcareous beds with traces of distal turbidites gradually become laminated calcareous shale or marls upsection (Cobiella-Reguera, 1996b). Pszczolkowski (1978) reported horizons with folds and submarine slide breccias, and layers of detrital limestones with shallow-water elements (e.g., ooids, *Favreina*) intercalated in pelagic deposits. The upper Artemisa Formation (Figure 6) is characterized by cherty horizons with abundant radiolaria, which are used to identify the Sumidero member. Pszczolkowski (1978, 1999) reported pink and light-brown limestones in this upper member, but coloration may be caused by alteration by groundwater, and the original sediments might well be dark, like the remaining rocks in the formation.

Paleoenvironments in the lower La Zarza beds are little studied. Pszczolkowski (1999) proposed that dolomitized limestones in the lower La Zarza member might have accumulated in partially restricted coastal lagoons, whereas the upper part of this unit was deposited on the outer shelf. For the Sumidero member, this author interpreted a pelagic deposit accumulated below the aragonite compensation depth. However, the presence of sediments with the above-described turbiditic features in upper La Zarza member beds and the occasional occurrence of synsedimentary breccias and folds (Pszczolkowski, 1978; Myczyñski and Pszczolkowski, 1994; Cobiella-Reguera, 1996b) suggest basin and probably local slope conditions late in the Tithonian, according to the Wilson (1980) model. The context suggests external zones of a carbonate ramp that deepened during the Late Jurassic and earliest Cretaceous. Indeed, as usually interpreted, the La Zarza member contains late Oxfordian sediments (but see below), albeit little biostratigraphic data below the Tithonian exist. Thus, lacking significant data on facies changes in the La Zarza member, we can only identify deepening that provided a favorable environment for ammonites that accumulated in deposits close to the Kimmeridgian–Tithonian boundary in the eastern Guaniguanico Cordillera. Assuming deepening during the Late Jurassic–Early Cretaceous would then be appropriate. The undisturbed horizontal laminations in many beds and the lack of bioturbation and benthic autocthonous fauna, together with the dark color in nonweathered rock surfaces (high values of TOC are reported by Moretti et al., 2003, for the Sumidero member), clearly point to deposition in poorly ventilated, dyssaerobic to anoxic bottoms, although no conclusive interpretation of water depth is available.

The basal Artemisa Formation contains late Oxfordian ammonites (Kutek et al., 1976), found only at the type locality of the Francisco Formation, near Cinco Pesos, Sierra del Rosario. In two beds of marly nodular limestones, separated by a thin layer of marly shales, and lying on Francisco Formation shales, Kutek et al. (1976) reported a poorly preserved ammonite assemblage containing *Mirosphinctes* and *Cubaspidoceras*, for which they envisaged not to be post-Bimammatum age. The age of this ammonite assemblage was interpreted as middle Bimammatum chron by Myczyñski et al. (1998). According to literal translation of Kutek et al. (1976, p. 305), "...most of the ammonites are preserved as incomplete moulds and imprints, which suggest solution and mechanical destruction of shells in deposits. Voids originating from solution of primary organic structures are sometimes filled with coarse secondary calcite." In one locality (ibid. p. 306) "...a layer of the ammonite bearing limestones is directly overlain by a dolomite sediment yielding tubular phosphatic pellets, a few mm in size, possibly of organic origin (?coprolites)." A little higher in the section, these authors quoted "...a layer... with numerous fissures...infilled with sparry dolomite with numerous phosphatic structures... and small borings, possibly made by polychaetes." "A bed with numerous fissures, infilled with sparry dolomite, with numerous phosphatic structures and small borings, possibly made by polychaetes." Kutek et al. (1976) concluded that the observed phenomena are evidence of starved deposition, including breaks in sedimentation. Also near Cinco Pesos, in beds possibly only a little higher in the section, one of us (J. L. Cobiella-Reguera) sampled the following assemblage (identified by J. Fernández-Carmona and S. Blanco-Bustamante): *Colomisphaera nagyi*, *Stomiosphaera moluccana*, *Globochaete alpina*, *Cadosina aff. parvula*, and *Cadosina fusca*. Fernández-Carmona and Blanco-Bustamante interpreted this association as Kimmeridgian, although according to Fernández Carmona and Pendás Amador (1998), *C. fusca* is more likely a Tithonian species. These findings suggest that additional research is necessary to establish the age of beds in the lowermost La Zarza member, perhaps entailing a redefinition of the lower boundary of the unit, excluding the two basal limestone beds with ammonites, which could be assigned to the Francisco Formation, because of their unconformable contact with the overlying beds (Figure 3).

In the upper La Zarza member, successive assemblages of ammonites are characterized by *Parakeratinites*, *Hybonoticeras*, *Pseudolissoceras*, *Lytohoplites*, and *Micracanthoceras-Corongoceras* or *Durangites-Himalayites-Salinites*, as well as tintinnoids (*Chitinoidella* and *Crassicollaria*), identified and correlated by Myczyñski and Pszczolkowski (1994) and Myczyñski (1999). These authors interpreted the upper La Zarza member as Tithonian and the overlying cherty deposits of the Sumidero member as Berriasian. Similarly, Pszczolkowski (1999) placed the boundary between the La Zarza and Sumidero members, which marks the onset of siliceous sedimentation, at the Jurassic–Cretaceous (Tithonian–Berriasian) boundary. Previously, however, Myczyñski (1989, his figure 2) and Myczyñski and Pszczolkowski (1990, their table 1) suggested that the La Zarza member and the El Americano member (see below) might include Berriasian rocks. According to our comments on the El Americano member in the Guasasa Formation (Sierra de los Organos) (see also Pszczolkowski et al., 2005), which were based on Tavera et al. (1994), Olóriz et al. (1995), and an updated interpretation of calpionellids (Myczyñski, 1989; Myczyñski and Pszczolkowski, 1994), we agree with the assignment of an early Berriasian age (see below) in the La Zarza member. The finding of *Stomiosphaerina proxima* and *Calpionella elliptica* in the upper La Zarza member, near Soroa, is also significant (Figure 7). This could extend the age to the middle Berriasian if the citation of *C. elliptica* did not involve known homeomorphs from the late Tithonian and earliest Berriasian. In addition, no precise calpionellid biostratigraphy from the Sumidero member has been published. Preliminary data compiled by Pszczolkowski (1978, his table 6) and data in Figure 7 indicate that the late middle Berriasian (uppermost B to C calpionellid zones) is recognized from lower but not the lowermost horizons in this member containing *C. elliptica*, probable *Remaniella*, and *Calpionellopsis*. Although Pszczolkowski (1999) interpreted the youngest Sumidero deposits as Valanginian in age, the precise identification of the first intercalated sandstones of the overlying Polier Formation in the eastern Sierra del Rosario (Figure 7) suggests that they did not range above the Berriasian. In the senior author's experience, the Valanginian sections at Sierra del Rosario contain significant amounts of sandstones and therefore belong to the Polier Formation.

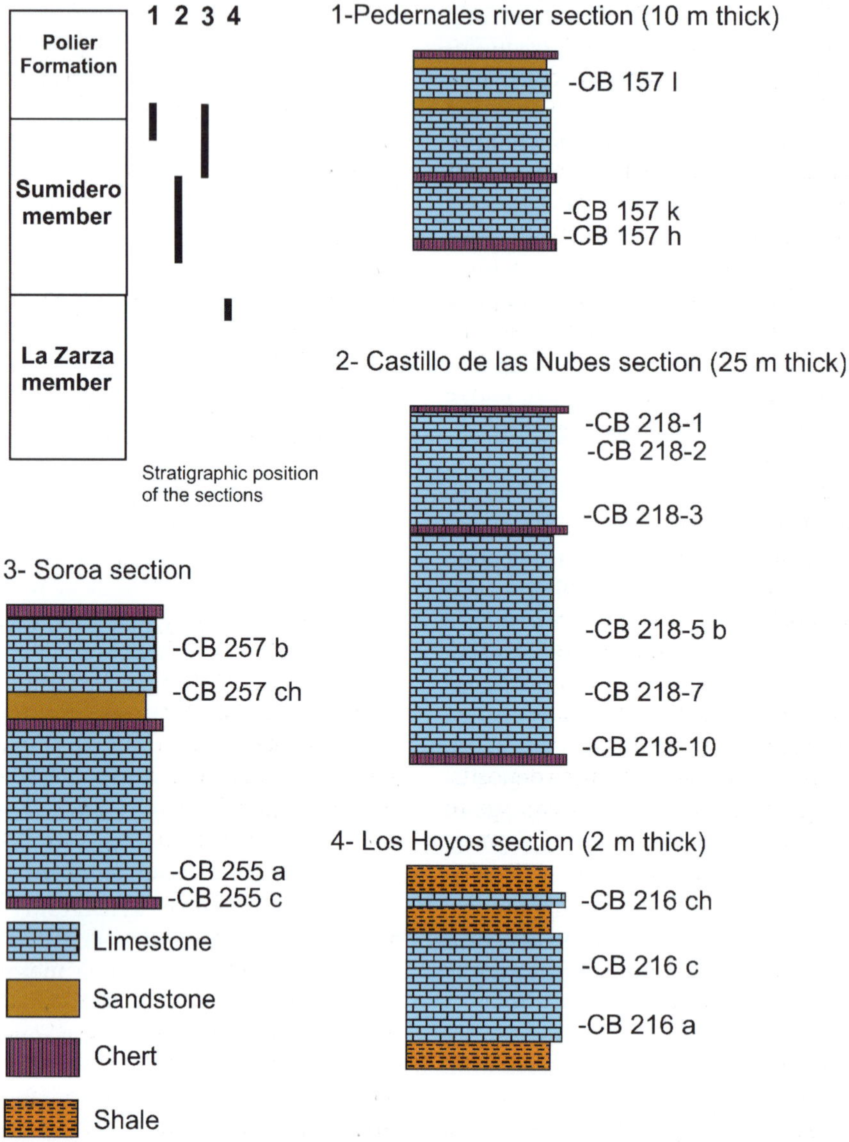

FIGURE 7. Schematic stratigraphic columns in upper Tithonian–Berriasian sections around Soroa (Figure 2) in southeastern Sierra del Rosario, Guaniguanico Cordillera, western Cuba. Black bars beside the general column represent the stratigraphic thickness of each minor column. Samples (CB). Pedernales river section: Artemisa Formation (Sumidero member): CB 157 h = *Tintinopsella carpathica, Calpionella alpina, Stomiosphaerina proxima, Crassicollaria* sp.?; CB 157 k = *Calpionella* sp., calcareous calpionellids; Polier Formation: CB 157 l = *Cadosina fusca, Stomiosphaerina proxima, Calpionella alpina, Tintinopsella* sp., *Calpionella cf. C. elliptica*. Castillo de las Nubes section: Artemisa Formation (Sumidero member): CB 218-10 = *Tintinopsella longa, Calpionella alpina, Stomiosphaerina proxima*, calcareous calpionellids; CB 218-7 = *Tintinopsella carpathica, Calpionella alpina, Stomiosphaerina proxima*, calcareous calpionellids; CB 218-5b = *Calpionella alpina* (small forms), *Tintinopsella carpathica, Calpionella elliptica, Nannoconus* sp.; CB 218-3 = *Calpionella elliptica*, calcareous calpionellids, *Nannoconus* spp.; CB 218-2 = *Tintinopsella longa, Calpionella alpina* (small forms), *Stomiosphaerina proxima*, calcareous calpionellids, *Nannoconus* spp.; CB 218-1 = *Tintinopsella longa, Calpionella alpina* (small forms), *Stomiosphaerina proxima*. Soroa section: Artemisa Formation (Sumidero member): CB 255 c = *Tintinopsella* sp., *Calpionella* sp., calcareous calpionellids; CB 255 a = *Calpionella* sp., calcareous calpionellids; CB 257 ch = *Calpionella alpina, Tintinopsella* sp.; Polier Formation: CB 257 b = *Tintinopsella carpathica, Calpionella alpina, C.* sp., *Colomisphaera* sp., *Crassicollaria* sp., *Crustacadosina* sp. Los Hoyos section: Artemisa Formation (La Zarza member): CB 216 a = *Colomisphaera carpathica, Cadosina fusca, Didemnoides moreti, Saccocoma* sp.; CB 216 c = *Saccocoma* spp., *Colomisphaera* sp., *Cadosina* sp.; CB 216 ch = *Stomiosphaerina proxima, Calpionella cf. C. elliptica*. Paleontologist: José Fernández-Carmona.

Polier Formation

Conformably overlying the Artemisa Formation, the Polier Formation comprises a succession of well-bedded, commonly laminated dark limestones with intercalations of sandstones, siltstones, and claystones (Figure 8) (Cobiella-Reguera et al., 1997; Pszczolkowski 1999). Limestones include common radiolarian-rich horizons, as well as turbiditic ones with bioclasts derived from shallow-water environments. Thin cherty horizons occur at the bottom of the unit and, locally, at the top. Terrigenous beds are silica-rich distal turbidites (Cobiella-Reguera et al., 1997; Pszczolkowski, 1999), which are much more frequent in the northern sections in the Sierra del Rosario and its western prolongation (Esperanza zone). As these thrust sheets contain sections originally located to the south (Cobiella-Reguera et al., 2000), a source of terrigenous continental deposits was probably active from the south. This contrasts with the southward dominant trend in paleocurrents reported by Pszczolkowski (1978) and Cobiella-Reguera et al. (1997). The Polier Formation represents deposition in a deep-water carbonate environment above the calcite compensation depth, receiving siliciclastic turbidites more or less mixed with shallow-water carbonates (Pszczolkowski, 1982, 1999; Cobiella-Reguera et al., 1997). The fine undisturbed laminations, dark color, and absence of autocthonous benthic fauna all clearly point to accumulation on dysaerobic to anoxic bottoms, but conclusive data about water depth are unavailable for these open-sea deposits. Pszczolkowski (1999) attributed a late Berriasian age to the lowermost Polier Formation on the basis of previously reported calpionellids (Pszczolkowski, 1978, his table 6). In horizons close to the base, Fernández-Carmona and Blanco-Bustamante identified *Tintinnopsella carpathica*, *Calpionella alpina*, *C. elliptica*, *Crassicollaria* sp., and *Calpionellopsis simplex* (Figure 7). Reference to these species does not include data on the relative abundance or possible species successions, so evaluating their interpretations is difficult. We cannot rule out the possibility that horizons belonging to the latest–middle Berriasian may be represented, too. In addition, a revision of data available to the senior author indicates that the latest–middle Berriasian and mainly the early–late Berriasian are recognized in the lowermost Polier Formation (Figure 7). In fact, our most recent revision of calpionellids indicates latest–middle Berriasian (upper C zone) and mainly early–late Berriasian (D1 and undifferentiated D zones) for the lowermost Polier Formation based on the first appearance datum (FAD) of *Tintinnopsella longa* together with *C. simplex*, *Calpionellopsis* sp., *Remaniella* sp., and *Calpionella* sp. cf. *elliptica*. Pszczolkowski (1999) held the lower boundary of Polier Formation to be diachronous, but no biostratigraphic data were given to support this claim. Pszczolkowski and Myczyñski (1999, 2003) considered the top of the Polier Formation to be Aptian in age.

Figure 8. Detail of Polier Formation outcrop in the upper San Juan River Valley, north of Las Terrazas, easternmost Sierra del Rosario (Figure 2). The white composite bed is a slightly weathered limestone. The upper laminated part shows flat isolated ripple marks and is separated from the lower structureless and radiolaria-nannoconid-rich mudstone by a stylolitic suture, most probably related to an erosion surface. Overlying dark rocks are sandstones and shales (turbidites). Coordinates: 22°51′54.5″N; 82°56′44.5″W.

Jagua Formation

This unit, in the Sierra de los Organos, contains a 160-m-thick (525-ft-thick) carbonate succession of dark (TOC average 0.9%) (Moretti et al., 2003) wackestones and packstones with a variable amount of clayey sediments (Figure 3). The formation is commonly subdivided as follows: at the base, the Pan de Azúcar member (coquinas and bioclastic or black oolitic limestones) and the Zacarías member (argillites and black siltstones with scarce ammonites); in the middle, the Jagua Vieja member (claystones and marly limestones with calcareous concretions ["quesos"] containing a variety of ammonites and marine vertebrates); and at the top, the Pimienta member (well-stratified micritic limestone with some layers of claystones) (Pszczolkowski, 1978). Judoley and Furrazola-Bermúdez (1965) recognized conglomerates and breccias at the bottom of their sequences, overlying upper Oxfordian deposits dated with ammonites. Some authors have interpreted breccias atop the Jagua Formation, at the boundary with the Guasasa Formation (Figure 9), as evidence of synsedimentary tectonic activity of variable intensity (Hatten, 1967; Judoley and Furrazola-Bermúdez, 1968; Khudoley and Meyerhoff, 1971), whereas others evoke a tectonic origin (Knipper and Puig, 1967). Pszczolkowski (1999) found evidence of local erosion prior to deposition of shallow carbonates at the base of the Guasasa Formation (San Vicente member). The Jagua Formation is assumed to be a neritic deposit (Pszczolkowski, 1978, 1999) in which the Zacarías member accumulated under deltaic influence, whereas the remaining members might have accumulated on a carbonate ramp (Cobiella-Reguera, 1996b), from very

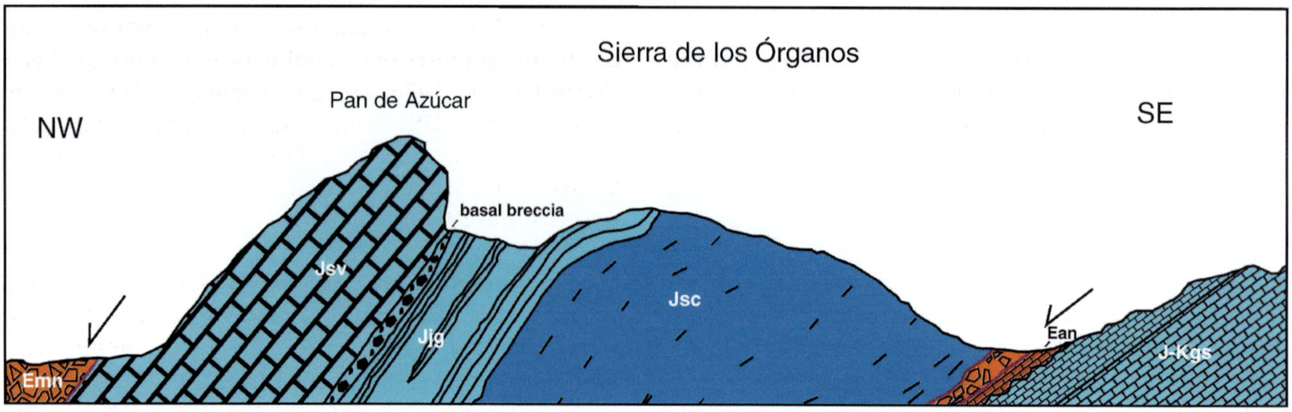

FIGURE 9. Profile across Pan de Azúcar area in Sierra de los Organos (modified from Hatten, 1967). The names of the stratigraphic units are used in the present chapter. Jsc = San Cayetano Formation; Jjg = Jagua Formation; Jsv = San Vicente member (lower Guasasa Formation); J-Kgs = Guasasa Formation; Ean = Ancón Formation (upper Paleocene); Emn = Manacas Formation (upper Paleocene?–lower Eocene). Observe the breccias at the base of San Vicente member. Not to scale. Section length: about 2 km (1.2 mi). The location is in Figure 2.

shallow waters near carbonate shoals for the Pan de Azúcar member to somewhat deeper water for the Jagua Vieja member and middle to outer shelf deposits for the Pimienta member (Pszczolkowski, 1978, 1999). All the members were deposited on poorly oxygenated sea floors. Similar to the Francisco Formation in the Sierra del Rosario, but thicker, the Jagua Formation represents the transition from terrigenous successions of the San Cayetano member to San Vicente member carbonates in the Guasasa Formation (Sierra de los Organos). Metamorphic equivalents of the Jagua Formation, with ammonites, have been reported from the Cangre belt (Somin and Millán, 1981).

The age of the Jagua Formation has been estimated on the basis of ammonites, except for the Pan de Azúcar member, which has traditionally been correlated with the Zacarías member. Ammonites have been collected since the beginning of the 20th century (e.g., Judoley and Furrazola-Bermúdez, 1968; Myczyñski, 1976a, b; Wierzbowski, 1976, and references therein). The classic article by Judoley and Furrazola-Bermúdez (1968) does not include precise biostratigraphy. Myczyñski (1976a, 1994) and Wierzbowski (1976) provided biostratigraphic interpretations based on their correlation with the European standard. These authors attributed a middle Oxfordian age to assemblages containing mainly *Vinalesphinctes* and *Cubasphinctes* (in the Jagua Vieja and the Zacarías members, according to Wierzbowski, 1976, his table 2) and late Oxfordian (especially in the Pimienta member) to those dominated by *Mirosphinctes* and *Cubaspidoceras*. On the basis of a precise correlation with data obtained from Chile, southern North America, and the western Tethys, Myczyñski et al. (1998) concluded that ammonites from the Jagua Formation range from the Bifurcatus to pre-latest Bimammatum chrons. The Pan de Azucar member has been stratigraphically correlated with the Zacarias member (early Bifurcatus age, according to Myczynski et al., 1998) but no ammonites have been found, and outcrops limitations exists (tectonics; to compare De La Nuez, 1972, 1974, and Wierzbowski, 1976). In addition, the bioclastic, oolitic and ammonites-free Pan de Azúcar member could either mask hiatuses or include late Transversarium chron deposits (Figure 3); alternatively, it could not include sediments younger than this age (e.g., stratigraphical gap at the top). The latter interpretation (not included in Figure 3) provides an alternative to traditional ones, and is based on the fact that occurrences of Zacarias Mb seem to be related to thinner Jagua Vieja member deposits (e.g., Myczyński, 1976; Figures 5–6, p. 264)-i.e., The Pan de Azúcar member seems to be unrelated to stratigraphic changes in Zacarias and Jagua Vieja members and, therefore, unconnected to environmental-depositional dynamics during the Bifurcatus Chron ammonite-bearing deposits. Whatever the case, and before detailed sedimentology of Pan de Azúcar member be available, the lack of ammonites in the Pan de Azúcar member and uppermost San Cayetano Formation impedes approaching more conclusive interpretations.

The correlation of the upper boundary of the Jagua Formation, i.e., the top of the Pimienta member, cannot be established accurately because of the absence of ammonites in the uppermost Pimienta member. The interpretation is further complicated by the fact that the overlying facies (San Vicente member in the Guasasa Formation) is devoid of ammonites. Therefore, we can only establish the age of the youngest ammonite-bearing horizons in the Jagua Formation (pre-uppermost Pimienta member), which Myczyñski et al. (1998) established as within-Bimammatum chron (probably Bimammatum subchron). The age of the uppermost Jagua Formation could then be younger than Bimammatum subchron.

However, the absence of Planula zone ammonites (uppermost Oxfordian or lowermost Kimmeridgian) and the supporting occurrence of conglomerates and breccias in many basal outcrops of the overlying San Vicente member would indicate erosion or nondeposition of latest Oxfordian beds in the Pimienta member of the Jagua Formation We envisage a correlation between this erosional (or non depositional) episode and the interval of starved (or condensed) sedimentation in the lowermost Artemisa Formation (see above). In Figure 3, an unconformity is assumed between the Pimienta and San Vicente members, together with a late, but not latest, Oxfordian age (Planula chron) for the youngest beds of the Pimienta member in the Jagua Formation.

Guasasa Formation

In the Sierra de los Organos, Upper Jurassic and Lower Cretaceous carbonates overlying the Jagua Formation are known as the Guasasa Formation, the lower part of which is the San Vicente member (Figure 3). In the original interpretation (Pszczolkowski, 1978), the Guasasa Formation comprised a thick, densely stratified lower section, deposited in very shallow waters (San Vicente member), and overlying, well-stratified rocks deposited in deeper waters (El Americano, Tumbadero, Tumbitas, and Infierno members). As implicitly recognized by Pszczolkowski (1999), the Infierno member is indistinguishable from the Pons Formation, which has priority in regional stratigraphic nomenclature.

San Vicente Member

Lithology, considerable thickness (up to 650 m [2132 ft]), and widespread geographic distribution support the interpretation of this unit at the formation level; however, local stratigraphic terminology is herein maintained. The San Vicente member includes thick-bedded to massive gray-black limestones, frequently dolomitized and with occasional chert nodules and lenses (Pszczolkowski, 1999, and references therein). No wave-resistant reef buildups have been reported. The San Vicente member has been interpreted as a carbonate shelf deposit (Hatten, 1967; Khudoley and Meyerhoff, 1971; Pszczolkowski, 1978, 1981, 1999; Myczyński and Pszczolkowski, 1990, and references therein) accumulated on a well-oxygenated bottom, with abundant micrites containing *Favreina* (Crustacea fecal pellets) in the lower part and calcarenites (intraoosparites and intrabiosparites) in the upper part. This facies change suggests a shallowing trend of the carbonate bank (Pszczolkowski, 1981). Breccias have been observed at the base of the unit (Figure 9) (Judoley and Furrazola-Bermúdez, 1965; Hatten, 1967; Pszczolkowski, 1978; see comments on the Jagua Formation). Bedding is thinner in horizons close to the overlying El Americano member, and micrite and calcilutite beds intercalated with detrital limestones indicate initial pulses of drowning of the carbonate bank (Pszczolkowski, 1978, Cobiella-Reguera, 1996a, b) at the Kimmeridgian–Tithonian boundary. Somin and Millán (1981) reported tectonically thinned metamorphic equivalents to the San Vicente member in the Cangre belt (Piotrowska, 1978).

Fossils of stratigraphic interest are scarce in the San Vicente member (gastropods, echinoderms, and algae in Judoley et al., 1988; *Favreina salevensis* and *G. alpina* in Fernández Carmona, 1998; bivalves, algal debris, echinoid spines, and benthic forams in Pszczolkowski, 1978). The age of the San Vicente member can be interpreted with some confidence as it rests on upper Oxfordian (Bimammatum chron?) beds of the Jagua Formation. Sedimentary breccias are frequently reported from basal outcrops of the unit (e.g., Judoley and Furrazola-Bermúdez, 1965; Hatten, 1967; Myczyński, 1976a, b; Pszczolkowski, 1999), but Knipper and Puig (1967) considered them to be of tectonic origin because of competence differences between the stratified (Pimienta member) and the massive or thick bedded (San Vicente) limestones. The abrupt Jagua-San Vicente facies change suggests a disconformity and/or unconformity, probably located at the same stratigraphic level as the Francisco-Artemisa disconformity and/or unconformity in Sierra del Rosario (Figure 3). The unit is conformably covered by lower Tithonian limestones of the El Americano member. The transition is rather abrupt, probably because of a rapid relative sea floor deepening.

On the basis of the occurrence of Hybonoticeras near the base of the overlying El Americano member (8 m [26 ft] above the top of the San Vicente member) and the biochronostratigraphic interpretation made by Myczyński (1999), we assume that the San Vicente member did not range into the Tithonian. Nevertheless, the base could include some upper Oxfordian rocks, according to Pszczolkowski (1999, and references therein). Obviously, the age of the lowermost San Vicente beds is unresolved. This interpretation suggests that San Vicente deposition occurred during an unfavorable time for ammonites in the Guaniguanico Cordillera, both in the Sierra de los Organos (uppermost Pimienta + San Vicente member) and in the Sierra del Rosario (lower La Zarza member, as reinterpreted in this chapter). This is consistent with the close resemblance of the main stratigraphic features of the Upper Jurassic in both regions, as recognized by Pszczolkowski (1994a). No evidence of significant, widespread synsedimentary instability pointing to an abrupt morphological jump in the basin floor has been found in the transitional San Vicente-La Zarza beds in Sierra del Rosario (Pszczolkowski, 1978), neither within the San Vicente and El Americano nor within any other Kimmeridgian–Tithonian lithostratigraphic unit in western Cuba. Assuming continuous deposition within this interval, we envisage a continuous

FIGURE 10. Well bedded (10–40 cm [4–16 in.] thick) dark limestone strata of El Americano member containing thin shaly intercalations. Rancho San Vicente, Viñales, Pinar del Rio province (Figure 2). Coordinates: 22°40′09.2″N; 83°42′33.7″W.

lateral progressive deepening from shallow-water banks (San Vicente member) to deeper neritic settings (lower La Zarza member of Artemisa Formation), with an overall carbonate ramp morphology. Meanwhile, local return to shallow-water carbonates occurred in Sierra del Rosario; the trend towards deepening continued in the Sierra de los Organos. However, lacking outcrops for transitional sections between these two ranges impedes a more integrated and conclusive interpretation.

El Americano Member

Like the San Vicente member, this unit has only been described in the Sierra de los Organos. The El Americano member consists of well-stratified gray-black limestones, with occasional fine clayey intercalations (Figures 3, 10). Maximum thickness is about 45 m (148 ft). Based on the microfacies with Saccocoma under the *Colomisphaera* sp.-*Cadosina parvula* assemblage zone (Myczyñski and Pszczolkowski, 1990) or the *Saccocoma*-Didemnidae microfacies (Myczyñski and Pszczolkowski, 1994), Pszczolkowski (1999, and references therein) interpreted these sediments as having accumulated under deep-neritic shelf conditions. This author (ibid.) interpreted microfacies with tintinnoids (Myczyñski and Pszczolkowski, 1990) and especially with radiolaria (Myczyñski and Pszczolkowski, 1994) as evidence for outer shelf or even bathyal environments at the end of the Tithonian. However, Myczyñski (1989) interpreted shallowing during the late Tithonian. As usual, it is unclear whether the evolution of microfacies was mainly controlled by depth or, instead, by combined factors such as distance from shore, hydrodynamics, and ecological conditions. This latter combination of factors is difficult to reconstruct with certainty in paleomargins undergoing deformation near oceanic accretion zones, as was the North American paleomargin in western Cuba during the Tithonian and Early Cretaceous (e.g., Marton and Buffler, 1999).

Ammonites from the El Americano member have long been known. Myczyñski and Pszczolkowski (1990, 1994) and Myczyñski (1999, and references therein) reported successive assemblages of ammonites and/or calpionellids. In recent studies, ammonite assemblages are characterized by *Mazapilites* with or without *Hybonoticeras*, *Pseudolissoceras* commonly associated with the so-called "*Virgatosphinctes*", *Paralytohoplites* (previously *Lytohoplites*), and the *Proniceras-Durangites-Kossmatia-Corongoceras* and *Protancyloceras-Vinalesites*; the latter two assemblages were recently replaced by the *Durangites-Himalayites-Salinites* assemblage. Pszczolkowski (1999) concluded that the El Americano member was Tithonian in age. Nevertheless, no conclusive interpretation of the upper Tithonian based on ammonites in Mesoamerica exists, as remarked by Villaseñor et al. (2000) for Mexico, and an updated analysis of combined ammonite and calpionellid ranges is required. Recently, Pszczolkowski and Myczyñski (2003) interpreted the topmost horizons of the El Americano member to be early Berriasian in age, and Pszczolkowski et al. (2005) placed the Jurassic/Cretaceous (J/C) boundary in the youngest beds of the El Americano member based on the combined record of calpionellids, radiolaria, nannoconids, and ammonites. Yet calpionellids and/or ammonites in Tavera et al. (1994) and Olóriz et al. (1995) allow upper horizons from the El Americano member previously considered as part of the Crassicollaria and the Calpionella zones to be reinterpreted as early to middle Berriasian (e.g., Pop, 1976; Judoley et al., 1988; Myczyñski, 1989; Myczyñski and Pszczolkowski, 1990; see table 5 of Pszczolkowski, 1978). Significantly, calpionellids reported from horizons within the upper El Americano member and interpreted as belonging to the upper Crassicollaria and/or Calpionella zones (Pop, 1976; Judoley et al., 1988; Myczyñski, 1989; Myczyñski and Pszczolkowski, 1990; see table 5 of Pszczolkowski, 1978) indicate early to early–middle Berriasian.

Tumbadero Member

The Tumbadero member includes well-stratified, commonly laminated mudstones and wackestones (Figure 3). Abundant calpionellids and radiolarians are observed; the latter typically associated with intercalations of black, cherty horizons. Thickness is up to 50 m (164 ft). Based on progressive deepening interpreted for the underlying El Americano member proposed by Pszczolkowski (1999), sediments of the Tumbadero member have been assumed to indicate accumulation in bathyal waters, below the aragonite compensation depth. Pszczolkowski et al. (2005) proposed generalized

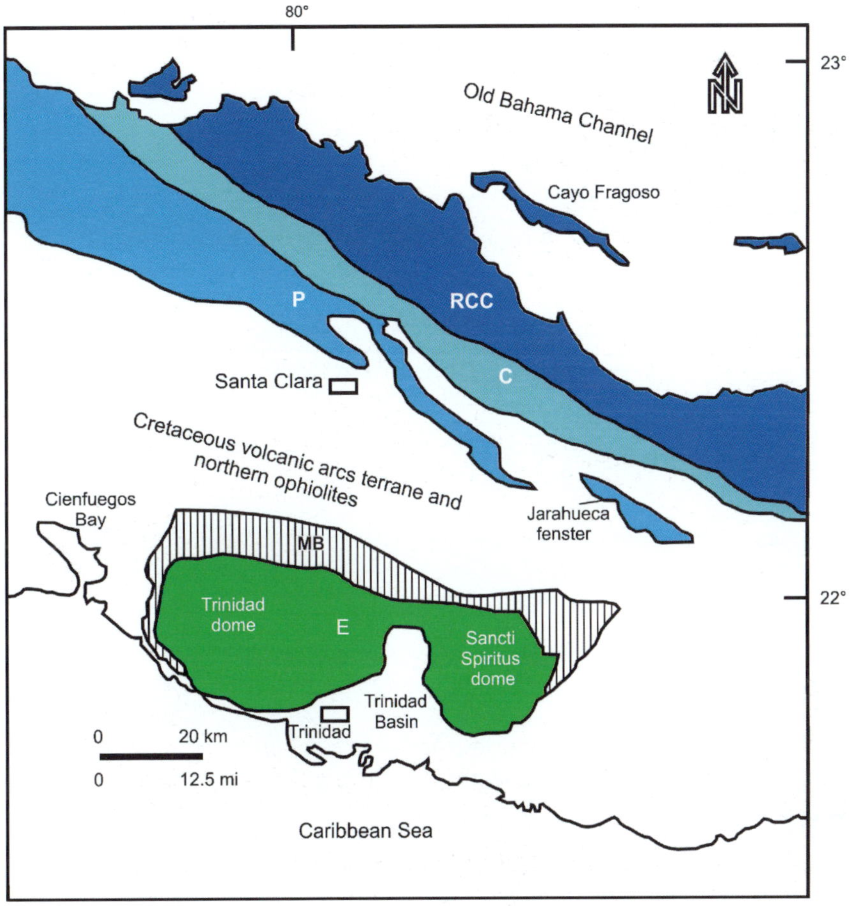

FIGURE 11. Main Upper Jurassic–Neocomian paleotectonic units in central Cuba. RCC = Remedios-Cayo Coco-Canal Viejo zone (interpreted as the southern fringe of the Bahamas platform); C = Camajuaní zone; P = Placetas zone; MB = Mabujina complex (metamorphic basement of the volcanic terrane); E = Escambray massif. Contacts between all these units are tectonic, resting ophiolites and the volcanic terrane on passive continental margin sequences.

poor oxygenation for deep waters in the proto-Caribbean Basin at the time. As in the case of the El Americano member, the criteria employed to reconstruct paleodepth are not conclusive, although Tumbadero member deposits undoubtedly accumulated offshore.

Ammonites from the Tumbadero member are scarce and difficult to identify (e.g., Myczyñski and Pszczolkowski, 1990; Myczyñski, 1994), making calpionellid data of special value. Following Tavera et al. (1994) and Olóriz et al. (1995), the age of the Tumbadero member is early to middle or early–late Berriasian (Figure 3) in view of data obtained by (1) Pop (1976) from cherty horizons in his Artemisa Formation in the Sierra de los Organos; (2) Pszczolkowski (1978, table 5), who indicated the dominance of *Calpionella* spp. and reported *Calpionellopsis* spp. from the uppermost 10 m (33 ft) in this member; and (3) Myczyñski (1989) and Myczyñski and Pszczolkowski (1990). Pszczolkowski (1978, 1999) assumed a Berriasian age for the Tumbadero member, which Pszczolkowski and Myczyñski (2003) estimated at most from earliest to early–late Berrriasian. Recently, Pszczolkowski et al. (2005) interpreted lower Tumbadero deposits to be *C. alpina* subchron in age. The lower cherty levels of the Tumbadero member must correlate with the Sumidero member of the Artemisa Formation.

Tumbitas Member

According to Pszczolkowski (1999), the Tumbitas member contains thick bedded light gray biomicrites with a rich pelagic fossil association (calpionellids, radiolarian, and nannoconids). The thickness is around 40 m (131 ft) but in some sections attains 80 m (262 ft). The unit is upper Berriasian–lower Valanginian, according to calpionellids in Pszczolkowski (1978, 1999).

Upper Jurassic in the North American Paleomargin in North-Central Cuba

The second large outcrop of the North American Mesozoic paleomargin in Cuba is located along its northern fringe, from northeastern Havana province, in the west, to Holguin, in the east (Figures 1, 11). As in western Cuba, alpine tectonic style is recognizable (Figure 12); but in contrast to western Cuba, where extensive Jurassic outcrops are seen, in north-central Cuba, these are limited and Cretaceous rocks predominate. The overwhelming majority of outcropping Jurassic rocks belong to the uppermost Jurassic because of detachments coincident with this stratigraphic level. Although numerous wells exist, relatively few data have been published.

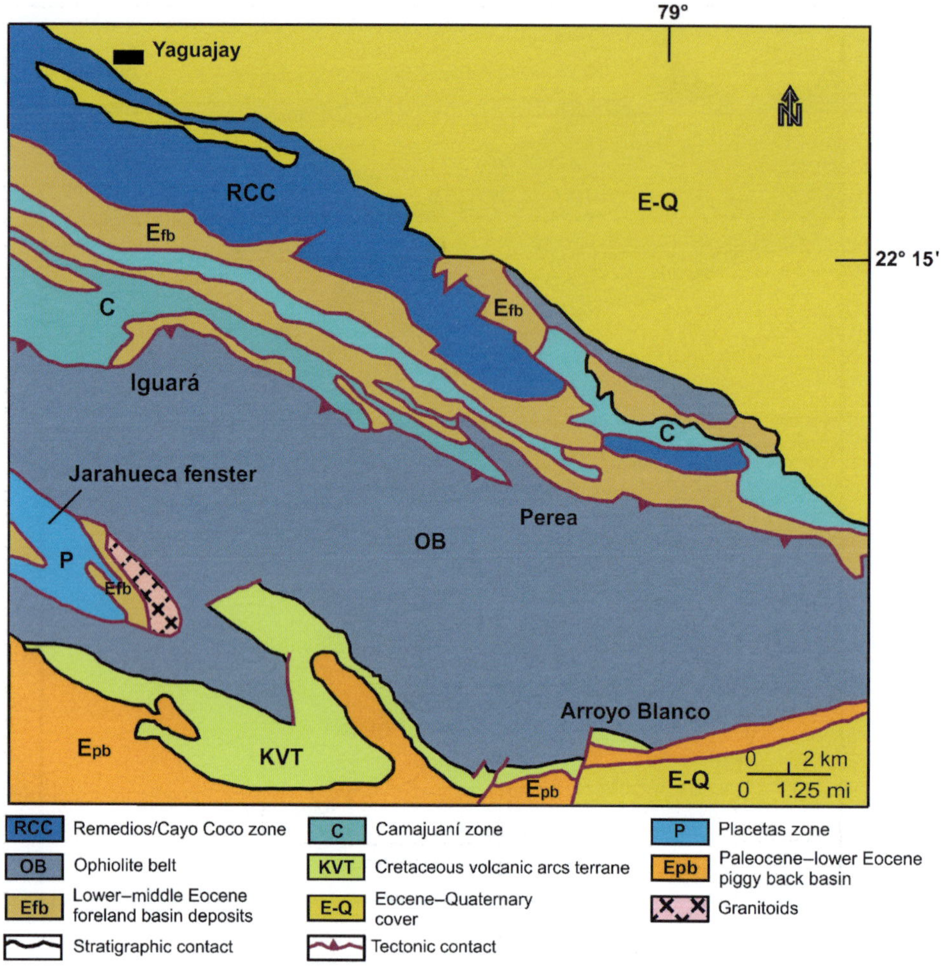

FIGURE 12. Tectonic map of northern Sancti Spiritus province (data from Pushcharovsky, 1988, and other sources). The map shows the well-developed northwest–southeast trend of the continental paleomargin zones. See Figure 1 for the location.

The quality and precision of these data, generally poorer than for western Cuba, impede precise correlation between the two regions.

From north to south, and slightly modifying the nomenclature preferred by Cuban specialists (see below), three stratigraphic successions may be distinguished in the north-central Cuba late Jurassic–Berriasian paleomargin: (1) The Remedios-Cayo Coco-Canal Viejo, (2) the Camajuaní, and (3) the Placetas zones (Figures 11, 13). This stratigraphic scheme follows guidelines established by field and oil geologists over more than 50 years and is supported by new data obtained from intensive oil and gas exploration conducted during the last 40 years (e.g., Meyerhoff and Hatten, 1968; Pardo, 1975; Kuznetsov et al., 1985; Díaz Otero et al., 1992, 1997; Alvarez-Castro et al., 1998; Blanco González, 1998; Segura-Soto and Blanco-Bustamante, 1998; Pszczolkowski and Myczyñski, 2003). Structural and stratigraphic interpretations envisage increasing relative autochthony northward across these structural zones (e.g., Pindell et al., 2006). Detailed stratigraphic research on the Camajuaní and Placetas zones clearly shows their original spatial relationships with the Florida-Bahamas megabank (e.g., Cobiella-Reguera, 2000; Pszczolkowski and Myczyñski, 2003; Pindell et al., 2006).

Remedios-Cayo Coco-Canal Viejo Zone

This zone corresponds to the southern boundary of the Bahamas platform (Figures 11, 12) (Díaz Otero et al., 1997). Data on the Jurassic successions are obtained almost exclusively from subsurface geology. The Jurassic section is divided into two large successions (Figure 13). The lower, the Punta Alegre Formation (Meyerhoff and Hatten, 1968), is found in three large diapirs and some wells. Punta Alegre Formation is made up of evaporites (anhydrites, gypsum and halites) with rare shales. Because of diapirism, exotic blocks of limestones and dolomites of variable size lie among the evaporites. The unit seems to lie beneath the upper Jurassic–lower Cretaceous Cayo Coco Formation. Salvador (1987) remarked on the possible correlation of the Punta Alegre Formation with Callovian evaporites from the rim of the Gulf of Mexico, an opinion shared by Alvarez Castro et al. (1998). In both cases, the age is assumed almost exclusively on the basis of the stratigraphic position, using poor biostratigraphic

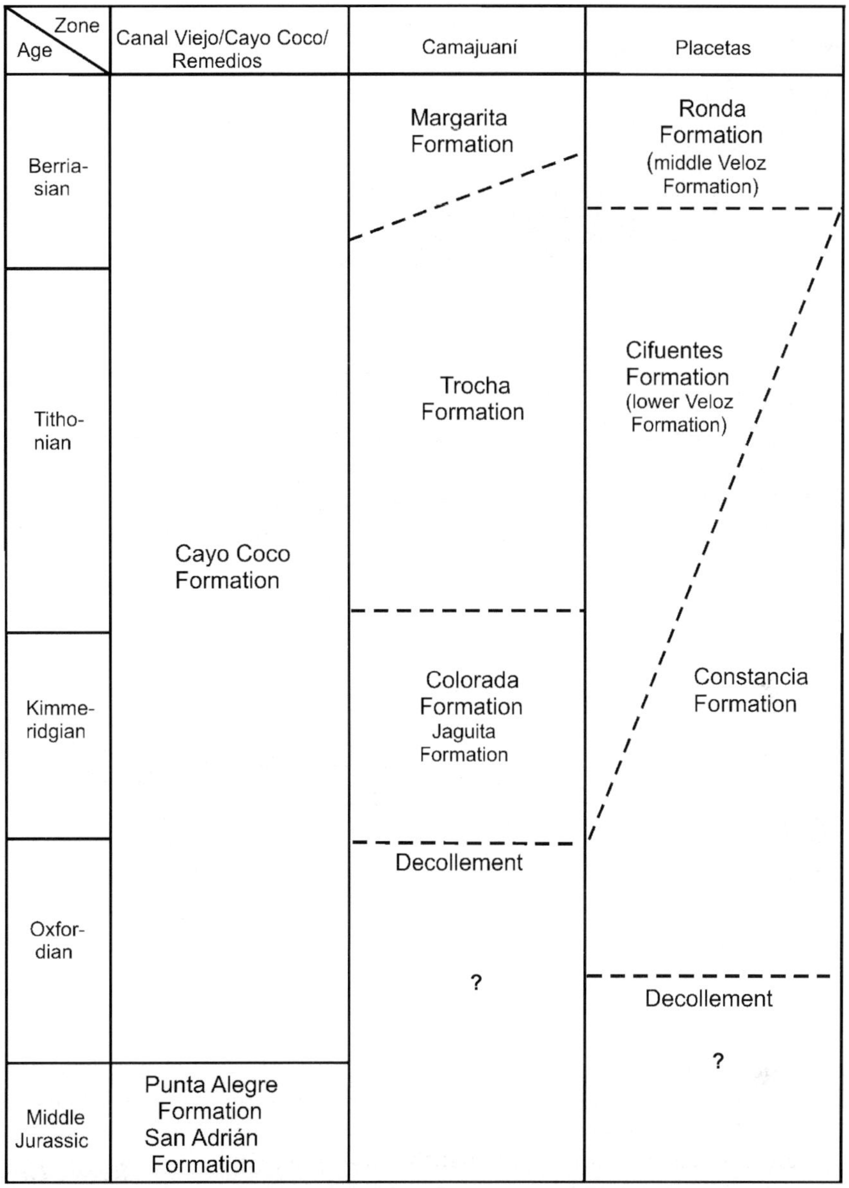

FIGURE 13. Schematic correlation chart of Upper Jurassic–Neocomian deposits in the north-central Cuban Mesozoic paleomargin.

data. Southwest of the city of Matanzas (100 km [62 mi] east of Havana), salt deposits (gypsum) in small tectonic sheets (Figure 14, San Adrián Formation) could correlate with the Punta Alegre Formation, although these deposits contain more xenoliths and clayey intercalations (Meyerhoff and Hatten, 1968; Albear-Fránquiz and Piotrowski, 1984).

The upper succession is represented by the Cayo Coco Formation (Figure 13), comprising about 1800 m (5905 ft) of carbonates (limestones and dolomites) and anhydrites. No anhydrites are present in southernmost sections (Perros Formation in some stratigraphic schemes). The presence of evaporites in this unit and in the underlying Punta Alegre Formation suggests sedimentary continuity, but available biostratigraphy does not allow for greater precision. Both the Punta Alegre Formation and the San Adrián Formation are shallow-to-very-shallow sediments (continental to inner shelf transitional environments). The Cayo Coco Formation was interpreted as deposited on a shallow submarine plateau, near the Bahamas-Florida block. (e.g., Meyerhoff and Hatten, 1968; Schlager et al., 1984), and at the edge of the Bahamas platform (Pindell et al., 2006).

In the Remedios-Cayo Coco-Canal Viejo zone, subsurface data are insufficient for a precise identification of horizons belonging to the upper Jurassic in the evaporitic sediments of the Punta Alegre Formation (if indeed such horizons exist). In the context of the regional interpretation favored in this study, we envisage a possible correlation with evaporites of the Gulf of Mexico Basin (see above), meaning that the carbonates of the Cayo Coco Formation may have initiated late Jurassic

Figure 14. Severely deformed rocks in San Adrián diapir (see Figure 1). Folded recrystallized gypsum locally contains horizons with abundant distorted shaly clasts (arrow). According to Albear-Fránquiz and Piotrowski, 1984, these clasts probably were interbeds originally. Coordinates: 23°04′45.0″N; 81°41′22.5″N.

deposition (Figure 11). Nevertheless, microfossils identified to date do not allow for the involvement of Cayo Coco Formation carbonates in the correlations here considered.

Camajuaní Zone

Upper Jurassic and Berriasian deposits are better represented on the surface than the Remedios-Cayo Coco-Canal Viejo zone (Figure 11). The Camajuaní zone mainly comprises carbonate sediments, frequently dolomitized, showing intercalations with clastic textures, including breccias (Figure 13) (e.g., Meyerhoff and Hatten, 1968; Shopov, 1982) and abundant biogenic-pelagic limestones. Terrigenous sediments are scarce. Cherty horizons appear within the Berriasian. The oldest layers in the Camajuaní zone date from the Kimmeridgian, and their basal contact is a tectonic one (Figure 13). According to Cuban oil geologists, the Kimmeridgian and the Tithonian are represented by the Jagüita and Colorada formations (Alvarez Castro et al., 1998; Sánchez et al., 1998; Segura-Soto and Blanco-Bustamante, 1998), which are composed of grayish to brownish micritic, pelagic limestones (*Saccocoma*, ammonite embryos, Cadosinidae, and radiolaria) alternating with detrital limestones (finer-grained in the Jagüita Formation) that commonly show graded bedding. In the Colorada Formation, clastic limestones contain abundant oolites and common minor irregularities at the base of the strata related to erosion of the underlying sediments (turbidites?). Intercalations of shales and siltstones exist. Pszczolkowski and Myczyński (2003) considered the Tithonian and, locally, most of the Berriasian beds to constitute an independent unit, the Trocha Formation (Figure 13). The Berriasian is represented by the Margarita Formation, in which Hatten et al. (1988) reported microbioclastic, oolitic limestones with nannoconids intercalated in mudstones and wackestones with calpionellids and radiolaria. Segura-Soto and Blanco-Bustamante (1998) hold the clastic limestones to be turbiditic deposits and believe that the Camajuaní zone represents sedimentation at the base of the slope corresponding to banks of the Remedios-Cayo Coco-Canal Viejo zone. Schlager et al. (1984) considered the Camajuaní zone to continue westward in the eastern Gulf of Mexico. Pindell et al. (2006) interpreted Camajuaní successions to represent the southern Bahamas continental slope and rise (Khudoley and Meyerhoff, 1971; Meyerhoff and Hatten, 1974; Pardo, 1975, among others). In this context, the Camajuaní zone could correspond to slope deposits accumulated immediately south and southwest of the Bahamas-Florida megabank during the late Jurassic and early Cretaceous (a subject which will be revisited below) on relatively stable grounds because no signs of synsedimentary instability, typical in slope deposits, have been identified. In fact, outer-shelf deposition could also be envisaged for Camajuani zone deposits during the Jurassic and the early Cretaceous.

The following fossils have been identified from the Colorada Formation: *Dingodinium tuberosum*, *Crustocadosina semiradiata*, *Commitosphaera pulla*, *Guttilina* sp., *Favreina* sp., *Saccocoma* sp., *Perisphinctaceae* sp. and *Pseudolissoceras* sp., which Sánchez et al. (1998) interpreted as Kimmeridgian–Tithonian in age (at least the early Tithonian is present with *Pseudolissoceras*). In the Margarita Formation, Hatten et al. (1958) found *C. elliptica*, *C. carpathica* (=*T. carpathica*), *C. oblonga* (=*Calpionellopsis oblonga*) and *C. darderi* (=*Calpionellites darderi*), indicating the middle Berriasian to upper Valanginian (Figure 13). Cherty sediments in the Margarita Formation, Camajuaní zone, date at least from the middle Berriasian according to calpionellids.

Placetas Zone

The Placetas zone is the southernmost recognizable belt of the Jurassic North American paleomargin in central Cuba (Figure 11). Grenvillian metamorphic rocks form a basement intruded by Jurassic (172 Ma old) granites (Somin and Millán, 1981; Renne et al. 1989). Two main lithologic successions can be distinguished in most of the area. The lower one is mainly terrigenous, and the upper one is almost entirely made up of carbonates (Alvarez Castro et al., 1998; Pszczolkowski and Myczyński, 2003, and references therein) (Figure 11). Pardo (1975) and Pszczolkowski (1986b) considered the terrigenous section to rest on granites with basal arkoses of granitic clasts derived from its substrate (? regolith). Above, and nearly always through tectonic contact, lie more varied arkosic sediments (Constancia Formation) showing a gradual transition to the overlying carbonates (Figure 13) (Somin and Millán, 1981; Renne et al., 1989; Pszczolkowski and Myczyñski, 2003).

Until recently, the age of the basal terrigenous succession of the Placetas zone was estimated to be Tithonian or Berriasian because the overlying limestones were considered to belong to the Berriasian. Pszczolkowski (1986b) reported *C. elliptica* (Cadisch), *C. alpina* (Lorenz), *Tintinopsella carpathica* (*Murgeanu et Filipescu*), and *Remaniella dadayi* (*Knaue*) in limestone intercalations of the upper Constancia Formation. However, oil wells in the north of the province of Matanzas show limestone intercalations in the upper part of the basal terrigenous succession (also identified as the Constancia Formation), which yielded *Globuligerina oxfordiana*, *Globuligerina* spp., *Lenticulina* sp., *Orbiculiforma* sp., and *Cenosphaera* sp. (Díaz-Colell, 1996; Alvarez Castro et al., 1998; Blanco González, 1998). Blanco González (1998) also recorded spores and concluded that the Constancia Formation (as defined by Cuban oil geologists) is Oxfordian–early Kimmeridgian in age (cf. Alvarez Castro et al., 1998). The recent observation of *Globuligerina oxfordiana* in the Constancia Formation can be considered indicative of the Oxfordian, although detailed paleontological analyses are necessary to obtain a more precise biochronostratigraphy (see Grigelis and Gorbatchik, 1980; Banner, 1982; Bignot and Janin, 1984; Giovanelli and Schiavinotto, 1986; Samson et al., 1992, for extended treatment of Jurassic Conoglobigerinidae and Globuligerinidae). Current data for foraminifers and spores seem to suggest that the changeover of siliciclastic sedimentation (Constancia Formation) to carbonates (named Cifuentes Formation by Cuban oil geologists) probably occurred during the latest Oxfordian or earliest Kimmeridgian (Figure 13), but precise data are not available. Recently, Pszczolkowski and Myczyñski (2003) revisited sections from the lower Placetas zone and interpret the Constancia (terrigenous sequence)-Veloz-Cifuentes (carbonate sequence) Formation to be time transgressive (diachronous), migrating from Kimmeridgian? to upper Berriasian as one moves from west (northern Matanzas) to east (northern central Cuba) (Figure 13). According to the preceding data, age interpretations of carbonate intercalations in the upper Constancia Formation cannot be considered conclusive.

Pszczolkowski and Myczyñski (2003) concluded that the Constancia Formation strata were probably deposited in grabens, in a tectonic setting similar to the pre-Cretaceous siliciclastic sections of the southeastern Gulf of Mexico grabens shown by Marton and Buffler (1994, 1999).

The Constancia Formation resembles the Bacunayagua Formation, which is found as slivers between tectonic sheets of the Placetas sequence. The Constancia Formation is found in deep wells (sometimes called the A horizon) of northern Havana and Matanzas provinces. The Bacunayagua Formation also occurs in rare and small outcrops. The arkosic sandstones and conglomerates of the Bacunayagua Formation have been considered Campanian–Maastrichtian and even early Tertiary in age (Albear-Fránquiz and Iturralde-Vinent, 1985), but the mineral composition, which is rich in clasts of sialic provenance, is very different from the Upper Cretaceous or early Paleogene volcaniclastic sandstones of Cuban oil fields.

According to Alvarez Castro et al. (1998), Blanco González (1998), and Sánchez et al. (1998), the overlying Kimmeridgian–Tithonian Cifuentes Formation is composed of dark gray to black, well-stratified biomicrites and mudstones with fine terrigenous intercalations (Fernández Carmona, 1998). Kimmeridgian deposits in this formation, with *Favreina salevensis*, *Didemnoides moreti*, and algae, accumulated in very shallow waters, whereas the Tithonian ones did so in outer shelf or even bathyal conditions (Fernández Carmona, 1998). The lower part of the Cifuentes Formation (lowermost Veloz Formation according to Pszczolkowski and Myczyñski, 2003) contains *Globochaete alpina* and *Didemnoides moreti*, whereas *Favreina salevensis*, *Globochaete alpina*, *Saccocoma*, *Chitinoidella* spp, *Cadosinidae*, and aptychi have been found in the middle and upper parts. *Crassicollaria* spp. and *Calpionella alpina* (large forms), of Tithonian age, appear in the uppermost layers of the lower Cifuentes Formation. Therefore, shallow carbonate sedimentation during the Kimmeridgian was followed by open-marine sedimentation during the Tithonian. This trend is similar to that identified in coeval deposits in the Guaniguanico Mountains.

Biomicrites with radiolaria and cherty and dolomitic intercalations of the Ronda Formation (middle part of Veloz Formation) (Pszczolkowski and Myczyñski, 2003) overlie the Cifuentes Formation (Fernández Carmona, 1998; Sánchez et al., 1998). Pop (1976) interpreted lower Ronda Formation beds as being of late Berriasian to probable Valanginian age. Calpionellid assemblages identified by Pop (1976) in deposits with cherty intercalations in the former province of Las Villas and the lithostratigraphic scheme allow us to deduce that Ronda Formation deposition may have begun during the middle to late Berriasian (Figure 13).

Further east, in northern Camaguey province, the Constancia Formation has not been identified. Beneath the carbonate sections, reported as the Veloz Formation (Iturralde-Vinent, 1988a, b), pillow basalts, brecciated hyaloclastites with fine intercalations of micritic and biomicritic limestones with radiolaria, calpionellids, *Saccocoma*, *Cadosina* sp., and ammonite embryos have been found. At the base of the Veloz Formation, Iturralde-Vinent (1988a, b) reported remains of *Saccocoma*, *Cadosina* sp., *Chitinoidella cubensis*, and ammonites. These data are interpreted as Tithonian. We cannot exclude, however, the possibility of older upper Jurassic horizons (e.g., Kimmeridgian, at least in part) for tholeitic rocks of the Nueva Maria Formation if *Saccocoma* sp. and *Cadosina* sp. come from levels below those in which calpionellids were found. Pszczolkowski and Myczyñski (2003) assumed an early Tithonian age for the Nueva Maria Formation, whereas the Veloz Formation would span from upper lower Tithonian to lower Cretaceous in northern Camaguey. Iturralde-Vinent (1997) envisaged the Nueva Maria Formation as part of the proto-Caribbean oceanic crust, but data supporting this inference are insufficient.

From the preceding comments, several general features of the upper Jurassic–Cretaceous stratigraphy of the Sierra del Rosario and Placetas zone are obviously very similar and possibly belong to the same hemipelagic tectonopaleogeographic domain, as envisaged by Cobiella-Reguera (2000) and recently proposed by Pszczolkowski and Myczyñski (2003). The latter authors used ammonite paleobiogeography to reinforce close relationships during Tithonian to early Cretaceous among the Guaniguanico sections, and the central Cuba Placetas and Camajuani zones. In between Guaniguanico and central Cuba, upper Jurassic–lower Cretaceous Placetas zone sections (and less frequently, beds from the Camajuani zone) have been recorded in many deep wells in northern Havana and Matanzas provinces (Kuznetsov et al., 1985). These units also crop out occasionally in northern Matanzas (Veloz Formation in Pushcharovsky, 1988) and northern Havana (Albear-Fránquiz and Piotrowski, 1984). However, greater bio- and lithostratigraphic precision is required to establish a reliable correlation between upper Jurassic–lower Cretaceous sections of the Guaniguanico Cordillera and north-central Cuba.

Other proposals for correlating Placetas and Guaniguanico deposits have been made. Schlager et al. (1984) envisaged regional correlation for the Placetas deposits and considered leg 77 sites 537 and 538 to represent the western prolongation of the Placetas zone. However, lower Cretaceous carbonate pelagic deposits drilled in both sites were on isolated knolls, whereas Placetas-type pelagic and hemipelagic sediments accumulated in deeper waters southward from the Bahamas platform. In fact, deep-water Mesozoic deposits of the southeastern Gulf of Mexico (Schlager et al., 1984) are very similar to Placetas-Sierra del Rosario sections. The most significant difference is the occurrence of terrigenous sediments in some Cuban sections, particularly in western Cuba, and their virtual absence in the southeastern Gulf. In addition, Schafhauser et al. (2003) reported Berriasian and Aptian deep-water carbonate deposits in Belize at the southern extension of the Guaniguanico Basin. These beds would have been thrusted from the east on to Cretaceous neritic carbonate sections (Coban Formation) and lower Tertiary turbidites (Sepur Formation), probably during the early Tertiary. If this interpretation is correct, an earliest Cretaceous deep-water basin extended along the southern fringe of the North American plate from Camaguey (east-central Cuba) to the present northwestern Caribbean, east from Belize (i.e., the proto-Caribbean or westernmost extreme of the Tethys). North of the deep-water seaway, extensive carbonate banks developed on the Bahamas-Florida platform and surrounding the emergent Yucatan Platform.

Upper Jurassic–Lowermost Cretaceous in the North American Paleomargin in Eastern Cuba (Maisí)

In easternmost Cuba, a small area in the Maisí region (Figure 1) shows metasedimentary upper Jurassic and Cretaceous rocks attributed to a continental margin (Cobiella-Reguera, 1983, 2000; Iturralde-Vinent, 1996a). Contacts are tectonic, except with the overlying upper Tertiary sedimentary deposits, which most probably are erosional across the region. Iturralde-Vinent (1996a) considered these rocks to belong to the Asunción terrane, but their tectonic setting below the ophiolite belt favors their interpretation as part of the paleomargin (Cobiella-Reguera, 2000, 2005a).

Phyllites and dark-colored shales, with quartzitic metasandstones, marbles, and metabasalts, are found in the lowest structural unit (Sierra Verde Formation). Millán et al. (1985) reported poorly preserved calpionellids and *Nannoconus* (*sensu lato*) that suggest a Tithonian to earliest Cretaceous age for a sample gathered from an isolated outcrop of unclear stratigraphic-structural setting (G. Millán, 1985, personal communication). Overlying the lower structural unit are black marbles, occasionally dolomitic, and calcareous and calcareous-micaceous schists, containing episodic intercalations of metamorphic cherts (La Asunción Formation) (Cobiella-Reguera, 1983). According to Millán et al. (1985), *Ophthalmidium* sp., *Spirillina* sp., *Chitinoidella* sp. and *Miliolidae* were identified in some horizons. The presence of *Chitinoidella* in the sample suggests an age ranging from late–early Tithonian to early–late Tithonian. Cobiella-Reguera (1983) and Cobiella et al. (1984) describe the striking resemblance between the Sierra Verde Formation and Jurassic terrigenous successions cropping out in western Cuba (metavolcanics included), but Millán et al. (1985) attributed a Cretaceous age to the Sierra Verde Formation. The metacarbonates (Chafarina or La Asunción Formation) are Tithonian in age, at least in part. Structural complexity and the poor information available do not allow the correlation of Maisí outcrops with those of the North American paleomargin in western and central Cuba.

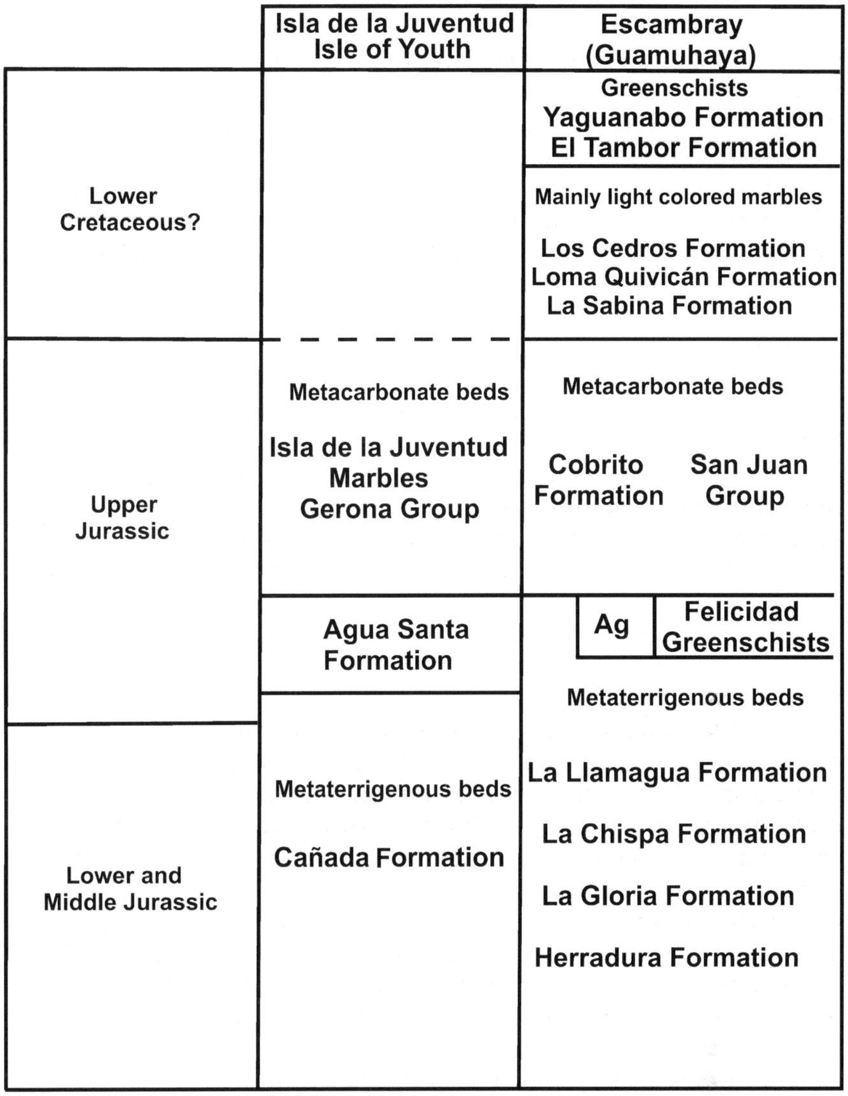

FIGURE 15. Correlation chart of the southern metamorphic terranes (data from Millán Trujillo, 1997a, b, c, and references therein). Ag = Algarrobo schists.

Passive Paleomargin Sequences in the Metamorphic Massifs of Southern Cuba (Isla de la Juventud and Escambray)

Metamorphic rocks belonging to passive continental margin sequences crop out in two large areas of southern Cuba: the metamorphic massifs of Isla de la Juventud and the Escambray (Guamuhaya). Both regions are extensive tectonic windows beneath large nappes formed by Cretaceous volcanic materials and their metamorphic basement (Figures 1, 11). Since 1970, southern Cuban metamorphic sections have been studied in some detail (e.g., Somin and Millán, 1981; Millán et al., 1985; Millán Trujillo, 1997a, b, c; Stanek et al., 2006). In both these massifs, the succession attributed to the Jurassic represents deposition on a passive paleomargin, with a lower metaterrigenous sequence, followed by metacarbonates (Millán Trujillo, 1997a) (Figure 15). However, great tectonic complexity affecting Jurassic rocks (tectonic sheets) and deformation caused by metamorphism make stratigraphic interpretation difficult (Stanek et al., 2006).

In Isla de la Juventud (Isle of Youth or Pinos; Figures 1, 15), the lower metaterrigenous, quartzose succession (Cañada Formation) shows abundant intercalations of micaceous and graphitic schists and provided psilate trilete spores that suggest a Mesozoic age (Millán Trujillo, 1997a). A transitional upper section of schists with marble intercalations has been identified (Agua Santa Formation). The upper part of the geologic section is made up of marbles (sometimes dolomitic) of the Gerona Group. In the lower part of the last unit, benthic foraminifers (*Spirillina* sp., *Ophthalmidium* [*Spirophthalmidium*] sp.) of Mesozoic age (middle Triassic–late Jurassic) have been found, together with unidentified cephalopods (Millán Trujillo 1997a). Overlying them are thick marbles barren in fossils, and layers of metacherts (Isla de la Juventud marbles). Although biostratigraphy is relatively imprecise, the similarity of the major lithologic features to those identified in the Guaniguanico Cordillera has led almost all researchers to accept a correlation between these two domains (Millán Trujillo, 1997a, b).

As in the Isla de la Juventud, the oldest beds in the Escambray (Guamuhaya) Massif (Figures 1, 15) are quartzose, metaterrigenous beds, showing locally variable metamorphism (La Chispa, La Llamagua, La Gloria, and Herradura formations in Millán Trujillo, 1997c). Transitional sections between the metaterrigenous and the upper metacarbonates (upper La Chispa Formation, Figure 15) also contain local mafic metavolcanic rocks (Felicidad Greenschists). In the lower part of the metacarbonate successions (San Juan Group), poorly preserved ammonoid remains of late Oxfordian age (*Perisphinctes* and *Mirosphinctes*) have been found (Millán and Myczyñski, 1978, 1979, who interpreted middle Oxfordian). These authors also described a single, ex-situ specimen as Perisphinctidae, probably of Tithonian age. At the middle-upper part of this unit (Mayarí Formation), Millán Trujillo (1997a) reported remains of Perisphinctidae showing morphology comparable to that of Tithonian forms, in addition to *Cadosina* sp. and forms similar to *C. carpathica* (Borza) (Millán Trujillo, 1997a, c). The latter species is interpreted as Kimmeridgian–Tithonian by Fernández Carmona and Pendás Amador (1998). Above the black marbles section, unfossiliferous pale metacarbonates were tentatively interpreted by Millán Trujillo (1997a) as Cretaceous.

Iturralde-Vinent (1996a, 2006) joined the Mesozoic sections of Escambray, Isla de la Juventud, and Guaniguanico mountains in his Southwestern terranes. Indeed, Jurassic sections in the Guaniguanico and Escambray mountains are similar. According to Cobiella-Reguera (1996a), they were probably deposited in the same basin, although their present respective location with regard to ophiolites and the Cretaceous volcanic arcs terrane is different. Regional geology suggests that rocks in Escambray and Isla de la Juventud were always located south of the Cretaceous volcanic arcs, whereas upper Jurassic–Cretaceous sections belonging to the Guaniguanico Cordillera were deposited along the North American margin (Cobiella-Reguera 1996a). Data from the southeastern Gulf of Mexico by Marton and Buffler (1999) and Moretti et al. (2003), and also from northern central Cuba (Cobiella-Reguera, 2000; Pszczolkowski and Myczyñski 2003), clearly show close stratigraphic relationships with Mesozoic sedimentary sections in Guaniguanico. Therefore, it seems unlikely that the latter are tectonostratigraphic terranes (e.g., as envisaged by Iturralde-Vinent, 1996a, 2003, 2006; Pszczolkowski, 1999; and partially Pindell et al., 2006) because they do not represent blocks exotic to the North American Mesozoic margin.

Pindell et al. (2006) described the Escambray terrane as a "...tectonically-unroofed, deep level of the Great Arc's forearc where passive margin strata has been subducted and subcreted in the Aptian-Albian... p. 323" The Escambray-Cretaceous volcanic arc terrane (represented by the Mabujina complex) tectonic contact is crossed by 88–80 Ma pegmatites (Stanek et al., 2006), showing that terrane welding in central Cuba is probably a pre-Coniacian event. According to Pindell et al. (2006), the southern metamorphic terranes and Guaniguanico were torn from the eastern or southern Yucatan margin during the Campanian and Maastrichtian, when the western-leading edge of the Great Caribbean Arc slid along the Yucatan border. In that case, at least some remains of Jurassic passive margin sections recording a middle Cretaceous high-pressure metamorphic event or structures pointing to pre-Coniacian thrusting should be present in the eastern Yucatan margin. However, no evidence for such events in southeastern Yucatan is reported in the geological literature (e.g., Morán-Zenteno, 1994). Other authors have evoked a Yucatan (Maya block) origin for the southern metamorphic terranes and Guaniguanico (Iturralde-Vinent, 1996a; Kerr et al., 1999; Schaffhauser et al., 2003). However, no sign of such an extraordinary event (i.e., terrane emplacement) has been reported, and the correlation between the Cuban sections (Guaniguanico Cordillera, Isla de la Juventud and Escambray Massif) and southeastern Yucatan has only been attempted to date by the latter authors. According to Schaffhauser et al. (2003), the lower Cretaceous Guaniguanico-like sections in western Belize were emplaced as thrust sheets between the Maastrichtian and the Eocene, several million years after the Pindell et al. (2006) proposed strike slip and collision of the Great Caribbean Arc with southeastern Yucatan. In the authors' opinion, no solid tectonic or stratigraphic evidence pointing to an original position for Cuba's southern metamorphic terranes along the Yucatan southeastern margin exists; this is compatible with the interpretation of Escambray as a Mesozoic polygenetic unit (terrane) amalgamated into a subduction zone (e.g., Cobiella-Reguera, 2000; Garcia-Casco et al., 2006; Stanek et al., 2006).

Final Remarks on Paleogeography and Tectonic Evolution. General Correlation with Late Jurassic–Berriasian Events in the Southeastern Gulf of Mexico

Advances in the knowledge of the regional geology during the last 30 years in Cuba and its surroundings, especially the southeastern Gulf of Mexico, allow a regional correlation of the upper Jurassic–Berriasian sequences and events. The passive margin type sections in the Guaniguanico mountains, north-central Cuba, Maisí (easternmost Cuba), and the southern metamorphic massifs record part of the early stages of the breakup of Pangea in Mesoamerica and the birth and early development of the southeastern Gulf of Mexico and the westernmost Tethys (proto-Caribbean).

According to Marton and Buffler (1999) (see also Salvador, 1987; Pindell, 1994, among others), the oldest oceanic crust in the Gulf of Mexico is probably of Callovian age and was formed in the central part of the basin. Later, in the Oxfordian, the process extended to the present eastern Gulf of Mexico, moving south until the Berriasian. The process was characterized by extensive rifting in the southeastern Gulf, which was deformed as grabens and semigrabens separated by uplifted blocks (Marton and Buffler, 1999; Moretti et al., 2003). Some intrusive mafic magmatism accompanied this movement (Marton and Buffler, 1994, 1999; Schlager, Buffler et al., 1984). This tectonic setting was related to the drifting and rotation of the Yucatan (Maya) block, which attained its present position in the Berriasian (Marton and Buffler, 1994, 1999).

Marked changes in paleogeography in western Cuba during the latest-middle to late Jurassic (especially the Oxfordian) were involved in the establishment of a deep seaway between the proto-Gulf of Mexico and the central North Atlantic. These events favored the entrance of middle–late Oxfordian ammonites into the basin where Sierra de los Organos and Sierra del Rosario (Guaniguanico Cordillera) sediments accumulated. Magmatism recorded in the El Sábalo Formation and in small isolated diabase bodies in the upper Artemisa Formation (Sumidero member) is possibly related to this process (Cobiella-Reguera, 1996a).

A clear correlation of the total stratigraphic range extending from horizons with carbonate concretions in the upper San Cayetano Formation to the top of the Jagua Formation in western Cuba is observed, with the initial third-order tectonoeustatic sequence (3rd-TES-I) recognized at the northern rim of the Gulf of Mexico Basin and north-central Mexico (Olóriz et al., 2003).

Marton and Buffler (1994, 1999) (see also Schlager, Buffler et al., 1984; Moretti et al., 2003) divided the supposed upper Jurassic sections in the southeastern Gulf in two main sequences: (1) a lower terrigenous one and (2) upper marine carbonates. This is the same general trend as the upper Jurassic–Berriasian North American paleomargin deposits in Cuba (Cobiella-Reguera, 2000; Pszczolkowski and Myczyñski, 2003). The transition from terrigenous to carbonate sedimentation occurred in the late middle Oxfordian in western Cuba (and possibly in the Escambray mountains of central Cuba, Figure 15). Possibly, the event was related to the opening process of a narrow oceanic depression in the southeastern Gulf and the origin of the oceanic proto-Caribbean (westernmost Tethys) Basin. In the central Cuba Placetas zone, the terrigenous-carbonate transition spans Kimmeridgian to upper Berriasian, according to Pszczolkowski and Myczyñski (2003). In this region, the lower terrigenous arkosic sequence (Constancia Formation) probably accumulated in small grabens, a tectonic scenario similar to that present in the southeastern Gulf (Pszczolkowski and Myczyñski, 2003). In the southeastern Gulf, Marton and Buffler (1999) dated the change from terrigenous to carbonate sediments as Kimmeridgian based on their interpretation of the Guaniguanico sections. However, they did not consider that, however related to the Gulf, in western Cuba the tectonic setting was different and upper Jurassic–Berriasian beds accumulated on a recently formed passive margin (Haczewski, 1976). Upper Jurassic beds from the Guaniguanico Cordillera, in the west, to Maisí in the east (Figure 1), record the first sediments deposited along the northern margin of the young proto-Caribbean Basin (Figures 3, 16) (Marton and Buffler, 1999). A complicated paleogeographic and paleotectonic evolution existed during the Kimmeridgian. According to Olóriz et al. (2003), a tilt-reversal affected the northern rim of the Gulf of Mexico Basin during the Kimmeridgian, forced terrigenous input in the southeastern United States, and was related to the assumed beginning of the differential subsidence of the central Gulf. This stratigraphic interval more or less coincided with deposition of the shallow-water carbonates of the San Vicente member. Block tectonics presumably affected western Cuba (Pszczolkowski, 1978) with local transgressions and regressions in a background of subsidence, whereas north-central Cuba underwent relative transgression (Placetas zone). Differential tectonoeustasy conditions terminated at the latest Kimmeridgian-earliest Tithonian when generalized transgression occurred throughout western and north-central Cuba. This comparatively sudden change in depositional conditions close to the Kimmeridgian–Tithonian boundary correlates with that registered at the northern rim of the Gulf of Mexico Basin, where Cotton Valley deposits rest on the youngest carbonates of the Haynesville Formation (Olóriz et al., 2003). Thus, the initiation of generalized transgressive conditions in the North American paleomargin of the proto-Caribbean Basin in Cuba contrasted with the opposite trend at the northern rim of the Gulf of Mexico Basin: an isochronous (in geologic, biostratigraphic sense) but differential response to a tectonic pulse affected the northern Gulf as opposed to the southern margins of the present Gulf (former North America-proto-Caribbean passive margin) (Olóriz et al., 2003).

Deeper depositional conditions started during the Tithonian along the paleomargin in northern Cuba. Possibly, this event can be correlated with the beginning of the drowning and stepback of the small carbonate platforms developed along the southeastern Gulf (Marton and Buffler, 1999; Moretti et al., 2003). A similar trend has been identified at the northern rim of the Gulf of Mexico Basin (Olóriz et al., 2003).

In the earliest Cretaceous, spreading and rifting ceased in the southeastern Gulf of Mexico (Marton and Buffler, 1994, 1999; Pindell, 1994, among others) when the Yucatan block reached its present position. Rapid

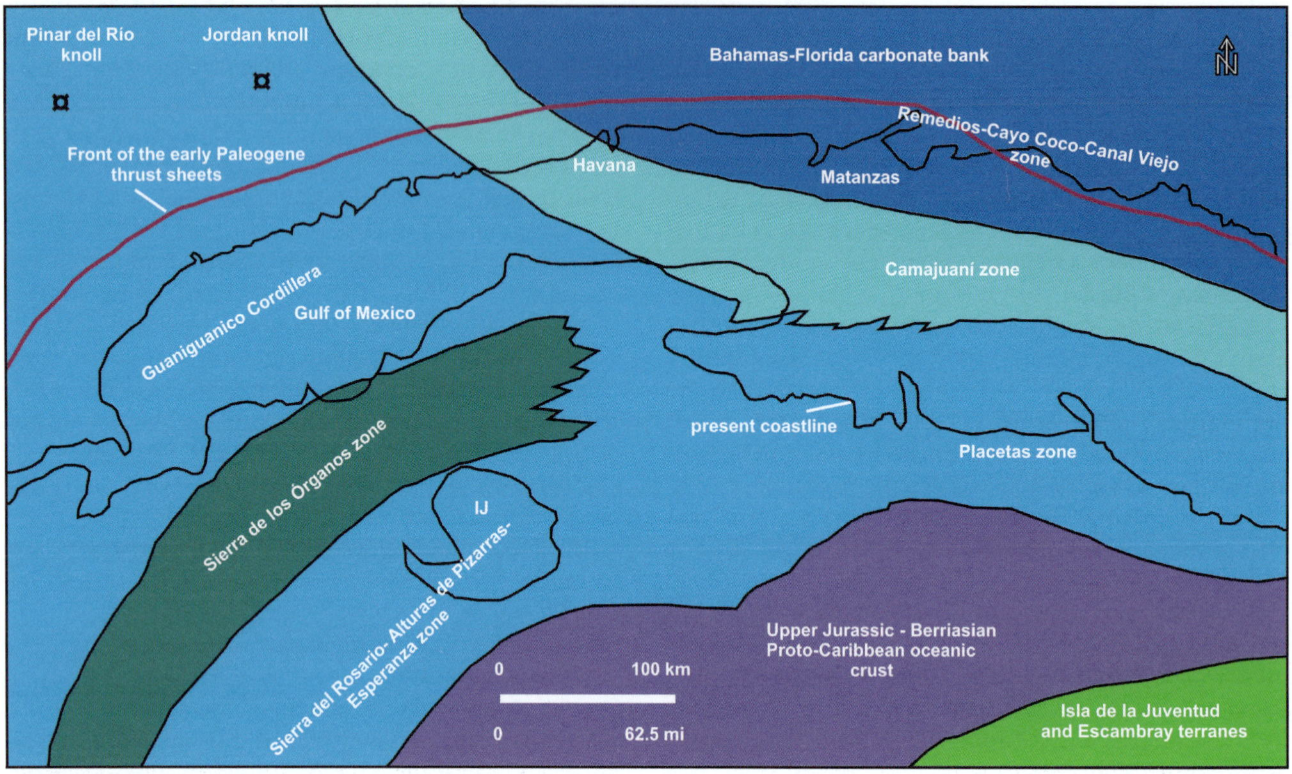

FIGURE 16. Berriasian palinspastic map of western and central Cuba and its surroundings. The figure includes some younger regional features: the overthrust front of the Cuban orogeny (late Paleocene–middle Eocene), the Pinar fault, and the Gulf of Mexico knolls. Except for the proto-Caribbean, the remaining units are blocks with thinned to normal continental crust. The western proto-Caribbean is interpreted as the oceanic crust created during the Oxfordian (or later)–Berriasian interval caused by separation from the North American paleomargin and southward drifting of a microcontinent whose margin consists of metasediments of the Isla de la Juventud and Escambray massifs (microcontinent?) (Cobiella-Reguera 2000). Estimates of horizontal displacements caused by the Cuban orogeny are from Pszczolkowski (1983) and Cobiella-Reguera (2005b).

thermal subsidence commenced in the entire area, and a marine transgression finally submerged the previously emergent blocks surrounding the Yucatan Platform. The unconformity that records the change in tectonic setting was called the late Berriasian surface (Marton and Buffler, 1999).

The updated interpretation of calpionellid biostratigraphy from leg 77 allows the correlation of events in Guaniguanico with the restructuring of the southeastern Gulf and adjacent areas during the latest Jurassic to earliest Cretaceous. In fact, no autochthonous sediments older than the Cretaceous have been recovered from leg 77, and they overlay fluvial and nearshore arkoses of probable Berriasian age (Watkins and McNulty, 1984). Our reinterpretation indicates that the age of generalized flooding in the area was middle Berriasian and not late Berriasian as stated by Marton and Buffler (1999). Their calpionellids in the basal marine beds indicate calpionellid zone C (i.e., late–middle Berriasian). Hence, basement highs in the southeastern Gulf (holes 536, 537) were at or near sea level during the early Berriasian, then drowned during the middle Berriasian. Basement areas close to western Cuba (hole 538 at the Catoche Knoll) might be considered to have a minimum age of late–middle Berriasian, although they were interpreted by Buffler, Schlager et al. (1984) and Schlager, Buffler et al. (1984) as including early? to late Berriasian rocks resting on an igneous basement in 538A. The lowest of these rocks, unit V, is oolitic, skeletal, oncolitic, and has minor pelagic limestones containing zone C calpionellids (bottom sample 538A-30-1) (Premoli-Silva and McNulty, 1984). Moreover, reworked *Crassicollaria brevis* in samples 538A-25 and 538A-26 could be early Berriasian instead of late Tithonian, as suggested by Premoli-Silva and McNulty (1984), on the basis of the relative abundance of this calpionellid species during these ages. These reinterpreted data correspond to a late–middle Berriasian restructuring event in the southeastern Gulf. This event was represented in the North American paleomargin in Cuba (northern proto-Caribbean area) by initiation of siliceous deposition (Figures 3, 13, Sumidero and Tumbadero members in Guaniguanico highlands and, possibly, Margarita and Veloz [Ronda] formations in north-central Cuba) and resulted in a deepening phase. The early–middle Berriasian crisis interpreted by Marques et al. (1991) and widely recognized in the central North Atlantic Basin and the northern rim of the Gulf of Mexico Basin by Olóriz et al. (2003) correlates with this event, thus giving a regional context to the geodynamic significance of this tectonoeustatic event.

As Figure 16 depicts, in middle and late Berriasian, a belt of deeper water carbonates (Camajuaní, Placetas, Sierra del Rosario, and Sierra de los Órganos zones) surrounded the northern margin of the proto-Caribbean (western Tethys) at least from the present southern east-central Cuba to the northwestern Caribbean Sea and southeastern Gulf of Mexico.

CONCLUSIONS

Jurassic–Cretaceous rocks from the North American passive continental paleomargin crop out along northern Cuba. Coeval sections exist in metamorphic terranes at Escambray (central Cuba) and the Isla de La Juventud (Isle of Youth, southern Cuba).

Together with sedimentary records from the southeastern Gulf of Mexico, these sections reveal early stages of Pangea's breakup in Mesoamerica and concomitant early structuring in both the southeastern Gulf of Mexico and the proto-Caribbean (westernmost Tethys) areas.

The generalized occurrence of a lower terrigenous and an upper marine carbonate sequence reveals a major phase in paleomargin structuring. Shifting from terrigenous to carbonate sedimentation occurred during the late middle Oxfordian in western Cuba, and possibly in the Escambray mountains (central Cuba) area. In the Placetas zone of north-central Cuba, turnover from terrigenous to carbonate deposition progressed throughout the Kimmeridgian to Berriasian. All along northern Cuba, generalized transgression occurred close to the Kimmeridgian–Tithonian boundary.

The Yucatan block reached its present position early in the Cretaceous, and spreading and rifting ceased in the southeastern Gulf of Mexico. A middle Berriasian unconformity records this major paleogeographic change, which could relate to widespread cherty sedimentation across northern Cuba. Deep-water sedimentation along the southern paleomargin of North America and extensive carbonate banks on the Bahamas-Florida platform and surrounding the emergent Yucatan Platform developed during the middle and late Berriasian.

Correlation with major lithofacies shifts in both Mexico and the northern rim of the Gulf of Mexico indicates phase structuring affecting the whole Mexico-Caribbean area with differential regional expression.

ACKNOWLEDGMENTS

This research was supported by Cuban and Spanish projects (CGL2005-01316, MCYT/MICINN, Spain) and the Research Group RNM-178 (Junta de Andalucia, Spain). The authors are indebted to suggestions made by the reviewers and the editor Claudio Bartolini on an early draft of this contribution. We are grateful to Dr. Gary Proust for his detailed and constructive review of our manuscript.

REFERENCES CITED

Albear-Fránquiz, J., and M. Iturralde-Vinent, 1985, Estratigrafía de las provincias de La Habana, *in* Instituto de Geologia y Paleontologia, ed., Contribución a la Geología de las provincias de La Habana y Ciudad de La Habana: La Habana, Editorial Científico-Técnica, p. 12–54.

Albear-Fránquiz, J., and J. Piotrowski, 1984, El enclave yesífero de San Adrian, Cuba. Observaciones sobre su ubicacion geólogo-tectónica: Ciencias de la Tierra y del Espacio, v. 9, p. 17–30.

Alvarez Castro, J., R. García Sánchez, R. Segura Soto, and S. Valladares Amaro, 1998, Historia geológica del desarrollo de las rocas del margen continental del dominio Las Villas, basada en la evolución sedimentaria de la paleocuenca, *in* Geología y Minería '98: Centro Nacional de Información Geológica Memorias 1, p. 20–23.

Andó, J., Sz. Harangi, L. Szakmany, and L. Dosztaly, 1996, Petrología de la asociación ofiolítica de Holguín, *in* M. Iturralde-Vinent, ed., Ofiolitas y arcos volcánicos de Cuba: International Geological Correlation Programe, Project 364, Geological Correlation of Ophiolites and Volcanic Arcs in the Circumcaribbean Realm, Miami, Florida, p. 154–176.

Areces-Mallea, A., 1991, Consideraciones paleobiogeográficas sobre la presencia de *Piazopteris branneri* (Pterophyta) en el Jurásico de Cuba: Revista Española de Paleontología, v. 6, no. 2, p. 126–134.

Banner, F. T., 1982, A classification and introduction to the Globigerinacea, *in* F. T. Banner and A. R. Lord, eds., Aspects of micropaleontology: London, Allen and Unwin, p. 142–239.

Bignot, G., and M.-C. Janin, 1984, Découverte de *Globuligerina oxfordiana* (Foraminifère planctonique) dans le Bajocien stratotypique de la Falaise des Hachettes (Sainte-Honorine-des-pertes, Calvados, France): Compte Rendus de la Académie des Sciences de Paris, v. 298, serie II, no. 17, p. 751–756.

Blanco González, M., 1998, Palinología del Jurásico Superior en Cuba central, *in* Geología y Minería '98: Centro Nacional de Información Geológica Memorias 2, p. 271–273.

Blein, O., S. Guillot, H. Lapierre, B. Mercier de Lepinay, J. Lardeaux, G. Millán-Trujillo, M. Campos, A. García, 2003, Geochemistry of the Mabujina complex, Central Cuba: Implications on the Cuban Cretaceous Arc Rocks, The Journal of Geology, v. 111, p. 89–101.

Buffler, R., and W. Thomas, 1994, Crustal structure and evolution of the southeastern margin of North America and the Gulf of Mexico Basin, *in* R. Speed, ed., Phanerozoic evolution of North American continent-ocean transitions: Boulder, Colorado, Geological Society of America, Decade of North American Geology, Continent-Ocean Transect Volume, p. 219–264.

Buffler, R. T., et al., 1984, Part II. Site chapters, *in* R. T. Buffler et al., 1984, Initial reports of the DSDP 77: Washington, D.C., U.S. Government Printing Office, p. 25–339.

Cobiella, J., F. Quintas, M. Campos, and M. Hernández, 1984, Geología de la región central y suroriental de la provincia de Guantánamo: Santiago de Cuba, Editorial Oriente, 125 p.

Cobiella-Reguera, J., 1983, Sobre la posición tectónica de la secuencia metaterrígena- carbonatada de la región de Maisí: Minería y Geología, v. 1, p. 71–82.

Cobiella-Reguera, J., 1996a, El magmatismo Jurásico (Caloviano?–Oxfordiano) de Cuba occidental: Ambiente de formación e implicaciones regionales: Revista de la Asociación Geológica Argentina, v. 51, no. 1, p. 15–28.

Cobiella-Reguera, J., 1996b, Estratigrafía y eventos Jurásicos en la cordillera de Guaniguanico, Cuba occidental: Minería y Geología, v. 13, no. 3, p. 11–25.

Cobiella-Reguera, J., 1998, The Cretaceous system in Cuba—An overview: Zentralblatt für Geologie und Paläontologie, v. 1, no. 3–6, p. 431–440.

Cobiella-Reguera, J., 2000, Jurassic and Cretaceous geological history of Cuba: International Geology Review, v. 42, p. 594–616.

Cobiella-Reguera, J., 2005a, Emplacement of Cuban ophiolites: Geologica Acta, v. 3, p. 273–294.

Cobiella-Reguera, J., 2005b, Relaciones originales Cordillera de Guaniguanico, SE Golfo de Mexico, Bloque Maya: Memories First Cuban Convention on Earth Sciences, 6th Geological Congress, GEO7-1, p. 1–4, CD-ROM.

Cobiella-Reguera, J., A. Hernández-Escobar, N. Díaz-Díaz, and P. Obregón-Pérez, 1997, Estudio de algunas areniscas de las formaciones San Cayetano y Polier, Sierra del Rosario, Cuba occidental: Minería y Geología, v. 14, no. 3, p. 59–68.

Cobiella-Reguera, J., A. Hernández- Escobar, N. Díaz-Díaz, and S. Gil-González, 2000, Estratigrafía y tectónica de la Sierra del Rosario, Cordillera de Guaniguanico, Cuba occidental: Minería y Geología, v. XVIII(1), p. 5–15.

De La Nuez, M. L., 1972, Sobre la edad de los esquistos arcillosos en los alrededores del Mogote Zacarías, provincia de Pinar del río: Actas de la Academia de Ciencias de Cuba, La Habana, Instituto de Geología, 1, p. 19–20.

De La Nuez, M. L., 1974, Sobre la edad de los esquistos arcillosos en los alrededores del Mogote Zacarías, provincia de Pinar del Río: Contribución a la Geología de Cuba, La Habana, Publicación Especial, 2, p. 141–156.

Díaz-Colell, M., 1996, Sobre la fauna de globuligerinas en pozos de la region petrolífera Varadero-Bahía de Cárdenas: Importancia de su presencia en la Formación Constancia, *in* Resúmenes Primer Taller Internacional sobre la Geología de Cuba, el Golfo de México y el Caribe Noroccidental: Universidad de Pinar del Río, p. 27–28.

Díaz-Otero, C., G. Furrazola-Bermúdez, and M. Iturralde-Vinent, 1992, Estratigrafia del banco carbonatado cretácico "Remedios" del área Cuba norte— Las Bahamas: Minería y Geología, v. 3, no. 3, p. 19–32.

Díaz Otero, C., G. Furrazola Bermúdez, and M. Iturralde-Vinent, 1997, Estratigrafía de la zona Remedios, *in* G. Furrazola-Bermúdez and K. Nuñez Cambra, comp., Estudios sobre Geología de Cuba: La Habana, Instituto de Geología y Paleontología, p. 221–242.

Draper, G., and J. Barros, 1994, Cuba, *in* S. Donovan and T. Jackson, eds., Caribbean geology, an introduction: Kingston, University of the West Indies Publishers' Association, p. 65–83.

Echevarría-Rodríguez, G., G. Hernández-Pérez, J. López-Quintero, J. López-Rivera, R. Rodríguez-Hernández, J. Sánchez-Arango, R. Socorro-Trujillo, R. Tenreyro-Pérez, and J. Yparraguire-Pena, 1991, Oil and gas exploration in Cuba: Journal of Petroleum Geology, v. 14, no. 3, p. 259–274.

Fernández Carmona, J., 1998, Bioestratigrafía del Jurásico Superior-Cretácico Inferior Neocomiano de Cuba occidental y su aplicación en la exploración petrolera: Ph.D. thesis, Centro de Investigaciones del Petróleo, La Habana, 90 p.

Fernández Carmona, J., and M. Pendás Amador, 1998, Familias Stomiosphaeridae y Cadosinidae: Nueva utilización de la micropaleontología aplicada a la exploración petrolera, *in* Geología y Minería '98: Centro Nacional de Información Geológica Memorias 2, p. 287–289.

Furrazola-Bermúdez, G., C. Judoley, M. Mijailovskaya, Y. Miroliubov, Y. Novojatsky, A. Nuñez-Jimenez, and J. Solsona, 1964, Geología de Cuba: Editorial Universitaria, La Habana, 239 p.

Garcia-Casco, A., R. Torres-Roldan, M. Iturralde-Vinent, G. Millan, K. Nunez Cambra, C. Lazaro Calisalvo, and A. Rodríguez Vega, 2006, High-pressure metamorphism of ophiolites in Cuba: Geologica Acta, v. 4, p. 63–88.

Giovanelli, M. C., and F. Schiavinotto, 1986, Indagini biometriche su foraminiferi globigeriformi del Dogger-Malm umbro-marchigiano: Indicazioni preliminari, *in* G. Pallini, ed., Commemorazione di Raffaele Piccinini, Atti I Convengo Fossili, Evoluzione, Ambiente, Pergola, Comitato Centenario Raffaele Piccinni, Beli/Peraso, 1984, p. 131–134.

Gordon, M., P. Mann, D. Cáceres, and R. Flores, 1997, Cenozoic tectonic history of the Northamerica-Caribbean plate boundary in western Cuba: Journal of Geophysical Research, v. 102, p. 10,055–10,082.

Grafe, F., K. Stanek, A. Baumann, W. Maresch, W. Hames, C. Grevel, and G. Millan, 2001, Rb/Sr and $^{40}Ar/^{39}Ar$ mineral ages of granitoid intrusives in the Mabujina unit, central Cuba: Thermal exhumation history of the Escambray massif: The Journal of Geology, v. 109, p. 615–631.

Grigelis, A.A., and T. N. Gorbatchik, 1980, Morphology

and taxonomy of Jurassic and Early Cretaceous representatives of the superfamily Globigerinacea (Favusellidae): Journal of Foraminiferal Research, v. 10, no. 3, p. 180–190.

Gygi, R. A., and A. von Hillebrandt, 1991, Ammonites (mainly Gregoryceras) of the Oxfordian (Late Jurassic) in northern Chile and time-correlation with Europe: Schweizerisches Paläontologisches Abhandlungen, v. 113, p. 137–167.

Haczewski, G., 1976, Sedimentological reconnaissance of the San Cayetano Formation: An accumulative continental margin in the Jurassic of western Cuba: Acta Geologica Polonica, v. 26, no. 2, p. 331–353.

Hardenbol, J., J. Thierry, M. B. Farley, T. Jacquin, P.-C. de Graciansky, and P. R. Vail, 1998, Mesozoic and Cenozoic sequence chronostratigraphic framework of European Basins, jurassic charts, *in* P.-C. de Graciansky, J. Hardenbol, J. Jacquin, and P. R. Vail, eds., Mesozoic and Cenozoic Sequence Stratigraphy of European Basins: SEPM Special Publication, 60, p. 3–13.

Hatten, C., 1967, Principal features of Cuban geology: Discussion: AAPG Bulletin, v. 51, p. 780–789.

Hatten, C., O. Schooler, N. Giedt, and A. Meyerhoff, 1958, Geology of central Cuba. Eastern Las Villas and western Camaguey Provinces (unpublished report): La Habana, Oficina Nacional de Recursos Minerales, 250 p.

Hatten, C., M. Somin, G. Millán, P. Renné, R. Kistler, and J. Mattinson, 1988, Tectonostratigraphic units of central Cuba, *in* Transactions of the 11th Caribbean Geological Conference, Barbados, p. 1–14.

Haydoutov, I., I. Boyanov, and G. Millán, 1989, Nuevos aspectos de la génesis del protolito del complejo anfibolitico Mabujina, sur de Cuba central, *in* Resúmenes y Programa: La Habana, Primer Congreso Cubano de Geología, p. 97–98.

Hayes, C., T. Vaughan, and A. Spencer, 1901, Report on a geological reconnaissance of Cuba: Washington, D.C., U.S. Government Printing Office, p. 1–134.

Hemleben, C., M. Spindler, and O. Anderson, 1989, Modern planktonic foraminifera: Berlin, Springer-Verlag, 365 p.

Hutson, F., P. Mann, and P. Renne, 1998, $^{40}Ar/^{39}Ar$ dating of single muscovite grains in Jurassic siliciclastic rocks (San Cayetano Formation): Constrains in the paleoposition of western Cuba: Geology, v. 26, no. 1, p. 83–86.

Iturralde-Vinent, M., 1988a, Consideraciones generales sobre el magmatismo de margen continental de Cuba: Revista Tecnológica, v. 18, no. 4, p. 17–24.

Iturralde-Vinent, M., 1988b, Composición y edad de los depósitos del fondo oceánico (asociación ofiolítica) del Mesozoico de Cuba, en el ejemplo de Camaguey: Revista Tecnológica, v. 18, no. 3, p. 13–25.

Iturralde-Vinent, M., 1996a, Introduction to Cuban geology and geophysics, *in* M. Iturralde-Vinent, ed., Ofiolitas y arcos volcánicos de Cuba: International Geological Correlation Programe Project 364, Geological correlation of ophiolites and volcanic arcs in the Circumcaribbean realm, Miami, Florida, p. 3–35.

Iturralde-Vinent, M., 1996b, Geología de las ofiolitas de Cuba, *in* M. Iturralde-Vinent, ed., Ofiolitas y arcos volcánicos de Cuba: International Geological Correlation Programe Project 364, Geological correlation of ophiolites and volcanic arcs in the Circumcaribbean realm, Miami, Florida, p. 83–120.

Iturralde-Vinent, M., 1996c, Magmatismo de margen continental de Cuba, *in* M. Iturralde-Vinent, ed., Ofiolitas y arcos volcánicos de Cuba: International Geological Correlation Programe Project 364, Geological correlation of ophiolites and volcanic arcs in the Circumcaribbean realm, Miami, Florida, p. 121–130.

Iturralde-Vinent, M., 1997, Introducción a la geología de Cuba, *in* G. Furrazola-Bermúdez and K. Nuñez Cambra, comp., Estudios sobre la geología de Cuba: La Habana, Centro Nacional de Información Geológica, p. 35–68.

Iturralde-Vinent, M., 2003, The conflicting Paleontologic versus stratigraphic record of the formation of the Caribbean seaway, *in* C. Bartolini, R. Buffler, and J. Blickwelde, eds., The circum-Gulf of Mexico and the Caribbean: Hydrocarbon habitats, basin formation, and plate tectonics: AAPG Memoir 79, p. 75–88.

Iturralde-Vinent, M., 2006, Mesocenozoic Caribbean paleogeography: Implications for the historical biogeography of the region: International Geology Review, v. 48, p. 791–827.

Iturralde-Vinent, M. A., and E. G. Lidiak, 2006, Caribbean tectonic, magmatic, metamorphic and stratigraphic events. Implications for plate tectonics (United Nations Educational, Scientific and Cultural Organization [UNESCO]/International Union of Geological Sciences [IUGS] International Geoscience Programme project 433 "Caribbean plate tectonics"): Geologica Acta, v. 4, no. 1–2, p. 1–5.

Judoley, C. M., and G. Furrazola-Bermúdez, 1968, Estratigrafía y Fauna del Jurásico de Cuba, Instituto Cubano de Recursos minerales: Departamento Científico de Geología: La Habana, Academia de Ciencias de Cuba, Departamento de Geología, 126 p.

Judoley, K., and G. Furrazola-Bermúdez, 1965, Estratigrafía del Jurásico Superior de Cuba: Ministerio de Industrias, Instituto Cubano de Recursos Minerales, Publicación Especial, 3, p. 1–32.

Judoley, K., G. Furrazola-Bermúdez, E. Linares, and S. Gil, 1988, Bioestratigrafía y paleogeografía del Jurásico de Cuba: Instituto de Geología y Paleontología, La Habana, 50 p. (unpublished).

Kerr, A., M. Iturralde-Vinent, A. Saunders, T. Babbs, and J. Tarney, 1999, A new plate tectonics model of the Caribbean: Implications from a geochemical reconnaissance of Cuban Mesozoic rocks: Geological Society of America Bulletin, v. 111, p. 1581–1599.

Khudoley, K., and A. Meyerhoff, 1971, Paleogeography and geological history of Greater Antilles, Boulder, Colorado: Geological Society of America Memoir 129, 199 p.

Knipper, A., and R. Cabrera, 1974, Tectónica y geología histórica de la zona de articulación entre el mio y el eugeosinclinal de Cuba y del cinturón hiperbasítico de Cuba, Contribución a la geología de Cuba, Instituto de Geología: La Habana, Academia de Ciencias de Cuba, p. 15–77.

Knipper, A., and M. Puig, 1967, Estructura tectonica de las montanas de Sierra de los Órganos en la zona del pueblo de Viñales y situación en ella de los cuerpos de serpentinitas. Revista de Geología: Academia de Ciencias de Cuba, v. 1, no. 1, p. 122–137.

Kutek, J. A., A. Pszczolkowski, and A. Wierzbowski, 1976, The Francisco Formation and an Oxfordian ammonite faunule from the Artemisa Formation, Sierra del Rosario, western Cuba: Acta Geologica Polonica, v. 26, no. 2, p. 299–319.

Kuznetsov, V., J. Sánchez, G. Furrazola, and R. García, 1985, Nuevos datos sobre la estratigrafía de los mantos tectónicos de la costa norte de Cuba: Centro de Investigaciones Geológicas: Serie Geológica, v. 2, no. 85, p. 106–118.

Ladd, J., and R. Sheridan, 1987, Seismic stratigraphy of the Bahamas: AAPG Bulletin, v. 71, p. 719–736.

Lewis, J., and G. Draper, 1990, Geology and tectonic evolution of the northern Caribbean margin, *in* G. Dengo and J. Case, eds., The Caribbean region: Boulder, Colorado, Geological Society of America, The Geology of North America, v. H, p. 77–140.

Lewis, J., G. Draper, J. Proenza, J. Espaillat, and J. Jimenez, 2006, Ophiolite-related ultramafic rocks (serpentinites) in the Caribbean region: A review of their occurrence, composition, origin, emplacement and Ni-laterite soil formation: Geologica Acta, v. 4, no. 1–2, p. 237–263.

Linares Cala, E., 1999, Manifestaciones superficiales y someras de hidrocarburos y aguas sulfurosas en Cuba: Relaciones con unidades tectonoestratigraficas y yacimientos gasopetroliferos: Mineria y Geologia, v. 16, no. 2, p. 39–45.

Llanes Castro, A., D. García Delgado, and D. Meyerhoff, 1998, Hallazgo de fauna jurásica (Tithoniano) en ofiolitas de Cuba central, *in* Geología y Minería '98: Centro Nacional de Información Geológica Memorias 2, p. 241–244.

López-Palomino, R., A. Villasenor-Martínez, and F. Olóriz-Sáez, 2006, Primer registro del genero *Vinalesphinctes* (Ammonitina) en el Oxfordiano de Mexico: Significación bioestratigrafica y consideraciones paleobiogeograficas en el Jurasico Superior de America: Revista Mexicana de Ciencias Geologicas, v. 23, p. 162–183.

Marques, B., F. Olóriz, and F. J. Rodríguez-Tovar, 1991, Interactions between tectonics and eustasy during the Upper Jurassic and lowermost Cretaceous. Examples from the South of Iberia: Bulletin de la Société Géologique de France, v. 162, no. 6, p. 1109–1124.

Marton, G., and R. Buffler, 1994, Jurassic reconstruction of the Gulf of Mexico Basin: International Geology Review, v. 36, p. 545–586.

Marton, G., and R. Buffler, 1999, Jurassic–Early Cretaceous tectono-paleogeographic evolution of the southeastern Gulf of Mexico Basin, *in* P. Mann, ed., Caribbean basins: Sedimentary basins of the World, v. 4, p. 63–91.

McFarlan, E., and L. Silvio Menes, 1991, Lower Cretaceous, *in* A. Salvador, ed., The Gulf of Mexico Basin: Boulder, Colorado, Geological Society of America, The Geology of North America, v. J, p. 181–204.

Meyerhoff, A., and C. Hatten, 1968, Diapiric structures in central Cuba, *in* J. Braustein and G. D. Brien, eds., Diapirism and diapirs: AAPG Memoir 8, p. 315–357.

Meyerhoff, A., and C. Hatten, 1974, Bahamas salient of North America: Tectonic framework, stratigraphy and petroleum potential: AAPG Bulletin, v. 58, no. 6, p. 1201–1239.

Millán, G., and R. Myczyñski, 1978, Fauna jurásica y consideraciones sobre la edad de las secuencias metamórficas del Escambray: La Habana, Academia de Ciencias de Cuba, Informe Científico-Técnico, v. 80, p. 3–14.

Millán, G., and R. Myczyñski, 1979, Jurassic ammonite fauna and age of metamorphic sequences of Escambray: Bulletin de l'Académie Polonaise des Sciences, Série des Sciences de la Terre, v. 27, no. 1 and 2, p. 37–47.

Millán, G., M. Somin, and C. Díaz, 1985, Nuevos datos sobre la geología del macizo montañoso de Sierra del Purial, Cuba oriental, *in* G. Millán and M. Somin, eds., Contribución al conocimiento geológico de las metamorfitas del Escambray y Purial: Reporte de Investigacion del Instituto de Geologia y Paleontologia, Academia de Ciencias de Cuba, La Habana, p. 52–69.

Millán Trujillo, G., 1996, Metamorfitas de la asociación ofiolítica de Cuba, *in* M. A. Iturralde-Vinent, ed., Ofiolitas y arcos volcanicos de Cuba: International Geological Correlation Programe Project 364, Geological correlation of ophiolites and volcanic arcs in the Circumcaribbean realm, Miami, Florida, U.S.A., Special Contribution 1, p. 131–146.

Millán Trujillo, G., 1997a, Posición estratigráfica de las metamorfitas cubanas, *in* G. Furrazola-Bermúdez and K. Nuñez Cambra, comp., Estudios sobre Geología de Cuba: La Habana, Centro Nacional de Información Geológica, p. 251–258.

Millán Trujillo, G., 1997b, Geología del macizo metamórfico Isla de la Juventud, *in* G. Furrazola-Bermúdez and K. Nuñez Cambra, comp., Estudios sobre Geología de Cuba: La Habana, Centro Nacional de Información Geológica, p. 259–270.

Millán Trujillo, G., 1997c, Geología del macizo metamórfico Escambray, *in* G. Furrazola-Bermúdez and K. Nuñez Cambra, comp., Estudios sobre Geología de Cuba: La Habana, Centro Nacional de Información Geológica, p. 271–289.

Montgomery, H., E. Pessagno, and I. Muñoz, 1992, Jurassic (Tithonian) radiolaria from La Desiderade (La Desiderade) (Lesser Antilles): Preliminary paleontological and tectonic implications: Tectonics, v. 11, no. 6, p. 1426–1432.

Morán-Zenteno, D., 1994, Geology of the Mexican Republic: AAPG Studies in Geology 39: AAPG, Tulsa, 160 p.

Moretti, I., R. Tenreyro, E. Linares, J. G. Lopez, J. Letouzey, C. Magnier, F. Gaumet, J. Lecomte, J. O. Lopez, and S. Zimine, 2003, Petroleum systems of the Cuban northwest offshore zone, *in* C. Bartolini, R. Buffler, and J. Blickwelde, eds., The circum-Gulf of Mexico and the Caribbean: Hydrocarbon habitats, basin formation, and plate tectonics: AAPG Memoir 79, p. 675–696.

Myczyñski, R., 1976a, A new ammonite fauna from the Oxfordian of the Pinar del Río province, western Cuba: Acta Geologica Polonica, v. 26, no. 2, p. 261–297.

Myczyñski, R., 1976b, *Organoceras* gen. n. (Ammonoidea)

from the Oxfordian of Cuba: Acta Geologica Polonica, v. 21, no. 4, p. 391–394.

Myczyñski, R., 1989, Ammonite biostratigraphy of the Tithonian of western Cuba: Annales Societatis Geologorum Poloniae, v. 59, no. 1 and 2, p. 43–125.

Myczyñski, R., 1994, Caribbean ammonite assemblages from Upper Jurassic–Lower Cretaceous sequences of Cuba: Studia Geologica Polonica, v. 105, p. 91–107.

Myczyñski, R., 1999, Some ammonite genera from the Tithonian of western Cuba and their paleobiogeographic importance: Studia Geologica Polonica, v. 114, p. 91–112.

Myczyñski, R., and A. Pszczolkowski, 1976, The ammonites and age of the San Cayetano Formation from the Sierra del Rosario, western Cuba: Acta Geologica Polonica, v. 26, no. 2, p. 321–329.

Myczyñski, R., and A. Pszczolkowski, 1990, Tithonian stratigraphy in the Sierra de los Organos, western Cuba: Correlation of the ammonite and microfossil zones, *in* G. Pallini, F. Cecca, S. Cresta, and M. Santantonio, coords., Atti II Convengo Internazionale, Fossili, Eoluzione, Ambiente, Pergola 1987, Comitato Centenario Raffaele Piccinini, eds., p. 405–415.

Myczyñski, R., and A. Pszczolkowski, 1994, Tithonian stratigraphy and microfacies in the Sierra del Rosario, western Cuba: Studia Geologica Polonica, v. 105, p. 7–38.

Myczyñski, R., F. Olóriz, and A. B. Villaseñor, 1998, Revised biostratigraphy and correlations of the middle–upper Oxfordian in the Americas (southern U.S.A., Mexico, Cuba and northern Chile): Neues Jahrbuch für Geologie und Paläontologie Abhandlungen, v. 207, no. 2, p. 185–206.

Olóriz, F., J. E. Caracuel, B. Marques, and F. J. Rodríguez-Tovar, 1995, Asociaciones de tintinnoides en facies ammonitico rosso de la Sierra Norte (Mallorca): Revista Española de Paleontología, No. Homenaje al Dr. Guillermo Colom, p. 77–93.

Olóriz, F., F. Villaseñor, and C. González-Arreola, 2003, Major lithostratigraphic units in land-outcrops in north-central Mexico and the subsurface of the northern rim of the Gulf of Mexico Basin (Upper Jurassic–lowermost Cretaceous): A proposal for correlation of tectono-eustatic sequences: Journal of South American Earth Sciences, v. 16, no. 3, p. 119–142.

Pardo, G., 1975, Geology of Cuba, *in* A. Nairn and F. Stehli, eds., The ocean basins and margins: New York, Plenum Press, v. 3, p. 553–613.

Pessagno Jr., E. A., and C. Martin, 2003, Tectonostratigraphic evidence for the origin of the Gulf of Mexico, *in* C. Bartolini, R. T. Buffler, and J. Blickwede, eds., The circum-Gulf of Mexico and the Caribbean: Hydrocarbon habitats, basin formation, and plate tectonics: AAPG Memoir 79, p. 46–74.

Pessagno Jr., E. A., D. M. Hull, C. B. Martin, X. Meng, H. Montgomery, J. G. Ogg, J. Urrutia-Fucugauchi, 1999, Stratigraphic evidence for northwest to southeast tectonic transport of Jurassic terranes in Central Mexico and the Caribbean (western Cuba), *in* P. Mann, ed., (K. J. Hsu, series ed.) Caribbean sedimentary basins: Sedimentary basins of the World: Amsterdam, Elsevier, v. 4, p. 123–150.

Pindell, J., 1994, Evolution of the Gulf of Mexico and the Caribbean, *in* S. Donovan and T. Jackson, eds., Caribbean geology. An introduction: Kingston, Jamaica, The University of the West Indies Publishers' Association, p. 13–40.

Pindell, J., and J. Barrett, 1990, Geological evolution of the Caribbean Region: A plate tectonics perspective, *in* G. Dengo and J. Case, eds., The Caribbean region: Boulder, Colorado, Geological Society of America, The Geology of North America, v. H, p. 405–432.

Pindell, J., L. Kennan, K. Stanek, W. Maresch, and G. Draper, 2006, Foundations of Gulf of Mexico and Caribbean evolution: Eight controversies resolved: Geologica Acta, v. 4, p. 303–341.

Piotrowska, K., 1978, Nappe structures in the Sierra de los Organos, western Cuba: Acta Geologica Polonica, v. 28, no. 1, p. 97–170.

Pop, G., 1976, Tithonian–Valanginian calpionellid zones from Cuba: Dări de Seamă ale Şedinţelor, v. 62, p. 237–266.

Premoli-Silva, I., and C. L. McNulty, 1984, Planktonic foraminifera and calpionellids from Gulf of Mexico sites, Deep Sea Drilling Project leg 77, *in* R. T. Buffler et al., eds., Initial reports of the DSDP 77: Washington, D.C., U.S. Government Printing Office, p. 547–584.

Pszczolkowski, A., 1978, Geosynclinal sequences of the Cordillera de Guaniguanico in western Cuba: Their lithostratigraphy, facies development and paleogeography: Acta Geologica Polonica, v. 28, no. 1, p. 1–96.

Pszczolkowski, A., 1981, El banco carbonatado jurásico de la sierra de los Organos, Provincia de Pinar del Rio; su desarrollo y situación paleotectónica: Ciencias de la Tierra y del Espacio, v. 3, p. 37–50.

Pszczolkowski, A., 1982, Cretaceous sediments and paleogeography in the western part of the Cuban miogeosyncline: Acta Geologica Polonica, v. 32, no. 1–2, p. 135–161.

Pszczolkowski, A., 1983, Tectonica del miogeosinclinal cubano en el area limitrofe de las Provincias de Matanzas y Villaclara: Ciencias de La Tierra y del Espacio, v. 6, p. 53–61.

Pszczolkowski, A., 1986a, Composición del material clástico de las arenitas de la Formación San Cayetano en la Sierra de los Organos (Provincia de Pinar del Río, Cuba): Ciencias de la Tierra y del Espacio, v. 14, p. 71–79.

Pszczolkowski, A., 1986b, Secuencia estratigráfica de Placetas en el área limítrofe de las provincias de Matanzas y Villaclara (Cuba): Bulletin of the Polish Academy of Sciences, v. 34, no. 1, p. 67–79.

Pszczolkowski, A., 1994a, Lithostratigraphy of Mesozoic and Paleogene rocks of Sierra del Rosario, western Cuba: Studia Geologica Polonica, v. 105, p. 39–66.

Pszczolkowski, A., 1994b, Geological cross-sections through the Sierra del Rosario thrust belt, western Cuba: Studia Geologica Polonica, v. 105, p. 67–90.

Pszczolkowski, A., 1994c, Interrelationship of the terranes in western and central Cuba—Comment: Tectonophysics, v. 234, p. 339–344.

Pszczolkowski, A., 1999, The exposed passive margin of

North America in western Cuba, *in* P. Mann, ed., Caribbean basins. Sedimentary basins of the World, 4 (series editor: K.J. Hsu): Amsterdam, Elsevier, p. 93–121.

Pszczolkowski, A., and J. de Albear, 1983, La secuencia vulcanógeno-sedimentaria de Sierra del Rosario: Ciencias de la Tierra y del Espacio, v. 6, p. 41–52.

Pszczolkowski, A., and R. Myczyñski, 1999, Nannocoid assemblages in upper Hauterivian–lower Aptian limestones of Cuba: Their correlation with ammonites and some planktonic foraminifers: Studia Geologica Polonica, v. 114, no. 1, p. 35–75.

Pszczolkowski, A., and R. Myczyñski, 2003, Stratigraphic constraints on the Late Jurassic–Cretaceous paleotectonic interpretations of the Placetas belt in Cuba, *in* C. Bartolini, R. Buffler, and J. Blickwede, eds., The circum-Gulf of Mexico and the Caribbean: Hydrocarbon habitats, basin formation and plate tectonics: AAPG Memoir 79, p. 545–581.

Pszczolkowski, A., D. García Delgado, and S. Gil González, 2005, Calpionellid and nannoconid stratigraphy and microfacies of limestones at the Tithonian–Berriasian boundary in the Sierra del Infierno (western Cuba): Annales Societatis Geologorum Poloniae, v. 75, no. 1, p. 1–16.

Pushcharovsky, Y., ed., 1988, Mapa geológico de la República de Cuba: Academia de Ciencias de Cuba, Academia de Ciencias de la URSS, escala 1:250,000, 42 sheets.

Renne, P., J. Mattinson, C. Hatten, M. Somin, G. Millán-Trujillo, and E. Linares-Cala, 1989, Confirmation of Late Proterozoic age for the Socorro complex of north central Cuba from $^{40}Ar/^{39}Ar$ and U-Pb dating, *in* Resúmenes y Programa del Primer Congreso Cubano de Geología: Sociedad Cubana de Geología, p. 118.

Rigassi-Studer, D., 1963, Sur la géologie de la Sierra de los Organos, Cuba: Archives des Sciences, Société de Physique et d' Histoire Naturelle, Génève, v. 16, no. 2, p. 339–350.

Rojas-Agramonte, Y., A. Kröner, J. Pindell, A. García-Casco, D. García-Delgado, D. Liu, and Y. Wang, 2008, Detrital zircon geochronology of Jurassic sandstones of western cuba (San Cayetano Formation): Implications for the Jurassic paleogeography of the NW proto-Caribbean, American Journal of Science, v. 308, p. 639–656.

Salvador, A., 1987, Jurassic paleogeography and origin of the Gulf of Mexico Basin: AAPG Bulletin, v. 71, p. 419–451.

Samson, Y., M.-C. Janin, G. Bignot, J. Guyader, and G. Breton, 1992, Les globuligerines (Foraminifères planctoniques) de l'Oxfordien inferieur de Villers-sur-mer (Calvados, France) dans leur gisement: Revue de Paléobiologie, v. 11, no. 2, p. 409–431.

Sánchez, J. R., R. Tenreyro, R. Socorro, S. Blanco, and D. Brey, 1998, Relaciones cuenca-plataforma en el paleomargen continental Bahamas-Cuba: Estratigrafía, sedimentogénesis y paleogeografía, *in* Geología y Minería '98: Centro Nacional de Información Geológica Memorias 2, p. 312–314.

Schafhauser, A., W. Stinnesbeck, B. Holland, T. Adatte, and J. Remane, 2003, Lower Cretaceous pelagic limestones in southern Belize: Protocaribbean deposits in the southeastern Maya block, *in* C. Bartolini, R. Buffler, and J. Blickwelde, eds., The circum-Gulf of Mexico and the Caribbean: Hydrocarbon habitats, basin formation, and plate tectonics: AAPG Memoir 79, p. 624–637.

Schlager, W., R. Buffler, D. Angstadt, and R. Phair, 1984a, Geologic history of the southeastern Gulf of Mexico, *in* R. T. Buffler et al., eds., Initial reports of the Deep Sea Drilling Project 77: Washington, D.C., U.S. Government Printing Office, p. 715–738.

Schlager, W., et al., 1984b, Deep Sea Drilling Project, leg 77, southeastern Gulf of Mexico: Geological Society of America Bulletin, v. 95, p. 226–236.

Segura-Soto, and Blanco-Bustamante, 1998, La secuencia de Camajuaní: Un sector del margen divergente del continente norteamericano en continua sumersión desde el Tithoniano hasta el Turoniano, *in* Geología y Minería '98: Centro Nacional de Información Geológica, La Habana, Memorias 2, p. 315–317.

Shopov, V., 1982, Estratigrafía y subdivisión de las zonas de Placetas y Camajuaní en la antigua provincia de Las Villas (Cuba central): Ciencias de La Tierra y del Espacio, v. 4, p. 40–46.

Somin, M., and G. Millán, 1981, Geology of the metamorphic complexes of Cuba (in Russian): Moscow, Nauka, 219 p.

Stanek, K., W. Maresch, F. Grafe, Ch. Grevel, and A. Baumann, 2006, Structure, tectonics and metamorphic development of the Sancta Spiritus Dome (eastern Escambray massif, central Cuba): Geologica Acta, v. 4, p. 151–170.

Stephan, J., et al., 1990, Paleogeodynamic maps of the Caribbean: 14 steps from Lias to present: Bulletin de la Société géologique de France, v. 6, no. 6, p. 915–919.

Tavera, J. M., R. Aguado, M. Company, and F. Olóriz, 1994, Integrated biostratigraphy of the Durangites and Jacobi zones (J/K boundary) at the Puerto Escaño section in southern Spain (Province of Cordoba): Geobios, Mémoire Spécial 17, p. 469–476.

United Nations Educational, Scientific and Cultural Organization (UNESCO) and International Union of Geological Sciences (IUGS), 2000, International stratigraphic chart: Division of Earth Sciences, United Nations Educational, Scientific and Cultural Organization (UNESCO), 16 p.

Valdés-Nodarse, E., 1998, Pb-Zn "sedex" deposits and their cooper stockworkroots, western Cuba: Mineralium Deposita, v. 33, p. 560–567.

Villaseñor, A., F. Olóriz, and I. López-Palomino, 2004, Inner whorls of Gregoryceras (Ammonitina, Peltoceratinae) as the first occurrence of the genus in Mexico: Rivista Italiana di Paleontologiae Stratigrafia, v. 110, no. 1, p. 249–254.

Villaseñor, A. B., F. Olóriz, and C. González-Arreola, 2000, Recent advances in Upper Jurassic (Kimmeridgian–Tithonian) ammonite biostratigraphy of north-central Mexico based on recently collected ammonite assemblages: GeoResearch Forum, v. 6, p. 249–262.

Watkins, D. K., and C. L. McNulty, 1984, Paleontological synthesis, leg 77, *in* R. T. Buffler et al., eds., Initial reports of the Deep Sea Drilling Project 77: Washington, D.C., U.S. Government Printing Office, p. 703–714.

Wierzbowski, A., 1976, Oxfordian ammonites of the Pinar

del Río province (western Cuba); their revision and stratigraphical significance: Acta Geologica Polonica, v. 26, no. 2, p. 137–260.

Wilson, J., 1980, Carbonate facies in geological history (Russian translation): Moscow, Nedra, 463 p.

Winker, C., and R. Buffler, 1988, Paleogeographic evolution of early deep-water Gulf of Mexico and margins, Jurassic to Middle Cretaceous (Comanchean): AAPG Bulletin, v. 72, p. 318–346.

Witrock, R. B., A. R. Friedmann, J. J. Galluzzo, L. D. Nixon, P. J. Post, and K. M. Ross, 2003, Biostratigraphic chart of the Gulf of Mexico off-shore region, Jurassic to Quaternary, U.S. Department of the Interior, Minerals Management Service, New Orleans.

Young, F., and F. Oloriz, 1993, Ammonites from the Smackover limestone, Cotton Valley field, Webster Parish, Louisiana, U.S.A, *in* S. Elmi, Ch. Mangold, and Y. Alméras, eds., 3ème Symposium International Céphalopodes Actuels et Fossiles (Symposium F. Roman): Geobios, M.S., v. 15, p. 401–409.

21

Moreno, M., C. Gonzalez, A. Cruz, J. Meneses, M. Araujo, T. Hernández, and F. Audemard, 2009, The deeper exploration of the Southern Gulf of Mexico, Le Acach-Chuktah, *in* C. Bartolini and J. R. Román Ramos, eds., Petroleum systems in the southern Gulf of Mexico: AAPG Memoir 90, p. 453–458.

The Deeper Exploration of the Southern Gulf of Mexico, Le Acach-Chuktah

Madain Moreno, Carlos Gonzalez, Alicia Cruz, and Javier Meneses
PEMEX Exploration and Production

Miguel Araujo, Tito Hernández, and Felipe Audemard
Schlumberger (DCS)

The Le Acach-Chuktah area belongs to the northern exploration frontier of the Southern Gulf of Mexico. A campaign carried out by Pemex in the last 7 seven years where 6 wildcats testing the most relevant structures were drilled in an area >3000 km^2 at water depths ranging from 250 to 400 m. All the wells were drilled at a depth >4000 m and they were targeted to reach the northwards limits of the known Fractured Cretaceous and the Kimmeridgian play. In addition to these targets, a Middle Miocene sand 30 m thick has been cut in all the wells. This deepwater sand body constitutes a widespread reservoir containing heavy oil, commonly associated with seismic attributes of fluid contacts close to the structural tops. At deeper levels, oil has also been tested in Oxfordian red beds in 2 wells.

A reassessment of the area allowed the identification of several new locations, four of which are illustrated here. They are primarily oriented to evaluate the known targets, considering as secondary objectives the Miocene and deeper Jurassic. The preliminary drilling in the area indicates that the Upper Cretaceous section is faulted out or thinner when compared to the known southern sections. At deeper water columns it will also be possible to penetrate older sedimentary sections probably associated with the Lower Jurassic and Late Paleozoic. The structures appear primarily resulting from a deeply rooted compressional regime probably initiated by a phase of half-graben inversions, overprinted by salt tectonics, which subsequently triggered gravitational structures developed from five shallow decoupling surfaces located within the Tertiary sections.

The results of this study were generated from the integration of multiple disciplines and tools in order to unravel additional potential for this area. From the original 2800 km^2 3D seismic we only use the northern 1000 km^2. The suites of logs from the 6 wells already drilled in the area and vicinities were used. All the 6 sonic logs, checkshots and available VSP were combined in order to generate a new vintage of homogenized synthetics. These showed a very good calibration with the seismic. The latest 3D seismic processing was available. In parallel the data from cores and cuttings were also integrated with all the recorded logs and a series of ELANs were created using the same reference frame for all the wells.

The test results were also considered within the context of the logs in order to calibrate reservoir quality, porosity distribution, lateral continuity, thickness change and oil quality. Figure 1 outlines the study area, the relative location with respect to the southern Gulf of Mexico coastline, the wells distribution as well as the water depths. Diagram including several key parameters, such as facies changes interpreted from logs, cores, cuttings and seismic, the porosity values measured, structural controls, depths, oil type and quality as well as source bed distributions were generated for the Upper-Mid Mesozoic section of the study area.

Reprinted from Extended Abstracts of the AAPG International Conference, October 24–27, 2004, Cancun, Mexico.
DOI:10.1306/13191095M903340

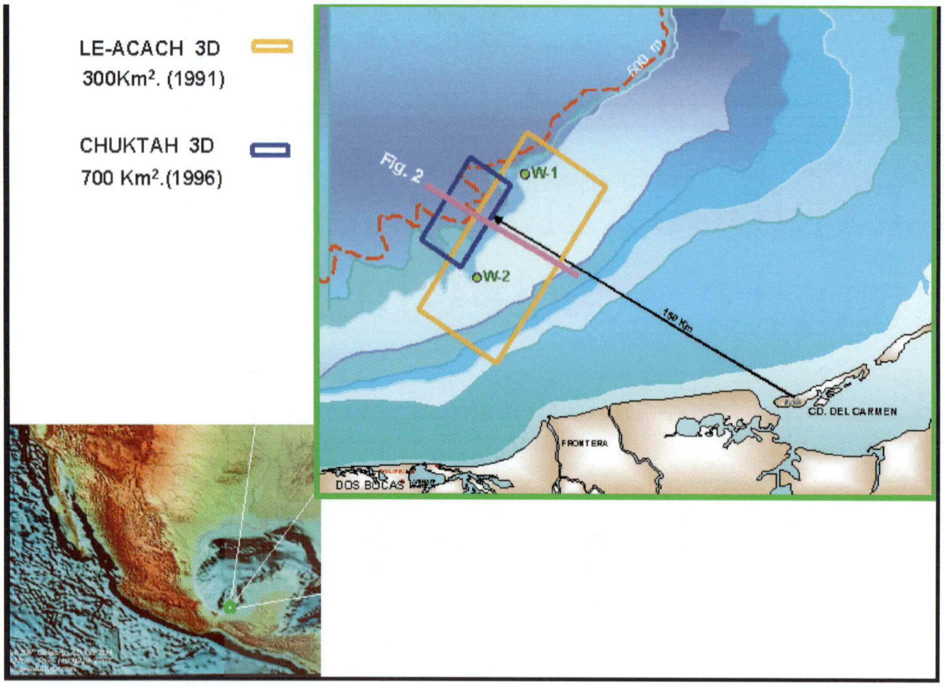

FIGURE 1. Relative location of the study area.

Key markers as the Top Eocene, Top Cretaceous, Top Kimmeridgian and Top Oxfordian were mapped in time. The uppermost level represents the seal for the underlying reservoirs. The lower three constitute the reservoirs targeted in this region, the fractured Cretaceous carbonates and the porous oolitic section from the Kimmeridgian. The Oxfordian being the deepest potential reservoir is the less predictable due to its restricted areal distribution. It is usually associated with shallow marine to continental facies of variable thickness. It has been interpreted to be developed as the sedimentary infill of half-grabens controlled by transfer zones.

It is very important to recall upon the Kimmeridgian marker. It also defines the base of the regionally distributed source beds for this region. Recently, a new source bed has been documented within the Oxfordian and needs to be considered in any evaluation of the petroleum system. This finding helps visualize alternate models for charge and maturity aspects. It certainly reduces concerns on specific areas where the overburden is thinning seaward.

Special attention is devoted to the middle Miocene at the top of a *Chaotic unit* where a regional plastic band is found (Figure 3). This section is internally deformed and averages 300 milliseconds in thickness and many of the overlying normal fault systems are rooted in and several salt piercements were confined to this level. The band constitutes in essence a major decoupling surface for the Southern Gulf of Mexico, right on top of the upper detachment shown in Figure 3. The upper limit of the unit is bounded by a regional sandstone body. This several meters thick, coarse grained deepwater unit has proven to host heavy crudes, observed during the drilling. In seismic sections the interval reflects the presence of hydrocarbon indicators when structures are not affected by salt intrusions. This promotes the idea of options to consider an additional shallower reservoir residing in this clastic section. The porosity, permeability values and fluid characteristics could be reevaluated from logs in order to combine it with the evidence from the seismic to develop a new play.

In this area there are at least two distinct salt intervals below the Kimmeridgian, The thickness of the sedimentary section where they are hosted is in the order of several kilometers equivalent to more than two seconds in time from the reflection seismic data. The ages for these units are not defined yet. An educated guess extrapolated from the broader knowledge around this area, points out to be Permian in age, the San Andresa Formation from the Permian Basin to the north of Mexico-Southwest Texas.

Five shallower markers were also mapped in order to control the stratigraphic distribution of this major sedimentary package and better resolve the Neogene distribution changes; due to thickness variation and growth regulated by extensional tectonics, which is in turn floating on top of the underlying Mid-Miocene regional decoupling surface. This later surface defines the boundary between two domains, the prevalent deeper compressional regime and the self adjusting gravitational regime gliding on top, which constantly readjusts due to contractional effect of salt tectonics and mud diapirism associated with over-pressuring of the 2 km thick upper Miocene-Pliocene section (Figure 3).

Depth conversion for the targeted reservoirs required a sophisticated approach due to lateral velocity changes

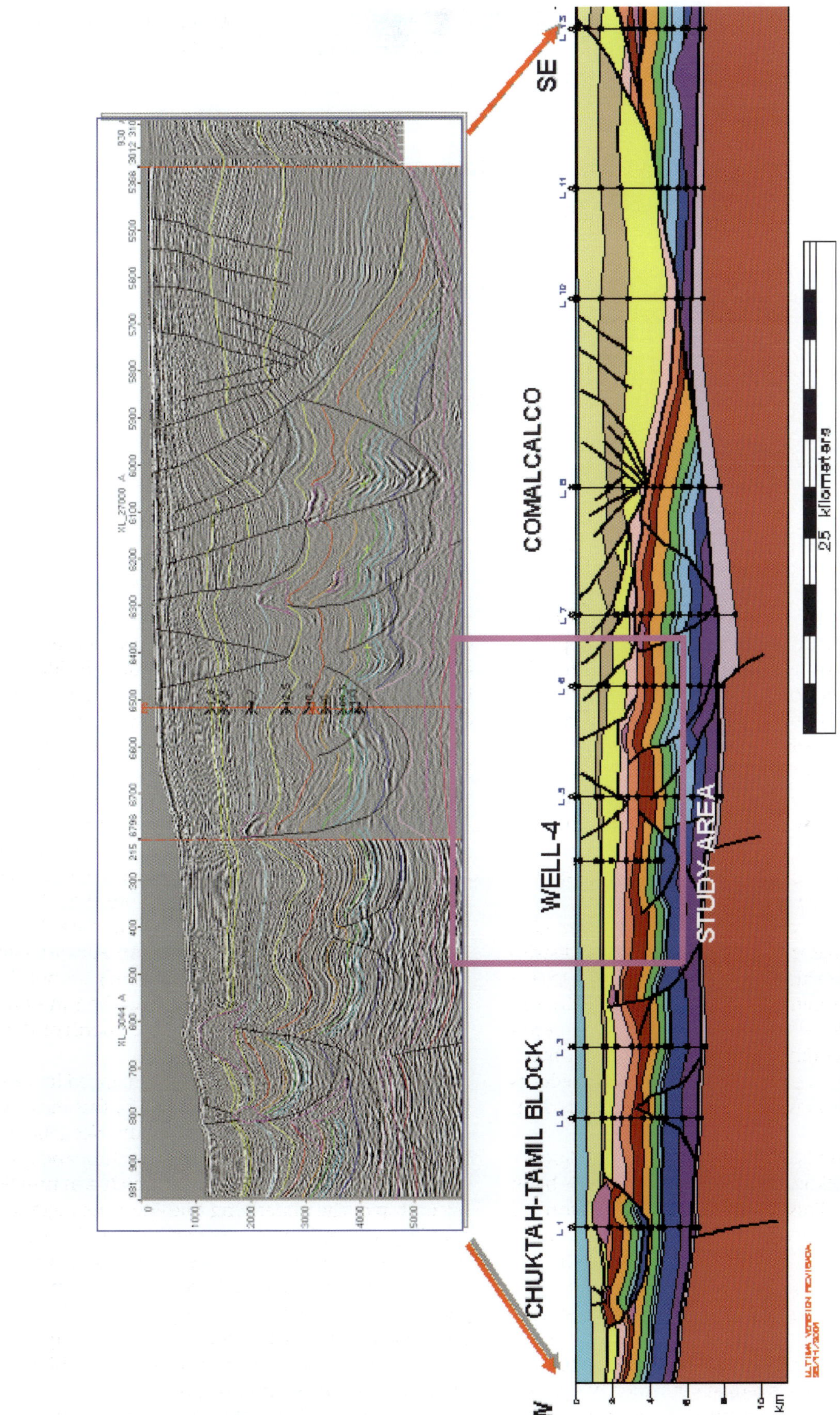

Figure 2. A regional NW-SE cross-section . The upper 6 units represent the Tertiary. The Upper and Lower cretaceous are green and the Jurassic is color coded in blue.

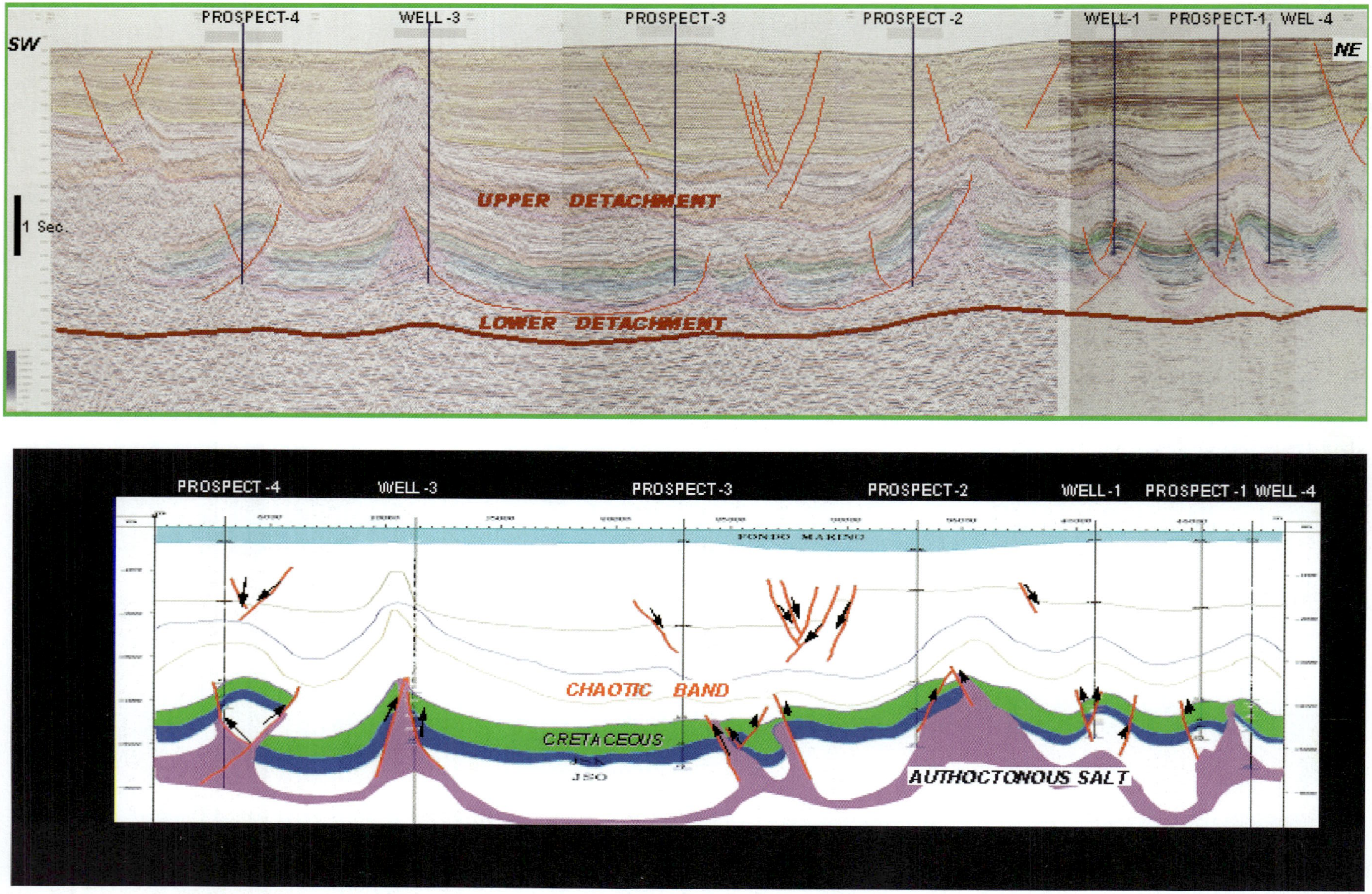

FIGURE 3. A section showing the four prospects, the two major decoupling surfaces and the uppermost autochthonous salt unit (Callovian).

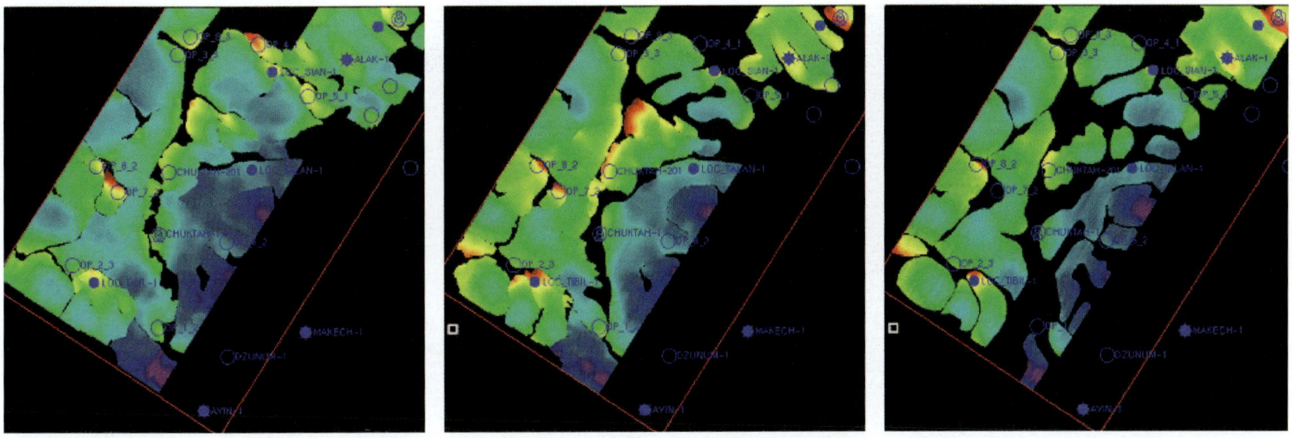

FIGURE 4. The three structural maps of the targeted reservoirs; Cretaceous (left), Kimmeridgian (center) and Oxfordian (right) show the different opportunities to be evaluated within the limits of the study area.

in the area. These are attributed to salt emplacement or to overpressure zones developed at the base of the extensional system decoupled in the Oligocene-Lower Miocene interval. These zones appear similar in shapes to the well-known roller structures, but in this case they are associated with low velocity shale intervals well developed at the base of the footwalls of the normal faults. These effects of low velocities zones are highlighted in sonic logs and evident when constructing the Time-Depth curves for all the wells drilled in this region.

The depth converted maps after considering these velocity problems allowed the identification of four structurally controlled prospects. Four of those are shown here in cross section and map views (Figures 4, 5). Another nine opportunities were identify for further documentation pending on the outcome of the proposed wells for this campaign. It is extremely important to undertake the detailed velocity analysis throughout the trajectory of exploration wells and careful well design will certainly minimize time and drilling costs in the region.

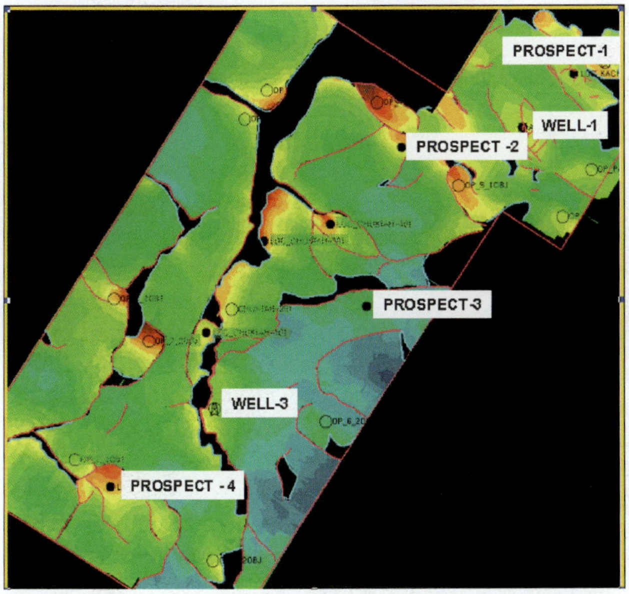

FIGURE 5. Relative location of the main prospects defined.

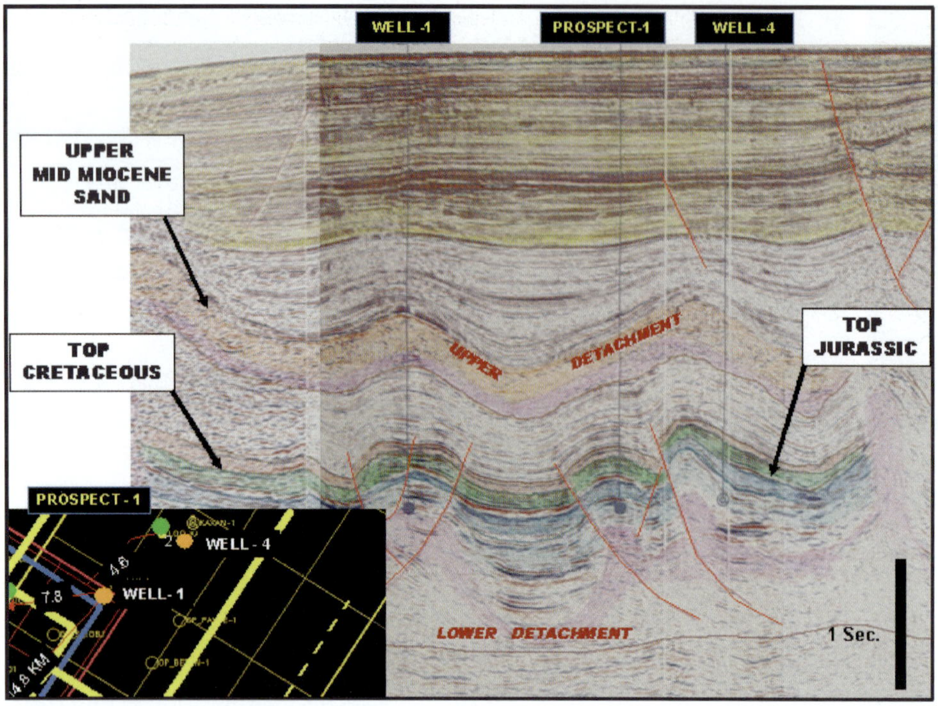

FIGURE 6. Prospect 1 represent a typical Mesozoic configuration with double objectives (Cretaceous and Jurassic). Wells on both sides were discoveries.

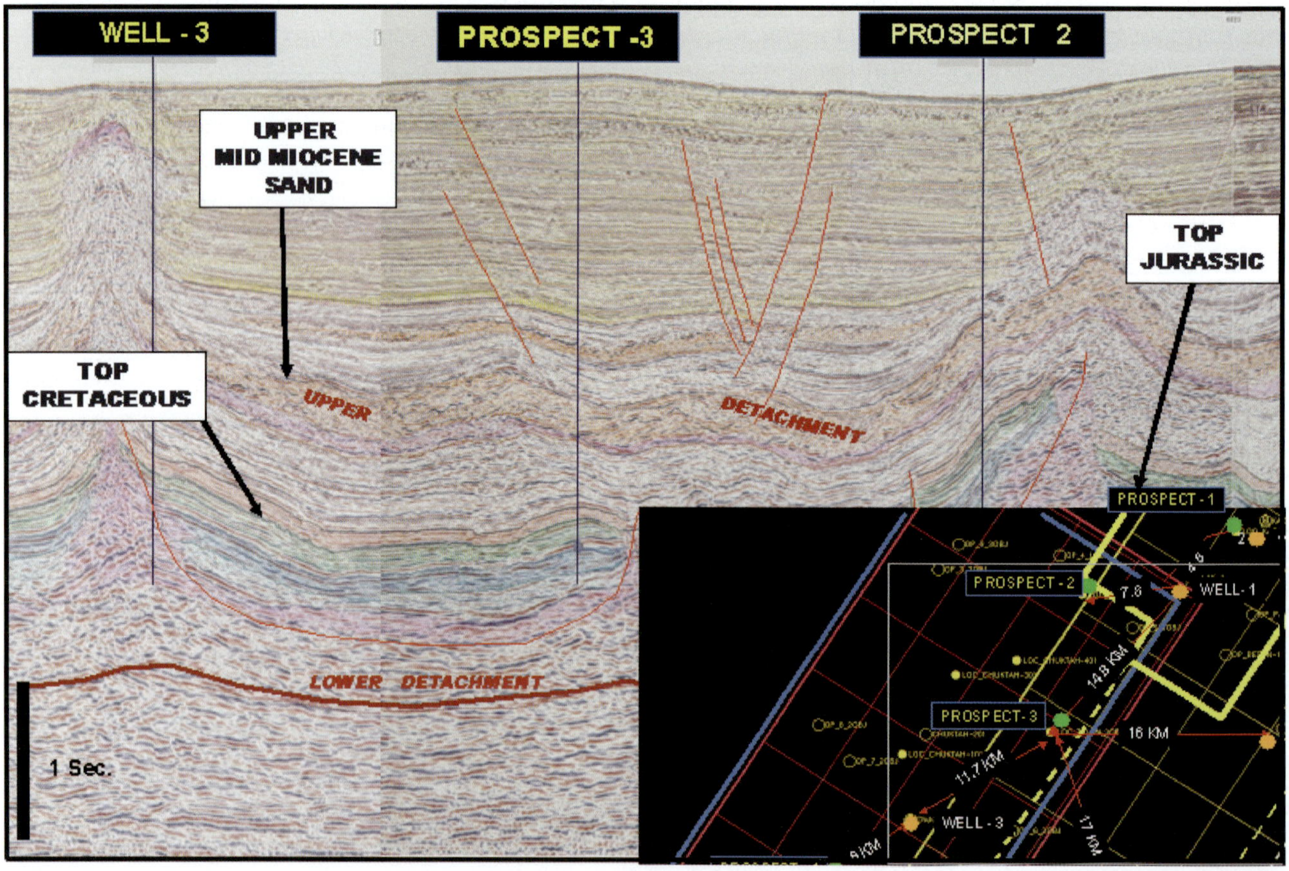

FIGURE 7. This section shows the prospect 3 as a structure less influenced by salt intrusions where reservoir integrity seems better preserved compared to the previously drilled structures on the sides.

22

Gonzalez, N., F. Angeles, J. Reyes, J. Meneses, C. Puerto, D. Balasejus, A. Suter, Y. Helfenbein, and F. Audemard, 2009, Choch Mulix a prolific area on the western margin of the Akal Horst, southern Gulf of Mexico, *in* C. Bartolini and J. R. Román Ramos, eds., Petroleum systems in the southern Gulf of Mexico: AAPG Memoir 90, p. 459–464.

Choch Mulix a Prolific Area on the Western Margin of the Akal Horst, Southern Gulf of Mexico

Nicolas Gonzalez, Francisco Angeles, Javier Reyes, Javier Meneses, and Carlos Puerto

PEMEX Exploration and Production

Detlef Balasejus, Andreas Suter, Yves Helfenbein, and Felipe Audemard

Schlumberger (DCS)

In less than a decade of intense activity this area is already a case history. Several discoveries in different Mesozoic intervals have already been made; the delineation phase is ongoing for three of these, where recoverable reserves are greater than 750 MMB. At the same time, four wildcats are on stream. We are presenting here some of the results associated with the recent and future exploration phase.

The area includes 9 wells drilled in less than 600 km^2. They reach depths >5000 m, all are producing with a productivity range from 1000 to 10000 barrels/d. The Upper Cretaceous fractured limestones constitute the upper reservoir whereas the Kimmeridgian is associated with oolitic limestone and dolomitic facies. These producing levels reside on top and below the Tithonian source beds. The area is deformed, therefore making it difficult to image in 3D seismic, where a deeper compressional regime is responsible for the antiformal structures, associated with salt tectonics and they have in turn induced gravitational gliding for most of the Tertiary section. The major regimes are separated by a pressurized zone from where most of the shale diapirs are rooted in.

The presented examples highlight the methodology used to document the new prospects, the spill point maps, the upside potential in the tertiary, as well as the one proposed for deeper sections. The integrated study considers most of the relevant variables concerning the geometry, petrophysics, fluids and facies distributions. They all together are tuned into the understanding of the petroleum system along with different depth conversions performed to honor the data.

The results of this study were generated from the integration of multiple disciplines and tools in order to unravel additional potential for this area. From the original 1100 km^2 of the 3D seismic, only the newly processed northern 700 km^2 were used. The suites of logs from the 9 wells already drilled in the area and vicinities were used. The 9 sonic logs, checkshots and available VSP were combined in order to generate a new vintage of homogenized synthetics. These showed a very good calibration with the seismic. The data from cores and cuttings were also integrated with all the recorded logs and a series of ELANs were created using the same reference frame for all wells.

The test results were also considered within the context of the logs in order to calibrate reservoir quality, porosity distribution, lateral continuity, thickness change and oil quality. Figure 1 shows a work flow diagram on how the entire data set was integrated.

Reprinted from Extended Abstracts of the AAPG International Conference, October 24–27, 2004, Cancun, Mexico.

DOI:10.1306/13191096M903340

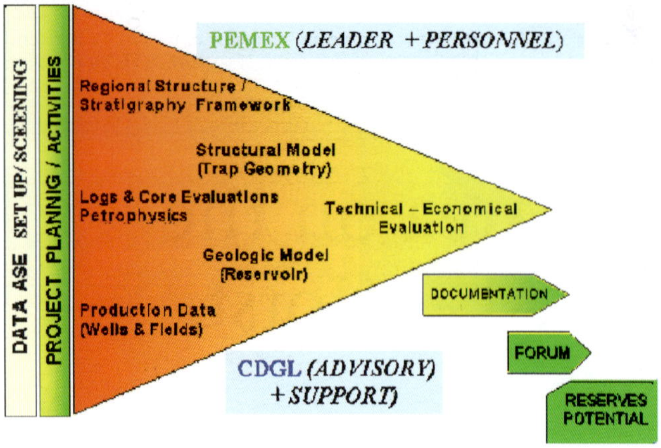

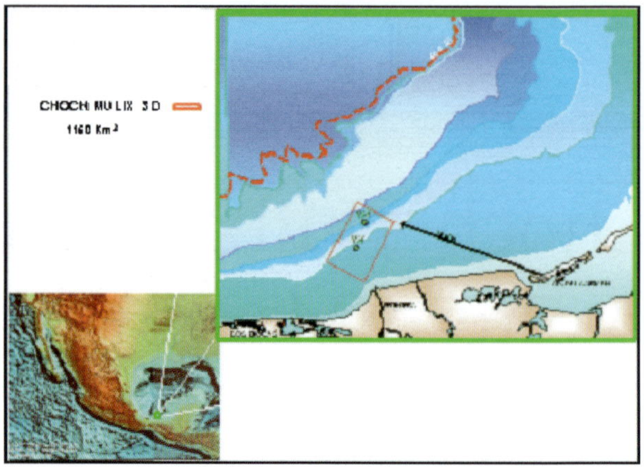

FIGURES 1 AND 2. The flow diagram used for the Choch Mulix study (left) Relative location of the area to the Southern Gulf of Mexico (right).

The outline of the study area, relative location with regards to the southern Gulf of Mexico coastline, wells distribution as well as the water depths are shown in Figure 2.

A fence diagram including several key parameters, such as facies changes interpreted from logs, cores, cuttings and seismic, porosity measurements, structural controls, depths, oil type and quality including source bed distributions were generated for the Upper-Mid Mesozoic section of the study area (Figure 3).

Key markers as the Top Eocene, Top Cretaceous, Top Kimmeridgian were digitally mapped in time. The uppermost level represents the seal for the underlying reservoirs. The lower two constitute the main reservoirs targeted in this region (the fractured Cretaceous carbonates and the porous oolitic section from the Kimmeridgian). The lowermost marker also defines the base of the most regional source beds commonly identified for this region (Figure 4). Four other overlaying markers were also mapped in order to control stratigraphic distribution of major sedimentary packages and better outline the Mio-Pliocene distribution, due to thickness changes and growth regulated by extensional tectonics, which is in turn floating on top of the underlying regional decoupling surface. The basal marker of the Oligocene-Miocene defines the boundary between two domains: the prevalent deeper contractional regime and the self adjusting gravitational gliding regime floating on top. The upper interval constantly readjusts due to the compressional effect of salt tectonics and mud diapirism associated with the over-pressuring of the more than 2 km thick upper Miocene-Pliocene section (Figure 5).

Depth conversion for the targeted reservoirs required a more sophisticated approach due to considerable lateral velocity changes in the area. These are attributed to overpressure zones developed at the base of the extensional system decoupled in the Oligocene–Lower Miocene interval. These zones show geometrical similarities to the

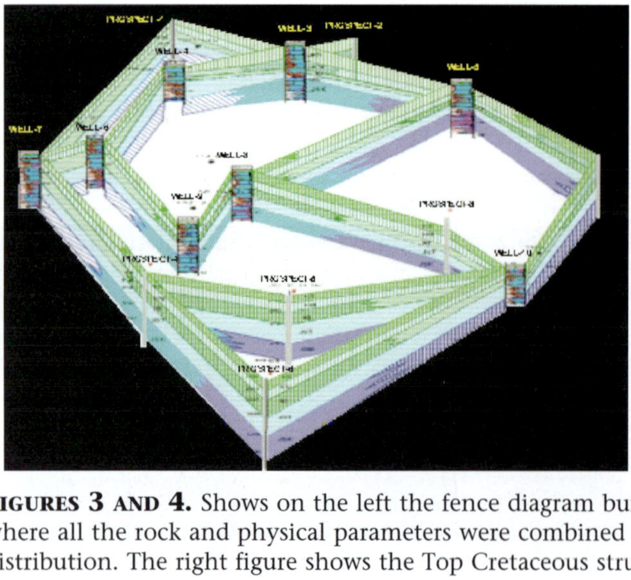

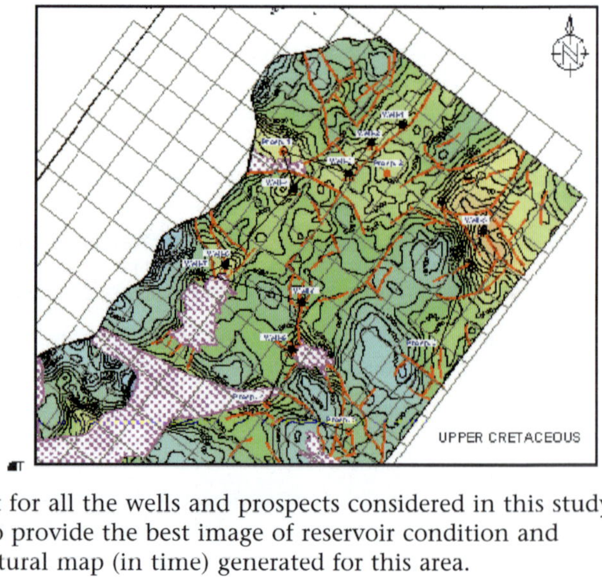

FIGURES 3 AND 4. Shows on the left the fence diagram built for all the wells and prospects considered in this study where all the rock and physical parameters were combined to provide the best image of reservoir condition and distribution. The right figure shows the Top Cretaceous structural map (in time) generated for this area.

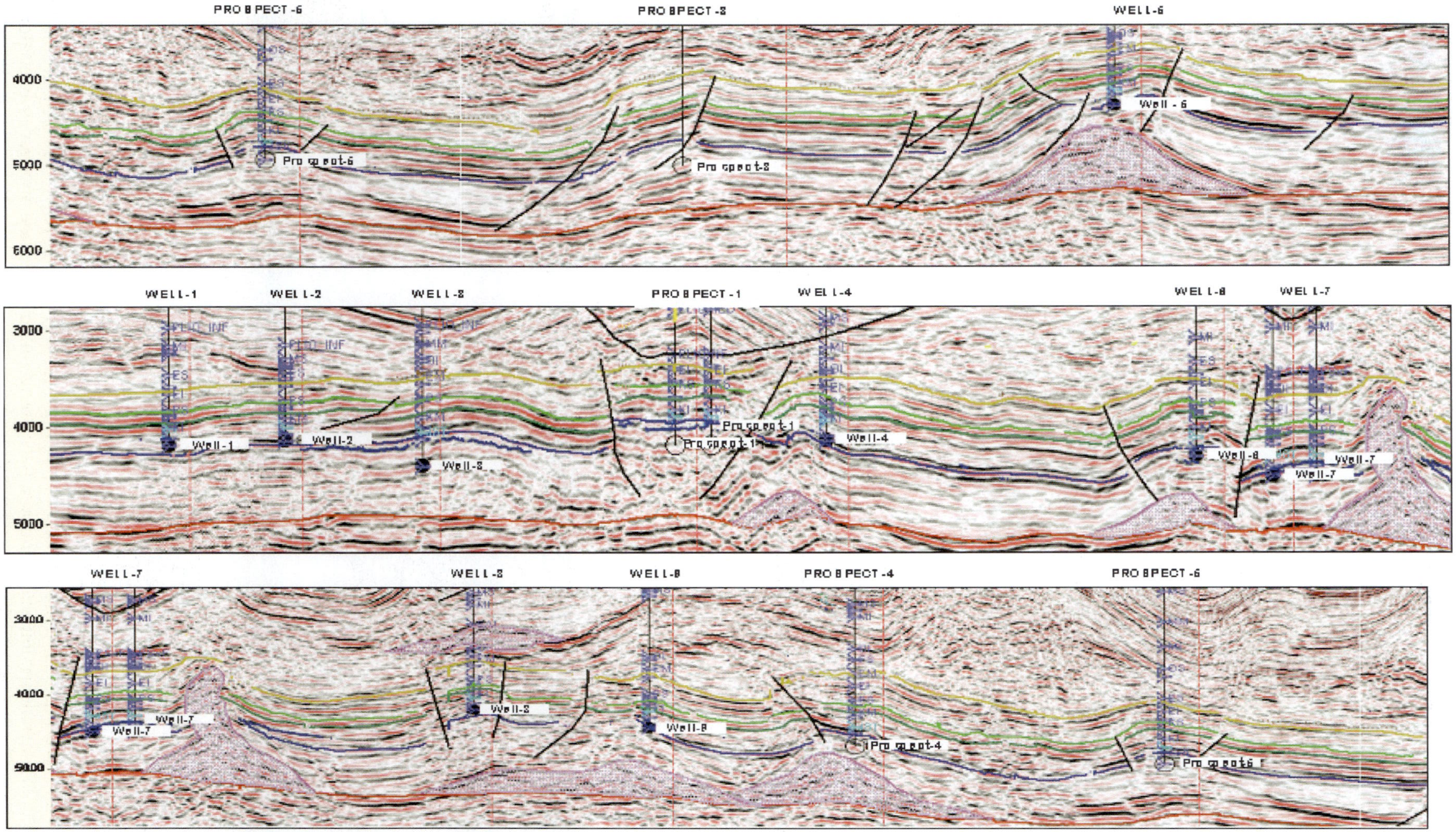

FIGURE 5. A panel including all the wells and prospects showing the consistent calibration of the seismic and well data. The views are focused to the Mesozoic intervals.

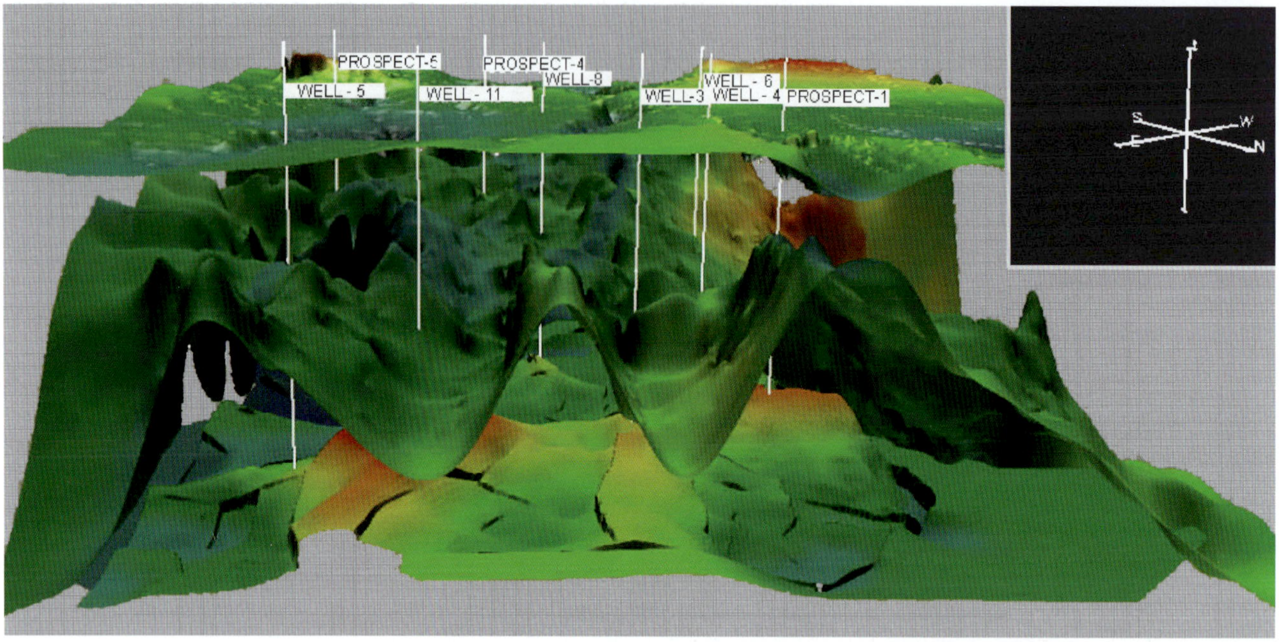

Figure 6. A 3D view of the level of complexities for the velocity model considered in the depth conversion used for this study.

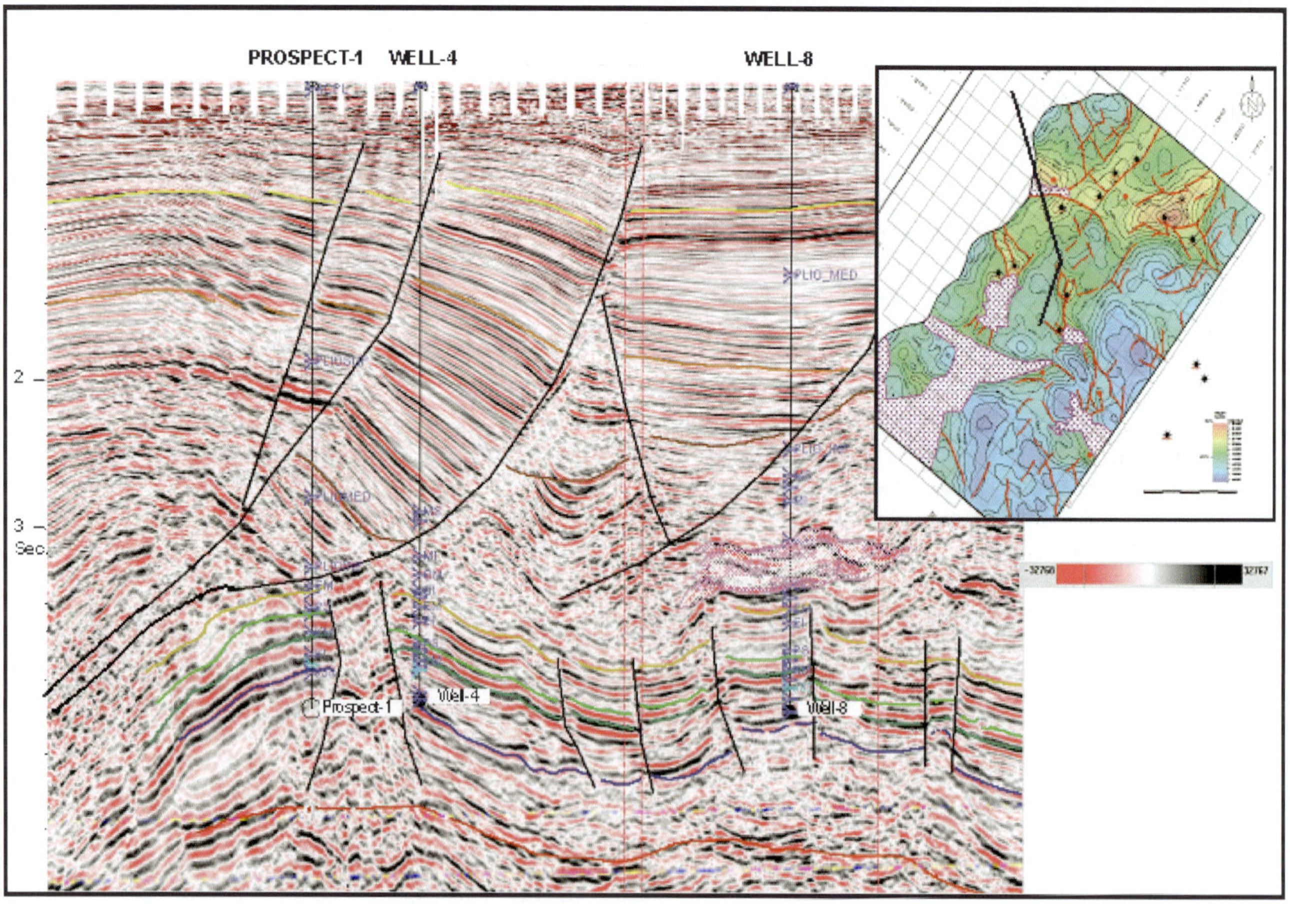

Figure 7. A NNW-SSE seismic section across Prospect 1. The wells on both sides are discoveries associated with productive intervals of light oil tested within the Mesozoic.

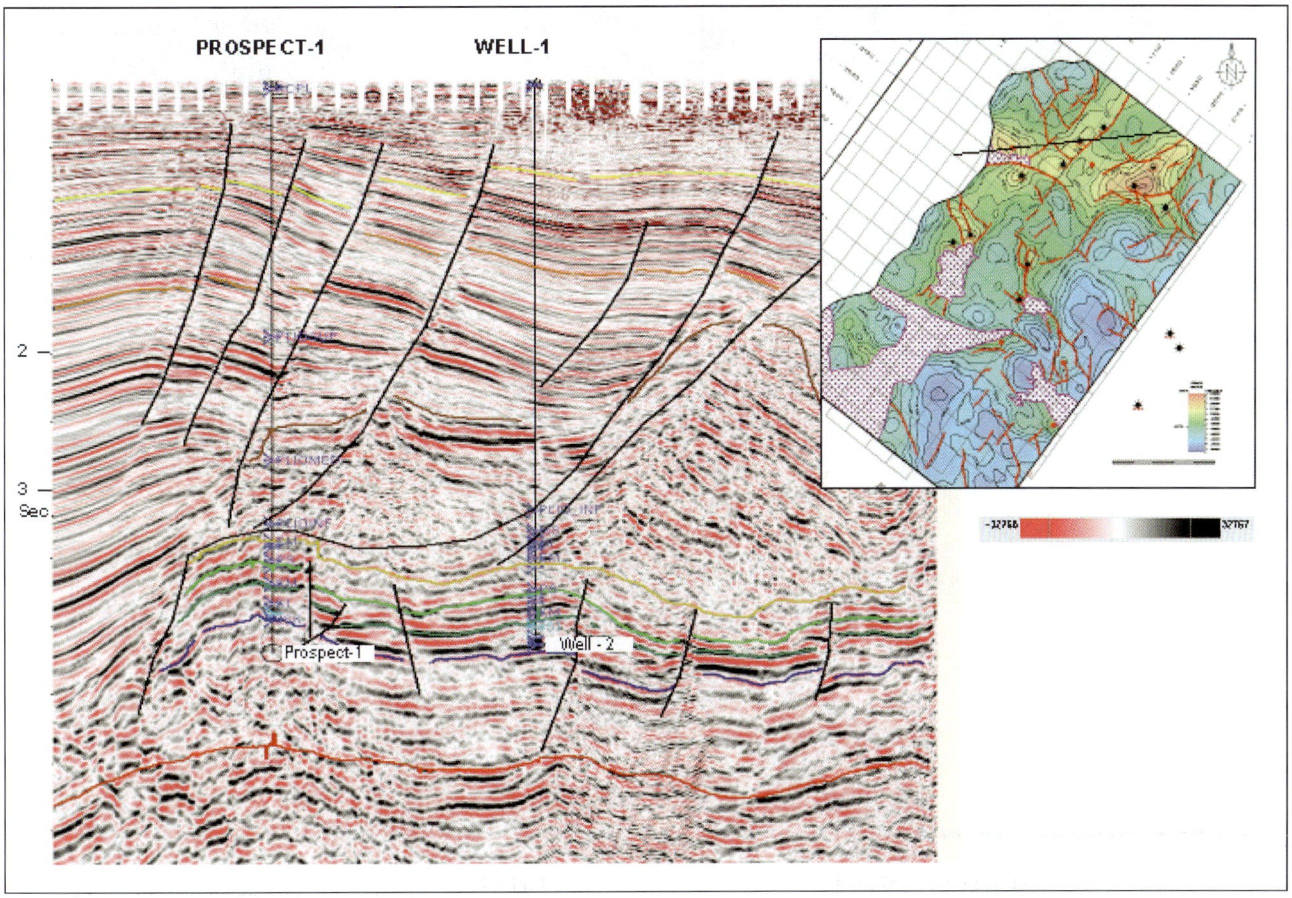

FIGURE 8. An E-W seismic section showing the image of Prospect 1.

well known roller structures of salt tectonics, however in this case they are associated with low velocity intervals. In the area the over-pressure zones are hosted in shale diapirs, which are well developed at the base of the footwalls of the normal faults. These are highly deformed overpressured shales, which have been confined within. These effects of low velocities zones are evident in sonic logs and are self-explanatory when constructing the Time–Depth curves. A 3D volume of the outlining of the Upper and Lower limits of these velocities anomalies are shown on Figure 6.

The depth conversion maps were created after taking into account the velocity problem, with a result of three independent structurally controlled prospects documented. One of those is shown in orthogonal cross section and the structural map in depth for the top Cretaceous (Figure 7).

Additionally, other prospects were identified for further documentation when evaluating the resulting spill-point map for the area. This map also allowed the visualization of relocations for appraisal and development wells. The spill-point map is shown on Figure 9, where tentative limits are outlined to highlight the extent of the proven areas for all these structures previously drilled and how much additional potential remains.

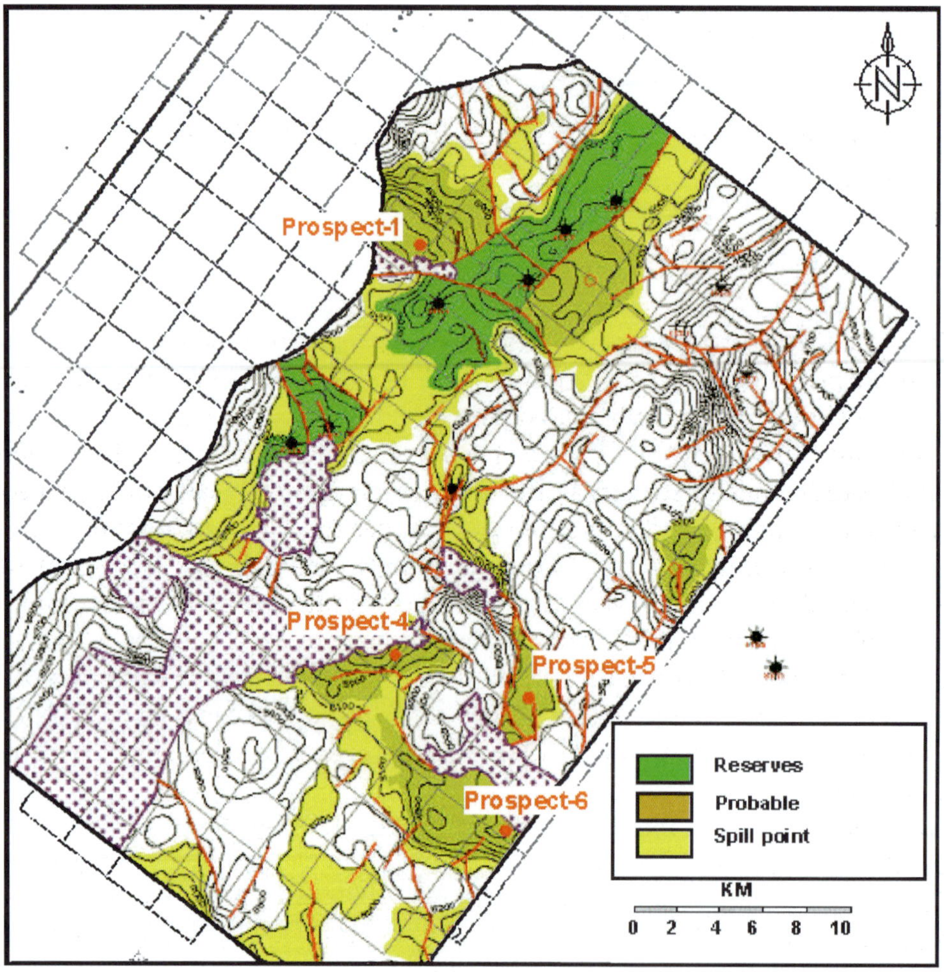
Prospect-1
Prospect-4
Prospect-5
Prospect-6
Reserves
Probable
Spill point
KM
0 2 4 6 8 10

23

Soto-Cuervo, A., V. Ortega-González, and G. Mora-Oropeza, 2009, Present and future of the Salina del Istmo Basin and its offshore extension into the Gulf of Mexico, *in* C. Bartolini and J. R. Román Ramos, eds., Petroleum systems in the southern Gulf of Mexico: AAPG Memoir 90, p. 465–470.

Present and Future of the Salina del Istmo Basin and its Offshore Extension into the Gulf of Mexico

A. Soto-Cuervo, V. Ortega-González, and G. Mora-Oropeza
PEMEX Exploration & Production

ABSTRACT

The Salina del Istmo Basin is a geographic region, which includes onshore and offshore prospects. The basin comprehends three main areas named: the Catemaco, Agua Dulce and Marbella folded belts; the Pescadores and omalcalco withdrawal salt basins; and the "Sal Somera" area. Three generic play types have been identified: Neogene's sandstones in structural and combined traps over salt bodies; Paleogene's sandstones in structural and combined traps under the salt; and Mesozoic structured carbonates in compressional traps under the salt. In the first play (onshore) 50 fields were discovered in the last 50 years and have a cumulative production with over than 2 billions bbl of equivalent oil. In the offshore portion 6 oil and gas fields were discovered, with 106 millions of barrels of original reserves. The second and third plays are in the early exploration stage, and are expected to provide better geological conditions. Higher volumes and better hydrocarbons quality characterize these plays, however with a higher exploration risk due to prospects depth.

More than 300 exploration opportunities are identify, in the Salina del Itsmo Basin, with a considerable volume of hydrocarbons and an average exploration risk of 1:5. Therefore, these traps offer an attractive perspective in the short and medium term. The best short term development possibilities are on the coast, near the giant land fields, where oil production facilities currently exist.

STUDY AREA

The Salina del Istmo Basin (SIB) is located in the south of Gulf of Mexico covering an area from Veracruz and Tabasco states to extending about 100, 000 km^2 offshore towards the Gulf of Mexico center (Figure 1).

BACKGROUND

The biggest three oil and gas fields, with 37 million bbl of oil accumulated production and 30 bcf of gas, were discovered in the 50s and 60s. However new exploration studies were carried out in the 90s. In 1998 the Tabscoob

Reprinted from Extended Abstracts of the AAPG International Conference, October 24–27, 2004, Cancun, Mexico.
DOI:10.1306/13191097M903340

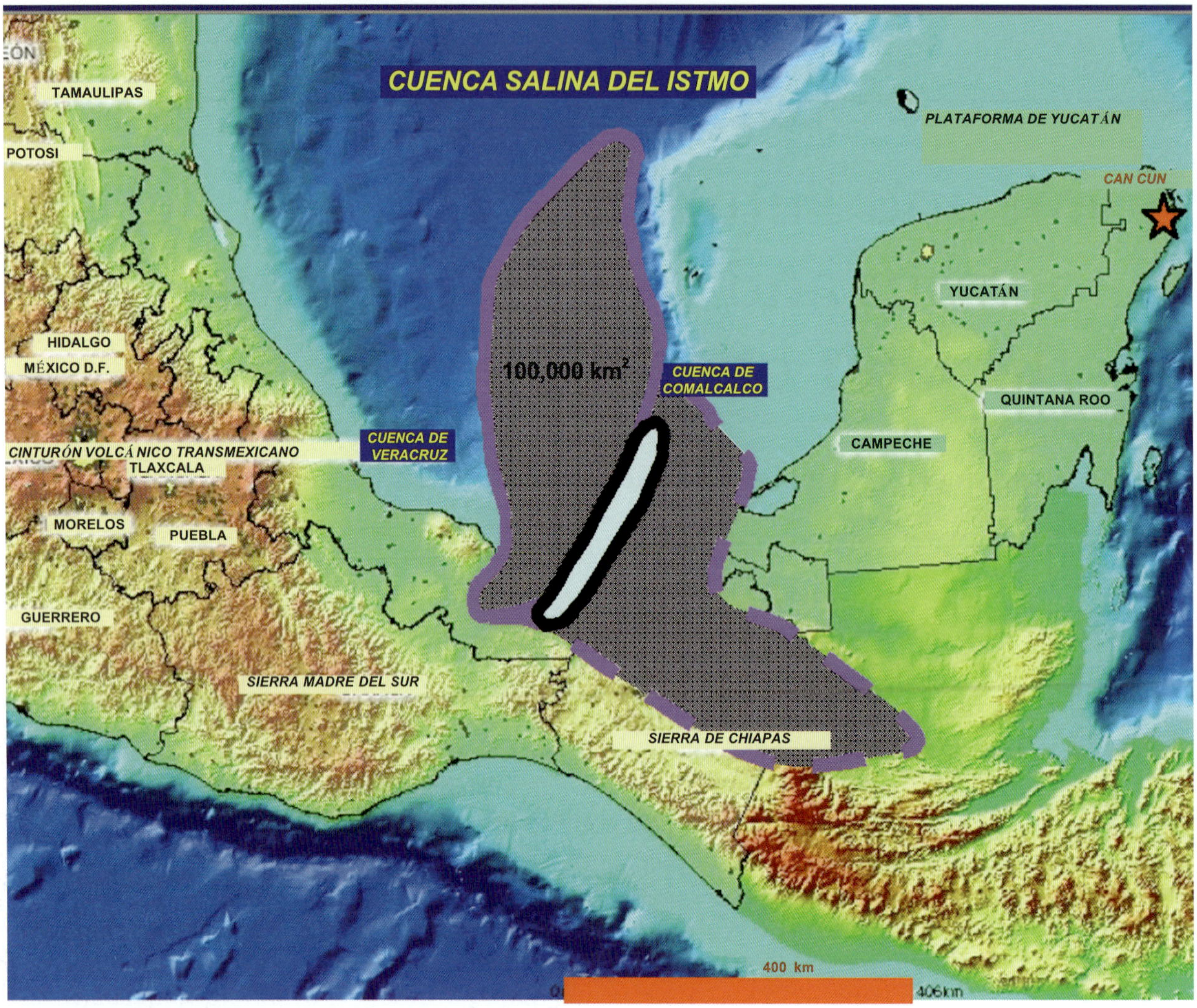

FIGURE 1. Location map.

oil and gas field was discovered. Beginning with 2003, 12 exploration wells were drilled and 22 reservoirs were discovered: 12 of oil, 5 of associated gas and 5 of dry gas, distributed in 8 fields, with a 62% of geological successful. About 116 million bbl of equivalent oil, ranging from 8 to 28 degree API, had been certified. Low API values are due to biodegradation of shallow reservoirs.

Currently, there are more than 300 exploration opportunities identified, where the main targets are concentrated in shallow tertiary, sub salt and Mesozoic plays. Seal and reservoir rock quality are the main risks in open marine facies. New discoveries results suggest that the best exploration strategy is to focus on sub salt and Mesozoic plays, with high possibilities to find bigger oil volume fields. Exploration in the SIB has been developed more widely onshore, and it has focused to the tertiary sands, while in offshore areas the exploration is just beginning. To date, the SIB has all the facilities for oil and gas production and export.

PRODUCTION HISTORY

The SIB has a long and prolific history of oil production and continues to provide exploration targets. The exploration initiated in 1902 with the drilling of the Amesquite-1 well. In 1906 the San Cristóbal-1 well discovered the first field, and the production began in 1910, with the discovery of the Ixhuatlán field. In the 60s the Cinco Presidentes field was discovered and the production reached 150 000 bbl/d.

Currently, Cinco Presidentes, Ogarrio, Cuichapa, Blasillo, Magallanes, El Plan and Bacal are the most important fields in the SIB. The production in the basin is 39,000 bbl/d of oil and 56 million cfpd of gas. The historical exploration success of 35% allowed the discovery of 48 tertiary fields, with an accumulated production of 2.044 billion bbl and an average recovery factor of 22% (Figure 2). The remnant reserves of these fields ascend to 788 million bbl of equivalent oil. From 1976 to 2002 reserves had not been incorporated in the basin.

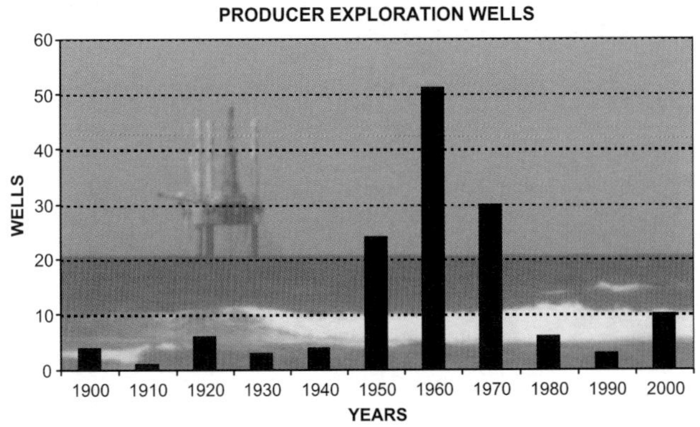

FIGURE 2. Exploration in the Salina del Istmo Basin.

The second stage of exploration in the SIB, initiates in 2003 with the discovery of 8 fields; 6 in the offshore and 2 in the onshore portion, with reserves of 116 million bbl of equivalent oil. Reached the 16% of the total reserves incorporated in Mexico during 2003. (Table 1).

GEOLOGY

The SIB shares the geological history of the rest of the Gulf of Mexico that initiates with the rifting in the Middle Jurassic, evolving from oceanic opening to a passive margin basin (Figure 3). During the first stage of the basin, thick packages of salt were deposited in an extensive area. Later, the drift of the Yucatan block towards the south allowed that great mass of salt was separated, remaining one part in the north of Gulf in USA and the second bordering the western side of the present Yucatan Platform. The mass of salt was buried by sediments from the Late Jurassic, and during the Mesozoic and the Tertiary; the salt has suffered episodes of mobility that deformed the sedimentary sequence in different ways.

The Catemaco, Agua Dulce and Marbella belts represent the effects of tectonic during the Middle Miocene to Late Miocene in the region. These folded belts are characterized by high depth decollement, and they are located over the autochthonous salt and involve also the Mesozoic's basins and platforms. Seismic and well information suggest the structures have been formed around 12.5 million years ago and were accompanied by the emplacement of the canopy of "Sal Somera" towards the north.

During the Pliocene the clastic platform margin progrades towards the north over an extensive body of salt developing withdrawal basins, associated to the counter regional faults, such as the Comalcalco and Pescadores basins.

SOURCE AND PETROLEUM CHARGE

Despite of 58 fields discovered, in Agua Dulce area, source rocks are unknown, and therefore their characteristics have been defined indirectly. Oil analyses indicate that the source rocks of this area are Upper Jurassic Tithonian (152–144 Ma) in age, constituted of shaly limestones and bituminous shales deposited in open marine

TABLE 1. Summary of reserves incorporated in the SIB during 2003.

POZO	*1P*		*2P*		*3P*		*PCE (mmb)*
	Aceite (mmb)	*Gas Natural (mmmpc)*	*Aceite (mmb)*	*Gas Natural (mmmpc)*	*Aceite (mmb)*	*Gas Natural (mmmpc)*	
Amoca-1 (1, 2)	26.8	10.2	31.8	12.1	69.3	26.3	75.6
Centli-1 (1, 2)						3.1	0.6
Namaca-1 (1, 2)					1.5	13.6	4.3
Xaxamani-1 (2)		8.3	5.6	21.9	13.2	28.1	18.9
Xicope-1 (1)		10.1		10.1		17.6	3.4
Yetic-1 (2)					1.7	6.7	3
Guaricho-1 (3)	2.9	2.1	7.6	5.4	7.6	5.4	8.8
Gubicha-1 (4)	0.7	1.2	0.7	1.2	0.7	1.2	1
TOTAL	30.4	31.9	45.7	50.7	94	102	115.6

708.8	Total Nacional			
380.6	Marinas y Sur	54%	(1) Play Orca	(3) Play Bacal
115.6	Cuenca Salina	16%	(2) Play 5 Presidentes	(4) Play Magallanes

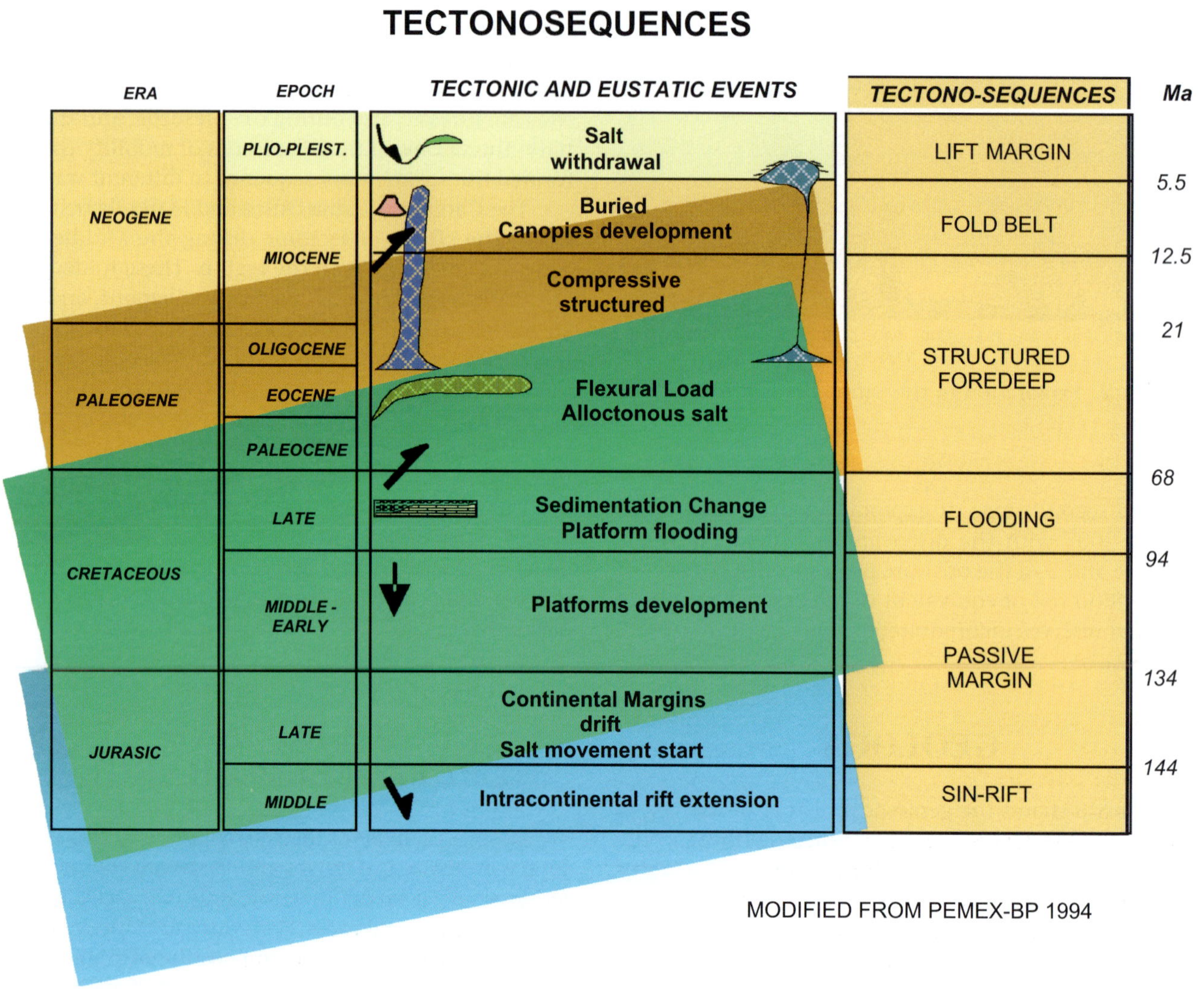

FIGURE 3. Tectonosequences Table.

conditions with regional distribution. Hydrocarbons geochemical analyses correlation and oceanic floor samples indicate that they have similarity with the oils from fields near the coast and the land fields neighbors. Both of them show a similar geological generation conditions. Based on the characteristics of heavy, light oils (27° to 43° API) and gas in this area, source rocks probably are currently in the generation window. In the onshore portion the SIB produces light oils and gas (Figure 4).

ECONOMIC PLAY ANALYSIS

Three generic types of plays are identified in the SIB Upper Jurassic–Tertiary Petroleum System: 1. Neogene's Sandstones in structural and combined traps over salt bodies. 2. Paleogene's Sandstones in structural and combined traps under the salt and 3.Mesozoic structured carbonates in compressional traps under the salt. In the first play (onshore) 50 fields were discovered in the last 50 years and have a cumulative production with over than 2 billions bbl of equivalent oil. In the offshore portion 6 oil and gas fields were discovered, with 106 millions of barrels of original reserves. The second and third plays are in the early exploration stage, and are expected to provide better geological conditions. Higher volumes and better hydrocarbons quality characterize these plays, however with a higher exploration risk due to prospects depth. Since economic point of view, these plays are important and they change the exploration strategy in the SIB. They offer attractive perspective to explore in the short and medium term.

The next definitions of plays were modified from Pemex-BP, 1994, and adapted here to be used in the SIB.

1. The Neogene's Sandstones in structural and combined traps over salt bodies.

Brujas Play (Pleistocene–Recent 1.7-0 Ma).–This play has not yet been proven in the SIB. It is formed by a

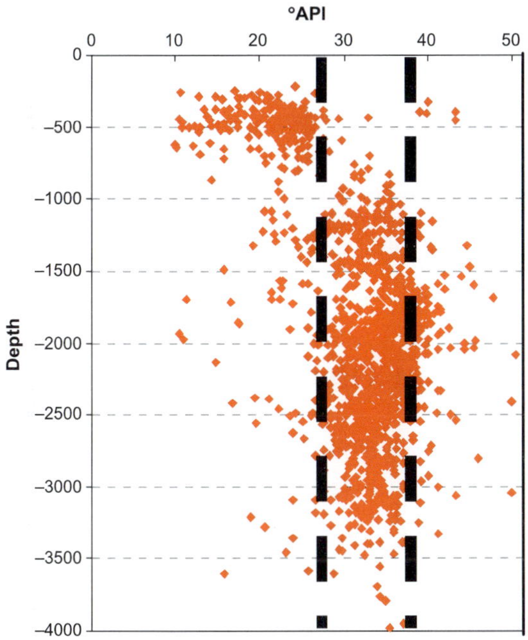

FIGURE 4. Light oils (27° to 38° API) located in the SIB.

series of turbidite sandstones deposited in the slope and quickly overlapped by progradant deposits of platform. The intraformational shale provides vertical and horizontal seal. Anticline structures turtle type that contains this play, are rollovers related to growth faults and anticline structures made by withdrawal salt.

Orca Play (Pliocene Late 3.0-1.7 Ma).–It is associated to turbidity sands against normal faults. The seal is formed by intra formation shales. The trap is made by pinchouts of sands, or can appear against growth faults caused by salt withdrawal, the deposit is contemporary with the age of the structures.

This play is divided in Upper Orca Play (1.7–2.4 Ma) and Lower Orca Play (2.4–3.0 Ma); both have fluvial deltaic, shelf and basinal facies. In the offshore portion, the Upper Orca Play (1.7–2.4 Ma) was proven in the Tecolli-1 well. The oil discovered has less that 8° API. The Lower Orca Play (2.4–3.0 Ma) produces in the wells Tabscoob-1 (44° API), Amoca-1 (18° API); and Nemiti-1 (8° API) and is also producer of dry gas in Xicope-1, Centli-1 and Namaca-1 wells.

Cinco Presidentes Play (Pliocene Early 5.5–3.0 Ma).–Associated to "turbidite sands" pinching out against the salt or overlapping it. The play is divided in Upper Cinco Presidentes Play (3.0–3.8 Ma) and Lower Cinco Presidentes Play (3.8–5.5 Ma). The progradation of this clastic wedge caused great mobilization of the Miocene's canopies and the development of extensive withdrawal salt basins with a Plio-Pleistocene thick fill. To the offshore there is a transition of fluvial deltaic and shelf facies including basinal turbidities.

The Cinco Presidentes Play has been proved in several wells. In this play were discovered 6 reservoirs with the follow distribution; In the Lower Cinco Presidentes Play (3,8–5,5 Ma) twice oil reservoirs in the Amoca-1 well (27–28° API). In the Santa Ana Field this play produces oil of 32° API, while in Marbella-1 well this oil is the 26° API. In the onshore portion the Guáricho-1 well produced oil of 39° API.

The Upper Cinco Presidentes Play (3.0–3.8 Ma) produces oil of 26° API and gas from Yetic-1 and Namaca-1 wells; 25° API from Xaxamani-1; 32° API from Santa Ana field, and dry gas in Campa-1 well.

Magallanes Play. (Middle-Late Miocene (12.5–5.5 m.y.). This play is constituted by pinchouts of turbidite sandstones against compressional structures. The Play is distributed in the different bands of the folded belts. The origin of the play is associated to "piggy back" basins. The turbidite sands were transported by the submarine canyons among compressional structures; these sands filled up the synclinals that generated conditions of pinchouts towards the top of the structures. Seal rocks are present in many intervals of the sequence.

The deformation of the SIB, begins during the Miocene (12.5 my), and stop in 5.5 m.y., which is also the age of this play. This play has just been proven onshore in the Gubicha-1 well, where 36° API oil was recovered.

Blasillo Play. Early Miocene-Middle Miocene (21–12,5 Ma). The Blasillo Play is composed of shallow marine and turbidite sands, later deformed into large anticlines. The sands were deposited in a shallow marine shelf and on the slope base as lobules of submarine fans. The seal is present as intraformational shales. In the 5.5 m.y. limit, when the compressive deformation finished, the area already structured was covered by the Sal Somera canopy. The Blasillo Play Sub salt varies from 21 to 12.5 Ma in age, and is composed by turbidite sands structured by folding under the salt canopy. This Play has been proved without success, in the marine area, by the wells Tascoob-1, Chichini-1, Xicope-1 and Xaxamani-1. Nevertheless, this play has good potential in the onshore portion.

2. Paleogene's Sandstones in structural and combined traps under the salt.

The Blasillo Play Subsalt had been drilled in the Salsomera-1 well and in land wells near of the coast such as Tonalá-902, Gurumal-2, Rabasa-1 and Palotada-1, and Yaxche-1 in offshore. This play is a target in a short and medium term in the SIB.

3. Mesozoic carbonates structured on compressional traps under the salt.

In the SIB, this play had not yet been proven. However, based on production from near fields, can be a potential target. Yagual-1 and Yaxche-1 oil and gas fields are example of oil accumulation in this type of play.

Therefore, this carbonates are strategic targets in a short and medium term.

The plays described ahead, are important plays to be explored, and they are the result of data analyses made by PEMEX-BP (1994) and PEMEX (2002).

Play Yum (Late Cretaceous, 94-66.5 Ma). This play is composed by open marine fractured carbonates, in great thrusted structures underneath the tertiary fields. The prospects are in open marine fractured rocks, involved in compressional structures. Late Cretaceous shaly carbonates and Paleogene shales form the seal rocks.

Níspero Play (Early Cretaceous-Middle, 134-94 Ma). This play is formed by Middle Cretaceous local deposits of breccias, product of the erosion of the carbonated platforms. Potential methods and seismic information indicate basement high areas in the SIB, where carbonated platforms were developed. Compressional structures are the main type of traps. Late Cretaceous open marine facies constitute the seal for this play.

Jujito Play (Late Jurassic (144–134 Ma). It is a conceptual play, represented by oolitic deposits, Late Jurassic in age, in thrusted structures. The definition of this play is based on the identification of high basements areas, where isolated platforms were developed. Tithonian's shales and Kimmeridgian open marine shaly facies form the seal for his play.

STRATEGY AND RESULTS

1. The exploratory wells were drilled in the crest of the salt cored anticline structures, were the sands are less developed and with low reservoir quality.
2. The best reservoir quality sands with better thickness are located in the flank of the structures such as was found in the Amoca-1 well.
3. The volumes of hydrocarbons found are not indicative of the volumes that we expect in the structure.
4. Directional and horizontal wells are the best technologies we have to use in the development of the reservoirs.

CONCLUSIONS

1. The SIB is a geologic province with high hydrocarbon traps potential.
2. Blasillo, Magallanes, Orca and Cinco Presidentes are plays already tested, while the Mesozoic, Blasillo Subsalt, and Brujas are hypothetical plays. From 2002 to the 2004, 25 reservoirs have been discovered: 15 of oil, 5 of humid gas and 5 of dry gas.
3. The Cinco Presidentes Play is the best play discovered in the SIB.
4. Shallow traps contained biodegrade hydrocarbons.
5. Historical records of the reservoirs in the SIB show us that the production is mainly light oil.

REFERENCES CITED

PEMEX, 2002, Proyecto de Sistemas Petroleros.

Pemex-BP, 1994, Estudio Cuencas Terciarias del Sureste y Área Marbella.

24

Fouad, K., L. F. Brown Jr., W. A. Ambrose, D. Dunlap, M. Aranda-Garcia, and U. Hernández-Romano, 2009, Genetic facies analysis using seismic geomorphology and seismic attributes in the continental shelf of eastern Mexico, *in* C. Bartolini and J. R. Román Ramos, eds., Petroleum systems in the southern Gulf of Mexico: AAPG Memoir 90, p. 471–476.

Genetic Facies Analysis Using Seismic Geomorphology and Seismic Attributes in the Continental Shelf of Eastern Mexico

Khaled Fouad, L. Frank Brown Jr., William A. Ambrose, and Dallas Dunlap

Bureau of Economic Geology, Jackson School of Geosciences, The University of Texas at Austin

Mario Aranda-Garcia and Ulises Hernández-Romano

PEMEX Exploración y Producción, Mexico

ABSTRACT

Methods for generating seismic facies maps have developed significantly in the last decade. These methods include facies mapping based on seismic geomorphology and sequence attribute extractions (Posamentier and Kolla, 2003). The results of these methods are used to define depositional systems, erosion, paleotopography, and inferred lithology estimation.

The seismic geomorphology technique is similar to back-stripping of a geologic sequence that represents a certain time-stratigraphic surface. We analyze depositional features from timeslices and windowed attribute extractions in a flattened volume within a stratigraphic sequence. Facies can be inferred from the resulting interpreted maps, which basically confine the variation in seismic reflections. Such variation, caused by geology change within seismic sequences and systems tracts, is expressed by change in reflection pattern, amplitude, and frequency (i.e., chaotic, hummocky, continuous). This technique produces quick and efficient results that capture the lateral changes in reflection pattern geometries.

In this article, we demonstrate how we deployed seismic geomorphology and attribute extraction to build out the facies in a third-order sequence. We show an example in the Laguna Madre area, a 3-D cube that extends over 2000 km^2 southeast of the Tuxpan Platform. This research is part of a study done by the Bureau of Economic Geology to investigate Neogene hydrocarbon plays in the Tuxpan area.

Reprinted from Extended Abstracts of the AAPG International Conference, October 24–27, 2004, Cancun, Mexico.

DOI:10.1306/13191098M903340

GEOLOGIC SETTING AND SEQUENCE STRATIGRAPHY

The Neogene sediments in the Laguna Madre area provide an excellent opportunity for studying lithofacies where they are well expressed seismically along the shelf margin of eastern Mexico. The lower and middle Miocene rocks comprise mostly lowstand prograding deltaic wedge, slope-fans, and basin-floor-fan deposits. In contrast, the upper Miocene and lower Pliocene rocks record more landward systems tracts. Landward sediments are characterized by cyclic on-shelf highstand and transgressive tracts to prograding wedges in the proximal parts.

Sequence and facies recognition was interpreted on the basis of the integration of well logs and cores, seismic data, and volume attributes (Risch et al., 1996). Robust sequence stratigraphic interpretation was performed to define the main third-order depositional cycles.

Multiple east-dipping, synthetic growth faults that sole into a single detachment are the principal structural features. The growth faults are commonly inferred to be the result of depositional loading triggering subsequent gravitational sliding of strata basinward.

Sequences in the Laguna Madre–Tuxpan (LM–T) area are bounded by unconformities and defined in the manner described by Van Wagoner et al. (1990). The timing of principal drivers of these unconformities is mainly ascribed to relative fall of sea level enhanced by local tectonic activity.

In this study we will show selected examples from shallow-water facies in the upper Miocene to lower Pliocene and deepwater facies in the middle Miocene deposits.

During the early Pliocene to the late Miocene an abrupt increase in sediment supply and accommodation rates occurred that resulted in remarkable sigmoidal clinoforms (Figure 1). The thick aggrading sediments and the prograding wedges suggest that the sedimentation rate slightly exceeded the accommodation space during this time. The upper Miocene section as defined in this report is enveloped by two surfaces, a regional sequence boundary (SB43) and a maximum flooding surface (FS40). This interval comprises several smaller depositional cycles defined by the stratigraphic marker beds FS42 and SB41, recognized from seismic data in the study area. The highamplitude packages above unconformities SB43 and SB41 are represented on well logs as sand-rich intervals. These two unconformities exhibit very little or no detectable relief on the shelf. For example, the SB43 unconformity is expressed in seismic data with virtually undetectable erosional relief in the area of Well-1, and the SB41 unconformity is expressed in seismic as a bright amplitude having 10 to 25 ms of relief, representing limited incision of valley-fill systems on the shelf.

The middle Miocene sediments are complex deepwater fan systems. The transgressive-regressive cycle of the middle Miocene is defined by two surfaces. The first is an erosional surface, SB55, underlying the highamplitude single- to multiple-cycle basin-floor-fan deposits. The second is a maximum flooding surface, FS50, representing termination of a transgressive episode that marks a rapid decrease in sediment supply.

SEQUENCES AND LITHOFACIES

Three cores from the interval between FS40 and SB43, were analyzed. Dominant composition is massively burrowed siltstone and very fine grained sandstone in slightly upward coarsening successions. Sandstones typically are poorly sorted owing to significant amounts of muddy matrix introduced by burrowing. Sandstones in the FS40 to SB43 interval may easily be interpreted as shelf bars because they are intensely burrowed and have a finegrained, slightly upward coarsening grain-size profile. However, a shelf-bar genesis for these sand bodies on the outer shelf is problematic, considering that the shelf is typically a muddy, sediment-starved system (Kulm et al., 1975). A hypothesized origin for these "shelf" sandstones is that they are reworked transgressive deposits from earlier highstand and lowstand progradational episodes. During highstand before the creation of the SB43 unconformity, sandy sediment was delivered to the outer shelf. These deltaic and shoreface deposits were then reworked and subsequently stranded on the shelf during transgression, after which they were thoroughly burrowed by organisms.

LOGS AND SEISMIC PATTERNS

The Pliocene and upper Miocene sections are penetrated in a number of wells in both the inner- and outer-shelf areas. This section exhibits cyclic on-shelf highstand and transgressive systems tracts, which are characterized by strong parallel seismic reflections dipping slightly seaward. Many incised valleys mark the top of the highstand systems tracts, particularly in the Pliocene section, typically represented by muddy transparent amplitude reflections. In some instances, the localized valley fills encompass single to multiple cycles of high amplitudes commonly interpreted to represent sandy sediments.

Wireline logs of the Pliocene and upper Miocene sections display thick coarsening and upward-shallowing highstand sandstone and siltstone intervals alternating with thin intervals of shaly transgressive upward-fining (backstepping) successions. The highstand sediments in

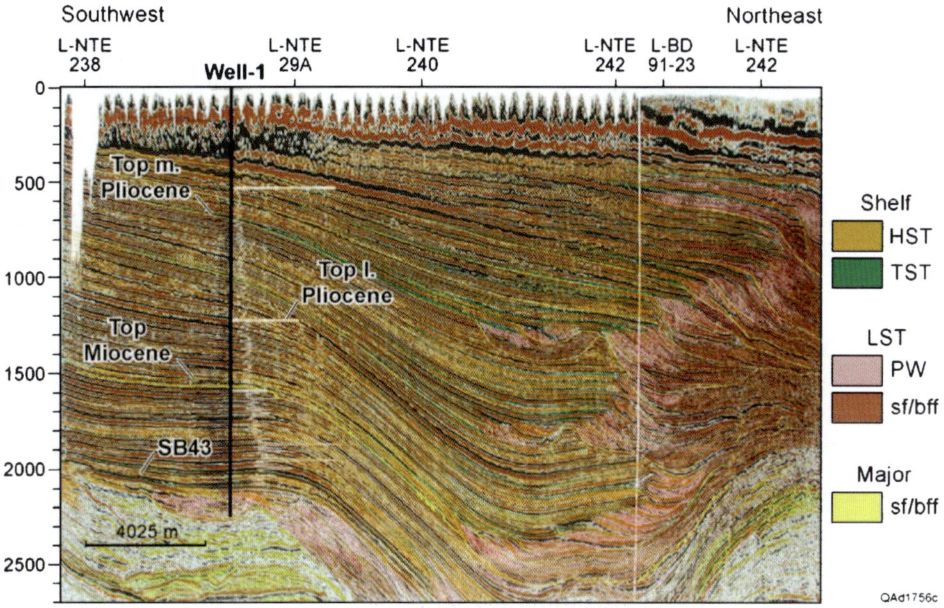

FIGURE 1. Arbitrary line from the Lankahuasa 3-D survey, showing sequence-stratigraphic framework of the upper Miocene FS40 to SB43 and Pliocene. Abbreviations are HST = highstand systems tract, TST = transgressive systems tract, PW = prograding wedge, sf = slope fan, and bff = basin-floor fan. Prograding wedges mark shelf breaks downdip of shelf systems.

this section are characterized on seismic by moderate-strength parallel reflections that range from continuous to discontinuous. Transgressive facies, however, are represented by thinner, more continuous parallel seismic reflections. Major flooding surface FS40 in Well-1 is a high-gamma ray, low-resistivity shaly zone near the top of the sandy succession overlying the SB43 unconformity. Seismic reflection patterns of the FS40 to SB43 show the presence of multiple transgressive-regressive depositional cycles on a ramplike shelf. Flooding surfaces in the FS40 to SB43 interval occur within thin (commonly <100 m), lowamplitude condensed sections that cap high-amplitude zones.

Basinward, the lower and upper Miocene strata are lowstand prograding wedges. These wedges are mostly muddy and are characterized by sigmoidal shapes with weak progradational seismic reflections and transparent amplitudes. Basin-floor-fan deposits are also spread at the toe of these clinoforms, where they are commonly characterized on seismic by bidirectional downlap reflections.

The middle Miocene deepwater facies are characterized by high-amplitude continuous and discontinuous reflections. These fan deposits can be classified as either confined (canyon, channels) or unconfined (sheet, lobes) geometrical bodies. Map patterns of the extracted amplitudes suggest that the fans are highimpedance sediments.

STRATAL SURFACE METHODS

Stratal surface interpretation techniques were used to determine facies distribution. Stratal surfaces are sheetlike rock units whose seismic signatures reflect the acoustic impedance boundary associated with surrounding strata. Maximum flooding surfaces (MFS), pervasive throughout the area, served as useful surfaces for flattening the seismic volume. These were subsequently used to define upper and lower limits of the stratigraphic cycle. The analysis carried out from horizon slicing and attribute extraction in the 3-D survey revealed a number of possible sand geometries in the Neogene section. Facies maps that were generated at specific intervals represent a specific systems tract. Also, facies-sensitive seismic attributes were extracted from the flattened volume within a narrow window (10 ms) within a specified vertical (geologic time) zone of interest (Taner et al., 1994). The extracted attributes were then draped on top of the structure horizons in 3-D perspective.

RESERVOIR OCCURRENCES AND SAND GEOMETRIES

Map patterns of extracted amplitudes selected at specific zones of the upper Miocene and lower Pliocene exhibit a wide variety of on-shelf depositional systems ranging from wave- and fluvial-dominated deltas to lagoonal and transgressive-shoal systems. Amplitude patterns selected within the middle Miocene section define the shape of a toe-slope-fan complex.

Growth faulting resulted in a narrow linear intraslope subbasin that was filled during the late Miocene and early Pliocene. A stratal slice map at SB43 displays two principal pre-SB43 sources of fluvial sediment and a wavedominated delta system with strong strike-oriented (northwest-southeast-trending) coastal sandstone facies of strandplain or barrier-bar systems (Figure 2). A river-dominated delta system to the south was inferred from

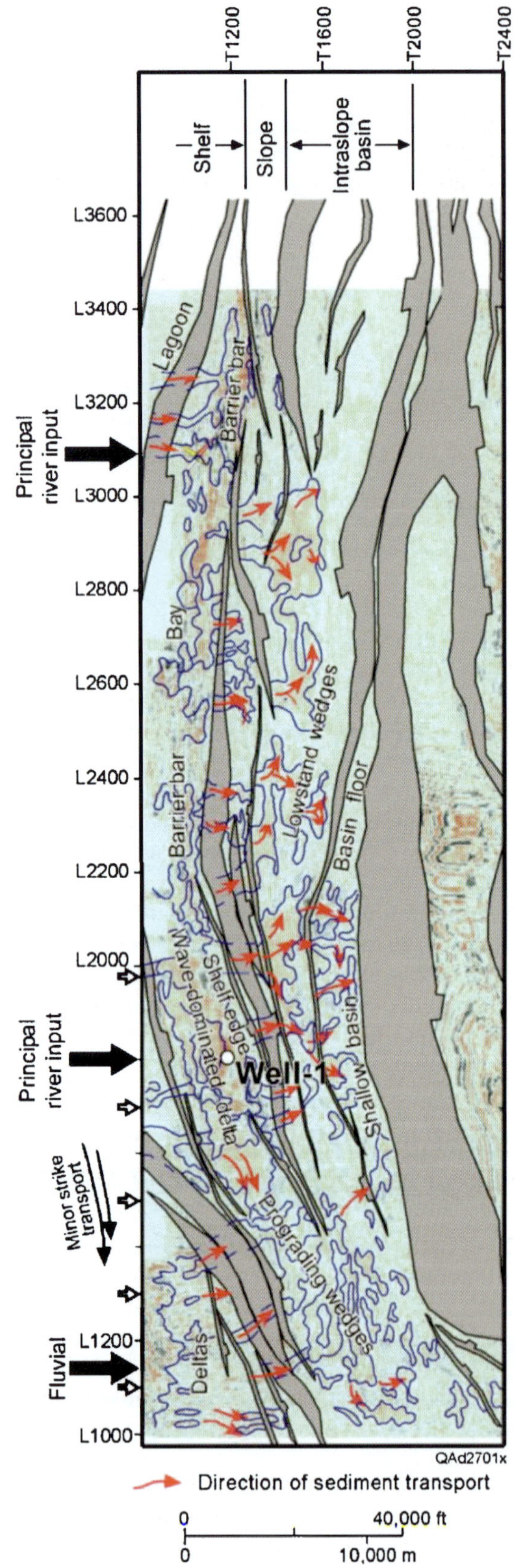

FIGURE 2. Amplitude-slice map on near top of SB43 (5.73 Ma) showing highstand shelf sediments and postunconformity prograding wedges and basin-floor-fan deposits.

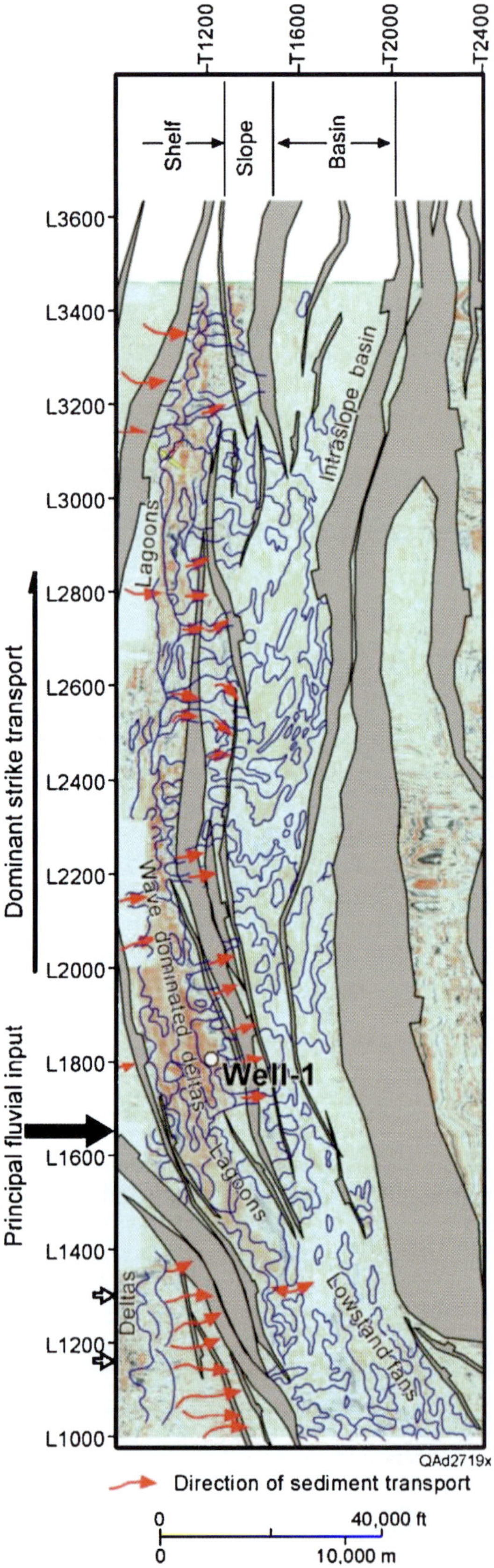

FIGURE 3. Amplitude-slice map on near top of SB41 (5.5 Ma) showing highstand shelf sediments and postunconformity prograding wedges and basin-floor-fan deposits.

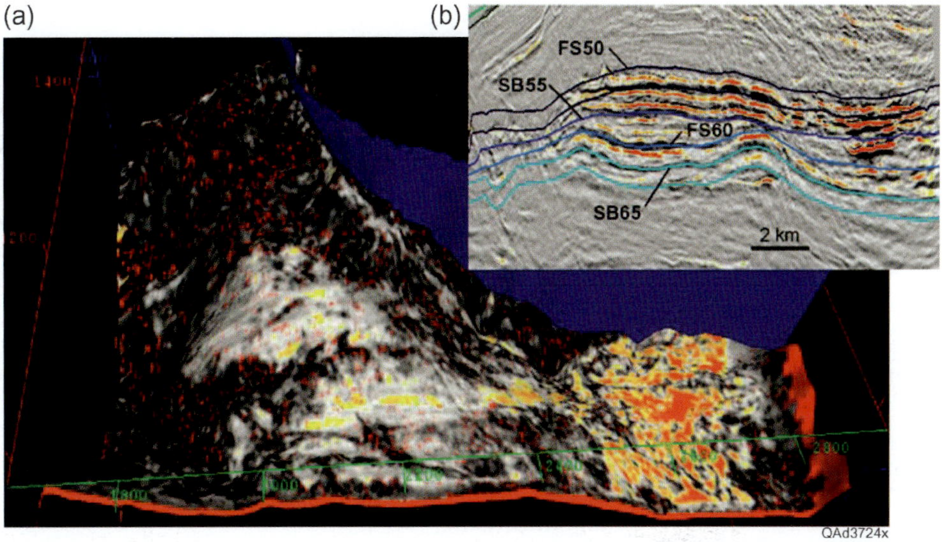

Figure 4. (a) Absolute-amplitude map extracted from flattened volume within the middle Miocene between SB65 and FS50, showing limits of fan at south part of Lankahuasa survey. (b) Seismic line 1100, showing fan deposits.

easttrending, narrow amplitude patterns. The continental margin north of Well-1 and Well-2 was intensively faulted, but to the south, where a river-dominated delta system existed, it was a structurally stable ramp area. Along the more intensively faulted shelf margin, lowstand delta systems and basin-floor fans were linked to wavedominated delta systems and strike-aligned barriers.

Transgression that followed SB43 significantly redistributed the sandstone facies of the wave-dominated delta system. A large lagoon formed in the area of Well-1 and Well-2 landward of transgressive barrier islands.

An amplitude-slice map at the SB41 unconformity is shown in Figure 3. This interval is restricted mostly in the south part of the study area. Seismic facies and wireline-log patterns resemble those described earlier for the FS42 to SB43 interval. Figure 3 displays the distribution of lowstand systems deposited on SB41, in addition to the underlying eroded highstand tract below the unconformity.

Another amplitude-slice map higher in the FS40 to SB43 interval is from the FS40 to SB41 interval. This interval displays parallel, mixed high- and low-amplitude reflections in shelf areas and progradational clinoform facies basinward of its associated growth fault. There is also a linear intraslope subbasin that was the site of a basinfloor- fan system and lowstand deltaic deposition. A stratal slice slightly higher in the section at the FS40 maximum flooding surface documents the starved basin and flooded, on-shelf coastal systems.

A similar procedure of attribute extraction from multihorizon flattening of seismic volume was applied to the middle Miocene interval between FS50 and SB55. The amplitude-slice map shows a diverse spectrum of bright amplitudes, continuous and discontinuous. Map patterns define the geometry of a complex fan that was cut from the north by the major growth fault (Figure 4). Because the area is in the vicinity of the Mexican Volcanic Belt, it is possible that these infill sediments of the fan, characterized by bright amplitudes, are related to volcanic materials. However, if these deepwater sediments have some porosity, these facies could represent potential reservoirs. Detailed analysis of these amplitude anomalies needs to be performed, in conjunction with our detailed facies maps, to test the probability of hydrocarbon occurrence.

CONCLUSIONS

We demonstrated an integrated approach using well logs, cores, and seismic geomorphology to illuminate spatial distribution of important facies in a selected 3-D area southeast of the Tuxpan Basin. Marker events at the inner shelf sites represented by the aggradational systems tract provide key horizons for mapping depositional packages. These surfaces are flat and continuous, and they overlie sigmoidal facies. Therefore, they are excellent candidates for stratigraphically flattening seismic data volumes. The facies map images depositional systems and reflects the preserved geometry of the systems. Relative impedance contrasts assist in discriminating sandstone and shale sediments.

ACKNOWLEDGMENTS

The authors wish to thank PEMEX for support of this research and for permission to publish these data. We also acknowledge support of this research by Landmark Graphics Corporation via the Landmark University Grant Program. We thank Eric Potter and Luis Sánchez-Barreda for reviewing the manuscript. Finally, we thank Susie Doenges for editing the manuscript. Publication was authorized by the Director, Bureau of Economic Geology, The University of Texas at Austin.

REFERENCES CITED

Kulm, L. D., R. C. Rousch, J. C. Harlett, R. H. Neudeck, D. M. Chambers, and E. J. Runge, 1975, Oregon continental shelf sedimentation: Interrelationships of facies distribution and sedimentary processes: Journal of Geology, v. 83, p. 145–176.

Posamentier, H. W., and V. Kolla, 2003, Seismic geomorphology and stratigraphy of depositional elements in deep-water settings: Journal of Sedimentary Research, v. 73, p. 367–388.

Risch, D. L., B. E. Donaldson, and C. K. Taylor, 1996, Deep-water facies analysis using 3-D seismic sequence stratigraphy and workstation techniques: An example from Plio-Pleistocene strata, Northern Gulf of Mexico, *in* P. Weimer and T. L. Davis, eds., Applications of 3-D Seismic Data to Exploration and Production: AAPG Studies in Geology No. 42 and Society of Exploration Geophysicists Geophysical Developments Series No. 5, AAPG/SEG, Tulsa, p. 45–56.

Taner, M. T., J. S. Schuelke, R. O'Doherty, and E. Baysal, 1994, Seismic attributes revisited (abs.): Society of Exploration Geophysicists 64th Annual International Meeting, Expanded Abstracts, p. 1104–1106.

Van Wagoner, J. C., R. M. Mitchum, K. M. Campion, and V. D. Rahmanian, 1990, Siliciclastic sequence stratigraphy in well logs, cores, and outcrops: AAPG Methods in Exploration Series, v. 7, p. 1–55.

25

Ambrose, W. A., K. Fouad, S. Sakurai, D. C. Jennette, L. F. Brown Jr., E. H. Guevara, D. B. Dunlap, T. F. Wawrzyniec, S. C. Talukdar, M. A. Garcia, U. H. Romano, H. R. Ruiz, R. C. Hernández, J. A. Vega, and E. M. Zamora, 2009, Neogene tectonic, stratigraphic, and play framework of the southern Laguna Madre–Tuxpan continental shelf, Gulf of Mexico, *in* C. Bartolini and J. R. Román Ramos, eds., Petroleum systems in the southern Gulf of Mexico: AAPG Memoir 90, p. 477–502.

Neogene Tectonic, Stratigraphic, and Play Framework of the Southern Laguna Madre–Tuxpan Continental Shelf, Gulf of Mexico

William A. Ambrose, Khaled Fouad, Shinichi Sakurai, David C. Jennette, L. F. Brown Jr., Edgar H. Guevara, and Dallas B. Dunlap

Bureau of Economic Geology, John A. and Katherine G. Jackson School of Geosciences, the University of Texas at Austin, Austin, Texas

Tim F. Wawrzyniec

Department of Earth and Planetary Sciences, 1 University of New Mexico, Albuquerque, New Mexico

Suhas C. Talukdar

Consultant The Woodlands, Texas

Mario Aranda Garcia, Ulises Hernández Romano, Hector Ruiz Ruiz, and Ramón Cárdenas Hernández

Petroleos Mexicanos Exploración y Producción, Edificio de Exploración Int. Campo Pemex (Anexo al Activo de Prod. Poza Rica) Poza Rica, Veracruz, Mexico

Juan Alvarado Vega and Eduardo Macías Zamora

Petroleos Mexicanos Exploración y Producción, Tampico, Tamaulipas, Mexico

ABSTRACT

Neogene shelf, slope, canyon, and slope-to-basin-floor transition plays in the southern Laguna Madre–Tuxpan (LM-T) continental shelf reflect a variety of structural and stratigraphic controls, including gravity sliding and extension, compression, salt evacuation, and lowstand canyon and fan systems. The Neogene in the LM-T area was deposited along narrow shelves associated with a tectonically active coast affected by significant uplift and erosion of carbonate and volcanic terrains. This study characterizes 4 structurally defined trends and 32 Neogene plays in a more than 50,000-km^2 (19,300-mi^2) area linking the Veracruz and Burgos basins.

Reprinted from AAPG Bulletin, v. 89, no. 6, (June 2005), pp. 725–751.
DOI:10.1306/13191099M903340

The Cañonero trend in the southern part of the LM-T area contains deep-seated basement faults caused by Laramide compression. Many of these faults are directly linked to the interpreted Mesozoic source rocks, providing potential pathways for vertically migrating hydrocarbons. In contrast, the Lankahuasa trend, north of the Cañonero trend, contains listric faults, which detach into a shallow horizon. This trend is associated with thick Pliocene shelf depocenters. The dominant plays in the Faja de Oro–Náyade trend in the central part of the LM-T area contain thick lower and middle Miocene successions of steeply dipping slope deposits, reflecting significant uplift and erosion of the carbonate Tuxpan platform. These slope plays consist of narrow channel-fill and levee sandstones encased in siltstones and mudstones. Plays in the north end of the LM-T area, in the southern part of the Burgos basin, contain intensely deformed strata linked to salt and shale diapirism. Outer-shelf, slope, and proximal basin-floor plays in the Lamprea trend are internally complex and contain muddy debris-flow and slump deposits.

Risk factors and the relative importance of play elements vary greatly among LM-T plays. Reservoir quality is a critical limiting play element in many plays, especially those in the Cañonero trend directly downdip from the trans-Mexican volcanic belt, as well as carbonate-rich slope plays adjacent to the Tuxpan platform. In contrast, trap and source are low-risk play elements in the LM-T area because of the abundance of large three-way and four-way closures and the widespread distribution of organic-rich Upper Jurassic Tithonian-age source rock. The potential for hydrocarbon migration in LM-T plays is a function of the distribution of deep-seated faults inferred to intersect the primary Mesozoic source. Their distribution is problematic for the Lankahuasa trend, where listric faults sole out into the Paleocene. Seal is poorly documented for LM-T plays, although the presence of overpressured zones and thick bathyal shales is favorable for seal development in middle and lower Miocene basin and slope plays.

OBJECTIVE, DATABASE, AND METHODS

The objective of this study was to provide a geologic framework for defining and delineating Miocene and Pliocene gas plays in the southern Laguna Madre–Tuxpan (LM-T) continental shelf and adjacent onshore areas (Figure 1). The study was conducted jointly by the Bureau of Economic Geology and Petroleos Mexicanos Exploración y Producción.

The Neogene trend in the LM-T continental shelf is a frontier area for oil and gas exploration in the Gulf Coast. The only offshore Neogene discoveries at the time of this writing occur in upper Miocene shelf and lower Pliocene slope sandstones in the Lankahuasa trend. Production in the eastern Mexico continental shelf has historically been from Mesozoic carbonate reservoirs in the area of the Faja de Oro and Sardina three-dimensional (3-D) surveys (Figure 1). Although approximately 20 wells are present in the area of these two 3-D surveys, the Neogene section in most of these wells has neither been completely logged nor fully evaluated. Moreover, most of the 32 Neogene hydrocarbon plays defined and mapped in this study have yet to be fully tested with exploratory wells. This integrated geologic study delineates the major tectonic and structural features, maps the principal paleogeographic elements and facies tracts, and characterizes Neogene plays in terms of their principal play elements, including reservoir quality and presence, trap, seal, source, migration, and timing.

The study area encompasses more than 50,000 km^2 (19,300 mi^2) and includes the offshore continental shelf between the Veracruz and Burgos basins to the 500-m (1600-ft) isobath, plus a small part of onshore Mexico (Figure 1). The eastern limit of the study area encompasses the Neogene paleoshelf, slope, and slope-to-basin-floor transition, with abyssal plain areas lying east of the mapped area. The database consists of six 3-D seismic surveys, collectively covering more than 8000 km^2 (3088 mi^2), as well as more than 14,000 linear km of offshore and onshore two-dimensional (2-D) lines (not shown). Twenty-seven offshore and eighteen onshore wells are included in the study. Four offshore wells are in the Lankahuasa 3-D survey area, ten are within the area of the Faja de Oro 3-D survey, ten are in the Sardina 3-D survey area, two are inside the Náyade 3-D survey area, and one is in the Lamprea 3-D survey area. Twelve

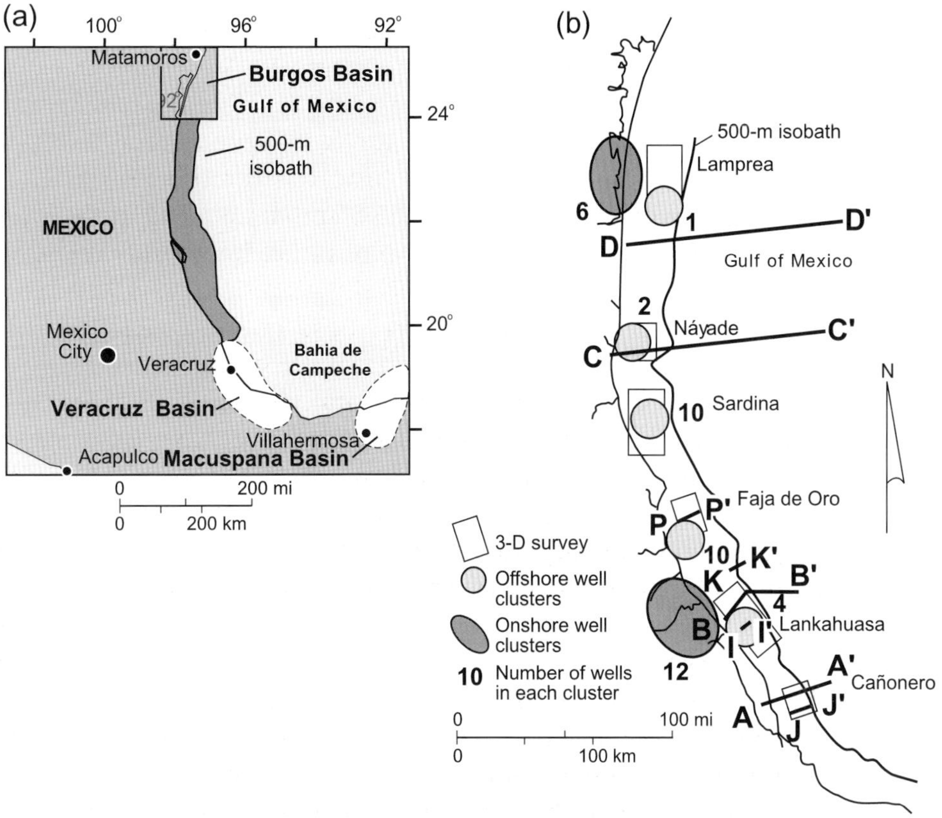

FIGURE 1. (a) Location of the southern LM-T continental shelf and adjacent Veracruz and Burgos basins. (b) Distribution of 3-D seismic surveys used in this study. Seismic sections AA′ to DD′ are shown in Figure 4. Seismic section II′ is shown in Figure 7a. Seismic section JJ′ is shown in Figure 7b. Seismic section KK′ is shown in Figure 8. Seismic section PP′ is shown in Figure 9b.

onshore wells that occur in clusters penetrate the middle Miocene, and deeper sections are west of the Lankahuasa 3-D survey. Six onshore wells are from onshore areas west of the Lamprea 3-D survey (Figure 1b). Of these onshore wells, approximately one-half contain usable logs and information from cuttings.

The study area was divided into four structural trends (Cañonero, Lankahuasa, Faja de Oro–Náyade, and Lamprea [Figure 2]). Each trend, described in more detail in the section on structural framework, was delineated on the basis of variations in timing and style of structural elements and served as a framework for evaluating trap, migration, and timing for each play.

The tops of the Mesozoic (structural basement) and regional detachments were mapped to document the structural elevation of inferred Mesozoic potential source rocks and to highlight areas of greater fault connectivity to this interpreted source interval. Mapped closures were evaluated from structure maps to define trap styles and to quantify trap density.

The Neogene section was divided into eight principal stratigraphic units bounded by regional unconformities, flooding, and fan-abandonment surfaces, which, from oldest to youngest, are (1) lower Miocene_1, (2) lower Miocene_2, (3) middle Miocene, (4) upper Miocene_1, (5) upper Miocene_2, (6) lower Pliocene, (7) middle Pliocene, and (8) upper Pliocene (Figure 3). Many of the regional stratal surfaces in the LM-T area were extrapolated from those recognized and correlated in the Bureau of Economic Geology–Petroleos Mexicanos regional study of the adjacent Veracruz basin (Jennette et al., 2003). An informal, two-part stratigraphic nomenclature scheme used for the Miocene section in this study (for example, FS40 and SB43) was adopted from Jennette et al. (2003). Biostratigraphic data from both onshore and offshore wells were used to constrain the age and paleobathymetry of flooding surfaces and fan-abandonment surfaces. The age of intervals defined in the study was determined in studies of foraminifera and nannoplankton assemblages conducted by paleontologists at the Petroleos Mexicanos and Instituto Mexicano del Petróleo.

Petrophysical models were constructed for wells with the best rock and log data. The latter typically consisted of gamma-ray, resistivity, density, sonic, and neutron curves. These models provided computations of water saturation (S_w), shale volume (V_{sh}), porosity (ϕ), and net pay (oil- or gas-bearing reservoir), parameters used in assessing reservoir quality and reservoir presence.

A geochemical study was conducted to define the source of known oil and gas occurrences, to identify oil- and gas-generating source rocks and recognize their relative importance, to define hydrocarbon-generating areas and timing of oil and gas generation and migration from different source rocks, and to infer which hydrocarbon types may be available for Neogene plays in different parts of the study area and evaluate the risk

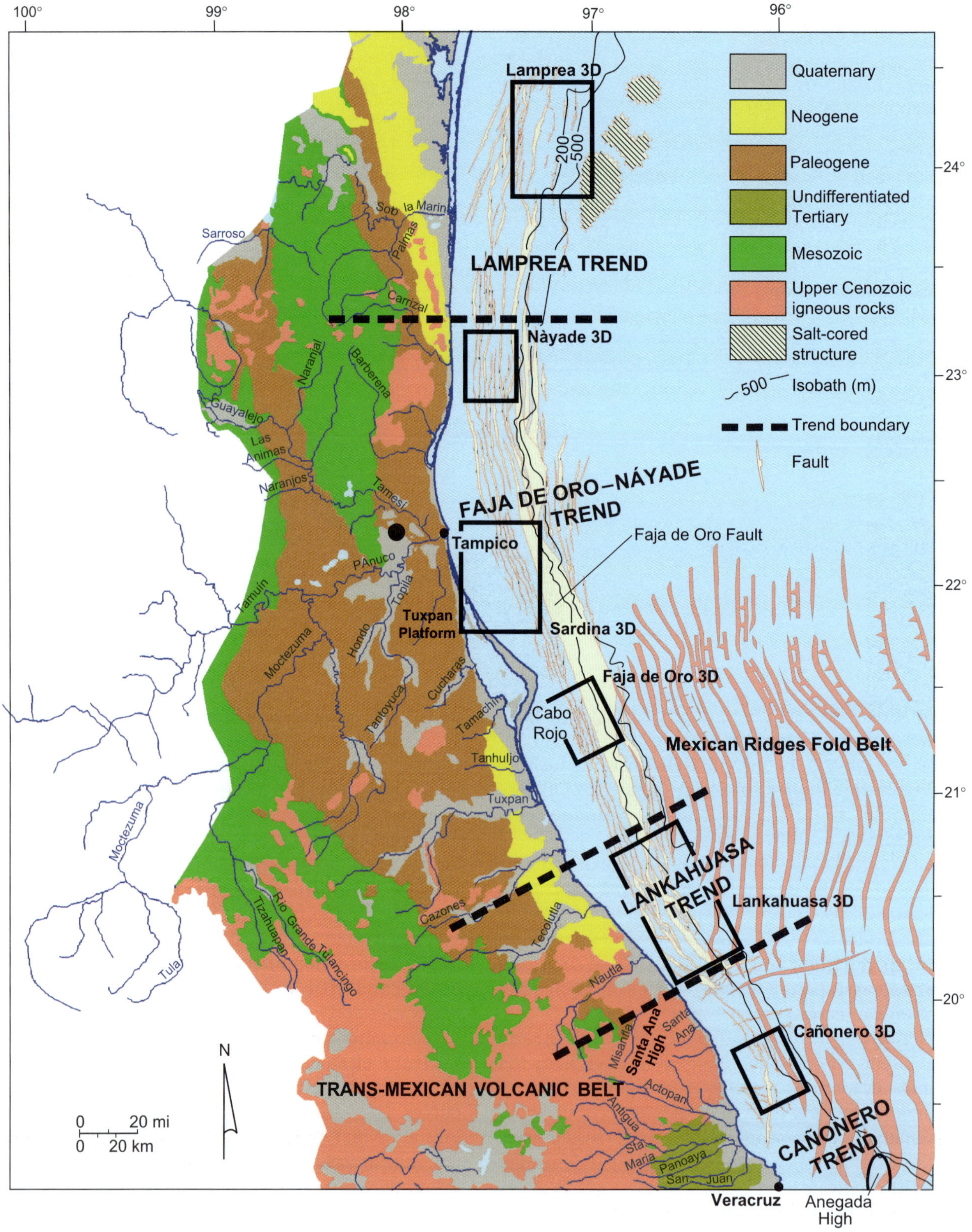

FIGURE 2. Physiographic provinces, structural trends, and 3-D seismic surveys in the southern LM-T continental shelf and adjacent areas.

factors for hydrocarbon charge. The geochemistry database included total organic carbon (TOC) and pyrolysis data on source rocks, vitrinite reflectance (R_o) data, gas-composition data, carbon-isotope data from sea-bottom sediment cores, bottomhole-temperature data, and heat-flow data from eight offshore locations. Data on gases,

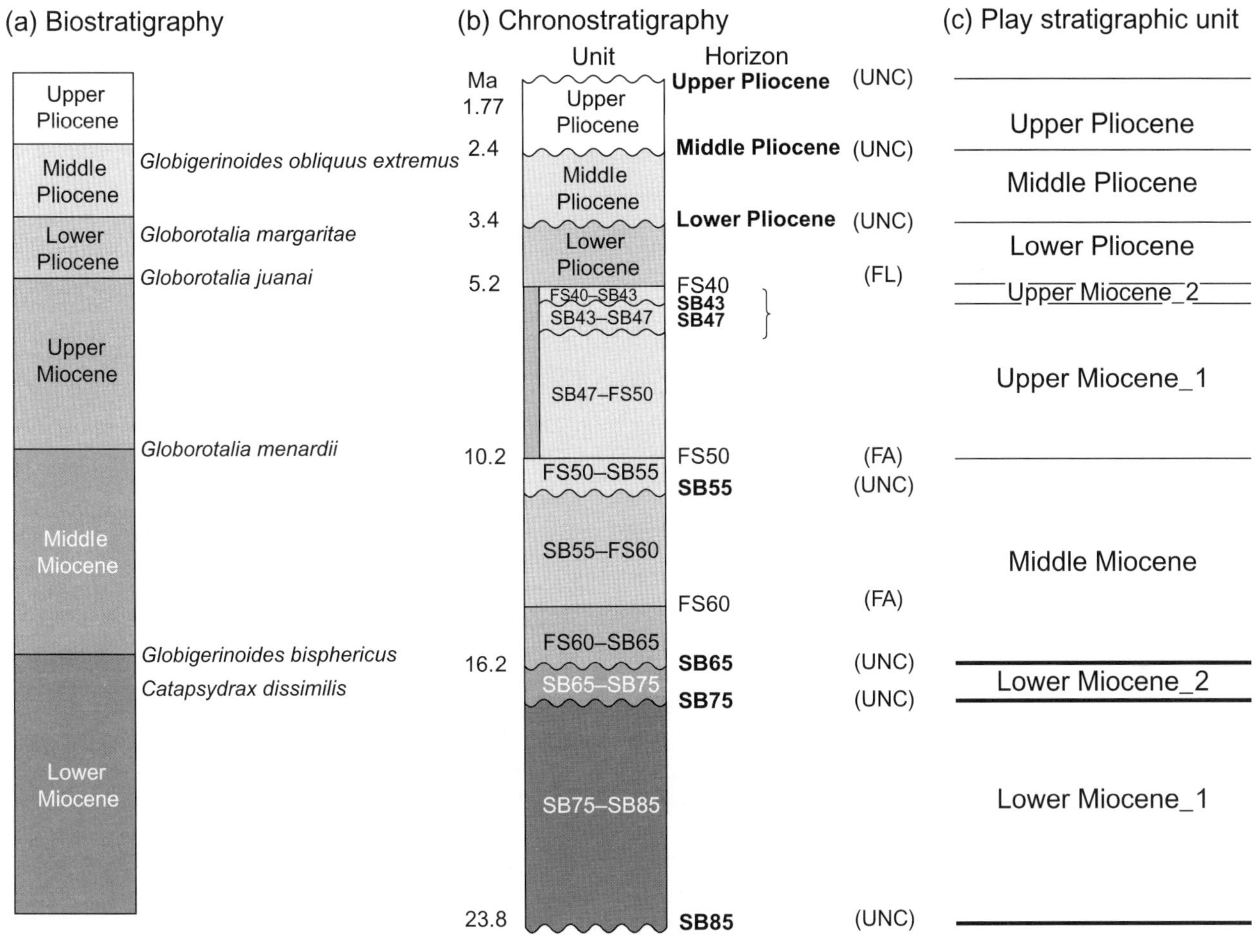

FIGURE 3. Laguna Madre–Tuxpan Neogene stratigraphic columns. (a) Major biostratigraphic horizons. (b) Chronostratigraphy and informal stratigraphic units correlated and mapped in this study, with stratigraphic occurrence of major sequence boundaries (unconformities) and flooding surfaces. (c) Play stratigraphic units defined and mapped in this study. Abbreviations: UNC = unconformity; FL and FS = flooding surface; FA = fan-abandonment surface; SB = sequence boundary. Biostratigraphic nomenclature and ages are based on Petroleos Mexicanos data.

oils, and source rocks were also used for interpretation, and other data were available from Mello et al. (1996), Román-Ramos et al. (1996), Guzmán-Vega and Mello (1999), Goldhammer and Johnson (2001), González and Holguín (2001), Guzmán-Vega et al. (2001), Magoon et al. (2001), Román-Ramos and Holguín-Quiñones (2001), and Jordan and Wilson (2003).

The 32 Neogene plays were defined using a two-stage classification scheme, consisting of (1) geologic age, with eight divisions in the Miocene and Pliocene, and (2) principal paleogeographic type, including shelf, slope, slope-to-basin-floor transition, and channelized facies composed of submarine-canyon deposits. Paleogeographic types were inferred from geometry and character of seismic reflectors in regional dip lines, supplemented with core and log data, as well as from isochron and amplitude-extraction maps. Isochron maps were particularly effective in delineating the regional distribution of basin-floor fans, where a strong relation was inferred to exist between fan depocenters and interval thickness.

Plays were characterized using a variety of play elements, including reservoir presence and quality, trap, seal, source, migration, and timing. Criteria were established for assessing play elements for each play. For example, reservoir quality was inferred directly from core and petrographic data and indirectly from log computations. For those play elements for which data were sparse or absent in the study area, analogs from adjacent and nearby basins in the Gulf Coast were used. Criteria for all play elements evaluated in this study are described in the section on plays. Play ranking is not addressed in this paper, as results are confidential. However, general comments are provided for the relative adequacy of each major play element for LM-T plays (see Table 1 and the section on play elements).

Table 1. Play-element adequacy for the Laguna Madre–Tuxpan plays in the Cañonero, Lankahuasa, Faja de Oro–Náyade trends.*

Trend	*Play*	*Reservoir Presence and Quality*	*Trap*	*Seal*	*Source*	*Timing and Migration*
Cañonero	shelf	untested	untested	untested	*favorable*	*favorable*
	slope	untested	untested	*favorable*	*favorable*	*favorable*
	slope-to-basin-floor transition	*unfavorable*	favorable	*favorable*	*favorable*	*favorable*
	canyon	unfavorable	favorable	*favorable*	*favorable*	*favorable*
Lankahuasa	shelf	proven	proven	proven	proven	proven
	slope	proven	proven	proven	proven	proven
	slope-to-basin-floor transition	*favorable*	favorable	*favorable*	*favorable*	untested
	canyon	*unfavorable*	untested	*favorable*	*favorable*	*favorable*
Faja de Oro–Náyade	shelf	untested	*unfavorable*	untested	*favorable*	untested
	slope	proven	proven	proven	proven	proven
	slope-to-basin-floor transition	*favorable*	favorable	*favorable*	*favorable*	untested
	canyon	untested	untested	untested	*favorable*	untested
Lamprea	shelf	favorable	*favorable*	untested	*favorable*	*favorable*
	slope	unfavorable	*unfavorable*	unfavorable	*favorable*	untested
	slope-to-basin-floor transition	untested	favorable	untested	*favorable*	untested
	canyon	untested	untested	untested	*favorable*	untested

**Proven = plays with established production or favorable production tests. Favorable = production not yet established, but existing data indicate adequacy. Unfavorable = production not yet established, and existing data suggest inadequacy. Untested = production not yet established, and insufficient data exist to evaluate adequacy. Italics = production not established, and although insufficient play-element data are present, adequacy is inferred from facies and play analogs from the neighboring Veracruz and Burgos basins. Data types and criteria indicating adequacy are discussed in the text section Play Elements.*

STRUCTURAL FRAMEWORK

Regional Tectonic Setting

The LM-T continental shelf was initially established as a series of sediment-starved, Jurassic carbonate platforms that formed on a relatively thick, extended, and modified continental crust (Feng et al., 1994). Pindell (1994) and Pindell et al. (2000) associated this modified crust with back-arc, south-southeast–directed extension and the opening of the Areperos Ocean, which is thought to be a back-arc extension related to the northwest-directed compression along the west coast of North America. Alternatively, others contended that the extended crust of the early LM-T shelf represents an attenuated crust directly associated with the south-southeast–directed opening of the Gulf of Mexico (Buffler, 1983; Buffler and Sawyer, 1985; Marton and Buffler, 1994; Wawrzyniec et al., 2003).

Following the opening of the Gulf of Mexico, the early, western shelf of the Gulf was a relatively stable, clastic-sediment-starved, carbonate platform (Buffler, 1983; Watkins and Buffler, 1996). The first major influx of siliciclastic sediment on the shelf occurred during the Paleocene. The onset of the Sevier-Laramide orogeny resulted in differential landward uplift and subsidence of the Cretaceous Tuxpan carbonate platform (Winker, 1984; Galloway et al., 2000). By the late Paleocene to early Eocene, the shelf was dominated largely by erosion and sediment bypass. However, by the end of the Eocene, fluvial-dominated deltas prograded eastward, a consequence of increased sediment supply and uplift in the Sevier-Laramide hinterland. By the Oligocene, the Tuxpan platform was buried by siliciclastic sediments. Throughout the Neogene and Quaternary, the shelf was dominated by siliciclastic deposits, recording a complex interplay between tectonic uplift, subsidence, and eustasy.

Structures recognized in the LM-T shelf include multiple east-dipping, synthetic faults that sole into a single detachment (Feng et al., 1994; Trudgill et al., 1999) and associated hanging-wall monoclines. These structures are the direct result of depositional loading and subsequent basinward gravitational sliding of strata (Buffler, 1983). Extensional grabens along the shelf were mechanically linked to compression in the Mexican

Ridges fold belt (Figure 2) and gravitational gliding along a weak detachment horizon in the shaly Oligocene section (Buffler, 1983; Feng et al., 1994; Trudgill et al., 1999; Wawrzyniec et al., 2003).

Local Structural Trends and Physiographic Setting

The LM-T area is divided from south to north into four structural trends: the Cañonero, Lankahuasa, Faja de Oro–Náyade, and Lamprea (Figure 2). Each trend is delineated on the basis of variations in timing and style of structural elements. Other important physiographic elements and structural features include the trans-Mexican volcanic belt, Santa Ana high, Mexican Ridges fold belt, and Tuxpan platform (Figure 2).

The trans-Mexican volcanic belt is related to subduction along the Mid-American trench and is about 200 km (124 mi) from the boundary between the Cocos and North American plates (Pindell, 1994; Ferrari et al., 1999). Volcanism in the Cañonero and southern part of the Lankahuasa trends is recorded seismically in the form of bright-amplitude, high-impedance middle and upper Miocene intrusive bodies and tuffaceous debris-flow deposits. The Santa Ana high is a basement uplift located in the trans-Mexican volcanic belt. It is also associated with a basinward deflection of basement structures and serves as the southern boundary of the growth-fault-dominated Lankahuasa trend (Figure 2). The Mexican Ridges fold belt, extending from the north end of the Veracruz basin to the deep offshore, more than 150 km (93 mi) east of Cabo Rojo, is a series of subparallel, north- and northwest-trending folds related to downdip compression and shortening at the distal end of major detachment faults (Pew, 1982) (Figure 2). The Tuxpan platform, located south of Tampico, is a Cretaceous carbonate platform that forms the western boundary of the Faja de Oro fault system. This fault system commonly offsets lower Miocene strata by hundreds of meters.

Cañonero Trend

The Cañonero trend, located downdip (eastward) of the trans-Mexican volcanic belt, is the northern continuation of the Veracruz basin. It contains deep-seated basement faults associated with compression and strike-slip motion. Strike-slip trends along the Anegada high continue along strike until they terminate along the east end of the Santa Ana high. Growth in Neogene strata along major east-dipping faults is relatively minor in lower and middle Miocene strata and more significant in upper Miocene and Pliocene strata in the Cañonero trend (Figure 4a). Anticlinal structures east of these main east-dipping faults in the Cañonero trend commonly contain four-way closures associated with middle Miocene to Pliocene inversion. Middle Miocene folds in the Cañonero trend may have had a bathymetric expression because they coincide with the basinward limit of many lower to middle Miocene lowstand basin-floor-fan complexes. The folds appear to have been initiated at the same time, with the more eastward folds continuing to actively deform to the present day. This deformation is evident in the bathymetric expression of normal faults that appear to accommodate crestal collapse of these roughly coast-parallel folds.

Lankahuasa Trend

The Lankahuasa trend is dominated by gravity-sliding tectonics and extensional grabens, where most of the fault displacement is accommodated in the Pliocene strata (Figure 4b). The major faults in the Lankahuasa trend are a series of shallow-detachment listric faults associated with shelf depocenters of upper Miocene and Pliocene age. The Lankahuasa trend contains numerous synthetic faults associated with primary or multiple antithetic faults, all of which appear to be synkinematic throughout the Neogene. The result was the formation of multiple wine-glass-shaped extensional grabens that grew younger and more developed basinward. Where these grabens are positioned over basement highs, strata in them form rollover structures. These grabens are associated with numerous, small-scale, three-way closures and areally limited, fault-bounded, low-impedance amplitude anomalies.

Faja de Oro–Náyade Trend

The Faja de Oro–Náyade trend, north of the Lankahuasa trend, occurs where several large listric faults take a basinward step and merge into a primary synthetic listric fault (Faja de Oro fault) that extends along the coast for nearly 200 km (124 mi). Updip of this fault is a system of synthetic, listric faults whose offset appears to have been arrested at the time that the main fault formed (Figures 2, 4c). The net result is a series of synthetic normal faults that merge into a single detachment, with the most basinward fault having a pronounced rollover, hanging-wall fold, and net horizontal offset of 12–22 km (7.5–13.6 mi). The primary detachment appears to extend as a continuous subhorizontal plane that can be subdivided into two segments. The shallow segment overlies the buried carbonate shelf and occurs at a depth of 2–3 s (two-way traveltime [TWT]), directly below the base of the Miocene section. This segment terminates at the east edge of the platform, where it steps down into the deeper part of the basin. The deep part of the detachment is thought to be located at the top of regional geopressure, which is interpreted to lie within upper Cretaceous or lower Tertiary shales (Watkins and Buffler, 1996).

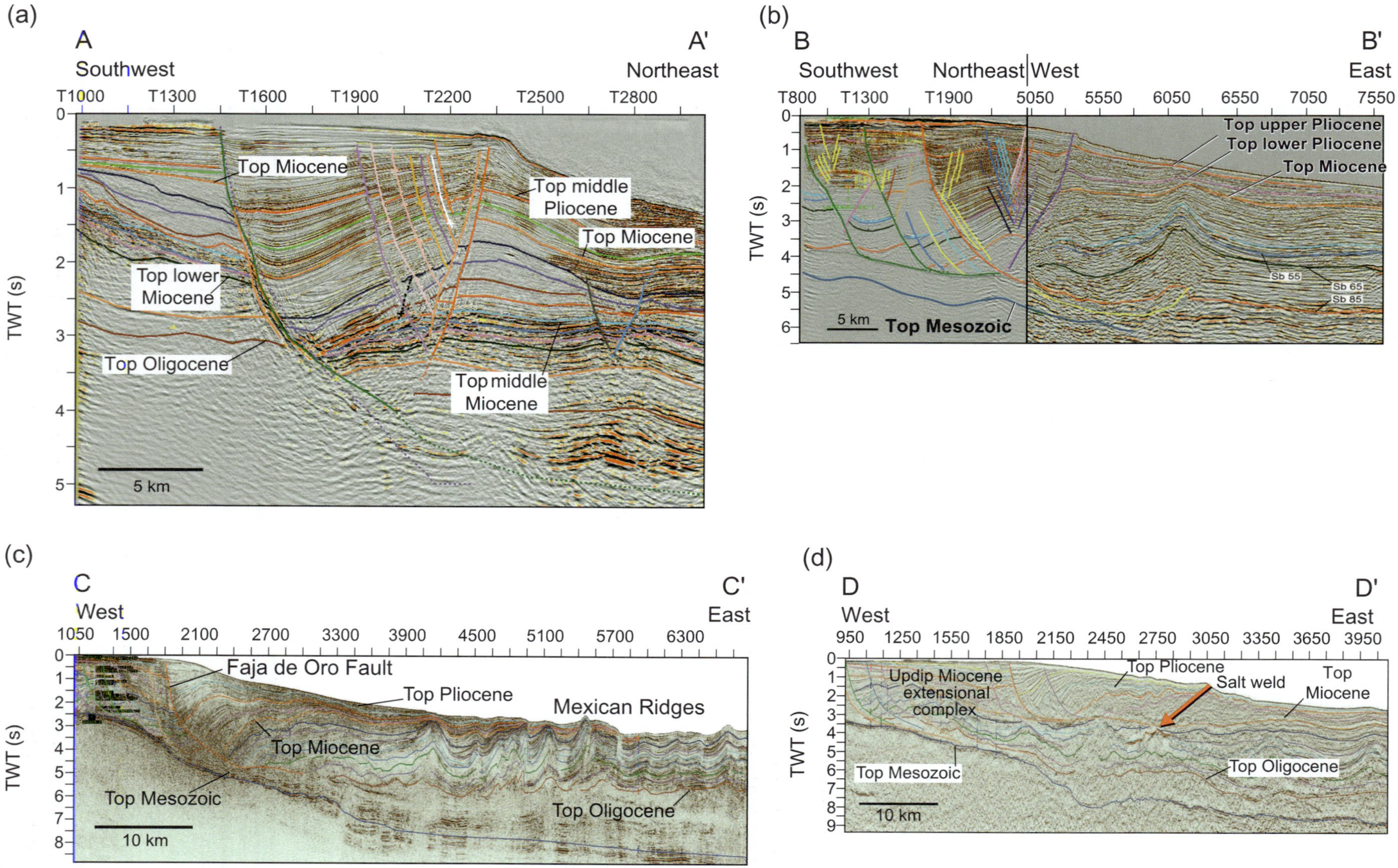

FIGURE 4. Seismic cross sections from the LM-T area. (a) Three-dimensional line AA′ from the south part of the Cañonero trend showing deeply penetrating, east-dipping master fault, associated with complex antithetic fault system. Vertical scale is TWT (ms). (b) Composite 3-D and 2-D line BB′ from the central part of the Lankahuasa trend showing highly extended Neogene section. (c) Two-dimensional line CC′ from the northern part of the Faja de Oro–Náyade trend, showing compressional Mexican Ridges fold belt. Vertical scale is TWT (ms). (d) Two-dimensional line DD′ from the Lamprea trend, illustrating an updip, upper Miocene extensional complex, a thick Pliocene section associated with growth faults, and salt welds. Vertical scale is TWT (seconds). Lines of section are shown in Figure 1.

Lamprea Trend

The southern boundary of the Lamprea trend is defined where greatly deformed strata are associated with salt and shale diapirism and salt withdrawal. The Lamprea trend is transitional northward to the Burgos basin (Figure 4d). This area contains Pliocene minibasins and slope-derived slump and debris-flow deposits, commonly developed above upper Miocene age salt welds. At least two distinct detachment levels are associated with structural styles consistent with halokinetic processes (Figure 4d). An updip Miocene extensional complex lies above the Cretaceous carbonate shelf and consists of synthetic, listric faults that sole into a detachment in the upper Paleocene section. This detachment appears to continue into the deep basin. However, a bright reflection at about 3.5 s TWT, located in the middle of the line, is interpreted to be a salt weld or a collapsed salt diapir. Furthermore, the position of this feature below a fault-bend-fold complex that modifies a lower Pliocene basin-floor-fan depocenter indicates that the salt extruded from this weld has since been displaced by the collapse of upper Pliocene inner slope depocenters. The second detachment became active in the lower Pliocene and soles into the base of the Pliocene, indicating that extension and gravitational collapse began as soon as the lower Pliocene strata began to blanket the extruded salt.

STRATIGRAPHIC FRAMEWORK

Unconformities

Sequences in the LM-T area are bounded by unconformities and defined in the manner described by Van Wagoner et al. (1990). Although some of these unconformities occur during periods of major eustatic sea level fall (Haq et al., 1987; Vail, 1987), in the LM-T area, they are thought to have been greatly influenced by local tectonic activity along the Sierra Madre Oriental orogenic belt. For example, the Veracruz basin, structurally continuous with the southern part of the LM-T area (Cañonero trend), was tectonically active throughout much of the Neogene, and major unconformities in this basin were produced through intermittent uplift along the Sierra Madre Oriental mountain belt (Jennette et al., 2002, 2003). Uplift in the western margin of the Veracruz basin resulted in major incision along a narrow shelf and delivery of large volumes of sediment to the basin floor. These deposits commonly directly overlie major unconformities and typically consist of submarine-canyon, channelized-fan, and sandy to conglomeratic basin-floor-fan facies.

Unconformities in LM-T area are inferred in seismic sections where there are abrupt, erosional contacts between units of different geometry and amplitude character. In the western (updip) part of the LM-T area, these unconformities are interpreted to be bypass surfaces associated with deep incision and canyons having hundreds of meters of erosional relief (Ambrose et al., 2003a) (Figure 5a). These canyons are interpreted to be 3–5 km (1.8–3.1 mi) wide, merging downdip (eastward) into a single, large canyon projected to the east edge of the Cañonero survey (Figures 5b, 6a). The canyon, varying in width from 5 to 8 km (3.1 to 5 mi), occurs as a single lenticular feature in strike section, associated with 400–550 ms (480–660 m; 1574–2165 ft) of erosion. The canyon-fill facies are differentiated into proximal- and distal-fill types (Figure 5b, c). The proximal canyon fill is defined by a relatively narrow (<5-km; <3.1-mi), asymmetrical strike profile, with a thick (>500-ms [>600-m; >1968-ft]) succession of nearly flat-lying, moderately bright-amplitude reflectors that lap onto the third-order basal SB43 unconformity (Figure 5b). In contrast, the distal canyon fill has a greater width-to-depth ratio and is more symmetrical in strike profile (Figure 5c). Downdip of the canyon system, the SB43 unconformity is nearly concordant with underlying reflectors and underlies a thick, bright-amplitude succession of basin-floor-fan deposits (Figure 5d).

Flooding and Fan-Abandonment Surfaces

Neogene flooding surfaces and fan-abandonment surfaces in the LM-T area commonly consist of dim amplitudes that form the upper bounding surface of sandy, bright-amplitude intervals (Figure 7). Flooding surfaces in the upper Miocene_2 (FS40 to SB43) interval in the Lankahuasa 3-D survey occur within thin (commonly <100-m [<330-ft]), dim-amplitude sections (Figure 7a). Two bright-amplitude successions in the upper Miocene_2 interval directly overlie sequence boundaries SB41 and SB43, respectively. These unconformities exhibit minor relief on the shelf. For example, the SB43 unconformity has virtually no detectable erosional relief in the western part of the Lankahuasa 3-D survey, and the SB41 unconformity is expressed in seismic lines as a bright-amplitude reflector having 10–25 ms (15–35 m; 49–114 ft) of relief, representing limited incision of valley-fill systems on the shelf (Figure 7a).

Fan-abandonment surfaces are also key stratigraphic horizons in the LM-T area, particularly those associated with thick, bright-amplitude lower and middle Miocene successions. For example, the fan-abandonment surface at the top of the middle Miocene (FS50) occurs at the base of a thick (>500-ms TWT [>600-m; >1968-ft]) section of dim-amplitude, muddy slope deposits (Figure 7b) and represents a regional chronostratigraphic surface, as well as a regionally extensive, potential seal above the middle Miocene section.

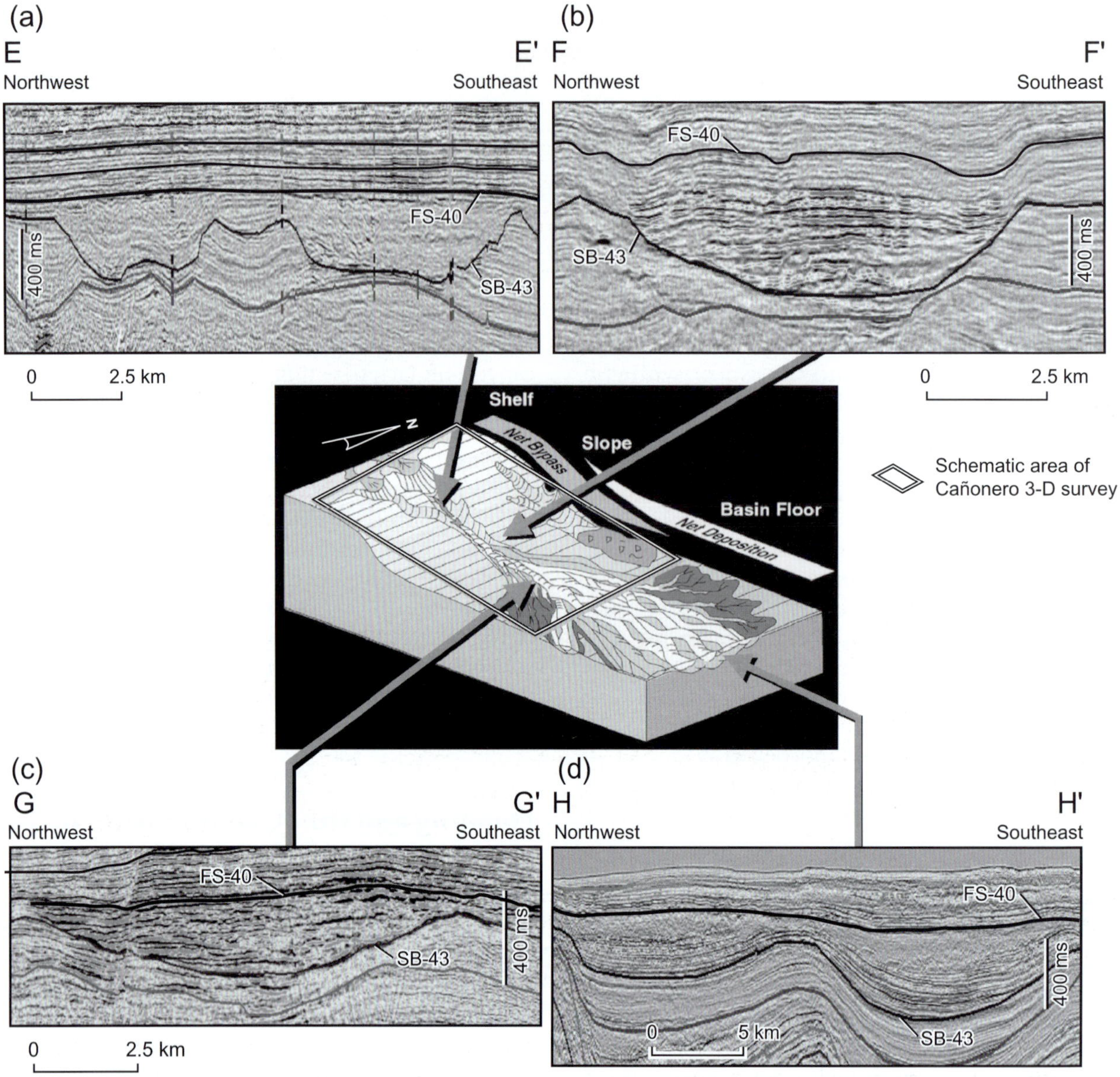

FIGURE 5. Schematic lowstand model depicting incised-slope, proximal-submarine-canyon, distal-submarine-canyon, and basin-floor-fan depositional systems, with cross-sectional insets providing examples from the upper Miocene_2 in the Cañonero trend. (a) EE′, dim-amplitude, shale-filled incised slope. (b) FF′, proximal-canyon fill. (c) GG′, distal-canyon fill. (d) HH′, basin-floor fan. Locations of sections (a–c) are shown in Figure 6a. Section (d) is approximately 10 km (6 mi) east of the Cañonero 3-D survey, and its location is shown in Figure 6b. Model modified from Beaubouef et al. (1999).

Paleogeography

Seismic-based paleogeographic maps, supplemented by log and core data, provide the foundation for defining and mapping LM-T plays. These maps are based on regional systematic changes in the geometry and amplitude character of reflectors observed in seismic sections (Figure 8).

Shelf seismic facies in this study are defined as sets of parallel reflectors updip (westward) of clinoforms. Although the overall geometry of reflectors in the shelf seismic facies is parallel, some individual reflectors are wavy and subparallel and are interpreted to represent small-scale valley fills notched into the shelf.

Slope seismic facies consist of clinoforms. The internal architecture of the slope facies is typically complex, with downlaps indicating prograding wedges and multiple erosional surfaces marking the bases of lowstand systems tracts.

Slope-to-basin-floor transition seismic facies are subdivided into two types, proximal and distal. The proximal facies are composed of wavy, subparallel sets of reflectors with moderate to bright amplitudes. These

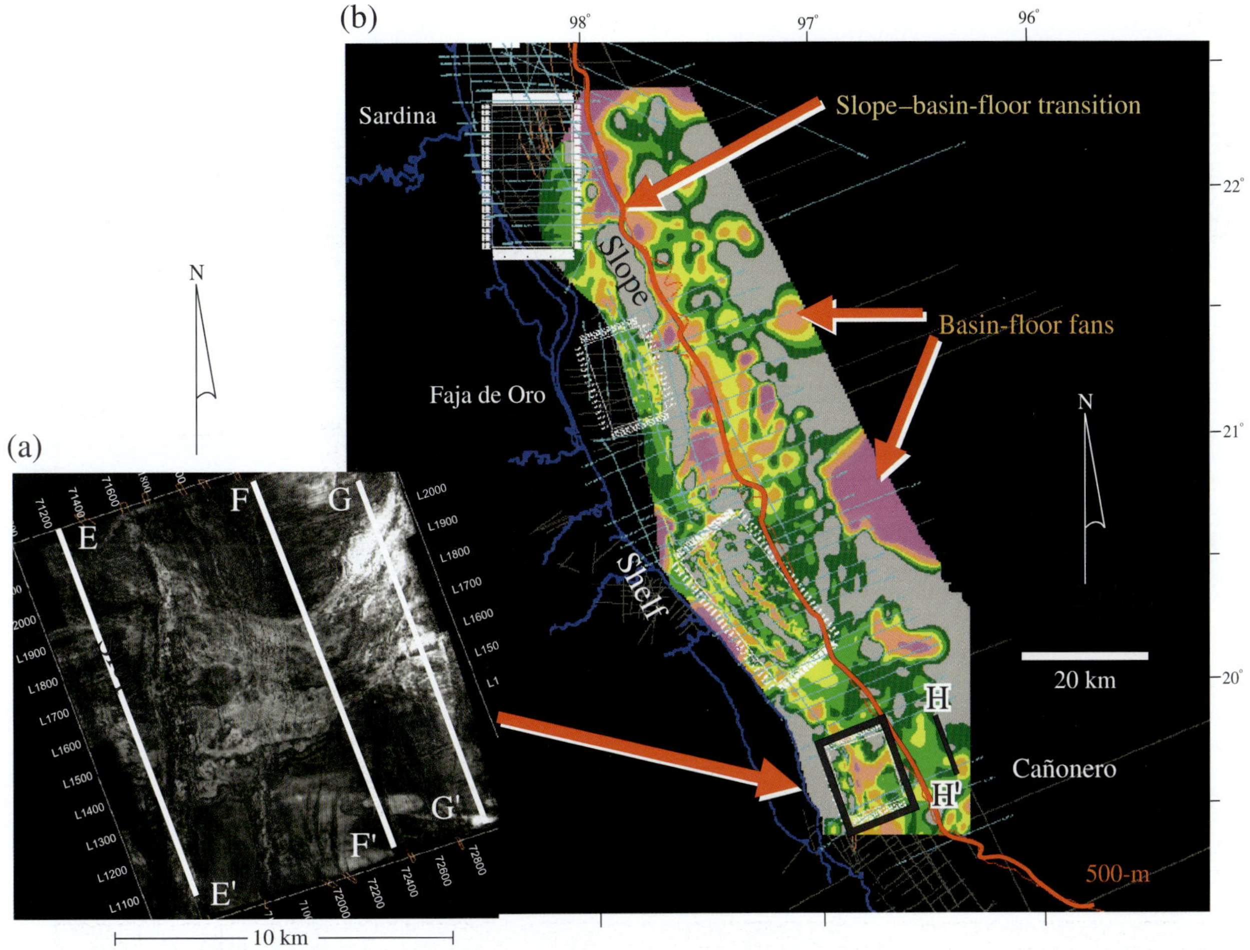

FIGURE 6. Upper Miocene_2 canyon systems. (a) Reflection-coefficient attribute map in the Cañonero 3-D survey, illustrating updip canyons merging eastward into a single, wider canyon. (b) Upper Miocene_2 isochron map, displaying regional context of the Cañonero canyon system. Large-scale, lobate, thick isochron features in the deep offshore (>500-m [>1600-ft] isobath) are interpreted to be basin-floor fans. Sections EE′ to HH′ are shown in Figure 5.

amplitudes in the Cañonero trend may represent thick accumulations of sandstone and/or conglomeratic sandstone, which are similar in seismic response to that of sandy and conglomeratic fan deposits in the adjacent Veracruz basin (Jennette et al., 2003). However, interpretation of recent data from similar deposits in the southern part of the Lankahuasa 3-D survey indicates the presence of volcanogenic conglomerates. Distal slope-to-basin-floor transition seismic facies are composed of parallel, dim-amplitude reflectors. Although the distal basin-floor seismic facies is composed of relatively dim amplitudes with respect to the proximal basin-floor facies, it may also be locally sandy, typically containing one or more thin zones of bright-amplitude reflectors.

The canyon seismic facies are interpreted to be major incision features. They are inferred from truncation of older reflectors by units with a lenticular cross-sectional geometry in strike view (Figure 5a–c) and are typically narrow and dip elongate in geometry (Figure 6a).

The Neogene in the LM-T area is an overall progradational, offlapping succession (Figures 9–11). The lower Miocene_1 (SB 75 to SB85) shelf edge is inferred to be west (onshore) of the modern Gulf of Mexico shoreline, and the slope-to-basin-floor transition is interpreted to be along the modern shoreline between the Lamprea and Sardina 3-D surveys (Figure 9a). The lower Miocene_1 slope in the Faja de Oro 3-D survey is a thick succession of steeply dipping wedges and clinoforms (Figure 9b). This section typically contains thin sandstones deposited in narrow, slightly sinuous, and dip-elongate channel and levee complexes (Figure 9c). These channel complexes are 1–2 km (0.6–1.2 mi) wide and are composed of amalgamated and multistoried channel complexes (commonly 10–30 m [33–100 ft] thick, interpreted from well data) and are associated with thin overbank sandstones.

The upper Miocene_2 (FS40 to SB43) shelf edge advanced eastward to the modern offshore (Figure 10a). In the central part of the Lamprea 3-D survey, the upper Miocene_2 stratigraphic unit is composed of high-impedance, slope mudstones (Figure 10b). Upper Miocene_2 cyclic shelf deposits (Figure 7a) are widely distributed in the western part of the Lankahuasa 3-D survey (Figure 10c).

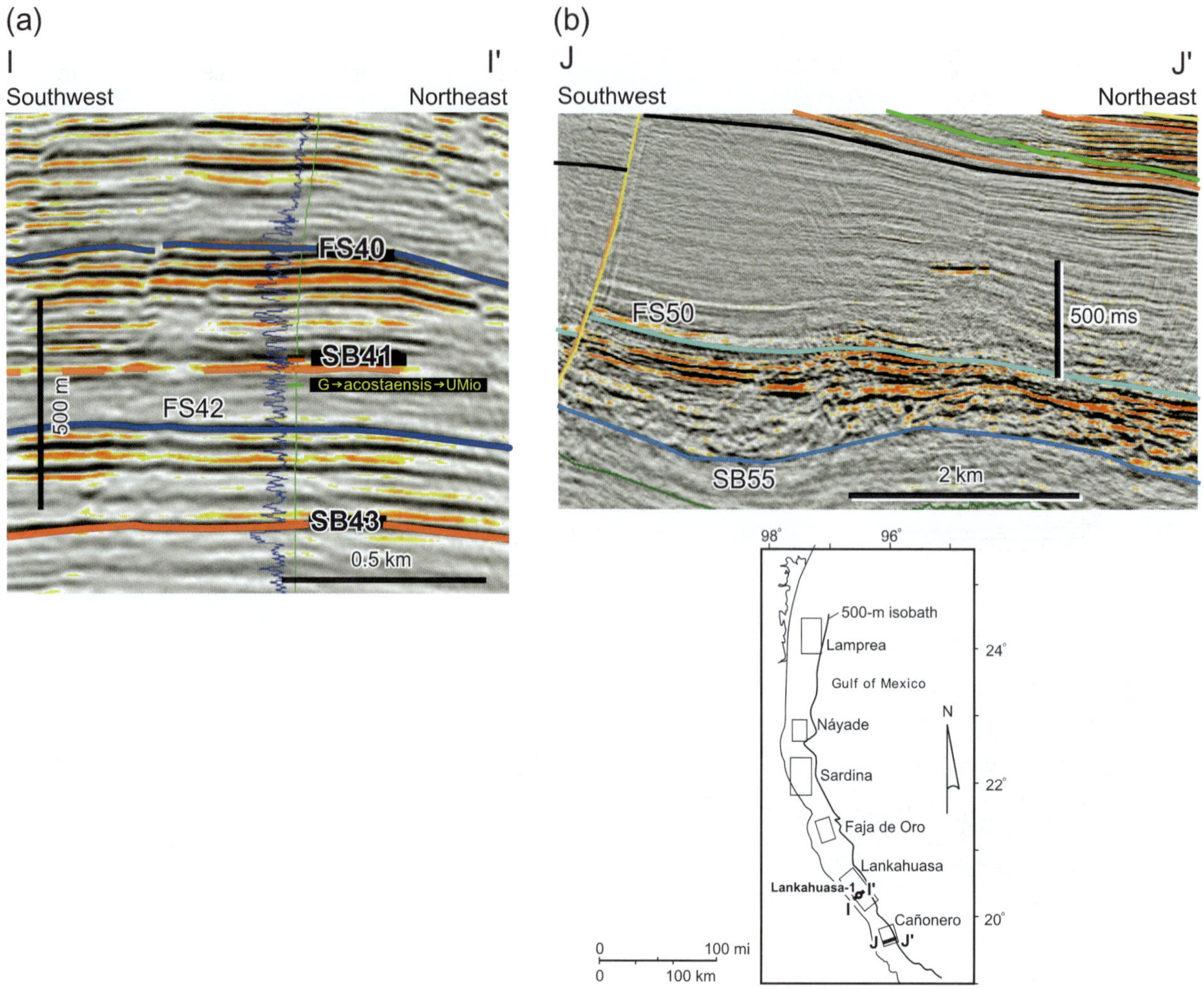

FIGURE 7. (a) Seismic section II′, showing log and seismic response of sequence boundaries (SB41, SB43) and flooding surfaces (FS40, FS42) in upper Miocene_2 shelf successions in the Lankahuasa-1 well. (b) Seismic dip section JJ′ in the Cañonero 3-D survey, showing seismic character of major fan-abandonment surface (FS50) at the top of the middle Miocene in the Cañonero 3-D survey. The Pemex 1 Lankahuasa is located in Figure 12. Cross section II′ is located in Figures 1 and 12a. Cross section JJ′ is located in Figure 1.

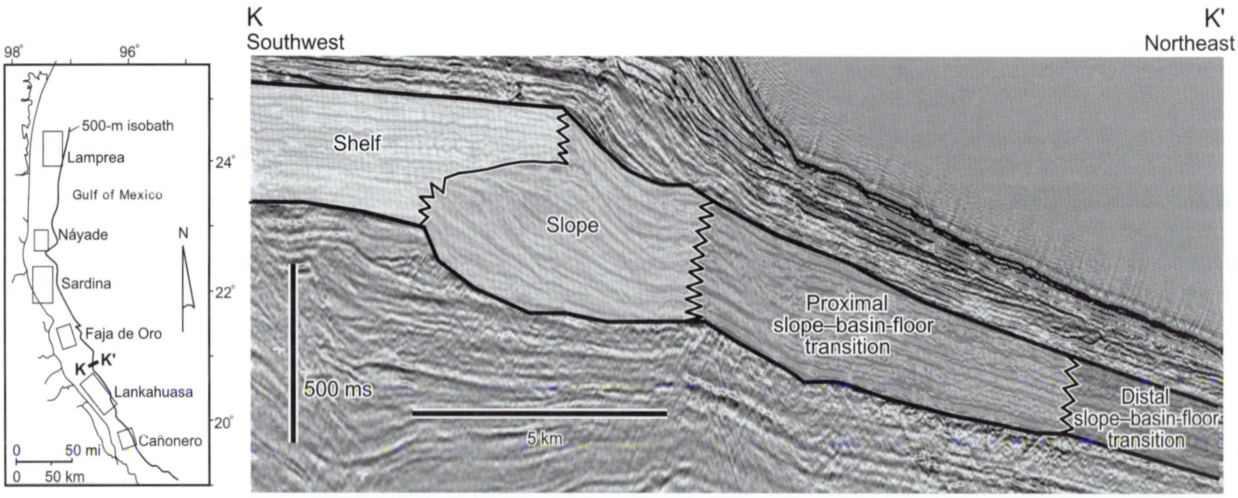

FIGURE 8. Two-dimensional seismic dip section KK′, showing seismic recognition criteria for major paleogeographic types in the LM-T area. Section is located in Figure 1.

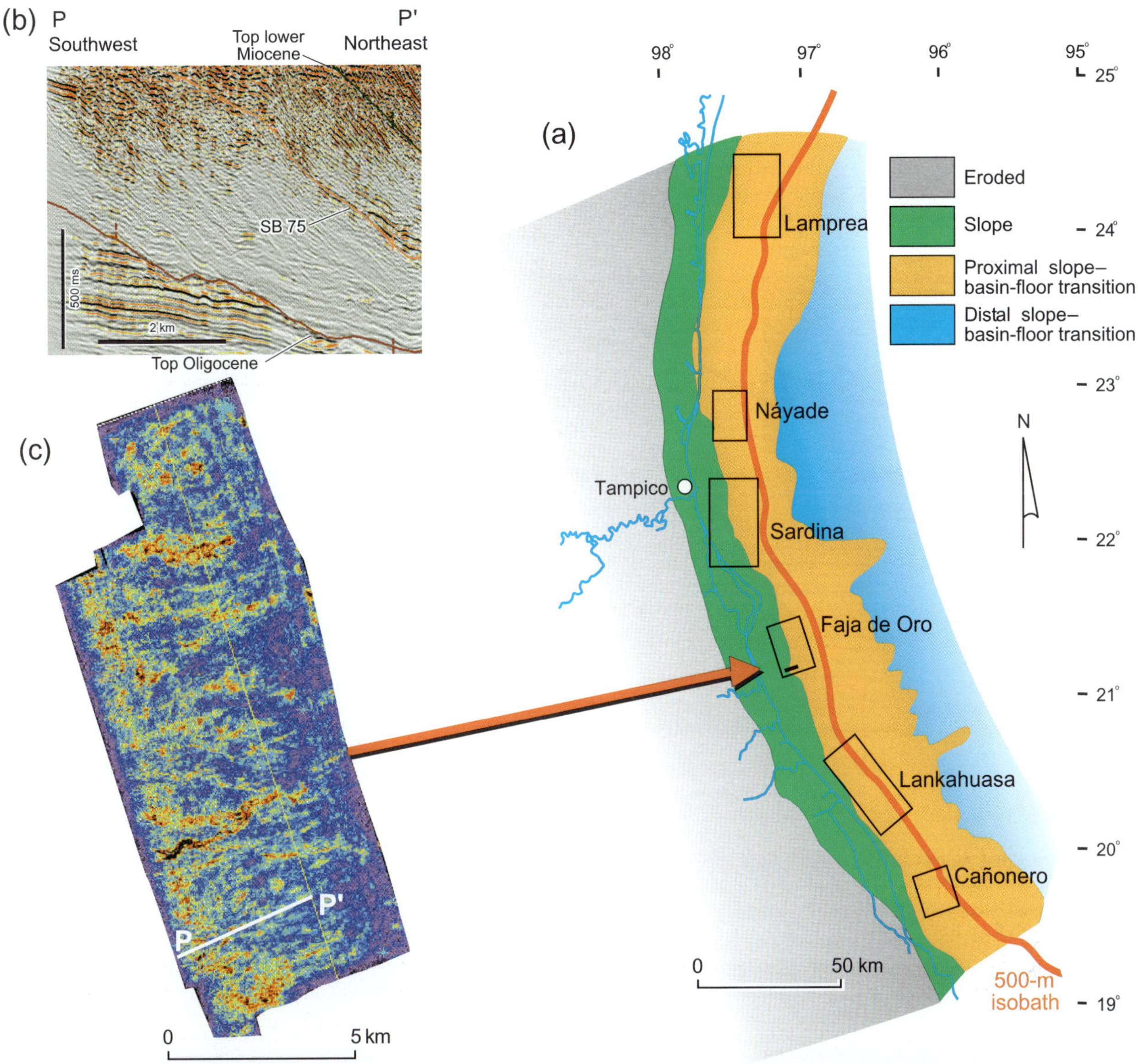

FIGURE 9. (a) Lower Miocene_1 paleogeography with slope and shelf deposits distributed updip (westward) of the 500-m (1600-ft) isobath. (b) Seismic dip section PP′ in the Faja de Oro 3-D survey, illustrating steeply dipping lower Miocene_1 slope deposits. Section is located in (c). The vertical exaggeration is approximately 3:1. (c) Maximum-amplitude map of the lower 80-ms interval in the lower Miocene_1 slope play in the Faja de Oro 3-D survey, illustrating narrow and dip-elongate slope-channel deposits.

However, south of the Lankahuasa 3-D survey, these shelf deposits are deeply dissected (locally >400 m [>1300 ft]) by a major lowstand canyon system in the area of Cañonero 3-D survey (Figure 10d).

Upper Pliocene slope deposits are located close to the modern 500-m (1600-ft) offshore isobath (Figure 11a). The upper Pliocene is interpreted to be eroded in an extensive area in the offshore from the Faja de Oro to the Sardina 3-D survey. The upper Pliocene section is complex in the Lamprea 3-D survey, where it consists of several shelf successions intercalated with inner slope slump deposits (Figure 11b).

PLAYS

Thirty-two plays mapped in the LM-T study are defined by eight age divisions and four types of paleogeographic settings. The eight age divisions, based on Petroleos Mexicanos biostratigraphic data and shown in Figure 3c, are (1) lower Miocene_1, (2) lower Miocene_2, (3) middle Miocene, (4) upper Miocene_1, (5) upper Miocene_2, (6) lower Pliocene, (7) middle Pliocene, and (8) upper Pliocene. The four main paleogeographic types (shelf, slope, canyon, and slope-to-basin-floor transition) in the LM-T area are discussed separately.

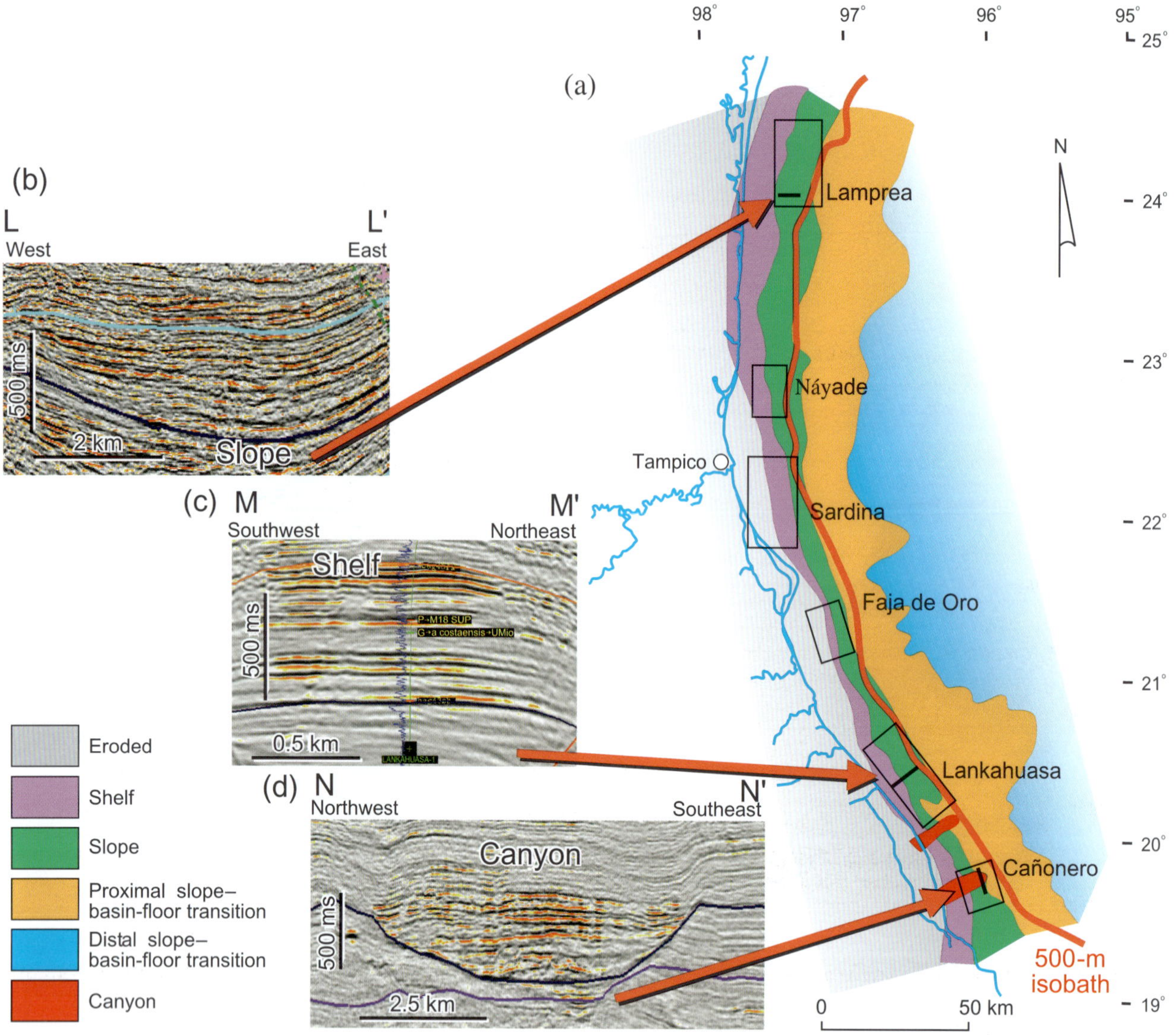

FIGURE 10. (a) Upper Miocene_2 paleogeography. (b) Three-dimensional seismic section LL′, showing mud-dominated, high-impedance slope deposits in the Lamprea 3-D survey. (c) Three-dimensional seismic section MM′, illustrating shelf successions in the Lankahuasa 3-D survey. (d) Three-dimensional seismic section NN′, portraying canyon-fill deposits in the Cañonero 3-D survey.

Shelf

Shelf plays defined in this study encompass a wide range of depositional environments, including coastal plain, shore zone, and submerged continental platform to the shelf edge. Shelf plays, as exemplified by the upper Miocene_2 shelf play in the Lankahuasa 3-D survey and trend, are typified by successions of thin sandstones with predominantly serrate log patterns (Figure 7a). Gas-bearing sandstones in the upper Miocene_2 shelf play in the Lankahuasa trend are common in transgressive systems tracts. These transgressive deposits are composed of strike-oriented isochron and amplitude patterns, indicating destruction and wave reworking of older deltaic and shoreface deposits (Figure 12a, b). Reservoir quality in these deposits is variable, where thin, burrowed sandstones (Figure 12c) pinch out into marine siltstone and mudstone. These sandstones are typically poorly sorted because of high amounts of muddy matrix introduced by burrowing. Common ichnofauna in these cores are Teichichnus, Rhizocorallium, Terebellina, and minor Thalassinoides. In contrast, Ophiomorpha and Skolithos are absent. Teichichnus, Rhizocorallium, Terebellina, and Thalassinoides can occur in a wide variety of water depths, but the absence of Ophiomorpha and Skolithos, common in sublittoral environments (Seilacher, 1967), indicates that the cored intervals are most likely from shelf, distal-deltaic, or lower shoreface settings.

A possible interpretation of upper Miocene_2 sandstones in the Lankahuasa trend includes shelf bars

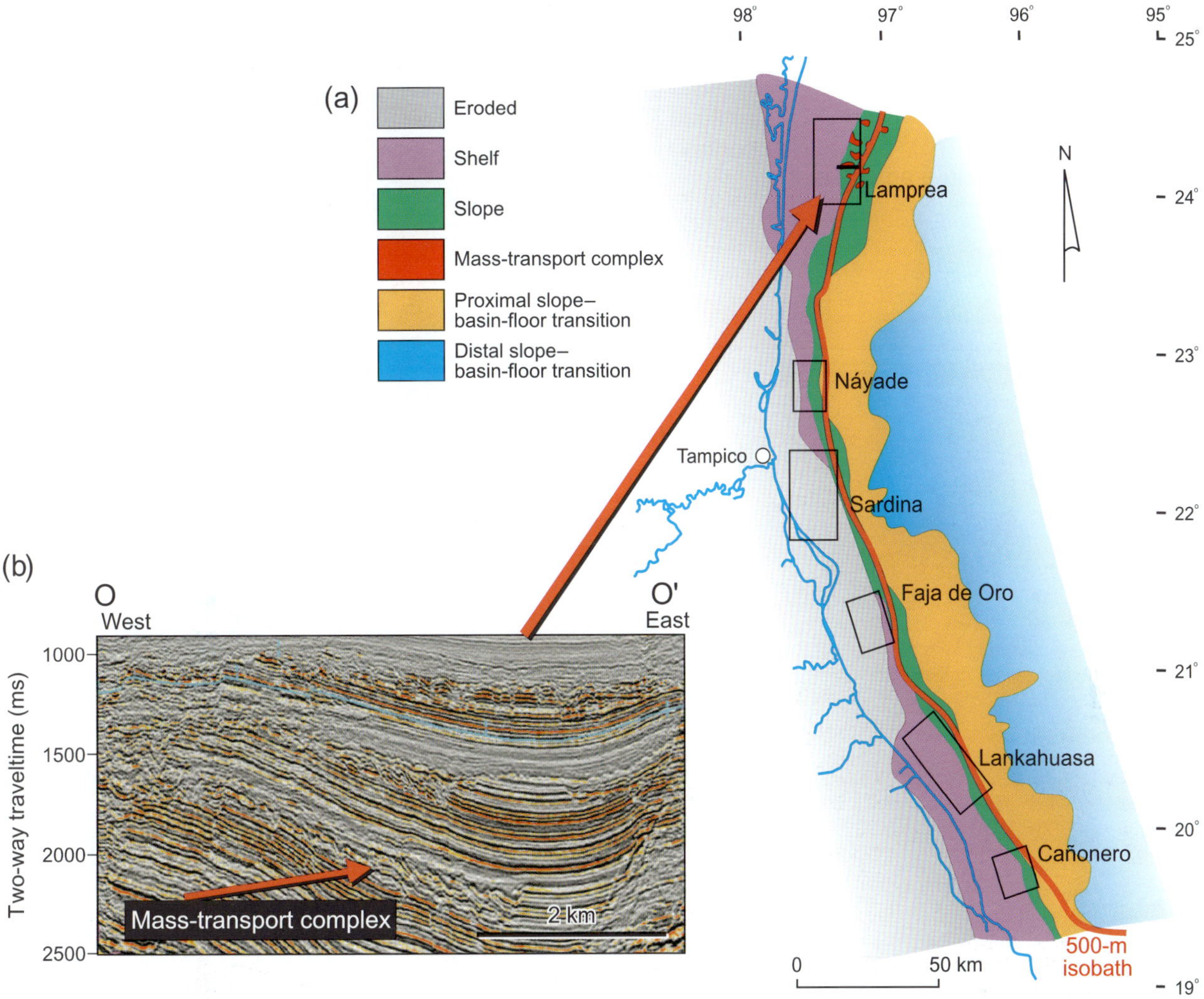

FIGURE 11. (a) Upper Pliocene paleogeography. (b) Three-dimensional seismic section OO′, showing inner slope slump deposits in the Lamprea 3-D survey.

because they are intensely burrowed and encased in marine mudstone and siltstone. However, these shelf sandstones may be reworked deposits from earlier lowstand progradational episodes (Figure 12d). During the SB43 lowstand, sandy deltaic sediments were inferred to have been delivered to an exposed outer shelf. These lowstand deltaic deposits were then reworked, burrowed by open-marine fauna, and subsequently stranded on the shelf during transgression in a sequence of events similar to those described by Plint (1988).

The upper Miocene outer shelf in the Lankahuasa trend is inferred from core and log data to have been a muddy, sediment-starved system of the type described by Kulm et al. (1975), despite the fact that the shelf was narrow. Logs of the Pliocene section in the Lankahuasa 3-D survey are dominated by successions of thin sandstones with serrate log responses, and percent-sandstone values computed from these logs are commonly less than 20%; thick, aggradational sandstones with blocky and blocky-to-upward-coarsening log signatures are absent. The muddy nature of the upper Miocene shelf in the Lankahuasa trend may have been the result of proximity to the trans-Mexican volcanic belt as well as the lack of large-scale depocenters. Additional investigations will be necessary to understand the influence of these factors on the upper Miocene shelf in this part of the LM-T area.

Slope

Slope plays in the LM-T area are typically mud dominated. Reservoir geometries consist of narrow, slightly sinuous, and dip-elongate channel and levee complexes. Reservoir quality is a major limiting factor in the LM-T slope plays, especially in the Tuxpan platform and adjacent areas, where many lower Miocene slope-channel sandstones are calcareous and with low porosity (Fouad et al., 2003a). These plays have a potential for stratigraphic traps, although potential reservoir size is limited.

The lower Miocene_1 slope play in the Faja de Oro 3-D survey is typical of other slope plays in the LM-T

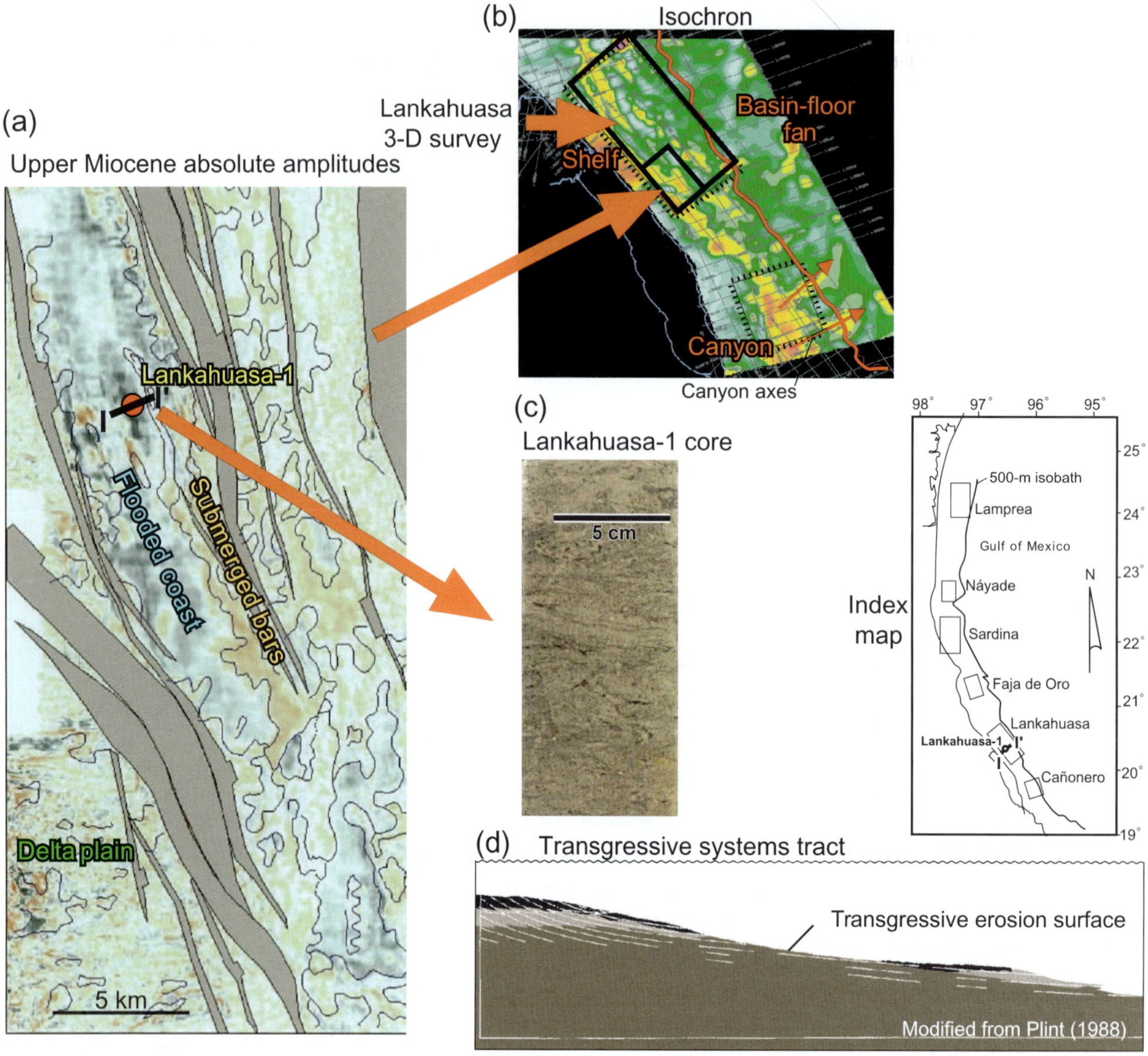

FIGURE 12. Upper Miocene_2 shelf systems in the Lankahuasa 3-D survey. (a) Absolute amplitude map of upper Miocene_2 transgressive deposits 40 ms below the FS40 maximum flooding surface, showing strike-parallel system of flooded coast and offshore submerged-bar deposits. (b) Isochron map of the upper Miocene_2 stratigraphic unit, with bright colors indicating thicker values. (c) Core from Pemex 1 Lankahuasa, showing heavily burrowed, fine-grained sandstones. (d) Model from Plint (1988), with in-place drowning and submergence of coastal deposits during transgression. Seismic section II′, indicated in (a), is shown in Figures 1 and 7a.

area. It consists of steeply dipping wedges and clinoforms above a major unconformity at the top of the Oligocene (Figure 9b). These strata commonly dip eastward at 10–15°, as measured on dip sections without vertical exaggeration. The internal architecture of the lower Miocene section is complex, consisting of numerous downlapping and crosscutting wedges.

A maximum-amplitude map of the interval 80 ms (~394 ft [~120 m]) above the top-of-Oligocene unconformity indicates that the bright-amplitude bodies are quite narrow (commonly only 0.3 mi [0.5 km] wide) and dip elongate, extending downdip (eastward) at least 3 mi (5 km) (Figure 9c). Eastward-merging tributary patterns are common, as well as pinch-outs downdip (basinward). A minor number of the bright-amplitude trends are sinuous, especially in the southern part of the Faja de Oro 3-D survey.

Canyon

Submarine canyon plays in the LM-T area are well developed in the upper Miocene_2 in the Cañonero trend (Figures 5, 6), where they are inferred to be related to periods of lowstand and associated with sediment bypass

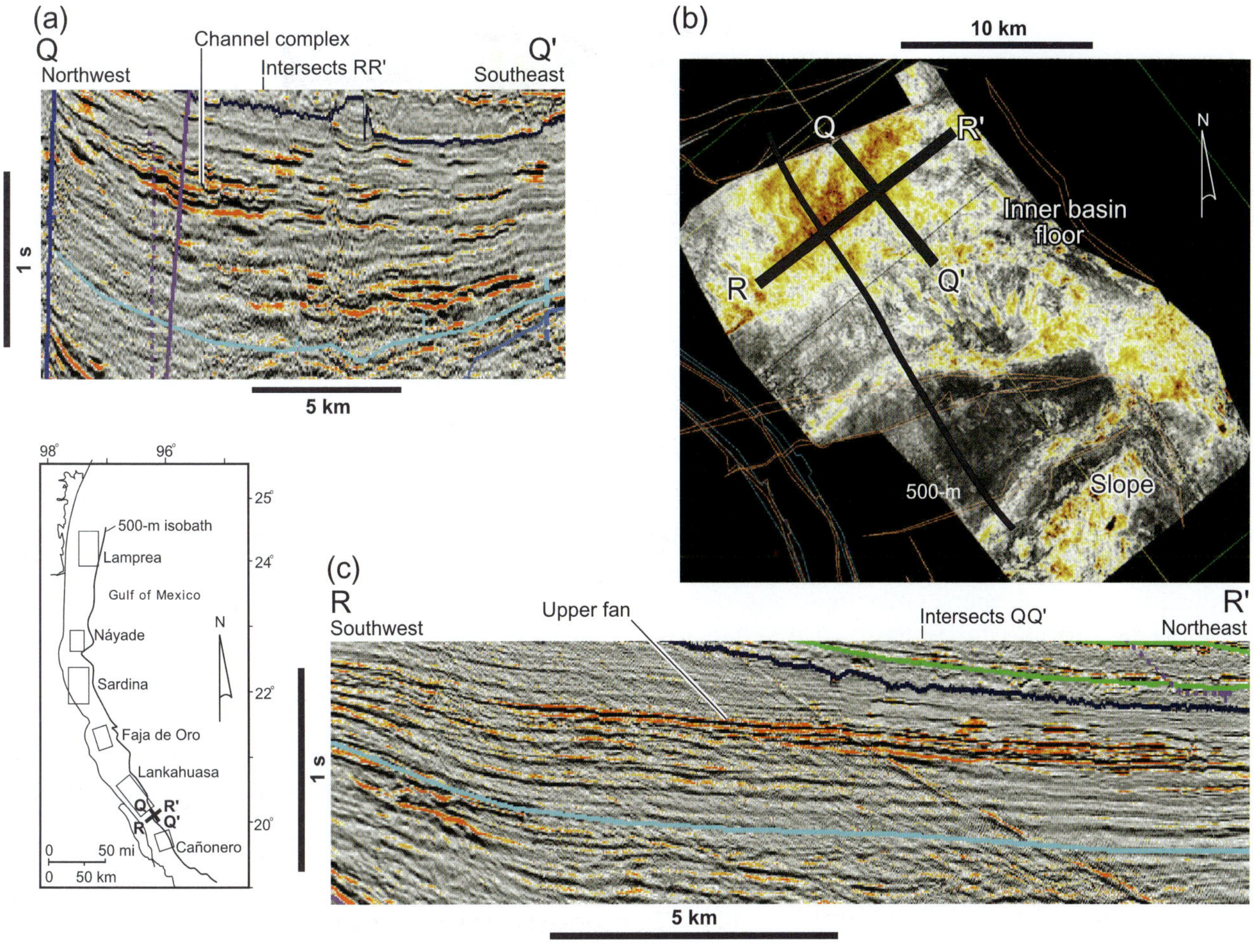

FIGURE 13. Upper Miocene_1 slope-to-basin-floor transition. (a) Three-dimensional seismic strike section QQ′, showing lenticular, bright-amplitude, and low-impedance slope-channel deposits. (b) Maximum peak amplitude map of slope-to-basin-floor transition above a local unconformity. (c) Three-dimensional seismic dip section RR′, displaying continuity of proximal-fan sheet deposits.

onto the basin floor (Ambrose et al., 2003a; Fouad et al., 2003b). They are defined as a single lenticular feature in strike section, varying in width from 5 to 8 km (3.1 to 5 mi), and are associated with 400–550 ms (480–660 m; 1574–2165 ft) of downcutting of older upper Miocene strata (Figure 5b, c). This canyon system has a wide variety of seismic responses and geometries that vary systematically from updip to downdip. For example, the updip trend is inferred to be mud prone, characterized by dim amplitudes, with locally chaotic reflectors along the margins possibly indicating slump facies (Figure 5a). In contrast, the downdip trend is composed of bright-amplitude strata that onlap onto the basal canyon unconformity (Figure 5b, c). The canyon system grades distally into basin-floor-fan deposits (Figure 6b), characterized by a dim- to moderate-amplitude set of parallel reflectors with minor onlap onto the unconformity (Figure 5d). Amplitudes in the Cañonero submarine canyon trend are incompletely understood because of only one well penetration at the time of this writing. Although the canyon fill has abundant bright-amplitude surfaces, these amplitudes may not indicate sandstone. Instead, they may indicate high-impedance tuffaceous deposits from the trans-Mexican volcanic belt.

Slope-to-Basin-Floor Transition

Slope-to-basin-floor transition plays in the LM-T area are developed mainly in the deeper offshore, east of the 500-m (1600-ft) isobath (Figures 6b, 13). These plays are inferred to contain thick (several hundred meters) channelized-lobe and extensive fan-sheet deposits. They commonly overlie regional unconformities and consist of bright-amplitude zones of wavy, subparallel reflectors capped by dim-amplitude flooding or fan-abandonment surfaces. These regional unconformities define individual episodes of lowstand-derived, submarine-canyon, and basin-floor-fan development that are interpreted to be tectonically driven, related to uplift in the Sierra Madre Oriental mountain belt. Reservoir quality in this play is expected to be highly variable, consisting of a variety of architectural elements, including fan lobes and sheets, channel complexes and associated overbank deposits, and muddy interchannel areas (Figure 13).

Fan lobes and sheets are volumetrically the most significant component of the basin-floor plays, judging by analogy from similar plays in the adjacent Veracruz basin (Jennette et al., 2003). Trap sizes in these plays are large, especially in downdip, four-way closures associated with compressional toe thrusts at the distal end of growth faults. This combination of fan lobes and sheets, as well as larger trap sizes, makes this play one of the most favorable in the LM-T area, although reservoir quality in the Cañonero trend is inferred to be poor because of proximity to the trans-Mexican volcanic belt.

Play Elements

Neogene plays in the LM-T area are characterized by a variety of play elements, including reservoir presence and quality, trap, seal, source, migration, and timing, which together provide constraints on a play's chance of success. Criteria were developed separately for evaluating data from each play element, and data for assessing play elements were used from a wide variety of sources. The relative importance and adequacy of play elements vary greatly for LM-T plays (Table 1). For example, considerable differences exist in inferred reservoir quality among the plays because of diverse source terrains and facies variability. Great variations are also present in trap styles and trap density, reflecting different tectonic settings in each structural trend, ranging from compressional in the Cañonero trend to purely extensional with gravity-slide tectonics in the Lankahuasa and Faja de Oro trends, as well as halokinesis in the Lamprea trend. The migration potential, a function of the presence and distribution of deep-seated faults cutting into the primary Mesozoic source rocks, also varies greatly throughout the eastern Mexico Gulf Coast.

Reservoir Presence and Quality

Reservoir presence in the LM-T area was inferred from log response and observations of impedance character of reflectors from 3-D seismic data. Where these data were lacking or ambiguous, analogous data were used from the neighboring Veracruz basin. At the time of this writing, reservoir presence was documented for only two Neogene plays in the LM-T area: the upper Miocene_2 shelf in the Lankahuasa trend and the middle Miocene slope play in the Faja de Oro–Náyade trend. However, recent drilling activity in the Lankahuasa 3-D survey also indicates the presence of hydrocarbons in middle and lower Pliocene slope and shelf plays (Cuevas et al., 2004).

Reservoir quality in the LM-T plays was assessed directly from core data and indirectly from petrophysical calculations of average V_{sh} and average porosity from wells. Where these data were absent, additional inferences of reservoir quality were based on facies analogs from the Veracruz and Macuspana basins (Jennette et al., 2002, 2003; Ambrose et al., 2003b). Log computations were made over the entire play interval instead of individual sandstone beds to best represent the entire play interval. However, many of the well-based results of reservoir quality were deemed not to be representative of the plays because too few wells were present to adequately test the plays (Table 1). This is especially the case for middle and lower Miocene plays, in which the only available data were from the Sardina and Náyade 3-D survey areas.

Slope-to-basin-floor transition plays in the Cañonero trend and the Veracruz basin are inferred to contain thick, sandy, and conglomeratic successions in fan sheets and lobes, although reservoir quality may be limited by the presence of volcanic clays and muddy debris-flow deposits. Slope plays are inferred to be relatively muddier than the basin-floor plays, judging by abundant sidewall core data.

Trap

Traps in the LM-T plays were evaluated from size range and geographic distribution of closures mapped from seismic data. Plays with a small numbers of traps or limited trap density were deemed to have an unfavorable trap potential (Table 1). Many plays are currently inadequately tested for trap because of limited numbers of wells. Most closures west of the 500-m (1600-ft) isobath are three-way and fault bounded, ranging in size from less than 5 to more than 50 km^2 (1.9 to more than 19 mi^2). For example, the upper Miocene_2 gas-bearing zones in the western part of the Lankahuasa 3-D survey occur in a large ($\sim$60-km^2; $\sim$23-mi^2) closure, bounded updip (westward) by a major listric fault. Many of the four-way closures in the LM-T area occur along or downdip (east) of the 500-m (1600-ft) isobath associated with the Mexican Ridges fold belt.

Trap styles differ between each major LM-T structural trend. For example, many closures in the Cañonero trend are composed of three-way and small four-way traps associated with antithetic faults and inverted structures, respectively (Figure 4a). The most common closures in the Lankahuasa trend are three-way, fault-bounded traps, particularly in complexly faulted grabens (Figure 4b). Narrow, three-way, fault-bounded closures are common in the Faja de Oro–Náyade trend (Figure 4c). Closures in the Lamprea trend are a combination of three-way and four-way types, reflecting the presence of mobile shale and salt structures (Figure 4d).

Although delineation of stratigraphic traps was beyond the scope of this regional project, there is a potential for such traps in the LM-T area, especially in the deep-water slope, canyon, and slope-to-basin-floor transition plays. The sandstone architecture of the slope plays, typified by the lower Miocene_1 slope play in the Faja de Oro and Sardina 3-D surveys, is complex, consisting of a dip-elongate network of narrow channel-fill and levee deposits pinching out into low-permeability mudstones and siltstone and which exhibit multiple pinch-outs (Fouad et al., 2003a) (Figure 13). Similarly, pinch-outs of canyon-fill deposits in the Cañonero trend are common, although the reservoir quality of these canyon-fill deposits may be limited by volcanic detritus from the updip trans-Mexican volcanic belt. The LM-T slope-to-basin-floor transition plays in the Cañonero trend are hypothesized to have a good potential for stratigraphic traps, particularly in the proximal facies tract, with stratigraphic compartmentalization related to complex channelized-fan deposits (Jennette et al., 2003).

Seal

Data from wells are insufficient to test many LM-T plays (Table 1). However, overpressure data from a few wells indicate that most overpressured zones occur in the middle and lower Miocene. Moreover, there may be differences in potential top seal among LM-T plays, inferred from variations in thickness of shales above flooding surfaces and fan-abandonment surfaces. For example, the middle Miocene is overlain by a thick succession of slope shales that should provide an excellent opportunity for vertical seal (Figure 7b). In contrast, the Pliocene section, especially in the Lankahuasa trend, is composed of sandy shelf and inner slope systems, with thin, intervening intervals of shale that may provide poor top seal (Figure 4b).

Seal in LM-T plays could potentially be compromised in areas intersected by major deep-seated faults. Many of these deep-seated faults are associated with current seafloor gas and oil seeps. In addition, intensely faulted zones that may contain numerous fractures, for example, the heavily faulted grabens in the east-central Lankahuasa 3-D survey (Figure 4b), may have a seal risk.

Source

Source, inferred from distribution, thickness, and thermal maturity of Tithonian deep-water shales in 46 wells in onshore and nearshore areas of the eastern Mexico Gulf Coast, is deemed adequate for all LM-T Neogene plays. Calcareous shales and argillaceous limestones of the three Upper Jurassic units (Oxfordian, Kimmeridgian, and Tithonian) contain abundant good to excellent type II oil-prone source rocks with TOC of 2% and higher and hydrogen index (HI) between 400 and 600 mg HC/g TOC. The Upper Jurassic source rocks are mature to overmature in the offshore LM-T area and are inferred to have generated and expelled large quantities of oil and gas. The Tithonian is one of the most important source rock units of the Upper Jurassic, based on its source characteristics, net thickness, and regional distribution, as well as for its widespread contribution to oil accumulations (Jordan and Wilson, 2003). Upper Jurassic samples from the onshore areas are immature to mature. We have estimated an original (i.e., prior to maturation) TOC of 3% and an original HI of 600 mg HC/g TOC for the Tithonian type II source rocks, which are used in maturity and hydrocarbon generation models.

Greater thicknesses of the Upper Jurassic were mapped in 12 grabens in the Tampico-Misantla basin by Román-Ramos and Holguín-Quiñones (2001). They estimated an average TOC of 2.2%, HI of 500 mg HC/g TOC, and thickness of 550 m (1800 ft) for the Upper Jurassic. Tithonian and Oxfordian units are interpreted to contain the richest source rock intervals in the Upper Jurassic, and the Kimmeridgian is a less important source rock unit.

Román-Ramos and Holguín-Quiñones (2001) mapped TOC enrichments in the Poza Rica area, the Tuxpan platform, and Arenque area in the Sardina 3-D survey. Jordan and Wilson (2003) evaluated TOC and pyrolysis data of approximately 800 samples from 100 wells in the Burgos, Tampico-Misantla, and Sureste basins of Mexico. They found that source rock TOC averaged as much as 3–4% in these basins. Based on estimations of the vertical extent of this richness by log-derived TOC calculations, they suggested that most of the Tithonian section is of source rock richness, except for some transitional beds at the base of the section.

Forty oil shows with similar gas-composition characteristics were identified in this study from ocean-bottom sediment cores from the Lankahuasa and Lamprea 3-D surveys and from areas north of Lankahuasa 3-D survey. Four of these oil shows have been correlated in this study, based on biomarkers (terpanes and steranes) and carbon isotope data with Tithonian-sourced oils from the Tampico-Misantla basin described by Guzmán-Vega et al. (2001). This oil show–source correlation indicates the presence of oil-prone Tithonian source rocks in the subsurface and the existence of an active Upper Jurassic petroleum system in the LM-T offshore area. A maturity reconstruction of the Upper Jurassic in the offshore LM-T study area suggests that the Tithonian source rocks are presently mature to overmature (Figure 14). Assuming a source quality similar to that of the onshore areas (which is quite likely), it can be inferred that the Tithonian sources have generated and expelled large quantities of oil and gas in the offshore areas. Gases produced from upper Miocene sandstones in the Lankahuasa 3-D survey and several gas shows from north of Lankahuasa have been identified in this study as dry, very mature thermogenic gases (R_o about 2.0% and higher) based on molecular and carbon isotope compositions.

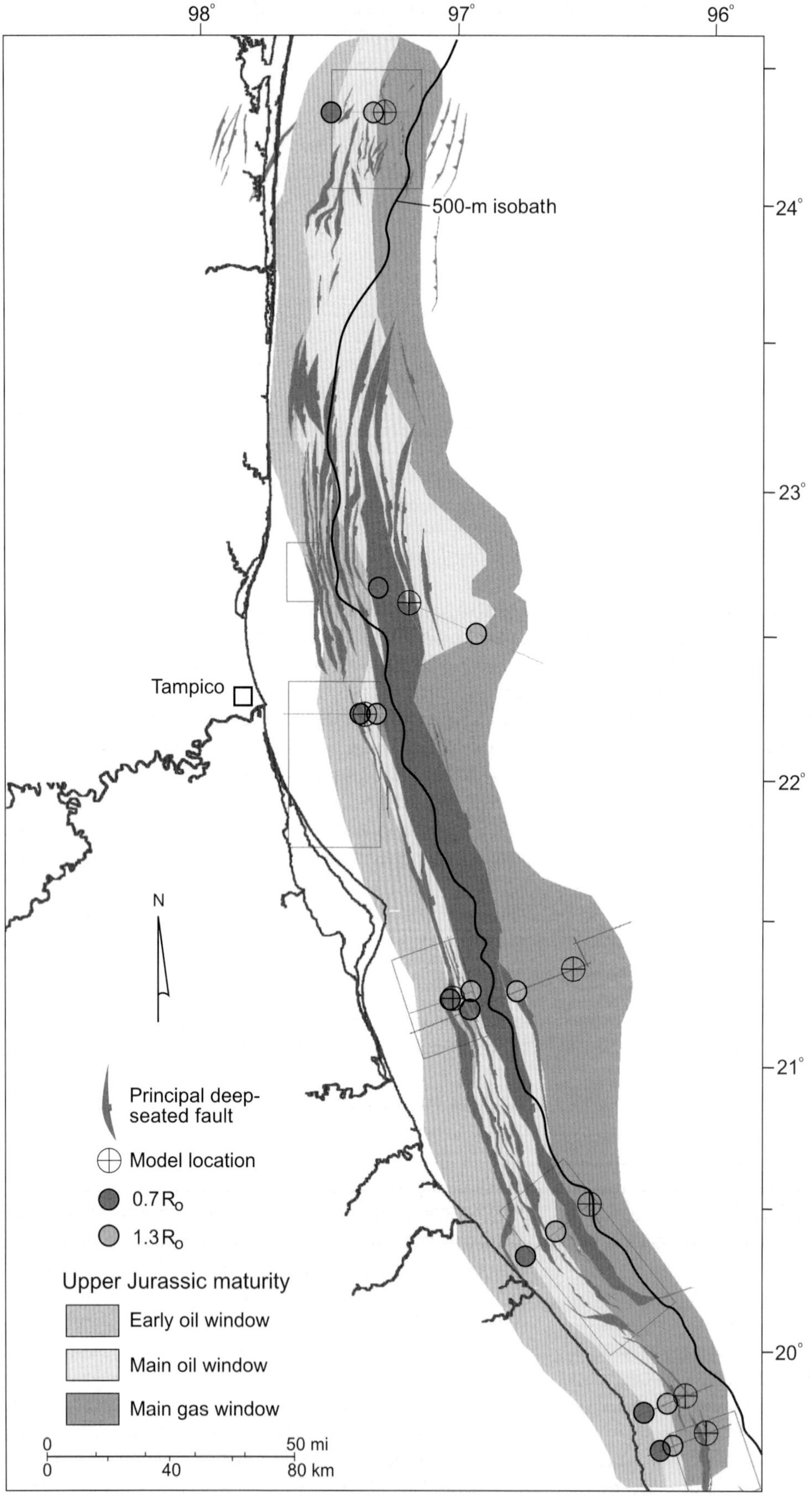

FIGURE 14. Upper Jurassic maturity in the LM-T area, based on calculations of vitrinite reflectance (R_o) in eight model locations.

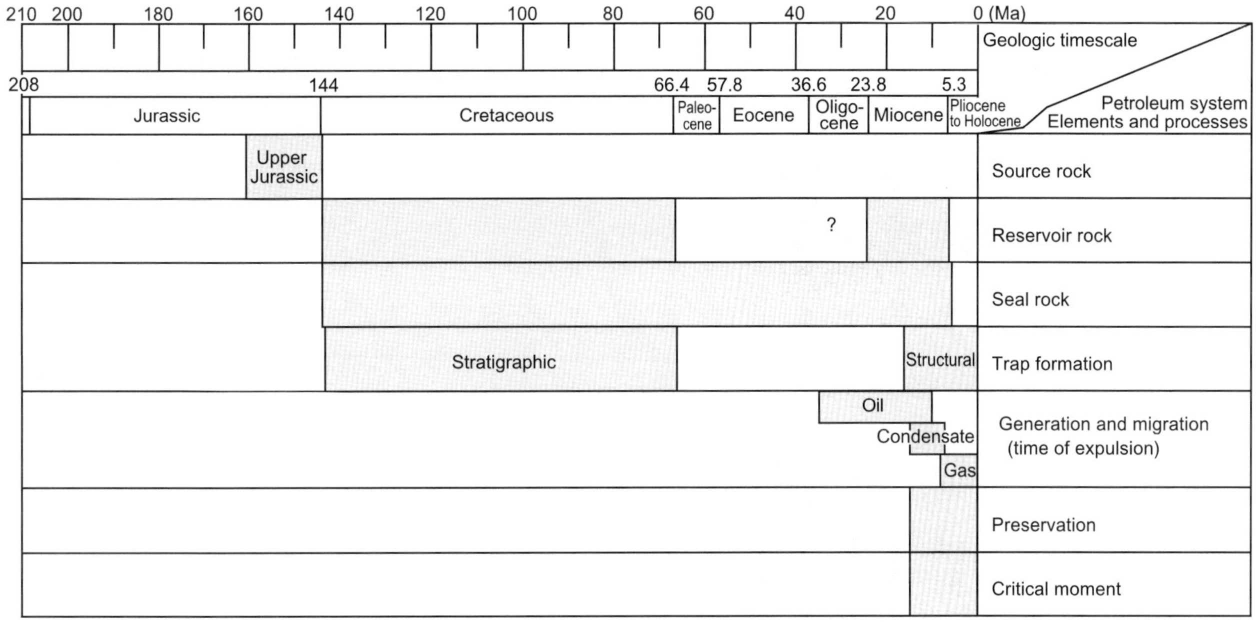

FIGURE 15. Upper Jurassic petroleum system chart for the Lankahuasa trend.

In contrast with the Tithonian section, Cretaceous and Tertiary strata have only a limited potential for providing hydrocarbon sources. The Cretaceous succession contains mostly carbonate reservoir rocks with minor intervals of calcareous shales and argillaceous limestones. Evaluation of these source rocks shows fair to good TOC contents (mostly between 0.50 and 1.5%, average 0.95%) and mostly type II oil-prone and type II/III oil- and gas-prone kerogens (HI 400–600 and 200–400 mg HC/g TOC, respectively). Potential Cretaceous source rocks are mostly immature in the onshore areas. Paleogene shales evaluated from the onshore areas show only immature, poor gas source rocks. Neogene (Miocene and Pliocene) shales are identified as minor biogenic gas source rocks.

Migration and Timing

Source rock evaluation suggests that the possibility of hydrocarbon charge from the deep Upper Jurassic source is a necessary condition for large gas accumulations in LM-T Neogene plays. Therefore, it is critical that vertical migration paths existed and large quantities of gas were available for migration when Neogene traps formed.

Variations in migration potential, inferred from the geometry of deep-seated faults likely to tap the predominant Upper Jurassic source, exist among the various LM-T trends. For example, the Cañonero trend contains numerous basement-involved faults inferred to provide direct upward-migration pathways from the Mesozoic to the Neogene. In contrast, areas north of the Cañonero trend and downdip of principal listric faults are problematic for upward migration of Mesozoic-sourced hydrocarbons. These listric faults have a nearly horizontal detachment surface above the top of the Mesozoic, soling out into Oligocene shales. Areas updip of the large-scale listric fault systems contain plays that are intersected by major faults cutting the ancestral Cretaceous shelf edge and that are directly connected to Mesozoic source rock intervals.

Timing was assessed from hydrocarbon modeling from several locations throughout the LM-T area, primarily in the shallow offshore. Input parameters for these models included stratigraphic and lithologic data, temperature and thermal-history data, and geochemical parameters, such as thermal-maturity data, bottomhole temperature and heat-flow data, HI and primary oil and gas potential, kerogen-to-oil and oil-to-gas kinetics, thermal-maturity data, hydrocarbon type, and isotopic data. Ages and lithologic compositions for the Paleogene, Cretaceous, and Upper Jurassic intervals were obtained from publications and from Petroleos Mexicanos data.

Timing is not a limiting play element for the LM-T plays because nearly all areas have experienced recent (5.0 Ma and younger) expulsion of hydrocarbons. However, some variations are present in the timing of hydrocarbon expulsion and trap formation between different trends in the LM-T area. These variations do not affect play possibilities but may have an impact on the probability of success of each prospect.

The Upper Jurassic petroleum system chart for the Lankahuasa trend is given as an example, where oil, condensate, and main gas generation and migration from the Upper Jurassic oil source (oil-prone type II) occurred

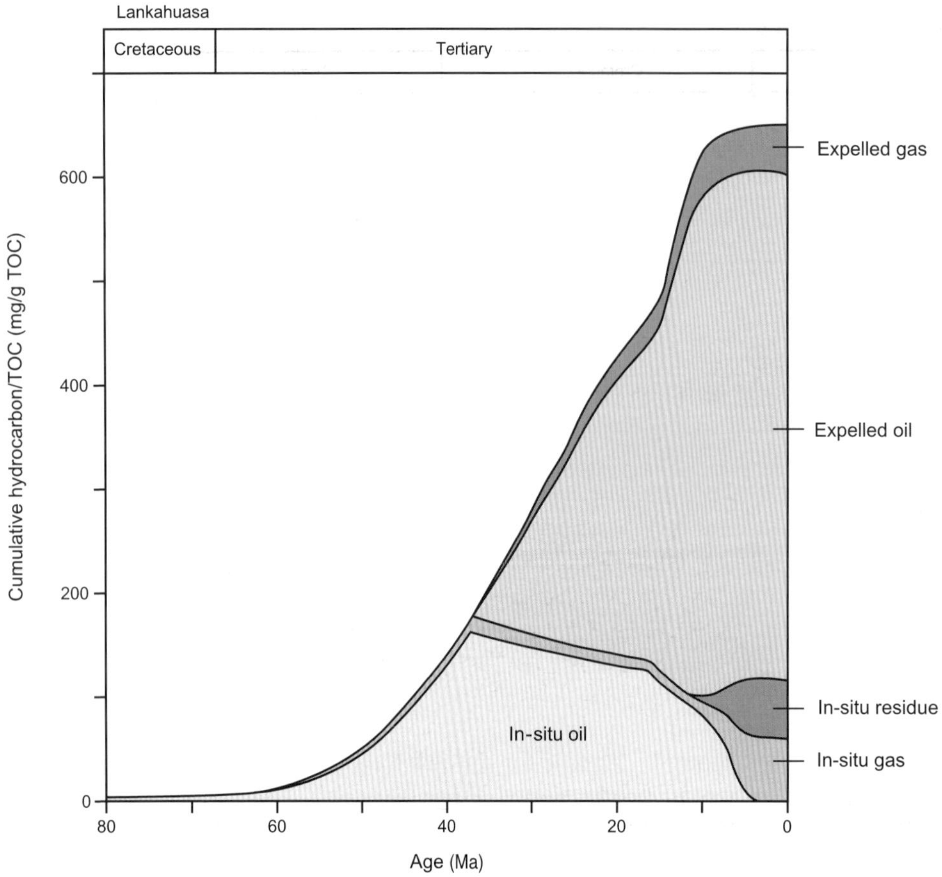

FIGURE 16. Cumulative oil and gas generation and expulsion vs. geologic time for the upper Jurassic source in the Lankahuasa trend.

during the Oligocene to middle Miocene (35.0–11.5 Ma), middle Miocene to late Miocene (13.0–7.8 Ma), and late Miocene to Holocene (9.6–0 Ma), respectively (Figure 15). Hydrocarbon types that could migrate from the source into Neogene plays during the development of traps and migration paths (16.4–0 Ma) are oil, condensate, and gas. During intense early Pliocene faulting, only gas could migrate from the source into Neogene plays. Listric faults in this trend detach above the source rock interval. As a result, vertical, cross-stratal, fault-related migration conduits connecting Jurassic source rocks with Neogene reservoirs and traps are interpreted to be a significant risk in this trend.

Most of the oil in the Lankahuasa trend is inferred to have been generated and expelled during the 35.0–16.4 Ma period prior to Neogene trap formation (Figure 16). Therefore, the quantity of hydrocarbons (oil, condensate, and gas) that could migrate directly from the source during Neogene trap formation was relatively much less important compared to the quantity of oil expelled from the source prior to Neogene trap formation (Figure 16), when the source was at about 0.70–0.85% R_o maturity and was in the main oil window (Figure 17).

In the other LM-T trends, starting in the south with the Cañonero trend, oil, condensate, and main gas generation and expulsion from the Upper Jurassic source occurred during middle Eocene to late Oligocene (47.0–27.0 Ma), early Oligocene to early Miocene (31.5–18.0 Ma), and early Miocene to Holocene (22.0–0 Ma), respectively. Hydrocarbon types that could migrate into Neogene plays during trap formation in the late Miocene to Holocene (10.2–0 Ma) were gas from the Upper Jurassic source rock intervals.

In the Faja de Oro–Náyade trend, oil generation and expulsion from the Upper Jurassic source occurred during middle Miocene to Holocene (10.2–0 Ma). In the Faja de Oro 3-D survey, formation of Neogene traps occurred during the middle Miocene to middle Pliocene (16.4–3.4 Ma). Therefore, oil could have migrated into Neogene traps. In the Sardina 3-D survey area, oil generation and expulsion from the Upper Jurassic source occurred during the late Miocene to Holocene (6.5–0 Ma). Traps in the Neogene intervals formed during the middle Miocene (16.4–10.2 Ma). In the Náyade 3-D survey area, oil generation and expulsion from the Upper Jurassic source occurred during the late Miocene to Holocene (11.0–0 Ma). Neogene traps formed during the middle Miocene–Holocene (16.4–0 Ma).

In the Lamprea trend, oil, condensate, and main gas generation and expulsion from the Upper Jurassic source occurred during the early Miocene to middle late Miocene (18.0–5.3 Ma), the late Miocene to late Pliocene (7.2–2.0 Ma), and the middle Pliocene to Holocene

FIGURE 17. Burial-history diagram with thermal-maturity windows for the Lankahuasa trend.

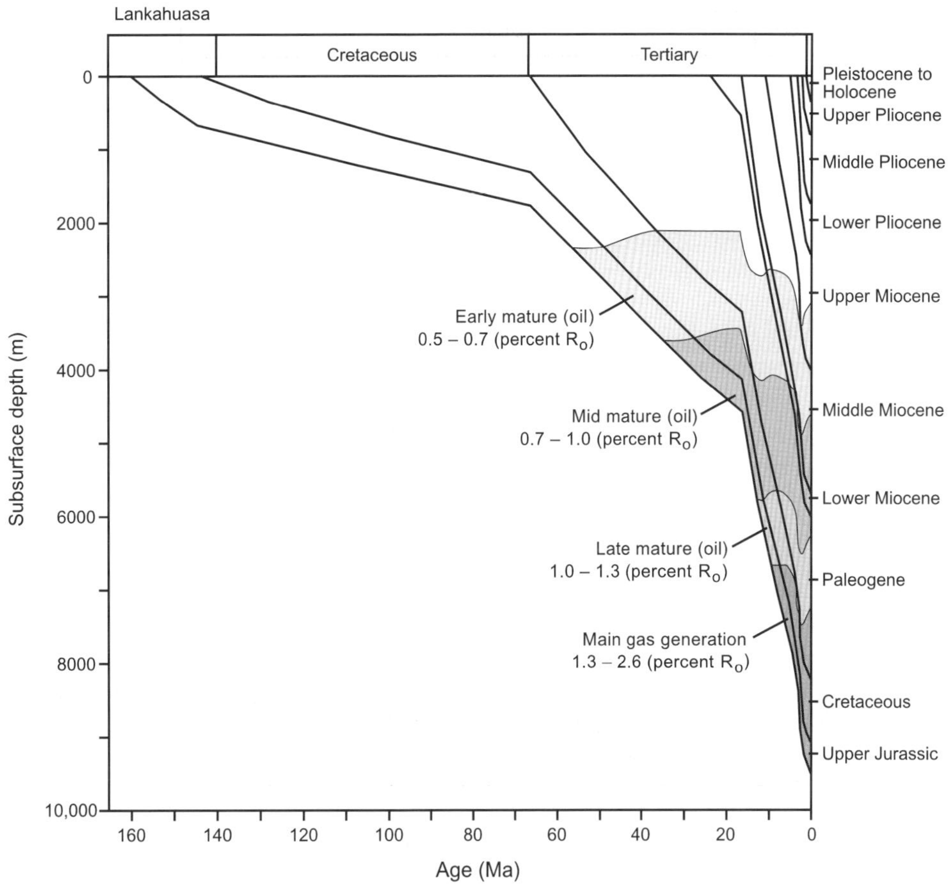

(3.3–0 Ma), respectively. If fault-related migration paths were available, the hydrocarbon types that could migrate from the Upper Jurassic source intervals into Neogene plays during trap formation in the late Miocene to late Pliocene (6.0–1.8 Ma) would consist of light oil, condensate, and gas.

SUMMARY

Neogene shelf, slope, canyon, and slope-to-basin-floor transition plays in the LM-T continental shelf are controlled by a variety of structural and stratigraphic features, including gravity sliding and extension, compression and strike-slip motion, salt evacuation, and formation of lowstand canyon and fan systems associated with major uplift and incision of the Sierra Madre Oriental mountain belt. The Lankahuasa trend, discussed first in this summary, has the greatest overall potential of all the LM-T trends, considering it contains a greater number of already proven plays, as well as an array of more favorable play elements (Table 1). The other trends (Cañonero, Faja de Oro–Náyade, and Lamprea) have a similar overall but lesser potential than the Lankahuasa trend and are discussed in geographical order from south to north.

At the time of this writing, the Lankahuasa trend has two proven plays, the upper Miocene_2 shelf and the lower Pliocene slope. Reservoir quality in the upper Miocene_2 shelf play is adequate, although individual sandstones in the play are thin and contain a wide range of porosity values (10–25%). However, these data are currently unavailable for the lower Pliocene slope play. Trap types for both plays are similar, being three-way and fault bounded above a system of major shallow-detachment listric faults. Seal is provided by laterally continuous flooding surfaces and slope mudstones, in combination with fault offsets, locally in excess of 100 m (330 ft). However, seal may be compromised locally in intensely faulted and fractured grabens in the eastern part of the Lankahuasa 3-D survey (Figure 4b). Migration in the Lankahuasa trend is adequate because of the existence of proven plays and the widespread presence of oil and gas seeps in the area.

Plays in the Cañonero trend contain a wide variety of favorable, unfavorable, and untested play elements (Table 1). Reservoir presence and quality have yet to be demonstrated for all plays in the Cañonero trend, and recent drilling activity in the upper Miocene_2 canyon play suggests the presence of volcanogenic deposits. Although trap is currently untested in the shelf and slope plays, abundant three-way and areally limited four-way closures are hypothesized to exist in the shelf-to-basin-floor transition and canyon plays. Seal is expected to be favorable in the slope-to-basin-floor transition and canyon plays, where thick, dim-amplitude successions of shale are inferred to overlie flooding and fan-abandonment surfaces (Figures 5, 7b). Source, migration, and timing are deemed to be favorable play elements in the Cañonero trend, where deep-seated basement faults are inferred to have provided upward migration for hydrocarbons.

The Faja de Oro–Náyade trend contains only one Neogene play where reservoir presence has been demonstrated, occurring in thin, silty sandstones in the middle Miocene slope play (Table 1). The western part of the Faja de Oro–Náyade trend is dominated by thick successions of steeply dipping lower and middle Miocene slope plays containing narrow channel and levee complexes (Figure 9). Traps in these slope plays are defined by narrow, three-way closures against faults (Fouad et al., 2003a). The source in the western part of the Faja de Oro–Náyade trend is expected to be predominantly oil, and migration pathways are inferred to be provided by major fault systems developed along the ancestral Cretaceous carbonate shelf margin, west of the Faja de Oro fault (Figure 4c). Major listric faults in the downdip Faja de Oro–Náyade trend are interpreted to sole out into Oligocene shales, above the primary Mesozoic source, and therefore, migration in this part of the trend is problematic and untested.

The Lamprea trend, like the Faja de Oro–Náyade trend, contains a great number of untested play elements (Table 1). Reservoir quality is deemed to be adequate in the shallow, uncompacted shelf plays, but a recently drilled well in more deeply buried plays, such as the middle Miocene slope play, indicates low porosity values because of high siltstone content, calcite cement, and compaction. A great variety of trap types exists in the Lamprea trend, consisting of three-way, fault-dependent traps in the updip Miocene extensional complex (Figure 4d), sparse four-way traps in the downdip, eastern part of the trend, and areally limited traps associated with diapirs. Seal is untested for almost all of the plays, except for the middle Miocene slope play (Table 1). Source is inferred to be favorable, but potential migration pathways may be compromised by impermeable salt and mud diapirs, as well as salt welds.

ACKNOWLEDGMENTS

The Bureau of Economic Geology, John A. and Katherine G. Jackson School of Geosciences, University of Texas at Austin, acknowledges support of this research by Landmark Graphics Corporation via the Landmark University Grant Program. Partial support for this publication was received from the John A. and Katherine G. Jackson School of Geosciences and the Geology of Foundation of the University of Texas at Austin.

This report was funded by Petroleos Mexicanos Exploración y Producción under Contract No. PEP-RN-RDIE-072/02-P. The Bureau of Economic Geology appreciates the opportunity to contribute to Petroleos Mexicanos' assessment of the hydrocarbon resources of the southern Laguna Madre–Tuxpan continental shelf and the continuous help and project participation by management and geoscientists of the Coordinación de Exploración at Poza Rica and the Asset Office at Tampico. Significant coordination between Petroleos Mexicanos and the Bureau of Economic Geology was facilitated by Luis Sanchez-Barreda.

The manuscript benefited greatly from the reviews of Robert M. Mitchum, William E. Galloway, Paul Post, and Eric Potter. Manuscript editing was by Lana Dieterich. Pat Alfano prepared the illustrations under the direction of Joel Lardon, manager, Media Information Technology. Publication was authorized by the director of the Bureau of Economic Geology.

REFERENCES CITED

Ambrose, W. A., et al., 2003a, Miocene valley-fill, slope, and submarine-canyon systems in the Laguna Madre–Tuxpan area, Mexico: Gulf Coast Association of Geological Societies Transactions, v. 53, p. 11–18.

Ambrose, W. A., et al., 2003b, Geologic framework of upper Miocene and Pliocene gas plays of the Macuspana basin, southeastern Mexico: AAPG Bulletin, v. 87, no. 9, p. 1411–1435.

Beaubouef, R. T., C. R. Rossen, F. B. Zelt, M. D. Sullivan, D. C. Mohrig, D. C. Jennette, J. A. Bellian, S. J. Friedman, R. W. Lovell, and D. S. Shannon, 1999, Deep-water sandstones, Brushy Canyon Formation, west Texas: AAPG Continuing Education Course Notes Series 40, variously paginated.

Buffler, R. T., 1983, Structure of the Mexican Ridges fold belt, southwest Gulf of Mexico, *in* A. W. Bally, ed., Seismic expression of structural styles; a picture and work atlas; v. 2: AAPG Studies in Geology 15, p. 2.3.3-16–2.3.3-21.

Buffler, R. T., and D. S. Sawyer, 1985, Distribution of crust and early history, Gulf of Mexico basin: Gulf Coast Association of Geological Societies Transactions, v. 35, p. 333–344.

Cuevas, J. A., M. Aranda Garcia, J. R. Román-Ramos, and J. Morales-Marin, 2004, Historical synopsis and geological framework of hydrocarbon plays, continental shelf, southwestern Gulf of Mexico (abs.): AAPG Annual Meeting Program, v. 13, p. A30.

Feng, J., R. T. Buffler, and M. A. Kominz, 1994, Laramide orogenic influence on late Mesozoic–Cenozoic subsidence history, western deep Gulf of Mexico basin: Geology, v. 22, p. 359–362.

Ferrari, L., M. López-Martínez, G. Aguirre-Diaz, and G. Carrasco-Nuñez, 1999, Space-time patterns of Cenozoic arc volcanism in central Mexico: From the Sierra Madre Occidental to the Mexican volcanic belt: Geology, v. 27, no. 4, p. 303–306.

Fouad, K., W. A. Ambrose, S. Sakurai, and D. Jennette, 2003a, Wave-shape classification and attribute analysis of the lower Miocene deep-water reservoirs, Madre basin, offshore Mexico (ext. abs.): Society of Exploration Geophysics Transactions, Seventy-Third Annual Meeting and International Exposition, CD-ROM, 4 p.

Fouad, K., W. A. Ambrose, F. Brown, and D. Jennette, 2003b, Seismic facies an attribute analysis of the Miocene incised-valley and submarine-canyon systems in Tuxpan basin, offshore Mexico (ext. abs.): Society of Exploration Geophysicists Seventy-Third Annual Meeting and International Exposition, CD-ROM, 4 p.

Galloway, W. E., P. E. Ganey-Curry, X. Li, and R. T. Buffler, 2000, Cenozoic depositional history of the Gulf of Mexico basin: AAPG Bulletin, v. 84, p. 1743–1774.

Goldhammer, R. K., and C. A. Johnson, 2001, Middle Jurassic–Upper Cretaceous paleogeographic evolution and sequence-stratigraphic framework of the northwest Gulf of Mexico rim, *in* C. Bartolini, R. T. Buffler, and A. Cantu-Chapa, eds., The western Gulf of Mexico basin: Tectonics, sedimentary basins, and petroleum systems: AAPG Memoir 75, p. 45–81.

González, R., and N. Holguín, 2001, Las rocas generadoras de Mexico: Asociación Mexicana de Geólogos Petroleros, v. 49, no. 1–2, p. 16–30.

Guzmán-Vega, M. A., and M. R. Mello, 1999, Origin of oil in the Sureste basin, Mexico: AAPG Bulletin, v. 83, no. 7, p. 1068–1095.

Guzmán-Vega, M. A., L. Castro Ortiz, J. R. Román-Ramos, L. Medrano-Morales, L. C. Valdez, E. Váquez-Covarrubias, and G. Ziga-Rodríguez, 2001, Classification and origin of petroleum in the Mexican Gulf Coast basin: An overview, *in* C. Bartolini, R. T. Buffler, and A. Cantu Chapa, eds., The western Gulf of Mexico basin: Tectonics, sedimentary basins, and petroleum systems: AAPG Memoir 75, p. 127–142.

Haq, B. U., J. Hardenbol, and P. R. Vail, 1987, Chronology of fluctuating sea levels since the Triassic: Science, v. 235, p. 1156–1166.

Jennette, D., T. Wawrzyniec, K. Fouad, D. Dunlap, J. Meneses-Rocha, F. Grimaldo, R. Muñoz, D. Barrera, C. Williams-Rojas, and A. Escamilla-Herrera, 2003, Traps and turbidite reservoir characteristics from a complex and evolving tectonic setting, Veracruz basin, southeastern Mexico: AAPG Bulletin, v. 87, no. 10, p. 1599–1622.

Jennette, D. C., et al., 2002, Play-element characterization of the Miocene and Pliocene Veracruz basin, southeastern Mexico: Gulf Coast Association of Geological Societies Transactions, v. 52, p. 441–454.

Jordan, C. L., and J. L. Wilson, 2003, Geologic controls on the organic richness of Tithonian (Upper Jurassic) source rocks in the Burgos, Tampico-Misantla, and Sureste basins of Mexico, *in* N. E. Rosen, ed., Structure and stratigraphy of south Texas and northeast Mexico: Applications to exploration: Gulf Coast Section SEPM Foundation and South Texas Geological Society, Special Publication, CD-ROM, p. 22–24.

Kulm, L. D., R. C. Rousch, J. C. Harlett, R. H. Neudeck, D. M. Chambers, and E. J. Runge, 1975, Oregon continental shelf sedimentation: Interrelationships of facies distribution and sedimentary processes: Journal of Geology, v. 83, p. 145–176.

Magoon, L. B., T. L. Hudson, and H. E. Cook, 2001, Pimienta-Tamabra, a giant supercharged petroleum system in the southern Gulf of Mexico, onshore and offshore Mexico, *in* C. Bartolini, R. T. Buffler, and A. Cantú-Chapa, eds., The western Gulf of Mexico basin: Tectonics, sedimentary basins, and petroleum systems: AAPG Memoir 75, p. 83–125.

Marton, G., and R. T. Buffler, 1994, Jurassic reconstruction of the Gulf of Mexico basin: International Geology Review, v. 36, p. 545–586.

Mello, M. R., et al., 1996, Geochemical characterization of oils from the Gulf of Mexico and southeastern Mexican basins: A petroleum system approach to predict deep water probes, *in* E. Gómez-Luna and A. M. Cortez, eds., 5th Latin American Congress of Organic Geochemistry, Cancún, Mexico, Memoir, Abstracts Volume, p. 106–109.

Pew, E., 1982, Seismic structural analysis of deformation in the southern Mexican ridges: M.A. Thesis, University of Texas at Austin, Austin, 101 p.

Pindell, J. L., 1994, Evolution of the Gulf of Mexico and the Caribbean, *in* S. K. Donovan and T. A. Jackson, eds., Caribbean geology: An introduction: Kingston, Jamaica, UWI Publishers, p. 12–39.

Pindell, J. L., L. Kennan, and S. Barrett, 2000, Putting it all together again: AAPG Explorer, v. 21, p. 58–62.

Plint, A. G., 1988, Sharp-based shoreface sequences and "offshore bars" in the Cardium Formation of Alberta: Their relationship to relative changes in sea level, *in* C. K. Wilgus, B. S. Hastings, H. Posamentier, J. Van Wagoner, C. A. Ross, and C. G. St. C. Kendall, eds., Sea level changes, an integrated approach: SEPM Special Publication 42, p. 357–370.

Román-Ramos, J., R. E. Mena-Sánchez, and L. Bernal-Vargas, 1996, Evaluación geoquímica de los recursos petrolíferos de la Cuenca Tampico-Misantla, Mexico, *in* M. E. Gómez-Luna and A. M. Cortez, eds., 5th Latin American Congress of Organic Geochemistry, Cancún, Mexico, Memoir, Abstracts Volume, p. 112–114.

Román-Ramos, J. R., and L. N. Holguín-Quiñones, 2001, Subsistemas generadores de la región norte de Mexico: Boletín Asociación Mexicana de Geólogos Petroleros, v. 49, no. 1–2, p. 68–84.

Seilacher, A., 1967, Bathymetry of trace fossils: Marine Geology, v. 5, p. 413–428.

Trudgill, B. D., M. G. Rowan, J. C. Fiduk, P. Weimer, P. E. Gale, B. E. Korn, R. L. Phair, W. T. Gafford, G. R. Roberts, and S. W. Dobbs, 1999, The Perdido fold belt, northwestern deep Gulf of Mexico: Part 1. Structural geometry, evolution and regional implications: AAPG Bulletin, v. 83, p. 88–113.

Vail, P. R., 1987, Seismic stratigraphy interpretation using sequence stratigraphy: Part I. Seismic stratigraphy interpretation procedure, *in* A. W. Bally, ed., Atlas of seismic stratigraphy, v. 1: AAPG Studies in Geology 27, p. 1–10.

Van Wagoner, J. C., R. M. Mitchum, K. M. Campion, and V. D. Rahmanian, 1990, Siliciclastic sequence stratigraphy in well logs, cores, and outcrops: AAPG Methods in Exploration Series 7, p. 1–55.

Watkins, J. S., and R. T. Buffler, 1996, Deepwater frontier exploration potential: Gulf Coast Association of Geological Societies Transactions, v. 46, p. 79–94.

Wawrzyniec, T., et al., 2003, Cenozoic deformational styles of the Laguna-Madre–Tuxpan shelf and Mexican Ridges fold belt, Mexico: Gulf Coast Association of Geological Societies Transactions, v. 53, p. 843–854.

Winker, C. D., 1984, Clastic shelf margins of the Post-Comanchean Gulf of Mexico: Implications for deep-water sedimentation, *in* Characteristics of gulf basin deep-water sediments and their exploration potential: Annual Research Conference, Gulf Coast Section SEPM Program and Abstracts, p. 109–120.